Bertram O. Fraser-Reid • Kuniaki Tatsuta • Joachim Thiem (Eds.)

Glycoscience
Chemistry and Chemical Biology

Section Editors: G. Coté, S. Flitsch, Y. Ito, H. Kondo, S.-I. Nishimura, B. Yu

With contributions by numerous experts

Second Edition

With 638 Figures and 159 Tables

Volume 1

 Springer

Prof. Dr. Bertram O. Fraser-Reid
Natural Products and Glycotechnology
Research Institute Inc.
North Caroline State University
4118 Swarthmore Road
Durham, NC 27707
USA
dglucose@aol.com

Prof. Dr. Kuniaki Tatsuta
University Graduate School
of Science and Engineering
3-4-1 Ohkubo, Shinjuku
Tokyo
Japan
tatsuta@waseda.jp

Prof. Dr. Joachim Thiem
Faculty of Science
Department of Chemistry
University of Hamburg
Martin-Luther-King-Platz 6
20146 Hamburg
Germany
thiem@chemie.uni-hamburg.de

Library of Congress Control Number: 2008921863

ISBN: 978-3-540-36154-1

This publication is available also as:
Electronic publication under ISBN: 978-3-540-30429-6 and
Print and electronic bundle under ISBN: 978-0-387-36157-2
DOI: 10-1007/978-3-540-30429-6

Springer is part of Springer Science+Business Media

springer.com

Editor: Marion Hertel, Heidelberg, Germany
Development Editor: Sylvia Blago, Heidelberg, Germany
Typesetting and Production: le-tex Jelonek, Schmidt & Vöckler GbR, Leipzig, Germany
Cover Design: Frido Steinen-Broo, Girona, Spain

Printed on acid-free paper   SPIN: 11539346   2109 letex — 5 4 3 2 1 0

Preface to the Second Edition

All of the commitments that we expressed in the preface of the 2001 first edition still hold true. The fact that we now introduce a second edition, seven years later, reflects the burgeoning research developments in glycoscience, and the broad impact these developments are having on contemporary biological science. The term "Glycoscience", selected first for a comprehensive collection of reviews a decade ago in Topics in Current Chemistry, has become the generally accepted language to describe the interests of scientists at all levels who are involved in research associated with chemistry, chemical biology and the biology of saccharides.

Readers of Glycoscience 2nd Edition will immediately notice that the editorial board has been supplemented with six younger colleagues of top scientific rank in the glycoscience area. We are happy to have been able to convince:

G. Coté, National Center for Agricultural Utilization and Research, Agricultural Research Service, Peoria, USA;

S. Flitsch, University of Manchester, UK;

Y. Ito, Institute of Physical and Chemical Research (RIKEN), Saitama, Japan;

H. Kondo, Shionogi & Co., Ltd., Osaka, Japan;

S.-I. Nishimura, Hokkaido University, Sapporo, Japan;

B. Yu, Shanghai Institute of Organic Chemistry, Chinese Academy of Sciences, Shanghai, China

to serve as Section Editors.

Their dedicated work in modifying and extending the coverage of the first edition, while maintaining its focus, has resulted in a contemporary appearance. This is particularly true of chapters which cover the bridges that link chemistry to biological chemistry to medicinal chemistry.

On behalf of the Editors and the Section Editors it is our pleasure to extend our gratitude to our dedicated expert colleagues, for their prodigious and scholarly work in updating the chapters. Finally, we are very pleased to note the constant and dedicated support and productive collaboration that we enjoyed with Drs. Marion Hertel and Sylvia Blago, and their highly dedicated editorial staff from Springer, Heidelberg.

January 2008

Bertram O. Fraser-Reid
Kuniaki Tatsuta
Joachim Thiem

Preface to the First Edition

Attempts at understanding of the structure and function of sugars date back to the work of eminent scientists active at the turn of the 19th/20th century. This field of research has developed into today's flourishing area of chemistry and biochemistry of carbohydrates. Over the years major scientific contributions have been associated with this research area and the results have had a strong impact on general chemistry.

At the end of the last century the significance of glycoscience, covering all aspects of the chemistry and chemical biology of carbohydrates and glycoconjugates, increased enormously owing to the challenge of interpreting complex natural processes at the molecular level.

Thus, the purpose of this comprehensive approach is to address and assist a broad readership ranging from graduate students to research scientists involved in glycochemistry and glycobiology, working in areas such as organic chemistry, biochemistry, molecular biology, immunology, and microbiology, as well as pharmaceutical, medicinal, agricultural, food chemistry, etc. The special design of this timely and major modern work on glycoscience is intended to go beyond the collection of specialized reviews. Accordingly, the first general chapters are focussed on principles and methods, giving detailed insight into the current status of structure analysis, synthesis and mechanistic interpretation. Subsequent chapters deal with mono-, oligo-, polysaccharides as well as with the chemistry and chemical biology of glycolipids, glycoproteins, other glycoconjugates and glycomimetics, and they always involve in depth discussion of the biochemical, biological and biomedical background. Because of this presentation, the impact of this collection should be attractive for both specialists, and non-specialists, as well as for newcomers to glycoscience from borderline research areas.

For the editors the task to put together such a comprehensive treatise was feasible only after they could convince the leaders in the fields to contribute and write their chapters with absolute authority. If this work does fulfil the desired purpose of becoming a useful, comprehensive collection in the area of chemistry and chemical biology of glycoscience, many thanks will be due to all of the expert authors.

July 2001

Bertram O. Fraser-Reid
Kuniaki Tatsuta
Joachim Thiem

Table of Contents

Volume 1

Volume 3

Editors-in-Chief

Prof. Dr. Bertram O. Fraser-Reid
Natural Products and Glycotechnology
Research Institute Inc.
North Caroline State University
4118 Swarthmore Road
Durham, NC 27707
USA
dglucose@aol.com

Prof. Dr. Kuniaki Tatsuta
University Graduate School
of Science and Engineering
3-4-1 Ohkubo, Shinjuku
Tokyo
tatsuta@waseda.jp

Prof. Dr. Joachim Thiem
Faculty of Science
Department of Chemistry
University of Hamburg
Martin-Luther-King-Platz 6
20146 Hamburg
Germany
thiem@chemie.uni-hamburg.de

Section Editors

Gregory L. Coté
National Center for Agricultural Utilization
U.S. Department Agriculture
Bioproducts Research Unit
Peoria, IL, USA
greg.cote@ars.usda.gov

Sections:
1. General Principles
4. Monosaccharides
5. Oligosaccharides
6. Complex Polysaccharides

Sabine Flitsch
School of Chemistry
University of Manchester
Manchester, UK
Sabine.Flitsch@manchester.ac.uk

Sections:
11. Biosynthesis and Degradation
12. Glycomedicine

Yukishige Ito
Institute of Physical and Chemical Research
(RIKEN)
Saitama, Japan
yukito@riken.jp

Sections:
3. Chemical Glycosylation Reactions
5. Oligosaccharides

Hirosato Kondo
Discovery Research Laboratories
Shionogi & Col, Ltd.
Osaka, Japan
hirosato.kondou@shionogi.co.jp

Sections:
9. Glycomimetics
10. Key Technologies and Tools
*   for Functional Glycobiology*

Shin-Ichiro Nishimura
Laboratory of Advanced Chemical Biology
Graduate School of Advanced Life Science
Hokkaido University
Sapporo, Japan
shin@glyco.sci.hokudai.ac.jp

Sections:
7. Glycolipids
8. Glycoproteins

Biao Yu
Chinese Academy of Sciences
Shanghai Institute of Organic Chemistry
Shanghai, China
byu@mail.sioc.ac.cn

Sections:
2. General Synthetic Methods
4. Monosaccharides

List of Contributors

Adachi, Masaatsu
Graduate School of Bioagricultural Sciences
Nagoya University
464-8601 Nagoya
Japan
madachi@agr.nagoya-u.ac.jp

Aich, Udayanath
Department of Biomedical Engineering
The Johns Hopkins University
Baltimore, MD 21218
USA

Allscher, Thorsten
Department Chemie und Biochemie
Ludwig-Maximilians-Universität München
Butenandtstr. 9
81377 München
Germany

Anderson, Kevin
Division of Medicinal and Natural Products
Chemistry, College of Pharmacy
University of Iowa
115 South Grand Avenue
Iowa City, IA 52242-1112
USA

Ando, Hiromune
Division of Instrumental Analysis
Life Science Research Center
Gifu University
Gifu-shi
Gifu 501–1193
Japan
hando@gifu-u.ac.jp

Ariga, Toshio
Institute of Molecular Medicine and
Genetics, Institute of Neuroscience
Medical College of Georgia
Augusta, GA 30912–2697
USA

Asano, Naoki
Faculty of Pharmaceutical Science
Hokuriku University
920-1181 Kanazawa
Japan
n-asano@hokuriku-u.ac.jp

Balbin, Yury Valdés
Center for Synthetic Antigens
Faculty of Chemistry
University of Havana
Ciudad Habana
Cuba 10400
yury@fq.uh.cu

Beau, Jean-Marie
Laboratoire de Synthèse de Biomolécules
Institut de Chimie Moléculaire
et des Matériaux associé au CNRS
Université Paris-Sud
91405 Orsay Cedex
France
jmbeau@icmo.u-psud.fr

BeMiller, James N.
Department of Food Science
Whistler Center for Carbohydrate Research
Purdue University
West Lafayette, IN 47909–2009
USA
bemiller@purdue.edu

Bennett, Clay S.
Chemistry Department
The Scripps Research Institute
La Jolla, CA 92037
USA

Bezouska, Karel
Department of Biochemistry
Faculty of Science
Charles University Prague
Hlavova 8
12840 Praha 2
Czech Republic
bezouska@biomed.cas.cz

Bishop, Lee
Department of Chemistry
Wayne State University
5101 Cass Avenue
Detroit, MI 48202
USA

Blixt, Ola
Department of Cellular and Molecular
Medicine, Faculty of Health Sciences
University of Copenhagen
2200 Copenhagen N
Denmark
olablixt@imbg.ku.dk

Breton, Christelle
Molecular Glycobiology
CERMAV-CNRS
(affiliated with Université Joseph Fourier)
Grenoble 38041
France
christelle.breton@cermav.cnrs.fr

Brown, Stephanie
Department of Microbiology
and Molecular Genetics
Harvard Medical School
Boston, MA 2115
USA
Department of Chemistry
and Chemical Biology
Harvard University
Cambridge, MA 02138
USA

Calderón, Janoi Chang
Center for Synthetic Antigens
Faculty of Chemistry
University of Havana
Ciudad Habana
Cuba 10400
janoi@fq.uh.cu

Ceroni, Alessio
Imperial College London
Division of Molecular Biosciences
Biopolymer Mass Spectrometry Group
London SW7 2AZ
UK
a.ceroni@imperial.ac.uk

Chen, Qi
The State Key Laboratory of Environmental
Chemistry and Ecotoxicology
Research Center for Eco-Environmental
Sciences
Chinese Academy of Sciences
Beijing 100085
China

Collins, Michelle E.
Department of Food Biosciences
The University of Reading
Reading RG6 6AP
UK
m.e.collins@reading.ac.uk

Dell, Anne
Biopolymer Mass Spectrometry Group
Division of Molecular Biosciences
Imperial College London
SW7 2AZ London
UK
a.dell@imperial.ac.uk

Doisneau, Gilles
Laboratoire de Synthèse de Biomolécules
Institut de Chimie Moléculaire
et des Matériaux associé au CNRS
Université Paris-Sud
91405 Orsay Cedex
France
gdoisneau@icmo.u-psud.fr

Du, Yuguo
The State Key Laboratory of Environmental
Chemistry and Ecotoxicology
Research Center for Eco-Environmental
Sciences
Chinese Academy of Sciences
Beijing 100085
China
duyuguo@rcees.ac.cn

Eggleston, Gillian
United States Department of Agriculture
SRRC-ARS-USDA
1100 Robert E. Lee Boulevard
New Orleans, LA 70124
USA
gillian@srrc.ars.usda.gov

Evans, Philip G.
School of Pharmacy
University of Reading
Whiteknights
Reading, Berkshire RG6 6AD
UK

Evers, David
Division of Medicinal and Natural Products
Chemistry, College of Pharmacy
University of Iowa
115 South Grand Avenue
Iowa City, IA 52242-1112
USA

Feizi, Ten
Glycosciences Laboratory
Faculty of Medicine
Imperial College London
Northwick Park and St Mark's Campus
Harrow, Middlesex HA1 3UJ
UK
t.feizi@imeprial.ac.uk

Fernandes, Daryl
Ludger Ltd.
Culham Science Centre
Abingdon, Oxfordshire OX14 3EB
UK
daryl.fernandes@ludger.com

Fokt, Izabela
M. D. Anderson Cancer Center
The University of Texas
Houston, TX 77030
USA

Friedman, Robert B.
Friedmann Associates
6654 North Mozart Street
Chicago, IL 60645
USA
bobbf@juno.com

Fujimoto, Yukari
Department of Chemistry
Graduate School of Science
Osaka University
Toyonaka, Osaka 560–0043
Japan

Fukase, Koichi
Department of Chemistry
Graduate School of Science
Osaka University
Osaka 560-0043
Japan
koichi@chem.sci.osaka-u.ac.jp

Ginsberg, Cynthia
Department of Microbiology
and Molecular Genetics
Harvard Medical School
Boston, MA 2115
USA
Department of Chemistry
and Chemical Biology
Harvard University
Cambridge, MA 02138
USA

Glasser, Wolfgang G.
Dept. Wood Science and Forest Products
Virginia Tech
Blacksburg, VA 24061
USA
wglasser@vt.edu

Gómez, Ana M.
Instituto de Química Orgánica General
(CSIC)
28006 Madrid
Spain
iqog106@iqog.csic.es

Grindley, T. Bruce
Department of Chemistry
Dalhousie University
Halifax, NS B3H 4J3
Canada
Bruce.Grindley@Dal.Ca

Gruber, Todd D.
Department of Chemistry
University of Wisconsin
1101 University Avenue
Madison, WI 53706
USA

Grynkiewicz, Grzegorz
Pharmaceutical Research Institute
Rydgiera 8
01-793 Warsaw
Poland
g.grynkiewicz@ifarm.waw.pl

Guo, Zhongwu
Department of Chemistry
Wayne State University
5101 Cass Avenue
Detroit, MI 48202
USA
zwguo@chem.wayne.edu

Haslam, Stuart M.
Biopolymer Mass Spectrometry Group
Division of Molecular Biosciences
Imperial College London
SW7 2AZ London
UK

Holst, Otto
Structural Biochemistry
Research Center Borstel
23485 Borstel
Germany
oholst@fz-borstel.de

Icart, Luis Peña
Center for Synthetic Antigens
Faculty of Chemistry
University of Havana
Ciudad Habana
Cuba 10400
luisp@fq.uh.cu

Imberty, Anne
Molecular Glycobiology
CERMAV-CNRS
(affiliated with Université Joseph Fourier)
Grenoble 38041
France
anne.imberty@cermav.cnrs.fr

Ishiwata, Akihiro
RIKEN (The Institute of Physical
and Chemical Research)
Saitama 351–0198
Japan

Isobe, Minoru
Graduate School of Bioagricultural Sciences
Nagoya University
464-8601 Nagoya
Japan
isobem@agr.nagoya-u.ac.jp

Ito, Yukishige
RIKEN (The Institute of Physical
and Chemical Research)
Saitama 351–0198
Japan
yukito@riken.jp

Jang-Lee, Jihye
Biopolymer Mass Spectrometry Group
Division of Molecular Biosciences
Imperial College London
SW7 2AZ London
UK

Jarosz, Sławomir
Institute of Organic Chemistry
Polish Academy of Sciences
01–224 Warsaw
Poland
sljar@icho.edu.pl

Jeon, Heung Bae
Center for Bioactive Molecular Hybrids
and Department of Chemistry
Yonsei University
Seoul 120–749
Korea

Joshi, Hiren J.
German Cancer Research Center
Central Spectroscopy B090
69120 Heidelberg
Germany
h.joshi@dkfz.de

Kiessling, Laura L.
Department of Chemistry
University of Wisconsin
1101 University Avenue
Madison, WI 53706
USA
kiessling@chem.wisc.edu

Kim, Kwan Soo
Center for Bioactive Molecular Hybrids
and Department of Chemistry
Yonsei University
Seoul 120–749
Korea
kwan@yonsei.ac.kr

Kiso, Makoto
Department of Applied
Bioorganic Chemistry
Faculty of Applied Biological Sciences
Gifu University
Gifu-shi
Gifu 501–1193
Japan
kiso@cc.gifu-u.ac.jp

Klüfers, Peter
Department Chemie und Biochemie
Ludwig-Maximilians-Universität München
Butenandtstr. 9
81377 München
Germany
kluef@cup.uni-muenchen.de

Kobayashi, Yoshiyuki
Daiichi Sankyo Research Institute
4250 Executive Square
La Jolla, California 92037
USA
ykobayashi@daiichisankyo-us.com

Koeller, Kathryn M.
Chemistry Department
The Scripps Research Institute
La Jolla, CA 92037
USA

Kondo, Hirosato
Discovery Research Laboratories
Shionogi & Co., Ltd.
12-4, Sagisu 5-chome
Fukushima-ku 553-0002 Osaka
Japan
hirosato.kondou@shionogi.co.jp

Křen, Vladimír
Institute of Microbiology
Academy of Sciences of the Czech Republic
Centre of Biocatalysis and Biotransformation
Vídeňská 1083
142 20 Prague
Czech Republic
kren@biomed.cas.cz

Lieth, Claus-W. von der
(deceased)

Lindhorst, Thisbe K.
Otto Diels-Institut für Organische Chemie
Christian-Albrechts-Universität zu Kiel
24098 Kiel
Germany
tklind@oc.uni-kiel.de

Liu, Jun
The State Key Laboratory of Environmental
Chemistry and Ecotoxicology
Research Center for Eco-Environmental
Sciences
Chinese Academy of Sciences
Beijing 100085
China

Liu, Yan
Glycosciences Laboratory
Faculty of Medicine
Imperial College London
Northwick Park and St Mark's Campus
Harrow, Middlesex HA1 3UJ
UK

López, J. Cristóbal
Instituto de Química Orgánica General
CSIC
Juan de la Cierva 3
28006 Madrid
Spain
clopez@iqog.csic.es

Lucas, Susana Dias
Pharmaceutical Research
Discovery Chemistry
F. Hoffmann-La Roche Ltd.
4070 Basel
Switzerland

Maaß, Kai
Institute of Biochemistry
Faculty of Medicine
University of Giessen
35392 Giessen
Germany
Kai.maass@biochemie.med.uni-giessen.de

Macmillan, Derek
Department of Chemistry
University College London
20 Gordon Street
WC1H 0AJ London
UK
d.macmillan@ucl.ac.uk

Madsen, Robert
Department of Chemistry
Center for Sustainable and Green Chemistry
Technical University of Denmark
Lyngby 2800
Denmark
rm@kemi.dtu.dk

Mayer, Peter
Department Chemie und Biochemie
Ludwig-Maximilians-Universität München
Butenandtstr. 9
81377 München
Germany

McGill, Nathan W.
School of Chemistry
The University of Melbourne
Parkville, VIC 3052
Australia
n.mcgill@pgrad.unimelb.edu.au

Mori, Yasutaka
Department of Chemistry
Graduate School of Science
Osaka University
Toyonaka, Osaka 560–0043
Japan

Morris, Howard R.
Biopolymer Mass Spectrometry Group
Division of Molecular Biosciences
Imperial College London
SW7 2AZ London
UK

Mortell, Kathleen H.
Department of Chemistry
University of Wisconsin
1101 University Avenue
Madison, WI 53706
USA

Murphy, Paul V.
Centre for Synthesis and Chemical Biology
School of Chemistry
and Chemical Biology
University College Dublin
Belfield
Dublin 4
Ireland
paul.v.murphy@ucd.ie

Nakata, Masaya
Department of Applied Chemistry
Faculty of Science and Technology
Keio University
Yokohama 223-8522
Japan
msynktxa@applc.keio.ac.jp

Nishikawa, Toshio
Graduate School of Bioagricultural Sciences
Nagoya University
464-8601 Nagoya
Japan
nisikawa@agr.nagoya-u.ac.jp

Nishimura, Shin-Ichiro
Laboratory of Advanced Chemical Biology
Graduate School of Advanced Life Science
Frontier Research Center for the
Post-Genome Science and Technology
Hokkaido University and Drug-Seeds
Discovery Research Laboratory
National Institute of Advanced Industrial
Science and Technology
Sapporo
Japan
shin@glyco.sci.hokudai.ac.jp

Nowogródzki, Marcin
Institute of Organic Chemistry
Polish Academy of Sciences
01–224 Warsaw
Poland

Osborn, Helen M. I.
School of Pharmacy
University of Reading
Whiteknights
Reading, Berkshire RG6 6AD
UK
h.m.i.osborn@reading.ac.uk

Oscarson, Stefan
Department of Organic Chemistry
Arrhenius Laboratory
Stockholm University
Stockholm 106 91
Sweden
stefan.oscarson@ucd.ie

Pang, Poh-Choo
Biopolymer Mass Spectrometry Group
Division of Molecular Biosciences
Imperial College London
SW7 2AZ London
UK

Panico, Maria
Biopolymer Mass Spectrometry Group
Division of Molecular Biosciences
Imperial College London
SW7 2AZ London
UK

Parry, Simon
Biopolymer Mass Spectrometry Group
Division of Molecular Biosciences
Imperial College London
SW7 2AZ London
UK

Payne, Richard J.
School of Chemistry
The University of Sydney
Sydney, NSW
Australia

Polt, Robin
University of Arizona
Tucson, AZ 85721-0041
USA
polt@u.arizona.edu

Priebe, Waldemar
M. D. Anderson Cancer Center
The University of Texas
Houston, TX 77030
USA
wpriebe@mdanderson.org, wp@wt.net

Ranzinger, René
German Cancer Research Center
Central Spectroscopy B090
69120 Heidelberg
Germany
r.ranzinger@dkfz.de

Rastall, Robert A.
Department of Food Biosciences
The University of Reading
Reading RG6 6AP
UK
r.a.rastall@reading.ac.uk

Razi, Nahid
Department of Molecular Biology
Glycan Array Synthesis Core D
Consortium for Functional Glycomics
The Scripps Research Institute
La Jolla, CA 92037
USA
nrazi@scripps.edu

Rice, Kevin G.
Division of Medicinal and Natural Products
Chemistry, College of Pharmacy
University of Iowa
115 South Grand Avenue
Iowa City, IA 52242-1112
USA
kevin-rice@uiowa.edu

Richardson, Jonathan P.
Department of Chemistry
University College London
20 Gordon Street
WC1H 0AJ London
UK

Robina, Inmaculada
Departamento de Química Orgánica
Universidad de Sevilla
41071 Sevilla
Spain
robina@us.es

Robyt, John F.
Laboratory of Carbohydrate Chemistry and
Enzymology, Department of Biochemistry,
Biophysics, and Molecular Biology
Iowa State University
Ames, IA 50011
USA
jrobyt@iastate.edu

Sadamoto, Reiko
Graduate School of Advanced Life Science
Hokkaido University
Sapporo 001–0021
Japan
reikosd@glyco.sci.hokudai.ac.jp

Santana, Violeta Fernández
Center for Synthetic Antigens
Faculty of Chemistry
University of Havana
Ciudad Habana
Cuba 10400
violeta@fq.uh.cu

Schmidt, Richard R.
Fachbereich Chemie, Fach M 725
Universität Konstanz
78457 Konstanz
Germany
richard.schmidt@uni-konstanz.de

Spencer, Daniel
Ludger Ltd.
Culham Science Centre
Abingdon, Oxfordshire OX14 3EB
UK
daniel.spencer@ludger.com

Steiner, Andreas
Glycogroup, Institut für Organische Chemie
Technische Universität Graz
8010 Graz
Austria

Stone, Bruce A.
Department of Biochemistry
La Trobe University
Bundoora, VIC 3083
Australia
b.stone@latrobe.edu.au

Stütz, Arnold
Glycogroup, Institut für Organische Chemie
Technische Universität Graz
8010 Graz
Austria
stuetz@tugraz.at

Suda, Yasuo
Department of Nanostructure
and Advanced Materials
Graduate School of Science and Engineering
Kagoshima University
890-0065 Kagoshima
Japan
ysuda@eng.kagoshima-u.ac.jp

Sutton-Smith, Mark
Biopolymer Mass Spectrometry Group
Division of Molecular Biosciences
Imperial College London
SW7 2AZ London
UK

Svensson, Birte
Department of Chemistry
Carlsberg Laboratory
2500 Valby
Denmark

Tadano, Kin-ichi
Department of Applied Chemistry
Keio University
Hiyoshi, Kohoku-ku
223-8522 Yokohama
Japan
tadano@applc.keio.ac.jp

Tanaka, Katsunori
Department of Chemistry
Graduate School of Science
Osaka University
Osaka 560-0043
Japan

Tanaka, Shin-ichi
Department of Chemistry
Graduate School of Science
Osaka University
Toyonaka, Osaka 560–0043
Japan

Thiem, Joachim
Department of Chemistry
Faculty of Science
University of Hamburg
20146 Hamburg
Germany
joachim.thiem@chemie.uni-hamburg.de

Thimm, Julian
Department of Chemistry
Faculty of Science
University of Hamburg
20146 Hamburg
Germany

Tissot, Berangere
Biopolymer Mass Spectrometry Group
Division of Molecular Biosciences
Imperial College London
SW7 2AZ London
UK

Toshima, Kazunobu
Department of Applied Chemistry
Faculty of Science and Technology
Keio University
Yokohama 223–8522
Japan
toshima@applc.keio.ac.jp

Totani, Kiichiro
RIKEN (The Institute of Physical
and Chemical Research)
2-1 Hirosawa, Wako-shi
351-0198 Saitama
Japan
totani@riken.jp

Tvaroška, Igor
Institute of Chemistry, Centre for Glycomics
Slovak Academy of Sciences
845 38 Bratislava
Slovak Republic
chemitsa@savba.sk

Urban, Dominique
Laboratoire de Synthèse de Biomolécules
Institut de Chimie Moléculaire
et des Matériaux associé au CNRS
Université Paris-Sud
91405 Orsay Cedex
France
domurban@icmo.u-psud.fr

Vauzeilles, Boris
Laboratoire de Synthèse de Biomolécules
Institut de Chimie Moléculaire
et des Matériaux associé au CNRS
Université Paris-Sud
91405 Orsay Cedex
France
bvauzeil@icmo.u-psud.fr

Velasco-Torrijos, Trinidad
Centre for Synthesis and Chemical Biology
School of Chemistry
and Chemical Biology
University College Dublin
Belfield
Dublin 4
Ireland
Current address:
Department of Chemistry
National University of Ireland Maynooth
Co. Kildare
Ireland
trinidad.velascotorrijos@nuim.ie

Verez-Bencomo, Vicente
Center for Synthetic Antigens
Faculty of Chemistry
University of Havana
Ciudad Habana
Cuba 10400
vicente@fq.uh.cu

Vogel, Pierre
Laboratoire de glycochimie
et de synthèse asymétrique
Ecole Polytechnique Fédérale de Lausanne
(EPFL), BCH
1015 Lausanne-Dorigny
Switzerland
pierre.vogel@epfl.ch

Wakao, Masahiro
Department of Nanostructure
and Advanced Materials
Graduate School of Science and Engineering
Kagoshima University
890-0065 Kagoshima
Japan
wakao@eng.kagoshima-u.ac.jp

Walker, Suzanne
Department of Microbiology
and Molecular Genetics
Harvard Medical School
Boston, MA 2115
USA
Department of Chemistry
and Chemical Biology
Harvard University
Cambridge, MA 02138
USA
suzanne_walker@hms.harvard.edu

Wang, Yuhang
The State Key Laboratory of Natural
and Biomimetic Drugs
School of Pharmaceutical Sciences
Peking University
Xue Yuan Rd #38
Beijing 100083
China

Wessel, Hans Peter
Pharmaceutical Research
Discovery Chemistry
F. Hoffmann-La Roche Ltd.
4070 Basel
Switzerland
hans_p.wessel@roche.com

Williams, Spencer J.
School of Chemistry
The University of Melbourne
Parkville, VIC 3052
Australia
sjwill@unimelb.edu.au

Wilson, Iain B. H.
Department für Chemie
Universität für Bodenkultur
(University of Natural Resources and
Applied Life Sciences)
Muthgasse 18
1190 Wien
Austria
iain.wilson@boku.ac.at

Witczak, Zbigniew J.
Department of Pharmaceutical Sciences
Nesbitt School of Pharmacy
Wilkes University
Wilkes-Barre, PA 18766
USA
zbigniew.witczak@wilkes.edu

Wittmann, Valentin
Fachbereich Chemie
Universität Konstanz
78457 Konstanz
Germany
mail@valentin-wittmann.de

Wong, Chi-Huey
Chemistry Department
The Scripps Research Institute
La Jolla, CA 92037
USA
wong@scripps.edu

Wrodnigg, Tanja
Glycogroup, Institut für Organische Chemie
Technische Universität Graz
8010 Graz
Austria

Yanagisawa, Makoto
Institute of Molecular Medicine and
Genetics, Institute of Neuroscience
Medical College of Georgia
Augusta, GA 30912–2697
USA

Yarema, Kevin J.
Department of Biomedical Engineering
The Johns Hopkins University
Baltimore, MD 21218
USA
kyarema1@jhu.edu

Ye, Xin-Shan
The State Key Laboratory of Natural
and Biomimetic Drugs
School of Pharmaceutical Sciences
Peking University
Xue Yuan Rd #38
Beijing 100083
China
xinshan@bjmu.edu.cn

Young, Travis
Department of Chemistry
University of Wisconsin
1101 University Avenue
Madison, WI 53706
USA

Yu, Robert K.
Institute of Molecular Medicine and
Genetics, Institute of Neuroscience
Medical College of Georgia
Augusta, GA 30912–2697
USA
ryu@mcg.edu

Zeng, Guichao
Institute of Molecular Medicine and
Genetics, Institute of Neuroscience
Medical College of Georgia
Augusta, GA 30912–2697
USA

Zhang, Jianbo
Department of Chemistry
East China Normal University
200062 Shanghai
China
jbzhang@chem.ecnu.edu.cn

Zhu, Xiangming
School of Chemistry
and Chemical Biology
University College Dublin
Belfield
Dublin 4
Ireland
xiangming@ucd.ie

Part 1
General Principles

1.1 Structure and Conformation of Carbohydrates

T. Bruce Grindley
Department of Chemistry, Dalhousie University,
Halifax, NS B3H 4J3, Canada
Bruce.Grindley@Dal.Ca

Abstract

The conformational analysis of monosaccharides, disaccharides, and oligosaccharides is reviewed. Conformational terms are introduced through examination of the conformations of cyclohexane and cyclopentane then applied to the pyranose, furanose, and septanose rings. Concepts such as the anomeric effect are discussed. Topics of current interest, such as hydroxymethyl group and hydroxyl group rotation and disaccharide conformations are summarized. Physical methods for studying conformation are outlined.

In: *Glycoscience*. Fraser-Reid B, Tatsuta K, Thiem J (eds)
Chapter-DOI 10-1007/978-3-540-30429-6_1: © Springer-Verlag Berlin Heidelberg 2008

Keywords

Conformation; Chair; Anomeric effect; Hydroxymethyl; Monosaccharides; Disaccharides; Oligosaccharides; Coupling constants; Molecular mechanics; Molecular dynamics

Abbreviations

A-value	free energy cost for a substituent on a chair conformation of a cyclohexane ring to change from an equatorial to axial orientation
B	boat
B3LYP	the Becke-3 Lee–Yang–Parr density functional
C	chair
CD	circular dichroism
DFT	density functional theory
E	envelope conformation of a five-membered ring
G	Gaussian
gg, gt, tg	conformations of hydroxymethyl or other side-chains on pyranose or furanose rings
GVB	generalized valence bond
H	half chair
HOMO	highest energy occupied molecular orbital
KDO	3-deoxy-D-*manno*-2-octulosonic acid
MP	Moeller–Plesset
NANA	5-acetamido-3,5-dideoxy-D-glycero-D-galacto-2-nonulosonic acid
nOe	nuclear Overhauser enhancement
ORD	optical rotatory dispersion
ROA	resonant Raman optical activity
S	skew conformation of a six-membered ring
T	twist conformation of a five-membered ring
TB	twist boat
TC	twist chair

1 Conformational Analysis

1.1 Introduction

The shapes or conformations of carbohydrates strongly influence both their reactivity and their physical properties. Several reviews of various aspects of the conformational analysis of carbohydrates have been published [1,2,3,4,5,6,7].

Monosaccharides exist chiefly in the six-membered pyranose form and in the five-membered furanose form, both of which adopt puckered shapes to minimize eclipsing interactions. These puckered shapes are most commonly described in terms of the shapes of the symmetrical conformations adopted by the analogous hydrocarbons, cyclohexane and cyclopentane [8,9,10,11], but may also be described in terms of puckering parameters [12,13,14]. Minima on the conformational potential energy surface are called conformers.

1.2 Conformations of Cycloalkanes and Heterocycles

1.2.1 Conformations of Cyclohexanes

The most stable conformer of cyclohexane is the D_{3d} symmetric chair (C) (**1**) conformation, in which orientations parallel to the C_3 axis are termed axial while those roughly perpendicular to the axis are termed equatorial. Chair conformers invert or undergo ring reversal via the C_2 symmetric half-chair (H) saddle point (**2**) and intermediate D_2 twist-boat, or, as used for carbohydrates, skew (S) conformer (**3**). The barrier to ring reversal ($\Delta G^{\neq}$) is about 43 kJ mol^{-1}; values of $\Delta H^{\neq}$ and $\Delta S^{\neq}$ obtained for cyclohexane-d_{11} [15] and cyclohexane-1,1,2,2,3,3,4,4-d_8 [16] were 44.8 and 48.2 kJ mol^{-1} and 9.2 and 19.2 J mol^{-1} K^{-1}, respectively. The skew conformer is flexible, exchanging atoms above and below the plane defined by the two carbons on the C_2 axis and the two adjacent to one of these, via the C_{2v} symmetric boat (B) (**4**) conformation, a saddle point on the conformational potential energy surface. Rapid cooling of a mixture of cyclohexane and argon from 800 °C to 20 K isolated a conformational mixture containing some skew form. The rate of conversion to the chair was used to give a $\Delta H^{\neq}$ value for this process, which, combined with the known chair to chair barrier, gave a value of 23 kJ mol^{-1} for the stability of the S form relative to that of the C [17].

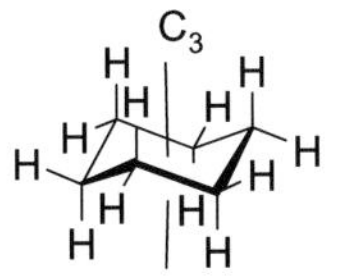

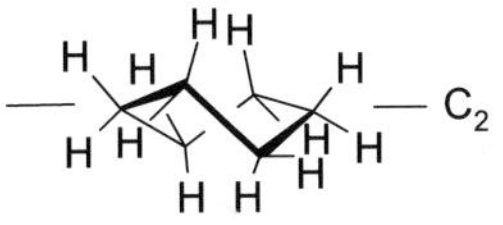

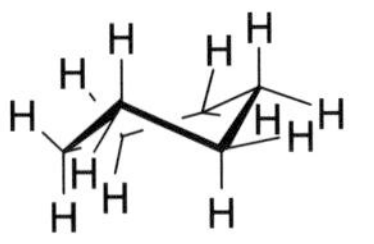

A substituent on a cyclohexane ring can assume either an equatorial or an axial orientation and the free energy preference for the equatorial conformer is termed the A value for the substituent (❷ *Fig. 1*). Tables of A values are available [8,18]. The values (at 300 K) most relevant for the current topic are those for methyl (7.31 [19] or 7.61 kJ mol^{-1} [20]), for hydroxymethyl (7.36 kJ mol^{-1} [21]), for hydroxyl (2.5 to 4.6 kJ mol^{-1} depending on sol-

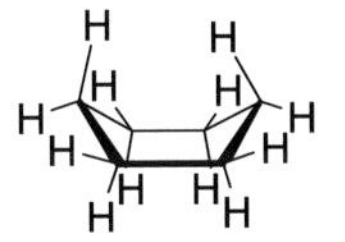

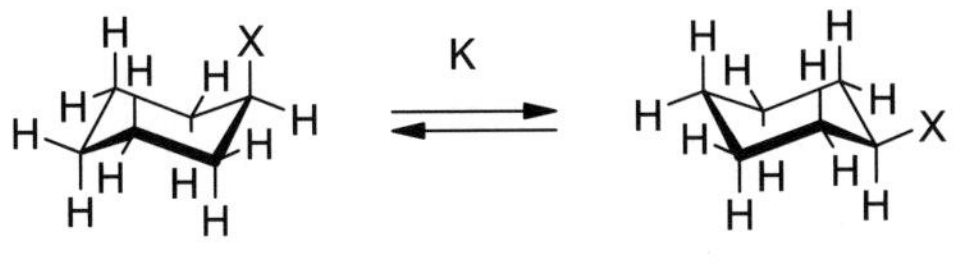

❏ **Figure 1**
Derivation of A values for substituted cyclohexanes: $A = RT \ln K = -\Delta G°$

☐ Figure 2
The 1,3-diaxial interaction in *cis*-1,3-disubstituted cyclohexanes

vent [8,22,23]), for methoxyl (2.3–3.1 kJ mol^{-1} [23,24]), for carboxyl (5.7 kJ mol^{-1} [25]), for carboxylate (7.9–8.2 kJ mol^{-1} [25,26]), and for methoxycarbonyl (4.6–5.5 kJ mol^{-1} [24,25, 26,27]). On the basis of the lack of change in geometry of the *syn* axial CH bonds on addition of an axial methyl, it was concluded [20] that the cause of the equatorial preference is repulsive steric interactions between the axial group and the ring carbons including the *gauche* torsional interaction rather than the traditional [8] explanation of steric interactions with the 1,3-related axial hydrogens (synaxial interaction). However, see below.

The interactions between substituents are important for the stabilities of the conformations of carbohydrates. Non-geminally disubstituted cyclohexanes can exist as *cis*- and *trans*-isomers. The relative stabilities of *cis*- and *trans*-1,4-disubstituted cyclohexanes or the relative stabilities of the diaxial versus the diequatorial conformers of *trans*-1,4-disubstituted cyclohexanes can be predicted successfully from the *A* values of the two substituents [8]. Another factor arises when *cis*-1,3-disubstituted-1,3-cyclohexanes are considered: the interaction between *syn*-1,3-diaxially related substituents, as shown in ❷ *Fig. 2*. Corey and Feiner summarized values for this interaction [28]. For carbohydrates, the most important of those available are: the OH / OH interaction (7.9 kJ mol^{-1} [29]), the CH$_3$ / OH interaction (7.9–11.3 kJ mol^{-1} [30,31,32]) and the OAc / OAc interaction (8.4 kJ mol^{-1}[33]).

The two chair conformers of *cis*-1,2-disubstituted cyclohexanes both have additional *gauche* interactions between substituents (5). One might expect that the stability differences between the diequatorial and diaxial conformers of the *trans*-isomer would be the sum of the *A*-values for the axial substituents minus the value of the *gauche* interaction for the equatorial substituents (6). For methyl substituents, where the *gauche* interaction is that of butane, this has been shown to be approximately true [34,35]. Corey and Feiner [28] found that *gauche* interactions between substituents other than methyl could not be estimated by averaging 1/2 the substituent's *A* values but have tabulated these *gauche* interaction values. Those involving polar groups are highly dependent on solvent because the diaxial and diequatorial conformers have very different dipole moments. Direct determination by measurement of the difference in stability between the diequatorial and diaxial conformers requires the subtraction of the substituent *A* values from the total stability difference. These latter values may be uncertain or have large solvent dependencies, leading to large uncertainties in the interaction energies. The most relevant values are the OH:OH interaction (1.5 kJ mol^{-1} in water [29], the OH:OMe interaction (2.7 kJ mol^{-1} in carbon disulfide [36], 1.9 kJ mol^{-1} in pentane [37], 1.9 kJ mol^{-1} in methanol [37]), the OMe:OMe interaction (5.3 kJ mol^{-1} in pentane, 2.3 kJ mol^{-1} in methanol [37]), and the OH:CH$_3$ interaction (1.6 kJ mol^{-1} [38]).

5

6

The relative stabilities of the two chair conformers of 1,1-geminally disubstituted cyclohexanes are additive based on A values [39,40] unless the populations of rotamers present for one of the two substituents are different to when in the monosubstituted cyclohexane [8,41] or unless there is differential solvation of one of the substituents in this compound as opposed to the cyclohexane derivative [39]. The latter situation applies with 1-methylcyclohexanol (**7**), where ΔG is 1.3 kJ mol^{-1} in carbon disulfide in favor of the methyl equatorial conformation, whereas subtraction of A values gives 3.2 kJ mol^{-1} [39]. In contrast, for 1-methoxy-1-methylcyclohexane, the observed and calculated ΔG values are 3.1 and 3.7 kJ mol^{-1}, respectively [40], a difference that will not have large conformational consequences. The difference between observed and calculated ΔGs for **7** presumably arises because the difference in the aggregation of the equatorial and axial oriented hydroxyl conformers through hydrogen bonding is much larger for the tertiary alcohol **7** than for the corresponding conformers of cyclohexanol. This effect could be important for conformational equilibria of ketoses or branched chain sugars but is probably negligible in hydroxylic solvents.

7

1.2.2 Conformations of Tetrahydropyran Derivatives

The replacement of a CH$_2$ group by an oxygen atom to give tetrahydropyran or oxane alters the shape of the chair conformation slightly by replacing two 153 pm C–C bonds by two 142 pm C–O bonds [42,43]. This has little effect on the barrier to ring reversal [8,44]. However, the substituent A-values now depend on position. The A-values for a methyl group at positions 2, 3, and 4 are 12.0, 5.98, and 8.16 kJ mol^{-1}, respectively [45], compared to ~7.5 kJ mol^{-1} for methylcyclohexane. The value at position 2 is significantly increased relative to that for methylcyclohexane because the shorter bond lengths increase steric repulsion, while that at position-3 is decreased because one of the CH$_3$/H synaxial interactions is replaced by a CH$_3$/lone pair interaction (❯ *Fig. 3*).

A methoxy group at position-2 of tetrahydropyran is much more stable in the axial orientation than would be expected based on its A value, a manifestation of the anomeric effect [46,47,48,49,50,51,52,53], originally observed with carbohydrate derivatives by

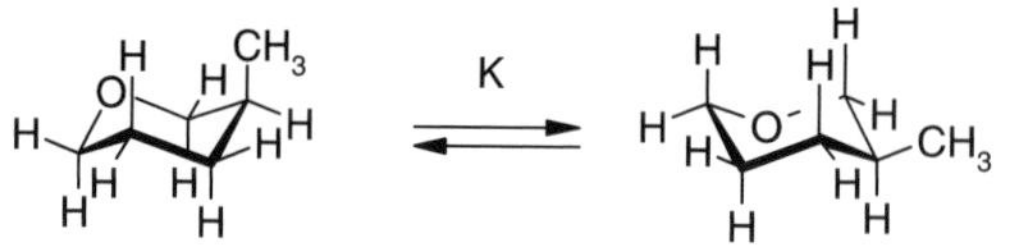

◘ Figure 3
The A value for a methyl group at position 3 of tetrahydropyran is less than that of cyclohexane, 6.0 vs 7.5 kJ mol^{-1}

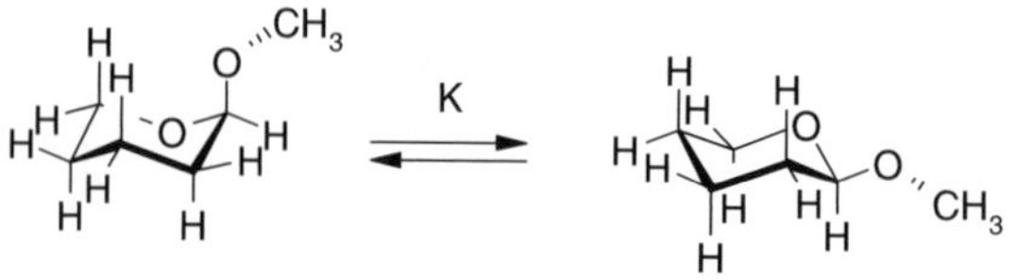

◘ Figure 4
The anomeric effect: the axial conformer is more stable than predicted based on the A value of the methoxy group. The exoanomeric effect: the preferred rotamers about the exocyclic C–O bond are those with the methyl *gauche* to the endocyclic oxygen atom and to H-1

Edwards [54] and named by Lemieux [33]. This effect is due to electronic interactions between the exocyclic and endocyclic oxygen atoms and their C–O bonds (see later). It had been recognized that cyclohexane A values were not appropriate as reference values for the size of the anomeric effect because of the different geometry in the heterocycle [55]. Franck [56] suggested that the cyclohexane A values for polar substituents be scaled to obtain reference heterocyclic A values by using a line defined by the relationship of heterocyclic to cyclohexane A values for non-polar substituents such as methyl and using hydrogen as the zero point. The resulting equation is $\Delta G(2\text{-thp}) = 1.53 \times \Delta G(\text{Cyhx}) + 0.04\,\text{kJ mol}^{-1}$. This suggestion has now been generally accepted [50,53]. Thus, an observed equatorial to axial free energy difference for a methoxy group at position-2 in tetrahydropyran of -3.3 kJ mol^{-1} in a non-polar solvent [57] and the cyclohexane A value of 2.7 kJ mol^{-1} scaled to 4.2 kJ mol^{-1} (see above) gives an anomeric effect for methoxy of 7.5 kJ mol^{-1} in tetrahydropyran (❷ Fig. 4).
Two explanations have been advanced to account for the anomeric effect. The first involves the stabilizing effect of bonding interactions between n electrons on one oxygen atom and the σ* orbital of the bond connecting the other oxygen atom and the central anomeric carbon atom [58]. The second involves destabilizing dipole-dipole repulsion between the two oxygen atoms and their lone pairs [54]. The relative importance of these factors has been difficult to establish.
Complicating the discussion are the two different descriptions of the nature of the lone pairs involved. In the traditional view, the lone pairs on oxygen are depicted as sp^3 hybridized. However, these lone pairs would be equivalent for water, which is not compatible with the photoelectron spectra, where the lone pair orbitals are very different in energy [59,60]. In the correct description, the two orbitals differ in their extent of p character; the higher energy orbital of HOMO, n$_p$ is close to being a pure p orbital, while the lower energy orbital, n$_\sigma$ has much more a s character, resulting in a picture that resembles sp^2 hybridization, with the n$_p$ orbital perpendicular to the plane containing the two substituents bonded to the oxygen

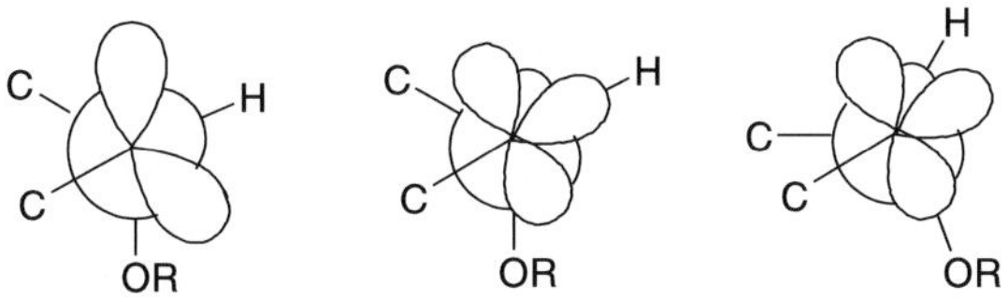

□ **Figure 5**
Newman projections of lone pair orbitals on one oxygen atom in a COCOR unit. *Left:* the geometry of an axial OR group in a tetrahydropyran ring with sp³ orbitals on the endocyclic oxygen. *Center:* the geometry of an axial OR group in a tetrahydropyran ring with n_p and n_σ orbitals on the endocyclic oxygen. *Right:* the geometry has been altered to show the best overlap of the n_p orbital with the $\sigma *$ orbital of the OR bond. This results in the COCC torsional angle closing from about staggered to ~30°

atom [48,50,52]. This description has somewhat different stereochemical consequences for n → σ* overlap (❷ *Fig. 5*). In the traditional description, the most favorable overlap from the n orbital of O1 occurs when the C–O1–C–O2 torsional angle is 60°, which matches that present for the axial OR group in a 2-substituted tetrahydropyran. The correct description leads to the most favorable overlap at a torsional angle of 90° where the intra-ring COCC torsional angle has been reduced to 30°. Considerable overlap is still present at the tetrahydropyran intra-ring torsional angle but Dubois et al. have suggested that this type of overlap explains the larger anomeric effect present in furanoses [61]. The recent conclusion that acyclic acetals $RCH(OR_1)(OR_2)$ increasingly adopt conformations with the smaller of R_1 or R_2 eclipsed with the acetal H as the two other R groups increase in size is in accord with this description of the oxygen lone pairs [62].

Evidence for the importance of the first explanation includes changes in bond lengths about the anomeric center. In the axial conformer, a lone pair on the endocyclic oxygen atom is aligned with the exocyclic C–O bond leading to orbital interaction and bond shortening and concomitant bond lengthening of the exocyclic C–O bond, the endo anomeric effect. In both conformers, if the methoxy methyl is *gauche* to the endocyclic C–O bond and *anti* to the C1–C2 bond, a lone pair will interact with the aligned endocyclic C–O bond leading to exocyclic C–O bond shortening. The preference for these conformers is termed the exo anomeric effect. This bond shortening has been clearly observed for the central C–O bonds in the favored *+gauche,+gauche* (+g,+g) or *-gauche,-gauche* (−g,−g) (**8**) conformers of dimethoxymethane, both in the gas phase [63] and recently in the solid [64]. In this conformation, there are two possible n → σ* interactions that compete. The C–O bond arrangements in this conformation correspond to those in the axial conformer of 2-methoxytetrahydropyran while those in the *gauche,anti* (g,a) conformer of dimethoxymethane (**9**) are similar to those in the equatorial conformer of 2-methoxytetrahydropyran. Where only one oxygen atom is the source of the n electrons, it is calculated using high level ab initio methods that greater bond shortening of the *a* C–O bond occurs, as in the g,a conformer [65,66]. These calculations also support the contention that n → σ* interactions are very important for the anomeric effect. Similar bond shortening is observed for the appropriate conformers of $ClCH_2OH$ [67]. Statistical analyses of bond length data from X-ray diffraction studies support the bond shortening due to n → σ* interactions [61,68,69] and are in agreement with greater bond shortening in the a,p conformer. In addition, electron withdrawing groups in 2-alkoxytetrahydropyrans [70,71] or

2-phenoxytetrahydropyrans [72,73,74], which lower the energy of the σ^* orbital and increase the interaction, were found to increase the preference for the axial conformer. A similar effect was seen in substituted 2,2-diphenyl-1,3-dioxanes where the phenyl rings bearing electron-withdrawing substituents preferred the axial orientation and vice versa [75]. The smaller magnitude of the $^1J_{C,H}$ value at anomeric centers for axial CH bonds than for equatorial CH bonds has been interpreted in terms of $n \rightarrow \sigma^*$ overlap [76,77], but recently it has been shown that the calculated dependence of the size of this value on the HCOC torsional angle is incompatible with this explanation [78]. The effects of replacing hydrogen by deuterium atoms on the positions of anomeric equilibria for the D-glucopyranoses were interpreted in terms of $n \rightarrow \sigma^*$ overlap [79].

The alternative explanation for the anomeric effect involves destabilizing dipole-dipole repulsion between the two oxygen atoms and their lone pairs [33,54] (see ❷ Fig. 6). This explanation was reexamined by Box [80,81], who suggested that the bond shortening evidence can be explained by n orbital repulsion. Persuasive support for this position has come from ab initio calculations on 2-methoxytetrahydropyran conformers by da Silva and coworkers, performed using the generalized valence bond-perfect wave function at the GVB-PP/6–31G(d,p) level [52]. This approach only uses a localized description of bonding which inherently excludes $n \rightarrow \sigma^*$ overlap. These calculations were able to fully account for the relative conformer energies and bond length shortening and lengthening previously explained by $n \rightarrow \sigma^*$ overlap. The changes in bond lengths with geometry are caused by differences in the % s character in the local orbitals and the contribution of $n \rightarrow \sigma^*$ overlap was considered to be insignificant [52]. In agreement, Perrin et al. showed how dipole-dipole repulsion could also lead to bond length alteration [82]. Wiberg and Marquez demonstrated that there was a significant solvent effect on the axial-equatorial equilibrium of 4,6-dimethyl-2-methoxytetrahydropyran and suggested that the reduction of electrostatic interaction between dipoles with increasing solvent polarity

◻ Figure 6
Approximate representation of the geometries of the n_p and n_s orbitals on oxygen atoms at the anomeric center. In the equatorial conformer, the n_s orbital on the exocyclic oxygen atom is aligned with one lobe of the n_p orbital on the endocyclic oxygen atom and one lobe of the n_p orbital on the exocyclic oxygen atom is aligned with the n_p orbital on the endocyclic oxygen atom, leading to repulsive destabilization. In the axial conformer, only one pair of orbitals is aligned, leading to less repulsion

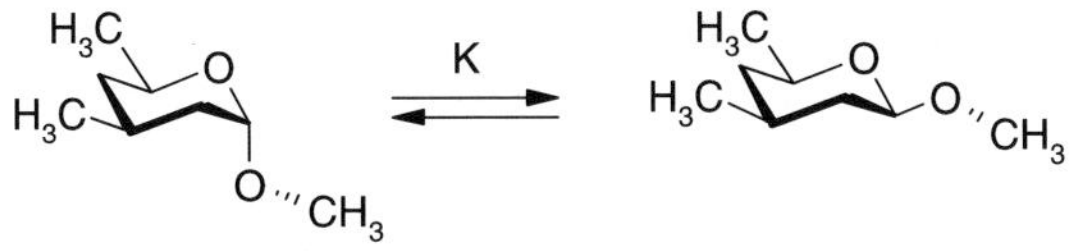

Figure 7
ΔG for the equilibrium ranged from 4.2 kJ/mol in benzene to 1.8 kJ/mol in acetonitrile [83]

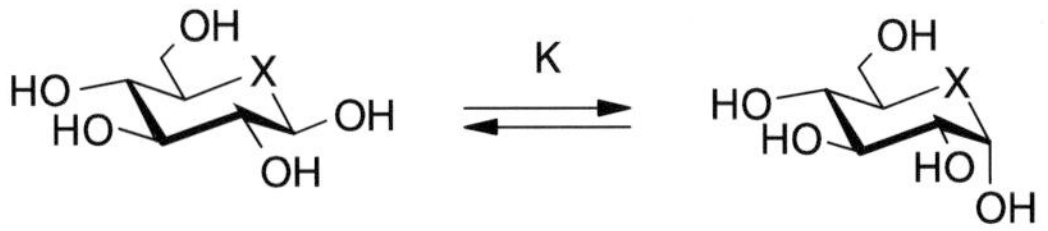

Figure 8
For X = O, K is 0.56 at 60 °C in water; for X = NH, K is 1.7 at 50 °C in water [87]

was responsible for this change [83] (❯ *Fig. 7*). Lemieux and coworkers had shown earlier that solvent effects influenced equilibria in 2-alkoxy-substituted tetrahydropyrans [71,84] but they interpreted the solvent dependency in terms of specific solvent molecule interactions. Jorgenson et al. pointed out that the solvent effect reflects the fact that the equatorial conformer has a larger dipole moment and hence is better solvated in more polar solvents [85]. Perrin et al. stated that the fact that the positions of the axial equatorial equilibria were almost identical in 2-methoxy-1,3-dimethylhexahydropyrimidine and 2-methoxy-1,3-dioxane indicated that electrostatic repulsion was dominant in determining the anomeric effect [77]. If n → σ* interactions were more important, the nitrogen donor would be expected to have a much larger axial stabilizing effect which was not observed. However, interpretation of the results of this study was complicated by the steric effects of the methyl groups on nitrogen and its conclusion was disputed by Salzner [86]. In addition, for norjirimycin, 5-amino-5-deoxy-D-glucopyranose-glucopyranose, which exists entirely in the piperidine ring form, the anomeric effect is 3.0 kJ mol^{-1} larger than for D-glucopyranose [87], consistent with n → σ* interactions (❯ *Fig. 8*). However, on balance, the calculations of da Silva's group indicate that n → σ* interactions are of minor importance in explaining the anomeric effect.

1.2.3 Conformations of Cyclopentanes and Tetrahydrofurans

The symmetric puckered conformations of cyclopentane are the C_s symmetric envelope (E) (**10**) with four carbon atoms in a plane and the C_2 symmetric twist (T) (**11**) with three carbon atoms in a plane [88]. Unlike cyclohexane, these conformations are of almost equal energy and are separated by barriers of about RT or less [89]. There are ten envelope conformations, each with one of the five carbon atoms out of the plane in one of the two directions, and ten corresponding twist conformations. The individual conformations freely exchange which atom or atoms are out of the plane, a process termed pseudorotation, and the whole sequence of conformations is called the pseudorotational itinerary (❯ *Fig. 9*).

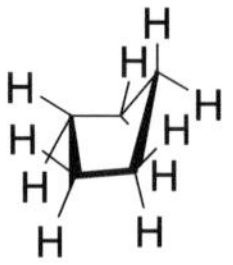

10 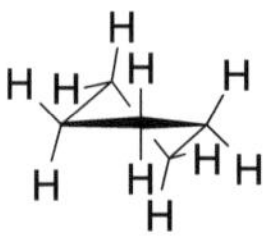   **11**

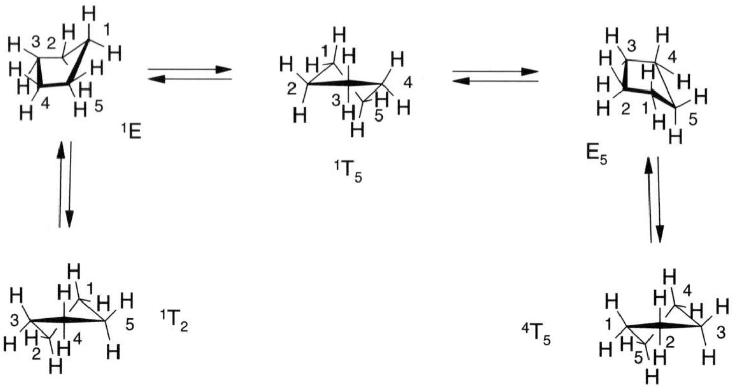

■ Figure 9
Part of the pseudorotational itinerary of cyclopentane

Substituents cause parts of the pseudorotational itinerary to become more populated to avoid eclipsing and 1,3-diaxial interactions. In the envelope conformation, the ring atom out of the plane has pseudoequatorial (**12**) and pseudoaxial positions (**13**) as do the atoms next to it that partially avoid eclipsing interactions. Substituents on the remaining two atoms are eclipsed. In the twist conformation, the atom on the C_2 axis has two identical isoclinal positions that also partially avoid eclipsing interactions (**14**). Thus, the barrier to pseudorotation for methyl-cyclopentane, which corresponds to having the methyl group in the least stable site, is larger, 14.2 kJ mol^{-1} [88].

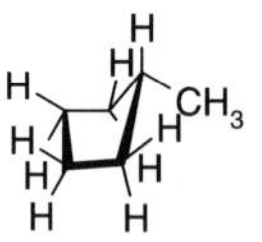

12 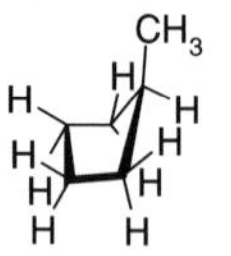   **13** 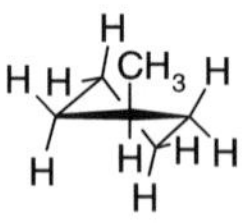   **14**

Microwave [90,91,92] and far-IR [93,94] spectra indicate that the barrier to pseudorotation in tetrahydrofuran is small (0.8 to 2.0 kJ mol^{-1}). Modeling the potential energy surface has proven to be difficult. The current theoretical view, obtained using calculations at the MP2/cc-pV5ZMP2/cc-pVTZ level, is that there are two minima, the most stable being the C_s symmetric envelope conformer (^1E) (**15**), the next the C_2 symmetric twist conformer ($^2T_3$) (**16**), 0.55 kJ mol^{-1} higher, with the saddle point on the pseudorotational itinerary being a C_1 symmetric conformation 0.78 kJ mol^{-1} above the global minimum [95]. Older methods obtain the twist form as more stable [96]. In the solid state, the twist conformer is observed in both 104 and 140 K X-ray [97] and 5 K high-resolution neutron powder [98] diffraction studies. In solid tetrahydrofuran, the pseudorotational motion is a large-amplitude ring deformation vibration with an amplitude of about 140 cm^{-1} [99]. The pseudorotational barrier increases in water by about 1.0 kJ mol^{-1} to about 2.1 ± 0.8 kJ mol^{-1} [100]. When substituents are present, their requirements become more important than the inherent tetrahydrofuran preferences. For instance, the microwave spectrum of 3-hydroxytetrahydrofuran shows that it exists in a ^2E conformation having an axial hydroxyl group hydrogen bonded to the ring oxygen (**17**), with no evidence for pseudorotation both in the gas phase [101] and in aqueous solution [102].

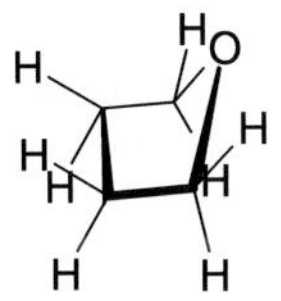

15 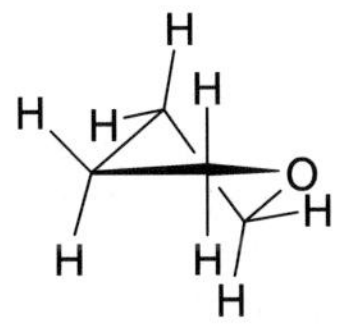   **16**

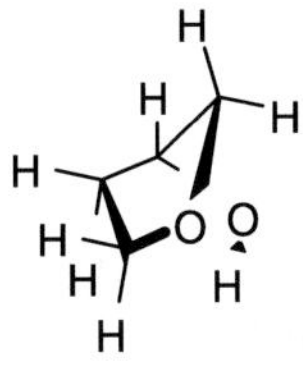

17

1.3 Conformations of Monosaccharides

1.3.1 Conformations of Acyclic Carbohydrates

In the solid state, most acyclic carbohydrate derivatives adopt a conformation having the carbon atoms in an extended, planar zig-zag arrangement, unless there are parallel, 1,3-steric interactions between oxygen atoms (Hassel–Ottar effect), written in abbreviated form as O//O interactions [103,104,105,106] (❏ *Fig. 10*). In the last few years, a number of examples have been observed where either O//O or CO interactions are present, particularly for compounds with chains longer than five carbons [107,108,109,110,111,112,113,114]. It is clear that the magnitude of the destabilizing effect associated with these interactions is less than in six-membered rings where torsional and bond angle relaxation is more difficult energetically [108,112]. *O*-Acetyl derivatives are more likely to adopt conformations with O//O or CO interactions than unsubstituted compounds [115].

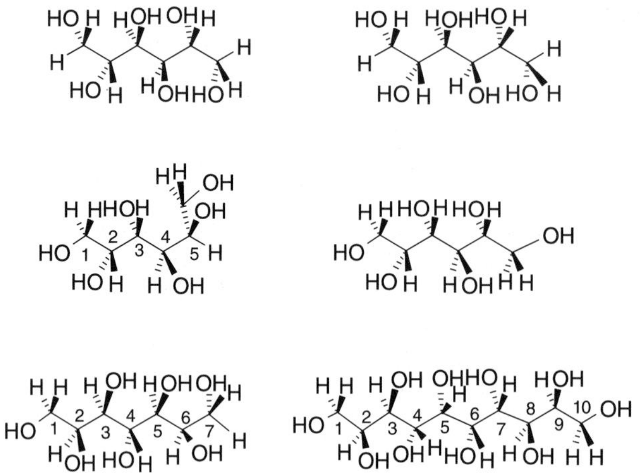

◘ Figure 10
The conformations in the solid state of top left : galactitol, planar zig-zag with extended oxygen atoms [94]; top right, D-mannitol, planar zig-zag with *gauche* oxygen atoms [95]; center left, D-altritol, the $_4G^-$ conformation adopted in the solid-state with a CO interaction [105]; center right, D-altritol, showing the planar zig-zag conformation avoided; bottom left, *meso*-D-*glycero*-L-*altro*-heptitol, with a O//O interaction between O-3 and O-5 [100]; bottom right, L-*galacto*-D-*galacto*-decitol, with a O//O interaction between O-5 and O-7 [98]

In solution, the same considerations apply [116]. The ^{1}H NMR spectra of aqueous solutions of all of the alditols from four to seven carbons have been analyzed [117,118,119] and the coupling constants have been used to determine the most populated conformers, with the assumption that all rotamers have ideal staggered torsional angles. Similar but not as precise information has been obtained from ^{13}C NMR chemical shifts [108,120,121]. It was concluded that the order of the magnitudes of repulsive interactions is the same as for six-membered rings but the actual sizes of the repulsive interactions were smaller. The chain adopts a planar, zig-zag conformation, except where O//O interactions are present, when it twists to replace an O//O by a C//H interaction, or where more than one O//O interactions are present, the chain may twist to give a *gauche* conformation that contains a C//O interaction. Nomenclature for *gauche* conformations was developed by Horton and Wander [116]; a $_2G^-$ conformation is obtained from the planar zig-zag conformation by a 120° clockwise rotation of the remote atom along the C-2–C-3 bond; a $_3G^+$ conformation is obtained from the planar zig-zag conformation by a 120° counterclockwise rotation of the remote atom along the C-3–C-4 bond. Chain preference falls in the order: planar, single twist, double twist. Chain twisting generally results in an oxygen atom extending the chain rather than a hydrogen atom. The preferred rotamer of the hydroxymethyl groups has the oxygen atom extending the chain [119] although significant amounts of all hydroxymethyl rotamers are observed. On the basis of studies of alditols in a variety of solvents [122,123], it was concluded that as solvents change from low polar to protic to very polar aprotic, the preference of C–O bonds in 2,3-butanediol units changes from *gauche* to *anti* [123].

1.3.2 Conformations of Pyranoses

Ring Conformations   The two chair conformations of pyranose sugars are named by defining a reference plane that contains four ring atoms and has the lowest numbered carbon atom out of the plane [124]. In the chair, abbreviated C, the positions of the atoms above and below the plane are given as superscripts and subscripts, respectively. To remove ambiguity, it is also necessary to stipulate that the atom written as a superscript is on the side of the reference plane from which the numbering of the remaining atoms appears clockwise (❷ *Fig. 11*). The abbreviations for boat, skew or twist-boat, and half-chair conformations are B, S, and H, respectively. Only one reference plane is possible for the boat but the skew is named so that the selected reference plane, which contains three adjacent atoms and one other, has the atom with the lowest possible number exocyclic (❷ *Fig. 12*) [124].

Pyranose derivatives adopt chair conformations unless an unusual combination of destabilizing interactions is present. Angyal developed a set of destabilizing interactions that can be used to estimate the relative stabilities of the two chair conformers in aqueous solution [125,126]. These values were determined before many of the A-values discussed above for cyclohexane and tetrahydropyran derivatives were measured and are formulated in terms of 1,3-diaxial

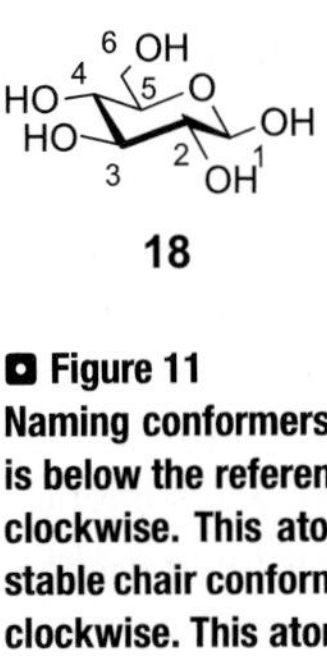

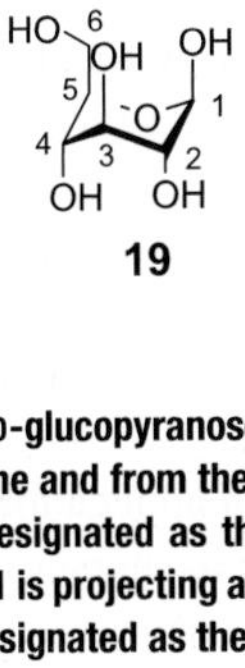

■ Figure 11
Naming conformers of β-D-glucopyranose. For 18, the more stable chair conformer, the lowest number C, C-1 is below the reference plane and from the atom above the plane, C-4, the order of the remaining atoms appears clockwise. This atom is designated as the superscript, that is, this is the $^4C_1$ conformation. For 19, the less stable chair conformer, C-1 is projecting above the reference plane and the order of the remaining atoms appears clockwise. This atom is designated as the superscript, that is, this is the $^1C_4$ conformation

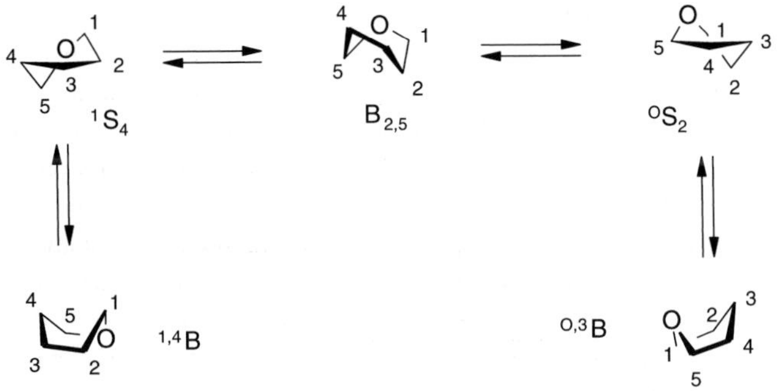

■ Figure 12
Part of the skew-boat pseudorotational itinerary for aldopyranoses to illustrate conformational nomenclature

◘ Table 1
Destabilizing effects in pyranose and cyclitol rings in aqueous solution at room temperature in kJ/mol (22 °C or 25 °C) [126]

1,3-Diaxial interactions		Gauche interactions			Anomeric effects	
C:O	10.4	C/O	1.9	for OH	if O-2 is equatorial	2.3
O:O	6.3	O/O	1.5		if O-2 is axial	4.2
C:H	3.8	C/C	3.8		if O-2 and O-3 are axial	3.6
O:H	1.9				if 2-deoxy	3.6

and *gauche* interactions and the anomeric effect (see ❷ *Table 1*). In general, these terms are remarkably successful for predicting which chair conformer is most populated in aqueous solution and also for predicting anomer populations. For D-aldopyranoses, the group on the pyranose ring with the largest A value is the hydroxymethyl group and this group causes most derivatives to adopt the $^4C_1$ conformation. In the $^1C_4$ conformation, this group is axial and has 1,3-diaxial interactions with substituents or hydrogen atoms on C-1 or C-3. Augé and David [127] noted that the value adopted by Angyal for the 1,3-diaxial interaction of a methyl and a hydrogen ($3.76\,\mathrm{kJ\,mol^{-1}}$) gave an equatorial preference for the hydroxymethyl group of $7.5\,\mathrm{kJ\,mol^{-1}}$, similar to the cyclohexane methyl A value of $7.4\,\mathrm{kJ\,mol^{-1}}$ [19,20]. However, Eliel et al. [45] found that the A value for a hydroxymethyl group at C-2 of tetrahydropyran was $12.1\,\mathrm{kJ\,mol^{-1}}$, about $4.6\,\mathrm{kJ\,mol^{-1}}$ greater than its cyclohexane value. Therefore, Augé and David [127] proposed a correction termed the "proximity" correction of $4.6\,\mathrm{kJ\,mol^{-1}}$ for pyranose conformations with axial hydroxymethyl or methyl substituents at C-5. This correction only increases the proportion of the already dominant $^4C_1$ conformation of most sugars and gave improved agreement for data from many substituted idopyranosyl and altropyranosyl derivatives.

The idopyranose anomers will be considered as examples of applications of Angyal's interaction energies. They are particularly interesting because the α-anomer is one of three α-aldohexopyranoses where both chair conformers are populated significantly (α-altropyranose and α-gulopyranose are the others) and also because the conformations of L-iduronic acid have significance for the biological properties of heparin and other glycosylaminoglycans [128]. For β-D-idopyranose, estimation of the destabilizing interactions by Angyal's method [125,126] gives: for the $^4C_1$ conformer, $2 \times 1.9\,(\mathrm{O:H}) + 6.3\,(\mathrm{O:O}) + 1.9\,(\mathrm{C/O}) + 1.5\,(\mathrm{O/O}) + 3.6$ (anomeric effect) $= 17.1\,\mathrm{kJ\,mol^{-1}}$; for the $^1C_4$ conformer, $1.9\,(\mathrm{O:H}) + 3.8\,(\mathrm{C:H}) + 10.4\,(\mathrm{C:O}) + 1.9\,(\mathrm{C/O}) + 3 \times 1.5\,(\mathrm{O/O}) = 22.5\,\mathrm{kJ\,mol^{-1}}$. Populations of conformers can be estimated from average NMR coupling constants between vicinal hydrogen atoms, $^3J_{\mathrm{H,H}}$, if the values for the individual conformers are known or can be estimated or calculated. These coupling constants are known for most aldoses and ketoses [129,130,131,132]. The ΔG value, $5.4\,\mathrm{kJ\,mol^{-1}}$ predicts $K(^1C_4/^4C_1) = 0.1$ at 298 K; the value derived from observed coupling constants and estimated values for $J_{2,3}$ and $J_{3,4}$ in pure conformers [131] is $K = 0.33$, which corresponds to ΔG of $2.7\,\mathrm{kJ\,mol^{-1}}$. The difference between the observed and calculated equilibrium constant is larger than for most pyranoses but the $^4C_1$ conformer (**20**) is correctly predicted to be more stable than the $^1C_4$ conformer (**21**).

20   **21**

The conformational situation for α-D-idopyranose is more complicated. Estimation of the destabilizing interactions by Angyal's method [125,126] gives: for the $^4C_1$ conformer (**22**), 2×1.9 (O:H) $+ 2 \times 6.3$ (O:O) $+ 1.9$ (C/O) $= 18.3$ kJ mol^{-1}; for the $^1C_4$ conformer (**23**), 2×3.8 (C:H) $+ 1.9$ (C/O) $+ 3 \times 1.5$ (O/O) $+ 2.3$ (anomeric effect) $= 16.3$ kJ mol^{-1}. Considering the uncertainties involved in the method, the difference of 2.0 kJ mol^{-1} is too small to be confident that only the $^1C_4$ conformer is populated significantly [$K(^4C_1/^1C_4) = 0.4$ at 298 K]. Synder and Serianni calculated $K = 0.25$ from the observed values of $J_{2,3}$ and $J_{3,4}$ using standard values for $J_{a,a}$ and $J_{e,e}$ of 9.5 and 2.0 Hz, respectively, which predict the $^1C_4$ conformer being more stable by 3.4 kJ mol^{-1} [131]. A complication is that the $J_{4,5}$ value observed, 5.0 Hz, is too large for an ideal $^1C_4$ conformer, where $J_{a,e}$ should be 3–4 Hz, as it also should be in the $^4C_1$ conformer. It is possible that a skew conformer, the $^OS_2$ conformer (**24**), which would have similar J values to the $^1C_4$ for H-1 to H-4, but a larger $J_{4,5}$, contributes, perhaps significantly [127,133,134]. Alternatively, the axial hydroxymethyl group in the $^1C_4$ conformer may be bent away from the ring resulting in a smaller H-4 H-5 torsional angle than normal and a bigger $J_{4,5}$ value [131]. For both anomers, the $^4C_1$ anomer was observed to be less stable than calculated using Angyal's method. On the basis of these equilibria in aqueous media, it does not appear that Angyal's values need to be corrected as suggested by Augé and David [127]. Recent ab initio calculations for α-D-idopyranose in the gas phase suggest that the conformational mixture present includes the B$_{3,O}$ conformer as well as those mentioned above [135].

22   **23**   **24**

The aldopentopyranoses are not constrained to the $^4C_1$ conformation by the presence of a hydroxymethyl group. As a result, the equilibria between the $^4C_1$ and $^1C_4$ conformers are more closely balanced, with the D-xylopyranoses present in aqueous solution mainly as $^4C_1$ conformers (**25**), the D-arabinopyranoses mainly as $^1C_4$ conformers (**26**), and the others as mixtures [136,137,138].

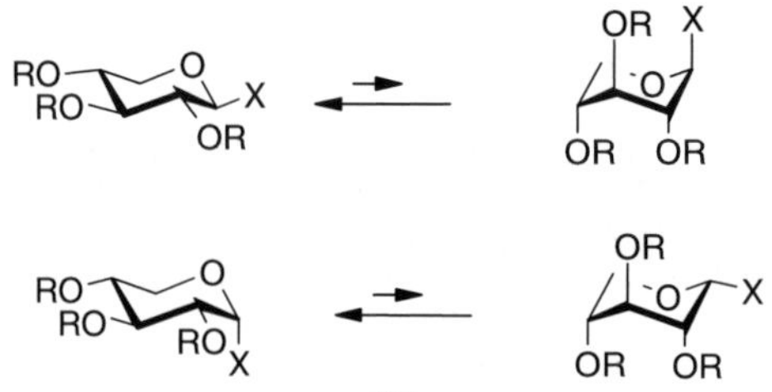

25

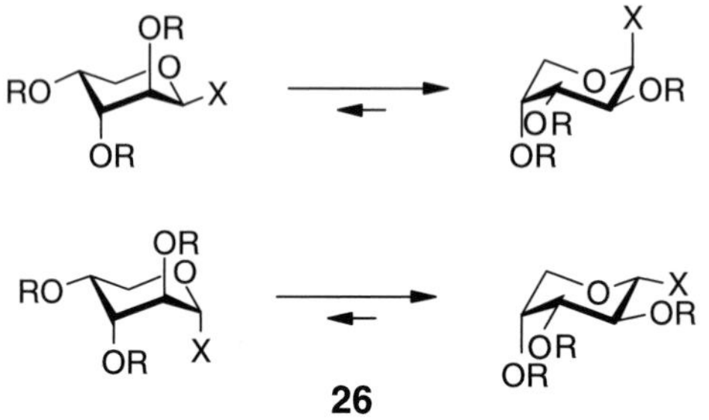

26

Comparison of the populated conformers of substituted and non-substituted aldopentopyranoses illustrates how various types of substitution influence the equilibria (see ❯ *Table 3*). It
should be noted that Angyal's interaction energies [125,126] predict the relative stabilities of
the $^4C_1$ and $^1C_4$ conformers of the unsubstituted aldopentopyranoses remarkably well, except
for those of arabinose, where the $^1C_4$ conformer is more stable than expected. It may be that
the interaction energies overestimate the effect of having an axial hydroxyl group at C-2 or
C-4, where there is a *gauche* interaction with the ring oxygen [139]. The acetylated derivatives
generally agree with these relative stabilities as well [140], once allowance is made for the
greater stabilization of the $^4C_1$ conformer of the α-anomer and the greater destabilization of
the $^4C_1$ conformer of the β-anomer by the larger anomeric effects (❯ *Table 2*) associated with
methoxy, acetoxy, and chloro groups than with the hydroxyl group. It is interesting that tri-
O-acetyl-β-D-xylopyranosyl fluoride exists entirely in a conformation with the fluoride group
axial [141]. It is likely that 1,3-synaxial repulsive interactions for two acetates are much less
than expected based on hydroxyl or methoxyl values and the same is true to a greater extent
for 1,3-diaxially related sulfates [142].

❏ **Table 2**
Other anomeric effects (kJ/mol)

Br[a]	> 13.4	Cl[a]	11.1		for OMe[c]	if O-2 is equatorial	5.6[d,e]
			12.8 if scaled[b]			if O-2 is axial	9.0[e]
OAc[f]	5.9						

[a]From tetrahydropyran equilibria [143]. [b]By the method of Franck [56]. [c]In methanol [2].
[d][144]. [e][145]. [f][146]

□ **Table 3**
Percentages of selected D-pentopyranose derivatives that exist as $^4C_1$ conformers

Configuration	Predicted[a]	Unsubstituted in water-d_2	Methyl 2,3,4-tri-O-acetyl[b]	1,2,3,4-Tetra-O-acetyl[b]	2,3,4-Tri-O-acetyl chloride[c]
α-ribo	54	41[d]	65[e]	77[e]	
β-ribo	73	74[f]	39[e]	43[e]	6[e]
α-arabino	13	~0[d]	17[c,e]	21[c,e]	
β-arabino	30	~2[d]	3[e]	4[e]	2[e]
α-xylo	94	~100[d]	>98[e]	>98[e]	>98[e]
β-xylo	98	~100[d]	81[e]	72[e]	21[e]
α-lyxo	72	56[d]	83[e]	71[e]	91[e]
β-lyxo	85	86[d]	58[e]	39[e]	

[140] [a]Calculated from conformer energy differences using Angyal's method [125]. [b]In acetone-d_6 unless otherwise specified. [c]In chloroform-d. [d]Calculated from data in [130] using $J_{1a,2a} = 7.8$ Hz, $J_{1e,2e} = 1.2$ Hz. If H-1 and H-2 are not *trans*, the values of $^3J_{H,H}$ for all pairs of *trans* non-anomeric vicinal protons were averaged, with $^3J_{a,a} = 9.5$ Hz, $^3J_{e,e} = 1.8$ Hz for $J_{2,3}$ and $J_{3,4}$, but $^3J_{a,a} = 10.3$ Hz, $^3J_{e,e} = 1.8$ Hz for $J_{4,5}$. [e][140]. [f]Calculated from data in [138]

The origin of the anomeric effect was discussed above in connection with tetrahydropyran conformations. The term "reverse anomeric effect" was coined to describe the tendency of substituents with atoms with formal positive charges attached to C-1, initially pyridinium ions, to adopt equatorial rather than axial orientations [147]. This effect, if real, has great significance for reactions at the anomeric center because many proceed via intermediates that involve increase in positive charge density on atoms attached to the anomeric center. Amongst the pieces of evidence in support of this effect were the surprising observations that in solution the N-(2,3,4,6-tetra-O-acetyl-α-D-glucopyranosyl)-4-methyl-pyridinium ion adopts a boat conformation (**27**) [148] and the N-(2,3,4-tri-O-acetyl-α-D-xylopyranosyl)pyridinium ion adopts a $^1C_4$ conformation (**28**) [149]. However, the steric effects of these groups are large, similar to phenyl, which has an A value of 9.2 kJ mol^{-1}, magnified to 14.6 kJ mol^{-1} at position-2 of a tetrahydropyran derivative [51]. The strongest evidence for this effect was the observation that the position of the equilibrium between the $^4C_1$ and $^1C_4$ conformers of N-(2,3,4-tri-O-acetyl-α-D-xylopyranosyl)imidazole moves toward the latter on protonation in chloroform-d as observed through the considerable changes in the average vicinal coupling constants [149,150]. Perrin and coworkers have reinvestigated this phenomenon [151,152,153] by studying the titration of glycosylimidazole anomers and other compounds and have concluded that the reverse anomeric effect does not exist. The most significant evidence is that glycosyl imidazole groups are more basic when axial than equatorial, contrary to prediction based on the reverse anomeric effect. Other groups have now come to the same conclusion [154,155,156]. Ammonium ions do not show this effect and it now seems clear that the observations that led to the concept of the reverse anomeric effect were due to unexpectedly large steric effects and perhaps to particular electrostatic interactions with imidazolium ions [157].

27

28

Biologically important monosaccharides contain a number of functional groups different to those considered thus far. Aldohexopyranuronic acids favor the $^4C_1$ conformation to about the same extent as do aldopyranoses [1]. Only α-L-iduronic and α-alturonic acids contain significant amounts of other conformations and the former is particularly complicated because the ^2S_O conformation is also populated [158]. Interestingly, the 2-sulfate of methyl α-L-iduronate exists predominantly in the $^1C_4$ conformation [159]. Methyl 2-acetamido-2-deoxy-α-L-altropyranuronic acid exists as a 65:35 $^4C_1$:$^1C_4$ mixture (**29**). The sodium salt and the methyl ester are present mainly as the $^4C_1$ conformer and entirely as the $^1C_4$ conformers, respectively, suggesting that charge on C-5 stabilizes the $^4C_1$ conformer. Interestingly, when this acid is linked at O-1 and O-4 as part of the O-specific polysaccharide of *S. sonnei*, it is only present in the $^4C_1$ conformation [160].

29

N-Acetyl-D-neuraminic acid (5-acetamido-3,5-dideoxy-D-*glycero*-D-*galacto*-2-nonulosonic acid, NANA) is a critical part of cell surface oligosaccharides. Under different conditions, it crystallizes as a dihydrate and in the anhydrous form. In both crystals, NANA is in the β-pyranose form with the ring in the $^2C_5$ conformation [161,162] that has the side chain and the carboxyl group equatorial and the glycosyl hydroxyl axial (**30**). In aqueous solution, the β-pyranose isomer is accompanied by a small amount (about 8%) of the α-pyranose form [163,164,165]. The $^2C_5$ conformation is maintained in solution for α- and β-pyranose forms, for their alkyl esters and for their glycosides [165,166,167,168]. The conformation of the side chain will be discussed below.

30

3-Deoxy-D-*manno*-2-octulosonic acid (KDO) is present in the inner core region of gram negative bacterial lipopolysaccharides [169,170] and in a green algae [171,172] as the α-pyranoside. In most bacterial exopolysaccharides, it is in the β-pyranoside form [173]. In X-ray structures, the pyranose rings of ammonium 3-deoxy-α-D-*manno*-2-octulopyranosidonate [174,175], of methyl (methyl 4,5,7,8-tetra-O-acetyl-3-deoxy-α-D-*manno*-2-octulopyranosid)onate [176], and of an α-(2→4)-linked disaccharide [177] all adopt the $^5C_2$

conformers that have the side chain and the carboxyl group equatorial (**31**). Solutions of free KDO show a much more complex tautomeric mixture than NANA, containing 60–65% α-pyranose, 2–11% β-pyranose, 20–25% α-furanose, and 8–9% β-furanose [178,179]. The ammonium salts of the α- and β-pyranose methyl glycosides are also present in solution as $^{5}C_{2}$ conformers but these compounds differ in their C-7-C-8 rotameric populations [174].

31

Non-chair conformations are often populated for anhydro sugars. Depending on the type of substitution, these sugars can exist entirely in boat conformations as in 2,6-anhydro [180] or 3,6-anhydro derivatives or to a small extent as for 1,6-anhydro-β-D-glucopyranose (❷ *Fig. 13*). 1,6-Anhydro-β-D-hexopyranoses are in equilibrium with the D-hexopyranoses and water when the latter compounds are heated in aqueous acid [181]. On formation of a 1,6-anhydro ring, the normal $^{4}C_{1}$ conformer is forced into a $^{1}C_{4}$ conformation. The amount of the 1,6-anhydro-β-D-hexopyranose present at equilibrium at 100 °C in aqueous acid can be predicted quite successfully from Angyal's interaction energies with a value of $\Delta G°$ for anhydro ring formation of -11.7 kJ mol^{-1} [181]. The $^{3}J_{H2,H3}$ and $^{3}J_{H3,H4}$ values observed for most 1,6-anhydro-β-D-glucopyranose derivatives are small, 1–2 Hz, consistent with a $^{1}C_{4}$ conformation and all of the derivatives that have been studied by X-ray diffraction adopt this conformer [182,183]. It was therefore surprising that the $^{3}J_{H2,H3}$ and $^{3}J_{H3,H4}$ values for 3-amino-3-deoxy-1,6-anhydro-β-D-glucopyranose were both 5.5 Hz in dimethyl sulfoxide-d_{6} [184], even though this compound adopted the $^{1}C_{4}$ conformation in the solid state [185], and $^{3}J_{H2,H3}$ and $^{3}J_{H3,H4}$ values for 3-amino-2-O-benzyl-3-deoxy-1,6-anhydro-β-D-glucopyranose were both < 2.0 Hz in chloroform-d [183]. It was concluded that the all axial substituent orientations in 1,6-anhydro-β-D-glucopyranose derivatives bring the energy of the $^{1}C_{4}$ conformation (**32**) close to a boat conformation, the $B_{O,3}$ (**33**). The more polar $B_{O,3}$ conformer is favored by more polar solvents and even the parent compound is present in this conformer to an extent of about 20% in water or dimethyl sulfoxide [183]. Two large or very polar groups on C-2 and C-4 also move the equilibrium toward the boat [186,187,188,189] as can unusual hydrogen bonding situations [190].

32   33

Exocyclic Groups   The conformations adopted by the exocyclic groups will be considered in three sections: the hydroxymethyl group and longer side chains, the hydroxyl groups, and the anomeric group.

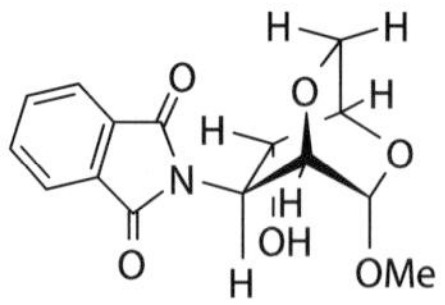

Figure 13
A boat conformation of a 2,6-anhydro derivative recently observed [165]

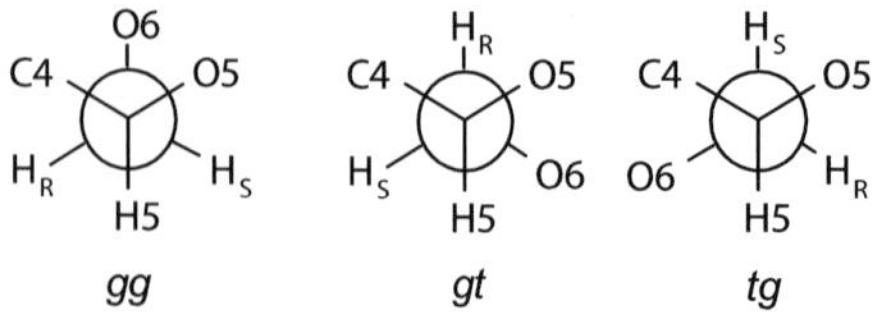

gg   *gt*   *tg*

Figure 14
Nomenclature for hydroxymethyl rotamers in a D-aldopyranose

The conformations of the hydroxymethyl group are normally discussed in terms of the three staggered rotameric conformations (❷ *Fig. 14*), termed *gg*, *gt*, and *tg*. The two letters indicate the orientation (*gauche* or *trans*) of O-6 with respect to the ring oxygen, then with respect to C-4. This topic was reviewed by Bock and Duus in 1994 [191]. The *tg* orientation has recently been shown to destabilize anomeric oxacarbenium ions significantly more than the other two orientations giving this topic significance for anomeric reactivity [192,193]. The preferences have been determined mainly from X-ray data [194,195], from NMR studies using average $^3J_{H,H}$ values [191], from $^2J_{H,H}$, $^2J_{C,H}$, $^3J_{C,H}$, and $^4J_{C,H}$ values [196,197,198], from chiroptical methods [199,200], and recently by resonant two-photon ionization and resonant ion-dip infrared spectroscopy [201,202]. The most important factor in determining the relative populations of the three rotamers is the orientation of the C-4 substituent with respect to the hydroxymethyl group. For the D-pyranoses, sugars with the *trans* orientation (*gluco-* or *manno-* configurations) have approximate populations *gg:gt:tg* of 6:4:0 while those with the *cis* orientation (*galacto*-configuration) have *gg:gt:tg* of 2:6:2. The configuration at C-1 [191,203], substitution [204,205], and solvent effects [204,205,206,207,208,209] also influence rotamer populations. For instance, the rotamer populations are *gg:gt:tg* 40:53:7 for methyl α-D-glucopyranoside but 31:61:8 for methyl β-D-glucopyranoside [197]. In aqueous solutions for glucopyranosides, hydrogen bonding to O-4 in the *tg* rotamer is weakened, leading to the observed predominance of the *gg* and *gt* rotamers [207]. Thus, the major effects influencing these populations are O//O interactions with the substituents on O-4, the *gauche* effect, hydrogen bonding and solvation [191]. Interestingly, the ratio *gg:gt:tg* only changes to 45:54:1 when the 4-hydroxy group is removed from methyl α-D-glucopyranoside (or galactopyranoside) or to 31:65:4 when removed from the β-anomer [191]. The *gauche* effect obviously has considerable importance in determining the position of these equilibria. Considerable advances in computational methods for the incorporation of solvation have been achieved recently that now allow the position of this equilibrium to be predicted accurately [207,210,211,212,213,214,215,216] and the pathway for rotamer interconversion to be studied [216].

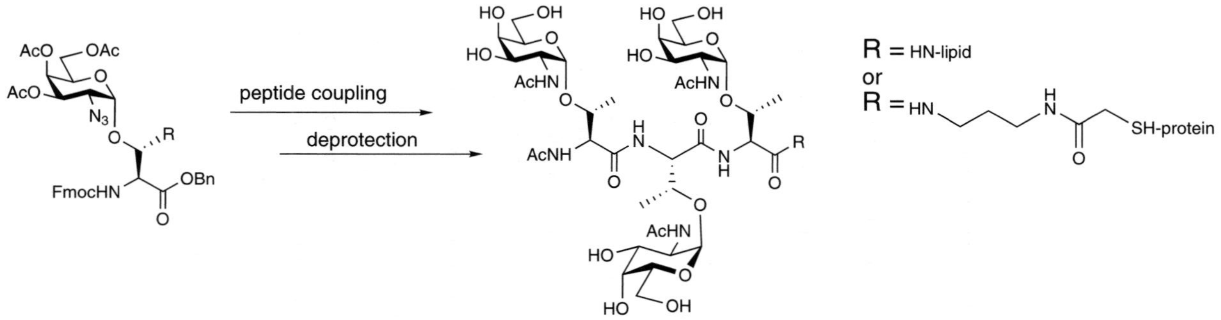

⬛ **Figure 15**
Possible side-chain conformations of *N*-acetyl-D-neuraminic acid

The conformations of the side-chains of sialic acid and KDO can be predicted using the principles outlined above. For *N*-acetyl-D-neuraminic acid (❷ *Fig. 15*), the conformation of the side chain can be termed *gg*, *gt*, or *tg*, using the orientations of C-8 with respect to O-6 and C-5 to define the conformers, as for hydroxymethyl conformers. The *gt* conformer has three minor O//H interactions, an N//H interaction, two C/O and an O/O interaction, the *gg* conformer has C//H and the major C//O interaction plus two C/O and a C/C interaction, while the *tg* conformer has major C//N and C//O interactions plus C/C and O/O interactions. Not surprisingly, the *gt* conformer is preferred as shown by small $J_{6,7}$ values both for the parent compound (0.8 Hz [166]) and for α-(268) linked oligomers (0.5 to 1.0 Hz [217]). The planar extended conformation about the C-7 – C-8 bond avoids the O//O and C//O interactions present in the other rotamers and is strongly preferred ($J_{7,8}$ 9.2 Hz) for the parent compound [166,218]. The $J_{8,9}$ values are 6.2 and 2.5 Hz, consistent with a mixture of the OH extended conformer (❷ *Fig. 15* bottom center) and the *gauche* conformer which avoids an O//O interaction (bottom right). The NMR relaxation times for C-7 and C-8 are similar to those of the secondary ring suggesting that there is only one populated conformer for the C-7 C-8 portion of the side chain [167,168].

However, in the α-(268) linked sialic acid oligomers, the OH-8 is replaced by OR-8, where R is the bulky quaternary center C-2' of the next sialic acid. Not surprisingly, in the α-(268) linked sialic acid trisaccharide and in the α-(268) linked sialic acid polymer, colominic acid, mixtures of rotamers about the C-7–C-8 bond are adopted [219,220] that have been interpreted for the polymer as giving a helix [221] or a short-lived helix [219], consistent with this polymer being a conformational epitope for serogroup B of *Neisseria meningitidis*. More recent detailed NMR relaxation studies have concluded that the polymer is a random coil with no helix present for any extended period and that the polymer has considerable flexibility about the exocyclic torsional angles [222].

The axial hydroxyl group at C-5 and the *R*-configuration at C-7 of 3-deoxy-α-D-*manno*-2-octulopyranosidonic acid result in the *gt* conformer about the C-6–C-7 bond being strongly preferred to avoid C//O and O//O interactions [174] (❷ *Fig. 16*). Interestingly, the populations of the hydroxymethyl group rotamers change when the anomeric configuration of the methyl glycoside of the ammonium salt is converted from α to β. The *a* conformer (defined by the relationship of C-6 and O-8) should be more stable; it has only a O/O interaction, while the *g*– conformer has an O//O interaction and the *g*+ conformer has O/O and O//O interactions. The *a* conformer fits the *J* values for the α anomer [174] and is present in the crystal structure [174,175]. The β anomer has *J* values closer to that expected for the *g*+ conformer; it seems reasonable that a postulated [223] hydrogen bond from the axial carboxylate anion stabilizes this rotamer (**34**). However, the peracetate of the α-anomer methyl ester also adopts this conformer, both in the crystal and in solution [176], as does the inner KDO in an α(268) linked dimer [177].

34

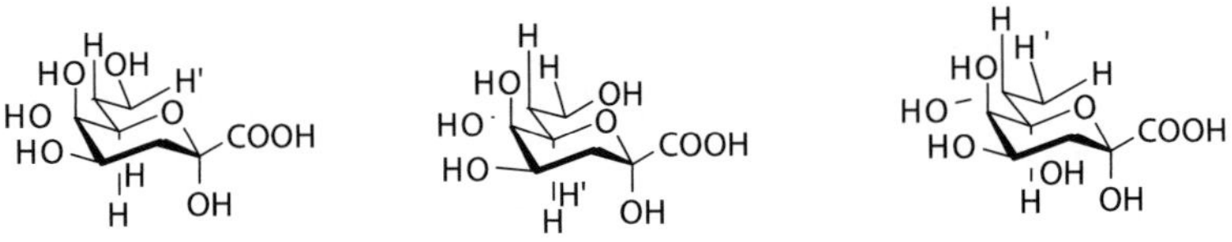

a   g-   g+

□ Figure 16
Possible side-chain conformations of 3-deoxy-α-D-*manno*-2-octulopyranosidonic acid

When considering the orientations of the hydroxyl groups in carbohydrates, it should be noted that the rotational barrier for methanol, $4.48\,\text{kJ mol}^{-1}$ [224] is about one-third the value in ethane or dimethyl ether [225]. Thus, preferences for staggered conformers in HCOH units are considerably smaller than for hydroxymethyl groups. Information about the orientations of OH groups has come from theoretical studies, from ^{1}H NMR studies in dimethyl sulfoxide, where exchange is slow, in cooled water or water/acetone mixtures, and recently from low temperature gas phase IR spectra. In the gas phase, calculations indicate that the secondary hydroxyl groups are all oriented either clockwise or counterclockwise because this sets up semicircular intramolecular hydrogen bonding networks [210,226,227,228].

A recently developed method, resonance ion-dip infrared spectroscopy, has provided experimental support for this conclusion about the orientations of hydroxyl groups of carbohydrates in the gas phase [229,230]. In this technique, a sample heated in a controlled manner in an oven to between 100 and 230° under an Ar pressure of 4 to 5 bar is allowed to escape through a nozzle. The expansion causes the jet to cool rapidly to 5 to 10 K. IR and UV spectra are recorded using pulsed tunable lasers and the spectra of individual conformers can be selected by UV hole-burning using a high power laser tuned to a band from that conformer and identified by comparison with spectra calculated using DFT theory. The compound being studied must contain a chromaphore and phenyl glycosides have been used. Surprisingly, only one to three conformers are observed for monosaccharides and disaccharides with clockwise or counterclockwise circular hydrogen bonding networks [231,232,233]. Monohydrates can be studied adding water to the Ar atmosphere and identified by mass spectrometry. It was found that molecules of water insert into the H-bonding circuits where the hydrogen bonds are weakest and the hydrated conformers can be the same as the unhydrated ones, minor unhydrated structures or structures that had negligible populations when unhydrated [233] (see ❷ *Fig. 17* for examples). Recently, this technique has been combined with resonant Raman optical activity (ROA) [234] spectra of aqueous solutions at room temperature to suggest that in most cases, the same conformations are populated under these conditions as in the hydrated Ar jet at 5 to 10 K [235], although benzyl β-D-lactoside was found to change to the conformation found for lactose by NMR spectroscopy [236].

In dimethyl sulfoxide, exchange of hydroxyl protons is slow enough that hydroxyl protons are observed separately coupled to adjacent CH protons in the ^{1}H NMR spectra of carbohydrates [237,238,239]. Hydroxyl protons involved in intramolecular hydrogen bonds are shielded in comparison to those involved in intermolecular hydrogen bonds to dimethyl sulfoxide [240,241]. The coupling constants can be used in conjunction with Karplus-type relationships [242,243] to identify the orientation of the OH groups. The J values observed for

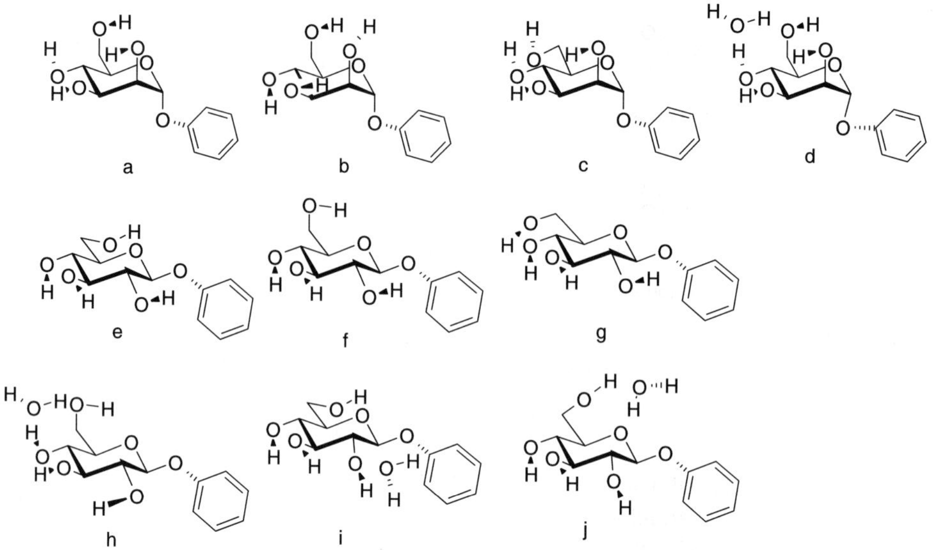

1

□ **Figure 17**

Hydroxy group conformers observed in the gas phase at 5–10 K by ion-dip IR spectroscopy: a–c, phenyl α-D-mannopyranoside, with stability decreasing from a to c; d, the only conformer observed for phenyl α-D-mannopyranoside hydrate; e to f, phenyl β-D-glucopyranoside, with stability decreasing from e to f; g to i, monohydrates of phenyl β-D-glucopyranoside, with stability decreasing from g to i. Note that the most stable conformer, g, was not populated for the anhydrous compound and was calculated to be 10 kJ mol^{-1} less stable than e

fully solvated equatorial hydroxyls are consistent with equal occupancy of all three rotamers (4.5–5.5 Hz) [240,241]. Fully solvated axial hydroxyl groups have slightly smaller J values (4.2–4.4 Hz), consistent with lower occupancy of the rotamer with the hydrogen atom on the interior of the ring. Values outside these ranges indicate that the OH group is an intramolecular H-bond donor [240]. Partial labeling with deuterium can be used to identify which hydroxyl groups are involved in hydrogen bonds by noting which hydroxyl signals show splitting due to the isotopic effect of the hydrogen-bonded hydrogen isotopes (SIMPLE NMR) [244]. The signs of the shift changes provide structural information [245]. Alternatively, the temperature dependence ($\Delta\delta/\Delta T$) of OH chemical shifts can be used; fully solvated OH groups show large $\Delta\delta/\Delta T$ while OH groups that are intramolecularly H-bonded show small $\Delta\delta/\Delta T$ [246]. All the hydroxyl groups in α-D-glucopyranose are involved in weak intramolecular hydrogen bonding. However, since this does not affect the magnitudes of their coupling constants, individual hydroxyl groups must be hydrogen bonded only a small proportion of the time. *Syn*-1,3-diaxial hydroxyl groups form stronger intramolecular hydrogen bonds that do affect J values [238].

In the more polar solvent water, there should be less tendency for ordering through intramolecular hydrogen bonding. Thus, hydroxyl orientations are mostly randomly distributed in water for monosaccharides, as indicated by low temperature NMR experiments [247,248]. As in DMSO solutions, equatorial hydroxyl groups flanked by axial oxygen atoms have slightly larger coupling constants [247]. More recently, information about the orientations of OH

eq-exo-syn   eq-non-exo   eq-exo-anti

ax-exo-syn   ax-non-exo   ax-exo-anti

☐ **Figure 18**
Representation of the three basic conformations about the anomeric CO bond: top, for β-D-pyranoses; bottom, for α-D-pyranoses

groups has been obtained in a wider range of NMR data. Chemical shifts are particular informative [249,250,251], but information can also be obtained from temperature coefficients of chemical shifts, rates of exchange with water, and nOe and NOESY experiments [252,253]. Some strong hydrogen bonds that are present in DMSO are not present in aqueous solutions [254,255] and sometimes different hydrogen bonds form [256]. In other situations, the same hydrogen bonds are present in water as in DMSO [257,258]. A variety of types of evidence has shown that certain intraresidue hydrogen bonds persist in aqueous solutions of oligosaccharides [252,259,260,261,262].

The orientations about the anomeric CO bond can be described as shown in ❷ *Fig. 18* [261]. This nomenclature was chosen over the traditional *ag*$^+$ nomenclature because the latter changes when the ring chair is inverted. The *exo-syn* and *exo-anti* conformations for both α and β anomers are favored by the exo-anomeric effect [49,52,71,263], which minimizes lone pair repulsions. The *exo-anti* conformations are disfavored by the two *gauche* interactions with O-5 and C-2. As a result, the *exo-syn* conformations are the only conformations populated in crystal structures of carbohydrates [69,264]. Theoretical calculations support the significance of the exo-anomeric effect [52,265,266]. Comparison of the conformations of C-glycosides, such as C-lactose and related compounds, which cannot have an exo-anomeric effect, with those of their *O*-containing analogs, decisively demonstrate the significance of the exo-anomeric effect [6,267,268].

It should also be noted that the non-exo conformations are disfavored by equatorial substituents at C-2, due to the O//O interaction, thus sterically increasing the preference for the eq-*exo-syn* conformation [269] (❷ *Fig. 19*). However, even without the exo-anomeric effect and without

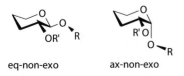

eq-non-exo   ax-non-exo

☐ **Figure 19**
Steric destabilization of non-exo conformations by equatorial substituents on C-2

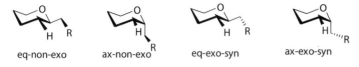

eq-non-exo   ax-non-exo   eq-exo-syn   ax-exo-syn

◻ **Figure 20**
Steric destabilization of non-exo conformations by H-2

substituents at O-2, there is a substantial preference for the *exo-syn* conformations as observed for 2-deoxy-C-glycosides [270] and calculated for 2-ethyltetrahydropyran [269]. The preference arises because the CH_2R group avoids the RH interaction with the hydrogen on C-2 by being *syn* to the oxygen atom (❂ *Fig. 20*).

1.3.3 Conformations of Furanoses

Because pseudorotation is so facile for cyclopentane and tetrahydrofuran derivatives and because substituent conformational preferences are not large, furanose derivatives are normally present as mixtures of conformations dominated by the interactions between substituents. These can be described as mixtures of ideal twist and envelope conformations but this description is often inadequate for intermediate conformations. The alternative description in terms of the pseudorotational itinerary is more precise [12,271,272,273] but less easy to visualize. Two different formalisms are used. The Altona–Sundaralingam (AS) system [272] described here is related to the more general Cremer–Pople (CP) system [12] by subtracting 90° from the CP phase angle. The CP puckering amplitude can be converted to the AS amplitude by dividing by 100 [7]. In the AS system, the infinite number of conformations on the pseudorotational itinerary are described in terms of the maximum torsion angle, θ_m, and the pseudorotation phase angle P. The pseudorotation phase angle P is calculated from the endocyclic torsional angles, $\theta_0, \theta_1, \theta_2, \theta_3$, and θ_4 according to Eq. (1) [272,273,274]:

$$\tan P = \frac{[(\theta_4 + \theta_1) - (\theta_3 + \theta_0)]}{[2\,\theta_2(\sin 36° + \sin 72°)]}. \tag{1}$$

The phase angle P is defined to be 0° when θ_2 has a value that is maximally positive, corresponding to the conformation $^3T_2$ (❂ *Fig. 21*) and returns to the same point at $P = 360°$. From the phase angle P, the five torsion angles are related by:

$$\theta_j = \theta_m \cos(P + j\delta) \tag{2}$$

where $j = 0$ to 4 and $\delta = 720°/5 = 144°$. The maximum torsion angle, θ_m, is derived by setting $j = 0$:

$$\theta_m = \frac{\theta_0}{\cos P}. \tag{3}$$

In the pseudorotation cycle (❂ *Fig. 21*), a change of P by 180° reverses the signs of all torsion angles. At every phase angle P, the sum of the torsion angles is 0°. Envelope and twist conformations alternate every 18° and T conformations are found at even multiples of 18°. The section of the pseudorotation cycle with phase angles of $0 \pm 90°$ is referred to as the

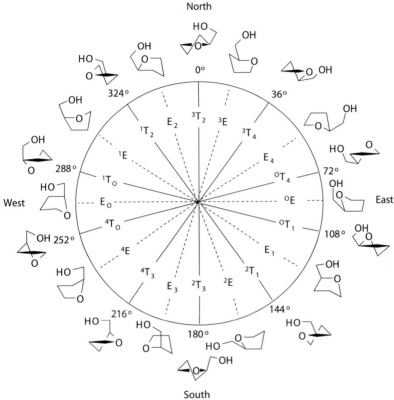

North

West

East

South

☐ Figure 21
Pseudorotational itinerary for a D-aldopentofuranose

northern half or N while that with phase angles of $180 \pm 90°$ (the lower half of the pseudoro-tational itinerary) is referred to as the southern half or S. For nucleotides and nucleosides, the preferred conformations lie in part of the N section with P angles of 0 to 36° and in the S section with phase angles of 144 to 190° [272,275,276,277,278,279,280]. In the N section, these correspond to the three conformations with C-3 above the plane of the ring (C3-endo) while the favored conformations in the S section have C-2 above the plane of the ring (C2-endo). Interconversion of N and S conformers is rapid, on the same time-scale as molecular reorientation [281]. Alteration of substitution and regiochemistry markedly alters the populat-ed conformations [276,280,282,283]. Conformational studies on nucleosides and nucleotides have shown that the N Ω S equilibrium is biased towards S in 2-deoxyribofuranose derivatives but is approximately 1:1 for ribofuranoses [273,284].

Lowary, Haddad and coworkers studied methyl 3-O-methyl-α-D-arabinofuranoside computa-tionally at the B3LYP/6–31G* level unsolvated and in aqueous solution using the B3LYP/6–31+G**SM5.42/BPW91/631G* method and obtained conformer distributions close to those derived from $^3J_{H,H}$ values, mainly a mixture of the $^oT_4$ or E_4 conformers (N) and the $^2T_1$ or ^2E_o conformers (S) [285]. The same group studied all of the methyl aldofuranosides using the same methods except that only idealized envelope conformers were considered [286]. In most cases, the change in the regions of the pseudorotational itineraries that were populated on moving to the solution were calculated to be small. Agreement with results from detailed analyses of $^3J_{H,H}$ and $^3J_{COCH}$ data [287] was good as was agreement with the conformations present in the crystal [288]. For individual compounds, the conformational situations varied widely; methyl α-D-xylofuranoside appears to exist as close to a single E_1 conformer in solution with most conformers $> 12 \, \text{kJ} \, \text{mol}^{-1}$ higher in energy while the conformers of methyl β-D-xylofuranoside lie on a fairly flat potential energy surface. β-Ribofuranose has also been studied using an ab initio molecular dynamics method that concluded that the ring conformer and the hydroxymethyl group rotamer populated were linked as well as emphasizing the importance of solvation [289,290]. The former observation had been made earlier [291,292]. β-D-Galactofuranose derivatives exist preferentially in the southwestern part of the pseudorotational cycle, in the ^4E, ^4T_O, E_O, and ^1T_O range, as shown by analysis of coupling constants and molecular mechanics calculations using the PIMM91 force field [229]. Methyl α-D-mannofuranoside is present in an E_3 conformation in the crystal [293].

Anomeric effects were first demonstrated in furanosides by means of X-ray diffraction [61,68, 294]. The importance of the anomeric effect for furanosides in solution was shown by studies of fused ring systems [295,296], from studies of nucleosides [297,298], and by comparing C-, N-, and O-furanosyl glycosides in the solid state and solution [299]. Interestingly, the magnitude of the anomeric effect for nucleosides is pH dependent with the largest effect being observed under the most acidic conditions [300].

1.3.4 Conformations of Septanoses

There have been recent indications that septanose sugars can have biological activity [301,302, 303]. Stoddard outlined clearly how the conformational properties of cycloheptanes [304,305] can be applied to oxepanes and septanoses [2] and recent studies [306,307] have used the nomenclature scheme introduced by Stoddart that is analogous to those used for conformations of furanoses and pyranoses [124]. For cycloheptane, the C_2-symmetric twist-chair (TC) (35) is the most stable conformation as indicated by molecular mechanics calculations [304,308,309] and high level ab initio studies [310,311], and confirmed by spectroscopic information [310,312,313], electron diffraction [314], and inelastic neutron scattering [315]. The TC conformer pseudorotates through a C_s-symmetric barrier, the chair (C) conformation (36), calculated to be about 4 to $6 \, \text{kJ} \, \text{mol}^{-1}$ above the TC [310,311] at room temperature. Another family of pseudorotating conformations includes the boat (B) and twist-boat (TB) where the B is minimum, about $12 \, \text{kJ} \, \text{mol}^{-1}$ above the TC [311]. Virtually all conformational information on septanoses has been interpreted in terms of TC conformers. In the TC conformation, the C_2 axis runs through one carbon atom and the center of the opposing bond giving four different types of carbon positions, numbered 1 to 4 starting with the atom on the axis (35). Positions 2–4 have axial and equatorial substituent positions. Hendricksen calculated with an early forcefield that it is much more unfavorable to have methyl groups on axial positions at positions 2 and 3 than in cyclohexane [305]. Position 1 has two identical substituent positions, termed isoclinal that were calculated to be similar to equatorial positions [305].

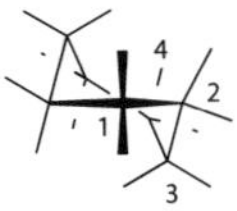

35 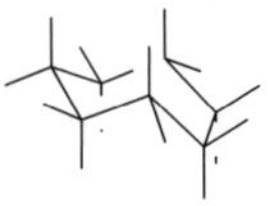   36

For oxepane, there are four different types of TC conformations, with the O atom on the axis of symmetry or at positions 2, 3, or 4 with respect to a pseudoaxis of symmetry, termed TC_D, TC_C, TC_B, and TC_A, respectively [313] (❂ *Fig. 22*). Molecular mechanics calculations indicate that the C_2-symmetric conformer is less stable [313,316] and this is supported by analysis of IR and Raman spectra [312]. The IR and Raman spectra were interpreted in terms of only two of the remaining TC conformations being populated, TC_C and TC_B [313]. The full TC/C pseudorotational itinerary includes 14 TC conformations, arising from pseudoaxes of symmetry going through each heavy atom and with right or left-handed twists of the remaining heavy atoms. In the TC conformation, the atom on the C_2 axis and the two flanking atoms lie in a plane with two adjacent remaining atoms above the plane and two below. The numbers of these out-of-plane atoms are used for naming TC conformations of septanoses [2], e. g., the most stable conformer of methyl α-D-*glycero*-D-idoseptanoside is $^{3,4}TC_{5,6}$ (**37**).

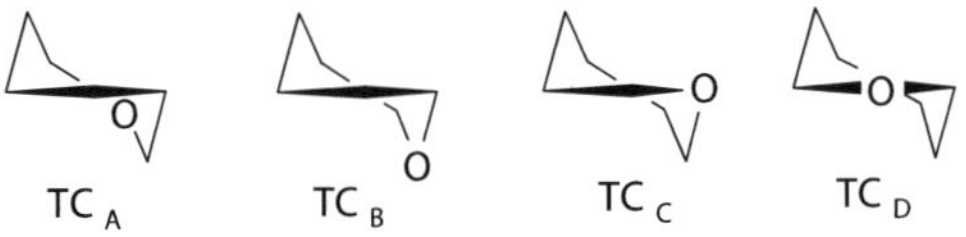

TC$_A$ TC$_B$ TC$_C$ TC$_D$

❏ **Figure 22**
The four possible TC conformations of oxepane with the nomenclature of Bocian and Strauss [313]

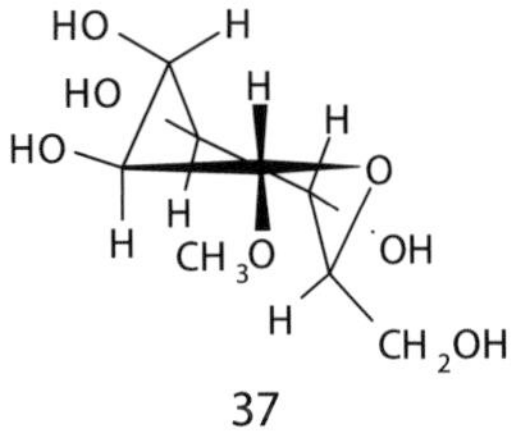

37

The conformational principles developed earlier appear to control septanose conformations. In both solution and in crystals, TC conformations are preferred that have hydroxymethyl, alkoxy groups, and hydroxyl groups in axial or isoclinal positions [307,317,318,319,320,321,322,323, 324]. The favored TC conformations have oxepane conformations of type TC$_B$ or TC$_C$. The equilibria between conformers are often closely balanced as indicated by the significant change in conformers populated when O-5 of methyl β-D-*glycero*-D-guloseptanoside was methylated [319].

1

1.4 Conformations of Disaccharides, Trisaccharides, and Oligosaccharides

1.4.1 Conformations of Disaccharides

The conformations adopted by disaccharides are defined in terms of the torsional angles across the glycosidic linkage, starting from the anomeric center. The angle Φ is defined by the torsional angle H1–C1–O1–Ci, where i is the number of the carbon atom in the aglycone. In virtually all structures determined by X-ray crystallography that were not constrained to have other values, Φ lies in the range expected on the basis of the exo-anomeric effect [69]. In solution, some glycosidic linkages have been observed to have minor populations of non-exo conformers [325,326,327] and one branched oligosaccharide was observed to have a considerable population of non-exo conformers about a β-ribofuranosyl linkage [328]. For glycosides of 2-uloses, the definition of this torsional angle is changed to C1–C2–O2–Ci. The second torsional angle, Ψ, is defined as C1–O1–Ci–Hi, where i is defined as above. If the first carbon atom in the aglycone is exocyclic, it is necessary to define a third torsional angle, ω, O1–Ci–Cj–Hj and Ψ now becomes C1–O1–Ci–Cj (see ❯ Fig. 23 and ❯ Fig. 24). In crystal structures, these torsional angles are often defined in terms of the heavy atoms, that is, C1–C1–O1–Ci or

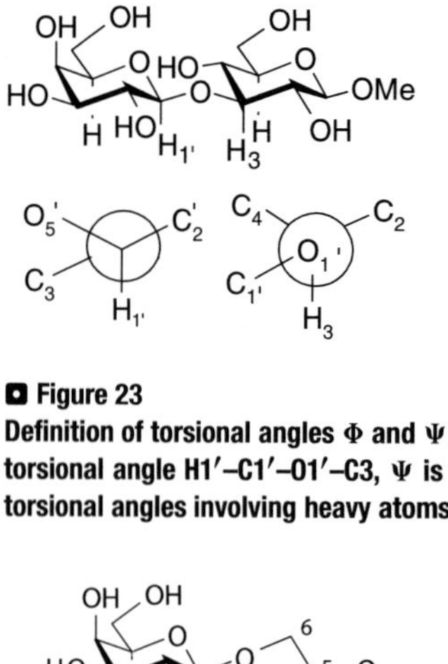

◻ **Figure 23**
Definition of torsional angles Φ and Ψ: for methyl β-D-galactopyranosyl-$(1\rightarrow3)$-β-D-glucopyranoside, Φ is the torsional angle H1'–C1'–O1'–C3, Ψ is the torsional angle C1'–O1'–C3–H3. The Newman projections show how torsional angles involving heavy atoms from X-ray data correspond to angles defined as above

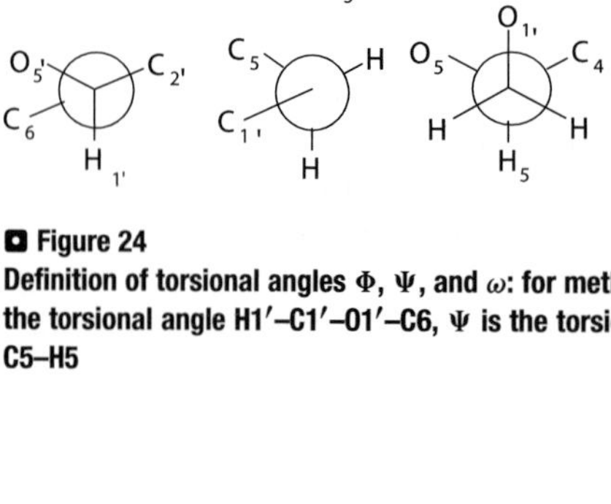

◻ **Figure 24**
Definition of torsional angles Φ, Ψ, and ω: for methyl β-D-galactopyranosyl-$(1\rightarrow6)$-β-D-glucopyranoside, Φ is the torsional angle H1'–C1'–O1'–C6, Ψ is the torsional angle C1'–O1'–C6–C5, ω is the torsional angle O1'–C6–C5–H5

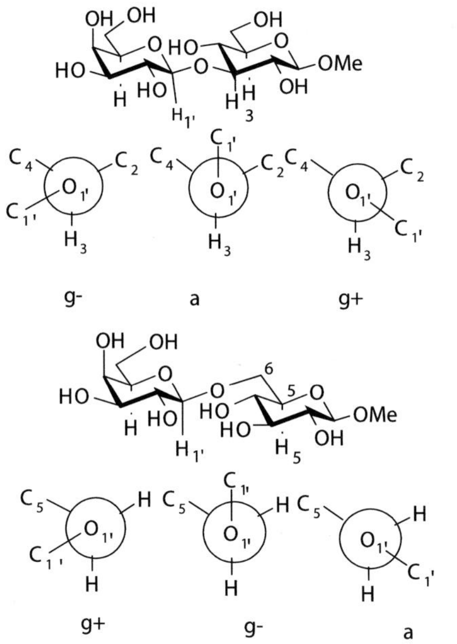

Figure 25
Designations of rotamers about the Ψ angle in disaccharides

O5–C1–O1–Ci, rather than H1–C1–O1–Ci. Approximate conversions can be made by adding + or −120°, as appropriate (see ❯ *Fig. 23* and ❯ *Fig. 24*) [1]. The book by Rao et al. [1] contains an extensive compilation of X-ray structures of disaccharides.

Unlike the torsional angle Φ, the angle Ψ commonly adopts a variety of values. It is convenient to be able to designate these values in terms of rotamers and a scheme for doing this is shown in ❯ *Fig. 25*. Conformational analysis of disaccharides can proceed by using a theoretical approach to estimate the variation in energy with Φ and Ψ. These results are usually presented in a two dimensional plot of Φ against Ψ, representing energy changes as contours, called a Ramachandran plot [329]. Analysis of CC, CO, CH and OH steric interactions in the different rotamers about the Ψ angle can simplify the evaluation of experimental data [330,331]. This procedure is illustrated in ❯ *Fig. 26* for β-(1→3) linked disaccharides. It can be seen that the *a* conformer is disfavored because having H-3 *anti* to C-1′ results in the remaining two carbons being *gauche* to C-1′. Indeed, no disaccharides linked through secondary oxygen atoms listed in the Tables in the book by Rao et al. [1] adopt this conformation in the solid state. For the particular linkage shown in ❯ *Fig. 26*, the g+ conformer is favored over the g− conformer (see ❯ *Table 4*) , because in the latter, C-4 has a 1,3-diaxial interaction with O-5, whereas in the former, the 1,3-diaxial interaction of the carbon atom that is *gauche* is that of C-2 with H-1′. In the solid state, both g− and g+ rotamers are observed but the values of the Ψ angle are normally much less than staggered [1], presumably because it is energetically advantageous to decrease the interactions of the *gauche* carbon at the expense of increasing the non-bonded interactions involving the aglycone H at the linkage center. In agreement with this statement, Lemieux and Koto concluded from hard-sphere calculations

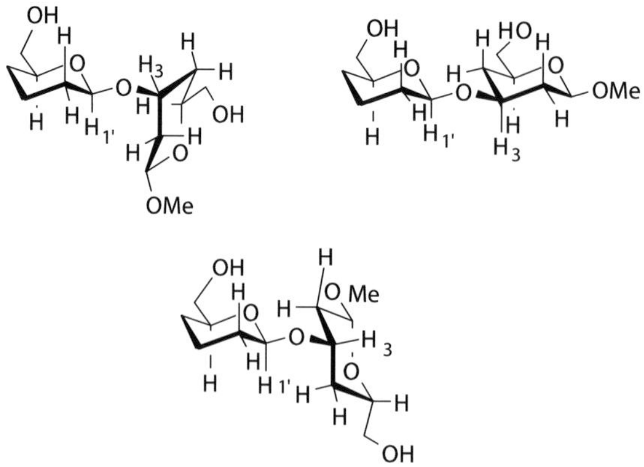

Figure 26
Estimation of steric energies in disaccharides, for methyl 2,3,4-trideoxy-β-D-*glycero*-aldohexopyranosyl-(1→3)-2,4-dideoxy-β-D-*threo*-aldohexopyranoside: top left, *g+* conformer; top right, *g−−* conformer; bottom, *a* conformer

Table 4
Steric interactions in methyl 2,3,4-trideoxy-β-D-*glycero*-aldohexopyranosyl-(1→3)-2,4-dideoxy-β-D-*threo*-aldohexopyranoside

Conformer	Interaction	Value (kJ mol^{-1})
g+	H-30-5′	1.9
	C-2H-1′	3.8
	H-2axC-1′	3.8
	Total	9.5
g−	C-40-5′	10.4
	H-4eC-1′	3.8
	Total	14.2
	C-4H-1′	3.8
	C-20-5′	10.4
	C-1′H-2ax	3.8
	C-1′H-4ax	3.8
	Total	21.8

on both cyclohexyl D-glucopyranoside anomers that the Ψ angle should have values close to 0° [332], but the calculated preference is much too large based on observed conformer mixtures for disaccharides [326,333,334,335]. Anderson has provided evidence that eclipsed conformations of this type are also important for acetals [62]. High-level ab initio calculations on 2-cyclohexyloxytetrahydropyran indicate that the potential energy surface is fairly flat at Ψ angles of $0 \pm 50°$, and the *anti* conformer is much less stable (~16 kJ mol^{-1}) [336]. Thus, it is perhaps more appropriate to refer to a conformer with a C1–O1–Ci–Hi torsional angle close to 0° as a *syn* conformer.

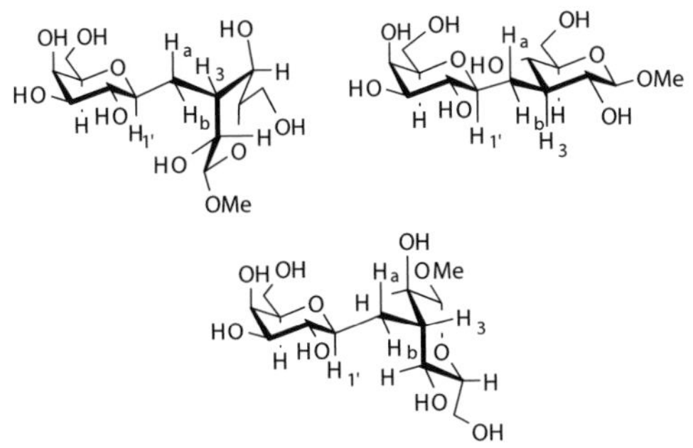

◘ Figure 27
Estimation of steric energies in C- and O-disaccharides, for methyl β-D-galactopyranosyl-(1→3)-β-D-glucopyranoside and its C-disaccharide analogue: top left, *g+* conformer; top right, *g−* conformer; bottom, *a* conformer

Substituents can markedly influence the inherent stabilities as shown in ❯ *Fig. 27* where the effects of adding equatorial hydroxyl groups for both O- and C-glycosides to give a typical 163-linked disaccharide are illustrated, with the values given in ❯ *Table 5*. This O-linked disaccharide has been observed by nOe measurements to have a small proportion of the *a* conformer along with the major *syn* and *g−* conformers [325]. In β-(164) linked derivatives, such as lactose or cellobiose, the interactions of the adjacent CH_2OH group become important and, in solution, the *a* conformer can become comparable in stability to, or even more favorable than, a *g* conformer [6,331,337,338]. Although a number of disaccharides and oligosaccharides have been found to have intermolecular hydrogen bonds in water, water/acetone, or dimethyl sulfoxide solutions by careful NMR measurements [241,252,259,262,326], no general criteria have yet been developed to predict their influence on disaccharide conformations. Although usually the most populated conformations are those bound by proteins [331,339,340,341], some oligosaccharides bind in conformations that contain a glycosidic linkage in an *a* conformer [338] or other conformations that are not highly populated in the free state [342,343]. A tethered disaccharide has been synthesized that is constrained to this conformation [344].

Calculation of the stabilities and geometries of conformers using the force fields of molecular mechanics or molecular dynamics programs provides an outline of the conformational possibilities that can be used to help interpret the experimental measurements [4,207,333,339,345, 346]. It was concluded that most of the current force fields agree on the geometries of the lower energy conformers of disaccharides although there is disagreement about the relative stabilities of these minima [347].

Recently, Almond performed molecular dynamics calculations on a number of disaccharides and oligosaccharides in water [348] using the CHARMm forcefield modified for carbohydrates [349]. He observed a number of persistent intersaccharide hydrogen bonds that influenced the mixture of conformers predicted to be present. On the basis of these calculations, he made the important suggestion that disaccharides linked through α-linkages will be flexible with many hydrogen bonds to water, while disaccharides linked through β-linkages will be involved in intersaccharide hydrogen bonds in water and will be relatively inflexible [348].

■ **Table 5**
Steric interactions for methyl β-D-galactopyranosyl-$(1\rightarrow3)$-β-D-glucopyranoside and its C-disaccharide analogue

Conformer	Interaction	Value for C-glycoside (kJ mol^{-1})	Value for O-glycoside (kJ mol^{-1})
$g+$	O-2C1'	10.4	10.4
	H-3O-5'	1.9	1.9
	C-2H-1'	3.8	3.8
	O-2'Hb	1.9	
	O-4Ha	1.9	
	C-1Hb	3.8	
	Total	23.7	16.1
$g-$	C-4O-5'	10.4	10.4
	O-4C-1'	10.4	10.4
	O-2'Hb	1.9	
	O-2Hb	1.9	
	Total	24.6	20.8
a	C-4H-1'	3.8	3.8
	C-2O-5'	10.4	10.4
	C-1'H-2	3.8	3.8
	C-1'H-4	3.8	3.8
	O-2Ha	1.9	
	O-2'Hb	1.9	
	O-4Hb	1.9	
	Total	27.5	21.8

1.4.2 Conformations of Trisaccharides and Oligosaccharides

No new factors arise when the conformations of trisaccharides or oligosaccharides are considered but intermonosaccharide interactions can become more important. For reviews, see [350,351,352]. However, discussion of these conformations using the terms employed with disaccharides is much more complicated because of the multiple $\Phi\Psi$ maps and because of the possibility of coordinated motion. An alternative description uses a generalized order parameter, S^2, which is related to the spatial restriction of internal reorientation, with values ranging from 0 to 1 [353,354,355,356,357]. The dynamics of isotropically tumbling molecules are described in terms of overall and internal correlation times. The generalized order parameter [353] has been extended to include two distinct correlation times [358]. When the oligosaccharide is relatively rigid, S^2 for all segments is close to 1. Terminal monosaccharides in linear oligosaccharides often have lower values of S^2 than the central regions consistent with independent motion [359,360], as do hydroxymethyl groups [361].

2 Physical Methods

2.1 Introduction

The physical methods used for the analysis of carbohydrate conformations are becoming increasingly sophisticated as the molecules studied and the questions asked become larger and more complicated. X-ray crystallography provides a reference point, the conformations of compounds in the solid state. In solution, molecular modeling supplies a framework which can be used to interpret the results of the vast array of NMR spectral observations and those from other methods. For reviews, see [2,354].

2.2 X-Ray Crystallography

The determination of crystal structures of small molecules has now become routine provided that suitable crystals are available [362,363,364,365]. The advent of higher power X-ray sources, more synchronized light, and better recording devices has meant that smaller and smaller crystals can be solved by diffraction methods. At the same time, the analysis of patterns from larger and larger molecules has become easier so that the limit on the use of X-ray crystallography is the production of crystals.

The crystal structures of a large number of carbohydrate derivatives have been solved [1,106, 366]. Crystal structures provide critical data on bond lengths, bond angles, torsional angles, intramolecular distances, etc., and this data has played and is continuing to play an important role in the discussion of the factors that influence conformation. Consideration of conformations of individual molecules using X-ray data must take into account intermolecular forces within the crystal lattice. Analysis of large sets of X-ray data from the Cambridge Data File avoids this difficulty and this method has influenced discussion of the anomeric effect [61,62,69]. Although most molecules crystallize in the conformation that is most highly populated in solution, this is not always true. For instance, 3-ammonio-3-deoxy-1,6-anhydro-β-D-glucopyranose is present in solution mostly in a boat conformation [184,367] but crystallizes in a chair [367]. Of increasing use is the determination of crystal structures of carbohydrates bound to proteins [368,369]. This allows analysis of the conformation of the bound carbohydrate and of the binding factors, both of which allow the synthesis of better binding molecules that may have medicinal applications.

2.3 NMR Spectroscopy

NMR spectroscopy is the most important technique for the examination of structure and conformation in solution [370,371,372,373]. Many aspects of NMR spectroscopy can be employed for the study of carbohydrates [130,351,374,375,376,377]. Assignment techniques are discussed in many books and will not be considered here. Chapters by Widmalm [354] and Serianni [376] include sections on assignment of carbohydrate spectra.

2.3.1 Chemical Shifts

^{1}H and ^{13}C NMR chemical shifts for unsubstituted and simple substituted carbohydrate derivatives and the effects of substituents on these shifts have been summarized [130,350,351,378, 379,380,381]. The bulk of carbohydrate protons are observed between 3.2 and 4.2 ppm with equatorial protons being less shielded than axial and primary less shielded than secondary [130,375,382,383]. Anomeric protons are more deshielded (~4.3 to 5.5 ppm) and similarly anomeric carbons (~ 90 to 110 ppm) are deshielded with respect to the bulk of the secondary (~ 67 to 82 ppm) and primary carbons (~60–65 ppm). The factors that influence hydroxyl proton chemical shifts have been summarized and evaluated computationally [239,251].

2.3.2 Scalar Coupling

One-Bond Coupling The magnitudes of $^1J_{C,H}$ are about 125 Hz in saturated alkanes but increase in the presence of electronegative substituents [384] and are typically 140–150 Hz for non-anomeric carbons. The sizes of anomeric $^1J_{C,H}$ are related to the relative orientations of the bonds at the anomeric center of pyranoses, which are in turn related to CH bond lengths [76]; values are ~170–175 Hz if H-1 is equatorial (typically α-D anomers in $^4C_1$ conformations) but ~160–165 Hz if H-1 is axial (typically β-D anomers in $^4C_1$ conformations) [385,386,387]. Values from both furanoside anomers are normally similar, > 170 Hz [388,389] although conformational restriction makes the $^1J_{C,H}$ values for 2,3-anhydrofuranosides diagnostic of configuration [390]. For the same reasons, $^1J_{C,H}$ values for the anomeric CH units of septanoses are unreliable indicators of configuration [307]. Additional electronegative substituents increase the magnitude of $^1J_{C,H}$ again to 176–185 Hz in orthoesters [391]. In oligosaccharides, the sizes of anomeric $^1J_{C,H}$ are also influenced by the $\Phi\Psi$ angles in predictable ways [387].

Two-Bond Coupling A number of different two-bond couplings can provide useful structural and conformational information. The magnitudes of H–C–H coupling constants in saturated systems range from about 0 to −15 Hz and adjacent electronegative atoms cause them to increase algebraically in an orientationally dependent fashion [392], for instance, $^2J_{H,H}$ is ~−6 Hz in 1,3-dioxanes but > −2 Hz in 1,3-dioxolanes. $^2J_{H,H}$ has been used for conformational studies [196,393] and its magnitude at C-6 of pyranose sugars is related to both the C-5 C-6 torsional angle as well as the C-6 O-6 torsional angle [196]. It could be utilized more, particularly in the study of 1,6-linked oligosaccharides.
The magnitude of $^2J_{C,H}$ depends on substitution, electronegativity, and bond angle and can be positive or negative, increasing with increased numbers of electronegative substituents [394]. Values range from about −8 to +10 Hz in carbohydrates with larger values if one of the carbons is the anomeric carbon [395,396,397]. Rules that relate the size of $^2J_{C,H}$ to the orientation of the oxygen atoms have been formulated [395,397] and used to assign the configuration in *N*-acetylneuraminic acid and derivatives [398,399] and in oligosaccharides [400]. More recently, conformational information has been obtained by calculating these values for all possible conformational minima for comparison with experimental values for hydroxymethyl rotamers [197].

The effects on the size of $^2J_{C,C}$ across both oxygen and carbon atoms have been investigated experimentally and theoretically [401,402]. Across the anomeric oxygen, the magnitude of $^2J_{C,C}$ depends on the size of the Φ torsional angle much more than on the Ψ torsional angle but also depends on the C–O–C bond angle [402], making its use more difficult.

Three-Bond Coupling   The magnitude of the coupling constant between protons on adjacent carbons, $^3J_{H,H}$, is related to the torsional angle between the protons [403] by the Karplus equation [404]. Several groups have modified the original equation to encode the effects of the orientations and electronegativities of electronegative substituents on the magnitude of $^3J_{H,H}$ but the equation of Haasnoot et al. [405], derived using $^3J_{H,H}$ values in six-membered rings, has seen the most use. The electronegativity values used with it have been modified and it has been reparameterized [406,407] to be:

$$^3J_{H,H} = 14.63\cos^2(\phi) - 0.78\cos(\phi) + \sum_i \lambda i \left\{ 0.34 - 2.31\cos^2[s_i(\phi) + 18.4|\lambda i|] \right\} \quad (4)$$

where Φ is the torsional angle, 8_i are the modified electronegativities, and s_i is the sign factor, either +1 or −1, defined according to the sign of the torsional angle [406]. Altona et al. have noted that use of this equation is limited to unstrained saturated systems but molecular orbital calculations have suggested that bond angle variation along the coupling path also influence the size of $^3J_{H,H}$ [408,409,410,411]. This latter factor may make Eq. (4) less accurate when extended to furanosides or acyclic systems.

$^3J_{H,H}$ values have been used extensively to evaluate which conformation is present. When conformational mixtures are present, the coupling constants for contributing conformers can be measured below the temperature of coalescence or estimated from the Haasnoot–Altona equation, Eq. (4). The position of the equilibrium is then evaluated by using the weighted average of the values for the contributors. This technique has been employed by many researchers [119,130,140,191,205,412], largely because it is accepted that the methods available for estimating values for contributors are reasonably accurate. It has been justified theoretically [413].

Karplus-type equations have been developed for many other systems of interest, most notably H–C–O–H [242,243], H–C–O–C [387,414,415], H–C–C [197,387], and C–O–C–C [416, 417]. It should be noted that coupling to hydroxyl protons can be observed in water or water/acetone or water/DMSO mixtures if the NMR tubes are rinsed in phosphate buffers and the carbohydrate derivatives are thoroughly deionized [239,249]. If the magnitudes of the coupling constants to carbon are determined on natural abundance material, considerable amounts are required. Alternatively, ^{13}C labeled material can be used [416,417,418,419] and starting materials with varieties of labels are available.

Long-Range Coupling   Long range coupling constants ($^4J_{H,H}$ and $^5J_{H,H}$) were examined in the early literature on NMR of carbohydrates [420,421]. In saturated systems, they are largest over a W coupling pathway ($<2\,Hz$) and thus have some potential to reveal conformational information. They can be obtained conveniently using gradient-enhanced, two-dimensional homonuclear correlation techniques [422]. Long-range CH couplings ($^4J_{COCCH}$ and $^4J_{CCCCH}$) to enriched ^{13}C atoms have also proven useful for conformational studies [198].

2.3.3 Dipolar Coupling

In normal liquids, most molecules move isotropically and dipolar couplings are averaged to 0. In static solids and in liquid crystals, molecular motion is non-existent or anisotropic and dipolar couplings are observed. In solids, dipolar couplings are obscured by chemical shift anisotropy but in liquid crystals, they are observed and can be large, up to the kHz range [423]. The size of the direct dipolar coupling (D_{ij}) between nuclei i and j is:

$$D_{ij} = \frac{-K_{ij}S_{ij}}{r_{ij}^3} \tag{5}$$

where K_{ij} is equal to a constant times the product of γ_i and γ_j and S_{ij} is a traceless symmetric tensor, the order tensor, indicating how the spins are oriented with respect to the liquid crystal axis and the magnetic field [424]. Analysis of these spectra is complex since every spin 1/2 nucleus is coupled to every other nucleus. However, Tjandra and Bax showed that partial orientation of small proteins can be achieved in phospholipid bilayers known as bicelles [425]. The value of the order parameter increases with increasing bicelle concentrations allowing separation of dipolar and scalar couplings. This technique has been applied to carbohydrates [236,333,426,427,428,429,430,431]. Measurements of dipolar couplings provide a valuable complement to NOE measurements to define internuclear distances since the distance dependence of this interaction is r^{-3}, rather than r^{-6}. A concern is that interaction of the carbohydrate with the liquid crystal may alter the populations of individual conformers [432].

2.3.4 Nuclear Overhauser Effect

Techniques based on the nuclear Overhauser effect (nOe) are now the most important methods used for defining three-dimensional structures of carbohydrates. There is a recent edition [433] of earlier texts [434,435]. Overhauser originally predicted that saturation of electrons in a metal would polarize the metal nuclear spins [436] and the term nOe refers to the change in intensity of a signal resulting upon irradiation of another signal. For a two-spin system, *IS*, the nOe effect $f_I(S)$ is defined as the fractional change in the intensity of *I* on saturating *S*:

$$f_I(S) = \frac{(I - I^\circ)}{I^\circ} \tag{6}$$

where I° is the equilibrium intensity of *I* [435]. ➲ *Figure 28* shows a diagram of transition probabilities (W_{0IS}, W_{1I}, W_{1S}, and W_{2IS}) and spin states for the two-spin system. Solomon [437] showed that on saturating *S*:

$$f_I(S) = \frac{\gamma_S}{\gamma_I}\frac{W_{2IS} - W_{0IS}}{W_{0IS} + 2W_{1I} + W_{2IS}} = \frac{\gamma_S}{\gamma_I}\frac{\sigma_{IS}}{\Delta_{IS}} \tag{7}$$

where γ_I and γ_S are the magnetogyric ratios for the two nuclei, σ_{IS} is the cross-relaxation rate, and Δ_{IS} is the longitudinal relaxation rate. Relaxation for ^1H and ^{13}C nuclei normally occurs via the dipole-dipole mechanism where the motion of nearby magnets, usually ^1H nuclei that move with the molecular motion, creates a magnetic field that has a frequency component that matches the transition frequency. For small molecules in solution at room temperature, the

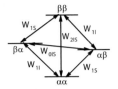

◻ Figure 28
Energy level diagram for a two-spin system, IS, showing definitions of transition probabilities and spin states

average correlation time, τ_C, the time in which a molecule rotates one radian, is on the order of 10^{-11} s, giving a tumbling rate of about 600,000 MHz. For protons in a 500-MHz spectrometer, a tumbling rate of 500 MHz would lead to the most efficient W_1 transitions, and of 1000 MHz would lead to the most efficient W_2 transitions. The W_{0IS} transition corresponds to the frequency difference between the two nuclei which can be 0 to a few kHz for homonuclear cases but will be large for heteronuclear cases, comparable to W_1.

For relaxation in the spin pair IS by the dipolar mechanism in a molecule rotating equally in all directions (isotropically), it can be shown [433,438] that:

$$W_{0IS} = \frac{1/10 K^2 \tau_C}{1 + (\omega_I - \omega_S)^2 \tau_C^2}, \qquad W_{1I} = \frac{3/20 K^2 \tau_C}{1 + \omega_I^2 \tau_C^2},$$

$$W_{1S} = \frac{3/20 K^2 \tau_C}{1 + \omega_S^2 \tau_C^2}, \qquad W_{2IS} = \frac{3/5 K^2 \tau_C}{1 + (\omega_I + \omega_S)^2 \tau_C^2}, \qquad (8)$$

where ω is the angular frequency, $2\pi\nu$, and $K = (\mu_o/4\pi)(h/2\pi)\gamma_I\gamma_S r_{IS}^{-3}$. When $\omega\tau_C$ is $\ll$ 1, log W increases linearly with log τ_C (extreme narrowing). Small molecules in non-viscous solvents generally have correlation times in this region. For homonuclear spin pairs, $\omega_I - \omega_S$ is small and log W_{0IS} increases linearly with log τ_C to large values of τ_C. Log W_1 and log W_{2IS} increase linearly with log τ_C until $\omega\tau_C \approx 1$ when the slope of the curve changes sign to be linear with $-\log \tau_C$. The curves bend when molecular weights are in the 1000–2000 range, the molecular weight range of about pentasaccharides to decasaccharides. For homonuclear pairs, the maximum nOe, η_{max}, is 0.5 under extreme narrowing conditions but becomes -1 when tumbling is slow, typical of polysaccharides. Near $\omega\tau_C = 1$, η_{max} becomes 0 and nOes are small in this tumbling regime. This change over occurs in the tumbling regime where $\omega\tau_C$ changes from about 10^{-1} to 10. The difficulty with employment of nOes in this region can be overcome by using a spin-locking field to lock the spins in the rotating frame. Under these conditions, called a ROESY experiment in the 2D version, extreme narrowing enhancements are observed [439].

When more than two spins are present, the relationships for steady-state nOe become complicated by transfer of magnetization. However, the initial rate of buildup of nOe in multispin systems depends only on σ_{IS}, which for homonuclear relaxation is proportional to r_{IS}^{-6}. It is convenient to compare the initial σ_{IS} for a pair of protons of interest with those for a reference pair where the internuclear distance is known [433,440,441]. Then,

$$\frac{f_I(S)}{f_{ref}} = \frac{\sigma_{IS}}{\sigma_{ref}} = \left(\frac{r_{IS}}{r_{ref}}\right)^{-6}. \qquad (9)$$

Interpretation of nOe results must be performed with caution because a number of difficulties, both technical, such as spin diffusion, and theoretical, such as the use of oversimplified or incorrect models can lead to incorrect interpretations.

2.4 Circular Dichroism and Optical Rotatory Dispersion

Only chiral molecules exhibit circular dichroism (CD) or optical rotatory dispersion (ORD) spectra. ORD is circular birefringence spectroscopy, that is, the difference in refractive index for circularly polarized light as a function of wavelength, or simply, the measure of optical rotation as a function of wavelength. Each individual UV absorption appears as an S-shaped or mirror-image S-shaped curve that is null at 8_{max}. CD is the difference in absorption of right and left circularly polarized light and appears as a Gaussian peak centered on 8_{max} for the UV absorption [442,443]. CD has been employed more for carbohydrates than ORD [444,445]. Since absorption maxima for most carbohydrates occur below 185 nm, vacuum CD must be used [444]. Considerable conformational information can be obtained because the appearance of the curves is very sensitive to the three-dimensional orientations of groups with respect to the chromophores. The most important factors in carbohydrate CD [445] are the anomeric configuration and the orientation about the Φ angle, normally exo-syn. Simple monosaccharides and some disaccharides have been studied [446] and conformational conclusions have been drawn for disaccharides and polymers [447,448,449]. Compounds containing other longer wavelength chromophores have also been studied [450,451].

Alternatively, chromophores can be added to carbohydrates through the formation of *para*-substituted benzoate esters or other functional groups containing chromophores. The high intensities of the CD spectra of these derivatives make this technique quite sensitive [452]. The orientations of the chromophores with respect to the sugar can be estimated and the resulting CD curves are very sensitive to substituent orientations. Thus, this technique is also useful for conformational studies [199,200,203,453].

2.5 Molecular Modeling

Molecular modeling has become an essential technique to outline the possibilities inherent for a particular system. The experimental observations, which are usually weighted averages, can then be interpreted in terms of these possibilities. Information about structure, conformation, and dynamics can be obtained. It is impossible to interpret complex sets of experimental observations, such as the multiple sets of average internuclear distances provided by nOe experiments, without knowledge about the geometries of two, three or more potential energy minima and without having some information about molecular dynamics. Methods for performing molecular modeling studies of carbohydrates are well established [329,347,454,455]. Initial studies used the hard sphere method to estimate non-bonded interactions [456] calculated with the potential of Kitaigorodsky [457]:

$$V = 3.5 \left[-0.04(r_o/r)^6 + 8.6 \cdot 10^3 \exp\left[-13 \left(\frac{r_o}{r} \right) \right] \right]. \tag{10}$$

Addition of terms to account for the exoanomeric effect in α and β glycosides gave the Hard Sphere Exo-Anomeric (HSEA) force field [458,459], which was very successful in predicting conformations of the blood group oligosaccharides [460]. Force-fields have continued to develop [4,195,347,461,462,463,464,465,466,467]. Most can be summarized as follows:

$$E_{potential} = E_{bond} + E_{angle} + E_{dihedral} + E_{non\text{-}bonded} + E_{electrostatic} + \text{cross-terms} \quad (11)$$

where each term describes the increase in energy associated with change in geometry from an equilibrium value for a particular structural feature. For instance, bond stretching can be represented by a simple harmonic [454]:

$$E_{bond} = \frac{1}{2}K_b(l - l_o)^2 \quad (12)$$

where K_b is the force-constant for bond-stretching and l and l_o are the actual and the equilibrium bond lengths, respectively. The energy cost of deviations from equilibrium is more severe at greater deviations and higher order terms can be added to represent this [468]:

$$E_{bond} = \frac{1}{2}K_b(l - l_o)^2[1 - c_1(l - l_o) + c_2(l - l_o)^2] \quad (13)$$

where c_1 and c_2 are additional empirical parameters, or a Morse curve can be used [469]. Force-fields that only include harmonic terms and explicit diagonal elements in the force constant matrix are termed class 1 force fields. Class 2 force-fields add cubic and higher terms and contain off-diagonal elements in the force-constant matrix, that is, terms such as stretch-bend interactions, which detail how the effect of bond-angle bending is changed as the bonds are stretched [470].

Further steps in evaluation of conformational contributors include minimization of the energy from an initial geometry. Techniques for performing this are embedded in all modern force fields. The potential energy surface for carbohydrates contains many minima due to the exocyclic hydroxy, hydroxymethyl, and anomeric groups. Torsional angle driving, where one torsional angle is given an arbitrarily high force-constant, can be used in manual or automatic fashion. Once all minima have been identified, the contribution of individual conformers to the conformational ensemble can be calculated using the Boltzmann distribution.

Dihedral angle driving becomes extremely tedious beyond the monosaccharide stage [471]. Random methods can be used. In stochastic searching, a minimization is performed on the initial geometry, then every Cartesian coordinate of every atom is altered a random amount, then another minimization is performed. The second minimum is compared with the first and saved if it is different. Multiple repetitions allow a library to be built up [472]. Similar methods can be used with torsional angles rather than Cartesian coordinates. Monte Carlo searching imposes an energy test on whether the new conformation is saved and used as a new starting point [473,474,475]. These methods are useful if conformational minima only are required.

Solvent effects can be incorporated in various ways. The simplest is to treat the solvent as a bulk dielectric medium and adjust the equation used to evaluate electrostatic interactions for the dielectric constant [454]. A more sophisticated theory, the reaction-field model, is based on the theory of interactions of a multipole solute molecule in a polarizable continuum. The polar molecule induces a reaction field in the solvent which decreases the energy [476]. An alternative approach is to describe the effect of solvation as follows:

$$\Delta G^{\circ}_{solvation} = \Delta E_{internal} + G_P + G_{CDS} \quad (14)$$

where G_P accounts for the favorable effects of mutual electrostatic polarization of the solute and the solvent minus the cost of distorting the solvent, $\Delta E_{internal}$ accounts for the cost of internal distortion of the solute and G_{CDS}, which depends on the solvent-accessible surface area, includes the effects of interactions dominated by the first solvation shell, such as cavitation [211,477,478]. Another approach is to use Langevin dynamics [333,479,480]. This area is under active development [478,481,482,483,484,485].

Molecular dynamics allows the examination of the time dependence of a system that includes a number of solute and solvent molecules in a cell [486,487,488]. The system of particles is termed the ensemble and the number of particles, the volume and either the energy or the temperature are kept constant. The time evolution of the ensemble is obtained from Newton's second law of motion:

$$\frac{dv}{dt} = \frac{F}{m} \tag{15}$$

where the acceleration is dv/dt, F is the force, and m is the atomic mass. Very small time steps keep the energy constant. The force is obtained from the molecular mechanics force field,

$$\frac{dE_{pot}}{dr} = -F. \tag{16}$$

First derivatives of the potential energy with respect to the Cartesian coordinates are determined as part of the minimization process. The total energy of the system is the sum of the potential and kinetic energies. The kinetic energy is proportional to the temperature:

$$E_{kinetic} = \frac{1}{2}mv^2 = \frac{3}{2}kT \tag{17}$$

and this can be used to define the temperature of the ensemble. A difficulty is handling the borders of the cell and various approaches have been used, including periodic border conditions, where the cell is symmetrically replicated in all directions.

The necessity of using small time steps means that many steps must be calculated in order to adequately sample conformational space. Therefore, molecular dynamics calculations are normally performed with class 1 force fields. Molecular dynamics simulations are analyzed by following the trajectory of a variable with time, often $\Phi\Psi$ angles for oligosaccharides. Most modeling programs now can perform molecular dynamics calculations [347,462,466,489].

Quantum mechanical calculations [490,491,492,493] are now fast and accurate enough that they have become the method of choice for studying conformations of individual monosaccharides and disaccharides [231,494,495,496,497]. The use of density functional theory [498] allows the study of much larger systems. High-level methods have to be employed to yield accurate results in terms of energies [499]. The continuing improvements in computational power will result in acceleration of the use of these methods.

Acknowledgement

I would like to thank Al French for comments on this chapter from the previous edition.

References

1. Rao VSR, Qasba PK, Balaji PV, Chandrasekaran R (1998) Conformation of Carbohydrates. Harwood Academic Publishers, Amsterdam
2. Stoddart JF (1971) Stereochemistry of Carbohydrates. Wiley-Interscience, New York
3. Dais P (1995) Adv Carbohydr Chem Biochem 51:63
4. Vliegenthart JFG, Woods RJ (2006) NMR Spectroscopy and Computer Modeling of Carbohydrates: Recent Advances. American Chemical Society, Washington, DC
5. Imberty A, Pérez S (2000) Chem Rev 100:4567
6. Jiménez-Barbero J, Espinosa JF, Asensio JL, Cañada FJ, Poveda A (2001) Adv Carbohydr Chem Biochem 56:235
7. French AD (1998) Carbohydrates. Wiley, New York, p 233
8. Eliel EL, Wilen SH, Mander LN (1994) Stereochemistry of Organic Compounds. Wiley, New York
9. Juaristi E (1995) Conformational Behaviour of Six-Membered Rings. VCH Publishers, New York
10. Moss GP (1996) Pure Appl Chem 68:2193
11. Nasipuri D (1991) Stereochemistry of Organic Compounds: Principles and Applications. Wiley, New York
12. Cremer D, Pople JA (1975) J Am Chem Soc 97:1354
13. Cremer D, Szabo KJ (1995) Ab Initio Studies of Six-membered Rings: Present Status and Future Developments. In: Juaristi E (ed) Conformational Behaviour of Six-Membered Rings. VCH Publishers, New York, p 59
14. Haasnoot CAG (1992) J Am Chem Soc 114:882
15. Anet FAL, Anet R (1975) Conformational Processes in Rings. In: Jackman LM, Cotton FA (eds) Dynamic Nuclear Magnetic Resonance Spectroscopy. Academic Press, New York, p 543
16. Höfner D, Lesko SA, Binsch G (1978) Org Magn Reson 11:179
17. Squillacote M, Sheridan RS, Chapman OL, Anet FAL (1975) J Am Chem Soc 97:3244
18. Bushweller CH (1995) Stereodynamics of Cyclohexane and Substituted Cyclohexanes. Substituent A Values. In: Juaristi E (ed) Conformational Behaviour of Six-Membered Rings. VCH Publishers, New York, p 25
19. Booth H, Everett JR (1980) J Chem Soc, Perkin Trans 2 255
20. Wiberg KB, Hammer JD, Castejon H, Bailey WF, DeLeon EL, Jarret RM (1999) J Org Chem 64:2085
21. Kitching W, Olszowy HA, Adcock W (1981) Org Magn Reson 15:230
22. Eliel EL, Gilbert EC (1969) J Am Chem Soc 91:5487
23. Schneider H-J, Hoppen V (1978) J Org Chem 43:3866
24. Jensen FR, Bushweller CH, Beck BH (1969) J Am Chem Soc 91:344
25. Eliel EL, Reese MC (1968) J Am Chem Soc 90:1560
26. Tichý M, Sipos F, Sicher J (1966) Coll Czech Chem Commun 31:2889
27. Booth H, Dixon JM, Khedhair KA (1992) Tetrahedron 48:6161
28. Corey EJ, Feiner NF (1980) J Org Chem 45:765
29. Angyal SJ, Mchugh DJ (1956) Chem Ind 1147
30. Eliel EL, Haubenstock H (1961) J Org Chem 26:3504
31. Tichý M, Orahovats A, Sicher J (1970) Coll Czech Chem Commun 35:459
32. Tavernier D, De Pessemier F, Anteunis M (1975) Bull Soc Chim Belg 84:333
33. Lemieux RU, Chu P (1958) ACS Meeting, San Francisco, CA
34. Manoharan M, Eliel EL (1983) Tetrahedron Lett 24:453
35. Booth H, Grindley TB (1983) J Chem Soc, Chem Commun 1013
36. Subbotin OA, Sergeev NM, Zefirov NS, Gurvich LG (1975) J Org Chem USSR 11:2233
37. Rockwell GD, Grindley TB (1996) Austral J Chem 49:379
38. Tichý M, Sicher J (1967) Coll Czech Chem Commun 32:3687
39. Subbotin OA, Sergeev NM (1978) J Org Chem USSR 14:1388
40. Jordan EA, Thorne MP (1986) Tetrahedron 42:93
41. Wiberg KB, Castejon H, Bailey WF, Ochterski J (2000) J Org Chem 65:1181
42. Riddell FG (1980) The Conformational Analysis of Heterocyclic Compounds. Academic Press, London
43. Kleinpeter E (1998) Adv Het Chem 69:217
44. Lambert JB, Mixan CE, Johnson DH (1973) J Am Chem Soc 95:4634

45. Eliel EL, Hargrave KD, Pietrusiewicz KM, Manoharan M (1982) J Am Chem Soc 104:3635
46. Szarek WA, Horton D (1979) Anomeric Effect: Origin and Consequences. ACS, Washington, DC
47. Thatcher GRJ (1993) Anomeric Effect and Associated Stereoelectronic Effects. ACS, Washington, DC
48. Kirby AJ (1983) The Anomeric Effect and Related Stereoelectronic Effects at Oxygen. Springer, Berlin Heidelberg New York
49. Tvaroška I, Bleha T (1989) Adv Carbohydr Chem Biochem 47:45
50. Juaristi E, Cuevas G (1994) The Anomeric Effect. CRC Press, Boca Raton, FL
51. Franck RW (1995) Stereoelectronic Effects in Six-membered Rings. In: Juaristi E (ed) Conformational Behaviour of Six-Membered Rings. VCH Publishers, New York, p 159
52. Bitzer RS, Barbosa AGH, da Silva CO, Nascimento MAC (2005) Carbohydr Res 340:2171
53. Graczyk PP, Mikolajczyk M (1994) Topics Stereochem 21:159
54. Edwards JT (1955) Chem Ind 1102
55. Eliel EL, Giza CA (1968) J Org Chem 33:3754
56. Franck RW (1983) Tetrahedron 39:3251
57. Booth H, Dixon JM, Readshaw SA (1992) Tetrahedron 48:6151
58. Romers C, Altona C, Buys HR, Havinga E (1969) Topics Stereochem 4:39
59. Brundle CR, Turner DW (1968) Proc Roy Soc A 307:27
60. Laing M (1987) J Chem Ed 64:124
61. Cossé-Barbi A, Watson DG, Dubois J-E (1989) Tetrahedron Lett 30:163
62. Anderson JE (2000) J Org Chem 65:748
63. Astrup EE (1973) Acta Chem Scand 27:327
64. Yokoyama Y, Ohashi Y (1999) Bull Chem Soc Jpn 72:2183
65. Petillo PA, Lerner LE (1993) Origin and Quantitative Modeling of the Anomeric Effect. In: Thatcher GRJ (ed) Anomeric Effect and Associated Stereoelectronic Effects. ACS, Washington, DC, p 156
66. Salzner U, von Rague Schleyer P (1994) J Org Chem 59:2138
67. Omoto K, Marusaki K, Hirao H, Imade M, Fujimoto H (2000) J Phys Chem A 104:6499
68. Cossé-Barbi A, Dubois J-E (1987) J Am Chem Soc 109:1503
69. Schleifer L, Senderowitz H, Aped P, Tartakovsky E, Fuchs B (1990) Carbohydr Res 206:21
70. Pierson GO, Runquist OA (1968) J Org Chem 33:2572
71. Praly J-P, Lemieux RU (1987) Can J Chem 65:213
72. Ouedrago A, Lessard J (1991) Can J Chem 69:474
73. Kirby AJ, Williams NH (1992) J Chem Soc, Chem Commun 1286
74. Briggs AJ, Glenn R, Jones PG, Kirby AJ, Ramaswamy P (1984) J Am Chem Soc 106:6200
75. Uehara F, Sato M, Kaneko C, Kurihara H (1999) J Org Chem 64:1436
76. Wolfe S, Pinto BM, Varma V, Leung RYN (1990) Can J Chem 68:1051
77. Juaristi E, Cuevas G (1992) Tetrahedron 48:5019
78. Cuevas G, Martínez-Mayorga K, Fernandez-Alonso MD, Jiménez-Barbero J, Perrin CL, Juaristi E, López-Mora NS (2005) Angew Chem Int Ed 44:2360
79. Lewis BE, Schramm VL (2001) J Am Chem Soc 123:1327
80. Box VGS (1990) Heterocycles 31:1157
81. Box VGS (1998) Heterocycles 48:2389
82. Perrin CL, Armstrong KB, Fabian MA (1994) J Am Chem Soc 116:715
83. Wiberg KB, Marquez M (1994) J Am Chem Soc 116:2197
84. Lemieux RU, Pavia AA, Martin JC, Watanabe KA (1969) Can J Chem 47:4427
85. Jorgensen WL, de Tirago PIM, Severance DL (1994) J Am Chem Soc 116:2199
86. Salzner U (1995) J Org Chem 60:986
87. Pinto BM, Wolfe S (1982) Tetrahedron Lett 23:3687
88. Fuchs B (1978) Topics Stereochem 19:1
89. Pitzer KS, Donath WE (1959) J Am Chem Soc 81:3213
90. Engerholm GR, Luntz AC, Gwinn WD, Harris DO (1969) J Chem Phys 50:2446
91. Meyer R, Lopez JC, Alonso JL, Melandri S, Favero PG, Caminati W (1999) J Chem Phys 111:7871
92. Mamleev AH, Gunderova LN, Galeev RV (2001) J Struct Chem 42:365
93. Greenhouse JA, Strauss HL (1969) J Chem Phys 50:124
94. Davidson R, Warsop PA (1972) J Chem Soc, Faraday Trans 2 1875
95. Rayón VM, Sordo JA (2005) J Chem Phys 122:204303
96. Wu AA, Cremer D (2003) Int J Mol Sci 4:158
97. Luger P, Buschmann J (1983) Angew Chem Int Ed Eng 22:410

98. David WIF, Ibberson RM (1992) Acta Crystallogr C48:301
99. Cadioli B, Gallinella E, Coulombeau C, Jobic H, Berthier G (1993) J Phys Chem 97:7844
100. Strajbl M, Baumruk V, Florián J (1998) J Phys Chem B 102:1314
101. Lavrich RJ, Rhea RL, McCargar JW, Tubergen MJ (2000) J Mol Spectrosc 199:138
102. Lavrich RJ, Torok CR, Tubergen MJ (2001) J Phys Chem A 105:8317
103. Berman HM, Rosenstein RD (1968) Acta Crystallogr B24:435
104. Kim HS, Jeffrey GA, Rosenstein RD (1968) Acta Crystallogr B24:1449
105. Jeffrey GA, Kim HS (1970) Carbohydr Res 14:207
106. Jeffrey GA (1990) Acta Crystallogr B46:89
107. Köll P, Kopf J, Morf M, Zimmer B, Brimacombe JS (1992) Carbohydr Res 237:289
108. Angyal SJ, Saunders JK, Grainger CT, Le Fur R, Williams PG (1986) Carbohydr Res 150:7
109. Kopf J, Morf M, Hagen B, Bischoff M, Köll P (1994) Carbohydr Res 262:9
110. Köll P, Bischoff M, Bretzke C, Kopf J (1994) Carbohydr Res 262:1
111. Köll P, Bruns R, Kopf J (1997) Carbohydr Res 305:147
112. Köll P, Kopf J (1996) Austral J Chem 49:391
113. Schouten A, Kanters JA, Kroon J, Comini S, Looten P, Mathlouthi M (1998) Carbohydr Res 312:131
114. Kopf J, Bischoff M, Köll P (1991) Carbohydr Res 217:1
115. Kopf J, Morf M, Zimmer B, Köll P (1992) Carbohydr Res 233:35
116. Horton D, Wander JD (1974) J Org Chem 39:1859
117. Hawkes GE, Lewis D (1984) J Chem Soc, Perkin Trans 2 2073
118. Lewis D (1986) J Chem Soc, Perkin Trans 2 467
119. Lewis D, Angyal SJ (1989) J Chem Soc, Perkin Trans 2 1763
120. Angyal SJ, Le Fur R (1980) Carbohydr Res 84:201
121. Schnarr GW, Vyas DM, Szarek WA (1979) J Chem Soc, Perkin Trans 1 496
122. Franks F, Kay RL, Dadok J (1988) J Chem Soc, Faraday Trans 1 84:2595
123. Gallwey FB, Hawkes JE, Haycock P, Lewis D (1990) J Chem Soc, Perkin Trans 2 1979
124. IUPAC-IUB Joint Commission on Biochemical Nomenclature (1981) Pure Appl Chem 1902
125. Angyal SJ (1968) Austral J Chem 21:2737
126. Angyal SJ (1969) Angew Chem Int Ed Eng 8:157
127. Augé J, David S (1984) Tetrahedron 40:2101
128. Casu B, Petitou M, Provasoli M, Sinaÿ P (1988) TIBS 13:221
129. Angyal SJ, Pickles VA (1972) Austral J Chem 25:1695
130. Bock K, Thøgersen H (1982) Ann Rep NMR Spectrosc 13:1
131. Snyder JR, Serianni AS (1986) J Org Chem 51:2694
132. Snyder JR, Johnston ER, Serianni AS (1989) J Am Chem Soc 111:2681
133. Dowd MK, French AD, Reilly PJ (1994) Carbohydr Res 264:1
134. Tobiason FL, Swank DD, Vergoten G, Legrande P (2000) J Carbohydr Chem 19:959
135. Kurihara Y, Ueda K (2006) Carbohydr Res 341:2565
136. Rudrum M, Shaw DF (1965) J Chem Soc 52
137. Lemieux RU, Stevens JD (1966) Can J Chem 44:249
138. Franks F, Lillford PJ, Robinson G (1989) J Chem Soc, Faraday Trans 1 85:2417
139. Srivastava RM, Pavão AC, Seabra GM, Brown RK (1997) J Mol Struct 412:51
140. Durette PL, Horton D (1971) Adv Carbohydr Chem Biochem 26:49
141. Hall LD, Manville JF (1969) Can J Chem 47:19
142. Probst KC, Wessel HP (2001) J Carbohydr Chem 20:549
143. Anderson CB, Sepp DT (1967) J Org Chem 32:607
144. Bishop CT, Cooper FP (1963) Can J Chem 41:2743
145. Smirnyagin V, Bishop CT (1968) Can J Chem 46:3085
146. Lemieux RU (1964) In: de Mayo P (ed) Molecular Rearrangements, Part 2. Wiley, New York, p 735
147. Lemieux RU, Morgan AR (1965) Can J Chem 43:2205
148. Lemieux RU (1971) Pure Appl Chem 25:527
149. Paulsen H, Györgydeák Z, Friedmann M (1974) Chem Ber 107:1590
150. Vaino AR, Chan SSC, Szarek WA, Thatcher GRJ (1996) J Org Chem 61:4514
151. Perrin CL (1995) Tetrahedron 51:11901
152. Perrin CL, Fabian MA, Brunckova J, Ohta BK (1999) J Am Chem Soc 121:6911
153. Perrin CL, Kuperman J (2003) J Am Chem Soc 125:8846

154. Randell KD, Johnston BD, Green DF, Pinto BM (2000) J Org Chem 65:220
155. Jones PG, Kirby AJ, Komarov IV, Wothers PD (1998) Chem Commun 1695
156. Vaino AR, Szarek WA (2001) J Org Chem 66:1097
157. Chan SSC, Szarek WA, Thatcher GRJ (1995) J Chem Soc, Perkin Trans 2 45
158. Ernst S, Venkataraman G, Sasisekharan V, Langer R, Cooney CL, Sasisekharan R (1998) J Am Chem Soc 120:2099
159. Hricovíni M (2006) Carbohydr Res 341:2575
160. Batta G, Lipták A, Schneerson R, Pozsgay V (1997) Carbohydr Res 305:93
161. Flippen JL (1973) Acta Crystallogr B29:1881
162. Ogura H, Furuhata K, Saitô H, Izumi G, Itoh M, Shitori Y (1984) Chem Lett 1003
163. Jaques LW, Brown EB, Barrett JM, Brey WS Jr, Weltner W Jr (1977) J Biol Chem 252:4533
164. Dabrowski U, Friebolin H, Brossmer R, Supp M (1979) Tetrahedron Lett 4637
165. Beau J-M, Schauer R, Haverkamp J, Dorland L, Vliegenthart JFG (1980) Carbohydr Res 82:125
166. Brown EB, Brey WS Jr, Weltner W Jr (1975) Biochim Biophys Acta 399:124
167. Czarniecki MF, Thornton ER (1977) J Am Chem Soc 99:8273
168. Batta G, Gervay J (1995) J Am Chem Soc 117:368
169. Unger FM (1981) Adv Carbohydr Chem Biochem 38:324
170. Wilkinson SG (1996) Prog Lipid Res 35:283
171. Becker B, Lommerse JPM, Melkonian M, Kamerling JP, Vliegenthart JFG (1995) Carbohydr Res 267:313
172. Becker B, Melkonian M, Kamerling JP (1998) J Phycol 34:779
173. Jann K (1983) Exopolysaccharides of E. Coli. In: Anderson L, Unger FM (eds) Bacterial Polysaccharides: Structure, Synthesis, and Biological Activities. ACS, Washington, DC, p 171
174. Birnbaum GI, Roy R, Brisson JR, Jennings HJ (1987) J Carbohydr Chem 6:17
175. Sengupta D, van Derveer D (1985) Ind J Chem 24B:1268
176. Kratky C, Stix D, Unger FM (1981) Carbohydr Res 92:299
177. Mikol V, Kosma P, Brade H (1994) Carbohydr Res 263:35
178. Cherniak R, Jones RG, Gupta DS (1979) Carbohydr Res 75:39
179. Brade H, Zähringer U, Rietschel ET, Christian R, Schulz G, Unger FM (1984) Carbohydr Res 134:157
180. Coxon B (1999) Carbohydr Res 322:120
181. Angyal SJ, Dawes K (1968) Austral J Chem 21:2747
182. Cerný M, Stanek J Jr (1977) Adv Carbohydr Chem Biochem 34:24
183. Grindley TB, Cude A, Kralovec J, Thangarasa R (1994) The Chair-Boat Equilibrium of 1,6-Anhydro-β-D-glucopyranose and Derivatives: An NMR and Molecular Mechanics Study. In: Witczak ZJ (ed) Levoglucosenone and Levoglucosans, Chemistry and Applications. ATL Press, Mount Prospect, ILL, p 147
184. Trnka T, Cerný M, Budesínský M, Pacák J (1975) Coll Czech Chem Commun 40:3038
185. Noordik JH, Jeffrey GA (1977) Acta Crystallogr B33:403
186. Paulsen H, Koebernick H (1976) Chem Ber 109:104
187. van Rijsbergen R, Anteunis MJO, De Bruyn A (1982) Bull Soc Chim Belg 91:297
188. Dais P, Shing TKM, Perlin AS (1984) J Am Chem Soc 106:3082
189. Li C, Bernet B, Vasella A (1991) Carbohydr Res 216:149
190. Wessel HP (1992) J Carbohydr Chem 11:1039
191. Bock K, Duus JØ (1994) J Carbohydr Chem 13:513
192. Jensen HH, Nordstrom LU, Bols M (2004) J Am Chem Soc 126:9205
193. Jensen HH, Bols M (2006) Acc Chem Res 39:259
194. Marchessault RH, Pérez S (1979) Biopolymers 18:2369
195. Kouwijzer MLCE, Grootenhuis PDJ (1995) J Phys Chem 99:13426
196. Stenutz R, Carmichael I, Widmalm G, Serianni AS (2002) J Org Chem 67:949
197. Thibaudeau C, Stenutz R, Hertz B, Klepach T, Zhao S, Wu QQ, Carmichael I, Serianni AS (2004) J Am Chem Soc 126:15668
198. Pan QF, Klepach T, Carmichael I, Reed M, Serianni AS (2005) J Org Chem 70:7542
199. Nobrega C, Vázquez JT (2003) Tetrahedron: Asym 14:2793
200. Roen A, Padron JI, Vázquez JT (2003) J Org Chem 68:4615
201. Jockusch RA, Kroemer RT, Talbot FO, Simons JP (2003) J Phys Chem A 107:10725
202. Jockusch RA, Talbot FO, Simons JP (2003) Phys Chem Chem Phys 5:1502

203. Padrón JI, Morales EQ, Vázquez JT (1998) J Org Chem 63:8247

204. de Vries NK, Buck HM (1987) Carbohydr Res 165:1

205. Rockwell GD, Grindley TB (1998) J Am Chem Soc 120:10953

206. de Vries NK, Buck HM (1987) Recl Trav Chim Pays-Bas 106:453

207. Gonzalez-Outeiriño J, Kirschner KN, Thobhani S, Woods RJ (2006) Can J Chem 84:569

208. Kirschner KN, Woods RJ (2001) Proc Nat Acad Sci 98:10541

209. Tvaroska I, Taravel FR, Utille JP, Carver JP (2002) Carbohydr Res 337:353

210. Barrows SE, Storer JW, Cramer CJ, French AD, Truhlar DG (1998) J Comp Chem 19:1111

211. Cramer CJ, Truhlar DG (1999) Chem Rev 99:2161

212. Dolney DM, Hawkins GD, Winget P, Liotard DA, Cramer CJ, Truhlar DG (2000) J Comp Chem 21:340

213. Momany FA, Appell M, Willett JL, Bosma WB (2005) Carbohydr Res 340:1638

214. Momany FA, Appell M, Willett JL, Schnupf U, Bosma WB (2006) Carbohydr Res 341:525

215. Tvaroška I, Taravel FR, Utille JP, Carver JP (2002) Carbohydr Res 337:353

216. Suzuki T, Kawashima H, Sota T (2006) J Phys Chem B 110:2405

217. Baumann H, Brisson JR, Michon F, Pon R, Jennings HJ (1993) Biochemistry 32:4007

218. Sawada T, Hashimoto T, Nakano H, Shigematsu M, Ishida H, Kiso M (2006) J Carbohydr Chem 25:387

219. Brisson JR, Baumann H, Imberty A, Pérez S, Jennings HJ (1992) Biochemistry 31:4996

220. Vasudevan SV, Balaji PV (2002) Biopolymers 63:168

221. Yamasaki R, Bacon B (1991) Biochemistry 30:851

222. Henderson TJ, Venable RM, Egan W (2003) J Am Chem Soc 125:2930

223. Bhattacharjee AK, Jennings HJ, Kenny CP (1978) Biochemistry 17:645

224. Lees RM, Baker JG (1968) J Chem Phys 48:5299

225. Lowe JP (1968) Prog Phys Org Chem 6:1

226. Barrows SE, Dulles FJ, Cramer CJ, French AD, Truhlar DG (1995) Carbohydr Res 276:219

227. Gregurick SK, Kafafi SA (1999) J Carbohydr Chem 18:867

228. Suzuki T, Sota T (2005) J Phys Chem B 109:12603

229. Talbot FO, Simons JP (2002) Phys Chem Chem Phys 4:3562

230. Simons JP, Jockusch RA, Carcabal P, Hung I, Kroemer RT, Macleod NA, Snoek LC (2005) Int Rev Phys Chem 24:489

231. Carcabal P, Patsias T, Hunig I, Liu B, Kaposta C, Snoek LC, Gamblin DP, Davis BG, Simons JP (2006) Phys Chem Chem Phys 8:129

232. Carcabal P, Hunig I, Gamblin DP, Liu B, Jockusch RA, Kroemer RT, Snoek LC, Fairbanks AJ, Davis BG, Simons JP (2006) J Am Chem Soc 128:1976

233. Carcabal P, Jockusch RA, Hunig I, Snoek LC, Kroemer RT, Davis BG, Gamblin DP, Compagnon I, Oomens J, Simons JP (2005) J Am Chem Soc 127:11414

234. Barron LD, Hecht L, Mccoll IH, Blanch EW (2004) Mol Phys 102:731

235. Macleod NA, Johannessen C, Hecht L, Barron LD, Simons JP (2006) Int J Mass Spect 253:193

236. Martin-Pastor M, Canales A, Corzana F, Asensio JL, Jiménez-Barbero J (2005) J Am Chem Soc 127:3589

237. Gillet B, Nicole D, Delpuech J-J, Gross B (1981) Org Magn Reson 17:28

238. Angyal SJ, Christofides JC (1996) J Chem Soc, Perkin Trans 2 1485

239. Sandström C, Kenne L (2006) Hydroxy Protons in Structural Studies of Carbohydrates by NMR Spectroscopy. In: Vliegenthart JFG, Woods RJ (eds) NMR Spectroscopy and Computer Modeling of Carbohydrates: Recent Advances. ACS, Washington, DC, p 114

240. Bernet B, Vasella A (2000) Helv Chim Acta 83:995

241. Bernet B, Vasella A (2000) Helv Chim Acta 83:2055

242. Fraser RR, Kaufman M, Morand P, Govil G (1969) Can J Chem 47:403

243. Rader CP (1969) J Am Chem Soc 91:3248

244. Christofides JC, Davies DB (1987) J Chem Soc, Perkin Trans 2 97

245. Vasquez TE, Bergset JM, Fierman MB, Nelson A, Roth J, Khan SI, O'Leary DJ (2002) J Am Chem Soc 124:2931

246. Muddasani PR, Bozó E, Bernet B, Vasella A (1994) Helv Chim Acta 77:257

247. Adams B, Lerner LE (1994) Magn Reson Chem 32:225

248. Batta G, Kövér KE (1999) Carbohydr Res 320:267

249. Sandström C, Baumann H, Kenne L (1998) J Chem Soc, Perkin Trans 2 809

250. Sandström C, Baumann H, Kenne L (1998) J Chem Soc, Perkin Trans 2 2385

251. Bekiroglu S, Sandström A, Kenne L, Sandström C (2004) Org Biomol Chem 2:200

252. Bekiroglu S, Kenne L, Sandström C (2004) Carbohydr Res 339:2465

253. Siebert HC, Andre S, Vliegenthart JFG, Gabius HJ, Minch MJ (2003) J Biomol NMR 25:197

254. Adams B, Lerner LE (1992) J Am Chem Soc 114:4827

255. Leeflang BR, Vliegenthart JFG, Kroon-Batenburg LMJ, van Eijck BP, Kroon J (1992) Carbohydr Res 230:41

256. Bock K, Frejd T, Kihlberg J, Magnusson G (1988) Carbohydr Res 176:253

257. Poppe L, van Halbeek H (1991) J Am Chem Soc 113:363

258. Sandström C, Magnusson G, Nilsson U, Kenne L (1999) Carbohydr Res 322:46

259. Bekiroglu S, Kenne L, Sandström C (2003) J Org Chem 68:1671

260. Hakkarainen B, Fujita K, Immel S, Kenne L, Sandström C (2005) Carbohydr Res 340:1539

261. Asensio JL, Cañada FJ, Garcia-Herrero A, Murillo MT, Fernández-Mayoralas A, Johns BA, Kozak J, Zhu ZZ, Johnson CR, Jiménez-Barbero J (1999) J Am Chem Soc 121:11318

262. Ivarsson I, Sandström C, Sandström A, Kenne L (2000) J Chem Soc, Perkin Trans 2 2147

263. Lemieux RU, Koto S, Voisin D (1979) The Exoanomeric Effect. In: Szarek WA, Horton D (eds) Anomeric Effect, Origin and Consequences. ACS, Washington, DC, p 17

264. Marchessault RH, Pérez S (1978) Carbohydr Res 65:114–120

265. Cramer CJ, Truhlar DG, French AD (1997) Carbohydr Res 298:1

266. Tvaroška I, Carver JP (1998) Carbohydr Res 309:1

267. Espinosa JF, Cañada FJ, Asensio JL, Martín-Pastor M, Dietrich H, Martín-Lomas M, Schmidt RR, Jiménez-Barbero J (1996) J Am Chem Soc 118:10862

268. Asensio JL, Cañada FJ, Chen XH, Khan N, Mootoo DR, Jiménez-Barbero J (2000) Chem Eur J 6:1035

269. Houk KN, Eksterowicz JE, Wu Y-D, Fuglesang CD, Mitchell DB (1993) J Am Chem Soc 115:4170

270. Wu T-C, Goekjian PG, Kishi Y (1987) J Org Chem 52:4819

271. Hall LD, Steiner PR, Pedersen C (1970) Can J Chem 48:1155

272. Altona C, Sundaralingam M (1972) J Am Chem Soc 94:8205

273. Saenger W (1984) Principles of Nucleic Acid Structure. Springer, Berlin Heidelberg New York

274. Altona C, Geise HJ, Romers C (1968) Tetrahedron 24:13

275. Gelbin A, Schneider B, Clowney L, Hsieh SH, Olson WK, Berman HM (1996) J Am Chem Soc 118:519

276. Plavec J, Thibaudeau C, Chattopadhyaya J (1996) Pure Appl Chem 68:2137

277. Podlasek CA, Stripe WA, Carmichael I, Shang MY, Basu B, Serianni AS (1996) J Am Chem Soc 118:1413

278. Felli IC, Richter C, Griesinger C, Schwalbe H (1999) J Am Chem Soc 121:1956

279. Polak M, Seley KL, Plavec J (2004) J Am Chem Soc 126:8159

280. Plevnik M, Crnugelj M, Stimac A, Kobe J, Plavec J (2001) J Chem Soc, Perkin Trans 2 1433

281. Plavec J, Roselt P, Földesi A, Chattopadhyaya J (1998) Magn Reson Chem 36:732

282. Polak M, Mohar B, Kobe J, Plavec J (1998) J Am Chem Soc 120:2508

283. Verberckmoes F, Esmans EL (1995) Spectrochim Acta A 51:153

284. Olson WK, Sussman JL (1982) J Am Chem Soc 104:270

285. Houseknecht JB, McCarren PR, Lowary TL, Hadad CM (2001) J Am Chem Soc 123:8811

286. Houseknecht JB, Lowary TL, Hadad CM (2003) J Phys Chem A 107:5763

287. Houseknecht JB, Lowary TL, Hadad CM (2003) J Phys Chem A 107:372

288. Evdokimov A, Gilboa AJ, Koetzle TF, Klooster WT, Schultz AJ, Mason SA, Albinati A, Frolow F (2001) Acta Crystallogr B57:213

289. Suzuki T, Sota T (2005) J Phys Chem B 109:12603

290. Suzuki T, Kawashima H, Kotoku H, Sota T (2005) J Phys Chem B 109:12997

291. Gordon MT, Lowary TL, Hadad CM (1999) J Am Chem Soc 121:9682

292. Gordon MT, Lowary TL, Hadad CM (2000) J Org Chem 65:4954

293. Temeriusz A, nulewicz-Ostrowska R, Paradowska K, Wawer I (2003) J Carbohydr Chem 22:593

294. Kopf J, Köll P (1984) Carbohydr Res 135:29

295. Ellervik U, Magnusson G (1994) J Am Chem Soc 116:2340

296. Grundberg H, Eriksson-Bajtner J, Bergquist KE, Sundin A, Ellervik U (2006) J Org Chem 71:5892

297. Thibaudeau C, Földesi A, Chattopadhyaya J (1998) Tetrahedron 54:1867

298. Luyten I, Thibaudeau C, Sandström A, Chattopadhyaya J (1997) Tetrahedron 53:6433

299. Oleary DJ, Kishi Y (1994) J Org Chem 59:6629

300. Luyten I, Thibaudeau C, Chattopadhyaya J (1997) J Org Chem 62:8800

301. Castro S, Duff M, Snyder NL, Morton M, Kumar CV, Peczuh MW (2005) Org Biomol Chem 3:3869

302. Bozo E, Gati T, Demeter A, Kuszmann H (2002) Carbohydr Res 337:1351

303. Bozo E, Medgyes A, Boros S, Kuszmann J (2000) Carbohydr Res 329:25

304. Hendrickson JB (1967) J Am Chem Soc 89:7036

305. Hendrickson JB (1967) J Am Chem Soc 89:7043

306. DeMatteo MP, Snyder NL, Morton M, Baldisseri DM, Hadad CM, Peczuh MW (2005) J Org Chem 70:24

307. DeMatteo MP, Mei S, Fenton R, Morton M, Baldisseri DM, Hadad CM, Peczuh MW (2006) Carbohydr Res 341:2927

308. Entrena A, Campos JM, Gallo MA, Espinosa A (2005) Arkivoc 88

309. Entrena A, Campos J, Gómez JA, Gallo MA, Espinosa A (1997) J Org Chem 62:337

310. Anconi CPA, Nascimento CS, Dos Santos HF, De Almeida WB (2006) Chem Phys Lett 418:459

311. Wiberg KB (2003) J Org Chem 68:9322

312. Bocian DF, Strauss HL (1977) J Am Chem Soc 99:2866

313. Bocian DF, Strauss HL (1977) J Am Chem Soc 99:2876

314. Dillen J, Geise HJ (1979) J Chem Phys 70:425

315. Verdal N, Wilke JJ, Hudson BS (2006) J Phys Chem A 110:2639

316. Espinosa A, Gallo MA, Entrena A, Gómez JA (1994) J Mol Struct 323:247

317. Pakulski Z (1996) Pol J Chem 70:667

318. Pakulski Z (2006) Pol J Chem 80:1293

319. DeMatteo MP, Snyder NL, Morton M, Baldisseri DM, Hadad CM, Peczuh MW (2005) J Org Chem 70:24

320. Ng CJ, Craig DC, Stevens JD (1996) Carbohydr Res 284:249

321. Tran TQ, Stevens JD (2002) Austral J Chem 55:171

322. Driver GE, Stevens JD (2001) Carbohydr Res 334:81

323. Choong W, McConnell JF, Stephenson NC, Stevens JD (1980) Austral J Chem 33:979

324. James VJ, Stevens JD (1982) Cryst Struct Commun 11:79

325. Dabrowski J, Kozar T, Grosskurth H, Nifant'ev NE (1995) J Am Chem Soc 117:5534

326. Landersjö C, Stenutz R, Widmalm G (1997) J Am Chem Soc 119:8695

327. Eklund R, Lycknert K, Söderman P, Widmalm G (2005) J Phys Chem B 109:19936

328. Asensio JL, Hidalgo A, Cuesta I, Gonzalez C, Cañada J, Vicent C, Chiara JL, Cuevas G, Jiménez-Barbero J (2002) Chem Eur J 8:5228

329. French AD, Brady JW (1990) Computer Modeling of Carbohydrates, An Introduction. In: French AD, Brady JW (eds) Computer Modeling of Carbohydrate Molecules. ACS, Washington, DC, p 1

330. Babirad SA, Wang Y, Goekjian PG, Kishi Y (1987) J Org Chem 52:4825

331. Ravishankar R, Surolia A, Vijayan M, Lim S, Kishi Y (1998) J Am Chem Soc 120:11297

332. Lemieux RU, Koto S (1974) Tetrahedron 30:1933

333. Landersjö C, Stevensson B, Eklund R, Östervall J, Söderman P, Widmalm G, Maliniak A (2006) J Biomol NMR 35:89

334. Garcia-Aparicio V, Fernandez-Alonso MDC, Angulo J, Asensio JL, Cañada FJ, Jiménez-Barbero J, Mootoo DR, Cheng XH (2005) Tetrahedron: Asym 16:519

335. Cheetham NWH, Dasgupta P, Ball GE (2003) Carbohydr Res 338:955

336. Odelius M, Laaksonen A, Widmalm G (1995) J Phys Chem 99:12686

337. Asensio JL, Espinosa JF, Dietrich H, Cañada FJ, Schmidt RR, Martín-Lomas M, Andre S, Gabius HJ, Jiménez-Barbero J (1999) J Am Chem Soc 121:8995

338. Milton MJ, Bundle DR (1998) J Am Chem Soc 120:10547

339. Harris R, Kiddle GR, Field RA, Milton MJ, Ernst B, Magnani JL, Homans SW (1999) J Am Chem Soc 121:2546

340. Poppe L, Brown GS, Philo JS, Nikrad PV, Shah BH (1997) J Am Chem Soc 119:1727

341. Fernandez-Alonso MD, Cañada FJ, Solis D, Cheng XH, Kumaran G, Andre S, Siebert HC, Mootoo DR, Gabius HJ, Jiménez-Barbero J (2004) Eur J Org Chem 1604

342. Haselhorst T, Espinosa JF, Jiménez-Barbero J, Sokolowski T, Kosma P, Brade H, Brade L, Peters T (1999) Biochemistry 38:6449

343. Haselhorst T, Weimar T, Peters T (2001) J Am Chem Soc 123:10705

344. Geyer A, Müller M, Schmidt RR (1999) J Am Chem Soc 121:6312
345. Siebert HC, Jiménez-Barbero J, Andre S, Kaltner H, Gabius HJ (2003) Describing Topology of Bound Ligand by Transferred Nuclear Overhauser Effect Spectroscopy and Molecular Modeling. In: Lee YC, Lee RT (eds) Recognition of Carbohydrates in Biological Systems Pt A: General Procedures. Academic Press, San Diego, p 417
346. Eklund R, Widmalm G (2003) Carbohydr Res 338:393
347. Pérez S, Imberty A, Engelsen SB, Gruza J, Mazeau K, Jiménez-Barbero J, Poveda A, Espinosa JF, van Eyck BP, Johnson G, French AD, Louise M, Kouwijzer CE, Grootenuis PDJ, Bernardi A, Raimondi L, Senderowitz H, Durier V, Vergoten G, Rasmussen K (1998) Carbohydr Res 314:141
348. Almond A (2005) Carbohydr Res 340:907
349. Woods RJ, Dwek RA, Edge CJ, Fraser-Reid B (1995) J Phys Chem 99:3832
350. Hounsell EF (1994) Adv Carbohydr Chem Biochem 50:311
351. Hounsell EF (1995) Prog Nuc Magn Reson Spectrosc 27:445
352. Bush CA, Martín-Pastor M, Imberty A (1999) Ann Rev Biophys Biomol Struct 28:269
353. Lipari G, Szabo A (1982) J Am Chem Soc 104:4559
354. Widmalm G (1998) Physical Methods in Carbohydrate Research. In: Boons G-J (ed) Carbohydrate Chemistry. Blackie, London, p 448
355. Lycknert K, Rundlof T, Widmalm G (2002) J Phys Chem B 106:5275
356. Lycknert K, Widmalm G (2004) Biomacromolecules 5:1015
357. Andersson A, Ahl A, Eklund R, Widmalm G, Maler L (2005) J Biomol NMR 31:311
358. Clore GM, Szabo A, Bax A, Kay LE, Driscoll PC, Gronenborn AM (1990) J Am Chem Soc 112:4989
359. Kjellberg A, Rundlöf T, Kowalewski J, Widmalm G (1998) J Phys Chem B 102:1013
360. Kjellberg A, Widmalm G (1999) Biopolymers 50:391
361. Mäler L, Lang J, Widmalm G, Kowalewski J (1995) Magn Reson Chem 33:541
362. Glusker JP, Lewis M, Rossi M (1994) Crystal Structure Analysis for Chemists and Biologists. Wiley-VCH, New York
363. Hammond C (1997) The Basics of Crystallography and Diffraction. Oxford University Press, New York
364. Ladd MFC, Palmer RA (2003) Structure determination by X-ray crystallography. 4th edn. Kluwer Academic/Plenum Press, New York
365. Hammond C (2001) The Basics of Crystallography and Diffraction. 2nd edn. Oxford University Press, New York
366. Jeffrey GA, Sundaralingam M (1985) Adv Carbohydr Chem Biochem 43:203
367. Maluszynska H, Takagi S, Jeffrey GA (1977) Acta Crystallogr B33:1792
368. Ravishankar R, Suguna K, Surolia A, Vijayan M (1999) Acta Crystallogr D55:1375
369. Alibes R, Bundle DR (1998) J Org Chem 63:6288
370. Friebolin H (2005) Basic One- and Two-Dimensional NMR Spectroscopy. 4th edn. Wiley-VCH, New York
371. Becker ED (2000) High Resolution NMR : Theory and Chemical Applications. 3rd edn. Academic Press, San Diego, CA
372. Braun S, Berger S (2004) 200 and more Basic NMR Experiments: A Practical Course. 3rd edn. Wiley-VCH, New York
373. Homans SW (1992) A Dictionary of Concepts in NMR. Rev. edn. Clarendon, Oxford
374. Vliegenthart JFG, Dorland L, van Halbeek H (1983) Adv Carbohydr Chem Biochem 41:209
375. van Halbeek H (1996) Carbohydrates and Glycoconjugates. In: Grant DM, Harris RK (eds) Encyclopedia of Nuclear Magnetic Resonance. Wiley, New York, p 1107
376. Serianni AS (2000) Carbohydrate Structure, Conformation, and Reactivity: NMR Studies with Stable Isotopes. In: Hecht SM (ed) Bioorganic Chemistry: Carbohydrates. Oxford University Press, New York, p 244
377. Jiménez-Barbero, J. and Peters, T.(2003) NMR Spectroscopy of Glycoconjugates. Wiley-VCH, Weinheim
378. Bock K, Pedersen C (1983) Adv Carbohydr Chem Biochem 41:27
379. Bock K, Pedersen C, Pedersen H (1984) Adv Carbohydr Chem Biochem 42:193
380. Shashkov AS, Nifant'ev ÉE, Amochaeva VY, Kochetkov NK (1993) Magn Reson Chem 31:599
381. Hobley P, Howarth O, Ibbett RN (1996) Magn Reson Chem 34:755
382. Lemieux RU, Stevens JD (1965) Can J Chem 43:2059

383. Jansson P-E, Kenne L, Widmalm G (1989) Carbohydr Res 188:169
384. Marshall JL (1983) Carbon-Carbon and Carbon-Proton NMR Couplings : Applications to Organic Stereochemistry and Conformational Analysis. Verlag Chemie International, Deerfield Beach, FL
385. Perlin AS, Casu B (1969) Tetrahedron Lett 2921
386. Bock K, Pedersen C (1974) J Chem Soc, Perkin Trans 2 293
387. Tvaroška I, Taravel FR (1995) Adv Carbohydr Chem Biochem 51:15
388. Cyr N, Perlin AS (1979) Can J Chem 57:2504
389. Mizutani K, Kasai R, Nakamura M, Tanaka O, Matsuura H (1989) Carbohydr Res 185:27
390. Callam CS, Gadikota RR, Lowary TL (2001) J Org Chem 66:4549
391. Hällgren C (1992) J Carbohydr Chem 11:527
392. Cahill R, Cookson RC, Crabb TA (1969) Tetrahedron 25:4681
393. Grindley TB, Szarek WA (1974) Can J Chem 52:4062
394. Hansen PE (1981) Prog Nuc Magn Reson Spectrosc 14:175
395. Cyr N, Hamer GK, Perlin AS (1978) Can J Chem 56:297
396. Schwarcz JA, Cyr N, Perlin AS (1975) Can J Chem 53:1872
397. Bock K, Pedersen C (1977) Acta Chem Scand B 31:354
398. Prytulla S, Lambert J, Lauterwein J, Klessinger M, Thiem J (1990) Magn Reson Chem 28:888
399. Staaf M, Weintraub A, Widmalm G (1999) Eur J Biochem 263:656
400. Pachler KGR (1996) Magn Reson Chem 34:711
401. Church T, Carmichael I, Serianni AS (1996) Carbohydr Res 280:177
402. Cloran F, Carmichael I, Serianni AS (2000) J Am Chem Soc 122:396
403. Lemieux RU, Kullnig RK, Bernstein HJ, Schneider WG (1958) J Am Chem Soc 80:6098
404. Karplus M (1963) J Am Chem Soc 85:2870
405. Haasnoot CAG, DeLeeuw FAAM, Altona C (1980) Tetrahedron 36:2783
406. Altona C, Francke R, de Haan R, Ippel JH, Daalmans GJ, Hoekzema AJAW, van Wijk J (1994) Magn Reson Chem 32:670
407. Altona C (1996) Vicinal Coupling Constants and Conformation of Biomolecules. In: Grant DM, Harris RK (eds) Encyclopedia of Nuclear Magnetic Resonance. Wiley, New York, p 4909
408. Barfield M, Smith WB (1992) J Am Chem Soc 114:1574
409. Barfield M, Smith WB (1993) Magn Reson Chem 31:696
410. Imai K, Osawa E (1990) Magn Reson Chem 28:668
411. Osawa E, Ouchi T, Saito N, Yamoto M, Lee OS, Seo MK (1992) Magn Reson Chem 30:1104
412. Marino JP, Schwalbe H, Griesinger C (1999) Acc Chem Res 32:614
413. Anet FAL, Freedberg DI (1993) Chem Phys Lett 208:187
414. Mulloy B, Frenkiel TA, Davies DB (1988) Carbohydr Res 184:39
415. Tvaroška I, Gajdos J (1995) Carbohydr Res 271:151
416. Milton MJ, Harris R, Probert MA, Field RA, Homans SW (1998) Glycobiology 8:147
417. Bose B, Zhao S, Stenutz R, Cloran F, Bondo PB, Bondo G, Hertz B, Carmichael I, Serianni AS (1998) J Am Chem Soc 120:11158
418. Zhu YP, Pan QF, Thibaudeau C, Zhao SK, Carmichael I, Serianni AS (2006) J Org Chem 71:466
419. Coxon B, Sari N, Batta G, Pozsgay V (2000) Carbohydr Res 324:53
420. Coxon B (1972) Conformational Analysis via Nuclear Magnetic resonance Spectroscopy. In: Whistler RL, BeMiller JN (eds) Methods in Carbohydrate Chemistry. Academic Press, New York, p 513
421. Kotowycz G, Lemieux RU (1973) Chem Rev 73:669
422. Otter A, Bundle DR (1995) J Magn Reson B 109:194
423. Emsley JW, Lindon JC (1975) NMR Spectroscopy using Liquid Crystal Solvents. Pergamon Press, Oxford
424. Diehl P (1996) Structure of Rigid Molecules Dissolved in Liquid Crystalline Solvents. In: Grant DM, Harris RK (eds) Encyclopedia of Nuclear Magnetic Resonance. Wiley, New York, p 4591
425. Tjandra N, Bax A (1997) Science 278:1697
426. Bolon PJ, Prestegard JH (1998) J Am Chem Soc 120:9366
427. Kiddle GR, Homans SW (1998) FEBS Lett 436:128
428. Rundlöf T, Landersjö C, Lycknert K, Maliniak A, Widmalm G (1998) Magn Reson Chem 36:773
429. Venable RM, Delaglio F, Norris SE, Freedberg DI (2005) Carbohydr Res 340:863
430. Zhuang TD, Leffler H, Prestegard JH (2006) Protein Science 15:1780

431. Bush CA (2003) Origins of Flexibility in Complex Polysaccharides. In: Cheng HN, English AD (eds) NMR Spectroscopy of Polymers in Solution and in the Solid State. ACS, Washington, DC, p 272

432. Berthault P, Jeannerat D, Camerel F, Salgado FA, Boulard Y, Gabriel JCP, Desvaux H (2003) Carbohydr Res 338:1771

433. Neuhaus D, Williamson MP (2000) The Nuclear Overhauser Effect in Structural and Conformational Analysis. 2nd edn. Wiley-VCH, New York

434. Neuhaus D, Williamson MP (1989) The Nuclear Overhauser Effect in Structural and Conformational Analysis. VCH Publishers, New York

435. Noggle JH, Schirmer RE (1971) The Nuclear Overhauser Effect. Academic Press, New York

436. Overhauser AW (1953) Phys Rev 92:411

437. Solomon I (1955) Phys Rev 99:959

438. Krishna NR, Agresti DG, Glickson JD, Walter R (1978) Biophys J 24:791

439. Malliavin TE, Desvaux H, Delsuc MA (1998) Magn Reson Chem 36:801

440. Keepers JW, James TL (1984) J Magn Reson 57:404

441. Donati A, Rossi C, Martini S, Ulyanov NB, James TL (1998) App Magn Reson 15:401

442. Nakanishi K (1994) Circular Dichroism : Principles and Applications. VCH, New York

443. Woody RW (1996) Theory of Circular Dichroism of Proteins. In: Fasman GD (ed) Circular Dichroism and the Conformational Analysis of Biomolecules. Plenum Press, New York, p 25

444. Johnson WCJr (1987) Adv Carbohydr Chem Biochem 45:73

445. Stevens ES (1996) Carbohydrates. In: Fasman GD (ed) Circular Dichroism and the Conformational Analysis of Biomolecules. Plenum Press, New York, p 501

446. Arndt ER, Stevens ES (1993) J Am Chem Soc 115:7849

447. Stroyan EP, Stevens ES (2000) Carbohydr Res 327:447

448. Arndt ER, Stevens ES (1997) Carbohydr Res 303:73

449. Schafer SE, Stevens ES (1996) Carbohydr Polym 31:19

450. Szabo L, Smith BL, McReynolds KD, Parrill AL, Morris ER, Gervay J (1998) J Org Chem 63:1074

451. Andersson M, Kenne L, Stenutz R, Widmalm G (1994) Carbohydr Res 254:35

452. Wiesler WT, Vázquez JT, Nakanishi K (1987) J Am Chem Soc 109:5586

453. Morales EQ, Padrón JI, Trujillo M, Vázquez JT (1995) J Org Chem 60:2537

454. Burkert U, Allinger NL (1982) Molecular Mechanics. ACS, Washington

455. Woods RJ (1996) The Application of Molecular Modeling Techniques to the Determination of Oligosaccharide Solution Conformations. In: Lipkowitz KB, Boyd DB (eds) Reviews in Computational Chemistry. VCH Publishers, New York, p 129

456. Rao VSR, Sundararajan PR, Ramakrishnan C, Ramachandran GN (1967) In: Ramachandran GN (ed) Conformations of Biopolymers. Academic Press, New York, p 721

457. Kitaigorodsky AI (1978) Chem Soc Rev 7: 133

458. Lemieux RU, Bock K, Delbaere LTJ, Koto S, Rao VS (1980) Can J Chem 44:631

459. Stuike-Prill R, Meyer B (1990) Eur J Biochem 194:903

460. Thøgersen H, Lemieux RU, Bock K, Meyer B (1982) Can J Chem 44:44

461. Kuttel M, Brady JW, Naidoo KJ (2002) J Comp Chem 23:1236

462. Case DA, Cheatham TE, III, Darden T, Gohlke H, Luo R, Merz KM Jr, Onufriev A, Simmerling C, Wang B, Woods RJ (2005) J Comp Chem 26:1668

463. Lii JH, Chen KH, Johnson GP, French AD, Allinger NL (2005) Carbohydr Res 340:853

464. Kony D, Damm W, Stoll S, Hunenberger PH (2004) J Phys Chem B 108:5815

465. Lins RD, Hunenberger PH (2005) J Comp Chem 26:1400

466. Ewig CS, Berry R, Dinur U, Hill JR, Hwang MJ, Li HY, Liang C, Maple J, Peng ZW, Stockfisch TP, Thacher TS, Yan L, Ni XS, Hagler AT (2001) J Comp Chem 22:1782

467. Stortz CA (2006) Carbohydr Res 341:663

468. Allinger NL, Yuh YH, Lii J-H (1989) J Am Chem Soc 111:8551

469. Engelsen SB, Rasmussen K (1997) J Carbohydr Chem 16:751

470. Hwang MJ, Stockfisch TP, Hagler AT (1994) J Am Chem Soc 116:2515

471. Stortz CA (1999) Carbohydr Res 322:77

472. Saunders M (1987) J Am Chem Soc 109:3150

473. Peters T, Meyer B, Stuike-Prill R, Somorjai R, Brisson JR (1993) Carbohydr Res 238:49

474. von der Lieth CW, Kozar T, Hull WE (1997) Theochem 395:225

475. Brocca P, Bernardi A, Raimondi L, Sonnino S (2000) Glycoconjugate J 17:283

476. Abraham RJ, Bretschneider E (1974) In: Orville-Thomas WF (ed) Internal Rotation in Molecules. Wiley-Interscience, New York, p 481

477. Tomasi J, Mennucci B, Cammi R (2005) Chem Rev 105:2999

478. Chamberlin AC, Cramer CJ, Truhlar DG (2006) J Phys Chem B 110:5665

479. Shen MY, Freed KF (2005) J Comp Chem 26:691

480. Dixon AM, Venable R, Widmalm G, Bull TE, Pastor RW (2003) Biopolymers 69:448

481. Kelly CP, Cramer CJ, Truhlar DG (2006) J Phys Chem A 110:2493

482. Curutchet C, Orozco M, Luque FJ, Mennucci B, Tomasi J (2006) J Comp Chem 27:1769

483. Chen JH, Im WP, Brooks CL (2006) J Am Chem Soc 128:3728

484. Lin ST, Hsieh CM (2006) J Chem Phys 125:

485. Wang ML, Wong CF (2006) J Phys Chem A 110:4873

486. Haille JM (1992) Molecular Dynamics Simulation. Elementary Methods. Wiley, New York

487. Balbuena PB, Seminario JM (1999) Molecular Dynamics: from Classical to Quantum Methods. Elsevier, Amsterdam

488. Allen MP, Tildesley DJ (1987) Computer Simulation of Liquids. Clarendon Press, Oxford

489. Spieser SAH, van Kuik JA, Kroon-Batenburg LMJ, Kroon J (1999) Carbohydr Res 322:264

490. Levine IN (2000) Quantum Chemistry. 5th edn. Prentice Hall, Upper Saddle River, NJ

491. Lowe JP, Peterson KA (2006) Quantum Chemistry. 3rd edn. Elsevier Academic Press, Burlington, MA

492. Lewars E (2003) Computational Chemistry: Introduction to the Theory and Applications of Molecular and Quantum Mechanics. Kluwer Academic Publishers, Boston, MA

493. Cramer CJ (2004) Essentials of Computational Chemistry: Theories and Models. 2nd edn. Wiley, Chichester, UK

494. Schnupf U, Willett JL, Bosma WB, Momany FA (2007) Carbohydr Res 342:196

495. French AD, Johnson GP (2006) Can J Chem 84:603

496. Hricovíni M (2006) Carbohydr Res 341:2575

497. Miura N, Taniguchi T, Monde K, Nishimura SI (2006) Chem Phys Lett 419:326

498. Becke AD (1993) J Chem Phys 98:5648

499. Lii JH, Ma BY, Allinger NL (1999) J Comp Chem 20:1593

1.2 General Properties, Occurrence, and Preparation of Carbohydrates

John F. Robyt
Laboratory of Carbohydrate Chemistry and Enzymology, Department
of Biochemistry, Biophysics, and Molecular Biology, 4252 Molecular Biology
Building, Iowa State University, Ames, IA 50011, USA
jrobyt@iastate.edu

In: *Glycoscience*. Fraser-Reid B, Tatsuta K, Thiem J (eds)
Chapter-DOI 10-1007/978-3-540-30429-6_2: © Springer-Verlag Berlin Heidelberg 2008

Abstract

D-Glucose and its derivatives and analogues, N-acetyl-D-glucosamine, N-acetyl-D-muramic acid, D-glucopyranosyl uronic acid, and D-glucitol represent 99.9% of the carbohydrates on the earth. D-Glucose is found in the free state in human blood and in the combined state in disaccharides, sucrose, lactose, and α,α-trehalose, in cyclic dextrins, and in polysaccharides, starch, glycogen, cellulose, dextrans; N-acetyl-D-glucosamine and an analogue N-acetyl-D-muramic acid are found in bacterial cell wall polysaccharide, murein, along with teichoic acids made up of poly-glycerol or -ribitol phosphodiesters. Other carbohydrates, D-mannose, D-mannuronic acid, D-galactose, N-acetyl-D-galactosamine, D-galacturonic acid, L-iduronic acid, L-guluronic acid, L-rhamnose, L-fucose, D-xylose, and N-acetyl-D-neuraminic acid are found in glycoproteins, hemicelluloses, glycosaminoglycans, and polysaccharides of plant exudates, bacterial capsules, alginates, and heparin. D-Ribofuranose-5-phosphate is found in many coenzymes and is the backbone of RNAs (ribonucleic acid), and 2-deoxy-D-ribofuranose-5-phosphate is the backbone of DNA (deoxyribonucleic acid). D-Fructofuranose is found in sucrose, inulin, and levan. The general properties and occurrence of these carbohydrates and general methods of isolation and preparation of carbohydrates are presented.

Keywords

D-Glucose; D-Fructose; Sucrose; Lactose; α,α-Trehalose; Starch; Glycogen; Cyclodextrins; Dextrans; Alternan

Abbreviations

ADPGlc	adenosine-diphospho-glucose
ATP	adenosine triphosphate
CGTase	cyclomaltodextrin glucanyltransferase
d.s.	degree of substitution
DNA	deoxyribonucleic acid
FACE	fluorophore-assisted capillary electrophoresis
FAD	oxidized flavin adenine dinucleotide
HPLC	high pressure liquid chromatography
MALDI-TOF MS	matrix-assisted laser desorption ionization–time of flight mass spectrometry
NAD$^+$	oxidized nicotinamide adenine dinucleotide
NADH	reduced nicotinamide adenine dinucleotide
NADPH	reduced nicotinamide adenine dinucleotide phosphate
ORD	optical rotatory dispersion
RNA	ribonucleic acid
NADPH	nicotinamide adenine dinucleotide-phosphate
NAG	N-acetyl-D-glucosamine
NAM	N-acetyl-D-muramic acid
TLC	thin-layer chromatography
TFMS	trifluoromethane sulfonic acid
UDPGlc	uridine-diphospho-glucose

1 General Properties and Occurrence of Carbohydrates

Carbohydrates have the following major properties: (1) they are polyhydroxy aldehydes or ketones; (2) they have chiral or asymmetric carbons that are generally manifested by the rotation of plane polarized light; (3) they have the ability to form multiple hydrogen bonds, generally giving them the property of being water-soluble, but they also can be water-insoluble when they form intermolecular hydrogen bonds with each other to give crystals or large, high molecular weight, insoluble crystalline aggregates, granules, or fibers; (4) many have reactivities of aldehydes that can be oxidized to acids by reagents that are thereby reduced (e. g., reducing an oxidizing agent such as an alkaline solution of copper(II) or ferricyanide/cyanide), and they, hence, are considered to be reducing sugars, or they can themselves be reduced by reducing reagents, such as $NaBH_4$, to give sugar alcohols; (5) the aldehyde or ketone groups in carbohydrates with five or more carbons will react with intramolecular alcohol groups to form cyclic structures with hemiacetal and hemiketal hydroxyl groups; (6) the hemiacetal or hemiketal hydroxyls are more reactive than the alcohols and can react intermolecularly with alcohols and amines to give acetals or ketals (glycosidic bonds) that are fairly stable; (7) they have two kinds of alcohol groups, secondary and primary, that can undergo the usual reactions of alcohols to give esters and ethers and can be replaced, for example, by hydrogen, halogens (F, Cl, Br, and I), amino groups, N-acetyl amino groups, and sulfhydryl groups; (8) they are generally, although not all of them, sweet-tasting (for example, D-glucose, D-glucitol, D-fructose, D-xylose, D-xylitol and sucrose are sweet-tasting) by forming specific hydrogen and hydrophobic bonds with the sweet-taste receptors on the tongue; and (9) when attached to proteins or cell surfaces, the structural diversity of oligosaccharides mediate a large number of biochemical and biological processes.

In the 19th century, several naturally occurring carbohydrates were known, such as glucose (then called dextrose), fructose (then called levulose), mannose, galactose, sucrose, lactose, starch, and cellulose. Some of these had been known for thousands of years, for example, sucrose, starch, and cellulose. Also in the 19th century, the empirical formula for all of these materials was found to be $C_n(H_2O)_n$ and they were originally thought to be hydrates of carbon, hence the name carbohydrates.

Carbohydrates are now more completely defined as polyhydroxy aldehydes or ketones and compounds that can be derived from them by reduction to give sugar alcohols, oxidation to give sugar acids, substitution of hydroxyl group(s) by hydrogen to give deoxy sugars or by amino or N-acetyl amino groups to give deoxy-amino sugars, derivatization of a hydroxyl group by phosphate or sulfate to give sugar phosphates or sugar sulfates, and by condensation reactions of a hydroxyl group of one sugar with the hemiacetal group of another sugar to give disaccharides, trisaccharides, oligosaccharides, and polysaccharides.

2 Carbohydrate Property of Optical Rotation of Plane Polarized Light

An important property of carbohydrates that was recognized in the 19th century was that they generally, but not always, rotated plane polarized light and that this was specific for each carbohydrate. This property is due to the presence of asymmetric or chiral carbons that have four different groups attached to the carbons. Those carbohydrates that rotate plane polarized light are said to be optically active. It was also recognized that the optical rotation was dependent on several factors: (1) the structure of the substance; (2) the length of the cell; (3) the concentration of the substance; (4) the wavelength of the plane polarized light; and (5) the temperature. The following relationship was derived to encompass these variables:

$$\alpha_{obs} = [\alpha]_\lambda^t \, l c$$

where α_{obs} = the observed optical rotation in degrees, l = the length of the cell holding the compound in dm (decimeter), c = the concentration of the sample in $g \, mL^{-1}$, usually in water, $[\alpha]_\lambda^t$ = the specific optical rotation constant of the substance at temperature, t, and wavelength, λ. Most polarimetric measurements are made with the D-line from a sodium lamp and each carbohydrate has a characteristic $[\alpha]_D^t$, although the optical rotation can also be measured continuously as a as a function of the wavelength (i. e., optical rotatory dispersion, ORD). Carbohydrate molecules with two-fold symmetry about a central point or plane do not rotate plane polarized light and are said to have a meso-structure.

3 The Structures of Carbohydrates

3.1 The Simplest Carbohydrates

There are three carbohydrates that are the simplest carbohydrates that fulfill the definition given above. They are the following:

L-glyceraldehyde dihydroxyacetone D-glyceraldehyde

$[\alpha]_D^{25} = -8.7°$ $[\alpha]_D^{25} = +8.7°$

◻ Scheme 1
The simplest carbohydrates

There are two forms for glyceraldehyde that are distinct and cannot be superimposed onto each other. Prof. Fischer defined the one with the chiral hydroxyl group to the right as D-glyceraldehyde, where the "D" indicates that the hydroxyl group is to the right or *dextro* and the one with the chiral hydroxyl group to the left as L-glyceraldehyde where the "L" indicates that the hydroxyl group is to the left or *levo*. It just so happened that for D-glyceraldehyde, plane polarized light was rotated to the right and L-glyceraldehyde rotated plane polarized light to the left. This is not always the case. Some carbohydrates with the D-configuration rotate plane polarized light to the left and some carbohydrates with the L-configuration rotate plane polarized light to the right.
A large majority of the carbohydrates found on the earth belong to the D-family of structural isomers. In the course of evolution, the reason that the D-family of structural isomers was

selected over those of the L-family of structural isomers is not clear. It, however, is not likely that it was a matter of chance. It has been known for many years that irradiation of a racemic mixture of D- and L- isomers with circularly polarized light will selectively destroy one of the two isomers, leaving the other more or less intact [1]. Circularly polarized light has been observed when there is high sunspot activity. High levels of circularly polarized light have also been observed coming from the *Orion nebula* [2]. The selection of D-carbohydrates could have occurred by this type of irradiation when carbohydrates were first being formed on the earth.

3.2 Analogues of D-Glyceraldehyde

D-Glyceraldehyde has some derived analogues, such as the reduced sugar alcohol, glycerol, its oxidized product, D-glyceric acid, and their phosphorylated analogues, 1-phospho-D-glycerol, 3-phospho-D-glyceraldehyde, and 3-phospho-D-glyceric acid, whose structures are shown as:

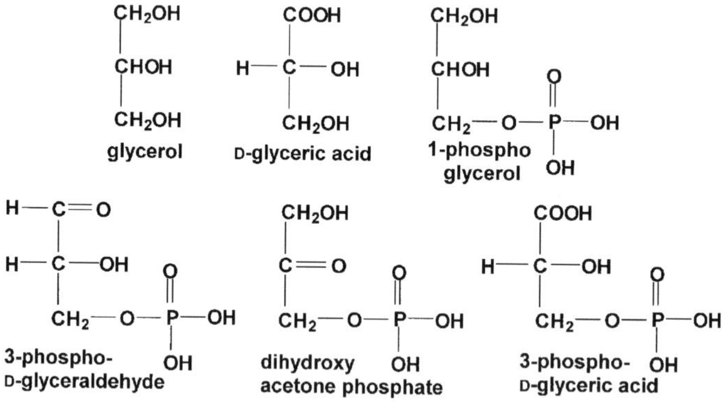

□ Scheme 2
Analogues and derivatives of the naturally occurring three-carbon carbohydrates

Glycerol is found as the backbone compound that is esterified by fatty acids to give a class of lipids known as triacyl glycerols (glycerides), and glycerol-1-phosphate is the backbone of a major class of phospholipids. 3-Phospho-D-glyceraldehyde, dihydroxy acetone phosphate, and 3-phospho-D-glyceric acid are all found in both the reactions of photosynthesis and in the degradative reactions of glycolysis.

3.3 The Formation of Carbohydrates Containing More than Three Carbons

From a theoretical stand point, Professor Emil Fischer showed that by adding a new chiral carbon between the aldehyde group and the asymmetric carbon of D-glyceraldehyde, a chiral pair of 4-carbon D-tetraoses would be obtained, namely, D-erythrose and D-threose. Adding another set of similar chiral carbons to each of the two D-tetraoses, gives four 5-carbon D-pentoses, and likewise adding a similar set of chiral carbons to each of the four D-pentoses, gives eight 6-carbon D-hexoses. It should be noted that it is only the configuration of the last asymmetric

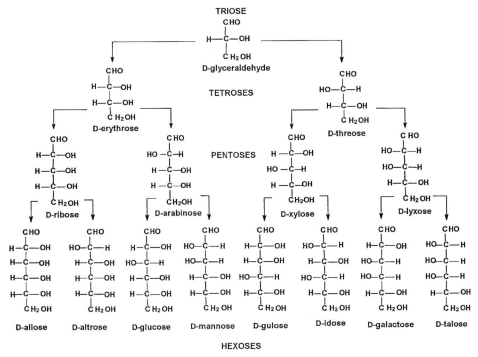

□ Figure 1
Structural family of D-carbohydrates, from triose to hexoses, with their names

carbon with the hydroxyl group to the right that makes it a "D" carbohydrate and the entire mirror image of the D-carbohydrates gives the L-carbohydrate. See ❯ Fig. 1 for the names and structures of the family of D-carbohydrates.

3.4 Special Properties of Pentoses and Hexoses

The pentoses and hexoses have a propensity for forming six-membered rings in which one of their hydroxyl groups reacts intramolecularly with the aldehyde group to form a cyclic hemiacetal. This reaction creates a new asymmetric center on the aldehyde carbon to give two isomers, called alpha (α) and beta (β). Two five-membered rings are also formed, but in much smaller amounts, as the six-membered cyclic structures are much more thermodynamically stable (i. e., less strained) than are the five-membered rings. The six-membered rings are called pyranoses and the five-membered rings are called furanoses. For D-glucose at 20 °C and equilibrium, there are five compounds: 0.004% the open aldehyde chain, 66% β-D-glucopyranose, 33% α-D-glucopyranose, and 0.498% each of β-D-glucofuranose and α-D-glucofuranose. See ❯ Fig. 2 for the structures of the five forms of D-glucose in equilibrium.

If one starts with α-D-glucopyranose, $[\alpha\text{-}]_D^{25} = +112°$, the optical rotation drops to +52° and if one starts with β-D-glucopyranose, $[\alpha\text{-}]_D^{25} = +19°$, the optical rotation increases and becomes constant at +52°, which is the optical rotation for an equilibrium mixture of the five structural

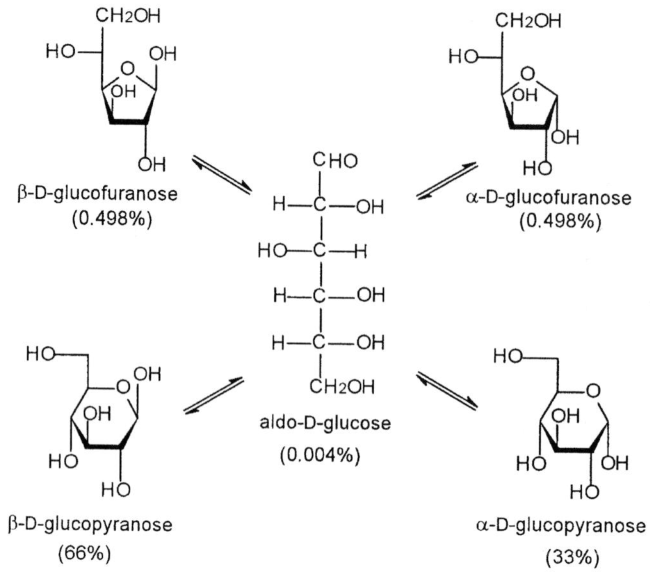

■ Figure 2

The five structural forms of D-glucose at equilibrium in aqueous solution at 20 °C

forms of D-glucose. The process is known as mutarotation and is relatively slow at pH 7 and 20 °C. It can be accelerated by catalysis with either acid or base or by adding an enzyme, known as mutarotase. Dilute base (pH 10) is a better catalyst by a factor of 5,000 than dilute acid (pH 4). Mutarotase acts as an acid–base catalyst and catalyzes the reactions 4–5 orders of magnitude faster than base.

3.5 D-Glucose: the Most Prominent Carbohydrate on the Earth

Of the 15 possible D-carbohydrates in ❿ *Fig. 1*, only a handful occurs in nature to any extent. By far, D-glucose and its analogues are the most prominent and represent 99.9% of the carbohydrates on the earth. Why is this? While D-glucose forms the six-membered cyclic structure and has the Haworth structure as shown in ❿ *Fig. 3*, the ring actually has a three-dimensional chair conformation, with two kinds of geometric bonds around the carbons, those that are within the plane of the ring (called equatorial bonds) and those that are perpendicular to the ring (called axial bonds). D-Glucose can exist in two chair conformations, the C1 or $^4C_1$ chair and the 1C or $_4C^1$ chair (see ❿ *Fig. 3* for the structures). In the C1 or $^4C_1$ conformation, all of the hydroxyl or bulkiest groups for β-D-glucopyranose are attached to the ring by equatorial bonds that put the hydroxyl or bulkiest groups as far apart as possible from each other, giving the most thermodynamically stable structure possible. If β-D-glucopyranose is in the other chair conformation, 1C or $_4C^1$, all of the hydroxyl groups are axial and are placed as close together as possible, giving the most thermodynamically unstable structure possible. Thus, β-D-glucopyranose exists primarily in the C1-conformation. α-D-Glucopyranose also exists

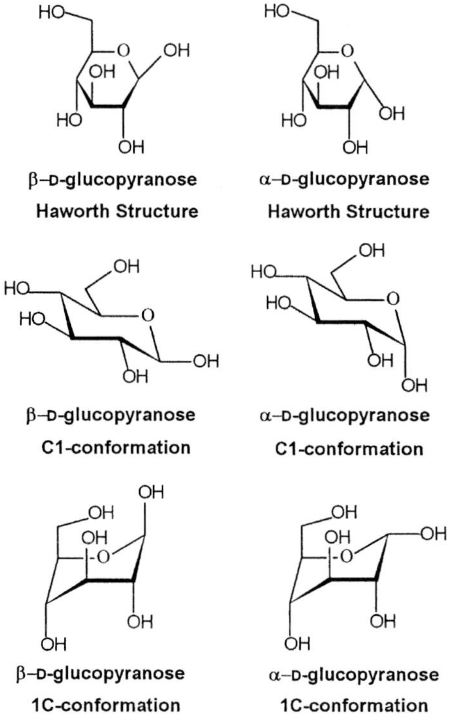

□ **Figure 3**
Haworth structures for α- and β-D-glucopyranose and their C1 and 1C conformations

in the C1-conformation, with only the hemiacetal hydroxyl group in the axial position. This is the most likely reason that D-glucose is the predominant carbohydrate on the earth, as it is the only D-hexose that can have all of its hydroxyl groups (exclusive of the hemiacetal hydroxyl group) equatorial, and at equilibrium in solution β-D-glucopyranose has a ratio of ∼2:1 to α-D-glucopyranose, which has its hemiacetal hydroxyl group axial. D-Xylose, a pentose, can also form a six-membered ring by its terminal hydroxyl group reacting with the aldehyde group and form a C1 conformation and place all of its bulky hydroxyl groups equatorial, but because it has five carbons, it would have to be split into a 2-carbon fragment and a 3-carbon fragment for metabolism and would require two separate pathways for further metabolism. D-Glucose has 6-carbons and is split into two 3-carbon fragments, D-glyceraldehyde-3-phosphate and dihydroxy acetone-phosphate that are interconvertible and requires only a single metabolic pathway for further metabolism. This is an additional plausible reason that D-glucopyranose is the predominant carbohydrate on the earth.

The D-pentoses will also exist in solution as the six-membered ring structure, but both D-xylose and D-ribose often have their C-5 hydroxyl groups phosphorylated and the C-5 hydroxyl group cannot react with the aldehyde group to form the six-membered ring, and therefore they do the next best thing, with the C-4 hydroxyl group reacting with the aldehyde group to form the five-membered, furanose ring, hemiacetal structure.

3.6 Occurrence of D-Erythrose, D-Ribose, and D-Xylose

Of the two D-tetroses, only 4-phospho-D-erythrose is found in any quantity, as an intermediate in the photosynthetic reactions. Of the four pentoses, only D-ribose and D-xylose occur to any extent. The phospho-*aldo*-D-pentoses, D-ribose-5-phosphate and D-xylose-5-phosphate, are found in the photosynthetic reactions. Three phospho-*keto*-D-pentoses are also found as intermediates in the photosynthetic reactions: D-ribulose-5-phosphate, D-xylulose-5-phosphate, and D-ribulose-1,5-bis-phosphate, the latter carbohydrate being directly involved in the fixing of CO_2 in photosynthesis.

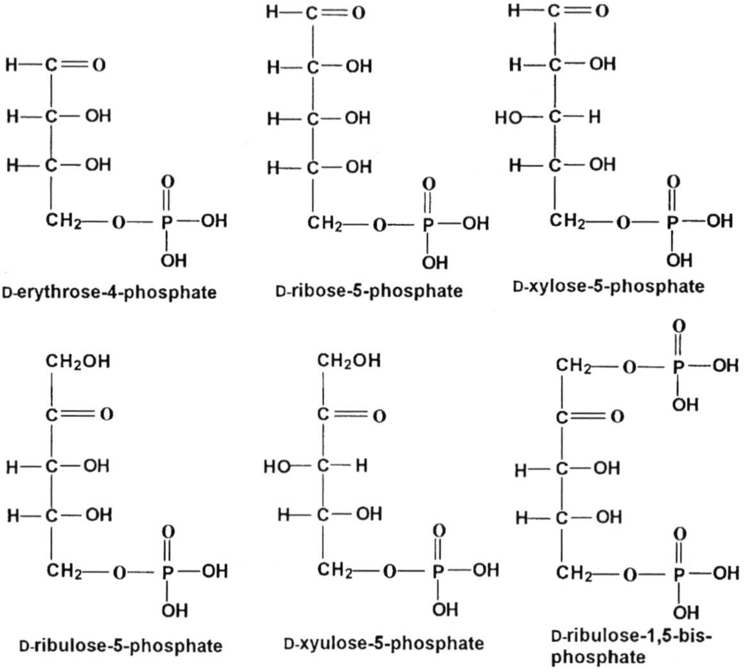

D-erythrose-4-phosphate D-ribose-5-phosphate D-xylose-5-phosphate

D-ribulose-5-phosphate D-xylulose-5-phosphate D-ribulose-1,5-bis-phosphate

☐ **Scheme 3**
Important naturally occurring four- and five-carbon sugar phosphates

Nicotinamide adenine dinucleotide (NAD$^+$ and NADH), coenzymes containing D-ribofuranose-diphosphate, are involved in many oxidation and reduction reactions, respectively, of carbohydrate metabolism. D-Ribofuranose is also the main component of the universal energy donor and energy carrier, adenosine triphosphate (ATP), which is one of the primary products of the light reactions of photosynthesis and is responsible for providing the energy for the formation of the carbon–carbon bond in the fixation of CO_2. It is also important in the transfer and utilization of energy in the metabolism of nonphotosynthesizing organisms. Another important D-ribofuranose coenzyme that is formed as a primary product of the light reactions of photosynthesis is the reducing coenzyme, nicotinamide adenine dinucleotide-phosphate,

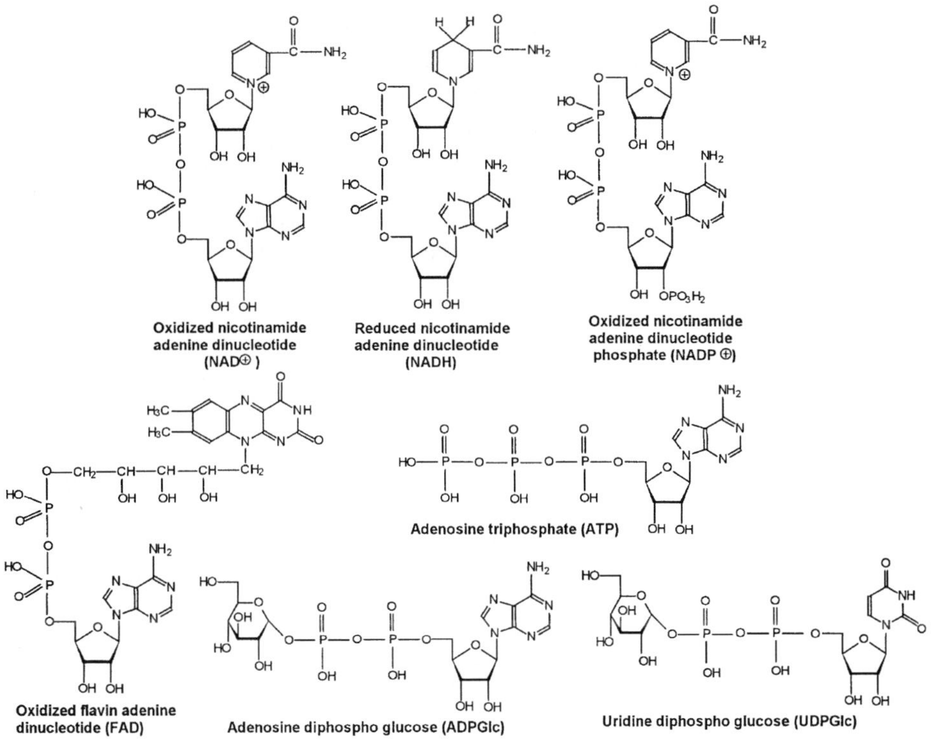

▣ Figure 4
Structures of the D-ribofuranose-5-phosphate nucleotide coenzymes that are important in biochemical metabolism

(NADPH), which is similar to the coenzymes, NAD$^+$ and NADH, mentioned above, but with an additional phosphate attached to the 2-position of the ribose unit. See ❷ *Fig. 4* for the structures of these coenzymes. NADPH is responsible for reducing the carbon–carbon bond that is formed in fixing CO_2 in photosynthesis.

ATP is the universal energy carrier and source of energy in biochemical systems; NAD$^+$ and NADH, oxidation and reduction coenzymes; FAD, oxidative coenzyme, containing ribitol-5-phosphate; ADPGlc, high-energy glucose donor, involved in starch biosynthesis; and UDPGlc, another high-energy glucose donor, involved in cellulose, glycogen, and sucrose biosyntheses, and in the enzymatic conversions of D-glucopyranose to many other sugars, such as, D-galactopyranose, D-glucopyranouronic acid, D-xylopyranose, and L-arabinopyranose.

D-Ribofuranose-5-phosphate occurs as the backbone component of the ribonucleic acids, RNA, that are involved in the biosynthesis of proteins. There are three kinds of RNA's: a small RNA, transfer-RNA that forms a high-energy, amino-acid covalent compound that transfers individual amino acids to the ribosome to be incorporated into proteins; an intermediate sized RNA, messenger-RNA that carries the codon or genetic information of a protein to the ribosome where the code is read and the peptide bonds of the protein are synthesized; and the largest sized RNA, ribosomal-RNA that composes the ribosome, the organelle where proteins are synthesized and assembled. 2-Deoxy-D-ribofuranose-5-phosphates are the backbone components of deoxyribonucleotides (DNA), which primarily act as the carrier of the genetic information, necessary for the formation of proteins involved in life processes. Like RNA, these molecules have two purines (adenine and guanine) and two pyrimidines (uracil and cytosine for RNA and thymidine and cytosine for DNA) that are linked β to carbon-1 of the D-ribofuranose-phosphate units to give N-glycosides.

D-Xylopyranose occurs as one of the major components in the hemicelluloses (see ❷ *Sect. 6* on cellulose and ❷ *Sect. 7* on hemicelluloses and ❷ Chap. 6.3 on cellulose).

3.7 Occurrence of Hexoses

Of the eight D-hexoses in ❷ *Fig. 1*, only three occur to any extent: D-glucose, its 2-isomer, D-mannose, and its 4-isomer, D-galactose. Another hexose that is found is D-fructose, which is a keto-sugar that can be derived from dihydroxy acetone, as the aldehyde carbohydrates were derived from D-glyceraldehyde. D-Fructose is formed when D-glucose or D-mannose are treated with alkali, which isomerizes carbons 1 and 2 of the two D-hexoses [3]. The diphosphate of D-fructose, D-fructose-1,6-bis-phosphate is the first hexose that is formed in the photosynthetic process and it is rapidly converted into D-fructose-6-phosphate, and then into D-glucose-6-phosphate.

4 Properties and Occurrence of D-Glucose

4.1 D-Glucose in the Free State

Free D-glucose occurs primarily in the blood of many higher animals, where it serves as an immediate source of energy and as a stabilizer of the osmotic pressure, and a precursor for the formation of glycogen and fat in muscle tissue. In normal humans, the concentration of blood glucose is 80–120 mg/100 mL^{-1} or 5–7 mM. This can increase to 200–300 mg/100 mL^{-1} or 11–20 mM after a high-carbohydrate meal, but then is relatively rapidly decreased by the action of insulin and often goes below 80 mg/mL^{-1} and then goes slowly up to normal levels. In uncontrolled diabetics the glucose often goes much higher, to 140–1100 mg/100 mL^{-1} or 8–60 mM. In controlled diabetics, the glucose will often be slightly higher than normal, for example, 120–140 mg/100 mL^{-1}.

D-Glucose is also found in the free state in honey, grapes, and raisins. It is produced by the action of glucoamylase (amyloglucosidase) on starch, where glucoamylase hydrolyzes the α-(1→4) linkage of the glucose units at the nonreducing-ends of the starch chains, giving inversion of the configuration, forming β-D-glucose. Glucoamylase will also hydrolyze α-(1→6) branch linkages, although at a rate about 0.1 that of the α-(1→4) linkage. Eventually glucoamylase will completely convert all of the starch into D-glucose.

4.2 D-Glucose in the Combined State

D-Glucose is found in a combined form in which its hemiacetal group has reacted with a hydroxyl on another glucose or another carbohydrate to form an acetal or glucosidic linkage and form such carbohydrate substances as, starch, dextran, cellulose, sucrose, lactose, α,α-trehalose.

4.2.1 Occurrence of D-Glucose Combined with D-Fructose, D-Galactose, and D-Glucose, and High-Energy D-Glucose Donors

There are three major, naturally occurring disaccharides: sucrose, lactose, and α,α-trehalose and D-glucopyranose is found in all three. Sucrose is a nonreducing disaccharide composed of D-glucopyranose joined to D-fructofuranose $(1\rightarrow2)$ to give α-D-glucopyranosyl-$(1\rightarrow2)$-β-D-fructofuranoside. Sucrose is widely distributed in plants, primarily as 6^{Fru}-phosphosucrose, which is the transport form of carbohydrates in plants. When it reaches certain parts of the plant, the glucose moiety is converted into starch by a series of reactions that give adenosine-diphospho-glucose (ADPGlc) or uridine-diphospho-glucose (UDPGlc), high-energy donors of glucose for the biosynthesis of starch and cellulose, respectively. Relatively large amounts (15–20% by weight) of free sucrose are found in the stems and tubers of sugar cane and the tubers of sugar beets. Free sucrose is also found in many other plants but in lower amounts, such as their fruits. Sucrose is the predominant sugar found in honey, produced by bees, in the sap of maple trees, giving maple syrup, and in sorghum, and in dates. See ❷ *Fig. 5* for the structures of sucrose, lactose, and α,α-trehalose and ❷ *Fig. 4* for the structures of ADPGlc and UDPGlc.

4.2.2 Properties and Occurrence of Sucrose and Sucrose Oligosaccharides Containing D-Galactose

Sucrose has five important properties: (a) the linkage between D-glucopyranose and D-fructofuranose is of high energy being an acetal–ketal linkage of a six-membered ring attached to a five-membered ring, making sucrose a nonreducing sugar. (b) The acetal–ketal linkage is

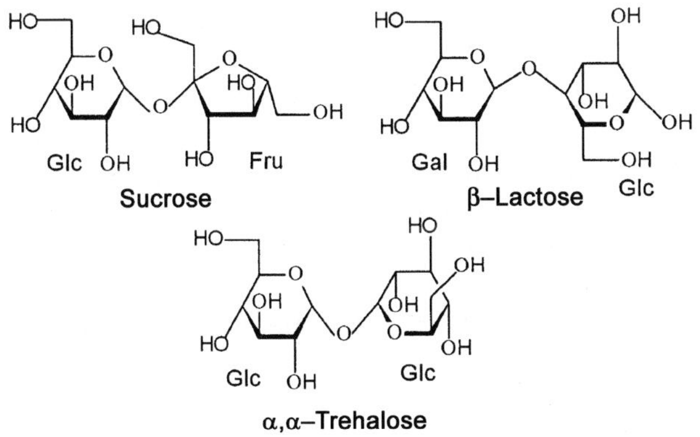

◻ Figure 5
Structures of the three naturally occurring disaccharides: sucrose, lactose, and α-α-trehalose, containing α-D-glucopyranose combined with β-D-fructofuranose, β-D-galactopyranose, and α-D-glucopyranose, respectively

relatively labile and is hydrolyzed by mild acid (pH 4) and is the donor of D-glucopyranose for the biosynthesis of dextrans and related polysaccharides by glucansucrases and the donor of D-fructofruanose for the biosynthesis of levan and inulin (see ❍ *Sect. 11*), fructofuranose polysaccharides by levansucrase and inulinsucrase; (c) it is the sugar of commerce because of the ease of obtaining it in large quantities in a pure state from sugar cane and sugar beets; (d) it crystallizes relatively easily; and (e) it has a pleasant sweet taste and has been recognized by humans for over 10,000 years as a sweet food and a natural sweetening agent. See ❍ *Fig. 5* for the structure of sucrose.

The origin of sucrose is thought to have been in the Indus Valley, where many woody, wild sugar cane plants that have the fundamental characteristics of the modern cultivated strains can still be found growing today. Sugar cane grows well in a warm, humid, tropical or semi-tropical climate. In the late 18th century on the European continent, the sugar beet was found to be an alternative source of sucrose that did not require a tropical or semi-tropical climate for growth. Sucrose is hydrolyzed into its component sugars (D-glucose and D-fructose) by the action of the enzyme, invertase, a β-fructofuranosidase, and by mild acid. In this form it is known as invert sugar, due to the fact that the direction of rotation of polarized light is inverted from dextrorotatory to levorotatory on hydrolysis. Honey is usually a mixture of sucrose and invert sugar. Yeasts also have invertase and can hydrolyze sucrose and then ferment the component sugars into ethyl alcohol.

In addition to sucrose, several plants also form a series of sucrose-based oligosaccharides with chains of α-1→6 D-galactopyranose units linked to the D-glucose moiety of sucrose [4]. The first in the series is the trisaccharide, raffinose, in which D-galactopyranose is linked α-(1→6) to sucrose; the second is a tetrasaccharide, stachyose, in which D-galactopyranose is linked α-(1→6) to the D-galactopyranose unit of raffinose. The next is a pentasaccharide, verbascose, with D-galactopyranose linked α-(1→6) to the terminal D-galactose unit of stachyose, and the next is a hexasaccharides, ajugose, with D-galactopyranose linked α-(1→6) to the terminal D-galactose unit of verbascose.

These D-galactopyranosyl sucrose oligosaccharides are particularly found in the tubers and seeds of legumes. Raffinose is found in cottonseeds and in sugar beets. Although sugar beets only contain about 0.05% by weight raffinose as compared with 16–18% sucrose, it has been isolated and crystallized with a purity of better than 99% from sugar beet syrup, where it accumulates during the processing of sucrose.

Soybeans are a good source of stachyose, where it is found to the extent of 2–3% by weight. In general, legume seeds and the mullein root are sources of verbascose. The enzyme invertase and mild acid specifically hydrolyze the oligosaccharides to give D-fructose and the corresponding reducing oligosaccharides that are terminated at the reducing-end with D-glucose. For example, raffinose is hydrolyzed to give D-fructose and the reducing disaccharide, melibiose [α-D-galactopyranosyl-(1→6)-D-glucose].

Another series of galacto-sucrose oligosaccharides involves the attachment to the D-fructofuranose moiety of sucrose [4]. The attachment of α-D-galactopyranose 1→6 to the fructose moiety gives the nonreducing trisaccharide, planteose. It is found primarily in the seeds of the *Plantago* family of plants, for example, the common weed and herb, plantain. Mild acid hydrolysis gives D-glucose and the reducing keto-disaccharide, planteobiose [α-D-galactopyranosyl-(1→6)-D-fructose]. Another nonreducing trisaccharide, melezitose, has α-D-glucopyranosyl linked (1→3) to the D-fructofuranose moiety of sucrose. It is found in the sweet exudates of many trees, such as larch, Douglas fir, Virginia pine, and poplars. Mild acid hydrolysis of melezitose gives D-glucose and the reducing disaccharide, turanose, [α-D-glucopyranosyl-(1→3)-D-fructose].

Several sucrose analogues have been enzymatically synthesized in the laboratory. Levansucrase can transfer a D-fructofuransoyl unit from raffinose to D-xylose, giving a nonreducing sucrose disaccharide analogue, xylsucrose [α-D-xylopyranosyl-(1→2)-D-fructofuranoside] and the reducing disaccharide, melibiose [α-D-galactopyranosyl-(1→6)-D-glucose [5]. A similar reaction of levansucrase with raffinose and D-galactose gives galactosucrose [α-D-galactopyranosyl-(1→2)-D-fructofuranoside] (also referred to as galsucrose) and melibiose [6]. Reaction of sucrose and lactose with levansucrase gives D-glucose and the nonreducing trisaccharide, lactosucrose [4^{Glc}-β-D-galactopyranosyl sucrose] [7]. Reaction of dextransucrase with sucrose and D-fructose gives an unusual reducing disaccharide, leucrose, containing an α-(1→5) linkage of D-glucopyranosyl linked (1→5) to D-fructopyranose [α-D-glucopyranosyl-(1→5)-D-fructopyranose] [8,9].

A relatively large number of sucrose derivatives have been chemically synthesized [10]. Some notable chloro derivatives have been obtained by the reaction of sucrose with sulfuryl chloride in pyridine/chloroform at low temperatures, for example 4,6,1′,6′-tetrachloro-4,6,1′,6′-tetradeoxy-glactosucrose, 4,6,6′-trichloro-4,6,6′-trideoxy-sucrose, and many others were formed [11,12,13,14,15]. These chloro-compounds of sucrose were 10–100 times sweeter than sucrose. One of them, Sucralose (4,1′,6′-trichloro-4,1′,6′-trideoxy-galactosucrose), was 650-times sweeter than sucrose, with no after-taste, and a sweet-taste identical to sucrose. It is used commercially as a noncariogenic and noncaloric sweetener in soft drinks, candies, cookies, jellies, and many other prepared foods, as well as a general substitute for table sugar. Sucralose is enzymatically inert and passes through the human body without being metabolized or absorbed.

4.2.3 Properties and Occurrence of D-Glucose Combined with D-Galactose to Give Lactose and Higher Oligosaccharides

Lactose is a disaccharide composed of β-D-galactopyranose linked (1→4) to D-glucose and is found in the milk of mammals, where it serves as a source of energy and nourishment for the newborn. Lactose is a reducing disaccharide because the D-glucopyranose residue has a free hemiacetal group at C1. Human milk contains 85 g L^{-1} lactose and cow's milk contains about 50 g L^{-1}.

Human milk also contains lactose oligosaccharides in which various different monosaccharide residues are attached to the D-galactopyranosyl residue. α-L-Fucose (6-deoxy-L-galactose) is attached (1→2) to the galactose moiety [16], α-N-acetyl-D-neuraminic acid is attached (2→3) to the galactose moiety [16,17], and the β-N-acetyl-D-glucosamine residue is attached either (1→3) or (1→6) to the galactose moiety [18]. The latter serves to produce a core structure that can be further extended by the addition of β-D-galactopyranose residues linked either (1→2) or (1→4) [19]. The β-D-galactopyranosyl-β-N-acetyl-D-glucosamine disaccharide is often added in multiples to give a repeated core structure to which α-N-acetyl-D-neuraminic acid and α-L-fucose residues are added to the ends of the oligosaccharides. The so called human blood group determinants (see ❷ Sect. 14) have structural similarities to the human milk oligosaccharides [19] and it is thought that through this relationship the milk oligosaccharides impart some form of early immunological protection to the newborn. See ❷ *Fig. 5* for the structure of lactose.

4.2.4 Properties and Occurrence of α-D-Glucose Combined with α-D-Glucose to Give α,α-Trehalose

α,α-Trehalose is also one of the three naturally occurring disaccharides. It is a nonreducing disaccharide with two D-glucopyranose residues joined together in an α-1↔α-1 acetal–acetal linkage and is of relatively high energy, like sucrose. Unlike sucrose, however, the acetal–acetal linkage is quite stable and is one of the most difficult linkages to be hydrolyzed by acid. The reason for this is not absolutely clear, but it has been hypothesized to be due to the stabilization of the molecule by intramolecular hydrogen bonds between the D-glucose residues. α,α-Trehalose is found in insect lymph fluid ("insect blood") where it acts as a source of chemical energy [19] and it is also found in mushrooms, honey, yeast, fungi, lobster, and shrimp as a source of energy [20,21] all of which have the enzyme, trehalase, that hydrolyzes α,α-trehalose to give two molecules of D-glucose that can be used for energy.

Besides being used as a source of energy, some plants and animals also use α,α-trehalose as a stabilizing agent during extreme conditions [22]. High concentrations of α,α-trehalose in the tissues of certain insects and in desert plants allows them to survive in a state of suspended animation under conditions of water deficiency. α,α-Trehalose helps frogs to survive in a frozen state and it helps to protect the DNA of salmon sperm from dehydration. α,α-Trehalose has also found applications in the preservation of organs taken for use in organ transplants [23,24]. See ❷ *Fig. 5* for the structure of α,α-trehalose.

5 Properties and Occurrence of D-Glucose in Polysaccharides and Cyclodextrins

5.1 Properties and Occurrence of Starch

Starch is an abundant polysaccharide composed of D-glucose residues. It is found in the green leaves, stems, roots, seeds, fruits, tubers, and bulbs of most plants, where it serves as the storage of chemical energy obtained from the energy of the sun light in the process of photosynthesis. Starch also serves as the major source of chemical energy for most nonphotosynthesizing organisms such as bacteria, fungi, insects, and animals. It is found in relatively large amounts in the major food crops of the world. Starch is present 80% by weight in the rice kernel, 78% in the potato tuber, 75% in green bananas, 73% in the maize kernel, 68% in wheat flour, and 60% in rye and lentils. Starch provides about 65% of the dietary calories in the human diet.

Starch occurs in plants as water-insoluble granules produced in plant organelles, plastids (chloroplasts and amyloplasts). The granules have specific shapes and sizes that are characteristic of their botanical source [25]. Most starches are composed of a mixture of two types of polysaccharides, a linear polysaccharide, consisting of α-(1→4) linked D-glucopyranose residues, called amylose, and a branched polysaccharide of α-(1→4) linked D-glucopyranose residues with 5–6% α-(1→6) branch linkages, called amylopectin. Amylose has an average of 500 to 5,000 D-glucopyranose residues per molecule, depending on the source; amylopectin is much larger and has an average of 100,000 to 1,000,000 D-glucopyranose residues per molecule [26,27]. When at equilibrium with its surroundings, starch granules will contain 10–15% w/w water. The amylose and amylopectin molecules in the granules can be solubilized by heating the granules in water, where they swell and eventually burst, releasing the individual molecules. Starch granules can also be dissolved in 9:1 dimethyl sulfoxide/water solutions [28]. See ❷ Fig. 1 in ❷ Chap. 6.2 for the structures of segments of amylose and amylopectin.

The amounts of amylose and amylopectin differ for starches from different botanical sources. Most so-called normal starches have 20–30% amylose and 80–70% amylopectins, respectively [29,30]. There are mutant varieties, such as waxy maize, waxy rice, and waxy potato, that are composed of 100% amylopectin. There also are the high amylose varieties, such as amylomaize-V that consists of 53% amylose and 47% amylopectin and amylomaize-VII that is 70% amylose and 30% amylopectin, just the reverse of the "normal" starches. Many of the "normal" starches have been found to have an intermediate component that is slightly branched amylose with 0.5–3% α-(1→6) branch linkages [26,27,29,30].

All starches can be completely converted into D-glucose by acid hydrolysis at high temperatures (100 °C) and by the action of the enzyme, glucoamylase, at lower temperatures (20–40 °C), when the granules are solubilized. Humans and other organisms can completely convert solubilized starches into D-glucose by the combined action of several enzymes, such as α-amylases found in saliva and in the small intestine, and α-(1→6)-glucosidase and α-(1→4)-glucosidase that are secreted by special cells in the lining of the small intestines.

Starches have been chemically modified to improve their solution and gelling characteristics for food applications. Common modifications involve the cross linking of the starch chains, formation of esters and ethers, and partial depolymerization. Chemical modifications that have been approved in the United States for food use, involve esterification with acetic anhydride, succinic anhydride, mixed acid anhydrides of acetic and adipic acids, and 1-octenylsuccinic anhydride to give low degrees of substitution (d.s.), such as 0.09 [31]. Phosphate starch esters have been prepared by reaction with phosphorus oxychloride, sodium trimetaphosphate, and sodium tripolyphosphate; the maximum phosphate d.s. permitted in the US is 0.002. Starch ethers, approved for food use, have been prepared by reaction with propylene oxide to give hydroxypropyl derivatives [31].

The solubility of the starch granules has been increased by reaction of starch granules in water with 7% hydrochloric acid for one week at 20 °C to give "Lintner soluble starch". Recent modifications to increase the solubility of starch granules have involved the reaction of the starch granules with hydrochloric acid in anhydrous alcohols, such as methanol, ethanol, 2-propanol, and 1-butanol to give a new class of limit dextrins whose average degree of polymerization can be controlled between 1800 and 30 [32,33,34]. Enzymatic conversions of starches into mixtures of maltodextrins are used in food preparations. Starch is the major source for the commercial preparation of D-glucose and D-fructose. Starches have been modified to give tertiary amino alkyl ethers, quaternary ammonium ethers, amino ethylated ethers, cyanamide ethers, starch anthranilates, cationic dialdehyde starch, carboxymethyl ethers, and carboxy starch for various applications in the sizing of paper, formation of coatings, sizing of textiles, flocculation, and emulsification technologies [35].

5.2 Properties and Occurrence of Glycogen

Glycogen is an α-glucan that is widely distributed in mammals in the liver, muscle, and brain and in fish, insects, and some species of bacteria, fungi, protozoa, and yeasts, as a reserve form of chemical energy. It is a high molecular weight polysaccharide (1×10^6 to 2×10^9 Da) composed of D-glucopyranose residues linked together by α-(1→4) glycosidic linkages with 10–12% α-(1→6) branch linkages [36]. It has been compared with amylopectin and called "animal starch." But, it is quite different from amylopectin in that it has over twice as many α-(1→6) branch linkages per molecule, giving the many chains an average chain length of 8 to 10 D-glucopyranose residues compared to 20 for amylopectin. Further, the branch linkages do not occur in clusters, as they do in amylopectin, and are randomly distributed, giving glycogen different chemical and physical properties from amylopectin. Glycogen does occur in particles or granules of about 25 nm, called β-particles [37]. The β-particles are further combined into a larger mass, called α-particles, which consists of approximately 100 β-particles. Nevertheless, in contrast to starch granules, the glycogen particles are quite water-soluble, because of the relatively high percent of branch linkages and the absence of intermolecular bonding, giving the absence of crystallinity. Glycogen reacts poorly with triiodide, giving a light brown color or no color, but never the blue color given by starch granules and amylose nor the maroon color given by amylopectin.

Glycogen has a specific function in the liver of mammals, where its primary role is to maintain the normal concentration of D-glucose in the blood. In humans, it can provide 100–150 mg of glucose per minute over a sustained period of 12 h, if necessary [38]. In skeletal muscle, its primary function is to provide immediate energy for muscle movement by being converted into α-glucopyranose-1-phosphate and in the human brain, where glycogen normally provides about 100 g of α-glucose-1-phosphate per day for energy used by the brain [38].

Blue-green algae, which are photosynthetic bacteria (cyanobacteria) and not eukaryotic algae, synthesize glycogen as a reserve energy storage polysaccharide instead of synthesizing starch [39]. Glycogen is also synthesized by nonphotosynthetic bacteria, such as *Escherichia coli* that synthesizes it intracellularly from UDPGlc [39] and *Neisseria perflava* that synthesizes it extracellularly from sucrose by the enzyme, amylosucrase [40]. The function of glycogen for these bacteria has been postulated to provide reserve energy in times of the absence of nutrients and as a source of energy for the formation of spores [39].

5.3 Properties and Occurrence
of Dextrans, Alternan, Mutan, and Pullulan

Dextrans are a large family of bacterial polysaccharides that have a contiguous series of D-glucopyranose residues linked α-(1→6) to each other [41]. Over 100 strains of *Leuconostoc mesenteroides* [42], *Streptococcus mutans*, *S. sobrinus*, and *S. salivarius* produce specific enzymes, dextransucrases that synthesizes dextrans from sucrose [41]. All of the dextrans are branched, primarily by α-(1→3) glycosidic linkages, but also by α-(1→2) and α-(1→4) linkages in specific *L. mesenteroides* strains. Differences in the number and arrangements of these branches, such as having single glucose branches or long α-(1→6) linked branch chains, and the order and frequencies of the branches, impart differences in the structures and properties [41,42]. The classic prototypical dextran is the commercial product synthesized by dextransucrase from *L. mesenteroides* NRRL B-512F. It has 95% α-(1→6) linkages with 5% α-(1→3) branch linkages that have branches that are both single glucose residues and long α-(1→6) linked chains and contains ~10^6 to 10^8 D-glucopyranose residues. Other strains make a wide variety of dextrans with various degrees of branching, not only through α-(1→3) linkages, but through α-(1→2) and/or α-(1→4) linkages as well. In some cases, the degree of branching is as low as 3%, and in other instances, virtually every glucose residue in the backbone may be substituted with a branch linkage [41].

Some of these bacteria also elaborate glucansucrases that synthesize polysaccharides that are not considered dextrans because they do not have contiguous α-(1→6) linked main chains. *L. mesenteroides* NRRL B-1355 secretes a dextransucrase that synthesizes a B-512F-type dextran and another enzyme, alternansucrase, that synthesizes an α-glucan from sucrose that has alternating α-(1→6) and α-(1→3) linked glucose residues in the main chains with 7–11% α-(1→3) branch chains of alternating α-(1→6) and α-(1→3) linked glucose residues [41]. This α-glucan is called alternan. Another glucansucrase, mutansucrase, secreted by *Streptococcus mutans*, synthesizes a linear glucan in which the D-glucopyranose residues are linked α-(1→3). It is particularly characterized by being extremely water-insoluble in contrast to the dextrans and alternan that are highly water-soluble [41].

Pullulan is a polysaccharide that is elaborated by several species of the fungus, *Aureobasidium*, particularly *A. pullulans*. This fungus is typified by the presence of black pigments and is sometimes called "black yeast" [43]. Pullulan is a water-soluble, linear polysaccharide of D-glucopyranose residues joined together by a repeating sequence of two α-(1→4) and one α-(1→6) linkages. The structure is that of a polymer of maltotriose units joined together end to end by α-(1→6) linkages [44,45,46]. In addition to maltotriose units, it also has ~5–7% maltotetraose units located in the interior of the polysaccharide chain [47].

These bacterial polysaccharides have been considered to be "slimes"; they are often in reality loose capsules that are produced extracellularly by the bacteria. It was found that low molecular weight *L. mesenteroides* NRRL B-512F dextran could be used as a blood plasma extender and was produced on a relatively large scale during the "cold war", but also found uses as a gel-filtration material when cross-linked by epichlorohydrin to give a family of cross-linked dextrans [41].

5.4 Properties and Occurrence of D-Glucose in Cyclic Dextrins

A number of different kinds of nonreducing cyclic dextrins containing D-glucopyranose residues occur. The first to be observed were the cyclomaltodextrins (sometimes referred to in the older literature as cyclodextrins or Schardinger dextrins), which have been known for over 100 years. They were first found in rotting vegetables and then in the fermentation of starch by a heat-resistant microorganism called *Bacillus macerans*. The compounds were crystallized from alcohol solutions and shown to be α-(1→4) linked, nonreducing, cyclic dextrins composed of six, seven, and eight D-glucopyranose residues, named cyclomaltohexaose, cyclomaltoheptaose, and cyclomaltooctaose or α-CD, β-CD, and γ-CD [48]. These cyclomaltodextrins are formed from starch by the enzyme, cyclomaltodextrin glucanyltransferase (CGTase). *Bac. macerans* CGTase primarily forms α-CD; other bacteria, for example *Bac. circulans* elaborates a CGTase that primarily forms β-CD and *Brevibacterium sp.* elaborates a CGTase that primarily forms γ-CD [49]. Larger cyclomaltodextrins, having 9, 10, 11, and

12 D-glucopyranose residues were later obtained in relatively small quantities, and even later, cyclomaltodextrins having as many as 25 glucose residues were obtained [50]. The internal cavity of the cyclomaltodextrins is relatively hydrophobic, giving them the property of forming complexes with a wide variety of organic molecules [51].

Cycloisomaltodextrins, linked α-(1→6) containing seven, eight, and nine D-glucopyranose residues have been found to be formed by a bacterial cycloisomaltodextrin dextran-glucanyl-transferase, acting on B-512F dextran [52]. A cyclic tetrasaccharide, containing four glucose residues with alternating α-(1→6) and α-(1→3) linkages has been obtained from the reaction of a bacterial enzyme on alternan [53]. This enzyme, 3-α-isomaltosyltransferase, is part of a 2-enzyme system that converts starch to the cyclic tetrasaccharide [54,55]. Also, *Bacillus stearothermophilus* starch branching enzyme catalyzed a reaction with amylose to give macrocyclic dextrins with one α-(1→6) linkage at the site of transglycosylation coupling [56].

Another group of cyclic dextrins is the cyclosophorans, which consist of 17–40 D-glucopyranose residues linked β-(1→2). They are produced by *Rhizobium* species involved in nitrogen-fixing nodules on the roots of legumes [59] and are also found in plant crown galls, produced by *Agrobacterium tumefaciens* [58]. The sizes of the cyclosophorans vary depending on the particular species of *Rhizobium*, which are also specific for the particular type of legume that they associate with to form the nodules. There is some evidence that the cyclosophorans play a role in the formation of the nodules and the crown galls [59].

Bradyrhizobium species synthesize a related cyclic dextrin that contains 12 D-glucopyranose residues linked by a repeating sequence of three contiguous β-(1→6) linkages followed by three contiguous β-(1→3) linkages. One of the β-(1→3) sequences has a single branched D-glucopyranose residue substituted β-(1→6) onto the center D-glucopyranose residue [60]. A cyclodextrin containing only β-(1→3) linkages (cyclolaminarinose) has been found to be elaborated by a recombinant strain of *Rhizobium meliloti* TY7 mutant that is deficient in forming cyclosophoran, but carrying the genetic locus of *Bradyrhizobium japonicum* USDA 110. The cyclolaminarinose dextrin has 10 D-glucopyranose residues, with a single laminaribiose disaccharide substituted β-(1→6) onto the ring [61].

5.5 Properties and Occurrence of Cellulose

Cellulose is usually considered the most abundant carbohydrate on the earth, occurring in all plant cell walls to the extent of approximately 50% by weight; 20–40% of the cell wall is made up of hemicelluloses, and the remaining 10–30% is the noncarbohydrate, lignin, which acts as a cross-linking and cementing agent in the plant cell wall, covalently attached to the hemicelluloses [62]. Hemicelluloses are a family of polysaccharides, with a structure similar to cellulose, but besides D-glucopyranose residues, they contain several other monosaccharide residues, such as D-xylopyranose, D-mannopyranose, D-galactopyranose, D-glucopyranose uronic acid, and L-arabinofuranose residues (see ❷ *Sect. 6*).

Cellulose is a very large, linear polysaccharide of $\sim 10^6$ to 10^8 D-glucopyranose residues, linked β-(1→4) to each other. Because of its high water-insolubility, its actual size has never been accurately determined. It is a β-glucan with a very tight helical structure in which the individual glucose residues are oriented 180°to each other [63]. Because of this conformation and the β-linkages, cellulose chains readily form intermolecular hydrogen bonds, giving multiple chains associated together in 3-dimensional bundles that further associate with other bundles to form micelles or fibers [63,64]. These micelles make a very tough and resistant material that gives shape, strength, and water and substance impermeability to the plant cell

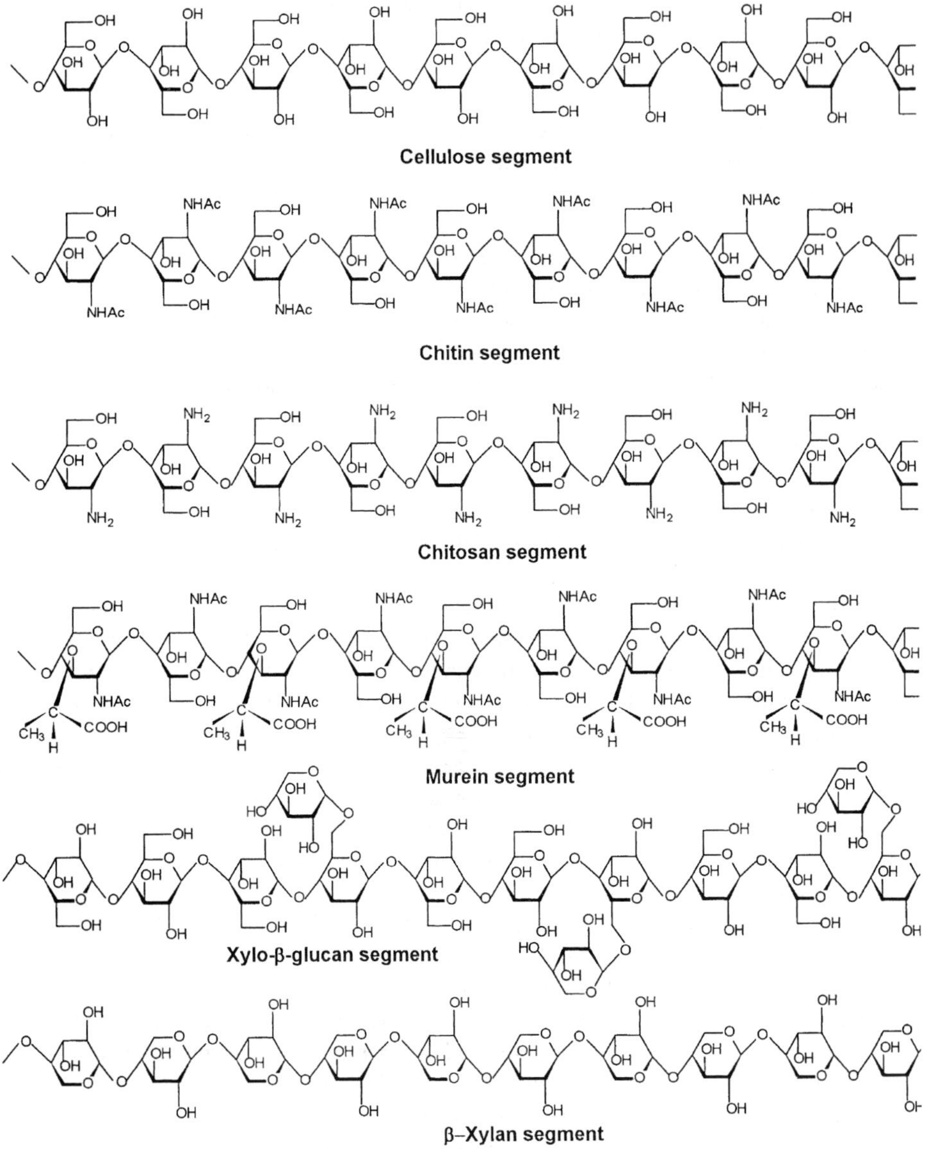

Cellulose segment

Chitin segment

Chitosan segment

Murein segment

Xylo-β-glucan segment

β–Xylan segment

◻ Figure 6
Haworth structures of β-(1→4) linked segments of cellulose, chitin, chitosan, murein, xylo-β-glucan, and xylan

wall. The fibers are also quite resistant to chemical and enzymatic attack. Very pure cellulose is found in the cotton boll and is also synthesized in a relatively pure form by some species of bacteria, such as *Acetobacter xylinum*, *Agrobacterium tumefaciens*, and related bacteria [65]. When one thinks of cellulose, it is usually in connection with trees and wood that come to mind. Cellulose is a major component of flax, comprising 80% (w/w), and jute, comprising 60–70% (w/w). Grasses, such a papyrus and bamboo have been important sources of cellulose going back into ancient times. Papyrus was used as an early form of paper, made from the pith of the papyrus plant, a wetlands sedge that grows to 5 meters (∼16 ft) in height and was once abundant in the Nile Delta of Egypt. It was first known to have been used as a writing material in ancient Egypt (at least as far back as the First Dynasty, 3,000 BC) but it was also widely used throughout the Mediterranean, as well as in Europe and Southwest Asia, until about the 11th Century AD. Papyrus was prepared as a thin film from the outer bark that was glued together with starch paste to give it body and the ability to hold ink. Bamboo also served man from very early times, and continues to do so, as a building material to form houses, roofs, furniture, and so forth. Paper today is manufactured from several cellulose sources, such as wood chips and sawdust, the fibers of the sugar cane plant (called bagasse), maize (corn) stalks, and the straws of rye, oats, and rice. The β-(1→4) glycosidic linkage of cellulose is more resistant toward acid hydrolysis than the α-(1→6) linkage of amylopectin. Cellulose is slowly hydrolyzed by 1-M HCl at 100 °C. See ❷ *Fig. 6* for the structure of a segment of the cellulose molecule.

6 Properties and Occurrence of Hemicelluloses

Hemicelluloses are a family of four basic types of polysaccharides, composed of two or more monosaccharide residues. All have structural features similar to cellulose in that they have their main chains that are β-(1→4) linked, with the exception of the arabinoglactans that are β-(1→3) linked. The main chains are homopolysaccharides composed of a single monosaccharide residue, but they are highly branched by one or two different kinds of monosaccharides that are linked for the most part to give single monosaccharide branches.

As previously mentioned, the cell walls of most plants contain 40–60% cellulose. The remaining carbohydrate, representing 40–50% (w/w) of the cell wall is composed of hemicelluloses. The composition of the hemicelluloses varies from one plant type to another [66]. The four basic types are

1. Xyloglucans composed of α-D-xylopyranose linked 1→6 to approximately every third D-glucose residue of cellulose [66,67,68];
2. Xylan composed of D-xylopyranose linked β-(1→4) and glucurono-arabino-xylan, which is composed of β-(1→4) D-xylopyranose chain with 4-*O*-methyl-α-D-glucopyranosyluronic acid linked 1→2 and α-L-arabinofuranosyl linked 1→3 to the xylan chain [66,67,68,69];
3. Mannan is composed of β-(1→4)-D-mannopyranose chains [70]; another type of D-mannan is galactomannan that has D-galactopyranose linked α-(1→6) to the D-mannan chain [71].
4. Arabinogalactan is composed of β-(1→3) linked D-galactopyranose chain with β-(1→6) linked D-galactopyranose branches and to a lesser degree a L-arabinofuranose disaccharide,

β-L-arabinofuranosyl-(1→3)-α-L-arabinopyranosyl linked 1→6 to the D-galactopyranose chain [72,73].

Xyloglucans and xylans are widely distributed, found in most plant cell walls, see ❷ *Fig. 6* for the structures of these hemicelluloses. Glucurono-arabino-xylan is also widely distributed, especially in soft-wood trees [74]. Glucuronoxylan is prevalent in hard-wood trees [74]. Hemicellulose composed exclusively of D-mannopyranose is found in palm seed endosperm [70], where it is known as vegetable ivory or ivory nut mannan. Galactomannan are particularly found in soft-wood trees, but also are found in the seedpods of the locust bean. In particular, a galactomannan known as guar gum has been obtained from the seeds of the legume, guar, grown in the semiarid regions of India. It has found widespread use as a thickening agent in food products. Arabinogalactans are prevalent in soft woods such as, larch, black spruce, Douglas fir, cedar, and juniper [74]. Segments of the structures of D-xyloglucan and D-xylan are shown in ❷ *Fig. 6*.

6.1 Properties and Occurrence of Pectin

Pectins are related to hemicelluloses and occur in the plant cell wall in low amounts of 1–5% (w/w). They are more prevalent in fruits, for example, apple pulp (10–15%) and in orange and lemon rinds (20–30%). Pectins do have a number of chemical and physical properties that differ from the hemicelluloses. They are composed of D-galactopyranosyl uronic acids linked α-(1→4) and have a relatively wide percentage of the carboxyl groups esterified with methyl groups. There also is a small amount (1 in ~25 uronic acid residues) with L-rhamnopyranosyl (6-deoxy-L-mannopyranosyl) residues linked α-(1→2) to the D-galactopyranosyl uronic acid residues. In addition, pectins also have 2-*O*-acetyl or 3-*O*-acetyl ester groups attached to the uronic acids [75]. The average molecular weight can vary from 20,000 to 400,000 Da, with a typical average molecular weight of 100,000 Da. Pectins act in plants as an intercellular cementing agent that provides body to fruits, and of course, it is used in foods as a gelling agent, especially in the preparation of jellies and confections.

7 Cellulose-like Polysaccharides Containing *N*-Acetyl-D-Glucosamine and D-Glucosamine

There are several polysaccharides containing the β-(1→4) structure, but with monomer residues other than D-glucopyranose, such as *N*-acetyl-D-glucosamine, D-glucosamine, *N*-acetyl-D-muramic acid.

7.1 Properties and Occurrence of Chitin

Chitin is a polysaccharides with the exact same structure as cellulose but containing *N*-acetyl-D-glucosamine or D-glucosamine and are fairly widely distributed [76] (see ❷ *Fig. 6* for the structure of a segment of chitin). It is a structural polysaccharide that forms fibers, is water impermeable, and replaces cellulose in the cell walls of many species of lower organisms, such as fungi, yeasts, green algae, and brown and red seaweeds. It also comprises the major component of the exoskeleton of insects, where it makes a hard shell-like material that is quite strong. Chitin is also found in the cuticles of worms and in the shells of crustaceans, such as mollusks, shrimps, crabs, and lobsters [77].

Chitin, like cellulose, has a highly ordered, crystalline structure in which the chains are inter-molecularly hydrogen bonded in an antiparallel arrangement, a parallel arrangement, and a mixed arrangement of two parallel and one antiparallel repeating arrangement [78]. Also like cellulose, it is very insoluble in water and most other solvents. In arthropods, the chitinous shell, or exoskeleton, does not grow, and is periodically cast off or molted. After the old shell is shed, a new, larger shell is produced, providing room for further growth. Chitin is very rigid, except between some body segments and joints, where it is much thinner and allows movement of the various parts.

7.2 Properties and Occurrence of Chitosan

Chitosan is a polysaccharide very similar to chitin, except that the N-acetyl-D-glucosamine is replaced by D-glucosamine in which the N-acetyl group is removed (see ❷ *Fig. 6* for the structure of a segment of chitosan). Chitosan is found occurring naturally mixed with chitin in the cell walls of some fungi and seaweeds. It is, however, primarily produced chemically by treating chitin with strong alkali to deacetylate the N-acetyl-amino group [79]. The degree of deacetylation can range from 60 to 100%, giving a family of chitosans. The free amino group of chitosan has a pK_a value of ~6.5 and it can be protonated in mildly acidic solutions, giving a positive charge to the glucosamine residues. The positive charges on chitosan produce very different physical and chemical properties from chitin. Because of the repulsion of the positive charges, chitosan chains do not line up and associate to form micelles and fibers, as does cellulose and chitin. Chitosan, thus, is water-soluble at acidic pH values.

Because of the positive charges on chitosan, it has found a number of applications. It binds to negatively charged surfaces, such as mucosal membranes, and has been used as a bandage material for wounds that is biocompatible and biodegradable [80,81]. Positively charged chitosan enhances the transport of polar drugs across epithelial tissues and is used to transport drugs in humans [82]. It has been used as an enhancer for plant growth, and as an aid in the defense of plants against fungal infections. Chitosan is used in water purification, as a material in a sand filtration system where it binds fine sediment particles during filtration, greatly aiding the removal of turbidity; it also removes phosphates by ion exchange, heavy metals by chelation, and oils by hydrophobic adsorption from water [81]. Chitosan has also been found useful for the immobilization of enzymes and cells [81,82,83,84].

7.3 Properties and Occurrence of *N*-Acetyl-D-Glucosamine and *N*-Acetyl-D-Muramic Acid in Murein – The Bacterial Cell Wall

The major component of all known bacterial cell walls is a polysaccharide composed of N-acetyl-D-glucosamine (NAG) linked together by β-$(1{\rightarrow}4)$ glycosidic bonds, as in chitin, but with every other NAG residue substituted at C-3 by an ether linkage to the hydroxyl group of L-lactic acid to give N-acetyl-D-muramic acid (NAM) [85,86,87,88]. This results in a nine-carbon N-acetyl-amino-sugar acid, with a repeating β-(1→4)-NAG-NAM sequence of 40–150 residues, giving a polysaccharide, called murein [89].

A pentapeptide is attached to the carboxyl group of the L-lactic acid by an amino group that forms a peptide (amide) bond [90]. Attached to this pentapeptide is a pentaglycine linked to the ε-amino group of an L-lysine by a carboxyl group. The glycine end forms a cross-link to another decapeptide [90]. Using slightly different amino acids in both the pentapeptide and the cross-linking peptide gives different peptides that are genus-dependent.

The murein-peptidoglycan gives rigidity and different specific shapes, such as rods, spheres, or spirals to bacterial cells. Because of the cross-linking of the murein chains, the peptidoglycan is considered one giant, bag-shaped macromolecule [91]. The structures of segments of chitin, chitosan, and murein are shown in ❷ *Fig. 6*.

7.4 Properties and Occurrence of Glycosaminoglycans Composed of Amino Sugars and Uronic Acids

Glycosaminoglycans make up a group of polysaccharides that are found in animal tissues. They are composed of repeating disaccharides units of N-acetyl-D-glucosamine or N-acetyl-D-galactosamine and D-glucopyranosyluronic acid residues. The linkages are primarily β at positions 3 and 4. They are most often attached to protein backbones, forming what is called a proteoglycan [92].

7.4.1 Hyaluronic Acid

Hyaluronic acid consists of repeating disaccharides of β-D-glucopyranosyluronic acid-(1→3)-N-acetyl-D-galactosamine linked β-(1→4) to the next disaccharide. This proteoglycan can have between 500 and 50,000 residues per chain [92]. Hyuronic acid is found widely distributed in mammalian cells and tissues, where it is found in synovial fluid that lubricates the joints, in the vitreous humor of the eye, and in connective tissue, such as the umbilical cord, the dermis, and the arterial wall. It also occurs as a capsular polysaccharide around certain bacteria, such as pathogenic streptococci [92].

7.4.2 Chondroitin Sulfate

Chondroitin sulfate also consists of a repeating disaccharide of β-D-glucopyranosyluronic acid-(1→3)-N-acetyl-D-galactosamine linked β-(1→4) to the next disaccharide unit, with sulfate groups attached to C-4 or C-6 of the N-acetyl-D-galactosamine residue. It occurs as a major component of cartilage found in the cornea of the eye, the aorta, the skin, and lung tissues, where it is located between fibrous protein molecules. Chondroitin sulfate provides a soft and pliable texture to these tissues [92].

7.4.3 Dermatan Sulfate

Dermatan sulfate is derived from chondroitin 4-sulfate by the action of a C-5 epimerase that inverts the carboxyl group of the β-D-glucuronic acid, giving the very rare sugar, α-L-idopyranosyl uronic acid (α-L-iduronic acid). Some of the L-iduronic acids are sulfated at C-2. Dermatan sulfate is found primarily in the skin [93].

7.4.4 Keratan Sulfate

Keratan sulfate consists of the disaccharide, N-acetyl-lactosamine, linked β-(1→3) to the D-galactopyranose residue of the next N-acetyl-lactosamine unit. Keratan sulfate is the most heterogeneous of the glycosaminoglycans, with variable sulfate content linked to C-4 or C-6 of the D-galactopyranose residue in lactosamine, and small amounts of L-fucose (6-deoxy-L-galactose), D-mannose, and N-acetyl-neuraminic acid residues [93] (also see ❷ *Sects. 10.2* and ❷ *14* on the occurrence of N-acetyl-D-neuraminic acid and L-fucose in other systems). Keratan sulfate is found in the cornea, on the surfaces of erythrocytes, in cartilage, and in bone.

7.4.5 Heparan Sulfate

Heparan sulfate consists of the repeating disaccharide β-D-glucopyranosyluronic acid-(1→4)-N-sulfato-2-amino-2-deoxy-α-D-glucopyranosyl linked (1→4). This polysaccharide is linked to a core protein to give a proteoglycan that is found as a matrix component of arterial wall, lung, heart, liver, and skin [93].

8 Polysaccharides Containing Uronic Acids That Have Some of Their Carboxyl Groups Inverted by a C-5 Epimerase to Give New Polysaccharides with New Properties

8.1 Heparin Sulfate

Heparin sulfate is formed from heparan sulfate by the action of an enzyme, C-5 epimerase that inverts the carboxylate group attached to C-5 of the D-glucopyranosyl uronic acid residues to give the rare and unusual sugar, α-L-idopyranosyl uronic acid. Heparin sulfate is released from the heparin of proteoglycans of mast cells into the blood stream when there is an injury to blood vessels, in the heart, liver, lungs, and skin. The release of heparin near the site of the injury acts as an anti-coagulating agent, preventing massive clotting of the blood and, hence, preventing run-away clot formation [93].

8.2 Alginates

Alginates are found primarily in brown seaweeds in amounts of 18–40% by weight of the plant. The majority is extracellular, being located between the cells [94]. One of the major species of seaweeds that contains alginates is the giant kelp, *Macrocyctis pyrifera*. It grows along the California coast of the US, the northwestern and southwestern coasts of South America, and the southeastern coasts of Australia and New Zealand [95].

Alginates are formed from poly-D-mannopyranosyluronic acid by the action of a C-5 epimerase that inverts the C-5 carboxyl group to give approximately 33% (w/w) of the

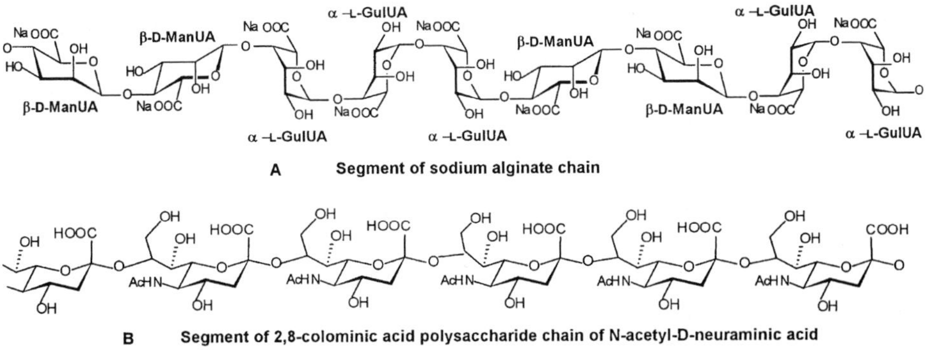

Figure 7
Conformational structures of segments of sodium alginate, containing β-D-mannopyranosyl uronic acid, and α-L-gulopyranosyl uronic acid; and 2,8-α-colominic acid, containing N-acetyl-α-D-neuraminic acid, linked (2→8)

rare and unusual sugar α-L-gulopyranosyl uronic acid. The ratio of the two uronic acids varies with different species of seaweed, type of tissue, and age of the plant [96]. The two kinds of uronic acids are combined together in blocks of variable numbers and also as alternating residues. In the brown seaweed, alginates most frequently exists as the calcium salt and are converted to the sodium salt when isolated.

The biological role that alginates play in seaweeds is that of a protective agent against desiccation during low tide. An unusual and useful property of sodium alginate is the ability to instantly form gels when in contact with divalent metal ions, such as calcium, barium, strontium, copper, cobalt, nickel, and so forth [97]. The strength and firmness of the gels are proportional to the amount of α-L-gulopyranosyluronic acid present in the alginate [98]. The strength and firmness is also dependent on the starting concentration of the alginate, the higher the concentration, for example 5% (w/v), gives very strong, firm gels, even though it is only 5% calcium alginate and 95% water.

Sodium alginate is used in food preparations as a thickening agent, a stabilizer, and an emulsifier in ice cream, cream cheese, salad dressings, frozen foods, pharmaceuticals, and so forth. Calcium alginate is the major ingredient in the "pimento" found in stuffed olives. A very important use of calcium alginate is the formation of gels that are used to encapsulate enzymes, hormones, drugs, and whole cells for carrying out various processes while being immobilized [99]. See ❯ *Fig. 7* for the conformational structure of a segment of alginate.

Alginates are also produced extracellularly by some bacteria, such as *Pseudomonas aeruginosa* and *Azotobacter vinelandii* [100], where they are believed to play a role in biofilm formation, pathogenesis, and soil aggregation. Their gels, however, are inferior to seaweed alginates, because of the presence of *O*-acetylation, which inhibits gel formation.

9 Occurrence and Properties of Plant Exudate Polysaccharides

Several complex polysaccharides are secreted by plants to seal wounds. Gum arabic is a complex material, containing protein, lipid, and carbohydrate, produced by Acacia trees found in the arid regions of Africa, in Nigeria, Mauritania, Senegal, and the Republic of Sudan. The structure of the polysaccharide portion has a main chain of D-galactopyranose residues linked β-(1→3) and D-glucopyranosyl uronic acid linked β-(1→6). The main chains have branch chains of two to five residues, consisting of α-L-arabinofuranosyl, α-L-rhamnopyranosyl, β-D-glucopyranosyl uronic acid, and 4-O-methyl-β-D-glucopyranosyl uronic acid [101]. The latter two uronic acids occur most frequently at terminal ends of the branched chains.

Another plant exudate is gum ghatti or Indian gum that can be obtained from a large tree grown in the deciduous forests of India and Sri Lanka. Gum ghatti is composed of L-arabinofuranose, D-galactopyranose, D-mannopyranose, D-glucopyranose uronic acid, and D-xylopyranose in approximately the molar ratios of 10:6:2:2:1 [102]. A third exudate gum is gum tragacanth that is primarily obtained from trees growing in Iran, Syria, and Turkey. It is a highly branched arabinogalactan with α-D-xylopyranose and α-L-fucopyranose branch residues [103]. These gums are primarily used to increase viscosity, provide body, stabilize emulsions, and suspend other materials and have been used for thousands of years in confectioneries, cosmetics, textiles, coatings, paints, pastes, and polishes.

10 Occurrence of Carbohydrates in Bacterial Polysaccharides

A large number of bacterial polysaccharides are known [104]. The major structural component of the bacterial cell wall is a polysaccharide, known as murein and composed of a repeating unit of one N-acetyl-D-glucosamine and an O-lactyl substituted N-acetyl-D-glucosamine (N-acetyl-D-muramic acid) see ❯ Sect. 7.3.

10.1 Xanthan, a Water-Soluble Bacterial Polysaccharide

In the 1950s, the US Department of Agriculture's Northern Regional Research Laboratories in Peoria, Illinois screened bacterial cultures to obtain a replacement for the plant exudates, which had become rare and expensive. They found that *Xanthomonas campestris*, when grown on D-glucose in an aerobic submerged fermentation, produces xanthan, a water-soluble polysaccharide gum [105,106]. It has a cellulose backbone of β-(1→4) linked D-glucopyranose residues with a trisaccharide of D-mannopyranose linked β-(1→4) to D-glucopyranosyl uronic acid linked β-(1→2) to a D-mannopyranosyl [β-D-Man p-(1→3)-β-D-Glc pUA-(1→2)-α-D-Man p-(1→3)-] attached to every other D-glucose residue in the cellulose chain by an α-(1→3) linkage [107,108]. Some of the nonreducing terminal D-mannopyranose residues of the trisaccharide have a cyclic six-membered pyruvic acid ketal attached to C4 and C6, and some of the inner D-mannopyranose units are acetylated at C6 [109].

The branching of the cellulose chain by the trisaccharide makes the otherwise insoluble cellulose molecule water-soluble. At low concentrations, xanthan produces high viscosities at low temperatures. These properties provide a number of uses as a thickener and bulking agent for prepared foods, such as salad dressings, syrups, toppings, relishes, ice cream, and baked goods. It is also used as a carrier and emulsifying agent in cosmetics and pharmaceuticals [110].

There are other bacterial gel polysaccharides with different properties, composed of D-glucopyranose, L-rhamnopyranose, D-glucuronic acid, and L-mannopyranose that are obtained from *Pseudomonas elodea* (syn. *Sphingomonas elodea*), which produces gellan, and also species of *Alcaligenes* that produce welan and rhamsan [111].

10.2 Pathogenic Bacterial Capsular Polysaccharides

Salmonella species have an O-antigen that is a heteropolysaccharide, imparting pathogenicity to the organism. It is composed of a repeating tetrasaccharide unit, made up of a sequence of D-mannopyranose, L-rhamnopyranose, and D-galactopyranose, with a variable 3,6-dideoxy-D- or L-hexose linked to the D-mannose residue as a branch residue [112]. *Salmonella* easily mutates and over 100 different kinds of capsular polysaccharides have been identified for various species and mutants. The polysaccharides vary according to the linkage positions and the α- or β-configurations [112]. They also vary with the nature of the attachment of four different 3,6-dideoxy carbohydrate residues: D-paratose (3,6-dideoxy-D-glucopyranose), D-tyvelose (3,6-dideoxy-D-mannopyranose), D-abequose (3,6-dideoxy-D-galactopyranose) and L-colitose (3.6-dideoxy-L-galactose) [113]. *Salmonella* readily mutates the structure of this polysaccharide, giving very wide diversity of structures and a basis of avoiding antibody neutralization.

α- Abe*p*
1
$\downarrow$
3
$[\rightarrow 4\text{-Man}p\text{-}\alpha\text{-}(1\rightarrow 3)\text{-Gal}p\text{-}\alpha\text{-}(1\rightarrow 4)\text{-}\alpha\text{-L-Rha}p\text{-}\alpha\text{-}(1\rightarrow]_n$

☐ **Scheme 4**
A typical *O*-antigen *Salmonella* capsule polysaccharide

Streptococcus pneumoniae strains constitute a large group of pathogens, responsible for bacterial pneumonia. All virulent strains have a voluminous capsule that is responsible for their pathogenicity. The capsules are all relatively complex heteropolysaccharides with diverse structures. The monosaccharide residues contain D-glucopyranose, D-glucopyranose uronic acids, L-rhamnopyranose, and *N*-acetyl-D-glucosamine [113]. There are some capsules with unusual carbohydrates, such as sugar alcohols (glycerol, erythritol, D-threitol, and ribitol), amino-sugars (*N*-acetyl-L-fucosamine, *N*-acetyl-D-mannosamine, *N*-acetyl-2-amino-2,6-dideoxy-L-talose [commonly called L-pneumosamine]), as well as D-galactofuranose and phosphodiesters. The structures are repeating tetra-, penta-, and hexa-saccharides. Like *Salmonella* O-antigens, the repeating units have permuted glycosidic linkages at different positions and with either α- or β-configurations to give a wide diversity of structures, with over 120 different known structures [114].

An unusual acidic polysaccharide capsule is produced by the Gram-negative pathogens, *Neisseria meningitides* and *Escherichia coli*. These polysaccharides contain the unusual nine-carbon sugar, *N*-acetyl-D-neuraminic acid, which is formed by an enzyme catalyzed aldo-condensation between the methyl group of pyruvic acid and the aldehyde group of *N*-acetyl-D-mannosamine, followed by the formation of a six-membered ring with a three hydroxy-carbon side chain. The sugar acid is linked α-(2→8) or α-(2→9) with itself to give a linear polysaccharide, called colominic acid [115,116] (see ❷ *Fig. 7* for the structure of α-(2→8) colominic acid). An interesting variation is the colominic acid produced by *E. coli* Bos-2 that has the alternating sequence of α-(2→8) and α-(2→9) linkages [117].

11 Properties and Occurrence of D-Fructose in Polysaccharides

Polysaccharides that exclusively contain D-fructose are known as fructans and there are two known kinds, inulin and levan. Inulin is a polysaccharide containing β-D-fructofuranose linked (2→1) [118]. Inulins are found in the roots and tubers of the family of plants known as the Compositae, which includes asters, dandelions, dahlias, cosmos, burdock, goldenrod, chicory, lettuce, and Jerusalem artichokes. Other sources are from the Liliacae family, which includes lily bulbs, onion, hyacinth, and tulip bulbs. Inulins are also produced by certain species of algae [119]. Several bacterial strains of *Streptococcus mutans* also produce an extracellular inulin from sucrose [120].

Levan is a polysaccharide containing β-D-fructofuranose residues linked (2→6) with (2→1) branch linkages. They are primarily found in grasses [119] and are produced extracellularly by several bacterial strains of *Bacillus subtilis*, *Aerobacter levanicum* (syn. *Erwinia herbicola*) [121], and *Streptococcus salivarius* [122]. They are of higher molecular weight than the inulins, having 100–200 D-fructofuranose residues per molecule. The branch chains are relatively short, containing 2–4 D-fructofuranose residues.

12 Properties and Occurrence of Sugar Alcohols

12.1 Glycerol

When aldoses or ketoses are reduced, sugar alcohols are formed. For example, glycerol is a simple, three carbon sugar alcohol, formed by the reduction of glyceraldehydes. It is found as a major component in two types of lipids, triacylglycerol (triglyceride fats and oils) and phospholipids. In the former, the three hydroxyl groups of glycerol are esterified by fatty acids. In the latter, glycerol is esterified by two fatty acids at the first two carbons and by phosphoric acid at the third carbon. The phosphoric acid is further esterified by the hydroxy groups of ethanolamine, *N,N,N*-trimethyl ethanolamine (choline), or by the hydroxy group of L-serine. The triglycerides make up the well-known fat deposits found in adipose tissue and the phospholipids are major components found in the lipid bilayers of membranes of cells and organelles and play important roles in nerve transmission. Glycerol is also a common component in the teichoic acids (see ❷ *Sect. 12.3*).

Free glycerol is obtained from the saponification of fats and oils. It is a slightly sweet, highly water-soluble liquid. It has the ability to absorb water, making it a valuable humectant and an emollient for skin conditioners. It is also used as a plasticizer in the formation of polymeric materials and is used in the manufacture of pharmaceuticals and the explosive, trinitroglycerine.

12.2 Properties and Occurrence of Free Sugar Alcohols, D-Glucitol, D-Mannitol, Ribitol, Xylitol, and D-Arabinitol

D-glucose can be reduced either chemically or enzymatically to give D-glucitol (frequently called D-sorbitol). It was first obtained from the fresh juice of the berries of the mountain ash [123]. D-Glucitol occurs widely in plants, being found in algae and higher plants. It is especially prevalent in red seaweed, where it occurs to the extent of 10–14% by weight [124] and is found in relatively large amounts in pears, apples, cherries, prunes, peaches, and apricots, where it imparts a sweet taste to these fruits [125].

D-Mannitol is also widely distributed in plants and was the first crystalline sugar alcohol to be obtained from a natural source, the manna ash [126]. It is also found in large amounts (70–90% w/w) in the exudates of the olive and the plane trees [127]. D-Mannitol is found in relatively large amounts in seaweeds of *Laminaria* and *Mycrocystis* species [128]. Species of the mold *Aspergillus*, produce D-mannitol by fermentation, using D-glucose or acetate as carbon sources [129].

Ribitol, the sugar alcohol from the reduction of D-ribose, is found as a constituent of the vitamin riboflavin (vitamin B_2). It is also a constituent of the teichoic acids, see ❷ *Sect. 12.3*. The reduced product of D-xylose is xylitol, which has a very sweet taste and also imparts an unusual cooling sensation. It is found in several fruits, such as plums, raspberries, and strawberries, where it occurs to the extent of about 1% by weight and gives a distinctive and pleasant taste to these fruits. D-Arabinitol is found in mushrooms in amounts as high as 9–10% by weight [130], in lichens [131], and in avocado seeds [132]. D-Arabinitol is produced by some species of yeast (*Debaryomyces subglobosus* and *Endomycopsis chodati*) through fermentation of D-glucose, D-mannose, and sucrose [133].

12.3 Sugar Alcohols in Teichoic Acids

The teichoic acids are bacterial polymers of sugar alcohols (glycerol or ribitol) and phosphoric acid joined end to end by phosphodiester linkages to the primary alcohol groups. They are found in conjunction with the peptidoglycan of Gram-positive bacterial cell walls [134]. The C2 hydroxy group of glycerol is frequently acylated by D-alanine or glycosylated by N-acetyl-D-glucosamine or D-glucopyranose [135]. In some *Bacillus* species phosphodiester linkages join glycerol units between the C1 hydroxy of one unit to the C2 hydroxy of the next unit. Ribitol residues are joined together between the C1 hydroxyl of one unit to the C5 hydroxy of the adjoining unit. The C3 or C4 hydroxy groups can be acylated by D-alanine and the C2 hydroxyl group can be glycosylated by a number of different carbohydrate residues, for example, N-acetyl-D-glucosamine, D-glucopyranose, and di- or tri-saccharides of D-glucopyranose. More complex teichoic acids occur that have a repeating sequence of glycerol joined $(1 \rightarrow 4)$ to N-acetyl-D-glucosamine by a phosphodiester linkage and N-acetyl-D-glucosamine joined $(1 \rightarrow 3)$ by a phosphodiester linkage to the next glycerol unit [136].

13 Properties and Occurrence of Deoxy Sugars

The most abundant and probably best known deoxy sugar is 2-deoxy-D-ribofuranose, which is found as the carbohydrate component in the genetic polymer, deoxyribonucleic acid, the carrier of genes in the chromosomes of living organisms. Other deoxy sugars include 6-deoxy-L-mannose (L-rhamnose), which is found in glycosides and in *Salmonella* sp. O-antigen polysaccharides (see ❷ *Sect. 10.2*). The third deoxy sugar is 6-deoxy-L-galactose (L-fucose), found in glycocoproteins, such as the blood group substances (see ❷ *Sect. 14*). The fourth deoxy sugar, 6-deoxy-D-glucose (D-quinovose) is found in acarbose, the naturally occurring pseudotetrasaccharide, produced by *Actinoplanes* sp. fermentation. Acarbose is an inhibitor of α-glucosidase [137]. It also occurs in some of its analogues, such as

α-acarviosine-(1→6)-cellobiose and α-acarviosine-(1→6)-lactose, which act as inhibitors of β-glucosidases and β-galactosidase [138]; and α-maltohexaose-(1→4)-acarbose and α-maltododecaose-(1→4)-acarbose, which are potent inhibitors of α-amylases in the nM range [139]. Four naturally occurring 3,6-dideoxy sugars appear in the different *Salmonella* sp. *O*-antigen capsular polysaccharides, see ❷ *Sect. 10.2*, for their names and structures. Four 2,6-dideoxy sugars (2,6-dideoxy-D-ribo-hexaose [D-digitoxose]; 2,6-dideoxy-3-*O*-methyl-D-ribo-hexose [D-cymarose]; 2,6-dideoxy-D-xylo-hexose [D-boivinose]; and 2,6-dideoxy-3-*O*-methyl-D-xylo-hexose [D-sarmentose]) are found in a number of plants, as the carbohydrate component of the so-called cardiac glycosides [140].

14 Properties and Occurrence of Carbohydrates in Glycoproteins

Glycoproteins make up a large class of important biological compounds. It is estimated that over 75% of the known (~3,000) proteins are glycosylated. The carbohydrates are believed to mediate a number of biological functions: (1) the correct folding of a protein tertiary structure after biosynthesis, (2) establishment and stabilization of protein conformation, (3) secretion of proteins through membranes, (4) control of protein turnover, (5) protection of proteins from proteinase hydrolysis, (6) increase in protein water-solubility, (7) biological recognition involved in growth, cell differentiation, organ formation, fertilization, processes of bacterial and viral infections, formation of tumors, tumor metastasis, allergies, and autoimmune diseases.

Carbohydrates are primarily attached to proteins in two ways: (1) by linkage of C1 to the amide nitrogen of L-asparagine, giving *N*-linked carbohydrate proteins and (2) by formation of acetal linkages with the hydroxyl group of L-serine or L-threonine, giving *O*-linked carbohydrate proteins. The carbohydrate can be a single monosaccharide residue or it can be an oligosaccharide, containing several monosaccharide residues. There are six major carbohydrates involved in glycoproteins; they are *N*-acetyl-D-glucosamine, *N*-acetyl-D-galactosamine, D-mannopyranose, D-galactopyranose, L-fucose, and *N*-acetyl-D-neuraminic acid [141].

The attachment of carbohydrate to nitrogen is invariably by *N*-acetyl-D-glucosamine and the attachment to oxygen is invariably by *N*-acetyl-D-galactosamine. *N*-linked carbohydrates are invariably composed of a "core" pentasaccharide [142] of the following structure:

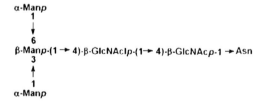

❑ Scheme 5
Core oligosaccharide for *N*-linked glycoproteins

The *N*-linked glycosides can be classified into three families that result from the further attachment of monosaccharides to the two branched D-mannopyranose residues at the nonreducing-end of the core pentasaccharide. These additional residues make up the variable regions of the oligosaccharides. The first family is the "high mannose family" that has additional

α-D-mannopyranose residues attached to the two terminal D-mannopyranose residues of the core pentasaccharide. These residues are linked α-(1$\rightarrow$3) and α-(1$\rightarrow$6). Substitution is terminated by α-(1$\rightarrow$2) linkages of the D-mannopyranose residues. An example of this type of N-glycoside is found attached to ovalbumin.

The second family of N-linked oligosaccharides are called the "lactosamine family" in which the D-mannose residues of the core pentasaccharide are substituted 1$\rightarrow$2 by lactosamine, which is a lactose analogue with N-acetyl-D-glucosamine substituted for D-glucopyranose at the reducing-end of lactose. The lactosamine is frequently substituted by N-acetyl-D-neuraminic acid (for the structure of N-acetyl-D-neuraminic acid, see the monomer residue in colominic acid, ➋ Sect. 8.2 and ➋ Fig. 7) linked 2$\rightarrow$3 or 2$\rightarrow$6 [142,143]. The third family has a mixed structure of the high mannose and lactosamine families [144,145,146,147].

The O-linked saccharides are not as common as the N-linked saccharides. A relatively simple saccharide is that attached to L-threonine of the highly glyosylated (1 out of every 3 amino acid residues), antifreeze protein found in the blood sera of fish living in the Arctic and Antarctic waters. This glycoside is a disaccharide, β-D-Galp-(1$\rightarrow$3)-α-D-GalNAc p [147].

Lactosamine and isolactosamine (lactosamine analogue with D-galactopyranose linked β-(1$\rightarrow$6) to N-acetyl-D-glucosamine) are a well characterized set of O-linked oligosaccharides that make up the ABO human blood group substances [148]. They are on the surface of erythrocytes and divide human blood into four distinct types. The following core structure makes up the O-blood type and is found in all four blood types:

```
α-L-Fucp
   1
   ↓
   2
β-Gal-(1→3)-GlcNAcp
            1
            ↓
            6
            Galp-(1→3)-β-GlcNAcp-(1→3)-β-GlcNAcp-(1→O-Ser /or The
            3
            ↑
            1
β-Gal-(1→4)-GlcNAcp
   2
   ↑
   1
α-L-Fucp
```

■ Scheme 6
Type 0 human blood group oligosaccharide

The core is composed of nine monosaccharide residues with two chains terminating in β-D-galactopyranose residues linked (1$\rightarrow$4) and (1$\rightarrow$3). The other three human blood groups have two additional monosaccharide residues added to the two β-D-galactopyranose residues of the two chains. A-type human blood group has two α-D-galactosamine residues, one each linked (1$\rightarrow$3) to the ends of the two chains; B-type has two α-D-galactopyranose residues, one each linked (1$\rightarrow$3) to the ends of the two chains; and AB-type has a mixture of α-D-galactosamine and α-D-galactopyranose residues, one each linked (1$\rightarrow$3) to the two chains. O-Type blood is the universal blood donor and can give blood to all four types, but can accept blood only from O-type donors; AB-type blood is the universal blood acceptor and can accept blood from all four types; A-type and B-type can accept blood from O-type donors or from donors with their own blood type, as they make antibodies against either the A- or B-types, and AB-type donors, precipitating the blood [148,149,150].

There are additional blood group variations. A common variation is an isomerization in which α-L-fucopyranose is moved from β-D-Galp to β-D-GlcNAcp and linked (1→4) to give the Lewis-a blood type. A second and related variation is the addition of another α-L-fucopyranose residue to β-D-GlcNAcp linked (1→4) to give two α-L-Fucp residues on the first chain, giving Lewis-b blood type. These kinds of variations can occur for each of the ABO blood types, giving O-type-Lewis-a, O-type-Lewis-b, A-type-Lewis-a, and so forth: O-Le^a, O-Le^b, A-Le^a, A-Le^b, B-Le^a, B-Le^b, AB-Le^a, AB-Le^b [150].

15 Separation and Purification of Carbohydrates

The source and the specific physical and chemical properties of carbohydrates determine the methods that are used for their separation and purification. Mono-, di-, tri- and sometimes higher-saccharides, for example maltodextrins, isomaltodextrins, and raffinose-sucrose dextrin series, are usually quite soluble in water. Carbohydrates, thus, are often obtained by the extraction of natural materials with hot water. As many impurities as possible are removed in an extraction mixture, such as salts, proteins, and lipids. Salts can be removed by precipitation and/or the use of ion exchangers. Lipids are removed with organic solvents, such as a 2:1 mixture of chloroform and methanol, and proteins are precipitated with acids and heat. High amounts of alkali and acid, however, should be avoided. Frequently, some of the last impurities in the aqueous extract, especially colored yellow to brown materials, can be removed by adding activated charcoal and filtering it out to give a clear solution before the extract is concentrated. The concentrated carbohydrate extract is obtained at an elevated temperature (50–60 °C) and an organic solvent such as methanol or ethanol is slowly added to the point where the clear solution just becomes cloudy. The solution is then cooled to ≈20 °C to give crystallization of the carbohydrate and then 4 °C to obtain additional crystals. Monosaccharides and disaccharides will often crystallize, while higher oligosaccharides are frequently obtained as amorphous precipitates that can be removed by centrifugation or filtration and dehydrated.

Many different chromatographic methods of separation (on charcoal, BioGel, silica gel, hydroxyapatite, paper) can be used on a preparative scale to give pure materials that can be studied and used even though they are not crystalline. Two typical examples are given for the isolation, purification, and crystallization of a monosaccharide, α-D-xylopyranose, and a disaccharide, lactose, from natural sources.

15.1 Isolation and Purification of α-D-Xylopyranose from Corn Cobs

Coarsely ground corn cobs or crude xylan can be used as starting materials. The xylan in either source is hydrolyzed with 7% (v/v) sulfuric acid by refluxing for 2.5 h. The mixture is filtered through cloth on a Büchner funnel with as much liquid as possible obtained by suction. The residue is washed with an equal volume of water by suspension as thin slurry and then filtered. A few drops of 1-octanol are added to the combined filtrates that are neutralized with barium carbonate. The solids (primarily barium sulfate) in the mixture are filtered and the residue washed by suspension in water and filtered. If corn cobs are used as the starting material, approximately one-sixth of a cake of baker's yeast is finely suspended in 10–15 mL of water and added to the clear filtrate that is covered with a cotton plug. Fermentation is allowed to go ≈15 h at 37 °C to remove D-glucose (if crude xylan is used, this step can be omitted). After removal of the yeast, activated charcoal is added with an equal weight of Celite 535 and the mixture is filtered by suction. The filtrate is concentrated to a syrup under reduced pressure at 50–60 °C, and three volumes of methanol are added with stirring. The solution is filtered and concentrated to syrup under reduced pressure. The syrup is dissolved in water and passed through a column (5.5 × 50 cm) of equal amounts of activated charcoal and Celite 535. The column is then washed with 6 L of water and the washings and the original filtrate are combined and concentrated under pressure to a syrup (~30 mL) that is filtered through a coarse sintered-glass filter. The filtrate is allowed to stand at 20 °C until crystallization of α-D-xylopyranose is complete. The crystals are removed by filtration and washed with cold (4 °C) 85% (v/v) aqueous methanol. A second crop of crystals are usually obtained by placing the supernatant at 4 °C [151].

15.2 Isolation and Purification of Lactose from Milk

Commercial skimmed (defatted) milk contains ~3% casein, 0.7% albumin, 4–5% lactose, and 1% minerals, along with small amounts of lactosamine and lactosamine oligosaccharides, with the remainder being water. The casein is first precipitated by warning to 40 °C and the addition of 1:10 (v/v) glacial acetic acid and water to 200 mL of milk, with continuous stirring. The dilute acetic acid is added until casein no longer separates. The precipitated casein is removed by centrifugation. Then 5 g of calcium carbonate is immediately added and stirred for ~5 min and then the solution is heated to boiling for ~10 min. This produces almost complete precipitation of the albumin, which is removed by vacuum filtration. The filtrate is concentrated by roto-vacuum evaporation to ~30 mL. Then 166 mL of hot ethanol is added, along with 5 g of activated charcoal; after it has been mixed well, the warm solution is filtered through a bed of Celite. The clear filtrate is allowed to stand 15–25 h at 20 °C or longer for crystallization. When crystallization is complete, the crystals are removed by filtration and a second crop of crystals are obtained by placing the clear solution at 4 °C.

15.3 Analysis, Isolation, and Purification of Monosaccharides and Oligosaccharides

Individual monosaccharides and their reduced sugar alcohols can be separated and analyzed by multiple ascent silica-gel, thin-layer chromatography [152], as well as the more complex mixtures of a series of homologous oligosaccharides, such as maltodextrins, isomaltodextrans, cellodextrins, chitosan- and chito-dextrins, cyclomaltodextrins, and the raffinose-sucrose dextrins can be quantitatively analyzed by multiple ascent silica-gel, thin-layer chromatography (TLC), followed by scanning densitometry [152,153,154]. See ❷ *Fig. 8* for a TLC separation of maltodextrins and cyclomaltodextrins. Pure individual oligosaccharides can be obtained in 50–200 mg amounts by preparative descending paper chromatography, using 70:30 (v/v) propanol-1/water solvent on 23 × 54 cm Whatman 3MM paper for 24–36 h on which the saccharides are separated, and detected by $AgNO_3/NaOH/Na_2S_2O_3$ development of a 1-cm strip

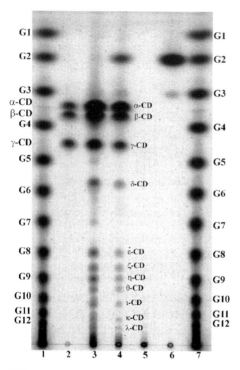

■ Figure 8

Thin-layer silica-gel chromatographic (TLC) separation of maltodextrins and cyclomaltodextrins and using What-man K5 silica gel plate, irrigated 18 cm three-times with 85:25:55:50 volume proportions of acetonitrile, ethyl acetate, propanol-1, water solvent. The carbohydrates were visualized on the plate by dipping it into a methanol solution, containing 0.3% (w/v) *N*-(1-naphthyl)ethylene diamine and 5% (v/v) sulfuric acid, dried, and heated at 120 °C for 10 min. The saccharides can be quantitated by scanning densitometry [152,153]. Lanes 1 and 7 are D-glucose (G1) and maltodextrins (G2 to G12); lane 2, cyclomaltohexaose (α-CD), cyclomaltoheptaose (β-CD), and cyclomaltooctaose (γ-CD); lane 3, a mixture of maltodextrins and cyclomaltodextrins; lane 4, cyclomal-todextrins (α-CD) to (λ-CD), with 6 to 16 D-glucopyranose residues; and lane 6 is maltose and maltotriose. From [166], reproduced by permission of the publisher, Elsevier Press

on each side of the paper. The paper is sectioned, and then the individual saccharides are elut-ed from the sectioned pieces of paper in pure form [155]. They can also be obtained in pure form in larger quantities by charcoal-Celite column chromatography: for example, cellodex-trins [156], isomaltodextrins [157], maltodextrins [158], and xylodextrins [159] have been prepared in this way. Sialyl oligosaccharides from human milk have been separated by ion-exchange chromatography [160] and maltodextrins have been separated by high performance liquid chromatography (HPLC) [161,162].

Capillary electrophoresis has been used to separate and analyze synthetically modified car-bohydrates in the nanogram to milligram range [163]. Fluorophore-assisted capillary elec-trophoresis (FACE) has successfully been used to separate nanogram amounts of maltodex-trins, containing 4–76 D-glucose residues [164,165,166], see ❷ *Fig. 9*. Many carbohydrates can be analytically separated by matrix-assisted, laser desorption, ionization-time of flight,

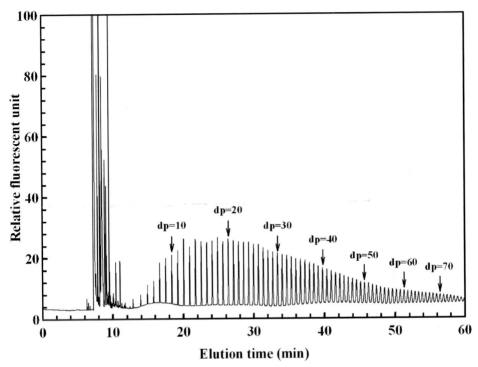

□ Figure 9
Fluorophore-assisted capillary electrophoresis (FACE) analysis of maltodextrins from G2 to G77. From [166], reproduced by permission of the publisher, Elsevier Press

mass spectrometry (MALDI-TOF MS). For example, cyclomaltodextrins, containing 6–25 D-glucose residues are readily analyzed by this technique [166] (see ❷ *Fig. 10*).

15.4 Separation and Purification of Water-Soluble Polysaccharides

Water soluble polysaccharides found in bacterial fermentations or produced enzymatically, such as the dextrans or xanthan, and so forth can be obtained and purified from the aqueous solutions by the addition of two volumes of ethanol, centrifugation, and the resulting pellet redissolved by slowly adding it to boiling water or by suspending it in water and autoclaving at 121 °C for 30 min and then reprecipitating it with two volumes of ethanol. The resulting precipitate can be obtained as a dry powder by treating it several times (5–10) with anhydrous acetone and then once with anhydrous ethanol, and dried in a vacuum oven at 40–50 °C for 12 h. The acetone removes the bulk of the water and the ethanol removes the last traces of water as the 95% azeotrope.

Dextrans with different structures synthesized by distinct dextransucrases that were elaborated by the same strains of *Leuconostoc mesenteroides* have been separated by differential ethanol precipitation, using different concentrations of ethanol, for example strains B-742,

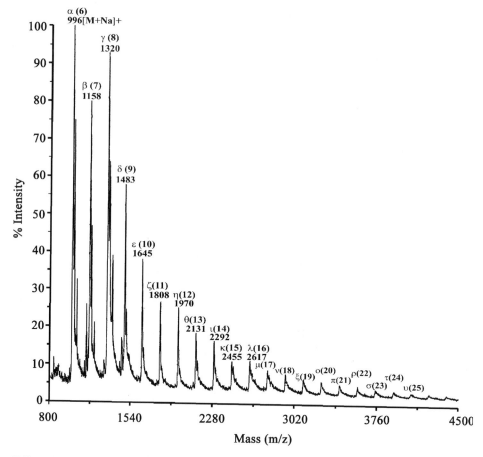

◼ Figure 10
Matrix-assisted laser desorption ionization-time of flight mass spectrometry (MALDI-TOF MS) analysis of cyclo-maltodextrins with 6 to 27 D-glucopyranose residues. The numbers in parentheses above the peaks are the number of D-glucopyranose residues in the cyclomaltodextrins. From [166], reproduced by permission of the publisher, Elsevier Press

B-1254, and B-1299 each gave two dextrans with different structures that were separated from each other; dextran and alternan, produced by strain B-1355, were also separated in this way [167].

15.5 Separation and Purification of Water-Insoluble Polysaccharides, Starch and Cellulose

The separation and purification of starch granules from plant extracts (see ❷ Sect. 2 in ❷ Chap. 6.2) and the fractionation of amylose and amylopectin is given in ❷ Sect. 5 in ❷ Chap. 6.2. The separation of cellulose and hemicelluloses from plant materials and from each other is given in ❷ Sect. 3.2 in ❷ Chap. 6.3.

15.6 Separation and Purification of Cyclomaltodextrins

The cyclomaltodextrins (α-CD, β-CD, and γ-CD) can be selectively obtained from a fermentation culture or an enzyme digest of cyclomaltodextrin glucanotransferase reaction with solubilized starch. The majority of the cyclomaltohexaose (α-CD) can be separated from cyclomaltoheptaose (β-CD) and γ-CD by their selective precipitation with p-cymene from the culture supernatant or from an enzyme digest [168]. The α-CD can then be precipitated from the supernatant with cyclohexene, which is extracted with acetone to remove the cyclohexene and the α-CD can be crystallized from water or a propanol-1/water solution [169]. The p-cymene precipitates of β-CD and γ-CD are put into a water solution and β-CD selectively precipitated from γ-CD with fluorobenzene. The γ-CD is then precipitated with anthracene saturated in diethyl ether. After the removal of the fluorobenzene from β-CD with acetone or ethanol extraction, β-CD can be crystallized from water, and after the removal of anthracene with acetone or ethanol extraction from γ-CD, it can also be crystallized from water [170,171]. The selective precipitations of the cyclomaltodextrins with various organic molecules is based on the selective formation of complexes of the organic molecules with the specific sizes of the cyclomaltodextrins and the relatively hydrophobic interior cavities of the cyclomaltodextrins [166,167,168].

15.7 Release of Oligosaccharides from Glycoproteins

The *O*-linked oligosaccharides are relatively easily released from the protein by a β-elimination reaction, using mild alkali (0.05–0.5-M NaOH) and temperatures of 0–45 °C for 15–216 h [172,173,174]. A standard procedure is 0.1-M NaOH at 37 °C for 48 h. Conditions, however, must be determined for each glycoprotein. The asparagine *N*-linked glycosides can be cleaved by hydrazinolysis [175]. The glycoprotein is heated at 100 °C with anhydrous hydrazine for 8–12 h in a sealed tube. Various endoglycosidases, such as endo-β-*N*-acetylglucosaminidase have also been used to release the oligosaccharides from glycoproteins [176,177]. A chemical method that releases both *O*- and *N*-linked oligosaccharides from glycoproteins involves trifluoromethane sulfonic acid (TFMS) [178]. TFMS reactions are performed at 0 °C for 0.5–2 h under nitrogen. After reaction, the mixture is cooled below −20 °C in a dry ice-ethanol bath and slowly neutralized by the addition of 60% (v/v) aqueous pyridine that is previously cooled to −20 °C. More information on the structure and analysis of glycoproteins may be found in Chap. 8.

References

1. Bonner WA (1991) Origins Life Evolut Biosph 21:72

2. Bailey J, Chrysostomou A, Hough JH, Gledhill TM, MeCall A, Clark S, Ménard F, Tamura M (1998) Science 281:672

3. Lobry de Bruyn CA, Alberda van Ekenstein W (1895) Rec Trav Chim 14:156; 203; (1896) 15:92; (1897) 16:241, 262, 274, 282; (1899) 18:147

4. French D (1954) Adv Carbohydr Chem 9:149

5. Avigad G, Feingold DS, Hestrin S (1956) Biochim Biophys Acta 20:129

6. Feingold DS, Avigad G, Hestrin S (1957) J Biol Chem 224:295

7. Avigad G (1957) J Biol Chem 229:121
8. Stodola FH, Sharpe ES, Koepsell HJ (1956) J Am Chem Soc 78:2514
9. Sharpe ES, Stodola FH, Koepsell HJ (1960) J Org Chem 25:1062
10. Khan R (1976) Adv Carbohydr Chem Biochem 33:236
11. Ballard JM, Hough L, Richardson AC, Fairclough PH (1973) J Chem Soc Perkin Trans I 1524
12. Hough L, Phadnis SP, Tarelli E (1975) Carbohydr Res 44:37
13. Parolis H (1976) Carbohydr Res 48:132
14. Hough L, Phadnis SP (1976) Nature 263:800
15. Hough L, Khan R (1978) Trends Biol Sci 3:61
16. Kuhn R, Gauhe A (1962) Chem Ber 95:518
17. Got R, Font J, Bourrillon R, Cornillot P (1963) Biochim Biophys Acta 74:247
18. Kuhn R, Ekong D (1963) Chem Ber 96: 683
19. Kalf GF, Rieder SV, J (1958) Biol Chem 230:691
20. Clegg JS, Filosa MF (1961) Nature 192:1077
21. Stewart LC, Richtmeyer NK, Hudson CS (1950) J Am Chem Soc 72:2059
22. Ingram J, Bartels D (1996) Ann Rev Plant Physiol Plant Molec Biol 47:377
23. Beattie GM, Leibowitz G, Lopez DA, Levine F, Hayek A (2000) Cell Transplant 9:91
24. Han B, Bischof N (2004) Cell Preserv Technol 2:91
25. Jane J-l, Kasemsuwan T, Leas S, Zobel H, Robyt JF (1994) Starch/Stärke 46:121
26. Hizukuri S, Takeda Y, Yasuda M, Szuki A (1981) Carbohydr Res 94:205
27. Hizukuri S (1991) Carbohydr Res 217:251
28. Leach HW, Schoch TJ (1962) Cereal Chem 39:318
29. Banks W, Greenwood CT (1975) Starch and its Components, Edinburgh University Press, Edinburgh, pp 15, 30
30. Whistler RJ, Daniel JR (1984) In: Whistler RJ, BeMiller JN, Paschall EF (eds) Starch: Chemistry and Technology. Academic Press, New York, pp 153–178
31. BeMiller JN, Whistler RL (1996) Starch Derivatives. In: Fennema OR (ed) Food Chemistry. Marcel Dekker, New York, pp 201–203
32. Fox JD, Robyt JF (1992) Carbohydr Res 227:163
33. Robyt JF, Choe J-Y, Hahn RS, Fuchs EB (1996) Carbohydr Res 281:203
34. Robyt JF, Choe J-Y, Fox JD, Hahn RS, Fuchs EB (1996) Carbohydr Res 283:141
35. Ruttenberg MW, Solarek D (1984) Starch Derivatives: Production and Uses. In: Whistler RJ, BeMiller JN, Paschall EF (eds) Starch: Chemistry and Technology, 2nd edn. Academic Press, San Diego, pp 311–388
36. Geddes R, Harvey JD, Wills PR (1977) Biochem J 30:257
37. Wanson J-C, Drochmans P (1968) J Cell Biol 38:130; (1972) 54:206
38. Geddes R (1985) Glycogen. In: Aspinall GO (ed) The Polysaccharides, vol. 3. Academic Press, San Diego, p 316
39. Preiss J, Walsh DA (1981) The Comparative Biochemistry of Glycogen and Starch. In: Ginsburg V, Robbins P (eds) Biology of Carbohydrates, vol. 1. Wiley, New York, p 203
40. Okada G, Hehre E (1973) Carbohydr Res 26:240; Tao BY, Reilly PJ, Robyt JF (1988) Carbohydr Res 181:163
41. Robyt JF (1995) Adv Carbohydr Chem Biochem 51:133
42. Jeanes A, Haynes WC, Wilham CA, Rankin JC, Melvin EH, Austin MJ, Cluskey JE, Fisher BE, Tsuchiya HM, Rist EE (1954) J Am Chem Soc 76:5041
43. Tsiyisaka Y, Mitsuhashi M (1993) Pullulan. In: Whistler RL, BeMiller JN (eds) Industrial Gums: Polysaccharides and Their Derivatives, 3rd edn. Academic Press, San Diego, pp 447–460
44. Bender H, Lehmann J, Wallenfels K (1959) Biochim Biophys Acta 36:309
45. Bouveng HO, Kiessling B, Lindberg B, McKay J (1962) Acta Chem Scand 16:615; (1963) 17:792
46. Wallenfels K, Keilich G, Bechtler G, Freudenberger D (1965) Biochem Z 341: 433
47. Catley BJ, Whelan WJ (1971) Arch Biochem Biophys 143:138
48. French D, Rundle RE (1942) J Am Chem Soc 64:1651; Freudenberg K, Cramer F (1948) Z Naturforsch 36:464; French D, Knapp DV, Pazur JH (1950) J Am Chem Soc 72:5150
49. Penninga D, Strokopytov B, Bozeboom HJ, Lawson CL, Dijkstra BW, de Vries GE, Bergsma J, Dijkhuizen L (1995) Biochemistry 34:3368; Mori S, Hirose, S, Oya T, Kitahata N (1994) Biosci Biotech Biochem 58:1968
50. Pulley AO, French D (1961) Biochem Biophys Res Commun 5:11; Endo T, Ueda H, Kobayashi S, Nagai T (1995) Carbohydr Res 269:369; Endo T, Nagase H, Ueda H, Kobayashi S, Nagai T (1997) Chem Pharm Bull 45:532

51. Szejtli J (1976) Stärke 29: 26; (1978) 30: 427; (1981) 33:387
52. Oguma T, Horiuchi T, Kobayashi M (1993) Biosci Biotech Biochem 57:1225
53. Côté G, Biely P (1994) Eur J Biochem 226:641
54. Aga H, Nishimoto T, Kuniyoshi M, Maruta K, Yamashita H, Higashiyama T, Nakada T, Kubota M, Fukuda S, Krimoto M, Tsujisaka Y (2003) J Biosci Bioeng 95:215
55. Kim Y-K, Kitaoka M, Hayashi K, Kim C-H, Côté GL (2003) Carbohydr Res 338:2213
56. Takaha T, Yanase M, Takata H, Okada S, Smith SM (1996) J Biol Chem 271:2902
57. Takata H, Takaha T, Okada S, Takagi M, Imanka T (1996) J Bacteriol 178:1600; Gorin PAF, Spencer JFT, Westlake DWS (1961) Can J Chem 39:1067; Zevenhuizen LPTM, Scholten-Koerselman HJ (1979) Antonie Leewenhoek 45:165; York WS, McNeil M, Darvill AG, Albersheim P (1980) J Bacteriol 142:243; Da Castro JM, Bruneteau M, Mutaftshiev S, Truchet G, Michel G (1983) FEMS Microbiol Lett 18:269
58. McIntire FC, Peterson WH, Riker AJ (1942) J Biol Chem 143:491
59. Dylan T, Helinski DR, Ditta GS (1990) J Bacteriol 172:1400; Cangelosi GA, Hung L, Pvanesarajah V, Stacey G, Ozga DA, Leigh JA, Nester EW (1987) J Bacteriol 169:2086
60. Miller KJ, Gore RS, Johnson R, Benesi AJ, Reinhold VN (1990) J Bacteriol 172:136; Rolin DB, Pfeffer PE, Osman SF, Szergold BS, Kappler F, Benesi AJ (1992) Biochim Biophys Acta 1116:215
61. Pfeffer PE, Osman SF, Hotchkiss A, Bhagwat AA, Keister DL, Valentine KM (1996) Carbohydr Res 296:23
62. O'Sullivan AC (1997) Cellulose 4:173
63. Marchessault RH, Sundararajan PR (1983) In: Aspinall GO (ed) The Polysaccharides, vol. 2. Academic Press, San Diego, pp 25–65
64. Hess K, Mahl G, Gütter E (1957) Kolloid Z 155:1
65. Ross P, Mayer R, Benziman M (1991) Microbiol Rev 55:35
66. Hus DS, Reeves RE (1967) Carbohydr Res 5:202
67. Aspinall GO, Krishnamurthy TN, Rosell K-G (1977) Carbohydr Res 55:11
68. Aspinall GO (1959) Adv Carbohydr Chem 14:429
69. Timell TE (1964) Adv Carbohydr Chem 19:247; (1965) 20:409
70. Aspinall GO, Molloy A, Craig JWT (1969) Can J Biochem 47:1063; Gould SEB, Rees DA, Wright NJ (1971) Biochem J 124:47
71. Bauer WD, Talmadge KW, Keestra K, Albersheim P (1973) Plant Physiol 51:174
72. Timell TE (1964) Adv Carbohydr Chem 19:247; (1965) 20:409
73. Stephen AM (1983) Other Plant Polysaccharides. In: Aspinall GO (ed) The Polysaccharides, vol. 2. Academic Press, San Diego, pp 166–169
74. Saka S (1990) Xyloglucans and Xylans. In: Hon DN-S, Shiraishi N (eds) Wood Cellulosic Chemistry. Marcel Dekker, New York, pp 59–88
75. Rolin C (1993) Pectin. In: Whistler RL, BeMiller JN (eds) Industrial Gums: Polysaccharides and Their Derivatives, 3rd edn. Academic Press, San Diego, pp 257–282
76. Karrer P, Francois G (1929) Helv Chem Acta 12:986
77. Ward K Jr, Seib PA (1970) Chitin. In: Pigman W, Horton D (eds) The Carbohydrates, vol. IIA. Academic Press, New York, pp 435–437
78. Rudall KM (1963) Adv Insect Physiol 1:257
79. Horton D, Lineback DR (1965) Methods Carbohydr Chem 5:403
80. Ueno H, Mori T, Fujinaga T (2001) Adv Drug Deliv Rev 52:105
81. Sanford PA (1989) Applications of Chitosan. In: Skjåk-Bræk G, Anthonsen T, Sanford PA (eds) Chitin and Chitosan. Elsevier Applied Science, London, pp 51–69
82. Vorlop KD, Klein J (1981) Biotech Lett 3:9
83. Jeon Y-J, Shahid F, Kim S-K (2000) Food Revs Int 16:159
84. Vandenberg GW, De La Noue J (2001) J Microencap 18:433
85. Jeanloz RW, Sharon N, Flowers HM (1963) Biochem Biophys Res Commun 13:20
86. Tipper DJ, Ghuysen J-M, Strominger JL (1965) Biochemistry 4:468
87. Sharon N, Osawa T, Flowers HM, Jeanloz RW (1966) J Biol Chem 242:223
88. Tipper DJ, Strominger JL (1966) Biochem Biophys Res Commun 22:48
89. Krulwich TA, Ensign JC, Tipper DJ, Strominger JL (1967) J Bacteriol 94:734
90. Ghuysen J-M (1968) Bacteriol Rev 32:425
91. Weidel W, Pelzer H (1964) Adv Enzymol 26:193
92. Fransson L-Å (1985) Mammalian Polysaccharides. In: Aspinall GO (ed) The Polysaccharides, vol. 3. Academic Press, San Diego, pp 338–386
93. Casu B (1985) Adv Carbohydr Chem Biochem 43:51

94. Painter TJ (1983) Algal Polysaccharides. In: Aspinall GO (ed) The Polysaccharides, vol. 2. Academic Press, San Diego, pp 263–264

95. Clare K (1993) Algin. In: Whistler RL, BeMiller JN (eds) Industrial Gums, Polysaccharides, and Their Derivatives, 3rd edn. Academic Press, San Diego, pp 108–118

96. Haug A, Larsen B (1962) Acta Chem Scand 16:1908

97. McNeely WH, Pettit DJ (1973) Algin. In: Whistler RJ, BeMiller JN (eds) Industrial Gums 2nd edn. Academic Press, San Diego, pp 74–75

98. Skjåk-Bræk G, Smidsrød O, Larsen B (1986) Int J Biol Macromol 8:330

99. Scott CD (1987) Enzyme Microb Technol 9:66

100. Linker A, Jones RS (1966) J Biol Chem 241:3845; Evans LR, Linker A (1973) J Bacteriol 116:915; Gorin PAJ, Spencer JFT (1966) Can J Chem 44:993; Pindar DF, Bucke CC (1975) Biochem J 152:617

101. Anderson DMW, Gill MCL, Jeffrey AM, McDougal FJ (1985) Phytochemistry 24:71

102. Aspinall GO, Hirst EL, Wickstrom A (1955) J Chem Soc 1160; Aspinall GO, Auret BJ, Hirst EL (1958) J Chem Soc 4408

103. Aspinall GO, Baille J (1963) J Chem Soc 1702

104. Sanford PA (1979) Adv Carbohydr Chem Biochem 36:266; Robyt JF (1998) Essentials of Carbohydrate Chemistry. Springer, Berlin, Heidelberg, New York, pp 193–218

105. Sloneker JH, Jeanes A (1962) Can J Chem 40:2066

106. Sloneker JH, Orentas DG (1962) Can J Chem 40:2188

107. Sloneker JH, Orentas DG, Jeanes A (1964) Can J Chem 42:1261

108. Lindberg B, Lorngren J, Thompson JF (1973) Carbohydr Res 28:351

109. Melton LD, Mindt L, Rees DA, Serson GR (1976) Carbohydr Res 46:245

110. Kang KS, Veeder GT, Mirrasoul PJ, Kaneko T, Cottrell IW (1982) Appl Environ Microbiol 43:1086

111. Jansson PE, Lindberg B, Widmalm G, Sanford PA (1984) Carbohydr Res 139:217; Kuo M-S, Mort AJ, Dell A (1986) Carbohydr Res 156:173; Jansson P-E, Lindberg B, Lindberg J, Maekawa E, Sanford PA (1986) Carbohydr Res 156:157

112. Lüderitz O, Staub AM, Westphal O (1966) Bacteriol Rev 30:193; Robbins PW, Uchida T (1962) Biochemistry 1:323

113. Bagdian G, Lüderitz O, Staub AM (1966) Ann NY Acad Sci 133:849

114. How MJ, Brimacombe JS, Stacey M (1964) Adv Carbohyd Chem 19:303; Larm O, Lindberg B (1976) Adv Carbohydr Chem Biochem 33:295; Lindberg B (1990) Adv Carbohydr Chem Biochem 48:279

115. McGuire EJ, Binkley SB (1964) Biochemistry 3:247

116. Bhattacharjee AK, Jennings HJ, Kenny CP, Martin A, Smith ICP (1975) J Biol Chem 250:1926

117. Egan W, Lui T-Y, Dorow D, Cohen JS, Robbins JD, Gotschlich EC, Robbins JB (1977) Biochemistry 16:3687

118. French AD, Waterhouse AL (1993) Structural Chemistry of Inulin. In: Suzuki M, Chatterton NJ (eds) Science Technology of Fructans. CRC Press, Boca Raton, FL, pp 41–82

119. Hendry GAF, Wallace RK (1993) Occurrence of Inulins and Levans. In: Suzuki M, Chatterton NJ (eds) Science Technology of Fructans. CRC Press, Boca Raton, FL, pp 119–140

120. Baird JK, Longyear VMC, Ellwood DC (1973) Microbios 8:143; Rossel K-G, Birkhead D (1974) Acta Chem Scand Ser B 28:589; Ebisu S, Kato K, Kotani S, Misaki A (1975) J Biochem (Tokyo) 78:879; Corrigan AJ, Robyt JF (1979) Infect Immun 26:387

121. Avigad G (1968) Bacterial Levans. In: Mark HF, Gaylard NG, Bikales NM (eds) Encyclopedia of Polymer Science Engineering, vol. 8. Wiley Interscience, New York, pp 71–78

122. Garzozynski SM, Edwards JR (1973) Arch Oral Biol 18:239; Eshrlich J, Stivala SS, Bahary WS, Garg SK, Long LW, Newbrun E (1975) J Dent Res 54:290; Marshall K, Weigel H (1976) Carbohydr Res 49:351

123. Boussingault J (1972) Compt Rend 74:939

124. Asahina Y, Shimoda H (1930) J Pharm Soc Japan 50:1; Haas P, Hill TG (1932) Biochem J 26:987

125. Strain HH (1937) J Am Chem Soc 56:2264

126. Proust M (1806) Ann Chim Phys 57:144

127. Jrier E (1893) Compt Rend 117:498

128. Bidwell RGS (1958) Can J Botany 36:337

129. Archibald AR, Baddiley J (1966) Adv Carbohydr Chem 21:354

130. Frèryacque M (1939) Compt Rend 208:1123

131. Lindberg B, Misiorny A, Wachtmeister CA (1953) Acta Chem Scand 7:591; Aghoramarthy K, Sarma KG, Seshadri TR (1961) Tetrahedron 12:173

132. Richtmyer NK (1970) Carbohydr Res 12:135

133. Anderson FB, Harris G (1963) J Gen Microbiol 33:137; Hajny GJ (1964) Appl Microbiol 12:87

134. Armstrong JJ, Baddiley J, Buchanan JG, Carss B, Greenberg GR (1958) J Chem Soc 4344

135. Archibald AR, Baddiley J (1966) Adv Carbohydr Chem 21:323

136. Archibald AR, Baddiley J, Burton D (1968) Biochem J 110:543; Archibald AR, Baddiley J, Heckels JE, Heptinstall S (1971) Biochem J 125:353

137. Truscheit E, Frommer W, Junge B, Muller L, Schmidt DD, Wingender W (1981) Angew Chem Int Ed Engl 20:744

138. Lee S-B, Park KH, Robyt JF (2001) Carbohydr Res 33:13

139. Yoon S-H, Robyt JF (2003) Carbohydr Res 338:1969

140. Courtois JÉ, Percheron F (1970) Phenanthrene Glycosides. In: Pigman W, Horton D (eds) The Carbohydrates, vol. IIA. Academic Press, New York, pp 216, 221–222

141. Montreuil J (1980) Adv Carbohydr Chem Biochem 37:158

142. Montreuil J (1975) Pure Appl Chem 42:431

143. Fournet B, Strecker G, Montreuil J, Dorl L, Haverkamp J, Vliegenthart JFG, Schmid K, Binette JP (1978) Biochemistry 17:5206

144. Nilsson B, Nordén NE, Svensson S (1979) J Biol Chem 254:4545

145. Tai T, Yamashita K, Setsuko I, Kobata A (1977) J Biol Chem 252:6687

146. Yamashita K, Tachibana Y, Kobata A (1978) J Biol Chem 253:3862

147. Feeney RE, Yeh Y (1978) Adv Prot Chem 32:191

148. Watkins WM (1966) Science 152:172

149. Lloyd KO, Kabat EA (1968) Proc Natl Acad Sci US 61:1470

150. Ginsburg V (1972) Adv Enzymol 36:131

151. Whistler RL, BeMiller JN (1962) Methods Carbohydr Chem 1:88

152. Robyt JF (2000) Thin-layer Chromatography of Carbohydrates. In: Wilson ID, Cooke M, Poole CF (eds) Encyclopedia of Separation Science, vol. 5. Academic Press, San Francisco, pp 2235–2244

153. Robyt JF, Mukerjea R (1994) Carbohydr Res 251:187

154. Han NS, Robyt JF (1998) Carbohydr Res 313:135

155. Robyt JF, White BJ (1987) Biochemical Techniques: Theory and Practice. Waveland Press, Prospect Heights, IL, pp 82–86

156. Miller GL, Dean J, Blum R (1960) Arch Biochem Biophys 91:21

157. Whelan WJ (1962) Methods Carbohydr Chem 1:321

158. French D, Robyt JF, Weintraub M, Knock P (1966) J Chromatog 24:68

159. Havlicek J, Samuelson O (1972) Carbohydr Res 22:307

160. Smith FD, Zopf DA, Ginsburg V (1978) Anal Biochem 85:602

161. Kainuma K, Nakakuki T, Ogawa T, (1981) J Chromatog 212:126

162. Ammeraal RN, Delgado GA, Tenbarge FL, Friedman RB (1991) Carbohydr Res 215:179

163. Kerns RJ, Vlahov IR, Lindhardt RJ (1995) Carbohydr Res 267:143

164. O'Shea MG, Sammuel MS, Konik CM, Morrell MK (1998) Carbohydr Res 307:1

165. Mukerjea Ru, Robyt JF (2003) Carbohydr Res 338:1811

166. Yoon S-H, Robyt JF (2002) Carbohydr Res 337:2245

167. Wilham CA, Alexander BH, Jeanes A (1955) Arch Biochem Biophys 59:61

168. Cramer F (1958) Chem Ber 91:308

169. French D (1957) Adv Carbohydr Chem 12:189

170. Thoma J, Stewart L (1965) Cycloamyloses. In Whistler RJ, Paschall EF, BeMiller JN, Roberts HJ (eds) Starch: Chemistry and Technology, vol. I. Academic Press, New York, pp 209–249

171. Anderson B, Seno N, Sampson P, Reilly JG, Hoffman P, Meyer K (1964) J Biol Chem 239:2716

172. Spiro RG, Bhoyroo VD (1971) Fed Proc Am Soc Exp Biol 30:1223

173. Spiro RG (1972) Methods Enzymol 28:35

174. Takasaki S, Mizuochi T, Kobata A (1982) Methods Enzymol 83:263

175. Muramatsu T (1978) Methods Enzymol 50:555

176. Kobata A (1978) Methods Enzymol 50:560

177. Tarentino AL, Trimble RB, Maley F (1978) Methods Enzymol 50:574

178. Sojar HT, Bahl OP (1987) Methods Enzymol 138:341

Part 2
General Synthetic Methods

2.1 Reactions at Oxygen Atoms

Ana M. Gómez
Instituto de Química Orgánica General, (CSIC),
28006 Madrid, Spain
iqog106@iqog.csic.es

In: *Glycoscience*. Fraser-Reid B, Tatsuta K, Thiem J (eds)
Chapter-DOI 10-1007/978-3-540-30429-6_3; © Springer-Verlag Berlin Heidelberg 2008

Abstract

Synthetic protocols based on carbohydrates require the differentiation of their abundant hydroxyl groups, by and large, in order to expose just one single hydroxyl group to the selected reagent. This differentiation is usually carried out with the assistance of protecting groups that block the rest of the hydroxyl groups while being compatible with the given reaction conditions. By corollary, the knowledge and apt choice of the appropriate protecting groups is a key factor in successful synthetic endeavors. In this chapter, an overview of the most commonly employed protecting groups in carbohydrate chemistry is given. Alkyl ethers, being robust protecting groups, have a long history in synthetic carbohydrate chemistry and in related structural studies of polysaccharides. Acetals and ketals, which are of fundamental importance in carbohydrate chemistry, are then discussed. Acyl and silyl protecting groups, which also play an important role in modern monosaccharide transformations, are also presented. Finally, recent blocking strategies are described, including orthogonal strategies, by which the protecting groups are harmoniously combined in modern carbohydrate chemistry.

Keywords

Protecting groups; Alkylation; Acetalation; Acylation; Carbonylation; Silylation; Phosphorylation

Abbreviations

ADMB	4-acetoxy-2,2-dimethylbutanoyl
All	allyl
Alloc or Aloc	allyloxycarbonyl
APAC	2-(allyloxy) phenyl acetyl
BDA	butane 2,3-diacetals
Bn	benzyl
Bocdene	2-(*tert*-butoxycarbonyl)-ethylidene
BOM	benzyloxymethyl ether
Bz	benzoyl
BzOBT	1-*N*-benzyloxy-1,2,3-benzotriazol

Cac or ClAc	chloroacetyl
CAN	ceric ammonium nitrate
CBz	benzyloxycarbonyl
CDA	cyclo-hexane-1,2-diacetal
CSA	camphorsulfonic acid
CCL	*Candida cylindracea* lipase
CVL	*Chromobacterium viscosum* lipase
DABCO	diazabicyclo[2.2.2]octane
DBMP	ditertbutylmethylpyridine
DBU	1,8-diazabicyclo[5,4,0]undec-7-ene
DCC	dicyclohexylcarbodiimide
DDQ	dichlorodicyanoquinone
DIB	(diacetoxyiodo)benzene
DISAL	3,5-dinitrosalicylate
Dispoke	dispiroketal
DMAP	4-(dimethylamino)pyridine
DME	dimethoxyethane
DMF	*N,N*-dimethylformamide
DMSO	dimethylsulfoxide
DMTr	dimethoxytriphenylmethyl
DMTST	dimethyl(methylthio)sulfonium trifluoromethane sulfonate
DTBMP	di-*tert*-butylmethylpyridine
DTBS	ditertbutylsilylene
EDAC	1-[3-(dimethylamino)propyl]-3-ethylcarbodiimide hydrochloride
Fmoc	fluoren-9-ylmethoxycarbonyl
HATU	2-(7-aza-1*H*-benzotriazole-1-yl)-1,1,3,3-tetramethyluronium hexafluorophosphate
IDCP	iodonium di-*sym*-collidine perchlorate
Lev	levulinoyl
MEM	methoxyethoxymethyl
MMTr	monomethoxytriphenylmethyl
Mocdene	2-(methoxycarbonyl)-ethylidene
MOM	methoxymethyl ether
NAP	2-naphthylmethyl
NBS	*N*-bromosuccinimide
NIS	*N*-iodosuccinimide
NMO	*N*-methylmorpholine-*N*-oxide
NMR	nuclear magnetic resonance
NPM	*p*-nitrophenylmethyl
PAB	pivaloylaminobenzyl
PBB	*p*-bromobenzyl
PCB	*p*-chlorobenzyl
PFL	*Pseudomonas fluorescens* lipase
Piv	pivaloyl
PIB	*p*-iodobenzyl

PLE	pig liver esterase
PMB	*p*-methoxybenzyl
PMBM	*p*-methoxybenzyloxymethyl
PN	protease *N*-neutral protease
Poc	propargyloxycarbonyl
PPL	lipase from porcine pancreas
PPTS	pyridinium *p*-toluenesulfonate
PSE	phenylsulfonylethylidene
RJL	*Rhizopus javanicus* lipase
SEE	1-[2-(trimethylsilyl)ethoxy]ethyl
SEM	trimethylsilylethoxymethyl ether
SET	single electron transfer
TBS or TBDMS	*tert*-butyldimethylsilyl
TBDPS	*tert*-butyldiphenylsilyl
TES	triethylsilyl
TFA	trifluoroacetic acid
TFAA	trifluoroacetic anhydride
THF	tetrahydrofurane
TIBAL	triisobutylaluminum
TIPDS	1,1,3,3-tetraisopropyldisiloxane
TIPS	trisopropylsilyl
TMEDA	tetramethylethylenediamine
TMS	trimethylsilyl
TMTr	trimethoxytriphenylmethyl
TPS	triphenylsilyl
Tr	trityl
Troc	2,2,2-trichloroethyloxycarbonyl
TBAF	tetrabutylammonium fluoride

1 Introduction

This chapter describes the chemical reactions at the oxygen atoms of carbohydrates along with some of their fundamental characteristics. The hydroxyl groups of carbohydrates display all the chemical properties associated with simple alcohols. The only difference is that carbohydrates contain many hydroxy groups with similar chemical character. Since the hydroxy groups in carbohydrates play different biological roles depending on their positions, the ability to perform chemical reactions on a particular hydroxy group is highly important. However, the regioselective transformation of one out of several hydroxy groups is far from being trivial. While the differentiation between the primary *versus* the secondary hydroxy groups is in general not too difficult, the discrimination between secondary hydroxy groups is a difficult task.

Usually, partially substituted derivatives are made with the aid of protecting groups. The protecting groups used in carbohydrate chemistry are the same as in any other area in organic chemistry [1,2,3,4]. In addition to this fact, it is important to point out that in carbohydrate derivatives protecting groups do more than protect; they also confer other effects to the molecule and can alter the course of a reaction. Important examples of such effects are the use of 2-*O*-participating groups in glycosyl donors [5] or the armed/disarmed concept for glycoside coupling [6].

This chapter aims to impart general synthetic strategies for most sugars and oligosaccharide structures through the use of some basic, well-proven protecting groups, coupled with general strategies towards regioselectivity. The discussion outlines frequently used protecting groups in carbohydrate chemistry, briefly surveying conditions for their introduction, stability, and removal. It should be noted at this stage that the hydroxyl group of the anomeric center, is unique in having two attached oxygen atoms and therefore it will be treated in a separated section.

2 Reactions at Non-Anomeric Hydroxyl Groups

2.1 Alkylation Reactions: Ether-Type Protecting Groups

Alkyl and aryl ethers are relatively stable to acids and bases due to the high C–O bond energy and it is difficult to recover the parent alcohols from them; therefore, most useful ether-type protections utilize resonance stabilization (by delocalization) of the benzylic-type cation or radical to facilitate the cleavage.

2.1.1 Methyl Ethers

Conversion to methyl ethers of non-anomeric hydroxyl groups is a long-established procedure used, in conjunction with ethylation and deutero-methylation, for the analysis of glycosides, oligosaccharides, and polysaccharides.

Methyl ethers are not normally regarded as protecting groups (though they may be considered in special cases [7]) because the removal is difficult requiring conditions not compatible with other functional groups. A recent study has demonstrated a wide range of susceptibilities to methylation of the hydroxyls in various methyl pyranosides using diazomethane together with transition-metal chlorides and boric acid [8]. On the other hand, the selective removal of an

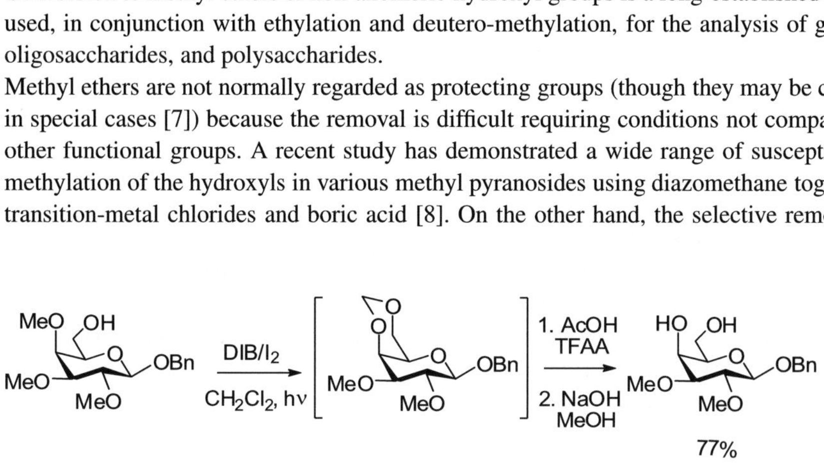

◙ Scheme 1
Selective removal of methoxy protecting groups

ether adjacent to a hydroxyl group in carbohydrate substrates was accomplished with (diacetoxyiodo)benzene (DIB) and I_2 under irradiative conditions (❷ *Scheme 1*). In this step, the methoxy protecting group was transformed into a mixture of acetals (methylenedioxy acetal or *O*-methyl acetate) which upon basic hydrolysis provides the diol [9,10].

2.1.2 Benzyl (Bn) Ethers

The classical permanent protecting group of carbohydrate hydroxyl functions is probably the benzyl ether. It is very stable and can be readily removed under essentially neutral conditions. For this reason, numerous benzylation and *O*-debenzylation procedures have been described. Benzyl ether formation is usually achieved by the reaction of alcohols and benzyl halides in the presence of a base such as sodium hydride in anhydrous DMF (❷ *Scheme 2*) [11], or a mild base (Ag$_2$O) in THF using a phase-transfer catalyst [12]. Benzylation can also be accomplished by the use of an acidic catalyst with benzyltrichloroacetimidate as the reagent [13]. A method using the reductive etherification of TMS ethers under non-basic conditions has also been reported [14].

❑ Scheme 2
Benzylation of methyl α-D-glucopyranoside

Benzyl ethers are highly stable to a wide range of reagents but are readily removed through catalytic reductive conditions [15]. Hydrogenolysis is commonly carried out using hydrogen gas with a palladium catalyst absorbed on charcoal although modifications involving hydrogen transfer have been used. A variety of alternative strategies include Na/liquid ammonia [16], anhydrous $FeCl_3$ [17,18], and $CrCl_2$/LiI [19].

Selective Benzylation   Selective benzylation of carbohydrate hydroxyl functions by direct one-step protection is difficult to achieve. Therefore, several techniques for the selective protection have been developed over the years and the most common are discussed below.

Reductive Opening of Benzylidene Acetals.   An attractive approach for the selective introduction of benzyl groups is provided by the regioselective opening of *O*-benzylidene acetals [20,21,22]. Generally, one of the two C–O bonds in benzylidene acetals can be selectively cleaved, and the direction of the cleavage is dependent on steric and electronic factors as well as, on the nature of the cleavage reagent. Reductive ring-opening of the 1,3-dioxane ring of 4,6-*O*-benzylidene-α-D-glucopyranosides gives the 6-*O*-benzyl and 4-*O*-benzyl ethers respectively in different ratios, depending on the combination of the reagent Lewis acid, solvent, and the substituent at C-3. Some examples are shown in ❷ *Table 1*.
4,6-*O*-Benzylidene-D-galactopyranosides behave in a similar manner to the D-gluco analogs in most cases [29,30]. In the ring opening of the dioxolane rings of 2,3-*O*-benzylidene-α-D-man-

■ Table 1

Reductive opening of 4,6-*O*-benzylidene-α-D-glucopyranoside

Entry	Reagent	Solvent and temperature	R	Yield (%) 6-*O*-Bn	4-*O*-Bn	References
1	NaBH$_3$CN/HCl	THF, 0 °C	Bn	81		[23]
2		THF, 0 °C	Bn	95		[23]
3		THF, rt	Bn (1-*O*-Allyl)	80		[24]
4		THF, rt	Bn (1-*O*-Allyl)	79	16	[25]
5	Me$_3$N·BH$_3$/AlCl$_3$	THF, rt	Bn	71		[26]
6		THF, rt	Bz	74		[26]
7		Toluene, rt	Bn		50	[26]
8		Toluene, rt	Bz		40	[26]
9	Me$_3$N·BH$_3$/BF$_3$·Et$_2$O	MeCN, 0 °C	Bn	30	55	[27]
10		CH$_2$Cl$_2$, 0 °C	Bn	3	73	[27]
11	Et$_3$SiH/CF$_3$CO$_2$H	CH$_2$Cl$_2$	Bn	81	55	[28]
12		CH$_2$Cl$_2$	Ac	98		[28]
13	BH$_3$·THF/Bu$_2$BOTf	CH$_2$Cl$_2$, 0 °C	Bn		87	[30]
14	BH$_3$·THF/Cu(OTf)$_2$	CH$_2$Cl$_2$, rt	Bn		94	[31]
15	Me$_2$EtSiH/Cu(OTf)$_2$	CH$_3$CN, 0 °C	Bn	84		[31]

noside derivatives, the directions of the reaction are determined by the configuration of the benzylidene carbon (❯ *Scheme 3a*). Regarding the reductive ring-opening of 1,2-*O*-benzylidene derivatives, in the case of *manno*-type derivative only a C–O1 bond was cleaved, whereas both the C–O1 and C–O2 bonds were cleaved in the case of the gluco-type compound (❯ *Scheme 3b*) [32].

Benzylidene acetals can also be opened under oxidative conditions, typically NBS in CCl$_4$, to give benzoyl ester protected halogen derivatives, thereby providing an entry into deoxycarbohydrate compounds [33].

Formation of Organotin Intermediates. Another method for selective benzylation refers to the activation of the hydroxyl groups of saccharides by the formation of organotin intermediates such as trialkylstannyl ethers or dialkylstannylene acetals [34]. When the substrate is treated with the tin reagent, one or two Sn–O bonds are formed, enhancing the nucleophilicity of the oxygen atom in the stannyl ether or stannylene acetal. This effect is not identical for the two oxygen atoms of a Sn-acetal, resulting in a differential increase of their nucleophilicity and an ensuing higher regioselectivity.

The activation is carried out by reaction of the polyol with bis(trialkyltin) oxide or a dialkyltin oxide with heating and can be performed in various solvents, the most common being methanol

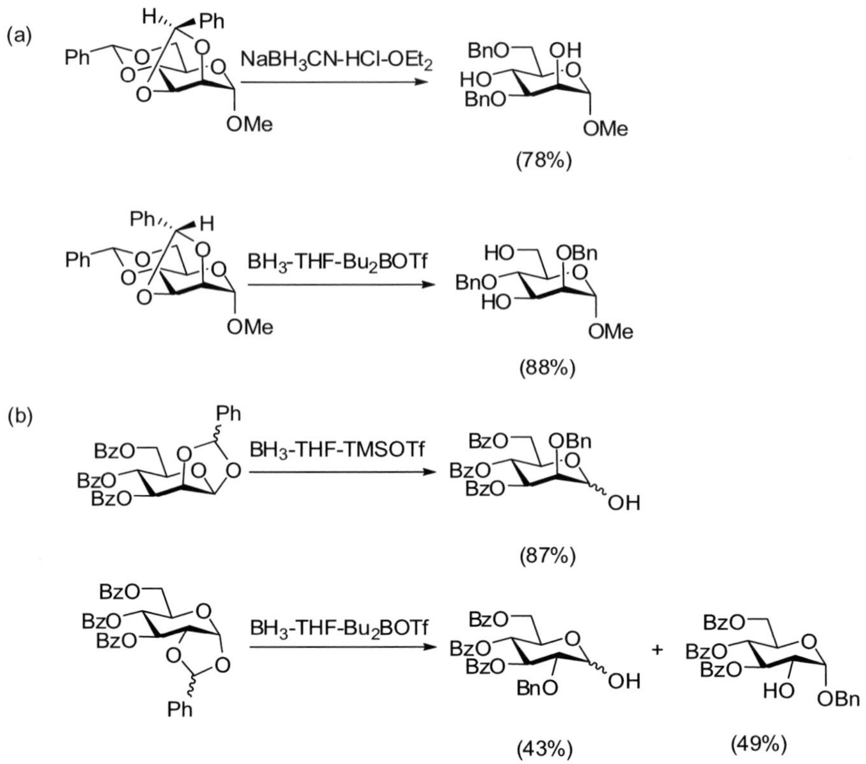

(a)

NaBH₃CN-HCl-OEt₂ → (78%)

BH₃-THF-Bu₂BOTf → (88%)

(b)

BH₃-THF-TMSOTf → (87%)

BH₃-THF-Bu₂BOTf → (43%) + (49%)

☐ Scheme 3
Regioselective benzylation by reductive ring opening of benzylidene acetals

or toluene. Subsequent treatment of the preformed tin intermediates dissolved in a polar aprotic solvent with alkyl halides in the presence of added nucleophiles such as tetrabutylammonium halides [35] or CsF [36] yields the corresponding alkyl (benzyl) ethers. Regarding regioselectivity, this is much the same irrespective of which type of tin derivative is used, the primary hydroxyl group and equatorial hydroxyl group in a vicinal *cis*-dioxygen configuration are preferentially benzylated. As exemplified in ❷ *Scheme 4*, this rule is generally correct, but the degree of selectivity is also dependent on structural features and other factors such as the presence of additives.

Regioselective de-*O*-benzylation.   An alternative strategy to partially benzylated carbohydrates has been accomplished by selective de-*O*-benzylation of easily available polybenzylated precursors. This has been achieved in limited cases by catalytic hydrogenolysis [37], catalytic hydrogen-transfer cleavage [38], acetolysis [39], hypoiodite fragmentation [40], iodine-mediated addition-elimination sequences [41], or use of Lewis acids [42].
Recently, isobutylalanes [43,44,45,46] or the combination CrCl₂/LiI [47] have been shown as efficient agents for the selective deprotection of poly-benzylated carbohydrates. The reaction with isobutylalanes is assumed to proceed through the formation of a penta-coordinated com-

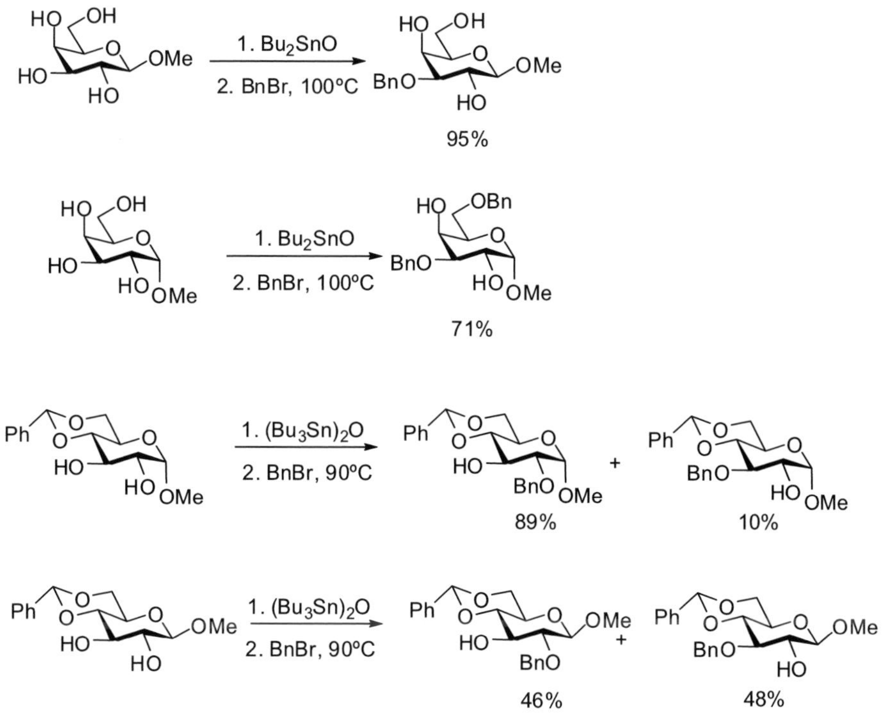

□ Scheme 4
Examples of stannyl mediated regioselective benzylation

plex between the aluminum reagent and a 1,2-*cis* oxygen pattern of the sugar. A second aluminum atom then selects the less hindered oxygen atom and directs the de-*O*-benzylation [46]. When one oxygen atom is clearly more accessible than the other the reaction is highly regioselective. In contrast, when the system CrCl$_2$/LiI is used, a three-point coordination model of the carbohydrate with Cr(II) or Cr(III) is needed for optimal selectivity [47] (❍ *Scheme 5*).

2.1.3 Substituted Benzyl Ethers

To increase the scope of available hydroxyl protecting groups substituted benzyl ethers, which can be selectively removed in the presence of unsubstituted benzyl ethers have been developed. These substituted benzyl ethers are generally less stable to different reaction conditions than unsubstituted benzyl ethers and therefore are used as temporary protecting groups.

p-Methoxy Benzyl (PMB) Ethers   Of the several benzyl ether-type protecting groups reported, *p*-methoxy benzyl (PMB) enjoys a unique position in carbohydrate chemistry due to the ease of its introduction and removal. PMB group demasking, in general, is mediated either by oxidizing agents or by Lewis acids. Thus, hydroxyl moieties protected as PMB ethers can be regenerated easily by oxidation with DDQ [48,49], DDQ-FeCl$_3$ [50], DDQ-Mn(OAc)$_3$ [51], or CAN [52,53]. In the case of DDQ, the *p*-methoxybenzyl group is cleaved selectively with-

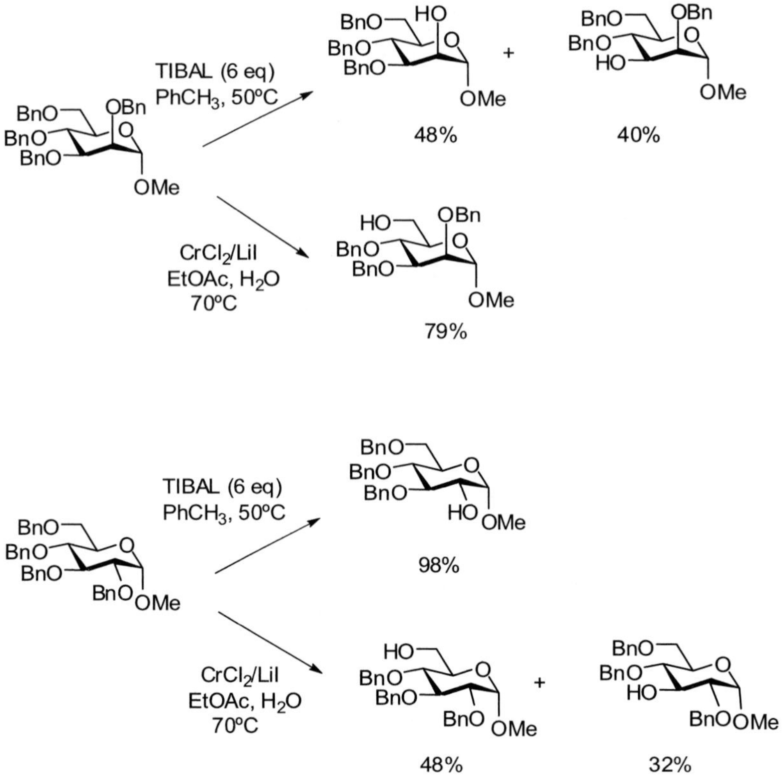

□ Scheme 5
Examples of regioselective de-*O*-benzylation

out affecting several other protecting groups, including benzyl ether. The reaction is assumed to proceed through an easy single electron transfer (SET) to DDQ to generate an oxonium ion which can be captured by water [48,49] (❷ *Scheme 6*).

Deprotection of polyhydroxylated carbohydrate PMB ethers can also be accomplished with SnCl$_4$ [54]. The reaction is compatible with benzyl, TMS, isopropylidene acetal, or methoxy protecting groups. Sometimes the reaction results in unusual regioselectivity and partial deprotection is observed. Preferential mono or bis cleavage of PMB ethers was achieved with careful control of the reaction conditions. However, the general conditions fail in the case of thioglycosides, and a combination of SnCl$_4$/PhSH needs to be used [55] (❷ *Scheme 7*). This combination is particularly useful in the cases where oxidative reagents such as DDQ or CAN need to be avoided. PMB ethers can also be cleaved with ZrCl$_4$ [56], SnCl$_2$/TMS-Cl/anisole [57], CF$_3$CO$_2$H in CH$_2$Cl$_2$ [58], CeCl$_3$·7H$_2$O/NaI [59], or I$_2$/MeOH [60]. The PMB group has also been used as an in situ-removable protecting group for a reactive hydroxyl group in a one-pot reaction involving two sequential glycosilations. The deprotection was performed with *N*-iodosuccinimide-trifluoromethanesulfonic acid at 0 °C and the procedure was used in the synthesis of the globotetraose (Gb4) tetrasaccharide [61].

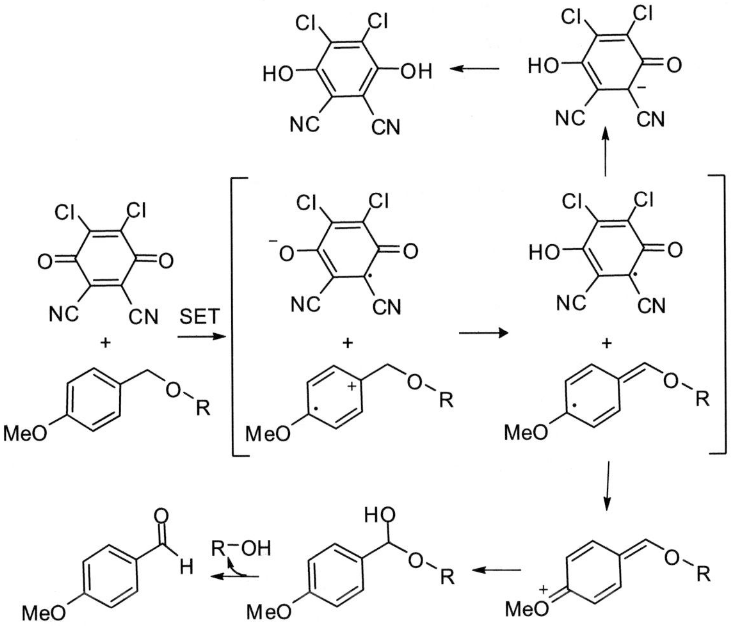

Scheme 6
Removal of *p*-methoxybenzyl ethers with DDQ

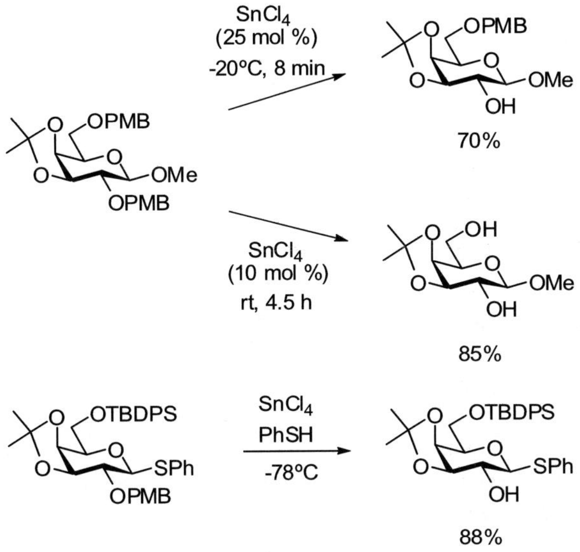

Scheme 7
Examples of deprotection of *p*-methoxybenzyl ethers with SnCl$_4$

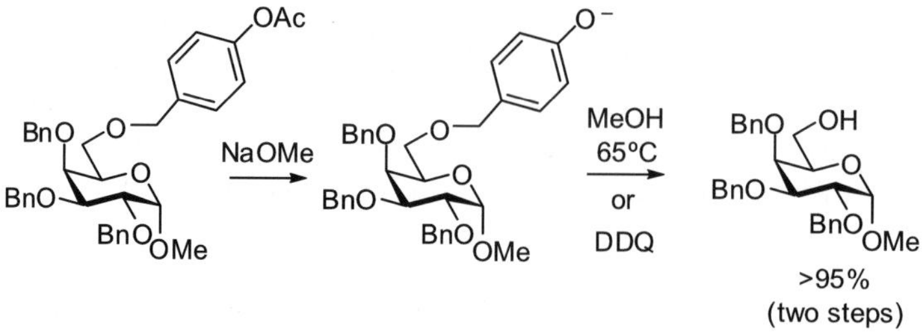

■ Scheme 8
Two-step deprotection of p-acetoxybenzyl protecting group

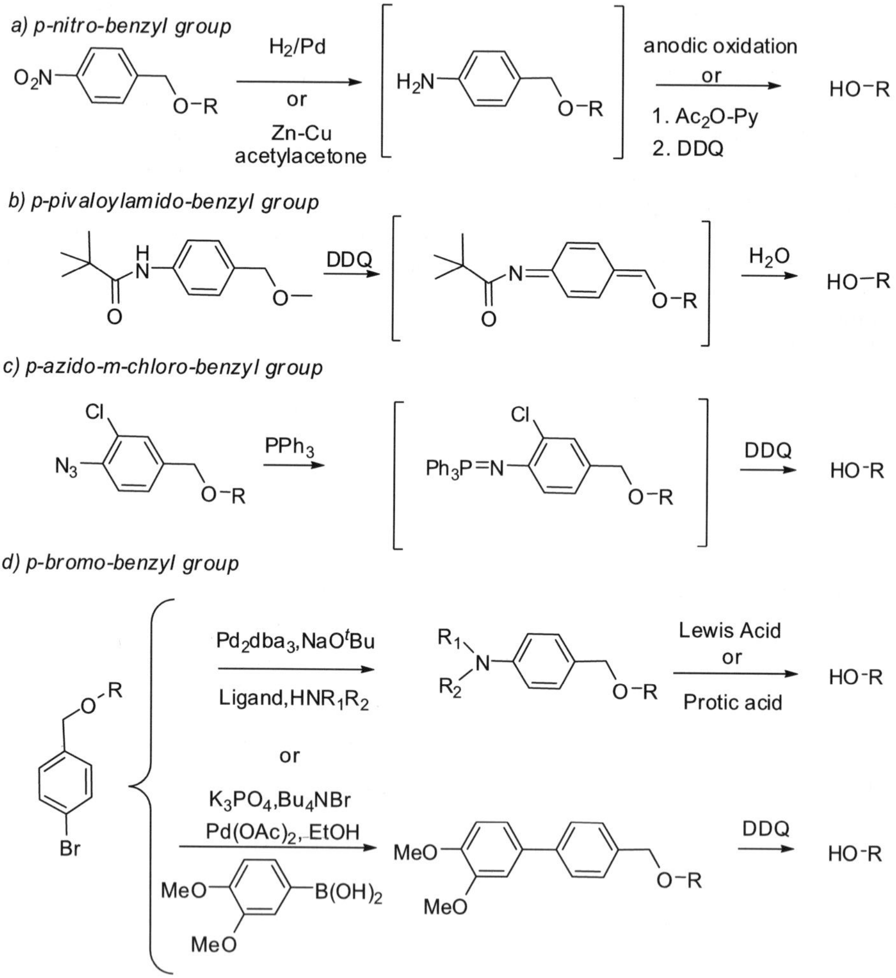

a) p-nitro-benzyl group

b) p-pivaloylamido-benzyl group

c) p-azido-m-chloro-benzyl group

d) p-bromo-benzyl group

■ Scheme 9
Examples of p-substituted benzyl-type protecting groups

Other *p*-hydroxybenzyl-derived protecting groups include *p*-acetoxybenzyl ether and 2-(tri-methylsilyl)ethoxymethoxybenzyl ether [62]. These groups require a two-stage deprotection strategy in which treatment with base or fluoride was followed by thermolysis or mild oxidation of the obtained *p*-phenoxybenzyl intermediate (❷ *Scheme 8*). The conditions are compatible with many of the standard manipulations of oligosaccharide synthesis and with the presence of benzyl or PMB ethers.

Other p-Substituted Benzyl-Type Ethers   Although the PMB protecting group has been extensively utilized in oligosaccharide synthesis, the acid sensitivity of this group sometimes restricts its synthetical application especially during glycosylations. Therefore, other *p*-substituted benzyl groups have been developed (❷ *Scheme 9*). The *p*-nitrobenzyl (or *p*-nitrophenylmethyl NPM) group, which is acid-stable, is readily cleaved via a two-stage procedure involving reduction to an *p*-amino-benzyl ether followed by mild anodic oxidation [63]. *p*-Acetamidobenzyl and *p*-pivaloylaminobenzyl (PAB) derivatives are also used as protecting groups for hydroxyl groups [64,65]. These ethers are much more stable under acidic conditions than a PMB ether, can be obtained by direct alkylation of the hydroxyl group or by acylation of the corresponding *p*-aminobenzylether, and are deprotected by treatment with DDQ. The oxidation occurs at a rate comparable to PMB ethers, so that no preferential cleavage could be achieved with DDQ between these two groups. *p*-Azido benzyl groups are also useful as protecting groups of hydroxyl moieties [66,67]. They can be removed much

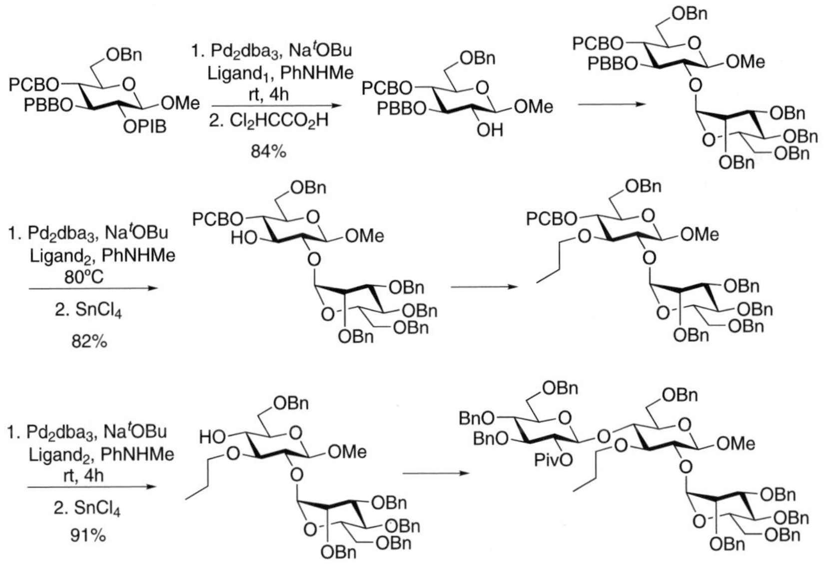

Ligand$_1$ = 1-(N,N-dimethylamino)-1´-(diciclohexylphosphino)biphenyl
Ligand$_2$ = (*o*-biphenyl)P(*t*Bu)$_2$

❏ Scheme 10
Iterative deprotection of *p*-halobenzyl ether protecting groups

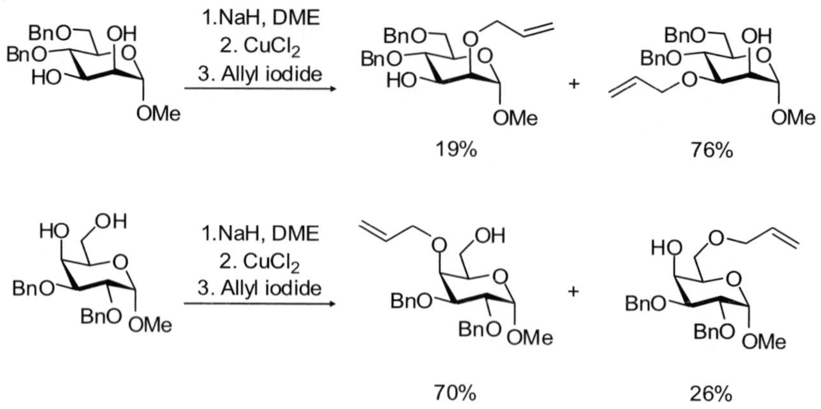

□ Scheme 11
Example of regioselective allylation from copper complexes

faster than the PMB group by DDQ oxidation after conversion of the azide group into the corresponding iminophosphorane. This group allowed for temporary protection of hydroxyl groups in solid-phase synthesis of oligosaccharides [68].

The chemically stable p-halobenzyl ethers (PIB = p-iodobenzyl, PBB = p-bromobenzyl; PCB = p-chlorobenzyl) are converted to labile arylamines via Pd-catalyzed amination [69]. Rapid deprotection of the amine benzyl ethers was observed under very mild Lewis acid conditions. Regarding compatibility of these novel protecting groups with others commonly used, selective cleavage was achieved in the presence of silyl ethers, PMB groups, and glycal double bonds. As shown in ❷ Scheme 10, the differences in the rates of reaction between aryl chlorides, bromides, and iodides in the Pd-catalyzed amination reactions allows for iterative deprotection. In a related method, the p-bromobenzyl group is converted to a DDQ-labile p-(3,4-dimethoxyphenyl)benzyl ether by a Suzuki–Miyaura coupling reaction. This protecting group played a key role in the access to the fully lipidated malarial GPI disaccharide [70].

2.1.4 Allyl and Related Ethers

The protection of alcohols with allyl [71] and related (prenyl, methylallyl, cinnamyl, homoallyl) groups is of great importance in carbohydrate synthesis due to their stability under the conditions required for glycoside formation. These groups are moderately stable to acids and bases, and offer the potential for selective dealkylation of differentially protected sites.

The most general method of preparing allyl ethers is to react the alcohol with allyl bromide or iodide in the presence of sodium hydride. The reaction is best carried out in a polar solvent, usually DMF [72]. Alcohols may also be alkylated after conversion to their barium salts. This technique is employed in the case of N-acyl derivatives of aminosugars to avoid any risk of alkylation at nitrogen which would accompany the use of sodium hydride as the base [73,74]. Conversion of alcohols to allyl carbonates, followed by palladium-catalyzed extrusion of CO_2, constitutes a milder alternative to the classical Williamson-type procedure [75,76]. Compounds containing base labile groups can also be allylated using allyl-trichloroacetimidate [77].

As in the case of benzylation, strategies used for selective allylation include prior conversion to organotin derivatives. When comparisons have been made, there does not seem to be significant differences in selectivities between benzylation and allylation [33,78]. Diols have also been selectively alkylated as their copper (II) salts. Under such conditions no disubstitution is observed and regioselectivity towards formation of a preferred monoallylated product (4,6-diols give mainly 4-substitution and 2,3-diols give mainly 3-substitution) is usually high (● *Scheme 11*) [79].

Common allyl deprotection methods are two-stage procedures that include isomerization to the more labile 1-propenyl group with a variety of agents (● *Scheme 12*). The most frequently employed conditions are treatment of the allyl ether with *t*-BuOK [80], Wilkinson catalyst [81], Pd/C [82], PdCl$_2$ [83], ruthenium(II) [84], and iridium(I) complexes [85] followed usually by acid hydrolysis or oxidation of the resulting enol ether. Also reported are methods including oxidative conditions such as DDQ [86], SeO$_2$ [87], NBS-*hv* [88], and OsO$_4$/NMO/NaIO$_4$ [89]. As a general rule substitution of the allylic framework either slows down or even inhibits the transition-metal catalyzed isomerization [71]. The crotyl and the prenyl groups are readily removed in DMSO/*t*-BuOK through γ-hydrogen elimination reactions. These processes are faster than the allyl to prop-1-enyl isomerization but the difference in rates does not appear to be sufficient to allow good selectivities [90]. Yb(OTf)$_3$ [91], I$_2$/CH$_2$Cl$_2$ [92], or DDQ [93] are mild and efficient methods to cleave prenyl ethers. Remarkable selectivity in the order methylprenyl > prenyl > mathallyl > allyl has been observed by using diphenyldisulfone in a sealed tube at 80 °C (● *Scheme 13*) [94]. These reactions are initiated by the benzene-sulfonyl radical formed from thermal homolysis of (PhSO$_2$)$_2$. The reaction conditions are compatible with the presence of other protecting groups such as acetals and allyl, benzyl, or silyl ethers.

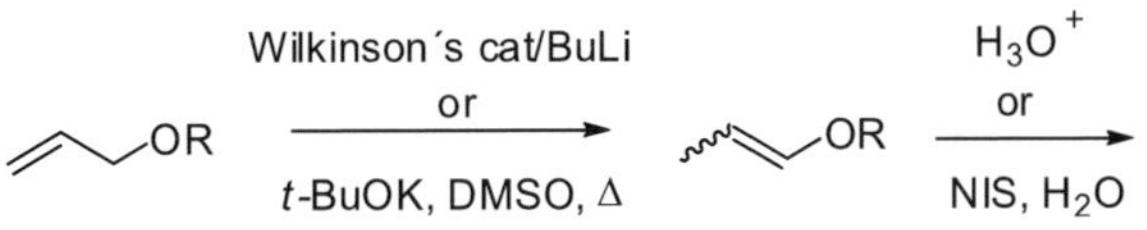

■ Scheme 12
Two-stage removal of allyl ether protecting groups

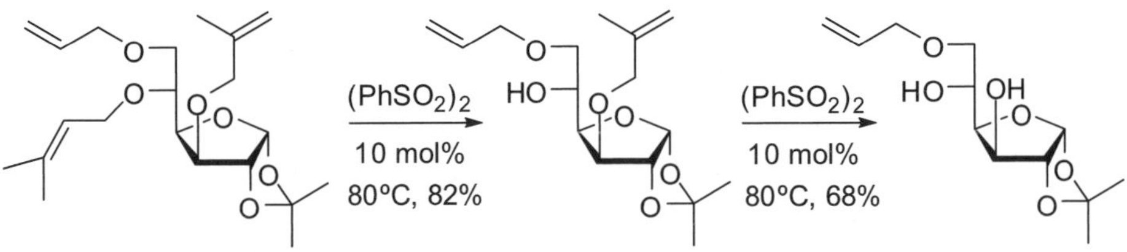

■ Scheme 13
Selective cleavage of branched allyl ethers

2.1.5 Trityl (Tr) Ethers

The usefulness of triphenylmethyl ethers as protecting groups in organic synthesis, in general, and in carbohydrate and nucleoside chemistry in particular is well documented. Its utility
is attributed to the ease in preparing and removing them as well as to the high selectivity
for primary positions observed in polyols. Tritylation of primary hydroxyl groups using trityl
chloride in pyridine is one of the oldest selective alkylation processes described in carbohydrate chemistry (◉ *Scheme 14*). Forcing conditions (trityl perchlorate and 2,4,6-tri-*tert*-butyl
pyridine in dichloromethane) may cause etherification of secondary hydroxyl groups [95].
Alcohols can also be protected as triphenylmethyl ethers by treatment with p-methoxybenzyl
trityl ether (PMBOTr) and DDQ under virtually neutral conditions [96].
Trityl ethers are generally cleaved under protic or Lewis acid conditions, such as formic
acid [97], trifluoroacetic acid [98], BCl_3 [99], $Yb(OTf)_3$ [100], and $VO(OTf)_2$ [101]. Recently, supported-acids [102,103] or Nafion-H [104] have been found to be useful reagents for the
removal of the triphenylmethyl group. Finally, trityl ethers are readily cleaved to the corresponding alcohols by using CBr_4/MeOH [105] or CBr_4-photoirradiation conditions [106].
Substituted trityl groups such as its mono- (MMTr), di- (DMTr) and trimethoxy- (TMTr)
derivatives are also used for the protection of primary hydroxyls (◉ *Fig. 1*). The MMTr and
DMTr groups can be cleaved [107,108] under much weaker acidic conditions than the parent
trityl ether due to the electron-releasing effect of their methoxy groups toward the benzene
ring. None of these trityl ethers is stable enough to survive under normal glycosylation conditions, and therefore they are only used as intermediates to construct building blocks in carbohydrate chemistry. However, the use of DMTr as a protecting group is extremely widespread
in oligonucleotide chemistry.

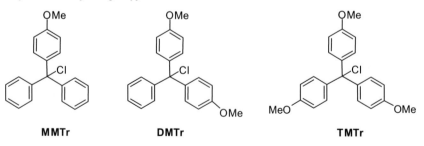

◻ Scheme 14
Tritylation of methyl α-D-glucopyranoside

MMTr   **DMTr**   **TMTr**

◻ Figure 1
Substituted trityl protecting groups

2.1.6 2-Naphthylmethyl (NAP) Ethers

The 2-naphthylmethyl (NAP) group was introduced by Esko et al. [109] and Spencer et al. [110] as a protecting group for polyhydroxy systems. It is stable under conditions normally used for glycoside formation and offers the potential for selective cleavage by hydrogenolysis even in the presence of benzyl groups [110] (❂ Scheme 15). Standard conditions for introduction of the NAP group are, the alkylation with naphthyl bromide [110] or the hydrogenolysis of dioxolane-type (2-naphthyl)methylene acetals [111,112].

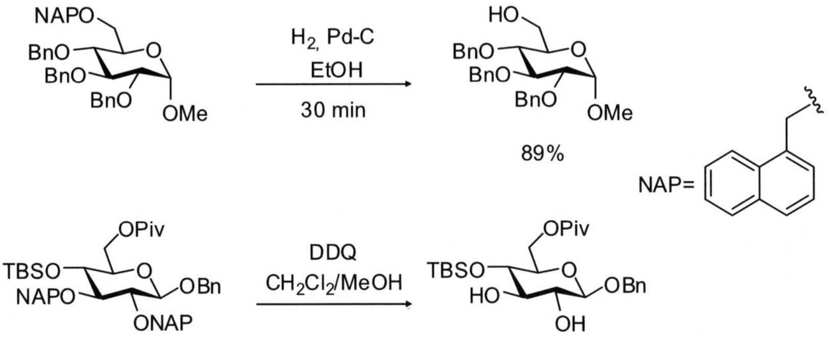

❏ Scheme 15
Selective removal of NAP ethers

The NAP ethers can easily be removed by hydrogenolysis [110] or by DDQ oxidation under conditions which other usual protecting groups like acetyl, pivaloyl, phthalimido, benzyl, and benzylidene survive [113,114]. Recently some successful applications of sugar NAP ethers in the synthesis of complex oligosaccharides have been reported [115].

2.1.7 Propargyl Ethers

Crich and coworkers, have successfully used propargyl ethers as protecting groups in oligosaccharide synthesis [116,117]. Propargyl ethers are readily introduced under standard techniques and are cleaved by a two-set deprotection protocol: an initial treatment with base followed by catalytic osmylation of the resulting allenyl ether (❂ Scheme 16). The use of the (1-naphthyl)-propargyl group opens the possibility for its one-step cleavage with DDQ [117]. These two protecting groups, because of their minimal steric character, are useful for improving the diastereoselectivity of β-mannosylation reactions [116,117,118].

2.1.8 o-Xylylene Ethers

In contrast to all ether-type protecting groups so far mentioned, the o-xylylene group is a bifunctional protecting group devised for simultaneous protection of two vicinal hydroxyl groups of a carbohydrate molecule [119]. The o-xylylene group can be easily introduced by direct alkylation of the diol with α,α'-dibromoxylene or by a two-step process involving an initial alkylation of one hydroxyl function followed by an intramolecular ring-closing reaction (❂ Scheme 17). The o-xylylene protecting group has been successfully used as an element of conformational control of remote stereochemistry in the synthesis of spiroketals [120].

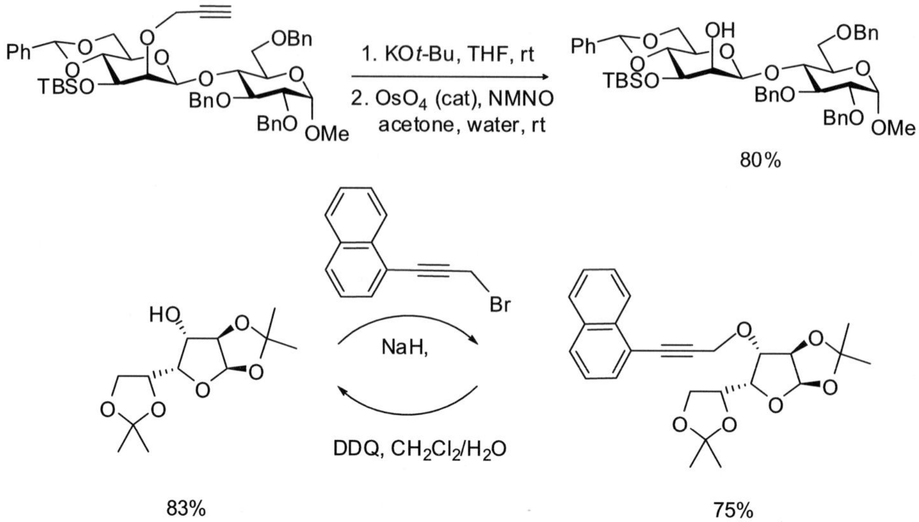

Ph
TBSO
BnO
OBn
BnO
OMe
1. KOt-Bu, THF, rt
2. OsO4 (cat), NMNO
acetone, water, rt
OH
80%
HO
Br
NaH,
DDQ, CH2Cl2/H2O
83%
75%

Scheme 16
Propargyl ethers as protecting groups

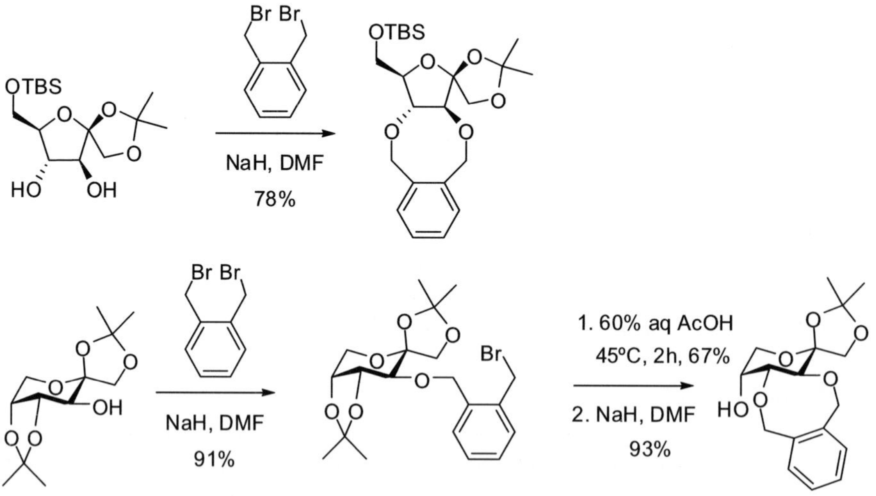

Br Br
OTBS
HO
OH
NaH, DMF
78%
OH
NaH, DMF
91%
Br
1. 60% aq AcOH
45ºC, 2h, 67%
2. NaH, DMF
93%
HO

Scheme 17
The *o*-xylylene protecting group

2.2 Acetalation Reactions: Acetal-Type Protecting Groups

Emil Fischer described as early as 1895 the formation of acetals of glycoses [121]. Since then, this type of protecting group has been extensively used in carbohydrate chemistry. Acetal protecting groups are readily available, easily introduced and removed, and stable to a good range of reactions. Standard conditions for the formation of acetals include treatment of a diol with a carbonyl reagent together with some acid catalyst.

Cyclic acetals such as benzylidene, isopropylidene, or 1,2-diacetals have been effectively used in the regioselective protection of diol systems. The ease of formation and the structures of products are a function of the regio- and stereochemistry of the hydroxyl groups and the properties of the employed carbonyl reagent. Acetals have been applied as protecting groups in many sugars including aminosugars and oligosaccharides. Exhaustive lists of catalyst and conditions can be found in reviews devoted to carbohydrates [122].

2.2.1 Cyclic Acetals

Isopropylidene (acetonides) and benzylidene derivatives are the most commonly used acetals for the simultaneous protection of 1,2- and 1,3-diols in carbohydrate and nucleoside chemistry [123]. Cyclohexylidene acetals are occasionally used, most often as an alternative to benzylidene acetals. Protection using cyclohexane-1,2-diacetals or the related butane-2,3-diacetals represents a new approach which has proved its value in complex oligosaccharide synthesis [124].

Besides being useful for selective protection of monosaccharides, cyclic acetals can display a number of interesting reactions, such as reductive or oxidative ring opening, that amplify the synthetic interest of these protecting groups [125].

Isopropylidene, Benzylidene, and Related Acetals   Isopropylidene or benzylidene acetals are formed either by direct condensation of the diol with the appropriate carbonyl compound (acetone or benzaldehyde, respectively) or by transacetalation with the corresponding dimethoxy acetal. Both processes are carried out in acidic conditions [123].

One advantage of these acetals is their regioselective introduction. Benzylidene derivatives are formed preferentially with 1,3-diols of which anomeric or primary hydroxyl groups are a part. Therefore, they are generally used for 4,6-*O*-protection of pyranoses forming either *cis*- or *trans*-fused 1,3-dioxane rings. In these six-membered rings only the thermodynamically more stable [126] equatorial phenyl-substituted derivatives are observed. Formation of benzylidene acetals has also been achieved under basic conditions using α,α-dihalotoluenes in refluxing pyridine [127].

In contrast, isopropylidene acetals are more stable as five-membered 1,3-dioxane rings formed on *cis*-1,2-diols. Practically all examples in the literature show that, the use of acetone for the acetonation of sugars, leads to 1,3-dioxane rings, which are thermodynamically favored [128]. If 2-alkoxypropene is used as reagent, a reversal on the regioselectivity is observed, and the kinetic products (4,6-*O*-isopropylidene acetals) are preferentially formed [129]. An intermediate behavior is observed for the transacetalation process involving 2,2-dimethoxypropane which gives results either similar to those obtained with acetone or similar to those obtained with enol ethers (❯ *Scheme 18*) [130].

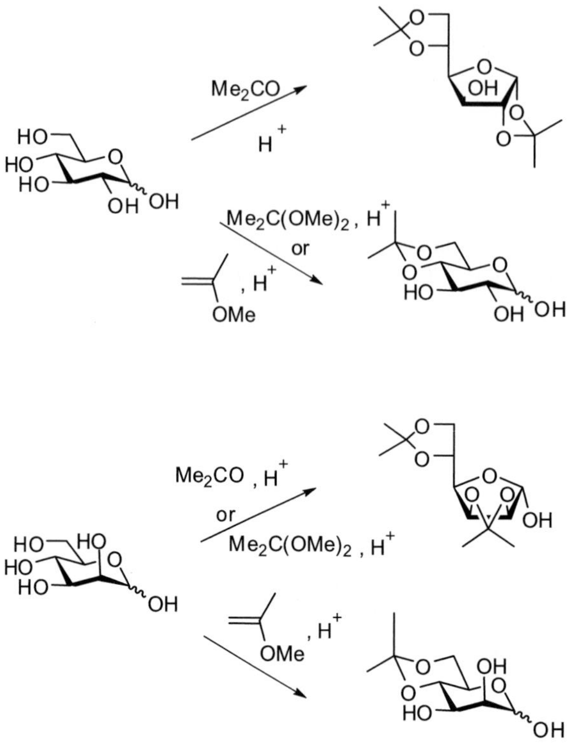

☐ **Scheme 18**
Examples of isopropylidene formation on hexoses

The most extensively adopted method for removal of these protecting groups involves the use of acidic conditions. There are various types of protic and Lewis acids that have been used for this purpose: aq. H_2SO_4 [131], Dowex acidic ion-exchange [132], trifluoroacetic acid [133], $Zn(NO_3)_2$ [134], supported $HClO_4 \cdot SiO_2$ [135] or recoverable $VO(OTf)_2$ [136]. Thiourea can also provoke the cleavage under essentially neutral conditions [137].

Relative to their deprotection, it should be emphasized, that the selective removal of one acetal in the presence of the same (or different) type of acetal, at distinct positions in the same molecule, is possible and has been observed quite often [138]. Several well-established observations can be summarized as follows (i) 1,3-dioxanes are hydrolyzed more easily than the corresponding 1,3-dioxolane, (ii) implication of the anomeric center renders the acetal function more stable, (iii) *cis*-fused 1,3-dioxolanes in a furanose or pyranose ring are more stable than the ones that involve a side chain, and (iv) *trans*-fused benzylidene acetals of hexopyranoses are hydrolyzed faster than the corresponding *cis*-fused acetals [123]. Some examples are shown in ❯ *Scheme 19*.

Some other alkyliden acetals with atypical properties have been used as protecting groups in carbohydrate chemistry. Thus, phenylsulfonylethylidene (PSE) acetals can be synthesized from glycosides under basic conditions. These derivatives are suitable for the protection of

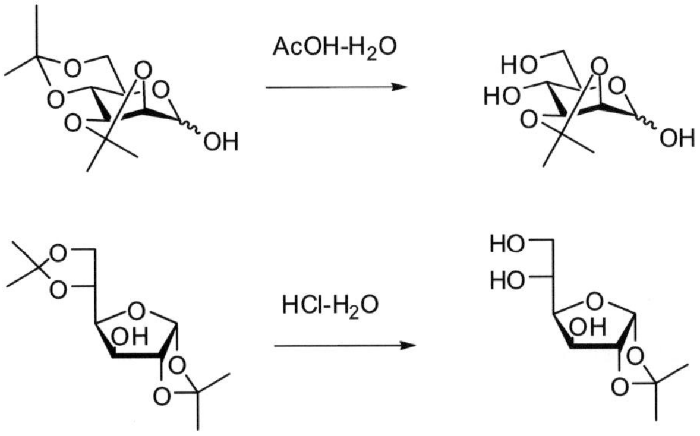

□ Scheme 19
Examples of selective removal of acetal protecting groups

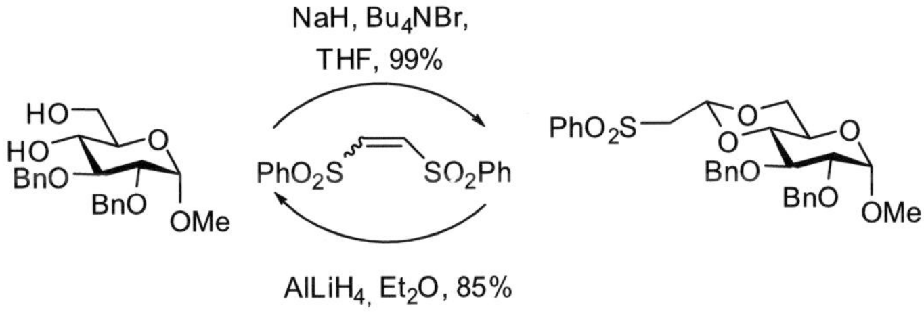

□ Scheme 20
Phenylsulfonylethylidene acetal as protecting group

1,2- and 1,3-diols. The equatorially configurated cyclic acetals are exclusively formed with 1,3-diols whereas the diastereoselectivity of the dioxolane-type acetals from 1,2-diols is quite poor. PSE acetals are deprotected to the corresponding diols under classical reductive conditions (LiAlH$_4$) (❍ *Scheme 20*) [139].

Vicinal diols in sugar substrates can also be protected as their 2-(*tert*-butoxycarbonyl)-ethylidene ("Bocdene") or 2-(methoxycarbonyl)-ethylidene ("Mocdene") derivatives in the reaction with *tert*-butyl or methyl propynoate. The acetal-like structures of these protecting groups is of interest because they are stable under acidic conditions, which allows their selective deprotection versus other acetals, and can be removed under basic conditions via an addition-elimination mechanism (❍ *Scheme 21*) [140]. The procedure is not suitable for 1,3- or 1,4 diols.

Diacetal Protecting Groups The pioneer work of Ley's group concerning the application of 1,2-diacetals such as the dispiroketal (dispoke) [141,142,143,144,145], the cyclo-hexane-1,2-diacetal (CDA) [146], and the butane 2,3-diacetals (BDA) [147] has found widespread application in carbohydrate chemistry (❍ *Fig. 2*) [148].

1,2-Diacetals are highly selective protecting agents which are able to discriminate di-equatorial diols in many carbohydrate derivatives [148]. These protecting groups are stable to functional group manipulation, glycosidation, and are easily removed at the end of a synthetic sequence

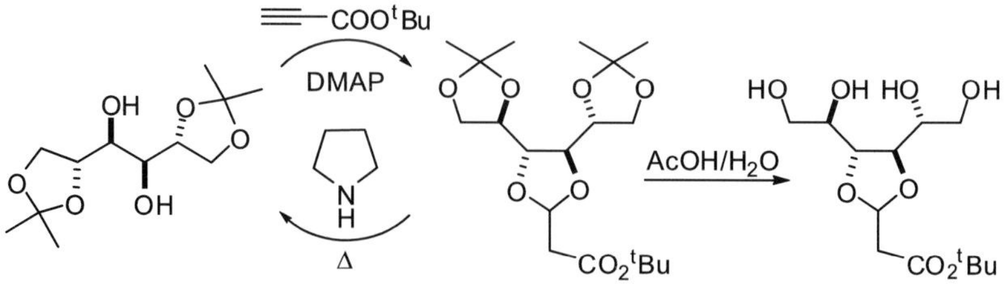

▣ Scheme 21
2-(*tert*-Butoxylcarbonyl)-ethylidene acetal as protecting group

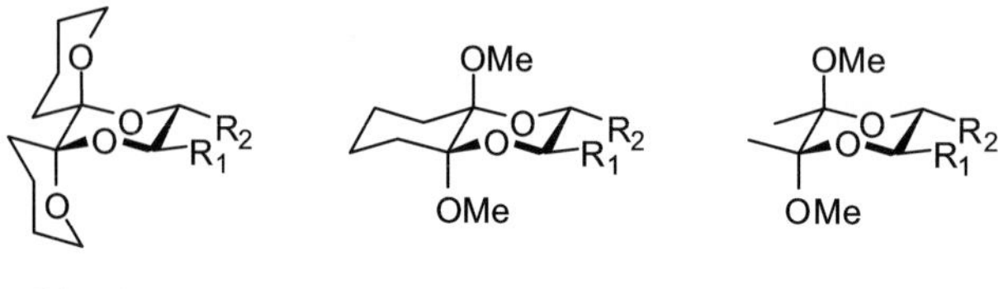

Dispoke   CDA   BDA

▣ Figure 2
Diacetal protecting groups

by aqueous trifluoroacetic acid [149]. Furthermore, 1,2-diacetal protected substrates present a rigid structural architecture that is able to effect reactivity-tuning during glycosidation reactions. This property has been successfully used in oligosaccharide synthesis [150].

The dispiroketal protection of monosaccharides is controlled by the stabilizing influence of multiple anomeric effects leading to a single diastereomeric derivative. In certain examples, where there is more than one diequatorial diol pair present in the molecule, as for example in D-glucose derivatives, reaction affords a mixture of diacetals (❷ *Scheme 22*). The reaction, often giving crystalline compounds, is carried out by treatment of the polyol with 3,4,3,′4′-tetrahydro-6,6′-bis-2H-pyran in chloroform at reflux in the presence of a catalytic amount of CSA [151].

Cyclohexane-1,2-diacetals (CDA) are, however, a better alternative for application to highly polar derivatives within the carbohydrate area. The cyclohexane-1,2-diacetals are formed by reacting 1,1,2,2-tetramethoxycyclohexane [146] or 1,2-cyclohexanedione [152] in boiling methanol containing a catalytic amount of CSA (❷ *Scheme 23*). The corresponding 1,4-dioxane products are formed with high stereoselective control owing to favorable anomeric effects and equatorial placement of functionality around the periphery of the 1,4-dioxane ring. The CDA derivatives are often highly crystalline and usually do not require chromatography. They are stable but can be deprotected readily. They are also able to withstand a wide variety of reaction types such as iodination, reduction, oxidation, Wittig coupling, silylation, and glycosidation reactions.

Likewise, butane-2,3-diacetals (BDA) are good protecting groups for vicinal diequatorial diols. They are prepared either from butane-2,3-dione [153] or from the tetramethoxy butane-2,3-diacetal [147] (❷ *Scheme 23*). BDA derivatives are usually isolated as solids rather than

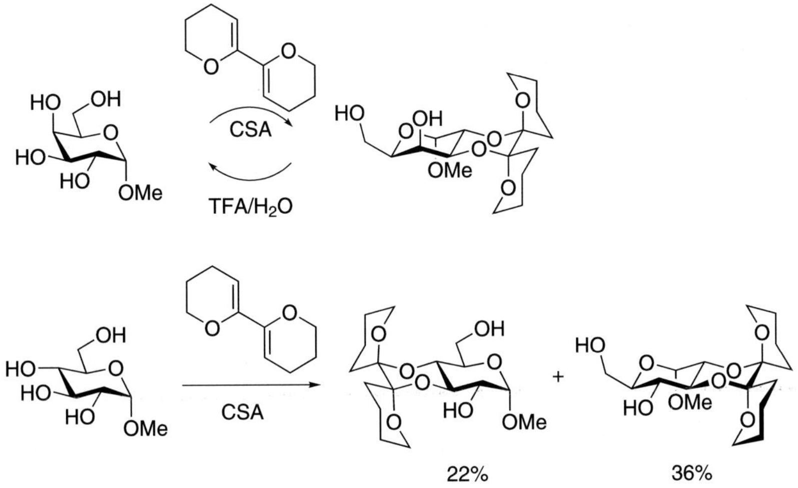

Scheme 22
Examples of protection of diequatorial vicinal diols with dispiroketals

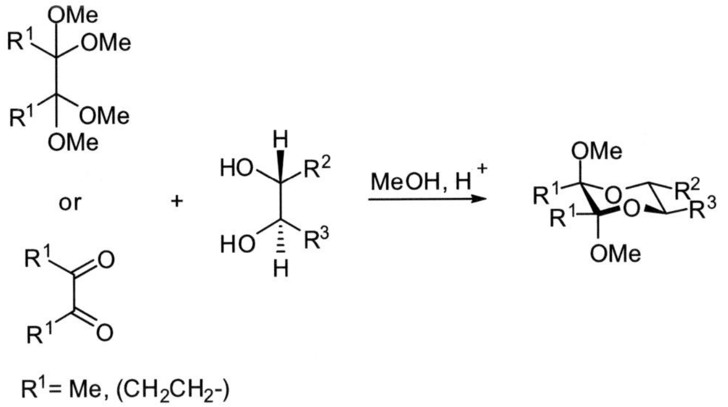

$R^1 = Me, (CH_2CH_2\text{-})$

Scheme 23
Acid-catalyzed formation of 1,2-diacetals from 1,2–*trans*-diols

highly crystalline materials but they have very desirable NMR features that help analysis of the products since the methyl groups act as useful diagnostic markers. As with CDA, BDA groups are readily removed at the end of the synthetic sequence.

Pyruvate Acetals Pyruvate ketals are present in many lipopolysaccharides of bacterial origin, in capsular polysaccharides (❯ *Fig. 3*) [154], and also in glycolipids isolated from fish nerve fibers [155]. As a result of the unique structural features including the presence of a negative charge on the carboxyl functional group and the chiral center, pyruvate ketals influence immunological specificity and patterns of immunological cross reactivity and therefore play an important role in cell–cell recognition processes [156]. Hexopyranosides containing pyru-

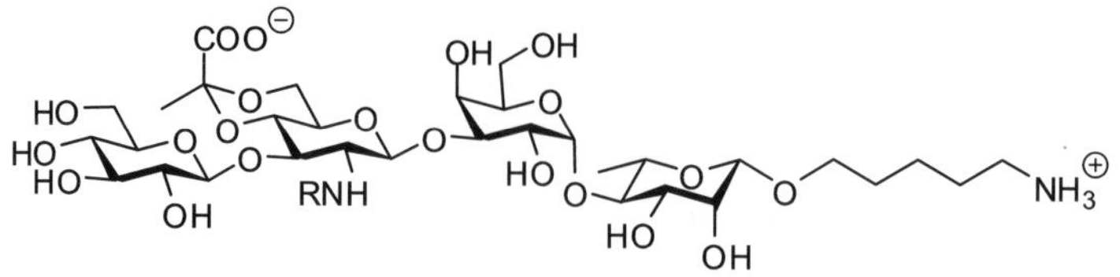

◻ Figure 3
Pyruvated tetrasaccharide related to *Streptococcus pneumoniae* Type 27

vic acid ketals are also useful tools for immunochemical studies of *Klebsiella* polysaccharides [157]. Most commonly, pyruvate ketals are present as 4,6-ketals of hexose residues [158]. However, they also occur as 1,4-dioxolanes formed either from *cis*-axial-equatorial [159] or *trans*-diequatorial hydroxyl groups [160].

Pyruvate ketals can be synthesized [161] by direct condensation of a pyruvate ester with a diol in the presence of a Lewis acid, but this is less preferred because of the electron-withdrawing effect of the adjacent carboxylate group [162,163]. Therefore, several indirect methods for the acetalization have been introduced including condensation with pyruvate derivatives [164,165] or generation of the carboxylate group by oxidation of a suitable precursor [166,167,168,169]. A more efficient route to pyruvic acid acetals starts from silylated diols [170] or by the reaction between diols and methyl pyruvate dialkyl dithioacetal [171,172] activated by methyl triflate, dimethyl(methylthio)sulfonium trifluoromethane sulfonate (DMTST), nitroso tetrafluoroborate (NOBF$_4$), SO$_2$Cl$_2$-trifluoromethanesulfonic acid, or *N*-Iodosuccinimide (NIS) and trifluoromethanesulfonic acid [173] (◉ *Scheme 24*).

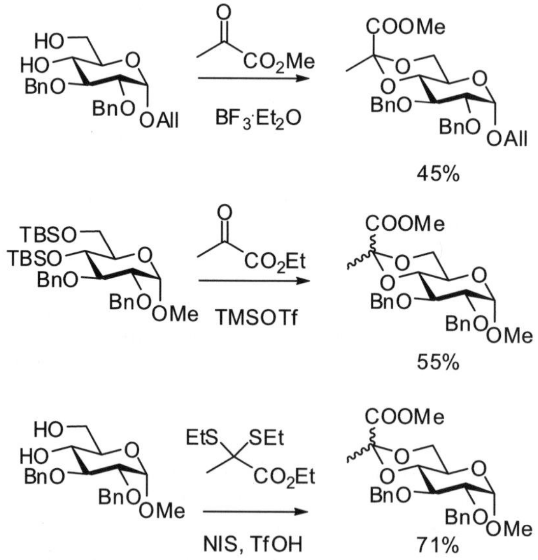

◻ Scheme 24
Examples of synthesis of pyruvate acetals of carbohydrates

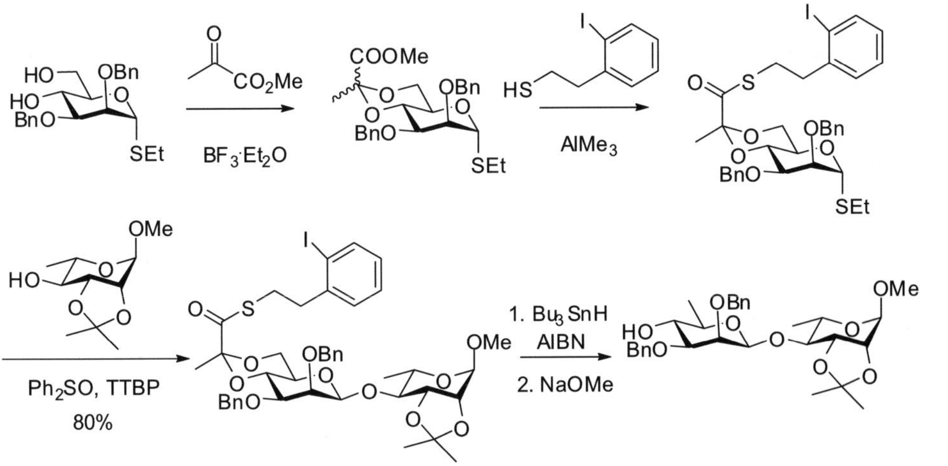

■ Scheme 25
Pyruvate acetals for stereoselective formation of β-D-mannopyranosides

Pyruvate-related acetals have been introduced by Crich as new protecting groups for carbo-hydrate thioglycoside donors [174]. The group conveys strong β-selectivity with thiomanno-side donors and undergoes a tin-mediated radical fragmentation to provide high yields of the synthetically challenging β-rhamnopyranosides (❷ Scheme 25) [174]. Besides this protect-ing group, it has been shown that this approach can also be applied to other related cyclic acetals [175,176,177].

2,2,2-Trihaloethylidene Acetals   In 1992 it was found that the reaction of hexafluoroace-tone or chloral and dicyclohexylcarbodiimide (DCC) with bis-vicinal triols having a *cis-trans* sequence of hydroxyl groups resulted in the formation of cyclic acetals in which the central carbon of the triol had the inverted configuration [178,179]. In this acetalization the oxygen atom of the carbonyl compound (but not that of the alcoholic component) is inserted into the acetal moiety. As shown in ❷ Scheme 26, this non-classical pathway involves the in situ for-mation of a cyclic imidocarbonic ester intermediate, followed by an intramolecular S_N2-attack by a deprotonated neighboring hemiacetal moiety [180,181]. The resulting cyclic acetals are acid-stable but can be converted into the acid-labile ethylidene acetals by treatment with Raney Ni or Bu_3SnH.

2.2.2 Acyclic Acetals

Acyclic *O,O*-acetals are used for the temporary protection of mono-alcohols. Most common-ly used are the tetrahydropyranyl (THP), the methoxymethyl (MOM), the benzyloxymethyl (BOM), or the methoxyethoxymethyl (MEM) protecting groups.

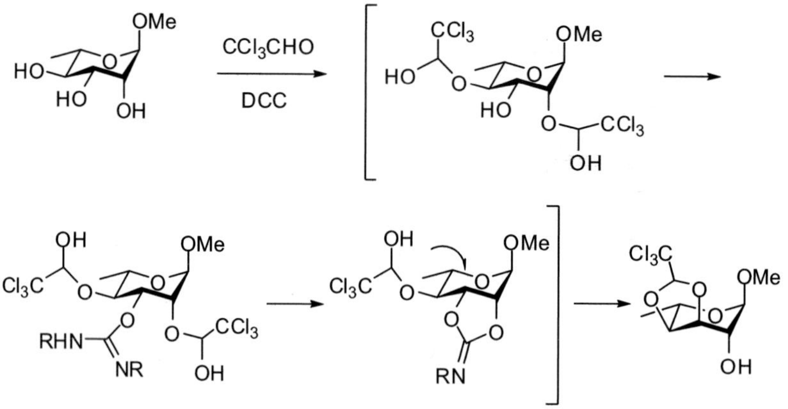

☐ Scheme 26
Example of epimerization by non-classical acetalization

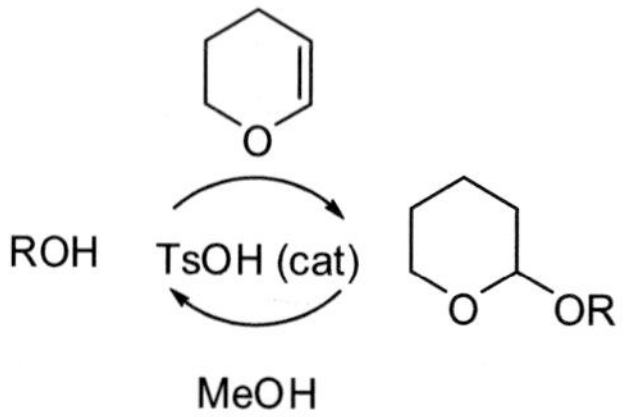

☐ Scheme 27
Tetrahydropyranyl protecting group

Tetrahydropyranyl Ether   The tetrahydropyranyl ether is one of the most frequently used protecting groups for alcohols during multi-step organic synthesis. It is usually introduced by treatment of the corresponding alcohol with 3,4-dihydro-2H-pyran in the presence of an acid catalyst (*p*-toluenesulfonic acid, PPTS, BF$_3$· OEt$_2$, or cation-exchange resins) (❷ *Scheme 27*). Several other methods aiming to introduce the THP group under neutral conditions have been reported [182,183,184,185,186,187]. The resulting tetrahydropyranyl ethers offer stability towards strongly basic reaction conditions, organometallics, hydrides, acylating reagents, and alkylation reagents and the deprotection is usually performed as an acidic hydrolysis or alcoholysis.

In spite of the stability of THP ethers, its use in carbohydrate chemistry is currently limited due to the resulting diastereomeric mixtures obtained by the introduction of an additional stereocenter. This fact makes chromatographic separation and characterization of the products difficult and therefore their use will probably decrease in the future.

Alkoxymethyl Ethers   The principal members of this set of protecting groups are: methoxymethyl ether (MOM) [188], methoxyethoxymethyl ether (MEM) [189], benzyloxymethyl ether (BOM) [190], *p*-methoxybenzyloxymethyl ether (PMBM) [191], and trimethylsilylethoxymethyl ether (SEM) [192] (❷ *Fig. 4*). Since these protecting units are devoid of chirality, their use introduces no stereochemical complications.

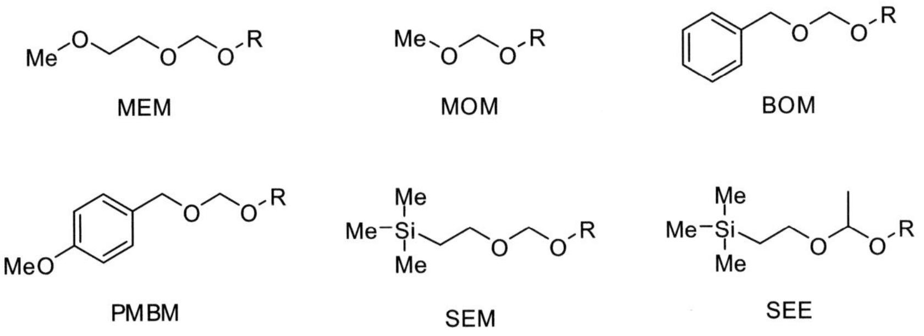

◘ Figure 4
Principal members of the alkoxymethyl ether family

Usually, the formation of the alkoxymethyl ethers is effected by reaction of the corresponding chloride with either the sodium or lithium salt of the alcohol to be protected. The resulting ethers are stable to a wide variety of conditions including many organometallic reagents, reducing conditions, oxidizing agents, and mild acids. Indeed a number of the alkoxymethyl ethers are orthogonal to each other and can be used strategically in multistep syntheses. For example, SEM, BOM, and PMBM ethers are cleaved using a fluoride source, hydrogenolysis and oxidation, respectively. The alkoxymethyl ethers also vary in their degree of lability to Brønsted acids with MOM ether being the most robust.

The SEM group can be removed under milder reaction conditions than the MEM or MOM analogs. As shown in ❷ *Scheme 28*, the compatibility of this group with the conditions required for other selective functional group transformation, together with its stability under glycosylation conditions, allowed the preparation of rhamnopyranosyl synthons for the elaboration of higher order oligosaccharides (❷ *Scheme 28*) [193].

The 1-[2-(trimethylsilyl)ethoxy]ethyl (SEE) group closely resembles the SEM group and can be introduced to alcohols with 2-(trimethylsilyl)ethyl vinyl ether in the presence of a catalytic amount of PPTS under neutral or slightly acidic conditions [194]. The SEE group can be removed with TBAF in THF (24 h, rt. to 45 °C).

2.3 Acylation Reactions: Ester-Type Protecting Groups

Acylation of hydroxyl groups of carbohydrates is one of the most commonly used functional group protection techniques in the synthesis of oligosaccharides. Acyl groups are readily introduced with many acylating agents of different reactivity. They are easily removed under basic (aqueous or non-aqueous) conditions, but are fairly stable under acidic conditions. The main drawback of esters as protecting groups is that they have a tendency to migrate (especially acetates), both under acidic and basic conditions. This is a concern in partially protected derivatives, and results in a mixture with the most stable compound preponderant. Thus, in *cis*-hydroxyls, there is normally a preferred migration from the axial position to the equatorial one and in 4,6-diols the migration goes from *O*-4 to *O*-6 preferentially [195].

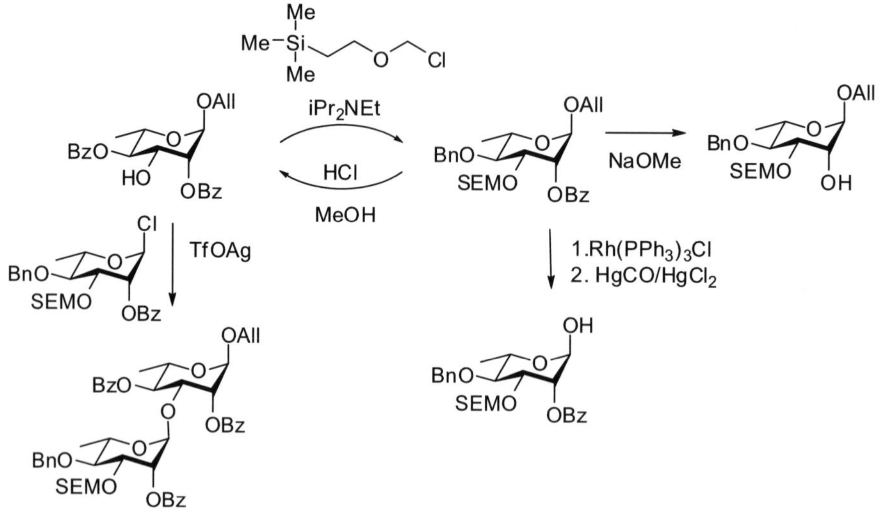

□ Scheme 28
Example of use of SEM protecting group

There is a very large number of different ester protecting groups available and only the more common representatives in carbohydrate chemistry will be treated here. This includes acetate and substituted acetates such as chloroacetate, pivaloate, and levulinate groups. Aroyl groups are frequently used, such as benzoyl and substituted benzoyl e. g. p-phenylbenzoyl and 2,4,6-trimethylbenzoyl groups.

2.3.1 Acetyl (Ac) and Benzoyl (Bz) Esters

General Aspects   In carbohydrate chemistry, per-O-acetylated sugars are inexpensive and useful intermediates for the synthesis of several natural products containing glycosides, oligosaccharides, and other glycoconjugates [196]. The acetylation reaction has also been employed for structural elucidation of many natural products containing carbohydrates. Acetylation of sugar alcohols is often carried out using a large excess of acetic anhydride or, more rarely, acetyl chloride, in the presence of pyridine (or other tertiary amine). Pyridine derivatives, such as 4-(dimethylamino)pyridine and 4-(pyrrolidino)pyridine have been added to the reaction as co-catalyst to speed up the acetylation reaction [197,198]. Similar considerations are valid for the O-benzoylation with the exception that benzoyl chloride rather than the anhydride is used.

Recently, imidazole has been successfully applied as a catalyst for the acetylation of carbohydrates in acetonitrile [199]. A variety of other catalysts in combination with excess of acetic anhydride and solvent includes sodium acetate [200], sulfuric acid [201], perchloric acid [202], and a number of Lewis acid catalysts such as, iodine [203], Sc(OTf)$_3$ [204], Cu(OTf)$_2$ [205], CoCl$_2$ [206], BiOCl-SOCl$_2$ [207], LiClO$_4$ [208], FeCl$_3$ [209], BiCl$_3$ [210], and a series of heterogeneous catalysts such as, montmorillonite K-10 [211], zeolites [212], nafion-H [213], HClO$_4$-SiO$_2$ [214], or molecular sieves [215]. Recently, a ZnCl$_2$–sodium acetate combination [216] or InCl$_3$ [217] with acetic anhydride under microwave conditions has been reported for the acetylation of carbohydrates. A few reports have also appeared on the acetylation of carbohydrates using ionic liquids as solvents and catalysts [218,219].

Zemplén deacylation is the most commonly used deblocking reaction for the removal of ester protecting groups [220]. Using this transesterification reaction, OH-functions can be regenerated under mild conditions, in methanol with a catalytic amount of sodium methoxide at room temperature. The difference in the rate of benzoate and acetate solvolysis is sufficient to enable removal of acetates in the presence of benzoates. Typical conditions for this selective cleavage include ammonia in MeOH.

Regioselective Acylation   Regioselective esterification of carbohydrates may be achieved in part by making use of the differing reactivities of hydroxyl groups. While the selective protection of primary hydroxy groups with sterically demanding acyl residues (e. g. by pivaloylation) is rather easily achieved, it is more difficult to protect one of a number of secondary hydroxy groups. Factors determining the regioselectivity of the acylation of secondary hydroxy groups in carbohydrates have been studied [221]; the most important ones being steric hindrance, intramolecular hydrogen bonding, and the configuration of the hydroxy groups. For instance, the presence of vicinal axial heteroatoms, such as O and S, enhances the nucleophilicity of the corresponding vicinal equatorial OH.

Regioselective acetylations have been promoted by different reagents such as alumina in refluxing ethyl acetate [222,223], silica gel-supported lanthanide chlorides and methylorthoacetate [224], a hindered base and acetyl chloride at low temperature [225], iminophosphorane bases and vinylacetate [226], NaH and 3-acetyl-thiazolidine-2-thiones [227], PPh$_3$/CBr$_4$ in ethyl acetate at high temperatures [228]. Recently, it has been shown that the rate and the selectivity of an acetylation reaction can be controlled by the counterion of the acetylating agent under nucleophilic catalysis. The team play of reagent, catalyst, and auxiliary base is responsible for the outcome of the reaction [229]. Thus, octyl β-D-glucopyranoside can be acetylated with high selectivity either on the primary or on secondary OH groups by using different acetylation agents under otherwise identical conditions (❧ *Scheme 29*).

The use of organotin reagents (❧ *Sect. 2.1.2* under ❧ *"Formation of Organotin Intermediates"*) provides a useful means of efficient regioselective acylations. There does not seem to be significant differences in selectivities between alkylation and acylation although forma-

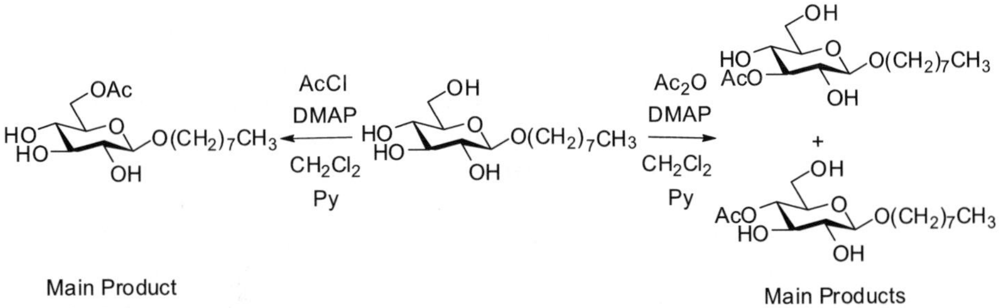

Main Product   Main Products

□ **Scheme 29**
Counterion-directed regioselective acetylation of octyl β-D-glucopyranoside

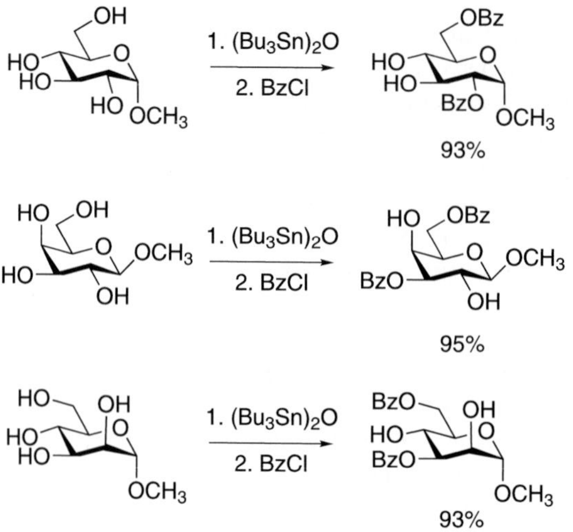

Scheme 30
Examples of stannyl mediated regioselective benzoylation

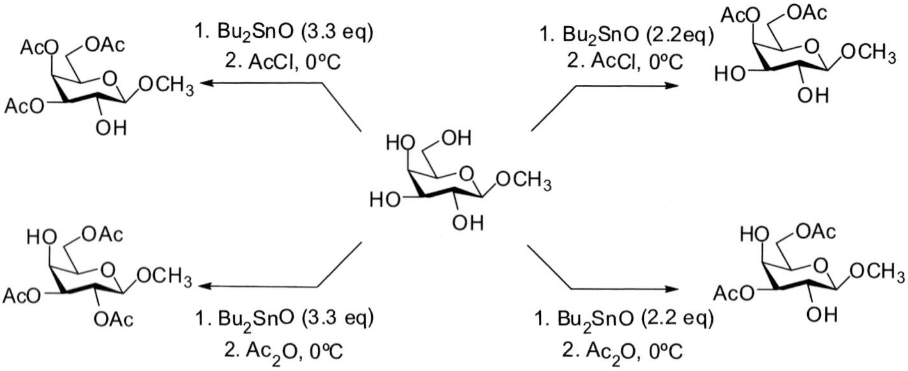

Scheme 31
Reagent-dependent multiprotection of methyl β-D-galactopyranoside

tion of O-acyl derivatives is a much faster reaction in any solvent, and it does not require heating or the presence of an additional nucleophile. Thus, treating methyl α-D-glucopyranoside with $(Bu_3Sn)_2O$ and then with BzCl gave the 2,6-di-O-benzoate, while the 3,6-di-O-benzoates were obtained upon analogous benzoylation of methyl β-D-galactopyranoside and methyl α-D-mannopyranoside, respectively (◉ Scheme 30) [230].

An extension of this tin chemistry to the regioselective acylation of unprotected sugars bound to a resin shows the possibility of using solid-phase techniques for the preparation of O-acyl derivatives of carbohydrates [231]. Very recently, it has been reported that organotin-mediated multiple carbohydrate esterifications can be controlled by the acylating reagent and the solvent polarity. When acetyl chloride is used, the reactions are under thermodynamic control, whereas when acetic anhydride is employed, kinetic control takes place (◉ Scheme 31) [232].

Enzymatic regioselective protection techniques are also an interesting and useful method for the protection of hydroxyl groups [233]. Such techniques are exclusively directed at regioselective acylation and deacylation, mostly by using different lipases [234] or proteases [235], which can catalyze acyl transfer reactions from activated esters to suitable acceptors. The most frequently used enzymes are *Porcine pancreatic lipase* (PPL), *Protease N-neutral protease* (PN), *Pseudomonas fluorescens lipase* (PFL), *Chromobacterium viscosum lipase* (CVL), and *Candida cylindracea lipase* (CCL). The results of the enzymatic acylation of several pyranoses and furanoses have been reviewed [236]. Almost all combinations of enzymes and substrates lead to acylation of the primary hydroxy group. The regioselectivities are usually higher than 70%, and the conversions between 40 and 100%. However, if the 6-OH groups are protected first or deoxygenated, in the corresponding enzymatic reactions, selectivities on the acylation of secondary hydroxyl groups are observed. An example is shown in ❷ *Scheme 32*, where enzymatic acyl transfer reactions turned out to be a viable method for the complete differentiation of the hydroxyl groups of glycal derivatives [237].

Enzymes are not only capable of introducing but also of removing acyl groups into carbohydrates [233]. For example, each of the three OH groups in 1,6-anhydroglucopyranose can be liberated selectively making use of enzymatic reactions (❷ *Scheme 33*) [238,239,240]. The lipase-mediated hydrolysis proceeds with higher velocity and, in many cases with better selectivity, if butanoates or pentanoates are employed as substrates instead of acetates. In all cases the reaction conditions are so mild that the acid sensitive structures remain unaffected.

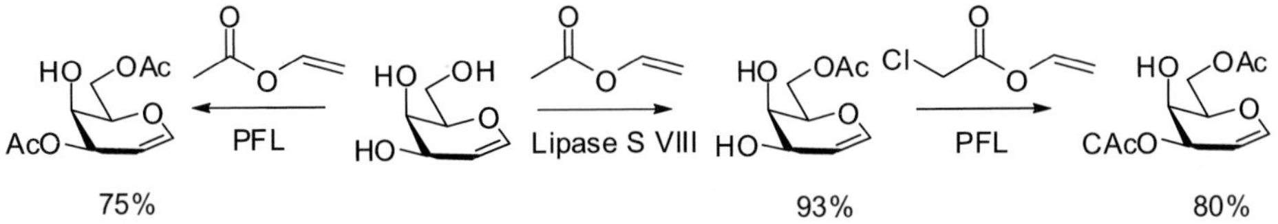

■ Scheme 32
Examples of enzymic regioselective acylation

PPL= Lipase from porcine pancreas
PLE= Pig liver esterase
RJL= Lipase from *Rhizopus javanicus*

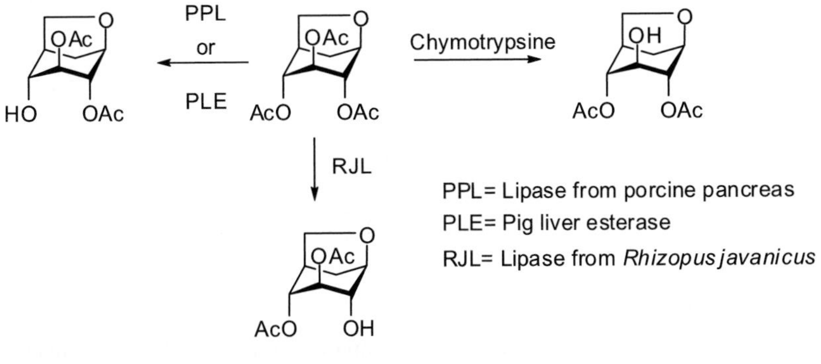

■ Scheme 33
Examples of enzymic regioselective deacylation

2.3.2 Substituted Acetyl Esters

The lability of acetates is enhanced by introducing chlorine atoms in the α-position. Thus, chloroacetates (Cac or ClAc) hydrolyze faster than acetates and trichloroacetates are so reactive that they are rarely used in synthesis. Thus far, thiourea [241], hydrazine dithiocarbonate [242], pyridine [243] and diazabicyclo[2.2.2] octane (DABCO) [244] are representative of dechloroacetylation reagents. 1-Selenocarbamoylpiperidine also deprotects the O-chloroacetyl group with high chemoselectivity in the presence of other acyl groups such as acetyl, pivaloyl, and Fmoc without the assistance of a base [245]. Thiourea, hydrazine dithiocarbonate, or 1-selenocarbamoylpiperidine are believed to deprotect the ClAc group by following a cyclization mechanism. This mechanism is illustrated in ➋ *Scheme 34*: the nucleophilic atom (X) of the reagent replaces the chlorine atom of the ClAc group, and then another nucleophilic atom (Y) attacks the carbonyl carbon to break the C–O bond, thereby resulting in the production of free hydroxyl. In contrast, tertiary-amine-containing reagents such as pyridine and DABCO presumably attack the α-carbon to form onium salt, which is then solvolyzed by water, MeOH, or EtOH to produce naked hydroxyl group.

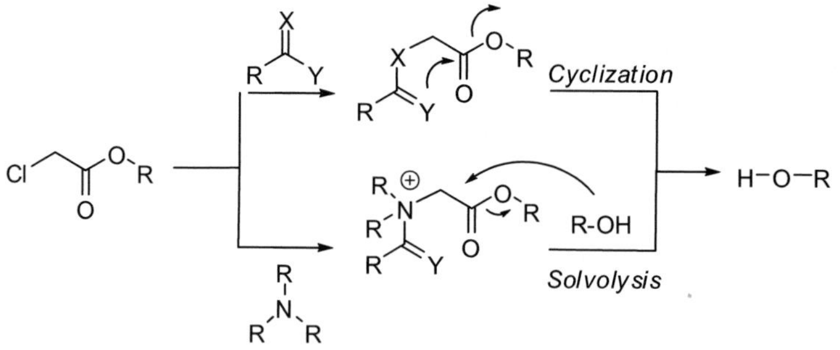

■ Scheme 34
Plausible mechanisms of removal of chloroacetyl groups

The chemoselective deprotection of the CAc group does not affect other protecting groups such as acyl derivatives (acetyl, benzoyl, or levulinoyl groups), carbonates, *p*-methoxybenzyl or silyl ethers and therefore has been included in sets of orthogonal protecting groups. However, the sensitivity of the chloroacetyl group may impose limitations for its application in the synthesis of complex oligosaccharides.

The 2-(allyloxy) phenyl acetyl (APAC) group has been proposed as a new robust acyl-type protecting group for hydroxyl groups [246]. It can be removed under mild conditions by relay deprotection whereby the phenolic allyl ether is cleaved by treatment with a transition metal followed by intramolecular ester cleavage by nucleophilic attack of the revealed hydroxyl (➋ *Scheme 35*). It is compatible with glycosylations and can perform efficiently neighboring group participation leading to the exclusive formation of 1,2-*trans* glycosides.

2.3.3 Pivaloyl (Piv) Esters

The bulky pivaloyl group has been used as a protecting group in the synthesis of acylated nucleosides [247], monosaccharides, and disaccharides [248]. The pivaloyl esters are usually highly crystalline compounds, its position in a molecule is easily detectable by ^{1}H NMR, and it can be

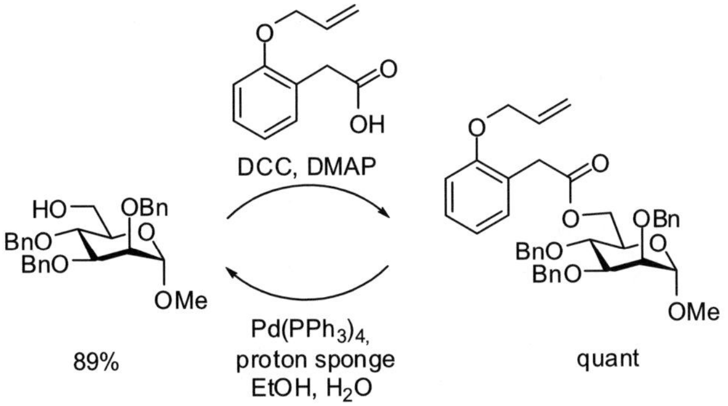

Scheme 35
The 2-(allyloxy)phenyl acetyl protecting group

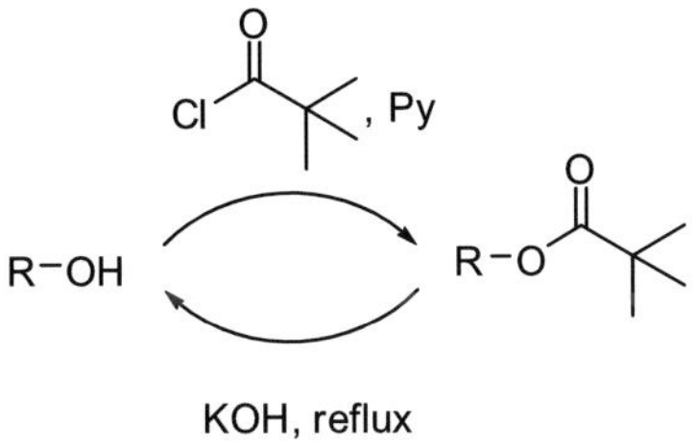

Scheme 36
The pivaloyl protecting group

removed totally or selectively by esterases from mammalian sera [249]. Furthermore, the use of pivaloate esters is advantageous in the stereoselective preparation of β-glycosidic linkages because they inhibit the sometimes competitive process of orthoester formation [250,251].

Pivaloylation of sugar alcohols is carried out using pivaloyl chloride in the presence of pyridine or with *N*-pivaloyl imidazole [252] in DMF. Pivaloyl esters are typically cleaved by base-catalyzed solvolysis (❷ *Scheme 36*). Their greater steric hindrance makes them react slower than other acyl groups and some selectivity in the hydrolysis of different ester protecting groups may be observed.

A systematic study in the selective pivaloylation of various pyranosides and oligosaccharides [253] has shown that, in the absence of adjacent axial alcoxy groups, pivaloylation preferentially occurs at the primary hydroxyl groups. However, in the presence of an adjacent axial function, the reactivity of the vicinal secondary hydroxyl group is as high as that of the primary group towards pivaloylation (❷ *Scheme 37*).

A limitation in the use of pivaloyl esters as protecting groups in a polyfunctional system is the harsh condition required for its cleavage (especially at sterically hindered secondary centers). The 4-acetoxy-2,2-dimethylbutanoyl (ADMB) esters have been proposed as an alternative because they are easily prepared, show similar reactivity in carbohydrate acylations, and are removed under much milder conditions (catalytic quantity of DBU at room temperature) [254].

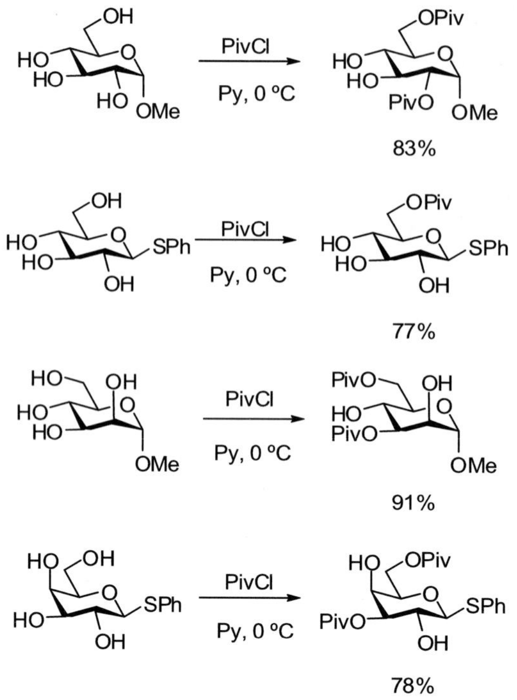

□ Scheme 37
Regioselective pivaloylation of hexopyranosides

2.3.4 Levulinoyl (Lev) Esters

The levulinoyl (4-oxopentanoyl) moiety is a very useful temporary protecting group in nucleotide, polysaccharide, and glycolipid synthesis.

The levulinoyl esters are prepared from the free hydroxyl group by treatment of levulinic acid with DCC [255,256] or 1-[3-(dimethylamino)propyl]-3-ethylcarbodiimide hydrochloride (EDAC) [257] in the presence of DMAP (❷ *Scheme 38*). Additionally, 3'- and 5'-*O*-levulinyl protected derivatives of 2'-deoxy nucleosides have been prepared by regioselective enzymatic acylation using a variety of lipases and acetonoxime levulinate as acylating agent [258]. In contrast to other ester substituents, the *O*-levulinoyl group is far less prone to migration [259].

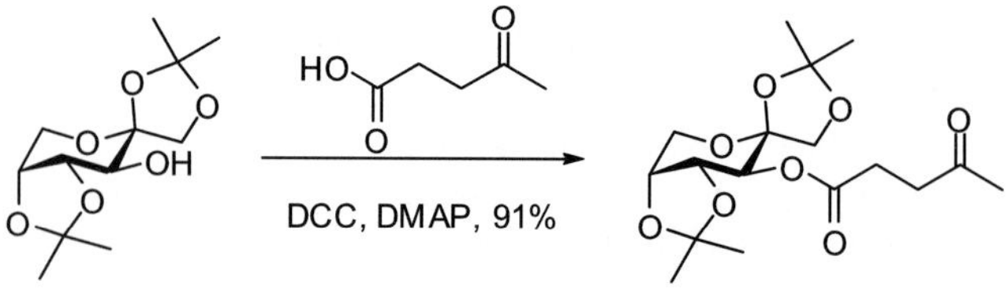

□ Scheme 38
Example of introduction of levulinoyl protecting group

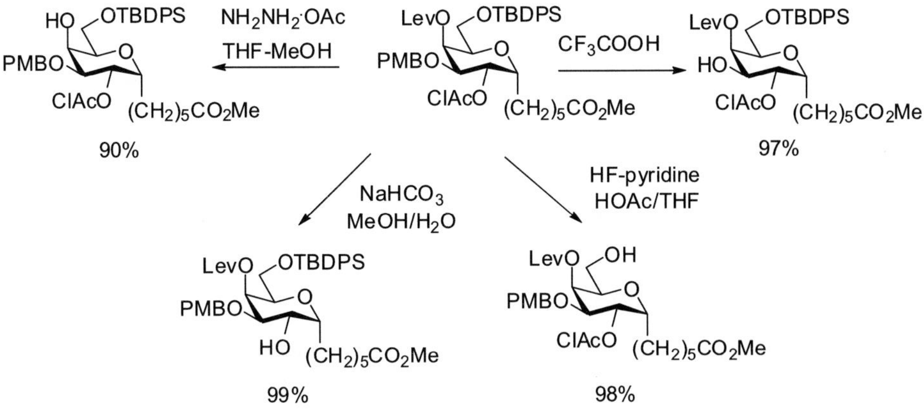

■ Scheme 39
Example of orthogonal hydroxyl protecting groups

Levulinates are stable during coupling reactions and can be selectively cleaved without affecting other protecting groups in the same molecule. Actually, levulinoyl esters can be removed with hydrazine acetate under conditions that do not harm other acyl groups such as acetates, benzoates, or even chloroacetates [255]. This selectivity has allowed the levulinoyl group to be included in a number of orthogonal sets designed for the synthesis of collections of oligosaccharides. For example, Wong and coworkers showed that chloroacetyl, p-methoxybenzyl, levulinoyl, and *tert*-butyldiphenylsilyl groups can each be removed selectively and the freed hydroxyl group employed in glycosylation reactions (● *Scheme 39*) [260]. Zhu and Boons showed that Fmoc, Lev, and diethylisopropylsilyl are another attractive set of orthogonal hydroxyl protecting groups for aminosugars [261].

2.4 Carbonylation Reactions: Carbonate-Type Protecting Groups

Carbonates represent an important family of protecting groups of hydroxyl groups. All of the members of the carbonate family are easy to introduce by reaction of the free alcohol with chloroformates or mixed carbonate esters. In general, carbonates are less reactive than esters towards basic hydrolysis owing to the reduced electrophilicity of the carbonyl afforded by the resonance deactivation by two oxygens. However, the conditions that attack esters may also attack carbonates.

Besides simple alkoxycarbonyl groups which are usually removed under basic hydrolysis, more sophisticated groups have been designed which are removed under milder and more specific conditions. In general, all the carbonate protecting groups are close relatives to the most important carboxyl protecting groups. The adaptation works because O-alkyl cleavage releases an unstable intermediate which decomposes with loss of carbon dioxide to give the free alcohol (● *Scheme 40*).

The most commonly installed carbonates on carbohydrate derivatives include CBz, Troc, Aloc, Poc, and Fmoc groups.

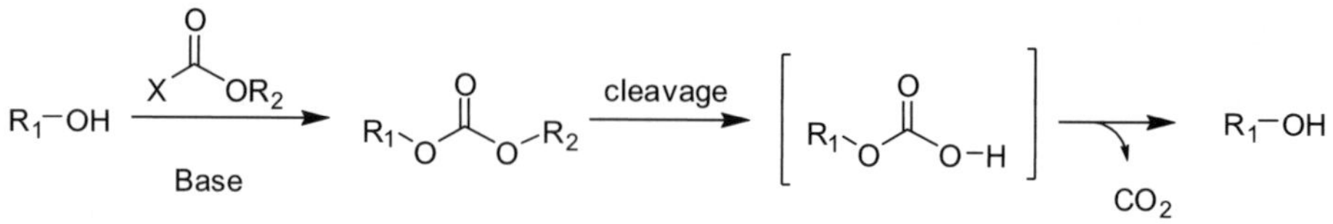

□ **Scheme 40**
Formation and cleavage of carbonate-type protecting groups

2.4.1 Benzyl Carbonates (Cbz)

The benzyloxycarbonyl group (Cbz or Z) is useful in carbohydrate synthesis, not only for *N*-protection of amino sugars, but also to protect alcohols [262,263]. The main advantage of this group is that it is cleaved by hydrogenolysis, and when compared to benzyl ethers, benzyl carbonates are not only removed more readily [264] but also allow hydroxyl group protection under softer conditions than those employed for benzylation.

The benzyloxycarbonates are usually prepared by treatment of the alcohol with benzyloxy-carbonyl chloride in the presence of a base (DMAP or *N*-ethyldiisopropylamine). Aqueous basic medium has to be avoided in polyol systems since these conditions favor the obtention of cyclic carbonates [265].

The benzyloxycarbonyl group has been used for the selective protection of monosaccharides. For instance, in the synthesis of a 1-*O*-carboxyalkyl GLA-60 analogue, a primary alcohol was selectively protected with benzylchloroformate and pyridine (❂ *Scheme 41*) [266]. Further-more, Gotor and Pulido showed that the reaction of D-glucose, D-mannose, and D-galactose with acetone *O*-(benzyloxycarbonyl)oxime in dioxane in the presence of a lipase from *Candida antarctica* allowed the selective benzyloxycarbonylation of the primary hydroxyl group [267]. On the other hand, the regioselective protection of secondary alcohols in pyranosides has also been achieved in high yields [268]. Thus, in the α-D-mannopyranoside series the 3-OH is the more reactive secondary alcohol whereas in the α-D-gluco and α-D-galacto series, the 2-OH is the more reactive group.

2.4.2 Allyl Carbonates (Aloc or Alloc)

The allyloxycarbonyl group [269] has shown a wide application in organic synthesis, especial-ly in the fields of peptides, nucleotides, and carbohydrates. Allyloxycarbonyl derivatives are more easily prepared than the corresponding allyl ethers and they are more stable than ester protecting groups which find frequent use in carbohydrate chemistry.

Allyloxycarbonyl groups have been conveniently installed on primary and secondary hydrox-yl groups of carbohydrate derivatives by reaction with allylchloroformate in the presence of TMEDA [270]. On the other hand, the Alloc group can be cleaved by transition-metal cata-lysts under conditions that are specific and with a high tolerance of other functional groups. As depicted in ❂ *Scheme 42*, allyl carbonates undergo facile oxidative addition with palla-dium(0) catalyst to afford π-allyl palladium complexes, which eject CO_2 to give, initially, allylpalladium alkoxides. Depending on the conditions, these intermediates either collapse to the allyl ether (❂ *Sect. 2.1.4*) or are intercepted by an external nucleophile to give the free alcohol [271].

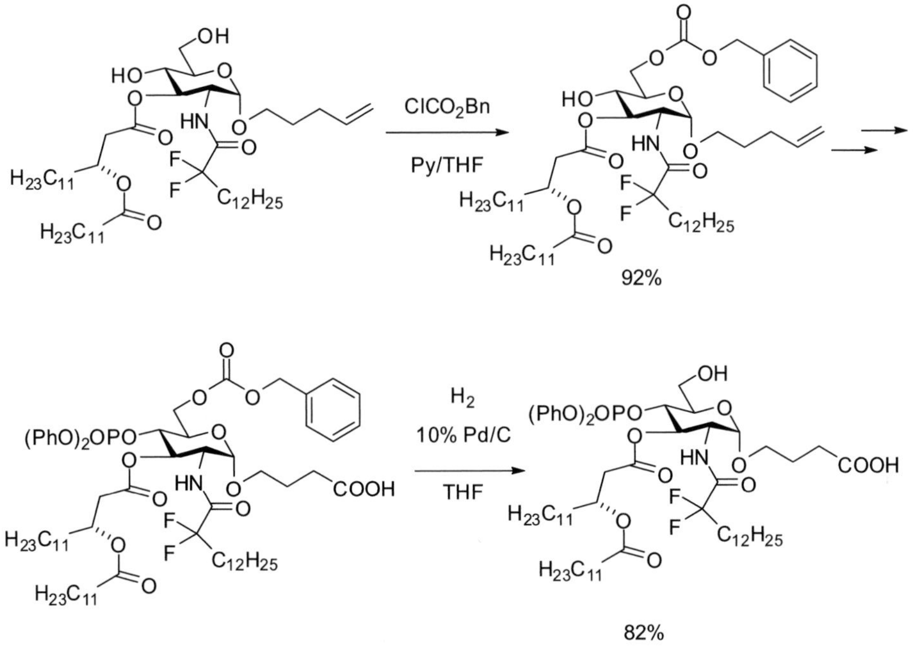

■ Scheme 41
Use of benzyloxycarbonyl group in the synthesis of a 1-*O*-carboxyalkyl GLA-60 analogue

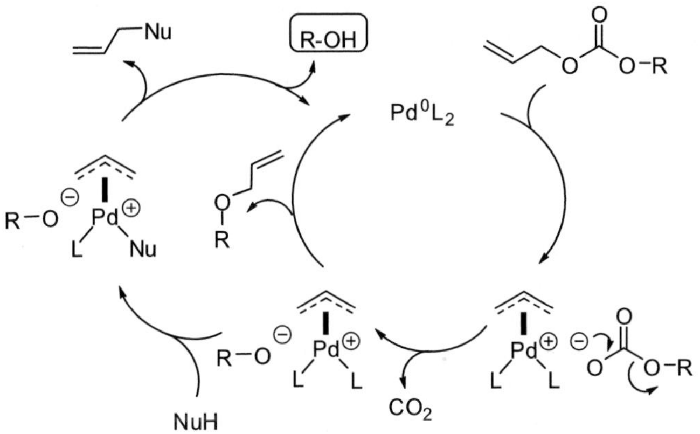

■ Scheme 42
Palladium (0)-mediated cleavage of allyloxycarbonyl groups

An impressive example of how this protecting group has become a powerful tool for the construction of glycopeptides and oligosaccharides comes from the first synthesis of the glycopeptide nephritogenoside [272]. Its structure shows a trisaccharide composed of three glucose moieties linked to a peptide of 21 aminoacids. Taking into account the instability of the

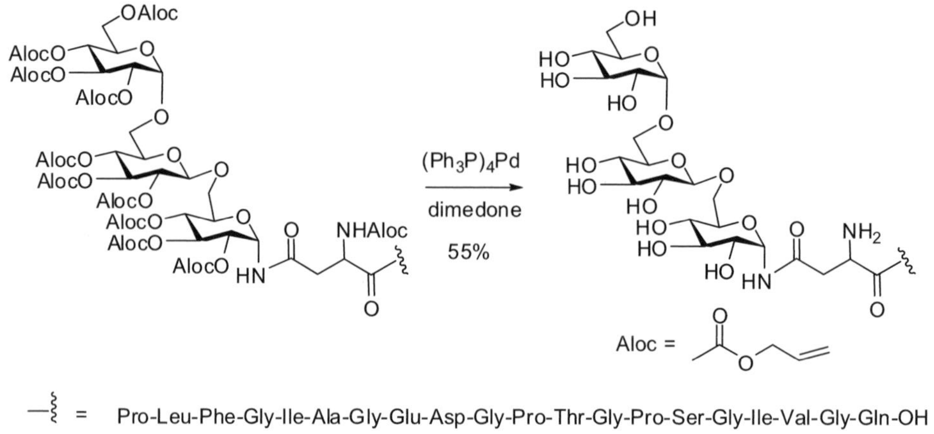

OAloc
AlocO
AlocO
AlocO
NHAloc
HN
(Ph3P)4Pd
dimedone
55%
OH
HO
NH2
Aloc =
= Pro-Leu-Phe-Gly-Ile-Ala-Gly-Glu-Asp-Gly-Pro-Thr-Gly-Pro-Ser-Gly-Ile-Val-Gly-Gln-OH

Scheme 43
Aloc as a powerful tool for the construction of glycopeptides

molecule under acidic and basic conditions, the Aloc group was chosen as the final protecting group of amino and hydroxyl groups. Removal of a total of 11 allyl carbonates was carried out in one single step by treatment with palladium(0) and dimedone to give free nephritogenoside (*Scheme 43*).

2.4.3 Propargyl Carbonates (POC)

Recently, it has been shown that the propargyloxycarbonyl (Poc) group can be used for the protection of the hydroxyl function in carbohydrates [273]. The protection is achieved by treating the alcohol with propargyloxycarbonyl chloride (PocCl) in the presence of a suitable base. The mild reaction conditions can be modulated to attain regioselective protections (*Scheme 44*). The resulting propargyl carbonates are compatible with acidic, basic, and also glycosylation conditions.

Propargyl esters are deprotected effectively using benzyltriethylammonium tetrathiomolybdate $[PhCH_2Net_3]_2MoS_4$. The deprotected products usually can be isolated by simple filtration. Under the conditions of deprotection benzylidene acetals, benzyl ethers, acetyl and levulinoyl esters, and allyl and benzyl carbonates are left untouched and therefore can be used effectively for orthogonal protection in carbohydrate chemistry.

The utility of propargyloxycarbonyl chloride in simultaneous protection of alcohols and amines has been explored and it is possible to deblock propargyl carbonates leaving propargyl carbamates untouched [274].

2.4.4 2,2,2-Trichloroethyl Carbonate (TrOC)

Although very popular as an amino protecting group in peptide and glycopeptide chemistry, only a few reports deal with the trichloroethoxycarbonyl group as a hydroxyl protecting group in carbohydrate derivatives. The selective deprotection is carried out by treatment with zinc in acetic acid to give 1,1-dichloroethylene [275] (*Scheme 45*).

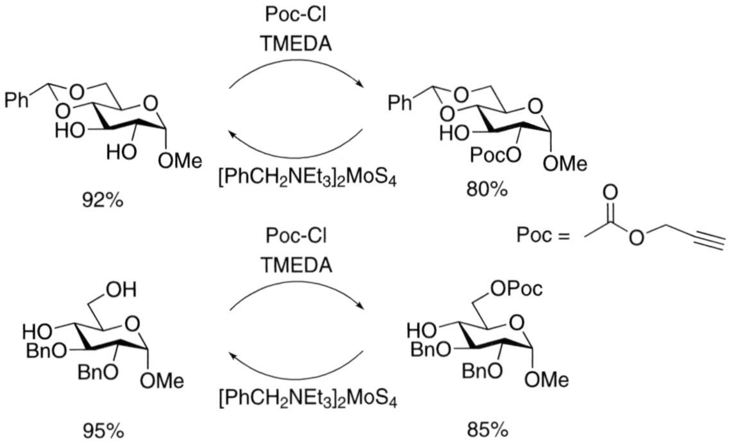

◻ Scheme 44
Propargyloxycarbonyl as a protective group in carbohydrates

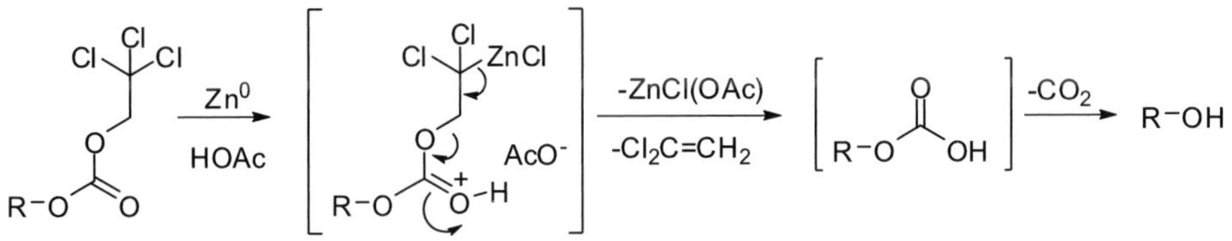

◻ Scheme 45
Selective deprotection of 2,2,2-trichloroethoxycarbonyl group

Trichloroethoxycarbonyl groups have been installed on primary and secondary hydroxyl groups of carbohydrate derivatives by standard coupling with 2,2,2-trichloroethyl chloroformate. It has been shown that a Troc group in the primary position of a glycosyl donor reduces its reactivity but enhances α-selectivity in glycosylation couplings (● *Scheme 46*) [276].

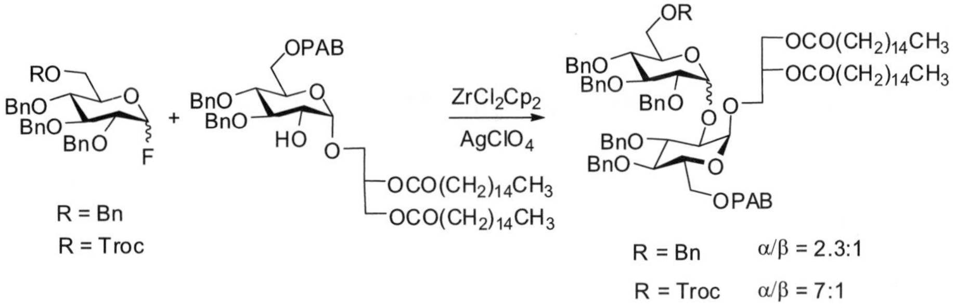

◻ Scheme 46
Example of the α-orienting effect of the 6-O-Troc group

2.4.5 Fluoren-9-ylmethoxycarbonyl (Fmoc) Group

The Fmoc group is a well-established amino-protecting group [277] often used in peptide synthesis, but only recently has been recognized as a temporary hydroxyl protecting group for oligosaccharide synthesis [277,278,279,280].

The Fmoc group is readily introduced under standard conditions using FmocCl and a catalytic amount of DMAP in pyridine. The resulting carbonates are exceptionally stable under acidic conditions and therefore survive glycosylation reactions.

The Fmoc group can be removed with mild bases such as ammonia, piperidine, or morpholine [281]. The cleavage goes through a rapid deprotonation of the fluorene group to generate an aromatic dibenzocyclopentadienide anion. In a subsequent slower step, elimination generates dibenzofulvene (itself an unstable species that rapidly adds nucleophiles) and a carbonate residue, which then decomposes with loss of carbon dioxide to release the free alcohol (❯ *Scheme 47*).

❑ Scheme 47
Selective deprotection of fluoren-9-ylmethoxycarbonyl group

Recently, several groups have reported the use of glycosyl donors bearing Fmoc-protected hydroxyl groups for the solid-phase synthesis of saccharide libraries. Thus, a lactosyl donor, bearing an Fmoc-protected hydroxyl group, has permitted the effective construction of lactose-containing oligosaccharides in a solid-phase system [282,283].

The Fmoc group has also been used as a temporary protecting group in the automated synthesis of Lewis antigens. The UV active dibenzofulvene moiety released after Fmoc cleavage allowed for real-time monitoring of the reaction progress and provided a qualitative assay for the efficiency of each glycosylation and deprotection cycle during automated assembly (❯ *Scheme 48*) [284].

2.4.6 2-[Dimethyl(2-naphthylmethyl)silyl]ethoxycarbonyl (NSEC) Group

The 2-[dimethyl(2-naphthylmethyl)silyl]ethoxycarbonyl group is a novel temporary protecting group to mask hydroxyl groups [285]. In an analogous manner to the Fmoc protection, this group may be particularly useful for the automated assembly of oligosaccharides, as its cleavage can be followed by UV.

The NSEC group can be introduced under standard conditions using NSECCl which is available in three steps from chlorodimethylvinylsilane and 2-(bromomethyl)naphthalene (❯ *Scheme 49*). The NSEC group may be difficult to introduce in sterically demanding positions.

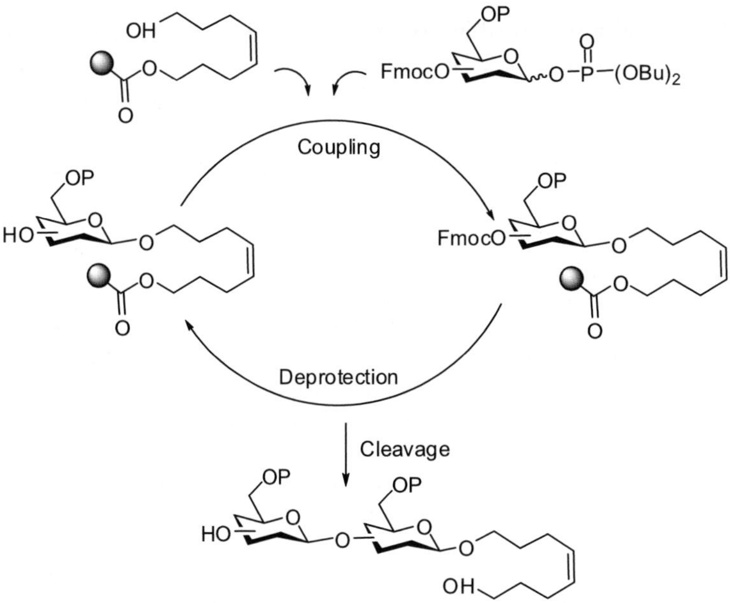

◻ Scheme 48
Fmoc as temporary protecting group in the automated synthesis of oligosaccharides

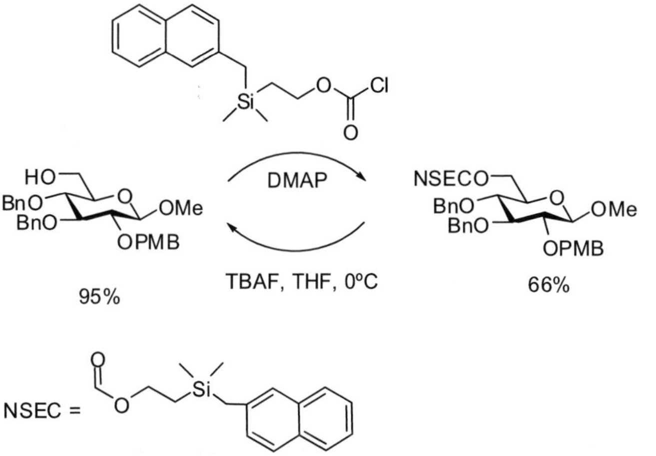

◻ Scheme 49
NSEC as a protective group in carbohydrates

The NSEC group is stable to glycosylation conditions using glycosyl phosphates, and it is not affected under deprotection conditions that facilitate the removal of Lev, Fmoc, allyl, and PMB groups. For ester-type protecting groups selective deprotection in the presence of NSEC derivatives is not possible.

The removal of NSEC carbonates is carried out by reaction with TBAF in the presence of esters including acetyl, levulinoyl, benzoyl and pivaloyl, but also allyl and PMB ethers are not affected. However, the NSEC group cannot be selectively removed in the presence of Fmoc protecting groups.

2.5 Silylation Reactions: Silyl-Type Protecting Groups

2.5.1 Silyl Ethers. General Aspects

Silyl ethers have become very important alcohol protecting groups, which are used, among other ways, routinely in carbohydrate chemistry [286]. Silyl derivatives are generally stable over a wide variety of reaction conditions and at the same time are selectively removable in the presence of other functional groups including other protecting groups.

Additionally, the ability to vary the organic groups on silicon introduces the potential to alter the R_3Si group in terms of both its steric and electronic characteristics and thereby influence the stability of the silylated species to a wide variety of reaction and deprotection conditions [287].

The synthetic potential of silyl ethers as protecting groups for the hydroxyl groups was appreciated in the early 1970s and now these derivatives are probably used more than any other protecting group in organic synthesis. ❷ *Figure 5* summarizes the structures, names, and abbreviations of the most commonly used silyl ethers.

The usual method for their introduction into sugars is the reaction of one or several hydroxyl functions in the sugar with a trialkylsilyl chloride in the presence of a base, such as pyridine and imidazole. The less sterically hindered the silyl group the easier it is to introduce. The introduction of the sterically unimpeded trimethylsilyl group to a primary, secondary, or tertiary alcohol is a straightforward process taking place with a variety of reagents under mild, high-yield reaction conditions. On the other hand, the introduction of the more steri-

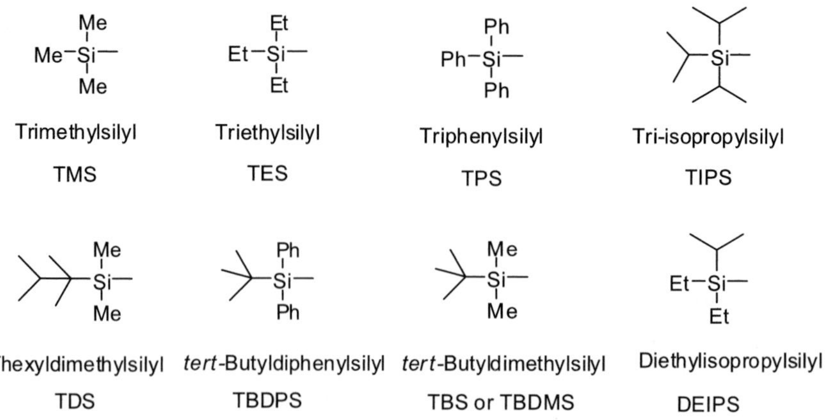

❷ **Figure 5**
Principal members of the silyl ether family

cally demanding *tert*-butyldimethylsilyl group requires reaction of the alcohol with *tert*-bu-tyldimethylchlorosilane in the presence of imidazole as a catalyst and the formation of the *tert*-butyldimethylsilyl ether of tertiary alcohols is very difficult. Alternatively, a more reactive form such as trialkylsilyl trifluoromethanesulfonates (R_3SiOTf) can be used.

As a rule, the bulkier the substituents, the greater the stability of the resulting silyl derivatives. However, stability is not only a function of steric bulk since electronic effects play a role as well, which can be exploited to differentiate stability under acidic or basic conditions. For example, phenyl-substituted silyl ethers are equal or more reactive than their trimethylsilyl counterparts under alkaline conditions, but less reactive under acidic conditions. In general terms, however, the relative stabilities of the silyl-protected functional groups will follow the order of: $^iPr_3Si > ThMe_2Si > {}^tBuPh_2Si > {}^tBuMe_2Si > {}^iPrMe_2Si > Et_3Si > Ph_2MeSi > Me_3Si$ [288].

On the other hand, silyl groups can migrate between different nucleophilic sites in a molecule under basic conditions. These migrations have to be considered as possible side reactions, and sometimes provide a valuable approach to interesting products that are not directly available [289]. For instance, in the synthesis of chemically modified cyclodextrines, the migration of the TBS groups from the 2-O to the 3-O on all the D-glucopyranose residues was observed during alkylation with sodium hydride in THF (❷ *Scheme 50*) [290].

In general, the size of the substituent on the silicon atom is directly related to the rate of deprotection with smaller silyl substituents being more easily cleaved under acidic conditions. Similarly, if the same protecting group is used to protect two or more hydroxyl groups, the silyl ether derived from the less sterically encumbered alcohol is usually the first to be deprotected [291]. On the other hand, removal of silicon protecting groups occurs under extremely mild and highly specific conditions using a fluorine source. In general, the order of cleavage of silyl ethers with basic fluoride reagents (such as TBAF) parallels the order found for basic hydrolysis; similarly, slightly acidic fluorine-based reagents such as HF-acetonitrile parallel the order found for acid hydrolysis.

❏ Scheme 50
Migrations of silyl groups in the synthesis of modified cyclodextrines

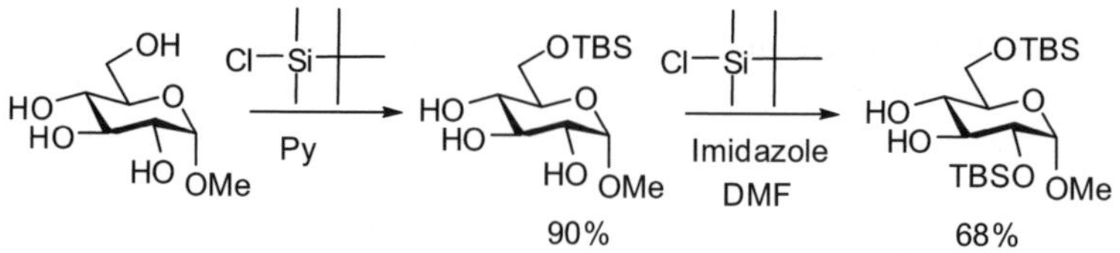

☐ **Scheme 51**
Selective silylation of methyl α-D-glucopyranoside

2.5.2 *tert*-Butyldimethylsilyl (TBS or TBDMS) Group

Since its introduction in 1972 [292], the *tert*-butyldimethylsilyl group has become the most popular of the general purpose silicon protecting groups. It can be easily installed in high yields under mild conditions and it is robust to a variety of reaction conditions. The TBS group is commonly introduced via *tert*-butyldimethylchlorosilane, in the presence of basic activators such as DMAP or imidazole in a dipolar aprotic solvent such as DMF (❯ *Scheme 51*). Hindered secondary alcohols can be silylated with TBSOTf using 2,6-lutidine as the base [293]. When the reaction is mediated by equimolecular amounts of dibutyltin oxide, the silylation with TBSCl gives the 6-monosilylated products in excellent yields [294].

N,O-Bis(*tert*-butyldimethylsilyl)acetamide silylates tertiary and hindered secondary alcohols in the presence of a catalytic amount of TBAF or another source of fluoride anion. Protection of primary hydroxyl groups in the presence of secondary ones is also possible (❯ *Scheme 52a*) [295].

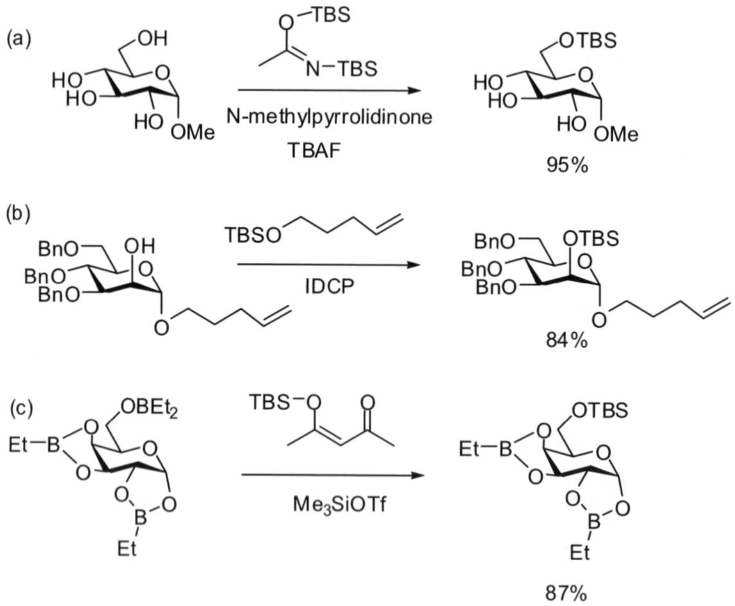

☐ **Scheme 52**
Alternative conditions for the *tert*-butyldimethylsilyl ether protection

tert-Butyldimethylsilyl pentenyl ether is also a suitable reagent for efficient silylation of primary and secondary hydroxyl groups (❷ *Scheme 52b*). Activation is carried out with iodonium di-*sym*-collidine perchlorate (IDCP) and this procedure can be applied even to pentenyl glycosides [296].

An unusual way for the preparation of TBS ethers involves the reaction of diethylboronyl ethers, obtained by the reaction of the corresponding alcohol with BEt$_3$, with the TBDMS-enolate of pentane-2,4-dione in the presence of a catalytic amount of TMSOTf (❷ *Scheme 52c*) [297].

The palladium(0) nanoparticle-catalyzed silylation of sugars by silane alcoholysis of *tert*-butyldimethylsilane has been proposed as an attractive alternative to the established silyl chloride method. The methodology gives convenient access to the 3,6-silylated methyl glycopyranosides as the dominant products rather than the 2,6-silylated glycosides typically obtained by the silyl chloride method [298]. Changing to homogeneous cationic catalysts of iridium and rhodium, 2,3,6- and 2,4,6-trisilylated derivatives are obtained in synthetically useful yields [299]. The TBDMS group has also been introduced [300] to alcohols or phenols by the Mitsunobu reaction (DEAD/PPh$_3$, THF, $-78\,°C$) using *tert*-butyldimethylsilanol.

Numerous methods are now available in the literature for the deprotection of TBS ethers under a variety of conditions. One of the most effective ways for the cleavage of silyl ethers is based on the exploitation of the high affinity of silicon towards fluoride ions. Thus, a number of reagents involving one form of fluoride or another, such as tetrabutylammonium fluoride [292], BF$_3$·Et$_2$O [301], hydrofluoric acid [302], fluorosilicic acid [303], ammonium fluoride [304], silicon fluoride [305], lithium tetrafluoroborate [306], and chlorotrimethylsilane/potassium fluoride dehydrate [307] have been developed for the deprotection of TBDMS ethers. Among these, TBAF is most frequently used but the strong basicity of the fluoride anion makes it inappropriate for base sensitive functionalities.

Similarly, acidic reagents such as HCl [308], H$_2$SO$_4$ [309], PPTS [310], TFA [311], TsOH [312] etc., have also been employed for this purpose but cannot be used in the presence of acid-sensitive functionalities. This has led to the development of several Lewis acids and other reagents including BF$_3$·OEt$_2$ [313], BCl$_3$ [314], Sc(OTf)$_3$ [315], Ce(OTf)$_4$ [316], InCl$_3$ [317], ZnBr$_2$ [318], Zn(BF$_4$)$_2$ [319], CeCl$_3$–NaI [320], BiBr$_3$ [321], BiOClO$_4$ [322], Cs$_2$CO$_3$ [323], CBr$_4$–MeOH [324], I$_2$ [325] and CAN [326] for desilylation.

Recently, an environmentally benign phosphomolybdic acid supported on silica gel has been used for the chemoselective deprotection of TBS ethers in carbohydrate derivatives (❷ *Scheme 53*). The mild conditions are compatible with the presence of other protecting groups such as isopropylidene acetal, OTBDPS, OTHP, OAllyl, OBn, OAc, OBz, *N*-BOc, *N*-CBz, and *N*-Fmoc which are stable under the reaction conditions. Another advantage of this procedure is that the catalyst can be readily recovered and recycled [327].

2.5.3 *tert*-Butyldiphenylsilyl (TBDPS) Group

The TBDPS group was introduced by Hanessian and Lavallee in 1975 [328]. The TBDPS group has greater steric demands than the TBS group and, therefore can result in much more selective protections of hydroxyl groups. The group is also less prone to migrate to proximate hydroxyl groups under neutral or acidic conditions than the TBS group but it may migrate under basic conditions [329].

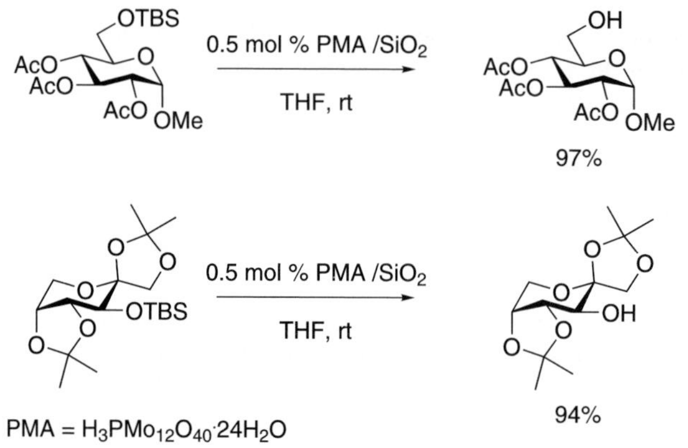

PMA = $H_3PMo_{12}O_{40}\cdot24H_2O$

■ **Scheme 53**
The ring oxygen is missing in the fructose derivatives

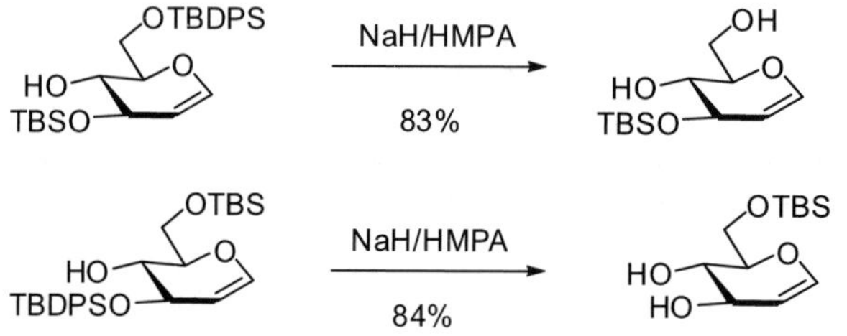

■ **Scheme 54**
Selective removal of TBDPS ethers in the presence of TBS ethers

The TBDPS ethers are prepared by treating alcohols with TBDPS-Cl in DMF in the presence of imidazole (the primary hydroxy group reacts faster than the secondary one). Tertiary alcohols do not silylate. The silylation of hindered alcohols is greatly accelerated with the aid of AgNO₃, NH₄NO₃, or NH₄ClO₄ [330].

TBDPS ethers are generally cleaved under the same acidic conditions as those used for TBS ethers but longer reaction times are necessary and consequently selective removal of TBS groups in the presence of TBDPS groups is very common [331,332]. The electron-withdrawing effect of the phenyl substituents enhances the electrophilicity of the silicon atom and therefore is more susceptible towards nucleophiles. For this reason it is possible to reverse the tendency of TBS ethers to cleave more easily than TBDPS ethers using ion fluoride or basic hydrolysis [333]. Some examples in glycal derivatives are shown in ❷ *Scheme 54*.

2.5.4 Triisopropylsilyl (TIPS) Group

The TIPS group [334] is one of the most sterically hindered silyl protecting groups, being removed only slowly under standard acid- or base-catalyzed hydrolysis conditions. The large steric bulk ensures high selectivity in the protection of primary hydroxyl groups over sec-

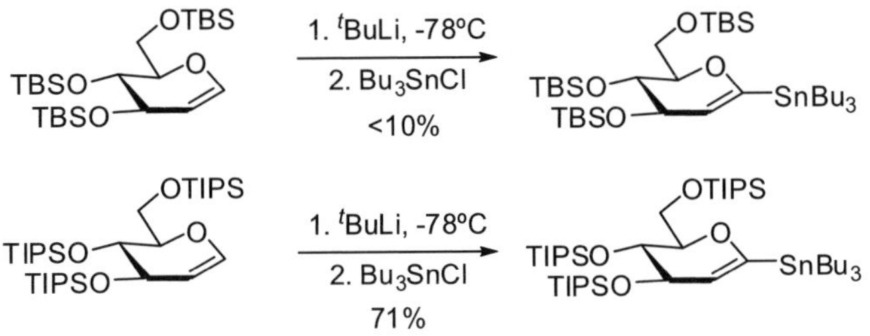

□ Scheme 55
TBS- vs. TIPS- ethers as protecting groups in the anomeric lithiation of glycals

ondary and valuable stability under a wide range of reaction conditions. It is noteworthy that TIPS groups are inert towards powerful bases such as *tert*-butyllithium, and therefore can be used as protecting groups in the anomeric lithiation of glycals (❂ *Scheme 55*) [335].

The TIPS group is usually introduced from triisopropylchlorosilane [336], but protection of hindered alcohols can be very slow in which case triisopropylsilyl triflate in the presence of 2,6-lutidine is used [293].

TIPS ethers are cleaved under the same conditions as those used for TBS ethers but longer reaction times are frequently necessary; consequently TBS ethers can be removed selectively in many cases.

2.5.5 1,1,3,3-Tetraisopropyldisiloxane (TIPDS) Group

The tetraisopropyldisiloxane-1,3-diyl group was introduced by Markiewicz et al. for simultaneous protection of the 3′- and 5′-hydroxy groups of ribonucleosides [337]. The group is usually introduced by the reaction of the bifunctional reagent 1,3-dichloro-1,1,3,3-tetraisopropyldisiloxane with the substrate in pyridine though imidazole in DMF solution can also be used. When applied to pyranoses, these conditions give the kinetic product, an 8-membered ring, which is formed by rapid reaction first at the least hindered hydroxyl group followed by a second intramolecular silylation with the next proximate hydroxyl at C-4.

Additionally, the eight-membered rings of TIPDS-acetals formed in this way can rearrange under the influence of acidic catalyst to the thermodynamically more stable seven-membered derivatives bridging two vicinal secondary hydroxyl functions [338].

The usefulness of the TIPDS protecting group in carbohydrate chemistry is well illustrated by the synthesis of a glyco(phospho)lipid of *Streptococci* cell membranes (❂ *Scheme 56*). Selective protection of the C-4 and C-6 hydroxyl groups of the pyranose was easily accomplished using the TIPDS group. Then the C-2 hydroxyl participated in a regioselective glycosylation under basic conditions to give the coupling product. At this stage of the synthesis the dynamic properties of the TIPDS group were exploited and the subsequent acid-catalyzed isomerization of the 4,6-O-disilyl-protected product results in the formation of the more stable 3,4-O-disiloxane. The freed primary hydroxyl function was then ready to be reacted with stearoyl chloride after which the naturally occurring glycolipid was eventually obtained [339].

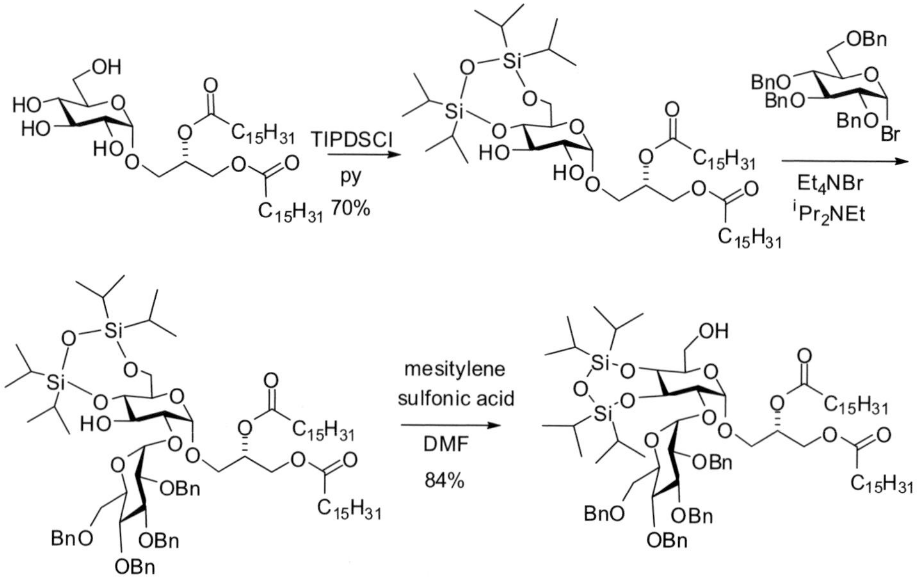

◻ Scheme 56
Application of TIPDS protecting group in the synthesis of a glyco(phospho)lipid

However, despite the attractive properties, which are inherent in the use of the TIPDS protecting group in sugar chemistry, its general applicability is limited, to some extension by the fact that this group can not withstand acidic conditions, which are commonly used in carbohydrate chemistry.

2.5.6 Di-*tert*-butylsilylene (DTBS) Group

The di*tert*butylsilylenediyl is a convenient and versatile protecting group introduced by Trost [340] and often used for the synthesis of anthracyclines [341] and nucleotide [342] derivatives. The DTBS group is not as robust as isopropylidene or benzylidene acetals and therefore its use is appropriate for systems requiring deprotection under very mild conditions. DTBS derivatives survive hydroboration, mild oxidation, Lewis acids, mild protic acids, and strong bases but hydrolysis occurs readily with HF in acetonitrile, HF·pyridine complex, or TBAF.

The formation of the silylene derivatives is effected by treatment of the diol with di*tert*butyl-dichlorosilane in the presence of 1-hydroxy-benzotriazole (HOBT) (❍ *Scheme 57*) [340]. Di*tert*butylsilyl ditriflate and 2,6-lutidine effects silylene formation faster and under milder conditions than the less-reactive dichloride [343]. When the silylation is carried out in pyranoside derivatives, the reaction proceeds selectively at 1,3-diol groups of C-4 and C-6 positions and the formation of the five-membered DTBS derivatives of 1,2 diols was not observed [344]. It has been shown that a DTBS group at the *O*4-*O*6 position of a galacto-type sugar directs α-predominant selective glycosylation in spite of the presence of a participatory group at C-2 such as benzoyl or Troc groups [345] (❍ *Scheme 58*). This new glycosylation method is a powerful strategy for the synthesis of α-galactosyl and galactosaminyl glycans [346,347].

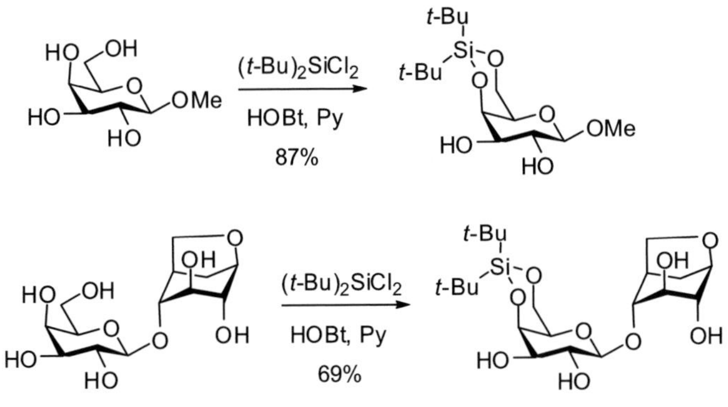

■ Scheme 57
Cyclic di-t-butylsilylene as protecting group

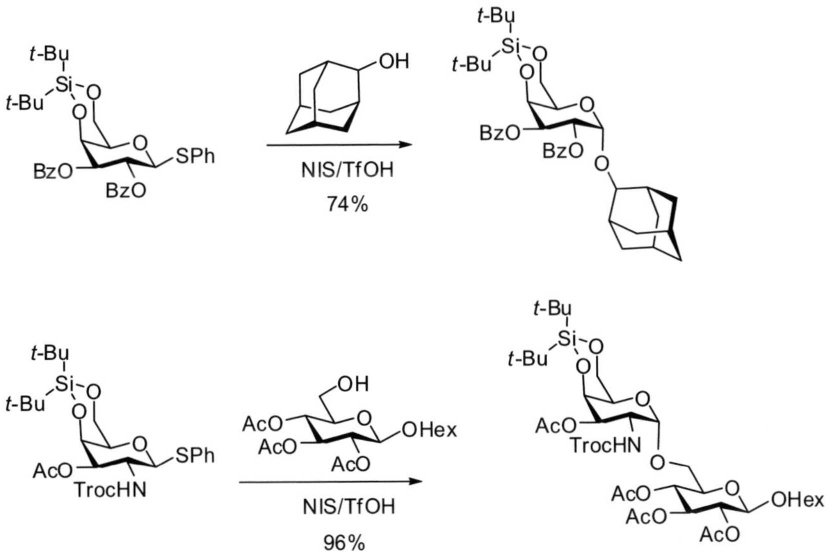

■ Scheme 58
Examples of di-t-butylsilylene-directed α-galactosylation

2.6 Phosphorylation Reactions

Phosphate esters play an important role in a wide variety of structurally diverse natural and biologically active compounds such as glycolipids, nucleic acids, nucleotides, proteins, coenzymes, steroids, and in particular carbohydrates. Introduction of a phosphate group essentially changes the physical and chemical properties of the parent molecule, resulting in changes to the polarization and intermolecular bonding characteristics of that molecule. Given the importance of this functional group it is not surprising that many methods have been developed for the phosphorylation of alcohol functions [348,349]. Both chemical and enzymic methods are available for the synthesis of specific phosphates.

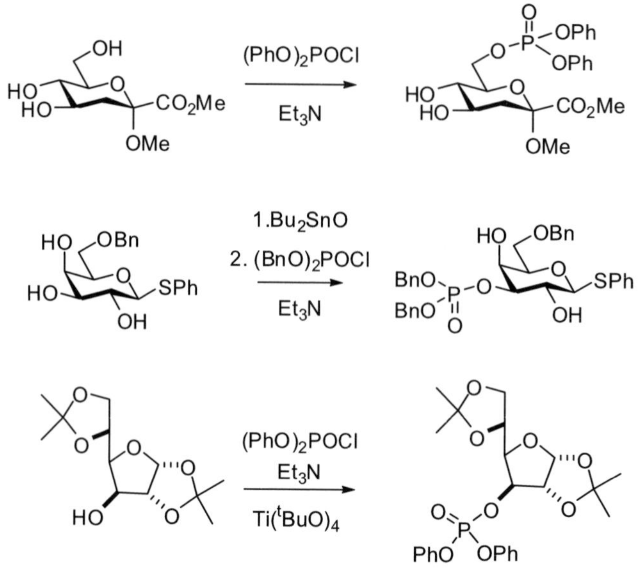

□ **Scheme 59**
Examples of chemical phosphorylation with phosphorous (V) reagents

Chemically, the most common phosphorylation reagents used are chlorophosphates [350]. These compounds are generally commercially available and are as stable as their routinely used acyl chloride counterparts to both air and moisture. The problem most commonly encountered with the use of such reagents is the conditions under which they will react. Phosphorylation is usually performed either through formation of the lithium [351] or thalium alkoxide [352], followed by the reaction with the chlorophosphate or simply by use of a proton scavenger such as pyridine [353] or Et$_3$N [354]. Alternatively, nucleophilic catalysis with DMAP [355] or tin-mediated phosphorylation [356] may be employed. A method for the phosphorylation of hydroxyl groups using a Lewis acid catalyst has been recently reported [357] (❷ *Scheme 59*). Solid-phase phosphorylating reagents have been used for the phosphorylation of unprotected nucleosides and carbohydrates [358,359]. These procedures exhibit high regioselectivity and only one monophosphorylated product is obtained. Carbohydrate and nucleoside diphosphates have also been synthesized by using solid-phase reagents [360].

Aside from P^V reagents, the most widely used and most successful of all chemical phosphorylation techniques is the use of reagents containing trivalent phosphorous which ensure the highest phosphorylation rates and permits one to avoid many side processes [361]. This methodology has been well developed and is used extensively in the construction of oligonucleotides. Phosphorous triamides phosphorylate efficiently monosaccharides whose molecules contain one free alcoholic hydroxyl [362] (❷ *Scheme 60*). Phosphorylation by dialkyl [363] or alkanediyl phosphoramidites [364], phosphonamidites [365] and phosphinamidites [366] follows a similar pathway. Treatment of monosaccharide derivatives whose molecules contain two closely located hydroxy groups with phosphamides results in cyclophosphorylation [367].

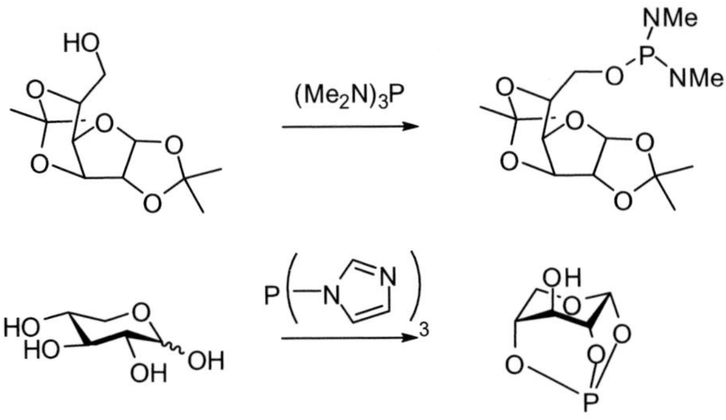

◼ Scheme 60
Examples of chemical phosphorylation with phosphorous (III) reagents

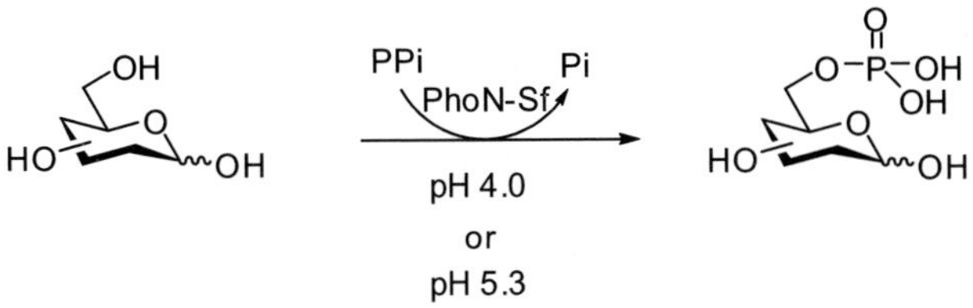

◼ Scheme 61
Example of enzymatic phosphorylation

A non-specific bacterial acid phosphatase from *Shigella flexneri* (PhoN-Sf) has been screened for regioselective phosphorylation of primary alcohol(s) of more than 20 different cyclic and acyclic monosaccharides using pyrophosphate as the phosphate donor (❯ *Scheme 61*) [368]. These studies have shown that PhoN-Sf is capable of phosphorylating a range of hexoses (D-glucose epimers, glycosides, and C-2 derivatives), pentoses, heptoses, ketoses, and acyclic carbohydrates.

3 Reactions at the Anomeric Hydroxyl

3.1 Alkylation Reactions

In general, reactions and conditions for the introduction of protecting groups in the anomeric hemiacetal group are the same as those mentioned previously for non-anomeric hydroxyl groups. Probably, the only exception relates to the alkylation reaction since alkyl ethers of the anomeric hydroxyl group, which are acetals rather than ethers, are normally formed under Fischer glycosylation conditions using the alcohol as the aglycon [369]. This process involves cleavage of the C-1–O-1 bond at the anomeric center and, therefore will not be treated here, but it is the method of choice for the preparation of alkyl glycosides.

3.1.1 Anomeric *O*-Alkylation and *O*-Arylation

The 1-*O*-alkylation of carbohydrates with simple alkylating agents, particularly methyl iodide and dimethyl sulfate, has long been known [370,371,372]. The reactivity of pyranoses and furanoses deprotonated at *O*-1 is, thus, analogous to that of alkoxides. Alkylation of fully protected pyranoses, due to the ring chain tautomerism between the two anomeric forms α and β and the open chain form (❖ *Scheme 62*), can take place at three different sites [373]. However, when the alkylation of 2,3,4,6-tetra-*O*-benzyl D-glucose is carried out in dioxane with sodium hydride and methyl triflate, the β-glucoside was obtained practically exclusively [374]. This selectivity has been explained on the basis of an enhanced nucleophilicity of the β-oxide atom which can be attributed to a steric effect in combination with a stereoelectronic effect resulting from repulsion of the lone electron pairs (kinetic anomeric effect) in the β-oxide [375] (❖ *Fig. 6*). Conversely, if the reaction is carried out at lower temperatures (−40 °C) the formation of α-anomer is preferred. Despite of the use of NaH, neither acyl migration nor orthoester formation occurred during the 1*O*-alkylation of acetyl-protected derivatives (❖ *Scheme 63*) [376,377]. The stereoelectronic effects in α- and β-furanosyl oxides should differ less for conformational reasons and the stereocontrol results primarily from steric and chelation effects.

The higher acidity of the 1-OH group of the hemiacetal (resulting from the indirect stabilization by the ring oxygen atom) allows for regioselective *O*-alkylation at this position regardless of the presence of other sugar hydroxy groups. Thus, as shown in ❖ *Scheme 64*, the alkyla-

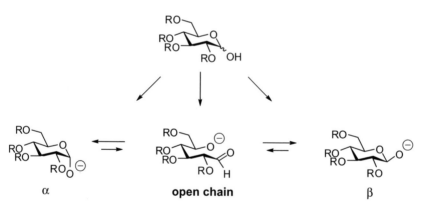

α   open chain   β

❖ Scheme 62
Ring chain tautomerism of fully protected pyranoses

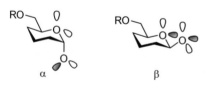

α   β

❖ Figure 6
Kinetic anomeric effect in the β-oxide

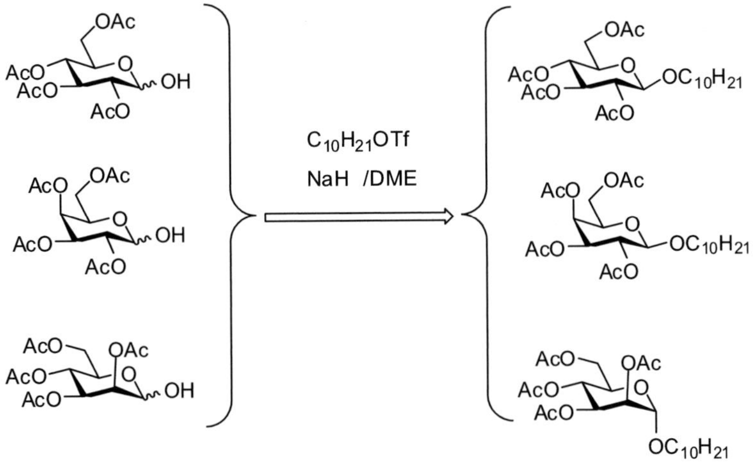

■ Scheme 63
Stereoselective anomeric O-alkylation of acyl-protected sugars

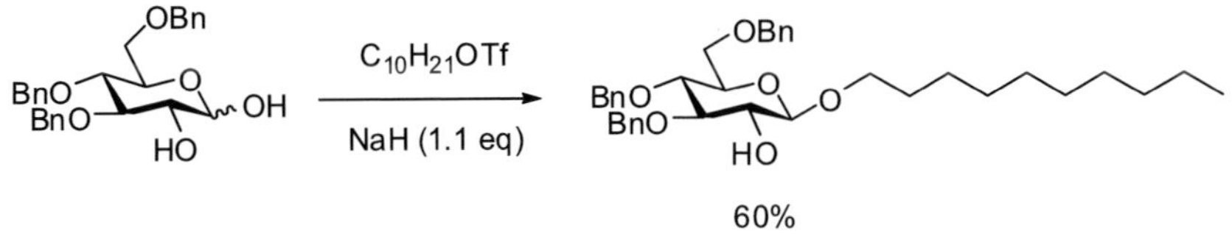

■ Scheme 64
Anomeric O-alkylation of partially protected pyranoses

tion of 2-O-unprotected tribenzylglucopyranose afforded exclusively the 1-O-alkylated product when the reaction was done with one equivalent of NaH, whereas the undesired 1,2-disubstituted isomer was also obtained when two equivalents of NaH were used [378].

Although ring-chain equilibration permits the formation of many products in fully unprotected monosaccharides, the regiocontrol in the per-O-benzylation towards uniform glycoside bond formation is generally very high [379,380] (❯ Scheme 65).

The 1-O-alkylation of pyranoses has also been used for glycosidic bond formation [381].

An alternative activation of the anomeric hydroxyl makes use of 1,2-O-dibutylstannylene acetals [382]. Thus, for instance condensation of the stannylene acetal of 3,4,6-tribenzylmannose with methyl iodide, allyl or benzyl bromide afforded the corresponding β-mannosides in almost quantitative yields (❯ Scheme 66).

On the other hand, carbohydrates carrying an aromatic aglycon are important natural products and therefore methods for the arylation of anomeric hemiacetals have also been developed. Both Mukaiyama [383] and Smith [384] have synthesized aryl glycosides by nucleophilic aromatic substitution for use as glycosyl donors. The method is quite efficient but requires activation by electron-withdrawing groups in the aromatic counterpart. Thus, direct reaction of 1-fluoro-2,4-dinitrobenzene with the 1-OH group of the hemiacetal gave 2,4-dinitroglycosides in excellent yields (❯ Scheme 67a). In the case of dinitrosalicylic (DISAL) acid

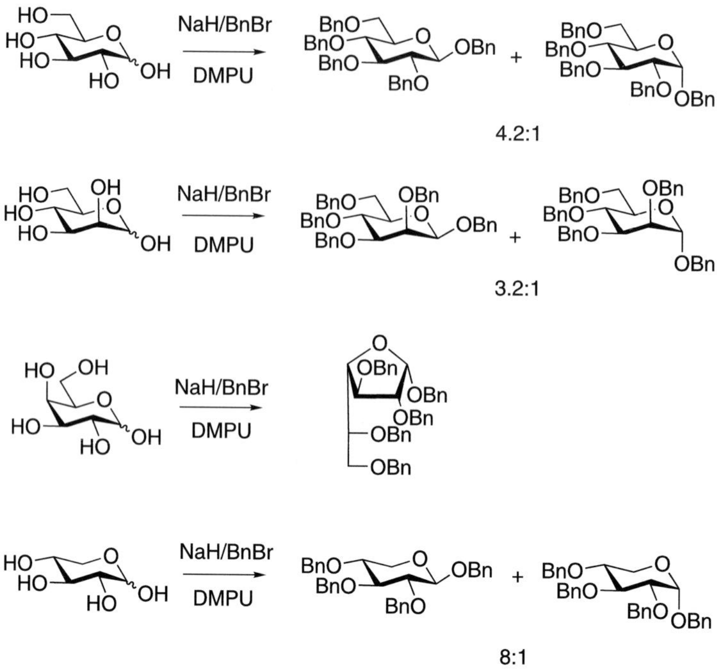

4.2:1

3.2:1

8:1

◻ **Scheme 65**
Two bonds are missing in the first compound

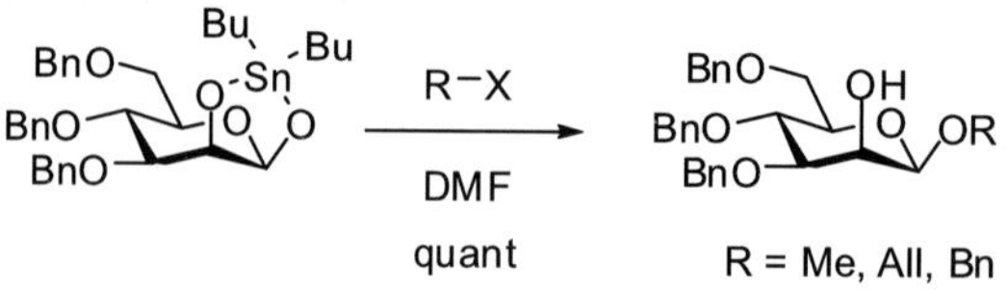

quant

R = Me, All, Bn

◻ **Scheme 66**
Anomeric *O*-alkylation via 1,2-*O*-stannylene acetals

derivatives, the use of DMAP as the base gave an α/β ratio similar to the starting 1-OH derivative. In contrast, formation of the β-anomer was favored using 1,4-dimethyl piperazine (❯ *Scheme 67b*) [385,386,387].

The reverse situation, in which the phenol acts as the nucleophile attacking activated carbohydrate hemiacetals, has also found several practical applications preparing *O*-aryl glycosides [388]. However, this situation implies an attack at the anomeric carbon and not at the anomeric oxygen.

3.1.2 Anomeric *O*-Dealkylation

On treatment with aqueous acid, glycosides are hydrolyzed to give the corresponding alcohol and the reducing sugar. Solvolysis of glycosidic bonds is one of the most general and important reactions in carbohydrate chemistry and so the literature on the acid hydrolysis of *O*-glycosidic

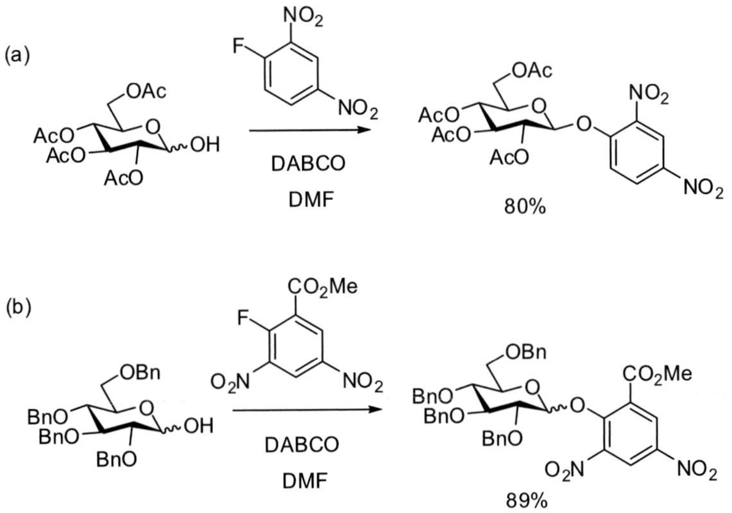

■ Scheme 67
Anomeric *O*-arylation via nucleophilic aromatic substitution

bonds covers thousands of titles. The data on more detailed studies of this reaction are covered by several reviews [389,390,391,392]. Although this process is formally the reverse direction of the alkylation reaction, the key step involves reaction at the anomeric carbon rather than at the anomeric oxygen and thus is beyond the scope of this chapter.

Nevertheless, a variety of protecting groups have been applied to the anomeric center, which are synthetically useful and provide alternative ways for the selective liberation of the anomeric oxygen. These include the following:

Benzyl Glycosides   In 1928, Freudenberg found that benzyl ethers of sugars were cleaved by hydrogenolysis with sodium amalgam and by catalytic hydrogenolysis that could be effected in acetic acid in the presence of platinum metals [393]. On palladium catalysis, hydrogen splits off the benzyl β-D-glycosides, at room temperature and atmospheric pressure to afford toluene and the reducing sugar (❂ *Scheme 68*) [394]. Hydrogenolysis is commonly carried out using hydrogen gas with a palladium catalyst absorbed on charcoal although modifications involving hydrogen transfer have been used.

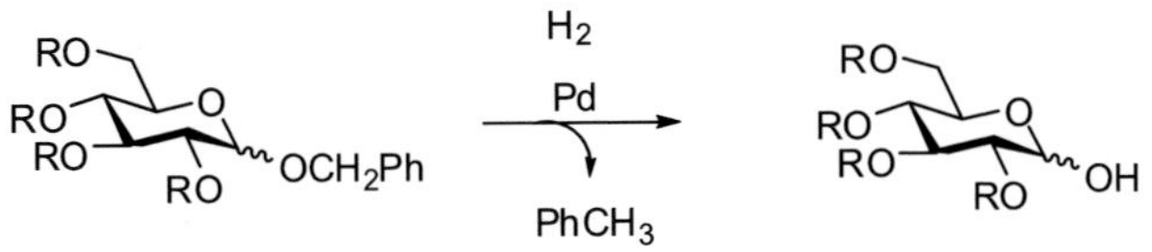

■ Scheme 68
Hydrogenolysis of benzyl glycosides

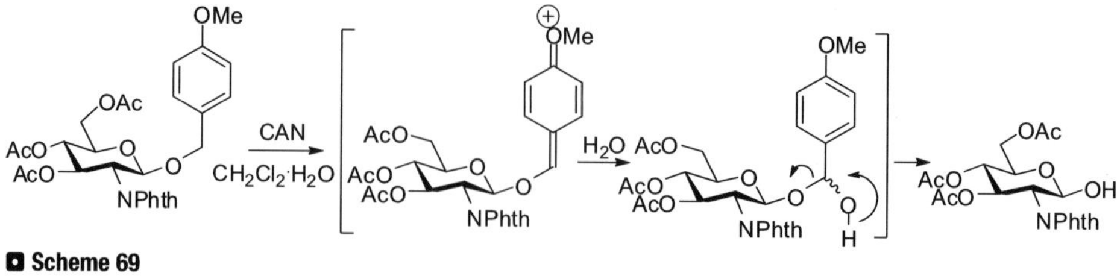

☐ **Scheme 69**
Oxidative deprotection of p-methoxybenzylglycoside

A remarkable rate difference in the hydrogenolysis of α- and β-benzyl D-gluco-and D-galac-topyranosides has been reported, with the β-anomers being more readily cleavable [395]. Ferric chloride has been employed for anomeric debenzylation in oligosaccharides [18].

Methoxy-substituted benzyl glycosides have been used as precursors for reducing sugars [396]. As mentioned in ❯ *Sect. 2.1.3* under ❯ *"p-Methoxy Benzyl (PMB) Ethers"*, their utility lies in the fact that they are more readily cleaved oxidatively than the unsubstituted benzyl ethers (❯ *Scheme 69*). All these transformations have great synthetic value although the process is not regioselective since it is operational for all the benzyloxy groups present in the sugar.

Particularly useful in protecting the anomeric hydroxyl are those substituted benzyl groups that are light-sensitive. Such groups are stable to a wide variety of chemical treatments and at the same time are sensitive to irradiation under conditions that leave other functional groups in the molecule unaffected. Thus, 2-nitro benzyl and 3,4-dimethoxy-6-nitrobenzyl (6-nitroveratryl) glycosides are more stable to acid hydrolysis than are the corresponding benzyl glycosides but are readily photolyzed at 320 nm to the reducing sugars in high yields (❯ *Scheme 70*) [397]. In this context photocleavable linkers for solid-phase synthesis, based on the lability of 2-nitro benzyl moieties under irradiation, have been applied to the liberation of the anomeric center in oligosaccharides [398].

Other substituted-benzyl glycosides that can be selectively removed in the presence of unsubstituted benzyl ethers have been developed. For example, 2-(hydroxycarbonyl)benzyl glycosides are easily sovolyzed by treatment with Tf₂O in the presence of di-*tert*-butylmethylpyridine (DBMP). The reaction implies anomeric C–O bond cleavage since it takes place by lactonization via the mixed anhydride to generate phthalide and the oxocarbenium ion (❯ *Scheme 71*) [399].

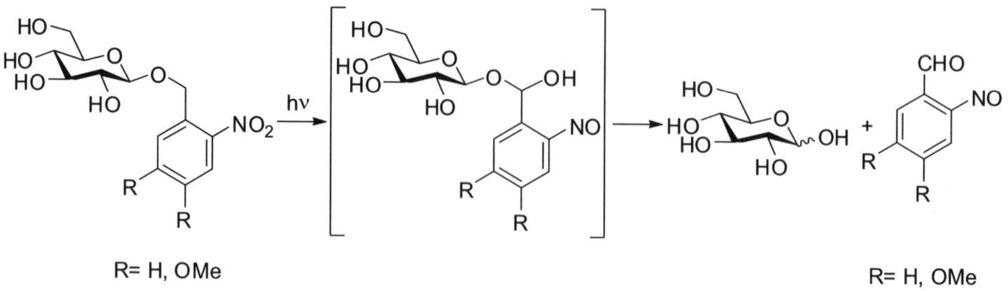

☐ **Scheme 70**
Example of light-sensitive glycosides: 2-nitrobenzyl and 6-nitroveratryl glycosides

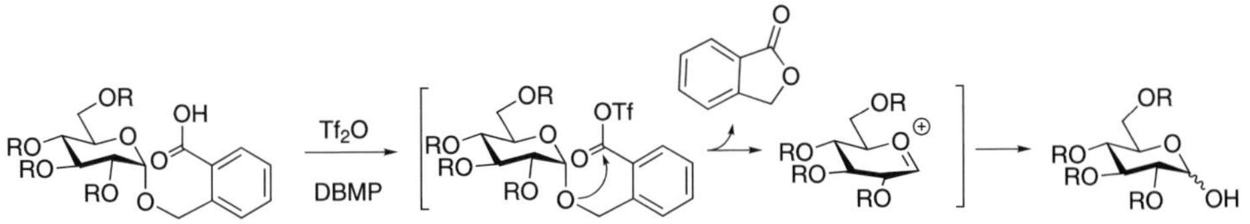

Scheme 71
Solvolysis of 2-(hydroxycarbonyl)benzyl glycosides

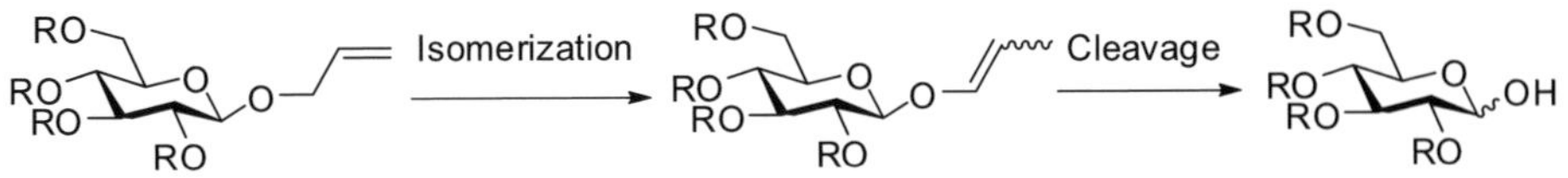

Scheme 72
Usual method for the deprotection of allyl glycosides

Allyl Glycosides   Another commonly used protecting group for the anomeric oxygen is the allyl group [400,401]. The most usual method for deblocking allyl glycosides involves the two-step reaction in which the allyl group is first converted into the more labile propenyl group and then is cleaved under mildly acidic conditions (see ➲ *Sect. 2.1.4*, ➲ *Scheme 72*).

Alternative methods for the deprotection of the allyl group at the anomeric position include Pd(PPh$_3$)$_4$/AcOH [402], in which the reaction proceeds by the formation of a π-allyl complex, or PdCl$_2$/CuCl/O$_2$ followed by photolysis in the presence of triethylamine [403]. Perfluoroalkylation with perfluoroalkyl iodide under sodium dithionite and sodium bicarbonate followed by elimination in the presence of zinc powder and ammonium chloride has also been disclosed as an efficient procedure for deprotection of the anomeric allyl group of carbohydrates (➲ *Scheme 73*) [404]. The reaction goes through the intermediacy of a radical addition of a perfluoroalkyl iodide to the double bond followed by Zn-mediated reductive β-elimination.

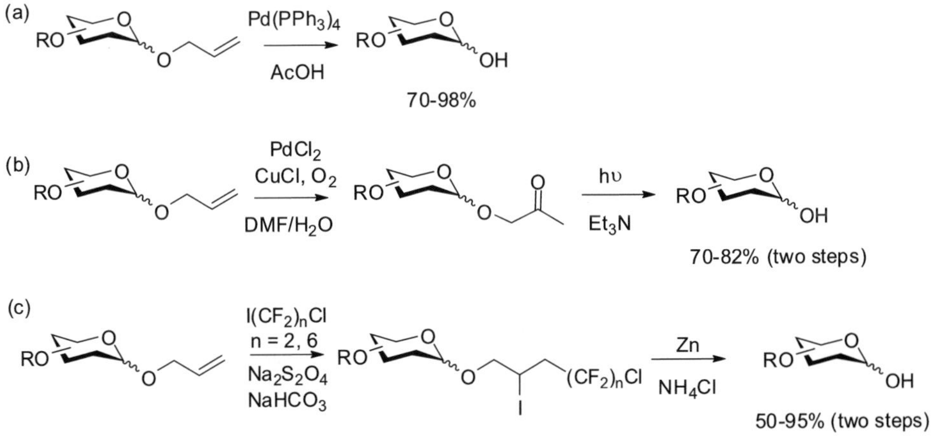

Scheme 73
Alternative methods for the deprotection of the allyl group at the anomeric position

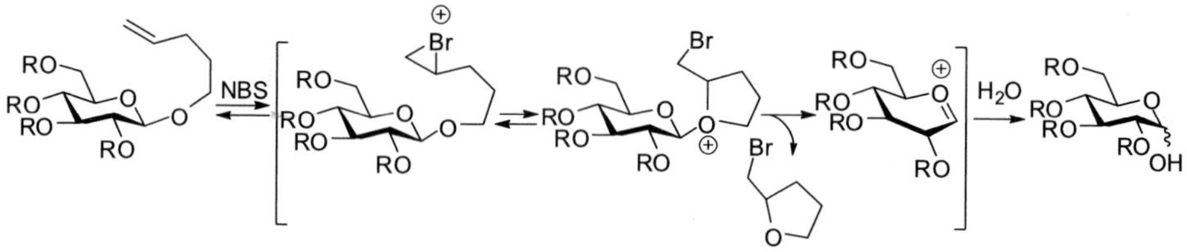

◻ Scheme 74
Deprotection of n-pentenyl glycosides

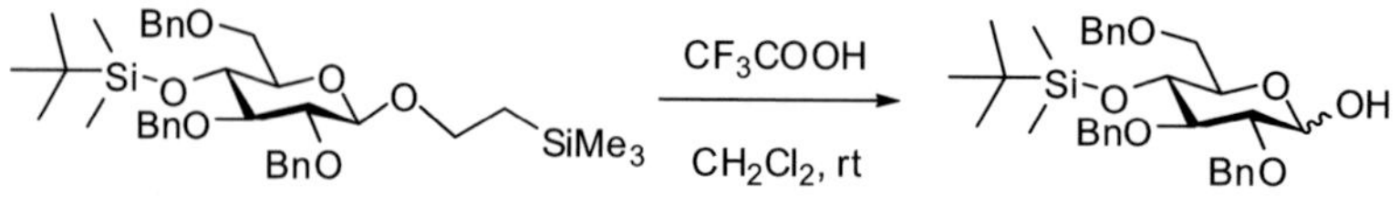

◻ Scheme 75
Deprotection of 2-(trimethylsilyl) ethyl glycosides

n-Pentenyl Glycosides   In 1988 Fraser-Reid and Mootoo reported the NBS-mediated reaction of *n*-pentenyl glycosides, in the presence of water to yield reducing monosaccharides (❯ *Scheme 74*) [405]. This transformation proved to be highly chemoselective leaving a wide variety of other functional groups unaffected.

The reaction takes place with cleavage of the anomeric C–O bond by electrophilic addition to the olefin followed by intramolecular displacement by the ring oxygen and eventual expulsion of the pentenyl chain, in the form of a halomethyltetrahydrofuran, to form an oxonium species (❯ *Scheme 74*). Trapping with water then leads to the reducing sugar. This transformation has also been extended to the use of *n*-pentenoyl esters [406,407].

2-(Trimethylsilyl) Ethyl Glycosides   An alternative procedure for protecting the anomeric center is based on the use of 2-(trimethylsilyl) ethyl glycosides [408,409]. Lipshutz et al. first found that LiBF$_4$ in CH$_3$CN caused the deblocking of the anomeric center [408], although extensive experimentation led Magnusson and coworkers to report on the use of trifluoroacetic acid in dichloromethane as the most effective reagent to carry out the same transformation (❯ *Scheme 75*) [409]. The reaction conditions are fully compatible with most of the normally used protecting groups, including silyl ethers and therefore this protecting group has found wide application in the synthesis of complex oligosaccharides.

3.2 Acylation Reaction

3.2.1 Anomeric *O*-Acylation

Free sugars, since they are polyhydroxy aldehydes or ketones, can be acylated through their hydroxyl groups (including the anomeric hydroxyl group) to give esters. However, the unusual property of the anomeric center to be a mixed function (ester and acetal) confers the glycosyl esters a special reactivity. Acetylation of unprotected sugars is complicated by the fact that they exist in solution as equilibrium mixtures of tautomers. The isomer obtained depends on

☐ Scheme 76
Acetylation of D-glucopyranose

the catalyst used and on the temperature. For example, acetylation of pure α- and β-D-glu-copyranoses with Ac$_2$O and pyridine at 0 °C occurs with retention of the configuration at the anomeric carbon [410], whereas acetylation of a mixture of α- and β-D-glucopyranoses in the presence of an acid catalyst (Ac$_2$O, ZnCl$_2$) takes place with predominant formation of the thermodynamically preferred (anomeric effect) α-D-glucopyranose pentaacetate, due to acid-induced anomerization. Conversely, β-D-glucopyranose pentaacetate, is formed preferentially when the acetylation is carried out in the presence of sodium acetate at higher temperatures (Ac$_2$O, NaOAc, Δ), a fact which has been explained on the basis of a lower rate for acetyla-tion when compared with mutarotation together with a preferential reactivity for the equatorial anomeric hydroxyl groups (❷ Scheme 76). For acetylation of ketoses low temperature acidic catalysts are preferred.

The ring size of the cyclic acetates formed under common acetylation procedures is normally pyranoid although sugars that form relatively stable furanose rings give more complex mix-tures. D-Galactose, for instance, in the presence of sodium acetate or pyridine at elevated temperatures gives appreciable amounts of furanose acetates [411].

Sugar benzoates have also been widely used since they are easy to prepare and more stable than the corresponding acetates. Benzoyl chloride in pyridine is the reagent of choice to carry out this transformation [412], and benzoylation in hot pyridine may lead to the isolation of glycofuranose benzoates [413]. More recently, a new method for the benzoylation of alcohols has been described using TMEDA as a base, which gave the expected benzoates in excellent yields [414]. In 2-N-protected 4,6-O-ketal derivatives of D-glucosamine a highly regio- and stereoselective acylation of the anomeric hydroxyl groups is possible using 1N-benzyloxy-1,2,3-benzotriazol (BzOBT) or benzoic anhydride and triethylamine as a base [415,416]. Because of the kinetic stereoelectronic effect or 1,3-diaxial repulsion, the O- is oriented in the equatorial position and only the β-anomer is formed (❷ Scheme 77).

When positions other than 1-OH are fully protected, anomeric acylation can be carried out by the usual methods for the esterification of alcohols (❷ Scheme 78). For example, car-bodiimide-mediated coupling [417,418] was the method used for the preparation of glycosyl benzyl phthalates [419] or n-pentenoyl esters [406,407] from the corresponding 1-OH sugars (❷ Scheme 78a). Combination of carbodiimides, active ester-forming reagents, and base cata-lysts have been studied for the selective acylation of monoprotected glucuronate esters and the uranium reagent HATU [420] has been found to be the reagent of choice [421].

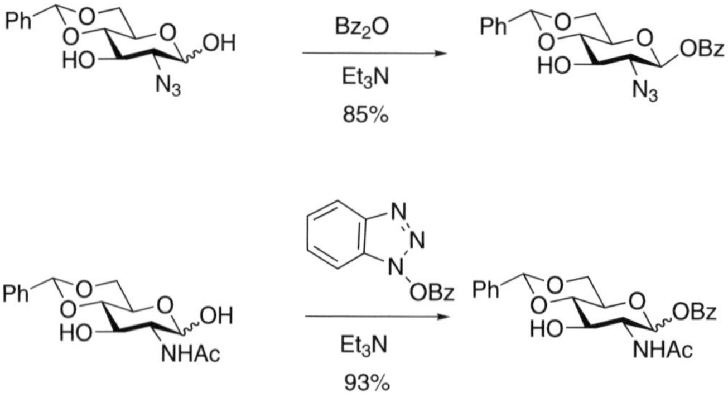

□ Scheme 77
The 1-OH shall be 1-OBz in the lower right structure

1-OH sugars react with acyl fluorides in the presence of cesium fluoride to furnish the corresponding glycosyl esters under essentially neutral conditions, with the α/β ratio being affected by changes in the order of addition of the reagents (❷ *Scheme 78b*) [422].

Acylation of the lithium salt of 1-OH sugars allows complete stereocontrol in the formation of glycosyl esters [423,424]. Metalation of 2,3,4,6-tetra-*O*-benzyl-α-D-glucopyranose, in tetrahydrofuran at −40 °C with 1.1 equiv. of *n*-BuLi, followed by acylation with acid chlorides produce mainly or exclusively α-glucosyl esters (❷ *Scheme 78c*). Increasing the reaction temperature and changing the solvent to benzene led to the preferred formation of β-glucosyl esters (❷ *Scheme 78d*). Analogously, tributylstannyl alkoxides can be used in place of the corresponding lithium salts [425].

2-Acylthio-3-nitropyridines, prepared from the corresponding carboxylic acids, have been used as acylating agents when the corresponding acid chlorides are unstable [426].

As an alternative to the direct esterification of the anomeric hydroxyl group, glycosyl esters have been prepared by displacement of a good leaving group at the anomeric position, although in that case the key step involves reaction at the anomeric carbon rather than at the anomeric oxygen. In this context, the direct glycosylation of trichloroacetimidates [427] with carboxylic acids is a particularly advantageous method (❷ *Scheme 78e*) [428]. This reaction, which involves inversion of the configuration at the anomeric center, is the method of choice for the stereoselective preparation of β-acyl-glycosides. The requisite trichloroacetimidates can be selectively produced from the corresponding hemiacetals under thermodynamically controlled conditions [429]. Analogously, 2-(trimethylsilyl) ethyl glycosides have also been transformed into the corresponding 1-*O*-acyl sugars by reaction with the appropriate anhydride in the presence of BF$_3$·Et$_2$O (❷ *Scheme 78f*) [430].

The Mitsunobu protocol has also been investigated in the stereocontrolled synthesis of glycosyl esters (❷ *Scheme 78g*) [431]. Complete stereochemical inversion at C-1 of the starting sugar is observed when the esterification is conducted with anomerically pure glycosyl hemiacetals. By corollary, complementary ratios of inverted products are formed when an anomeric mixture of sugars is esterified. The stereochemical outcome of the esterification is not affected

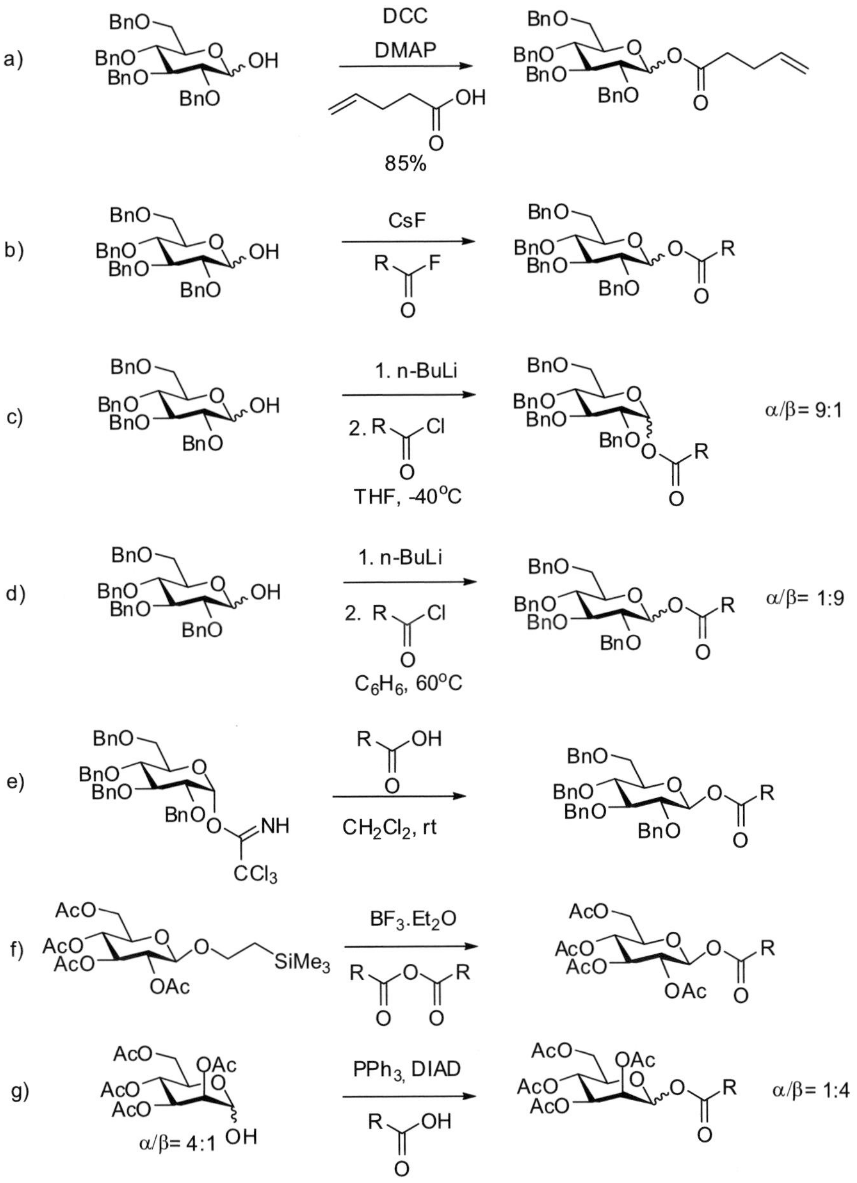

■ Scheme 78
Acylation reactions of fully protected pyranoses

by anchimeric assistance from acyl groups at C-2. Accordingly, 2,3,4,6-tetra-O-acetyl-D-mannose furnishes a mixture of 1-O-benzoates in which the β-anomer predominates, a result that is especially significant in view of the difficulties generally encountered in obtaining β-glycosides of D-mannose.

3.2.2 Anomeric *O*-Deacylation

Deprotection of the anomeric acyl group in acylated sugars can be effected in a number of manners including chemical and enzymatic methods.

Enzymic Deacylation   Both furanose and pyranose sugars can be efficiently deacetylated by suitable lipases under proper reaction conditions. The removal of the 1-*O*-acetyl group of glucose pentaacetate by *Aspergillus niger* lipase was reported after 20% conversion [432]. More recently it was found that the regioselectivity could be enhanced in the pyranose case by the presence of DMF [433]. Porcine pancreatic lipase in 10% DMF exclusively cleaved glucose pentaacetate ester at C-1 (70% isolated yield), and similar selectivities (and yields) were obtained for several peracetylated hexopyranoses. Peracetylated furanoses were deacylated at C-1 by the use of the lipase from *Aspergillus niger*. Finally, peracetylated reducing disaccharides have been specifically hydrolyzed at the anomeric center with a lipase from *Aspergillus niger* (Lipase A Amano 6) in a mixture of organic solvents and phosphate buffer (❷ *Scheme 79*) [434].

Chemical Deacylation   Several methods for the regioselective 1-*O*-deacylation of carbohydrates have been reported. Most of them involve regioselective nucleophilic attack upon the carbonyl group at *O*-1 thus liberating the anomeric oxygen. Nitrogen-containing nucleophiles have been widely used in this transformation: piperidine [435], hydrazine acetate [436], and hydrazine hydrate [437], have been reported to selectively hydrolyze anomeric acetates in peracetylated disaccharides (❷ *Scheme 80*). Hydrazine acetate in DMF [436], benzylamine in chloroform [438], hydrazine hydrate in pyridine [439], ammonia in an aprotic solvent [440], and 2-aminoethanol [441], have been used to regioselectively 1-*O*-deacylate per-*O*-

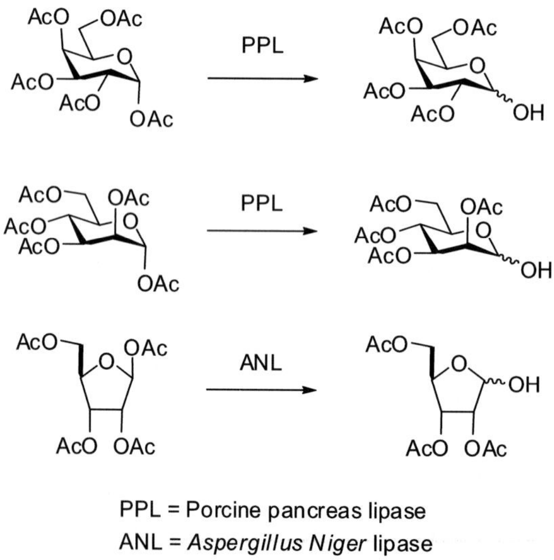

PPL = Porcine pancreas lipase
ANL = *Aspergillus Niger* lipase

❏ Scheme 79
Enzymic deacylation of acyl-glycosides

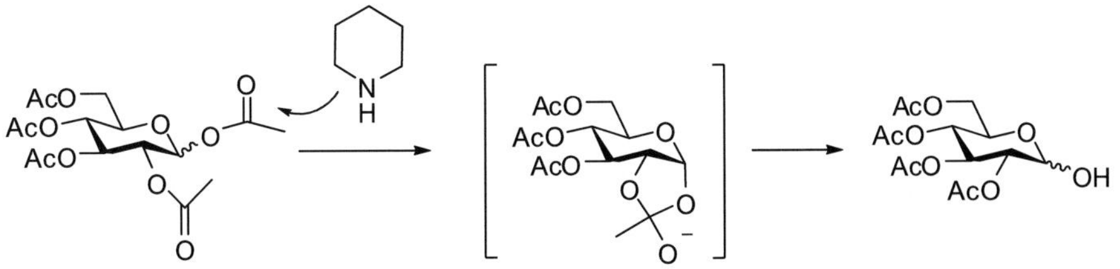

Scheme 80
Chemical deacylation of acyl-glycosides

acylaldoses. Other reagents used include potassium hydroxide [442], potassium cyanide [442], sodium methoxide [443], bis(tributyltin)oxide [442,444], tributyltin methoxide [444], ammonium carbonate [445], ammonium acetate [446], and mercuric chloride/mercuric oxide [447]. Heterogeneous anomeric deacetylation has also been reported by the use of magnesium oxide in methanol [448], or silica gel in methanol [449].

Finally, acid-catalyzed solvolysis of per-O-acyl hexopyranoses (SnCl$_4$, CH$_3$CN, H$_2$O) is an efficient method for removal of the anomeric acetyl group [450]. In this case the reaction takes place by cleavage of the C-1–OAc bond [451]. This reaction proceeds in 1 h at room temperature for sugars containing 1,2-trans-acetoxy groups and at 40 °C for 1,2-*cis* acylated pyranoses, and confirms the anchimeric assistance provided by the ester group at C-2.

3.3 Carbonylation and Thiocarbonylation of the Anomeric Hydroxyl

The reagents most commonly used for the preparation of sugar carbonates are phosgene, alkyl chloroformates, and diaryl carbonates. Phosgene reacts with free sugars giving rise to cyclic carbonates preferentially having five-membered rings. Depending on the sugar the anomeric position may be involved, for instance when D-glucose is treated with phosgene and pyridine a 1,2:5,6-diester derivative is obtained. Unprotected sugars also react with chloroformic esters in the presence of pyridine, although to yield alkoxycarbonyl compounds (▶ *Scheme 81a*) [452].

In the case of protected sugars, the anomeric hydroxy group reacts with chloroformic esters to give mixed esters (▶ *Scheme 81b*). Usually the coupling reaction is not stereoselective giving rise to an anomeric mixture of carbonates, although the α/β ratio can be influenced by the choice of the proper base (▶ *Scheme 81c*) [453]. Reaction of 2-thiopyridyl chloroformate with a glucose derivative results in an anomeric mixture (α:β, 1:2) (▶ *Scheme 81d*) whereas the use of bis(2-thiopyridyl)carbonate yields exclusively the β-anomer (▶ *Scheme 81e*) [454]. Very recently, a highly regio- and stereoselective reaction of D-glucopyranose 1,2-diols with allyl chloroformate or ethyl chloroformate has been reported [455].

Diaryl carbonates (e. g. carbonyl diimidazol, 4-nitrophenyl carbonate) can react sequentially with carbohydrate derivatives to furnish mixed sugar carbonates (▶ *Scheme 81f*) [456]. Although normally anomeric mixtures are generated the use of a succinimidyl group, in the presence of K$_2$CO$_3$, was effective for the synthesis of pure β-carbonates.

Anomeric alkyl xanthates are prepared by treatment of 1-OH sugars with sodium hydride in the presence of a catalytic amount of imidazole, carbon disulfide, and an alkyl halide (▶ *Scheme 81g*) [457].

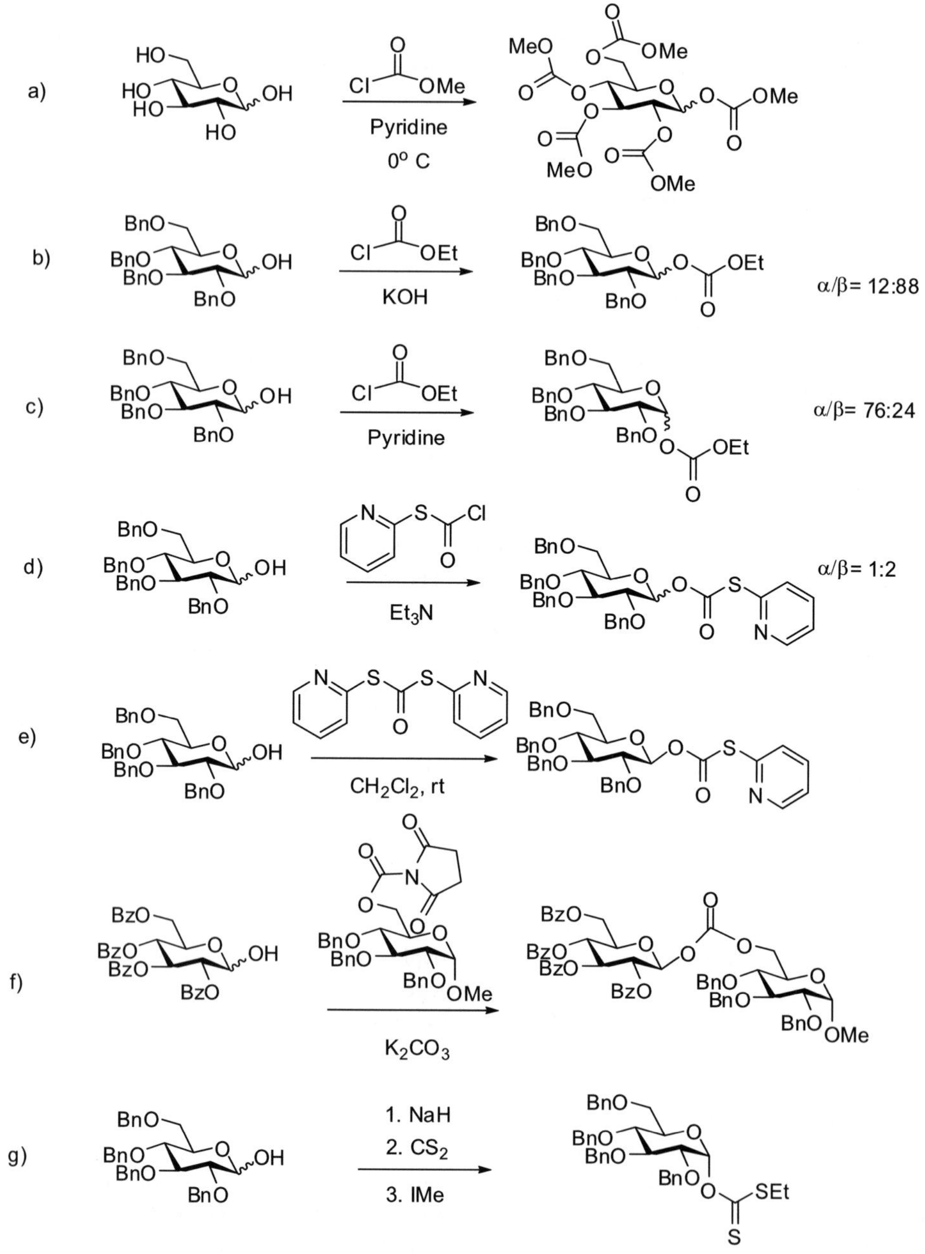

☐ Scheme 81
Preparation of sugar carbonates

3.4 Silylation

Anomeric silyl ethers have been prepared from 1-OH sugars and the corresponding silyl chloride in the presence of a base. When the hydroxyl group at C-2 is unprotected silyl group migrations away from the anomeric center have been observed [458].

3.5 Phosphorylation and Phosphitylation

Glycosyl phosphates are intermediates in biological glycosyl transfer and are constituents of cell membranes [459]. Both chemical and enzymic methods are available for the synthesis of specific phosphates. In the preparation of certain glycosyl phosphates, enzymic synthesis with the appropriate phosphorylase provides the simplest preparation. In this fashion, α-D-glucopyranosyl phosphate is readily prepared by the phosphorolysis of starch or glycogen [460].

Chemically, synthesis of glucosyl phosphates also may involve two different approaches based either on activation of the glycosyl oxygen or activation at the anomeric carbon. In the latter, glycosyl acetates [461], orthoesters [462], glycosyl halides [463], trichloroacetimidates [464], vinyl glycosides [465], glycals [466], or 1,2-orthoesters [467] are used as glycosyl donors and they are not the aim of this chapter. The above two-step procedures ensure in most cases the anomeric purity of the final glycosyl phosphates.

On the other hand, several alternative procedures for the synthesis of glycosyl phosphates involving 1-OH activation have been developed. The thallium salt of the anomeric hydroxyl readily undergoes substitution with a phosphochloridate in benzene or acetonitrile (❷ *Scheme 82*) [468]. The configuration of an organo-phosphate moiety introduced at the anomeric position is strongly influenced by the choice of solvent, so that a preponderance of either the α- or β-phosphate may be attained. These differences are reminiscent of solvent effects observed in syntheses of *O*-acyl esters and reflect differences in the anomeric composition in the reducing sugars as well as in the relative reactivities of the two anomers. Similarly, α-phosphates of *N*-acylglucosamine are prepared in high yields via the reaction of the corresponding 1-*O*-lithium salts with phosphorochloridate at low temperatures [469].

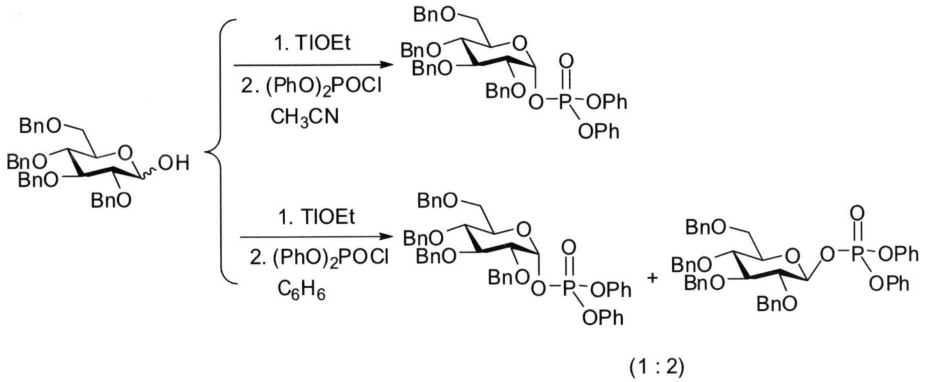

(1 : 2)

❏ **Scheme 82**
Preparation of aldosyl phosphates

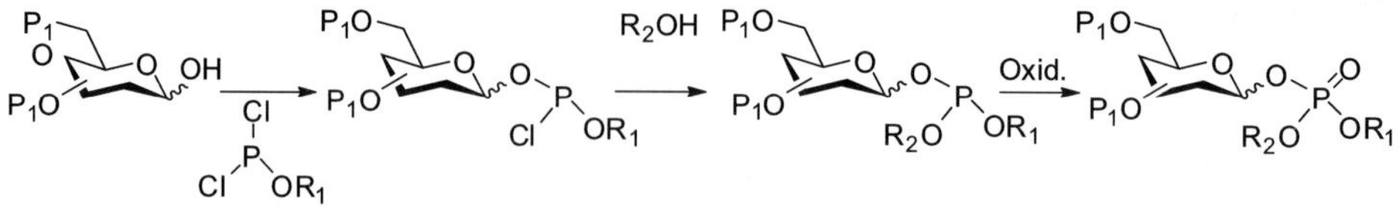

◘ Scheme 83
Preparation of aldosyl phosphates

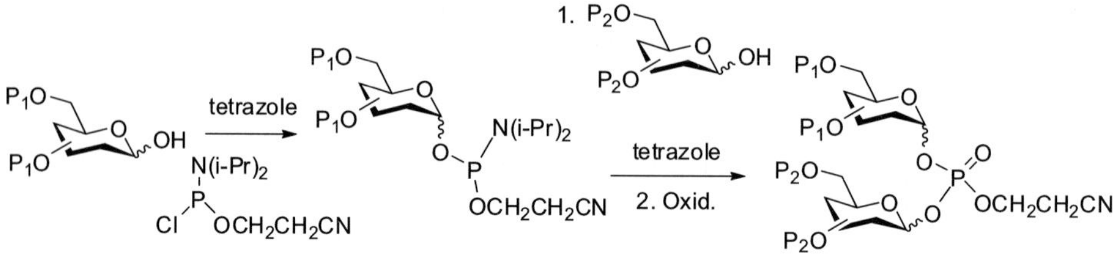

◘ Scheme 84
Preparation of glycosyl phosphodiesters

A different approach to the synthesis of aldosyl phosphates involves the intermediacy of aldosyl phosphites [470]. The reaction of the anomeric hydroxyl group, with a trivalent phophitylating reagent, furnished an anomeric phosphorochloridite, which is able to react with a hydroxyl-containing compound, to generate a phosphorous triester which, upon oxidation gives the corresponding aldosyl phosphoric triester (❷ Scheme 83).

More recently phophitylating reagents which, after oxidative transformation to the corresponding phosphates, allow removal of the protecting group at phosphorous (V) by mild base treatment, have been reported [471,472].

Recent studies have shown that stabilization in O-glycosyl phosphites can be achieved with the help of an electron-withdrawing O-alkyl group at the phosphite moiety (e. g. trichloroethyl vs. ethyl group) [473].

The phosphytilation approach has also been applied for the preparation of compounds in which two anomeric centers are part of a phosphodiester bond (❷ Scheme 84) [471].

Acknowledgement

A.M.G. is grateful to the Ministerio de Educación y Ciencia of Spain [CTQ2006–15279-CO3–02/BQU] for financial support.

References

1. Kocienski PJ (2005) Protecting Groups, 3rd edn. Georg Thieme Verlag, Stuttgart
2. Robertson J (2000) Protecting Group Chemistry. Oxford University Press, New York
3. Greene TWP, Wuts GM (1999) Protective Groups in Organic Synthesis, 3rd edn. Wiley, New York
4. Hanson JR (1999) Protecting Groups in Organic Synthesis. Sheffield Academic Press, New York
5. Goodman L (1967) Adv Chem and Biochem 22:109
6. Mootoo DR, Konradsson P, Udodong U, Fraser-Reid (1988) J Am Chem Soc 110:5583
7. Schürrle K, Beier B, Werbitzky O, Piepersberg W (1991) Carbohydr Res 212:321
8. Evtushenko EV (1999) Carbohydr Res 316:187
9. Boto A, Hernández D, Hernández R, Suárez E (2004) Org Lett 6:3785
10. Boto A, Hernández D, Hernández R, Suárez E (2006) J Org Chem 71:1938
11. Czerncki S, Georgoulis C, Provelenghiou C (1976) Tetrahedron Lett 39:3535
12. Hijfte LV, Little RD (1985) J Org Chem 50:3940
13. Wessel HP, Lersen T, Bundle DR (1985) J Chem Soc Perkin Trans I 2247
14. Hatakeyama S, Mori H, Kitano K, Yamada H, Nishizawa M (1994) Tetrahedron Lett 35:4367
15. Weissman SA, Zewge D (2005) Tetrahedron 61:7833
16. Iseloh U, Dudkin V, Wang Z G, Danishefsky S (2002) Tetrahedron Lett 43:7027
17. Park MH, Takeda R, Nakanishi K (1987) Tetrahedron Lett 28:3823
18. Rodebaugh R, Debenham JS, Fraser-Reid B (1996) Tetrahedron Lett 37:5477
19. Falck JR, Barma DK, Baati R, Mioskowski Ch (2001) Angew Chem Int Ed 40:1281
20. Bhattacharjee SS, Gorin PA (1969) J Can Chem 47:1195
21. Gelas J (1981) Adv Carbohydr Chem 39:71
22. Garegg PJ (1997) In: Hanessian S (ed) Preparative Carbohydrate Chemistry. Marcel Dekker, New York, p 53
23. Garegg PJ, Hultberg H (1981) Carbohydr Res 93:C10
24. Mani NS, Kanakamma PP (1994) Tetrahedron Lett 35:3629
25. Ernst A, Vasella A (1996) Helv Chim Acta 79:1279
26. Ek M, Garegg PJ, Hultberg H, Oscarson S (1983) J Carbohydr Chem 2:305
27. Oikawa M, Liu WC, Nakai Y, Koshida S, Fukase K, Kusumoto S (1996) Synlett 1179
28. DeNinno MP, Etienne JB, Duplantier KC (1995) Tetrahedron Lett 36:669
29. Jiang L, Chan TH (1998) Tetrahedron Lett 39:355
30. Garegg PJ, Hultberg H, Wallin S (1982) Carbohydr Res 108:97
31. Shie ChR, Tzeng ZH, Kulkarni SS, Uang BJ, Hsu ChY, Hung SCh (2005) Angew Chem Int Ed 44:1665
32. Suzuki K, Nonaka H, Yamaura M (2003) Tetrahedron Lett 44:1975
33. Hanessian S (1968) Adv Chem Ser 74:159
34. Grindley TB (1998) Adv Carbohydr Chem Biochem 53:17 and references therein
35. Alais J, Veyrières A (1987) J Chem Soc Perkin Trans I 377
36. Danishefsky SJ, Hungate R (1986) J Am Chem Soc 108:2486
37. Beauepère D, Boutbaiba I, Wadouachi A, Frechou C, Demailly G, Uzan R (1992) New J Chem 16:405
38. Cruzado C, Martín-Lomas M (1986) Tetrahedron Lett 27:2497
39. Yang G, Ding X, Kong F (1997) Tetrahedon Lett 38:6725
40. Madsen J, Viuf C, Bols M (2000) Chem Eur J 6:1140
41. Cipolla L, Lay L, Nicotra F (1997) J Org Chem 62:6678
42. Hori H, Nishida Y, Ohrui H, Meguro H (1989) J Org Chem 54:1346
43. Sollogoub M, Das SK, Mallet JM, Sinaÿ P (1999) C R Acad Sci Paris, t 2, Série IIc, 441
44. Chevalier-du Roizel B, Cabianca E, Rollin P, Sinaÿ P (2002) Tetrahedron 58:9579
45. Jia C, Zhang Y, Zhang LH, Sinaÿ P, Sollogoub M (2006) Carbohydr Res 341:2135
46. Lecourt T, Herault A, Pearce AJ, Sollogoub M, Sinaÿ P (2004) Chem Eur J 10:2960
47. Falck JR, Barma DK, Venkataraman SK, Baati R, Mioskowski C (2002) Tetrahedron Lett 43:963
48. Oikawa Y, Yoshioka T, Yonemitsu O (1982) Tetrahedron Lett 23:885
49. Horita K, Yoshioka T, Tanaka T, Oikawa Y, Yonemitsu O (1986) Tetrahedron 42:3021

50. Chandrasekhar S, Sumithra G, Yadav JS (1996) Tetrahedron Lett 37:1645
51. Sharma GVM, Lavanya B, Mahalingam AK, Radha Krishna P (2000) Tetrahedron Lett 41:10323
52. Johansson R, Samuelsson B (1984) J Chem Soc Perkin Trans 1 2371
53. Classon B, Garegg PJ, Samuelsson B (1984) Acta Chem Scand B 38:419
54. Kartha KPR, Kiso M, Hasegawa A, Jennings H (1998) J Carbohydr Chem 17:811
55. Yu W, Su M, Gao Z, Yang Z, Jin Z (2000) Tetrahedron Lett 41: 4015
56. Sharma GVM, Reddy ChG, Krishna PR (2003) J Org Chem 68:4574
57. Akiyama T, Shima H, Ozaki S (1992) Synlett 415
58. Yan L, Kahne D (1995) Synlett 523
59. Vaino AR, Szarek WA (1995) Synlett 1157
60. Cappa A, Marcantoni E, Torregiani E (1999) J Org Chem 64:5696
61. Bhattacharyya S, Magnusson BG, Wellmar U, Nilsson UJ (2001) J Chem Soc Perkin Trans 1, 2001, 886
62. Jobron L, Hindsgaul O (1999) J Am Chem Soc 121:5835
63. Fukase K, Tanaka H, Torii S, Kusumoto S (1990) Tetrahedron Lett 31:389
64. Fukase K, Yoshimura T, Hashida M, Kusumoto S (1991) Tetrahedron Lett 32: 4019
65. Fukase K, Egusa K, Nakai Y, Kusumoto S (1996) Molecular Diversity 182
66. Fukase K, Hashida M, Kusumoto S Tetrahedron Lett (1991) 32:3557
67. Egusa K, Fukase K, Kusumoto S (1997) Synlett 675
68. Egusa K, Kusumoto S, Fukase K (2003) Eur J Org Chem 3435
69. Plante OJ, Buchwald SL, Seeberger PH (2000) J Am Chem Soc 122:7148
70. Liu X, Seeberger PH (2004) Chem Commun 1708
71. Guibé F (1997) Tetrahedron 40:13509
72. Gill J, Gigg R, Payne S, Conant R (1987) J Chem Soc Perkin Trans I 423
73. Rollin P, Sinaÿ P (1977) J Chem Soc, Perkin Trans I 2513
74. Hindsgaul O, Norberg T, Le Pendu J, Lemieux RU (1982) Carbohydr Res 109:109
75. Oltvoort JJ, Klosterman M, Van Boom JH (1983) Recl Trav Chim Pays-Bas 102:501
76. Lakhmiri R, Lhoste P, Sinou D (1989) Tetrahedron Lett 30:4669
77. Iversen T, Bundle DR (1981) J Chem Soc, Chem Commun 1240
78. David S, Hanessian S (1985) Tetrahedron 41:643
79. Eby R, Schuerch C (1982) Carbohydr Res 100:C41
80. Gent PA, Gigg R (1976) Carbohydr Res 49:325
81. Warren CD, Jeanloz RW (1977) Carbohydr Res 53:67
82. Boss R, Scheffold R (1976) Angew Chem Int Ed Engl 15:558
83. Ogawa T, Nakabayashi S, Kitajima T (1983) Carbohydr Res 114:225
84. Nicolaou KC, Hummel CW, Bockovich NJ, Wong CH (1991) J Chem Soc Chem Commun 870
85. Oltvoort JJ, van Boeckel CAA, de Koning JH; van Boom JH (1981) Synthesis 305
86. Yadav JS, Chandrasekhar S, Sumithra G, Kache R (1996) Tetrahedron Lett 37:6603
87. Kariyone K, Yazawa H (1970) Tetrahedron Lett 11: 2885
88. Diaz RR, Melgarejo CR, Lopez-Espinosa MTP, Cubero II (1994) J Org Chem 59:7928
89. Kitov PI, Bundle DR (2001) Org Lett 3:2835
90. Gent PA, Gigg R, Conant R (1972) J Chem Soc Perkin Trans I 1535
91. Sharma GVM, Ilangovan A, Mahalingam AK (1998) J Org Chem 63:9103
92. Vatèle JM (2002) Synlett 507
93. Vatèle JM (2002) Tetrahedron 58:5689
94. Markovic D, Vogel P (2004) Org Lett 6:2693
95. Helferich B (1948) Adv Carbohydr Chem 3:79
96. Sharma GVM, Mahalingam AK, Prasad TR (2000) Synlett 1479
97. Bessodes M, Komiotis KA (1986) Tetrahedron Lett 27:579
98. MacCoss M, Cameroon DJ (1978) Carbohydr Res 60:206
99. Jones GB, Hynd G, Wright JM, Sharma A (2000) J Org Chem 65:263
100. Lu RJ, Liu D, Giese RW (2000) Tetrahedron Lett 41:2817
101. Yan MCh, Chen YN, Wu HT, Lin ChCh, Chen ChT, Lin ChCh (2007) J Org Chem 72:299
102. Pathak AK, Pathak V, Seitz LE, Tiwari KN, Katar MS, Reynolds RC (2001) Tetrahedron Lett 42:7755
103. Agarwal A, Vankar YD (2005) Carbohydr Res 340:1661
104. Rawal GK, Rani S, Kumar A, Vankar YD (2006) Tetrahedron Lett 47:9117

105. Yadav JS, Subba Reddy BV (2000) Carbohydr Res 329:885

106. Chen MY, Patkar LN, Lu KCh, Lee ASY, Lin ChCh (2004) Tetrahedron 60:11465

107. Schaller H, Weimann G, Lerch B, Khorana HG (1963) J Am Chem Soc 85:3821

108. Caruthers MH (1989) J Chem Ed 66:577

109. Sarkar AK, Rostand KS, Jain RK, Matta KL, Esko JD (1997) J Biol Chem 272:25608

110. Gaunt MJ, Yu J, Spencer JB (1998) J Org Chem 63:4172

111. Liptak A, Borbás A, Jánossy L, Szilágyi L (2000) Tetrahedron Lett 41:4949

112. Borbás A, Szabo ZB, Szilágyi L, Bényei A, Lipták A (2002) Carbohydr Res 337:1941

113. Kia J, Abbas SA, Locke RD, Piskorz CF, Alderfer JL, Matta KL (2000) Tetrahedron Lett 41:169

114. Liao W, Locke RD, Matta KL (2000) Chem Comm 369

115. Xia J, Alderfer JL, Piskorz CF, Matta KL (2001) Chem Eur J 7:356

116. Crich D, Jayalath P (2005) Org Lett 7:2277

117. Crich D, Wu B (2006) Org Lett 8:4879

118. Crich D, Jayalath P, Hutton TK (2006) J Org Chem 71:3064

119. García-Moreno MI, Aguilar M, Ortiz Mellet C, García Fernández JM (2006) Org Lett 8:297

120. Balbuena P, Rubio EM, Ortiz Mellet C, García Fernández JM (2006) Chem Commun 2610

121. Fisher E (1895) Berichte der Deutschen Chemischen Gesellschaft 28:1145

122. de Belder AN (1977) Adv Carbohydr Chem Biochem 34:179

123. Calinaud P, Gelas J (1997) In: Hanessian S (ed) Preparative Carbohydrate Chemistry. Marcel Dekker, New York, p 3

124. Ley SV, Baeschlin DK, Dixon DJ, Foster AC, Ince SJ, Priepke HWM, Reynolds DJ (2001) Chem Rev 101:53

125. Gelas G (1981) Adv Carbohydr Chem Biochem 39:71

126. Eliel EL (1971) Pure Appl Chem 25:509

127. Garegg PJ, Swahn CG (1980) Methods Carbohydr Chem 8:317

128. Foster AB (1972) In: Pigman W, Horton D (eds) The Carbohydrates: Chemistry, Biochemistry. Academic Press, New York, p 391

129. Gelas J, Horton D (1981) Heterocycles 16:1587

130. Clode DM (1979) Chem Rev 79:491

131. Manna S, Jacques YP, Falck JR (1986) Tetrahedron Lett 27:2679

132. Park KH, Yoon YJ, Lee SG (1994) Tetrahedron Lett 35:9737

133. Lablance Y, Fitzsimmons J, Adams EP, Rokacha J (1986) J Org Chem 51:789

134. Vijayasaradhi S, Singh J, Aidhan IS, (2000) Synlett 110

135. Agarwal A, Vankar YD (2005) Carbohydr Res 340:1661

136. Yan MCh, Chen YN, Wu HT, Lin ChCh, Chen ChT, Lin ChCh (2007) J Org Chem 72:299

137. Majumdar S, Bhattacharya A (1999) J Org Chem 64:5682

138. Haines AH (1981) Adv Carbohydr Chem Biochem 39:71

139. Chéry F, Rollin P, De Lucci O, Cossu S (2000) Tetrahedron Lett 41:2357

140. Ariza X, Costa AM, Faja M, Pineda O, Vilarrasa J (2000) Org Lett 2:2809

141. Ley SV, Woods M, Zanotti-Gerosa A (1992) Synthesis 52

142. Ley SV, Leslie R. Tiffin PD, Woods M (1992) Tetrahedron Lett 33:4767

143. Ley SV, Boons GJ, Leslie R, Woods M, Hollinshead DM (1993) Synthesis 689

144. Hughes AB, Ley SV, Priepke HWM, Woods M (1994) Tetrahedron Lett 35:773

145. Ley SV, Downham R, Edwards PJ, Innes JE, Woods M (1995) Contemp Org Synth 2:365

146. Ley SV, Priepke HWM, Warriner SL (1994) Angew Chem Int Ed Engl 33:2290

147. Montchamp JL, Tian F, Hart ME, Frost JW (1996) J Org Chem 61:3897

148. Ley SV, Baeschlin DK, Dixon DJ, Foster AC, Ince SJ, Priepke HWM, Reynolds DJ (2001) Chem Rev 101:53

149. Douglas NL, Ley SV, Osborn HMI, Owen DR, Priepke HWM, Warriner SL (1996) Synlett 793

150. Litjens REJN, van den Bos LJ, Codée JDC, Overkleeft HO, van der Marel GA (2007) Carbohydr Res 342:419

151. Ley SV, Downham R, Edwards PJ, Innes JE, Woods M (1995) Contemp Org Synth 2:365

152. Grice P, Ley SV, Pietruuszka J, Priepke HWM, Warriner SL (1997) J Chem Soc Perkin Trans I 351

153. Hense A, Ley SV, Osborn HMI, Owen DR, Poisson JF, Warriner SL, Wesson KE (1997) J Chem Soc Perkin Trans 1 2023

154. Lindberg B (1990) Adv Carbohydr Chem Biochem 48:279

155. Araki S, Abe S, Yamada S, Satake M, Fujiwara N, Kon K, Ando S (1992) J Biochem 112:461

156. Dudman WF, Heidelberger M (1969) Science 164:954

157. Thayer WR, Bazic CM, Camphausen RT, McNeil M (1990) J Clin Microbiol 28: 714
158. Hirase S (1957) Bull Chem Soc Jpn 30:68
159. Dutton GGS, Karunaratane DN (1984) Carbohydr Res 134:103
160. Gorin PAJ, Mazurek M, Duarte HS, Duarte JH (1981) Carbohydr Res 92:C1
161. Ziegler T (1997) Top Curr Chem 186:203
162. Liptak A, Szabo L (1989) J Carbohydr Chem 8:629
163. Ziegler T, Eckhardt E, Herold G (1992) Tetrahedron Lett 33:4413
164. Ziegler T, Eckhardt E, Herold G (1992) Liebigs Ann Chem 441
165. Ziegler T, Eckhardt E, Neumann K, Birault V (1992) Synthesis 1013
166. Gorin PAJ, Ishikawa T (1967) Can J Chem 45:521
167. Garegg PJ, Lindberg B, Kvarnstrom I (1979) Carbohydr Res 77:71
168. Collins PM, McKinnon C, Manro A (1989) Tetrahedron Lett 30:1399
169. Aspinell GO, Ibrahim IH, Khare NK (1990) Carbohydr Res 200:247
170. Hiruma K, Tamura J, Horito S, Yoshimura J, Hashimoto H (1994) Tetrahedron 50:12143
171. Liptak A, Szabo L (1988) Carbohydr Res 184:C5
172. Liptak A, Bajza I, Kerekgyarto J, Hajko J, Szilagyi L (1994) Carbohydr Res 253:111
173. Agnihotri G, Misra AK (2006) Tetrahedron Lett 47:8493
174. Crich D, Bowers AA (2006) J Org Chem 71:3452
175. Crich D, Banerjee A (2005) Org Lett 7:1395
176. Crich D, Yao Q (2004) J Am Chem Soc 126:8232
177. Crich D, Yao Q (2003) Org Lett 5:2189
178. Miethchen R, Rentsch D, Stroll N (1992) Tetrahedron 48:8393
179. Miethchen R (2003) J Carbohydr Chem 22:801
180. Miethchen R, Rentsch D, Frank M, Lipták A (1996) Carbohydr Res 281:61
181. Miethchen R, Rentsch D, Frank M (1996) J Carbohydr Chem 15:15
182. Kumar P, Dinesh CU, Reddy RS, Pandey B (1993) Synthesis 1069
183. Ranu BC, Saha M (1994) J Org Chem 59:8269
184. Bhalerao UT, Davis KJ, Rao BV (1996) Synth Commun 26:3081
185. Yadav JS, Srinivas D, Reddy GS (1998) Synth Commun 28:1399
186. Tanemura K, Horaguchi T, Suzuki T (1992) Bull Chem Soc Jpn 65:304
187. Maity G, Roy SC (1993) Synth Commun 23:1667
188. Kluge AF, Untch KG, Fried JH (1972) J Am Chem Soc 94:7827
189. Corey EJ, Gras J, Ulrich P (1976) Tetrahedron Lett, 11: 809
190. Stork G, Isobe M (1975) J Am Chem Soc 97:4745
191. Kozikowski AP, Wu J (1987) Tetrahedron Lett 28:5125
192. Lipshutz BH, Pegram JJ (1980) Tetrahedron Lett 21:3343
193. Hosoya T, Takashiro E, Matsumoto T, Suzuki K (1994) J Am Chem Soc 116:1004
194. Pinto BM, Buiting MMW, Reimer KB (1990) J Org Chem 55:2177
195. Haines AH (1976) Adv Carbohydr Chem Biochem 33:11
196. Garegg PJ (1992) Acc Chem Res 25:575
197. Hofle G, Steglich W, Vorbruggen H (1978) Angew Chem Int Ed Engl 17:569
198. Scriven EFV (1983) Chem Soc Rev 12:129
199. Tiwari P, Kumar R, Maulik PR, Misra AK (2005) Eur J Org Chem 4265
200. Wolfrom ML, Thompson A (1963) Methods Carbohydr Chem 2:211
201. Hyatt JA, Tindall GW (1993) Heterocycles 35:227
202. Binch H, Stangier K, Thiem J (1998) Carbohydr Res 306:409
203. Kartha KPR, Field RA (1997) Tetrahedron 53:11753
204. Lee JC, Tai CA, Hung SC (2002) Tetrahedron Lett 43:851
205. Tai AA, Kulkarni SS, Hung SC, (2003) J Org Chem 68:8719
206. Ahmad S, Iqbal J (1987) J Chem Soc Chem Commun 114
207. Ghosh R, Chakraborty A, Maiti S (2004) Tetrahedron Lett 45:9631
208. Lu KC, Hsieh SY, Patkar LN, Chen CT, Lin CC (2004) Tetrahedron 60:8967
209. Dasgupta F, Singh PP, Srivastava HC (1980) Carbohydr Res 80:346
210. Montero JL, Winum JY, Leydet A, Kamal M, Pavia A A, Roque JP (1997) Carbohydr Res 297:175
211. Bhaskar PM, Loganathan D (1998) Tetrahedron Lett 39:2215
212. Bhaskar PM, Loganathan D (1999) Synlett 129

213. Kumareswaran R, Pachamuthu K, Vankar YD (2000) Synlett 1652
214. Tiwari P, Misra AK (2006) Carbohydr Res 341:339
215. Adinolfi M, Barone G, Iadonisi A, Schiattarella M (2003) Tetrahedron Lett 44:4661
216. Limousin C, Cleophax J, Petit A, Loupy A, Lukacs G (1997) J Carbohydr Chem 16:327
217. Das SK, Reddy KA, Krovvidi VLNR, Mukkanti K (2005) Carbohydr Res 340:1387
218. Murugesan S, Karst N, Islam T, Wiencek JM, Linhardt RJ (2003) Synlett 1283
219. Forsyth SA, Macfarlane DR, Thomson RJ, von Itzstein M (2002) Chem Commun 714
220. Zemplén G, Pacsu E (1929) Ber Dtsch Chem Ges 62:1613
221. Kurahashi T, Mizutani T, Yoshida JI (1999) J Chem Soc Perkin Trans I 465
222. Posner GH, Oda M (1981) Tetrahedron Lett 22:5003
223. Rana SS, Barlow JJ, Matta KL (1981) Tetrahedron Lett 22:5007
224. Bianco A, Brufani M, Melchioni C, Romagnoli P (1997) Tetrahedron Lett 38:651
225. Ishihara K, Kurihara H, Yamamoto H (1993) J Org Chem 58:3791
226. Ilankumaran P, Verkade JG (1999) J Org Chem 64:9063
227. Yamada S (1992) J Org Chem 57:1591
228. Hagiwara H, Morohashi K, Sakai H, Suzuki T, Ando M (1998) Tetrahedron 54:5845
229. Kattnig E, Albert M (2004) Org Lett 6:945
230. Ogawa T, Matsui M (1981) Tetrahedron 37:2363
231. Peri F, Cipolla L, Nicotra F (2000) Tetrahedron Lett 41:8587
232. Dong H, Pei Z, Byström S, Ramström O (2007) J Org Chem 72:1499
233. Waldmann H, Sebastian D (1994) Chem Rev 94:911
234. Schmid RD, Verger R (1998) Angew Chem Int Ed Engl 37:1608
235. Bordusa F (2002) Chem Rev 102:4817
236. Drueckhammer DG, Hennen WJ, Pederson RL, Barbas III CF, Gautheron CM, Krach, T, Wong CH (1999) Synthesis 499
237. Holla EW (1989) Angew Chem Int Ed Engl 28:220
238. Kooeterman M, De Nijs MP, Weijnen JG, Schoemaker HE, Meijer EM (1989) J Carbohydr Chem 8:333
239. Zemek J, Kucar S, Anderle D (1987) Collect Czech Chem Commun 52:2347
240. Csuk R, Glhzer BJ (1988) Z Naturforsch 436:1355
241. Naruto M, Ohno K, Naruse N, Takeuchi H (1979) Tetrahedron Lett 20:251
242. van Boeckel CAA, Beetz T (1983) Tetrahedron Lett 24:3775
243. Johnson F, Starkovsky NA, Paton AC, Carlson AA (1964) J Am Chem Soc 86:118
244. Lefeber DJ, Kamerling JP, Vliegenthart FG (2000) Org Lett 2:701
245. Sogabe S, Ando H, Koketsu M, Ishihara H (2006) Tetrahedron Lett 47:6603
246. Arranz E, Boons GJ (2001) Tetrahedron Lett 42:6469
247. Robins MJ, Hawrelak SD, Kanai T, Siefert JM, Mengel R (1979) J Org Chem 44:1317
248. Tomic S, Petrovic V, Matanovic M (2003) Carbohydr Res 338:491
249. Petrovic V, Tomic S, Ljevakovic D, Tomasic J (1997) Carbohydr Res 302:13
250. Kunz H, Harreus A (1982) Liebigs Ann Chem 41
251. Marin J, Blaton MA, Briand JP, Chiocchia G, Fournier C, Guichard G (2005) ChemBioChem 6:1796
252. Santoyo-González F, Uriel C, Calvo-Asín JA (1998) Synthesis 1787
253. Jiang L, Chan TH (1998) J Org Chem 63:6035
254. Yu H, Williams DL, Ensley HE (2005) Tetrahedron Lett 46:3417
255. van Boom JJ, Burgers PMJ (1976) Tetrahedron Lett 52:4875
256. Chery F, Cronin L, O'Brien JL, Murphy PV (2004) Tetrahedron 60:6597
257. Adamo R, Kovác P (2006) Eur J Org Chem 2803
258. García J, Fernández S, Ferrero M, Sanghvi Y, Gotor V (2003) Tetrahedron: Asymmetry 14:3533
259. Rej RN, Glushka JN, Chef W, Perlin AS (1989) Carbohydr Res 189:135
260. Wong CH, Ye X, Zhang Z (1998) J Am Chem Soc 120:7173
261. Zhu T, Boons GJ (2000) J Am Chem Soc 122:10222
262. Morère A, Mouffouk F, Chavis C, Montero JL (1997) Tetrahedron Lett 38:7519
263. Morère A, Menut Ch, Vidil C, Skaanderup P, Thorsen J, Roque JP, Montero JL (1997) Carbohydr Res 3000:175
264. Mouffouk F, Morère A, Vidal S, Leydet A, Montero JL (2004) Synth Commun 34:303
265. Barker GR, Gillam IC, Lord PA, Douglas T, Spoors JW (1960) J Chem Soc 3885

266. Shiozaki M, Deguchi N, Macindoe WM, Arai M, Miyazaki H, Mochizuki T, Tatsuta T, Ogawa J, Maeda H, Kurakata SI (1996) Carbohydr Res 283:27

267. Pulido R, Gotor V (1993) J Chem Soc Perkin Trans I 589

268. Morère A, Mouffouk F, Jeanjean A, Leydet A, Montero JL (2003) Carbohydr Res 338:2409

269. Guibe F, Saint M'Leux Y (1981) Tetrahedron Lett 22:3591

270. Adinolfi M, Barone G, Guariniello L, Iadonisi A (2000) Tetrahedron Lett 41:9305

271. Kunz H, Unverzagt C (1984) Angew Chem Int Ed Engl 23:436

272. Teshima T, Nakajima K, Takahashi M, Shiba T (1992) Tetrahedron Lett 33:363

273. Sridhar PR, Chandrasekaran S (2002) Org Lett 4:4731

274. Ramesh R, Bhat RG, Chandrasekaran S (2005) J Org Chem 70:837

275. Woodward RB, Heusler K, Gosteli J, Naegeli P, Oppolzer W, Ramage R, Ranganathan S, Vorbruggen H (1966) J Am Chem Soc 88:852

276. Fukase K, Yoshimura T, Kotani S, Kusumoto S (1994) Bull Chem Soc Jpn 67:473

277. Carpino LA, Han GY (1970) J Am Chem Soc 92:574

278. Roussel F, Knerr L, Grathwohl M, Schmidt RR (2000) Org Lett 2:3043

279. Zhu T, Boons GJ (2000) Tetrahedron: Asymmetry 11:199

280. Freese SJ, Vann WF (1996) Carbohydr Res 281:313

281. Gioeli C, Chattopadhyaya JB (1982) J Chem Soc Chem Commun 672

282. Roussel F, Takhi M, Schmidt RR (2001) J Org Chem 66:8540

283. Roussel F, Knerr L, Schmidt RR (2001) Eur J Org Chem 2067

284. Love KR, Seeberger PH (2004) Angew Chem Int Ed 43:602

285. Bufali S, Höleman A, Seeberger PH (2005) J Carbohydr Chem 24:441

286. LaLonde M, Chan TH (1985) Synthesis 817

287. Hwu RJR, Tsay SC, Cheng BL (1998) In: Rappoport Z, Apeloig Y (eds) Chemistry of Organic Silicon Compounds, vol 2. Wiley, New York, p 431

288. Larso GL Silicon-Based Blocking Agents. In: Suplement to the Gelest-Catalog (ABCR) Silicon, Germanium & Tin Compounds, Metal Alkoxides and Metal Diketonates; See http://www.gelest.com

289. Arias-Pérez MS, López MS, Santos MJ (2002) J Chem Soc Perkin Trans 2:1459

290. Ashton PR, Boyd SE, Gattuso G, Hartwell EY, Königar R, Spencer N, Stoddart JF (1995) J Org Chem 60:3898

291. Nelson TD, Crouch RD (1996) Synthesis 1031

292. Corey EJ, Venkateswarlu A (1972) J Am Chem Soc 94:6190

293. Corey EJ, Cho H, Rücker C, Hua DH (1981) Tetrahedron Lett 22:3455

294. Bredenkamp MWS (1995) Afr J Chem 48:154

295. Johnson DA, Taubner LM (1996) Tetrahedron Lett 37:605

296. Colombier C, Skrydstrup T, Beau JM (1994) Tetrahedron Lett 44:8167

297. Dahlhoff WV, Taba KM (1986) Synthesis 561

298. Chung MK, Orlova G, Goddard JD, Schlaf M, Harris R, Beveridge TJ, White G, Hallett FR (2002) J Am Chem Soc 124:10508

299. Chung MK, Schlaf M (2005) J Am Chem Soc 127:18085

300. Clive DLJ, Kellner D (1991) Tetrahedron Lett 32:7159

301. Kelly DR, Roberts SM, Newton RF (1979) Synth Commun 9:295

302. Collington EW, Finch H, Smith IJ (1985) Tetrahedron Lett 26:681

303. Shimshock SJ, Waitermire RE, Deshong P (1991) J Am Chem Soc 113:8791

304. Zhang W, Robins MJ (1992) Tetrahedron Lett 33:1177

305. Corey EJ, Yi KY (1992) Tetrahedron Lett 33:2289

306. Metcalf BW, Burkhart JP, Jund K (1980) Tetrahedron Lett 21:35

307. Peng Y, Li WD (2006) Synlett 1165

308. Fukuda Y, Shindo M, Shishido K (2003) Org Lett 5:749

309. Nakamura T, Shiozaki M (2001) Tetrahedron Lett 42:2701

310. Corey EJ, Roberts BE (1997) J Am Chem Soc 119:12425

311. Furstner A, Albert M, Mlynarski J, Metheu M, DeClerq E (2003) J Am Chem Soc 125:13132

312. Shahid KA, Mursheda J, Okazaki M, Shuto Y, Goto F, Kiyooka S (2002) Tetrahedron Lett 43:6377

313. Jackson SR, Johnson MG, Mikami M, Shiokawa S, Carreira EM (2001) Angew Chem Int Ed 40:2694

314. Yang YY, Yang WB, Teo CF, Lin CH (2000) Synlett 1634–1636

315. Oriyama T, Kobayashi Y, Noda K (1998) Synlett 1047
316. Bartoli G, Cupone G, Dalpozzo R, Nino AD, Maiuolo L, Procopio A, Sambri L, Tagarelli A (2002) Tetrahedron Lett 43:5945
317. Jadav JS, Reddy BVS, Madan C (2000) New J Chem 24:853
318. Crouch RD, Polizzi JM, Cleiman RA, Yi J, Romany CA (2002) Tetrahedron Lett 43:7151
319. Ranu BC, Jana U, Majee A (1999) Tetrahedron Lett 40:1985
320. Bartoli G, Bosco M, Marcantoni E, Sambri L, Torregiani E (1998) Synlett 209
321. Bajwa JS, Vivelo J, Slade J, Repic O, Blacklock T (2000) Tetrahedron Lett 41:6021
322. Crouch RD, Romany CA, Kreshock AC, Menkoni KA, Zile JL (2004) Tetrahedron Lett 45:1279
323. Jang JY, Wang YG (2003) Tetrahedron Lett 44:3859
324. Lee A SY, Yeh HC, Shie JJ (1998) Tetrahedron Lett 39:5249
325. Lipshutz BH, Keith J (1998) Tetrahedron Lett 39:2495
326. Hwu JR, Jain ML, Tsai FY, Tsay SC, Balkumar A, Hakimelahi GH (2000) J Org Chem 65:5077
327. Kumar GDK, Baskaran S (2005) J Org Chem 70:4520
328. Hanessian S, Lavallee P (1975) Can J Chem 53:2975
329. Mulzer J, Schöllhorn B (1990) Angew Chem Int Ed Engl 29:431
330. Hardinger SA, Wijaya N (1993) Tetrahedron Lett 34:3821
331. Crouch RD (2004) Tetrahedron 60:5833
332. Prakash C, Saleh S, Blair IA (1989) Tetrahedron Lett 30:19
333. Shekhani MS, Khan KM, Mahmood K, Shah PM, Malik S (1990) Tetrahedron Lett 31:1669
334. Rucker C (1995) Chem Rev 95:1009
335. Friesen RW, Sturino CF, Daljeet AK, Kolaczewska A (1991) J Org Chem 56:1944
336. Bennett F, Knight DW, Fenton G (1991) J Chem Soc Perkin Trans I 1543
337. Markiewicz WT (1979) J Chem Res (S) 24
338. Verdegaal CHM, Jansse PL, de Rooij JFM, van Boom JH (1980) Tetrahedron Lett 21:1571
339. van Boeckel CAA, van Boom JH (1985) Tetrahedron 21:4575
340. Trost BM, Caldwell CG (1981) Tetrahedron Lett 22:4999
341. Trost BM, Caldwell CG, Murayama E, Heissler D (1983) J Org Chem 48:3252
342. Furusawa K, Katsura T (1985) Tetrahedron Lett 26: 887
343. Corey EJ, Hopkins PB (1982) Tetrahedron Lett 23:4871
344. Kumagai D, Miyazaki M, Nishimura SI (2001) Tetrahedron Lett 42:1953
345. Imamura A, Ando H, Korogi S, Tanabe G, Muraoka O, Ishida H, Kiso M (2003) Tetrahedron Lett 44:6725
346. Imamura A, Kimura A, Ando H, Ishida H, Kiso M (2006) Chem Eur J 12:8862
347. Imamura A, Ando H, Ishida H, Kiso M (2005) Org Lett 7:4415
348. Slotin LA (1977) Synthesis 737
349. Lemmen P, Richter W, Werner B, Karl R, Stumpf R, Ugi I (1993) Synthesis 1
350. Edmundson RS (1979) In: Barton D, Ollis WD (eds) Comprehensive Organic Chemistry. Pergamon Press, Oxford, UK, vol 2, p 1267
351. Ireland RE, Muchmore DC, Hengartner U (1972) J Am Chem Soc 94:5098
352. Granata A, Perlin AS (1981) Carbohydr Res 94:165
353. Mora, N; Lacombe JM (1993) Tetrahedron Lett 34:2461
354. Schlimbrene BR , Miller SJ (2001) J Am Chem Soc 123:10125
355. Sabesan S, Neira S (1992) Carbohydr Res 223:169
356. Manning DD, Bertozzi CR, Rosen SD, Kiessling LL (1996) Tetrahedron lett 37:1953
357. Jones S, Seltsianos D, Thompson KJ, Toms SM (2003) J Org Chem 68:5211
358. Parang K (2002) Bioorg Med Chem Lett 12:1863
359. Ahmadibeni Y, Parang K (2005) J Org Chem 70:1100
360. Ahmadibeni Y, Parang K (2005) Org Lett 7:5589
361. Nifantiev EE, Grachev MK, Burmistrov SY (2000) Chem Rev 100:3755
362. Kochetkov NK, Nifantiev EE, Gudkova IP, Ivanona NL, Leskin VA (1972) Zh Obshch Khim 42:450
363. Haines AH, Massy DJR (1996) Synthesis 1422
364. Nifantiev EE, Gudkova IP, Chan DD (1972) Zh Obshch Khim 42:506
365. Nifantiev EE, Tuseev AP, Tarasov VV (1966) Zh Obshch Khim 36:1124
366. Nifantiev EE, Schegolev AA (1965) Vestn Mosk Univ Ser II: Khim 80
367. Kochetkov NK, Nifantiev EE , Koroteev MP, Zhane ZK, Borisenko AA (1976) Carbohydr Res 47:221

368. van Herk T, Hartog AF, van der Burg AM, Wever R (2005) Adv Synth Catal 347:1155
369. Fischer E (1893) Chem Ber 26:2400
370. Purdie T, Irvine JC (1903) J Chem Soc 83:1021
371. Haworth WN (1919) J Chem Soc 107:8
372. Roth D, Pigman W (1960) J Am Chem Soc 82:4608
373. Schmidt RR (1986) Angew Chem Int Ed Engl 25:212
374. Schmidt RR, Reichrath M, Moering V (1984) J Carbohydr Chem 3:67
375. Box VGS (1982) Heterocycles 19:1939
376. Schmidt RR, Reichrath M, Moering U (1980) Tetrahedron Lett 21:3501
377. Klotz W, Schmidt RR (1994) J Carbohydr Chem 13:1093
378. Klotz W, Schmidt RR (1991) Synlett 168
379. Klotz W, Schmidt RR (1993) Liebigs Ann Chem 683
380. Lu W, Navidpour L, Taylor SD (2005) Carbohydr Res 340:1213
381. Esswein A, Rembold H, Schmidt RR (1990) Carbohydr Res 200:287
382. Srivastava VK, Schuerch C (1979) Tetrahedron Lett 35:3269
383. Mukaiyama T, Hashimoto Y, Hayashi Y, Shoda S (1984) Chem Lett 557
384. Huchel U, Schmidt C, Schmidt RR (1998) Eur J Org Chem
385. Sharma SK, Corrales G, Penadés S (1995) Tetrahedron Lett 36:5627
386. Koeners HJ, de Kok AJ, Romers C, van Boom JH (1980) Recl Trav Cim Pays-Bas 99:355
387. Peterson L, Jensen KJ (2001) J Org Chem 66:6268
388. Jacobsson M, Malmberg J, Ullervik U (2006) Carbohydr Res 341:1266
389. Capon B (1969) Chem Rev 69:407
390. BeMiller JM (1967) Adv Carbohydr Chem 22:25
391. Bochkov AF, Zaikov GE (1979) Chemistry of the O-glycosidic bond. Pergamon Press, Oxford
392. Garegg PJ (2004) Adv Carbohydr Chem Biochem 59:69
393. Freudenberg K, Dürr W, von Hochstetter H (1928) Chem Ber 61:1735
394. Richtmyer NK (1934) J Am Chem Soc 56:1633
395. Gómez AM, Danelón GO, Valverde S, López JC (1999) Carbohydr Res 320:138
396. Classon B, Garegg PJ, Samuelsson B (1984) Acta Chem Scand B 38:419
397. Zehavi U, Amit B, Patchornik A (1972) J Org Chem 37:2281
398. Rodebaugh R, Fraser-Reid B, Geysen HM (1997) Tetrahedron Lett 38:7653
399. Kim KS, Kim JH, Lee YJ, Lee YJ, Park J (2001) J Am Chem Soc 123 8477
400. Stanek JJr (1990) Topics Curr Chem 154: 209
401. Gigg J, Gigg R (1966) J Chem Soc C 82
402. Nakayama K, Uoto K, Higashi K, Soga T, Kusama T (1992) Chem Pharm Bull 40:1718
403. Lüning J, Möller U, Debski N, Welzel P (1993) Tetrahedron Lett 37:5871
404. Yu B, Zhang J, Lu S, Hui Y (1998) Synlett 29
405. MootooDR, Date V, Fraser-Reid B (1988) J Am Chem Soc 110:266
406. López JC, Fraser-Reid B (1991) J Chem Soc Chem Commun 159
407. Kunz H, Werning P, Schultz M (1990) Synlett 631
408. Lipshutz BH, Pegram JJ, Morey MC (1981) Tetrahedron Lett 22:4463
409. Jansson K, Ahlfors S, Fredj T, Kihlberg J, Magnusson G (1988) J Org Chem 53:5629
410. Behrend R, Roth P (1904) Liebigs Ann Chem 331:359
411. Schlubach HH, Prochownick V (1930) Chem Ber 63:2298
412. Levene PA, Meyer GM (1928) J Biol Chem 76:513
413. Fletcher HGJr (1953) J Am Chem Soc 75:2624
414. Sano T, Ohashi K, Oriyama T (1999) Synthesis 7:1141
415. Hung SCh, Thopate SR, Wang ChCh (2001) Carbohydr Res 330:177
416. Luo SY, Kulkarni SS, Chou ChH, Liao WM, Hung SCh (2006) J Org Chem 71:1226
417. Hassner A, Alexanian V (1978) Tetrahedron Lett 4475
418. Ziegler FE, Berger GD (1979) Synth Commun 9: 539
419. Kim KS, Lee YJ, Kim HY, Kang SS, Kwon SY (2004) Org Biomol Chem 2:2408
420. Carpino LA (1993) J Am Chem Soc 115:4397
421. Perrie JA, Harding JR, Holt DW, Johnston A, Meath P, Stachulski AV (2005) Org Lett 7:2591
422. Shoda S, Mukaiyama T (1982) Chem Lett 6:861
423. Pfeffer PE, Rothman ES, Moore GG (1970) J Org Chem 2925
424. Barrett AGM, Bezuidenhoudt BCB, Gasieki AF, Howell AR, Russel MA (1989) J Am Chem Soc 111:1392
425. Ogawa T, Nozaki M, Matsui M (1978) Carbohydr Res 60:C7
426. Barrett AGM, Bezoindehoudt BCB (1989) Heterocycles 28:209

427. Schmidt RR, Kinzy W (1994) Adv Carbohydr Chem Biochem 50:21
428. Schmidt RR, Michel J (1980) Angew Chem Int Ed Engl 731
429. Schmidt RR, Michel J (1984) Tetrahedron Lett 25:821
430. Ellervik U, Magnusson G (1993) Acta Chem Scand 47:826
431. Smith AB, Hale KJ, Rivero RA (1986) Tetrahedron Lett 27:5813
432. Shaw JF, Klibanov AM (1987) Biotech Bioeng 29:648
433. Hennen WJ, Sweers HM, Wang YF, Wong CH (1988) J Org Chem 53:4939
434. Khan R, Gropen L, Konowicz PA, Matulovà M, Paoletti S (1993) Tetrahedron Lett 34:7767
435. Rowell RM, Feather MS (1967) Carbohydr Res 4:486
436. Excoffier G, Gagnaire D, Utille JP (1975) Carbohydr Res 39:368
437. Khan R, Konowicz PA, Gardossi L, Matulovà M, Paoletti S (1994) 35:4247
438. Ferrer Salat C, Exero Agneseti P, Bemborad Caniato M (1976) Spanish Patent 430 636, Laboratorios Ferrer SL: CA (1977) 87:23683k
439. Ishido Y, Sakairi N, Sekiya M, Nakazaki N (1981) Carbohydr Res 97:51
440. Fiandor J, García López MT, de las Heras FG, Méndez Castrillón PP (1985) Synthesis 1121
441. Grynkiewicz G, Fokt I, Szeja W, Fitak H (1989) J Chem Res S 152
442. Watanabe K, Itoh K, Araki Y, Ishido Y (1986) Carbohydr Res 154:165
443. Itoh T, Takamura H, Watanabe K, Araki Y, Ishido Y (1986) Carbohydr Res 156:241
444. Nudelman A, Herzig J, Gottlieb HE, Keinan E, Sterling J (1987) Carbohydr Res 162:145
445. Mikamo M (1989) Carbohydr Res 191:150
446. Chittaboina S, Hodges B, Wang Q (2006) Let Organ Chem 3:35
447. Sambarah T, Fanwick PE, Cushman M (2001) Synthesis 1450
448. Herzig J, Nudelman A (1986) Carbohydr Res 153:162
449. Avalos M, Babiano R, Cintas P, Jiménez JL, Palacios JC, Valencia C (1993) Tetrahedron Lett 34:1359
450. Banaszek A, Bordas-Cornet X, Zamojski A (1985) Carbohydr Res 144:342
451. Lemieux RU, Brice C (1955) Can J Chem 33:109
452. Zemplén G, Laszlo ED (1915) Chem Ber 48:915
453. Boursier M, Descotes G (1989) C R Acad Sci 308:919
454. Lou B, Huynh HK, Hanessian S (1997) Oligosaccharide synthesis by remote activation: O-protected glycosyl 2-thiopyridylcarbonate donors. In Hanessian S (ed) Preparative carbohydrate chemistry. Marcel Dekker, New York, chap 19
455. Zhang J, Liang X, Wang D, Kong F (2007) Carbohydr Res 342:797
456. Iimori T, Shibazaki T, Ikegami S (1997) Tetrahedron Lett 38:2943
457. Pougny JR (1986) J Carbohydr Chem 5:529
458. Lassaletta JM, Schmidt RR (1995) Synlett 925
459. Nikolaev AV, Botvinko IV, Ross AJ (2007) Carbohydr Res 342:297
460. Putman EW (1963) Methods Carbohydr Chem 2:267
461. McDonald DL (1962) J Org Chem 27:1107
462. Volkova LV, Danilov LL, Evstigneeva RP (1974) Carbohydr Res 32:165
463. Putman EW (1963) Methods Carbohydr Chem 2:261
464. Schmidt RR, Stumpp M (1984) Liebigs Ann Chem 680
465. Boons GJ, Burton A, Wyatt P (1996) Synlett 310
466. Plante OJ, Palmacci ER, Seeberger PH (2001) Science 291:1523
467. Ravida A, Liu X, Kovacs L, Seeberger PH (2006) Org Lett 8:1815
468. Granata A, Perlin AS (1981) Carbohydr Res 94:165
469. Inage M, Chaki H, Kusumoto S, Shiba T (1982) Chem Lett 1281
470. Ogawa T, Seta A (1982) Carbohydr Res 110:C1
471. Westerdouin P, Veeneman GH, Marugg JE, van der Marel GA, van Boom JH (1986) Tetrahedron Lett 27:1211
472. Ichikawa Y, Sim MM, Wong CH (1992) J Org Chem 57:2943
473. Müller T, Hummel G, Schmidt RR (1994) Liebigs Ann Chem 325

2.2 Oxidation, Reduction, and Deoxygenation

Robert Madsen
Department of Chemistry, Center for Sustainable and Green Chemistry,
Technical University of Denmark, Lyngby 2800, Denmark
rm@kemi.dtu.dk

Abstract

In this chapter, methods for oxidation, reduction, and deoxygenation of carbohydrates are presented. In most cases, the reactions have been used on aldoses and their derivatives including glycosides, uronic acids, glycals, and other unsaturated monosaccharides. A number of reactions have also been applied to aldonolactones. The methods include both chemical and enzymatic procedures and some of these can be applied for regioselective transformation of unprotected or partially protected carbohydrates.

Keywords

Azidonitration; Deoxygenation; Dihydroxylation; Epoxidation; Hydrogenation; Oxidation; Photobromination; Reduction

In: *Glycoscience*. Fraser-Reid B, Tatsuta K, Thiem J (eds)
Chapter-DOI 10-1007/978-3-540-30429-6_4: © Springer-Verlag Berlin Heidelberg 2008

Abbreviations

AIBN	2,2′-azobisisobutyronitrile
BSTFA	bis(trimethylsilyl)trifluoroacetamide
CAN	ceric ammonium nitrate
DCC	N,N′-dicyclohexylcarbodiimide
DIBALH	diisobutylaluminum hydride
DMF	dimethylformamide
DMSO	dimethyl sulfoxide
HMPA	hexamethylphosphoric triamide
MS	molecular sieves
m-CPBA	m-chloroperoxybenzoic acid
NBS	N-bromosuccinimide
NIS	N-iodosuccinimide
NMO	N-methyl-morpholine N-oxide
PCC	pyridinium chlorochromate
PDC	pyridinium dichromate
TFAA	trifluoroacetic anhydride
TEMPO	2,2,6,6-tetramethyl-1-piperidinyloxy
TPAP	tetrapropylammonium perruthenate

1 Introduction

Carbohydrates have found valuable applications as enantiomerically pure starting materials for synthesis of non-carbohydrate compounds. Through chemical synthesis carbohydrates can be transformed into versatile synthetic intermediates that have functional groups and stereogenic centers structured in a framework found in many natural products, and the entire repertoire of common reactions in organic chemistry can be performed on carbohydrates [1]. However, carbohydrates are also densely functionalized molecules and as such constitute a special synthetic challenge. Regio- and chemoselectivity problems are frequently encountered which often makes it necessary to protect functionalities not involved in the desired transformation. In the following, oxidations, reductions, and deoxygenations will be discussed. An eight-volume comprehensive encyclopedia is recommended for further information on the different reagents for these reactions [2].

2 Oxidations

A large variety of oxidation procedures are available to the synthetic organic chemist [3]. As presented in this chapter, many of these methods can be applied to protected carbohydrates and some also for regioselective oxidation of unprotected or partially protected sugars.

2.1 Oxidation at the Anomeric Center

Unprotected aldoses can be selectively oxidized at the anomeric center to afford aldonic acids/aldonolactones. This oxidation can be achieved by chemical as well as by biochemical

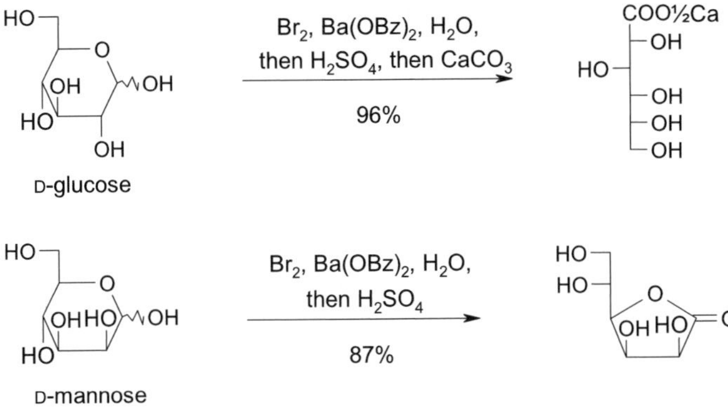

■ Scheme 1

methods. A common laboratory procedure for oxidation of aldoses uses 1.1 equiv. of bromine in an aqueous solution in the presence of an acid scavenger such as barium benzoate, barium carbonate, or calcium carbonate [4]. The scavenger is necessary since the liberated hydrobromic acid lowers the rate of the oxidation reaction. These conditions are very selective for the anomeric center and will not oxidize other hydroxy groups in the aldose. The product is typically isolated by crystallization either as the aldonolactone or as a salt of the aldonic acid [5]. It should be noted that aldonolactones usually exist as the five-membered 1,4-lactone contrary to aldoses which prefer the six-membered pyranose form. Gluconolactone is an important exception from the rule since it crystallizes as the 1,5-lactone. The oxidation with bromine takes place on the cyclic form of the aldose and not with the free aldehyde. Furthermore, the β-pyranose is oxidized faster than the α-pyranose for all the common aldoses [6]. Therefore, the initial product is the 1,5-lactone which will either ring-open to form a salt of the aldonic acid or rearrange to the thermodynamically more stable 1,4-lactone (● Scheme 1) [4]. The process can also be turned into a catalytic procedure by using an electrochemical oxidation of calcium bromide to generate bromine in a solution with the aldose and calcium carbonate [7].

Another catalytic method makes use of a homogeneous dehydrogenation catalyst in the presence of a hydrogen acceptor. The complex RhH(PPh$_3$)$_4$ catalyzes a clean dehydrogenation of unprotected aldoses into aldono-1,4-lactones in DMF [8]. Benzalacetone (PhCH=CHCOCH$_3$) serves as the hydrogen acceptor and is converted into 4-phenylbutan-2-one during the course of the reaction.

Although homogeneous catalysts are often used at the laboratory scale industrial applications usually prefer a heterogeneous catalyst due to the easy separation from the product and the recovery of the catalyst. Many heterogeneous catalysts have been studied for aerobic oxidation of unprotected aldoses. The favored catalysts are Pd/C and Au/C, which show very high selectivity for the hemiacetal function [9,10]. A drawback with Pd/C, however, is catalyst deactivation. This can be circumvented by promoting the catalyst with bismuth which seems to improve the catalyst performance by coordinating with the substrate [11]. Thus, aerobic oxidation of glucose over a Bi-Pd/C catalyst at pH 9 with continuous addition of sodium hydroxide gives rise to sodium gluconate in 99% yield (● Scheme 2) [9]. The catalyst can be

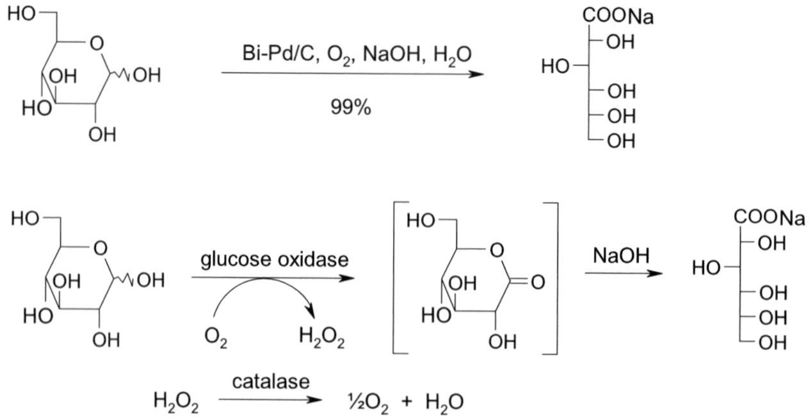

Scheme 2

recovered and reused five times without affecting the yield. The same oxidation can also be performed with an Au/C catalyst to afford sodium gluconate in a near quantitative yield, but in this case the reused catalyst reacts more slowly in the following runs [10]. When the oxidation of glucose is carried out with the parent Pd/C catalyst the yield drops to 78% due to incomplete conversion [9].

Despite the advances in heterogeneous catalysis the preferred industrial procedure for oxidation of glucose is still aerobic fermentation. The microbial process is usually performed with *Aspergillus niger* by the use of submerged fermentation [12]. The obtained glucono-lactone/gluconic acid will inhibit the fungal growth and is therefore neutralized with sodium hydroxide in order to maintain the pH of the growth medium around 6. The process is highly efficient and is able to oxidize glucose at a rate of 15 g/L per hour [12].

The oxidation can also be carried out by using an enzymatic reaction with the two enzymes that are responsible for the microbial oxidation: glucose oxidase and catalase (● *Scheme 2*) [13]. Glucose oxidase dehydrogenates glucose to gluconolactone by simultaneous reduction of dioxygen to hydrogen peroxide. The liberated peroxide is an inhibitor of glucose oxidase and must be removed in order to obtain a good conversion. This is achieved by decomposition with catalase and the conversion of glucose to gluconate is nearly quantitative under these conditions [12]. Glucose oxidase is very specific for D-glucose, but with a prolonged reaction time similar aldoses can also be oxidized in good yield including D-xylose, D-mannose, D-galactose, and D-glucosamine [14].

Protection of aldoses at the non-anomeric positions makes it possible to use many of the common procedures in organic chemistry for oxidizing lactols as shown with mannofuranose **1** and glucopyranose **3** (● *Table 1*). The reactions can be divided into three main categories: oxidations mediated by activated dimethyl sulfoxide (DMSO), oxidations with chromium(VI) oxides, and oxidations catalyzed by ruthenium oxides. The DMSO-mediated oxidations of alcohols can be promoted by several activators [27]. With the partially protected aldoses the activation has mainly been achieved with acetic anhydride and oxalyl chloride. Competing β-elimination does usually not occur unless the eliminating group is an ester, e. g., an acetate or a benzoate [27].

◻ **Table 1**
Oxidation of diisopropylidenemannofuranose 1 and tetrabenzylglucopyranose 3 to lactones 2 and 4

Substrate	Reagent	Solvent	Yield (%)	Reference
1	DMSO, Ac$_2$O	DMSO	79–96	[15,16]
3	DMSO, Ac$_2$O	DMSO	84	[17]
1	DMSO, (COCl)$_2$; Et$_3$N	CH$_2$Cl$_2$	82	[18]
3	DMSO, (COCl)$_2$; Et$_3$N	CH$_2$Cl$_2$	60	[19]
1	CrO$_3$·2C$_5$H$_5$N	CH$_2$Cl$_2$	80	[16]
1	CrO$_3$·2C$_5$H$_5$N, Ac$_2$O	CH$_2$Cl$_2$	97	[20]
3	PDC, 3 Å MS	CH$_2$Cl$_2$	86	[19]
1	PCC, 3 Å MS	CH$_2$Cl$_2$	86	[21]
3	PCC, 3 Å MS	CH$_2$Cl$_2$	95	[22]
1	RuO$_2$, NaIO$_4$	H$_2$O/CHCl$_3$/CCl$_4$	79	[23]
1	TPAP, NMO	MeCN	88	[24]
3	TPAP, NMO	CH$_2$Cl$_2$	94	[19]
1	Dess–Martin periodinane	CH$_2$Cl$_2$	95	[25]
1	RhH(PPh$_3$)$_4$, benzalacetone	DMF	93	[8]
1	NiO(OH)[a], K$_2$CO$_3$	H$_2$O	80	[26]

[a] Generated electrochemically

The chromium-mediated oxidations can be performed with a number of chromium(VI) reagents [28]. The Collins reagent (CrO$_3$.2C$_5$H$_5$N) is not very effective at oxidizing aldoses and 6 equiv. are needed in order to oxidize **1** in 30 min [16]. The reactivity of the reagent can be enhanced by adding acetic anhydride which makes it possible to oxidize **1** with 4 equiv. of the reagent in 5–10 min [20]. The same number of equivalents is usually required with pyridinium dichromate (PDC) and pyridinium chlorochromate (PCC) which are both commercially available [28]. In all reactions with chromium reagents the work-up is rather tedious and a significant amount of toxic waste is produced. Upon completion of the reaction most of the chromium byproducts are precipitated as a tarry mass which is then followed by purification of the products by silica gel flash chromatography.

In view of the purification and waste disposal problems with the chromium oxidations catalytic methods with ruthenium catalysts are more attractive. Ruthenium(VIII) oxide is a strong oxidant that will also oxidize alkenes, alkynes, sulfides, and in some cases benzyl ethers. The method is compatible with glycosidic linkages, esters and acetals, and is usually carried out in a biphasic solvent system consisting of water and a chlorinated solvent. Acetonitrile or a phase-transfer catalyst has been shown to further promote the oxidation [29,30]. Normally, a periodate or a hypochlorite salt serve as the stoichiometric oxidant generating ruthenium(VIII) oxide from either ruthenium(IV) oxide or ruthenium(III) chloride [30].

Another ruthenium-catalyzed oxidation uses tetrapropylammonium perruthenate (TPAP) [24]. Being a ruthenium(VII) oxide, the perruthenate ion is a less powerful oxidant than ruthenium(VIII) oxide and more functional groups are stable to the oxidation conditions, including alkenes, alkynes, amines, amides, benzyl, trityl and silyl ethers [24]. However, alcohols and lactols still undergo oxidations in high yield with N-methyl-morpholine N-oxide (NMO) as the stoichiometric oxidant. The reactions are usually carried out in dichloromethane, acetonitrile, or mixtures of both in the presence of molecular sieves [24].

A number of special oxidation methods have also been applied to partially protected aldoses. The Dess–Martin periodinane, 1,1,1-triacetoxy-1,1-dihydro-1,2-benziodoxol-3(1H)-one [31], is a mild and efficient oxidant compatible with carbohydrate protecting groups [25]. However, it is also quite expensive and should only be used if the other procedures fail. The hydrogen transfer reactions catalyzed by 5% of RhH(PPh$_3$)$_4$ can also be applied to partially protected aldoses [8].

In all the above methods for oxidizing carbohydrates a stoichiometric oxidant is added to the reaction mixture. This can be avoided by using an electrochemical oxidation. A nickel hydroxide electrode has been applied for oxidizing isopropylidene-protected carbohydrates in aqueous base [26]. While secondary hydroxy groups fail to react under these conditions, the hemiacetal at the anomeric center is oxidized to the lactone in good yield [26].

Besides aldoses methyl glycosides and glycals can also be oxidized at the anomeric center. Peracetylated methyl β-D-glucopyranoside 5 reacts with ozone at the anomeric center to give the corresponding open chain methyl ester [32]. When the reaction is performed with 2 equiv. of chromium(VI) oxide further oxidation occurs to give keto ester 6 in quantitative yield (❂ Scheme 3) [33]. Interestingly, the α-anomer of 5 does not react under these conditions. Ester-protected glycals or 2-hydroxyglycals react selectively with m-chloroperoxybenzoic acid (m-CPBA) in the presence of borontrifluoride etherate to afford α,β-unsaturated lactones (❂ Scheme 3) [34]. The Lewis acid mediates an allylic rearrangement which is followed by

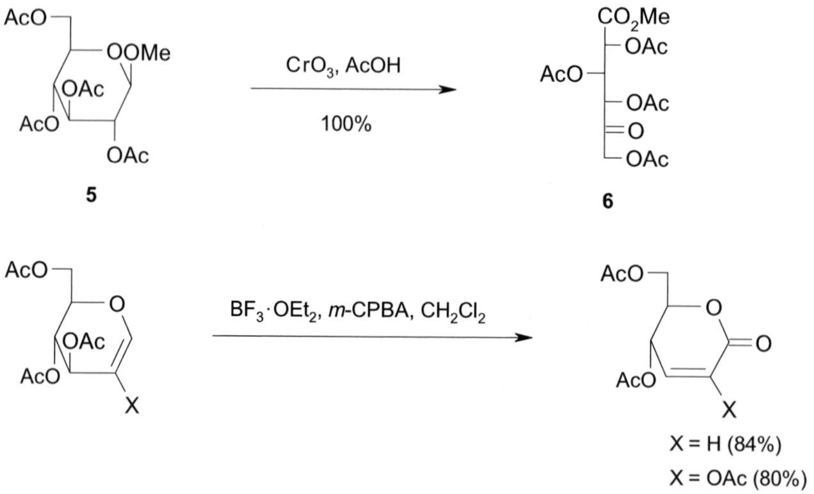

❑ Scheme 3

oxidation at the anomeric center by the peracid. The same transformation can be achieved with indium(III) chloride and 2-iodoxybenzoic acid [35].

2.2 Oxidation of Primary Alcohols to Aldehydes

An easily available protected carbohydrate containing a primary hydroxy group is diisopropy-lidenegalactopyranose 7. Oxidation of 7 to the corresponding aldehyde 8 illustrates very well the different reagents available for this transformation (❯ Table 2). Activated DMSO-mediated oxidations are usually the method of choice for converting a primary alcohol into an alde-hyde [27]. The conditions are mild and no overoxidation to the carboxylic acid occurs. The original Moffatt procedure uses N,N'-dicyclohexylcarbodiimide (DCC) in the presence of a proton source as activator. A good yield is obtained with alcohol 7, but with some alco-hols a significant amount of the corresponding methylthiomethyl ether is formed [27]. The urea byproduct can also be difficult to separate from a protected carbohydrate aldehyde. Acti-vation with sulfur trioxide or oxalyl chloride gives a more straightforward work-up and these procedures also give rise to very little of the methylthiomethyl ether byproduct 9 [27]. The Swern procedure (oxalyl chloride) is the most widely used protocol due to its combination of high reactivity with inexpensive reagents. To avoid the methylthiomethyl byproduct, the Swern oxidation is normally carried out at low temperature. A tertiary amine is added in the last step of the oxidation process. Triethylamine is most commonly used, but more hindered amines have been reported to give better yields in some cases [27]. These basic conditions can cause epimerization of the aldehyde or β-elimination. Although these side reactions have not been observed in the oxidation of 7, a similar substrate without the 1,2-isopropylidene group gave exclusively the β-elimination product when using the Swern procedure [42]. Acetic anhydride can also be used to activate DMSO, and this procedure has been used for a number of carbohy-drate alcohols [27,42]. However, the reaction is slow and yields often moderate due to signifi-

◻ Table 2
Oxidation of diisopropylidenegalactopyranose 7 to aldehyde 8

Reagent	Solvent	Yield (%)	Reference
DMSO, DCC, pyridine·HCl	DMSO	83–87	[36]
DMSO, SO$_3$·pyridine; Et$_3$N	DMSO	85	[37]
DMSO, (COCl)$_2$; Et$_3$N	CH$_2$Cl$_2$	82	[38]
CrO$_3$, pyridine, Ac$_2$O	CH$_2$Cl$_2$	84–93	[39,20]
PDC, Ac$_2$O	CH$_2$Cl$_2$/DMF	71	[40]
Pb(OAc)$_4$	Pyridine	74	[41]

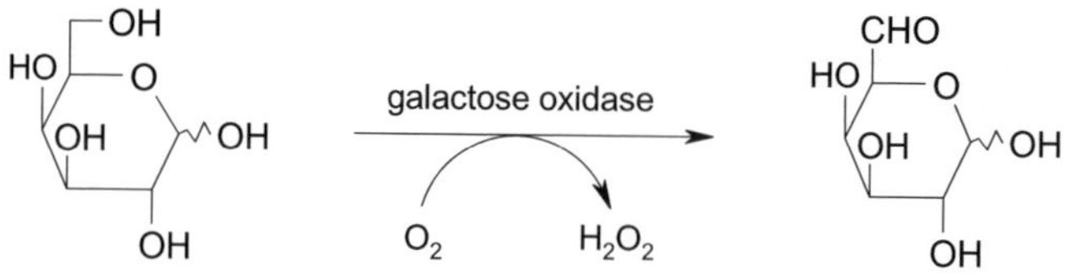

Scheme 4

cant formation of the methylthiomethyl ether and the acetate of the starting alcohol [27,42]. In fact, oxidation of **7** with DMSO/acetic anhydride gave methylthiomethyl ether **9** as the major product together with smaller amounts of the desired aldehyde **8** and the acetate of **7** [43].

Heavy metal reagents can also be used for oxidation of primary alcohols to aldehydes. Although experimentally less attractive than the Swern procedure, good yields can be obtained. Particularly, the chromium(VI)-based oxidants when activated by acetic anhydride have found use for carbohydrate alcohol oxidations even on large scale [20,39]. The major side reaction is overoxidation to the corresponding carboxylic acid. Contrary to these chromium reagents, lead(IV) acetate is generally not very reactive for oxidation of carbohydrate alcohols although oxidation of **7** to **8** has been achieved in a satisfactory yield [41].

The TPAP/NMO system [24] and the Dess–Martin periodinane [31] have been widely applied for oxidizing alcohols in complex natural product synthesis. Although both reagents are commercially available, they have so far found relatively little use in carbohydrate chemistry for oxidation of primary alcohols to aldehydes [44,45].

Unprotected or partially protected carbohydrates cannot generally be oxidized to the aldehyde using the chemical methods described above. Instead, an enzymatic oxidation can be used in some cases. Galactose oxidase catalyzes the oxidation of certain primary alcohols to aldehydes (⊙ *Scheme 4*) [46]. The enzyme is commercially available and is usually isolated from the fungus *Dactylium dendroides*. It is very specific for D-galactose and its derivatives including D-galactopyranosides, 2-deoxy-D-galactose, *N*-acetyl-D-galactosamine, D-talose, and a few alditols [46,47]. Other hexoses or pentoses do not undergo oxidation in the presence of galactose oxidase. The enzyme catalase is usually also added in the oxidation to decompose the hydrogen peroxide liberated during the reaction.

2.3 Oxidation of Primary Alcohols to Carboxylic Acids

Oxidation of primary alcohols with stronger oxidizing agents gives carboxylic acids. Several methods are illustrated for the oxidation of diisopropylidenegalactopyranose **7** and -sorbofuranose **11** (⊙ *Table 3*). Oxidation of the latter is an important step in the synthesis of vitamin C (ascorbic acid) from D-glucose [49]. The electrochemical oxidation of **11** with nickel(III) oxide hydroxide has been applied on an industrial scale using various nickel electrodes and chemical reactors [52]. On a laboratory scale, however, the same oxidation can be conveniently accomplished with a catalytic amount of nickel(II) chloride and sodium hypochlorite as the stoichiometric oxidant [53]. Potassium permanganate and ruthenium(VIII) oxide are also strong oxidants for converting a primary alcohol to the carboxylic acid [2]. Both reagents are rather non-selective and will also oxidize olefins, sulfides, and in some cases benzyl

□ **Table 3**
Oxidation of diisopropylidenegalactopyranose 7 and -sorbofuranose 11 to carboxylic acids 10 and 12

7: R = CH$_2$OH
10: R = COOH

11: R = CH$_2$OH
12: R = COOH

Substrate	Reagent	Solvent	Yield (%)	Reference
7	KMnO$_4$, NaOH, Bu$_4$NBr	H$_2$O/CH$_2$Cl$_2$	85	[48]
11	KMnO$_4$, KOH	H$_2$O	91	[49]
7	RuCl$_3$, NaIO$_4$	H$_2$O/MeCN/CHCl$_3$	82	[50]
7	RuO$_2$, NaOCl[a]	H$_2$O/CCl$_4$	86	[51]
11	RuO$_2$, NaOCl[a]	H$_2$O/CCl$_4$	83	[51]
7	NiO(OH)[a], KOH	H$_2$O	93	[26]
11	NiO(OH)[a], KOH	H$_2$O	90–96	[52]
11	NiCl$_2$, NaOCl, NaOH	H$_2$O	90	[53]

[a]Generated electrochemically

ethers. Ruthenium(VIII) oxide is used catalytically and a biphasic solvent system of carbon tetrachloride, acetonitrile, and water has proven to be beneficial for the oxidation [29,54] although successful ruthenium(VIII) oxide oxidations have also been accomplished in aqueous acetone [55].

When acid labile protecting groups are not present in the substrate, the Jones oxidation [chromium(VI) oxide, sulfuric acid] can be applied for preparation of uronic acids. Isopropylidene acetals are normally cleaved to some extent under these conditions [56]. The method usually requires an excess reagent (2–5 equiv.) to drive the oxidation to completion. For example, the Jones oxidation of methyl and allyl 2,3,4-tri-O-benzyl-α-D-glucopyranoside occurs with 2 equiv. of reagent to give the uronic acids in good yields [57]. Cleavage of acid labile protecting groups during the reaction can in some cases be an advantage. Worthy of note is the direct oxidation of trityl ether **13** to the uronic acid isolated as methyl ester **14** (❍ *Scheme 5*) [56]. The Jones oxidation can also be used on thioglycosides without concomitant oxidation at sulfur [58].

Another chromium(VI) oxidant for the preparation of uronic acids is PDC [28]. The oxidation is carried out in an aprotic solvent like dimethylformamide (DMF) or dichloromethane. Acetals and sulfides are stable under these conditions [59]. Although PDC is also used for oxidation of primary alcohols to aldehydes, use of a larger excess and/or a longer reaction time will give the carboxylic acid [60]. PDC can be further activated by addition of acetic anhydride which will shorten the reaction time [61]. If *tert*-butanol is added to the reaction, the *tert*-butyl ester can be obtained directly as shown by the conversion of **15** into **16** (❍ *Scheme 5*) [62]. Presumably, the intermediate aldehyde forms a hemiacetal with *tert*-butanol which is then further oxidized to the ester.

In some cases a two-step protocol is a milder procedure for oxidation of a primary alcohol to a carboxylic acid. The first step is then usually a Swern oxidation of the alcohol to the

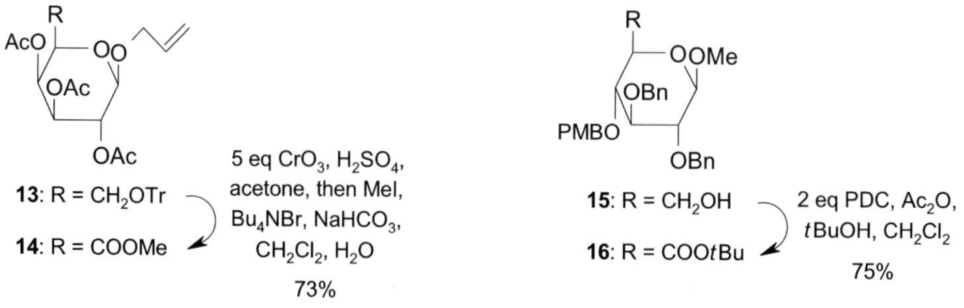

R
AcO ―OO
 OAc
  OAc
13: R = CH₂OTr
14: R = COOMe

5 eq CrO₃, H₂SO₄,
acetone, then MeI,
Bu₄NBr, NaHCO₃,
CH₂Cl₂, H₂O
73%

R
 ―OOMe
 OBn
PMBO
 OBn
15: R = CH₂OH
16: R = COOtBu

2 eq PDC, Ac₂O,
tBuOH, CH₂Cl₂
75%

◻ Scheme 5

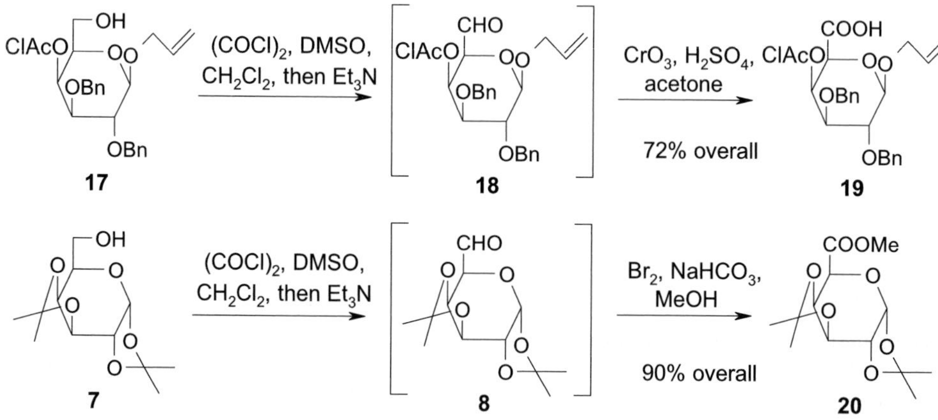

17: ClAcO ―OO, OBn, OBn, OH → (COCl)₂, DMSO, CH₂Cl₂, then Et₃N → [18: ClAcO, OBn, OBn, CHO] → CrO₃, H₂SO₄, acetone, 72% overall → [19: ClAcO, OBn, OBn, COOH]

7: O, O, O, O, OH → (COCl)₂, DMSO, CH₂Cl₂, then Et₃N → [8: CHO] → Br₂, NaHCO₃, MeOH, 90% overall → [20: COOMe]

◻ Scheme 6

aldehyde. This is not purified, but taken on directly to the next step. Oxidation of the aldehyde to the uronic acid is now an easier task than the direct oxidation of the starting alcohol. For example, Jones oxidation of alcohol 17 is a sluggish reaction accompanied by significant chloroacetyl migration [63]. However, Jones oxidation of aldehyde 18 proceeds readily to give uronic acid 19 in good overall yield (❍ Scheme 6) [63]. The latter aldehyde oxidation can also be achieved effectively with sodium chlorite [64] which has been applied in the oxidation of complex oligosaccharides [65]. By addition of bromine and methanol to the aldehyde, the methyl ester is obtained directly, e. g., 7 → 20 (❍ Scheme 6) [66]. The reaction proceeds through the hemiacetal which is more readily oxidized than the starting aldehyde or methanol. In this reaction bromine can be replaced with PDC which has been used for oxidation of thioglycosides that do not tolerate treatment with bromine [67].

Unprotected or partially protected glycosides cannot generally be oxidized to uronic acid by the above-described methods. However, for unprotected carbohydrates milder and more selective oxidants have been developed that take advantage of the primary alcohol function being more sterically accessible. An important reagent for this transformation is 2,2,6,6-tetramethyl-1-piperidinyloxy (TEMPO) which is a shelf-stable and commercially available nitrosyl radical

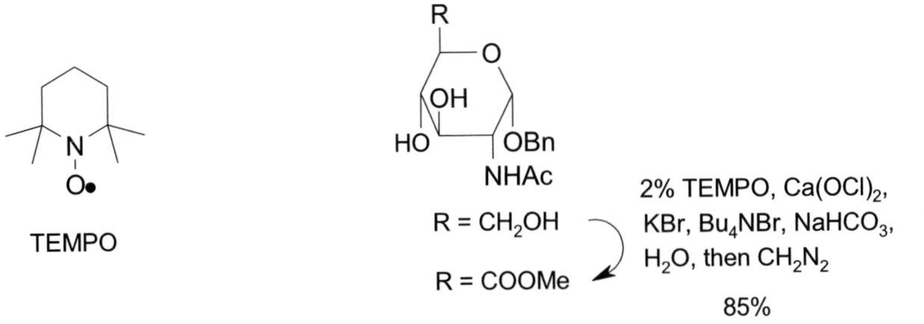

☐ Scheme 7

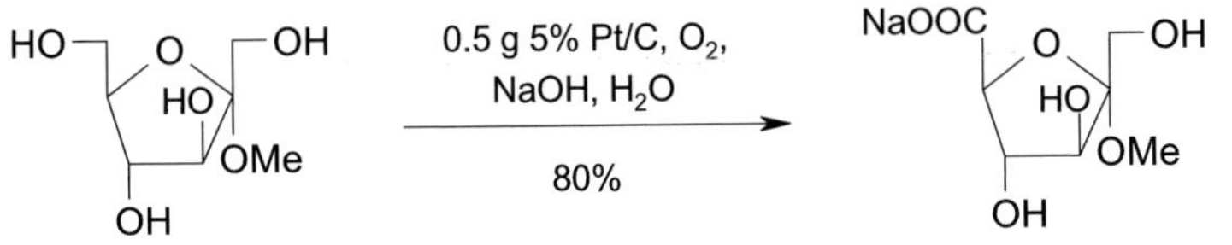

☐ Scheme 8

(❯ *Scheme 7*) [68,69]. It is soluble in water and used catalytically with sodium or calcium hypochlorite as the stoichiometric oxidant in the presence of bromide ions. Benzyl ethers, ester groups, olefins, and azides are stable to the TEMPO oxidation conditions [69,70] while thioglycosides may undergo oxidation at sulfur [58]. TEMPO is very regioselective for oxidation of the primary hydroxy group and has been successfully applied for oxidation of naturally occurring polysaccharides [71]. A drawback with the TEMPO procedure is the need for several inorganic salts in the reaction mixture that can be difficult to remove in the work-up. This can be circumvented by using (diacetoxyiodo)benzene as the stoichiometric oxidant in a dichloromethane/water mixture [72]. Under these conditions, thioglycosides can be converted into uronic acids without simultaneous oxidation to the sulfoxide/sulfone.

Another method for selective oxidation of carbohydrate primary alcohols involves platinum-catalyzed oxidation with oxygen in water [73]. Typically, Pt/C is used as the platinum source. The hemiacetal group at the anomeric center is preferentially oxidized, but when this is blocked the primary hydroxy group will undergo oxidation with very high selectivity over the secondary hydroxy groups [74]. Even if two primary hydroxy groups are present, as in methyl α-D-fructofuranoside **21**, the sterically most accessible group will undergo oxidation in high yield (❯ *Scheme 8*) [75]. Olefins are stable to the reaction conditions while amines and sulfides are catalyst poisons [73]. The oxidation is best carried out around neutral pH, and a base is usually added during the reaction to neutralize the acid as it is formed. The reaction is easily worked up as the catalyst is removed by filtration and no other inorganic salts are needed. However, the oxidation does require a large amount of platinum catalyst as is also evident in the oxidation of **21**. Even more problematic is the oxidation of more hindered hydroxy groups. Oxidation of L-sorbopyranose to L-*xylo*-hexulosonic acid (2-keto-L-gulonic acid) requires an average 1 g of platinum metal (i. e., 10 g of 10% Pt/C) to convert 0.3–8 g of L-sorbopyranose in 1 h [76]. Increased activity can be obtained by using bismuth- or lead-promoted platinum catalysts. The presence of the promoter seems to suppress poisoning of the catalyst caused by accumulation of oxygen on the metal surface. However, the promoters also change the selectivity profoundly [76], hence a general solution to the deactivation problem remains to be found.

When an aldose is not protected at C1, oxidation can occur both at C1 and at the primary position to give an aldaric acid. Strong nitric acid is the classical reagent for this oxidation. For example, treatment of D-glucose with concentrated nitric acid at 60 °C for 1 h affords D-glucaric acid isolated as the crystalline monopotassium salt in 41% yield [77]. Aldoses can also be oxidized to aldaric acids by the platinum-catalyzed oxidation with oxygen, but the need for relatively large amounts of platinum generally makes this procedure less attractive [78]. More recently, a modification of TEMPO was introduced for oxidation of aldoses to aldaric acids [79]. By this procedure glucose is converted into the monopotassium salt of glucaric acid in 85% yield [79].

2.4 Oxidation of Secondary Alcohols to Ketones

A commonly used, protected carbohydrate containing a secondary hydroxy group is diisopropylideneglucofuranose **23**. Oxidation to the corresponding ketone **24** illustrates some of the most widely applied methods for oxidation of secondary alcohols (❯ *Table 4*). Again, the reactions can be divided into three main categories: oxidations mediated by activated DMSO, oxidations with chromium(VI) oxides, and oxidations catalyzed by ruthenium oxides. For oxidations with activated DMSO the Swern procedure is the most widely used [27].

◻ Table 4
Oxidation of diisopropylideneglucofuranose 23 to ketone 24

Reagent	Solvent	Yield (%)	Reference
DMSO, (COCl)$_2$; Et$_3$N	CH$_2$Cl$_2$	92	[80]
DMSO, TFAA; Et$_3$N	CH$_2$Cl$_2$	85	[81]
DMSO, Ac$_2$O	DMSO	81	[82]
PCC, 3 Å MS	CH$_2$Cl$_2$	89	[83]
PDC, AcOH, 4 Å MS	CH$_2$Cl$_2$	98	[84]
PDC, Ac$_2$O	CH$_2$Cl$_2$	94	[85]
CrO$_3$·2C$_5$H$_5$N, Ac$_2$O	CH$_2$Cl$_2$	90	[20]
RuO$_2$, KIO$_4$	H$_2$O/CHCl$_3$	86	[86]
TEMPO, NaBr, NaOCl	H$_2$O/EtOAc	> 85	[87]
Dess–Martin periodinane	ClCH$_2$CH$_2$Cl	83	[88]

Trifluoroacetic anhydride (TFAA) is also a very potent activator for DMSO and concomitant trifluoroacetylation of the starting alcohol is usually not observed [27]. Both the Swern and the TFAA procedure are carried out at low temperature to prevent undesired side reactions, particularly formation of the methylthiomethyl ether. Before these two methods became developed, acetic anhydride was often used for DMSO activation. However, the oxidation under these conditions is slower and the methylthiomethyl ether byproduct is often observed [27].

Among the chromium(VI) oxides PCC [28,89] and PDC [28,40] are preferred for oxidation of carbohydrate secondary alcohols. The reaction is in both cases accelerated by molecular sieves [90] and anhydrous acetic acid [91]. Activation of PDC can also be achieved with acetic anhydride [40]. The most widely used procedure, however, seems to be PCC and 3 Å molecular sieves in dichloromethane [28,92]. PCC is mildly acidic, but acetal protecting groups remain stable to the oxidation conditions. Chromium(VI) oxide-pyridine complex is usually not a satisfactory oxidant for carbohydrate secondary alcohols. However, further activation by acetic anhydride gives good yields of ketones with 4 equiv. of reagent [20]. Secondary alcohols in thioglycosides can be oxidized under these conditions without accompanying oxidation at sulfur [93].

If a secondary alcohol is not easily oxidized by other methods the ruthenium(VIII) oxide catalyzed procedure is often recommended. As mentioned previously, this is a strong oxidation method which is not compatible with a number of functional groups. Sodium periodate usually serves as the stoichiometric oxidant, but sodium hypochlorite has also been used in the oxidation of secondary alcohols [94]. Because of the cheap oxidants and a straightforward work-up this reaction is well suited for large-scale oxidations [95]. The TEMPO procedure also employs a cheap stoichiometric oxidant and has been applied in the oxidation of **23** on a kilogram scale [87]. The TPAP-catalyzed method is a milder procedure and many functional groups are stable to these conditions. However, secondary alcohols are still oxidized to ketones in high yield with NMO as the co-oxidant [24].

The Dess–Martin periodinane [31] has also been used for oxidation of carbohydrate secondary alcohols [88,96]. Oxidations are usually carried out under neutral conditions in dichloromethane, chloroform, or acetonitrile. However, the Dess–Martin periodinane is used stoichiometrically and as such becomes a rather expensive oxidant. As a result, it is mostly recommended for special cases where the above-described procedures are insufficient.

Manganese(IV) oxide is very slow at oxidizing isolated secondary alcohols. However, if the alcohol is allylic or alpha to a lactone manganese(IV) oxide is the reagent of choice [3,97]. The reagent has to be activated for the oxidation and a commercial sample is usually not sufficient. Activated manganese(IV) oxide is prepared as a solid from potassium permanganate and manganese(II) sulfate [98]. Oxidations can be carried out in a variety of solvents, but ether, chloroform, or acetone are usually good choices. Hereby, allylic alcohol **25** and α-hydroxy-lactone **27** undergo oxidation in high yield (❷ *Scheme 9*) [99,100].

Manganese(IV) oxide will also oxidize the C3 hydroxy group in glycals [101]. However, a variety of other oxidants has also been applied for this special case (❷ *Table 5*). Silver carbonate on Celite is a mild and neutral oxidant that also gives good yields for allylic oxidations. Because of the heterogeneous reaction conditions, an excess (5 equiv. or more) of this reagent is needed [102]. Fully protected glycals can be oxidized directly with *N*-bromosuccinimide (NBS)/benzoyl peroxide [105] or with the Koser reagent (PhI(OH)OTs) [106].

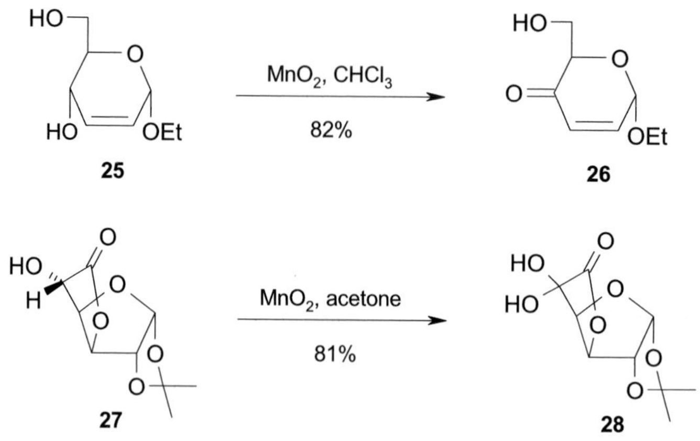

HO—／＼—O

MnO₂, CHCl₃ ... 82%

25 → 26

HO, H ... MnO₂, acetone ... 81%

27 → 28

■ Scheme 9

■ Table 5
Oxidation of unprotected and protected glucals at C3

R	R′	Reagent	Solvent	Yield (%)	Reference
H	TBS	MnO₂	CH₂Cl₂	72	[101]
H	TBS	Ag₂CO₃, Celite	Benzene	86	[102]
H	H	Ag₂CO₃, Celite	Benzene	70	[103]
H	H	(Bu₃Sn)₂O; NIS	Benzene	60	[104]
Ac	Ac	NBS, (BzO)₂, K₂CO₃	CCl₄	72	[105]
Ac	Ac	PhI(OH)OTs, 3 Å MS	MeCN	52	[106]

Secondary hydroxy groups in unprotected or partially protected carbohydrates are difficult to oxidize regioselectively by the above-described procedures. Instead, some special conditions have been developed in these cases [107]. Particularly noteworthy is the brominolysis of stannyl ethers and stannylene acetals. Because of the sensitivity of the Sn–O linkage, secondary hydroxy groups can be activated with tin for a variety of regioselective reactions including oxidation to the ketone [108]. The reaction is carried out by first forming the tin compound with either bis(tributyltin) oxide or dibutyltin oxide which is then subsequently treated with bromine in situ to affect the oxidation. Bis(tributyltin) oxide was found to be superior in an extensive study on oxidation of unprotected and partially protected methyl glycopyranosides [109,110]. Oxidation of methyl pento- and hexopyranosides with this reagent and bromine gives ketoglycosides in high yield (❷ Table 6). The regioselectivity is surprisingly high and is determined by the configuration of the hydroxy group being oxidized. An axial hydroxy group is always oxidized preferentially as compared to an equatorial hydroxy group. During the oxidation

☐ **Table 6**
Regioselective oxidation of unprotected and partially protected methyl glycopyranosides

$$\text{(Bu}_3\text{Sn)}_2\text{O, CHCl}_3\text{, 3 Å MS,}$$
$$\xrightarrow{\text{then Br}_2}$$

Substrate	Oxidized position	Yield (%)	Reference
Methyl α-D-glucopyranoside	C4	65[b]	[109]
Methyl β-D-glucopyranoside	C3	97	[109]
Methyl α-D-mannopyranoside	C2	53[b]	[110]
Methyl α-D-galactopyranoside	C4	70[b]	[109]
Methyl β-D-galactopyranoside	C3/C4[a]	67	[109]
Methyl α-D-allopyranoside	C3	91	[110]
Methyl β-D-allopyranoside	C3	84	[110]
Methyl α-D-altropyranoside	C3	84	[110]
Methyl β-D-altropyranoside	C3	82	[110]
Methyl α-D-xylopyranoside	C4	92	[109]
Methyl β-D-xylopyranoside	C3	93	[109]
Methyl β-L-arabinopyranoside	C4	93	[109]
Methyl 4,6-O-benzylidene-α-D-glucopyranoside	C2	95	[110]
Methyl 4,6-O-benzylidene-β-D-glucopyranoside	C3	98	[109]
Methyl 4,6-O-benzylidene-α-D-galactopyranoside	C3	59[b]	[110]
Methyl 4,6-O-benzylidene-β-D-galactopyranoside	C3	89	[110]
Methyl 4,6-O-benzylidene-α-D-altropyranoside	C3	76	[110]

[a] C3/C4 ratio 2/5
[b] Some unreacted starting material is also recovered

a proton is removed from the carbon bearing the secondary hydroxy group. In general, the more easily available this proton is, the more easily the secondary hydroxy group is oxidized to the corresponding ketone. This rule also applies to protected methyl glycosides, e. g., the 4,6-benzylidenehexopyranosides [109,110]. The products from these regioselective oxidations are hydroxyketones that often dimerize fairly rapidly [110].

Other partially protected carbohydrates also undergo very regioselective oxidation. Noteworthy is the oxidation of isopropylideneglucofuranose **29** to 5-ketofuranose **30** (❽ *Scheme 10*) [109]. For oxidation of the axial hydroxy group in *cis*-1,2 diols, the dibutylstannylene acetal method is often employed. Oxidation of methyl fucoside **31** with this procedure gives ketone **32** in good yield (❽ *Scheme 10*) [111].

The platinum-catalyzed oxidation with oxygen can also be applied for selective oxidation of secondary alcohols if no primary alcohol is present [73]. Like the tin-bromine method, axial secondary hydroxy groups will undergo preferential oxidation over equatorial hydroxy groups. However, as described above large amounts of platinum metal are required for these oxidations. Some improvement in catalyst activity has been achieved by promotion of platinum with bismuth or lead [76]. This also causes a change in selectivity and makes it possible in

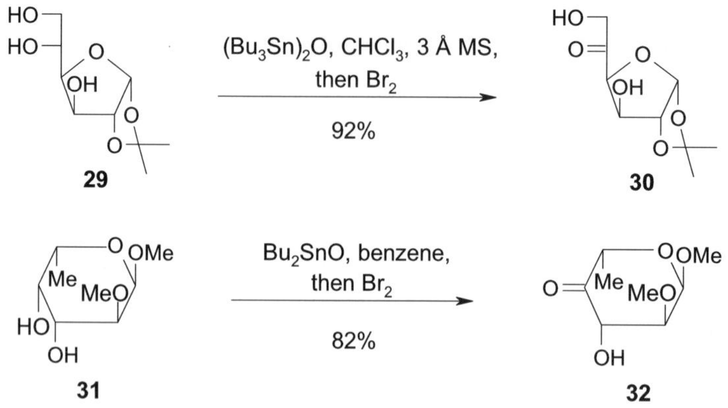

29 → (Bu₃Sn)₂O, CHCl₃, 3 Å MS, then Br₂ / 92% → 30

31 → Bu₂SnO, benzene, then Br₂ / 82% → 32

◻ Scheme 10

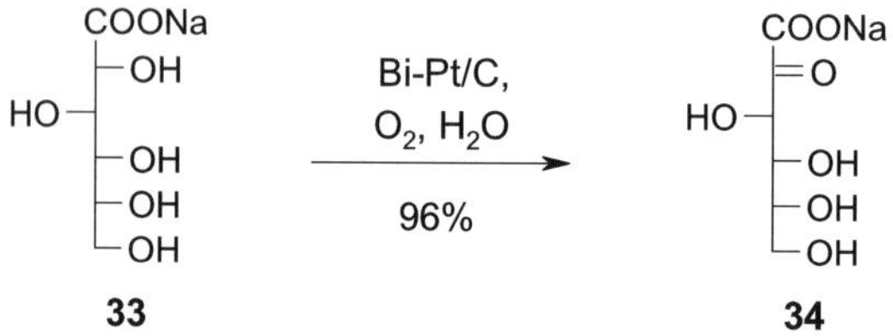

33 → Bi-Pt/C, O₂, H₂O / 96% → 34

◻ Scheme 11

some cases to oxidize a secondary alcohol in the presence of a primary alcohol. A remarkable example is the very regioselective oxidation at C2 in aldonic acids as shown with the conversion of sodium D-gluconate 33 into 2-keto-D-gluconate 34 (❍ Scheme 11) [112].

Efficient preparation of keto D-gluconates can also be achieved by fermentation. Microbial oxidation of D-glucose with various bacterial strains of the genus *Pseudomonas* produces 2-keto-D-gluconate which can be isolated by direct crystallization of the calcium salt 35 (❍ Scheme 12) [113]. The same product can be obtained by fermentation with *Gluconobacter* species [114]. In fact, with this genus both 2-keto-, 5-keto-, and 2,5-diketo-D-gluconates can be formed and, depending on the strain, good selectivity for either one of the three ketogluconates can be obtained [115]. For example, 5-ketogluconate 36 can be formed in yields up to 90% with *Gluconobacter suboxydans* (❍ Scheme 12) [115]. Besides oxidizing aldoses and aldonic acids *Gluconobacter* species are also known to mediate the oxidation of alditols [114]. Only alditols containing a D-*erythro* grouping adjacent to a primary alcohol will react with a reasonable growth rate. The oxidation occurs selectively at the secondary hydroxy group next to the primary alcohol (❍ Scheme 12) [116]. An example is the oxidation of D-glucitol (D-sorbitol) to L-sorbose [117] which is the first step in the classical route for production of vitamin C [49].

Some pyranosides can be oxidized at C3 using the bacterium *Agrobacterium tumefaciens*. This method has been particularly successful for oxidation of disaccharides. The conversion of sucrose into 3-keto-sucrose has been studied in detail (❍ Scheme 13) [118]. Lactose, maltose,

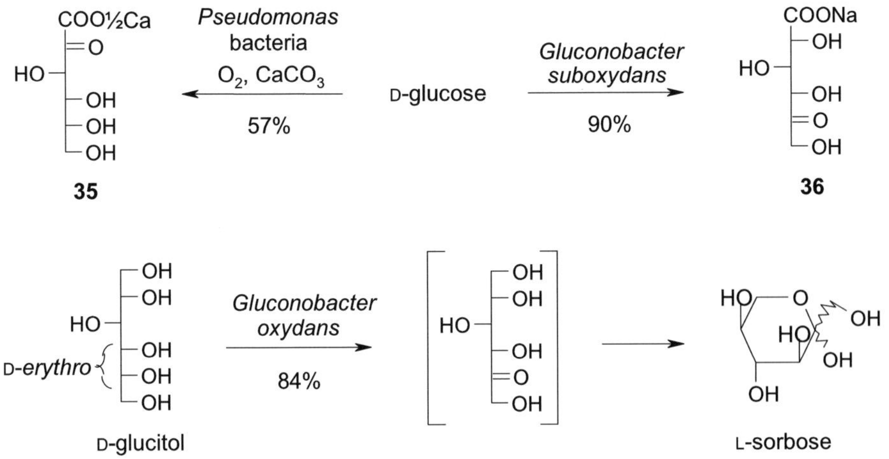

COO½Ca
=O
HO
─OH
─OH
─OH

35

Pseudomonas bacteria
O$_2$, CaCO$_3$
57%

← D-glucose →

Gluconobacter suboxydans
90%

COONa
─OH
HO
─OH
=O
─OH

36

─OH
─OH
HO
─OH
D-*erythro* { ─OH
─OH

D-glucitol

Gluconobacter oxydans
84%

─OH
─OH
HO
─OH
=O
─OH

L-sorbose

■ Scheme 12

sucrose

Agrobacterium tumefaciens
60%

3-keto-sucrose

pyranose 2-oxidase
O$_2$ H$_2$O$_2$

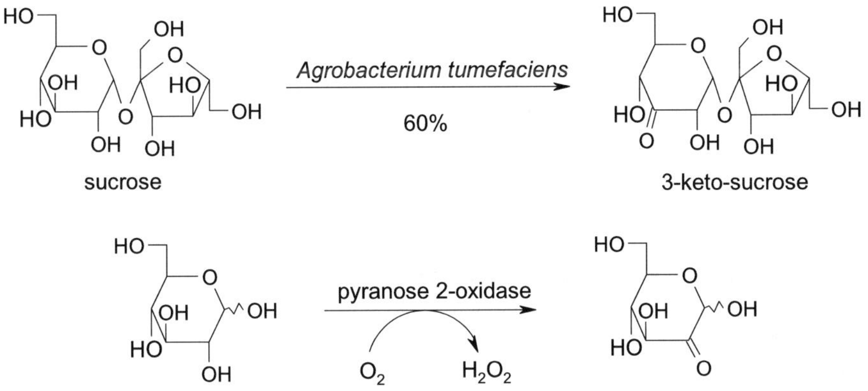

■ Scheme 13

and cellobiose also undergo this selective C3 oxidation [119]. The active enzyme, a 3-dehydro-genase, has been isolated and purified. In addition to disaccharides this enzyme also oxidizes D-glucose, D-galactose and their methyl glycosides at C3 [120].

Selective oxidation at C2 in pyranoses can in some cases be carried out with the enzyme pyranose 2-oxidase, which can be isolated from several wood-degrading fungi [121]. The natural substrate for the enzyme is D-glucose, but significant activity towards C2 oxidation in D-galactose, D-xylose, and D-allose has also been observed (❷ *Scheme 13*) [121]. In contrast, L-sorbose undergoes very selective oxidation at C5 [121]. In some cases, the formed 2-keto-D-aldose is not the oxidation end-product, but is accumulated in the mixture and then further converted into another product. In this way, D-glucose, D-galactose, and D-xylose have been converted into the corresponding 2,3-diketo-aldoses by oxidation with pyranose 2-oxidase from various sources [122].

2.5 Epoxidation, Dihydroxylation, and Azidonitration of Olefins

Some unsaturated carbohydrates are easily available especially glycals and products derived from Wittig-type olefinations. The epoxidation and dihydroxylation of these substrates constitutes an important reaction in the synthesis of more complex sugars. Stereocontrolled epoxidation of glycals gives 1,2-epoxy sugars that are useful glycosyl donors [123]. Peroxy acids cannot generally be used for the epoxidation because of the high reactivity of the 1,2-epoxide which will undergo ring-opening by the acid formed during the course of the epoxidation. An exception is the *m*-CPBA-KF mixture (❷ *Table 7*). The addition of potassium fluoride sufficiently reduces the solubility of *m*-chlorobenzoic acid to prevent the formation of products arising from epoxide ring-opening [124]. Strictly anhydrous conditions are necessary and commercial *m*-CPBA samples have to be further dried. Hereby, high yields of the desired 1,2-epoxy sugars can be obtained. The epoxidation occurs predominately from the α-face of the olefin placing the oxygen *anti* to the C3 substituent. Because of the axial C4 substituent, D-galactal derivatives show an even higher α-selectivity in the epoxidation than the D-glucal derivatives. Another reagent for the epoxidation is dimethyldioxirane [123], which has to be prepared from acetone and oxone. The byproduct of the epoxidation in this case is acetone which does not react with the product epoxide. As a result, near quantitative yields are obtained and dimethyldioxirane is often the reagent of choice for epoxidizing glycals [125].

Protected *C*-methylene pyranosides like **37a,b** can be obtained by Wittig methylenation from the corresponding ketones (❷ *Scheme 14*) [126]. Epoxidation to give the epoxy-branched sugars can now be carried out with a peroxy acid due to the increased stability of the formed epoxide. The electrophilic attack of the peroxy acid occurs from the less hindered face of the olefin. In this way, 2-*C*-methylene pyranosides **37a** and **37b** undergo epoxidation to give **38a** and **38b**, respectively, controlled by the axial anomeric methoxy group [126].

Hex-5-enopyranosides are available by various elimination reactions and can be epoxidized with *m*-CPBA and with a dioxirane. Like in the case with glycals the epoxides are quite sensitive to ring-opening reactions. Prolonged treatment with the epoxidizing agent in the presence of water gives hexos-5-uloses, which can be isolated in moderate to good yields [127]. For example, hex-5-enopyranoside **39** gives a 7:3 mixture of epoxides **40a,b** upon treatment with

❑ **Table 7**
Epoxidation of 3,4,6-tri-*O*-benzyl-D-glucal and -D-galactal

R¹	R²	Reagent	Yield (%)	α/β-ratio	Reference
H	OBn	*m*-CPBA, KF	95	9/1	[124]
OBn	H	*m*-CPBA, KF	95	20/1	[124]
H	OBn	Dimethyldioxirane	99	20/1	[123]
OBn	H	Dimethyldioxirane	99	only α	[125]

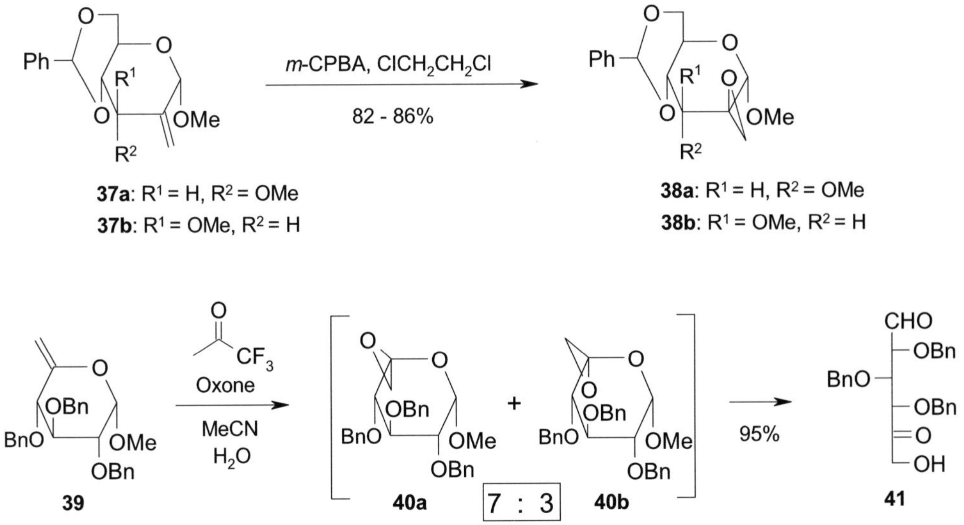

☐ Scheme 14

in situ generated methyl(trifluoromethyl)dioxirane while the reaction with m-CPBA furnishes **40a,b** in a ratio of 3:7 (❷ *Scheme 14*) [127]. Further reaction with water leads to the corresponding hexos-5-ulose which after loss of methanol affords **41** in 95% overall yield.

Dihydroxylation of olefins is typically carried out with a catalytic amount of osmium(VIII) oxide in the presence of NMO as co-oxidant. If the reaction proceeds slowly a tertiary amine is sometimes added to further accelerate the dihydroxylation [128]. In cyclic systems the dihydroxylation takes place from the sterically least encumbered face of the olefin. Sugars containing an exocyclic methylene group are thus converted into branched sugars, e. g., **42** → **43** (❷ *Scheme 15*) [129]. Aldonolactones containing a 2,3-double bond undergo dihydroxylation from the face opposite the side chain [55]. Glycals are dihydroxylated to give the 2,3-*trans* compound as the major product. For example, dihydroxylation of D-galactal with osmium(VIII) oxide gives a 4:1 mixture of D-galactose and D-talose [130]. A more interesting application is the dihydroxylation of sialic acid glycal **44** to give diol **45** in high yield (❷ *Scheme 15*) [131]. This diol can be converted into a special glycosyl donor for α-selective sialylation using neighboring group participation [131]. Dihydroxylation of unprotected glycals from the opposite face to give 2,3-*cis* products can be carried out with catalytic molybdenum(VI) oxide and hydrogen peroxide in water [130]. The reaction presumably involves epoxidation of the double bond directed by the C3 hydroxy group followed by epoxide ring-opening with water. Particularly attractive is the conversion of D-galactal into crystalline D-talose in high yield (❷ *Scheme 15*) [132].

Dihydroxylation of chain-extended unsaturated carbohydrates gives rise to higher sugars. Unsaturated ester **46** is available from glucose by a Wittig reaction in dioxane (❷ *Scheme 16*) [133]. Dihydroxylation of **46** affords a 5:1 mixture of two diastereomers and the major isomer is isolated by crystallization as the octonolactone **47** [134]. Interestingly, the two-step procedure can be converted into a one-pot transformation in dioxane by performing the

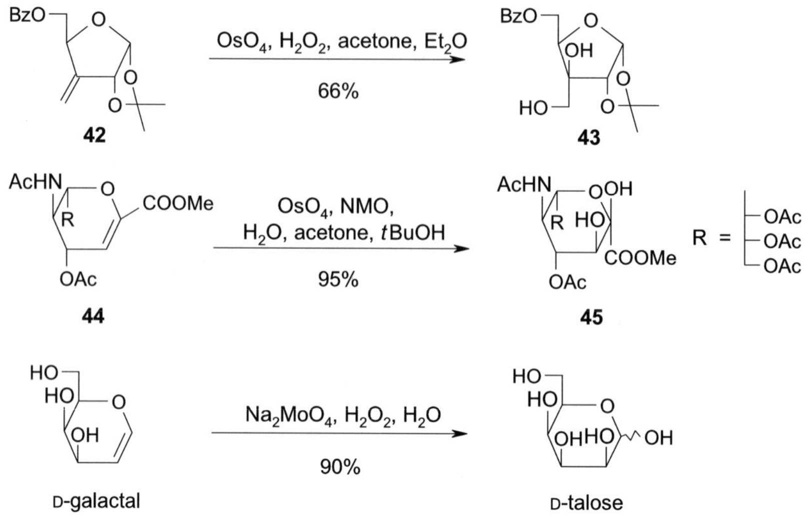

■ Scheme 15

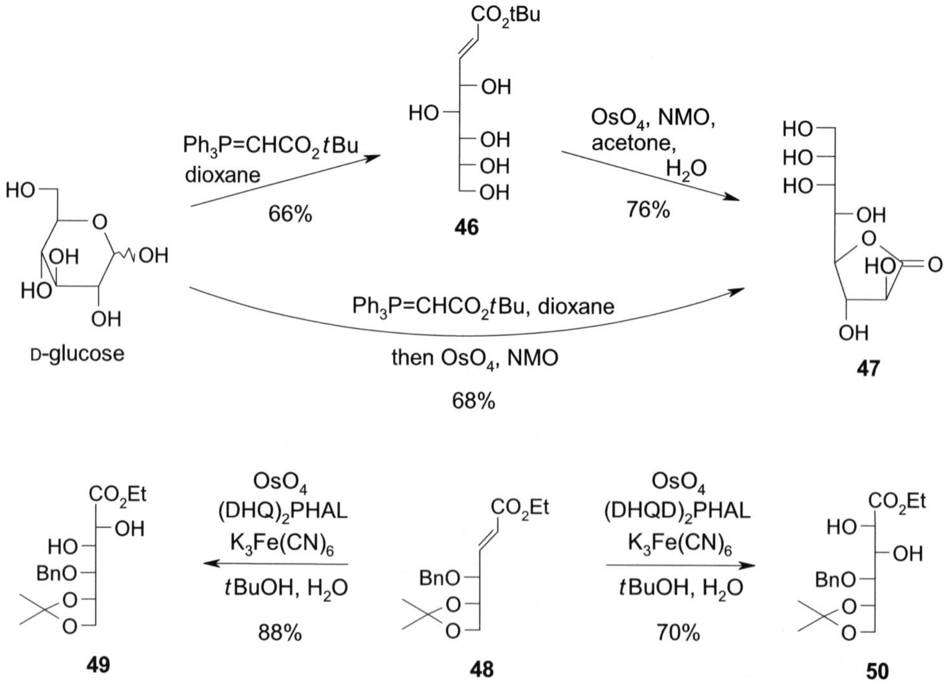

■ Scheme 16

dihydroxylation immediately after the Wittig reaction. In this way, lactone **47** is obtained in a better overall yield from glucose than by isolating the intermediate ester **46**. Similar yields and diastereoselectivities are observed when galactose, arabinose, and xylose are subjected to the one-pot Wittig-dihydroxylation reaction [134]. Mannose, ribose, and lyxose, on the other hand, give a lower diastereoselectivity in the dihydroxylation [134]. In all cases, the stereochemistry of the major product is in accordance with Kishi's empirical rule [135]. This predicts that for dihydroxylation of acyclic allylic alcohols (or protected alcohols), the relative stereochemistry between the preexisting hydroxy group (or protected hydroxy group) and the adjacent, newly introduced hydroxy group in the major product is *erythro*. Kishi's rule applies to dihydroxylation of a variety of carbohydrate allylic systems [136]. The selectivity in the dihydroxylation can be improved or inverted by using a chiral ligand for osmium [137]. This is demonstrated very effectively in a recent study where all six L-hexoses are prepared from L-ascorbic acid [138]. The latter is converted into L-erythrose and L-threose which are both subjected to a Wittig reaction to afford either the (*E*)- or the (*Z*)-unsaturated ester. Each Wittig adduct is then dihydroxylated into either the 3,4-*erythro* or the 3,4-*threo* product depending on the chiral ligand for osmium. Thus, dihydroxylation of (*E*)-olefin **48** in the presence of hydroquinine 1,4-phthalazinediyl diether ((DHQ)$_2$PHAL) gives exclusively the L-altro product **49** while the reaction with hydroquinidine 1,4-phthalazinediyl diether (DHQD)$_2$PHAL) affords only the opposite L-gluco isomer **50** (❷ *Scheme 16*). Similar results are obtained for the corresponding (*Z*)-olefin [138].

Although osmium(VIII) oxide continues to be the most popular reagent for dihydroxylation of olefins, it does have two major drawbacks: it is very expensive and very toxic. As a result, other reagents have been investigated. Potassium permanganate has been successfully applied for dihydroxylation of the electron-deficient double bond in 2,3-unsaturated aldonolactones and – lactams [139]. Ruthenium(VIII) oxide generated from catalytic ruthenium(III) chloride and stoichiometric sodium periodate has been applied for the dihydroxylation of various protected carbohydrates with very short reaction times [140]. Although good yields can be obtained with both potassium permanganate and ruthenium(VIII) oxide, these reagents are more powerful oxidants than osmium(VIII) oxide and less chemoselective. As a result, byproducts arising from overoxidation and oxidative fission are more common.

Osmium(VIII) oxide also catalyzes the aminohydroxylation of olefins using chloramine-T as a stoichiometric oxidant and nitrogen source. Aminohydroxylation has been performed on several unsaturated carbohydrates including glycals, hex-2-eno- and hex-3-enopyranosides [141]. The yields are typically in the 60–80% range, the major byproduct being the diol. However, the reaction suffers badly from poor regioselectivity and often large amounts of both isomers are obtained.

Azidonitration represents a special aminohydroxylation reaction because the azido group often serves as a masked amino group. Azidonitration is possible only on glycals where the azido group is introduced at C2 and the nitrate at C1. The reaction is typically carried out with 1.5 equiv. of sodium azide and 3 equiv. of ceric ammonium nitrate (CAN) in acetonitrile at about −15 °C [142,143,144]. In this way, a number of acetylated glycals **51a–e** undergo fairly selective azidonitration to give azidonitrates **52a–e** (❷ *Table 8*). The reaction works particularly well on derivatives of D-galactal [142,145,146]. In contrast, azidonitration of tri-acetyl D-glucal is not very selective under these standard conditions, but gives an almost equal amount of the corresponding 2-azido-D-glucose and –D-mannose products [142,147]. Howev-

◻ Table 8

Azidonitration of acetylated glycals

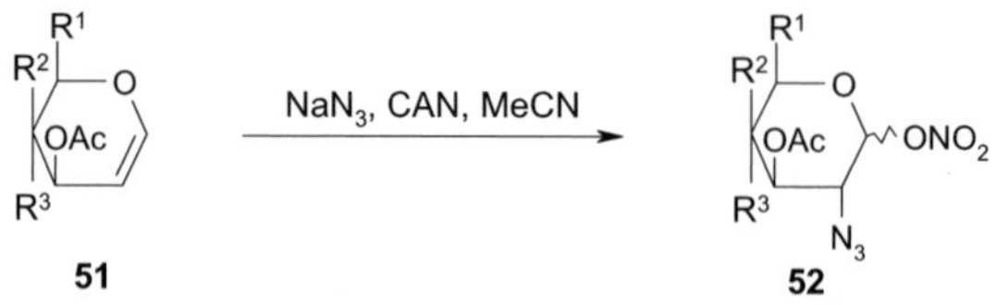

	R¹	R²	R³	Yield (%)	Reference
a	CH_2OAc	OAc	H	75	[142]
b	H	OAc	H	55	[143]
c	H	H	OAc	58[a]	[143]
d	COOMe	H	OAc	42[a]	[144]
e	COOMe	OAc	H	44[a]	[144]

[a]Product isolated as glycosyl acetate after treatment with NaOAc

er, increased selectivity for the 2-azido-D-mannose product can be obtained by lowering the reaction temperature to -40 °C and diluting the mixture with ethyl acetate [147]. On the other hand, use of the corresponding 4,6-O-benzylidene- or 4,6-O-isopropylidene-D-glucal show increased α-selectivity to give mainly the 2-azido-D-glucose product [148].

Although the yields obtained in the azidonitration reaction are sometimes moderate, the reaction continues to be quite important for preparation of 2-azidoglycosyl donors used in 2-amino-glycoside synthesis. In order to obtain a glycosyl donor the glycosyl nitrate can be directly converted into several glycosyl derivatives with the appropriate reagent. This includes conversion into the hemiacetal with hydrazine acetate [145], the glycosyl acetate with sodium acetate [142,143,144], the glycosyl chloride with tetraethylammonium chloride [142], or the glycosyl bromide with lithium bromide [142,146].

A closely related reaction to azidonitration is the azidophenylselenylation reaction that gives 2-azidophenylselenoglycosides from glycals. The reaction is carried out with sodium azide, diphenyl diselenide, and (diacetoxyiodo)benzene in dichloromethane at room temperature [149]. Typically, the yields for azidophenylselenylation of acetylated glycals are slightly higher than in the corresponding azidonitration reaction, e. g., **51a → 53** (❂ *Scheme 17*) [149]. In addition, only the α-selenoglycoside is formed. Selenoglycosides can serve directly as glycosyl donors in the preparation of O- and C-glycosides [150] or be hydrolyzed to the corresponding hemiacetal as shown for the conversion of **53** into **54** [149].

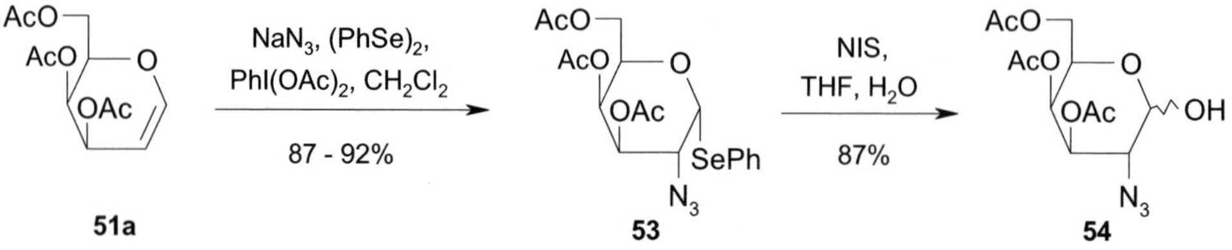

◻ Scheme 17

2.6 Bromination at Ring Positions

In tetrahydrofuran and –pyran systems homolytic hydrogen abstraction will occur at the ether carbon atoms due to the stabilization of the developing radical by one of the oxygen lone-pairs. In aldose derivatives this hydrogen abstraction can be directed towards the non-anomeric carbon adjacent to the ring oxygen if all hydroxy groups are protected with radical destabilizing protecting groups. In this way, bromine atoms can be substituted directly onto these ring positions by free radical photobromination [151]. The reaction is carried out with bromine or NBS in refluxing carbon tetrachloride under a tungsten or heat lamp. Bromine is the most reactive brominating agent, while NBS is more selective in some cases. In addition, bromine produces hydrogen bromide during the course of the reaction, which can lead to side reactions if not trapped with an acid scavenger. The hydroxy groups are most efficiently protected with acetyl or benzoyl groups. Sometimes side reactions can occur with acetyl groups, which can undergo α-bromination to the corresponding bromoacetate [151].

Using these photobromination conditions a number of ester-protected aldose derivatives undergo very selective bromination (❯ *Scheme 18*) [152]. Aldopyranosides (e. g., **55**) and methyl esters of uronic acids (e. g., **57**) are brominated at C5. The bromination is controlled by the anomeric effect to place bromine axial. In these D-glucopyranose derivatives the bromination occurs most readily for the β-anomers. In the corresponding α-anomers the axial anomeric protecting group makes the C5 proton less accessible, thus causing a more sluggish reaction. In 1,6-anhydropyranoses, however, the bromination occurs very selectively at C6 giving rise to the *exo*-bromide adduct, e. g., **60**.

The brominated sugars are usually quite stable and can be useful for a variety of purposes [151]. Substitution of bromine with deuterium can be used for preparation of labeled carbohydrates, while substitution with hydrogen sometimes can be used for inverting the stereochemistry. Noteworthy is the conversion of D-glucuronic acid derivative **61** into the corre-

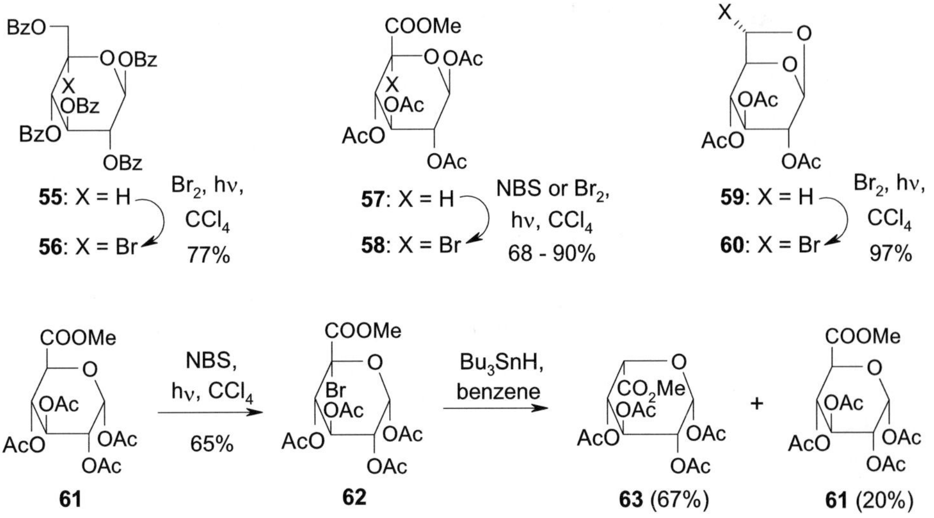

❒ Scheme 18

sponding L-iduronic acid **63** by bromination and subsequent debromination (❷ *Scheme 18*) [153]. It is important that the reaction is carried out with the α-anomer. If the corresponding β-anomer **58** is subjected to the same debromination conditions, the starting D-glucuronic acid **57** is obtained as the major product [151].

3 Reductions

Numerous reduction procedures are available in organic synthesis [154] and many of these can be applied for the reduction of carbohydrates.

3.1 Reduction at the Anomeric Center

Unprotected aldoses and ketoses can be reduced to afford alditols while aldonolactones can be reduced to give either aldoses or alditols. The reagent of choice for reduction to alditols is sodium borohydride since it is both cheap and convenient to use. The reduction is carried out under mild conditions at room temperature in an aqueous solution. Sodium borohydride is stable in water at pH 14 while it reacts with the solvent at neutral or slightly acidic pH, but at a slower rate than the rate of carbonyl reduction. In some cases, the product will form esters with the generated boric acid. These borate complexes can be decomposed by treatment with hydrochloric acid or a strongly acidic ion-exchange resin and the boric acid can be removed in the work-up as the low boiling trimethyl borate by repeated co-evaporation with methanol at acidic pH [155].

The reduction of aldoses/ketoses occurs readily with sodium borohydride and during the reaction the pH increases to about 9 (❷ *Scheme 19*) [155]. For the reduction of aldonolactones in water the first step of the reduction has to be carried out at a pH around 5 in order to avoid ring-opening of the lactone to the corresponding sodium salt which will not react with sodium borohydride. The pH control can be achieved by performing the reduction in the presence of an acidic ion-exchange resin, e. g., Amberlite IR-120 [156]. In this way, it is possible to stop the reduction at the aldose step. Alternatively, more sodium borohydride can be added and thereby increasing the pH to 9 by which the alditol is obtained (❷ *Scheme 19*). The reduction of aldonolactones to alditols can also be performed in anhydrous methanol or ethanol where hydrolysis of the lactone is not a side reaction [156].

Sodium borohydride only reduces aldonolactones when there is an electronegative substituent at C2. As a result, 2-deoxylactones are not reduced with this reagent, but can instead be reduced with disiamylborane in THF to the corresponding 2-deoxyaldose [157] or with calcium borohydride in ethanol to the alditol [158]. Disiamylborane is easily generated in situ by reacting borane-dimethylsulfide complex with 2-methyl-2-butene prior to addition of the lactone [157]. Sodium borohydride is not a useful reagent for large-scale industrial applications. In this case, catalytic hydrogenation in an aqueous solution over a heterogeneous catalyst is the preferred method for reducing aldoses and ketoses. The favored catalyst is Raney nickel [159] or a promoted Raney nickel [160]. The hydrogenations are typically carried out at high pressure and at temperatures around 120 °C. Lately, Ru/C and modifications thereof are gaining more attention due to a higher activity and fewer problems with metal leaching [161]. A particularly challenging example is the hydrogenation of D-fructose to afford D-mannitol where the best catalysts give D-mannitol:D-glucitol in a ratio of about 3:2 [162].

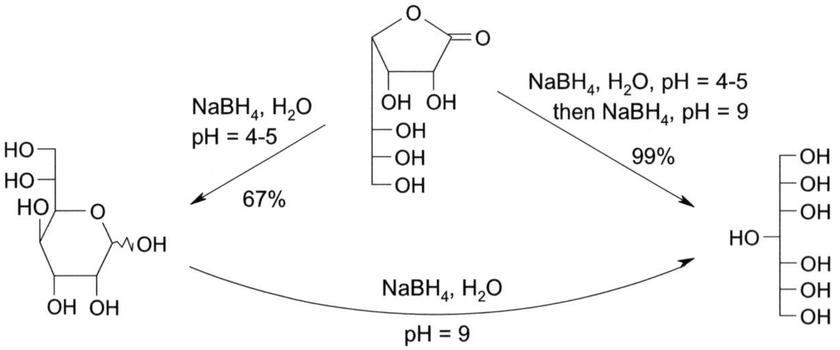

◘ Scheme 19

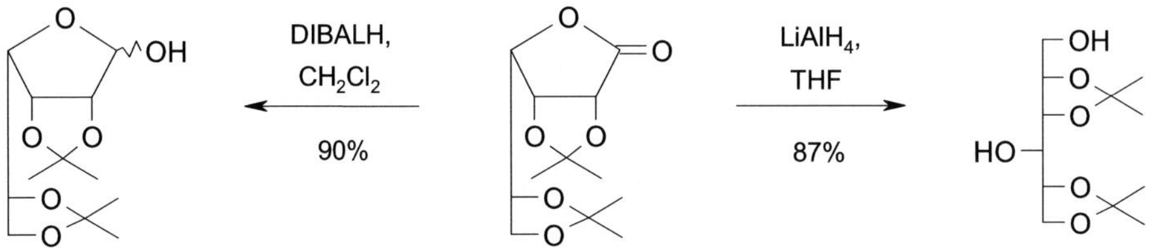

◘ Scheme 20

Partially protected aldoses and ketoses can also be reduced with sodium borohydride to the corresponding alditol. The reduction is typically carried out in an alcoholic solvent or in a mixture of THF and water [163]. The same reduction can be achieved in a non-protic solvent with lithium aluminum hydride and diisobutylaluminum hydride (DIBALH) which are more powerful reducing agents [164]. Protected aldonolactones can be reduced with sodium borohydride in a similar manner as described above in ❷ Scheme 19 [165]. In addition, protected aldonolactones can be reduced with DIBALH [166] and disiamylborane [167] to the corresponding aldose or with lithium aluminum hydride to the alditol (❷ Scheme 20) [168]. It should be noticed that ester-protected aldonolactones can be reduced with disiamylborane in high yield without reducing the ester groups [167] while the other reducing agents only tolerate ether and acetal protecting groups. The reduction of protected aldonolactones with sodium borohydride, DIBALH, or disiamylborane gives aldoses with the same ring size as the starting lactone and this method is particularly effective for synthesizing protected aldoses in the furanose form.

3.2 Reduction of Carboxylic Acids to Primary Alcohols

Uronic acids are important components in many naturally occurring polysaccharides. By chemical or enzymatic degradation of these polysaccharides, smaller uronic acid units ranging from monosaccharides to smaller oligosaccharides can be prepared. In this connection, reduction of the carboxyl group can serve both synthetic and analytical purposes.

Reduction of the free carboxylic acid in an otherwise fully protected uronic acid can be accomplished with borane in THF [169]. For example, the borane-THF complex reduces diisopropy-

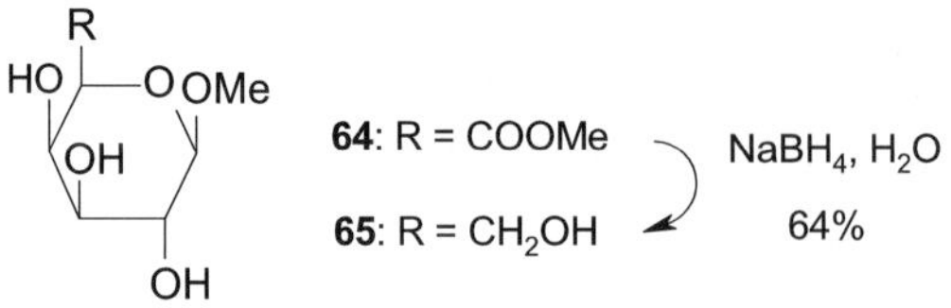

64: R = COOMe

65: R = CH₂OH

NaBH₄, H₂O

64%

□ **Scheme 21**

lidenegalacturonic acid **10** to diisopropylidenegalactose **7** (❂ *Table 3*) [170]. Esters of protected uronic acids are normally reduced to the corresponding alcohols with lithium aluminum hydride in ether or THF [171]. Unprotected glycosides of uronic acids are reduced with sodium borohydride in water. In this way, methyl galacturonate **64** is reduced to galactoside **65** which is isolated by crystallization after work-up with an ion-exchange resin (❂ *Scheme 21*) [171]. Free carboxylic acids, however, do not undergo direct reduction with sodium borohydride. Instead, an initial activation of the acid is necessary. This can be conveniently done in water with the water soluble carbodiimide, 1-(3-dimethylaminopropyl)-3-ethylcarbodiimide, which reacts selectively with the free carboxylic acid [172], which is thus sufficiently activated to be reduced with sodium borohydride. The entire procedure is carried out as a one-pot process in water and the method is well suited for analysis of uronic acids in polysaccharides [172].

3.3 Reduction of Ketones to Secondary Alcohols

Sodium borohydride is often the reagent of choice for the reduction of carbohydrate keto groups. The reduction is typically carried out in ethanol and the stereochemical outcome depends on steric and electronic factors in the substrate. It is important to note that sodium borohydride is a sterically undemanding reagent that is perfectly capable of approaching the ketone along the seemingly more hindered axial trajectory, thus leading to the equatorial alcohol. In fact, in simple cyclohexanones this axial hydride attack is favored for electronic reasons [154]. Sodium borohydride reductions of several unprotected methyl keto-glucopyranosides is shown in ❂ *Scheme 22* [109,110]. The ratio between the axial and the equatorial product alcohol is influenced by the stereochemistry at the anomeric center. For the methyl keto-α-glucosides the reduction always occurs from the face of the ketone opposite to the axial methoxy group at C1. For the keto-β-glucosides the stereochemical outcome is less predictable since the 2-ketoglucoside gives the axial alcohol while the 3-ketoglucoside affords the equatorial alcohol as the major product. With the 2-keto-β-glucoside the reduction has paved the way for one of the more reliable procedures for the preparation of β-mannopyranosides. The glycosylation is performed with a glucosyl donor to give a β-glucoside which is then oxidized at C2 and subsequently reduced to give the β-mannoside [173,174]. For keto-hexopyranosides derived from other aldoses the stereochemical outcome is highly dependent on the substrate and is often difficult to predict. In some cases, the reduction gives complete selectivity for one product while in other cases equal amounts of both diastereomers are obtained [109,110].

Protected ketosugars undergo reduction with sodium borohydride in a similar way. However, in some cases the steric or electronic nature of the protecting groups can have an additional influence on the selectivity for the reduction [175]. Protected ketosugars are normally prepared by direct oxidation of the corresponding hydroxysugar by one of the procedures mentioned

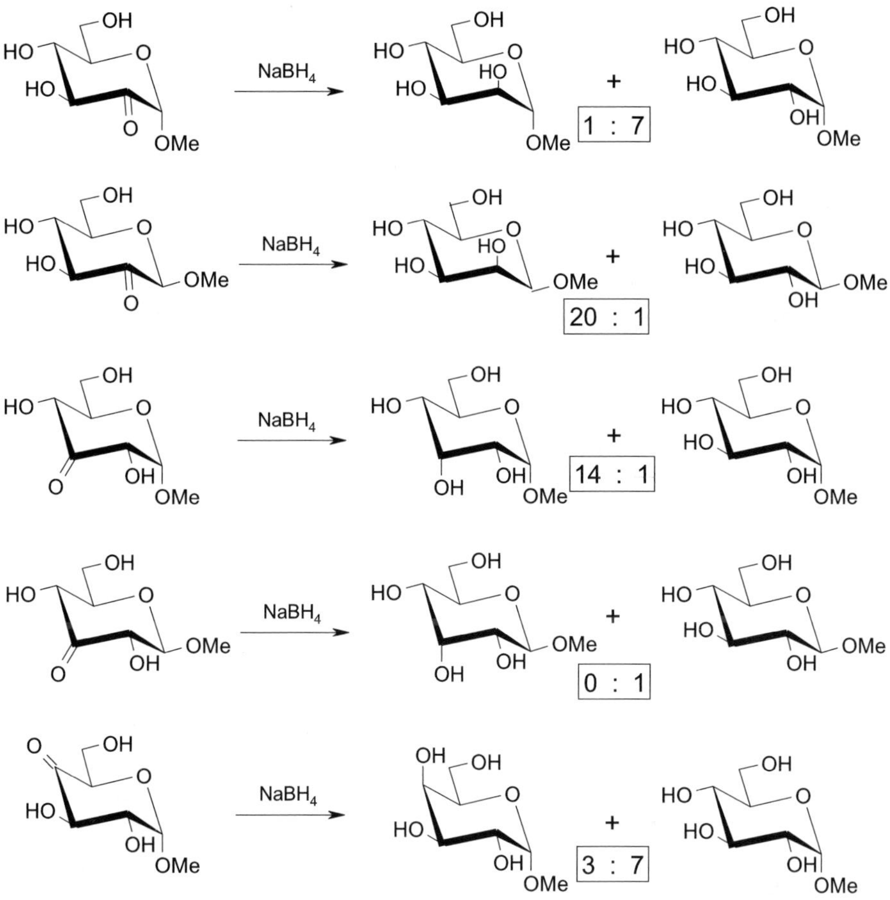

Scheme 22

previously. Subsequent reduction of the keto group then gives the possibility for inverting the stereochemistry. Several examples are illustrated in ❯ *Scheme 23* [80,95,176,177,178]. In the β-mannoside synthesis in the first example, the protective group at C3 plays a major role [176]. Additional studies have shown that high selectivity for the β-mannoside is obtained with a benzyl ether at C3 while an ester protective group gives more of the β-glucoside [179]. Inversion by oxidation-reduction can also be an efficient protocol for large-scale synthesis of some rare sugars. This is illustrated here by the preparation of diisopropylidene-protected psicose and allose from the corresponding fructose and glucose compounds [80,95].

Several other reagents have also been used for carbohydrate ketone reductions. Some of these can cause a dramatic change in the selectivity as compared to reduction with sodium borohydride (❯ *Table 9*) [180]. Borane is similar to sodium borohydride in the sense that it is sterically undemanding and also capable of approaching the ketone along the axial trajectory to give the equatorial alcohol. If sodium borohydride does not give a satisfactory yield of an equatorial alcohol, borane can in some cases be a better choice of reducing agent [181].

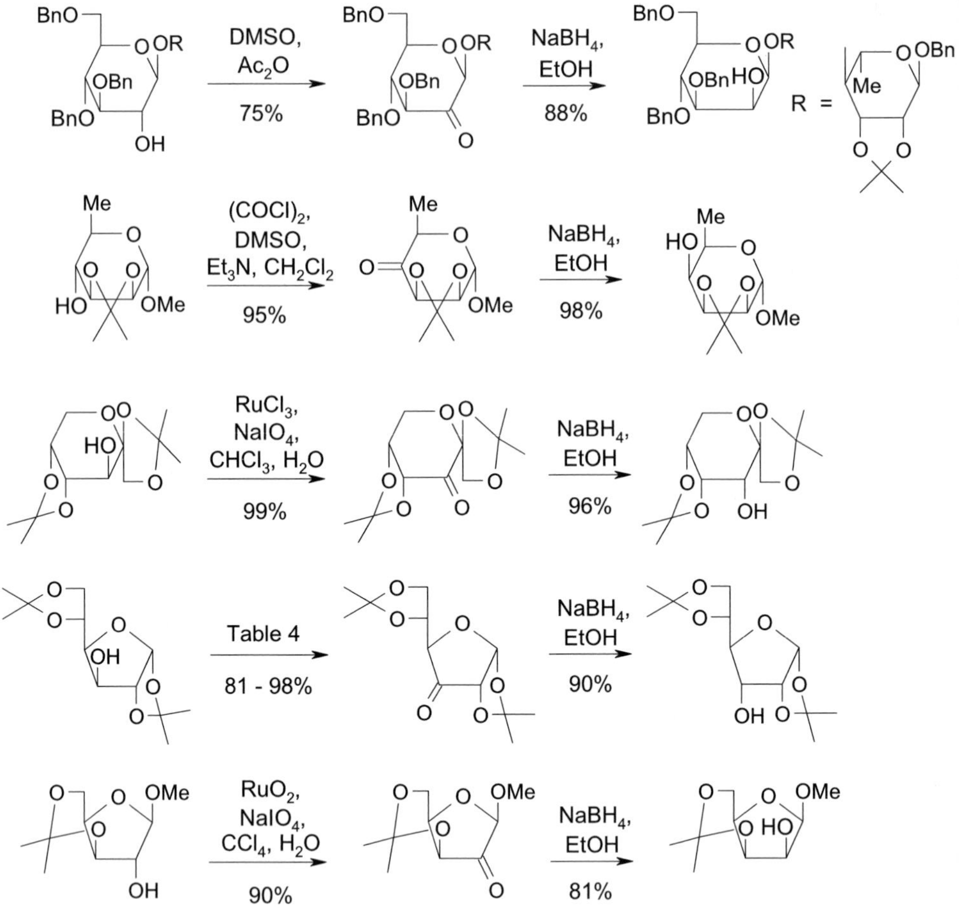

◻ Scheme 23

Preference for the axial alcohol, however, can be obtained by heterogeneous catalytic hydrogenation or by using a sterically hindered borohydride. Catalytic hydrogenation can also be performed with the Pt/C catalyst that has been used for reduction of 2-keto-β-glucosides to β-mannosides [174]. Benzyl ethers are not affected under these conditions. Lithium tri-(*sec*-butyl)borohydride (L-selectride) is a sterically very demanding borohydride which has also found successful use for reduction to β-mannosides [182]. Other ketoglycosides have also been shown to give more of the axial alcohol when reduced with L-selectride instead of sodium borohydride [183].

3.4 Reduction of Oximes to Primary Amines

Reduction of a ketoxime is a very useful procedure for introducing a primary amine. Many reducing agents will perform this reduction, including borane, lithium aluminum hydride, and

◻ **Table 9**
Reduction of 2-keto-α-D-glucoside

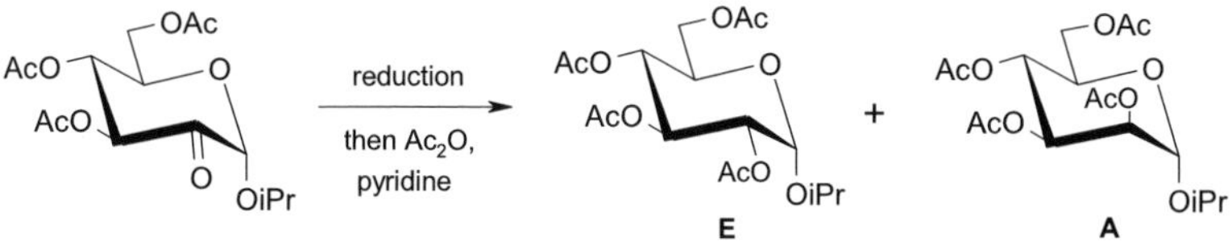

Reagent	Solvent	E:A
NaBH$_4$	H$_2$O/dioxane	96:4
(BH$_3$)$_2$	THF	94:6
H$_2$, Pd/C	EtOH	32:68
H$_2$, Rh/C	EtOH	20:80

◻ **Table 10**
Reduction of 2-ketoximes

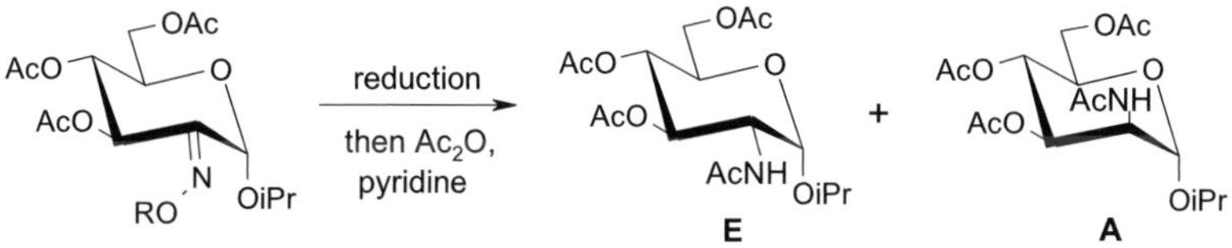

R	Reagent	Solvent	E:A
H	H$_2$, Pd/C, HCl	MeOH	1:1
H	LiAlH$_4$	THF	2:3
H	(BH$_3$)$_2$	THF	7:3
Ac	(BH$_3$)$_2$	THF	95:5

catalytic hydrogenation over various metals [154]. The oxime carbon is not as electrophilic as a ketone carbon, and as a result sodium borohydride alone does not reduce an oxime. However, in the presence of an additive such as nickel(II) chloride, titanium(III) chloride, or titanium(IV) chloride, borohydrides will also perform oxime reductions [2]. Like other oximes, carbohydrate oximes are also usually prepared from the corresponding ketone. Oximes at C2, however, can also be prepared by nitrosochlorination of glycals followed by reaction with alcohols [184]. The stereoselectivity in the oxime to amine reduction is often similar to the one obtained in the reduction of the corresponding ketone, which has been described in the previous section.

For reduction of 2-ketoximes the stereochemical outcome is determined by the reducing agent, the substituent on the oxime and the stereochemistry at the anomeric center (❯ *Table 10*) [185]. Borane gives the best selectivity for the 1,2-*cis* product and this selectivity can be dramatically enhanced by using the acetylated oxime. This turns out to be general for reduction of glucose derived 2-ketoxime esters (❯ *Scheme 24*) [186]. α-Glucosides of these 2-oxime esters give high yields of *N*-acetylglucosaminide derivatives (e. g., **67**) when reduced with borane followed by *N*-acetylation. On the other hand, β-glucosides give *N*-acetylmannosaminide

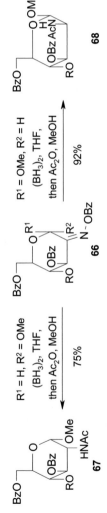

R = 2,3,4,6-tetra-O-benzoyl-β-D-galactopyranosyl

Scheme 24

Table 11
Reduction of 3-ketoximes

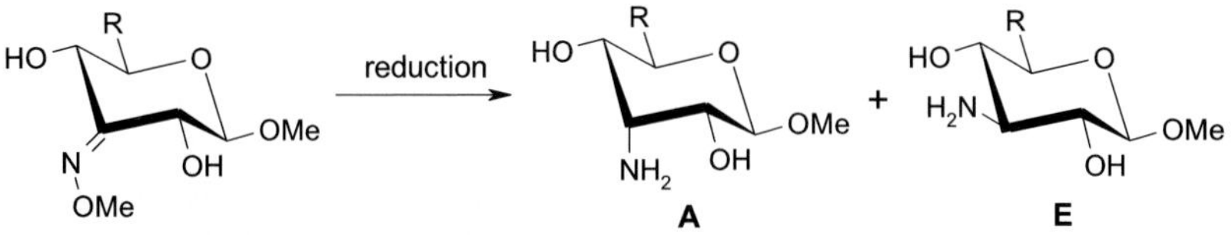
R
HO
O
OMe
N
OMe
OH
reduction
HO
NH2
OH
A
+
H2N
E

R	Reagent	Solvent	A:E	Yield (%)
H	AlH_3	THF	0:1	52
H	H_2, PtO_2	AcOH	7:3	87
CH_2OH	AlH_3	THF	1:2	61
CH_2OH	H_2, PtO_2	AcOH	7:1	67

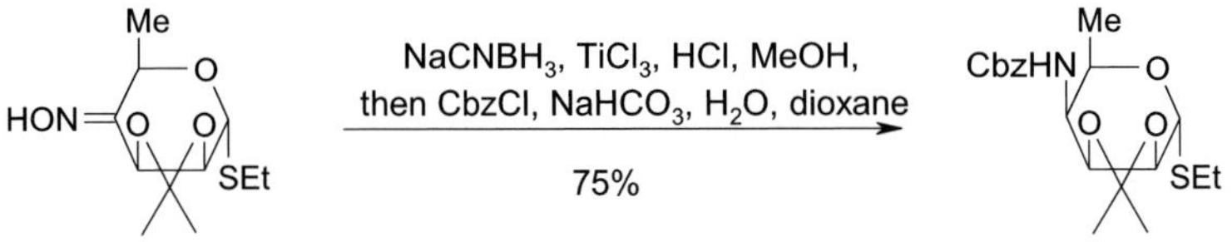
Me
HON
O
SEt
NaCNBH3, TiCl3, HCl, MeOH,
then CbzCl, NaHCO3, H2O, dioxane
75%
CbzHN

69

70

Scheme 25

residues (e. g., **68**) very stereoselectively [186]. Although borane-THF complex is normally the reagent of choice for this reduction of 2-ketoximes, sodium borohydride/nickel(II) chloride [187] and lithium borohydride/trimethylsilyl chloride [188] have also been shown to work well. The latter mixture presumably generates borane in situ.

A study on the reduction of 3-ketoximes shows that the stereochemical outcome depends on the reagent (*Table 11*) [189]. Catalytic hydrogenation favors the axial amine while the sterically less demanding alane gives more of the equatorial amine. For reduction of 4-ketoximes the same study shows that the axial product dominates regardless of the reducing agent [189]. This has also been observed in the reduction of **69** which gives the axial product **70** almost exclusively (*Scheme 25*) [190].

3.5 Hydrogenation of Olefins

Saturation of a carbohydrate double bond is almost always carried out by catalytic hydrogenation over a noble metal. The reaction takes place at the surface of the metal catalyst that absorbs both hydrogen and the organic molecule. The metal is often deposited onto a support, typically charcoal. Palladium is by far the most commonly used metal for catalytic hydrogenation of olefins. In special cases, more active (and more expensive) platinum and rhodium catalysts can also be used [154]. All these noble metal catalysts are deactivated by sulfur, except when sulfur is in the highest oxidation state (sulfuric and sulfonic acids/esters). The lower oxidation state sulfur compounds are almost always catalytic poisons for the metal catalyst and even minute traces may inhibit the hydrogenation very strongly [154]. Sometimes Raney nickel can be used to remove traces of sulfur impurities prior to the hydrogenation. Raney nickel can also be used on its own as a hydrogenation catalyst although it is less reactive than the noble metals and more catalyst is required [191].

Hydrogenation over noble metals is usually performed at room temperature and at 1–3 atmospheres of hydrogen. The progress of the reaction can be monitored by measuring the hydrogen uptake. Methanol, ethanol, or ethyl acetate are normally the solvents of choice. Catalytic hydrogenation is sensitive to the steric environment around the olefin, but less sensitive to electronic factors. The hydrogenation generally occurs from the sterically less demanding face of the olefin [154].

A number of carbohydrates contain an endocyclic olefin. If this is not further substituted, hydrogenation occurs readily as shown for hex-2-enopyranoside **71a** (❍ *Scheme 26*) [192]. A benzylidene protecting group is unaffected under these conditions. If the double bond is further substituted, as in hex-2-enopyranoside **71b**, the hydrogenation takes longer [193]. However, it occurs very selectively from the face of the olefin opposite to the axial anomeric methoxy group. Interestingly, when this methoxy group is absent, the facial selectivity reverses, as seen for hydrogenation of lactone **73** [194]. This is, however, in accordance with the general trend observed for 2,3-unsaturated aldonolactones, which preferentially undergo hydrogenation from the face of the olefin opposite to the side chain [195]. The elimination to form the unsaturated aldonolactone and the subsequent hydrogenation can be carried out as a one-pot procedure if a base is added to the reaction mixture. In this way, peracetylated aldonolactones can be hydrogenated in the presence of triethylamine to give 3-deoxylactones

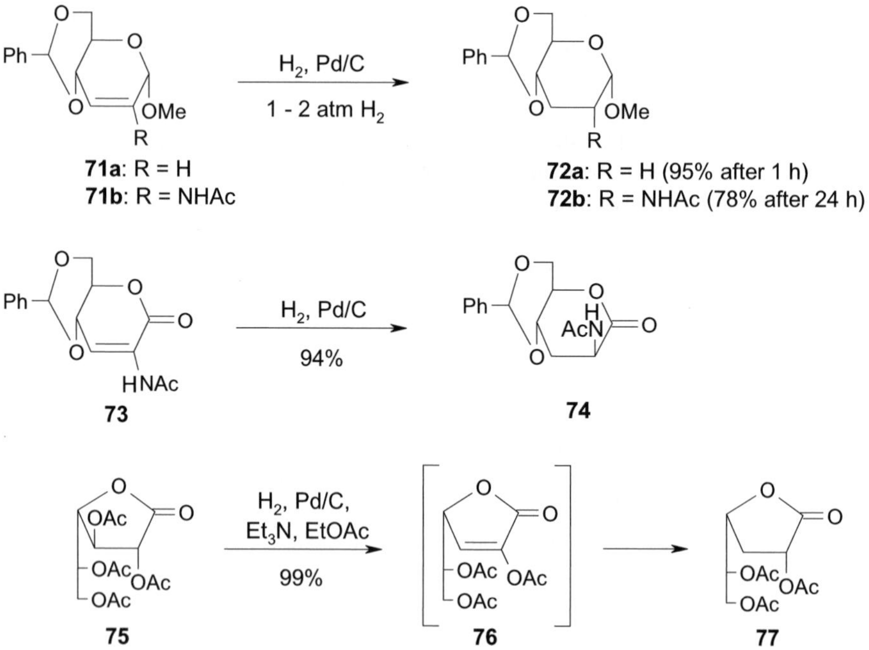

❑ Scheme 26

□ **Table 12**
Hydrogenation of ascorbic acid

78	Catalyst	H₂ Pressure (atm)	Temperature (°C)	Time	Yield (%)	Reference
23.1 g	2.2 g of 10% Pd/C	3.4	50	24 h	99	[198]
250.0 g	5 g of PdCl₂ on C	50.0	50	72 h	85	[199]
10.0 g	1 g of 5% Rh/C	3.7	rt	1.5 h	75–90	[200]

(❍ *Scheme 26*) [196]. Tetra-*O*-acetyl-D-galactono-1,4-lactone **75** yields 3-deoxylactone **77** through the unsaturated lactone **76**. Because of the facial selectivity in the hydrogenation, the 2-acetate group in the product is always *cis* to the side chain. 2,3-Unsaturated aldono-lactones can also be saturated by 1,4-reduction with tributyltin hydride, copper(I) iodide, and trimethylsilyl chloride [197]. Under these conditions other isolated olefins are not affected.

A special, but cheap, 2,3-unsaturated aldonolactone is ascorbic acid **78** which undergoes very selective hydrogenation to L-gulono-1,4-lactone **79** (❍ *Table 12*). However, ascorbic acid is very unreactive towards hydrogenation and a relatively large amount of catalyst is needed. If Pd/C is used as the catalyst, 100 g of ascorbic acid requires about 1 g of palladium metal for the hydrogenation [198,199]. If the more reactive and also more expensive Rh/C is used, about half of that amount is needed to hydrogenate the same amount of ascorbic acid [200]. As a result, the catalyst is the most expensive reactant for hydrogenation of ascorbic acid.

Another special substrate containing an endocyclic double bond is enol acetate **80** prepared from glucose isopropylidene ketone **24** (❍ *Scheme 27*) [201]. Hydrogenation occurs selective-ly from the face opposite to the 1,2-*O*-isopropylidene group to give gulofuranose **81** [85,202]. This reduction combined with the oxidation in ❍ *Table 4* can be used for conversion of D-glu-cose into D-gulose [201].

For carbohydrates containing an exocyclic double bond, hydrogenation will introduce a stere-ocenter in the ring. In this way, branched sugars can be obtained by hydrogenation of products derived from Wittig-type olefinations. For example, hydrogenation of **82**, also prepared from ketone **24**, gives the branched furanose **83** [203]. Again, the hydrogenation takes place from the face opposite to the 1,2-*O*-isopropylidene group. For hex-5-enopyranosides the hydrogena-tion gives 6-deoxypyranosides, e. g., hex-5-enopyranoside **84** gives rise to L-fucoside **85** as the major product (❍ *Scheme 27*) [204]. A smaller amount (17%) of the epimeric 6-deoxy-D-altro compound is also obtained in this reaction.

4 Deoxygenations

A hydroxy group can be removed by a number of methods that usually involve a two-step process where the hydroxy group is first converted into another functional group and this group is then subsequently replaced by hydrogen [205].

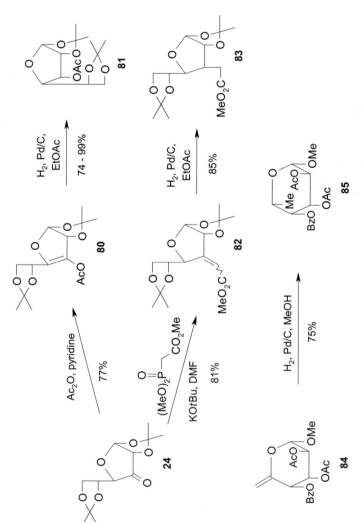

■ Scheme 27

4.1 Deoxygenation at the Anomeric Center

Deoxygenation of aldoses at C1 gives rise to 1,4- and 1,5-anhydroalditols which can be used as chiral scaffolds for further synthesis. One of the preferred methods for deoxygenating aldoses and their glycosides at C1 employs triethylsilane in the presence of trimethylsilyl triflate [206]. The strong Lewis acid mediates the formation of an oxocarbenium ion at C1 which is then reduced by the silane. Under these conditions several unprotected methyl aldopyranosides and aldofuranosides are converted into 1,5- and 1,4-anhydroalditols by persilylation with BSTFA [bis(trimethylsilyl)trifluoroacetamide] followed by treatment with triethylsilane and trimethylsilyl triflate in the same pot (❷ *Scheme 28*) [206]. Methyl hexopyranosides and methyl pentofuranosides give good yield of the corresponding 1,5-anhydrohexitols and 1,4-anhydropentitols, respectively. Some methyl pentopyranosides, on the other hand, undergo rearrangement into 1,4-anhydropentitols under the reaction conditions [206]. This rearrangement can be avoided by using the peracetylated methyl pentopyranosides or the peracetylated pentopyranoses as the starting material [207]. Besides ester-protected substrates the reaction can also be applied to ether-protected aldoses (❷ *Scheme 28*) [208]. Furthermore, a methoxy group and an acetate are not the only groups that can be reductively cleaved from C1. A hydroxy group can be removed in aldoses that are fully protected at all positions but the anomeric center [209]. In addition, 1,2-*O*-isopropylidenefuranoses undergo reductive cleavage of the acetal to afford 1,4-anhydroalditols if borontrifluoride etherate is used as the Lewis acid instead of trimethylsilyl triflate [210].

Another well-adapted method for removing the oxygen functionality at C1 uses a radical reduction of a protected pyranosyl halide (❷ *Scheme 28*) [211]. Formally, this is not a deoxygenation since the hydroxy group at C1 has already been replaced by a halide. However, glycosyl halides are easily available from aldoses by a one-pot procedure [212] and combined with the radical reduction this two-step route gives easy access to a number of 1,5-anhydroalditols. 1,4-Anhydroalditols, on the other hand, are more difficult to obtain by this method since the corresponding furanosyl halides require more steps for their preparation. The rad-

◨ Scheme 28

ical reduction is usually performed with tributyltin hydride and a catalytic amount of 2,2′-azobisisobutyronitrile (AIBN) and goes through the formation of a glycosyl radical which is stabilized by the endocyclic oxygen [213]. The radical reduction at C1 can also be carried out with titanocene borohydride which can be prepared from titanocene dichloride and sodium borohydride [214]. Furthermore, radical reductions can be achieved with protected aldoses containing a phenyl thionocarbonate at C1 [215]. Because of the toxicity of organotin compounds this reduction has been performed with a catalytic amount of tributyltin hydride in the presence of polymethylhydrosiloxane as the stoichiometric reductant [215].

Other methods for preparing 1,5-anhydroalditols employ lithium aluminum hydride reduction of protected 1,2-epoxy pyranoses and glycosyl halides. For example, the reductive ring-opening of the glucose-derived epoxide in ❷ *Table 7* affords 1,5-anhydro-3,4,6-tri-*O*-benzyl-D-glucitol in 74% yield [216] while the reduction of 2,3,4-tri-*O*-acetyl-α-L-rhamnopyranosyl bromide gives 1,5-anhydro-L-rhamnitol in 87% yield [217]. Additionally, anhydroalditols can also be prepared by Raney nickel-mediated desulfurization of thioglycosides [218] while the same reduction on aldose dialkyl dithioacetals gives rise to 1-deoxyalditols [219].

4.2 Deoxygenation of Primary Alcohols

Diisopropylidenegalactopyranose **7** has been deoxygenated at C6 under various conditions which illustrates some of the methods that are available for removing a primary alcohol (❷ *Table 13*). Sulfonates can be selectively introduced at the primary position in many carbohydrates and can be displaced by a hydride from either lithium aluminum hydride or sodium borohydride. Besides galactose the reduction has also been applied for removing primary sulfonates in glucose [230], mannose [231], and ribose [232]. Carboxylates, on the other hand, are not displaced by hydride, but can be removed by photolysis at 254 nm in an aqueous hexamethylphosphoric triamide (HMPA) solution [223]. The reaction goes through a radical mechanism and does not tolerate halides and other carbonyl groups in the substrate [233].

A more common radical reaction for deoxygenating alcohols is the Barton–McCombie reaction [234]. In this transformation the alcohol is converted into a thiocarbonyl derivative (xanthate, thionocarbonate, or thionocarbamate) which undergoes homolytic C-OCS cleavage upon treatment with tributyltin hydride and a radical initiator. In the original Barton–McCombie procedure secondary alcohols are treated with *N,N′*-thiocarbonyldiimidazole or carbon disulfide/methyl iodide/sodium hydride and the resulting thionocarbamate or xanthate is then reductively cleaved with tributyltin hydride [234]. However, these procedures proved inefficient for deoxygenating primary alcohols due to the slightly lower stability of a primary radical as compared to a secondary radical [234,235]. Instead, improved conditions for primary alcohols have been developed by acylation with 2,4,6-trichlorophenyl or 4-fluorophenyl chlorothionoformate followed by deoxygenation of the resulting thionocarbonate [224]. Although tributyltin hydride is an effective reducing agent which is compatible with esters, ethers, acetals, and olefins, organotin compounds are toxic and difficult to remove completely from the desired products. Furthermore, tributyltin hydride is rather expensive and has a limited shelf-stability. Therefore, alternative reagents have been investigated and particular attention has been given to hydrogen donors containing Si–H or P–H bonds [236]. So far, the most effective tin hydride

◻ Table 13

Deoxygenation of diisopropylidenegalactopyranose 7 to diisopropylidenefucopyranose 86

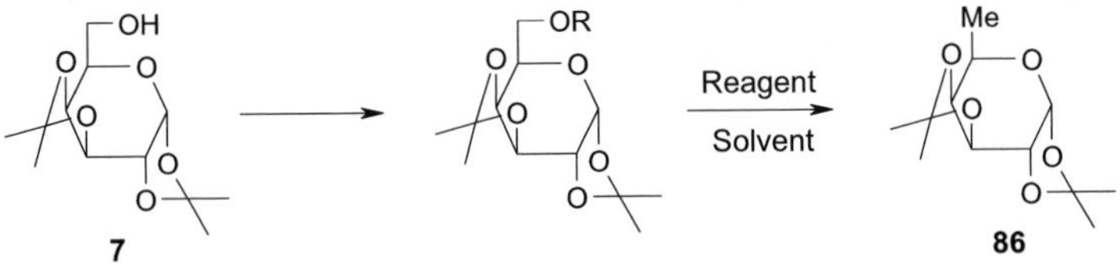

R	Reagent	Solvent	Yield (%)	Reference
Ts	LiAlH$_4$	Et$_2$O	59	[220]
Ts	NaBH$_4$	DMSO	88	[221]
Tf	NaBH$_4$	MeCN	92	[222]
Ac	hν	H$_2$O/HMPA	85	[223]
C=S(O-2,4,6-Cl$_3$Ph)	Bu$_3$SnH, AIBN	Toluene	91	[224]
C=S(O-4-FPh)	PhSiH$_3$, (BzO)$_2$	Toluene	88	[224]
C=S(O-4-FPh)	Ph$_3$SiH, (BzO)$_2$	Toluene	88	[225]
C=S(NHPh)	TMS$_3$SiH, AIBN	Benzene	85	[226]
C=S(O-4-FPh)	H$_3$PO$_2$, Et$_3$N, AIBN	Dioxane	91	[227]
C=S(O-4-FPh)	(MeO)$_2$PHO, (BzO)$_2$	Dioxane	90	[228]
C=S(SMe)	(Bu$_4$N)$_2$S$_2$O$_8$, HCO$_2$Na	DMF	86	[229]

substitute has been tris(trimethylsilyl)silane, but unfortunately this reagent is also quite expensive. Arylsilanes, hypophosphorous acid, and dialkyl phosphites are significantly less reactive than tributyltin hydride and require longer reaction times and larger amounts of the radical initiator. However, work-up and product purification with these alternative hydrogen donors is easy and particularly the P–H reagents are much cheaper than tributyltin hydride. More recently an interesting new procedure based on tetrabutylammonium peroxodisulfate and sodium formate has been published where a range of alcohols are deoxygenated in excellent yield in less than 1 h [229].

Another radical reaction for deoxygenating the C6 position in hexoses employs the corresponding 4,6-O-benzylidene derivative. These acetals undergo a thiol-catalyzed radical redox rearrangement to afford 6-deoxyhexoses with a benzoate at C4 [237]. The rearrangement is initiated by thermal decomposition of a peroxide which then reacts with the thiol to generate the reactive thiyl radical. The reaction works very well with *trans*-fused acetals, as in glucose, while the *cis*-fused acetals, as in galactose, give a poor regioselectivity resulting in deoxygenation at C4 and at C6. The rearrangement tolerates a range of functional groups and can even be

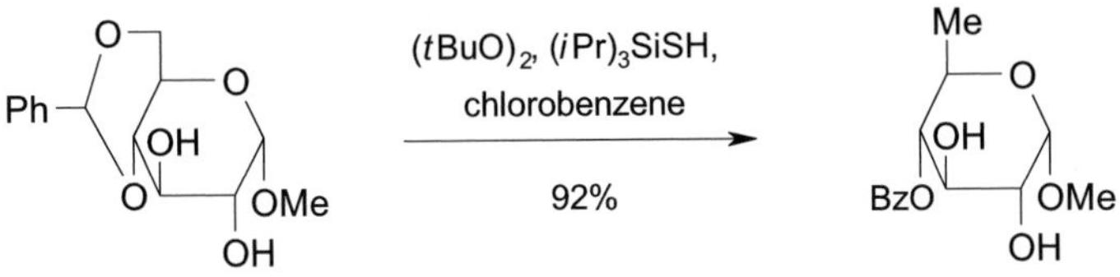

◻ Scheme 29

applied to a partially protected substrate (❂ *Scheme 29*) [237]. A similar radical deoxygenation at C6 can be achieved by using the more complex 4,6-*O*-[1-cyano-2-(2-iodophenyl)ethylidene] acetal [238]. In this case, the redox rearrangement is mediated by tributyltin hydride and AIBN. Again, the *trans*-fused acetals in glucose and mannose undergo a very regioselective fragmentation to yield the 6-deoxy compounds while the regioselectivity with the *cis*-fused acetal in galactose is poor [238].

4.3 Deoxygenation of Secondary Alcohols

The reduction of secondary sulfonates with lithium aluminum hydride or sodium borohydride is usually a poor reaction for deoxygenating secondary alcohols [220,222]. In most cases, the hydride attack will occur at sulfur and result in cleavage of the S–O bond to afford the starting secondary alcohol as the main product. An exception from this rule is observed when tetrabutylammonium borohydride is used for reduction of secondary triflates in refluxing benzene [239]. Under these conditions clean displacement with hydride occurs to give the corresponding deoxy compounds in good yield (❂ *Table 14*).

A much more common transformation for deoxygenating secondary alcohols is the Barton–McCombie reaction and various modifications of this method. Diisopropylideneglucofuranose **23** has served as a model compound for many of these deoxygenation reactions

■ Table 14

Deoxygenation of diisopropylideneglucofuranose 23 to 3-deoxyfuranose 87

R	Reagent	Solvent	Yield (%)	Reference
Tf	Bu$_4$NBH$_4$	Benzene	84	[239]
C=S(SMe)	Bu$_3$SnH	Toluene	75	[240]
C=S(imidazolide)	Bu$_3$SnH	Toluene	74	[235]
C=S(O-2,4,6-Cl$_3$Ph)	Bu$_3$SnH, AIBN	Benzene	100	[241]
C=S(O-4-FPh)	PhSiH$_3$, (BzO)$_2$	Toluene	100	[225]
C=S(SMe)	Ph$_2$SiH$_2$, AIBN	Toluene	92	[242]
C=S(SMe)	Ph$_3$SiH, (BzO)$_2$	Toluene	95	[225]
C=S(OPh)	TMS$_3$SiH, AIBN	Toluene	81	[243]
C=S(NHPh)	TMS$_3$SiH, AIBN	Benzene	99	[226]
C=S(SMe)	H$_3$PO$_2$, Et$_3$N, AIBN	Dioxane	91	[227]
C=S(SMe)	(MeO)$_2$PHO, (BzO)$_2$	Dioxane	97	[227]
C=S(SMe)	(Bu$_4$N)$_2$S$_2$O$_8$, HCO$_2$Na	DMF	98	[229]
Bz	Mg(ClO$_4$)$_2$, hν[a]	H$_2$O/iPrOH	86	[244]

[a]9-Ethyl-3,6-dimethylcarbazole is used as the photosensitizer

◻ **Scheme 30**

(❯ *Table 14*). The early procedures used tribultyltin hydride on the corresponding *S*-methyl xanthate or imidazoylthiocarbonyl derivative [234,235]. Later on it was shown that faster conversion and better yields could be obtained by using a radical initiator and a thiono-carbonate containing an electron-withdrawing aryl group [241]. AIBN is usually employed as the initiator and together with tributyltin hydride constitute the most widely employed reagent mixture for deoxygenating secondary alcohols (❯ *Scheme 30*) [245]. However, the toxicity of tin hydrides and the problems associated with the work-up has prompted a search for alternative reducing agents. In some applications catalytic amounts of tributyltin hydride or solid-supported tin hydrides have been used for deoxygenation of **23** [215,246]. In most cases, however, the tin hydride has been replaced by another reducing agent containing either a Si–H or a P–H bond. Several examples of these reagents are illustrated in ❯ *Table 14*. In addition, photoinduced deoxygenation of benzoates has been used for the removal of secondary alcohols [244]. The reaction is selective for benzoyl esters of secondary alcohols and cannot be used for deoxygenation of primary alcohols [247]. Recently, the photoinduced deoxygenation reaction has been applied for selective deoxygenation at C2 in aldonolactones [248].

The synthesis of 2-deoxyaldonolactones can also be achieved from the parent lactones if they contain a triflate or a tosylate at C2. The treatment of these sulfonated lactones with iodide [249], hydrazine [250], or by catalytic hydrogenolysis [251] leads to the corresponding 2-deoxyaldonolactones in good yields. When 2-bromo-2-deoxyaldonolactones are subjected to catalytic hydrogenolysis the reaction can give either the 2-deoxy- or the 2,3-dideoxylactone depending on the presence or absence of an acid scavenger (❯ *Scheme 31*) [252]. The debromination by hydrogenolysis in the presence of triethylamine is a well-established method for the synthesis of deoxysugars [253]. The formation of the dideoxylactone, however, is an unusual transformation that seems to proceed through the 2,3-unsaturated lactone.

2-Deoxyaldonolactones can also be prepared by a samarium(II) iodide-mediated deoxygenation reaction. By this procedure a range of protected and unprotected aldonolactones undergo selective reduction at C2 with 3 equiv. of the reagent (❯ *Scheme 31*) [254]. If the starting lactone contains an ester at C3 the reduction is accompanied by elimination to afford a 2,3-unsaturated aldonolactone [254].

Another radical reaction for preparation of 2-deoxysugars utilizes acylated glycosyl halides as the starting material. As mentioned previously, when these glycosyl halides are treated with tributyltin hydride and AIBN the initially formed glycosyl radical is reduced to give anhydroalditols [211]. However, if tributyltin hydride and AIBN are added very slowly over

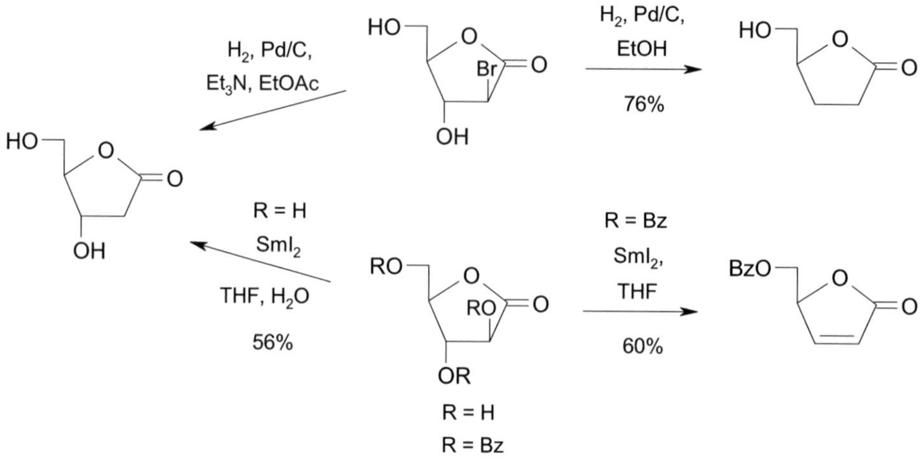

■ Scheme 31

several hours to a solution of the glycosyl halide, the concentration of the hydrogen donor will be low enough to ensure rearrangement of the glycosyl radical before the reduction takes place [213]. Under these conditions the acyloxy group at C2 will migrate to the anomeric center and the resulting C2 centered radical will then abstract hydrogen to afford a peracylated 2-deoxyaldose (❯ Scheme 32) [255]. The slow addition can be avoided if tributyltin hydride is replaced by tris(trimethylsilyl)silane since the latter is a less effective hydrogen donor and therefore allows the rearrangement to take place before the hydrogen abstraction [256].

2-Deoxyaldoses can also be prepared from partially protected glycals by addition of water or acetic acid [257]. A special reaction for synthesis of a 2-deoxysugar employs dibenzyliden-emannoside 88 (❯ Scheme 32) [258]. Reaction of 88 with butyl lithium leads to a selective

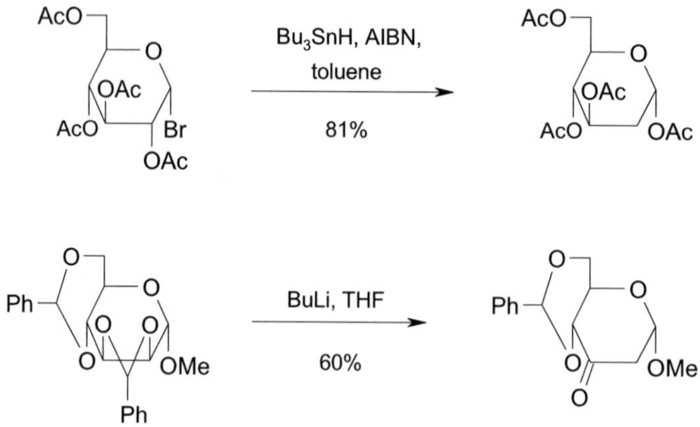

■ Scheme 32

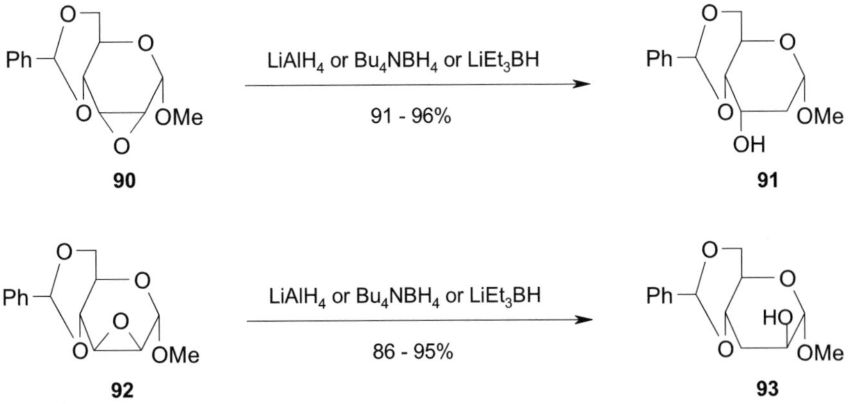

Scheme 33

deprotonation at C3 and subsequent elimination of benzaldehyde to give 2-deoxy-3-keto sugar **89**.

A more general procedure for synthesis of deoxysugars uses ring-opening of an epoxide with hydride. Monosubstituted epoxides will react at the primary position while 1,2-disubstituted epoxides can react at both secondary positions. In six-membered rings the epoxide opening is controlled by the Fürst–Plattner rule and gives rise to products with a *trans*-diaxial orientation between the secondary alcohol and the incorporated hydride. Lithium aluminum hydride is often used as the reducing agent [259], but the ring-opening can also be achieved with tetrabutylammonium borohydride [239], lithium triethylborohydride [260], and in situ generated borane [128]. The ring-opening is shown in ❷ *Scheme 33* with three different reagents on 2,3-epoxides **90** and **92** [239,259,260]. For both compounds, very regioselective ring-opening is observed to give 2-deoxyglycoside **91** from *allo* epoxide **90** and 3-deoxyglycoside **93** from *manno* epoxide **92**. It should be noticed that epoxides **90** and **92** are both easily prepared from tosylates of methyl 4,6-*O*-benzylidene-α-D-glucopyranoside [261]. The regioselective reduction can also be performed with 3,4-epoxides to give deoxysugars with an axial hydroxy group at C3 or C4 [262]. Furthermore, cyclic sulfates undergo regioselective ring-opening with tetrabutylammonium borohydride in the same way as the corresponding epoxides [263].

Acknowledgement

The author thanks the Lundbeck Foundation for financial support. The Center for Sustainable and Green Chemistry is sponsored by the Danish National Research Foundation.

References

1. Collins PM, Ferrier RJ (1995) Monosaccharides – Their Chemistry and Their Roles in Natural Products. Wiley, Chichester

2. Paquette LA (ed) (1995) Encyclopedia of Reagents for Organic Synthesis. Wiley, Chichester

3. Hudlický M (1990) Oxidations in Organic Chemistry, ACS Monograph Ser No 186. ACS, Washington, DC

4. Hudson CS, Isbell HS (1929) J Am Chem Soc 51:2225; Nelson WL, Cretcher LH (1930) J Am Chem Soc 52:403; Isbell HS (1963) Meth Carbohydr Chem 2:13

5. Isbell HS, Frush HL (1993) J Res Natl Bur Stand 11:649

6. Isbell HS, Pigman WW (1937) J Org Chem 1:505

7. Isbell HS, Frush HL (1931) J Res Natl Bur Stand 6:1145; Frush HL, Isbell HS (1963) Meth Carbohydr Chem 2:14; Zinner H, Voigt H, Voigt J (1968) Carbohydr Res 7:38

8. Isaac I, Stasik I, Beaupère D, Uzan R (1995) Tetrahedron Lett 36:383

9. Besson M, Lahmer F, Gallezot P, Fuertes P, Flèche G (1995) J Catal 152:116

10. Biella S, Prati L, Rossi M (2002) J Catal 206:242

11. Wenkin M, Ruiz P, Delmon B, Devillers M (2002) J Mol Catal A: Chem 180:141

12. Ramachandran S, Fontanille P, Pandey A, Larroche C (2006) Food Technol Biotechnol 44: 185

13. Beltrame P, Comotti M, Della Pina C, Rossi M (2004) J Catal 228:282

14. Pezzotti F, Therisod M (2006) Carbohydr Res 341:2290; Pezzotti F, Therisod H, Therisod M (2005) Carbohydr Res 340:139

15. Horton D, Jewell JS (1966) Carbohydr Res 2:251

16. Manna S, McAnalley BH, Ammon HL (1993) Carbohydr Res 243:11

17. Kuzuhara H, Fletcher HG (1967) J Org Chem 32:2531

18. Buchanan JG, Smith D, Wightman RH (1984) Tetrahedron 40:119

19. Benhaddou R, Czernecki S, Farid W, Ville G, Xie J, Zegar A (1994) Carbohydr Res 260:243

20. Garegg PJ, Samuelsson B (1978) Carbohydr Res 67:267

21. Shing TKM, Gillhouley JG (1994) Tetrahedron 50:8685

22. Hanessian S, Ugolini A (1984) Carbohydr Res 130:261

23. Hall RH, Bischofberger K (1978) Carbohydr Res 65:139

24. Ley SV, Norman J, Griffith WP, Marsden SP (1994) Synthesis 639

25. Csuk R, Dörr P (1995) J Carbohydr Chem 14:35

26. Schäfer HJ, Schneider R (1991) Tetrahedron 47:715

27. Tidwell TT (1990) Org React 39:297; Tidwell TT (1990) Synthesis 857

28. Luzzio FA (1998) Org React 53:1

29. Carlsen PHJ, Katsuki T, Martin VS, Sharpless KB (1981) J Org Chem 46:3936

30. Morris Jr PE, Kiely DE (1987) J Org Chem 52:1149

31. Dess DB, Martin JC (1991) J Am Chem Soc 113:7277; Meyer SD, Schreiber SL (1994) J Org Chem 59:7549

32. Deslongchamps P, Moreau C (1971) Can J Chem 49:2465

33. Pistia G, Hollingsworth RI (2000) Carbohydr Res 328:467

34. Lichtenthaler FW, Rönninger S, Jarglis P (1989) Liebigs Ann Chem 1153

35. Yadav JS, Subba Reddy BV, Suresh Reddy C (2004) Tetrahedron Lett 45:4583

36. Lee HH, Hodgson PG, Bernacki RJ, Korytnyk W, Sharma M (1988) Carbohydr Res 176:59; Howarth GB, Lance DG, Szarek WA, Jones JKN (1969) Can J Chem 47:75

37. Cree GM, Mackie DW, Perlin AS (1969) Can J Chem 47:511

38. Midland MM, Asirwatham G, Cheng JC, Miller JA, Morell LA (1994) J Org Chem 59:4438

39. Lough C, Hindsgaul O, Lemieux RU (1983) Carbohydr Res 120:43

40. Andersson F, Samuelsson B (1984) Carbohydr Res 129:C1

41. Ward DJ, Szarek WA, Jones JKN (1972) Carbohydr Res 21:305

42. Barili PL, Berti G, D'Andrea F, Bussolo VD, Granucci I (1992) Tetrahedron 48:6273

43. Goodman JL, Horton D (1968) Carbohydr Res 6:229

44. Cronin L, Murphy PV (2005) Org Lett 7:2691

45. Abe H, Terauchi M, Matsuda A, Shuto S (2003) J Org Chem 68:7439

46. Mazur AW (1991) In: Bednarski MD, Simon ES (eds) Enzymes in Carbohydrate Synthesis, ACS Symp Ser 466. ACS, Washington, DC, p 99

47. Andreana PR, Sanders T, Janczuk A, Warrick JI, Wang PG (2002) Tetrahedron Lett 43:6525; Root RL, Durrwachter JR, Wong C-H (1985) J Am Chem Soc 107:2997; Avigad G, Amaral D, Asensio C, Horecker BL (1962) J Biol Chem 237:2736

48. Vogel C, Jeschke U, Vill V, Fischer H (1992) Liebigs Ann Chem 1171

49. Reichstein T, Grüssner A (1934) Helv Chim Acta 17:311

50. Godage HY, Fairbanks AJ (2000) Tetrahedron Lett 41:7589

51. Torii S, Inokuchi T, Sugiura T (1986) J Org Chem 51:155

52. Lyazidi HA, Benabdallah MZ, Berlan J, Kot C, Fabre P-L, Mestre M, Fauvarque J-F (1996) Can J Chem Eng 74:405; Nanzer J, Langlois S, Cœuret F (1993) J Appl Electrochem 23:477; Robertson PM, Berg P, Reimann H, Schleich K, Seiler P (1983) J Electrochem Soc 130:591

53. Weijlard J (1945) J Am Chem Soc 67:1031

54. Hecker SJ, Minich ML (1990) J Org Chem 55:6051

55. Garner P, Park JM (1990) J Org Chem 55:3772

56. Steffan W, Vogel C, Kristen H (1990) Carbohydr Res 204:109; Betaneli VI, Ott AY, Brukhanova OV, Kochetkov NK (1988) Carbohydr Res 179:37

57. Jarosz S (1988) Carbohydr Res 183:201; van Boeckel CAA, Belbressine LPC, Kaspersen FM (1985) Recl Trav Chim Pays-Bas 104:259

58. Allanson NM, Liu D, Chi F, Jain RK, Chen A, Ghosh M, Hong L, Sofia MJ (1998) Tetrahedron Lett 39:1889

59. Westman J, Nilsson M (1995) J Carbohydr Chem 14:949

60. Vogel C, Gries P (1994) J Carbohydr Chem 13:37

61. Halkes KM, Slaghek TM, Hyppönen TK, Kruiskamp PH, Ogawa T, Kamerling JP, Vliegenthart JFG (1998) Carbohydr Res 309:161

62. Nilsson M, Svahn C-M, Westman J (1993) Carbohydr Res 246:161

63. Nakahara Y, Ogawa T (1988) Carbohydr Res 173:306

64. Chambers DJ, Evans GR, Fairbanks AJ (2005) Tetrahedron 61:7184

65. Clausen MH, Madsen R (2003) Chem Eur J 9:3821

66. Lichtenthaler FW, Jarglis P, Lorenz K (1988) Synthesis 790

67. Garegg PJ, Olsson L, Oscarson S (1995) J Org Chem 60:2200

68. De Souza MVN (2006) Mini-Rev Org Chem 3:155

69. Lin F, Peng W, Xu W, Han X, Yu B (2004) Carbohydr Res 339:1219

70. Davis NJ, Flitsch SL (1994) J Chem Soc Perkin Trans 1 359

71. Bragd PL, van Bekkum H, Besemer AC (2004) Top Catal 27:49

72. van den Bos LJ, Codée JDC, van der Toorn JC, Boltje TJ, van Boom JH, Overkleeft HS, van der Marel GA (2004) Org Lett 6:2165

73. Heyns K, Paulsen H (1962) Adv Carbohydr Chem 17:169

74. Fabre J, Betbeder D, Paul F, Monsan P (1993) Synth Commun 23:1357; Vleeming JH, Kuster BFM, Marin GB (1997) Carbohydr Res 303:175

75. Johnson L, Verraest DL, van Haveren J, Hakala K, Peters JA, van Bekkum H (1994) Tetrahedron Asymmetry 5:2475

76. Mallat T, Brönnimann C, Baiker A (1997) J Mol Cat A: Chem 117:425

77. Mehltretter CL (1963) Meth Carbohydr Chem 2:46

78. Venema FR, Peters JA, van Bekkum H (1992) J Mol Cat 77:75

79. Merbouh N, Bobbitt JM, Brückner C (2002) J Carbohydr Chem 21:65; Merbouh N, Thaburet JF, Ibert M, Marsais F, Bobbitt JM (2001) Carbohydr Res 336:75

80. Yoshikawa M, Okaichi Y, Cha BC, Kitagawa I (1990) Tetrahedron 46:7459

81. Yoshimura J, Sato K, Hashimoto H (1977) Chem Lett 1327

82. Mazur A, Tropp BE, Engel R (1984) Tetrahedron 40:3949

83. Lankin DC, Nugent ST, Rao SN (1993) Carbohydr Res 244:49

84. Shing TKM, Wong C-H, Yip T (1996) Tetrahedron Asymmetry 7:1323

85. Legler G, Pohl S (1986) Carbohydr Res 155:119

86. Baker DC, Horton D, Tindall Jr CG (1972) Carbohydr Res 24:192

87. Bio MM, Xu F, Waters M, Williams JM, Savary KA, Cowden CJ, Yang C, Buck E, Song ZJ, Tschaen DM, Volante RP, Reamer RA, Grabowski EJJ (2004) J Org Chem 69:6257

88. Rao HSP, Muralidharan P, Pria S (1997) Indian J Chem 36B:816

89. Hollenberg DH, Klein RS, Fox JJ (1978) Carbohydr Res 67:491

90. Herscovici J, Egron M-J, Antonakis K (1982) J Chem Soc Perkin Trans 1 1967

91. Agarwal S, Tiwari HP, Sharma JP (1990) Tetrahedron 46:4417; Czernecki S, Georgoulis C, Stevens CL, Vijayakumaran K (1985) Tetrahedron Lett 26:1699

92. Wang Z-X, Tu Y, Frohn M, Zhang J-R, Shi Y (1997) J Am Chem Soc 119:11224

93. Ichikawa S, Shuto S, Matsuda A (1998) Tetrahedron Lett 39:4525

94. Gonsalvi L, Arends IWCE, Sheldon RA (2002) Org Lett 4:1659

95. Mio S, Kumagawa Y, Sugai S (1991) Tetrahedron 47:2133

96. Skaanderup PR, Madsen R (2003) J Org Chem 68:2115; Cook GP, Greenberg MM (1994) J Org Chem 59:4704; Alonso RA, Burgey CS, Rao BV, Vite GD, Vollerthun R, Zottola MA, Fraser-Reid B (1993) J Am Chem Soc 115:6666

97. Gan L-X, Seib PA (1991) Carbohydr Res 220:117

98. Henbest HB, Jones ERH, Owen TC (1957) J Chem Soc 4909; Mancera O, Rosenkranz G, Sondheimer F (1953) J Chem Soc 2189

99. Fraser-Reid B, McLean A, Usherwood EW, Yunker M (1970) Can J Chem 48:2877

100. Kitahara T, Ogawa T, Naganuma T, Matsui M (1974) Agric Biol Chem 38:2189

101. Goodwin TE, Crowder CM, White RB, Swanson JS, Evans FE, Meyer WL (1983) J Org Chem 48:376

102. Dinh TN, Khac DD, Gandolfi I, Memoria Y, Fétizon M, Prangé T (1993) Bull Soc Chim Fr 130:287

103. Tronchet JMJ, Tronchet J, Birkhäuser A (1970) Helv Chim Acta 53:1489

104. Czernecki S, Leteux C, Veyrières A (1992) Tetrahedron Lett 33:221

105. Bouillot A, Khac DD, Fétizon M, Guir F, Memoria Y (1993) Synth Commun 23:2071

106. Kirschning A (1995) J Org Chem 60:1228; Kirschning A (1998) Eur J Org Chem 2267

107. Arterburn JB (2001) Tetrahedron 57:9765

108. Grindley TB (1994) In: Kovác P (ed) Synthetic Oligosaccharides – Indispensable Probes for the Life Sciences, ACS Symp Ser 560. ACS, Washington, DC, p 51

109. Tsuda Y, Hanajima M, Matsuhira N, Okuno Y, Kanemitsu K (1989) Chem Pharm Bull 37:2344

110. Liu H-M, Sato Y, Tsuda Y (1993) Chem Pharm Bull 41:491

111. Aspinall GO, Gammon DW, Sood RK, Chatterjee D, Rivoire B, Brennan PJ (1992) Carbohydr Res 237:57

112. Abbadi A, van Bekkum H (1995) Appl Catal A: General 124:409

113. Lockwood LB (1963) Meth Carbohydr Chem 2:51

114. Deppenmeier U, Hoffmeister M, Prust C (2002) Appl Microbiol Biotechnol 60:233

115. Silverbach M, Maier B, Zimmermann M, Büchs J (2003) Appl Microbiol Biotechnol 62:92; Weenk G, Olijve W, Harder W (1984) Appl Microbiol Biotechnol 20:400; Shinagawa E, Matsushita K, Adachi O, Ameyama M (1983) J Ferment Technol 61:359

116. Hann RM, Tilden EB, Hudson CS (1938) J Am Chem Soc 60:1201

117. Lockwood LB (1962) Meth Carbohydr Chem 1:151

118. Stoppok E, Walter J, Buchholz K (1995) Appl Microbiol Biotechnol 43:706; Stoppok E, Matalla K, Buchholz K (1992) Appl Microbiol Biotechnol 36:604

119. Maeda A, Adachi S, Matsuno R (2001) Biochem Eng J 8:217; Klekner V, Löbl V, Šímová E, Novák M (1989) Folia Microbiol 34:286

120. Van Beeumen J, De Ley J (1968) Eur J Biochem 6:331

121. Giffhorn F (2000) Appl Microbiol Biotechnol 54:727

122. Volc J, Sedmera P, Halada P, Dwivedi P, Costa-Ferreira M (2003) J Carbohydr Chem 22:207; Volc J, Sedmera P, Halada P, Prikrylová V, Haltrich D (2000) Carbohydr Res 329:219; Sedmera P, Volc J, Havlícek V, Pakhomova S, Jegorov A (1997) Carbohydr Res 297:375

123. Halcomb RL, Danishefsky SJ (1989) J Am Chem Soc 111:6661

124. Bellucci G, Catelani G, Chiappe C, D'Andrea F (1994) Tetrahedron Lett 35:8433

125. Cheshev P, Marra A, Dondoni A (2006) Carbohydr Res 341:2714

126. Yoshimura J, Sato K-i, Funabashi M (1979) Bull Chem Soc Jpn 52:2630

127. Enright PM, Tosin M, Nieuwenhuyzen M, Cronin L, Murphy PV (2002) J Org Chem 67:3733

128. Andresen TL, Skytte DM, Madsen R (2004) Org Biomol Chem 2:2951; Murphy PV, O'Brien JL, Smith III AB (2001) Carbohydr Res 334:327

129. Tronchet JMJ, Tronchet J (1977) Helv Chim Acta 60:1984

130. Bílik V, Kucár Š (1970) Carbohydr Res 13:311

131. Castro-Palomino JC, Tsvetkov YE, Schmidt RR (1998) J Am Chem Soc 120:5434

132. Bílik V (1972) Chem Zvesti 26:76

133. Railton CJ, Clive DLJ (1996) Carbohydr Res 281:69

134. Jørgensen M, Iversen EH, Madsen R (2001) J Org Chem 66:4625

135. Cha JK, Christ WJ, Kishi Y (1984) Tetrahedron 40:2247

136. Prenner RH, Binder WH, Schmid W (1994) Liebigs Ann Chem 73; Brimacombe JS, Kabir AKMS (1988) Carbohydr Res 179:21

137. Marshall JA, Beaudoin S (1994) J Org Chem 59:6614

138. Ermolenko L, Sasaki NA (2006) J Org Chem 71:693

139. López-Herrera FJ, Sarabia-García F, Pino-González MS, García-Aranda JF (1994) J Carbohydr Chem 13:767; Rassu G, Casiraghi G, Spannu P, Pinna L, Fava GG, Ferrari MB, Pelosi G (1992) Tetrahedron Asymmetry 3:1035

140. Tiwari P, Misra AK (2006) J Org Chem 71:2911; Shing TKM, Tam EKW, Tai VW-F, Chung IHF, Jiang Q (1996) Chem Eur J 2:50

141. Matsumoto K, Ebata T, Matsushita H (1995) Carbohydr Res 267:187; Schulte G, Meyer W, Starkloff A, Dyong I (1981) Chem Ber 114:1809; Dyong I, Schulte G, Lam-Chi Q, Friege H (1979) Carbohydr Res 68:257

142. Lemieux RU, Ratcliffe RM (1979) Can J Chem 57:1244

143. Hashimoto H, Araki K, Saito Y, Kawa M, Yoshimura J (1986) Bull Chem Soc Jpn 59:3131

144. Darakas E, Hultberg H, Leontein K, Lönngren J (1982) Carbohydr Res 103:176

145. Toyokuni T, Cai S, Dean B (1992) Synthesis 1236

146. Broddefalk J, Nilsson U, Kihlberg J (1994) J Carbohydr Chem 13:129

147. Paulsen H, Lorentzen JP, Kutschker W (1985) Carbohydr Res 136:153

148. Seeberger PH, Roehrig S, Schell P, Wang Y, Christ WJ (2000) Carbohydr Res 328:61

149. Mironov YV, Sherman AA, Nifantiev NE (2004) Tetrahedron Lett 45:9107; Czernecki S, Ayadi E (1995) Can J Chem 73:343; Santoyo-González F, Calvo-Flores FG, García-Mendoza P, Hernández-Mateo F, Isac-García J, Robles-Diás R (1993) J Org Chem 58:6122

150. Jiaang W-T, Chang M-Y, Tseng P-H, Chen S-T (2000) Tetrahedron Lett 41:3127; SanMartin R, Tavassoli B, Walsh KE, Walter DS, Gallagher T (2000) Org Lett 2:4051

151. Somsák L, Ferrier RJ (1991) Adv Carbohydr Chem Biochem 49:37

152. Vogel C, Liebelt B, Steffan W, Kristen H (1992) J Carbohydr Chem 11:287; Ferrier RJ, Tyler PC (1980) J Chem Soc Perkin Trans 1 1528; Blattner R, Ferrier RJ (1980) J Chem Soc Perkin Trans 1 1523; Ferrier RJ, Furneaux RH (1977) J Chem Soc Perkin Trans 1 1996

153. Medakovic D (1994) Carbohydr Res 253:299

154. Hudlický M (1996) Reductions in Organic Chemistry, 2nd edn, ACS Monograph Ser No 188. ACS, Washington, DC

155. Wolfrom ML, Thompson A (1963) Meth Carbohydr Chem 2:65; Hayward LD, Wright IG (1963) Meth Carbohydr Chem 2:258

156. Frush HL, Isbell HS (1956) J Am Chem Soc 78:2844

157. Bock K, Lundt I, Pedersen C (1981) Carbohydr Res 90:7

158. Johansen SK, Kornø HT, Lundt I (1999) Synthesis 171

159. Hough L, Theobald RS (1962) Meth Carbohydr Chem 1:94

160. Gallezot P, Cerino PJ, Blanc B, Fléche G, Fuertes P (1994) J Catal 146:93

161. Hoffer BW, Crezee E, Mooijman PRM, van Langeveld AD, Kapteijn F, Moulijn JA (2003) Catal Today 79–80:35; Kusserow B, Schimpf S, Claus P (2003) Adv Synth Catal 345:289

162. Heinen AW, Peters JA, van Bekkum H (2000) Carbohydr Res 328:449

163. Fleet GWJ, Son JC, Green DSC, di Bello IC, Winchester B (1988) Tetrahedron 44:2649; Helleur R, Rao VS, Perlin AS (1981) Carbohydr Res 89:83

164. Jiang S, Singh G, Wightman RH (1996) Chem Lett 67; Doane WM, Shasha BS, Russell CR, Rist CE (1967) J Org Chem 32:1080

165. Lerner LM, Kohn BD, Kohn P (1968) J Org Chem 33:1780

166. García-Moreno MI, Díaz-Pérez P, Mellet CO, Fernández JMG (2003) J Org Chem 68:8890; Rosen T, Taschner MJ, Heathcock CH (1984) J Org Chem 49:3994

167. Pedersen C, Jensen HS (1994) Acta Chem Scand 48:222; Kohn P, Samaritano RH, Lerner LM (1965) J Am Chem Soc 87:5475

168. Fleet GWJ, Son JC (1988) Tetrahedron 44:2637

169. Baer HH, Breton RL, Shen Y (1990) Carbohydr Res 200:377

170. Danishefsky SJ, Maring CJ (1989) J Am Chem Soc 111:2193

171. Lewis BA, Smith F, Stephen AM (1963) Meth Carbohydr Chem 2:68

172. Taylor RL, Shively JE, Conrad HE (1976) Meth Carbohydr Chem 7:149

173. Liu KK-C, Danishefsky SJ (1994) J Org Chem 59:1892

174. Garegg PJ (1992) Acc Chem Res 25:575

175. Chang C-WT, Hui Y, Elchert B (2001) Tetrahedron Lett 42:7019

176. Kochetkov NK, Dmitriev BA, Malysheva NN, Chernyak AY, Klimov EM, Bayramova NE, Torgov VI (1975) Carbohydr Res 45:283

177. Eis MJ, Ganem B (1988) Carbohydr Res 176:316
178. Bischofberger K, Brink AJ, de Villiers OG, Hall RH, Jordaan A (1977) J Chem Soc Perkin Trans 1 1472; Boeyens JCA, Brink AJ, Jordaan A (1978) J Chem Research (S) 187
179. Lichtenthaler FW, Schneider-Adams T (1994) J Org Chem 59:6728
180. Lemieux RU, James K, Nagabhushan TL (1973) Can J Chem 51:27
181. Borén HB, Ekborg G, Eklind K, Garegg PJ, Pilotti Å, Swahn C-G (1973) Acta Chem Scand 27:2639
182. Lichtenthaler FW, Lergenmüller M, Peters S, Varga Z (2003) Tetrahedron Asymmetry 14:727
183. Pelyvás I, Hasegawa A, Whistler RL (1986) Carbohydr Res 146:193
184. Lemieux RU, Ito Y, James K, Nagabhushan TL (1973) Can J Chem 51:7
185. Lemieux RU, James K, Nagabhushan TL, Ito Y (1973) Can J Chem 51:33; Lemieux RU, Gunner SW (1968) Can J Chem 46:397
186. Lichtenthaler FW, Kaji E, Weprek S (1985) J Org Chem 50:3505; Kaji E, Lichtenthaler FW (1995) J Carbohydr Chem 14:791
187. Smiatacz Z, Paszkiewicz E, Chrzczanowicz I (1991) J Carbohydr Chem 10:315
188. Karpiesiuk W, Babaszek A (1990) J Carbohydr Chem 9:909
189. Tsuda Y, Okuno Y, Iwaki M, Kanemitsu K (1989) Chem Pharm Bull 37:2673
190. Smid P, Jörning WPA, van Duuren AMG, Boons GJPH, van der Marel GA, van Boom JH (1992) J Carbohydr Chem 11:849
191. Thomas SS, Plenkiewicz J, Ison ER, Bols M, Zou W, Szarek WA, Kisilevsky R (1995) Biochim Biophys Acta 1272:37
192. Albano EL, Horton D (1969) J Org Chem 34:3519
193. Calvo-Mateo A, Camarasa MJ, De las Heras FG (1984) J Carbohydr Chem 3:461
194. Varela O, Nin AP, de Lederkremer RM (1994) Tetrahedron Lett 35:9359
195. Choquet-Farnier C, Stasik I, Beaupère D (1997) Carbohydr Res 303:185; Varela OJ, Cirelli AF, de Lederkremer RM (1979) Carbohydr Res 70:27
196. Bock K, Lundt I, Pedersen C (1981) Acta Chem Scand B35:155
197. Song J, Hollingsworth RI (2001) Tetrahedron Asymmetry 12:387
198. Andrews GC, Crawford TC, Bacon BE (1981) J Org Chem 46:2976
199. Czarnocki Z, Mieczkowski JB, Ziólkowski M (1996) Tetrahedron Asymmetry 7:2711
200. Soriano DS, Meserole CA, Mulcahy FM (1995) Synth Commun 25:3263
201. Meyer zu Reckendorf W (1972) Meth Carbohydr Chem 6:129
202. Lemieux RU, Stick RV (1975) Aust J Chem 28:1799; Tronchet JMJ, Bourgeois JM (1970) Helv Chim Acta 53:1463
203. Rosenthal A, Nguyen L (1969) J Org Chem 34:1029
204. Chiba T, Tejima S (1979) Chem Pharm Bull 27:2838
205. Hartwig W (1983) Tetrahedron 39:2609
206. Bennek JA, Gray GR (1987) J Org Chem 52:892
207. Jeffery A, Nair V (1995) Tetrahedron Lett 36:3627
208. Guo Z-W, Hui Y-Z (1996) Synth Commun 26:2067
209. Nicotra F, Panza L, Russo G, Zucchelli L (1992) J Org Chem 57:2154
210. Ewing GJ, Robins MJ (1999) Org Lett 1:635
211. Bamford MJ, Pichel JC, Husman W, Patel B, Storer R, Weir NG (1995) J Chem Soc Perkin Trans 1 1181; Kocienski P, Pant C (1982) Carbohydr Res 110:330
212. Kartha KPR, Jennings HJ (1990) J Carbohydr Chem 9:777
213. Praly J-P (2001) Adv Carbohydr Chem Biochem 56:65
214. Cavallaro CL, Schwartz J (1996) J Org Chem 61:3863
215. Tormo J, Fu GC (2002) Org Synth 78:239
216. Flaherty TM, Gervay J (1996) Tetrahedron Lett 37:961
217. Ness RK, Fletcher Jr HG, Hudson CS (1950) J Am Chem Soc 72:4547
218. Lemieux RU (1951) Can J Chem 29:1079
219. Sarbajna S, Das SK, Roy N (1995) Carbohydr Res 270:93
220. Schmid H, Karrer P (1949) Helv Chim Acta 32:1371
221. Thiem J, Meyer B (1980) Chem Ber 113:3067
222. Barrette E-P, Goodman L (1984) J Org Chem 49:176
223. Pete J-P, Portella C, Monneret C, Florent J-C, Khuong-Huu Q (1977) Synthesis 774
224. Barton DHR, Blundell P, Dorchak J, Jang DO, Jaszberenyi JC (1991) Tetrahedron 47:8969
225. Barton DHR, Jang DO, Jaszberenyi JC (1993) Tetrahedron 49:2793
226. Oba M, Nishiyama K (1994) Tetrahedron 50:10193

227. Barton DHR, Jang DO, Jaszberenyi JC (1993) J Org Chem 58:6838
228. Barton DHR, Jang DO, Jaszberenyi JC (1992) Tetrahedron Lett 33:2311
229. Park HS, Lee HY, Kim YH (2005) Org Lett 7:3187
230. Hanessian S (1972) Meth Carbohydr Chem 6:190; Schmidt OT (1962) Meth Carbohydr Chem 1:198
231. Nishio T, Miyake Y, Kubota K, Yamai M, Miki S, Ito T, Oku T (1996) Carbohydr Res 280:357
232. Sairam P, Puranik R, Rao BS, Swamy PV, Chandra S (2003) Carbohydr Res 338:303
233. Portella C, Deshayes H, Pete JP, Scholler D (1984) Tetrahedron 40:3635
234. Barton DHR, McCombie SW (1975) J Chem Soc Perkin Trans 1 1574
235. Rasmussen JR, Slinger CJ, Kordish RJ, Newman-Evans DD (1981) J Org Chem 46:4843
236. Studer A, Amrein S (2002) Synthesis 835
237. Dang H-S, Roberts BP, Sekhon J, Smits TM (2003) Org Biomol Chem 1:1330
238. Crich D, Bowers AA (2006) J Org Chem 71:3452
239. Sato K-i, Hoshi T, Kajihara Y (1992) Chem Lett 1469
240. Iacono S, Rasmussen JR (1986) Org Synth 64:57
241. Barton DHR, Jaszberenyi JC (1989) Tetrahedron Lett 30:2619
242. Barton DHR, Jang DO, Jaszberenyi JC (1993) Tetrahedron 49:7193
243. Schummer D, Höfle G (1990) Synlett 705
244. Buck JR, Park M, Wang Z, Prudhomme DR, Rizzo CJ (2000) Org Synth 77:153
245. Ruttens B, Kovác P (2004) Synthesis 2505
246. Boussaguet P, Delmond B, Dumartin G, Pereyre M (2000) Tetrahedron Lett 41:3377; Neumann WP, Peterseim M (1992) Synlett 801
247. Wang Z, Prudhomme DR, Buck JR, Park M, Rizzo CJ (2000) J Org Chem 65:5969
248. Bordoni A, de Lederkremer RM, Marino C (2006) Carbohydr Res 341:1788
249. Stewart AJ, Evans RM, Weymouth-Wilson AC, Cowley AR, Watkin DJ, Fleet GWJ (2002) Tetrahedron Asymmetry 13:2667
250. Malle BM, Lundt I, Furneaux RH (2000) J Carbohydr Chem 19:573
251. Kalwinsh I, Metten K-H, Brückner R (1995) Heterocycles 40:939
252. Lundt I, Pedersen C (1986) Synthesis 1052; Bock K, Lundt I, Pedersen C (1981) Carbohydr Res 90:17
253. Horton D, Cheung T-M, Weckerle W (1980) Meth Carbohydr Chem 8:195
254. Hanessian S, Girard C, Chiara JL (1992) Tetrahedron Lett 33:573; Hanessian S, Girard C (1994) Synlett 861
255. Giese B, Gröninger KS (1990) Org Synth 69:66
256. Giese B, Kopping B, Chatgilialoglu C (1989) Tetrahedron Lett 30:681
257. Lam SN, Gervay-Hague J (2003) Org Lett 5:4219; Costantino V, Imperatore C, Fattorusso E, Mangoni A (2000) Tetrahedron Lett 41:9177; Dupradeau F-Y, Hakomori S-i, Toyokuni T (1995) J Chem Soc Chem Commun 221
258. Testero SA, Spanevello RA (2006) Carbohydr Res 341:1057
259. Jones K, Wood WW (1988) J Chem Soc Perkin Trans 1 999; Prins DA (1948) J Am Chem Soc 70:3955
260. Baer HH, Hanna HR (1982) Carbohydr Res 110:19
261. Wiggins LF (1963) Meth Carbohydr Chem 2:188
262. Hedgley EJ, Overend WG, Rennie RAC (1963) J Chem Soc 4701
263. Zegelaar-Jaarsveld K, van der Plas SC, van der Marel GA, van Boom JH (1996) J Carbohydr Chem 15:665

2.3 Heteroatom Exchange

Yuhang Wang, Xin-Shan Ye
The State Key Laboratory of Natural and Biomimetic Drugs, School of
Pharmaceutical Sciences, Peking University, Xue Yuan Rd #38, Beijing
100083, China
xinshan@bjmu.edu.cn

In: *Glycoscience*. Fraser-Reid B, Tatsuta K, Thiem J (eds)
Chapter-DOI 10-1007/978-3-540-30429-6_5: © Springer-Verlag Berlin Heidelberg 2008

Abstract

Saccharides derived with a variety of functional groups are usually of great importance in many biological and chemical aspects. The chemical modifications of saccharides with heteroatoms at nonanomeric positions are reviewed in this chapter.

Keywords

Halogens; Nitrogen; Sulfur; Phosphorus; Nucleophilic substitution; Ring-opening; Addition reaction

Abbreviations

AIBN	azo-bis-isobutyronitrile
CAN	ceric ammonium nitrate
DAST	diethylaminosulfur trifluoride
DBU	1,8-diazabicyclo[5.4.0]undec-7-ene
DEAD	diethyl azodicarboxylate
DFMBA	N,N-diethyl-α,α-difluoro-(m-methylbenzyl)amine
DIAD	diisopropyl azodicarboxylate
dppb	1,4-bis(diphenylphosphino)butane
IDCP	iodonium di-*sym*-collidine perchlorate
IBX	1-hydroxy-1,2-benziodoxol-3(1H)-one 1-oxide
NBS	N-bromosuccinimide
PMB	p-methoxybenzyl
SMDA	sodium dihydrobis(2-methoxyethoxy)aluminate
TASF	tri(dimethylamino)-sulfonium difluorotrimethylsilicate
TFAA	trifluoroacetic anhydride

1 Introduction

The sugar molecule can be considered as a playground for exploratory organic synthesis, and many functional group manipulations can be carried out on the nonanomeric carbons of carbohydrates. Among these reactions replacements of nonanomeric oxygen atoms by heteroatoms (e. g. halogen, nitrogen, sulfur, phosphorus, etc.) afford valuable compounds having biological activities as well as synthetic usefulness. Halogeno sugars can serve as tools for studying carbohydrate-protein interaction, and useful intermediates in the synthesis of other important sugar derivatives. Amino sugars are widely distributed in living organisms and are essential units of various antibiotics such as amino glycosides. Thiosugars exhibit biologically important properties as the substrates for many carbohydrate-related enzymes. Phosphorus derivatives have been widely used as chiral ligands for asymmetric synthesis and also as biological intermediates. Here synthetic methodologies for the preparation of these specific classes of carbohydrates will be reviewed and discussed in this chapter, as the revised version of the corresponding chapter written by Boullanger and Descotes in the previous edition of this book.

1.1 Nucleophilic Substitution

The manipulation of appropriately activated hydroxyl groups by nucleophilic displacement reactions is an indispensable tool for the introduction of functionalities directly attached to the sugar framework.

Carbohydrates are excellent substrates to test nucleophilic displacement reactions with a variety of heteroatom-based nucleophiles. These reactions can be tried on primary and secondary hydroxyl groups at different sites of the sugar ring, and with different steric or stereoelectronic dispositions [1].

The substitution reactions at the anomeric carbon usually proceed easily and via a S_N1 mechanism in most cases. However, it is not the case in the nucleophilic substitutions at nonanomeric sites. Because of the presence of vicinal electron-withdrawing substituents (OR or NHR) which strongly destabilize the intermediate carbocations, S_N2 displacement reactions instead of S_N1 reactions are favored (❷ Scheme 1).

Besides displacement reactions, direct replacement of hydroxyl groups by heteroatoms is also an attractive general approach to afford heteroatom-substituted sugar derivatives. A general and important method is based on the reaction of an activated triphenyl phosphine derivative with a sugar alcohol to give an alkylphosphonium ion, which in turn is attacked by a nucleophile to give the substituted product with the configuration inversed. The plausible mechanism is depicted in ❷ Scheme 2 [2].

❏ Scheme 1

$$R_3P: \overset{\frown}{E-Y} \longrightarrow \left[\overset{\oplus}{E-PR_3Y} \right] \longrightarrow HE + \overset{\oplus}{R_3P-O-R^1} \longrightarrow R^1Nu + R_3P=O$$

$$R^1-\overset{..}{O}-H \qquad\qquad \underset{Nu}{}$$

☐ Scheme 2

In addition to the phosphorus-containing reagents, other reagents, such as the Vilsmeier–Haack reagent [$(Me_2N^+=CHX)X^-$] [3], sulfuryl chloride [4] and diethylaminosulfur trifluoride (DAST) [5] have also been employed for direct replacement of hydroxyl groups in carbohydrates.

Classical substitution reactions usually involve the introduction of leaving groups, among which sulfonic esters provide a versatile and simple method for activating hydroxyl groups of the carbohydrate for a bimolecular displacement reaction. The most common sulfonate leaving groups consist of mesylates and tosylates. The use of p-bromobenzenesulfonates (brosylates), which are ten times more reactive than tosylates, has been occasionally reported [6]. Triflates and imidazole-1-sulfonates (imidazylates) [7] are used in the substitutions at positions where other sulfonates are known to be ineffective. Halogens can also be used as leaving groups.

The efficiency of a displacement critically depends on the position of the leaving group and the chemical environment of the sugar ring (the steric- or stereoelectronic dispositions of substituents). Generally displacement at C-6 (primary position) can proceed under relatively milder conditions. Primary sulfonates of hexopyranosides are readily displaced by nucleophiles provided that the C-4 oxygen is in an equatorial orientation (e. g. 6-O-sulfonates of D-glucosides). And the analogous reactions in the D-galactopyranose series are particularly sluggish, presumably because of the polar, repulsive forces in the transition state involving lone pairs of electrons on the axial O-4 and the ring-oxygen atom. The examples are illustrated in ❿ Scheme 3 by the conversion of 1 and 3 to 2 and 4, respectively [8,9].

Displacements at C-3 or C-4 (secondary position) proceed with more difficulty than those at C-6. Much more drastic conditions (long reaction time, high temperatures, and polar aprotic solvents) are necessary. This mainly results from the influence of steric and polar factors from other groups in the carbohydrate ring. The difference of substitution reactivity may probably rely on the orientations of leaving groups. Axial leaving groups are substituted faster (about three times) than equatorial ones.

☐ Scheme 3

The displacements at C-2 are very difficult owing to the vicinal C-1 bearing two electron-with-drawing oxygens, which retard the departing of the leaving group. This substitution depends strongly on the orientations of the anomeric substituents. When C-3 also bears an electron-withdrawing group, the displacement at C-2 of α-glycosides can hardly proceed, while the corresponding reaction at C-2 of β-glycosides is much more facile.

Furthermore, S_N2 displacements in furanoid rings are easier in comparison with pyranoid rings, since the more flexible five-membered rings have a smaller increasing strain, favoring the formation of the transition state. In unsaturated rings, substituents at the allylic positions are particularly susceptible to nucleophilic displacement due to the presence of double bonds. For example, mesylate compound 5 can be converted to the corresponding azide 6 and iodide 7, when treated with NaN_3 and NaI, respectively (❷ Scheme 4). Both reactions are easily realized in acetone at room temperature [10,11].

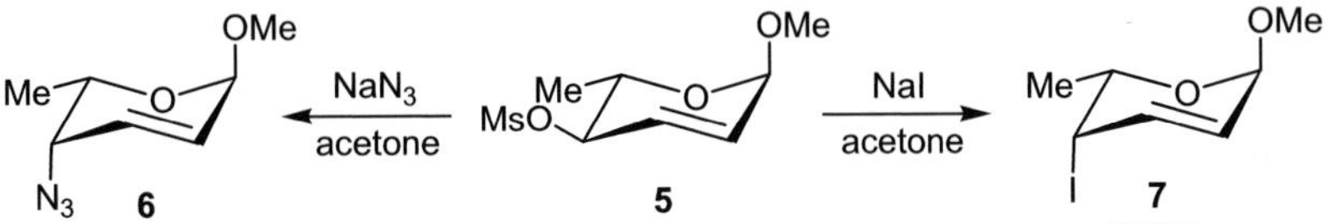

■ Scheme 4

Effective S_N2 substitution reactions are also greatly effected by anion nucleophilicity. An anion will be less nucleophilic when it is effectively solvated and when it is restricted by its counterion. This can be circumvented by selection of a tetraalkylammonium counterion, addition of a crown ether, and use of a solvent that effectively solvate cations.

Neighboring group participation involving acylamino or acyloxy groups is common in nucleophilic substitution. For example, in the reaction of methyl 4,6-O-benzylidene-2-deoxy-2-benzoylamino-3-O-mesyl-α-D-altro-pyranoside 8 with NaOEt, no 3-O-ethoxy-mannoside derivative was obtained. Instead, oxazoline 9 and epimine 10 were identified in this reaction (❷ Scheme 5) [12]. This results from the 1,2-trans-diaxial relationship between the leaving

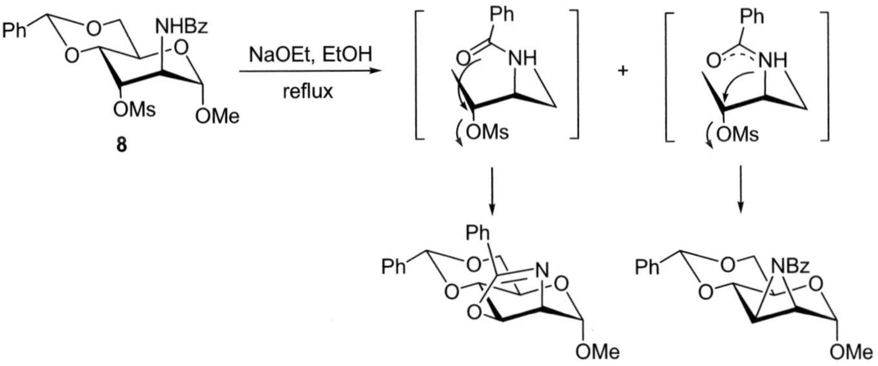

■ Scheme 5

☐ Scheme 6

group and the acylamino group, which is the ideal position for neighboring group participation. On the other hand, when the substituents in methyl 4,6-O-benzylidene-2-deoxy-2-
benzoylamino-3-O-mesyl-α-D-glucopyranoside **11** are in a 1,2-*trans*-diequatorial orientation,
the nucleophile competes with the acylamino group to form substituted products **12, 14**, and
oxazoline **13** (❍ *Scheme 6*) [13]. The participation of acyloxy groups is not very often encountered in pyranoid rings, even when the substituents are in suitable orientations. It seems that
acyloxy group participation is more facile in furanoid rings.

It has to be realized that an elimination reaction is the most frequent competing side reaction of
nucleophilic substitution. Strong bases and weak nucleophiles favor this reaction. And elimination reactions are more likely when the leaving group is in a *trans*-diaxial relationship with
a α-hydrogen atom at a sterically hindered position. Among halide ions, the fluoride ion is the
strongest base and the weakest nucleophile and, as a result, substitution by this anion is often
accompanied by elimination; sometimes, elimination products are the only ones formed (e. g.
15 to **16** in ❍ *Scheme 7*) [14]. To overcome elimination reactions, fluorides have been introduced by nucleophilic ring opening of epoxides, which is discussed in detail in ❍ *Sect. 1.2.*
The azide ion also is sufficiently basic to promote elimination reactions; however, because the
azide ion is an effective nucleophile, substitution is usually the dominant or exclusive reaction
(e. g. **17** to **18** in ❍ *Scheme 8*) [15].

In addition, another competing side reaction is molecular rearrangement of carbohydrate. For
example, attempted displacement of the 2-O-imidazyl ester of methyl 3,4,6-tri-O-methyl-β-D-

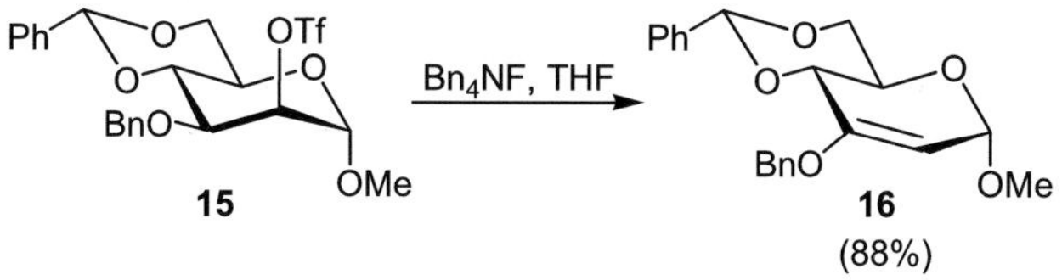

Scheme 7

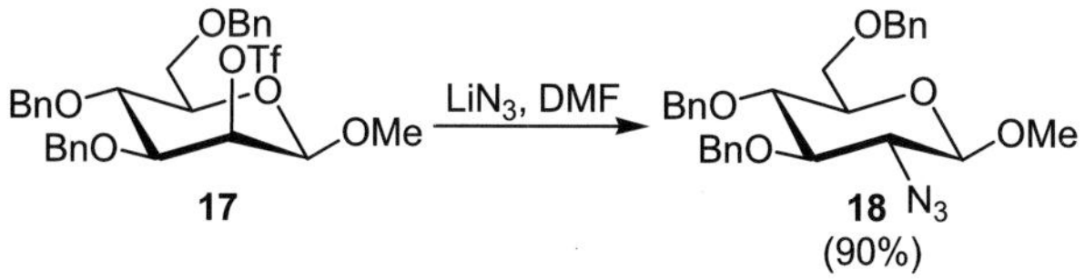

Scheme 8

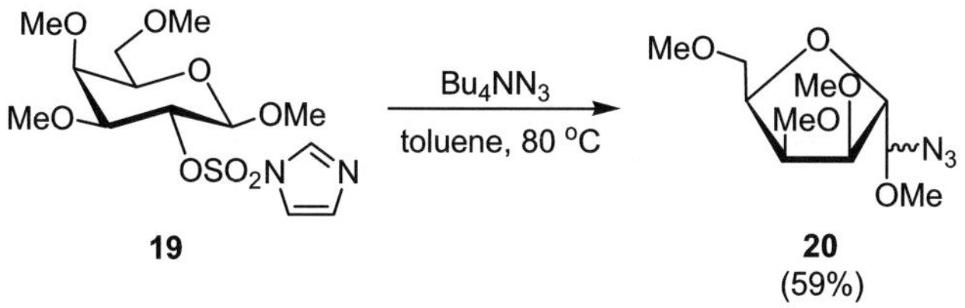

Scheme 9

galactopyranoside **19** with an azide ion led to the formation of the corresponding 2,5-anhydro sugar **20** (❖ *Scheme 9*) [16]. The predominance of a ring-contraction reaction over a substitution can be explained by a steric interaction between the incoming nucleophile and the axial C-4 substituent, and by the favored antiparallel disposition of the C-1-O-5 bond and the equatorial leaving groups at C-2.

1.2 Ring-Opening Reaction

1.2.1 Nucleophilic Ring Opening of Epoxides

Nucleophilic ring opening of epoxides with heteroatom nucleophiles is another valuable method for the synthesis of many heteroatom-modified carbohydrate derivatives. The cyclic nature of epoxides renders the competing elimination process stereoelectronically unfavorable. Analogous to the above-discussed S_N2 nucleophilic mechanism, nucleophiles can open epoxide rings, and give rise to Walden inversion at the attacked carbon, furnishing α-hydroxy derivatives as illustrated in ❖ *Scheme 10*.

In theory, nucleophiles can attack either of the two carbons on epoxide rings. Therefore, for asymmetric epoxides, two regioisomeric products will be formed. However, in many cases, the epoxide-opening reactions in carbohydrate rings can be highly regio- and stereoselective, even specific. For example, the *trans*-fused benzylidene D-manno compound **21**, when treated with NaN₃ gives the D-altro adduct **22** as the only product, while the same reaction of its D-allo epoxide **23** gives the D-altro adduct **24** preferentially (❖ *Scheme 11*) [17]. The propensity of the reaction to give *trans*-diaxial products possibly results from the transition state, in which the diaxial product adopts a chair conformation, whereas the diequatorial product

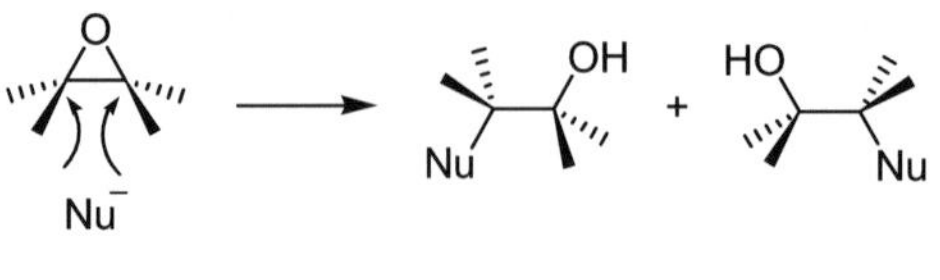

◻ Scheme 10

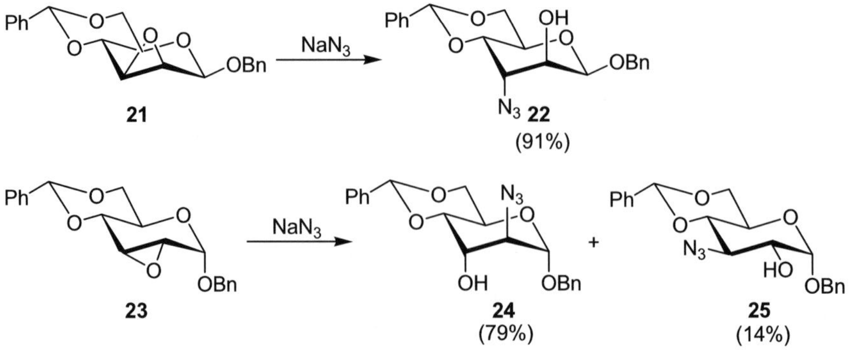

◻ Scheme 11

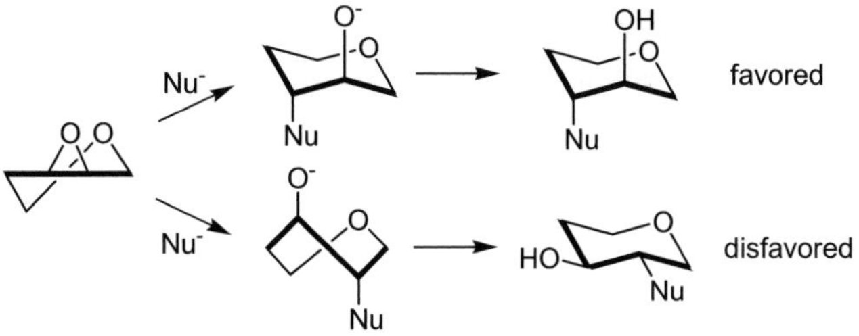

Scheme 12

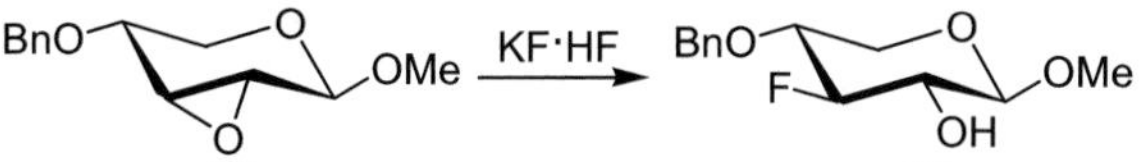

Scheme 13

adopts a twist-boat (❯ *Scheme 12*). Usually, the formation of a highly strained twist boat is energetically unfavorable and hence the diequatorial product is often not formed. In principle, the regioselectivity and stereoselectivity of carbohydrate epoxide rings can be improved by the introduction of additional rigidity in the starting material forcing the pyranose ring into one particular conformation. And for the less rigid conformations, it's difficult to predict the products.

Besides anionic nucleophiles (e. g. N_3^-), poor nucleophiles such as F^- can also be used in the ring-opening reactions in the presence of acids or Lewis acids as catalysts. For example, the reaction of methyl 2,3-anhydro-β-D-ribopyranoside **26** with F^- does not proceed except when acidic KF·HF is used to provide the nucleophile, yielding 3-deoxy-3-fluoroxylopyranoside **27** (❯ *Scheme 13*) [18].

Sometimes, intramolecular rearrangements involving neighboring group participation or epoxide-ring migration if free hydroxyl groups are suitably situated in the vicinity of the epoxide ring, can complicate the reaction, giving rearranged products together with expected epoxide ring-opening products (❯ *Scheme 14*) [19].

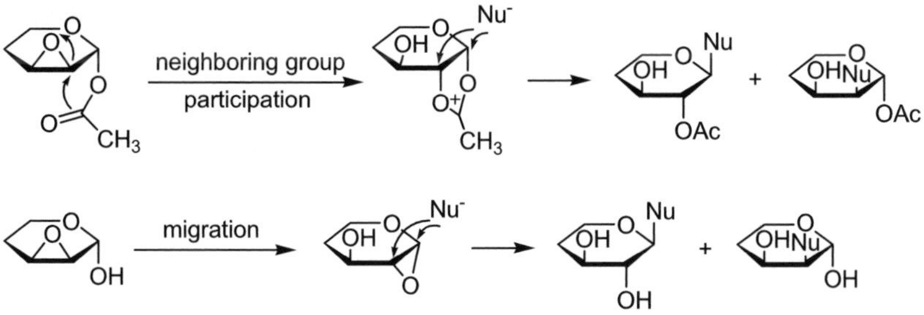

Scheme 14

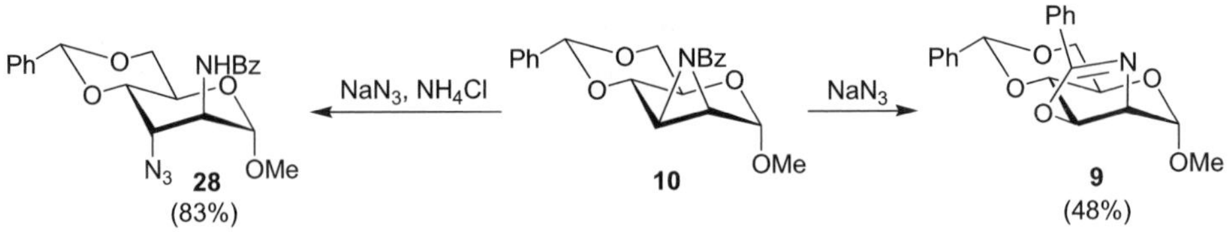

Scheme 15

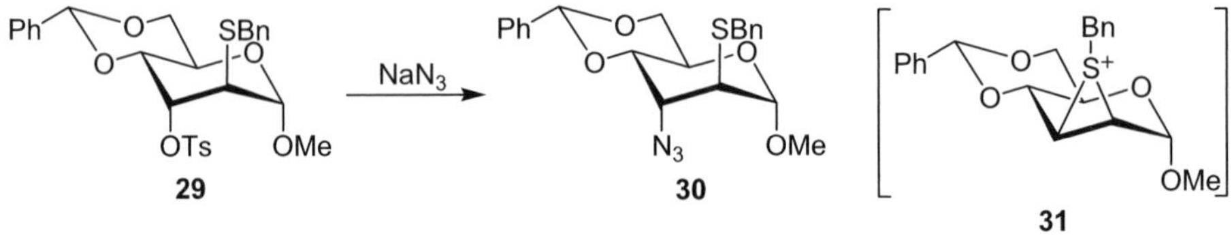

Scheme 16

Furthermore, the opening reactions of other three-membered rings such as aziridines and episulfides are similar to epoxides, except that they are more complicated. For example, the reaction of benzoylaziridine **10** with NaN$_3$ gives the *trans*-diaxial adduct **28** in the presence of NH$_4$Cl, whereas the oxazoline **9** is obtained as the major product in the absence of NH$_4$Cl (❯ *Scheme 15*) [20]. In another instance, the 2-benzylthio-2-deoxy-3-*O*-tosyl-α-D-altropyranoside **29** reacts with NaN$_3$ to afford the configuration-retained substitution derivative **30** (❯ *Scheme 16*), indicating the reaction occurs via the sulfonium salt intermediate **31** formed by intramolecular displacement from compound **29** [21].

1.2.2 Ring-Opening Reactions of Five/Six-Membered Rings

Five-membered ring compounds, such as cyclic sulfites, sulfates, sulfamidates, and thionocarbonates, can also be opened by treatment with nucleophiles in an analogous fashion to epoxide ring-opening. Nucleophilic ring-opening of cyclic sulfate **32** with NaN$_3$, for example, gives the azido-substituted product **33** (❯ *Scheme 17*) [22]. Noteworthy, these cyclic compounds are usually prepared from corresponding *cis* diols, whereas epoxides are synthesized from *trans* diols.

In some cases, six-membered rings can be involved in ring-opening reactions. For example, the reaction of 4,6-*O*-benzylidene glucopyranoside **34** with *N*-bromosuccinimide (NBS) in the presence of barium carbonate leads to the corresponding 4-*O*-benzoyl-6-bromo-6-deoxy

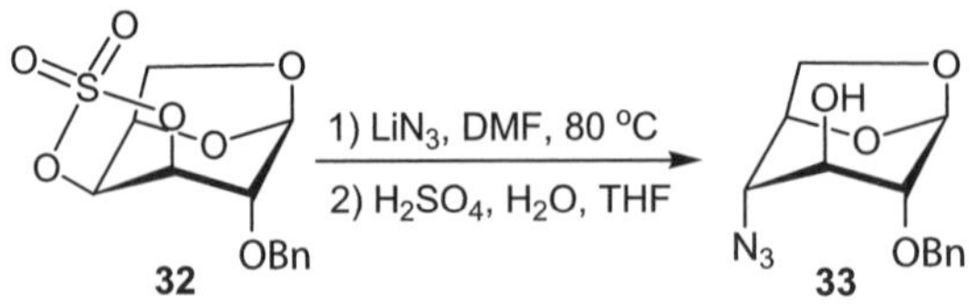

Scheme 17

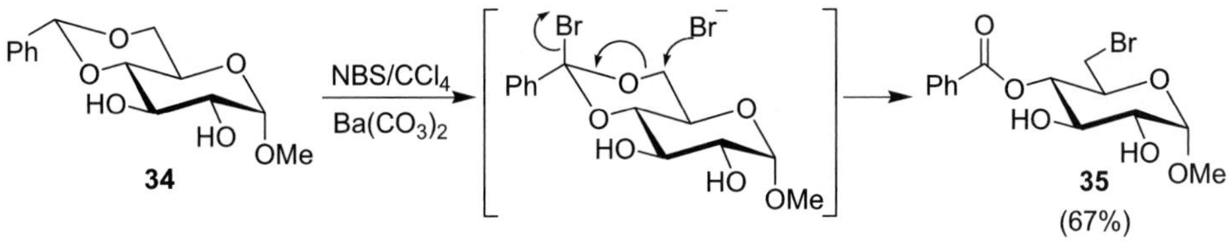

Scheme 18

glucopyranoside **35** (see ➤ *Scheme 18*) [23]. The reaction may probably undergo the radical bromination of the benzylic carbon atom followed by rearrangement to the 6-deoxy-6-bromo derivative.

1.3 Addition Reaction

Unlike the carbohydrates with double bonds at positions other than between C-1 and C-2 ('isolated alkenes'), which exhibit normal alkene chemistry, glycals are vinyl ethers and therefore undergo a number of highly selective addition reactions due to the strongly polarized double bonds and the presence of bulky substituents at the C-3 allylic centers. Straightforward addition reaction includes initial electrophilic addition at the double bond, followed by the addition of a nucleophile at C-1 to give the 1,2-*trans* adduct (➤ *Scheme 19*).

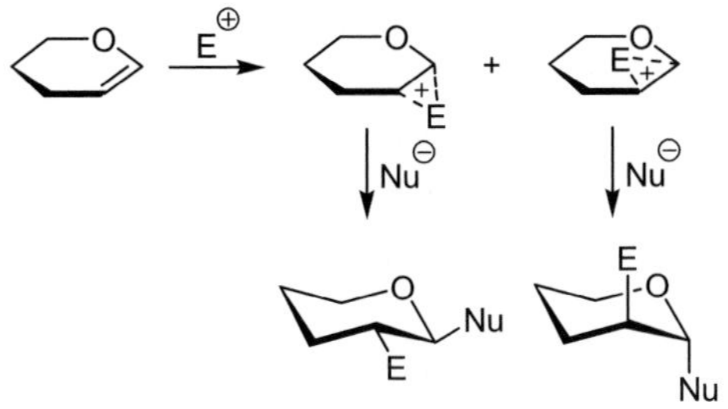

Scheme 19

The classical Ferrier rearrangements of acetyl glycals lead to the introduction of a nucleophilic group at C-1 (preferentially α-anomers) and the migration of the double bond to C-2 in the presence of acid catalysts (➤ *Scheme 20*) [24]. In some particular conditions (high temperature or stronger acid), 3-substituted isomeric products can be formed (➤ *Scheme 20*) [25].

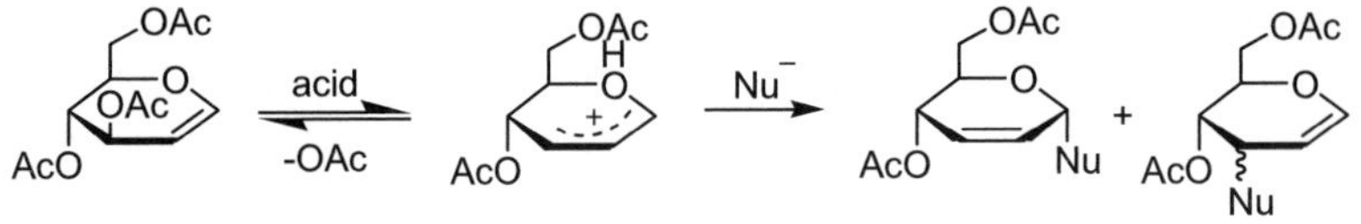

Scheme 20

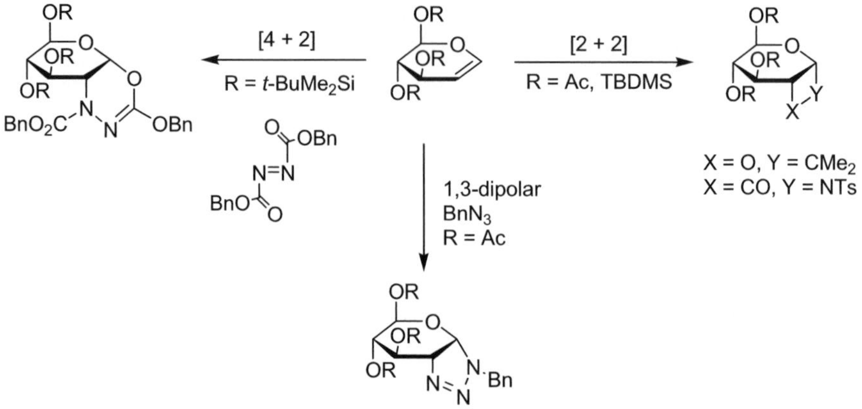

Scheme 21

Glycals can also undergo cycloaddition reactions and their derivatives are of interest for synthetic purposes. Diels–Alder [4+2], Paterno–Büchi [2+2], and 1,3-dipolar additions can be applied to the construction of fused cycloadducts (❯ *Scheme 21*) [26,27,28].

Radical additions can also be performed on the double bonds of unsaturated carbohydrates and regio- and stereoselectivities are high in some particular cases [29].

2 Introduction of Halogens

Deoxyhalogeno sugars are carbohydrate derivatives in which hydroxyl groups at positions other than the anomeric center have been replaced by halogen atoms. These compounds are susceptible to nucleophilic attack, leading to displacement, elimination, or anhydro-ring formation, therefore are useful intermediates in the synthesis of other uncommon sugars, such as deoxy and aminodeoxy sugars. Deoxyfluoro sugars are extensively applied to the study of carbohydrate metabolism and transport [30]. ^{18}F-labeled carbohydrates are employed for medical imaging [31]. Many chlorinated sugars have antibiotic activities and the trichloro sugar 'sucralose', 4,1′,6′-trichloro-4,1′,6′-trideoxy-*galacto*-sucrose, is known for its sweet profile 650 times stronger than sucrose [32,33]. The bromo and iodo derivatives have also been widely used, and have undergone a variety of nucleophilic substitution reactions [34].

2.1 Displacement of Sulfonic Esters

Sulfonate displacement has been well established as a common method for introducing halogen atoms into carbohydrates. The high reactivity and easy preparation of sulfonic esters contribute to their use in carbohydrate chemistry. Triflate displacements are popular for the preparation of secondary halogenated carbohydrates and often give satisfactory results when mesylates and tosylates fail to give products.

The efficiency of the displacement is to a large degree dependent upon the nucleophilicity of the halide. As illustrated in ❯ *Scheme 22*, for example, all four 2-deoxy-2-halo-α-D-arabino-

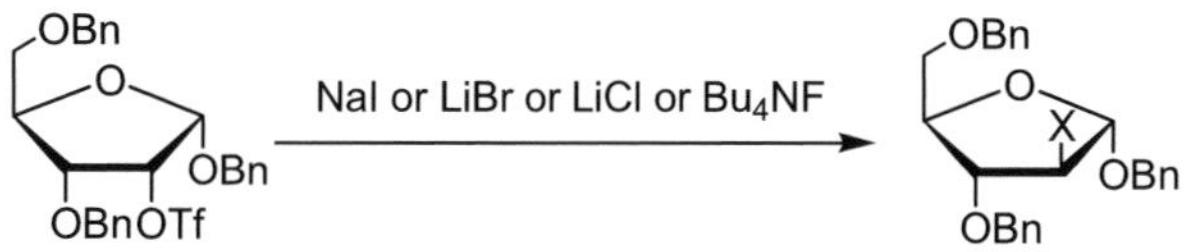

Scheme 22

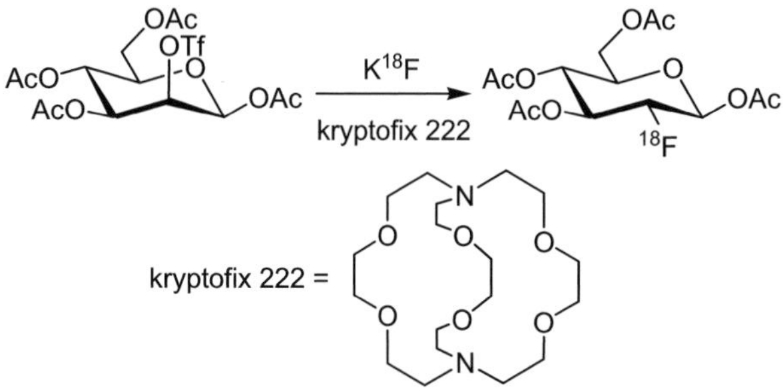

Scheme 23

furanosides **37** (X = F, Cl, Br, I) have been obtained from the corresponding 2-*O*-triflyl-α-D-ribofuranoside **36**, and the yields decrease in the order: I > Br > Cl > F [35].

Alkalimetal halides or the more highly soluble tetraalkylammonium salts are frequently used as the halide sources. The weakly nucleophilic character of the fluoride ion strongly affects its displacement efficiency. In some cases, a cryptand is used to increase its effective nucleophilicity (❷ *Scheme 23*) [36].

A variety of fluoride ion sources have been used in an effort to improve product yields in deoxyfluoro sugar synthesis. These salts include cesium fluoride, tetrabutylammonium fluoride, and tetrabutylammonium difluoride (Bu_4NHF_2) [37,38]. And currently, the most successful fluoride source is tri(dimethylamino)-sulfonium difluorotrimethylsilicate (TASF), which is soluble in many organic solvents and produces an anhydrous fluoride ion [39].

2.2 Direct Displacement of Hydroxyl Groups

2.2.1 Application of Alkoxyphosphonium Salts

Several reagents of this type are now available for replacing hydroxyl groups in carbohydrates by halogen atoms. The reaction of triphenylphosphine and tetrahalomethanes produces a halogenophosphonium ion in situ; this ion, when used in pyridine with unprotected sugars, can react with the hydroxymethyl groups to selectively give primary halo sugars. As shown in ❷ *Scheme 24*, treatment of methyl β-D-glucopyranoside **38** with triphenylphosphine and carbon tetrachloride, tetrabromide, or tetraiodide produces the corresponding 6-deoxy-6-haloglucosides **39** in almost quantitative yields [40,41].

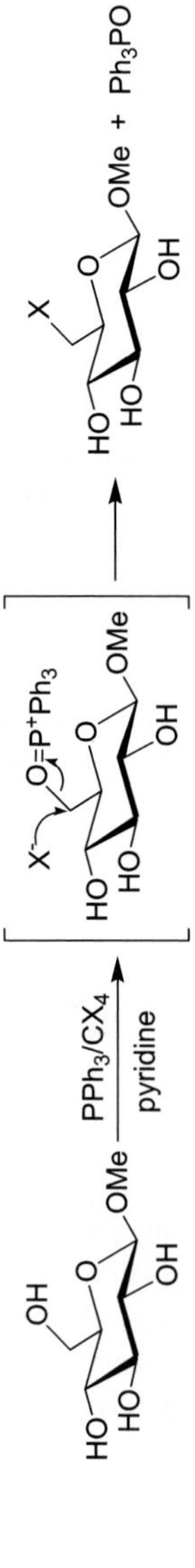

39 (X = Cl, Br, I)

38

□ Scheme 24

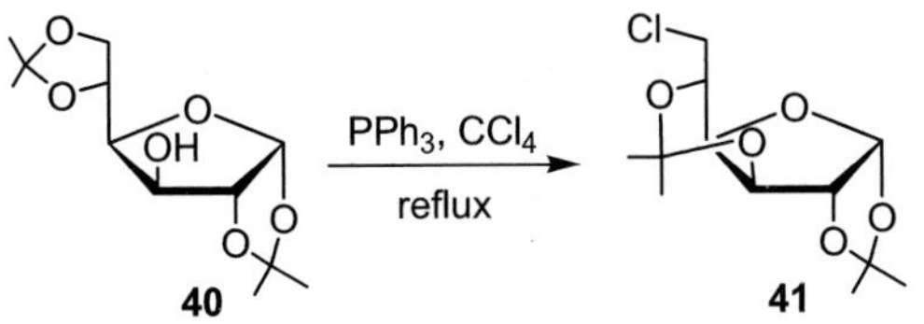

□ Scheme 25

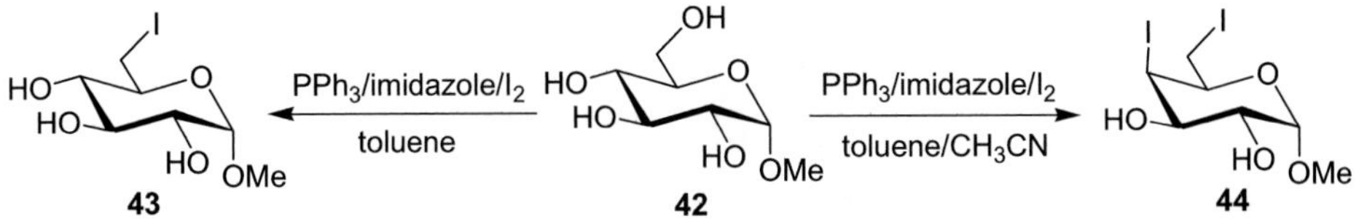

□ Scheme 26

However, some limitations in the application of this halogenation procedure have been noted. For example, 1,2;5,6-di-O-isopropylidene-α-D-glucofuranose **40**, in which C-3 is sterically hindered, gives the 5,6-acetal rearranged product **41** (❷ *Scheme 25*) [42].

Other combinations of reagents, such as triphenylphosphine/*N*-halosuccinimides [43], triphenylphosphine/imidazole/iodine, and triphenylphosphine/2,4,5-trihaloimidazole, react by a similar mechanistic pathway, providing a degree of regioselectivity. For example, methyl α-D-glucopyranoside **42**, when heated with PPh₃/imidazole/I₂ mixture in toluene, gives methyl 6-deoxy-6-iodo-α-D-glucoside **43**, whereas in toluene/acetonitrile, a solvent of increased polarity in which reactants are more soluble, displacements occur at both the 4- and 6-position to give methyl 4,6-dideoxy-4,6-diiodo-α-D-galactoside **44** (❷ *Scheme 26*) [44,45,46].

In another instance, when 2,4,5-tribromoimidazole is used, methyl 4,6-O-benzylidene-3-bromo-3-deoxy-α-D-allopyranoside **45** is selectively formed from the methyl glucoside **34** (❷ *Scheme 27*) [47]. This results from the greater nucleophilicity of C-3 hydroxyl, which can form alkoxyphosphonium ion easily.

The Mitsunobu procedure, which uses diethyl azodicarboxylate (DEAD) or diisopropyl azodicarboxylate (DIAD) to react with PPh₃ and alcohols providing configuration-inversed substitution products via alkoxytriphenylphosphonium ion intermediates (❷ *Scheme 28*), presents another route for introducing halogen atoms into carbohydrates [48].

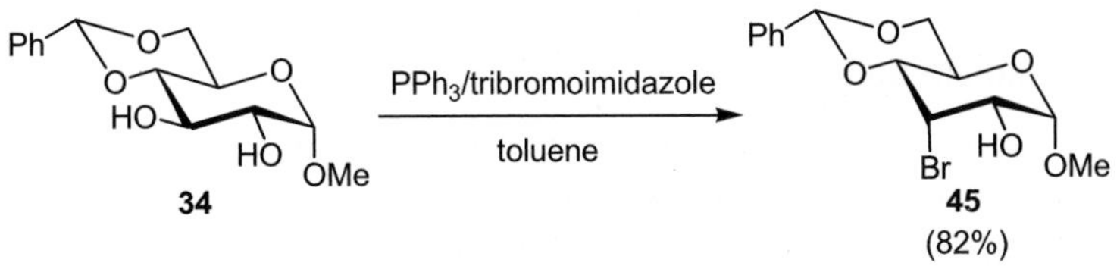

□ Scheme 27

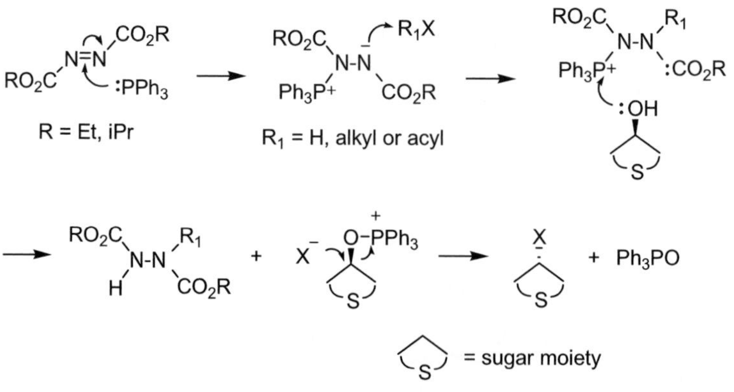

= sugar moiety

☐ Scheme 28

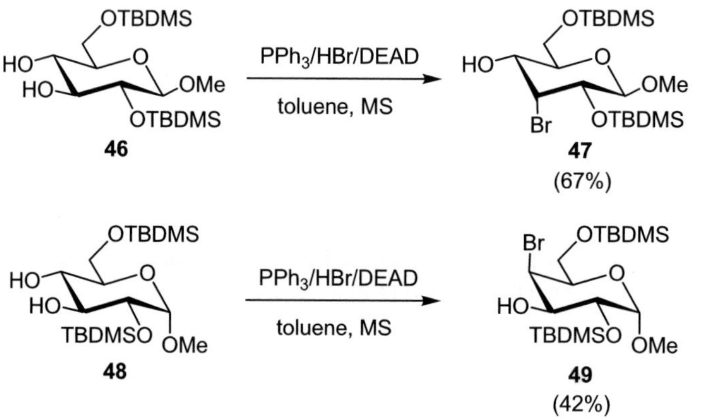

☐ Scheme 29

For example, methyl 2,6-di-*O*-*t*-butyldimethylsilyl-β-D-glucopyranoside **46** can be converted mainly to 3-deoxy-3-bromoalloside **47** when treated with PPh$_3$/HBr/DEAD, whereas the corresponding α-glucoside **48** reacts to give C-4 brominated product **49** preferentially under the same conditions (❖ *Scheme 29*) [49]

Triphenylphosphite methiodide [(PhO)$_3$P$^+$MeI$^-$] and dihalides [(PhO)$_3$P$^+$XX$^-$] (Rydon reagents), which are closely related to the reagents described above, have also been successfully applied to the synthesis of halogeno sugars [50].

2.2.2 Application of Iminium Salts

N,N-dimethylformamide reacts with chlorides of inorganic acids (phosgene, phosphoryl chloride, phosphorus trichloride, and thionyl chloride) to form an active salt: (chloromethylene)dimethyliminium chloride, which is also called Vilsmeier's imidoyl chloride reagent. This reagent, as shown in ❖ *Scheme 30*, reacts with the free hydroxyl group of the sugar derivative **50** to give an activated intermediate that can undergo nucleophilic substitution by

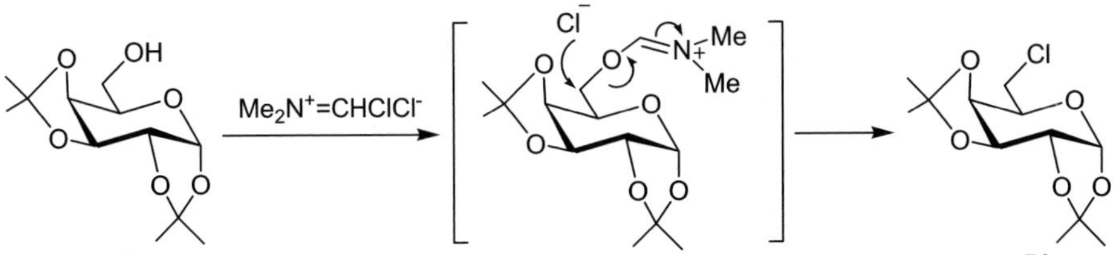

□ Scheme 30

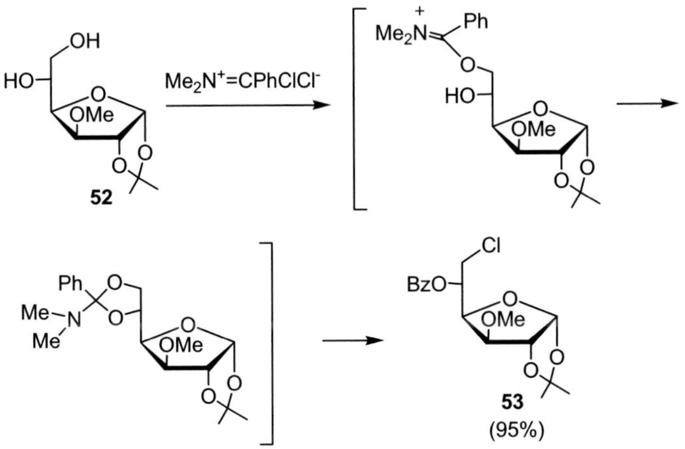

53
(95%)

□ Scheme 31

chloride ion with loss of dimethylformamide yielding the chlorinated derivative **51** in an excellent yield [51].

α-Chlorobenzylidene-*N*,*N*-dimethyliminium chloride reacts with the diol of sugar derivative **52** to form an addition product, which is then transformed into the 5-*O*-benzoyl-6-chloro-6-deoxy-glucose derivative **53** (❷ *Scheme 31*) [52].

2.2.3 Application of Chlorosulfates

The reaction of carbohydrates with sulfuryl chloride can form chlorosulfate groups initially, followed by S_N2 displacement of the liberated chloride ion. This provides another effective method for the preparation of chlorodeoxy sugars [34,53]. Those positions, where the steric and polar factors are favorable for a S_N2 reaction, are more susceptible to displacement. The chlorosulfates that have been substituted with a chloride can easily be cleaved by treatment with sodium iodide.

Thus, the representative reaction manner of this transformation is presented by the reactions of methyl α-D-glucopyranoside **42**, as outlined in ❷ *Scheme 32*. When a minimum proportion of pyridine is employed, the reaction of sulfuryl chloride with **42** produces 4,6-dichlorodeoxy sugar **54**. While in the presence of an excess of pyridine, 2,3-cyclic sulfate **55** can be formed from either compound **42** or **54**. And methyl α-D-glucopyranoside 2,3,4,6-tetra(chlorosulfate)

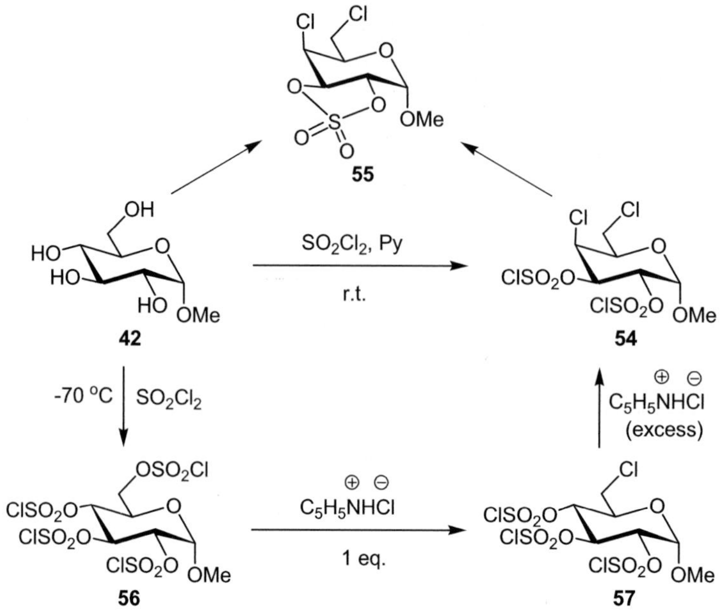

Scheme 32

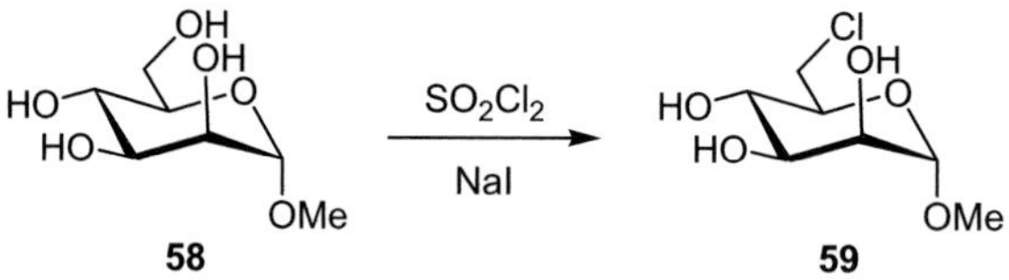

Scheme 33

56 is isolated when the reaction proceeds at $-70\,°C$. Compound **56** is further converted into the sugar derivative **54** via the formation of 6-chlorodeoxy sugar **57** on treatment with pyridine chloride [54,55,56,57].

In the transformation of **42** to **54**, the lack of substitution of the chlorosulfonyloxy group at C-3 is attributed to the presence of the β-*trans* axial methoxyl group at C-1; also, the chlorosulfonyloxy group at C-2 is deactivated to displacement by chloride ion. This 'β-*trans*-axial substituent effect' can also interpret the regioselectivity of the reaction of the α-D-mannopyranoside **58** to **59** (❂ *Scheme 33*) [57].

The chlorosulfated glycosyl chloride **60** is converted into 2-chloro-2-deoxy-α-D-lyxopyranosyl chloride **61** when treated with aluminum chloride (❂ *Scheme 34*) [58]. This reaction proceeds via an initial intramolecular displacement of chlorosulfate at C-2 by the anomeric chlorine atom, followed by the chloride ion attack at the more highly reactive center (C-1).

Pyranoid derivatives having a chlorosulfonyloxy group in a *trans*-diaxial relation with a ring proton, such as **62**, may undergo an elimination reaction to yield the unsaturated compound **63** (❂ *Scheme 35*) [59].

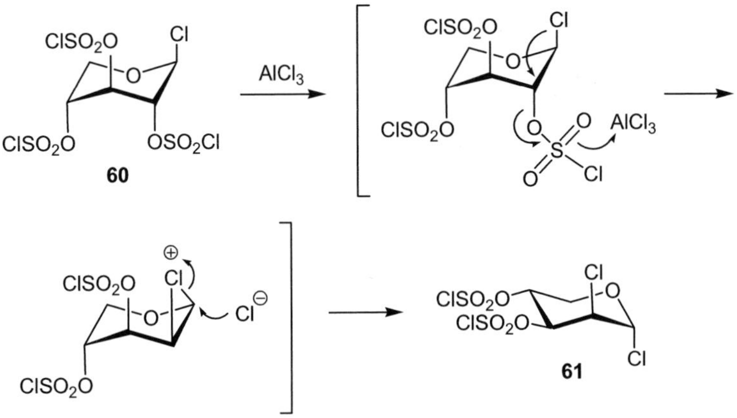

Scheme 34

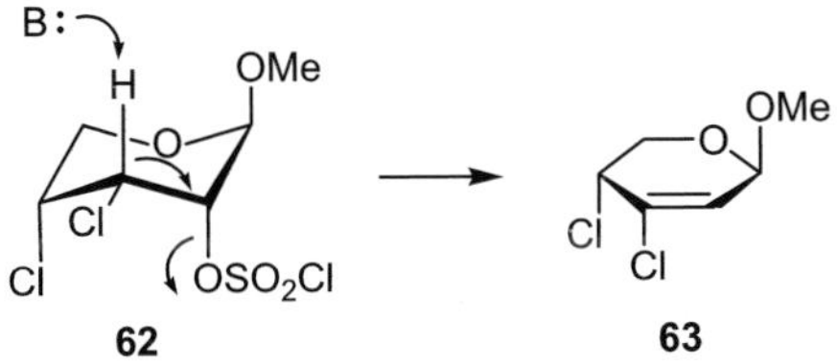

Scheme 35

2.2.4 Application of Diethylaminosulfur Trifluoride (DAST)

To date, diethylaminosulfur trifluoride (Et$_2$NSF$_3$) may be the most commonly applied reagent for direct fluorination. An alcohol displaces a fluoride of DAST resulting in an activated intermediate, which in turn is substituted by the liberated fluoride, as illustrated in the formation of **65** from **64** (● *Scheme 36*) [60].

Some selectivity has been observed with this reagent, and is similar to chlorosulfate displacement. Thus, due to the 'β-*trans*-axial substituent effect', methyl α-D-glucopyranoside **42** gives 4,6-difluorodeoxy sugar **66**, and methyl β-D-glucopyranoside **38** gives 3,6-difluorodeoxy sugar **67** (● *Scheme 37*) [61,62].

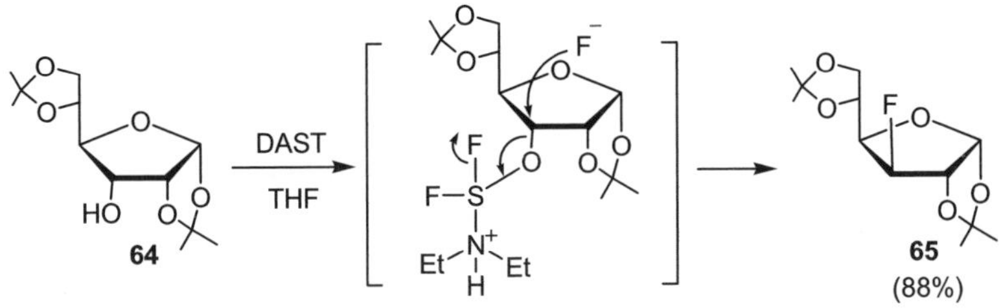

Scheme 36

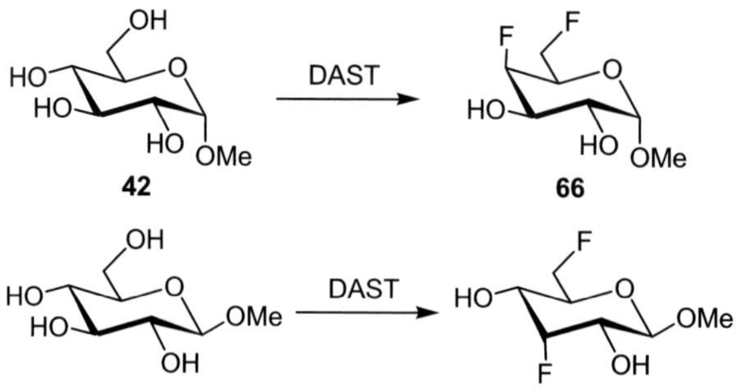

Scheme 37

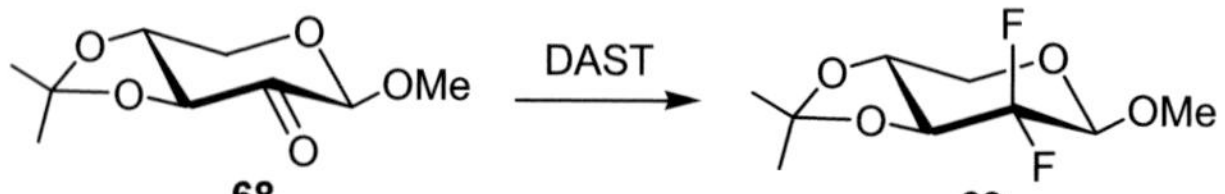

Scheme 38

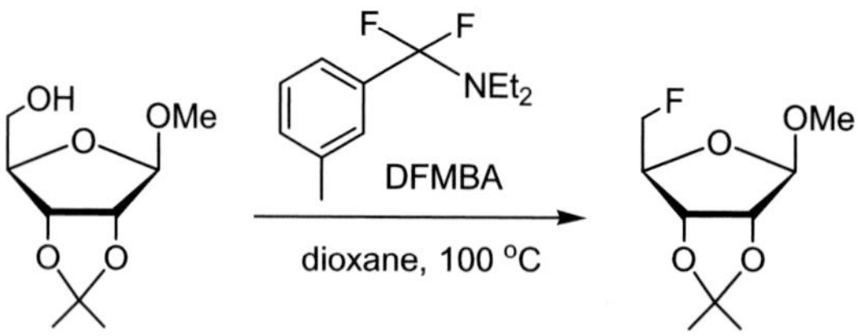

Scheme 39

A gem-difluoride is formed when a ketone or aldehyde is treated with DAST, as outlined in the reaction of **68** to **69** (● *Scheme 38*) [60].

For the synthesis of **71** from **70**, N,N-diethyl-α,α-difluoro-(m-methylbenzyl)amine (DFMBA) is used to avoid the unexpected migration of the methoxy group from the 1- to 5-position when DAST is used (● *Scheme 39*) [63].

2.3 Ring-Opening Reactions

2.3.1 Opening of Epoxide Rings

As an alternative practical method, the opening of epoxides by halide ions has often been used for introducing halogens, especially fluorine, into sugar molecules. Sodium iodide, magnesium bromide, and hydrogen halides (HCl, HF) are frequently used as halogen sources [64,65].

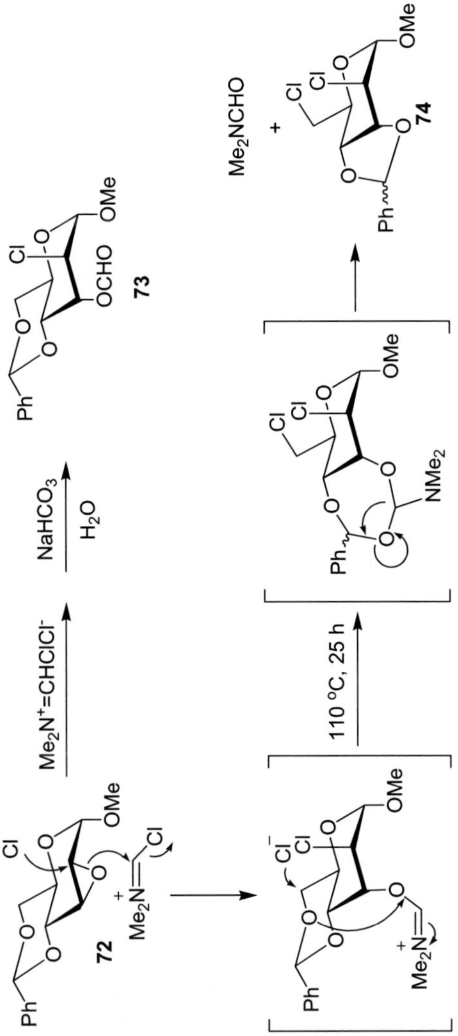

Scheme 40

A special example of epoxide opening reaction is outlined in ❷ *Scheme 40*. Methyl 2,3-anhydro-4,6-*O*-benzylidene-α-D-allopyranoside **72** gives the 2-chloro-2-deoxy-3-*O*-formyl-altroside **73** on treatment of (chloromethylene)dimethyliminium chloride at room temperature. Alternatively, when the reaction mixture is heated for a longer time, methyl 3,4-*O*-benzylidene-2,6-dichloro-2,6-dideoxy-α-D-altroside **74** is obtained by means of a rearrangement of the benzylidene group [51].

An interesting regioselectivity of epoxide opening occurs in the reaction of sugar **75** with HF·pyridine (❷ *Scheme 41*) [66]. Excellent selectivity of fluoride ion substitution at C-5 is obtained giving compound **76** after acetylation.

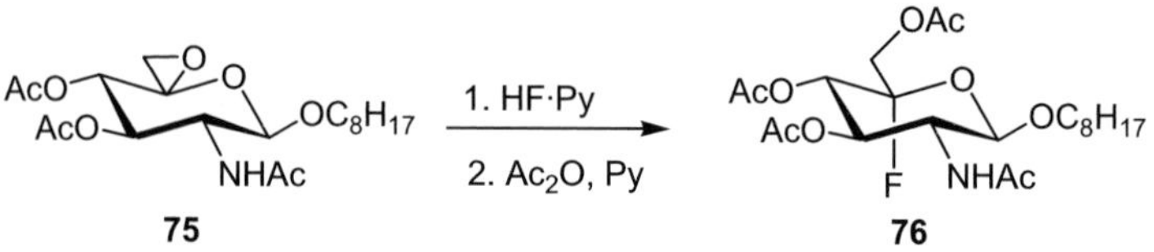

75   1. HF·Py   **76**
         2. Ac₂O, Py

◻ Scheme 41

2.3.2 Opening of Sulfur-Containing Rings

Five-membered ring compounds such as cyclic thionocarbonates can also be opened on treatment of methyl iodide. For example, thionocarbonate **77** gives regioselectively the ring-opening product **78** (❷ *Scheme 42*) [67]. Another regioselective ring opening has been observed in the transformation of fluorosugar **80** from the 5,6-cyclic sulfates **79** (❷ *Scheme 42*) [68,69].

77   MeI   **78**

79   1. TBAF   **80**
(R = Ac, Bn or Ms)   2. H₃O⁺   (85-90%)

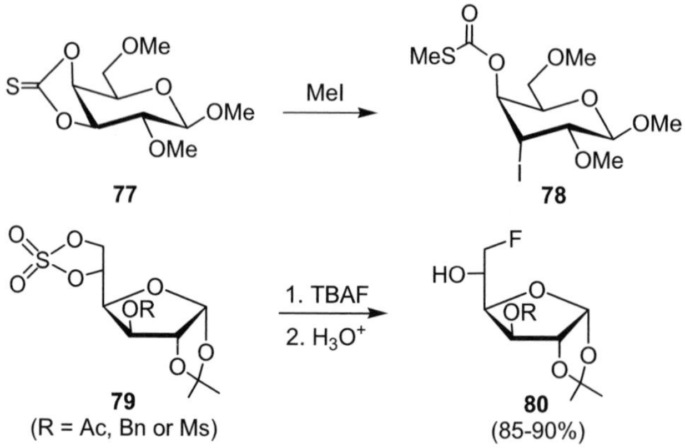

◻ Scheme 42

2.3.3 Opening of Benzylidene Acetals

N-Bromosuccinimide converts benzylidene acetals into bromodeoxy benzoates, and offers an excellent method for preparing 6-bromodeoxyhexoses (see ❷ *Sect. 1.2.2*). The reaction can also be applied to 2,3- and 3,4-benzylidene acetals of pyranosides and furanosides, and proceeds regioselectively [23]. When there are no stereochemical constrains to effect regioselectivity, mixtures of isomers are formed. And similar transformations can be performed using bromotrichloromethane as the bromine source [70].

2.4 Addition to Unsaturated Sugar Derivatives

Addition reactions of electrophilic halogenation reagents to unsaturated sugars constitute a valuable synthetic alternative to the S_N-strategy. 3,4,6-tri-O-acetyl-D-glucal **81** reacts with fluorine diluted with argon to give mixtures of 3,4,6-tri-O-acetyl-2-deoxy-2-fluoro-α-D-glucopyranosyl fluoride **82** and the corresponding β-D-*manno*-difluoride **83** [71]. An efficient variant of this reaction utilizes xenon difluoride additions catalyzed by boron trifluoride etherate, giving compound **82** in an increased selectivity (❷ *Scheme 43*) [72].

High regio- and stereoselectivity are obtained in the reaction given in ❷ *Scheme 44*. The reaction of glucal **81** with NBS mainly leads to 2-bromo-2-deoxy-α-D-mannopyranoside **84**, whereas 3,4-O-isopropylidene-D-galactal **85** affords only the β-D-galactopyranoside **86**. Thus, this reaction has made glycals very significant starting materials for the preparation of some important glycosides, especially 2-deoxy oligosaccharides after reduction of the halide groups [73,74].

Addition of iodine azide to glycals gives mostly the 1,2-*trans*-2-deoxy-2-iodoglycosyl azides [75]. In view of the hazardous nature of iodine azide, two mild methods were devel-

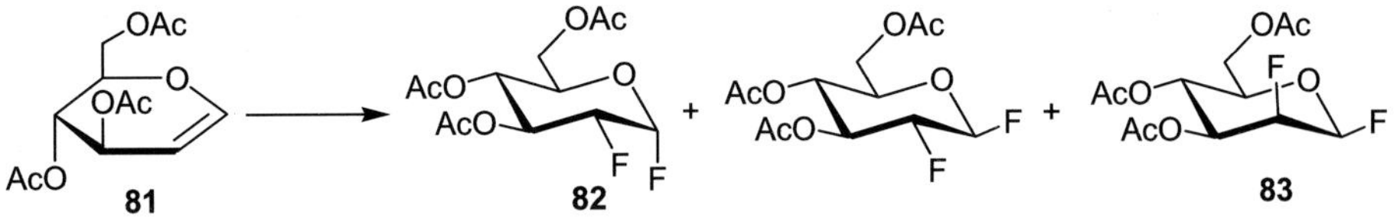

	82		83
F_2	35%	—	26%
XeF_2, $BF_3 \cdot Et_2O$	61%	12%	5%

❐ Scheme 43

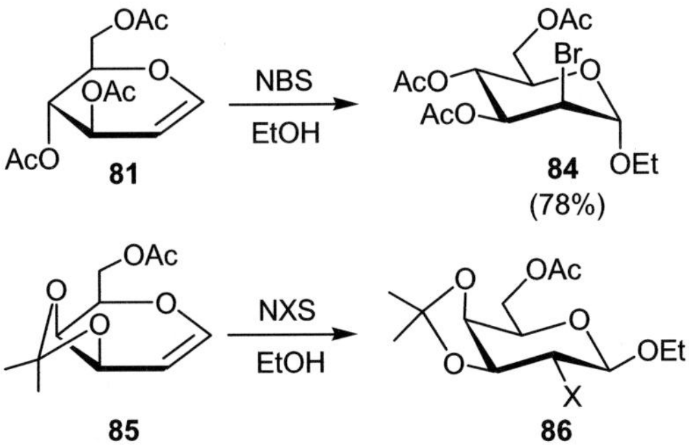

❐ Scheme 44

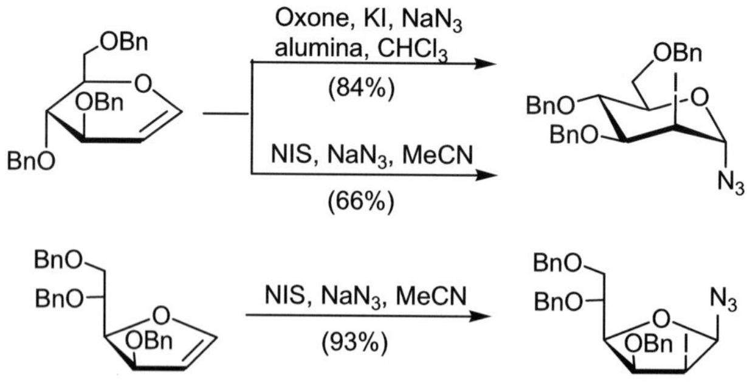

□ Scheme 45

oped for the preparation of 1,2-*trans*-2-deoxy-2-iodoglycosyl azides from glycals in good yields. The first method involves reaction of a glycal with oxone, potassium iodide, sodium azide, and neutral alumina in chloroform at room temperature (❷ *Scheme 45*). Whereas the second method involves its reaction with *N*-iodosuccinimide and sodium azide in acetonitrile at 0 °C; however, it is interesting to note that while pyran glycals give 1,2-*trans*-2-deoxy-2-iodoglycosyl azides, the furan glycals give exclusively the 1,2-*cis*-2-deoxy-2-iodoglycosyl azides (❷ *Scheme 45*) [76].

Under the conditions of the addition reaction glycals such as 2-fluoro-glycals can be converted into corresponding 2-deoxy-2-fluoro-2-iodo derivatives [77].

The isolated alkenes undergo normal additions, frequently with high regio- and stereoselectivity. For example, methyl 4,6-*O*-benzylidene-2,3-dideoxy-*α*-D-*erythro*-hex-2-enopyranoside **87** gives the 2,3-dibromo-D-*altro*-adduct **88** on treatment with bromine, and the 2-*O*-acetyl-3-bromo-3-deoxy adduct **89** is obtained when treated with acetyl hypobromite (❷ *Scheme 46*), indicating that the reactive bromonium ion intermediate is formed on the lower (*α*) face of the ring [78].

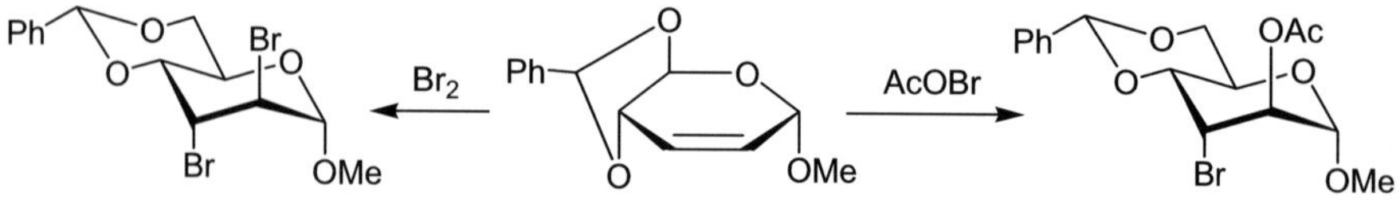

□ Scheme 46

2.5 Radical Bromination Reactions

Bromine can be introduced directly into particular ring positions of certain sugar derivatives by free radical processes which can show high selectivity and efficiency. For example, peracetylated glucose **90** is selectively brominated at the C-5 position by NBS yielding **91** however, in the case of glycosyl halides, such as tetra-*O*-acetyl-*β*-D-glucopyranosyl chloride **92**, anomeric substituted product **93** is mainly formed (❷ *Scheme 47*) [79,80].

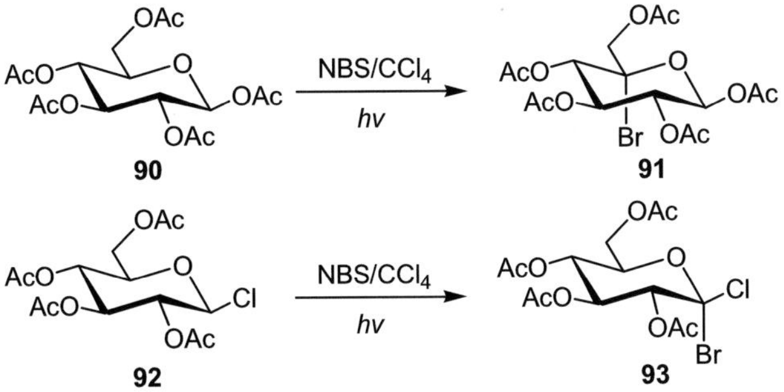

□ Scheme 47

3 Introduction of Nitrogen

Amino sugars are carbohydrate derivatives in which a hydroxyl group is replaced by an amino group at nonanomeric positions. 2-Amino-2-deoxy-D-glucose (D-glucosamine) is abundant in nature, appearing in particular in the polysaccharide chitin. 2-Amino-2-deoxy-D-galactose (D-galactosamine) is a constituent monosaccharide unit of dermatan and chondroitin sulfate found in mammalian tissue and cartilage. Nonulosaminic acids are 5-amino-5-deoxynonose derivatives usually found in combined form in mucolipids or mucopolysaccharides.

3.1 Nucleophilic Displacement Reactions

This substitution reaction can be performed using nucleophilic nitrogen-containing reagents such as ammonia, hydrazine, or sodium azide. Ammonia and hydrazine can overcome the dipolar repulsion against charged nucleophiles, but their products are still nucleophilic and can perform a second displacement. Azide ions are stable under many reaction conditions and can be converted into amines by a wide range of reducing agents. Sometimes, the addition of crown ether is required to increase the nucleophilicity of azide ions [7]. Phthalimide ions have also successfully been applied in substitution reactions to yield a protected amino sugar derivative [81,82].

Misunobu substitution conditions (PPh$_3$, DIAD) have also been successfully applied to the direct displacement of hydroxyl using zinc azide-pyridine complex [83].

Intramolecular substitutions provide a convenient and stereoselective approach for the preparation of vicinal *cis*-hydroxy amino derivatives. Thus, the sodium hydride-mediated cyclization of the D-*gluco* compound **94** leads to the formation of the oxazolidinone **95**, which is hydrolyzed by base affording the D-mannosamine **96** (❂ *Scheme 48*) [84].

Palladium-catalyzed azidation reaction is a novel regio- and stereoselective method for introducing azide groups into 2,3-dideoxyhex-2-enopyranosides. For example, ethyl 4,6-di-O-acetyl-α-D-*erythro*-hex-2-enopyranoside **97**, when treated with NaN$_3$ in the presence of Pd(PPh$_3$)$_4$ and dppb [1,4-bis(diphenylphosphino)butane], gives predominantly the 4-substituted product **98** with retention of configuration (❂ *Scheme 49*). The (π-allyl)palladium intermediate and the crowded C-2 by the aglycon moiety may contribute to the *trans* attack of N$_3^-$ at C-4 only [85].

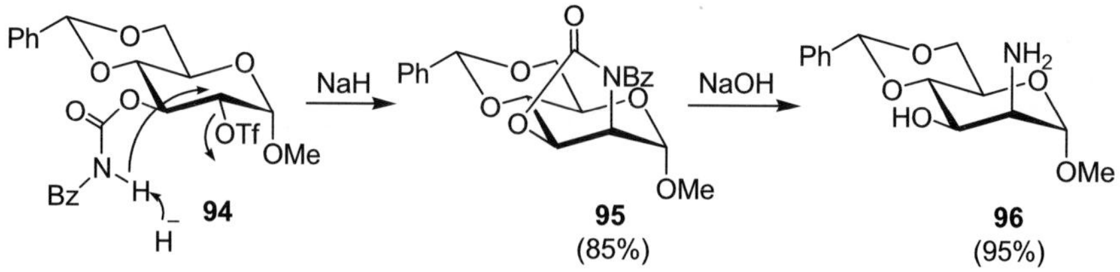

▣ Scheme 48

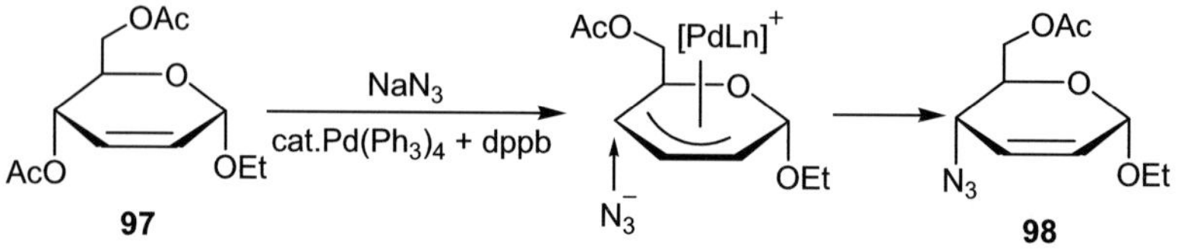

▣ Scheme 49

3.2 Opening of Epoxide Rings

Nucleophilic ring-opening of epoxides with nitrogen nucleophiles offers another route to amino sugars. Similar to the above-mentioned nucleophilic substitutions, epoxide openings can be conducted with ammonia, primary amines, guanidine, or azide ions [17,86]. Ammonium chloride is often used to neutralize the alkoxide produced during the opening of an epoxide.

3.3 Addition of Nitrogenous Reagents to Double Bonds

3.3.1 Addition to Glycals

Chloronitroxylation of glycal double bonds has been used as a means for synthesizing 2-amino-2-deoxy sugar derivatives. As illustrated in ❯ *Scheme 50*, the addition of nitrosyl chloride to tri-O-acetyl-D-glucal **81** gives a dimeric adduct which can be converted into the oxime **99** in high stereoselectivity with alcohols [87]. Reduction of the product provides corresponding amino sugar derivatives (see ❯ *Sect. 3.5*).

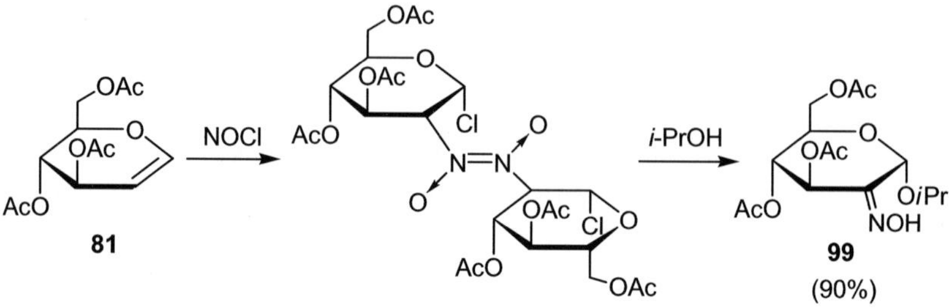

▣ Scheme 50

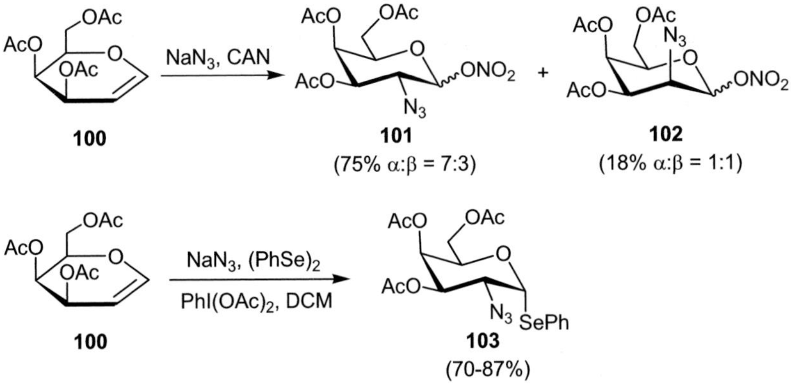

Scheme 51

Addition to the double bond can also be conducted by the so-called azidonitration reaction. This reaction occurs with sodium azide and ceric ammonium nitrate (CAN) resulting in a 2-azido-2-deoxyglycosyl nitrate via a radical azido addition, as illustrated in ❷ *Scheme 51* by the conversion of tri-*O*-acetylgalactal **100** into the D-*galacto* adduct **101** and a small amount of the *talo*-isomer **102**. The anomeric nitrate can be readily replaced by a halide, acetyl, or hydroxyl functionality [88]. In another similar reaction (❷ *Scheme 51*), sodium azide and diphenyl diselenide in the presence of (diacetoxy)iodobenzene react with the galactal **100** to give stereoselectively the α-phenylselenyl galactoside **103** (azidophenylselenylation) [89].

Halosulfonamidation of hexose-derived glycals followed by sulfonamide migration reaction provides also a useful approach for the synthesis of 2-amino sugar derivatives. For example, the reaction of tri-*O*-benzyl glucal **104** with iodonium di-*sym*-collidine perchlorate (IDCP) and benzenesulfonamide gives the *trans*-diaxial iodosulfonamide **105**, which undergoes sulfon-amide migration in the presence of lithium ethanethiolate yielding the 2-amino thioglycoside **106** (❷ *Scheme 52*) [90].

A transition metal-mediated approach to amidation of glycal substrates can lead to the forma-tion of 2-deoxy-2-trifluoroacetyl-amido derivatives. Thus, when a solution of (saltmen)Mn(N) is added to a mixture of trifluoroacetic anhydride (TFAA) and glycal **107** followed by sequen-tial treatment with thiophenol and BF$_3 \cdot$ Et$_2$O, the 2-*N*-trifluoroacetamido thioglycoside **108** can be obtained in good yield and excellent diastereoselectivity (❷ *Scheme 53*) [91].

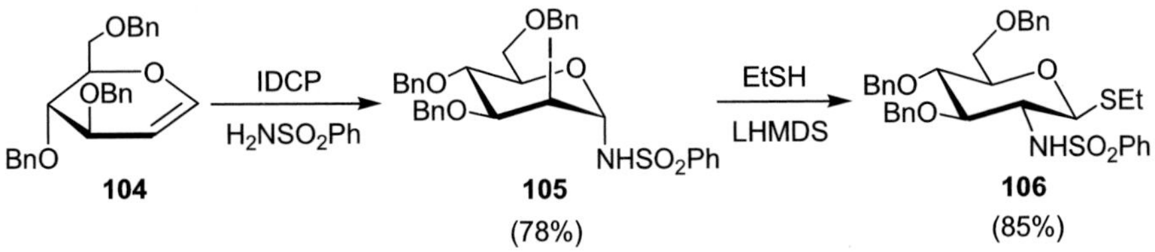

Scheme 52

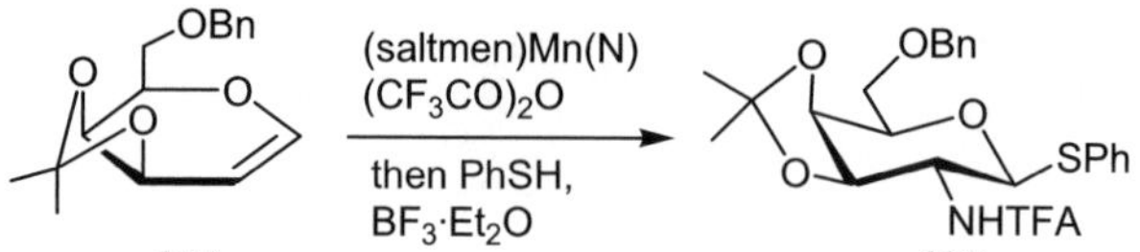

Scheme 53

3.3.2 Addition to Isolated Alkenes

The p-toluenesulfonylimidoosmium reagent, which may be generated from Chloramine-T and osmium tetraoxide, leads to cis-addition of OH and HNTs to the least-hindered face of double bonds. With the hex-2-enopyranoside **97**, the addition occurs to the top face of the double bond to give the regioisomers **109** and **110** (**◗** Scheme 54) [92].

Vinyl sulfone-modified carbohydrates, such as **111**, can be subjected to Michael addition reaction with primary or secondary amines to construct **112** bearing amino groups at the C-2 carbon in equatorial configurations (**◗** Scheme 55) [93].

An 1-hydroxy-1,2-benziodoxol-3(1H)-one 1-oxide (IBX)-mediated process provides another efficient and stereoselective preparative route for the synthesis of amino sugars. For example, reaction of the D-glucal derivative **113** with p-methoxybenzene isocyanate in the presence of a catalytic amount of 1,8-diazabicyclo[5.4.0]undec-7-ene (DBU) followed by treatment with IBX furnishes the cyclic carbonate **115**, through urethane **114**. Subsequent removal of the p-methoxybenzyl (PMB) protecting group by CAN affords the protected amino sugar **116** (**◗** Scheme 56) [94].

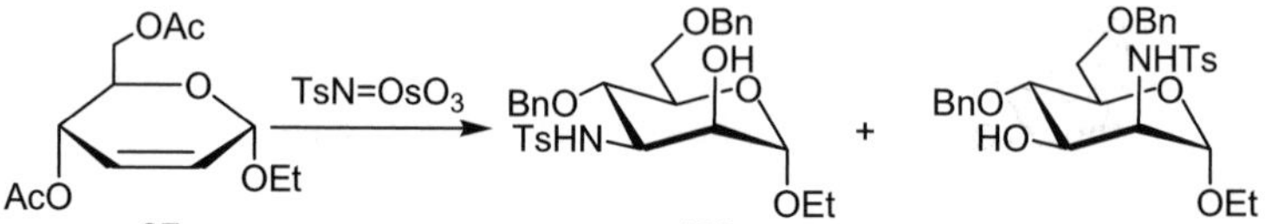

Scheme 54

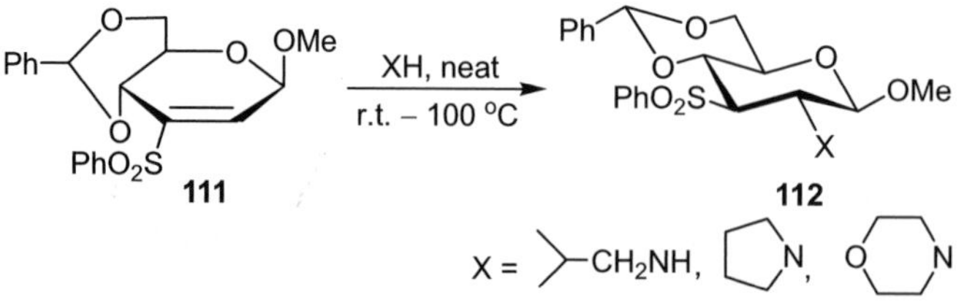

Scheme 55

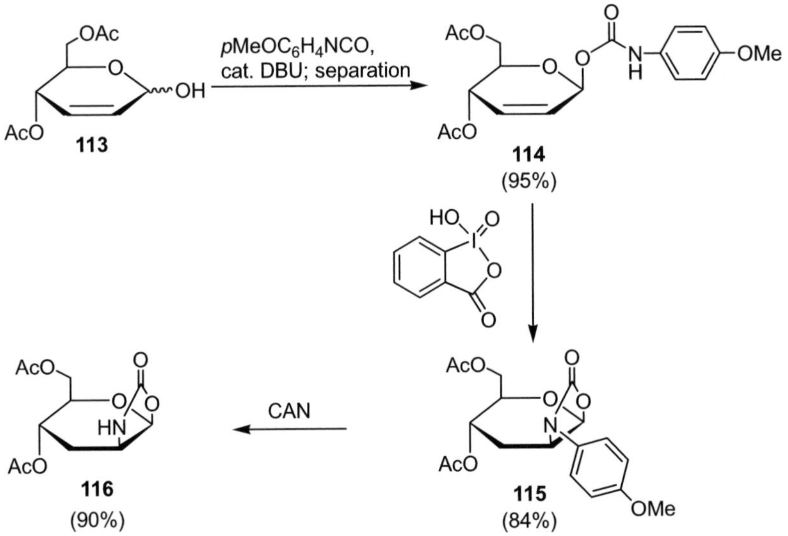

Scheme 56

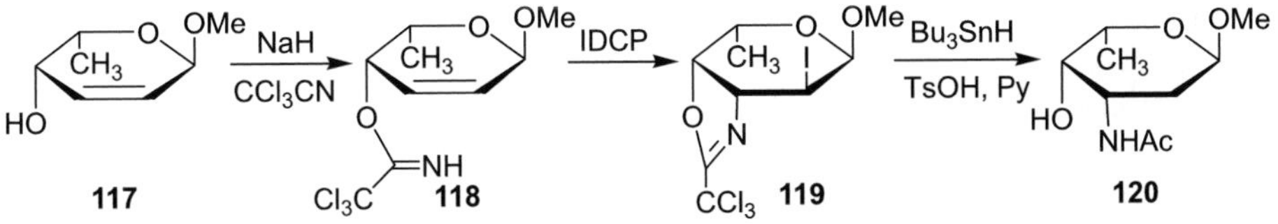

Scheme 57

Regio- and stereo-controlled iodocyclizations of allylic trichloroacetimidates provide a route to *cis*-hydroxyamino sugar from hexenopyranosides. For example, conversion of the hydroxyl of compound **117** into the trichloroacetimidate **118** followed by IDCP-mediated intramolecular cyclization, gives iodo-oxazoline derivative **119**, which is reduced (Bu₃SnH) and hydrolyzed (pyridine, TsOH) to afford *N*-acetyldaunosamine methyl glucoside **120** (❂ *Scheme 57*) [95].

3.4 Cyclization of Dialdehydes

Dialdehydes, obtained by periodate oxidation of appropriate cyclic polyhydroxy precursors, condensing with some nitrogen-containing reagents in basic solution has been developed into a useful synthesis of amino sugars. Thus, the dialdehyde **121**, on treatment with nitromethane in the presence of sodium methoxide produces a mixture of *aci*-nitro salt isomers **122**, which mainly give 3-amino-3-deoxyhexopyranosides **123** and **124** after reduction (❂ *Scheme 58*) [96].

In another instance, when the dialdehyde **125** is treated with phenylhydrazine, a cyclization occurs, probably via the monophenylhydrazone **126** to give the phenylazo derivative **127**,

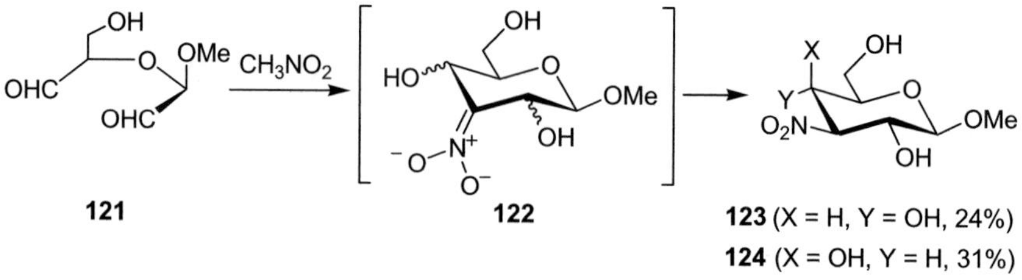

121   **122**   **123** (X = H, Y = OH, 24%)
                                             124 (X = OH, Y = H, 31%)

◻ Scheme 58

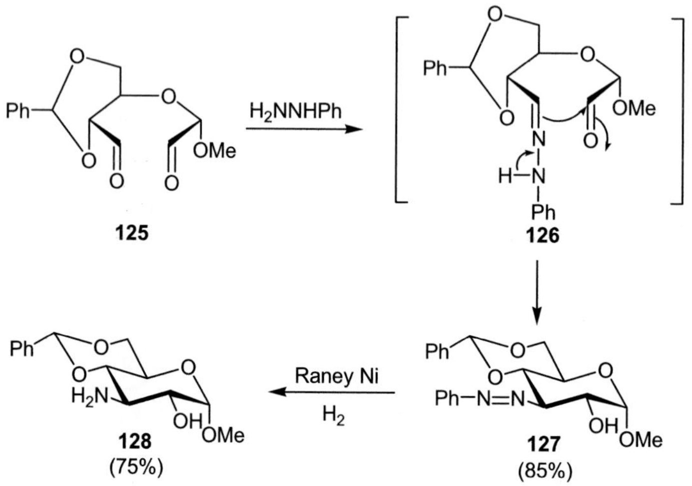

◻ Scheme 59

which can be hydrogenolyzed to the 3-amino-3-deoxy-α-D-glucopyranoside **128** with a good stereoselectivity (❂ *Scheme 59*) [97].

3.5 Reduction of Ulose Oximes

Oxime formation from uloses and subsequent reduction provides yet another route to amino sugars. The stereochemical outcome is dependent on the reducing agent, solvent, and protecting groups [98]. For example, PtO_2 reduction of the oxime **129** gives only the L-*ribo* derivative **130** with the amino group in an axial orientation, whereas borane reduction of acylated oxime **131** produces mainly the L-*arabino* derivative **132** with an equatorial amino substituent (❂ *Scheme 60*) [99].

3.6 Rearrangement Reactions

The rearrangement of allyl trichloroacetimidate into allyl trichloroacetamide (Overman rearrangement) has been used for the synthesis of amino sugars. For example, the trichloroace-

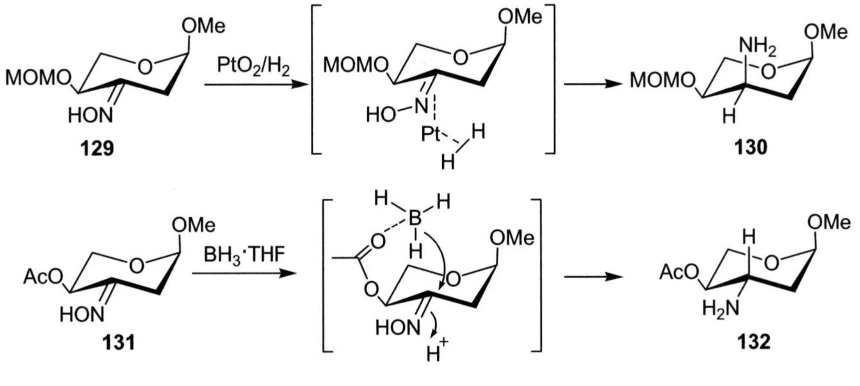

■ Scheme 60

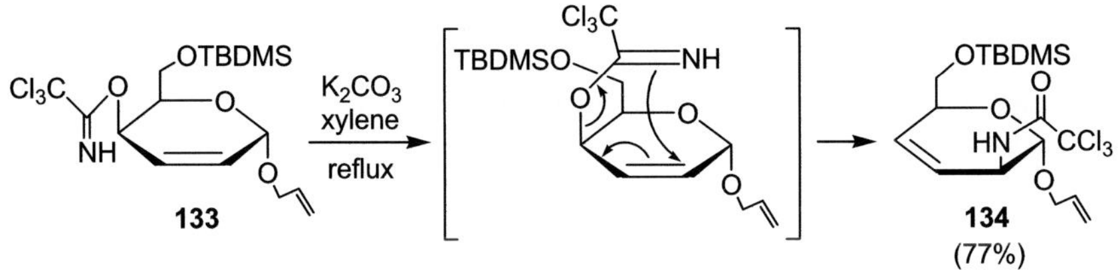

■ Scheme 61

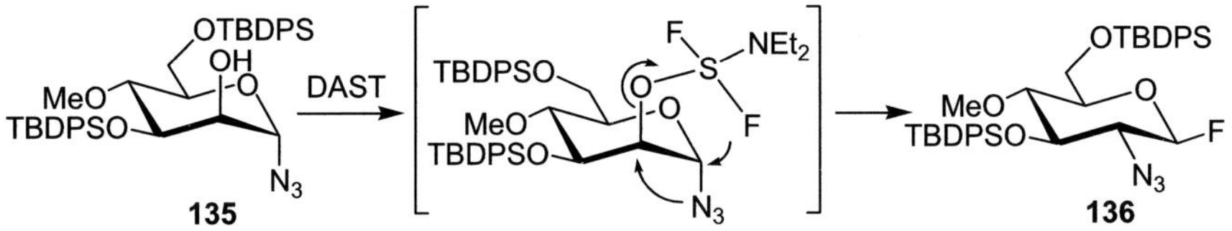

■ Scheme 62

timidate **133** undergoes reflux in xylene in the presence of potassium carbonate, giving the corresponding trichloroacetamide **134** with total stereoselectivity (❯ *Scheme 61*) [100].

In another interesting reaction, 2-azido-2-deoxypyranosyl fluoride **136** is obtained from the 2-OH unprotected mannosyl azide **135**, by DAST-mediated displacement involving neighboring group participation and migration of the anomeric azide group (❯ *Scheme 62*) [101].

3.7 Miscellaneous Methods

Under dehydrating conditions, D-fructose **137** will effectively react with amines to produce the corresponding imine product **138**, which can be transformed into the aldohexose **139** by using zinc halide as catalyst. Condensation of **139** with another equiv. of amine followed by ring-closure and hydrolysis gives the 2-alkylamino-2-deoxy-D-glucopyranose **140** (❯ *Scheme 63*) [102].

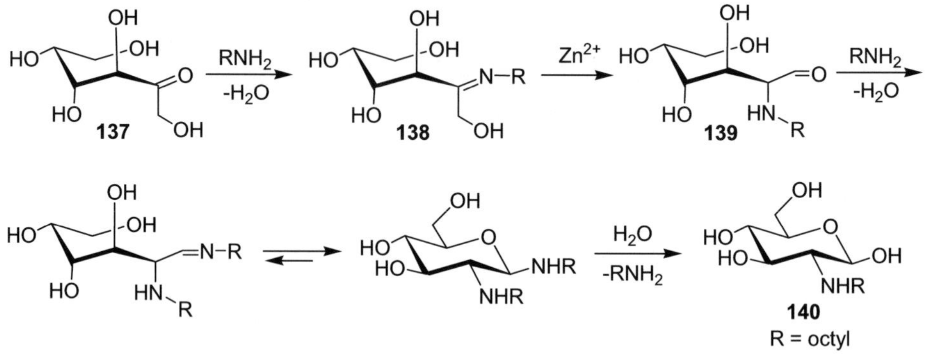

□ Scheme 63

A modification of the Kiliani–Fischer method for amino functionality introduction involves the condensation of free sugars with arylamines to give imines, which are treated directly with hydrogen cyanide to produce aminonitriles. Hydrogenolysis and hydrolysis of the resulting aminonitriles gives 2-amino-2-deoxyaldoses with one carbon extended [103].

4 Introduction of Sulfur and Selenium

4.1 Nucleophilic Substitutions

Thiosugars discussed here are the carbohydrates with oxygen replaced by sulfur at the nonanomeric positions. Normally these compounds can be obtained by the displacement of suitably positioned leaving groups using reagents containing nucleophilic sulfur such as thiocyanate ion ($^-$SCN), thiolacetate [Me(C=O)S$^-$], benzylthiolate (PhCH$_2$S$^-$) and ethyl xanthate [EtO(C=S)S$^-$] [104].

As expected, these substitutions occur readily at primary positions, as outlined by the thioacetate displacement applied to the 5-*O*-tosylxylose derivative **141** giving 5-thio-D-xylose **142** (**❷** *Scheme 64*). Similarly, the selenosugar derivative **143** is obtained from compound **141** by substitution using potassium selenocyanate (**❷** *Scheme 64*) [104]. And 6-*S*-phenyl-gluco/galactopyranosides are readily prepared from the corresponding 6-hydroxy-glycosides by direct substitution of the hydroxyl group using phenyl disulfide and tri-*n*-butylphosphine [105].

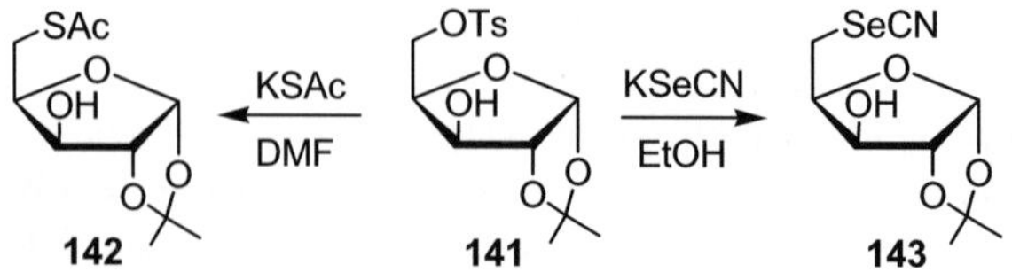

□ Scheme 64

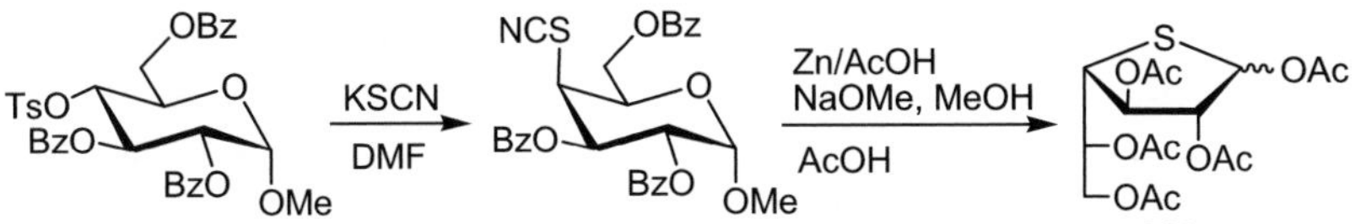

■ Scheme 65

A sluggish displacement by the thiocyanate ion occurs at C-4 in methyl 2,3,6-tri-*O*-benzoyl-4-*O*-tosyl-α-D-glucopyranoside **144** to give the 4-thiocyanogalactose derivative **145**. Subsequent reduction, deprotection and acetolysis yield preferentially the five-membered form **146** in which sulfur is in the ring (❷ *Scheme 65*) [106].

Phosphorus dithioacids [RR′P(S)SH] are effective sulfur nucleophiles for the synthesis of 3′-*S*- or 5′-*S*-nucleosides [107]. Furthermore, Mitsunobu reactions have also been used to convert primary sugar alcohols into corresponding thiosugar derivatives in one-pot manner [108].

4.2 Opening of Epoxide Rings

By the means of epoxide ring opening, thiosugars can also be obtained. Sometimes with nucleophiles such as thiocyanate or thiourea a more complex reaction occurs to convert epoxides directly into episulfides, as illustrated by the conversion of the 5,6-anhydro-L-*ido* compound **147** to the D-*gluco*-5,6-episulfide **148** (❷ *Scheme 66*). Also, episulfides can be ring opened with nucleophiles [109].

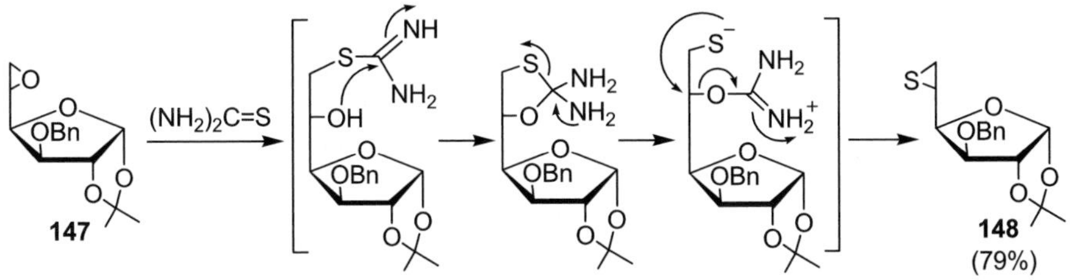

■ Scheme 66

4.3 Addition to Unsaturated Carbohydrates

Many known addition reactions of sulfur-containing reagents to unsaturated sugar derivatives involve free radical processes. For example, when triacetyl-D-glucal **81** is treated with thiolacetic acid in the presence of free radical initiators, the mixture of 2-thio-D-mannitol **149** and glucitol **150** is formed (❷ *Scheme 67*) [110].

In an analogous fashion, radical additions to exocyclic unsaturated sugars have also successfully been performed with high regio- and stereoselectivity. Thus, treatment of the compound **151** with benzylthio or thiolacetate radicals gives mainly **152** (❷ *Scheme 68*) [111].

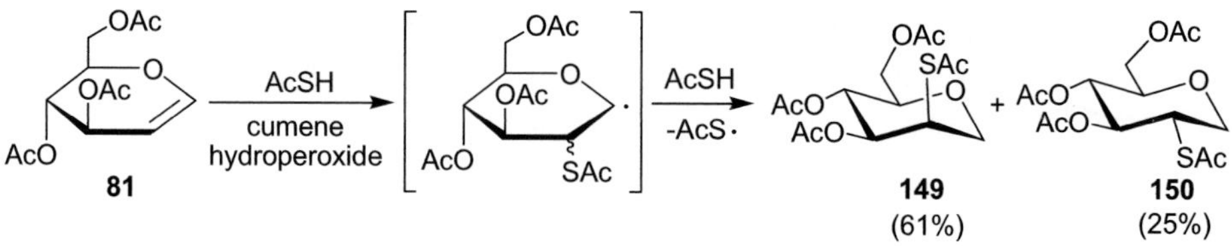

Scheme 67

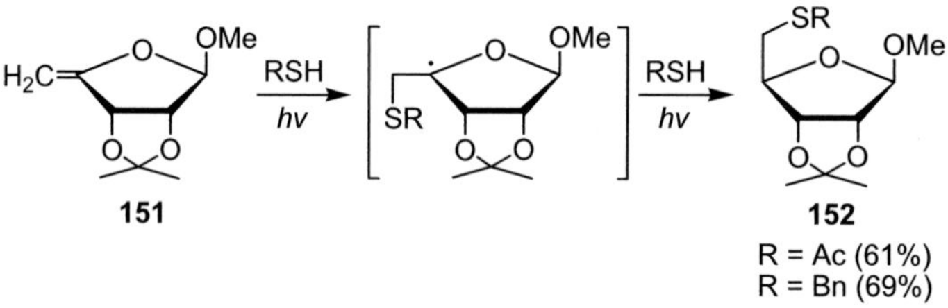

Scheme 68

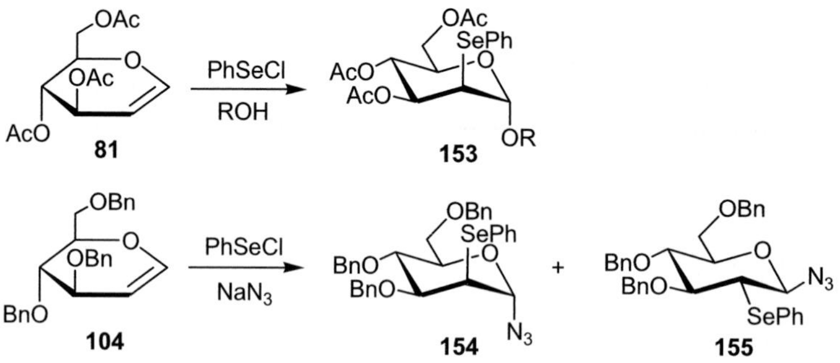

Scheme 69

As for the introduction of a selenium atom into carbohydrates, addition reactions to glycals are also effective. Thus, the glucal **81** can be treated with phenylselenyl chloride and gives mainly the *trans*-diaxial product **153** (❂ *Scheme 69*) [112]. Similarly, azidophenylselenylation of perbenzylated glucal **104** proceeds smoothly to give 2-*Se*-phenyl-2-selenoglycosylazides **154** and **155** (❂ *Scheme 69*) [113].

4.4 Rearrangement Reactions

4.4.1 1,2-Migrations

In 1,2-*trans*-thioglycosides, DAST can mediate a 1,2-migration of the anomeric thio functionality through an episulfonium ion to give the 1,2-*trans*-product, as illustrated by the conversion of **156** to **157** (❂ *Scheme 70*) [101]. And under the same conditions 2-hydroxy-1-seleno gly-

Scheme 70

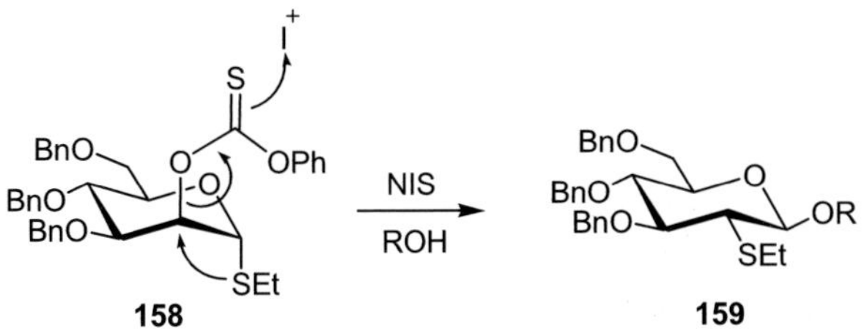

Scheme 71

cosides result in a similar 1,2-migration of the selenium group, with simultaneous installation of a fluoride group at C-1 [113].

In another similar reaction, activation of the phenoxythiocarbonyl ester on C-2 of the thiomannoside **158** by iodonium ions leads to the anomeric thioethyl group migration to the C-2 position, yielding the 1,2-*trans*-glucoside **159** (❖ *Scheme 71*) [114].

1,2-Migration and concurrent glycosidation of phenyl 2,3-*O*-thionocarbonyl-1-thio-α-L-rhamnopyranosides under the action of methyl trifluoromethanesulfonate (MeOTf) also give in high yields the 3-*O*-(methylthio) carbonyl-2-*S*-phenyl-2,6-dideoxy-β-L-glucopyranosides, which are ready precursors to the corresponding 2-deoxy-β-glycosides [115].

In the reduction of 2-azido-3-*O*-benzyl-4,6-*O*-bezylidene-2-deoxy-1-thio-β-D-mannopyranoside **160** with propane-1,3-dithiol and NEt$_3$ followed by the protection of the formed amino group, migration of the ethylthio group at C-1 and the amino group at C-2 occurs with the retention of configuration in both positions, to give 3-*O*-benzyl-4,6-*O*-bezylidene-2-*S*-ethyl-2-thio-β-D-mannopyranosyl-(2,2,2-trichloroethoxycarbonyl)amine **161** (❖ *Scheme 72*) [116].

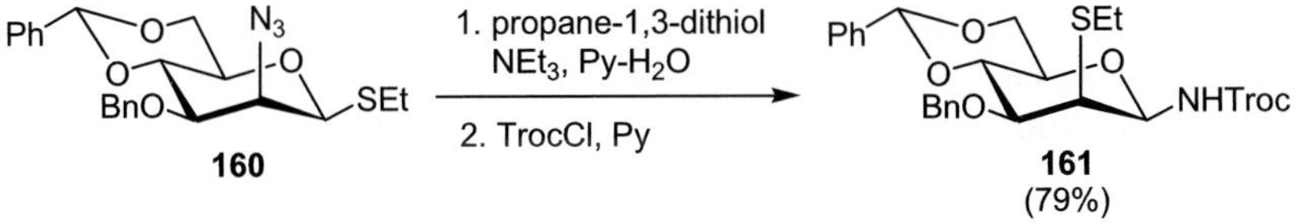

Scheme 72

4.4.2 Rearrangements

When 2-deoxy-2-iodoglycosylisocyanate **162** is exposed to ammonia, a long-distance rearrangement occurs to yield 1,2-fused aminothiazoline **163** (❖ *Scheme 73*) [117].

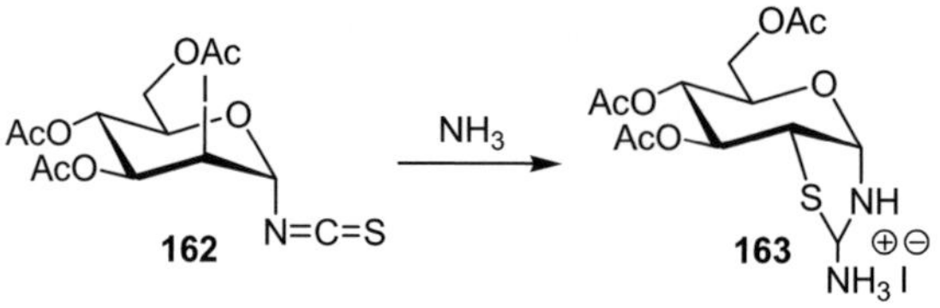

☐ Scheme 73

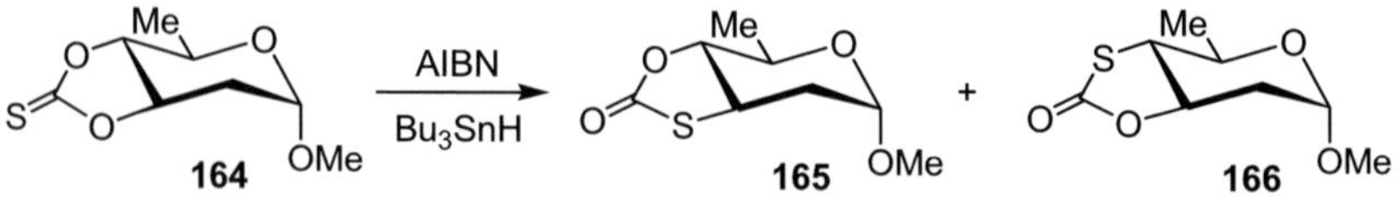

☐ Scheme 74

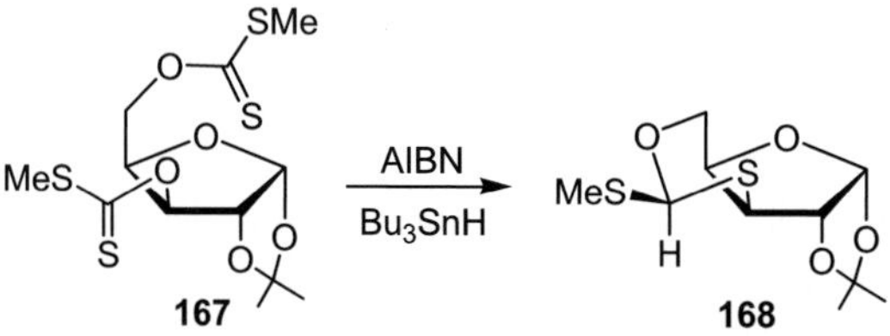

☐ Scheme 75

Thionocarbonates may proceed a radical-induced rearrangement to form thiosugars. Thus, treatment of 3,4-thionocarbonate **164** with azo-bis-isobutyronitrile (AIBN) and Bu₃SnH gives the 3- and 4-thio sugar derivatives **165** and **166** with retention of configuration (❯ *Scheme 74*) [118].

Dixanthates such as **167** also undergo rearrangement reactions under similar radical conditions to afford 3,5-cyclic dithioorthoformate **168** instead of the expected deoxygenation products (❯ *Scheme 75*) [119].

In the case of heating dixanthate **169** at 200 °C, orthotrithiocarbonates **170** and **171** are formed via a regioselective cyclization and with inversion of configuration. A similar procedure can also be applied to the 4,6-dixanthate **172** giving **173** and **174** (❯ *Scheme 76*) [120].

5 Introduction of Phosphorus

5.1 Nucleophilic Substitutions

The trivalent phosphorus atom bears a lone pair of electrons and therefore can be used as a nucleophilic reagent for substitution. Triphenylphosphine displacements on alkyl halides give phosphonium salts which, after the conversion into phosphorus ylides by strong bases,

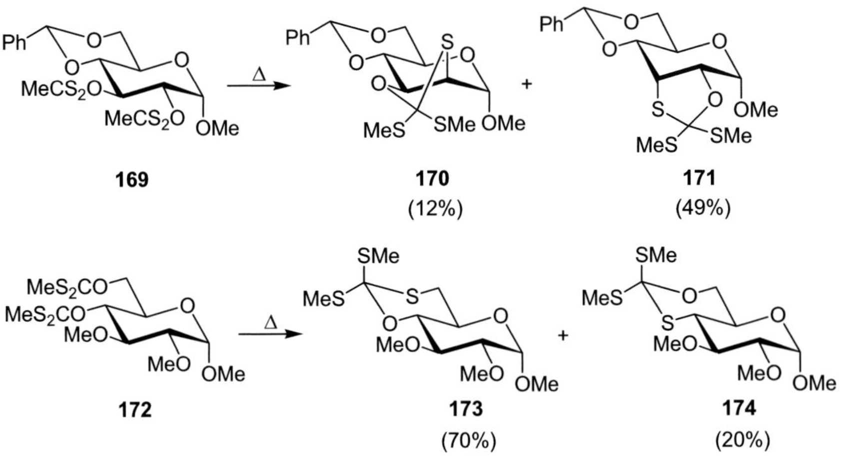

169 → **170** (12%) + **171** (49%)

172 → **173** (70%) + **174** (20%)

■ Scheme 76

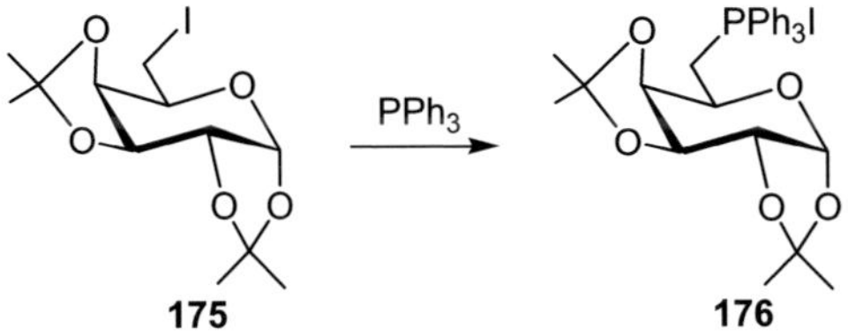

175 → **176**

■ Scheme 77

177 X = I
180 X = P(O)(OEt)$_2$
or X = P(O)(NEt)$_2$

178 X = Hal
181 X = P(O)(OEt)$_2$

179 X = I
182 X = P(O)(OEt)$_2$
or X = P(O)(OBu)$_2$

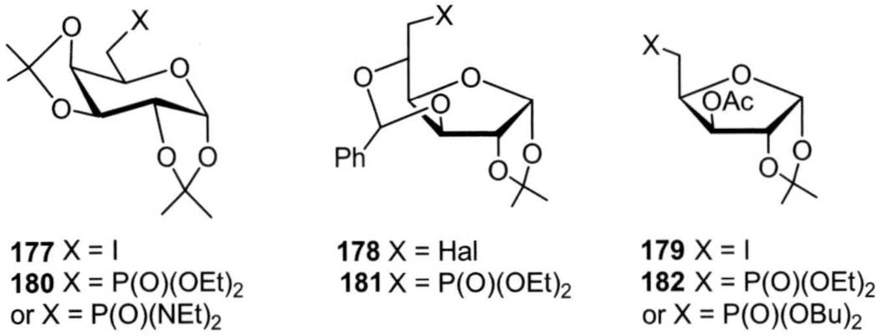

■ Scheme 78

are useful in Wittig alkene synthesis. A limited number of reactions of this type have been performed on primary sugar halides. For example, the reaction of 6-iodogalactose derivative **175** with PPh$_3$ gives the phosphonium iodide **176** (❷ *Scheme 77*) [121].

Other trivalent phosphorus-containing nucleophiles, such as trialkyl phosphites [P(OR$_3$)], dialkyl alkylphosphonites [R′P(OR)$_2$], or alkyl dialkylphosphinites [R′$_2$P(OR)], can convert primary carbohydrate halides **177**, **178**, **179** to corresponding phosphorus derivatives **180**, **181**, **182** via the Michaelis–Arbusov reaction (❷ *Scheme 78*) [122,123].

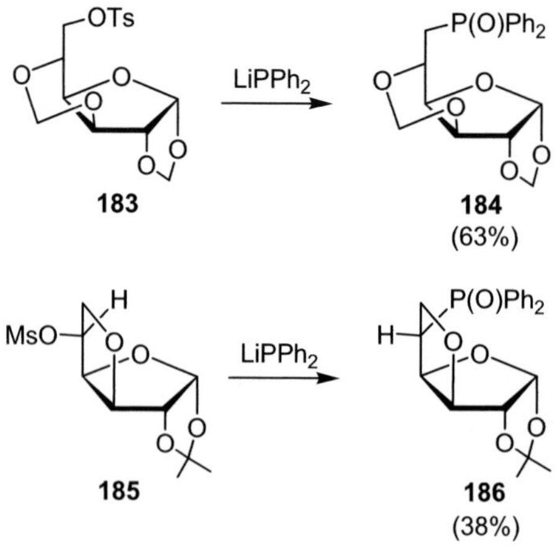

Scheme 79

Lithium diphenylphosphine (LiPR$_2$) has also been used to substitute sulfonates on sugar derivatives such as **183** and **185** to yield, after spontaneous oxidation, **184** and **186** (❯ *Scheme 79*) [124].

5.2 Ring-Opening Reactions

Epoxide ring opening is also an alternative route for the introduction of a phosphorus atom into carbohydrates. Thus, the D-*allo* epoxide **72** reacts with LiPPh$_2$, giving an intermediate phosphine which is spontaneously oxidized to the 2-phosphonate **187** [125], whereas the isomeric D-*manno* epoxide **188** affords the regioisomeric 3-phosphonate **189** (❯ *Scheme 80*) [124].

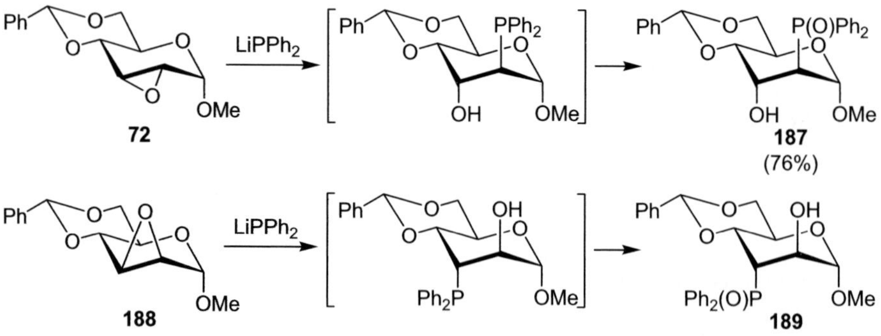

Scheme 80

A slight excess of potassium diphenylphosphine can open the oxetane ring of compound **190** to afford phosphine **191** (❯ *Scheme 81*) [126].

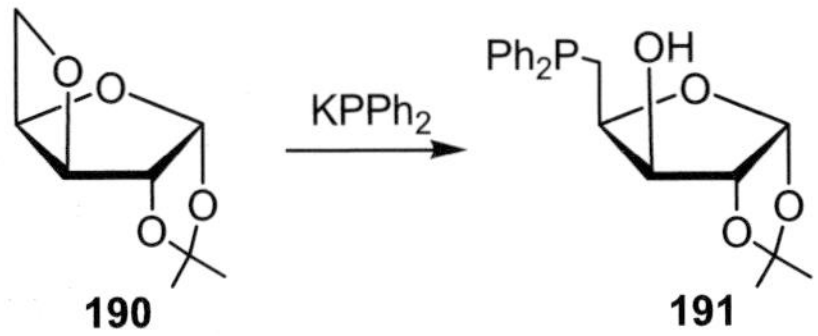

◻ Scheme 81

The 1,6-anhydro ring of 1,6-anhydro-D-glucopyranose can also be opened by using phosphinic acid (H_3PO_2) and sodium phosphinate (H_2PO_2Na) to give the 6-phosphate derivative [127].

5.3 Addition Reactions

5.3.1 Addition to Carbonyl Compounds

The addition of dialkyl phosphonates [$HP(O)(OR)_2$] to the carbonyl group of oxosugars gives rise to geminal phosphorus adducts. And the introduction orientation of phosphorus is effected by steric hindrance on sugar rings.

Thus, for example, the reaction of 1,6-anhydro-2-oxo-carbohydrate **192** with dimethyl phosphonate gives rise to the 2-phosphonate derivative **193** with a high stereoselectivity (❷ *Scheme 82*) [128]. And the addition of dimethyl phosphonate to the ulose **194** in the presence of DBU affords the (5*R*)-5-(dimethylphosphinyl)-D-*lyso*-hexofuranoside derivative **195a** and its (5*S*)-epimer **195b** (❷ *Scheme 82*) [129].

Tosylhydrazones can also undergo additions of pentavalent phosphorus derivatives, such as alkyl phenylphosphinate [Ph(H)P(O)OR]; dimethyl phosphonate [$HP(O)(OMe)_2$], or dimethyl phenylphosphonate [$PhP(O)(OMe)_2$], to introduce phosphorus atoms.

For example, compound **196** reacts with dimethyl phenylphosphonate to afford the adduct **197** and reductive removal of the tosylhydrazino group by $NaBH_4$ leads to the phosphi-

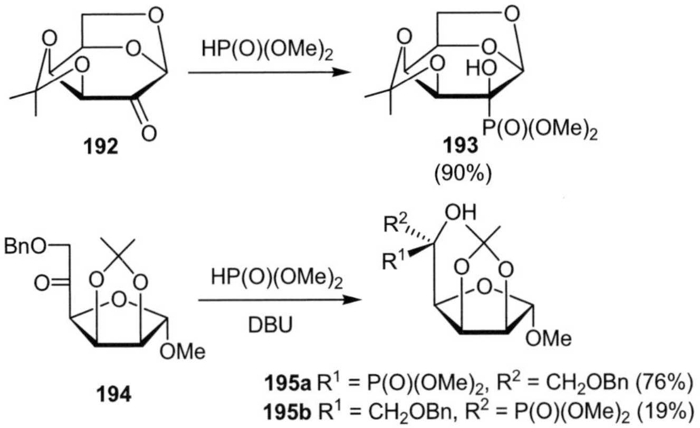

195a R^1 = P(O)(OMe)$_2$, R^2 = CH$_2$OBn (76%)
195b R^1 = CH$_2$OBn, R^2 = P(O)(OMe)$_2$ (19%)

◻ Scheme 82

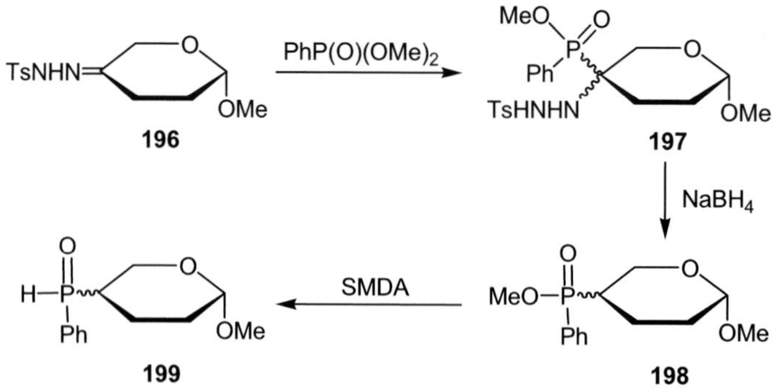

◼ Scheme 83

nate **198**, which can be further reduced to the phosphine oxide **199** by reaction with sodium dihydrobis(2-methoxy-ethoxy)aluminate (SMDA) (**◉** *Scheme 83*) [130].

5.3.2 Addition to Isolated Double Bond

Dialkyl phosphonates [e. g. HP(O)(OMe)$_2$] can be added to isolated double bonds of the 3-nitro derivative **200** to give the 3-nitro-2-phosphonate **201**. Also, phenylphosphine can react with the exocyclic double bond of compound **202** to afford the phosphine **203** in 63% yield together with a dimeric compound [131]. And compound **203** can be further oxidized to produce the phosphine oxide **204** quantitatively (**◉** *Scheme 84*) [132].

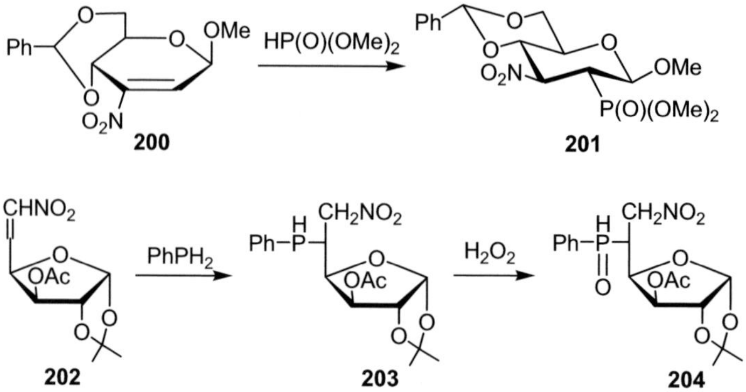

◼ Scheme 84

5.4 Coupling Reaction

A modified Arbuzov reaction has been applied to the coupling of the D-glucal/D-galactal derived vinyl halides **205** and triethyl phosphite at 150 °C in the presence of NiCl$_2$

as a catalyst, and 2-(diethoxyphosphoryl)hex-1-en-3-uloses **206** are obtained in fair yields (❯ *Scheme 85*) [133].

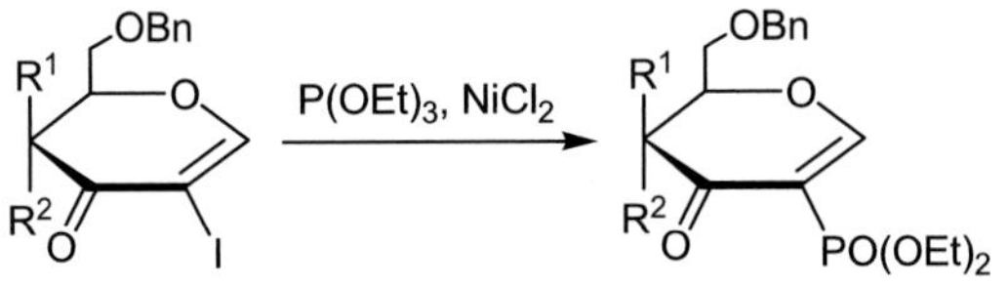

205a, 205b   **206a, 206b**

a: R^1 = H, R^2 = OBn (60%)
b: R^1 = OBn, R^2 = H (60%)

◻ Scheme 85

References

1. Hanessian S (1997) Selected reactions in carbohydrate chemistry. In: Hanessian S (Ed) Preparative carbohydrate chemistry. Dekker, New York, p 85
2. Boons GJ (2000) Functionalised saccharides. In: Boons GJ, Hale KJ (Eds) Organic synthesis with carbohydrates. Sheffield Academic Press, Sheffield, p 56
3. Szarek WA (1973) Adv Carbohydr Chem Biochem 28:250
4. Szarek WA (1973) Adv Carbohydr Chem Biochem 28:230
5. Card PJ (1985) J Carbohydr Chem 4:451
6. Wu MC, Anderson L, Slife TW, Jensen LJ (1974) J Org Chem 39:3014
7. Hanessian S, Vatèle JM (1981) Tetrahedron Lett 22:3579
8. Nadkarni S, Williams NR (1965) J Chem Soc 3496
9. Capon B (1969) Chem Rev 69:407
10. Brimacombe JS, Doner LW, Rollins AJ (1972) J Chem Soc Perkin Trans 1 2977
11. Brimacombe JS, Doner LW, Rollins AJ (1973) J Chem Soc Perkin Trans 1 1295
12. Buss DH, Hough L, Richardson AC (1963) J Chem Soc 5295
13. Zu Reckendorf WM, Bonner WA (1963) Tetrahedron 19:1711
14. Binkley RW, Hehemann DG (1979) Adv Carbohydr Chem 24:139
15. Vox JN, Van Boom JH, Van Boechel CAA, Beetz T (1984) 3:117
16. Richardson AC (1969) Carbohydr Res 10:395
17. Gnichtel H, Rebentisch D, Tompkins TC, Gross PH (1982) J Org Chem 47:2691
18. Wright JA, Taylor NF (1966–1967) Carbohydr Res 3:333
19. Buchanan JG, Schwarz JCP (1962) J Chem Soc 4770
20. Guthrie RD, Williams GJ (1976) J Chem Soc Perkin Trans 1 801
21. Christensen JE, Goodman L (1961) J Am Chem Soc 83:3827
22. Van Der Klein PAM, Filemon W, Veeneman GH, Van Der Marel GA, Van Boom JH (1992) J Carbohydr Chem 11:837
23. Hanessian S, Pleassas NR (1969) J Org Chem 34:1035, 1045 and 1053
24. Ferier RJ, Prasad N (1969) J Chem Soc 570
25. Ferier RJ, Ponpipom MJ (1971) J Chem Soc Chem Commun 553
26. Leblanc Y, Fitzsimmons BJ, Springer JP, Rokach J (1989) J Am Chem Soc 111:2995
27. Dios A, Geer A, Marzabadi CH, Franck RM (1998) J Org Chem 63:6673
28. Dahl RS, Finney NS (2004) J Am Chem Soc 126:8356
29. Giese B (1986) Radical in organic synthesis: Formation of carbon-carbon bonds. Pergamon Press, Oxford
30. Taylor NF (1988) Fluorinated carbohydrate. Chemical and biological aspects, ACS Symp Ser 374
31. Beuthien-Baumann B, Hamacher K, Oberdorfer F, Steinbach J (2000) Carbohydr Res 327:107

32. Tomas SS, Plenkiewicz J, Ison ER, Bols M, Zou W, Szarek WA, Kisilevsky R (1995) Biochim Biophys Acta 1272:37

33. Suami T, Hough L, Tsuboi M, Machinami T, Watanabe N (1994) J Carbohydr Chem 13:1079

34. Szarek WA (1973) Adv Carbohydr Chem Biochem 28:225

35. Su TL, Klein RS, Fox JJ (1982) J Org Chem 47:1506

36. Hamacher K, Coenen HH, Stocklin G (1986) J Nucl Med 27:235

37. Tsuchiya T, Takahashi Y, Endo M, Umezawa S, Umezawa H (1985) J Carbohydr Chem 4:587

38. Haradahira T, Maeda M, Kai Y, Omae H. Kojima M (1985) Chem Pharm Bull 33:165

39. Doboszewski B, Hay GW, Szarek WA (1987) Can J Chem 65:412

40. Anisuzzaman AKM, Whistler RL (1978) Carbohydr Res 61:511

41. Whistler RL, Anisuzzaman AKM (1980) Methods Carbohydr Chem 8:227

42. Haylock CR, Melton LD, Slessor KN, Tracey AS (1971) Carbohydr Res 16:375

43. Hanessian S, Ponpidom MM, Lavallee P (1972) Carbohydr Res 24:45

44. Garegg PJ, Samuelsson B (1980) J Chem Soc Perkin Trans 1 2866

45. Abram TS, Baker R, Exon CM (1982) J Chem Soc Perkin Trans 1 285

46. Garegg PJ (1984) Pure Appl Chem 56:845

47. Kielberg J, Frejd T, Jansson K, Sundin A, Magnusson G (1988) Carbohydr Res 176:271

48. Mitsunobu S (1981) Synthesis 1

49. Hannelore V, Brandstetter H, Zbiral E (1980) Helv Chim Acta 63:327

50. Verheyden JPH, Moffatt JG (1970) J Org Chem 35:2319

51. Hanessian S, Plessas NR (1969) J Org Chem 34:2163

52. Back TG, Barton DHR, Rao BL (1977) J Chem Soc Perkin Trans 1 1715

53. Hanessian S (1968) Adv Chem Ser 74:159

54. Bragg PD, Jones JKN, Turner JC (1959) Can J Chem 37:1412

55. Jones JKN, Perry MB, Turner JC (1960) Can J Chem 38:1122

56. Jennings HJ, Jones JKN (1963) Can J Chem 41:1151

57. Jennings HJ, Jones JKN (1965) Can J Chem 43:2372

58. Jennings HJ (1970) Can J Chem 48:1834

59. Jennings HJ, Jones JKN (1965) Can J Chem 43:3018

60. An SH, Bobek M (1986) Tetrahedron lett 27:3219

61. Card PJ (1983) J Org Chem 48:393

62. Card PJ (1985) J Carbohydr Chem 4:453

63. Kobayashi S, Yoneda A, Fukuhara T, Hara S (2004) Tetrahedron lett 45:1287

64. Penglis AAE (1981) Adv Carbohydr Chem Biochem 38:195

65. Tsuchiya T (1990) Adv Carbohydr Chem Biochem 48:91

66. Hartman MCT, Coward JK (2002) J Am Chem Soc 124:10036

67. Patroni JJ, Stick RV (1987) Aust J Chem 40:795

68. Fuentes J, Andulo M, Pradera MA (1998) Tetrahedron lett 39:7149

69. Fuentes J, Andulo M, Pradera MA (1999) Carbohydr Res 319:192

70. Chana JS, Collins PN, Farnia F, Peacock DJ (1988) J Chem Soc Chem Commun 94

71. Ido T, Wan CN, Fowler JS, Wolf AP (1977) J Org Chem 42:2341

72. Korytnyk W, Petrieó CR (1982) Tetrahedron 38:2547

73. Thiem J, Klaffke W (1992) Top Curr Chem 154:285

74. Horton D, Priebe W, Snaidman M (1990) Carbohydr Res 205:71

75. Lafont D, Descotes G (1987) Carbohydr Res 166:195

76. Rawal GK, Rani S, Madhusudanan KP, Vankar YD (2007) Synthesis 2:294

77. Mc Carter JD, Adam MJ, Wihters SG (1995) Carbohydr Res 266:273

78. Ansell MF (1983) Rodd's chemistry of carbon compounds, vol 1F, G Supplement. Elsevier, Amsterdam

79. Blattner R, Ferrier RJ (1980) J Chem Soc Perkin Trans 1 1523

80. Ferrier RJ, Haines SR, Gainsford GJ, Gabe EJ (1984) J Chem Soc Perkin Trans 1 1683

81. Karpeisiuk W, Banaszek A, Zamojski A (1989) Carbohydr Res 186:156

82. Kloosterman M, Westerduin P, Van Boom JH (1986) Recl Trav Chim Pays-Bas 105:136

83. Viand MC, Rollin P (1990) Synthesis 130

84. Knapp S, Kukkola PJ, Sharma S, Murali Dhar TG, Naughton ABJ (1990) J Org Chem 55:5700

85. De Oliveira RN, Cottier L, Sinou D, Srivastava RM (2005) Tetrahedron 61:8271

86. Williams NR (1970) Adv Carbohydr Chem Biochem 25:109

87. Lemieux RU, James K, Nagabhushan TL (1973) Can J Chem 51:48
88. Lemieux RU, Ratcliffe M (1979) Can J Chem 57:1244
89. Czernecki S, Ayadi E, Randriamandimby D (1994) J Org Chem 59:8256
90. Griffith DA, Danishefsky SJ (1990) J Am Chem Soc 112:5811
91. Du Bois J, Tomooka CS, Hong J, Carreira EM (1997) J Am Chem Soc 119:3179
92. Dyong I, Schilte G, Lam-Chi Q, Friege H (1979) Carbohydr Res 68:257
93. Ravindran B, Sakthivel K, Suresh CG, Pathak T (2000) J Org Chem 65:2637
94. Nicolaou KC, Baran PS, Zhong YL, Vega JA (2000) Angew Chem 112:2625
95. Paul HW, Fraser-Reid B (1986) Carbohydr Res 150:111
96. Baer HH (1972) Methods Carbohydr Chem 6:245
97. Guthrie RD, Johnson LF (1961) J Chem Soc 4166
98. Lemieux RU, James K, Nagabhushan TL, Ito Y (1973) Can J Chem 51:33
99. Pelyvas I, Hasegawa A, Whistler RL (1986) Carbohydr Res 146:193
100. Montero A, Mann E, Herradón B (2005) Tetrahedron Lett 46:401
101. Nicolaou KC, Ladduwahetty T, Randall JL, Chucholowski A (1986) J Am Chem Soc 108:2466
102. Piispanen PS, Norin T (2003) J Org Chem 68:628
103. Brossmer R (1962) Methods Carbohydr Chem 1:216
104. Trimnell D, Stout EI, Doane WM, Russel CR (1975) J Org Chem 40:1337
105. Yu B, Zhu X, Hui Y (2001) Tetrahedron 57:9403
106. Arela O, Cicero D, De Lederkremer RM (1989) J Org Chem 54:1884
107. Dabbkowski W, Michalska, M, Tworowska I (1998) J Chem Soc Chem Commun 427
108. von Itzstein M, Jenkins MJ, Mocerino M (1990) Carbohydr Res 208:287
109. Whistler RL, Lake WC (1972) Methods Carbohydr Chem 6:286
110. Igarashi K, Honma T (1970) J Org Chem 35:606
111. Matsuura K, Maeda S, Araki Y, Ishido Y (1970) Tetrahedron Lett 2869
112. Kaye A, Neidle S, Reese CB (1988) Tetrahedron Lett 29:2711
113. Nicolaou KC, Mitchell HJ, Fylaktakidou KC, Suzuki H, Rodríguez RM (2000) Angew Chem Int Ed 39:1089
114. Zuurmond HM, Van Deklein PAM, Van Der Marel GA, Van Boom JH (1993) Tetrahedron 49:6501
115. Yu B, Yang Z (2001) Org Lett 3:377
116. Veselý J, Rohlenová A, Džoganová M, Trnka T, Tišlerová I, Šaman D, Ledvina M (2006) Synthesis 4:699
117. Santoyo-Gonzales F, Garcia-Calvo-Flores F, Isaac-Garcia J, Hernandez-mateo P, Garcia-mendoza P, Robles-Diaz R (1994) Tetrahedron 50:2877
118. Somask L, Ferrier RJ (1991) Adv Carbohydr Chem Biochem 49:37
119. Herdewijn PAM, Van Aerschot A, Jie L, Esmans E, Peneau-Dupont J, Declerc JP (1991) J Chem Soc Perkin Trans 1 1729
120. Faure A, Kryczka B, Descotes G (1979) Carbohydr Res 74:127
121. Karpiesiuk W, Banaszek A (1994) Tetrahedron 50:2965
122. Seo K, Inokawa S (1970) Bull Chem Soc Jpn 43:3224
123. Seo K, Inokawa S (1975) Bull Chem Soc Jpn 48:1237
124. Hall LD, Steiner PR (1971) J Chem Soc Chem Commun 84
125. Brown MA, Cox PJ, Howie RA, Melvin OA, Taylor OJ, Wardell JL (1995) J Organomet Chem 498:275
126. Pàmies O, Diéguez M, Net G, Ruiz A, Claver C (2001) J Org Chem 66:8364
127. Nifant'ev EE, Gudkova IP, Kochetkov NK (1970) J Gen Chem USSR 40:425
128. Paulson H, Greve W (1973) Chem Ber 106:2124
129. Hanaya T, Yamamoto H (2002) Helv Chim Acta 85:2608
130. Yamashita M, Yoshikane M, Ogawa T, Inokawa S (1979) Tetrahedron 35:741
131. Paulson H, Greve W (1973) Chem Ber 106:2114
132. Takayanagi H, Yamashita M, Seo K, Yoshida H, Ogata T, Inokawa S (1974) Carbohydr Res 38:C19
133. Leonelli F, Capuzzi M, Calcagno V, Passacantilli P, Piancatelli G (2005) Eur J Org Chem 2671

2.4 Anhydrosugars

Sławomir Jarosz, Marcin Nowogródzki
Institute of Organic Chemistry, Polish Academy of Sciences,
01–224 Warsaw, Poland
sljar@icho.edu.pl

Abstract

The anhydrosugars are reviewed in this chapter. The emphasis is placed on the general methodology of their preparation, as well as, their application in stereocontrolled organic synthesis. The material is divided into two main parts: anomeric anhydrosugars and non-anomeric ones. In the first class, 1,2-sugar epoxides and 1,6-anhydrosugars are the most important since they offer significant synthetic potential in targeted synthesis of important compounds. From the second class, sugar epoxides (2,3- and 3,4-oxiranes) are particularly useful. The synthesis of both classes of anhydrosugars (anomeric and non-anomeric) is illustrated by selected examples including older examples (i. e. those described also in the previous edition of this monograph) and those published more recently. Various methods for the preparation of anhydrosugars are described in order to give the reader a general impression of the importance of such derivatives both as biologically active targets (illustrated by selected examples of anhydronucleosides) and optically pure building blocks in targeted synthesis. The material included in this chapter

In: *Glycoscience*. Fraser-Reid B, Tatsuta K, Thiem J (eds)
Chapter-DOI 10-1007/978-3-540-30429-6_6: © Springer-Verlag Berlin Heidelberg 2008

should help the reader solve any problems faced in their laboratories connected with planning and execution of the synthesis of optically pure targets.

Keywords

Anhydrosugars; Synthesis; Sugar oxiranes; 1,2-Sugar epoxides; 1,6-Anhydrosugars; Rearrangement

Abbreviations

AZT	3-azidothymidine
DET	diethyl tartrate
DIAD	di-isopropyl azodicarboxylate
DMDO	dimethyldioxirane
IDCP	iodonium dicollidine perchlorate
LNA	locked nucleic acids
MCPBA	*m*-chloroperbenzoic acid
OTs	toluene-*p*-sulfonate
RCM	ring closing metathesis
RNA	ribonucleic acid
THF	tetrahydrofuran
THP	tetrahydropyran
TPP	triphenylphosphine

1 Introduction

The material presented in this chapter describes the concise methodology of the preparation of anhydrosugars and their application in synthesis, updating the chapters published in the last edition [1,2]. However, basic information from the earlier edition is also included here allowing the reader to follow the present text more easily. Therefore, citation to this data refers (mostly) to the previously published chapters and not to original papers.

The anhydrosugars are an important class of saccharides, the synthesis and properties of which are described in many monographs [3,4,5,6,7]. Such derivatives are used for the preparation of modified carbohydrates (including *C*-glycosides) and complex enantiomerically pure products in which the chirality of the parent sugar is transferred to the target ('chiron approach' [8]). Many compounds from this class are also components of biologically active products. Anhydrosugars (also called 'intramolecular anhydrides') are derivatives that formally arise from elimination of the molecule of water from the parent carbohydrate. Various carbon atoms may be engaged in this formal process, thus providing different classes of anhydrosugars [3]. Generally, they belong to two main groups in which: (1) the anomeric carbon atom is involved in the anhydro structure and (2) the anhydro linkage is built between other carbon atoms of the sugar (❷ *Fig. 1*). These two classes of compounds will be described separately.

To construct the anhydro skeleton in the sugar molecule standard synthetic methods, used for the formation of 'normal' heterocyclic derivatives, are applicable [4,5]. Generally, one of the hydroxyl groups of the diol (from which the anhydro ring is formed) is activated (by

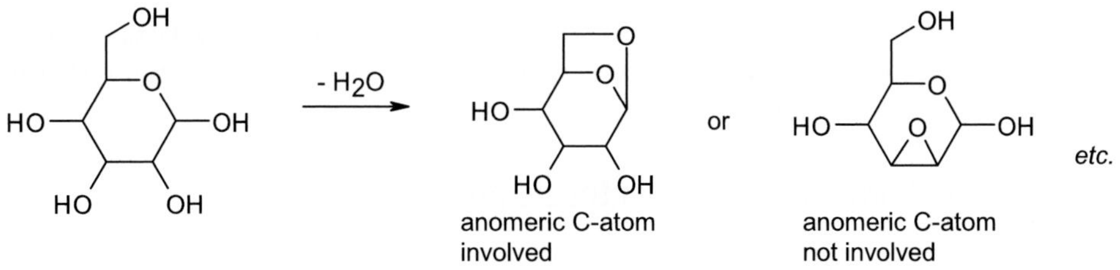

Figure 1
Examples of anhydrosugars derived from parent monosaccharides by formal elimination of the molecule of water

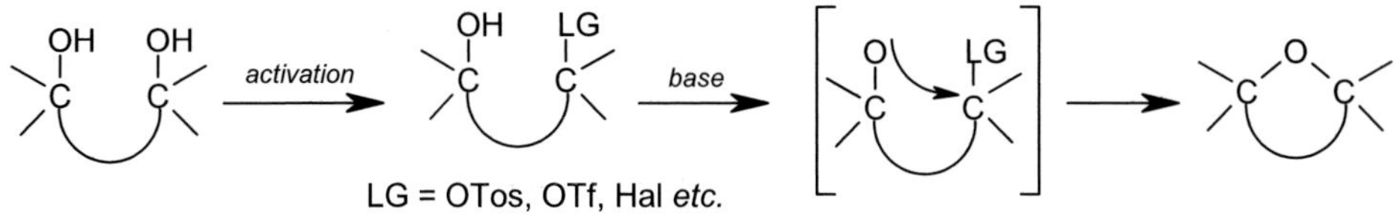

LG = OTos, OTf, Hal *etc.*

Figure 2
General method leading to anhydrosugars

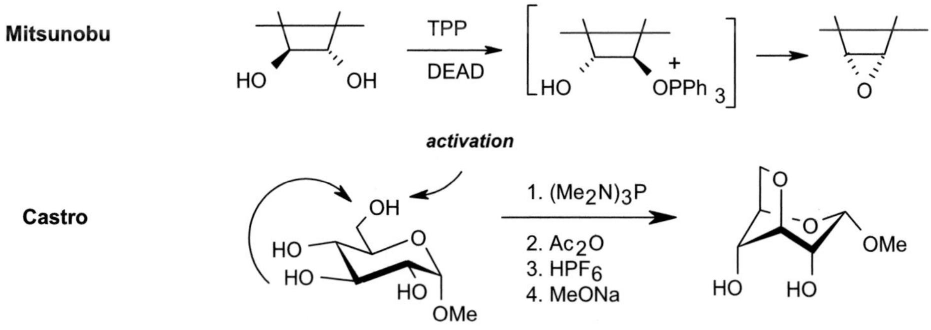

Figure 3
Synthesis of anhydrosugars by the Mitsunobu and Castro reaction

conversion into the leaving group such as tosylate, triflate, halogen etc.), while the second one acts as a nucleophile in basic media, which results in an intramolecular S_N2 closure of the anhydro ring (❷ *Fig. 2*).

Milder conditions, such as activation of one of the hydroxyl groups according either to Mitsunobu [9] or Castro [10] protocol, are also used for the preparation of anhydrosugars, which may be illustrated by examples shown in ❷ *Fig. 3*.

Compounds with the oxirane rings are special cases which, beside this general methodology, are also prepared by a number of other methods (direct epoxidation, reaction with sulfur ylides, Darzens' reaction, etc.) [1].

Anhydrosugars are also classified on the basis of the size of the heterocyclic anhydro ring. Thus, oxiranes (sugar epoxides), oxetanes, tetrahydrofuran, and tetrahydropyran derivatives are known. Reactivity of anhydrosugars is determined by the size of the heterocyclic anhydro ring. The most reactive are oxiranes and oxetanes, with oxolane (THF) and oxane (THP) derivatives being much less reactive. Selected examples of the representative derivatives: their synthesis and reactivity, from each group will be presented.

2 Anomeric Anhydrosugars: Synthesis and Reactions

Anomeric anhydrosugars represent a class of molecules that can be regarded as intramolecular glycosides. The glycosydic bond may engage the terminal alcohol of a sugar molecule (1,6-anhydropyranoses, 1,6-anhydrofuranoses, etc.) or any other atom (1,2-, 1,3-anhydrosugars etc.). The most representative are 1,6- and 1,2-anhydro-sugars; others are less common and have rather limited synthetic potential [3].

2.1 1,6-Anhydrosugars

2.1.1 1,6-Anhydrohexopyranoses

The most common 1,6-anhydrosugars are anhydro-aldopyranoses, which are formal derivatives of 6,8-dioxabicyclo[3.2.1]octane; less common are anhydro-aldofuranoses.

1,6-anhydrohexopyranose
6,8-dioxabicyclo[3.2.1]octane

1,6-anhydrohexofuranose
2,8-dioxabicyclo[3.2.1]octane

1,6-Anhydro-β-D-glucopyranose (levoglucosan), the most representative example of this class of compounds, was first isolated in pure form in 1894 by Tanret upon treatment of naturally occurring phenolic glycoside with barium hydroxide [2]. This compound is now conveniently obtained by pyrolysis of starch or cellulose and its production is covered by many patents. Other 1,6-anhydrosugars with different configurations, such as mannosan or galactosan, may be prepared by pyrolysis of mannan ivory nut meal or α-lactose [2,3].

OH
OH   OH
1,6-anhydroglucopyranose

OH
OH
1,6-anhydromannopyranose

OH
HO
OH
1,6-anhydrogalactopyranose

Generally, treatment of a sugar with a free hydroxyl group at the C-6 position possessing a good leaving group at the anomeric position (halogen, azide, tosylate etc.) with a strong base affords 1,6-anhydropyranoses [3]. Another method consists of a selective activation of a terminal hydroxyl group; this procedure is applicable for most 1,6-anhydro-pyranoses. Fraser-Reid and co-workers described a large-scale synthesis of 1,6-anhydro-D-*gluco*- and D-*manno*-pyranoses from free sugars by selective activation of the terminal position with tosyl chloride

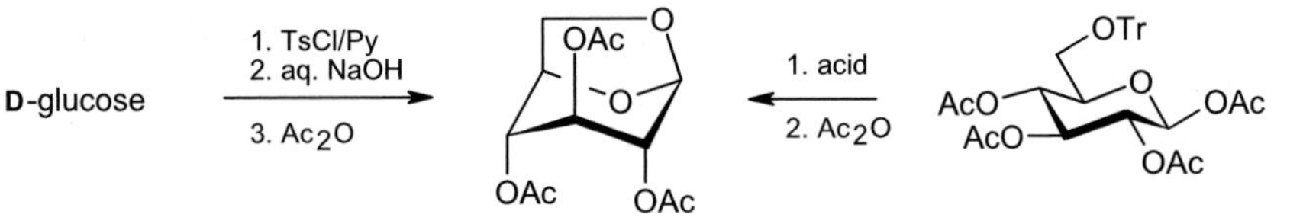

Scheme 1

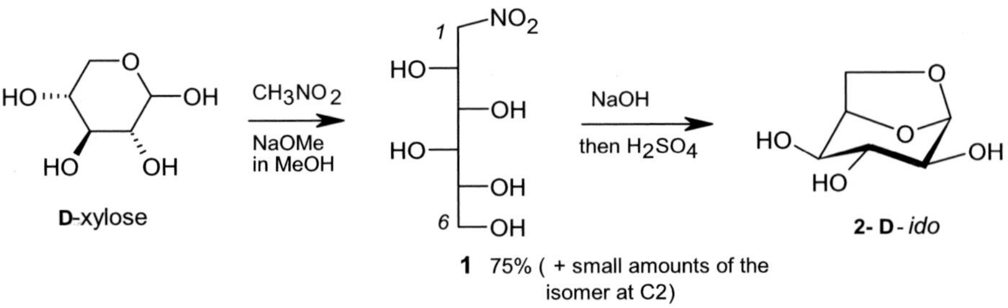

1   75% (+ small amounts of the
        isomer at C2)

Scheme 2

followed by cyclization of the intermediate in basic media [11]. This process can be facilitated by microwave irradiation [12]. Alternatively, the labile ethers (such as trityl) at the C6 position, while activated with Lewis acid, provide the anhydrosugars in excellent yield (❯ *Scheme 1*) [2].

In some cases special activation is not necessary. In acidic media, D-idose is easily converted into 1,6-anhydro-D-*ido*-pyranose, because the equatorial arrangement of all hydroxyl groups in the anhydro molecule facilitates this process. For example, starting from D-xylose, Köll prepared 1,6-anhydro-D-idose in good yield in a sequence of reaction shown in ❯ *Scheme 2* [13]. The free idose, obtained from 1-deoxy-1-nitro-D-iditol (**1**) by the Nef reaction, cyclized readily to the desired anhydrosugar [13]. During the synthesis of L-idose from the D-glucose [14] one may encounter the problem of the isolation of the free sugar, since it readily undergoes cyclization to the anhydro form.

Other methods of preparation of these valuable compounds are shown in ❯ *Scheme 3*. All of them are based on the intramolecular glycosylation of glycals by a terminal hydroxyl group. Treatment of the 3,4-di-*O*-benzoyl-D-glucal (**4**; R = Bz) with iodonium dicollidine perchlorate (IDCP) results in formation of the anhydro derivative **5a** (R = Bz), while activation of the hydroxyl group in a free glycal (**4**; R = H) with $(Bu_3Sn)_2O$ followed by iodocyclization pro-

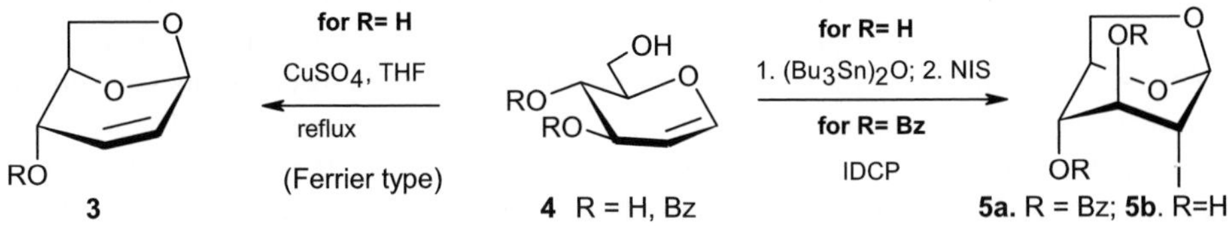

Scheme 3

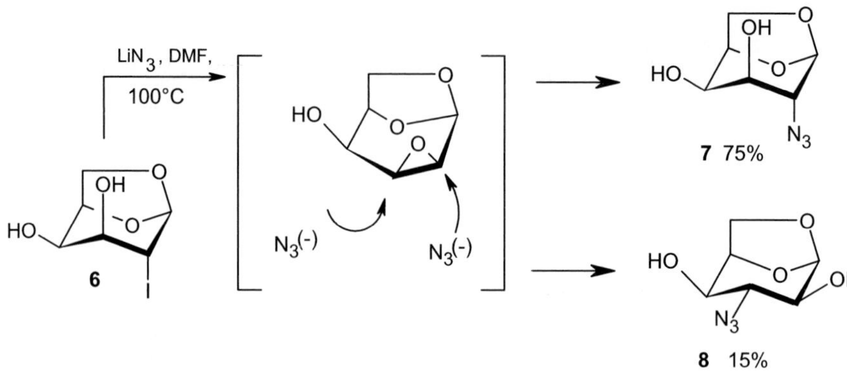

■ Scheme 4

vides the free compound **5b** (R = H). Internal Ferrier-type glycosylation leading to compound **3** was achieved by treatment of the free glycal with Lewis or protic acid (❯ *Scheme 3*) [2].

The presence of a leaving group in the anhydro skeleton opens up an easy path to modified anhydrosugars. For example, reaction of 2-iodo-2-deoxy-1,6-anhydrogalactose (**6**) with lithium azide provides the 2-azido-galactose derivative **7** together with small amounts of 3-azido-idose derivative **8** (❯ *Scheme 4*).

1,6-Anhydro-pyranoses with other than an oxygen heteroatom in the ring are known (although not common). Thiolevoglucosan has been known for many years [15]. An interesting example of the synthesis of 1,6-anhydrothiomannose (**9**) was presented recently [16].

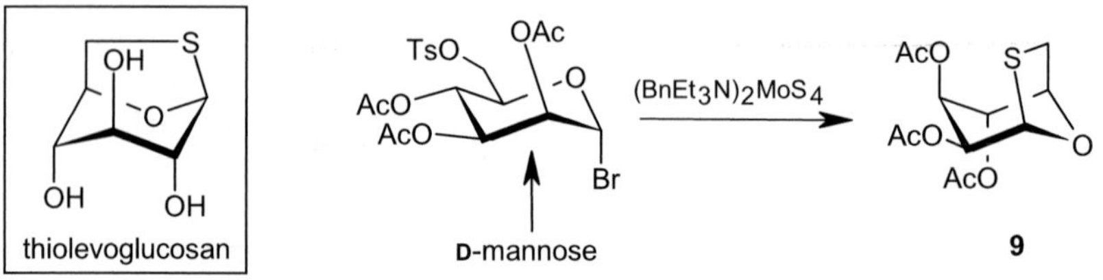

Such 'hetero' anhydrosugars may be used as convenient synthetic intermediates. For example, septanose iminosugars **11** were prepared via the aza-anhydrosugars **10** by Fuentes (❯ *Scheme 5*) [17].

The hydroxyl groups in anhydrosugars differ in steric orientation and, hence, their reactivity is also different [3]. This feature may be well illustrated by selective transformations of the hydroxyl groups in 1,6-anhydroglucose. Because of the $^1C_4$ conformation of its skeleton, all hydroxyl groups in 1,6-anhydro-D-glucose are placed at the axial positions. The hydroxyl group at the C-3 position is most hindered, thus less reactive than those at the C-2 and C-4 positions. This offers a great advantage for the preparation of many useful building blocks. For example, di-tosylation of 1,6-anhydro-D-glucopyranose leads to 2,4-di-*O*-tosyl derivative **12**, which readily cyclizes to the 'Cerny' epoxide **13** (❯ *Scheme 6*) [3,6].

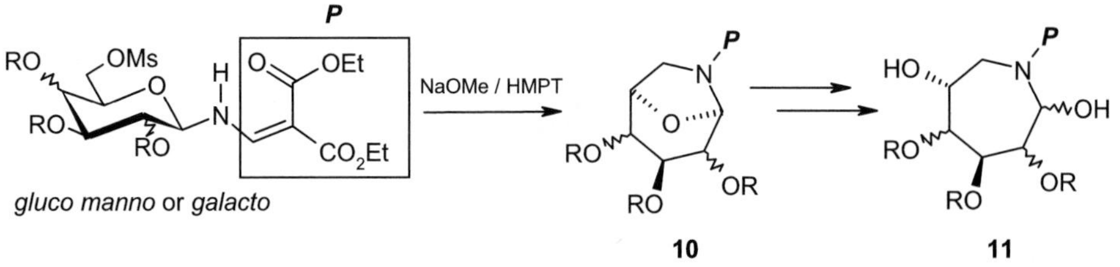

Scheme 5

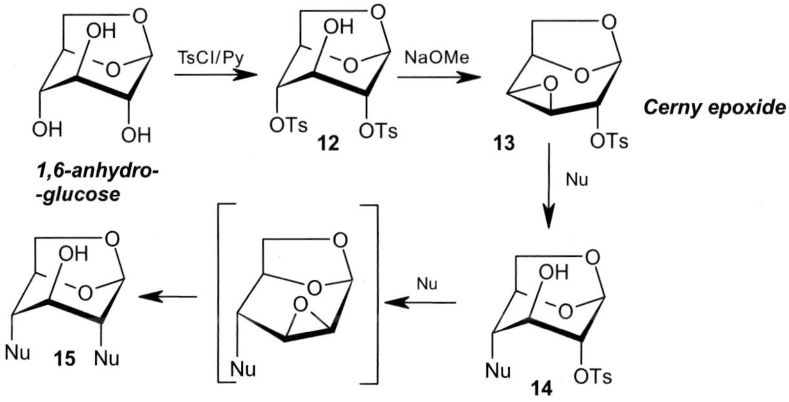

Scheme 6

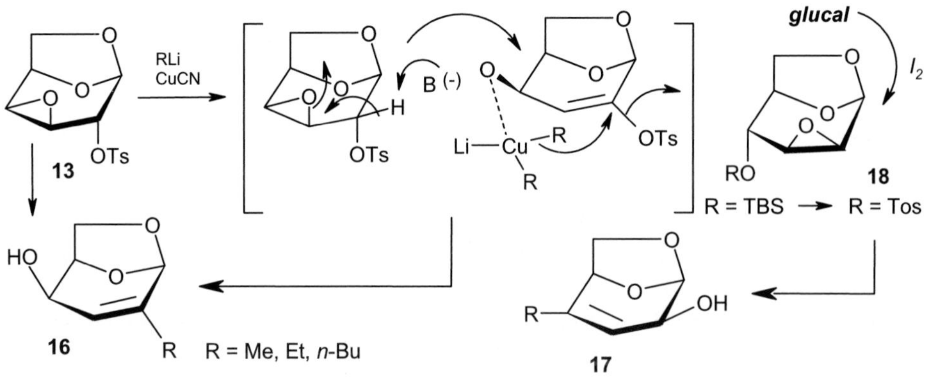

Scheme 7

Opening of the three-membered ring in **13** with various nucleophiles provides the corresponding mono- (**14**; at the C-2 position) and di-substituted (**15** at the C-2 and C-4 positions) derivatives (❂ *Scheme 6*) [3].

However, when organocuprate is used as a nucleophile, another product is formed. Instead of opening the oxirane ring, cuprate acts as a base which abstracts a proton from the C-2

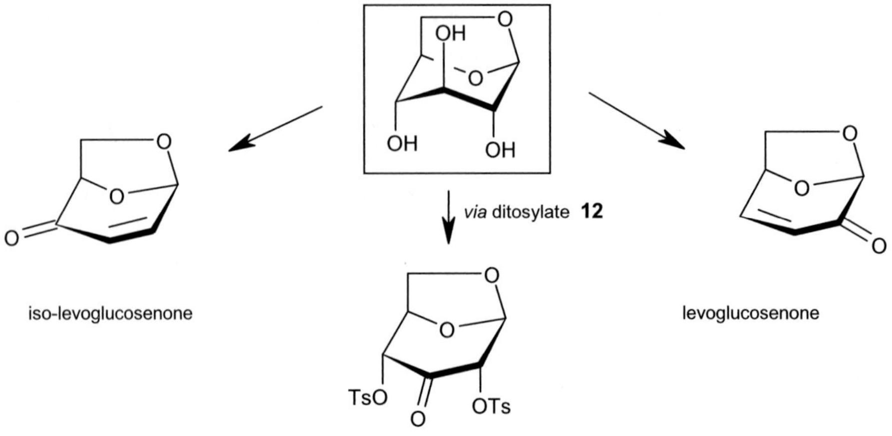

☐ Figure 4
Examples of keto-anhydrosugars prepared from 1,6-anhydroglucose

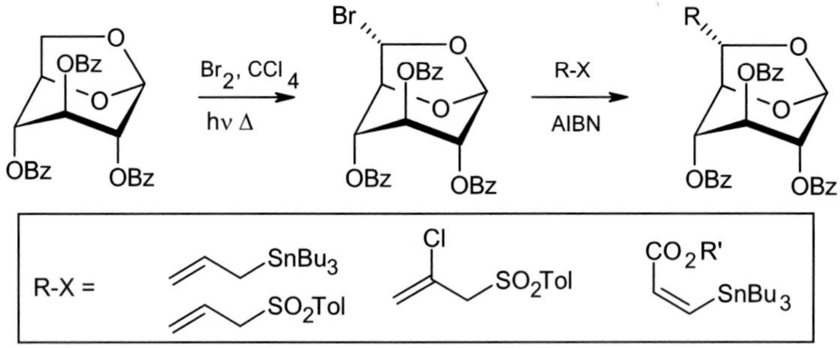

☐ Scheme 8

position inducing a cascade process leading finally to 2,3-unsaturated anhydrosugar **16** with a nucleophile placed at the C-2 position [18].

The alternative compound **17**, with the nucleophile placed at the C-4 position, was prepared by the same method [19] from regioisomeric (to Cerny epoxide) oxirane **18**, which was obtained from D-glucal by iodocyclization followed by standard reactions (❷ *Scheme 7*).

Many useful building blocks (besides those already presented) with preservation of the 1,6-anhydro skeleton such as levoglucosenone and its regioisomer *iso*-levoglucosenone can be prepared from 1,6-anhydroglucose [20]. The carbonyl group can be also introduced at the C-3 position by oxidation of the 2,4-di*O*-tosyl-1,6-anhydroglucose (**12**) [3] (❷ *Fig. 4*).

Besides the reactions involving the oxygen atoms, other processes directed to functionalization of the carbon skeleton are known. One of the most useful strategies is based on bromination of the C-6 position and further replacement of the bromine atom with a functional group under the radical conditions (❷ *Scheme 8*) [21].

Such derivatives, functionalized at the C-6 position, are convenient synthons for a variety of interesting compounds.

2.1.2 1,6-Anhydrofuranoses

1,6-Anhydrofuranoses derived from all eight diastereoisomeric aldohexoses were described in the literature [6]. These compounds are formed as side products in the synthesis of 1,6-anhydropyranoses from free (or partially protected) sugars. Treatment of free sugars with toluene-*p*-sulfonic acid in DMF solution affords the furanose and pyranose 1,6-anhydrides with the furanose form up to 33% for the *galacto-*, *allo-*, and *talo-* isomers [3]; the example is shown in ❷ *Scheme 9*.

A common procedure for the efficient synthesis of 1,6-anhydrofuranoses is based on cyclization of the 6-*O*-tosyl derivative of furanose with the free anomeric position and protected hydroxyl function at the C-5 position. Synthesis of 1,6-anhydro-D-mannofuranose is a good example of such a strategy (❷ *Scheme 10*) [22].

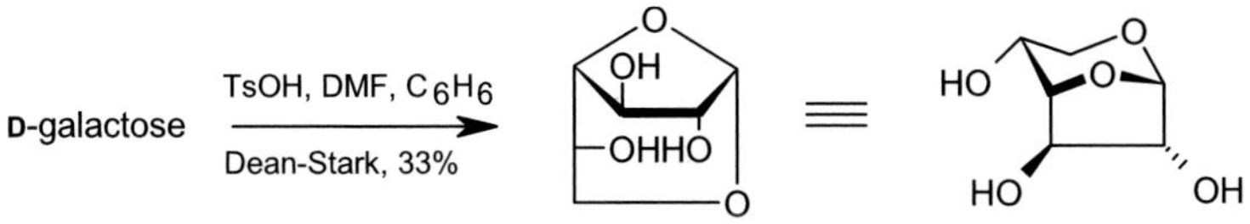

❑ Scheme 9

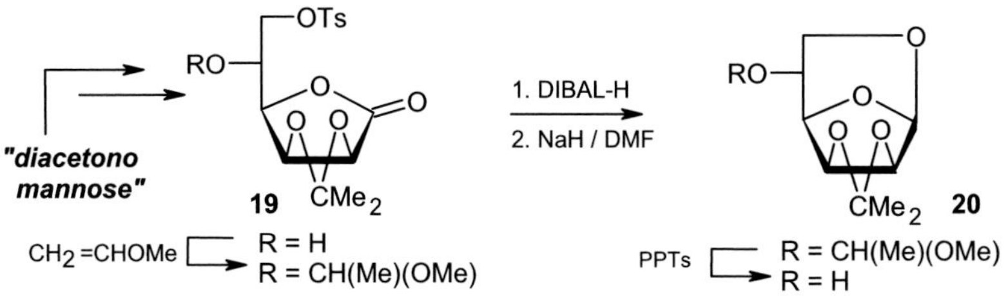

❑ Scheme 10

Standard transformation of the "diacetonomannose" provided 6-*O*-tosyl-5-*O*-protected lactone **19**, which was reduced to lactole with diisobutyl hydride. Cyclization under basic conditions afforded 1,6-anhydromannofuranose **20**, which can be deprotected easily at the C-5 position.

2.2 Higher Anhydroaldoses and Anhydroketoses

Higher sugars also can form the anhydrides in which the anomeric position and the terminal carbon atom are linked via a heterocyclic ring. For example, heptoses can form 1,7-anhydro derivatives although they are very rare. Only a few examples of such derivatives have been described in the literature (❷ *Fig. 5*) [3].

Anhydroketoses are derived mostly from 2-ketosugars and have similar properties to anhydroaldoses. Several such derivatives (e. g. **21**) were isolated from residues after pyrolysis of parent ketohexoses such as, for example, fructose [3,23]. The 3,8-anhydro derivative **24**, was isolated during the synthesis of D-*glycero*-D-*manno*-oct-3-ulose (**23**) by self-aldol condensation of D-erythrose (**22**) (❖ *Fig. 6*).

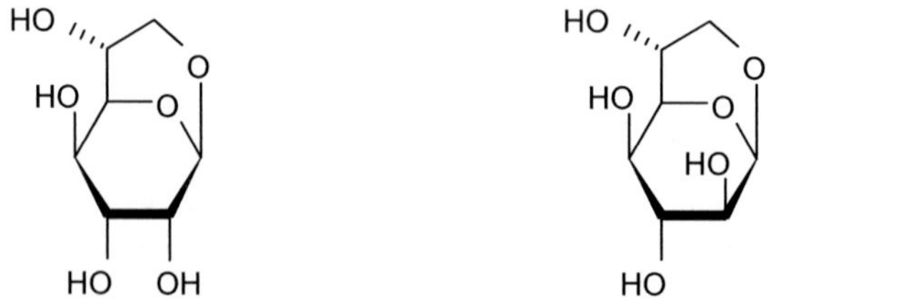

D-*glycero* -**D**-*gulo* -
1,7-anhydroheptopyranose

D-*glycero* -**D**-*ido* -
1,7-anhydroheptopyranose

❑ **Figure 5**
Examples of higher anomeric anhydrosugars

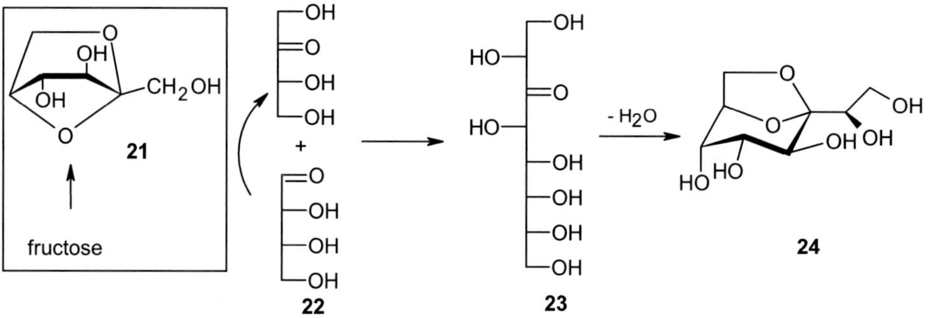

❑ **Figure 6**
Selected examples of anhydroketoses

2.3 1,2-Anhydrosugars

1,2-Anhydrosugars are an important class of compounds in which the three-membered oxirane ring is connected to the anomeric and the C-2 atoms of a sugar. The first compound of this class, the so-called Brigl's anhydride (**26**), was prepared by treatment of β-chloro-3,4,6-tri-*O*-acetyl-D-glucose (**25**) with ammonia (❖ *Scheme 11*) [3].

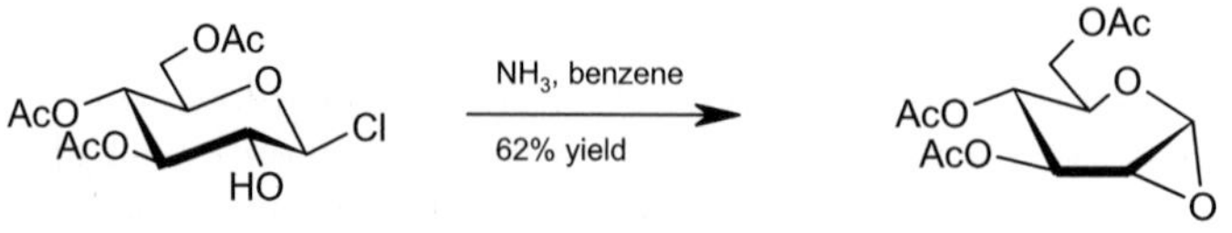

❑ **Scheme 11**

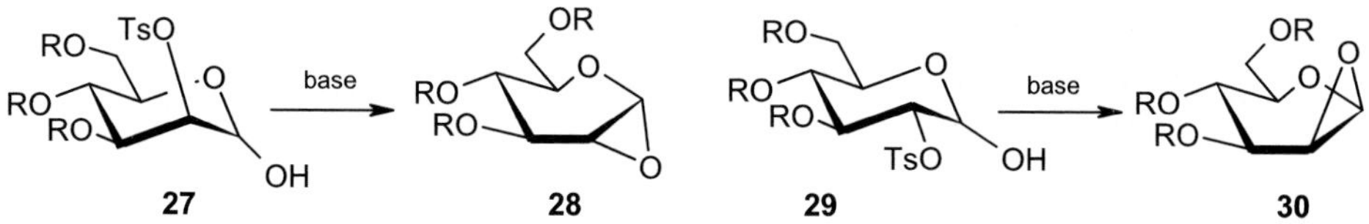

Figure 7
Synthesis of 1,2-anhydrosugars via activation of the 2-OH group

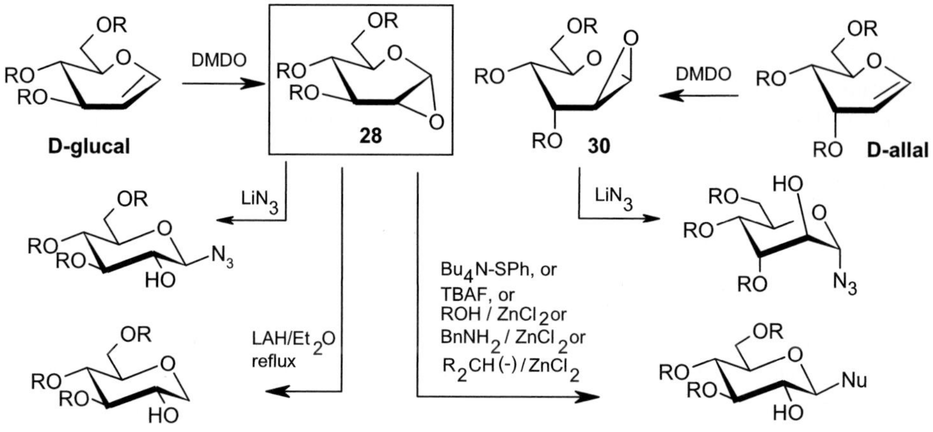

Scheme 12

There are three main routes leading to 1,2-anhydrosugars. The first one (similar to Brigl's original preparation) involves the attack of the hydroxyl group from the C-2 position on the anomeric center activated with a leaving group; both groups have to be in the *trans* arrangement. The second route is just the opposite: activated hydroxyl at the C-2 position (e. g. OTs) is attacked by the free anomeric hydroxyl with the inversion of the configuration at the C-2 stereogenic center [2,3] (❂ *Fig. 7*).

Treatment of the 2-OTs D-mannose derivative **27** with a base affords the 1,2-anhydro-glucose derivative **28**, while a similar reaction of the 2-OTs glucopyranose **29** yields the 1,2-anhydro-D-mannopyranose **30**.

The third method of synthesis of such epoxides, proposed by Danishefsky [24], involves a direct epoxidation of glycals with dimethyldioxirane (DMDO). The stereoselectivity of this reaction depends on the configuration at the C-3 position. If D-allal is used instead of the D-glucal the opposite epoxide is formed (❂ *Scheme 12*) [25]. Opening of the epoxide with azide (❂ *Scheme 12*) or oxygen nucleophiles proceeds with the inversion of the configuration at the anomeric center [25,26]. When such 1,2-epoxides are treated with a 'hydride' reagent smooth reduction is observed, which provides 1-deoxy-sugars in good yield (e. g. compound **31**) [27].

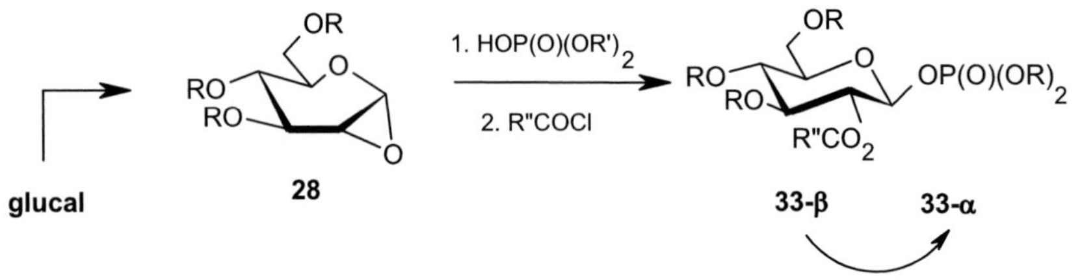

☐ Scheme 13

The 1,2-anhydrosugars are convenient synthons for the preparation of a variety of important derivatives. Generally, opening of the oxirane ring with various nucleophiles (alcoholates, amines, sulfides, carboanions, etc.) proceeds with the inversion of the configuration at the anomeric center [28,29,30].

Glycosyl phosphates, building blocks in automated solid-phase synthesis of complex oligo-saccharides, were prepared via a 1,2-epoxide (❷ *Scheme 13*). Oxidation of the appropriately substituted D-glucal **28** with DMDO followed by reaction with dialkyl phosphate provided the β-phosphate **33**, which anomerized further to the α-analog [31].

Reaction of 1,2-anhydro derivatives such as **34** with acetylene organometallics provides acetylenic *C*-glycoside **37** with the *retention* of the configuration at the anomeric center [29,32]. This can be explained by a complexation of the organometallic to the oxirane oxygen atom with a cleavage of the C1–O bond (**36**) and subsequent attack of the electrophile from the same (α) side. An interesting variation of this procedure with zirconium organometallics was proposed recently by Wipf [33]. This reaction provides also the α-anomers (**35**) substituted at the anomeric center with the *E*-olefin (❷ *Scheme 14*).

Diastereoselective synthesis of aryl *C*-glycosides of furanoses was realized in two ways. Oxidation of 1,2-unsaturated furanose **38** (obtained from thymidine according to the Danishefsky procedure) provides either D-*ribo*- (when the 2-OH is unprotected) or D-*arabino*- (the OH is protected with a bulky substituent) epoxides (**40** and **39**, respectively). The *trans* opening

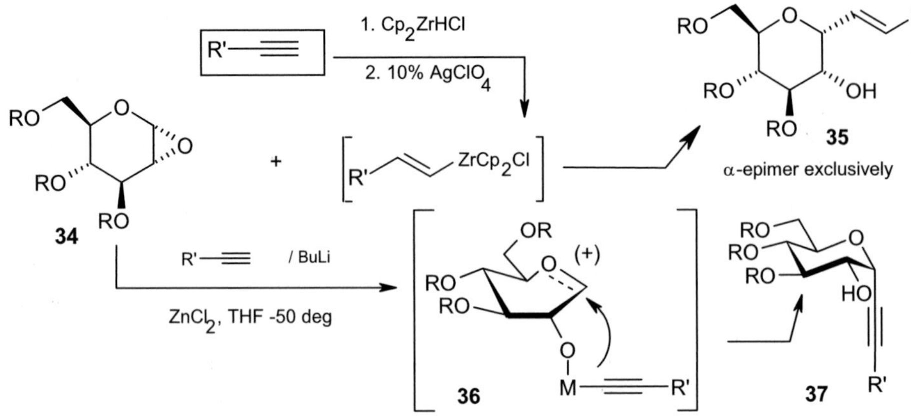

☐ Scheme 14

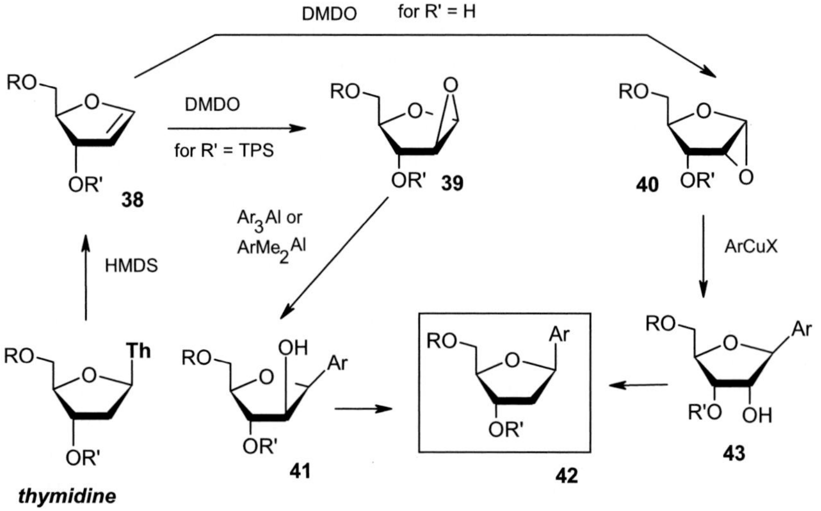

Scheme 15

of the oxirane ring [34] afforded the D-*ribo*-C-glycoside **43**, while the *cis*-opening with aluminum compounds provided the D-*arabino*-C-glycoside **41**. Both were further converted into the same 2-deoxy derivative **42** (◉ *Scheme 15*) [35].

2.4 1,3- and 1,4-Anhydrosugars

Such derivatives are obtained by reaction of partially protected glycosyl chlorides with a strong base (NaH, ᵗBuOK) in THF. The alternative version consists of treatment of the 4-*O*- or 5-*O*-

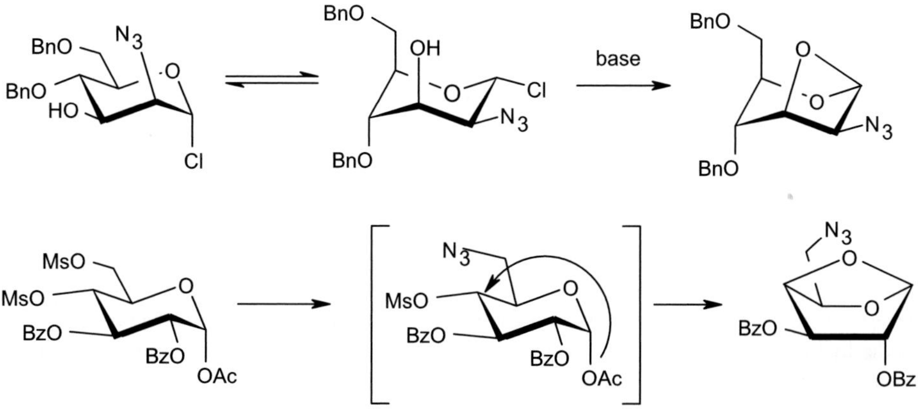

Figure 8
Examples of the 1,3- and 1,4-anhydropyranoses

sulfonylated partially protected sugars with a base or azide anion (NaN$_3$). Examples of the preparation of 1,3-anhydropyranose and 1,4-anhydropyranose are shown in ❷ *Fig. 8* [3]. Substituted 1,3-anhydroglucopyranoses, in the presence of the acidic catalysts, undergo regio-selective ring-opening polymerization providing (1→3)-β-D-glucopyranans. The 1,4-anhy-drosugars are stable in basic media, but are readily hydrolyzed with acid. 1,4-Anhydro-galac-tose is so unstable that it is already decomposed on silica gel.

3 Non-anomeric Anhydrosugars: Synthesis and Reactions

This group of compounds is characterized by the presence of a free (or protected as a *inter-molecular* glycoside) anomeric position. The anhydro function may be created between var-ious (except anomeric) carbon atoms, thus a large number of such derivatives are possible. The most convenient method of classification of such anhydrosugars is based on the size of the anhydro ring: sugar oxiranes, oxetanes, THF and THP derivatives are known. The latter are rather rare, however, interest in them has increased recently. Discussing the anhydrosug-ars one has to consider also two different positions of the anhydro ring, namely the *exo-* and *endo*-cyclic rings (❷ *Fig. 9*).

oxirane (with **endocyclic** ring)   oxirane (with **exocyclic** ring)   oxetane   THF derivative   THP derivative

❑ **Figure 9**
Different types of anhydrosugars with an anomeric carbon atom not involved in the anhydro ring

3.1 Sugar Oxiranes

Because of the well-pronounced differentiation of the reactivity of the hydroxyl groups in the sugar molecule, it is possible to prepare anhydrosugars differing in configuration from the same precursor. This strategy may be exemplified by preparation of either 5,6-anhydro-α-D-*gluco-* or 5,6-anhydro-β-L-*ido*-hexofuranoses (**46** and **45**, respectively) from 3-*O*-benzyl-1,2-*O*-isopropylidene-α-D-glucofuranose (**44**). Activation of the primary hydroxyl group (by selective tosylation) followed by cyclization in basic media affords the L-*ido*-isomer **46**. Alter-natively, protection of the most reactive group (6-OH) as benzoate followed by activation of the secondary one at the C5 position affords the 6-*O*-benzoyl-5-*O*-tosyl derivative; in basic media hydrolysis of the benzoate occurs readily and the anion generated at the oxygen atom from the C-6 position attacks the C-5 center with the inversion of the configuration providing the D-*gluco*-isomer **45** (❷ *Scheme 16*) [1].

A synthetically useful procedure for the one-pot conversion [1] of vicinal diols into epoxides involves selective mono-activation of one hydroxyl group (by reaction with 1 equiv. of tosyl

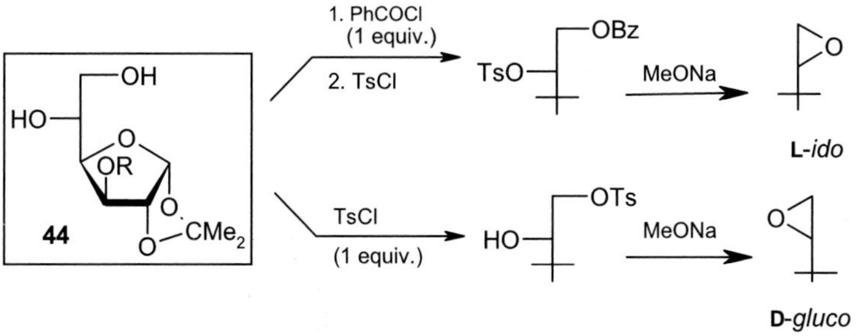

Scheme 16

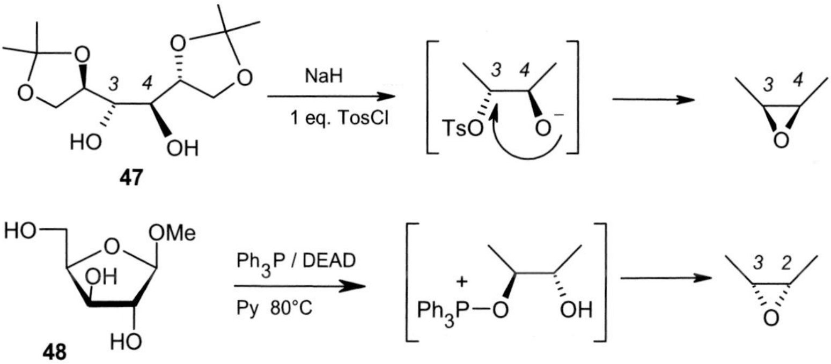

Scheme 17

chloride) followed by the attack of a second hydroxyl group with the closure of the oxirane ring. Thus, treatment of the diol (e. g. diacetonomannitol **47**) with two equivalents of sodium hydride generates the di-anion which reacts further with one equivalent of toluene-*p*-sulfonyl chloride. The intermediate thus formed, undergoes intramolecular S$_N$2 reaction leading to the appropriate oxirane (❍ *Scheme 16*). The activation can be performed also under much milder Mitsunobu conditions (e. g. compound **48** in ❍ *Scheme 17*) [36].

Synthesis of the smallest heterocyclic ring (the oxirane) is achieved conveniently also by other methods. The oxidation of a carbon–carbon double bond with hydrogen peroxide (alone or with organic nitriles: MeCN or PhCN), peracids (e. g. *m*-chloroperbenzoic), or organic peroxides\transition-metal catalysts [37] is applied to form sugar epoxides. Alternatively, transformation of a carbon–oxygen double bond into epoxides could be done by reaction of sugar aldehydes with: α-halogeno-acids (the Darzens' reaction), diazo-acids, or sulfonium ylides (❍ *Fig. 10*) [1].

This methodology may be illustrated by the synthesis of exocyclic epoxide **50** (a key compound for the preparation of sugar–lysine chimeras), which was realized by reaction of lactone **49** with methyl bromoacetate (❍ *Scheme 18*) [38].

The oxirane ring can be formed between C2-C3, or C3-C4, or other carbon atoms; the most common are 2,3-epoxides. Such compounds are important building blocks in the preparation of, for example, di- and oligosaccharides as well as modified monosaccharides. The sugars

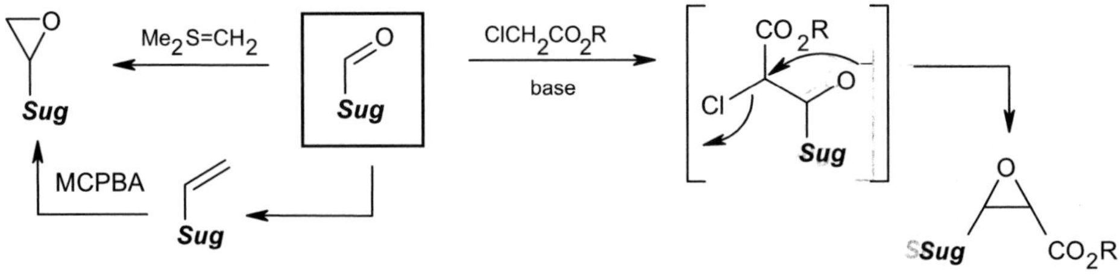

◻ Figure 10
Methods for the preparation of sugar epoxides

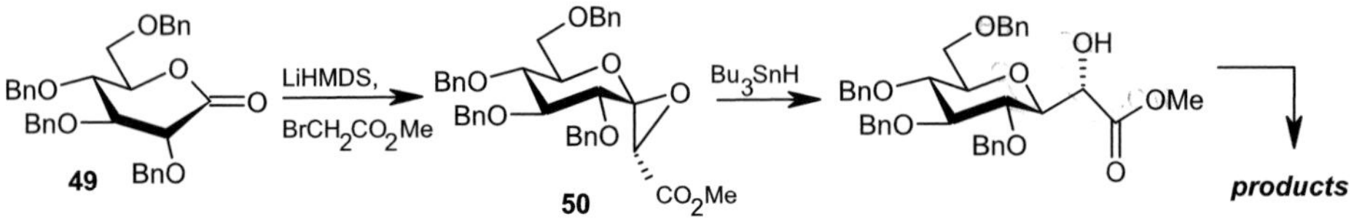

◻ Scheme 18

epoxides can be opened with a high degree of regiochemical control [39], which is governed by stereoelectronic factors. In some cases the regioselectivity of the opening of the oxirane ring may be inverted, if this process is done under chelating conditions. A detailed analysis of this process for several 6-deoxy-2,3- and 3,4-epoxysygars (prepared by the general methods already described within this chapter) was performed by Crotti [40]. The representative examples of this analysis are shown in ❷ *Fig. 11*. For the *cis*-relationship between the 3,4-epoxide ring and the methyl group at the C-6 position the opening of the ring was highly regioselective providing the C-3 product regardless of the reaction conditions (chelating or standard).
However, for the *trans*-relationship the regiochemistry of this process is slightly different. Opening of the 3,4-epoxide **51**-*trans* provides significant amounts of the C-4 product (❷ *Fig. 11*).

3.1.1 2,3-Anhydrosugars

These compounds are formed either by a nucleophilic displacement of appropriate leaving groups by a hydroxyl group placed at the C-α-position or by direct epoxidation of the C2–C3 double bond. Synthesis of 2,3-anhydro-derivatives of α-D-*manno*- and α-D-allopyranosides (**54** and **53**, respectively) from methyl 4,6-*O*-benzylidene-α-D-glucopyranoside (**52**) is a good example of the first method. Activation of the more acidic 2-OH group in α-methyl glucoside **52** with 1 equiv. of tosyl chloride affords the 2-*O*-Tos derivative which, upon treatment with base (NaOMe), cyclizes readily to the *manno*-epoxide **54** via nucleophilic attack of the anion generated from the hydroxyl group at the C-3 position. When, however, ditosylate is used under the same conditions, the stereoisomeric *allo*-epoxide **53** is formed, which results from the preferential hydrolysis of the tosylate from C-2 and subsequent attack of the 2-O anion on the C-3 position (❷ *Scheme 17*) [1,41]. Such sugar epoxides (e. g. **54**) were recently

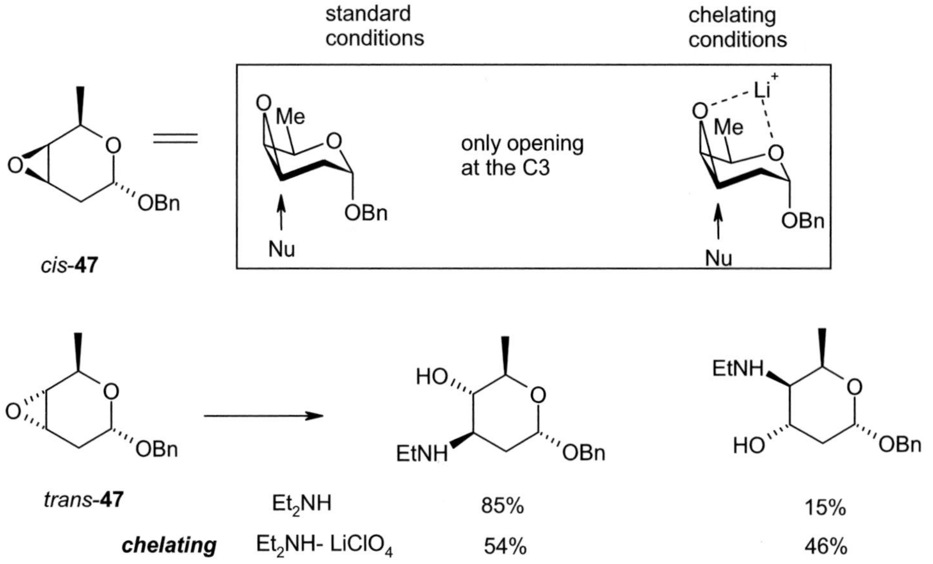

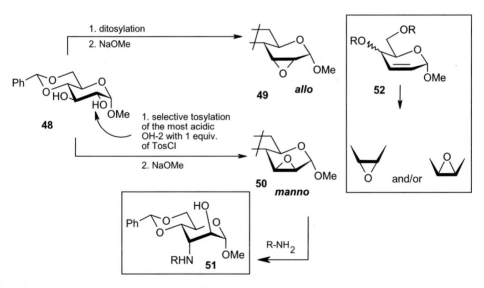

■ Figure 11
Examples of the opening of the epoxides with nucleophiles under chelating and standard conditions

■ Scheme 19

used for the preparation of aminosugars (e. g. **55**) (❖ *Scheme 19*) [42]. Epoxidation reaction of the double bond in monosaccharides (e. g. **56**) is used mainly for the preparation of 2,3-anhydrosugars or saccharides with the exocyclic oxirane ring.

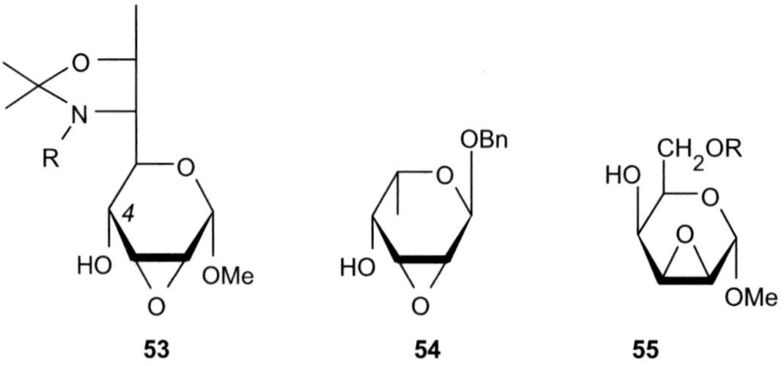

◻ Figure 12
Examples of 2,3-anhydrosugars obtained by epoxidation of the parent unsaturated derivatives

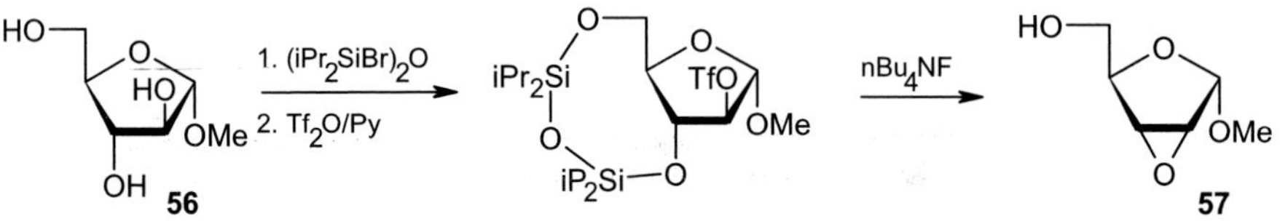

◻ Scheme 20

Stereoselectivity of the direct epoxidation of 2,3-unsaturated sugars with carboxylic or imino-peroxo acids (H_2O_2 + RCN) depends on the relative configuration of substituents at the adjacent centers, for example the configuration at C-1(OR) and C-4(OH). Since both such substituents may complex the epoxidizing agent by hydrogen bonding [43] they exhibit a directing effect in the oxidation process. Therefore, the highest selectivity of the epoxidation of 2,3-unsaturated sugars should be expected when both (oxy) groups at the C-1 and C-4 positions are in relative *cis* configuration. No wonder, therefore, that compound **57** was obtained as the only product by epoxidation of the corresponding unsaturated sugar, since both oxygen functions are in the *cis*-relation (❷ *Fig. 12*) [1].

The effect of a free OH group is usually much more pronounced than that of the protected one. For example, oxiranes **58** and **59** were obtained with high stereoselectivity (directed by the free OH), although both oxy-substituents act in the opposite directions [1,44] (❷ *Fig. 12*). When the epoxidation is performed not on allylic alcohols but allylic esters the selectivity of this process is low; both possible oxiranes are obtained in comparable amounts. The stereoselective formation of the 2–3-epoxides can be achieved also by the S_N2 process. Recently, Lowary proposed an efficient synthesis of 2,3-epoxy-*arabino*-furanoside **61** from the parent glycoside **60** in a sequence of reactions presented in ❷ *Scheme 20* [45]. Such anhydrosugars are convenient precursors for further functionalization at either the C-2 or C-3 position.

When the sulfur function is placed at the anomeric center (e. g. **62**), this molecule serves as a glycosyl donor in glycosylation reactions, thus it becomes a precursor of oligosaccharides (e. g. **63**). The epoxy ring located at the C2-C3 atoms may be properly functionalized, allowing us to prepare disaccharides of the desired stereochemistry (e. g. **64**, **65**; ❷ *Fig. 13*) [46,47,48]. By this methodology a number of di- and trisaccharides were obtained [49].

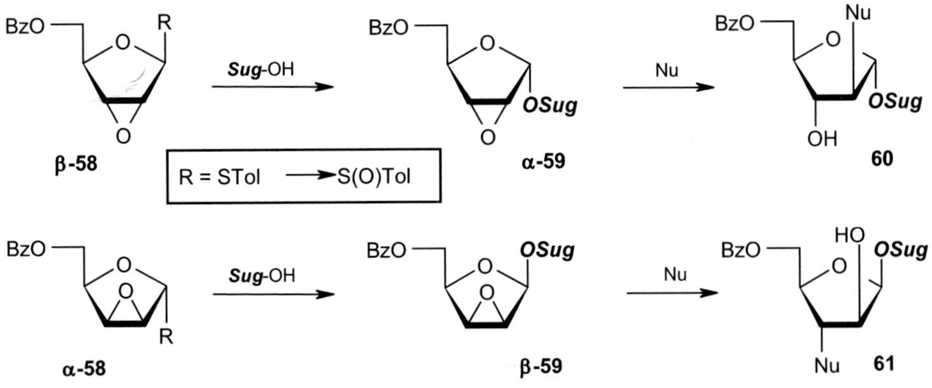

■ Figure 13
The concise methodology of the synthesis of di-saccharides from anhydrosugars

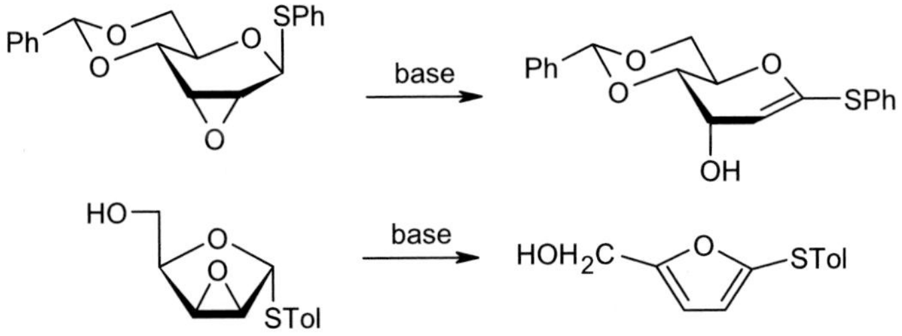

■ Figure 14
Behavior of 2,3-anhydro-thioglycosides under basic conditions

However, 2,3-anhydro-thioglycosides undergo—under basic conditions—other undesired reactions. If the anion is generated at the α-position to the anhydro ring, the β-elimination occurs readily providing the unsaturated carbohydrates [41]. Furanoses, under the same conditions undergo aromatization (❂ Fig. 14)

The carbon atoms in cyclic sulfates and sulfites are highly reactive towards nucleophilic reagents [50], which allows us to use such functionalities as the epoxide substitutes in many reactions. The 3,5-sulfites having the neighboring free hydroxyl group were used for preparation of anhydronucleosides. Thus, D-xylo-furanoside nucleoside 66 reacted with SOCl₂ in pyridine to afford the cyclic sulfite 67, which upon treatment with sodium bicarbonate furnished the 2,3-anhydro-nucleoside 68 (❂ Scheme 21).

However, synthesis of 2',3'-anhydro-cytidine 70 could not be realized in this way, since treatment of the corresponding cyclic sulfite 69 with bicarbonate afforded the D-ribo-derivative 72. The initially formed epoxide 70 was opened with the oxygen atom from the heterocyclic moiety and this anhydro derivative 71 upon another ring-opening process with a bicarbonate anion provided 72 (❂ Scheme 21) [51].

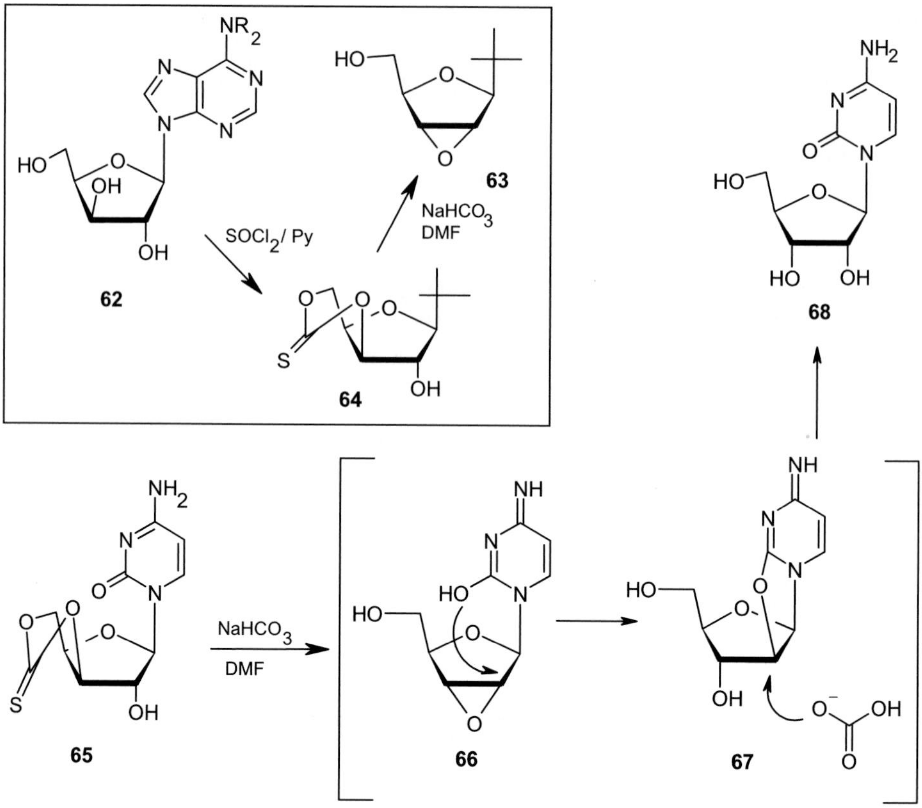

■ Scheme 21

3.1.2 Exocyclic Epoxides

These compounds are conveniently obtained by epoxidation of the exocyclic double bond in the sugar skeleton. Such oxidation with *m*-chloroperbenzoic acid (MCPBA) usually proceeds with moderate selectivity. Treatment of homologated galactose **73** with MCPBA afforded both epoxides **74** and **75** in a 3:1 ratio; these compounds are synthons for the preparation of antibiotic olguinine (❷ *Scheme 22*) [1].

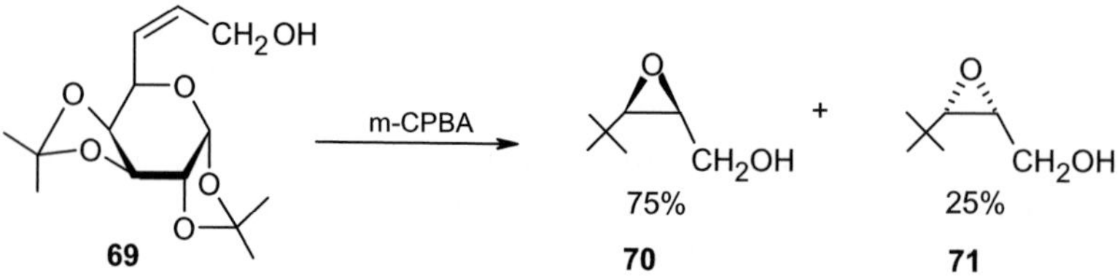

■ Scheme 22

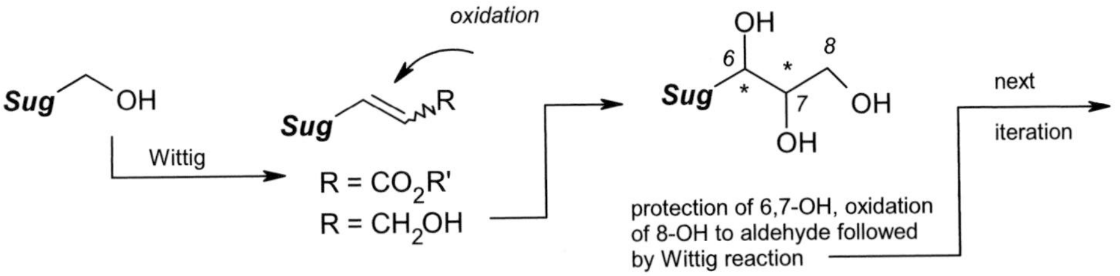

Figure 15
Brimacombe methodology of the synthesis of higher carbon sugars

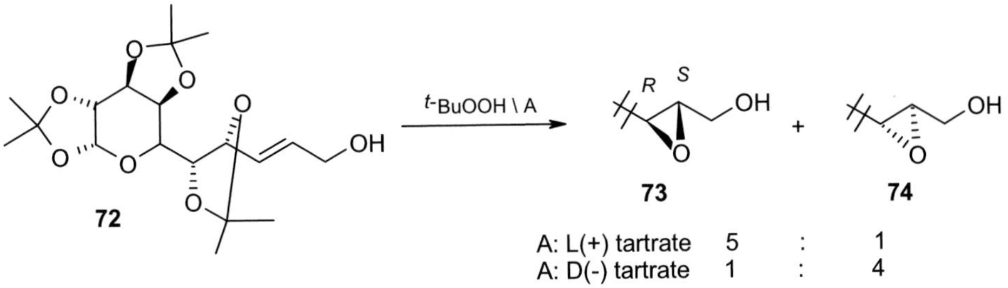

A: L(+) tartrate	5 :	1
A: D(-) tartrate	1 :	4

Scheme 23

One of the first applications of carbohydrate derivatives with an exocyclic oxirane ring in stereocontrolled synthesis was proposed by Brimacombe in his synthesis of higher sugars [52]. The general idea was based on the elongation of the parent monosaccharide at the terminal position by two carbon atoms using the Wittig methodology followed by functionalization of the resulting double bond, which was achieved either by osmylation or epoxidation. By this iterative elongation Brimacombe was able to prepare decoses (**◉** *Fig. 15.*)

One of the final steps of this synthesis involved epoxidation of the unsaturated decose (e. g. **76**), which was performed under the Sharpless conditions. When L-tartrate was used as chiral catalyst the 8(*R*),9(*S*) epoxide **77** was obtained as the main product, while with D-tartrate the opposite isomer **78** was formed (**◉** *Scheme 23*) [52].

The asymmetric Sharpless epoxidation allowed us to obtain the epoxide of desired stereochemistry by use of the proper catalyst, however, the selectivity was not high. Much more selective was the epoxidation process of a precursor of higher sugar pyranosidic nucleosides **79**, which provided only epoxide **80** with (–)-DET, while (+)-DET afforded exclusively the opposite stereoisomer **81** (**◉** *Scheme 24*) [1].

3.1.3 Rearrangement of Sugar Epoxides

In compounds with the free hydroxyl group placed at the α-position with respect to the oxirane ring the interconversion between epoxides may be noted. The first such rearrangement was observed by Lake and Peat already in 1939 and later by Buchanan [1]. Treatment of methyl 2,3-di-*O*-benzyl-4-*O*-tosyl-6-*O*-trityl-α-D-glucopyranoside with alkali resulted in formation of methyl 3,4-anhydro-α-D-galactoside together with the 2,3-anhydro-D-guloside. When the

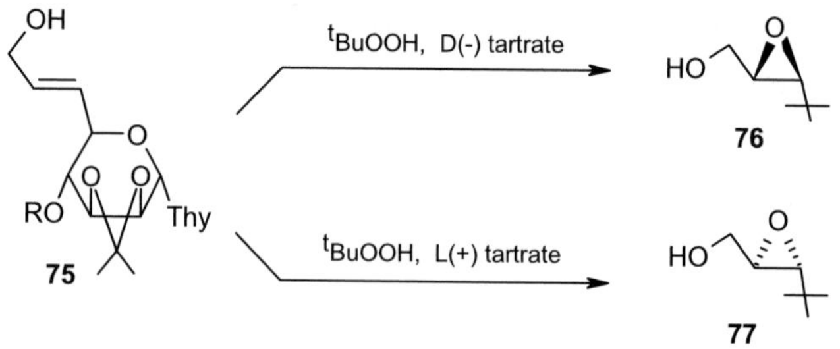

Scheme 24

hydroxyl group is placed at the β- or γ-positions with respect to the oxirane ring, formation of other anhydrosugars with a different size of heterocyclic ring might be observed [1,3]. The classical examples of such transformations are shown in ❷ *Fig. 16*.

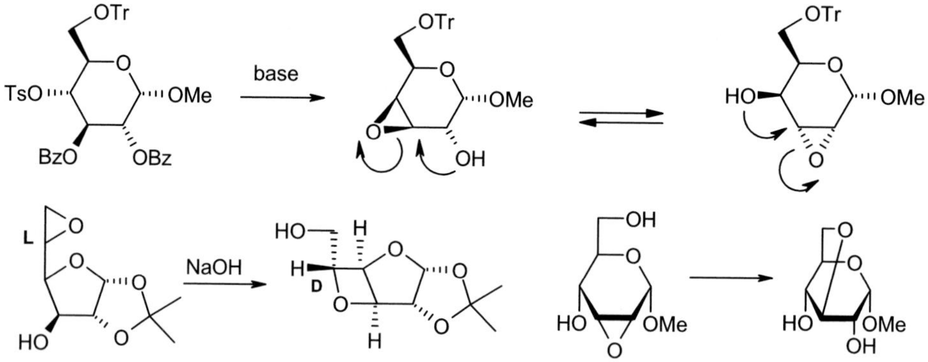

Figure 16
Selected examples of the rearrangement of sugar epoxides

3.2 Sugar Oxetanes, Oxolanes (THF), and THP Derivatives

These compounds are rather rare. One of the first examples was presented by Helferich who prepared 3,5-anhydro-1,2-*O*-isopropylidene-α-D-xylofuranose (**82**) from the appropriate 3,5-dimesyl derivative by treatment with ethanolic potassium hydroxide [1]. Another convenient approach to such derivatives utilized the reaction of 1,2-*O*-isopropylidene-5-*O*-triflate-α-D-glucofuranos-3,6-lactone (**79**) with base (K$_2$CO$_3$) in methanol, which produced the corresponding L-*ido*-anhydrosugar **80**. This was done by opening of the lactone ring with formation of the alcoholate at 3-OH, which substituted the triflate with inversion of the configuration at the C-5 atom (❷ *Fig. 17*).

The general methodology for the preparation of anhydropyranoses containing tetrahydrofuran and tetrahydropyran anhydro rings is described in the leading book for preparative carbohydrate chemistry: "*Methods in Carbohydrate Chemistry*" [53]. 3,6-Anhydrofuranoses

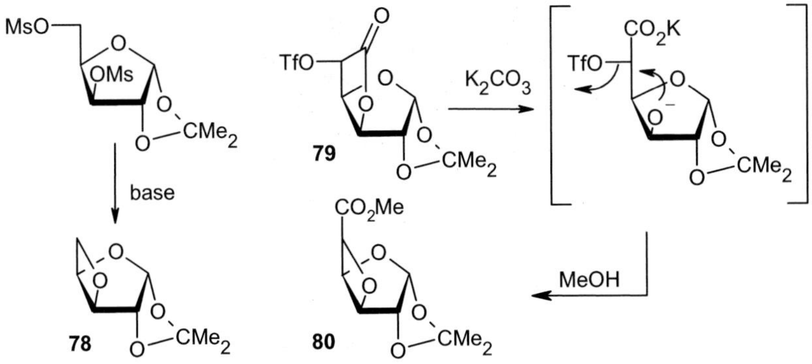

■ Figure 17
Formation of sugar oxetane derivatives

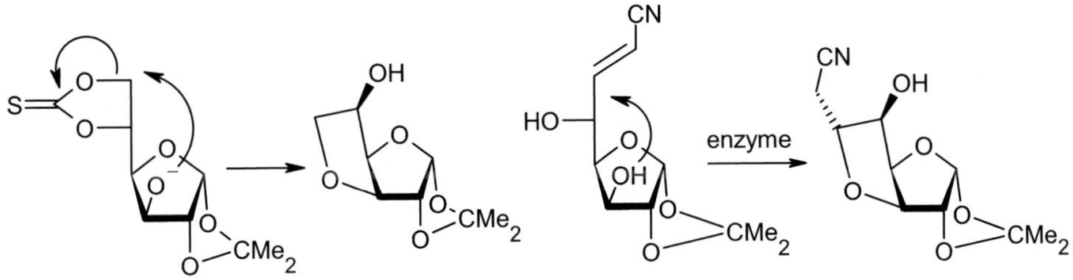

■ Figure 18
Selected examples of the preparation of oxolane anhydrosugars

are prepared similarly. The thiocarbonate, sulfate, or sulfite functionalities are used as leaving groups. The 1,4 addition of a sugar hydroxyl group to an activated electrophile is also used for the preparation of anhydrosugars. For example, treatment of 1,2-O-isopropylidene-α-D-glucofuranos-5,6-thiocarbonate with base caused the intramolecular attack of the 3-OH anion onto the C-6 atom with formation of 3,6-anhydro-glucofuranose.

No attack at C-5 leading to the oxetane (3,5-anhydro) has been noted [1]. Enzymatic cyclization of the α,β-unsaturated sugar nitrile provided the corresponding 3,6-anhydro-glucofuranose derivative in good yield [1] (❷ *Fig. 18*).

The most important compounds from this class are undoubtedly 3,6-anhydrofuranoses. Furanodictine A and B (produced by cellular slime mold *Dictyostelium discoideum*) showing neuronal differentiation activity [54] are good examples. These interesting derivatives may be conveniently obtained from the open-chain sugars. For example, synthesis of furanodictine A was realized from compound **85** obtained from D-arabinose in a few well-defined steps as shown in ❷ *Fig. 19* [55].

The tin methodology is particularly useful in organic chemistry. The tin moiety activates the allylic fragments; besides, the stannyl unit can be replaced with metal cations (generally lithium), thus organotin derivatives can be regarded as stable precursors of carbanions [56]. A highly oxygenated heterocycle isolated from natural sources—goniofurone, representative

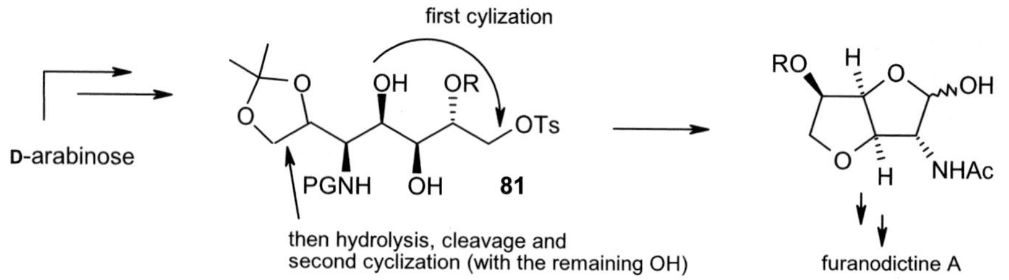

first cylization
D-arabinose
OH
OR
OTs
PGNH
OH
81
then hydrolysis, cleavage and
second cyclization (with the remaining OH)
RO
H
OH
NHAc
furanodictine A

Figure 19
Synthesis of a representative oxolane anhydrosugar

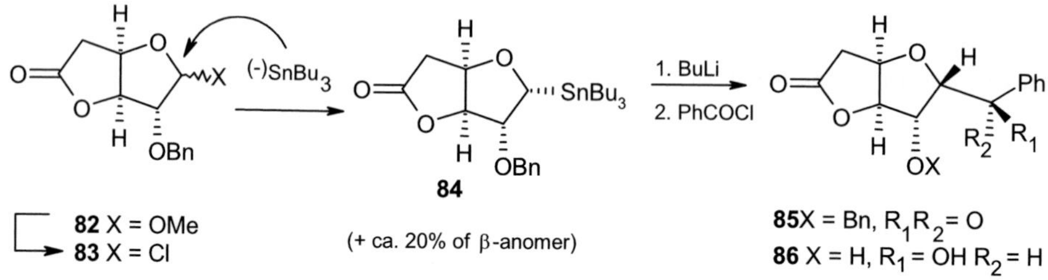

X
OBn
(-)SnBu3
SnBu3
84
1. BuLi
2. PhCOCl
Ph
OX
82 X = OMe
83 X = Cl
(+ ca. 20% of β-anomer)
85X = Bn, R1R2= O
86 X = H, R1= OH R2= H

Scheme 25

of 3,6-anhydrofuranoses—has been synthesized from the stannyl intermediate. Compound **86**, readily obtained in a few steps from D-glucurono-3,6-lactone, was converted into chloride **87**. The S_N2 displacement of the chlorine with the tin anion [$Bu_3Sn^{(-)}$ generated from $Bu_3SnSiMe_3$ by action of fluorides] afforded the stannyl derivative **88**, which reacted with benzoyl chloride to give ketone **89** finally transformed into **90** (➋ *Scheme 25*) [57].

Among tetrahydropyran anhydrosugar derivatives the most common are 3,6-anhydro-pyranoses. An example illustrating the synthesis of such types of compounds is the preparation of methyl 3,6-anhydro-*β*-D-glucoside from methyl 2,3,4-tri-*O*-acetyl-6-bromo-6-deoxy-*β*-D-gluco-pyranoside upon treatment with barium hydroxide. 3,6-Ahydro-D-galactose and 3,6-ahydro-D-mannose are prepared in a similar way (➋ *Fig. 20*).

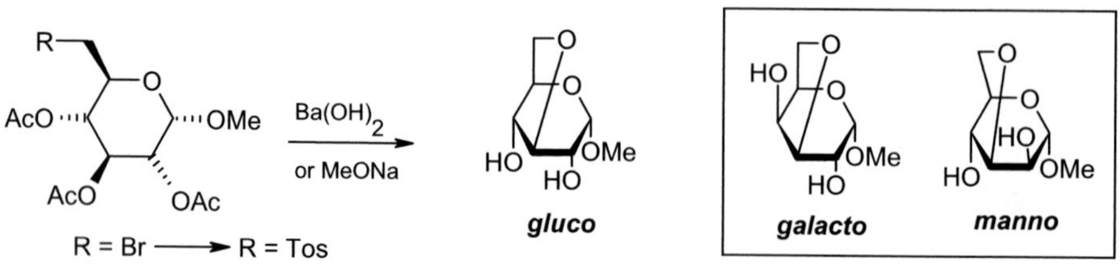

AcO
OMe
OAc
Ba(OH)2
or MeONa
R = Br → R = Tos
gluco
galacto
manno

Figure 20
Examples of 3,6-anhydropyranoses

4 Anhydronucleosides

Several anhydronucleosides such as oxiranes **91** [58] or oxetanes **92** [59] (❂ *Fig. 21*) inhibit HIV replication.

87   **88**

❏ **Figure 21**
Examples of anhydronucleosides

They are also useful intermediates in the preparation of biologically important compounds [60] including unsaturated nucleosides. Several representatives of the latter compounds such as 3′-azidothymidine (AZT), 2′,3′-dideoxyinosine (ddI), 2′,3′-dideoxycytosine (ddC), and 2′,3′-dideoxy-2′,3′-didehydro-thymidine (d4T) (see ❂ *Scheme 25*) are approved as *anti*-HIV drugs by the US Food and Drug Administration [1].

Nucleosides with conformationally restricted carbohydrate rings (locked nucleic acids—LNA) are RNA mimics containing the 2′-*O*,4′-*C*-methylene linkage. The bicyclic structure locks

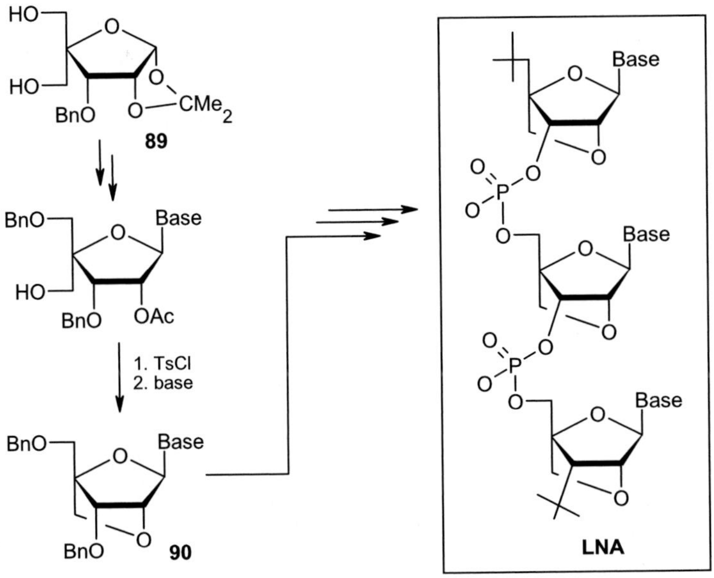

❏ **Figure 22**
Synthesis of locked nucleic acids (LNA)

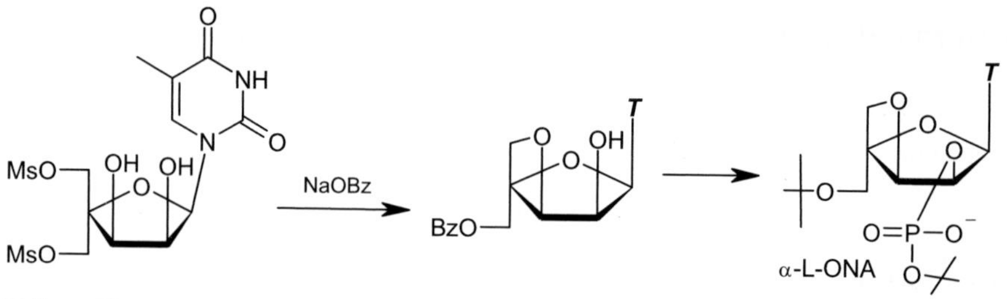

▢ Figure 23
Synthesis of 3,4-anhydronucleosides

nucleosides in the 3'-endo conformation, favorable for thermal stability of A-type duplexes. The key-compound in their synthesis is oxetane **94**, being prepared by standard reactions from **93** (❂ *Fig. 22*) [61].

Another example of monomers of this type, which have significant potential as building blocks for antigene and antisense strategies, was proposed recently by Wengel (❂ *Fig. 23*) [62].

Strained carbohydrates such as 2,3- and 3,5-anhydrofuranoses are important compounds, since the heterocyclic oxirane and oxetane rings can be cleaved by various reagents, thus giving rise to either synthetically useful intermediates or products possessing biological activity [63]. Recently, the synthesis of such derivatives from partially protected furanosides (or furanoses) has been reported [36].

2',3'-Anhydronucleosides are usually prepared from appropriate mono- or di-sulfonates [1]. 1-(2-Deoxy-3,5-anhydro)-β-D-*threo*-pentosylthymine (**96**), an example of the oxetane-type of anhydronucleoside, was prepared from dimesyl thymidine by the action of an alkali. The first synthesis of 2',3'-dideoxy-2',3'-didehydro-thymidine (d4T) was accomplished in 1966 by Horwitz et al. from this compound, which can be now prepared in quite large scale according to a slightly modified procedure proposed by the group from Bristol-Myers. The 5'-tritylated derivative of 3'-O-mesyl-thymidine was converted into 3',2-anhydroucleoside **95**, from which

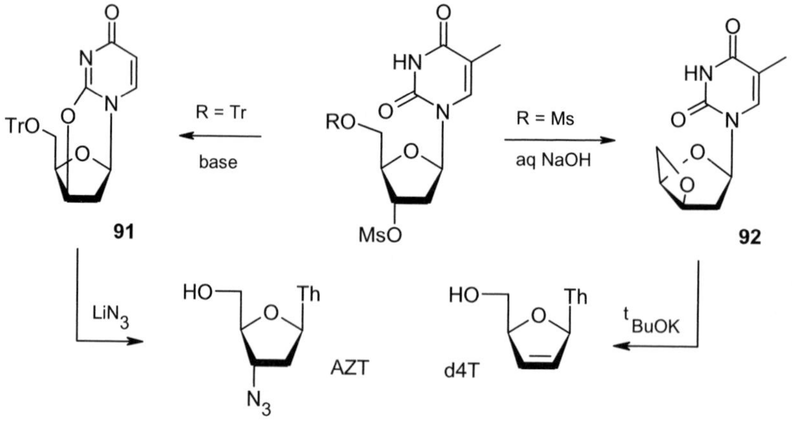

▢ Scheme 26

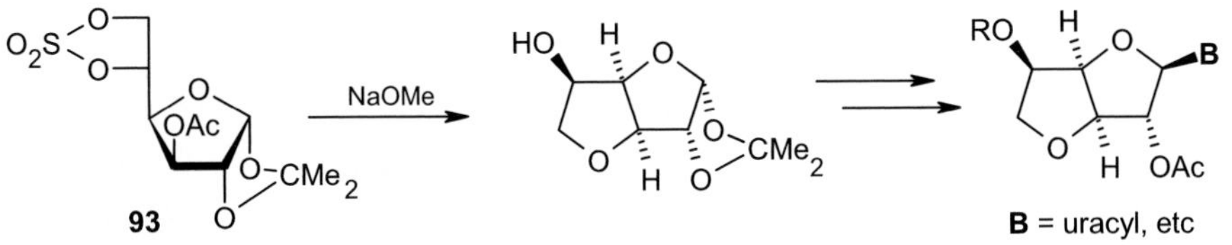

■ Figure 24
Synthesis of 3,6-anhydronucleosides

the 3-azido-thymidine (AZT) was prepared by an opening of the anhydro ring with lithium azide (❏ *Scheme 26*) [1]. Both compounds are used in the treatment of AIDS.

Oxolane anhydronucleosides were prepared recently from cyclic 5,6-sulfate **97** derived from diacetonoglucose. Under the basic conditions the hydroxyl group at the C-3 position was liberated and the anion thus formed attacked the C-6 position of the sugar providing the target anhydro derivative (❏ *Fig. 24*) [64].

This compound is a precursor of the corresponding anhydronucleosides.

5 Miscellaneous

Selected examples of the preparation of different types of anhydrosugars, which were not included in previous paragraphs, are presented in this Section. Many examples of such types of compounds are noted in the chemistry of sucrose. Sucrose is the most common disaccharide occurring in nature and 150 mln tons per year is produced. Intensive work is carried out to utilize this sugar also in markets other than food. The first example of anhydrosucrose was presented by Khan, who obtained the trianhydro-derivative **98** by treatment of tri-*O*-tosyl-sucrose with base [65]. Anhydro derivatives of sucrose are often obtained during Mitsunobu-type reactions [66,67], which are aimed at the synthesis of, for example, sucrose fatty esters. The esters at the terminal positions C-6 and C-6′ may be conveniently prepared under the Mitsunobu conditions (DIAD, TPP), when rather reactive acids are used [68]. However, when the acids are not reactive enough (or not present [67,69]), intramolecular etherification can compete with the desired intermolecular esterification, leading to anhydro-derivatives either at positions 3′,4′ or 3′,6′ (**99** and **100**, respectively) [70]. Treatment of free sucrose with phthalimide under the Mitsunobu conditions affords modified derivatives in which the primary 6-OH and 6′-OH groups are replaced with phthalimide while the secondary ones at the C-3′ and C-4′-positions were converted into the epoxide (**101**, ❏ *Scheme 27*) [71].

Reaction of the artificial sweetener sucralose (being 650 times sweeter than sucrose) with triphenylphosphine and diethyl azodicarboxylate afforded epoxide **102** from which the tetrachloro-derivative **103** was obtained (❏ *Fig. 25*) [72].

Triflation at *O*-2 in hepta-*O*-acetylsucrose **104**, followed by S$_N$2 displacement with amines led to the *C*-2 epimer of its deoxyamino analog, as well as 2,3-epoxide (**105**) and its ring-opening products (❏ *Scheme 28*) [73].

Sugar stannanes, convenient intermediates for the preparation of interesting enantiomerically pure compounds, are stable precursors of highly reactive carbanions. Such organostannanes may be easily obtained from anhydrosugars, as shown for example in ❏ *Fig. 26* [53].

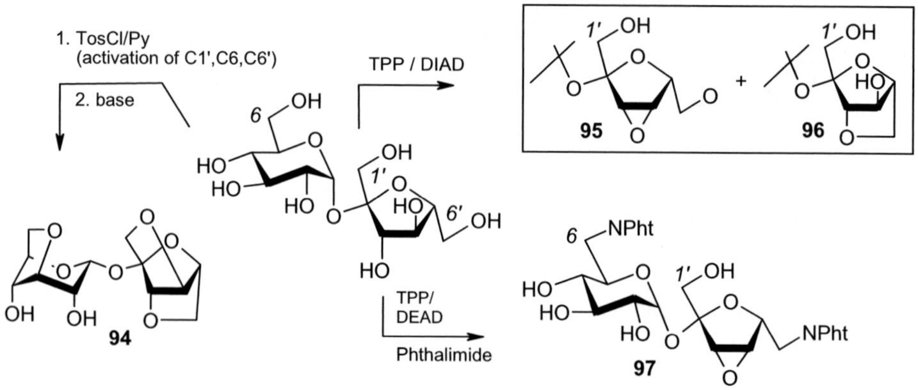

□ Scheme 27

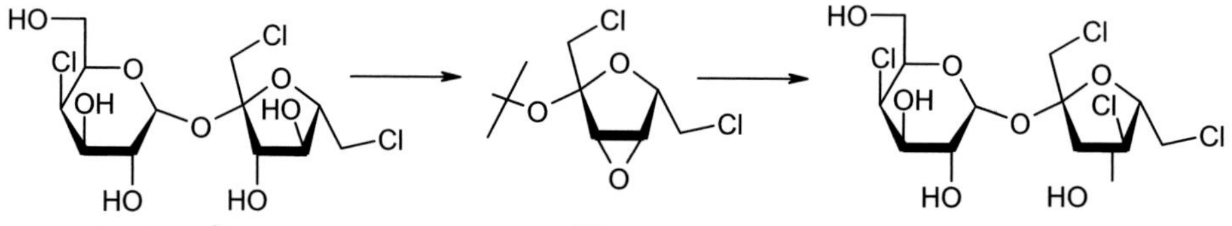

□ Figure 25
Selective transformation of sucralose

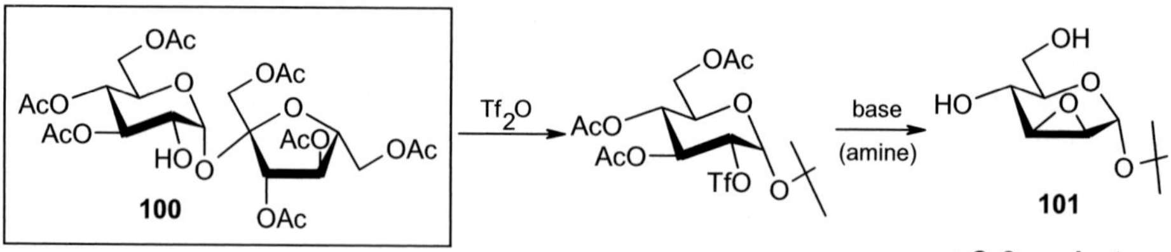

□ Scheme 28

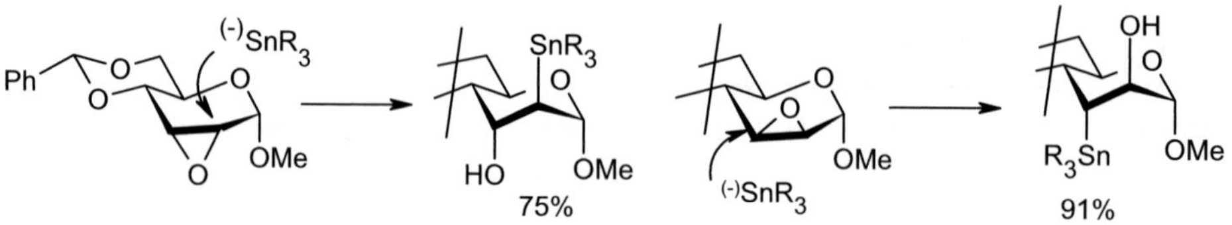

□ Figure 26
Opening of the ring of sugar epoxides with tin nucleophiles

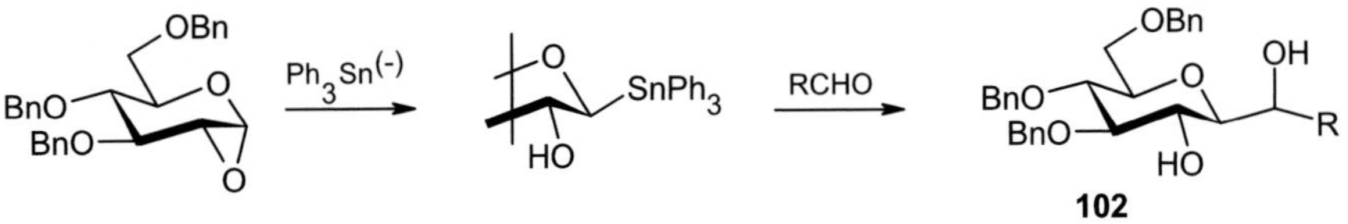

Scheme 29

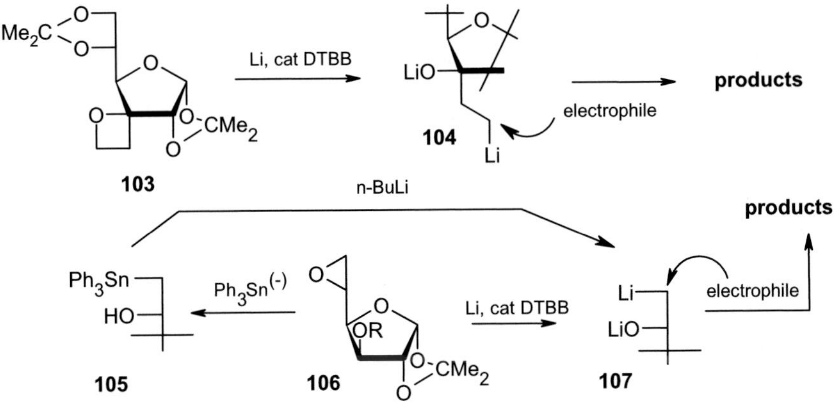

Scheme 30

If the 1,2-epoxide is used, the stannyl derivative obtained in this way may be applied in the synthesis of *C*-glycosides (e. g. **106**) with the retention of the configuration at the anomeric center (❂ *Scheme 29*) [53].

Reductive lithiation [74] of epoxides (**110**) or oxiranes (**107**) is a good method for the preparation of highly reactive carbohydrate anions (**111** or **108**). Such species may also be obtained in a two-step procedure involving the opening of the epoxide ring with a tin nucleophile (to **109**), which is replaced finally with lithium (❂ *Scheme 30*) [75].

Sugars with the 3-membered anhydro ring were used as convenient intermediates in the synthesis of biologically important compounds such as azasugars, which represent an important class of transition-state analogue inhibitors of glycosidases and glycosyl transferases. Compernolle applied sugar aziridines (e. g. **112**) for the preparation of heterocycles with a 6-membered ring [76], while sugar oxiranes (e. g. **113**) were used as substrates for the septanose-type azasugars [77] (❂ *Fig. 27*).

An interesting approach to azasugars (e. g. **116** in ❂ *Scheme 31*) from the epoxsides built at the anomeric centers of unprotected carbohydrates was presented recently [78]. Such epoxides (e. g. **115**) were obtained with very high selectivity by reaction of the corresponding hemiacetals (e. g. **114**) with sulfur ylid (❂ *Scheme 31*). This reaction is applicable for various sugar aldehydes [79].

The cascade reaction of symmetrical bis-epoxide **117** with azide provided the cyclopentane derivative **119** which was applied in the preparation of peptide-based drugs (❂ *Scheme 32*) [80].

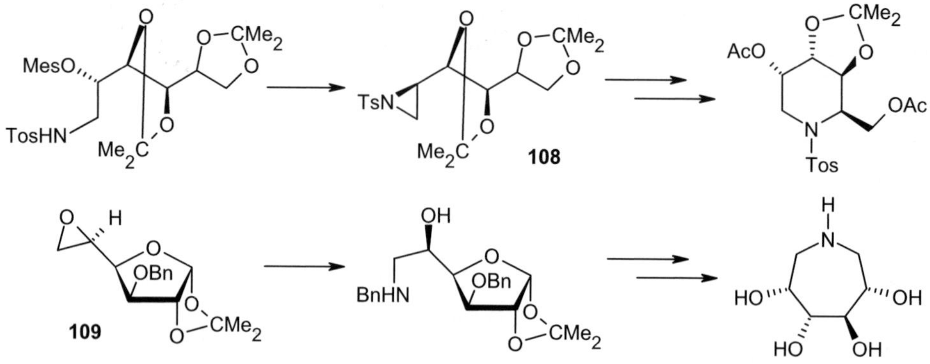

□ Figure 27
Examples of the synthesis of azasugars from anhydrosugar intermediates

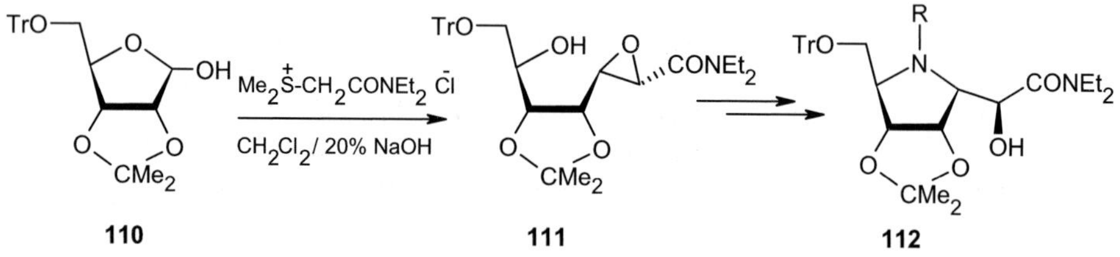

□ Scheme 31

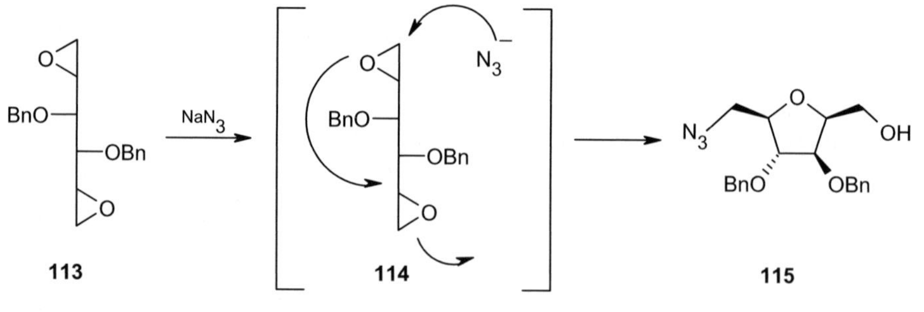

□ Scheme 32

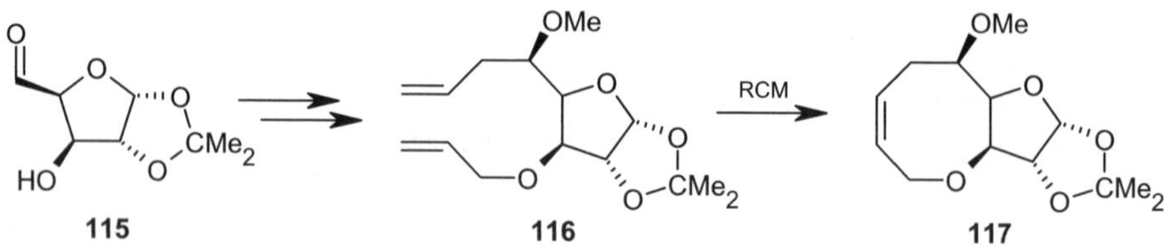

□ Figure 28
Methodology of the synthesis of anhydrosugars via the RCM process

Application of the ring-closing olefin metathesis reaction (RCM) for the preparation of carbocyclic rings from the appropriately functionalized sugars is now well documented [81]. Recently it was applied for the synthesis of compound **122**, which can be regarded as a (n,m)-anhydrosugar [81]. The intermediate diolefin **121** was prepared in a few standard steps from the corresponding dialdose **120** (❷ *Fig. 28*) [82].

6 Conclusion

This chapter briefly describes different classes of anhydrosugars, both anomeric and non-anomeric. Attention is paid mostly to anhydrosugars of great importance such as 1,6- and 1,2-anhydrosugars, and non-anomeric sugar epoxides. Several anhydrosugars (mostly the modified ones with the nucleoside structure) show interesting biological activity, the others are used as optically pure synthons in stereocontrolled organic synthesis. The material presented in this chapter comprises the basic chemistry of such derivatives, which is already described in many monographs and reviews (including the previous chapters from this series) and provides new applications of such important derivatives. The new data (updating previously published work) are not comprehensive, however, they illustrate the important aspects of the chemistry in which anhydrosugars are engaged. Therefore, only selected examples from the new literature are included here. This material should give the reader a general impression of the chemistry of such compounds and elucidate the modern aspects of anhydrosugar chemistry.

References

1. Jarosz S (2001) Glycoscience—Chemistry and Chemical Biology 1:291
2. Stick RV, Williams SJ (2001) Glycoscience—Chemistry and Chemical Biology 1:627
3. Cerny M (2003) Adv Carbohydr Chem Biochem 58:121
4. Ball DH, Parish FW (1969) Adv Carbohydr Chem Biochem 24:139
5. Williams NR (1970) Adv Carbohydr Chem Biochem 25:109
6. Cerny M, Stanek J Jr (1977) Adv Carbohydr Chem Biochem 34:23
7. Ferrier RJ, Middleton S (1993) Chem Rev 93:2779
8. Hanessian S (1983) Total Synthesis of Natural Products: The Chiron Approach. Pergamon Press, New York; Inch TD (1984) Tetrahedron 40:3161; Fraser-Reid B, Tsang R (1989) Strategies and Tactics in Organic Synthesis. Academic Press, New York
9. Mitsunobu O (1981) Synthesis 1
10. Castro BR (1983) Org React 29:1
11. Zottola MA, Alonso R, Vite GD, Fraser-Reid B (1989) J Org Chem 54:6123
12. Bailliez V, de Figuiredo RM, Olesker A, Cleophax J (2003) Synthesis 1015
13. Dromowicz M, Köll P (1998) Carbohydr Res 308:169
14. Wiggins LF (1963) Methods Carbohydr Chem 2:188; Whistler RL, Lake WC (1972) Methods Carbohydr Chem 6:286
15. Lundt I, Skelbaek-Pedrsen B Acta Chim Scand Ser B (1981) 35:637; Skelto B, Stick RV, Matthew D (2000) Austr J Chem 53:389; Sridhar P, Saravanan V, Chandrasekarn S (2005) Pure Appl Chem 77:145
16. Sivapriya K, Chandrasekaran S (2006) Carbohydr Res 341:2204
17. Fuenets J, Olano D, Pradera MA Tetrahedron Lett (1999) 40:4063
18. Krohn K, Gehle D, Flörke U (2005) Eur J Org Chem 2841
19. Krohn K, Gehle D, Flörke U (2005) Eur J Org Chem 4557
20. Witczak ZJ, Kaplon P, Kolodziej M (2002) J Carbohydr Chem 21:143
21. Kelly DR, Mahdi JG (2002) Tetrahedron Lett 43:511

22. Manna S, McAnalley BH, Ammon HL Carbohydr Res (1993) 243:11
23. Goursaud F, Peyrane F, Veyrieres A (2002) Tetrahedron 58:3629
24. Halcomb RL, Danishefsky SJ (1989) J Am Chem Soc 111:6661; Danishefsky SJ, Bilodeu MT (1996) Angew Chem Int Ed Engl 35:1380
25. Lee GS, Min HK, Chung BY (1999) Tetrahedron Lett 40:543
26. Timmers CM, van Starten NCR, van der Marel GA, van Boom JH J Carbohydr Chem (1998) 17:471
27. Flaherty TM, Gervay J (1996) Tetrahedron Lett 37:961
28. Du Y, Linhard RJ (1998) Tetrahedron 54:9913
29. Allwein SP, Cox JM, Howard BE, Johnson HWB, Rainier JD (2002) Tetrahedron 58:1997
30. Jacobsson M, Malmberg J, Ellervik U (2006) Carbohydr Res 341:1266
31. Seeberger PH, Haase W-CH (2000) Chem Rev 100:4349; Plante OJ, Palmacci ER, Andrade RB, Seeberger PH (2001) J Am Chem Soc 123:9545
32. Leeuwenburgh MA, van der Marel GA, Overkleeft HS, van Boom JH (2003) J Carbohydr Chem 22:549; Riseeuw DP, Grotenberg GM, Witte MD, Tuin AW, Leeuwenburgh MA, van der Marel GA, Overkleeft HS, Overhand M Eur (2006) J Org Chem 3877
33. Wipf P, Pierce JG, Zhuang N (2005) Org Lett 7:483
34. Chow K, Danishefsky S (1990) J Org Chem 55:4211
35. Singh I, Seitz O (2006) Org Lett 8:4319
36. Schultze O, Voss J, Adiwidjaja G (2001) Synthesis 229
37. Sharpless KB, Behrens CH, Katsuki T, Lee AWM, Martin VS, Takatani M, Viti SM, Walker FJ, Woodard SS (1983) Pure Appl Chem 55:589; Katsuki T, Martin VS (1996) Org Reactions 48:1
38. Zhang K, Wang J, Sun Z, Nguyen D.-H, Schweizer F (2007) Synlett 239; Knorr R, Treciak A, Bannwarth E, Gillesen D (1989) Tetrahedron Lett 30:1927
39. Zamojski A, Banaszek B, Grynkiewicz G (1982) Adv Carbohydr Chem Biochem 40:1
40. Crotti P, Di Bussolo V, Favero L, Macchia F, Pineschi M (2002) Tetrahedron 58:6069
41. Wang Y, Li Q, Cheng S, Wu Y, Guo D, Fan Q-H, Wang X, Zhang L-H, Ye X-S (2005) Org Lett 7:5577
42. Maxwell VL, Evinson EL, Emmerson DPG, Jenkins PR (2006) Org Biomol Chem 4:2724
43. Berti G (1973) Topics Stereochem 7:83
44. Coleman RS, Jones AB, Danishefsky SJ (1990) J Org Chem 55:2771; Horita K, Sakurai Y, Nagasawa M, Yonemitsu O (1997) Chem Pharm Bull 45:1558
45. Callam CS, Gadikota RR, Lowary TL Carbohydr Res (2001) 330:267; Bai Y, Lowary TL (2006) J Org Chem 71:9658
46. Gadikota RR, Callam CS, Wagner T, Del Fraino B, Lowary TL (2003) J Am Chem Soc 125:4155
47. Callam CS, Gadikota RR, Krein DM, Lowary TL (2003) J Am Chem Soc 125:13112
48. Tilekar JN, Lowary TL (2004) Carbohydr Res 339:2895
49. Cociorva OM, Lowary TL (2004) Tetrahedron 60:1481
50. Byun H-S, He L, Bittman R (2000) Tetrahedron 56:7051
51. Takatsuki K-I, Yamamoto M, Ohgushi S, Kohmoto S, Kishikawa K, Yamashita H (2004) Tetrahedron Lett 45:137
52. Brimacombe JS (1989) In: Atta-ur Rahman (ed) Studies in Natural Product Chemistry. Elsevier, Amsterdam, 4C:157
53. Lewis BA, Smith F, Stephen AM (1963) Methods Carbohydr Chem 2:172
54. Kikuchi H, Saito Y, Komiya J, Takaya Y, Honma S, Nahakata N, Ito A, Oshima Y (2001) J Org Chem 66:6982
55. Yoda H, Suzuki Y, Takabe K (2004) Tetrahedron Lett 45:1599
56. Jarosz S, Zamojski A (2003) Curr Org Chem 7:13
57. Ye J, Bhatt RK, Falck JR (1993) Tetrahedron Lett 34:8007
58. Webb TR, Mitsuya H, Broder S (1988) J Med Chem 31:1475
59. Counde O-Y, Kurz W, Eugui EM, Mc Roberts MJ, Verheyden JHP, Kurz LJ, Walker KAM (1992) Tetrahedron Lett 33:41
60. Huryn DM, Okabe M (1992) Chem Rev 92:1745
61. Rozners E (2006) Current Org Chem 10:675
62. Madsen AS, Hrdlicka PJ, Kumar TS, Wengel J (2006) Carbohydr Res 341:1398.
63. Unger FM, Christian R, Waldstätten P (1978) Carbohydr Res 67:257
64. Molas MP, Matheu MI, Castillon S, Isac-Garcia J, Hernandez-Mateo F, Calvo-Flores FG, Santoyo-Gonzalez F (1999) Tetrahedron 55:14649
65. Khan R (1972) Carbohydr Res 22:441
66. Descotes G, Mentech J, Veesler S (1989) Carbohydr Res 190:309.

67. Guthrie RD, Jenkins ID, Thang S, Yamasaki R (1983) Carbohydr Res 121:109; ibid. item. (1988) 176:306

68. Bottle S, Jenkins IA (1984) J Chem Soc Chem Commun 385; Abouhilale S, Greiner J, Riess JG (1991) Carbohydr Res 212:55; Baczco K, Nugier-Chauvin C, Banoub J, Thilbault P, Plusquellec D (1995) Carbohydr Res 269:79

69. Buchanan JG, Cummerson DA (1972) Carbohydr Res 21:293

70. Molinier V, Fitremann J, Bouchu A, Queneau Y (2004) Tetrahedron: Asymmetr 15:1753

71. Amariutei L, Descotes G, Kugel C, Maitre JP, Mentech J (1998) J Carbohydr Chem 7:21

72. Hough L (1991) In: Lichtenthaler FW (ed) Carbohydrates as Organic Raw Materials I. VCH, Weinheim, p 32

73. Lichtenthaler FW, Mondel S (1997) Carbohydr Res 303:293

74. Soler T, Bachki A, Falvello LR, Foubelo F, Yus M (2000) Tetrahedron: Asymmetr 11:493

75. Taylor OJ, Wardell JL (1988) Recl Trav Chem Pays-Bas 107:267; Cox PhJ, Doidge-Harrison SMSV, Howie RA, Nowell IW, Taylor OJ, Wardell JL (1989) J Chem Soc Perkin Trans 1 2017

76. Compernolle F, Joly GJ, Peeters K, Toppet S, Hoornaert GJ, Kilonda A, Babady-Bila (1997) Tetrahedron 53:12739; Pearson MSN, Mathe-Allainmat M, Ergeas V, Lebreton J (2005) Eur J Org Chem 2159

77. Tilekar JN, Patil NT, Jadhav HS, Dhavale DD (2003) Tetrahedron 59:1873

78. Pino-Gonzales MS, Assiego C (2005) Tetrahedron: Asymmetr 16:199

79. Lopez-Herrera FJ, Pino-Gonzalez MS, Sarabia Garcia FA, Heras Lopez, Ortega-Alcantara JJ, Pedraza-Cebrian MG (1996) Tetrahedron: Asymmetr 7:2065;

80. Poitout L, Le Merrer Y, Depezay JC (1995) Tetrahedron Lett 36:6887; Gruner SAW, Locardi E, Lohof E, Kessler H (2002) Chem Rev 102:491

81. Hansen FG, Bundgaard E, Mdsen R (2005) J Org Chem 70:10139

82. Kaliappan KP, Kumar N (2005) Tetrahedron 61:7461

2.5 C–C Bond Formation

Yuguo Du*, *Qi Chen*, *Jun Liu*
The State Key Laboratory of Environmental Chemistry and Ecotoxicology,
Research Center for Eco-Environmental Sciences, Chinese Academy of
Sciences, Beijing 100085, China
duyuguo@rcees.ac.cn

Abstract

In this chapter, synthetic methodologies for the preparation of *C*-branched carbohydrates, and nucleosides will be summarized and discussed.

Keywords

Chain extension; Aldol reactions; Branched-chain sugars; Radical reactions

Abbreviations

AIBN	azobisisobutyronitrile
CAN	ceric(IV) ammonium nitrate
CAMC	$3'$-β-carbamoylmethylcytidine

In: *Glycoscience*. Fraser-Reid B, Tatsuta K, Thiem J (eds)
Chapter-DOI 10-1007/978-3-540-30429-6_7: © Springer-Verlag Berlin Heidelberg 2008

COT	cyclooctatetraene
DLP	dilauoryl peroxide
DMF	dimethylformamide
HDA	hetero-Diels–Alder
HMDS	1,1,1,3,3,3-hexamethyldisilazane
MBH	Morita–Baylis–Hillman
MDA	methyl diazoacetate
MMTr	monomethoxytrityl
TBS	*tert*-butyldimethylsilyl
TBAF	tetra-*n*-butylammonium fluoride
TBAI	tetra-*n*-butylammonium iodide
TBDPS	tertbutyldiphenylsilyl
TEA	triethylamine
TEDMS	2-(trimethylsilyl)ethynyl]dimethylsilyl
THF	tetrahydrofuran
THP	tetrahydropyran

1 Introduction

The carbohydrates represent a structurally diverse group of compounds, which are usually derivatized with a variety of functional group modifications. *C*-branched sugar derivatives, in which a new C–C bond was formed at the non-anomeric center, can provide a versatile compound pool for drug screening. Generally speaking, any chemistry method (traditional or new) applied to C–C bond formation could be useful in *C*-branching sugar preparation. Among these methods are intramolecular alkylation and intramolecular condensation of aldehyde with enolates, phosphonates, and nitro-stabilized anions. Metal-mediated radical reactions, cycloadditions, and rearrangements have also been applied frequently.

2 C–C Bond Formation by Means of Nucleophilic Additions

2.1 Addition of Organometallics to Carbonyl Sugar

Organometallics, such as Grignard reagents and alkyllithium, are the most commonly used *C*-nucleophiles in C–C bond formation by addition to a carbonyl-functionalized sugar (❷ *Scheme 1*). Occasionally, organocopper [1] and alkylcerium reagents [2], prepared from alkyllithium and copper or cerium salts, are applied in this type of reaction. These nucleophilic additions, especially for Grignard reagents, are usually of high stereoselectivity due to the possible chelate-complexation between the metal ions and oxygenate group in the molecule [3]. The diastereofacial selectivity can be predicted and explained by different models [4]. The hydroxyl group generated in situ can also be further deoxylated or transformed into other functional groups.

For synthesizing the *C*-methyl-branched sugars [5] such as L-axenose, L-cladinose, L-arcannose, and L-vinelose, methylmagnesium bromide and methyllithium are often used to intro-

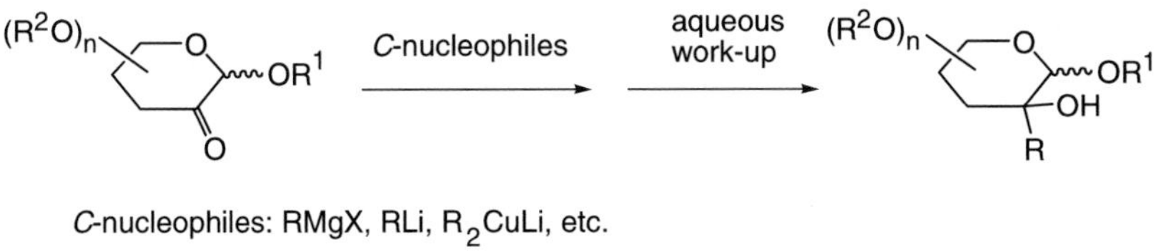

Scheme 1

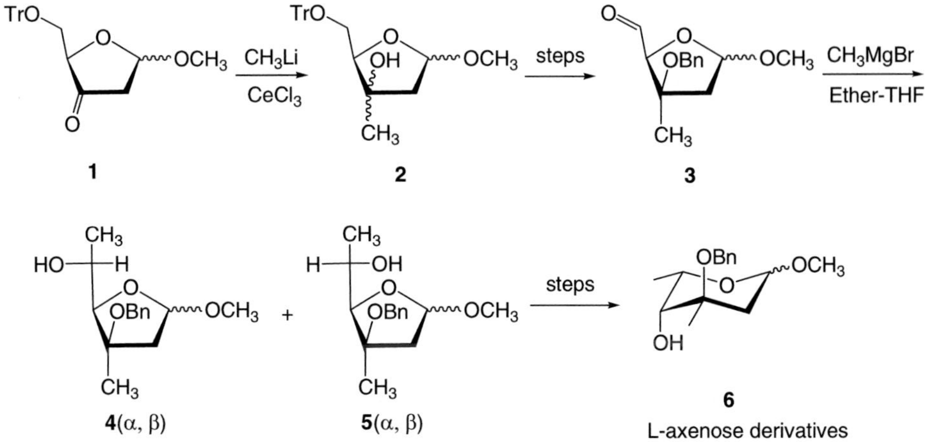

Scheme 2

duce a branching methyl group as well as a hydroxyl group with a certain configuration. Giuliano et al. [6] reacted 2-deoxy-D-glycero-pentofuranosid-3-ulose derivative **1** with methyl-cerium to give a 5:1 ratio of *threo* to *erythro* products **2** in the case of the α-anomer, and an exclusive *threo* product for the corresponding β-anomer. However, the same reaction gave unsaturated products using methyllithium as the nucleophile. When aldehyde **3** was treated with methylmagnesium bromide, a 1.3:1 ratio of products **4β** and **5α** favoring the D-arabino isomer **4β** in ether/tetrahydrofuran (THF) co-solvent were obtained. Equilibration of **5β** in methanolic hydrogen chloride gave pyranosides **6** as L-axenose derivatives (❷ *Scheme 2*).

Nielsen et al. [7] reported a synthesis of the [3.2.0] bicyclic β-nucleoside mimicking the anti-HIV drug AZT. Ketone **7** reacted with trichloromethyllithium to give the alcohol **8** with absolute stereoselectivity in a reasonable yield. Following the modified Corey–Link reaction [8], **8** was converted into the α-azido methyl ester **9**, taking advantage of in situ formation of a dichloroepoxide, ring opening by the azide ion, and methanolysis of the acyl chloride intermediate. The reaction was absolutely stereoselective, and only one compound was obtained (❷ *Scheme 3*). 2′-C-branched nucleosides have served as valuable probes to explore biomolecular structure and functions [9,10]. Piccirilli et al. [11] synthesized a series of ribonucleotides bearing substituents with increasing electron-withdrawing power (CH_3, CH_2F, CHF_2, CF_3) (❷ *Scheme 4*). Addition of difluoromethyl phenyl sulfone to ketoribose **10** in the presence of lithium hexamethyldisilazane in THF/HMPA gave sulfone **11** in good yield.

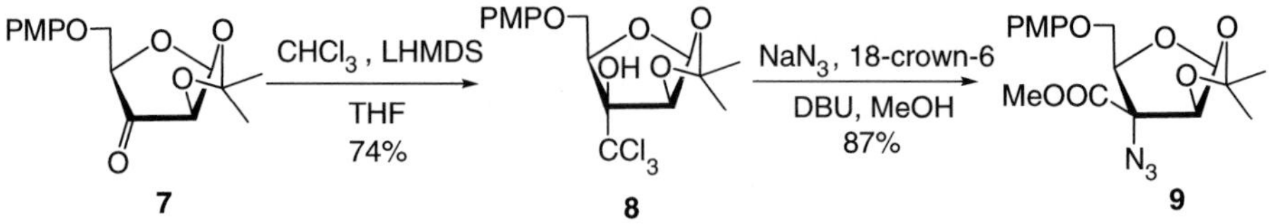

◻ Scheme 3

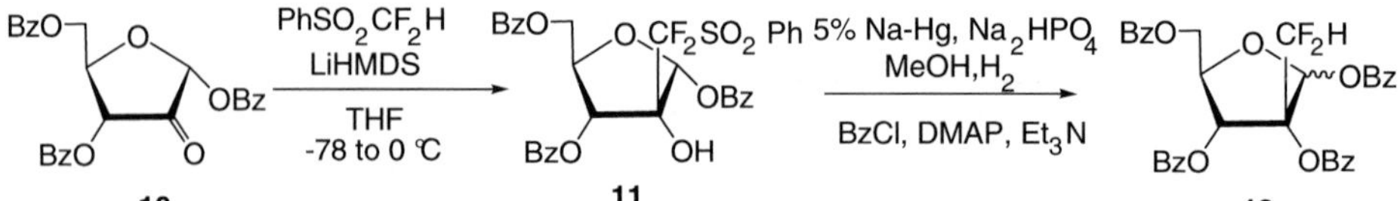

◻ Scheme 4

^{19}F-^{1}H NOE experiments indicated that the phenylsulfonyl difluoromethyl group attached stereoselectively to the β-face of the sugar. Single-electron-transfer reduction of **11** assisted by Na-Hg-MeOH-Na$_2$HPO$_4$ under hydrogen pressure, followed by benzoylation with BzCl, afforded perbenzoylated difluoromethylribose **12**.

A number of nucleoside analogues, either used clinically as anticancer drugs or evaluated in clinical studies, are of C-branched structures which can be prepared by introducing methyl [12], allyl [13], ethynyl [12a,14], trifluoromethyl [15], and the other groups [16] through addition of organometallics to sugar moieties. This will not be discussed in detail here.

2.2 Aldol-Type Condensations

The aldol condensation, as the most extensively studied reaction in C–C bond formation, is also commonly applied to the synthesis of C-branched sugars. In general, it involves the base-catalyzed addition of one molecule of carbonyl compound to a second molecule in such a way that the α-carbon of the first attached to the carbonyl carbon of the second to form a β-hydroxyl compound.

2.2.1 Aldol Reactions

Here, the name "aldol" refers to a β-hydroxycarbonyl compound which is derived from the nucleophilic additions between enol (or enolate) **13** and ketone (or aldehyde) **14** (❷ *Scheme 5*). Sugars, as the chiral polyhydroxy aldehyde or ketone compounds, are natural substrates for these reactions.

Serianni et al. [17] reported the synthesis of branched-chain aldotetrose **16** from lactol **15** and CH$_2$O under mild basic conditions through aldol reaction. In the same way, condensation of lactone **17** and acetone gave C-2 branched **18** using LiHMDS as a base at low temperature [18] (❷ *Scheme 6*).

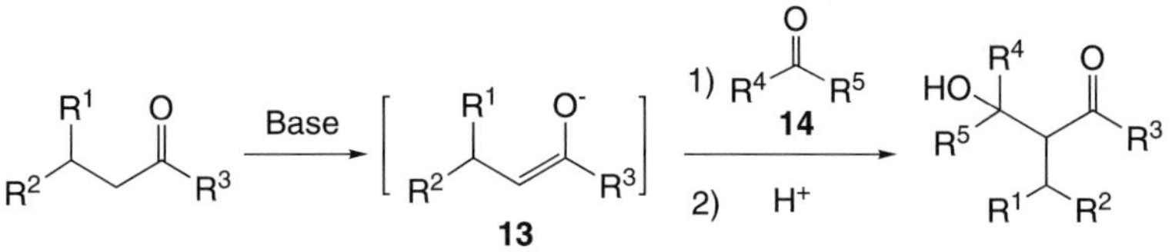

◻ Scheme 5

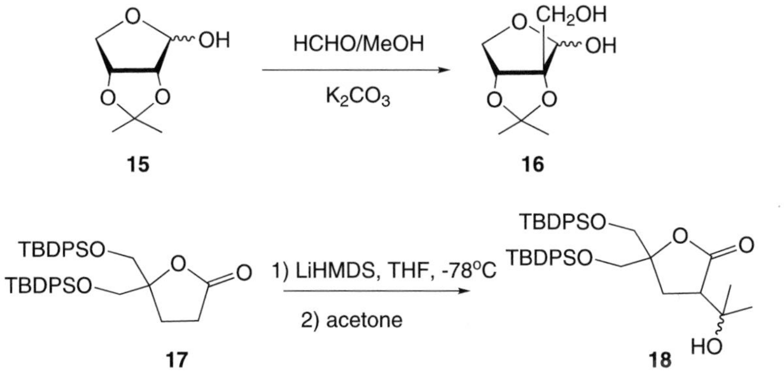

15 → **16**

17 → **18**

◻ Scheme 6

An interesting aldol condensation between 2-oxoglucopyranoside **19** and diethyl malonate has been carefully investigated and the reaction mechanism was illustrated in ❂ *Scheme 7* [19]. A butenolide-containing sugar **20**, available from the aldol condensation of methyl 4, 6-*O*-benzylidene-α-D-glucopyranosid-2-ulose **19**, and diethyl malonate, was autoxidized by air at the C-3 position affording α,β-unsaturated γ-lactone sugar **21**, which subsequently underwent 1,4-conjugate (Michael) addition of hydroxide ion (or water) leading to 2-*C*-branched-chain glycopyranosid-3-ulose **22**. The autoxidations could be performed in either weak basic, neutral, or weak acidic medium, respectively. When active methylene compound was introduced, a new type of *C*-branch sugar **23** was obtained [20].

2.2.2 Aldol-Cannizzaro Reactions

When aldehyde, carrying an active α-hydrogen, is coupled with formaldehyde, the product becomes a suitable substrate for Cannizzaro reaction, which can react with formaldehyde subsequently in one pot to give the corresponding alcohol and sodium formate. Sodium hydroxide is believed to be a good catalyst for this reaction (❂ *Scheme 8*).

An efficient method for large-scale preparation of apiose was developed by Koóš et al. [21], using an Aldol-Cannizzaro reaction as a key step. 2,3-*O*-isopropylidene-L-*threo*-tetrodialdose acetal **24** was reacted with excessive formaldehyde and gave 3-*C*-(hydroxymethyl)-2,3-*O*-isopropylidene-D-*glycero*-tetrose acetal **25**, which was deprotected to afford apios **26**. Besides, treatment of **27** with formaldehyde in the presence of sodium hydroxide yielded branched furanoside **28** under the same conditions [22] (❂ *Scheme 9*). It is worth noting that the prod-

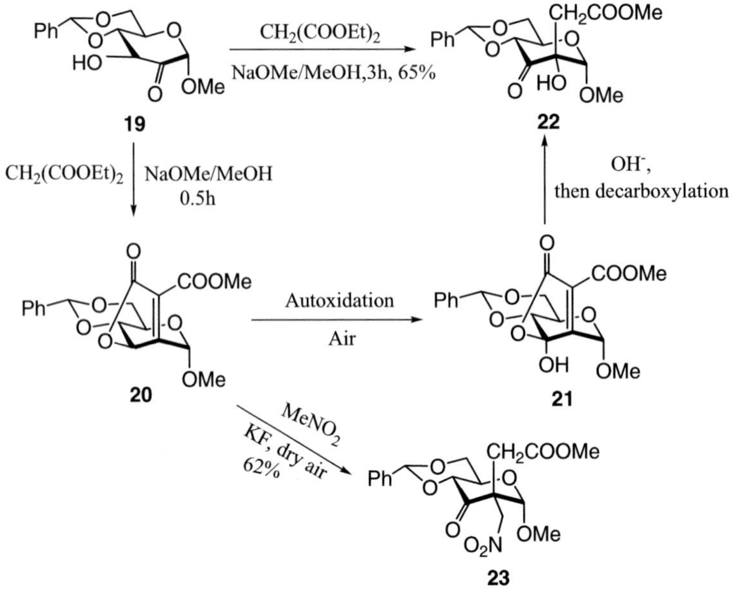

■ Scheme 7

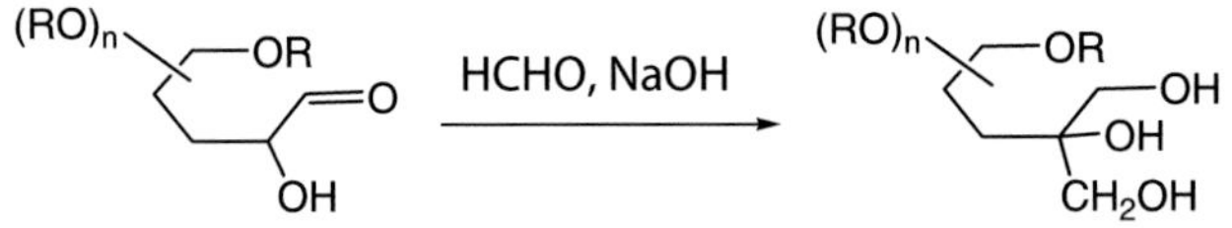

■ Scheme 8

uct of the Aldol-Cannizzaro reaction can be further transformed into locked nucleic acid (LNC) [23] containing other functional groups, such as ethenyl [24], ethynyl [25], and so on [26] (❷ Scheme 10).

2.2.3 Nitroaldol Condensations (Henry Reaction)

Condensation of aldehyde or ketone with nitroalkane, bearing an α-hydrogen, stereoselective-ly enables the formation of a new carbon–carbon bond under mild basic conditions with the generation of a β-nitroalcohol, which is called nitroaldol condensation or the Henry reaction (❷ Scheme 11). This is a good method for preparing the branch-chain sugars containing nitro-gen, since it provides a polyhydroxylated carbon framework with multiple stereogenic centers as well as the possibility for transformation of the nitro group to other functional groups. Many natural or unnatural azasugars have been obtained through this reaction for the development of new glycosidase inhibitors in recent years.

Bols [27] reported an improved synthesis of isofagomin **32** and noeuromycin **34** from D-ara-binose in six and seven steps, respectively. As shown in ❷ Scheme 12, Henry reaction of ketone sugar **29** gave the nitromethane adduct benzyl 4-deoxy-4-C-nitromethylene-D-arabi-

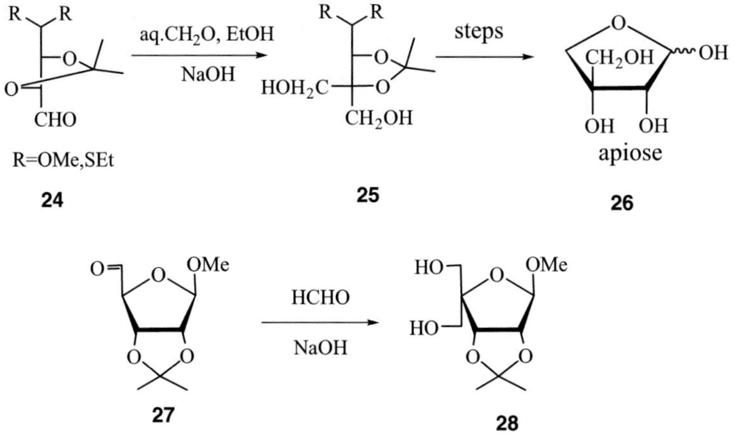

24 → 25 → 26

R=OMe,SEt

apiose

27 → 28

Scheme 9

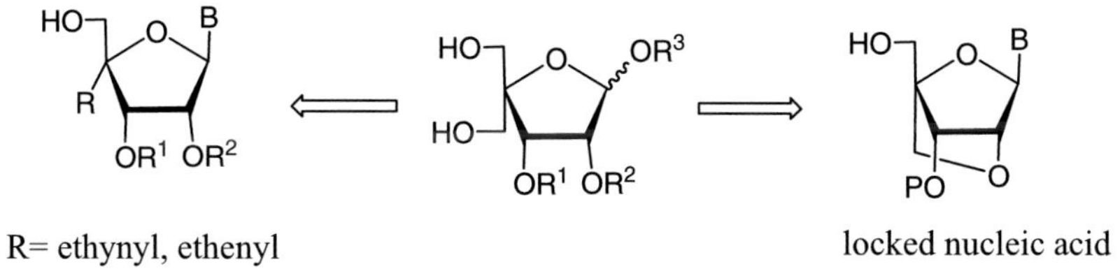

R= ethynyl, ethenyl

locked nucleic acid

Scheme 10

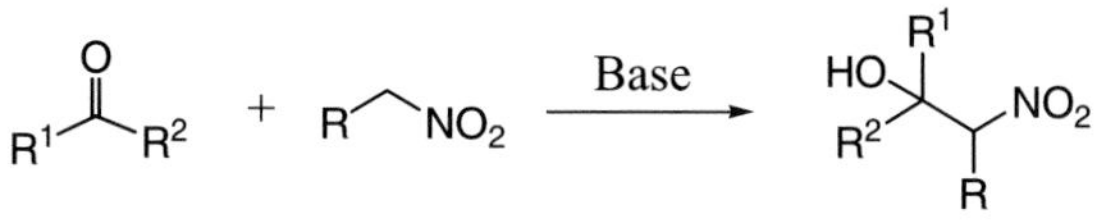

Scheme 11

no-pyranoside. After acetylation, fully protected **30** was obtained as a mixture of epimers which could be separated easily. Reductive elimination of **30** with NaBH$_4$ gave a 4R,4S mixture of benzyl 4-deoxy-4-C-nitromethylene-D-arabinopyranoside **31** in which the desired isofagomine/noeuromycin precursor (4S)-**31** was slightly favored (<2:1). Mixture **31** can be transformed into isofagomin **32** and noeuromycin **34**, along with their isomers **33** and **35**, respectively.

In the preparation of polyhydroxylated azepane as potential glycosidase inhibitors, Dhavale [28] described a short synthetic route utilizing the Henry approach. The nitroaldol reaction of 1,2-O-isopropylidene-3-O-benzyl-α-D-xylo-pentodialdose **36** and nitromethane in the presence of triethylamine at room temperature afforded α-D-gluco- and β-L-ido- nitroaldose **37**, the precursors to (2S, 3R, 4R, 5R) and (2S, 3R, 4R, 5S) tetrahydroxyazepanes **38** and **39**, in a 88:12 ratio in 95% yield (**Scheme 13**).

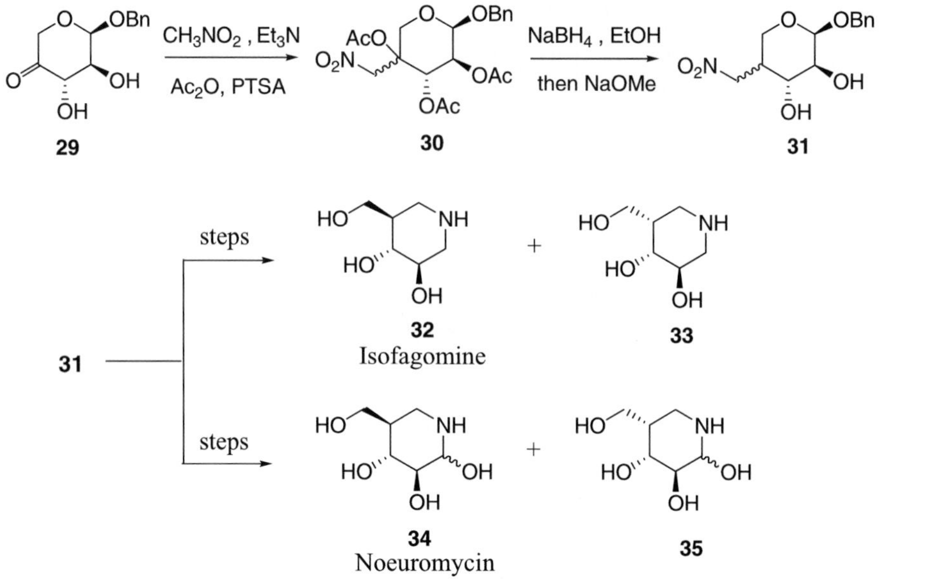

Scheme 12

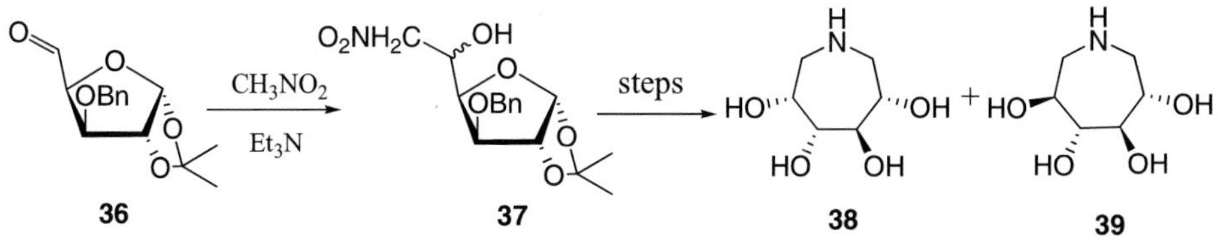

Scheme 13

Using the Henry reaction [29], addition of nitromethane to the carbonyl group of 5-O-benzoyl-1,2-O-isopropylidene-α-D-erythro-ketofuranos-3-ulose **40** took place stereoselectively to give ribo isomer **41** in almost qunatitative yield (❯ *Scheme 14*). KF was found to be better than any other bases for this reaction. The absolute configuration of all asymmetric carbon atoms in **41** was confirmed by single-crystal X-ray crystallographic analysis. The stereoselectivity probably resulted from the steric hindrance of the 1,2-O-isopropylidene group. Several new C-branched iminosugar derivatives bearing a pyrrole ring, such as **42**, were synthesized from **41**.

2.3 Wittig Reaction

Wittig reaction and its variants (e. g., Wadsworth–Emmons, Wittig–Horner, etc) are widely applied in the synthesis of C-branched sugar derivatives. The Wittig reagent reacts with aldehyde or ketone sugar affording the C=C bond as a potent functional group which could

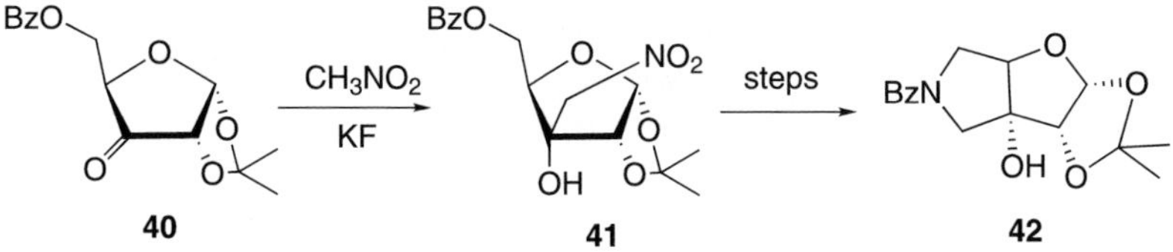

◻ Scheme 14

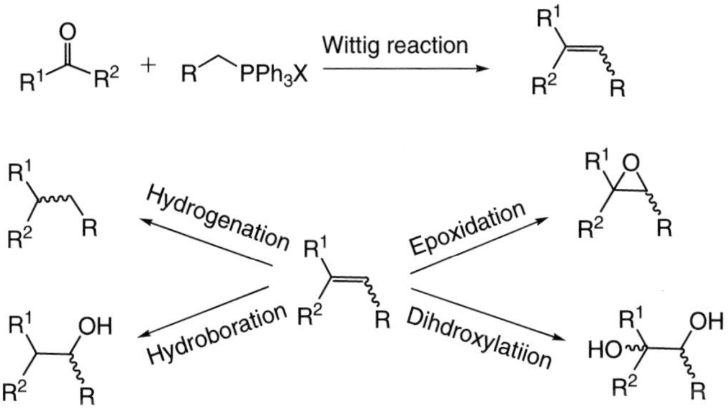

◻ Scheme 15

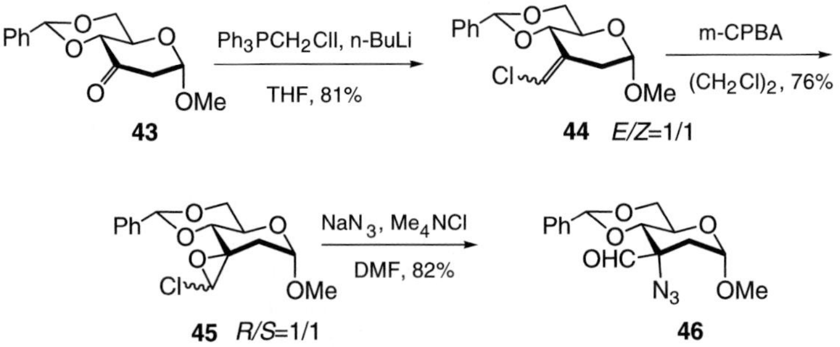

◻ Scheme 16

be converted into epoxides, alcohols, and other compounds via epoxidation, dihydroxylation, hydroboration, and catalytic hydrogenation, etc (**◉** *Scheme 15*).

A new synthetic method towards sugar aldehyde via α-chloroolefine and α-chloroepoxides was exploited by Sato et al. [30]. As shown in **◉** *Scheme 16*, Wittig reaction of **43** with $Ph_3P(I)CH_2Cl$ and n-BuLi in THF gave the corresponding chloroolefins **44** ($E/Z = 1/1$). The configurations of **44E** and **44Z** were determined by NMR spectrum (NOESY: H-4 and olefin proton). A mixture of these diastereomers was oxidized with *m*-chloroperbenzoic acid in 1,2-dichloroethane at 70 °C to give a *R,S*-mixture (1:1) of spiro α-chloroepoxide **45**, which was then treated with NaN_3 and Me_4NCl in dimethylformamide (DMF) at 80 °C furnishing the corresponding α-azidoaldehyde derivative **46** regioselectively.

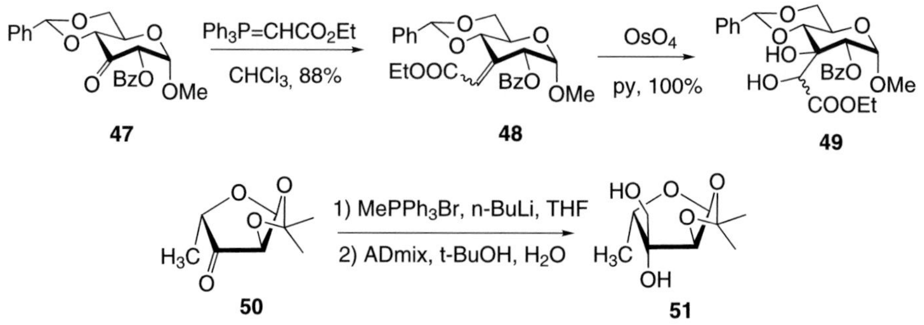

Scheme 17

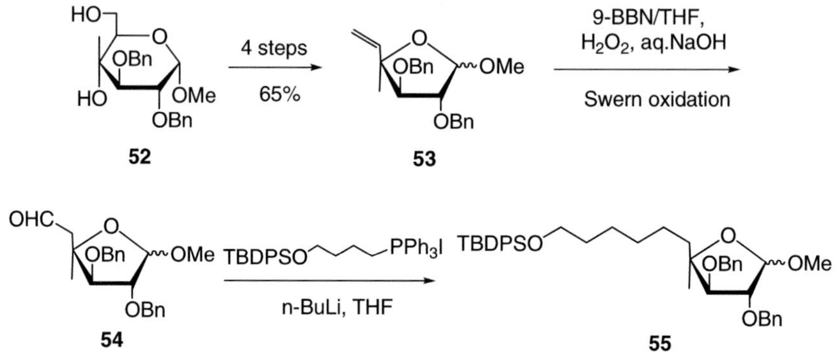

Scheme 18

In the synthesis of a miharamycin sugar moiety [31], a C-3-branched hexopyranoside was constructed through Wittig reaction and dihydroxylation. Olefination of the ketosugar **47** was performed with [(ethoxycarbonyl)-methylene]triphenylphosphorane affording **48** as a mixture of isomers ($Z:E = 7:3$). Oxidation of **48** with OsO_4 in pyridine led to a mixture of stereoisomers **49** in a quantitative yield, whereas a catalytic amount of OsO_4 and 4-methylmorpholine N-oxide in acetone/water gave only a 66% yield of **49**. Similarly, ketone sugar **50** was converted into the *exo*-methylene derivative by reaction with $H_2C=CHPPh_3$, a further Sharpless dihydroxylation using AD-mix (ether α or β) afforded the single stereoisomer [32] (**>** *Scheme 17*).

Asymmetric synthesis of macrolide antibiotic, ossamycin, was started from known **52**, which was further converted into furanoside **53** via zinc reduction and subsequent cyclization (**>** *Scheme 18*). Compound **53** was subjected to hydroboration, Swern oxidation, Wittig olefination, and hydrogenation furnished key intermediate **55** towards ossamycin [33].

Treatment of 5-O-TBS-1,2-O-isopropylidene-α-D-*erythro*-pentofuranos-3-ulose **56** with [(ethoxycarbonyl)-methylene]triphenylphosphorane gave (E/Z)-**57** (7:1; 90%) (**>** *Scheme 19*). Desilylation and hydrogenation of **57** at 25 psi H_2/Pd-C gave **58** with a trace amount of the over-reduced diol **59**. Formation of **59** was minimized at lower hydrogen pressure (5 psi). Reduction of **57** with $NaBH_4$/EtOH gave worse results. The 1,2-O-isopropylidene group is known to direct incoming reagents from the β-face of furanosyl derivatives and thus *ribo* diastereomers **58** (or **59**) were obtained with either catalytic hydrogenation or chemical reduction [34].

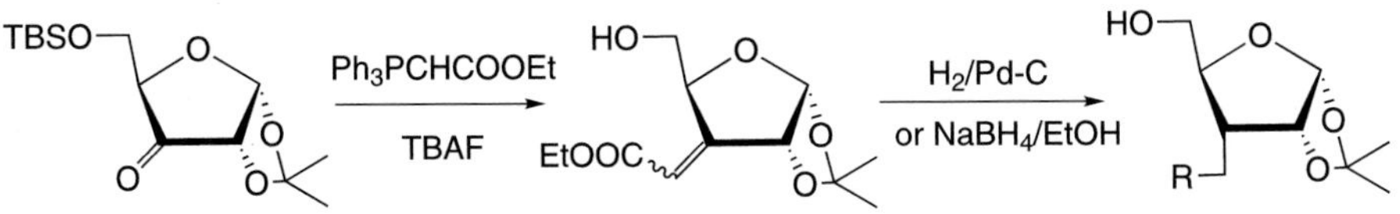

□ Scheme 19

2.4 Other Nucleophilic Additions

One carbon branch unit could be introduced into a sugar compound by addition of cyanide to sugar aldehyde or sugar ketone. As reported by San-Félix et al. [35], treatment of 3-ulose **60** with sodium cyanide afforded cyanohydrin **61**. The high stereoselectivity in the formation of **61** could be explained by the presence of conformationally rigid 1,2-O-isopropylidene functionality, which dictates the approach of the cyanide ion from the sterically less-hindered β-face of the ulose. Mesylation ($\rightarrow$**62**) and subsequent aldol-type cyclo-condensation afforded C-branched-3-spiro derivative **63**. A psicopyranose derivative containing 1,2,4-oxadiazole could be achieved [36] with the same method, and a potent functionality, such as cyanide, could also be introduced at the same time. Reaction of 1,2:4,5-di-O-isopropylidene-β-D-erythro-2-hexulopyranose-3-ulose **64** with NaCN under phase-transfer conditions yielded 1,2:4,5-di-O-isopropylidene-3-C-cyano-β-D-psicopyranose **65** in nearly quantitative yield and complete stereoselectivity. Refluxing of **65** with hydroxylamine in anhydrous methanol ($\rightarrow$**66**) followed by benzoyl chloride treatment expectantly gave 1,2,4-oxadiazole derivatives **67** (◗ *Scheme 20*).

Thiazole-based one carbon extension has been proved to be a useful tool in the synthesis of C-branched sugar derivatives [37]. To investigate an approach towards the synthesis of

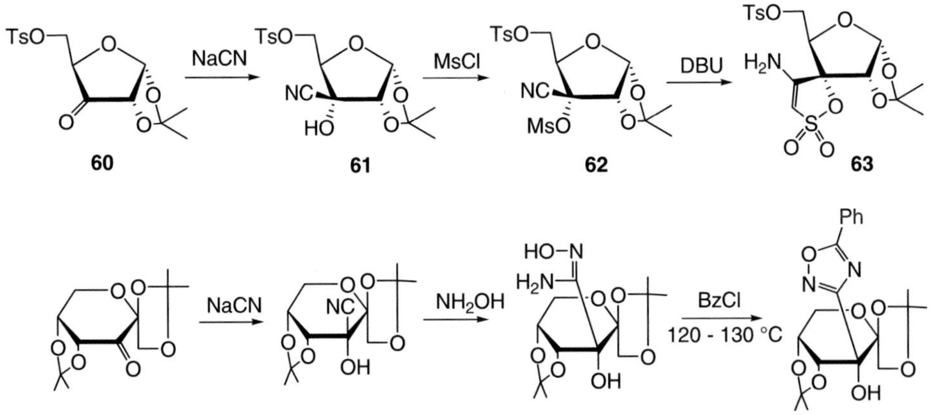

□ Scheme 20

C-branched sugar aceric acid [32] (❷ *Scheme 21*), 3-ulose **68** was treated with 2-(trimethylsilyl) thiazole (2-TST) in THF under thermodynamic control, followed by desilylation with tetra-*n*-butylammonium fluoride (TBAF), led to a 5:1 mixture of *endo-* and *exo*-adducts **69** and **71**, respectively. In contrast, addition of 2-thiazolyl lithium to compound **68** gave only the adduct **71**, arising from sterically controlled *exo*-addition to the trioxabicylo[3.3.0]octane ring system in an unoptimized 40% yield. Benzylation of isomeric adducts **69** and **71** afforded 3-*O*-benzyl derivatives **70** and **72**, respectively. The stereochemistry of thiazole addition to 3-ulose **68** was established by X-ray structural analyses of compounds **69** and **72**, which revealed that **72** has the required C-3 configuration for aceric acid synthesis.

The stereoselectivity in the reaction of 3-ketofuranose **68** and 2-thiazolyl lithium can be attributed to the kinetic control and steric hindrance from *endo*-face addition. The thiazolyl group of compounds **70** and **72** was converted into the formyl functionalized sugars **73** and **74** using essentially a one-pot, three-step literature method [38,39].

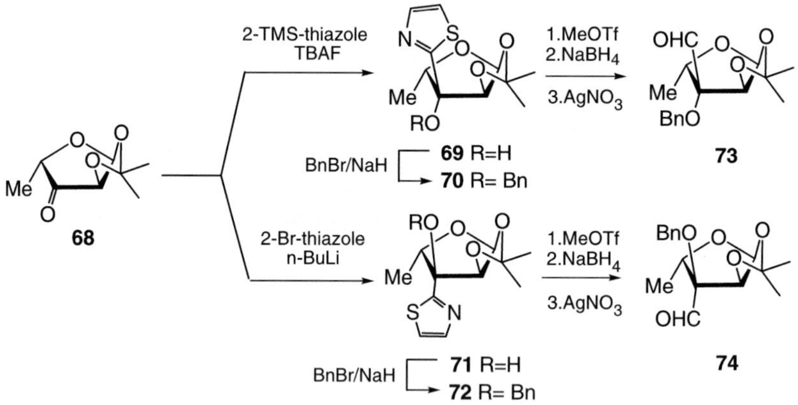

❑ Scheme 21

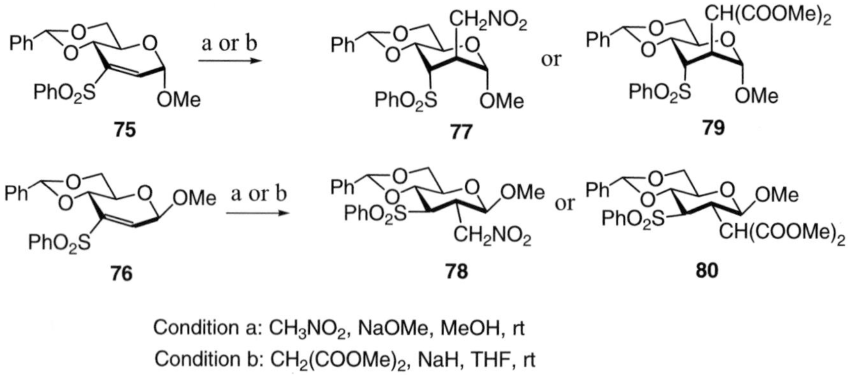

Condition a: CH₃NO₂, NaOMe, MeOH, rt
Condition b: CH₂(COOMe)₂, NaH, THF, rt

❑ Scheme 22

Suresh et al. [40] reported a diastereoselective addition of carbon nucleophiles to vinyl sulfone-modified carbohydrates, in which nucleophiles added to the C-2 position from a direction opposite to that of the anomeric methoxy group (❷ *Scheme 22*). The nucleophile, generated from CH_3NO_2 and NaOMe, reacted with **75** (or **76**) producing unique **77** (or **78**). Similarly, the nucleophile generated from dimethyl malonate and sodium hydride produced exclusively **79** and **80**, respectively. Interestingly, α-methyl glycoside **75** gave **77** and **79** having α-D-*altro* configuration, while β-methyl glycoside **76** produced β-D-gluco analogues **78** and **80** under the same reaction conditions. Crystal structures of **77/79** and **78/80** unambiguously established the absolute configurations at positions C-2 and C-3 of these compounds. Pentofuranosides presented the same results.

3 C–C Bond Formation by Metal or Metal Complex Mediated Reactions

Metal or metal complex-mediated reactions have been widely applied to *C*-glycoside synthesis [41]. Accordingly, various metal or metal complex catalysts such as rhodium, indium, samarium, palladium, and so forth have also been developed for the preparation of *C*-branched-chain sugars.

3.1 Rhodium Complex Catalyzed Reactions

Hydroformylation of alkenes by homogenous rhodium catalysts has attracted the attention of researchers for functionalization of complex molecules [42,43]. Al-Abed et al. [44] recently

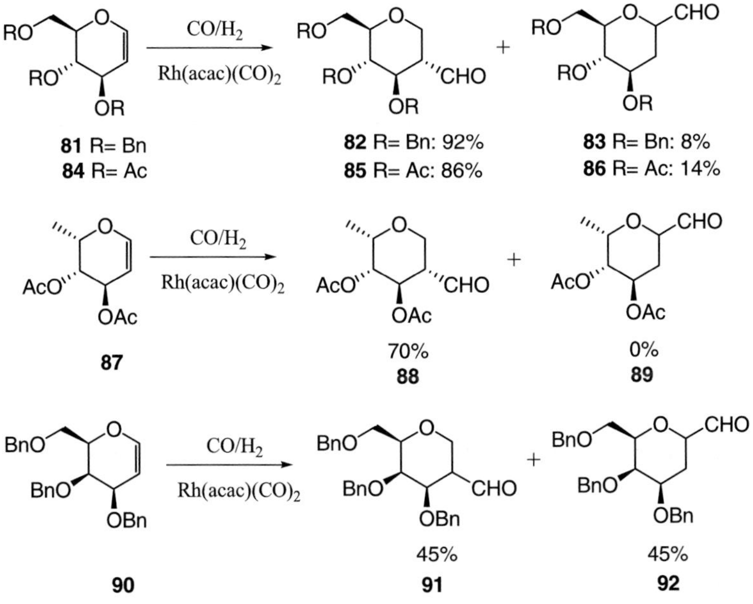

81 R= Bn
84 R= Ac

82 R= Bn: 92%
85 R= Ac: 86%

83 R= Bn: 8%
86 R= Ac: 14%

87

70%
88

0%
89

90

45%
91

45%
92

❑ **Scheme 23**

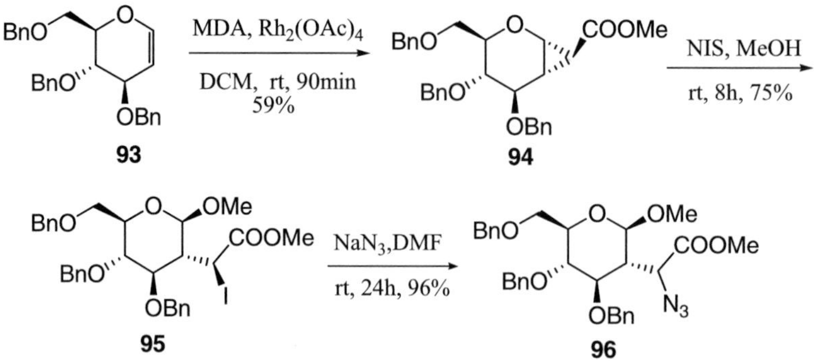

Scheme 24

reported hydroformylation of glycals utilizing the catalyst Rh(acac)(CO)$_2$ to give a mixture of the C-1 formyl 2-deoxy-C-glycoside and/or the C-2 formylpyran counterpart (❷ *Scheme 23*). Thus, hydroformylation of tri-O-benzyl-D-glucal **81** gave formyl adducts **82:83** in a ratio of 92:8, and the corresponding acetyl glucal **84** obtained similar results under these conditions. Unintelligibly, the 3,4-di-O-acetyl-6-deoxy idopyranose derivative **87** gave only the C-2 formyl product **88**, while tri-O-benzyl-D-galactal **90** gave a 1:1 mixture of formyl compounds **91** and **92**. The regioselectivity of formyl addition depends on polarization of the olefin, relative stability of the alkyl-metal complexes, the difficulty of β-elimination for conformationally rigid substrates, and the ratio of acyl-metal intermediates [45].

An efficient method for the synthesis of 2-C-branched glyco-amino acid derivatives by diastereoselective ring opening of carboxylated 1,2-cyclopropane sugars has been achieved using the stereocontrolled cyclopropanation of glycals mediated by rhodium acetate [46]. As shown in ❷ *Scheme 24*, tri-O-benzyl-D-glucal **93** was treated with methyl diazoacetate (MDA) in the presence of a catalytic amount of rhodium acetate furnishing 1,5-anhydro-2-deoxy-1,2-C-(*exo*-carbomethoxy methylene)-3,4,6-tri-O-benzyl-α-D-glucitol **94** in 59% yield. Treatment of **94** with NIS/MeOH afforded methyl-3,4,6-tri-O-benzyl-2-deoxy-2-C-(iodomethyl acetate)-β-D-glucopyranoside **95** as a single diastereomer in which two new stereocenters were introduced in a single reaction. Further reaction of **95** with NaN$_3$/DMF afforded azide **96** which could be converted into a glyco-amino acid ester after reduction.

3.2 Indium-Promoted Reactions

Organometallic reactions in aqueous media have been developed and their application in organic synthesis has been increasingly explored [47]. From this perspective, indium chemistry has captured much recent attention due to the comparable catalyzing abilities in aqueous media [48].

Lubineau reported a series of research works on the preparation of C-branched monosaccharides and C-disaccharides under indium promoted Barbier-type allylation in aqueous media [49]. In the case of substrate **97**, the reaction, which took place in H$_2$O/EtOH (1:2) at 50 °C, gave unique stereoisomer **98** with complete regio- and diastereoselectivity. From

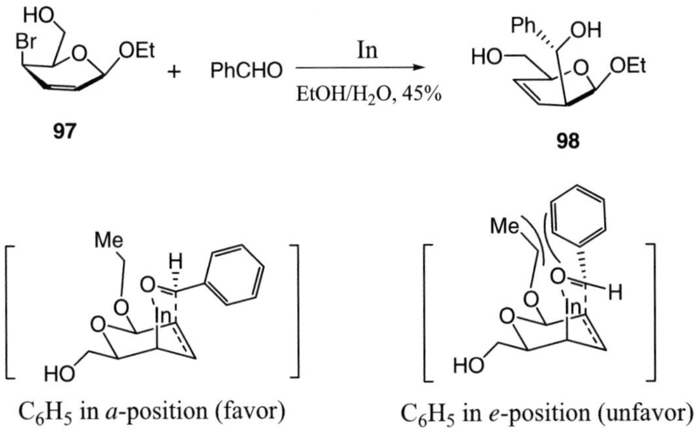

C$_6$H$_5$ in *a*-position (favor)   C$_6$H$_5$ in *e*-position (unfavor)

■ Scheme 25

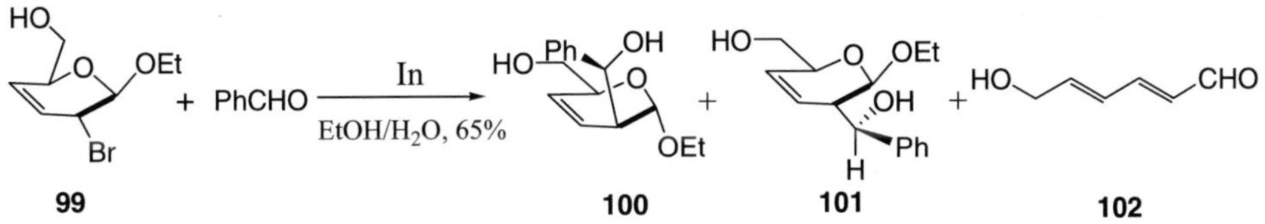

■ Scheme 26

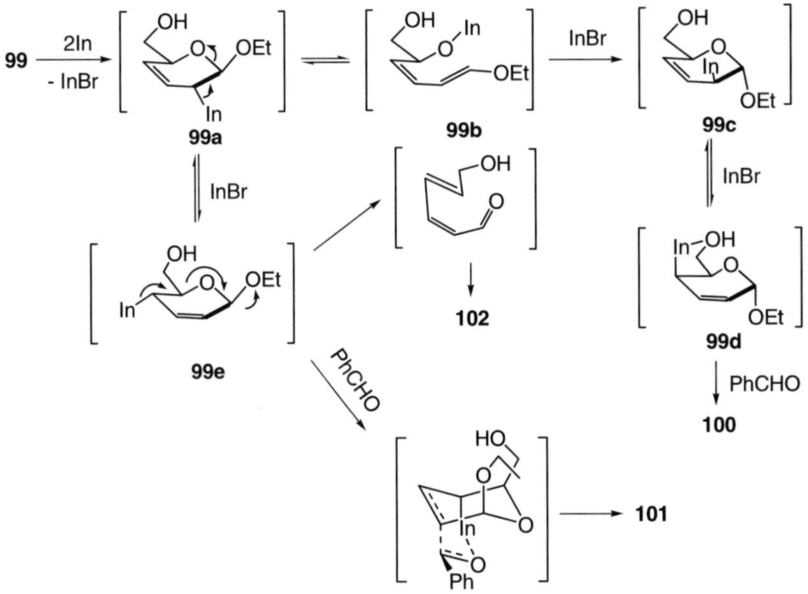

■ Scheme 27

a mechanistic point of view, the stereochemistry at C-7 can be explained by the formation of a six-membered cyclic transition state between the carbonyl and the allyl-indium complex moiety in which the phenyl group is in the axial position as depicted in ❯ *Scheme 25*. Indeed, molecular modeling showed clearly an unfavorable steric interaction between the equatorial phenyl substituent and the β-ethoxy group. When 2-bromo-4-enopyranoside **99** was employed in this reaction, a mixture of the C-2 axial product **100**, the C-2 equatorial adduct **101**, and the known aldehyde **102** was obtained in a 6/6/1 ratio (❯ *Scheme 26*), which could be rationalized by the mechanism given in ❯ *Scheme 27*. Firstly, an allylindium(I) species **99a** is formed which may undergo a ring opening through β-elimination. Secondly, indium(I) bromide formed in the reaction may act as a Lewis acid and cause the cyclization of the enol ether derivative **99b** to give **99c**, which will take a more favorable state **99d** via a stereospecific 1,3-allylindium migration. In the same manner, the species **99a** exists in equilibrium with its regioisomer **99e** in the presence of InBr. The desired products **100–102** were derived from these intermediates, respectively.

Starting from methyl 6-bromo-4,6-dideoxy-α-D-*threo*-4-enopyranoside **103**, 4-*C*-branched sugars have been prepared from various aldehydes with the same manner in a THF–phosphate buffer (0.11 M, pH 7.0) as the solvent (❯ *Scheme 28*). If the reaction is conducted in pure

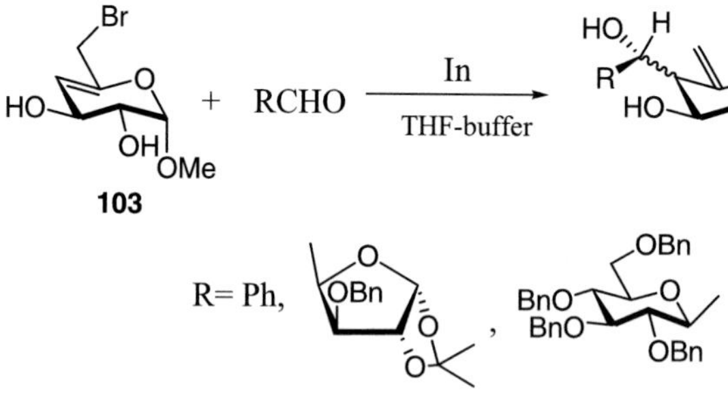

◻ Scheme 28

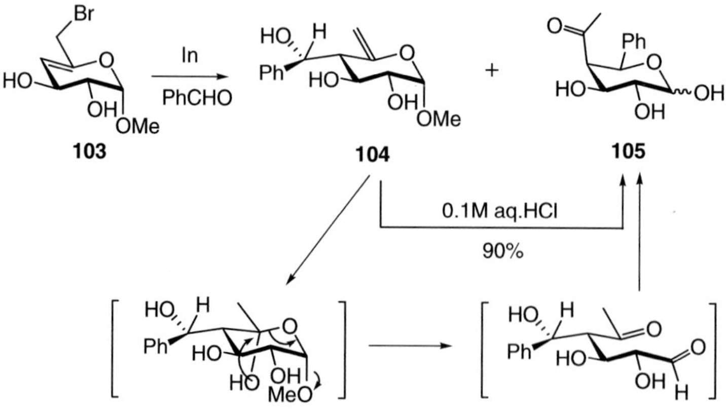

◻ Scheme 29

water, in the absence of the phosphate buffer, the 4-C-adduct **104** is slowly transformed (30% after 1.5 h) into an α,β mixture of **105**. In fact, **105** α, β can also be formed from the acid hydrolysis of the labile enol ether derivative **104** (❯ *Scheme 29*). After electrophilic addition of water, followed by pyran ring opening and elimination of the anomeric methoxy group, the cyclization of the C-7 hydroxyl group with the aldehyde led to the keto-derivatives **105** α, β. In order to confirm these results, compound **104** was treated with 0.1 M aq HCl affording **105** as a mixture of anomers (α/β, 1:3) in 90% yield.

3.3 Diiodosamarium-Mediated Reactions

Prandi [50] reported the preparation of methyl α-D-caryophylloside, a natural 4-C-branched sugar, in which the key step was diiodosamarium-promoted coupling reaction. As illustrated in ❯ *Scheme 30*, C–C bond formation between the crude acid chloride **106** and ketone **107** was mediated smoothly by SmI$_2$ in tetrahydropyran (THP). Expected products were isolated in 63% yield and in a 8:1 diastereoisomeric ratio. Reduction of the major diastereomer **108** with sodium borohydride in methanol at 0 °C was very slow, but the expected **109** was eventually obtained in 73% yield after 24 h at room temperature.

3′-β-Carbamoylmethylcytidine (CAMC) exhibits potent cytotoxicity against various human tumor cell lines and was synthesized using an intramolecular Reformatsky-type reaction promoted by SmI$_2$ as the key step [51]. The synthesis of the 3′-β-branched-chain sugar pyrimidine nucleosides is shown in ❯ *Scheme 31*. 2′-O-TBS-3′-ketouridine **110** was acylated with a bromoacetyl group to give the 5′-O-bromoacetyl derivative **111**, the precursor to the intramolecular Reformatsky-type reaction. Treatment of **111** with 2 equiv. of SmI$_2$ in THF at -78°C afforded the desired lactone **112** in good yield, while the Zn-promoted Reformatsky reaction did not obtain **112** under standard conditions. This is attributed to the strong chelating ability of the samarium enolate to the 3′-carbonyl oxygen to form a six-membered transition state. Ammonolysis of the lactone **112** at -70 °C gave the 3′-carbamoylmethyluridine derivative **113**.

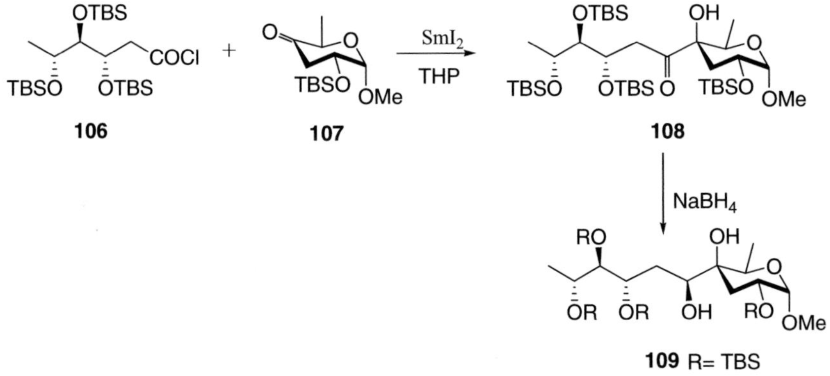

❒ Scheme 30

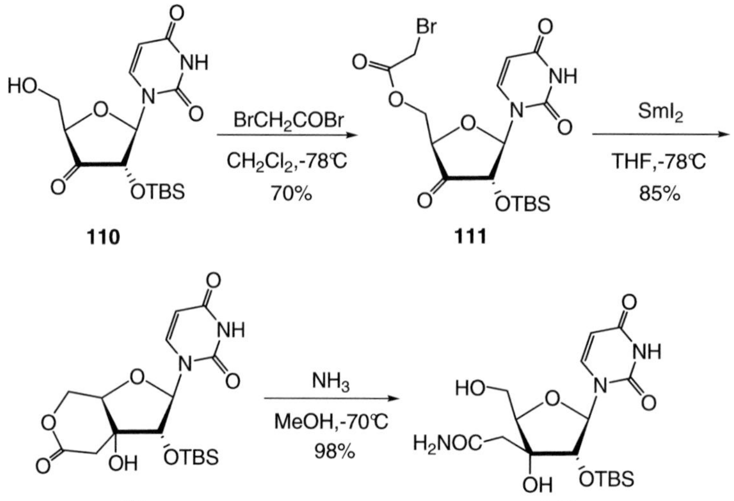

BrCH₂COBr
CH₂Cl₂,-78°C
70%

110

111

SmI₂
THF,-78°C
85%

NH₃
MeOH,-70°C
98%

112

113

■ Scheme 31

3.4 Other Metal-Catalyzed Reactions

Copper-catalyzed intramolecular cyclopropanation of glycal-derived diazoacetates has been well investigated by Pagenkopf [52]. The cyclopropanation was achieved by the addition of the diazoester to a refluxing solution of 5 mol% bis(N-tert-butylsalicylaldiminato)copper(II) **116** in either dichloromethane or toluene. Slow addition of the diazoester over 8–12 h was necessary to minimize the formation of dimeric fumarates and maleates. Catalyst **116** performed admirably in the cyclopropanation reactions in this study, whereas other catalysts, including Rh₂(OAc)₄, gave lower yields. The intramolecular cyclopropanation of glycals is compatible with a variety of protecting groups at the C-4 and C-6 positions, including cyclic silylene (**114a**), acetonide group (**114b**), and acyclic benzyl and TBS groups (**114c**). Cyclopropane **115a**, **115b**, or **115c** shown in ❷ Scheme 32 was formed as an exclusive stereoisomer in each case. As expected, the cyclopropanation is not limited to electron-rich olefins, and reaction of **117** proceeded with equal efficiency. In addition to the high diastereoselectivity inherent to this intramolecular reaction, the products present a parallel selectivity regardless of the protecting groups employed at C-4 and C-6.

Zeise's dimer [Pt(C₂H₄)Cl₂]₂ catalyzed ring opening of 1,2-cyclopropanate of sugars with O-nucleophiles generated 2-C-branched carbohydrates [53]. A number of O-nucleophiles can participate in the ring opening including alcohols, phenols, and water. A wide range of alcohols has been employed to give 2-C-branched glycosides ranging from simple methyl glycosides to complex disaccharides. A very high diastereoselectivity is obtained at the newly formed C-1 stereocenter. The α-glycoside, favored by the anomeric effect, is always the major product regardless of the stereochemistry of the starting cyclopropane (❷ Scheme 33).

Palladium-catalyzed carbonyl allylation [54] can be effectively applied to the regio- and diastereoselective synthesis of 2-C- and 4-C-branched sugars **121** and **122** from allylic esters

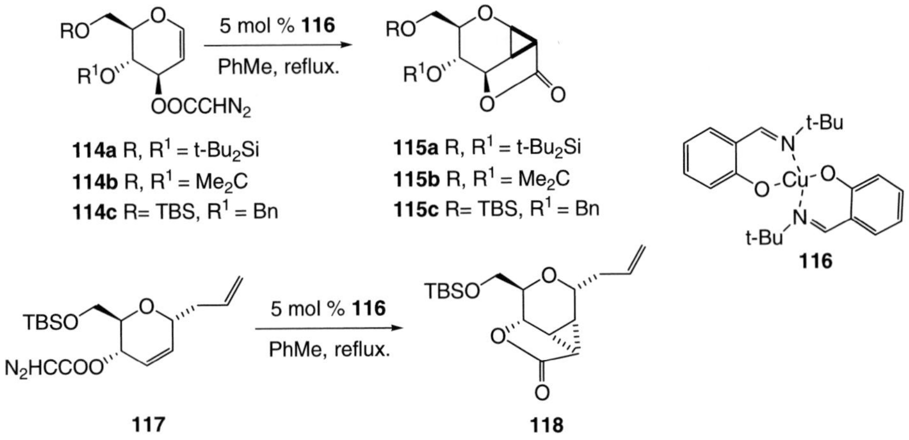

Scheme 32

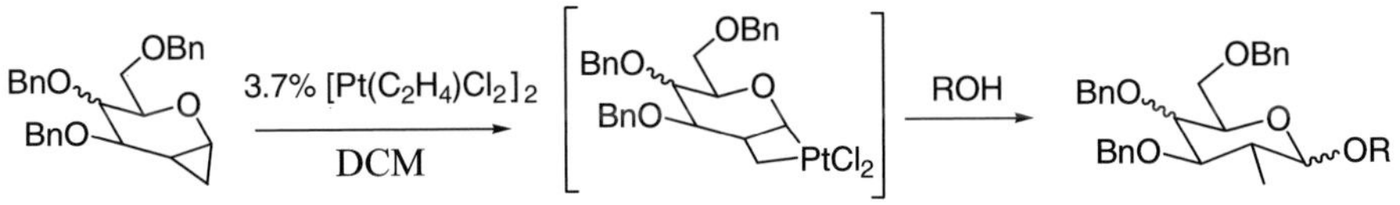

Scheme 33

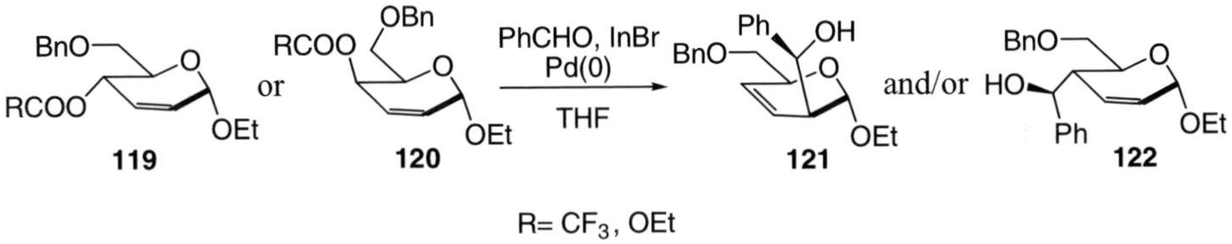

R= CF$_3$, OEt

Scheme 34

or carbonates, **119** and **120**, via formation of π-allylpalladium(II) intermediates and reductive transmetalation with indium(I) bromide (❷ *Scheme 34*). A postulated mechanism is depicted in ❷ *Scheme 35*. In the catalytic process, Pd(0) complexes may react with alkenes **119** or **120** to form π-allyl species with inversion of configuration of the carbon bearing the leaving group. The resulting π-allyl palladium(II) complexes **123** or **126** are then reductively transmetalated with indium(I) to give allylindium(III) species **124** and **125** (or **127** and **128**), followed by reacting with benzaldehyde to afford target compounds **121** and **122**, respectively. As was recently confirmed by Lubineau's group [55], InBr approaches the substrate from the same face as the palladium leading to a reductive transmetalation with retention of configuration.

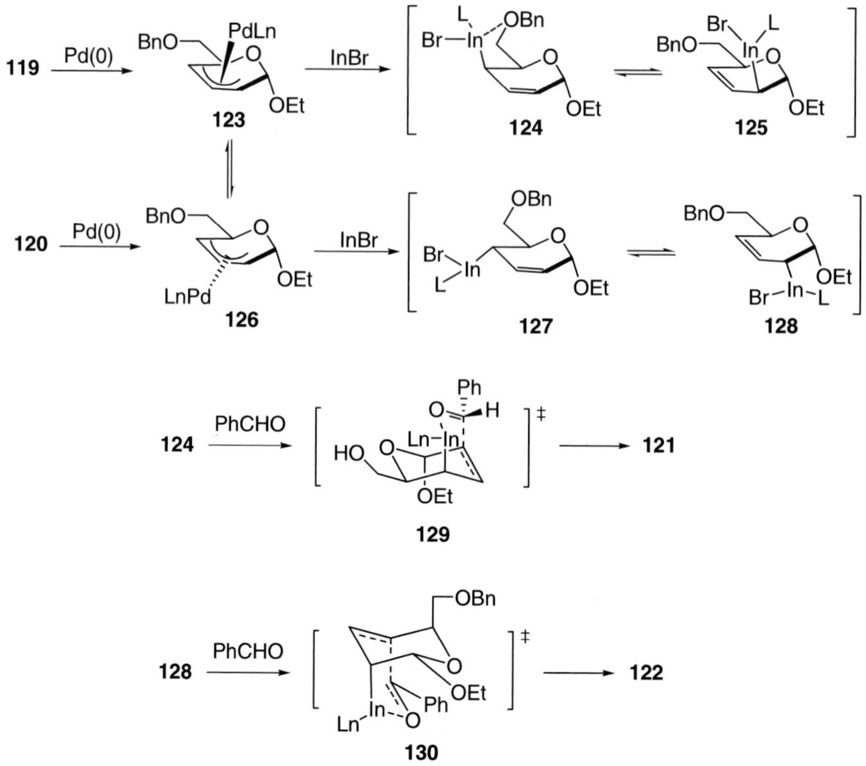

Scheme 35

4 Radical Cyclization

Over the past decades, radical chemistry has been developed into an important and integral part of organic chemistry. Radical cyclization becomes a facile and useful strategy for stereo- and regioselective C–C bond formation, affording useful chiral synthons for the synthesis of C-branched sugar derivatives [56,57]. The reactions in this section are divided into intramolecular and intermolecular free radical cyclization.

4.1 Intramolecular Free Radical Cyclization

The most representative examples of intramolecular free radical cyclization in carbohydrate chemistry are the syntheses of C-branched nucleosides derivatives. The key step in C-branched nucleoside preparation is the regio- and stereo-controlled formation of a new C–C bond at the branching point of the ribofuranose ring [58]. Among published reports, a temporary silicon connection is becoming a growing interest in the syntheses of C-branched nucleosides by intramolecular radical cyclization.

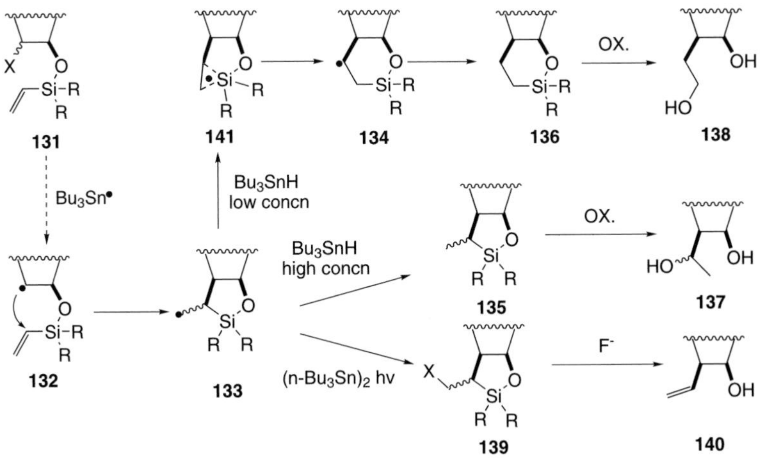

■ Scheme 36

Shuto's group has developed a highly versatile regio- and stereoselective method for introducing the C_2 substituent via an intramolecular radical cyclization reaction, in which a silicon-containing group was applied as a temporary radical acceptor tether [59,60], as summarized in
❷ *Scheme 36.*
The selective introduction of both 2-hydroxyethyl and 1-hydroxyethyl groups can be achieved, depending on the concentration of nBu$_3$SnH in the reaction system, via a 5-*exo*-cyclization intermediate **135** or a 6-*endo*-cyclization intermediate **136**, respectively. A vinyl group can also be introduced, to give **140**, by irradiation of the vinylsilyl ether in the presence of (n-Bu$_3$Sn)$_2$, followed by treatment of the resulting 5-*exo*-cyclization product **139** with a fluoride ion. The mechanism of radical cyclization is that the kinetically favored 5-*exo*-cyclized radical **133** was trapped by high concentration of n-Bu$_3$SnH to give **135**. At lower concentrations of n-Bu$_3$SnH and at higher reaction temperatures, the radical **133** rearranged into a more stable, ring-enlarged 4-oxa-3-silacyclohexyl radical **134** via a pentavalent-like silicon radical transition state **141**, which was then trapped with n-Bu$_3$SnH to give **136**. This method has been applied to the synthesis of many C-branched chain sugar nucleoside analogues.
The stereoselective introduction of an ethynyl group in various five- and six-membered iodohydrins has also been developed by the same group (❷ *Scheme 37*) [60]. Treatment of 3′-ethynyldimethylsilyl 2-deoxy-2-iodo-D-mannopyranoside **142** with Et$_3$B and TBAF in toluene furnished the methyl 2-deoxy-2-C-ethynyl-4,6-O-benzylidene-α-D-mannopyranoside **144** in 85% yield. However, a similar reaction of 3-iodo-D-ribosyl substrate **143** having the [2-(trimethylsilyl) ethynyl]dimethylsilyl at C-5 gave the desired product 5-O-acetyl-3-deoxy-3-C-ethynyl-1,2-O-(1-methyl ethylidene)-α-D-xylofuranose, in which the ethynyl group was introduced at the γ-*cis* position to the 5′-hydroxyl, in only 31% yield. This may be explained by an unfavored 6-*exo* radical cyclization.
2′-Deoxy-2′-iodo-3′-O-TEDMS uridine derivative **146**, readily available from uridine via 2,2′-anhydrouridine, was subjected to the above-mentioned procedure. After removal of the monomethoxytrityl (MMTr) group and subsequent acetylation with Ac$_2$O in pyridine,

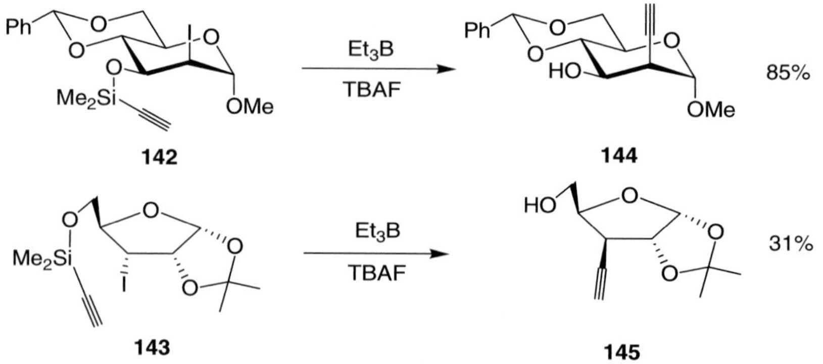

Scheme 37

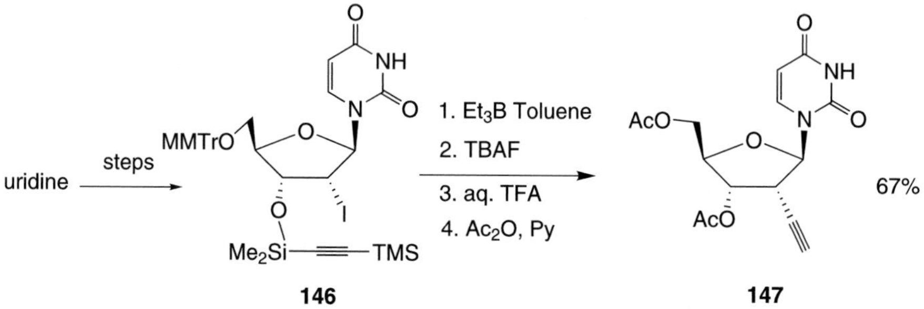

Scheme 38

3,′5′-di-*O*-acetyl-2′-deoxy-2′-*C*-ethynyluridine **147**, a key substrate designed for potential antimetabolites (● *Scheme 38*), was isolated in 67% yield.

The cyclization (5-*exo* and 6-*endo*) of radicals derived from 6-(bromomethyl)dimethylsilylated glycal uridine, was extensively investigated by Tanaka and coworkers [61] (● *Scheme 39*). Without the 2-substituent, 6-(bromomethyl)dimethylsilyl-1-[3,5-bis-*O*-TBDMS-2-deoxy-D-*erythro*-pent-1-enofuranosyl] uracil **148a** was subjected to a radical reaction with azobisisobutyronitrile (AIBN) and Bu₃SnH in refluxing benzene producing specifically 6-*endo*-cyclized products **149a** (58%) and **150a** (32%). No 5-*exo*-cyclized products were formed in this case. In the presence of the 2-methyl substituent, **148b** predominantly afforded 5-*exo* product **150b** under the same reaction conditions. However, the C2 radical intermediate was not stable and a substantial amount (29%) of rearranged product **150c** was formed through anomeric radical species. Cyclization of the other 2-substituted (R₂= CO₂Me, OBz, Cl) derivatives furnished exclusively 5-*exo* products.

Bromo-acetal **151** was subjected to tributyltin hydride promoted radical cyclization and afforded a stereoselective β-*C*-methyl derivative **152**. (● *Scheme 40*) Similarly, treatment of **151** under Keck's conditions [62] with allyltributyltin and AIBN resulted in the formation of a bis-*C*,*C*-glycoside **153** [63].

Kim et al. (● *Scheme 41*) developed a new method for the stereoselective syntheses of fused α-substituted γ-butyrolactone nucleosides via [1,5]-*C*,*H* insertion of α-diazo-γ- butyrolac-

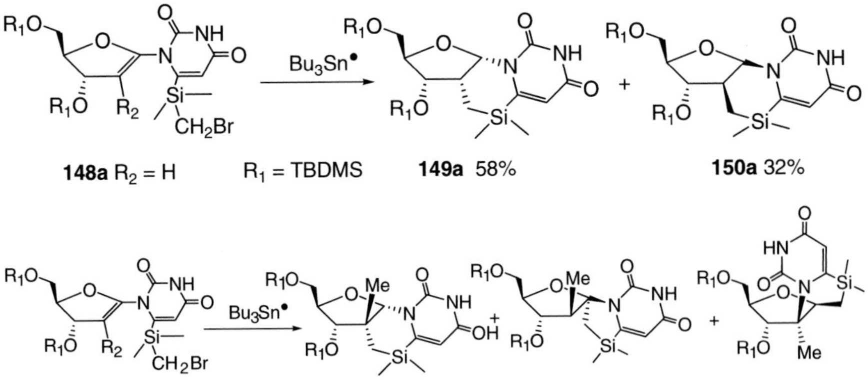

148a R_2 = H R_1 = TBDMS   **149a** 58%   **150a** 32%

148b R_2 = Me R_1 = TBDMS   **149b** 8%   **150b** 41%   **150c** 29%

◘ Scheme 39

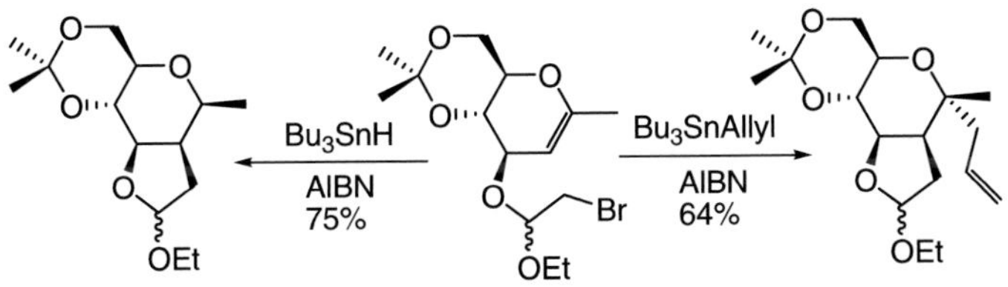

152   **151**   **153**

◘ Scheme 40

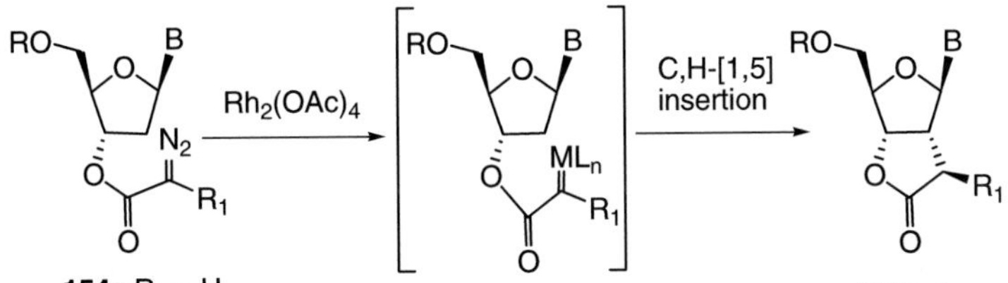

154a R_1 = H,
154b R_1 = COMe
154c R_1 = CO_2Me,
154d R_1 = CO_2Et

155a–d

◘ Scheme 41

tone nucleosides [58,64]. 2'-Deoxy-3'-diazoacetates of nucleosides **154a–d** (R_1=H, COMe, CO_2Me, CO_2Et) were chosen as templates for the insertion reactions. Initial studies on stereocontrolled *C,H*-insertion of 2'-deoxy-3'-α-diazoacetate nucleosides were performed in the presence of dirhodium tetraacetate (1.0 mol%) in dichloromethane at room temperature and only a trace amount of product was obtained. However, when the reaction mixture was refluxed, γ-butyrolactones **155a–d** were obtained with high diastereoselectivities in 58–80% yields.

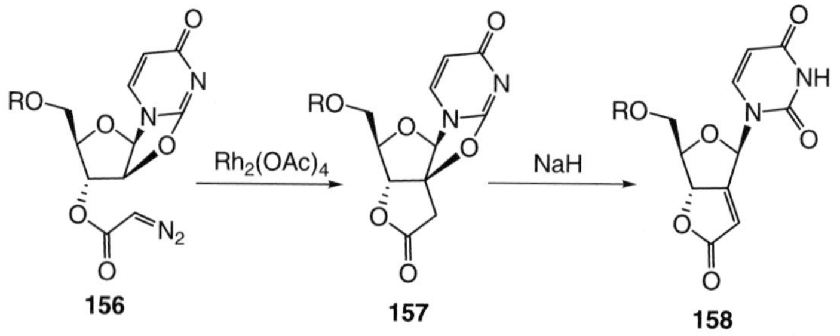

◘ Scheme 42

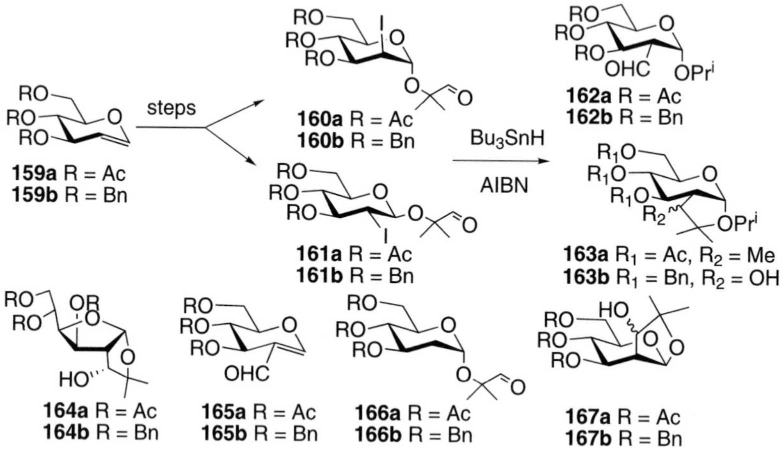

◘ Scheme 43

The same method was applied to the synthesis of biologically active nucleosides (*E*)-2′-deoxy-2′-(carboxymethylene)-5′-*O*-trityluridine-3′,2′-γ-lactone **158** (❷ *Scheme 42*). Exposure of 2′,5′-cyclouridine derivative to the House–Blankey protocol afforded the corresponding diazo compound **156**, which could be converted to the γ-butyrolactone of uridine **157** in 65% yield via [1,5]-*C,H* insertion. Lactone **158** could be smoothly obtained by an elimination reaction with sodium hydride in 85% yield.

A new three-step procedure of iodoetherification, ozonolysis, and radical cyclization-fragmentation, converting glucals into C-2 formyl pyranosides, was reported by Choe et al. [65]. The free-radical promoted cyclization of 1,1-dimethyl-2-oxoethyl-3,4,6-tri-*O*-acetyl-2-deoxy-2-iodo-α-D-mannopyranoside **160a**, prepared in 72% yield from commercially available tri-*O*-acetyl-D-glucal **159a** in two steps, produced a mixture of products in varying yields depending on the reaction conditions (❷ *Scheme 43*).

When the reaction was conducted in benzene with 1.0–1.5 equiv. of Bu₃SnH and 0.01–0.5 equiv. of AIBN under reflux, formyl-transfer product isopropyl 3,4,6-tri-*O*-acetyl-2-deoxy-2-*C*-formyl-α-D-glucopyranoside **162a** was isolated in 40% yield. The bicyclic alcohols

$R_1 = CHFCO_2Et, CF_3, CH_2CN; R_2 = Et, (CH_2)_2Ph$

◻ **Scheme 44**

163a (a 1:1 mixture of diastereomers), the C-2 formyl glycal **165a**, C-2 deoxysugar **166a**, and an unexpected furanoside product [1,2-bis-(acetyloxy)ethyl]hexahydro-2,2-dimethyl-furo-[2,3-b]furan-3,4-diol,4-acetate **164a** were also isolated from this reaction with different concentrations of **160a** and the reaction time. If the above radical reactions were promoted by tris(trimethylsilyl)silane, it did not form any of the furanoside **164a**. However, a 65% yield of the C-2 formyl glycal **165a** was isolated and the desired formyl transfer product **162a** was isolated in 41% yield under optimal conditions. This method can also be extended to synthesize C-2 branched galactosides and C-2 cyano glucopyranosides.

4.2 Intermolecular Free Radical Cyclization

A two-step preparation of 2-*C*-branched nucleoside derivatives through highly diastereoselective tandem xanthate radical addition–substitution reactions was investigated by Lequeux and coworkers (❷ *Scheme 44*) [66]. The addition was regioselective and the fluorocarboxylic ester functional group was unambiguously added onto C-2. Attempts to trap the radicals with ethyl acrylate were unsuccessful, and no addition to electron-poor alkenes was observed for this reaction because of the high electrophilic character of the carboxyfluoromethyl radical. The displacement of the resulting anomeric xanthates with various nucleophiles in the presence of Lewis acid allowed the formation of 2,3-disubstituted tetrahydrofuran derivatives.

In the above-mentioned reaction, the corresponding acetals or glycofuranoside derivatives can be prepared using *O*-nucleophiles. Treatment of a single diastereomer, ethyl (2-ethoxythio carbonylsulfanyl-tetrahydrofuran-3-yl)-fluoroacetate (**168**), with ethanol or neopentanol in the presence of silver triflate in toluene led to the formation of a *trans/cis* mixture of the corresponding acetals (❷ *Scheme 45*). The reactions gave low stereoselectivity, and the mixtures of 2,3-*trans* and -*cis* isomers ethyl (2-ethoxy-tetrahydrofuran-3-yl)fluoroacetate **169a** and ethyl[2-(2,2-dimethyl-propoxy)-tetrahydrofuran-3-yl] fluoroacetate **169b** were obtained in a 7:3 and 3:2 ratios, respectively.

Introduction of a *C*-nucleophile has been attempted by using organomagnesium reagents (PhMgBr, EtMgBr) in the presence of Lewis acid (AgOTf, BF$_3$·Et$_2$O, SnCl$_4$). Substitution of the anomeric xanthate to form *C*-glycoside was not observed in these cases. However, SnCl$_4$ promoted carbon–carbon bond formation can be performed from the *trans* ethyl (2-ethoxythio carbonylsulfanyl tetrahydrofuran-3-yl)-fluoroacetate **168** and Me$_3$SiCN at $-78\,°C$, resulting in 2-cyanotetrahydrofurans **169c** with a high yield (83%). This reaction was also attempted with nucleophile TMSAll at $0\,°C$ or $-78\,°C$ in the presence of AgOTf, Cu(OTf)$_2$, or SnCl$_4$, but the corresponding alkylated tetrahydrofuran derivatives were only detected as a minor component of the products.

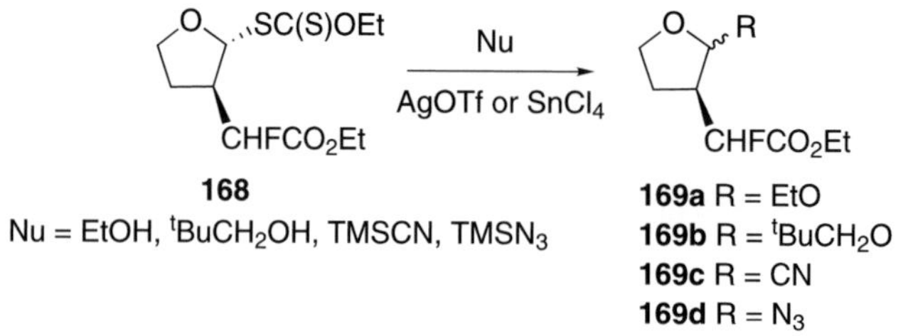

168

Nu = EtOH, ^{t}BuCH$_2$OH, TMSCN, TMSN$_3$

169a R = EtO
169b R = ^{t}BuCH$_2$O
169c R = CN
169d R = N$_3$

☐ Scheme 45

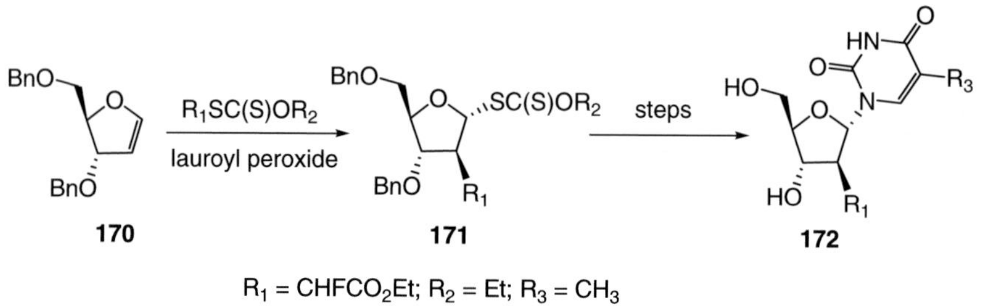

170   **171**   **172**

R$_1$ = CHFCO$_2$Et; R$_2$ = Et; R$_3$ = CH$_3$

☐ Scheme 46

A two-step synthesis of modified 2′-C-nucleoside precursor, ethyl [2-(5-methyl-2,4-dioxo-3,4-dihydro-2H-pyrimidin-1-yl)-4-hydroxyl-5-hydroxymethyltetra-hydrofuran-3-yl]fluoro-acetate **172**, from protected glycal **170** and xanthate has been developed following the same idea, and a diastereomeric 1:1 mixture of 2,3-*trans* product **171** was obtained in 57% yield (❷ *Scheme 46*). The use of triethylborane as a free-radical initiator was less successful and a longer reaction time was also required. Interestingly, introducing thymine at C-1 in the presence of silver triflate at 0 °C was highly stereoselective, and only a C1,C2-*trans* linked product was detected.

Dimethyl malonate in combination with ceric(IV) ammonium nitrate (CAN) has been used to generate electrophilic malonyl radicals [67]. Linker and coworkers extended this methodology with the addition of such radicals to substituted glycals [68] (❷ *Scheme 47*). The reactions proceed smoothly to afford the C-2 branched carbohydrates **174** and **175** with good yields and excellent regioselectivities. For unsubstituted galactal **173a** and carboxamide **173b**, only methyl glycosides **174a** and **174b** were obtained, respectively, whereas the nitrile **173d** afforded exclusively the ortho esters **175d**. On the other hand, the ester **173c** gave a mixture of both products **174c** and **175c**.

This result can be rationalized by the interaction between the SOMO of the electrophilic radical and the HOMO of the double bond. Furthermore, due to the steric shielding of the pseudo axial O-acetyl group, the radicals attack the double bond selectively from the α-face. The reaction mechanism is depicted in ❷ *Scheme 48*.

In an approach towards the synthesis of C-oligosaccharide containing α-D-Man-(1–4)-D-Man repeating units, C-4 allylated building block **183** was designed [69]. However, the allyla-tion of iodide **181** at C-4 with allyltributyltin and AIBN or dilauoryl peroxide (DLP) slow-

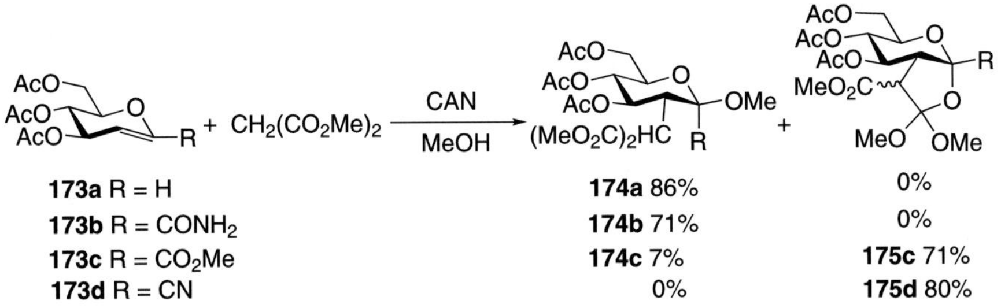

Scheme 47

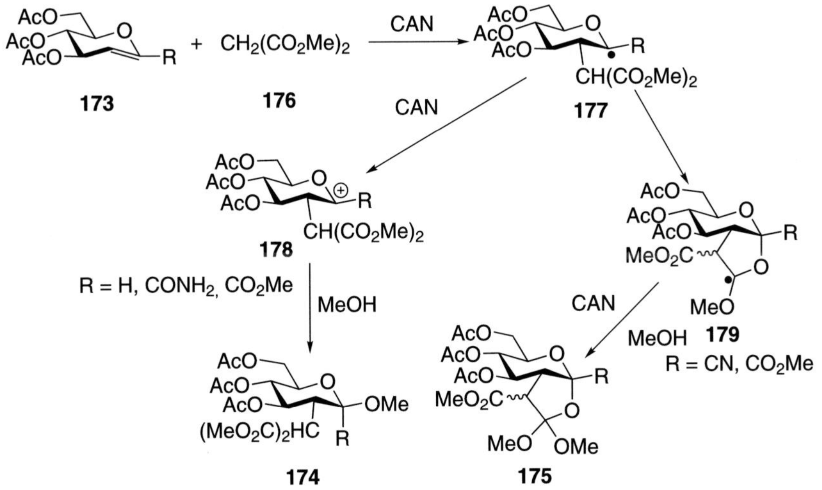

173a R = H   174a 86%   0%
173b R = CONH₂   174b 71%   0%
173c R = CO₂Me   174c 7%   175c 71%
173d R = CN   0%   175d 80%

Scheme 48

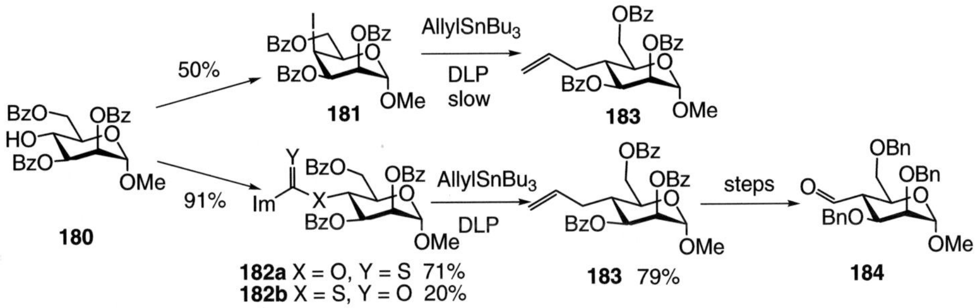

Scheme 49

ly gave desired **183**, together with considerable 4-deoxy byproduct due to 1,3-diaxial repulsion between iodide and the C_2-benzoate (**Scheme 49**). When thionocarbamate **182** was subjected to allyltributyltin and dilauoryl peroxide in refluxing benzene, only the equatorially C_4-allylated methyl 2,3,6-tri-O-benzoyl-4-deoxy-4-C-allyl-D-mannopyranoside **183** was obtained with 79% yield. After the transformation of protecting groups, double bond migration and olefin ozonolysis, formyl-branched monosaccharide **184** was achieved in high yield.

5 Rearrangement and Cycle Additions

Many of the 2-C-branched sugars are synthesized from glycals through 1,2-cyclopropanation and a subsequent selective ring opening via solvolysis in the presence of a stoichiometric amount of mercury(II) salts, strong acid, or halonium ions. In all cases an anomeric mixture of glycosides were often provided through an oxocarbonium-like intermediate, although α-glycosides are being favored due to the anomeric effect.

Zou recently reported a novel 1,2-migration of a 2′-oxoalkyl group via 1,2-cyclopropanate sugar derivatives [70]. Treatment of 2′-oxoalkyl 2-O-Ms-α-C-mannosides **185** with sodium methoxide in methanol produced 2-C-branched methyl β-glycoside **187** (40–50%) and a bicyclic derivative **188** (10–15%) (❷ *Scheme 50*). The proposed mechanism is that base-catalyzed enolate of 2′-oxoalkyl 2-O-Ms-α-C-mannosides **185** afforded 1,2-cyclopropanated intermediate **186** smoothly, which in turn underwent ring-opening with methoxide at the anomeric carbon to give 2-C-formylmethyl-2-deoxy-β-glycoside **187**. Presumably, an intramolecular rearrangement afforded bicyclic **188**. When **185** was treated with weaker bases, such as TEA in methanol or K_2CO_3 in acetonitrile-methanol co-solvent, **187** was obtained as the sole product, since the ring-opening enolation becomes less likely with a weaker base.

The ring opening of 1,2-cyclopropanated sugars, formed from 2′-aldehydo(acetonyl)-2-O-Ms(Ts)-α-C-glycosides, by nucleophiles such as alcohols, thiols, and azide under weak basic conditions, resulted in the formation of 2-C-branched β-glycosides and glycosyl azides in good to excellent yields. The best results were obtained with thiol nucleophiles and 1,2-*trans* 2-C-branched β-glycosides were always given (❷ *Scheme 51*).

In the development of new sugar amino acids and peptidomimetics, 6-allyloxy-2-[(*tert*-butyl diphenylsilanyloxy)methyl]-3,6-dihydro-2H-pyran-3-ol **189** was subjected to the refluxing

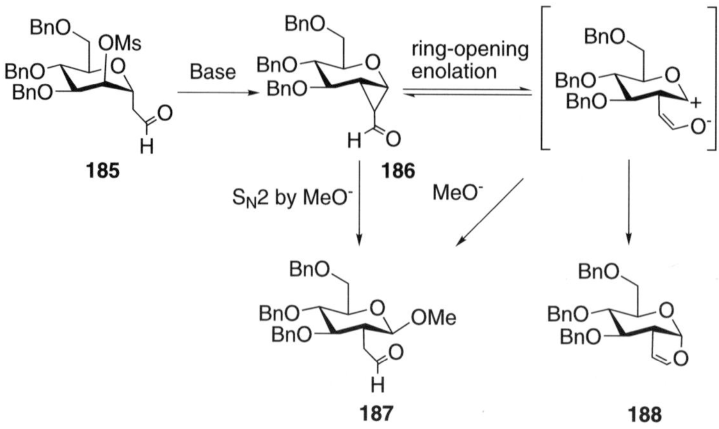

■ Scheme 50

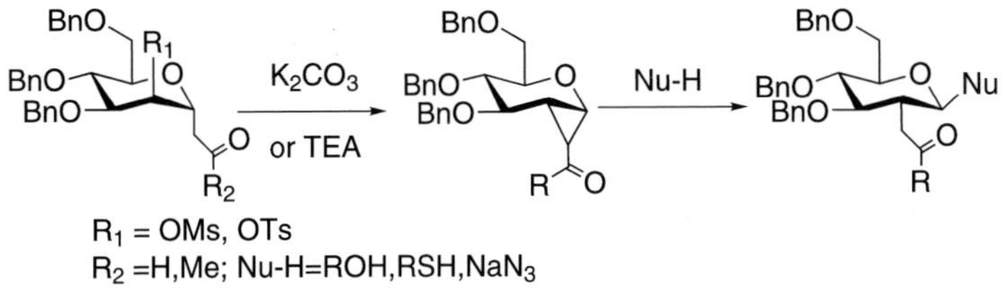

R₁ = OMs, OTs
R₂ =H,Me; Nu-H=ROH,RSH,NaN₃

□ **Scheme 51**

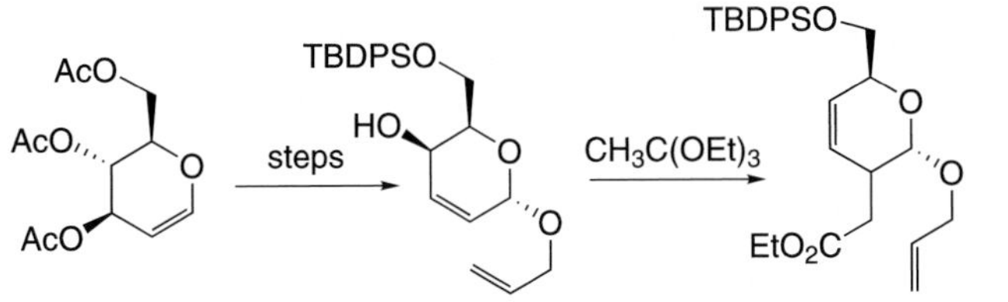

189   **190**

□ **Scheme 52**

triethyl orthoacetate in the presence of propanoic acid and hydroquinone, obtaining ethyl 2-{(2S,3S,6S)-2-allyloxy-6-[(*tert*-butyldiphenylsilanyloxy)-methyl]-3,6-dihydro-2*H*-pyran-3-yl} acetate **190** as the only product via Claisen–Johnson rearrangement [71] (❷ *Scheme 52*). Krohn reported a base-catalyzed vinyl tosylate-cuprate cross coupling to form regio- and stereodiverse *C*-branched 1,6-anhydrosugar building blocks [72]. As shown in ❷ *Scheme 53*, treatment of epoxide 1,6:3,4-dianhydro-2-*O*-tolylsulfonyl-β-D-galactopyranose **191** in THF with Gilman methyl cuprate afforded 1,6-anhydro-2,3-dideoxy-2-methyl-β-D-threo-hex-2-enopyranose **192** as the sole product. The reaction was initiated from the rearrangement of epoxide to allyl alcohol under basic conditions, and followed by a subsequent cross-coupling between vinyl tosylate and the cuprate. Under the same reaction conditions, 1,6:2,3-dianhydro-4-*O*-p-tolylsulfonyl-β-D-mannopyranose **193** and 1,6:3,4-dianhydro-2-*O*-p-toluolsulfonyl-β-D-altropyranose **195** was converted into 1,6-anhydro-3,4-dideoxy-4-methyl-β-D-threohex-3-enopyranose **194**, and 1,6-anhydro-2,3-dideoxy-2-methyl-β-D-erythrohex-2-enopyranose **196b**, respectively. Compound **195** was also reacted with the corresponding *n*-butyl Gilman cuprate to afford 1,6-anhydro-2-butyl-2,3-dideoxy-β-D-erythrohex-2-enopyranose **196c** in 68% yield, in addition to 22% of 1,6-anhydro-2,3-dideoxy-β-D-erythrohex-2-enopyranose **196a**. The product **196a** may be formed by hydrolysis of intermediates, such as metalated vinyl species.

An efficient route towards the syntheses of chiral 2-*C*-methylene-*O*-glycosides and chiral pyrano[2,3-b][1] benzopyrans using InCl₃ as the catalyst was reported [73]. Accordingly, reaction of 2-*C*-acetoxymethyl glycal derivatives and aliphatic (or aromatic) hydroxyl compounds in the presence of InCl₃ via Ferrier rearrangement furnished the corresponding 2-*C*-methylene glycosides (❷ *Scheme 54*). InCl₃, In(OTf)₃, and Yb(OTf)₃ were tested to catalyze the Ferrier rearrangement of 2-*C*-acetoxymethyl-3,4,6-tri-*O*-benzyl-D-glucal **197a** and the best result was obtained with 30 mol% InCl₃. The substrates **197a** and **197b**, on treatment with benzyl

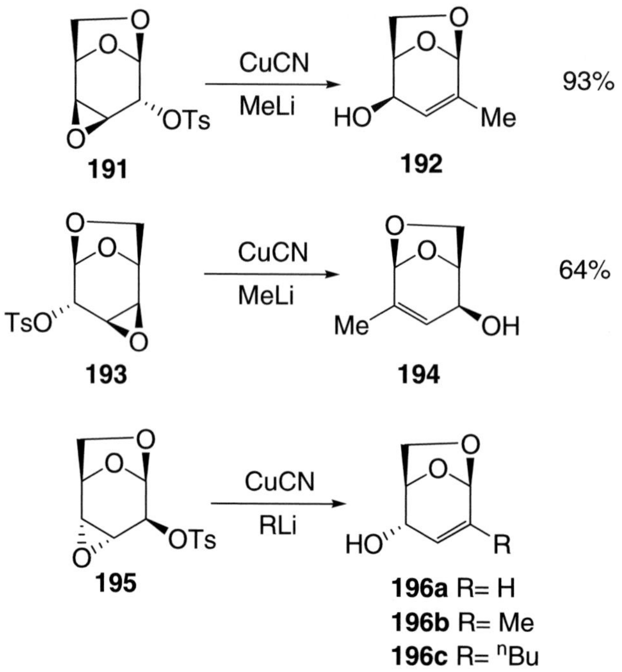

93%

64%

196a R= H
196b R= Me
196c R= ^{n}Bu

☐ Scheme 53

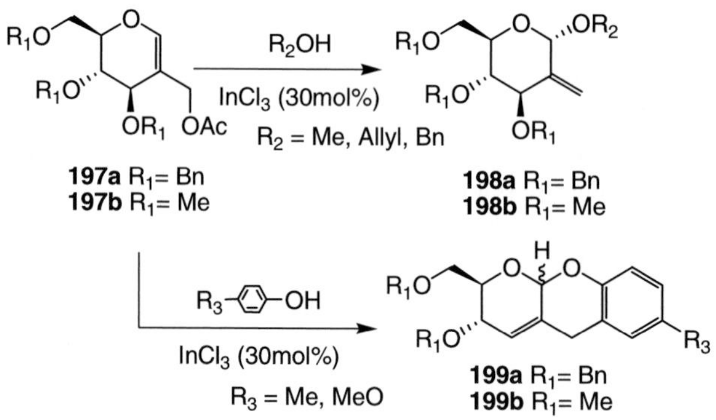

197a R_1= Bn
197b R_1= Me

198a R_1= Bn
198b R_1= Me

199a R_1= Bn
199b R_1= Me

☐ Scheme 54

alcohol, obtained the corresponding 2-C-methylene glycosides **198a** and **198b**, respectively, in good yields and exclusive α-selectivity. The excellent yields and exclusive α-selectivity were also obtained in the reactions of 2-C-acetoxymethyl-3,4,6-tri-O-methyl-D-galactal with benzyl, allyl, and *tert*-butyl alcohols. Interestingly, the reaction of 2-C-acetoxyglycals **197a** (or **197b**) with phenols formed the corresponding chiral carbohydrate-pyranobenzopyran derivatives **199a** (or **199b**) via Ferrier rearrangement and tandem cyclization in excellent yields and moderate to high stereoselectivities.

The reaction of dibenzylamine with glucose **200** in ethanol, in the presence of acetic acid, gave a high yield of Amadori ketose **201**, which was subsequently treated with calcium hydroxide in water and afforded, after acidic work-up, the branched ribono-1,4-lactone **202a**. No epimeric arabinonolactone **202b** was formed under these conditions [74] (❷ *Scheme 55*).

D-galactose **203** underwent the same Amadori rearrangement [75,76] to give crystal α-anomer of tagatosamine **204** (88% yield), which was subjected to CaO in H$_2$O and an acidic work-up, affording separable 2-*C*-methyl-branched lyxono-**205a** and xylono-**205b** [77] (❷ *Scheme 56*). Treatment of D-xylose **206** or L-arabinose **209** with the same procedure obtained 2-*C*-methyl-D-erythrono-**208a** and D-threono-**208b**, or L-threono-**211a** and L-erythrono-**211b**, respectively.

Several diastereoselective HDA (hetero-Diels–Alder) reactions of α,β-unsaturated carbonyl compounds and electron-rich alkenes have been exploited to gain carbohydrate derivatives with good diastereomeric excess. In HDA reaction, up to three chiral centers are formed with high stereoselectivity at each chiral carbon [78].

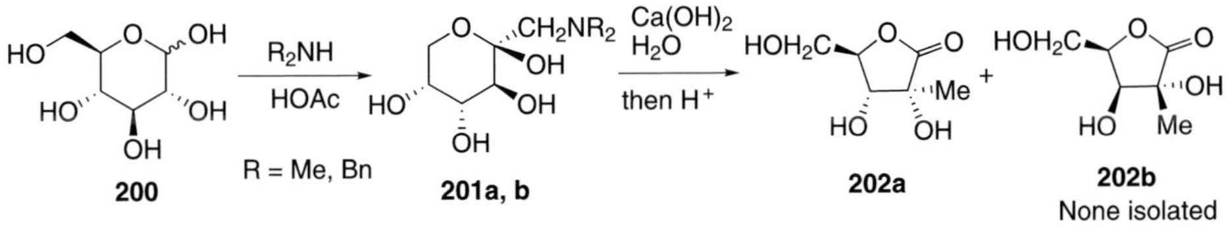

❑ Scheme 55

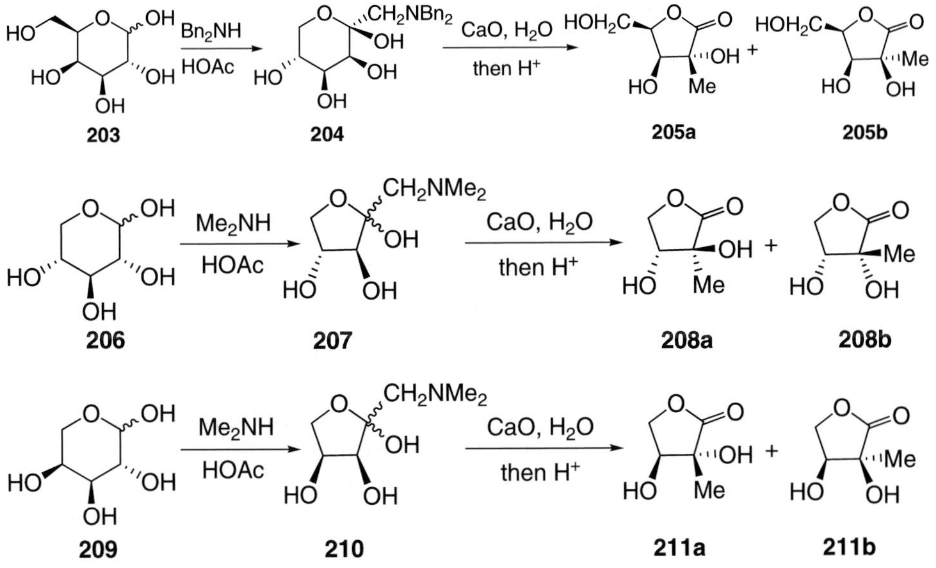

❑ Scheme 56

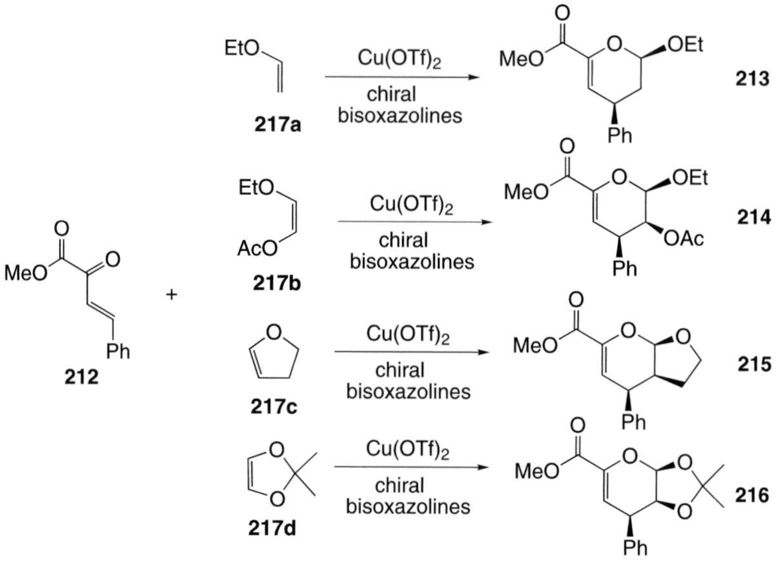

Scheme 57

Two independent groups demonstrated that the reaction between β,γ-unsaturated γ-keto esters and ethyl vinyl ether, in the presence of chiral bisoxazoline copper(II) complexes, led to enantiomerically enriched dihydropyrans which could be converted into attractive carbohydrate derivatives [79,80]. The representative reactions of (E)-2-oxo-4-phenylbut-3-enoic acid methyl ester 212 with ethyl vinyl ether 217a in the presence of different C_2-bisoxazoline ligands and copper(II) salts are presented in ❖ *Scheme 57*. The HDA reactions gave the dihydropyran 213 in very high yield (93–99%) with predominantly one diastereomer (de > 98% and ee > 99.5%). The same reaction with 212 and 217b–d was carried out smoothly to give dihydropyran moieties 214–216 with high diastereoselectivity (de > 95%) and enantioselectivity (up to 99.5% ee), respectively. These HDA products 213–216 are good synthons in the preparation of spirosugars and C-branched sugars.

6 Other Methods

Traditional syntheses of C-branched sugar analogs usually start from readily available chiral pool compounds, such as nucleosides or carbohydrates, by taking advantage of the already set or easily adjusted stereochemical relationships. However, such routes are frequently lengthy and laborious due to the multifunctional group sensitivities and extensive protecting group manipulations. In contrast, the asymmetric synthesis is extremely flexible for optimization. Because the upstream starting materials are usually small molecules, it is much easier to find a new reaction that provides a better way to a key intermediate [81].

Stereoselective iodolactonization of small achiral molecules is a very useful methodology to create a tetrahydrofuran framework leading to 3,'5'-C-branched carbohydrates [82]. Starting

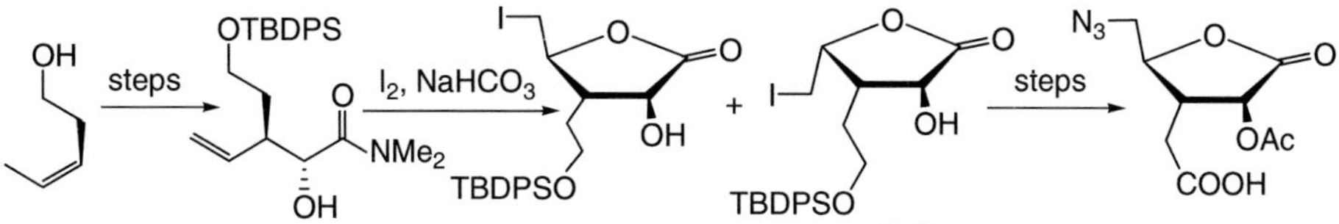

Scheme 58

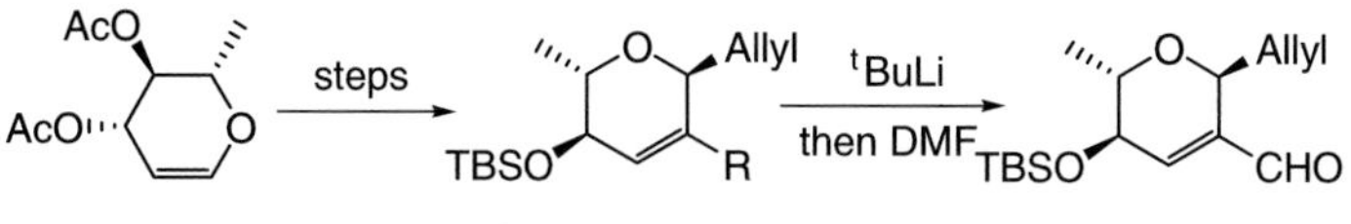

Scheme 59

from the commercially available (Z)-3-penten-1-ol **218**, iodolactonization of dimethyl amide **219** in THF in the presence of NaHCO$_3$ gave a mixture of 3,5-*trans* and *cis* iodolactones **220a** and **220b** in a ratio of 4:1. Attempts to optimize the reaction conditions, such as temperature, solvent, base, and amide substituents, did not improve the *trans*/*cis* ratio. Azide substitution of **220a** produced the enantiomerically pure ribonucleoside analog **221** (❷ *Scheme 58*).

Fraser-Reid [83] reported a two-step preparation of 2-*C*-branched sugar derivatives through halogen/metal exchange (❷ *Scheme 59*). 5-Bromo-3-(*tert*-butyldimethylsilyl)oxy-2,3-dihydro-2-methyl-6-(prop-2-en-1-yl)-6*H*-pyran **223a** was prepared in four steps from commercially available 3,4-di-*O*-acetyl-l-rhamnal **222**. All attempts to trap the vinyl lithium with a variety of halomethyl alkoxy electrophiles were unsuccessful. However, when a THF solution of **223a** was mixed with 5 equiv. of DMF and 6 equiv. of *tert*-butyllithium at −78 °C, the enal **224** was furnished quantitatively. It is worth mentioning that adding *tert*-butyllithium to the mixture, prior to mixing with DMF, led to low yields (33–69%) of enal **224**, with substantial formation of **223b**.

The employing of cyclic polyene, like cyclooctatetraene (COT) **225**, seems to be an esoteric and interesting synthetic approach aiming at the synthesis of hexoses and their branched analogues. Mehta reported the transformation of **225** into a rare sugar (DL)-β-allose and its C_2-branched sibling. Acetonide and TBS-protected cyclooctadienediol **226**, readily available from **225** in steps, was subjected to ozonolysis leading to the bicyclic hemiacetal **228** through the intermediacy **227**. Further modification of hydroxyl groups led to (DL)-methyl-2-deoxy-C_2-hydroxymethyl-β-allopyranoside **229** [84] (❷ *Scheme 60*).

Highly diastereoselective synthesis of C-3 branched deoxysugars has been studied by Shaw's group using Morita–Baylis–Hillman (MBH) reactions. The three-component reactions of aldehyde, sugarenone **230** (prepared from 3,4,6-tri-*O*-acetyl-D-glucal) and TiCl$_4$ were investigated under various reaction conditions. Gradually increasing reaction temperature from −78 °C to −30 °C, together with the adding of TBAI or Me$_2$S, obtained satisfactory yields and stereo outcomes [85].

Both aromatic and aliphatic aldehydes formed the products in good to excellent yields with almost complete diastereoselectivity (diastereoselectivity >99%). The adduct **231** underwent

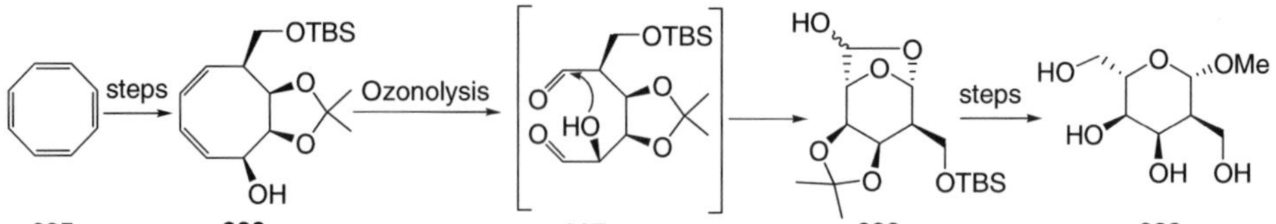

◻ Scheme 60

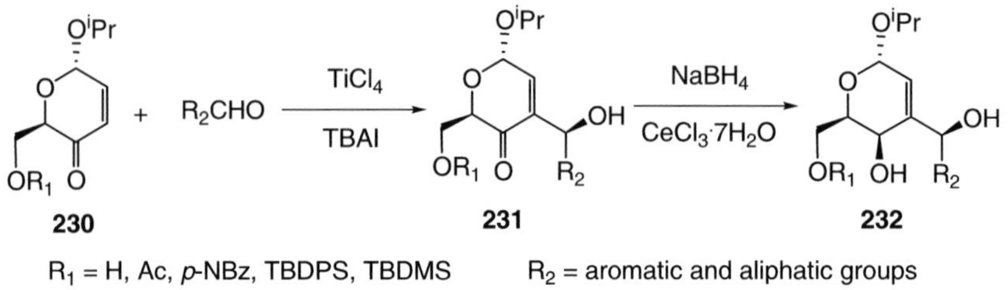

R₁ = H, Ac, p-NBz, TBDPS, TBDMS   R₂ = aromatic and aliphatic groups

R$_1$ = H, Ac, p-NBz, TBDPS, TBDMS   R$_2$ = aromatic and aliphatic groups

◻ Scheme 61

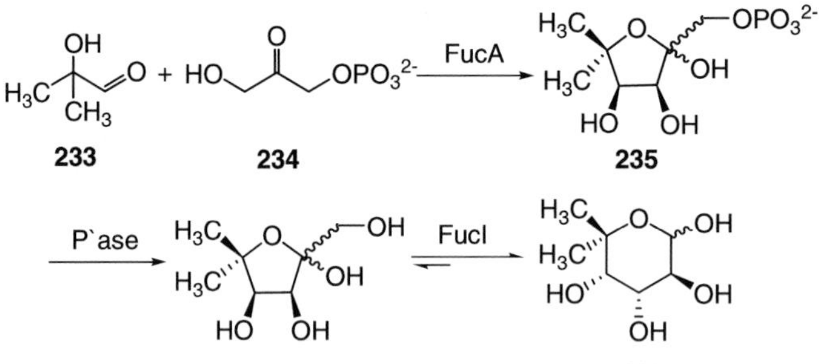

◻ Scheme 62

NaBH$_4$ reduction with the help of CeCl$_3$·7H$_2$O to obtain *threo* derivatives **232** (◉ *Scheme 61*). An enzymatic route for the synthesis of L-fucose analogs modified at the non-reducing end is reported by Fessner et al. [86]. Using 2-Hydroxy-2-methylpropanal **233** and dihydroxyacetone phosphate **234** as substrates, branched fucose derivative **237** has been prepared via recombinant L-fuculose 1-phosphate aldolase (FucA) and L-fucose ketol isomerase (FucI) in *E. coli* (◉ *Scheme 62*).

Zinc-mediated Barbier-type addition of **238**, followed by Luche's procedure obtained a mixture of the homoallylic alcohol diastereomers **239** and **240**. Alcohol **239** was carried through benzoylation, deketalization, silylation, and ozonolysis, to produce *C*-branched γ-lactone **241**. Benzoylation of **240** followed by hydroboration, PCC oxidation, debenzoylation, and alkaline-promoted cyclization directly formed *C*3-branching 2-deoxyfuranose **242** [87] (◉ *Scheme 63*). A novel method for stereoselective synthesis of 4'-α-carbon-substituted nucleosides, through epoxidation of 4',5'-unsaturated nucleosides and SnCl$_4$-promoted epoxy ring opening, was

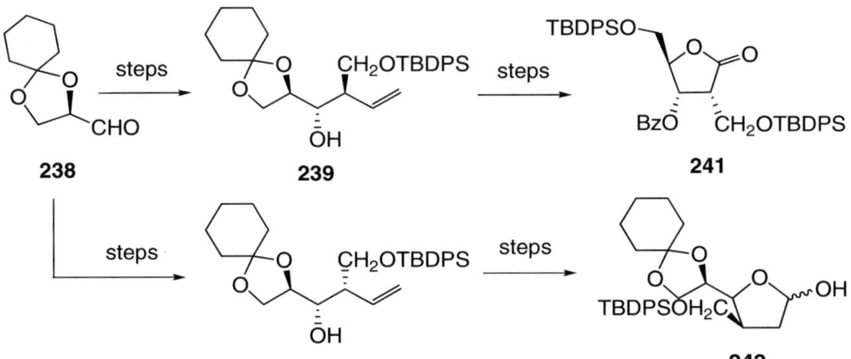

238 239 241 240 242

Scheme 63

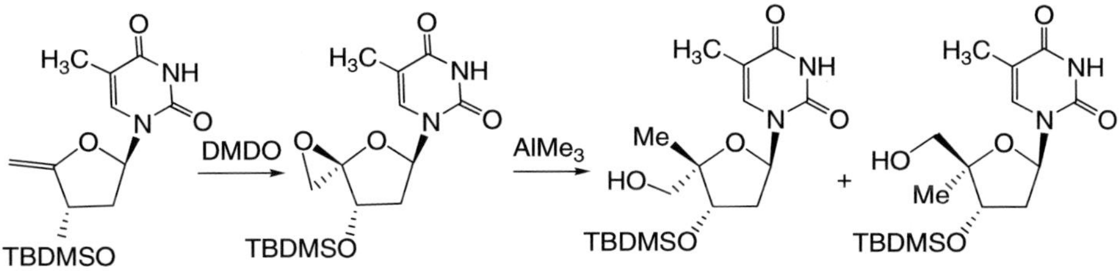

243 244 245 64% 246 5%

Scheme 64

reported by Haraguchi [88]. When 3-*O*-TBDMS-4,5-unsaturated thymidine **243** was treated with an acetone solution of DMDO (1.5 equiv.) at −30 °C, 4,5-epoxythymidine **244** was obtained as a single isomer. Ring-opening of **244** with Me_3Al (3 equiv.) afforded 4′-*β*-methyl isomer **245** (64%) and its isomer **246** (5%) (*Scheme 64*). However, reaction of **244** with allyltrimethylsilane (3 equiv.) and $SnCl_4$ (3 equiv.) gave a mixture of expected 4′-*β*-allylthymidine and byproduct 5′-*O*-trimethylsilyl derivative.

In conclusion, C–C bond formation on carbohydrates has absorbed lots of widely used methods from general synthetic organic chemistry. Typical among these methods are intramolecular alkylation and intramolecular condensation of aldehyde with enolates, phosphonates, and nitro-stabilized anions. Metal-mediated radical reactions, cycloadditions, and rearrangements have also been applied frequently.

Acknowledgement

Financial support from NNSFC (Project 20621703) is gratefully acknowledged.

References

1. Lipshutz BH, Sengupta S (1992) Org React 41:135

2. Giuliano RM, Villani FJ (1995) J Org Chem 60:202

3. Danishefsky SJ, Pearson WH, Harvey DF, Maring CJ, Springer JP (1985) J Am Chem Soc 107:1256

4. Gawley RE, Aubé J (1996) Principles of Asymmetric Synthesis. Pergamom Press, Oxford

5. Sato K, Yoshimura J (1982) Carbohydr Res 103:221

6. Smith GR, Villani FJ, Failli L, Giuliano RM (2000) Tetrahedron Asymmetry 11:139

7. Sørensen MH, Nielsen C, Nielsen P (2001) J Org Chem 66:4878

8. Dominguez C, Ezquerra J, Baker SR, Borrelly S, Prieto L, Espada M, Pedregal C (1998)Tetrahedron Lett 39:9305

9. Baker CH, Banzon J, Bollinger JM, Stubbe J, Samano V, Robins MJ, Lippert B, Jarvi E, Resvick R (1991) J Med Chem 34:1879

10. Ong SP, McFarlan SC, Hogenkamp HP (1993) Biochemistry 32:11397

11. Ye JD, Liao XG, Piccirilli JA (2005) J Org Chem 70:7902

12. (a) Girardet JL, Gunic E, Esler C, Cieslak D, Pietrzkowski Z, Wang G (2000) J Med Chem 43:3704;
    (b) Eldrup AB, Prhavc M, Brooks J, Bhat B, Prakash TP, Song Q, Bera S, Bhat N, Dande P, Cook PD, Bennett CF, Carroll SS (2004) J Med Chem 47:5284;
    (c) Franchetti P, Cappellacci L, Pasqualini M, Petrelli R, Vita P, Jayaram HN, Horvath Z, Szekeres T, Grifantini M, (2005) J Med Chem 48:4983

13. (a) Babu BR, Keinicke L, Petersen M, Nielsen C, Wengel J (2003) Org Biomol Chem 1:3514
    (b) Garcia I, Feist H, Cao R, Michalik M, Peseke K (2001) J Carbohydr Chem 20:681;
    (c) Herrera L, Feist H, Quincoces J, Michalik M, Peseke K (2003) J Carbohydr Chem 22:171;
    (d) Feist IOH, Herrera L, Michalik M, Quincocesd J, Pesekea K (2005) Carbohydr Res 340:547;
    (e) Li X, Uchiyama T, Raetz CRH, Hindsgaul O (2003) Org Lett 5:539

14. Nomura M, Sato T, Washinosu M, Tanaka M, Asao T, Shuto S, Matsuda A (2002) Tetrahedron 58:1279

15. Lena C, Mackenzie G (2006) Tetrahedron 62:9085

16. González Z, González A (2000) Carbohydr Res 329:901

17. Zhao S, Petrus L, Serianni AS (2001) Org Lett 3:3819

18. (a) Sigano DM, Peach ML, Nacro K, Choi Y, Lewin NE, Nicklaus MC, Blumberg PM, Marquez VE (2003) J Med Chem 46:1571;
    (b) Tamamura H, Bienfait B, Nacro K, Lewin NE, Blumberg PM, Marquez VE (2000) J Med Chem 43:3209

19. (a) Liu HM, Zhang F, Zhang J (2001) Carbohydr Res 334:323;
    (b) Liu HM, Zhang F, Zhang J, Li S (2003) Carbohydr Res 338:1737

20. Liu HM, Zhang F, Zou DP (2003) Chem Commun 2044

21. Koóš M, Mičová J, Steiner B, Alföldi J (2002) Tetrahedron Lett 43:5405

22. Gunic E, Girardet JL, Pietrzkowski Z, Eslerb C, Wang G (2001) Bioorgan Med Chem 9:163

23. (a) Håkansson AE, Koshkin AA, Sørensen MD, Wengel J (2000) J Org Chem 65:5161;
    (b) Koshkin AA, Fensholdt J, Pfundheller HM, Lomholt C (2001) J Org Chem 66:8504;
    (c) Koshkin AA (2004) J Org Chem 69:3711;
    (d) Meldgaard M, Hansen FG, Wengel J (2004) J Org Chem 69:6310

24. Montembault M, Bourgougnon N, Lebreton J (2002) Tetrahedron Lett 43:8091

25. Ohrui H, Kohgo S, Kitano K, Sakata S, Kodama E, Yoshimura K, Matsuoka M, Shigeta S, Mitsuya H (2000) J Med Chem 43:4516

26. Wu T, Nauwelaerts K, Aerschot AV, Froeyen M, Lescrinier E, Herdewijn P (2006) J Org Chem 71:5423

27. Andersch J, Bols M (2001) Chem Eur J 7:3744

28. Chakraborty C, Dhavale DD (2006) Carbohydr Res 341:912

29. Ji XM, Mo J, Liu HM, Sun HP (2006) Carbohydr Res 341:2312

30. (a) Sato K, Sekiguchi T, Hozumi T, Yamazaki T, Akai S (2002) Tetrahedron Lett 43:3087;
    (b) Sato K, Miyama D, Akai S (2004) Tetrahedron Lett 45:1523

31. Rauter A, Ferreira M, Borges C, Duarte T, Piedade F, Silva M, Santos H (2000) Carbohydr Res 325:1

32. Jones NA, Nepogodiev SA, MacDonald CJ, Hughes DL, Field RA (2005) J Org Chem 70:8556

33. Kutsumura N, Nishiyama S (2005) Tetrahedron Lett 46:5707

34. Robins MJ, Doboszewski B, Timoshchuk VA, Peterson MA (2000) J Org Chem 65:2939

35. Cordeiro A, Quesada E, Bonache MC, Velázquez S, Camarasa MJ, San-Félix A (2006) J Org Chem 71:7224

36. Yu J, Zhang S, Li Z, Lu W, Cai M (2005) Bioorg Med Chem 13:353

37. Hanessian S (1997) Preparative Carbohydrate Chemistry. Marcel Dekker, New York

38. Carcano M, Vasella A (1998) Helv Chim Acta 81:889

39. Dondoni A, Fogagnolo M, Medici A, Pedrini P (1985) Tetrahedron Lett 26:5477

40. Sanki AK, Suresh CG, Falgune UD, Pathak T (2003) Org Lett 5:1285

41. Du Y, Linhardt RJ, Vlahov IR (1998) Tetrahedron 54:9913

42. Horiuchi T, Ohta T, Shirakawa E, Nozaki K, Takaya H (1997) J Org Chem 62:4285

43. Kollar L, Sandor P (1993) J Organomet Chem 445:257 and references therein

44. Seepersaud M, Kettunen M, Abu-Surrah AS, Voelter W, Al-Abed Y (2002)Tetrahedron Lett 43:8607

45. Fernandez E, Ruiz A, Claver C, Castillon S (1998) Organometallics 17:2857

46. Sridhar PR, Ashalu KC, Chandrasekaran S (2004) Org Lett 6:1777

47. Li CJ (1993) Chem Rev, 93:2023 and references therein

48. (a) Li CJ, Chan TH (1999) Tetrahedron 55:11149;
(b) Chan TH, Isaac MB (1996) Pure Appl Chem 68:919

49. (a) Canac Y, Levoirier E, Lubineau A (2001) J Org Chem 66:3206
(b) Lubineau A, Canac Y, Goff N (2002) Adv Synth Catal 344:319;
(c) Levoirier E, Canac Y, Norsikian S, Lubineau A (2004) Carbohydr Res 339:2737

50. Prandi J (2001) Carbohydr Res 332:241

51. Ichikawa S, Minakawa N, Shuto S, Tanaka M, Sasaki T, Matsuda A (2006) Org Biomol Chem 4:1284

52. Yu M, Lynch V, Pagenkopf BL (2001) Org Lett 3:2563

53. Beyer J, Skaanderup PR, Madsen R (2000) J Am Chem Soc 122:9575

54. Norsikian S, Lubineau A (2005) Org Biomol Chem 3:4089

55. Fontana G, Lubineau A, Scherrmann MC (2005) Org Biomol Chem 3:1375

56. Lawrence AJ, Pavey JBJ, Chan MY, Fairhurst RA, Collingwood SP, Fisher J, Cosstick R, O'Neil IA (1997) J Chem Soc Perkin Trans 1:2761

57. Kittaka A, Tanaka H, Odanaka Y, Ohnuki K, Yamaguchi K, Miyasaka T (1994) J Org Chem 59:3636

58. Lim J, Choo DJ, Kim YH (2000) Chem Commun 553

59. (a) Kodama T, Shuto S, Nomura M, Matsuda A (2001) Chem Eur J 7:2332;
(b) Sukeda M, Shuto S, Sugimoto I, Ichikawa S, Matsuda A (2000) J Org Chem 65:8988

60. Sukeda M, Ichikawa S, Matsuda A, Shuto S (2003) J Org Chem 68:3465

61. (a) Ogalino J, Mizunuma H, Kumamoto H, Takeda S, Haraguchi K, Nakamura KT, Sugiyama H, Tanaka H (2005) J Org Chem 70:1684;
(b) Kumamoto H, Shindoh S, Tanaka H, Itoh Y, Haraguchi K, Gen E, Kittaka A, Miyasaka T, Kondo M, Nakamura KT (2000) Tetrahedron 56:5363

62. Keck GE, Enholm EJ, Yates BE, Wiley MR (1985) Tetrahedron 41:4079

63. Gomez AM, Casillas M, Valverde S, Lopez JC (2001) Tetrahedron Asymmetry 12:217

64. (a) Padwa A, Weingarten MD (1996) Chem Rev 96:223;
(b) Doyle MP, Forbes DC (1998) Chem Rev 98:911

65. Choe SWT, Jung ME (2000) Carbohydr Res 329:731

66. Baptiste LJ, Yemets S, Legay R, Lequeux T (2006) J Org Chem 71:2352

67. (a) Linker T, Hartmann K, Sommermann T, Scheutzow D, Ruckdeschel E (1996) Angew Chem Int Ed Engl 35:1730;
(b) Linker T, Sommermann T, Kahlenberg F (1997) J Am Chem Soc 119:9377

68. (a) Gyollai V, Schanzenbach D, Somsak L, Linker T (2002) Chem Commun 1294;
(b) Sommermann T, Kim BG, Peters K, Peters EM, Linker T (2004) Chem Commun 2624

69. Mikkelsen LM, Skrydstrup T (2002) J Org Chem 68:2123

70. (a) Shao HW, Ekthawatchai S, Chen CS, Wu SH, Zou W (2005) J Org Chem 70:4726;
(b) Shao HW, Ekthawatchai S, Wu SH, Zou W (2004) Org Lett 6:3497

71. Ana M, Enrique M, Bernardo H (2004) Eur J Org Chem 3063

72. (a) Krohn K, Gehle D, Flörke U (2005) Eur J Org Chem 2841;
    (b) Krohn K, Gehle D, Flörke U (2005) Eur J Org Chem 4557

73. Ghosh R, Chakraborty A, Maitia DK, Puranikb VG (2005) Tetrahedron Lett 46:8047

74. Hotchkiss DJ, Jenkinson SF, Storer R, Heinz T, Fleet GWJ (2006) Tetrahedron Lett 47:315

75. Hodge JE, Fisher BE (1963) Methods in Carbohydrate Chemistry, vol II. Academic Press, New York

76. Hou Y, Wu X, Xie W, Braunschweiger PG, Wang PG (2001) Tetrahedron Lett 42:825

77. Hotchkiss DJ, Soengas R, Booth KV, Weymouth-Wilson AC, Eastwich-Field V, Fleet GWJ (2007) Tetrahedron Lett 48:517

78. (a) Schmidt RR, Maier M, Tetrahedron Lett (1985) 26:2065;
    (b) Schmidt RR, Haag ZB, Hoch M (1988) Liebigs Ann Chem 885;
    (c) Dujardin G, Molato S, Brown E (1993) Tetrahedron Asymmetr 4:193
    (d) Tietze LF, Montenbruck A, Schneider C (1994) Synlett 509;
    (e) Tietze LF, Schneider C, Grote A (1994) Chem Eur J 2:139;
    (f) Dujardin G, Rossignol S, Brown E (1998) Synthesis 763

79. (a) Evans DA, Johnson JS (1998) J Am Chem Soc 120:4895;
    (b) Thorhauge J, Johannsen M, Jørgensen KA (1998) Angew Chem Int Ed 37:2404;
    (c) Evans DA, Olhava EJ, Johnson JS, Janey JM (1998) Angew Chem Int Ed Engl 37:3372;
    (d) Evans DA, Johnson JS, Olhava EJ, Janey JM (2000) J Am Chem Soc 122:1635

80. Audrain H, Thorhauge J, Hazell RG, Jørgensen KA (2000) J Org Chem 65:4487

81. (a) Robins MJ, Sarker S, Xie M, Zhang W, Peterson MA (1996) Tetrahedron Lett 37:3921;
    (b) Peterson MA, Nilsson BL, Sarker S, Doboszewski B, Zhang W, Robins MJ (1999) J Org Chem 64:8183;
    (c) Robins MJ, Doboszewski B, Timoshchuk VA, Peterson MA (2000) J Org Chem 65:2939

82. (a) Rozners E, Qun X (2003) Org Lett 5:3999;
    (b) Rozners E, Liu Y (2003) Org Lett 5:181;
    (c) Rozners E, Liu Y (2003) J Org Chem 70:9841;
    (d) Qun X, Rozners E (2005) Org Lett 7:2821

83. Fraser-Reid B, Chen XT, Haag D, Henry KJ, McPhail AT (2000) Chirality 12:488

84. Mehta G, Pallavi K (2004) Tetrahedron Lett 45:3865

85. Sagar R, Pant CS, Pathak R, Shaw AK (2004) Tetrahedron 60:11399

86. Fessner W-D, Goße C, Jaeschke G, Eyrisch O (2000) Eur J Org Chem 125

87. Chattopadhyay A, Goswami D, Dhotare B (2006) Tetrahedron Lett 47:4701

88. Haraguchi K, Takeda S, Tanaka H (2003) Org Lett 5:1399

2.6 C=C Bond Formation

Sławomir Jarosz, Marcin Nowogródzki
Institute of Organic Chemistry, Polish Academy of Sciences,
01–224 Warsaw, Poland
sljar@icho.edu.pl

Abstract

The material presented in this chapter describes the general methodology used for the preparation of unsaturated sugars. The 'older' methods (i. e. those being developed since at least the 1950s) which are still very useful and have general application are also presented but they are illustrated by newer examples. The direct formation of the double bond(s) is emphasized, but the methodology based on the rearrangement of unsaturated sugars into other olefinic carbohydrates is also reviewed.

The emphasis is placed on the general methods rather than synthesis of individuals, since this should give the reader a general view of the importance of unsaturated sugars and their application to stereocontrolled organic synthesis. All main classes of unsaturated sugars are described briefly. The anomeric unsaturated derivatives are excluded from this review unless they serve as starting materials in the preparation of other unsaturated carbohydrates via rearrangement reactions.

Keywords

Synthesis; Rearrangement; Unsaturated sugars; Sugar chirons; Targeted synthesis

In: *Glycoscience*. Fraser-Reid B, Tatsuta K, Thiem J (eds)
Chapter-DOI 10-1007/978-3-540-30429-6_8: © Springer-Verlag Berlin Heidelberg 2008

Abbreviations

DBU   1,8-diazabicyclo[5.4.0]undec-7-ene
DMSO   dimethyl sulfoxide
DMF   dimethylformamide
HCMV   human cytomegalovirus
NHDF   normal human fibroblasts
PTC   phase transfer catalysis
RCM   ring-closing metathesis
TIBAL tri-*iso*-butyl aluminum

1 Introduction

Sugar derivatives containing a double bond are an important class of compounds often used in stereocontrolled synthesis of optically pure targets. Their chemistry: preparation, reactions, and application in targeted synthesis is well developed and described in a number of monographs [1,2,3,4,5]. Therefore the classical methods, which have been introduced to the synthetic chemistry of sugars since (at least) the 1950s and are still very useful for their preparation will be presented in this report, which is an update of an earlier report [6]. The general methodology presented here will refer to the previous report rather than to original literature, but will be illustrated by more recent examples.

Only occasionally are unsaturated sugar derivatives such as blasticidin A (**1**), which inhibits blast disease in rice, found in nature. However, the non-natural unsaturated sugars often possess potent biological activity. The most representative example is, undoubtedly, 2,'3'-dideoxy-2,'3'-didehydro-thymidine d4T (Stavudine, Cerit®)—approved as the fourth *anti*-HIV drug by the US Food and Drug Administration—which exhibits an effect comparable to AZT in HIV-infected CEM cells (in vivo), is less toxic than AZT (the first drug used in an *anti*-HIV treatment) for bone marrow stem cells and was found to be less inhibitory to mitochondrial DNA replication [6,7]. Another example, neuraminic acid derivative **2**, is the anti-sialyl-inhibitory anti-influenza drug [1]. It is also known that the L-nucleosides are important components of antiviral agents; for example, L-2′3′-dideoxy-2′3′-didehydro-5-fluorocytidine (L-Fd4C) exhibits potent anti-HBV (antihepatitis B virus) activity (❷ *Fig. 1*) [8,9].

Generally two main classes of unsaturated sugars in which (1) the anomeric carbon atom is involved in a double bond (glycals) and (2) the unsaturation is placed between other carbon atoms are known. One has to discuss also derivatives with the *exo*- and *endo*-unsaturated bonds (either olefins/dienes or acetylenes). Another class of unsaturated monosaccharides is represented by the open-chain sugars with the double bond(s) and/or triple bond present in the molecule. The examples of different types of unsaturated sugars are shown in ❷ *Fig. 2*.

Compounds such as glycals will not be discussed here, unless they are used as substrates for the preparation of other, non-anomeric, unsaturated derivatives.

Unsaturated sugar chirons are versatile starting materials in organic synthesis. Many complex natural products whose structure incorporates five- and six-membered carbocyclic rings

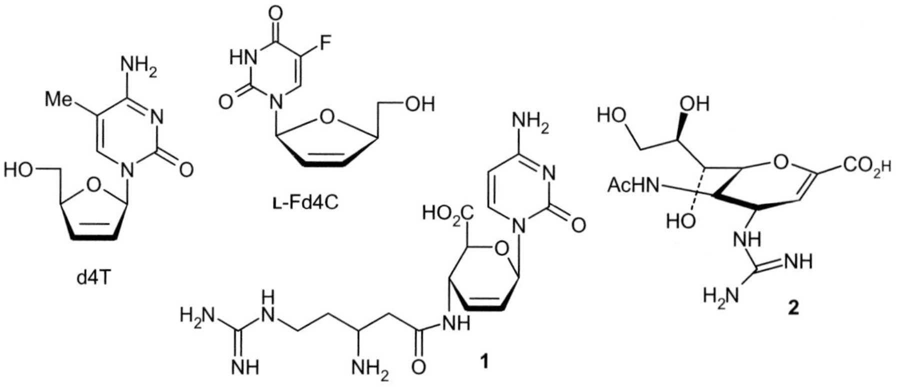

Figure 1
Examples of unsaturated sugars with potent biological activity

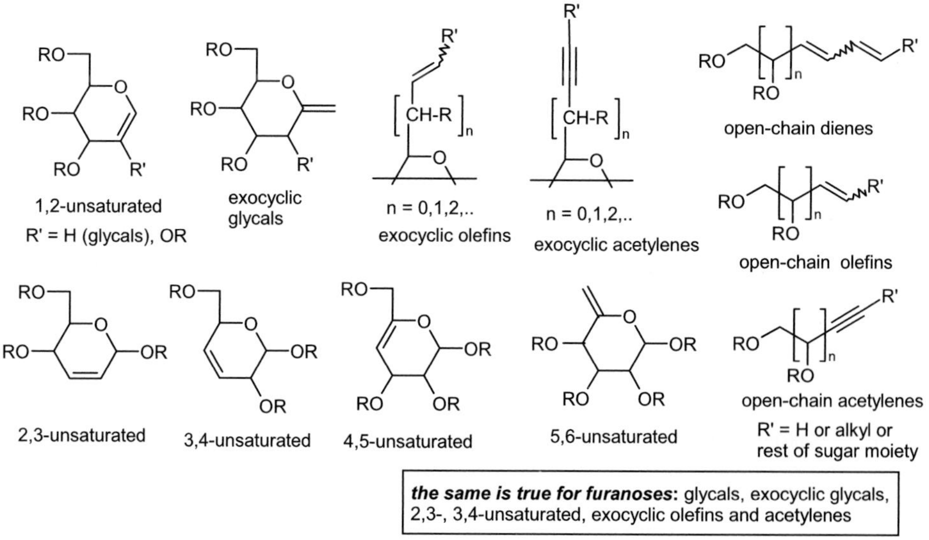

Figure 2
Different types of unsaturated sugars

have been synthesized from such readily available substrates [5,10]. The older methods, still used with great success in the synthesis of complex modified unsaturated carbohydrates, may be found in an excellent review by Ferrier [11]. For the newer ones see the recent reviews [1,2,4,6].

2 General Methods of the Formation of the Double Bond

The double bond may be introduced into the carbohydrate skeleton in several different ways. The methodology of the preparation of unsaturated sugars is well developed and almost any derivative can be prepared using classical methods described in a number of monographs and text-books. These methods include reduction of oxygen functions, elimination reactions, reactions of carbonyl compounds with the Wittig-type reagents, and rearrangements of a sugar skeleton (e. g. preparation of 2,3-unsaturated sugars from glycals). Carbohydrates are polyhydroxylated compounds and conversion of the *vic*-diol function into an olefin is the simplest and most obvious choice. Several different methods shown in ❯ *Scheme 1* were elaborated for such purposes. Direct conversion is achieved by treatment of a diol with triphenyl-phosphine and tri-iodoimidazole in refluxing toluene (Garegg's procedure, *route d*) [12]. Other common preparations require first activation of a diol followed by further transformation of such active intermediate(s) into olefins.

The di-*O*-tosylates (prepared by action of tosyl chloride in pyridine) are reduced with zinc (NaI/Zn; *route e*; Tipson–Cohen reaction) [13]. Cyclic ortho-esters (prepared by reaction of the diol with ethyl orthoformate) are transformed into olefins by simple heating in the presence of acids (Eastwood reaction, *route b*) [14]. Cyclic thiocarbonates (obtained by reaction of a diol with thiophosgene or *N,N'*-thiocarbonyl-di-imidazole) are reduced to olefin with trimethyl phosphite (Corey–Winter method, *route c*) [15]. Finally, reduction of vicinal di-xanthates with tri-*n*-butyltin hydride according to the Barton procedure [16] affords olefins via a reductive elimination process (*route a*). The Corey–Winter, Garegg, and Tipson–Cohen methods are most commonly applied for deoxygenation of sugar diols.

Epoxides and episulfides are also convenient precursors of unsaturated sugars. The first step involves nucleophilic opening of the three-membered ring to afford the intermediate from which the desired unsaturated derivative is obtained [1]. For example, conversion of the

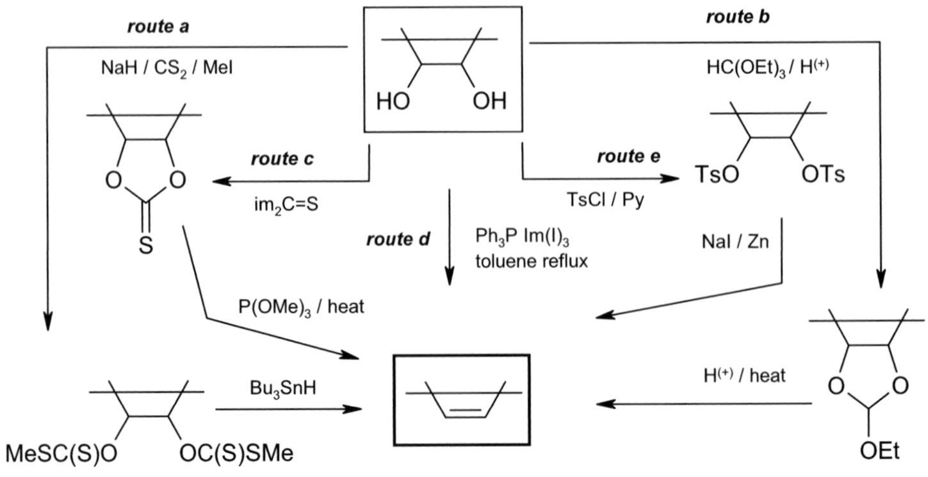

◻ Scheme 1

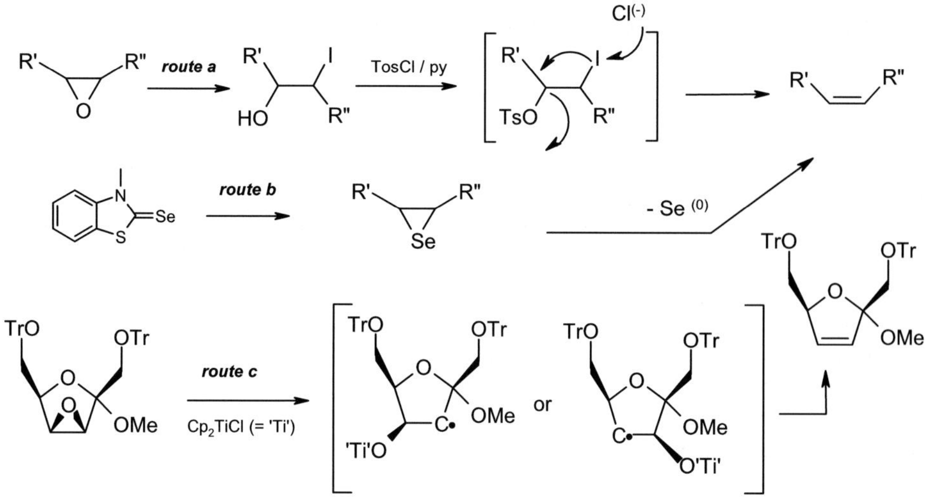

■ Scheme 2

epoxide into iodohydrin and subsequent reaction with pyridine\tosyl chloride affords appropriate olefinic sugars (*route a* ❷ *Scheme 2*). Reduction of oxiranes with selenium reagents, such as 3-methyl-2-selenoxo-1,3-benzothiazole, provides olefins in good yields (*route b*). Bis(cyclopentadienyl) titanium chloride induces a radical deoxygenation of epoxides; this titanium species is an extremely mild reducing agent, which may be illustrated by the fact that the olefin (obtained from the corresponding epoxide according to a mechanism shown in *route c*) could be isolated in good yield, although it is so unstable that even traces of acidic impurities present in CDCl$_3$ may cause its aromatization [6].

All these already presented general methods may be used for the construction of either *exo*- or *endo*-cyclic double bonds in a carbohydrate skeleton.

There are also special procedures which allow the preparation of sugars with only *endo*- or only *exo*-cyclic double bonds. The first group of compounds with *endo*-cyclic double bond may be prepared by total synthesis from non-carbohydrate precursors. The particularly useful hetero Diels–Alder reaction (❷ *Fig. 3*) allows one to obtain the dihydropyran skeleton either by reaction of a diene with a heterodienophile [3,17] or by reaction of a heterodiene with a 'normal' dienophile [18].

The approach proposed first by Zamojski was based on the thermal reaction of 1-methoxy-1,3-butadiene with a highly active heterodienophile (butyl glyoxylate), which provided a dihydropyran derivative—precursor of racemic monosaccharides [3,17].

Several years later Danishefsky introduced 1-methoxy-3-trimethylsilyloxy-1,3-butadiene, a highly reactive diene, which upon reaction with (not activated) aldehydes catalyzed with mild Lewis acids [(Eu(fod)$_3$] afforded cyclic α,β-unsaturated ketones [19]. Another method involved reaction of 1,4-di-alkoxy(acyloxy)-butadiene with an activated heterodienophile, which led to more functionalized derivatives (Schmidt) [20]. This methodology may be illustrated by the 'classical' synthesis of the precursor of purpurosamine B (**3**) and higher sugar

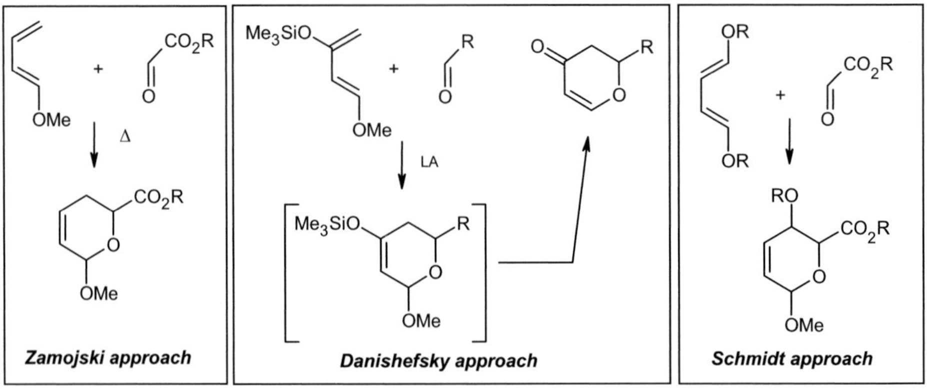

□ Figure 3
Synthesis of unsaturated sugars by a hetero Diels–Alder approach

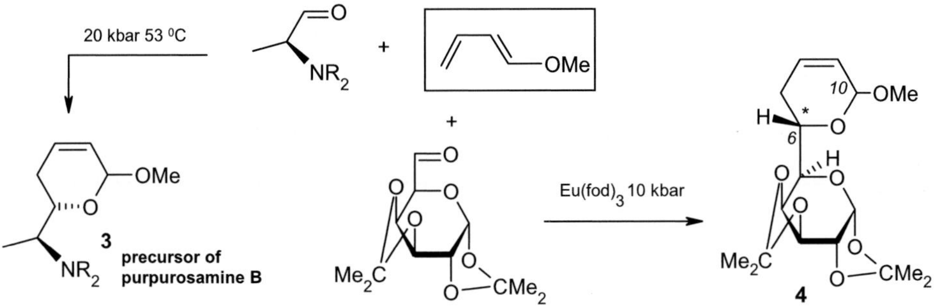

□ Scheme 3

dialdose **4** (❂ *Scheme 3*). The second reaction performed under 20 kbar pressure afforded the target with almost 100% diastereoselectivity. It was possible to lower the pressure (to 10 kbar) if the reaction was additionally catalyzed with mild Lewis acid [Eu(fod)$_3$] [6].

This process (hetero Diels–Alder reaction leading to a dihydropyran system) may be also conducted in an asymmetric version; application of chiral transition-metal catalysts based on BINOL, BINAP, bisoxazolines, etc. provides adducts in very high optical purity (*ee* up to 99%) [1,6]. In a series of papers Jurczak reported recently a highly enantioselective cycloaddition of 1-methoxy-1,3-butadiene and butyl glyoxylate catalyzed with chiral salen complexes [21].

The convenient precursors of unsaturated sugars, dihydropyran or dihydrofuran derivatives, can be also obtained from an appropriately functionalized furan molecule. Two main approaches are used. The first is based on the oxidative rearrangement of furfuryl alcohols (Achmatowicz reaction) [22], the second one involves an acid-catalyzed reaction of aldehydes with 2-trimethyl-silyloxyfuran (Casiraghi reaction) [23] (❂ *Scheme 4*).

In the original Achmatowicz approach, the (racemic) furfuryl carbinol is oxidized with bromine in the presence of methanol under weakly basic conditions to provide the Clauson-

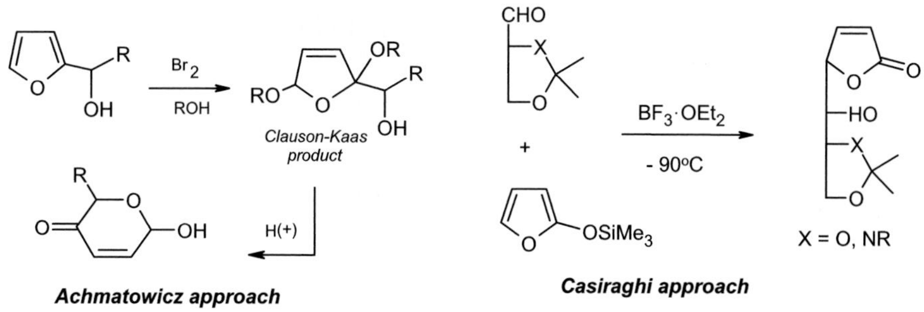

Achmatowicz approach

Casiraghi approach

X = O, NR

☐ **Scheme 4**

Kaas product, which is further hydrolyzed under rigorously controlled acidic conditions into the ald-2-enos-4-ulose derivative. Many other modifications [24] of the original Achmatowicz procedure such as oxidation of the furan ring with: *m*-CPBA, PCC, NBS, ^{t}BuOOH\VO(OAc)$_2$, or singlet oxygen are also used for this transformation [6].

Application of the Sharpless procedure for the kinetic resolution of racemic allyl alcohols [Ti(O*i*Pr)$_4$\$^t$BuOOH and L- or D-DET] to such a process provided an optically active dihydropyran together with enantiomerically pure unreacted furylcarbinol [6]. The Casiraghi approach leads to 2,3-unsaturated furanoses (or amino furanoses) by an acid-catalyzed reaction of 2-(trimethylsilyloxy)furan with sugar aldehydes or aminoaldehydes.

For introduction of the exocyclic double bond two methodologies are particularly useful. The first one is based on the reaction of the aldehydes (ketones) with the Wittig reagents (*route a*) [6]. In the second one (*route b*), an acetylenic functionality is introduced [25], which may be further converted either into the *Z*- or *E*-olefins by proper reducing reagents (H$_2$/ Lindlar catalyst or e. g. LiAlH$_4$, respectively) (❷ *Scheme 5*) [6].

All these general methods presented here will be illustrated by proper examples in the next sections of this chapter.

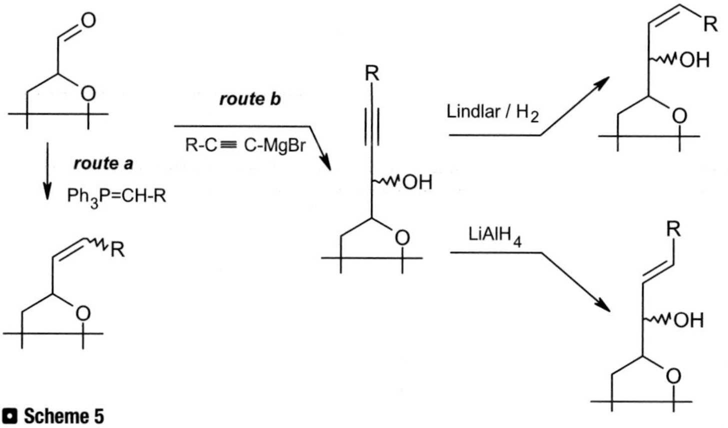

☐ **Scheme 5**

3 Monosaccharides with the Endocyclic Double Bond

In this section the selected and most representative examples (including also several older ones) of the preparation of unsaturated monosaccharides with the double bond present in the sugar ring will be described. The material will be divided into three main classes: 2,3-, 3,4-, and 4,5-unsaturated sugars in both pyranose and furanose form.

3.1 2,3-Unsaturated Monosaccharides

One of the most convenient methods for the synthesis of 2,3-unsaturated sugars is the so-called Ferrier-I rearrangement of glycals (❷ *Scheme 6*). This reaction allows one to obtain the unsaturated targets via the process in which the double bond migrates from the C1–C2 to the C2–C3 position [1,2,4,5,11].

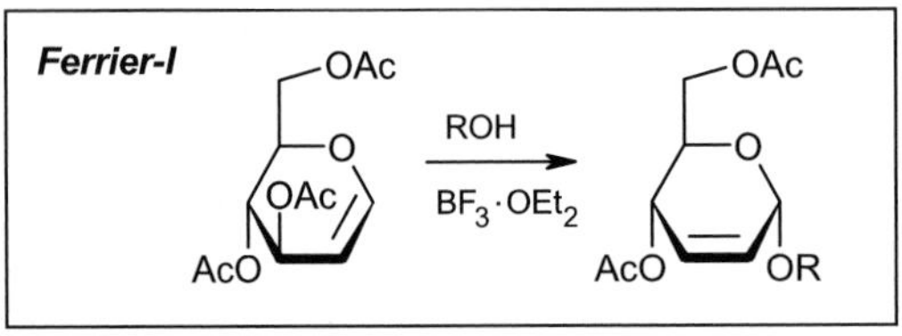

■ Scheme 6

It is not concerned, therefore, with the formation of the double bond and this process will be described in more detail in ❷ *Sect. 3.4 (rearrangement)*.

In this section only the direct synthesis of 2,3-unsaturated sugars, in which the double bond is created will be presented. Numerous syntheses of these valuable compounds have been described in the literature. One of the most important classes of such derivatives are 2,'3'-dideoxy-2,'3'-didehydronucleosides which possess a broad spectrum of biological activity. Two different approaches to 2,'3'-dideoxy-2,'3'-didehydro-thymidine (d4T) illustrate the general methodology of their preparation.

The first synthesis of d4T was accomplished in 1966 by Horwitz from 1-(2-deoxy-3,5-epoxy-β-D-*threo*-pentosyl)thymine; abstraction of a proton from the 2'-position by potassium ᵗbutoxide in DMSO resulted in an opening of the oxetane ring and formation of d4T. Other syntheses were based on the elimination of sulfoxide or the selenoxide moiety placed at the C-2' position in the nucleoside, which was prepared from cheap lactone **6** (❷ *Scheme 7*) [6].

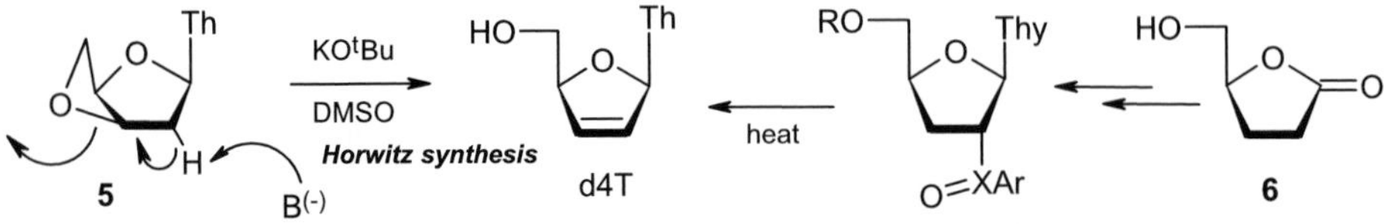

■ Scheme 7

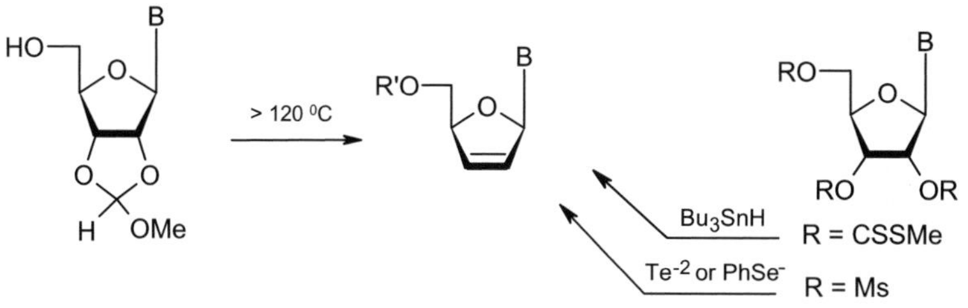

☐ Scheme 8

The 2,'3'-double bond in unsaturated nucleosides may be created also by thermal degradation of *ortho*-esters, reduction of appropriate di-xanthates with tri-*n*-butyltin hydride [26], or reductive elimination of dimesylates caused by telluride or selenide anions (❷ *Scheme 8*) [6,27]. The β-elimination reaction is a useful method for the preparation of unsaturated sugars. For example, treatment of the sugar lactone **7** (readily available from the corresponding glycal by the Ferrier-I rearrangement followed by standard transformations) with a strong base generates the anion from which the benzoate is eliminated providing the 2,3-unsaturated derivative **8** (❷ *Scheme 9*) [28].

Base-catalyzed isomerization of epoxides into allylic alcohols is a method of choice for the preparation of unsaturated sugars; olefin **10** was prepared by treatment of oxirane **9** with a strong base. This process is similar to the Horwitz synthesis of d4T presented in ❷ *Scheme 7*. The Tipson–Cohen methodology is frequently applied for the synthesis of 2,3-unsaturated sugars as shown in ❷ *Scheme 10*; such reduction might be performed also under microwave irradiation [6].

Formation of 2,3-unsaturated sugars was illustrated by the representative examples. Application of any other general methods which are described in ❷ *Sect. 2* should be considered when planning the synthesis of such derivatives.

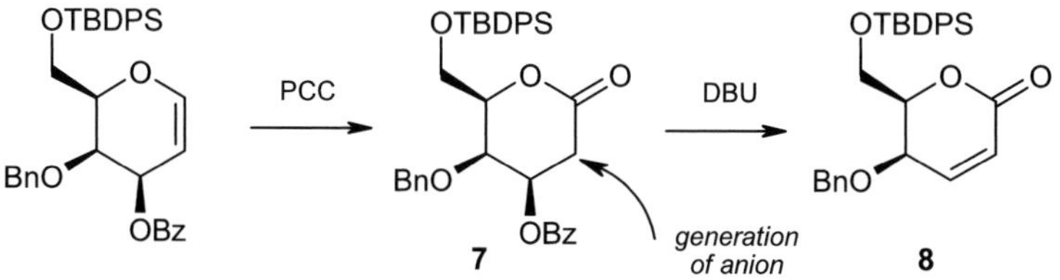

☐ Scheme 9

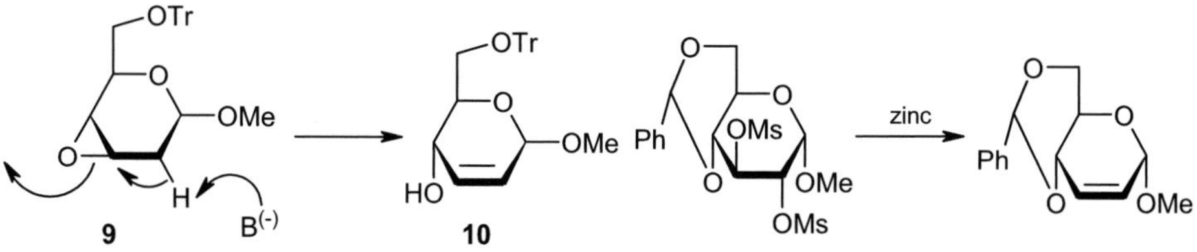

☐ Scheme 10

3.2 3,4-Unsaturated Monosaccharides

Synthesis of 3,4-dideoxy-3,4-didehydro-sugars can be conveniently performed by the Cohen–Tipson reduction of appropriate sulfonates. For example, methyl 3,4-dideoxy-3,4-didehydro-α-D-*erythro*-hex-3-enopyranoside is obtained from methyl α-D-glucopyranoside by selective protection of the 2- and 6-OH groups as benzoyl esters, subsequent tosylation (or mesylation) of the remaining 3,4-hydroxyl groups, and reduction of such di-sulfonic ester with zinc in refluxing DMF. The same target can also be obtained from methyl 2,6-di-*O*-benzoyl-3,4-di-*O*-mesyl-β-D-galactopyranoside; the latter reaction was performed under microwave irradiation to facilitate the process (❷ *Scheme 11*) [6].

Another example of application of the elimination process in the synthesis of these useful derivatives is shown in ❷ *Scheme 12*. Derivative **14** was obtained in high yield from 1,5-anhydro-D-fructose **13** via regioselective elimination of acetic acid [29].

Further illustration of this useful process is provided in ❷ *Scheme 13*. It was based on a different reactivity of the hydroxyl group in carbohydrates. Selective protection of the 3-OH in **15** leaves the 2-OH free. Oxidation of this hydroxyl with the Swern reagent affords a ketone from which the benzoic acid is eliminated leading to the 2,3-unsaturated compound **16**. Reduction of the carbonyl group is followed by spontaneous migration of the ester group from C-1, thus affording a free sugar **17** [1].

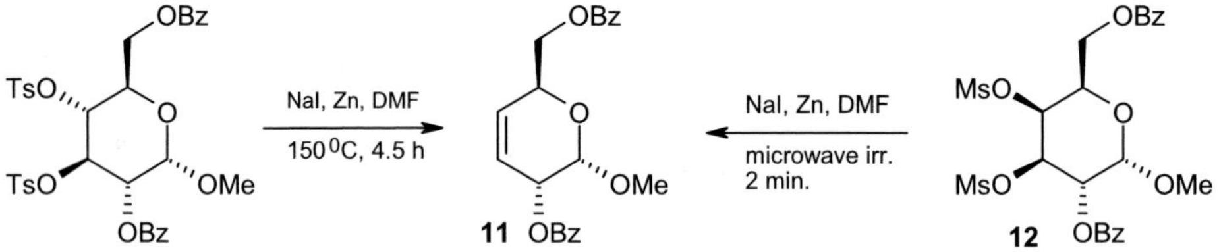

❏ Scheme 11

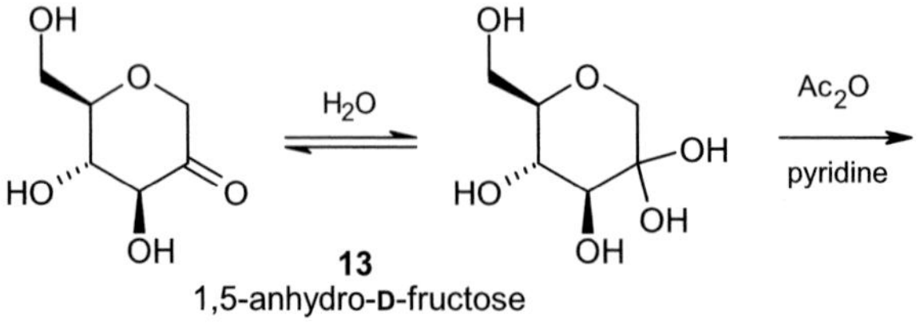

❏ Scheme 12

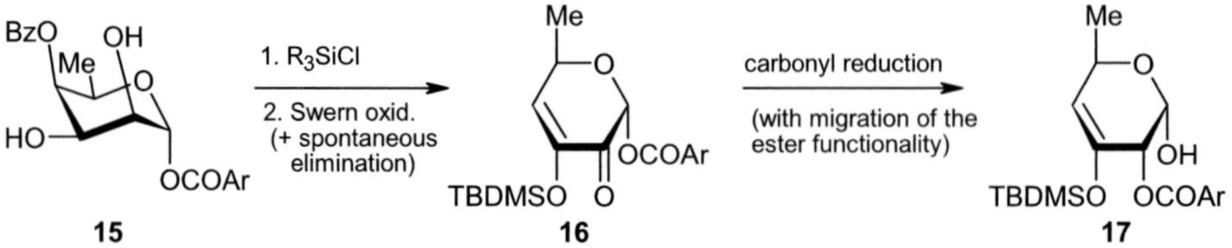

❏ Scheme 13

3.3 4,5-Unsaturated Monosaccharides

The 4,5-unsaturated pyranoses or 3,4-unsaturated furanoses are usually prepared by a base-catalyzed elimination of a leaving group such as halogen, sulfonyloxy etc. from appropriate sugar derivatives. The classical example is represented by the synthesis of compound **19** from 1,2:5,6-di-*O*-isopropylidene-α-D-glucofuranose **18** [1]. Analogous elimination performed for **20** led to derivative **21** from which capuramycin—a complex nucleoside antibiotic could be prepared readily (❷ *Scheme 14*) [30].

Elimination of the substituent from the β-position with respect to the electron-withdrawing group is often observed and leads to appropriate unsaturated products. Some natural glycopyranuronate conjugates with the characteristic 4-*O*-glycosidic bond are degraded to 4,5-unsaturated carbohydrates with liberation of sugar(s) sub-unit(s) by enzymes or in a proton-poor media (❷ *Fig. 4*) [6].

An example of chemically induced β-elimination is shown in ❷ *Scheme 15*. Treatment of the 3,4-*O*-isopropylidene-D-galactoside with a strong base induced the elimination of acetone with formation of the 3,4-unsaturated sugar [31].

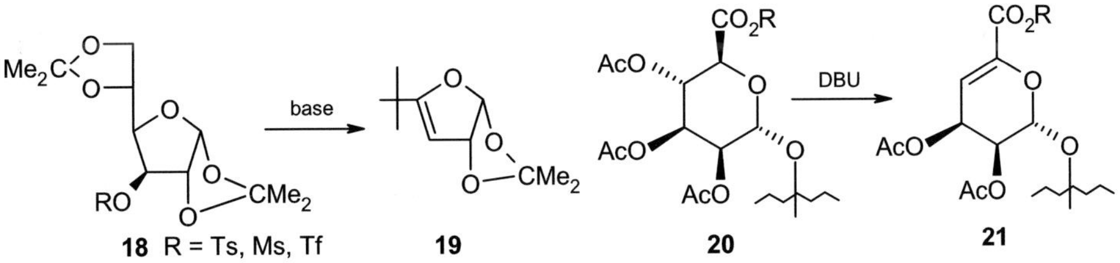

❏ Scheme 14

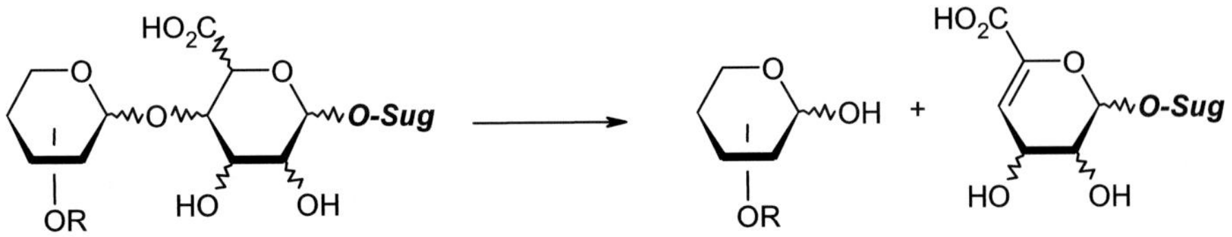

❏ Figure 4
Degradation of oligo- and polysaccharides by enzymes in proton-poor media

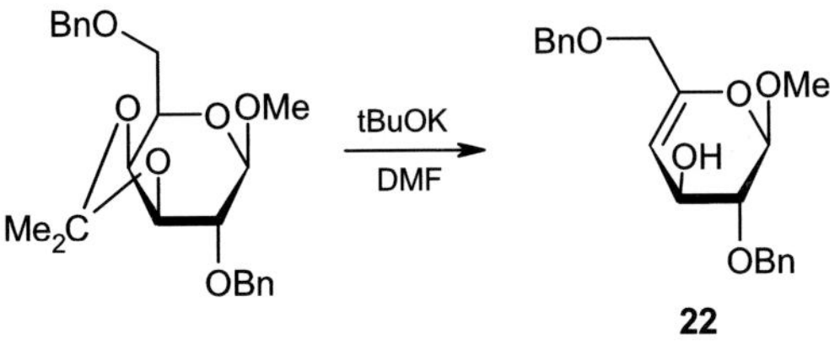

22

❏ Scheme 15

3.4 Rearrangement Reactions of Unsaturated Sugars

Rearrangements of unsaturated sugars are not concerned with a *direct formation* of a C=C bond, but with a migration of a double bond from one position to another. Surely, the most useful process of this type is the so-called Ferrier rearrangement [3,32], i. e. reaction of 1,2-unsaturated pyranoses or furanoses with nucleophiles in the presence of Lewis acids leading to 2,3-unsaturated sugars. In the original procedure, tri-*O*-acetyl-D-glucal (**23**; X = H) was treated with ethanol in the presence of boron trifluoride etherate which effected the addition of an alcohol at the C1 center with a simultaneous shift of the C1–C2 double bond to the C2–C3 position and elimination of acetic acid providing **24** (X = H). Other variants of this reaction, such as the Pd-catalyzed process [33] or microwave irradiation [34] allow one to obtain the desired products under milder reaction conditions. The 2-acyloxyglycals **23a** react with alcohols at elevated temperature to afford 2-substituted products **24** (X = Ac) [35]. When glycals are treated with alcohols in the presence of palladium chloride the corresponding 3,4-unsaturated sugars **25** are formed in good yield (❯ *Scheme 16*) [36].

The Ferrier-I rearrangement represents now a classical methodology which should be considered in planning the syntheses of optically pure targets from sugar chirons. Such a conclusion may be illustrated by the preparation of pyranosyl nucleosides obtained by this procedure.

For example, treatment of di-*O*-acetyl-L-rhamnal **26** with silylated pyrimidine or purine bases in the presence of a mild acidic catalyst (trityl perchlorate) afforded 2,3-unsaturated pyranosyl nucleosides, as a mixture of α,β-anomers **27** (with the β-anomer predominating) [37]

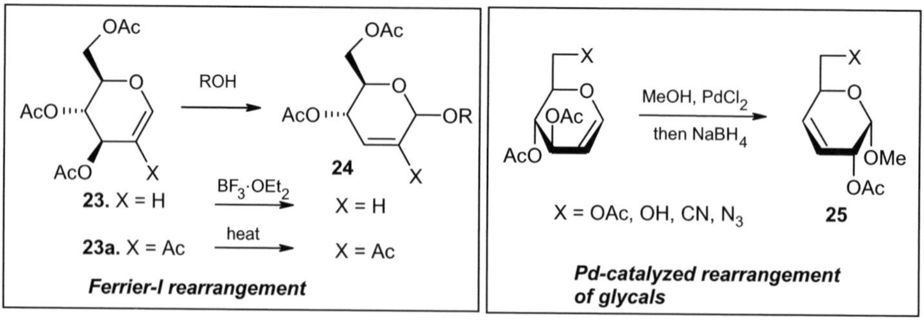

☐ Scheme 16

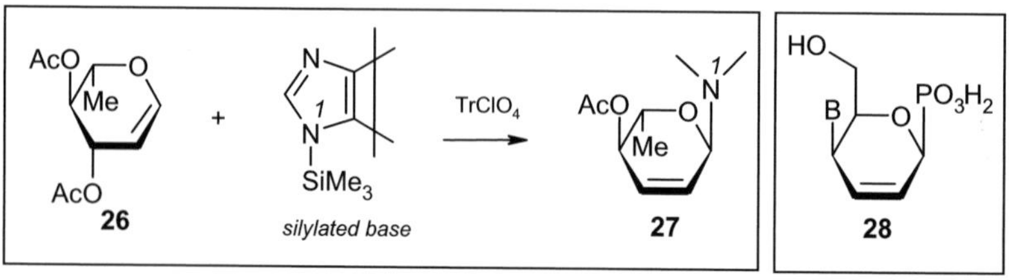

☐ Scheme 17

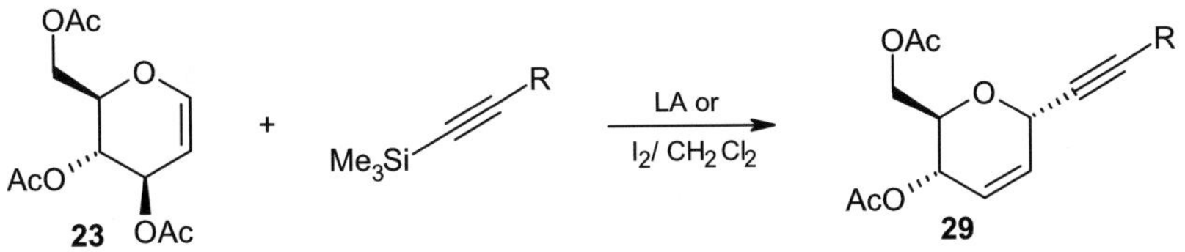

Scheme 18

(● *Scheme 17*). In the synthesis of phosphorylated pyranosyl nucleosides (which inhibit replication of HCMV in NHDF cells) such as **28**, the key step [38] consisted of the Ferrier-type coupling of glycals with alkyl phosphites catalyzed with boron trifluoride etherate.

Ferrier reaction is applicable also to many other nucleophiles (thiols, amines), which allow one to prepare a wide variety of useful 2,3-unsaturated products, many of which—especially 2,3-unsaturated furanoses—possess interesting biological properties. Introduction of the diphenylphosphine oxide group by Ferrier rearrangement of glycals was reported [39]. A particularly interesting process is represented by reaction of glycals with *C*-nucleophiles (such as e. g. allylsilanes, silylacetylenes, etc.), which directly provides the 2,3-unsaturated *C*-glycosides [1,2]. For example, addition of acetylenic species to glycal **23** (with its rearrangement) provides the 2,3-unsaturated-1-*C*-sugar acetylenes **29** in good yields (● *Scheme 18*) [40].

Another useful process applied in the synthesis of target unsaturated carbohydrates are sigmatropic rearrangements as for example the [3,3] rearrangement of **30** (prepared from appropriate alcohol and trichloroacetonitrile) leading to the amine **31** [41]. The [3,3] rearrangement of

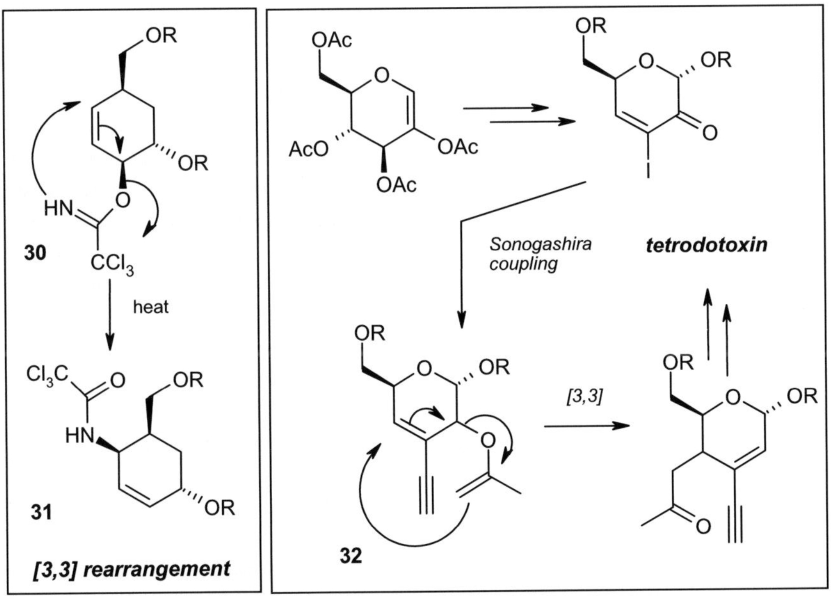

Scheme 19

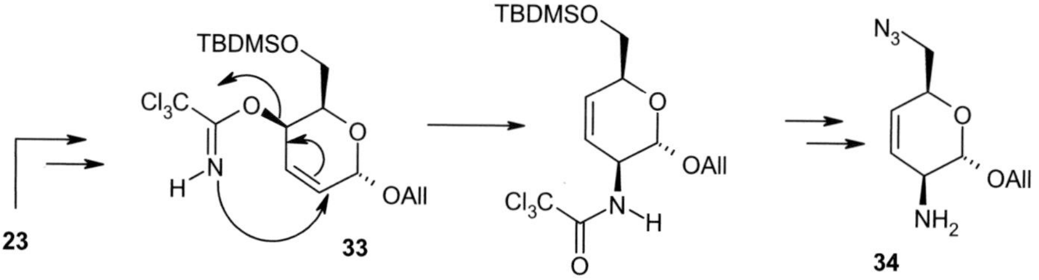

Figure 5
Synthesis of 3,4-unsaturated aminosugars via rearrangement of 2,3-unsaturated sugars

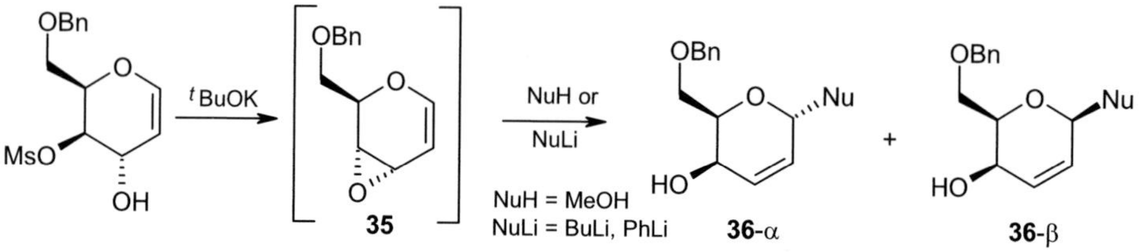

Scheme 20

unsaturated compound **32** was a key step in the synthesis of tertrodotoxin (a toxic principle of puffer fish poisoning) realized recently by Isobe (❯ *Scheme 19*) [42].

The key-step in the synthesis of a unsaturated sugar **34** bearing two amino groups (a scaffold which were used for the preparation of peptides with aromatic rings) was realized by the rearrangement of the adduct **33** readily prepared from D-glucal **23** (❯ *Fig. 5*) [43].

Another interesting example of the preparation of unsaturated *O*- and *C*-glycosides **36** was presented recently [44]. This was based on the S$_N$2′ attack of the nucleophile on unsaturated epoxide **35**. Substitution at C-1 proceeded with simultaneous migration of the double bond and opening of the three-member ring providing 2,3-unsaturated derivative **36** (❯ *Scheme 20*). Conversion of 4-deoxy-2,3-unsaturated sugars into 3,4-unsaturated ones reported by Banaszek and Zamojski already in 1972 [3] was one of the key-steps in total synthesis of monosaccharides by a hetero Diels–Alder approach.

Epoxide **37** (either of possible stereoisomers) was converted into the appropriate aminosugar (by reaction with aqueous dimethylamine), which was further oxidized with hydrogen peroxide to the *N*-oxide **38** and transformed into the 3,4-unsaturated sugar **39** via the thermal Cope-type rearrangement. A milder procedure for the preparation of derivative **39** was proposed by David [45]; after opening of the oxirane ring with a phenylselenide anion, the resulting sugar selenide was oxidized in situ with H$_2$O$_2$ to the selenoxide **40**, which underwent a smooth conversion into allylic alcohol **39** (R′ = H) (❯ *Scheme 21*).

Another example of the transposition of the C2–C3 double bond into the C3–C4 position was reported by Lubineau [46]. Tri-*O*-acetyl-D-glucal **23** was converted in three standard steps into 2,3-unsaturated sugar **41**. The modified Appel reaction provided bromide **42** which reacted with aldehyde in the presence of metallic indium affording the 3,4-unsaturated product **43** (❯ *Scheme 22*).

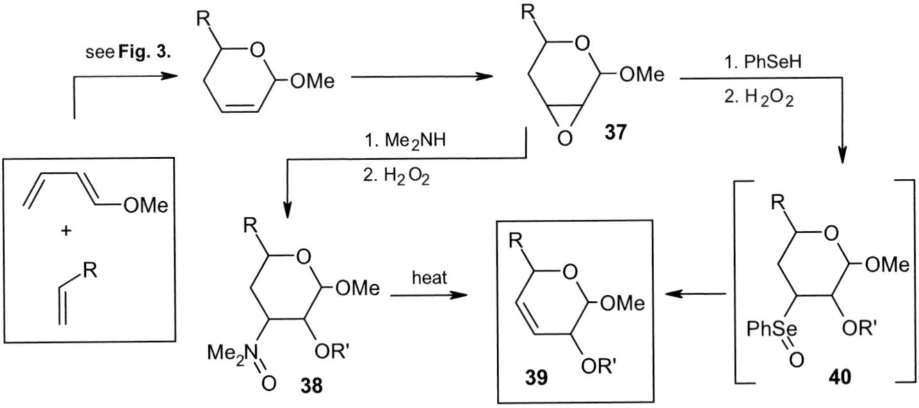

☐ Scheme 21

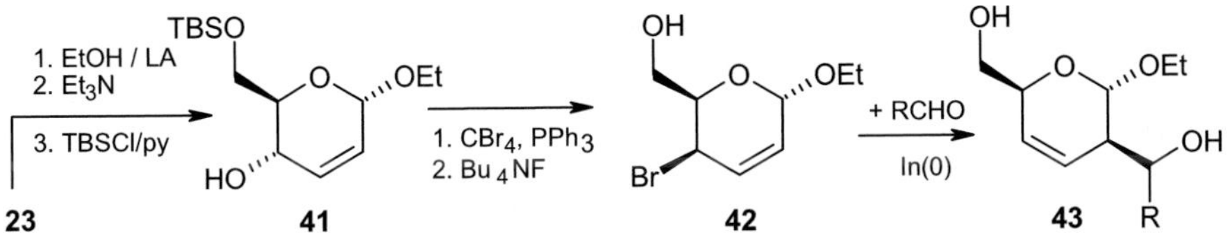

☐ Scheme 22

Synthesis of unsaturated sugars with the strongly electron-withdrawing trifluoromethyl group was initiated also from the tri-O-acetyl-D-glucal (**23**). Radical addition of the difluoro-rochloromethyl fragment to a double bond afforded two products **44** and **45**; reaction of the former with cesium fluoride furnished 2-trifluromethyl-2-eno-saccharide **46** (❷ *Scheme 23*) [47]. The transposition of the exocyclic double bond into the C2′–C3′ position leading to unsaturated nucleosides is presented also in ❷ *Scheme 24*. The S$_N$2′ reaction of **47** with lithium azide led to 2′-azidomethylene derivative **48**, while rearrangement of phosphinate **50** (prepared from **49** and Ph$_2$PCl) caused by iodine afforded **52** (via transition state **51**) [6].

The selected examples (both older and newer ones) shown in this part should emphasize to the reader the problem of planning and execution of the synthesis of unsaturated sugars via

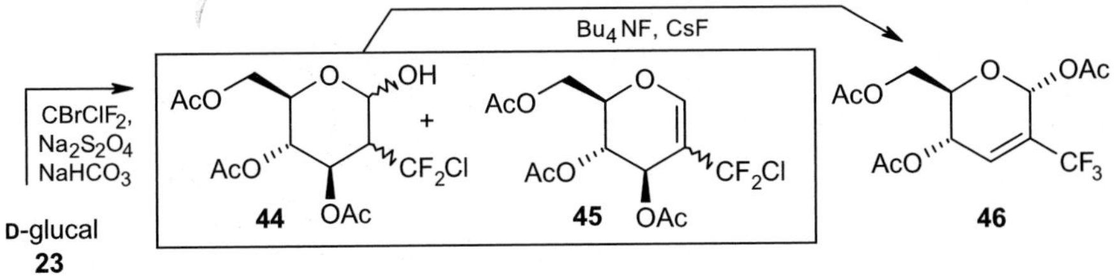

☐ Scheme 23

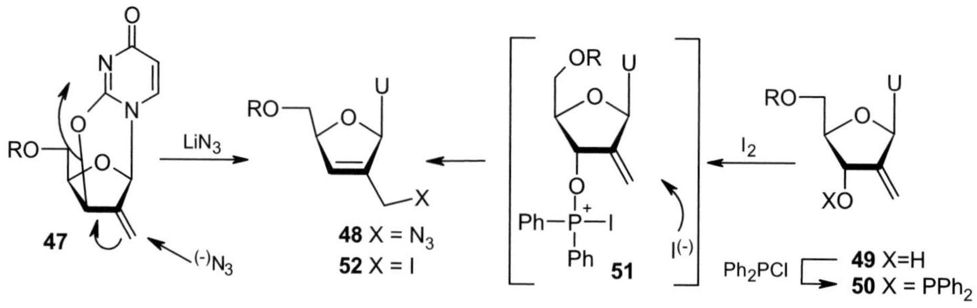

◻ Scheme 24

rearrangement reactions. This may also be illustrated by other syntheses and the reader can find more examples in the literature.

4 Monosaccharides with the Exocyclic Double Bond(s)

The Wittig (or Wittig-type) reaction is one of the most frequently used methods for the formation of a new carbon–carbon double bond. The terminal olefinic bond is usually constructed by reaction of aldehydes with methylenetriphenylphosphorane ($Ph_3P=CH_2$) or the dialkyl methyl-phosphonate anion. Both methods are applied for the preparation of unsaturated sugars [6]. However, serious problems are sometimes encountered using $Ph_3P=CH_2$, because of the rather high basicity of this reagent (which may cause undesired rearrangements of the parent carbonyl compounds) and moderate nucleophilicity. The ylids may be replaced with more reactive and less basic reagents such as Tebbe's reagent [48] or $CH_2Br_2/Zn/TiCl_4$ [49] (these reagents convert also esters into alkoxy-olefins). For the elongation of sugars by a C_2-unit appropriate aldehydes are treated with stabilized phosphoranes ($Ph_3P=CHCOR$), which leads to the α,β-unsaturated derivatives. Sugar ketones, being less reactive than aldehydes very often do not react with stabilized phosphoranes. More nucleophilic phosphonate anions might be, therefore, applied to overcome this problem as illustrated by synthesis of *exo*cyclic unsaturated compound **53**, an intermediate in the Hanessian synthesis of thromboxane (**❷ Scheme 25**) [6]. α-Alkoxy-α,β-unsaturated sugars are conveniently obtained by reaction of the corresponding aldehydes (ketones) with alkoxyphosphoranes [50] [$Ph_3P=C(OR)(CO_2R)$] or phosphonates [51] [$(MeO)_2P(O)CH(OR)CO_2R$].

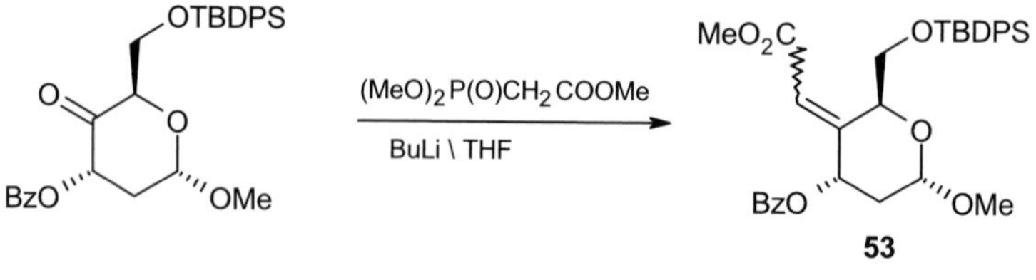

◻ Scheme 25

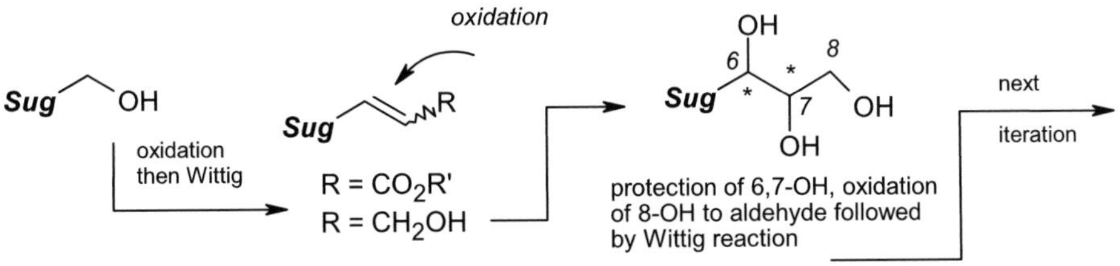

Figure 6
Brimacombe methodology of the synthesis of higher carbon sugars

The Wittig reaction is a common methodology applied in the preparation of higher carbon sugars. In the mid 1980s Brimacombe proposed a general synthesis of such compounds by a two carbon atom elongation of the parent monosaccharide, followed by functionalization of the resulting double bond [52]. By this iterative elongation Brimacombe was able to prepare decoses (● *Fig. 6*).

A more convenient method for higher sugar synthesis consists, undoubtedly, of the coupling of two properly activated monosaccharide subunits. Application of unstabilized sugar-derived phosphoranes was proposed by Secrist [53]. The phosphonium iodide upon treatment with a base should generate the ylid, which might either decompose in a β-elimination process or react with an aldehyde partner to afford the *cis*-Wittig adduct. Indeed, the precursor of a higher sugar antibiotic hikozamine was obtained according to this general idea under rigorously controlled conditions (● *Scheme 26*). This methodology was also applied for the synthesis of tunicamine [54], a backbone of another higher sugar antibiotic—tunicamycin.

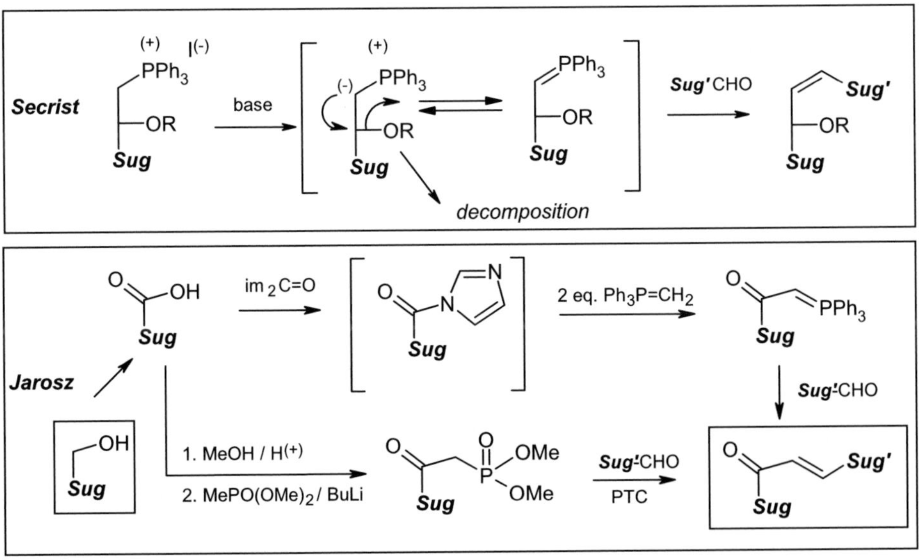

Scheme 26

A general methodology for the preparation of higher carbon sugars was proposed by Jarosz in the mid 1980s [6,55,56]. Protected monosaccharide with the terminal OH free was oxidized to an acid, which was further converted either into phosphorane or phosphonate. Upon reaction with another sugar synthon (aldehyde), α,β-unsaturated higher sugar enones were formed in good yields (➊ *Scheme 26*). The second approach is more versatile, since the phosphonates are more reactive and can be prepared in much higher yields than the corresponding phosphoranes [6,55].

This methodology may be also successfully applied to complex sugars. For example, selectively protected sucrose **54** with the 6-OH free was converted into the corresponding phosphonate, which upon reaction with '*diacetonogalactose*' aldehyde provided the higher sucrose precursor **55** in good yield (➊ *Fig. 7*) [57].

Another general approach leading to higher carbon sugars utilized acetylenic precursors [55, 58]. Conversion of an aldehyde into sugar acetylene was achieved using Corey's methodology [59] (Ph$_3$P/CBr$_4$, then a base). Treatment of the acetylene **56** with tributyltin hydride under the radical conditions afforded the *E*-vinyltin **57** (➊ *Scheme 27*).

Replacement of the tin moiety with lithium (this reaction proceeds with the retention of the configuration at the double bond) followed by reaction with sugar aldehyde afforded two diastereoisomeric higher sugar allylic alcohols **58** with the *E*-geometry across the double bond.

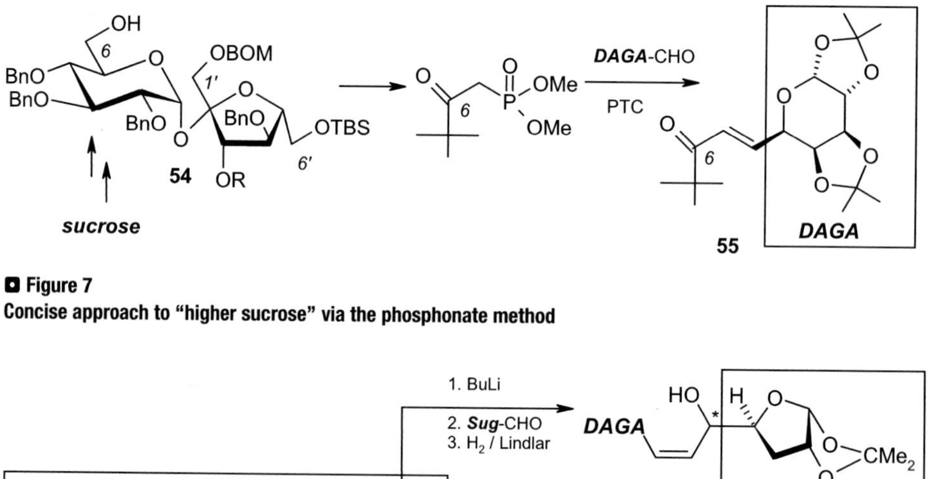

Figure 7
Concise approach to "higher sucrose" via the phosphonate method

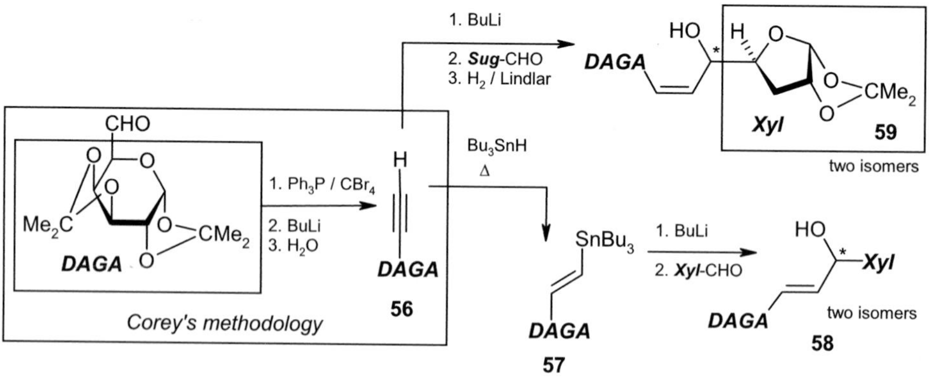

Scheme 27

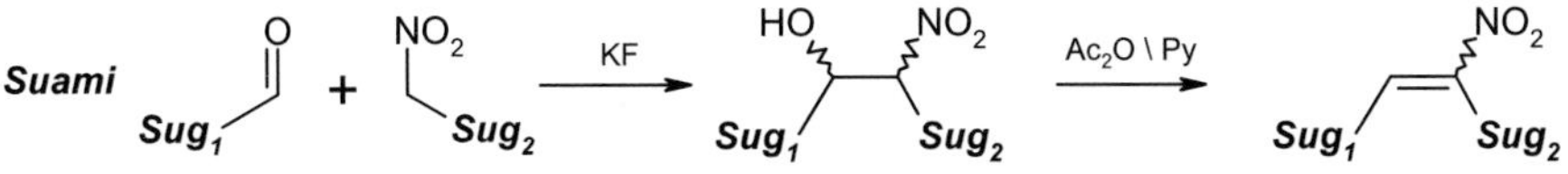

■ Scheme 28

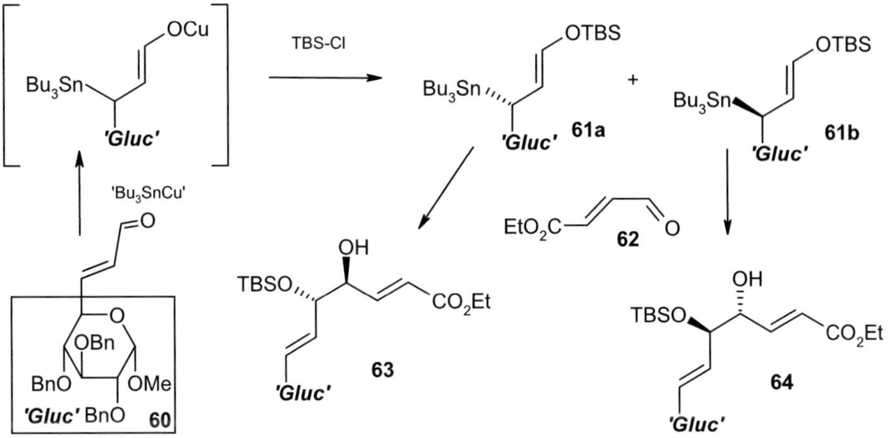

■ Scheme 29

Alternatively, the anion generated from the acetylene **56** reacted with sugar aldehydes providing two isomeric higher sugar propargylic alcohols, which were reduced (H$_2$/Lindlar catalyst) into the Z-olefins **59** (❷ *Scheme 27*) [58].

Other methods for coupling of monosaccharide sub-units leading finally to unsaturated higher sugars are also used, as for example the Henry reaction applied by Suami in his marvelous synthesis of tunicamine (❷ *Scheme 28*) [60].

Another interesting example consists of application of reactive (but at the same time stable) stannyl derivatives. Marshall prepared unsaturated precursors of dodecoses according to ❷ *Scheme 29* [61].

The synthesis was initiated from the easily available unsaturated aldehyde **60**, which reacted with tributyltin cuprate affording a 1,4-adduct (as a mixture of stereoisomers), trapped as trimethylsilyl ethers **61a** and **61b**. Reaction of these reactive species with synthon **62** afforded the unsaturated precursors of higher carbon sugars **63** and **64**.

4.1 5,6-Unsaturated Pyranosides (and 4,5-Unsaturated Furanosides)

This class of compounds needs special interest. Such derivatives are especially important, since they serve as extremely useful intermediates in the preparation of a wide variety of carbocyclic compounds with strong biological activity (e. g. inositols, carbasugars, etc.) [5,62].

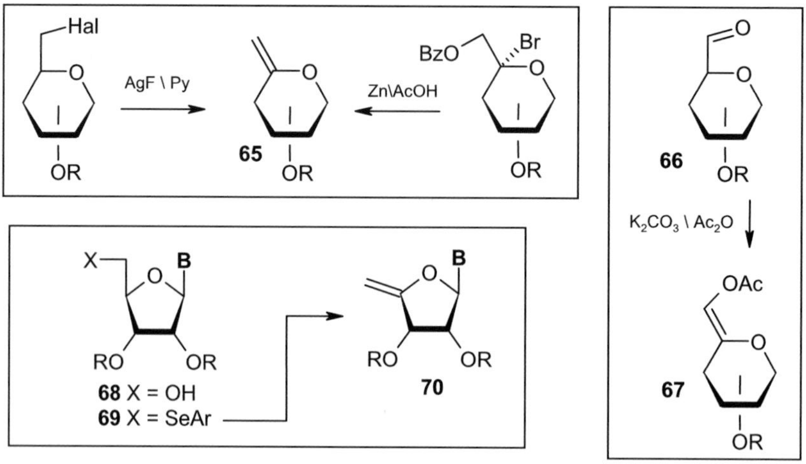

Scheme 30

5,6-Unsaturated pyranosides are easily converted into cyclohexane derivatives by a HgCl₂-catalyzed rearrangement discovered by Ferrier in 1979 (The Ferrier-II rearrangement) [5].

The first compound with the 5,6-unsaturation was synthesized already in 1930; the action of silver fluoride in pyridine on 6-halogenosugars caused elimination of a HX molecule with formation of compound **65** (❖ *Scheme 30*). This reaction soon became a standard method for the preparation of 5,6-unsaturated pyranosides [63]. Much cheaper than silver fluoride reagents such as DBU [64] or sodium hydride [65] may also be applied in such a process. Reduction of 5-bromo-sugars with zinc in acetic acid also affords the exocylic derivatives **65**, however, small amounts of the endocyclic (4,5-unsaturated) isomer are formed (❖ *Scheme 30*) [66].

Treatment of aldehydosugars **66** with potassium carbonate and acetic anhydride provides appropriate 6-acyloxy-5,6-unsaturated sugars **67** (❖ *Scheme 30*) [67]. An example of the preparation of 4,5-unsaturated furanose **70** (used for the synthesis of analogs of antibiotic sinefungin) is presented in ❖ *Scheme 30*. Alcohol **68** was converted into selenide **69** which was oxidized to selenoxide with NaIO₄ and further eliminated in the presence of triethylamine [68].

Recently the Ferrier-II rearrangement was used as a key step in the synthesis of unnatural (–) actinobolin. The 5,6-unsaturated compound **72** was prepared by a base-induced elimination of HI from the 6-deoxy-6-iodo-derivative **71** (❖ *Scheme 31*) [69].

The alternative reaction to the Ferrier-II rearrangement was proposed by Sinaÿ. Treatment of the 5,6-unsaturated glycoside with tri-*iso*-butyl aluminum (TIBAL) provides the corresponding carbocycles in high yields. The configuration at the anomeric center in the glycoside is preserved during the cyclization [70]. This reaction can be performed for *O*- and *C*-glycosides, disaccharides, thio- and selenyl-glycosides, etc. (❖ *Scheme 32*).

This reaction is complementary to Ferrier rearrangement. For example treatment of the thioglycoside **74** under the Ferrier conditions (HgCl₂) affords the ketone **73** with elimination of the sulfur moiety, while application of the Sinaÿ procedure leads to the cyclohexane derivative **75** with the sulfur atom present in the molecule (❖ *Scheme 33*).

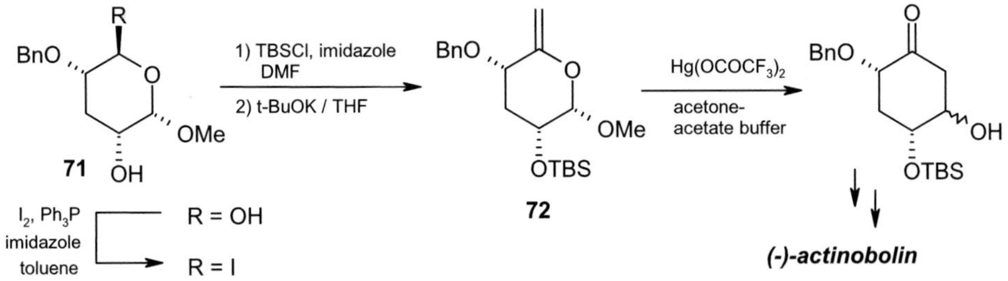

Scheme 31

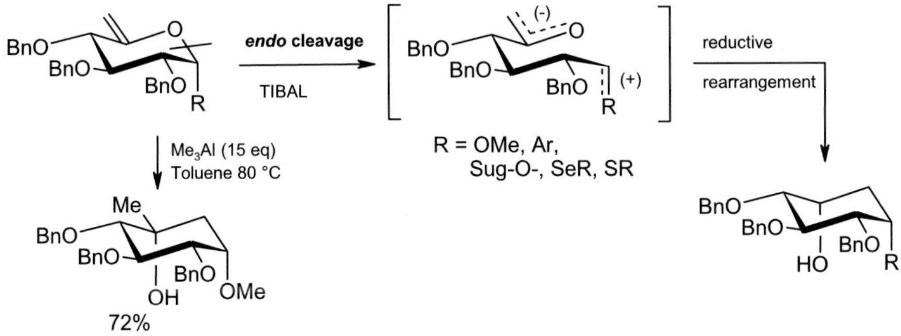

Scheme 32

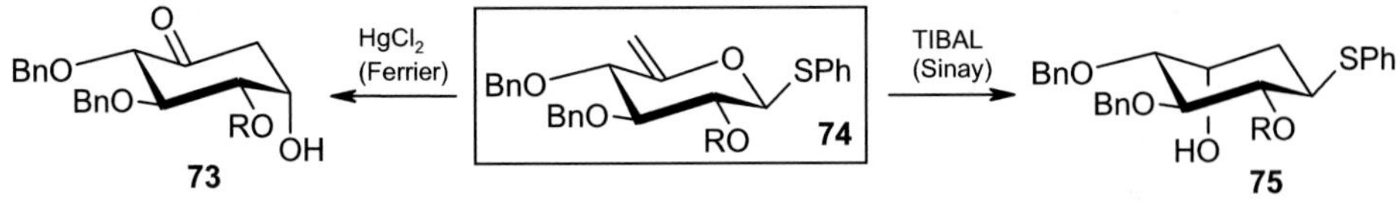

Scheme 33

5 Acyclic Unsaturated Sugars

One of the first unsaturated open-chain sugars (observed for the first time by Emil Fischer) was obtained by hydrolysis of tri-O-acetyl-D-glucal (**23**). Heating of this compound in water results in smooth formation of the E-α,β-unsaturated open-chain aldose **76** (❷ *Scheme 34*) [1]. The carbonyl group can be protected, which allows one to replace the acetate blocks into more stable benzyl groupings thus providing compound **77** (❷ *Scheme 34*) [71].

In 1979 Vasella introduced a convenient procedure for the preparation of unsaturated open-chain sugars by reductive fragmentation of 6-deoxy-6-halogenopyranosides (and 5-deoxy-5-halogeno-furanosides) with zinc in ethanol [72]; the same fragmentation is induced with n-butyl-lithium. Fürstner proposed modification of the original procedure and in his approach activated zinc on graphite was used [73]. The conditions of the Fürstner process are milder and, for example, unsaturated acid **79** can be prepared in very high yield from iodolactone **78** (❷ *Fig. 8*).

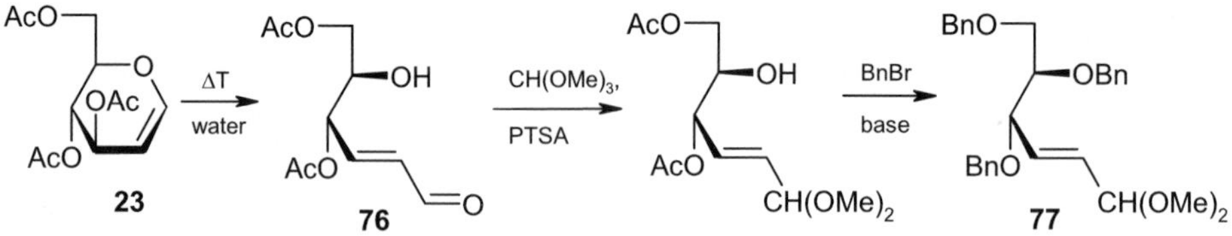

Scheme 34

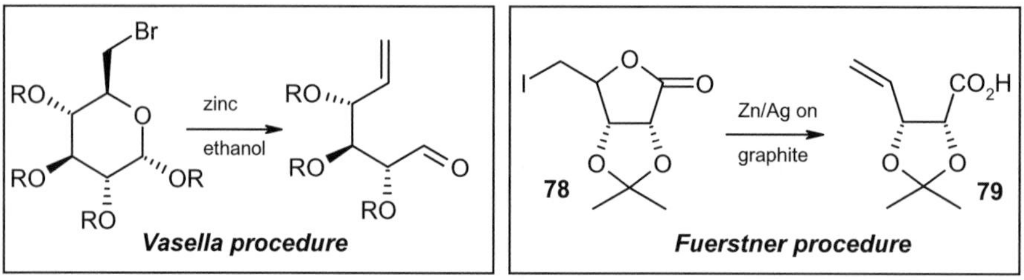

Vasella procedure   *Fuerstner procedure*

Figure 8
Synthesis of unsaturated open chain sugars by fragmentation reactions

Sugar allyltin derivatives are very useful synthetic intermediates. The most convenient and reliable method for their preparation is a so-called 'xanthate' procedure. The first compound of this class was prepared in 1988 by Mortlock and Thomas [74]. 1,2-*O*-isopropylidene-D-glyceraldehyde (**80**) was converted into allylic alcohol and further transformed into the corresponding xanthate **81**. This compound underwent thermal [3,3] rearrangement into the dithiocarbonate **81a**, which upon reaction with tri-*n*-butyltin hydride provided the final sugar allyltin **82** as a mixture of isomers with the *E*-one strongly predominating (**○** *Scheme 35*).

Application of other sugars (pentoses and hexoses) as starting materials allowed Jarosz to prepare a number of such useful compounds (such as **83** or **86** in Schemes 36 and 37) using the xanthate method [75,76,77].

It was observed that treatment of such allyltins with a Lewis acid (the best being zinc chloride) induced a controlled fragmentation leading to highly oxygenated dienoaldehydes with the *E*-configuration across the internal double bond. This procedure may be, therefore, a method of choice for the synthesis of unsaturated sugar dienes. For example sugar allyltin **83** (obtained in a few steps from 'diacetonoglucose') upon treatment with zinc chloride provided a dienoaldehyde **84**, which can be further elongated into the triene **85** by simple reaction with a stabilized Wittig reagent (**○** *Scheme 36*) [76].

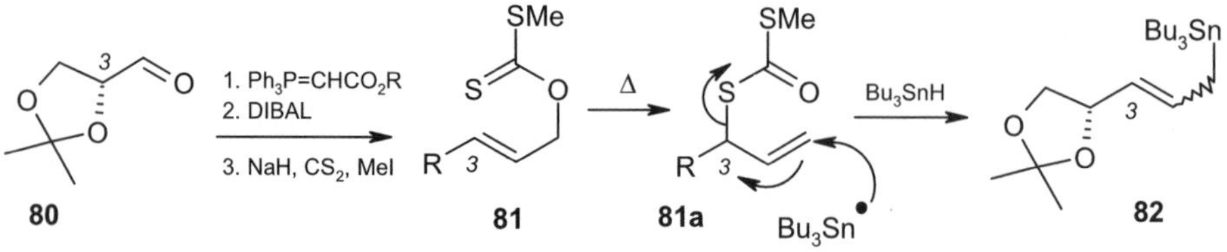

Scheme 35

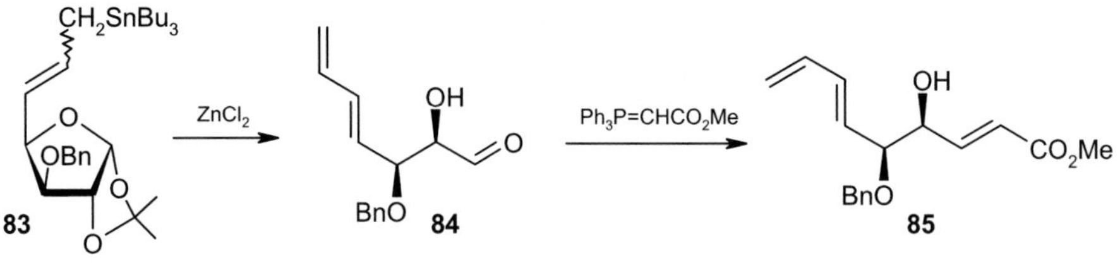

◻ Scheme 36

It was observed, that the hexose-derived allytins **86** are convenient precursors of highly oxygenated carbobicyclic derivatives. When such a compound was treated with ZnCl$_2$ the E-dienoaldehydes **87** were formed exclusively, which could be further transformed into the corresponding trienes (❷ *Scheme 37*) [76,77].

Reaction of dienoaldehyde with the stabilized Wittig reagent(s) provided a triene **88**, which upon high-pressure Diels–Alder cyclization furnished bicyclo[4.3.0]nonene [78]. Alternatively the aldehyde **87-E** was converted into the methyl ester **89**, which reacted with the dimethyl methylphosphonate anion affording the phosphonate **90**. Reaction of the latter with aldehydes under mild PTC conditions yielded a (regioisomeric to the previous one) triene **91**, which cyclized spontaneously to bicyclo[4.4.0]decene (❷ *Scheme 37*) [79].

The controlled fragmentation of the primary sugar allyltins is a convenient method for the preparation of dienoaldehydes with the E-configuration across the internal double bond. The Z-dienes are also available from sugar allyltins, but the secondary ones. When the sugar allylic mesylates (or chlorides) **86** reacted with a tin nucleophile, the S$_N$2' product **92** was obtained.

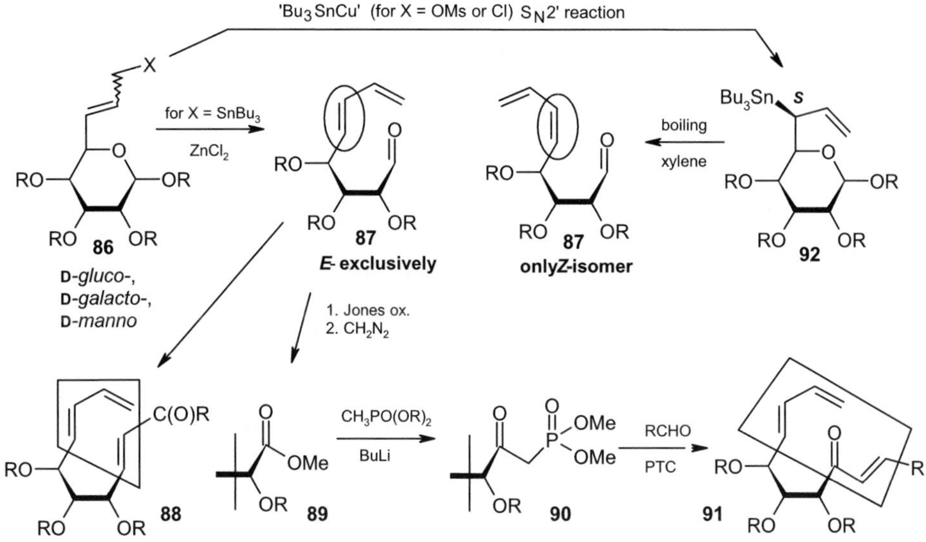

◻ Scheme 37

Moreover, only one stereoisomer was formed regardless of the geometry (*E*- or *Z*-) of the substrate, to which the *S*-configuration at the newly created stereogenic center was assigned [80]. It was found that such secondary allyltins decomposed thermally (the primary ones are stable up to at least 220 °C) into the dienoaldehydes **87** with the Z-geometry across the internal double bond. Therefore, both dienes are available with high stereoselectivity (❍ *Scheme 37*) [77].

6 Application of Unsaturated Sugars as Chirons

Azasugars [81], carbasugars [82], and *C*-glycosides [83] are important sugar mimics, which may be prepared by a variety of methods. In this section the selected methods for the preparation of such derivatives (and also other important compounds) from unsaturated sugars will be presented. Although this is not connected directly with the main subject of this chapter (which deals with the formation of a double bond in sugars), these important targets are prepared from unsaturated sugars, which are in turn synthesized by a number of methods described within this chapter. This would also show the very high synthetic potential of unsaturated sugars.

An interesting methodology to highly oxygenated pyrazolidines and indolizidines from 2,3-unsaturated sugar lactones was presented by Chmielewski [84]. The synthesis was initiated by a 1,4-addition of nitrogen nucleophiles to unsaturated lactones **93**, which resulted in formation of appropriate heterocyclic derivative **94** (❍ *Scheme 38*).

◻ Scheme 38

Further transformations allowed one to obtain a variety of natural and unnatural azasugars.

One of the most useful routes to highly oxygenated carbocycles consists of the transformation of monosaccharide into diolefin, which is further subjected to ring-closing metathesis reaction (*RCM*) with formation of an unsaturated carbocycle. The most convenient way to introduce a terminal double bond is provided by reductive dehalogenation of a terminal halogenosugar (the Vasella reaction, which was already presented in ❍ *Fig. 8*, ❍ *Sect. 5*) leading to the corresponding eno-aldehyde in good yield.

The unsaturation at C-1 may be introduced either by a simple Wittig (or Wittig-type) reaction (*C*$_1$ homologation; ❍ *Scheme 39; route a*) or by reaction with, for example, allylic building blocks (*C*$_3$ homologation); these two steps (reductive dehalogenation followed by a *C*$_3$ homologation) can be performed simultaneously (❍ *Scheme 39; route b*). Such prepared diolefins are then transformed into carbocycles [85] with the Grubbs' (or similar) catalysts. This versatile methodology allows one to prepare carbocycles with different sizes (5–8) of the ring, as shown in ❍ *Scheme 39* and ❍ *Scheme 40*.

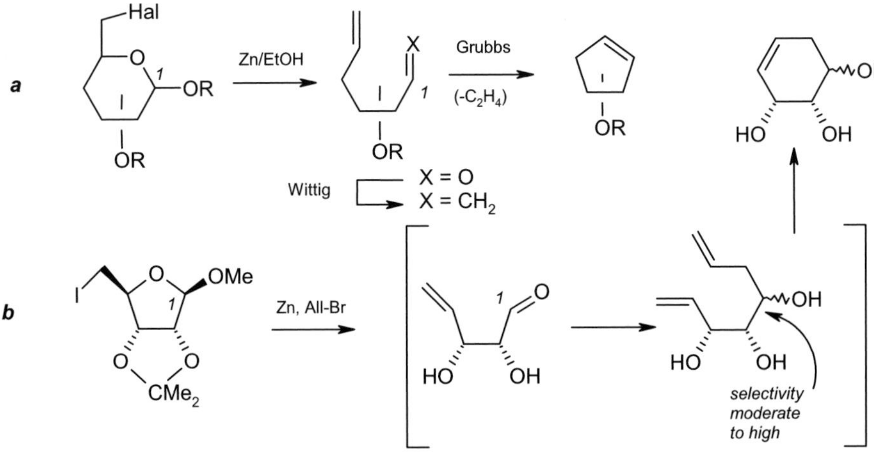

☐ Scheme 39

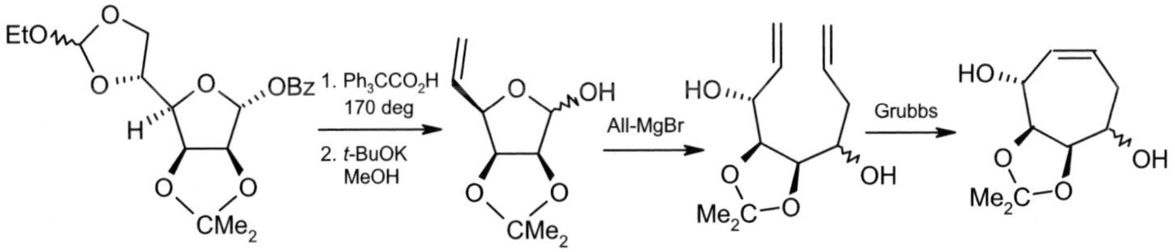

☐ Scheme 40

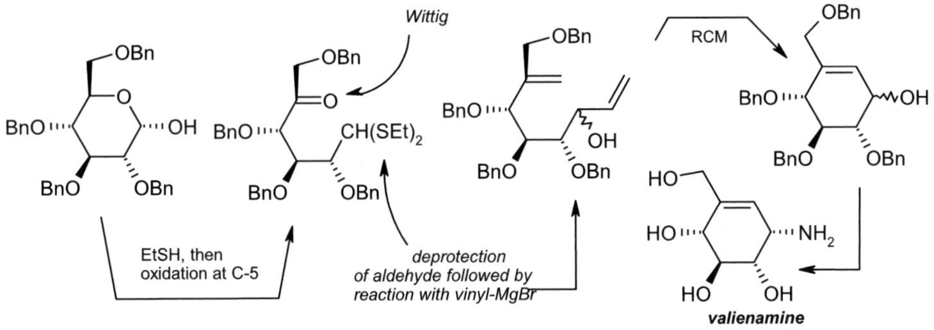

☐ Figure 9
Application of unsaturated sugars in the synthesis of valienamine

The carbocycles substituted at the double bond (with for example, an alkoxymethylene group) can also be prepared from the appropriately functionalized unsaturated sugars. Synthesis of α-glucosidase inhibitor—valienamine—was realized from unsaturated sugar chirons as shown in ❍ *Fig. 9* [86].

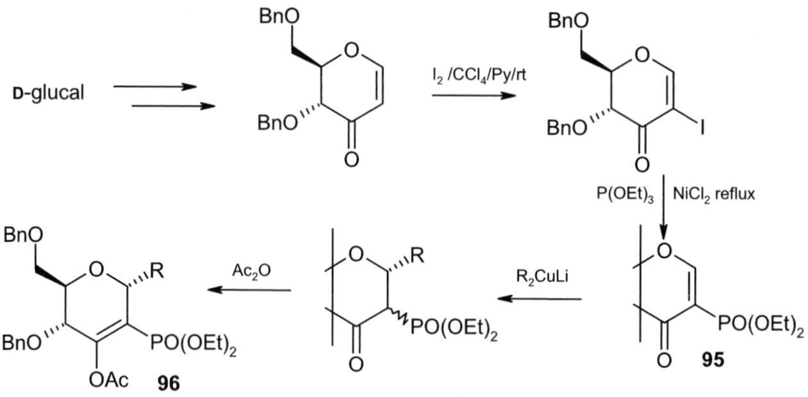

◘ Scheme 41

C-glycosyl phosphonates are important inhibitors of enzymes [87]. An interesting approach to such derivatives was proposed recently [88].

The synthesis was initiated from the unsaturated phosphonate **95** prepared in several steps from the D-glucal as shown in ❷ *Scheme 41*. The 1,4-addition of a soft nucleophile provided the adduct which, upon acetylation, was easily converted into the target **96**.

An interesting approach to symmetrical *C*-disaccharides was proposed recently by Lowary. 1-*O*-Acetyl-sugar **97** was converted into 1-*C*-allyl derivative **97a** by classical reaction with allyltrimethylsilane, and this derivative was 'dimerized' under the metathesis conditions with Grubbs' catalyst (❷ *Scheme 42*) [89].

Unsaturated sugars are used also as starting materials for the preparation of complex natural products and their analogs. For example, radical cyclization of **98** (readily prepared in well defined steps from D-glucal) led to synthon **99**, a precursor of antifungal agent GM222712 (a close analog of sordarin) [90]. The cascade reaction of **100** with the nucleophile catalyzed with gold-III provided also the bicyclic skeleton with a different arrangement of unsaturated bonds (❷ *Scheme 43*) [91].

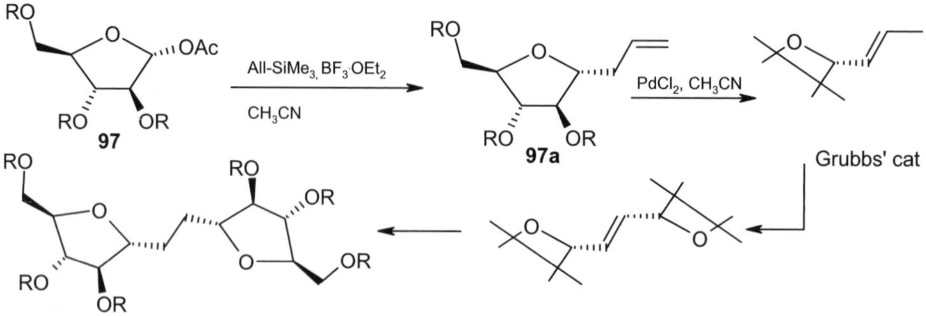

◘ Scheme 42

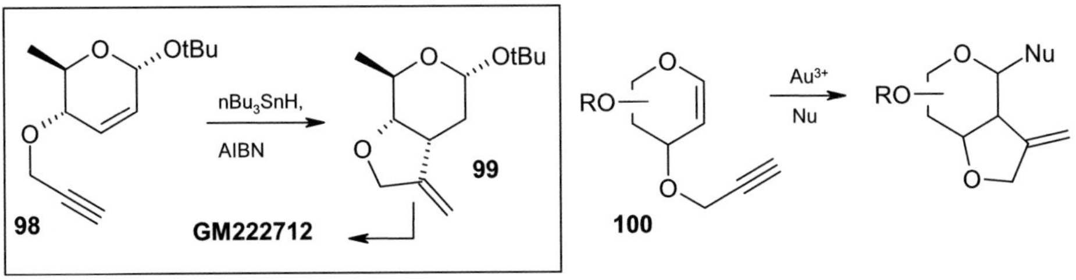

Scheme 43

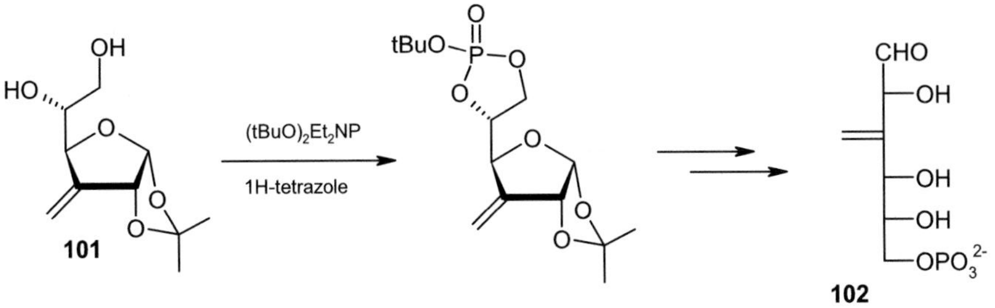

Scheme 44

The glucose analog, 3-deoxy-3-*C*-methylene-D-*ribo*-hexose (**102**) is a substrate of a D-xylose isomerase. The efficient synthesis from 3-deoxy-3-*C*-methylene-glucofuranose (**101** prepared by the classical method according to [92]) was described recently (● *Scheme 44*) [93].

The key step in the preparation of 5,6-dihydro-δ-pyrones with an incorporated 1,3-polyol system (an example, compound **105** possessing inhibitory activity against *C. cucumerinum* is shown in ● *Scheme 45*) consisted of formation of a 2,3-unsaturated furanose derivative **104**, which was prepared from **103**. Abstraction of an acidic proton from the C-2 induced β-elimination, providing the desired target **104** [94].

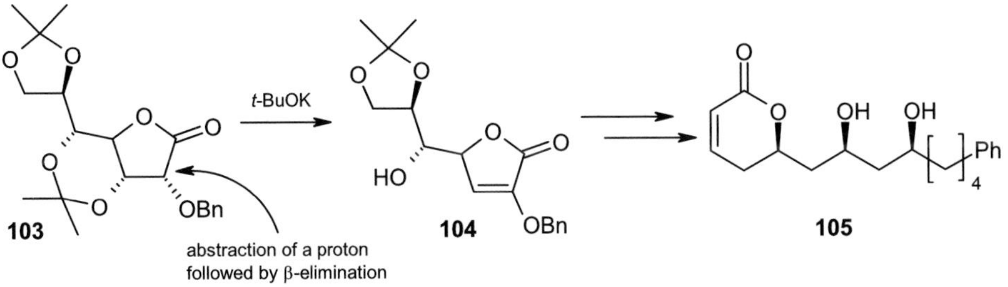

Scheme 45

7 Miscellaneous

α,β-Unsaturated γ-lactones occur widely in nature as simple metabolites of a broad class of natural products. Many of these compounds exhibit interesting biological activity such as, for example, antifungal, antibacterial, cytotoxic properties, etc. Because of their biological importance a number of methods for preparation of such compounds were described in the literature. An interesting method based on the elimination reaction was proposed by Rauter (❷ *Scheme 46*) [95].

The synthesis of higher carbon sugars by an Achmatowicz approach was realized by Indian scientists [96]. This example should have been presented in ❷ *Sect. 2* (describing general methods); however, since it illustrates a concise approach to the synthesis of unsaturated sugars using different methods it is included here.

Addition of allyl species to the free sugar **106** affords an open-chain unsaturated carbohydrate **107**. In a few well-defined steps the furyl derivative **108** is obtained, which upon oxidation with peroxide provides finally the higher carbon sugar **109** (❷ *Scheme 47*).

Another example of the usefulness of tri-*O*-acetyl-D-glucal in the preparation of a complex target in optically pure form is illustrated by the synthesis of iridoids.

Compound **110** (obtained by reaction of D-glucal with propargylic alcohol) was subjected to cyclization with Co$_2$(CO)$_8$, which led to a fused tricyclic derivative **111** (❷ *Scheme 48*) [97].

[2+2] Cycloaddition of carbenes to glycals is an efficient route leading to cyclopropanated sugars with high stereoselectivity [1]. Such cyclopropanes can be transformed into a number

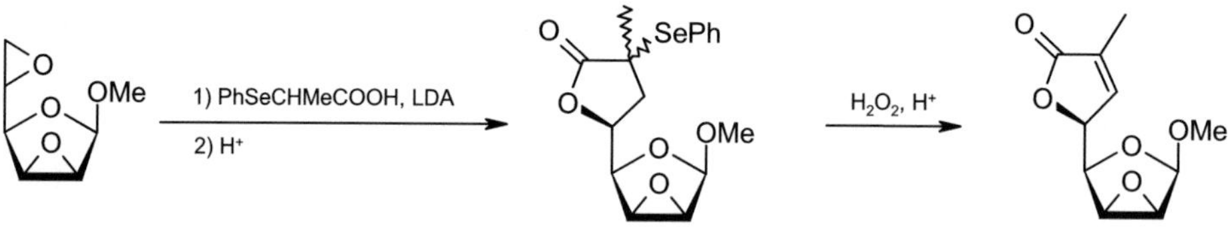

◻ Scheme 46

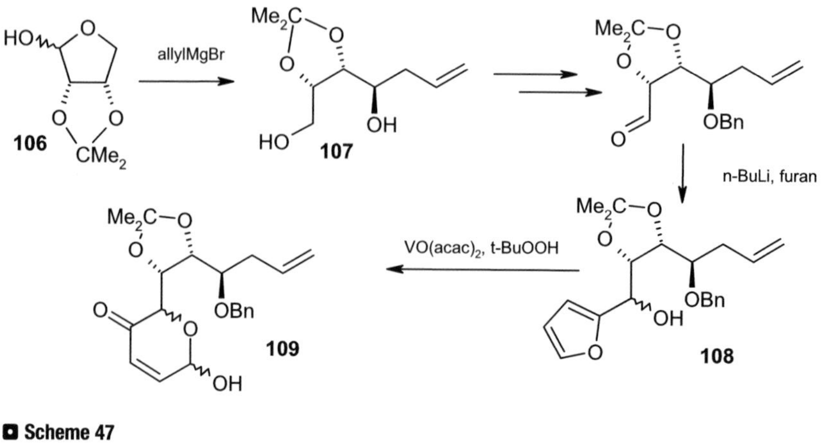

◻ Scheme 47

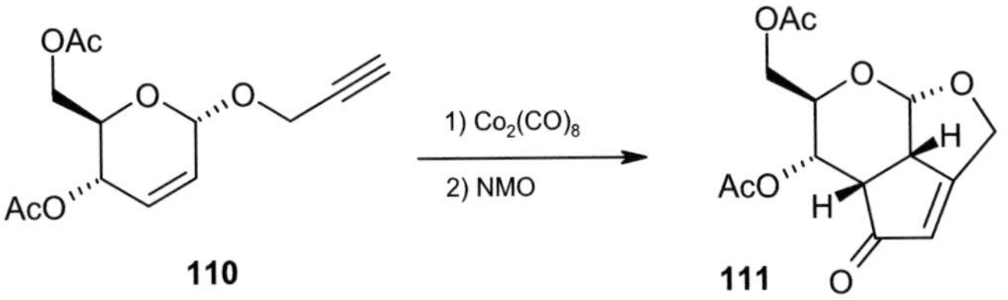

Scheme 48

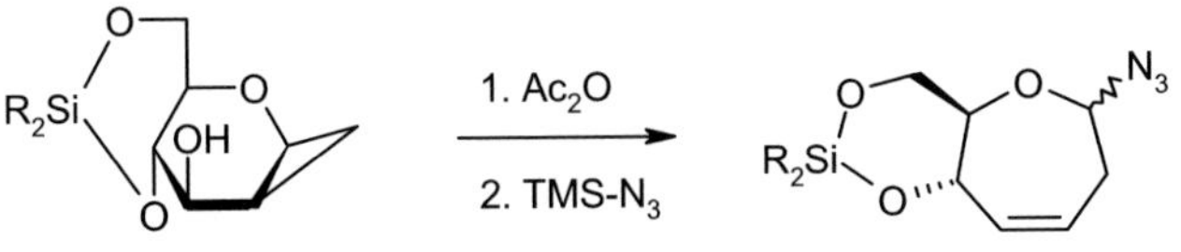

Scheme 49

of interesting compounds. The 3,4-unsaturated septanose was prepared from cyclopropane **112** upon reaction with trimethylsilyl azide (❷ *Scheme 49*) [1].

The palladium-catalyzed reaction of unsaturated sugars (already mentioned in ❷ *Sect. 3.4, rearrangements*) is a methodology that may also lead to 3,4-unsaturated sugars. A complex obtained from the 2,3-unsaturated derivative and Pd(0) species reacted with an azide anion providing either substituted 2,3- or 3,4-unsaturated sugar azides as shown in ❷ *Scheme 50* [98]. As shown in ❷ *Scheme 51* the outcome of the intramolecular reaction of unsaturated sugars depends on the configuration of the starting material. Cyclization of the 4(*S*) stereoisomer **113** induced by palladium species led to bicyclic unsaturated sugar **114**, while the same reaction performed for the 4(*R*) stereoisomer led to compound **115** with an opened sugar ring (❷ *Scheme 51*) [99].

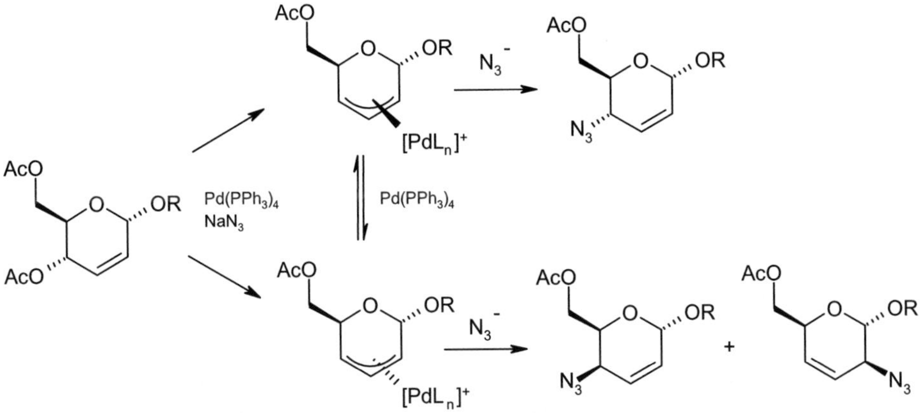

Scheme 50

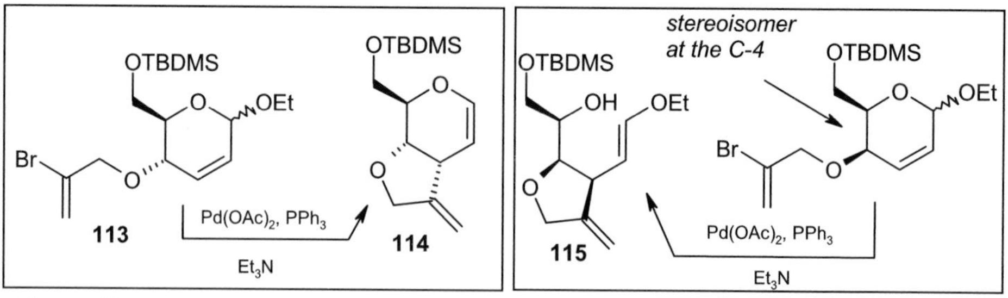

■ Scheme 51

8 Conclusions

Material presented in this chapter describes the methodology of the synthesis of unsaturated sugars (more precisely the formation of the C=C bond) by various methods. This is an updated version of the chapter that appeared in the 2001 edition of this book presenting the concise approach leading to such derivatives. The chemistry of simple (and also more complex) monosaccharides is very well developed, which allows one to apply this well-established methodology for the preparation of complex targets. Therefore, the older methods already described in the previous chapter are also included here, since they represent the basic knowledge of modern carbohydrate chemistry. The material is illustrated with examples of the synthesis of new targets, which have been prepared using methods developed in the last few decades.

Of course, this chapter does not pretend to be comprehensive. The idea behind this chapter was for the authors to show how to prepare the desired unsaturated sugars using proper method(s) and (eventually) how to apply these synthons in the synthesis of important targets.

In modern organic synthesis, monosaccharides have a very well-developed chemistry that allows us to prepare almost any derivative and they are convenient sources of chirality. We hope that the material presented in this chapter will help the reader to plan the synthesis of optically pure targets using the known and convenient carbohydrate approach.

References

1. Ferrier RJ, Hoberg JO (2003) Adv Carbohydr Chem Biochem 58:55
2. Ferrier RJ, Zubkov OA (2003) Org React 62:569
3. Zamojski A, Banaszek B, Grynkiewicz G (1982) Adv Carbohydr Chem Biochem 40:1
4. Ferrier RJ (2001) Topics Curr Chem 215:277
5. Ferrier RJ, Middleton S (1993) Chem Rev 93:2779
6. Jarosz S (2001) Glycoscience—Chemistry and Chemical Biology 1:365
7. Meier Ch, Lorey M, De Clercq E, Balzarini J (1997) Bioorg Med Chem Lett 7:99
8. Jeong L, Schinazi RF, Beach JW, Kim HO, Nampalli S, Shanmuganathan K, Alves AJ, McMillan A, Chu CK (1993) J Med Chem 36:181
9. Yun M, Moon HR, Kim HO, Choi WJ, Kim YCL, Park CCS, Jeong LS (2005) Tetrahedron Lett 46:5903 and references therein
10. Hanessian S (1983) Total Synthesis of Natural Products: The Chiron Approach. Pergamon Press, New York; Hanessian S, Franco J, Larouche B (1987) Pure Appl Chem 62:1990; Inch TD (1984) Tetrahedron 40:3161; Fraser-Reid B, Tsang R (1989) Strategies and Tactics in Organic Synthesis. Academic Press, New York:2; Fraser-Reid B (1996) Acc Chem Res 29:7
11. Ferrier RJ (1969) Adv Carbohydr Chem 24:199
12. Garegg PJ, Samuelsson B (1979) Synthesis 469; Liu Z, Classon B, Samuelsson B (1990) J Org Chem 55:4273

13. Tipson RS, Cohen A (1965–1966) Carbohydr Res 1:338

14. Josan JS, Eastwood FW (1968) Carbohydr Res 7:161

15. Corey EJ, Winter RAE (1963) J Am Chem Soc 85:2677; Horton D, Thomson JK, Tindall ChG Jr (1972) Methods Carbohydr Chem 6:297

16. Barrett AGM, Barton DHR, Bielski R (1979) J Chem Soc Perkin Trans 1 2378

17. Konowal A, Jurczak, J, Zamojski A (1968) Rocz Chem 42:2045

18. Schmidt RR (1986) Acc Chem Res 19:250

19. Danishefsky SJ, Pearson WH, Segmuller BE (1985) J Am Chem Soc 107:1281; Danishefsky SJ, DeNinno MP (1987) Angew Chem Int Ed Engl 26:15

20. Angerbauer R, Schmidt RR (1981) Carbohydr Res 89:173; Schmidt RR (1986) Acc Chem Res 19:250; Bataille C, Begin G, Guillam A, Lemiegre L, Lys C, Maddaluno J, Toupet L (2002) J Org Chem 67:8054

21. Kwiatkowski P, Asztemborska M, Caille JC, Jurczak J (2003) Adv Synth Catal 345:506; Kwiatkowski P, Asztemborska M, Jurczak J (2004) Synlett 1755; Kwiatkowski P, Asztemborska M, Jurczak J (2004) Tetrahedron: Asymmetr 15:3189; Chaladaj W, Kwiatkowski P, Jurczak J (2006) Synlett 3263

22. Achmatowicz O, Bukowski P, Zwierzchowska Z, Zamojski A (1971) Tetrahedron 27: 1973; Achmatowicz O (1981) In: Trost BM and Hutchison CR (eds) Organic Synthesis Today and Tomorrow 307. Pergamon Press, Oxford

23. Casiraghi G, Colombo L, Rassu G, Spanu P (1991) J Org Chem 56:2135 and 6523

24. Raczko J, Jurczak J (1995) In: Atta-ur Rahman (ed) Studies in Natural Product Chemistry. Elsevier, Amsterdam, 15:639 and literature cited therein

25. Horton D, Tsai JH (1979) Carbohydr Res 75:151

26. Doboszewski B, Chu CK, Van Halbeek H (1988) J Org Chem 53:2777

27. Clive DLJ, Wickens PL, Sgarbi PWM (1996) J Org Chem 61:7426; Clive DLJ, Sgarbi PWM, Wickens PL (1997) J Org Chem 62:3751

28. Baba T, Huang G, Isobe M (2003) Tetrahedron 59:6851

29. Deppe O, Glumer A, Yu S, Buchholz K (2004) Carbohydr Res 339:2077

30. Knapp S, Nandan SR (1994) J Org Chem 59:281

31. Capozzi G, Catelani G, D'Andrea F, Menichetti S, Nativi C (2003) Carbohydr Res 338:123

32. Ferrier RJ, Prasad N (1969) J Chem Soc C 570

33. Babu Satheesh R, O'Doherty GA (2003) J Am Chem Soc 125:12406; Babu Satheesh R, Zhou M, O'Doherty GA (2004) J Am Chem Soc 126:3428

34. Shanmugasundaram B, Bose AK, Balasubramian KK (2002) Tetrahedron Lett 43:6795

35. Ferrier RJ (1972) Methods Carbohydr Chem 6:307

36. Baer HH (1989) Pure Appl Chem 61:1217

37. Herscovici J, Montserret R, Antonakis K (1988) Carbohydr Res 176:219

38. Paulsen H, Thiem J (1973) Chem Ber 106:3850; Alexander P, Krishnamamurthy VV, Prisbe EJ (1996) J Med Chem 39:1321

39. Takano A, Fukuhara H, Ohno T, Kutsuma M, Fujimoto T, Shirai H, Iriye R, Kakehi A, yamamoto I (2003) J Carbohydr Chem 22:443

40. Hosokawa S, Kirschbaum B, Isobe M (1998) Tetrahedron Lett 39:1917; Saeeng R, Sirion U, Sahakitpichan P, Isobe M (2003) Tetrahedron Lett 44:6211

41. Ischikawa Y, Kobayashi Ch, Isobe M (1996) J Chem Soc Perkin Trans 1 377

42. Ohyabu N, Nishikawa T, Isobe M (2003) J Am Chem Soc 125:8798

43. Montero A, Mann E, Herradon B (2004) Tetrahedron Lett 46:401

44. Di Bussolo V, Caselli M, Romano MR, Pineschi M, Crotti P (2004) J Org Chem 69:7383

45. David S, Lubineau A, Vatele JM (1982) Carbohydr Res 104:41

46. Canac Y, Levoirier E, Lubineau A (2001) J Org Chem 66:3206

47. Wegert A, Reinke H, Miethchen R (2004) Carbohydrate Res 339:1833

48. Ali MH, Collins PM, Overend V G (1990) Carbohydr Res 205:428

49. Sharma M, Bobek M (1990) Tetrahedron Lett 31:5839

50. Frick W, Kruelle Th, Schmidt RR (1991) Liebigs Ann Chem 435

51. Itoh H, Kaneko T, Tanami K, Yoda T (1988) Bull Chem Soc Jpn 61:3356

52. Brimacombe JS (1989) In: Atta-ur Rahman (ed) Studies in Natural Product Chemistry. Elsevier, Amsterdam, 4C:157

53. Secrist JA III, Wu SR (1979) J Org Chem 44:1434; Secrist JA III, Barnes KD, Wu SR (1989) In: Horton ED, Hawking DL, McCorrey GJ (eds) Trends in Synthetic Carbohydrate Chemistry. ACS Symposium Series, 386:93

54. Karpiesiuk W, Banaszek A (1994) Bioorg Med Chem Lett 4:879

55. Jarosz S (2001) J Carbohydr Chem 20:93

56. Jarosz S, Skora S, Kosciolowska I (2003) Carbohydr Res 338:407
57. Mach M, Jarosz S (2001) J Carbohydr Chem 20:411; Jarosz S, Mach M (2002) Eur J Org Chem 769
58. Jarosz S (1993) J Carbohydr Chem 12:1149
59. Corey EJ, Hua DH, Pan BCh, Steitz SP (1982) J Am Chem Soc 104:6816
60. Suami T, Sasai H, Matsuno K (1983) Chem Lett 819
61. Marshall JA, Elliott, LM (1996) J Org Chem, 61:4611; Jarosz S (2003) Curr Org Chem 7:13
62. Ferrier RJ (1979) J Chem Soc Perkin Trans 1 1455
63. Blair MG (1963) Methods Carbohydr Chem 2:415
64. Mirza S, Molleyres LP, Vasella A (1985) Helv Chim Acta 68:988
65. Barton DHR, Auge-Dorey S, Camara J, Dalko P, Delaumeny JM, Gero SD, Quiclet-Sire B, Stuetz P (1990) Tetrahedron 46:215
66. Blattner R, Ferrier RJ, Tyler CP (1980) J Chem Soc Perkin Trans 1 1535
67. Prestwich GD (1996) Acc Chem Res 29:503
68. Meade EA, Krawczyk SH, Towsend LB (1988) Tetrahedron Lett 29:4073
69. Imuta S, Ochiai S, Kuribayashi M, Chida N (2003) Tetrahedron Lett 44:5047
70. Sollogub M, Millet JM, Sinaÿ P (2000) Angew Chem Int Ed 39:362; Pearce AJ, Sollogub M, Mallet JM, Sinaÿ P (1999) Eur J Org Chem 2103; Sollogoub M, Pearce AJ, Herault A, Sinaÿ P (2000) Tetrahedron: Asymmetr 11:283; Jia C, Pearce AJ, Bleriot Y, Zhang Y, Zhang LH, Sollogoub M, Sinaÿ P (2004) Tetrahedron: Asymmetr 15:699
71. Pathak R, Shaw AK, Bhaduri AP (2002) Tetrahedron 58:3535
72. Bernet B, Vasella A (1979) Helv Chim Acta 62:1990 and 2400
73. Fuerstner A (1993) Angew Chem Int Ed Engl 32:164
74. Mortlock SV, Thomas EJ (1988) Tetrahedron Lett 29:2479
75. Jarosz S, Fraser-Reid B (1989) J Org Chem 54:4011; Kozlowska E, Jarosz S (1994) J Carbohydr Chem 13:889
76. Jarosz S, Skóra S, Szewczyk K (2000) Tetrahedron: Asymmetr 11:1997
77. Jarosz S, Gawel A (2005) Eur J Org Chem 3415
78. Jarosz S, Skora S (2000) Tetrahedron: Asymmetr 11:1425
79. Jarosz S, Skora S (2000) Tetrahedron: Asymmetr 11:1433
80. Jarosz S, Szewczyk K, Zawisza A (2003) Tetrahedron: Asymmetr 14:1715
81. Afarinkia K, Bahar A (2005) Tetrahedron: Asymmetr 16:1239
82. Kobayashi Y (2001) Glycoscience: Chemistry and Chemical Biology III:2595; Ogawa S (1998) Carbohydrate Mimics: Concepts and Methods 87; Suami T (1990) Top Curr Chem 154:257; Suami T, Ogawa S (1990) Adv Carbohydr Chem Biochem 48:21.
83. Postema NHD, Piper JL, Betts RL (2005) Synlett 1345
84. Rabiczko J, Chmielewski M (1999) J Org Chem 64:1347; Panfil I, Lipkowska-Urbanczyk Z, Suwinska K, Solecka J, Chmielewski M (2002) Tetrahedron 58:1199
85. Jorgensen M, Hadwiger P, Madsen R, Stütz AE, Wrodnigg TM (2000) Curr Org Chem 4:565; Madsen R (2007) Eur J Org Chem 399
86. Chang YK, Lee BY, Kim DJ, Lee GS, Jeon HB, Kim KS (2005) J Org Chem 70:3299
87. Engel R (1977) Chem Rev 77:349; Dondoni A, Marra A (2000) Tetrahedron: Asymmetr 11:305; Orsini F, Caselli A (2002) Tetrahedron Lett 43:7259
88. Leonelli F, Capuzzi M, Calcano V, Passacantilli P, Piancatelli G (2005) Eur J Org Chem 2671
89. Chang GX, Lowary TL (2006) Tetrahedron Lett 47:4561
90. Bueno JM, Coteron JM, Chiara JL, Fernandez-Mayoralas A, Fiandor JM, Valle N (2000) Tetrahedron Lett 41:4379
91. Kashyap S, Hotha S (2006) Tetrahedron Lett 47:2021
92. Brimacombe JS, Miller JA, Zakir U (1976) Carbohydr Res 49:233
93. Burger A, Tritsch D, Biellmann JF (2001) Carbohydr Res 332:141
94. Ramana CV, Srinivas B, Puranik VG, Gurjar MK (2005) J Org Chem 70:8216
95. Rauter AP, Figueiredo J, Ismael M, Canda T, Font J, Figueredo M (2001) Tetrahedron: Asymmetr 12:1131
96. Krishna UM, Deodhar KD, Trivedi GK (2004) Tetrahedron 60:4829
97. Marco-Contelles J, Ruiz-Caro J (1999) J Org Chem 64:8302
98. de Oliveira RN, Cottier L, Sinou D, Srivastava RM (2005) Tetrahedron 61:8271
99. Bedjeguelal K, Joseph L, Bolitt V, Sinou D (1999) Tetrahedron Lett 40:87

2.7 Degradations and Rearrangement Reactions

Jianbo Zhang
Department of Chemistry, East China Normal University,
200062 Shanghai, China
jbzhang@chem.ecnu.edu.cn

In: *Glycoscience*. Fraser-Reid B, Tatsuta K, Thiem J (eds)
Chapter-DOI 10-1007/978-3-540-30429-6_9: © Springer-Verlag Berlin Heidelberg 2008

Abstract

This section deals with recent reports concerning degradation and rearrangement reactions of free sugars as well as some glycosides. The transformations are classified in chemical and enzymatic ways. In addition, the Maillard reaction will be discussed as an example of degradation and rearrangement transformation and its application in current research in the fields of chemistry and biology.

Keywords

Degradation; Rearrangement; Hydrolysis; Double bond shift; Ring transformation; Ring-contraction; Ring-expansion; Ferrier carbocyclization; Anomerization; Aromatization; Maillard reaction; Amadori reaction

Abbreviations

AGEs	advanced glycation end products
DAST	diethylaminosulfur trifluoride
DMDO	dimethyldioxirane
DMF	dimethylformamide
EFC	ethanol-from-cellulose
HFIP	1,1,1,3,3,3-hexafluoro-2-propanol
HMF	5-(hydroxymethyl)-2-furaldehyde, 5-hydroxymethylfuraldehyde
IDCP	iodonium dicollidine perchlorate
LTMP	lithium 2,2,6,6-tetramethylpiperidide
m-CPBA	3-chloroperoxybenzoic acid, *meta*-chloroperoxybenzoic acid
PTC	phase transfer catalysis
TASF	tris(dimethylamino)sulfonium difluorotrimethylsilicate
TMSCN	trimethylsilyl cyanide
TMSOTf	trimethylsilyl triflate
TMU	*N,N*-tetramethylurea

1 Overview

Degradation and rearrangement reactions in carbohydrate chemistry are described in the standard organic chemistry textbooks [1,2]. Recent books [3,4,5] contain informative surveys of developments in this area. This chapter dealing with degradation and rearrangement reactions of carbohydrate systems covers literature published in the last few years. Degradation reactions are classified into two main categories: hydrolysis from glycosides or polysaccharides into free sugars, and the degradations from free sugars into useful building blocks or chiral synthons for organic synthesis. The rearrangement reactions described herein are classified into four groups: double bond shifts, ring rearrangements associating with a double bond, ring isomerizations (contraction and expansion), and other processes. Current results on the Maillard reaction initiated by the Amadori reaction, which are so intimately associated with degradation and rearrangement reactions, are also discussed in this chapter.

2 Hydrolysis of Glycosides and Polysaccharides

Among all the degradation patterns for glycosides and polysaccharides, hydrolysis is the most important process in carbohydrate chemistry, either in nature or in biological systems. Before long, it became a process important in the food industry for production of free sugars and is now gaining more and more attention because of the present energy crisis. This is because petroleum is not an ideal chemical feedstock for industry, due to its intractability, while glycosides and polysaccharides—which are abundant and recyclable—can be utilized in the production of fuel and chiral synthons to be used instead of traditional petroleum [6].

2.1 Chemical Hydrolysis

Chemical hydrolysis is a very familiar reaction for the sugar industry. However, it may generate an array of possible degradation products. For example, very low rate constants for the spontaneous hydrolysis of nonactivated methyl β-D-glucopyranoside **1** have been determined at 220 °C [7], (❷ *Fig. 1*). At pH > 7, the rate constants approach a constant value. On hydrolysis at pH 10 in the presence of $H_2^{18}O$, the results show that the reaction occurs almost exclusively by cleavage of the C1/O1 bond. The β-anomer **1** is roughly twice as reactive as the α-anomer **2**, as are also the anomeric pair of methyl D-ribofuranosides **3** and **4**. Unlike the hydrolysis at pH < 7, the hydrolysis of **1** without catalysts proceeds with a negative entropy of activation. This is consistent with bimolecular attack of water on **1**.

Acid hydrolysis of isopropenyl α-D-glucopyranoside **5** at pH 3.0 and 25 °C occurs by *C*-protonation followed by cleavage of the alkenyl ether C/O bond. The α-anomer **5** is hydrolyzed 4.5-times faster that its β-anomer **6**. Spectroscopic evidence indicates greater conjugation of O1 with the double bond, and hence a greater basicity of the β-carbon of the double bond, in **5** compared to **6** [8].

An accelerating effect by the intramolecular nucleophilic catalysis of a phosphate anion upon hydrolysis of the phosphate **7** at 80TT °C and pH 6–9, in comparison with the unsubstituted **8** and its 2-*O*-methyl derivatives **9**, has been observed [9]. The reaction of **8** is base-catalyzed down almost to pH 7, while that of **9** is pH-independent up to pH 9–10 The hydrolysis of **7** proceeds about 100-times faster than that of **9** at pH ~9 and 80 °C. In comparison, the hydrolysis of **10** is pH-independent down to pH 7 and ~20-times slower than that of **7** at pH 9 and 80 °C [10].

A kinetic study of the acyl migration reaction of the 1-*O*-acyl β-D-glucopyranuronic acid **11**, a model drug ester glucuronide, employing a directly coupled stop-flow HPLC/600 MHz ^{1}H-NMR system at pH 7.4 and 25 °C, has been carried out [11]. The acyl migration rate of the β-1-*O*-acyl group of **11** is greater than any other regio-isomers. The simulating mutarotation rates for the 4-*O*-acyl isomers **12** are in good accord with the experimental values.

The fructofuranosyl cation **13** is the first formed product of the acid-catalyzed melt thermolyses of sucrose **14** (❷ *Scheme 1*). This reacts with hydroxy nucleophiles co-existing in the melt to give fructose-grafted products. Rigorous thermolysis of **14** itself at 170 °C furnishes a fructosylglucan with an average dp ~25 together with the known sucrose thermal oligosaccharides from **14**, such as **15** (3.9%) and **16** (4.1%) [12].

Mechanistic studies on acid hydrolysis of glycosides often encounter the *endo/exo*-cyclic cleavage problem [13]. For instance, the sulfuric acid (1%)-catalyzed acetolysis of the

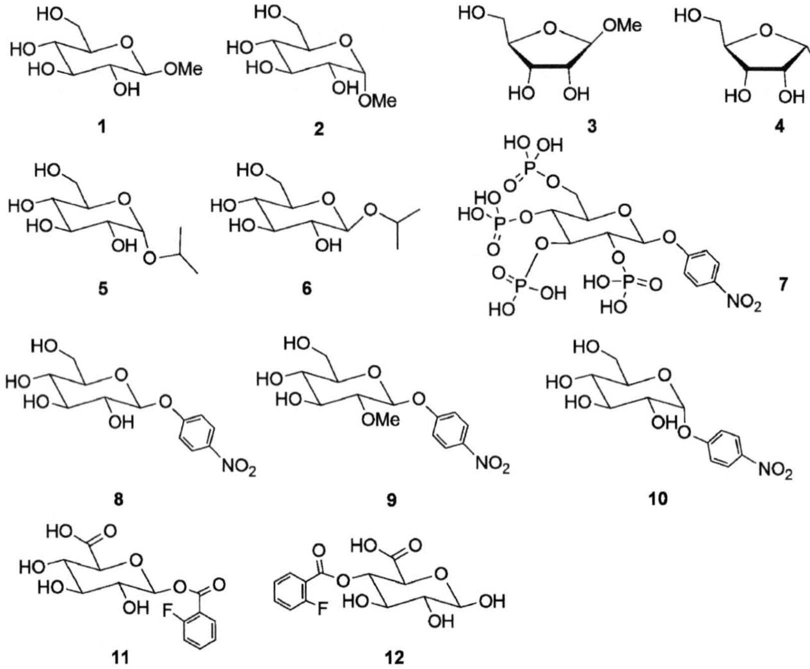

Figure 1
Glycosidesubstrates for nucleophilic hydrolysis

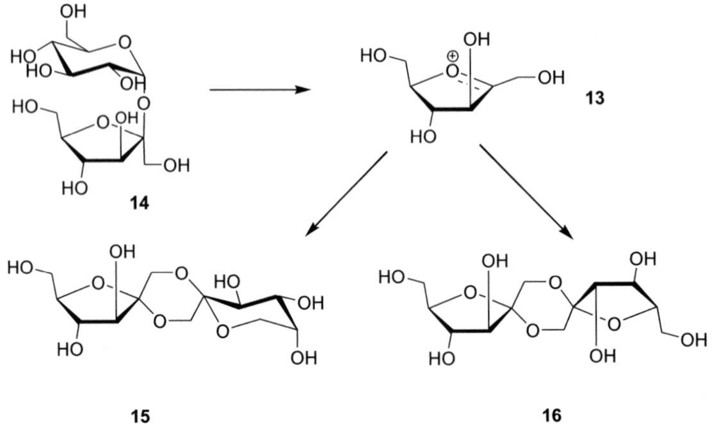

Scheme 1

anomeric ethyl glycoside derivatives **17** and **18** as well as the diastereoisomeric acetal **19** and **20** have been studied kinetically. The time-dependent distribution of the acetolysis products from **17** and **18** shows that their rapid mutual anomerization precedes their acetolysis to **21** and **22**, undoubtedly by way of the *endo*-cyclic cleavage product **23**, the precursor of **19** and **20** [14] (**❷** *Fig. 2*).

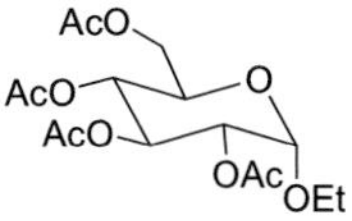

Figure 2
Substrates and intermediates in acetolysis of ethyl glycoside

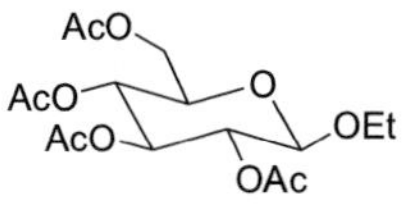

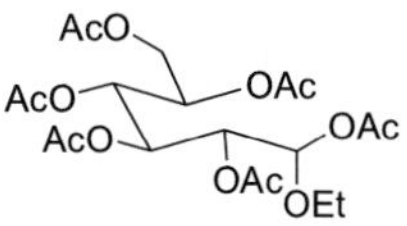

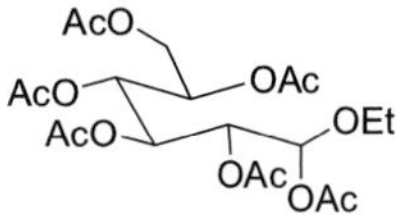

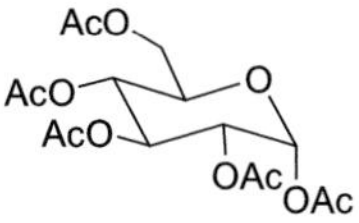

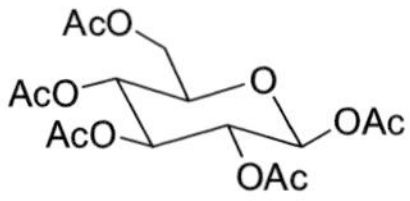

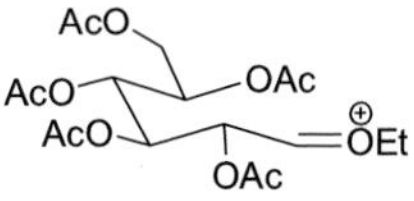

Sialosides have a distinct mechanism of hydrolysis for its unusual sugar structure of sialic acid. For example, the large β-dideuterium and small primary ^{14}C kinetic isotope effects observed at the anomeric carbon and the large secondary ^{14}C kinetic isotope effect observed at the carboxylate carbon in the acid-catalyzed solvolysis of CMP-N-acetyl neuraminate **24** support an oxocarbenium ion-like transition state **25** having the $_5$S conformation without nucleophilic participation of carboxylate and with the carboxylate anion in a looser environment than in the ground state [15] (**◉** *Fig. 3*). Such a zwitterion structure is consistent with the results from calculations using the COSMO-AM1 method for aqueous solutions [16].

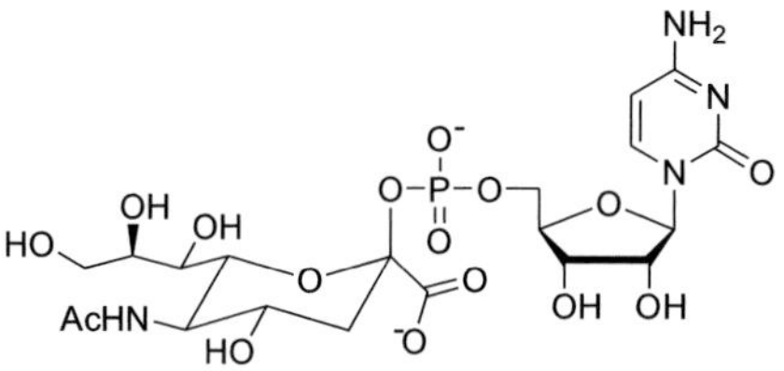

Figure 3
CMP-N-acetyl neuraminic acid and oxocarbenium ion-like transition state in sialoside hydrolysis

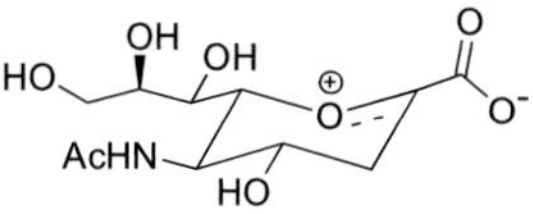

2.2 Enzymatic Hydrolysis

Glycoside hydrolase is one of the main categories of hydrolases in nature. Many references have suggested a distorted conformation for the substrate, which accelerated the hydrolytic process dramatically [17,18]. The influenza A sialidase hydrolyzes sialyl glycosides with retention of the anomeric configuration [16], whereas the *Salmomella typhimurium* sialidase works with inversion, although their protein folds and presumed active site residues are very similar [19,20]. Comparative studies using deuterium-labeled *p*-nitrophenyl N-acetyl-α-neuraminides **26–28** have postulated that the reactive substrate adopts a B$_{25}$ conformation with significant proton donation to the leaving group for the influenza virus enzyme, whereas the *S. typhimurium* enzyme works through a single chemical transition state derived from the ground state $^2C_5$ conformation with little proton donation to the leaving group [13]. The leaving group ^{18}O isotope effects are higher at pH 6.67 and 60 °C than at pH 2.69 and 50 °C in the nonenzymatic hydrolysis of **29** (❍ *Fig. 4*) [21,22]. This indicates that the C/O bond dissociation is complete at the transition state [21].

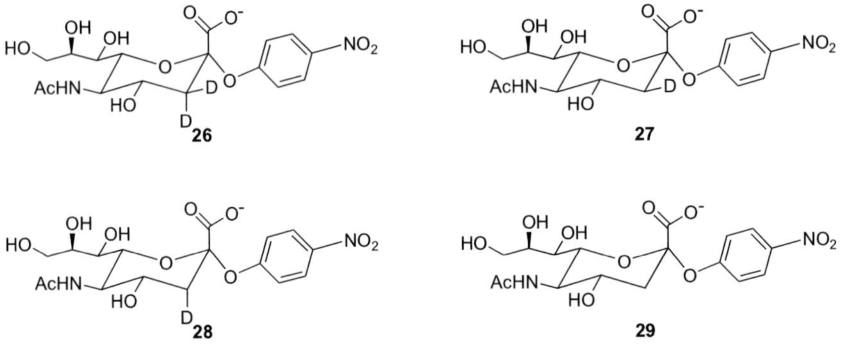

❑ Figure 4
Deuterium-labeled *p*-nitrophenyl *N*-acetyl-α-neuraminides

New analytic tools can be helpful in the research of enzymatic processes. For instance, time-course examination of enzymatic hydrolysis has recently been studied with ^1H-NMR spectroscopy. Thus, α-L-rhamnosyl and α-D-galactosyl hydrolysates from *Aspergillus* fungi have recently been found to be inverting hydrolyses [23,24].

With newly gained knowledge of hydrolytic mechanisms a novel artificial enzyme, the antibody enzyme AbZyme, was designed using the known mimics of the transition states [25]. Antibody Ab24, produced by in vitro immunization using the carrier-free hapten **30** and spleen cells in culture, catalyzes the hydrolysis of **8** with a k_{cat} of 0.02 h^{-1} and K_m of 160 μM ($k_{cat}/k_{uncat} = 2.2 \times 10^4$). Similarly, antibody Ab21 can catalyze the hydrolysis of galactoside **31** with a k_{cat} of 0.035 h^{-1} and K_m of 310 μM ($k_{cat}/k_{uncat} = 2.5 \times 10^4$) (❍ *Fig. 5*) [26].

At the same time, with the progress in development of new separation techniques and biotechnology, more and more enzymes have been found with interesting properties. For instance,

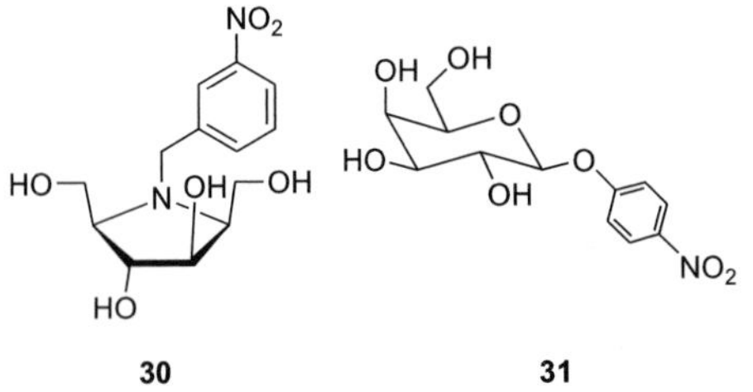

❑ Figure 5
Hapten and substrate for catalytic antibody Ab24

the α-1,4 glucan lyase (EC 4.2.2.-) from *Gracilariopsis lemeneiformis* is a new class of starch/glycogen degrading enzyme that digests the substrate from the nonreducing end while releasing 1,5-anhydro-D-fructose **32** successively, instead of the usual D-glucose **33** (❯ *Fig. 6*) [27,28,29].

32   **33**

❑ **Figure 6**
1,5-anhydro-D-fructose and D-glucose

Several recent books review the enzymes used in the conversion of renewable feedstocks such as starch and cellulose [30,31,32]. They provide many examples of the use of enzymes in the resource sector, specifically addressing their use in agriculture, forest products, and pulp and paper; they also address the greater use of agriculture and forestry residues and possible enzymatic modification. One recent example is the use of crude α-galactosidase from *Gibberella fujikuroi* to reduce the flatulence-inducing raffinose family sugars in chickpea flour. Crude enzyme treatment of chickpea flour resulted in complete hydrolysis of sugars of the raffinose family [33,34].

3 Degradation of Free Sugars

Although the free sugars are important industrial starting materials, they have not been focused upon until the recent findings of their degradation into useful organic resources [35,36].

3.1 Thermal Degradations

Thermal degradations in aqueous carbohydrate solutions are well documented [37,38,39]. Innovation of technology for the degradation of phytomass has focused upon the production of pyrolysis oil with high H/C and C/O ratios. Hydrothermal degradation appears to be attractive from this point of view [36,40,41,42]. Early in 1964, Qua and Fagerson tentatively identified furfural **34**, dihydroxyacetone **35**, glycolic acid **36**, glycolaldehyde **37**, and 5-(hydroxymethyl)-2-furaldehyde (HMF) **38**, and noted the presence of six additional volatile products from glucose **33** heated at 250 °C for 1 min in air. Recently, the scientists found that HMF **38** can be utilized as a very important intermediate for the petroleum industry [43,44,45].
For example, a problem is the formation of **35** during autoclave sterilization of various solutions for parenteral injection containing **33** as an excipient or a nutritional carbohydrate [46]. Similar degradations occur during food processing, especially in soft-drink production, with compound L-ascorbic acid **39** degrading into **34** and **35** [47,48]. Another example includes the roasting of coffee, during which there are many aphilic acids formed by carbohydrate degradation, which contribute to the smell and taste impact for coffee beans [49,50,51].

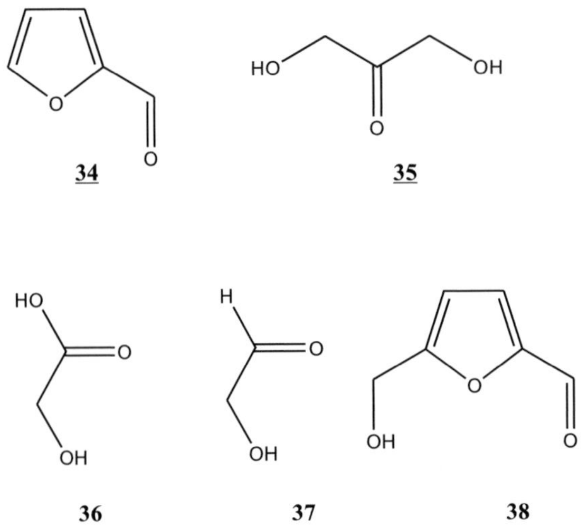

Figure 7
Main thermal degradation products from D-glucose

Ab initio molecular dynamics (MD) simulations were also applied in elucidation of xylose and glucose degradation pathways (❯ *Scheme 2*). In the case of D-xylose **39**, a 2,5-anhydride intermediate was observed leading to the formation of furfural **34** through elimination of water. This pathway agrees with one of the mechanisms proposed in the literature in that no open chain intermediates were found. In the case of D-glucose **33**, a series of intermediates were observed before forming the 2,5-anhydride intermediate that eventually leads to HMF **38** (❯ *Fig. 7*). One of these intermediates was a very short-lived open-chain form. Furthermore, two novel side-reaction pathways were identified, which lead to degradation products other than **38** [52].

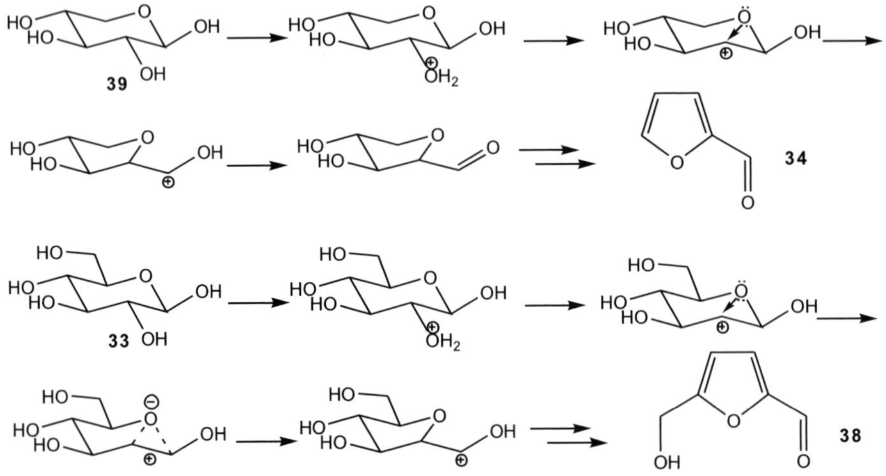

Scheme 2

3.2 Acidic Degradations

Alkyl glycosides are environmentally benign biosurfactants due to their biodegradability and low toxicity [53]. Usually they are produced through Fisher glycosylation using hydrophobic alcohols in acidic media. A practical problem is the control of the degradation reactions of starting free sugars in the acidic Fischer reaction. The situation is more serious in the case of D-Fructose **40**, which degrades into **38** [54,55,56] (❯ *Scheme 3*). In Fischer reactions of this type, silica-alumina cracking catalysts effectively catalyze reactions to give the glycosides **41**, **42**, and **43**, without formation of **38** [57]. These results convincingly indicate that both the glycosylation giving the furanosides **41** and **42** and the degradation to **38** proceed via the common cyclic intermediates **13** [54].

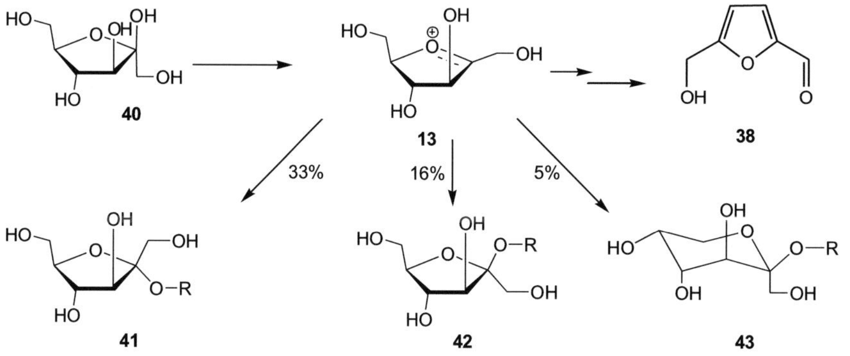

◻ Scheme 3

3.3 Alkaline Degradations

The alkaline degradation of reducing monosaccharides involves a series of consecutive reactions and gives many kinds of products [58]. For example, the alkaline degradation of **33** in aqueous calcium hydroxide at 100 °C results in a complex mixture of more than 50 compounds (❯ *Scheme 4*). Products obtained by the same degradation of **40** are similar to those from the

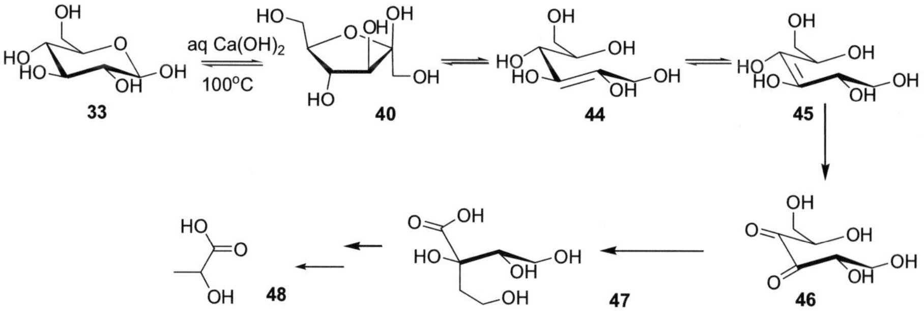

◻ Scheme 4

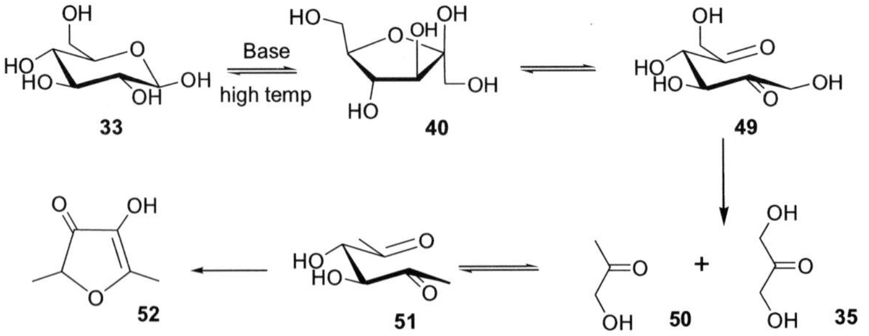

■ **Scheme 5**

reaction of **33**. Among the degradation products, lactic acid **44** is almost the sole major product in each case [59].

High-temperature alkaline degradation of **33** forms furaneol (**52**), an aroma compound, probably because of fragmentation of **49** into the C$_3$-fragments **35** and **50** (❖ *Scheme 5*). Fragment **50** dimerizes into the diketone **51**, the precursor of **52** [60].

3.4 Oxidative Degradations

Oxidative degradation reactions involving the anomeric center are classic processes and are well documented [61,62]. For example, lactose, maltose, cellobiose, and galactose can be degraded selectively in one step and in high yield into the corresponding next lower aldose and formic acid by H$_2$O$_2$ in the presence of borate. The selectivity further improves when a small amount of EDTA is added, in order to suppress the influence of transition metal ions, which catalyze the decomposition of H$_2$O$_2$ via radical pathways, leading to nonselective oxidative degradation of aldoses. The function of borate in the selective oxidative degradation of aldoses is two-fold: catalysis of the degradation of the starting aldose and protection of the next lower aldose against oxidation [63].

On alkaline oxidation of aldoses with (*N*-chloro-*p*-toluenesulfonamido) sodium (CAT), the monosaccharides **33**, **40**, D-mannose **54**, D-arabinose **55**, and D-ribose **56**, belonging to the 4,5- or 3,4-*ethythro*-series, afford the C$_4$-acids **59** and **60** in 35 to 49% yields while the yields of glyceric acid are low [64]. Thus, as illustrated in ❖ *Scheme 6*, hexoses are cleaved at the C1/C2 (a) and C2/C3 (b) bonds, whereas pentoses break at the C1/H1 (a) and C1/C2 (b) bonds.

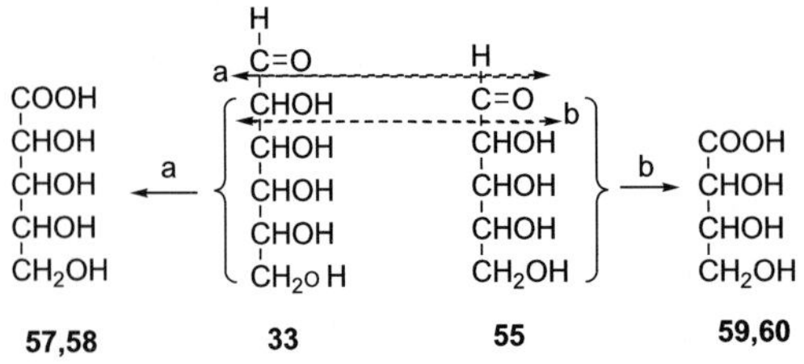

■ **Scheme 6**

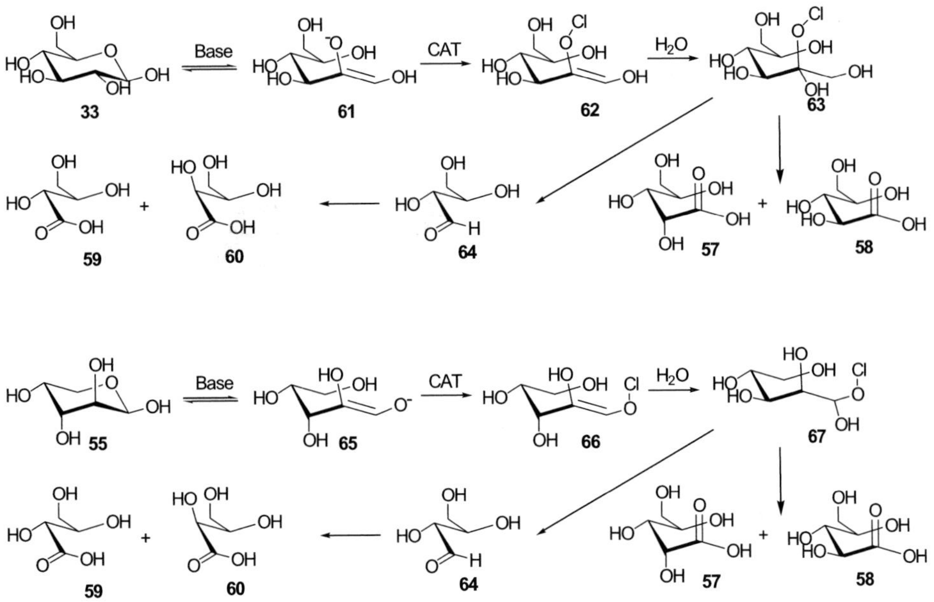

Scheme 7

These reactions are governed by the alkaline-induced slow equilibrium between hexoses and enediol anions and the irreversible, rate-determining formation of the intermediate **63** (❯ *Scheme 7*). The latter is transformed into **57** and **58** or **64**. In the case of pentose **55**, the intermediate **67** gives out **59** and **60** from the intermediate **64**.

(S)- and (R)-3-hydroxy-γ-butyrolactone (**68** and **69**, respectively) are two extremely flexible chiral synthons. They can be converted to an extremely large number of useful and important intermediates with a wide range of applications. Earlier synthetic routes to these compounds all relied on structural transformations or selective reductions of malic acid. They can now be obtained in high yield from several carbohydrate raw materials. For example, the (S)-lactone **68** can readily be prepared by the oxidation of 4-linked D-hexose sources such as cellobiose, lactose, maltose, maltodextrins, starch, etc., with hydrogen peroxide and an alkaline or alkaline-earth hydroxide. Treatment of a 4-linked hexose **70** with base leads to an isomerization to the 4-linked ketose, which readily undergoes β-elimination to form enone, which then tautomerizes to the diketone. The diketone is readily cleaved with hydrogen peroxide to give the salt of (S)-3,4-dihydroxybutyric acid and glycolic acid. Acidification and concentration yields the lactone **68** [65]. Similarly, the (R)-lactone **69** can be synthesized using a 4-linked L-hexose source since the chiral center in the product is derived from the 5-carbon of the hexose. The (R)-lactone was obtained in high yield from L-arabinose **71** by the simple strategy of functionalizing the 3-position by forming a 3,4-acetal and oxidizing it under similar conditions as those used for the preparation of the other isomer. This oxidation yields the dihydroxy acid and formic acid via the unsaturated aldehyde which tautomerizes to the R-dicarbonyl compound. The dihydroxy acid is then converted to the lactone **69** by acidification and concentration (❯ *Scheme 8*) [66].

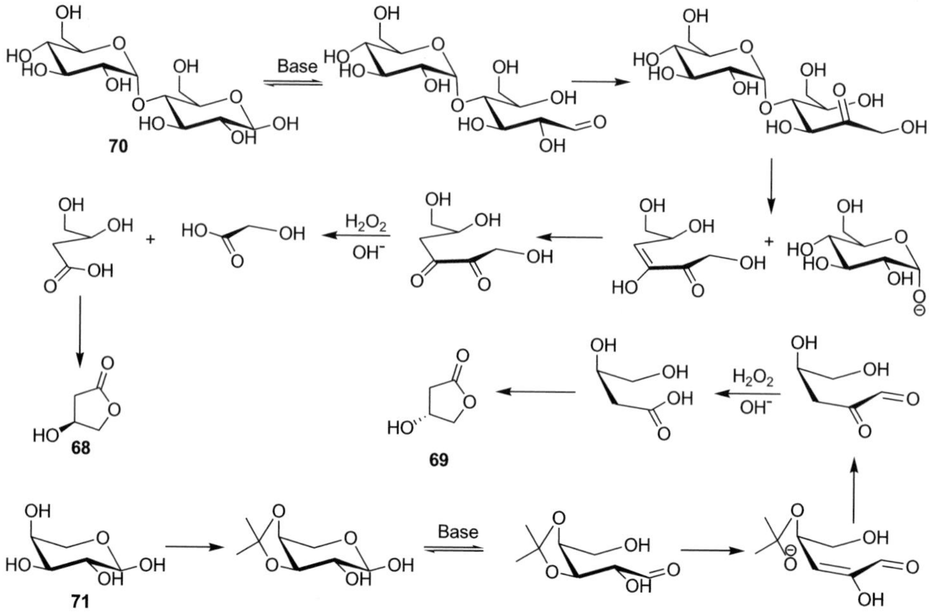

☐ Scheme 8

Titanium-containing zeolites, such as Ti-BEA, Ti-FAU, and TS-1 have been tested as catalysts for the Ruff oxidative degradation of calcium D-gluconate **72** to D-arabinose **55** using diluted hydrogen peroxide as the oxidant. Only large-pore zeolites Ti-BEA and Ti-FAU were found to be active. It was shown, in particular, that a very rapid leaching of titanium occurred and that the titanium species present in the solution were responsible for the catalytic activity observed [67,68].

Applying H_2O_2/CuO in alkaline solution, degradation of the carbohydrate-rich biomass residues results with formic, acetic and threonic acids as the main products. Gluconic acid was formed instead of glucaric acid throughout. Reaction of a 10% H_2O_2 solution with sugar beet molasses generated mainly formic and lactic acids. Important advantages of the microwave application were lower reaction times and reduced reagent demands [69].

3.5 Enzymatic Degradations

1,2,4-Butanetriol is an important intermediate in organic synthesis, for instance in the production of D,L-1,2,4-butanetriol trinitrate. Commercial synthesis of D,L-1,2,4-butanetriol employs $NaBH_4$ reduction of esterified D,L-malic acid. For every ton of 1,2,4-butanetriol synthesized, multiple tons of byproduct borates are generated. D,L-malic acid can also be hydrogenated over various catalysts (Cu–Cr, Cu–Al, Ru–Re) at 2900–5000 psi of H_2 and 60–160 °C reaction temperatures. Yields of 1,2,4-butanetriol range from 60 to 80%. A variety of byproducts are also formed during high-pressure hydrogenation. These byproducts are not generated when esterified malic acid is reduced using $NaBH_4$. D,L-malic acid is synthesized from the *n*-butane component of liquefiable petroleum gas via the intermediacy of maleic anhydride. The new synthesis of 1,2,4-butanetriol has been established with microbes. Enzymes from three different microbes are recruited to create biosynthetic pathways by which D-1,2,4-butanetriol **73** and L-1,2,4-butanetriol **74** are derived from D-xylose **39** and L-arabinose **55**, respectively [70] (❯ *Scheme 9*).

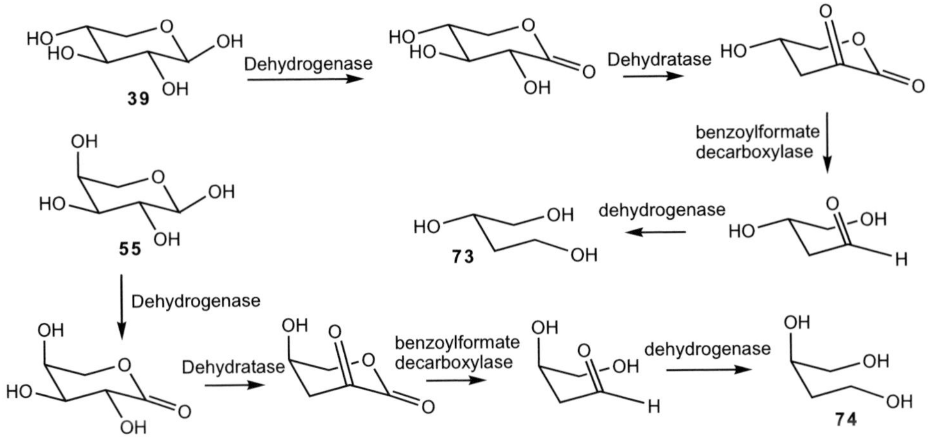

❑ Scheme 9

The use of ethanol as an alternative automobile fuel has been steadily increasing around the world for a number of reasons [71]. Domestic production and use of ethanol for fuel can decrease dependence on foreign oil, reduce trade deficits, create jobs in rural areas, reduce air pollution, and reduce global climate change carbon dioxide buildup. Ethanol, unlike gasoline, is an oxygenated fuel that contains 35% oxygen, which reduces particulate and NOx emissions from combustion. Ethanol can be made synthetically from petroleum or by microbial conversion of biomass materials through fermentation. In 1995, about 93% of the ethanol in the world was produced by the fermentation method and about 7% by the synthetic method. The fermentation method generally uses three steps: (1) the formation of a solution of fermentable sugars, (2) the fermentation of these sugars to ethanol, and (3) the separation and purification of the ethanol, usually by distillation. Ethanol-from-cellulose (EFC) holds great potential due to the widespread availability, abundance, and relatively low cost of cellulosic materials. However, although several EFC processes are technically feasible, only recently have cost-effective EFC technologies begun to emerge, which are quite important for rapidly developing countries such as China and Canada [72].

4 Rearrangement with Double Bond Shifts

4.1 [2,3]-Sigmatropic Rearrangements

Double bond rearrangements in carbohydrate systems lead to various kinds of sugar transformations. The [2,3]-Wittig rearrangements [73] initiated by deprotonation and followed by migration of an anionic substituent are illustrated in ❷ *Scheme 10*. The stereochemistry at C1 is well transformed to C3 and/or C4 [74,75]. The [1,2]-Wittig rearrangement without a double bond shift occurs in dependence on the conditions.

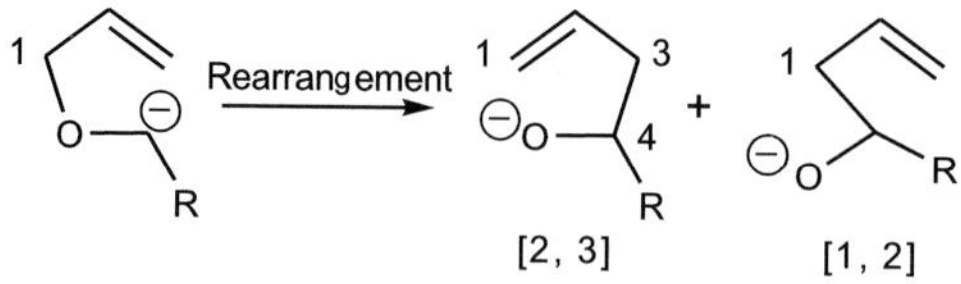

❑ Scheme 10

The [2,3]-Witting rearrangement has been employed in synthetic work on the tetrahydrofuran acetogenins from *Annonaceous* species starting from furanoid glycals [76]. The rearrangement of **75** is induced by a base to generate an anionic species, which rearranges into **76** and its epimer. In this case, erythro-2 predominates. Under the same conditions, **77** with a silyl-protecting group mainly gives the [1,2]-Witting rearrangement product **78** (❷ *Scheme 11*).

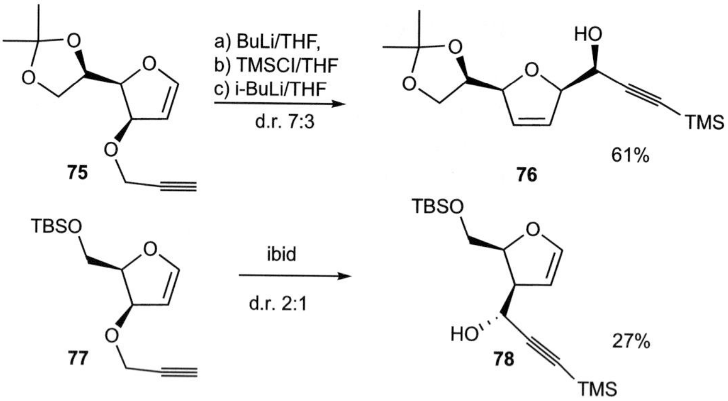

❑ Scheme 11

A [2,3]-sigmatropic rearrangement of a sulfoxide has been employed in the total synthesis of calicheamicin g [77]. The thioglycoside **79** is oxidized to give the sulfoxide intermediate, which spontaneously undergoes a suprafacial sigmatropic shift to the β-position to move the double bond towards the anomeric center. The resulting sulfinate is treated with a secondary amine to afford the desired rearranged glycal derivative **80** (❷ *Scheme 12*).

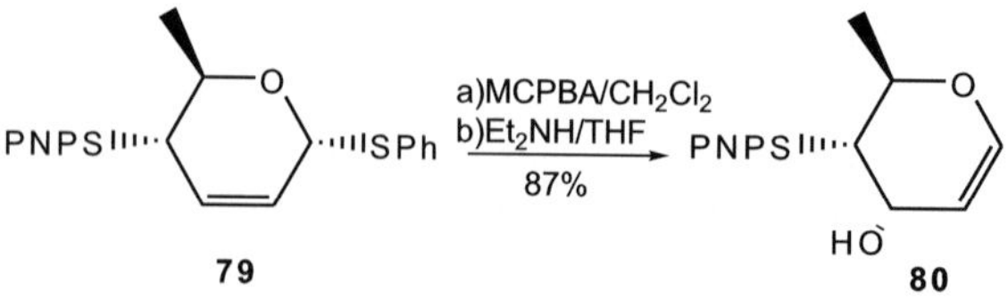

❑ Scheme 12

Recently, the *C*-analogue of sulfatide **83** was synthesized through a [2,3]-Wittig sigmatropic rearrangement (❯ *Scheme 13*) [78].

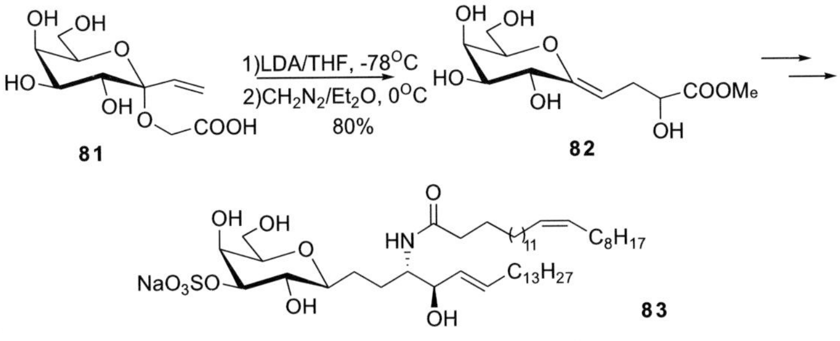

■ Scheme 13

4.2 [3,3]-Sigmatropic Rearrangements

4.2.1 Overman Rearrangement and Related Reactions

An allylic system that is easy to rearrange is a useful tool in a variety of synthetic methods. Allylic alcohol is readily converted into the corresponding trichloroacetimidate by brief treatment with trichloroacetonitrile in the presence of an appropriate base and usually results in high yields. Simple heating of the imidate of the allylic alcohol system induces the rearrangement reaction [79,80] (❯ *Scheme 14*).

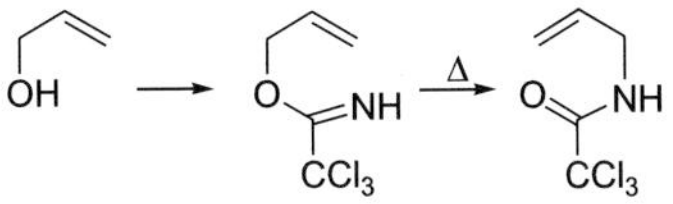

■ Scheme 14

For example, on heating at 160 °C in 1,2-dichlorobenzene, the allylic trichloroacetimidate **84** smoothly rearranges into the corresponding 2-amino-2-deoxy sugar **85** [81]. The suprafacial rearrangement from C2 to C4 is similarly performed to obtain the 4-amino-4-deoxy sugar derivative **87** [82] (❯ *Scheme 15*).

Under dehydrating conditions, the allylic carbamate **88** generates the allyl cyanate which, in turn, rearranges into the reactive allyl isocyanate and then reacts with nucleophiles [83] (❯ *Scheme 16*). The allylic carbamate is prepared by treatment of the allylic alcohol with trichloroacetyl isocyanate in dichloromethane at 0 °C and chemoselective removal of the alkali-labile trichloroacetyl group by mild reaction with cold methanolic potassium carbonate without affecting the carbamate linkage at all. The obtained carbamate **88** is dehydrated by the triphenyl phosphine/tetrabromomethane system under very mild conditions. This leads to the reactive isocyanate via spontaneous rearrangement. The isocyanate thus generated is trapped with a nucleophile such as pyrrolidine to furnish the aminosugar derivative **89**.

In the recent total synthesis of sphingofungin E (**90**), Overman rearrangement of an allylic trichloroacetimidate derived from diacetone-D-glucose **91** generated tetra-substituted carbon

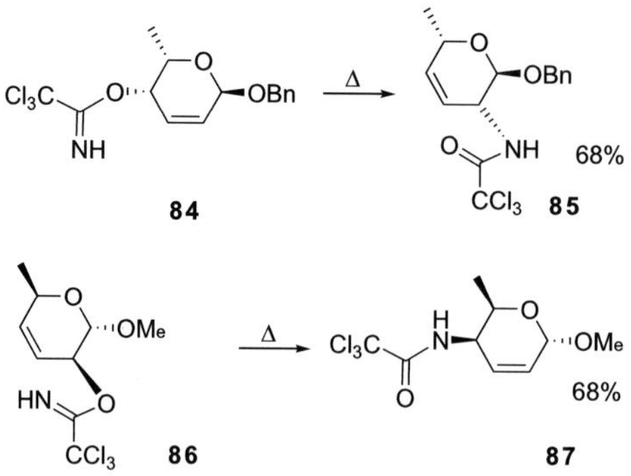

☐ **Scheme 15**

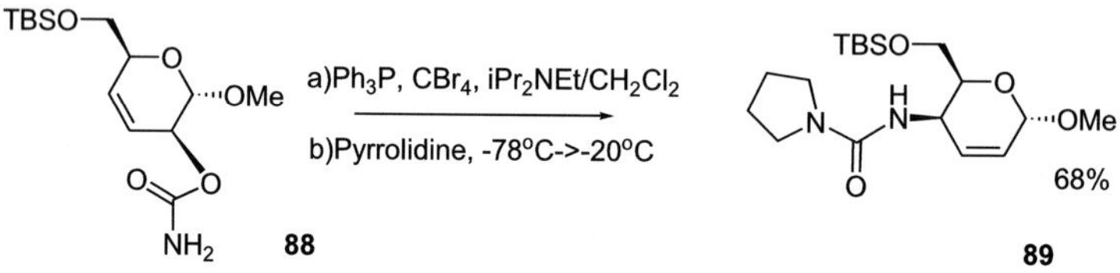

☐ **Scheme 16**

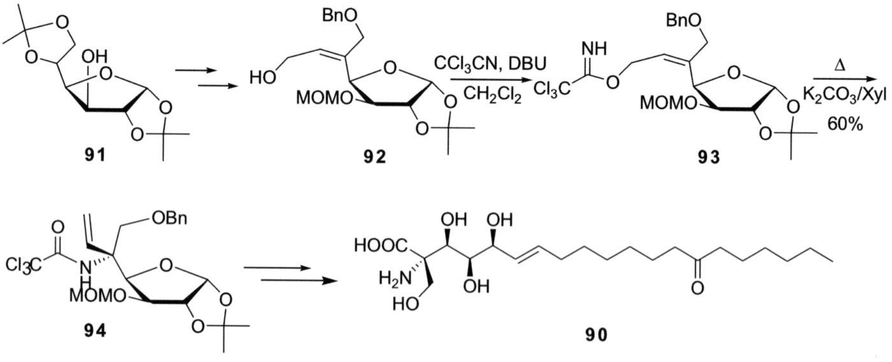

☐ **Scheme 17**

with nitrogen (**94**), and subsequent Wittig olefination afforded the highly functionalized part in sphingofungin E stereoselectively [84] (❷ *Scheme 17*).

4.2.2 Modified Claisen Rearrangements

Modifications of the Claisen rearrangement have been widely used in a variety of synthetic chemistry reactions [85] (❷ *Scheme 18*).

The simple Claisen rearrangement itself has been employed in the transformation of the vinylglycal **95** into carbocyclic compounds [86] (❷ *Scheme 19*). On heating at 240 °C in o-dichlorobenzene in a sealed tube for 1 h, the desired rearrangement of **95** proceeds in the

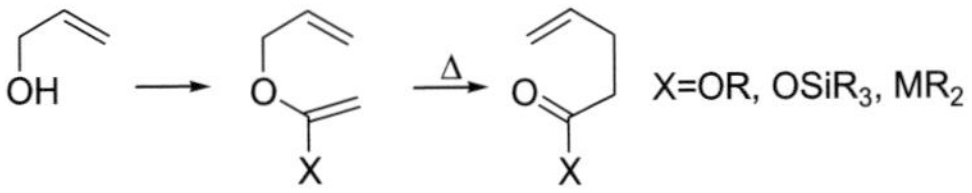

Scheme 18

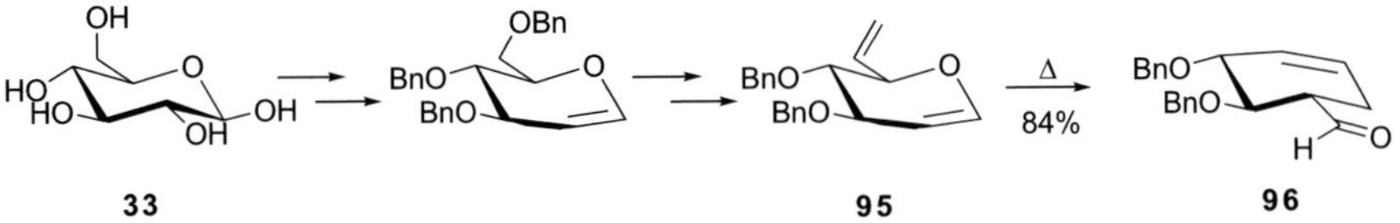

33   95   96

Scheme 19

expected direction to give the unsaturated carbocyclic system bearing an aldehyde function in 84% yield. This is a useful synthetic intermediate for a variety of pseudosugars.

Aromatic Claisen rearrangements in 2,3-unsaturated sugar systems are useful for the stereo-controlled synthesis of aryl-branched sugars [87] (❯ *Scheme 20*). The α-anomer **97** is much less reactive in comparison to the β-anomer **99**. This thermal rearrangement is carried out by refluxing in *N,N*-diethylaniline. The efficiency of the reaction is almost independent of the nature of the p-substituent in the phenyl group.

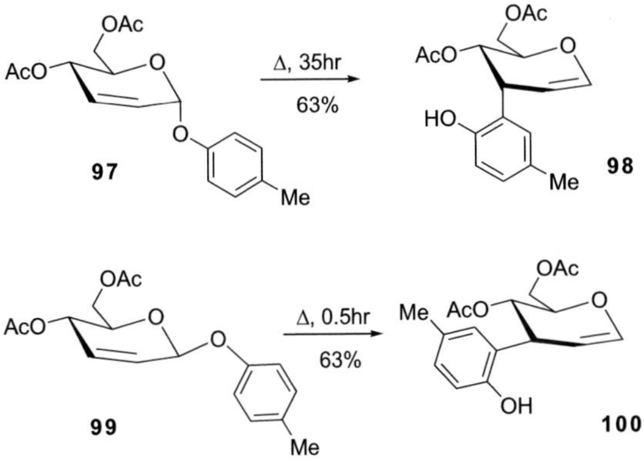

97   98

99   100

Scheme 20

4.2.3 Hetero-Cope Rearrangements

Cationic aza-Cope/Mannich tandem reactions [88,89,90,91] have been applied to the asymmetric synthesis of homochiral proline derivatives (azafuranosides) [92]. The β-amino alcohol **101** reacts with glyoxal at room temperature to generate a cyclic aminoacetal, which undergoes spontaneous dehydration to give rise to the ene-iminium intermediate (❯ *Scheme 21*). Through an aza-Cope reaction, this cationic species transforms into the bond-rearranged exomethylene intermediate. Then a Mannich-type cyclization takes place to give the homochiral proline derivative **102** quantitatively.

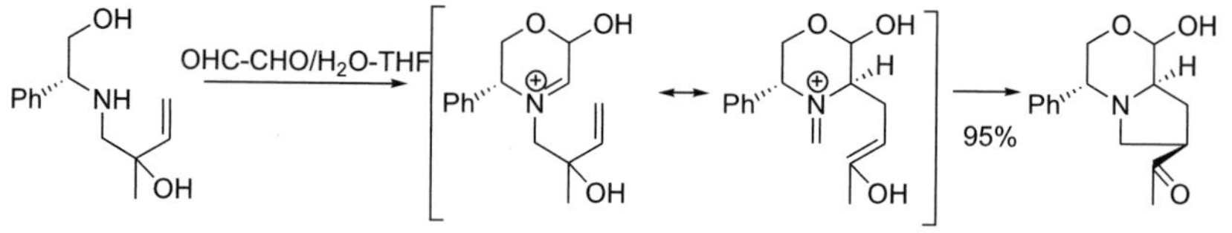

101 **102**

☐ Scheme 21

This protocol was later employed for the synthesis of $(-)$-α-allokainic acid [93]. Tandem aza-Cope/Mannich reactions of this type have also been employed to construct the framework of $(-)$-preussin [90]. On refluxing in trifluoroacetic acid, the protonated and functionalized oxazolidine **103** changes into the ene-iminium intermediate, which equilibrates with the bond-rearranged enol compound through an aza-Cope process (❷ Scheme 22). This is followed by cyclization through the Mannich reaction to give the functionalized pyrrole **104** in 78% yield with 86% ee. To proceed to $(-)$-preussin, a retro-Mannich fragmentation-Mannich cyclization of **104** is needed to establish the desired configuration of the pendants on the pyrrolidine ring.

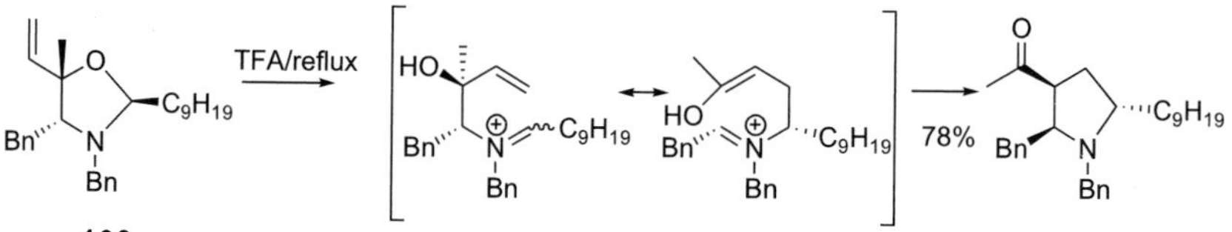

103 **104**

☐ Scheme 22

A neutral, metal-free rearrangement, formally a suprafacial [1,3]-sigmatropic migration, of the hydroxy group has been reported [94] (❷ Scheme 23). The direct migration of the hydroxy group is thermally forbidden. This rearrangement reaction probably proceeds by way of an intermediate formate, which undergoes an oxy-Cope rearrangement. Diethylaminosulfur trifluoride (DAST) is considered to react with the solvent DMF to generate the reactive quaternary amine salt, which rapidly converts **105** into the corresponding formate. The hypothetical 4-O-formate would then undergo acyloxy group migration accompanied by a double bond shift through an oxy-Cope rearrangement to give the 2-O-formate. Thus, the net results are suprafacial 1,3-shifts of the hydroxy group from the C4 to the C2 position. No substitution reaction of the hydroxy group with the fluoride ion seems to occur during this reaction.

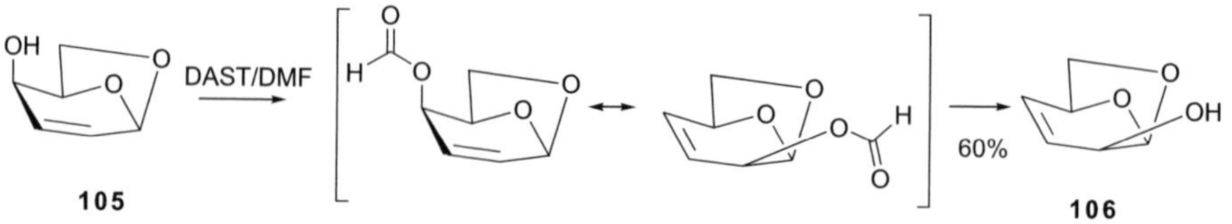

105 **106**

☐ Scheme 23

As an artificial enzyme, AbZyme was applied in a similar [3,3]-sigmatropic rearrangement [95,96]. The substrate **107** is prepared from a diacetone-D-mannitol through conventional synthetic transformations. On exposure to the artificial polychronal antibody in the presence of 2-(*N*-morpholino)ethanesulfonic acid and sodium chloride at 37 °C, with a molar ratio of 100:1 of the hexadiene **107** to the antibody, **107** is completely converted into the product **108** in 20 h (❷ *Scheme 24*).

107   **108**

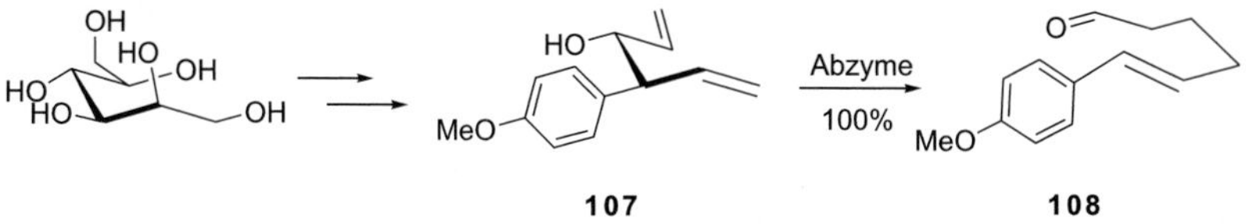

■ Scheme 24

4.3 Double Bond Inducing Ring-Closing Rearrangements

There are a variety of examples of ring-closing rearrangements with exhausting double and/or triple bonds and some recent examples are shown here. Pd(0)-complexes catalyze reactions of the unsaturated amine **109** to give the azasugar **110**, an intermediate in the synthesis of SS20846A **111** [97,98,99,100] (❷ *Scheme 25*).

109   **110**   **111**

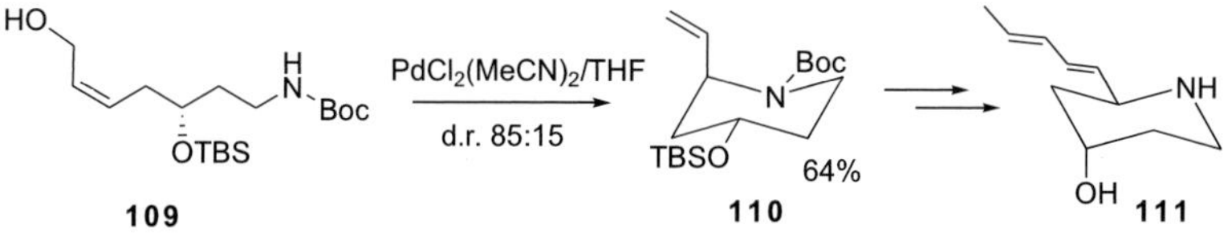

■ Scheme 25

Besides the conventional methods, the metallo-carbene route to access cyclic compounds has become a versatile tool in sugar chemistry. Synthesis of stavudine **112**, an antiviral nucleoside, from an allyl alcohol [101] is realized by a Mo(CO)₅-mediated cyclization reaction (❷ *Scheme 26*). Molybdenum hexacarbonyl smoothly reacts with the triple bond of **113** to generate the intermediate Mo-carbene, which undergoes a clean cyclorearrangement to yield the furanoid glycal **114**. Alkynol isomerization is effected by group-6 transition metal carbonyl complexes [102].

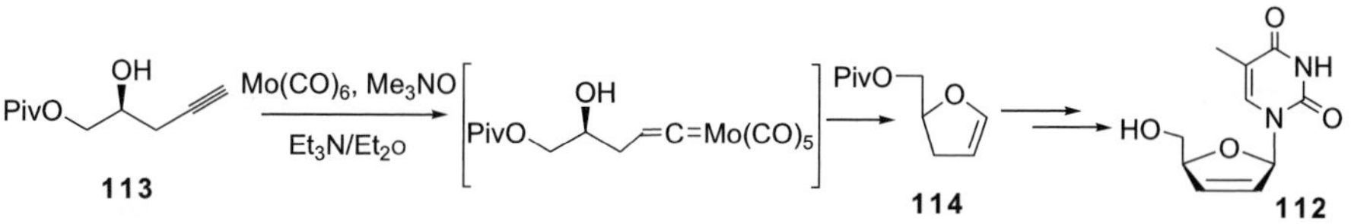

113   **114**   **112**

■ Scheme 26

5 Ring Isomerizations

Ring transformations are useful reactions in synthetic carbohydrate chemistry [103].

5.1 Ring Contractions

Nucleophilic displacement reactions of the sulfonyloxy group or its equivalents in the sugar ring are known to induce unexpected ring-contraction reactions [104]. However, the first target-oriented ring-contraction reaction of the sulfonate **115** (❷ *Scheme 27*) in the stereoselective total synthesis of (−)-rosmarinecine from D-glucosamine impressively demonstrated the novel utility of this kind of reaction [105]. Ring-contraction reactions of carbohydrates have now become a useful tool for syntheses of various types of compounds [103].

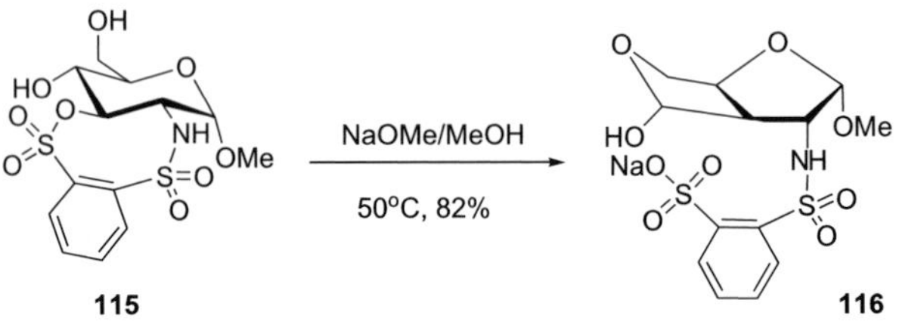

❑ Scheme 27

Epoxy sugars are good starting materials for the preparation of ring-contracted products. Various 2,3-epoxypyranosides such as **117** can be converted into furanosides directly by simple heating under reflux in toluene containing lithium bromide and *N*,*N*-tetramethylurea (TMU) [106] (❷ *Scheme 28*). Usually, the more stable 3-*C*-formyl derivatives are formed. In the case of the *C*-glycoside **119**, however, the mode of reaction changes to yield mainly the 2-*C*-formyl compound **120**.

Zr-mediated ring-contraction reactions using vinyl sugars are useful to synthesize carbocycles [107,108]. This method was later employed successfully for aza-sugar synthesis. The

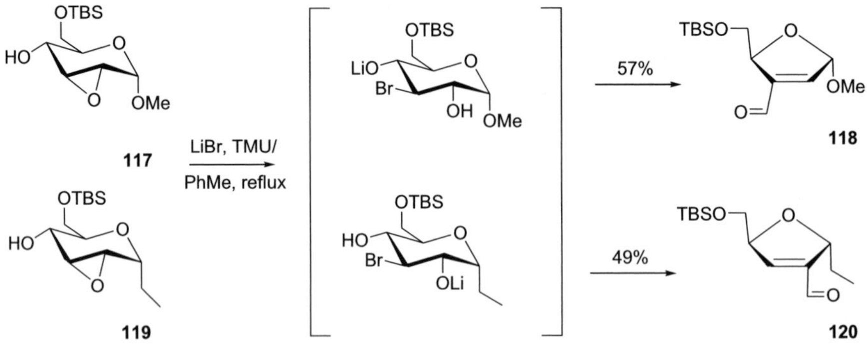

❑ Scheme 28

functionalized morpholine **121** is transformed into the pyrrolidine **122** with excellent stereose-lectivity. The stereochemistry at the junction of the main product is *cis*. This protocol has been applied to the synthesis of inositol phosphate analogs using the 5-*C*-vinyl glycoside deriva-tive **123** [109] (❯ *Scheme 29*).

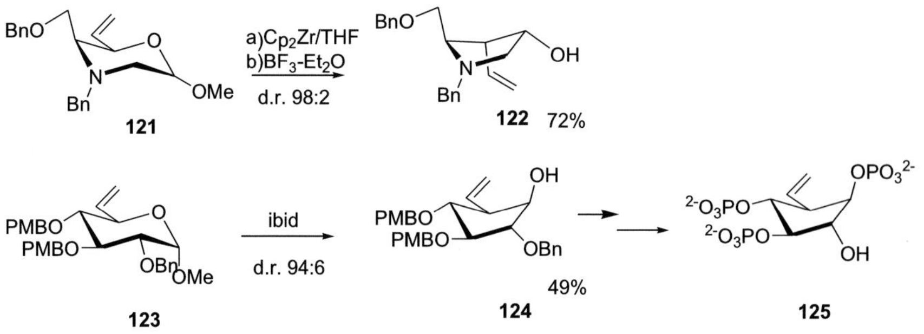

■ Scheme 29

The *O*-benzyl derivative of the glycal **126** undergoes stereoselective ring contractions on treat-ment with thallium(III) nitrate [110] (❯ *Scheme 30*).

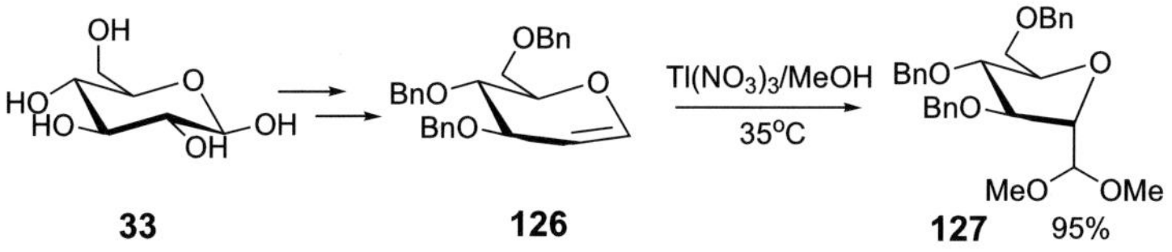

■ Scheme 30

Triflates of aldonolactones are a productive source of ring-contraction reactions. Com-pound **128** contracts its ring under acidic and basic conditions to give *C*-furanosides [111,112] (❯ *Scheme 31*). Triflates of glycosides occasionally yield ring-contracted products [113]. Another paper has provided an additional example of a ring-contraction reaction of a sug-ar triflate on reaction with tetrabutylammonium nitrite in moist toluene; the triflate yields a ring-contracted byproduct [114]. It has been found that the ring-contraction reactions of the triflate in the presence of pyridine depend on the acidity of the solvent; in the acidic solvent 1,1,1,3,3,3-hexafluoro-2-propanol (HFIP), ring-contraction of **130** proceeds smoothly. Oxygenophilic silyl protection of the 2-*O*-triflate of aldonolactone **131** prohibits ring contrac-tion so that the reaction results in the formation of the silyl-migrated epoxide **132** [115].

Mitsunobu conditions smoothly effect clean ring-contraction reactions of thiosugars [116] (❯ *Scheme 32*). From the thioheptanoid **134**, the thiopyranoid **135** is obtained. Mild sulfona-tion induces a spontaneous ring-contraction of the azaheptanoid **136** to afford the azapyra-noside **137** [117].

Some rare, four-membered sugar rings have been synthesized by ring-contraction reac-tions [103]. Similar to 2-*O*-triflates, which easily undergo ring-contraction reactions [118]. DAST-treatment of the thiopentofuranose derivative **138** affords the ring-contracted prod-uct **139** having a thietane framework (❯ *Scheme 33*). DAST-assisted ring-contraction has

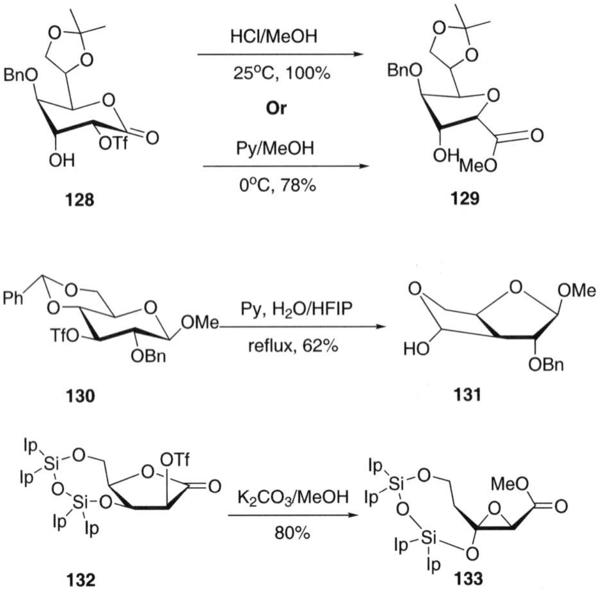

◘ Scheme 31

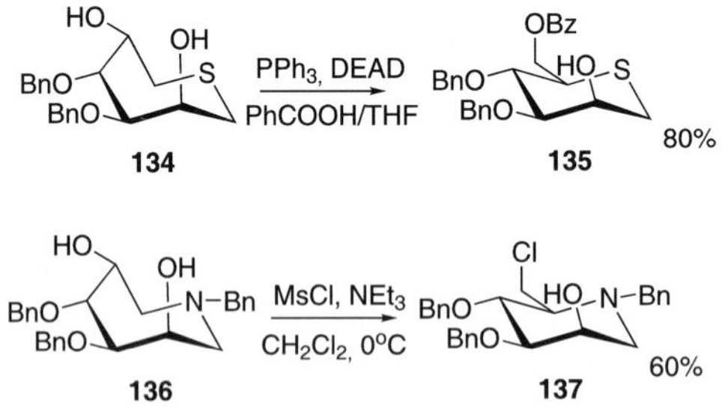

◘ Scheme 32

been found in the fluorination reaction of the thiosugar furanose **138** [119]. It is known that the sulfur(IV) fluoride/hydrogen fluoride system also promotes such ring-contraction reactions [120]. On Friedel–Craft reaction of the thiopentosyl bromide **140**, a ring-contraction process occurs [121]. The per-*O*-alkylated glycoside **142** is converted into the δ-lactone **143** with concomitant ring-contraction to furnish product **144** [122].

The Chan rearrangement was effectively used to build up the furanoid structure on the way to taxol [123,124] (**◉** *Scheme 34*).

In the syntheses of staurosporin congeners, ring-contraction reactions have been used effectively [125,126]. Novel stereoselective Beckmann-type rearrangement of TAN-1030A **147** produces the K-252 analog **148** via a hypothetical hemiacetal intermediate [127]. Oxidation of the model compound **149**, the staurosporin analog, results in ring contractive benzilic acid rearrangement to give a furanoid **150** possessing the framework of K252a [128] (**◉** *Scheme 35*).

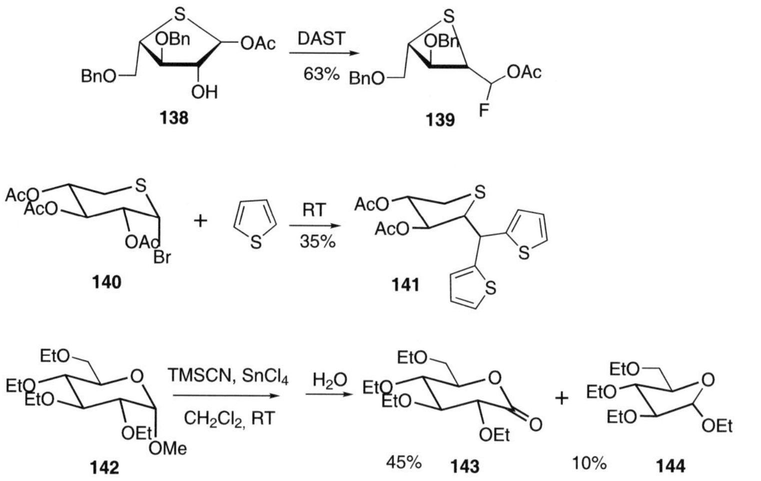

138   **139**

140   **141**

142   **143**   **144**

◻ Scheme 33

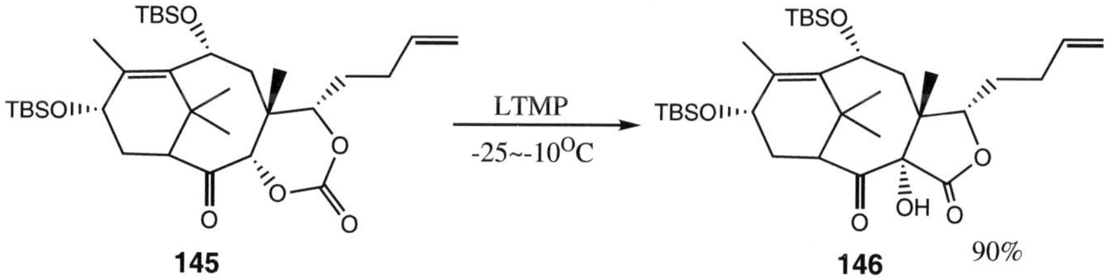

145   **146**   90%

◻ Scheme 34

149   **150**

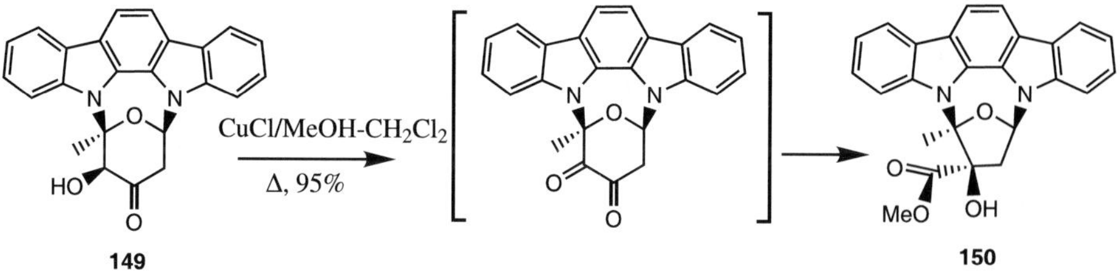

◻ Scheme 35

The protected oxyaminoglucoside **151** rearranges to the azafuranose form under deprotecting conditions [77] (❂ *Scheme 36*).

The mild reaction of the thioureido derivative **153** with methanol produces the compound **154** with migration of the acetyl group. On heating, this compound isomerizes into the *cis*-fused cyclic **155** [129] (❂ *Scheme 37*).

Epimerization at C2 of L-gulose **156** on reaction with KCN in buffered aqueous solution is thought to proceed by way of the open-chain intermediate. Free sugars produce cyclic products directly on reaction with the Wittig reagent. Thus, **159** is converted into **160** on prolonged heating with the reagent (❂ *Scheme 38*).

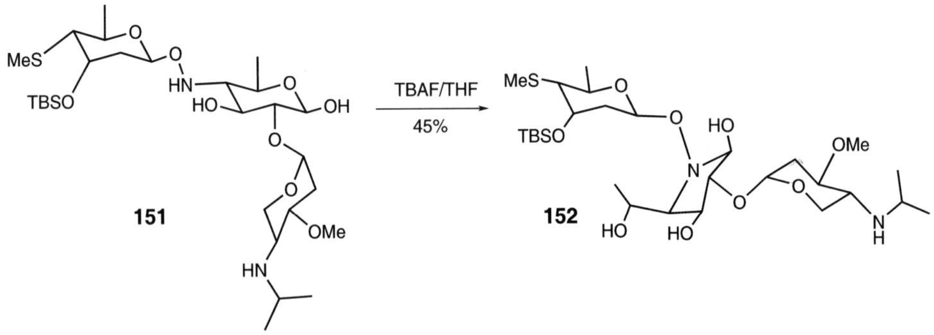

☐ Scheme 36

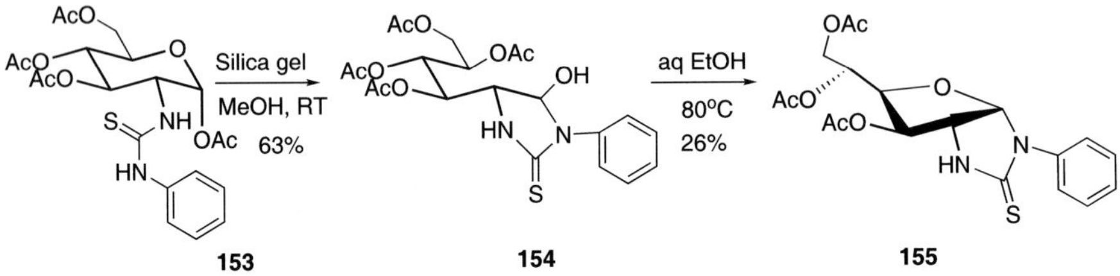

☐ Scheme 37

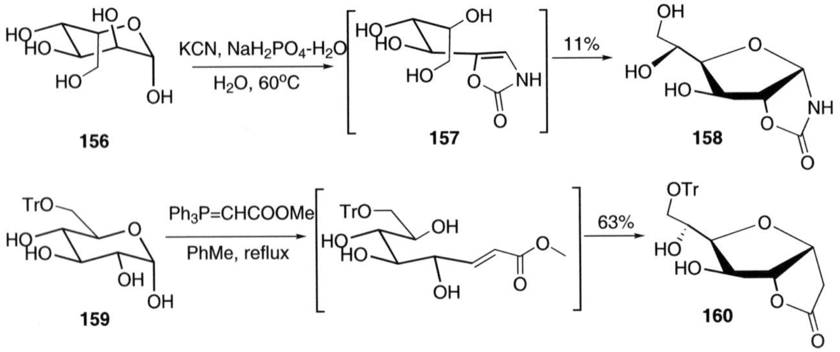

☐ Scheme 38

5.2 Ring Expansions

The cyclopropane system is a tool for inserting a methylene unit into a ring system to form a larger ring structure. Even densely functionalized pyranoids such as **161** [130,131] and cyclohexanes [132] expand into heptanoids and cycloheptanes, respectively. 1,2-*C*-dibromomethylene sugar **163** expands its pyranose ring to give oxepine **164** [133] (❯ *Scheme 39*).

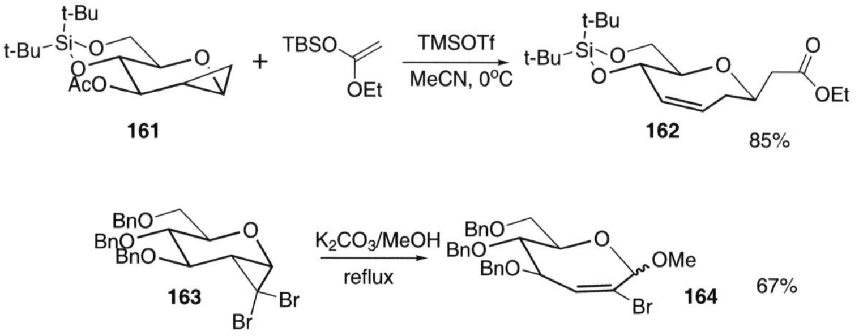

Scheme 39

α-Hydroxyfurans expand to pyranoids via sequential epoxyalcohol rearrangements. Epoxidation of the α-hydroxyfuran **165** with *meta*-chloroperoxybenzoic acid (m-CPBA) induces a cationic rearrangement, followed by dehydration, to form a pyranoid on the way to (+)-resineferatonin [134] (◗ *Scheme 40*). Dimethyldioxirane (DMDO), apparently more sensitive to the steric circumstances than m-CPBA, has been used for the selective epoxidation of the furan **167** [135]. The monoepoxide rearranges followed by hemiacetalization to afford the pyranoid intermediate **165** of the total synthesis of the eleuthesides. α-Aminofurans similarly expand into azapyranoids [136]. Racemic **169** is kinetically resolved to give (S)-**169** and the rearranged (2R, 6R)-**170** with modified Sharpless epoxidation. Compound (S)-**169** is transformed into (2S, 6S)-**170** on treatment with m-CPBA.

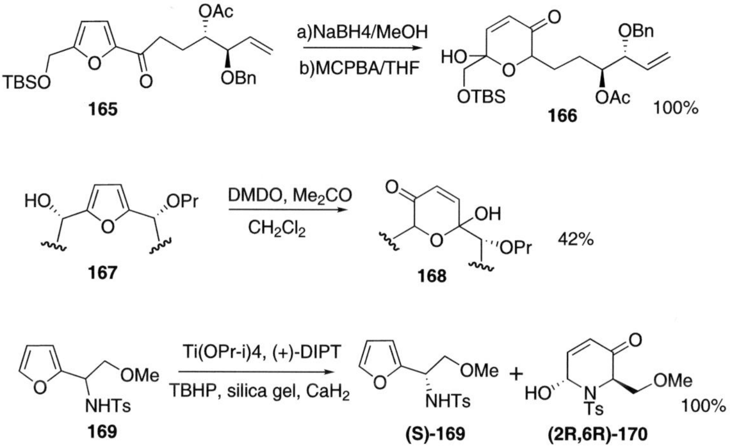

Scheme 40

The reaction of the bicyclic thiosugar **171** with N6-benzoyladenine in the presence of moist tin(IV) chloride furnishes the ring-rearranged nucleoside product **172** instead of the normal glycosylation product [137]. On heating mesylate **173** under reflux in the presence of a nucleophile, the thermal ring-expanding reaction occurs [138] (◗ *Scheme 41*).

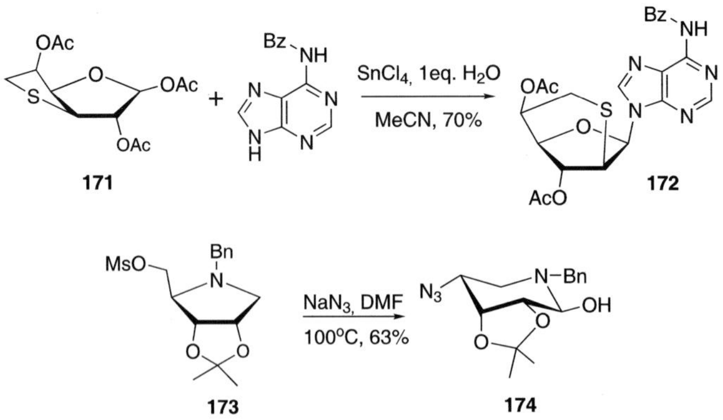

171 + **Bz–NH** $\xrightarrow[\text{MeCN, 70\%}]{\text{SnCl}_4, \text{1eq. H}_2\text{O}}$ **172**

173 $\xrightarrow[\text{100°C, 63\%}]{\text{NaN}_3, \text{DMF}}$ **174**

◻ Scheme 41

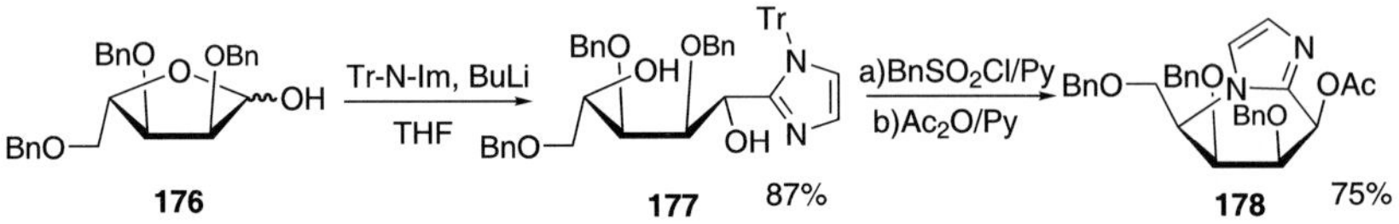

176 $\xrightarrow[\text{THF}]{\text{Tr-N-Im, BuLi}}$ **177** 87% $\xrightarrow{\text{a)BnSO}_2\text{Cl/Py} \atop \text{b)Ac}_2\text{O/Py}}$ **178** 75%

◻ Scheme 42

179 $\xrightarrow[\text{RT, 72\%}]{\text{BnNH}_2, \text{CH}_2\text{Cl}_2}$ **180**

181 $\xrightarrow{\text{a)TBAF/THF} \atop \text{b)H}_2, \text{Pd-C/EtOAc}}$ **182**

183 $\xrightarrow[\text{94\%}]{\text{H}_2, \text{Pd-C/EtOH}}$ **184**

185 $\xrightarrow[\text{94\%}]{\text{H}_2, \text{Pd-C/EtOH}}$ **186**   ⇌   **187**

188 $\xrightarrow{\text{a)H}_2, \text{Pd(OH)}_2\text{/MeOH} \atop \text{b)1N HCl}}$ **189**   49%

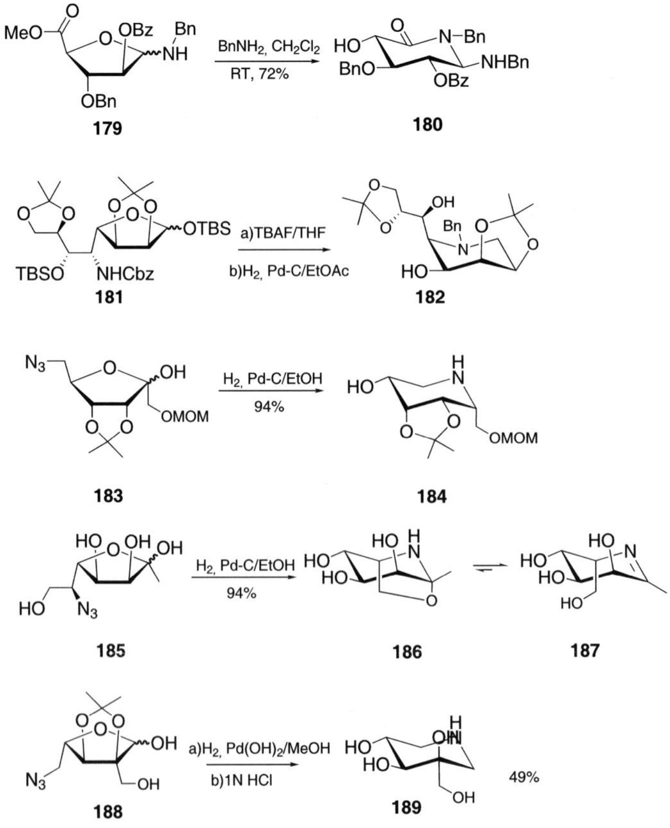

◻ Scheme 43

The key compound on the way to the debranched nagstatin **175** has been synthesized from the L-ribose derivative **176** by employing ring-chain interconversion involving addition of trityl imidazole, selective sulfonylation, and warm acetylation which causes detritylation with cyclization [139] (❯ *Scheme 42*).

The glycosylamine **179** transforms to the piperidinone **180** on reductive amination [140]. Sequential deprotection to regenerate hemiacetal OH and amino groups from **181** induces ring interconversion and reduction to give the azasugar **182** [141]. Similarly, the azidodeoxyketose derivative **183** can be converted to the piperidine derivative **184** by reductive aminocyclization [142]. Reduction of **185** affords the bicyclic azasugar **186** on intramolecular reductive cyclization, which is not a stable system and forms an equilibrium mixture with the monocyclic imine **187** [143]. Reactions of this type are also of use for the synthesis of the branched-chain 1-*N*-iminosugars such as **189**, which have been the subject of continuous attention as glycosidase inhibitors [144,145,146,147,148,149,150,151,152] (❯ *Scheme 43*).

On deprotection, followed by neutralization, the acetal **190** rearranges spontaneously to the azasugar **191** [153], an analog of the indolizine alkaloids for which synthetic approaches starting from carbohydrates [154] have recently been described [155] employing the olefin

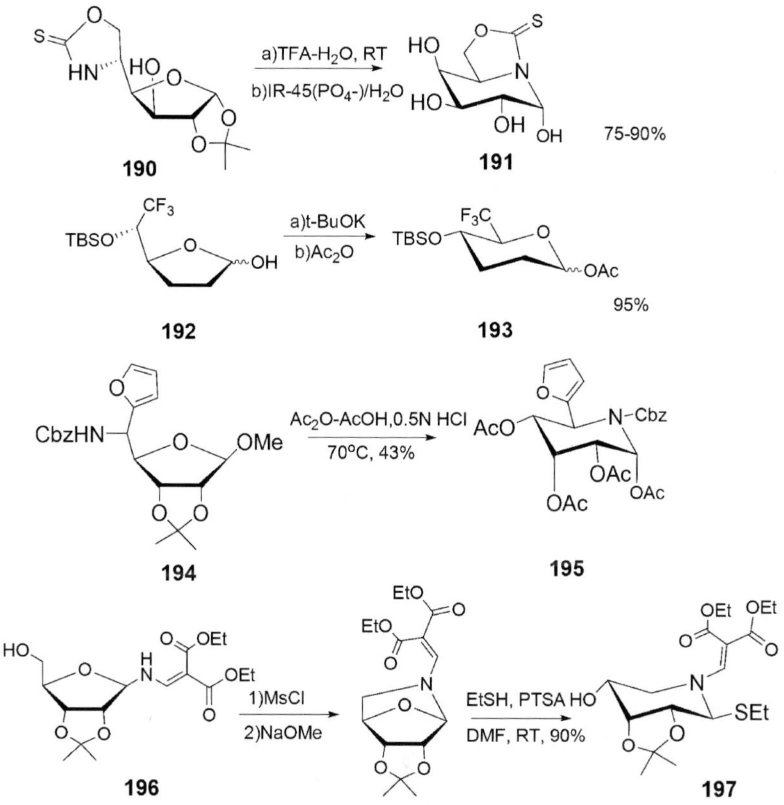

■ **Scheme 44**

metathesis protocol [156,157,158,159]. Under basic conditions, silyl-group shifts occur in a 6-deoxy-6,6,6-trifluorosugar **192** and form a pyranoside derivative [160]. Acetolysis of the methyl glycoside **194** mainly affords the piperidine **195** [161]. Azasugar ethyl thioglyco-side **197**, a new type of azasugar derivative, can be stereoselectively prepared from suitable glycosylenamine **196**, through anhydroazasugar derivatives. The thioethoxy group is introduced through a highly stereoselective substitution. The attack of EtSH was 100% stereose-lective [162,163] (❱ *Scheme 44*).

The protected 5-ulose derivative **198** can be converted into the piperidine **199** by reductive ami-nation [164]. The 5-*O*-sulfonyllactol **200** is reductively transformed into the azasugar **201** by way of oxime formation [165]. Reductive deprotection of the aminodeoxylactol derivative **202** affords the *N*-substituted piperidine **203** [166]. The unsaturated alcohol, readily obtained from the lactol **204** and Grignard reagent, cyclizes into the *C*-glycoside **206** [167]. The lactol **207** is converted into the unsaturated dithioacetal, which cyclizes slowly to give **208** on storage [168] (❱ *Scheme 45*).

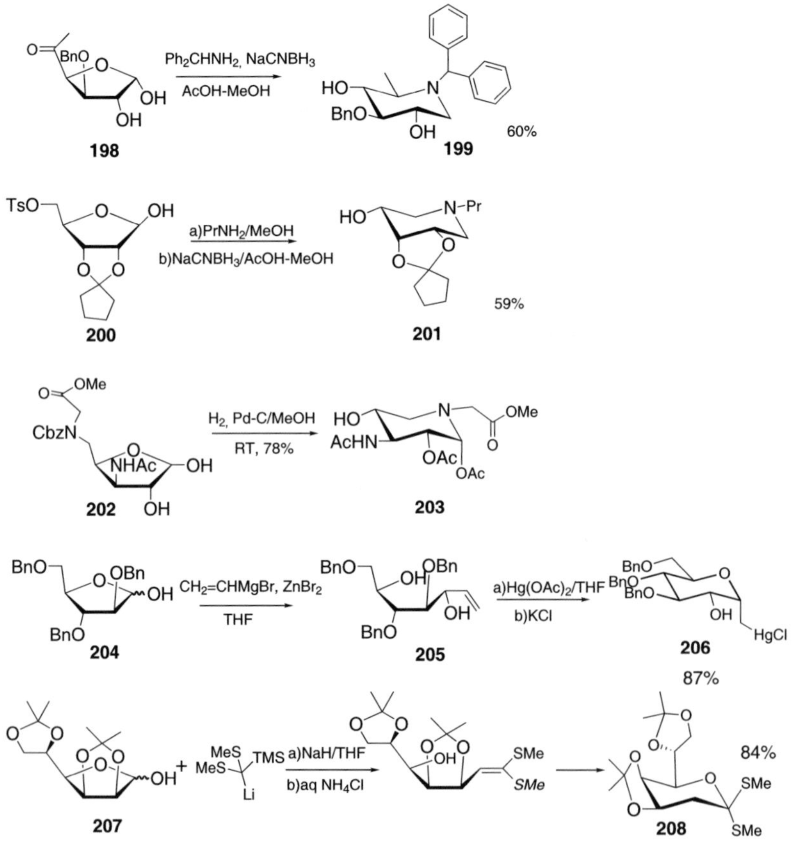

■ Scheme 45

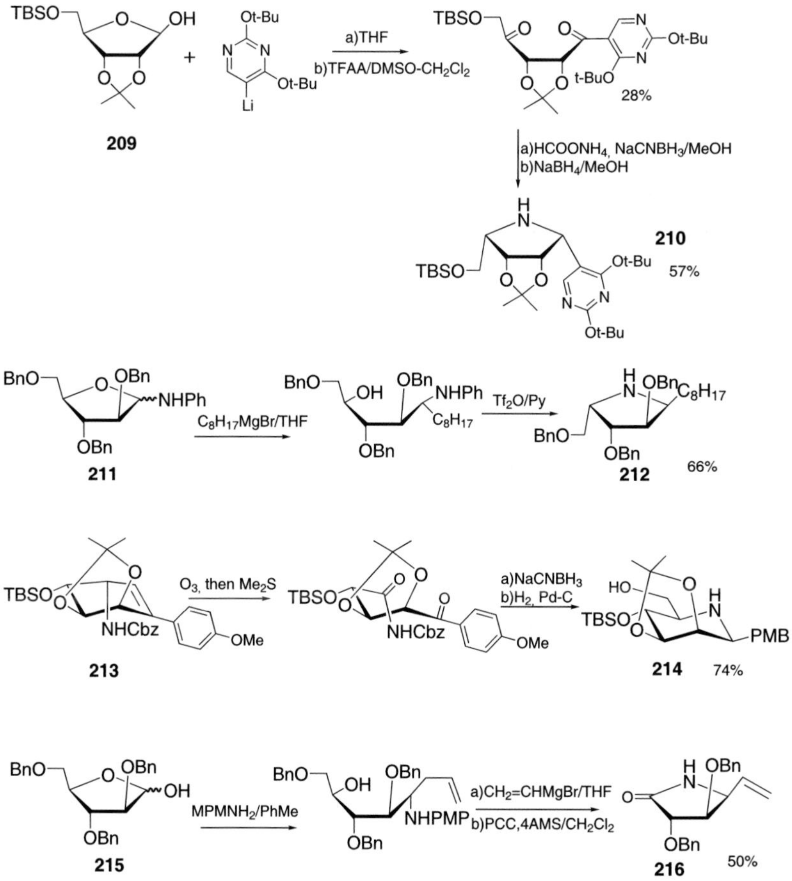

Scheme 46

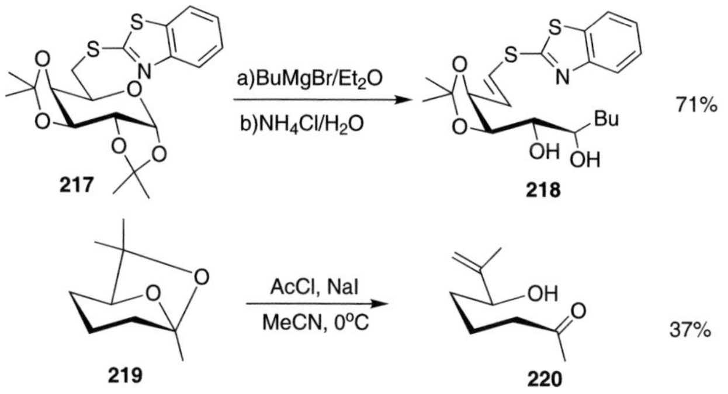

Scheme 47

5.3 Ring Transformation

Additive ring-opening of **209**, followed by Swern oxidation and aminocyclization, affords the aza-*C*-nucleoside **210** [169], belonging to an attractive class of *C*-glycosides [170,171,172,173, 174,175]. The glycosylamine **211** is converted to the azasugar **212** via an alkylative ring-opening reaction [176,177]. The aminoaldehyde derivative generated from the unsaturated aminocyclitol **213** cyclizes to give **214** [178] (❷ *Scheme 46*). Descending oxidative aminocyclization of **215** affords the lactam **216** [179].

5.4 Ring-Opening Rearrangements

A fragmentation ring-opening rearrangement reaction of **217** using a Grignard reagent has been reported [180]. The combined reagent acetyl chloride/sodium iodide induces a ring-opening rearrangement of the bicyclic ketal **219** [181]. The iodide ion serves to promote the reaction [182] (❷ *Scheme 47*).

6 Miscellaneous Reactions

6.1 Ferrier Carbocyclization and Related Reactions

The Ferrier II reaction, a carbocyclization reaction, is of widespread use as a tool for the conversion of glycosides into cyclitols [183,184,185]. Newer examples for the utilization of the reaction conducted under catalytic conditions [186,187] have appeared in the recent literature. Compound **221** is converted into cyclohexanone **222** on the way to (−)-mesembranol [188] (❷ *Scheme 48*). Compound **223** is transformed to the enone **224**, the precursor of several new cyclitol derivatives [189,190,191].

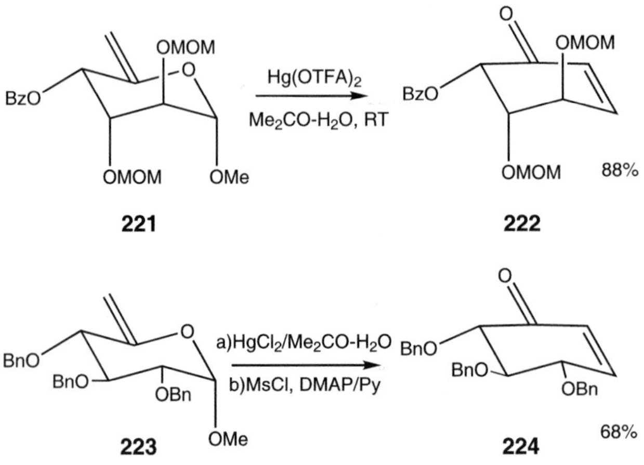

❏ Scheme 48

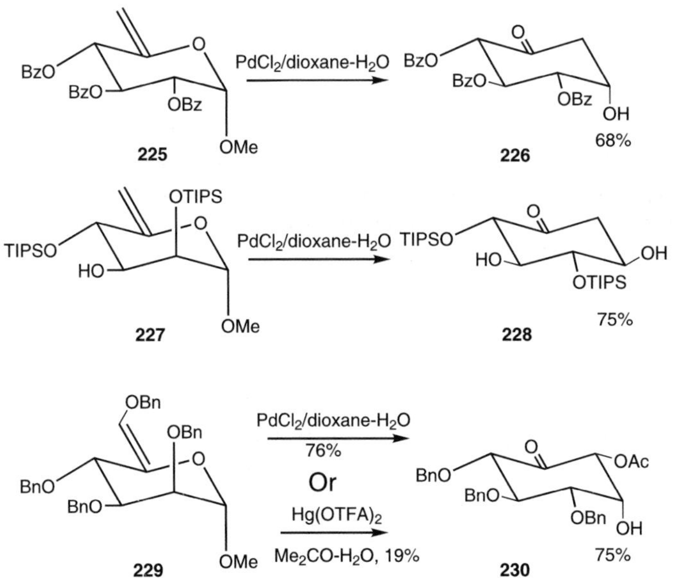

Scheme 49

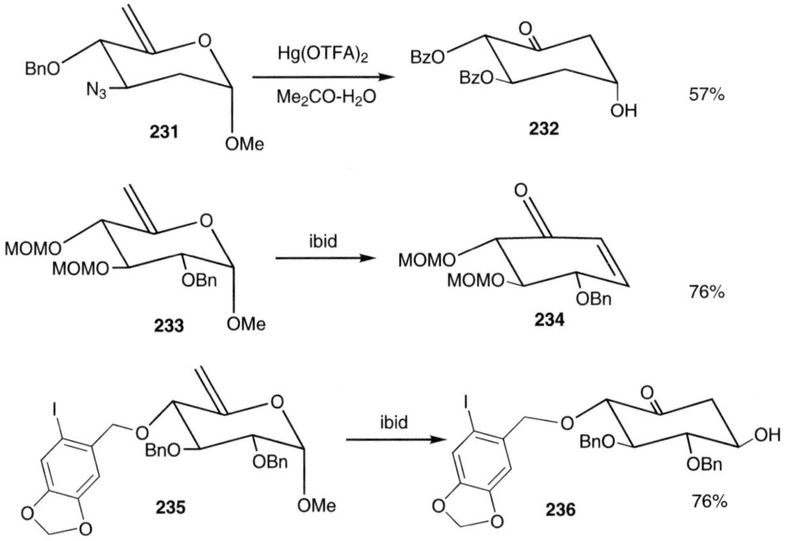

Scheme 50

The Pd-catalyzed carbocyclization affects good control on the orientation of the newly formed OH group [192] (**Scheme 49**). Thus, **225** and **227** afford the corresponding cyclitols with almost complete selectivity. In the rearrangement of **227**, the stereoselectivity is controlled by the bulky silyl ether-protecting group, which effects the conformational change.

This protocol can be applied to the 6-*O*-acetyl-5-enopyranoside **229** with good efficiency and the utility is well demonstrated by the synthesis of the D-myo-inositol phosphate, IP3 [193,194].

Compound **231** is converted to the Ferrier product **232**, the precursor of novel aminoglucosides [183] (❯ *Scheme 50*). Carbocyclization of the glycoside **233** gives the cyclohexane **234**, from which tetrazoline analogs can be synthesized [195]. The Ferrier cyclization found new utility in the synthetic chemistry of *Amaryllidaceae* alkaloids [196]. Thus, the glycoside **235** is transformed to the Ferrier-II product **236**, the logical intermediate to 7-deoxypancratistatin. A novel reductive carbocyclization of hex-5-enopyranosides retains the substituent at the anomeric center and the ring oxygen remains as the new hydroxy group [197]. The stereo-

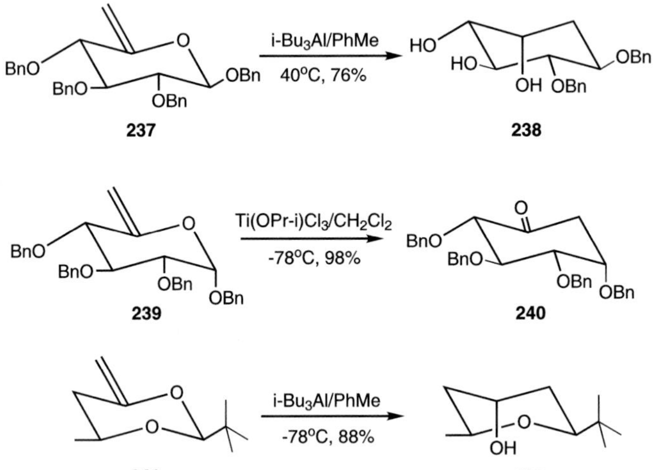

❑ Scheme 51

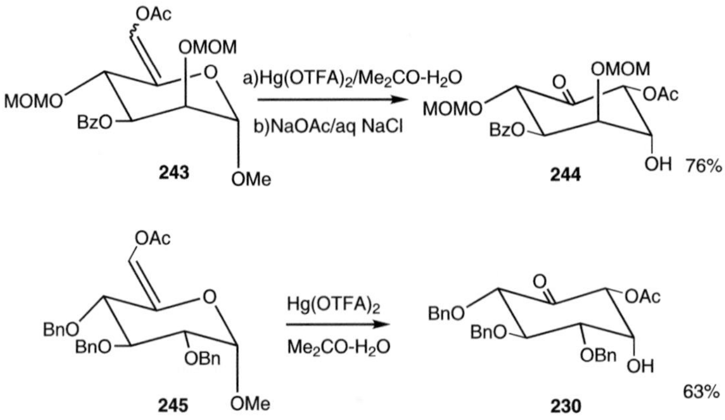

❑ Scheme 52

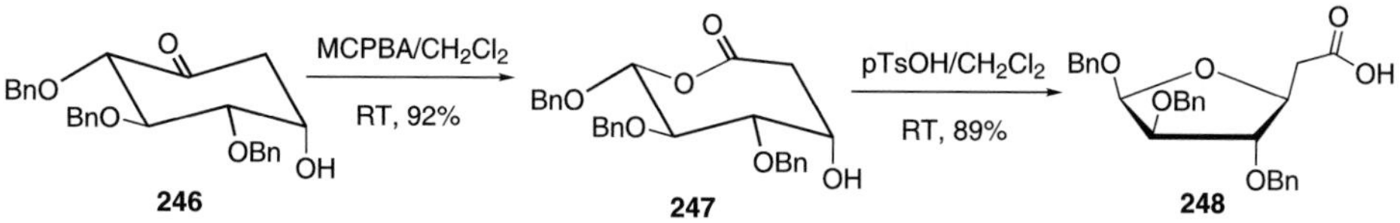

◻ Scheme 53

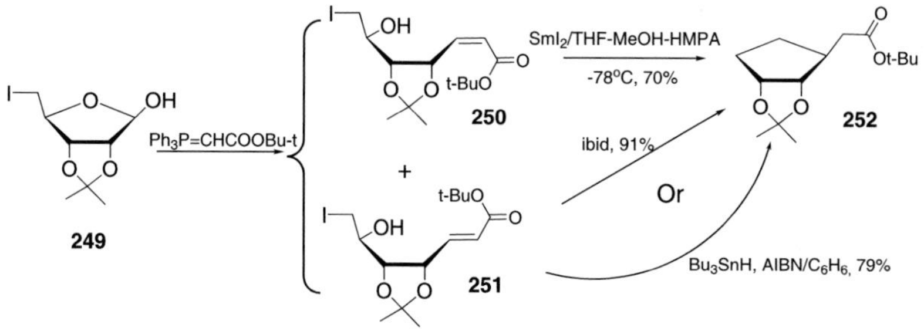

◻ Scheme 54

chemistry at the anomeric centers is retained as exemplified by the conversion of **237** to **238** (❯ *Scheme 51*). A more efficient cyclization also retains the aglycone; the glycoside **239** affords the cyclohexanone **240**. The cyclic acetal **241** is converted to the pyran **242** reductively [198,199,200].

The enol acetate **243** affords the Ferrier product **244**, a key compound to L-chiro-inositol polyphophates [201]. The Ferrier cyclization of **245** is useful for the preparation of the key intermediate to glycosylphophetidylinositols [202] (❯ *Scheme 52*).

Combination of the Ferrier-II and the Baeyer–Villiger reactions leads to the stereoselective synthesis of rare 5-deoxyfuranosiduronic acids [203]. As exemplified, the oxidation of the Ferrier II product **246**, followed by hydrolysis, gives the acid **248** [204] (❯ *Scheme 53*).

The evolution of SmI_2 as a reagent in synthesis has been one of the exciting recent developments in organic chemistry. The construction of highly functionalized carbocycles from carbohydrates promoted by SmI_2 is currently receiving significant interest and a series of carbocyclization strategies have been described in the literature. Treatment of the lactol **249** with the Wittig reagent readily gives the olefins, which undergo radical-induced cyclization [205]. Cyclization of the (Z)-isomer **250** under the action of SmI_2 is more stereoselective than that of the (E)-isomer **251** [206]. In the case of **252**, the diastereomeric excess of the products significantly depends on the choice of the reducing agents (❯ *Scheme 54*).

Stepwise conversion of the iodoglycoside **253** via Grob–Vasella fragmentation and cyclorearrangement induced by SmI_2 furnishes the carbocycles **255** and **256** with a trans-junction [207,208]. This reaction can be carried out in a one-pot manner whereby SmI_2 induces the fragmentation of the iodoglycoside [209]. While the iodoglycoside **258** mainly affords the carbocycle **259** with a cis-junction, the reaction of **253** only gives the quinovoside **257** (❯ *Scheme 55*).

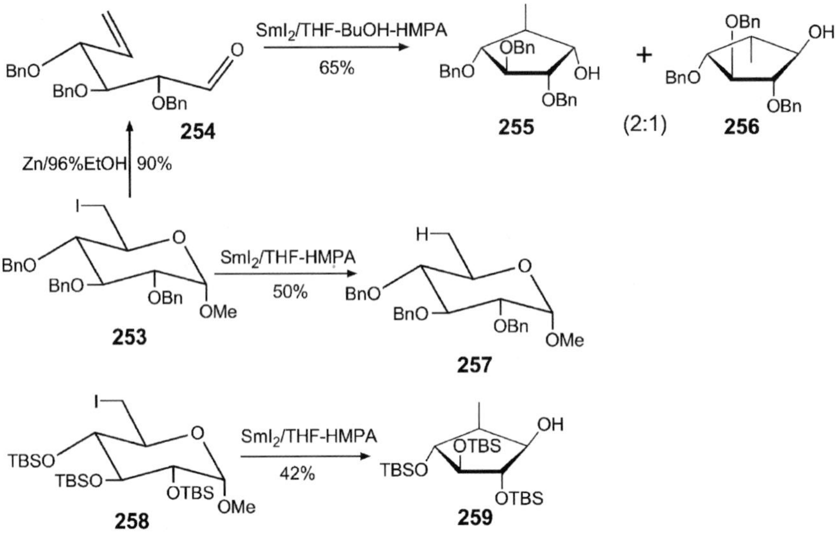

□ Scheme 55

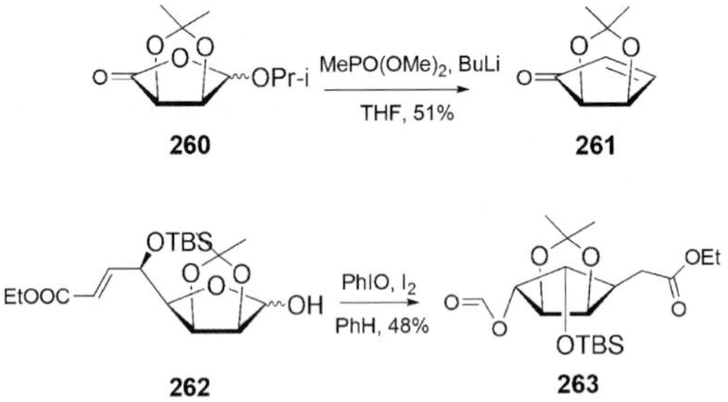

□ Scheme 56

The glycoside **260** is converted into the cyclopentenone **261** on reaction with dimethyl methanephosphonate and base [210] (❷ *Scheme 56*). The tandem β-fragmentation-cycloisomerization of the unsaturated lactol **262** gives the carbocycle **263** [211].

6.2 Anomerization and Related Rearrangements

Anomerization is a characteristic reaction of sugar [212,213]. The well-known reagent, Pascu's TiCl$_4$ for the anomerization of acetylated glycoside, rapidly anomerizes the benzyl-protected glucoside **264** [214,215] (❷ *Scheme 57*). The results from inhibition experiments indicate that TiCl$_4$ might coordinate with O5 and O6 to form a ring-opened intermediate. The use of catalytic amounts of TiBr$_4$ combined with MgBr$_2$·OEt$_2$ allows us to carry out longer reactions: the disaccharide **266** anomerizes to **267** completely. It has been reported that β-glycosides such as **268** anomerizes quantitatively [216]. Although the acetylated glycoside **270** is anomer-

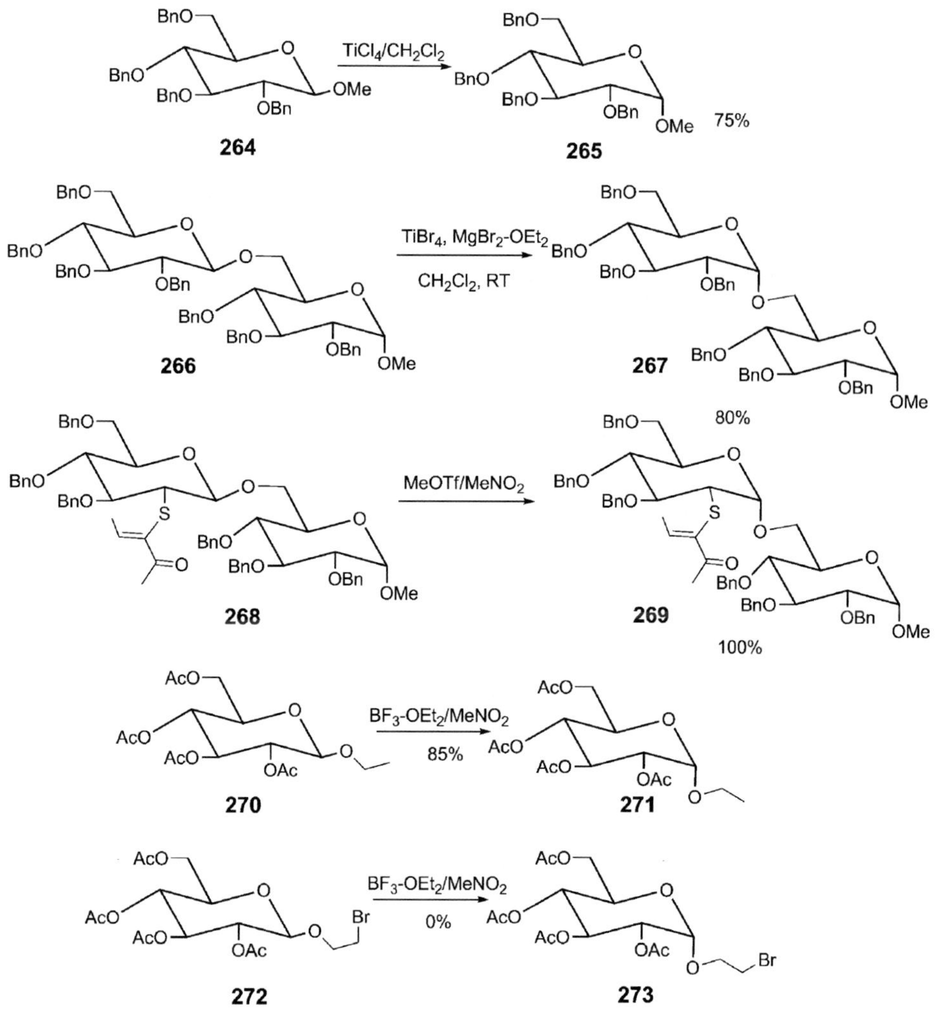

□ Scheme 57

ized in polar nitromethane containing BF$_3$·OEt$_2$, the bromide **272** is practically inert [217]
(❍ *Scheme 57*).

The silylated ketose **274** slowly anomerizes in the presence of TASF by way of the keto-form
[218]. The thioglycoside **276** anomerizes in the presence of a catalytic amount of IDCP [219].
Under PTC conditions, the β-chloride **278** also anomerizes [220]. Anomerization of **280**
to the α-form **281** via an open-chain zwitterionic intermediate has been suggested [221]
(❍ *Scheme 58*).

An investigation of the time course of the anomerization of β-iodide **282** has been carried
out using ^{1}H-NMR spectroscopy [222]. The NMR titration method to measure the shift of

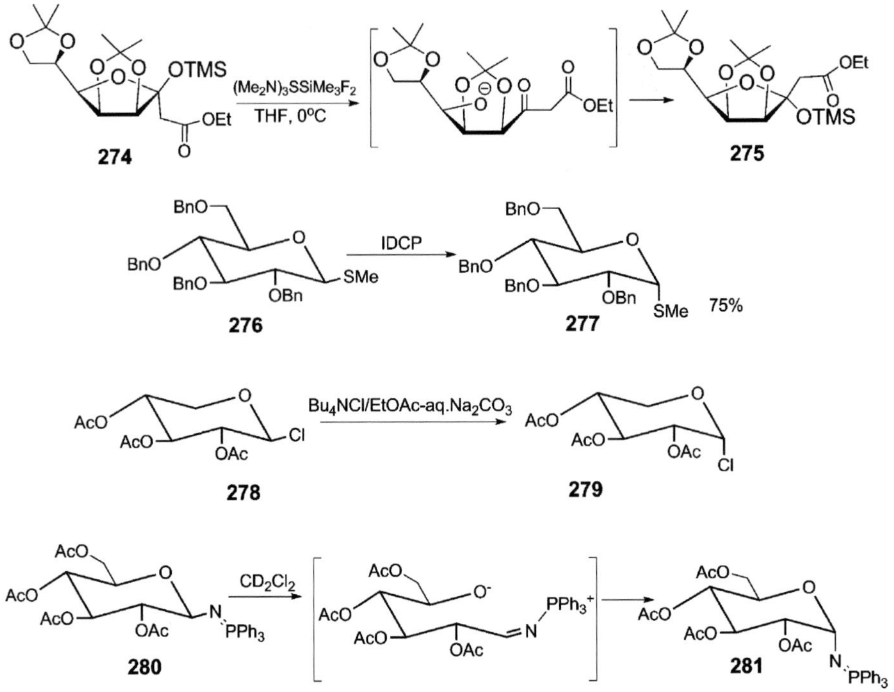

■ Scheme 58

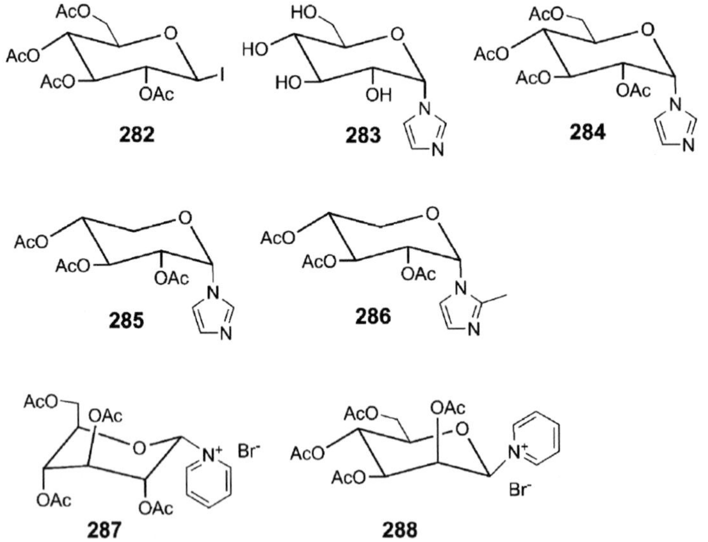

■ Figure 8
Anomeric equilibrium of glycosides and reverse anomeric effect

the anomeric equilibrium on protonation of **283** and **284** reveals that the protonated imidazoyl group has a small but distinct preference for the axial disposition than does the unprotonated group; which is the opposite of what the reverse anomeric effect predicts [223,224,225]. Compounds **285** and **286** increase the proportion of their $^1C_4$ conformers on *N*-protonation but not when the polarity of the solvent is increased as predicted by the reverse anomeric effect [226]. In solution, the α-glycosylpyridinium salt **287** adopts the $^1C_4$ conformation and the β-mannosyl compound **288** has the $^4C_1$ form; both of them have positively charged groups in an equatorial position at the anomeric center, indicating the manifestation of the reverse anomeric effect [227] (❖ *Fig. 8*).

Aryl *C*-glycosides such as **289** and **291** undergo α- to β-anomerization in the presence of an acid by way of open-chain intermediates [228,229]. Under basic conditions, **293** isomerizes into **295** through **294** [230] (❖ *Scheme 59*).

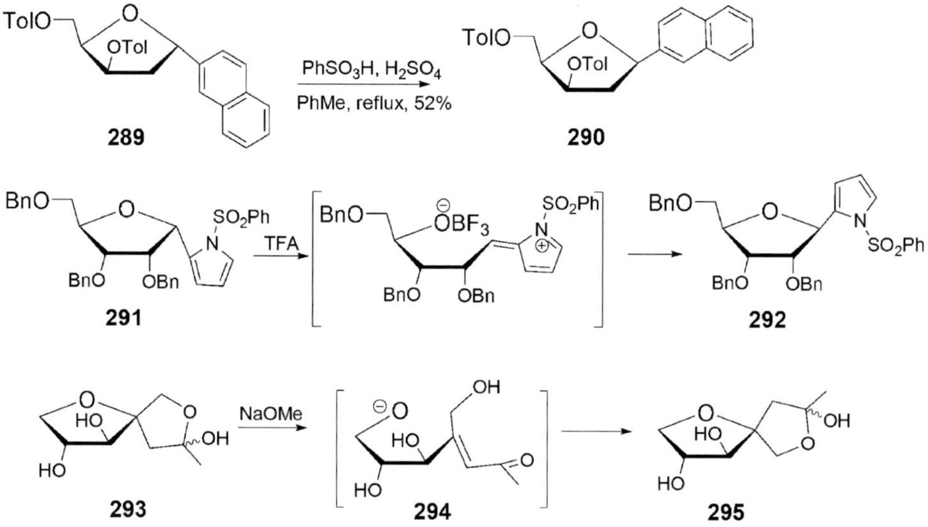

❏ Scheme 59

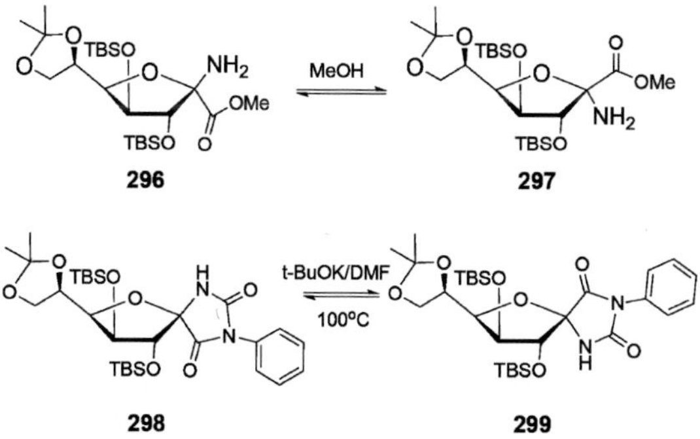

❏ Scheme 60

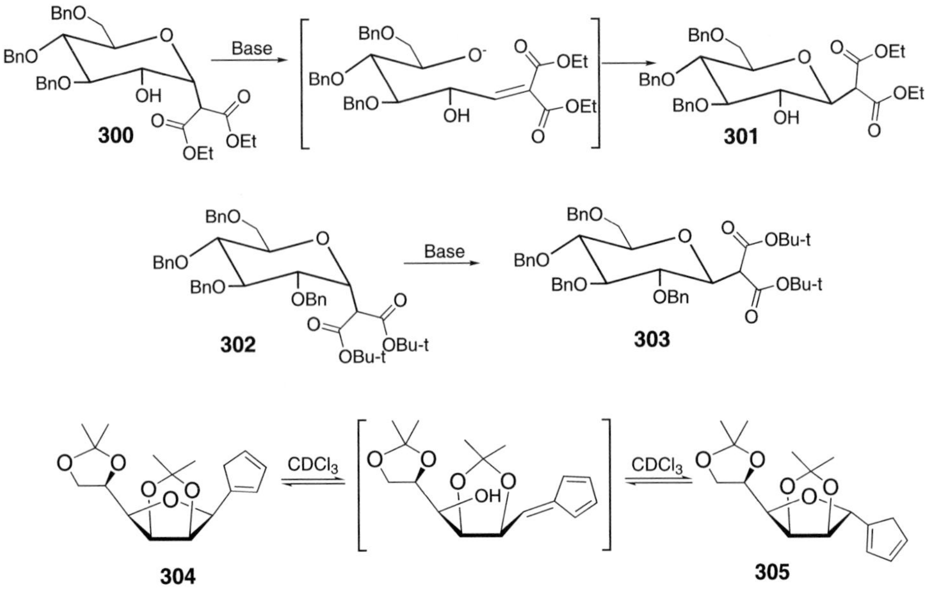

■ **Scheme 61**

The glycosylamine **296** anomerizes in methanolic solution. The spirohydantoins **298** and **299** form an equilibrium mixture under basic conditions [231] (❷ *Scheme 60.*

A possibility for the α- to β-anomerization of *C*-glycosides **300** and **302** by way of open-chain intermediates generated under basic conditions has been discussed [232,233]. Isomerization of pure **303** to **304** and vice versa probably occurs by way of open-chain isomerization through the linear intermediate [234] (❷ *Scheme 61*).

6.3 Aromatization of Sugars

Sequential elimination reactions, most of them being dehydration, involving the reaction at the anomeric center often produce various aromatic compounds [235] especially furans which have diverse use [236,237]. Explorations have been continued to open a new route to aromatics based on renewable biomass in place of fossilized material.

Oxidation of 5-hydroxymethylfuraldehyde **38** with hydrogen peroxide catalyzed by chloroperoxidase, a hemeperoxidase from *Caldariomyces fumago*, proceeds with good selectivity to furnish **306** [238] (❷ *Scheme 62*). The 6-aminodeoxyglycal derivative **307** is similarly converted into the furan **308** [239]. The dithiane **309** gives the oxacyclohexadiene **310**, on acid treatment [240]. Treatment of **311** with TMSOTf produces the pyrylium salt **312** [241].

The phenylosazone from D-xylose **39** can be converted into the pyrazoles **313** [242] . Isomaltulose **314** affords the glycosylated aromatic compound **315** [243]. On acidic acetylation the ulosonic acid ester **316** forms concomitantly the glycal derivative **317** and the furanoic acid derivative **318** [244] (❷ *Scheme 63*).

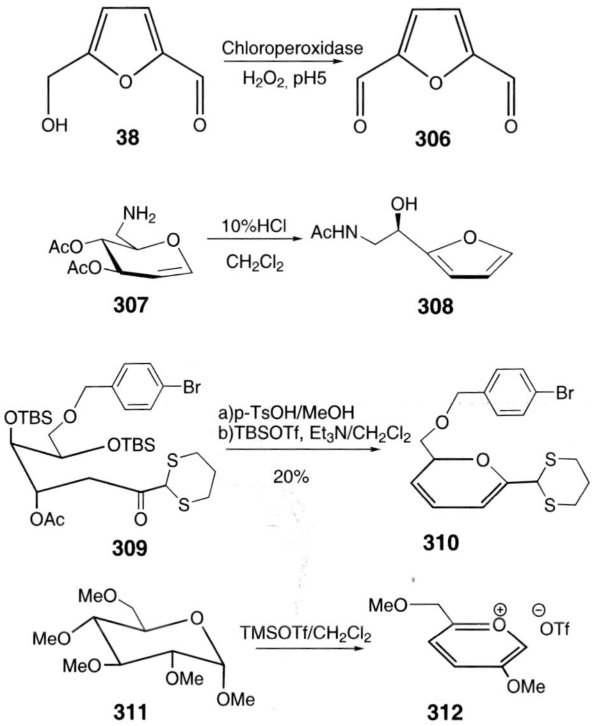

Scheme 62

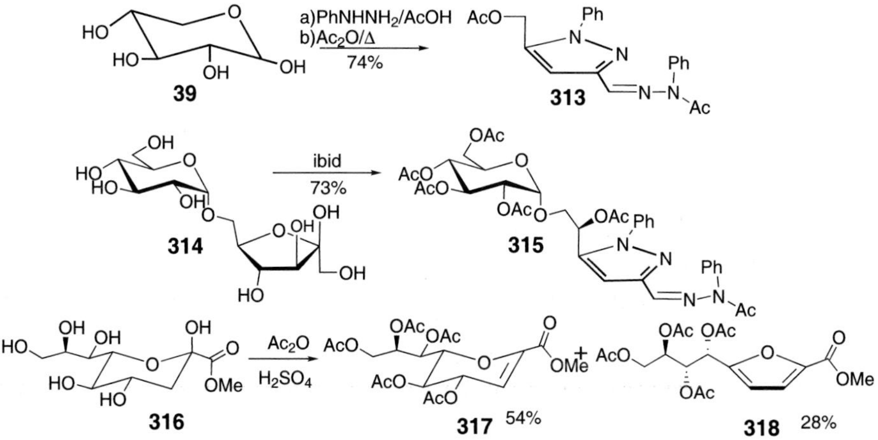

Scheme 63

Even on mild *C*-glycosylation using the TMSCN/TMSOTf system, a notable amount of the D-psicofuranose derivative **319** degrades to the furan **321** [245]. On reaction with Ph$_2$Hg, the chloride **322** gives the furan **323** exclusively [246]. The bromohydrin **324** degrades into furan **325** on heating with a base [247] (❷ *Scheme 64*).

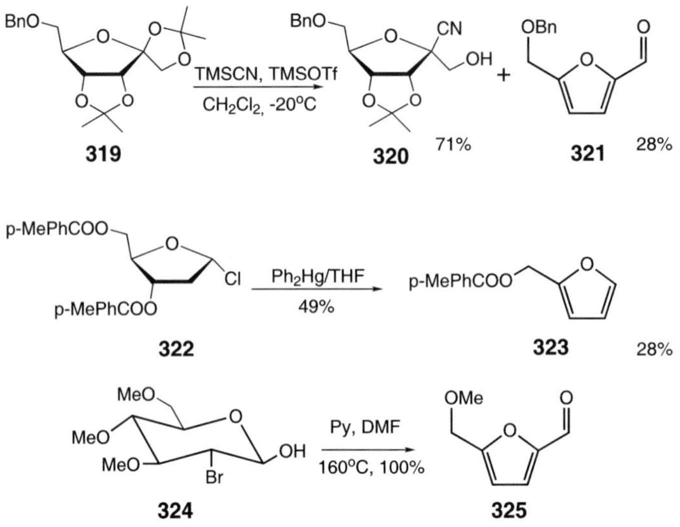

319   **320** 71%   **321** 28%

322   **323** 28%

324   **325**

◻ **Scheme 64**

7 The Maillard Reaction

The Maillard reaction is a complex group of degradation/rearrangement reactions initiated by reactions of free sugars and amines [248,249,250,251]. The reaction is of major interest for food processing [252,253,254,255,256,257] and life sciences [258,259,260,261,262,263,264, 265]. Degeneration of amine drugs in the presence of reducing sugars as excipients and deterioration of sugar artifacts are also related to the reaction [266,267].

The nonenzymatic reaction between reducing sugars and long-lived proteins *in vivo* results in the formation of glycation and advanced glycation end products, which alter the properties of proteins including charge, helicity, and their tendency to aggregate. Such protein modifications are linked with various pathologies associated with the general aging process such as Alzheimer disease and the long-term complications of diabetes. Although it has been suggested that glycation and advanced glycation end products altered protein structure and conformation, little structural data and information currently exist on whether or not glycation does indeed influence or change local protein secondary structure [268]. For example, in the blood, D-glucose can react with an NH_2 group of hemoglobin to form an imine that subsequently undergoes an irreversible rearrangement to a more stable a-aminoketone known as hemoglobin-AIc \index{hemoglobin-AIc}% [269].

Diabetes results when the body does not produce sufficient insulin or when the insulin it produces does not properly stimulate its target cells. Because insulin is the hormone that maintains the proper level of glucose in the blood, diabetics have increased blood glucose levels. The amount of hemoglobin-AIc formed is proportional to the concentration of glucose in the blood, so diabetics have a higher concentration of hemoglobin-AIc than nondiabetics. Thus, measuring the hemoglobin-AIc level is a way to determine whether the blood glucose level of a diabetic is being controlled [270,271].

Cataracts, a common complication in diabetics, are caused by the reaction of glucose with the group of proteins in the lens of the eye. It is thought that the arterial rigidity common in old age may be attributable to a similar reaction of glucose with the NH_2 group of proteins [250,272].

7.1 Mechanism of the Maillard Reaction

As illustrated in ❷ *Scheme 65*, **33** reacts with an amine to give an imine **326** that isomerizes into an aminoketose **327** (Amadori product), existing as an equilibrium mixture of cyclic hemiacetals, whereas **40** affords, by way of **328**, the hexosamine derivatives **329** and **330** (Heyns products), also in cyclic form. The Amadori–Heyns compounds \index{Amadori–Heyns compounds}% are at the head of the complex sequences of the Maillard reaction. The crystal structure of the Amadori product **331** between **33** and glycine has been determined more than three decades after the first proposal of its structure. Alternative preparations and X-ray analyses of Heyns products **332** and **333** have been reported [273,274].

The Amadori product from D-glucose **33** and L-proline decomposes at 130 °C in DMF to afford **33** and D-Mannose **54**, indicating the reversibility of the Amadori reaction. A kinetic study using **33** and phenylalanine indicates that the Schiff's base formation is the rate-determining step of the Maillard reaction [275,276].

Maltol **337** is one of the degradation products in monosaccharide solutions with amino acids forming Amadori compounds but not in the solution of monosaccharides alone. Heated solutions of monosaccharides yield **335**, the logical precursor of **337**, but not **337** itself. On the basis of the molecular mechanics calculation indicating that **335** adopts the conformation unfavorable for dehydration into **337**, a possible route via the dehydrated product **336**, an *ortho*-elimination product, has been postulated as a more favorable alternate reaction pathway [277].

7.2 Chemistry of Biologically Significant Maillard Products

In biological systems, Amadori products formed from aldoses and the amino group in peptides, decompose to release reactive sugar derivatives that are irreversibly consumed in the production of the advanced glycation end products (AGEs). In this sense, **338** is one of the key substances in the Maillard reaction [278]. A new specific assay of **338** has been developed using diaminonaphelene [279]. The dicarbonyl compounds **338** and **339**, the suggested intermediates in the degradation of the Amadori compound **331**, had been trapped with aminoguanidine [280,281]. The role of **338** generated in the Maillard cascade as a cross-linker of proteins has been emphasized [282]. Oxygen and metal cations accelerate the degradation of Amadori products to D-glucosone (**340**), a precursor of glyoxal **341** [283,284] (❷ *Scheme 66*). It is known that some Maillard products have strand-breaking activities to DNA. Many compounds found in foodstuffs are α,β-unsaturated ketones [285,286]. Compounds **53**, **342**, **343**, and **335** (❷ *Fig. 9*) cleave DNA single strands by generating hydroxyl radicals and other active oxygen radicals in the presence of Fe^{3+} and oxygen [287,288,289,290,291,292]. For example, the key hydrolyzate **344** generates hydroxyl radicals and the oxidation products **345** and **346**. An organic hydroperoxide **347**, presumably formed via direct oxidation of **339** or stepwise from **348**, the precursor of **335**, has been isolated [293,294,295,296,297] (❷ *Scheme 67*).

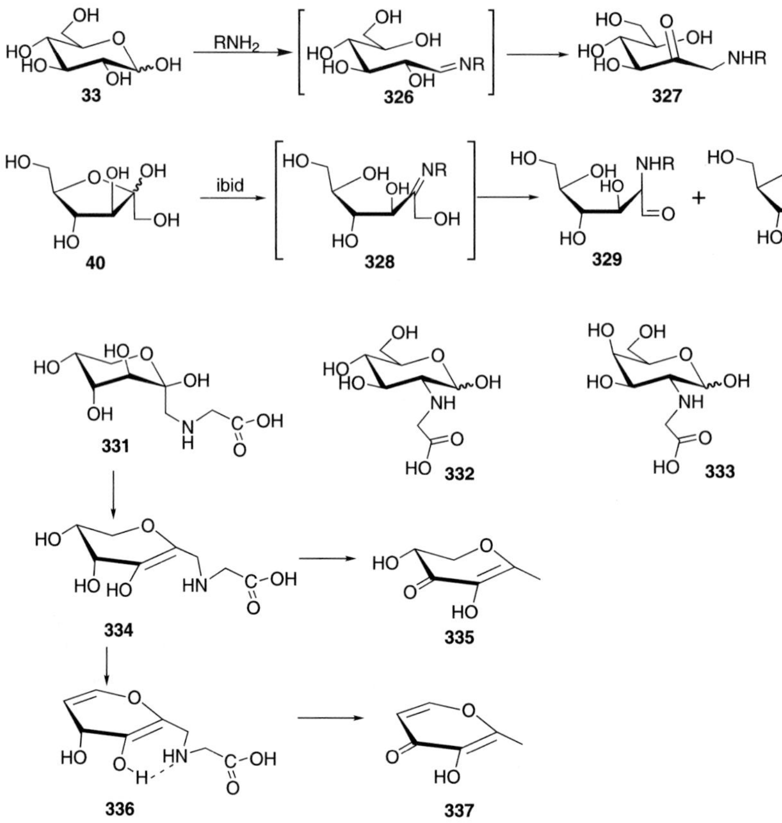

■ Scheme 65

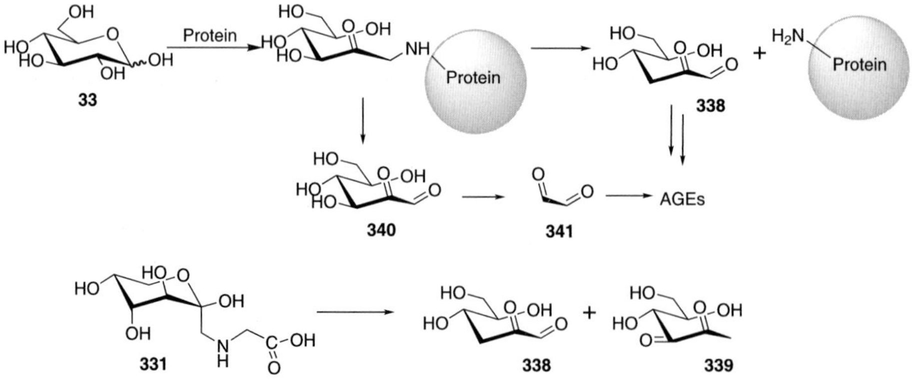

■ Scheme 66

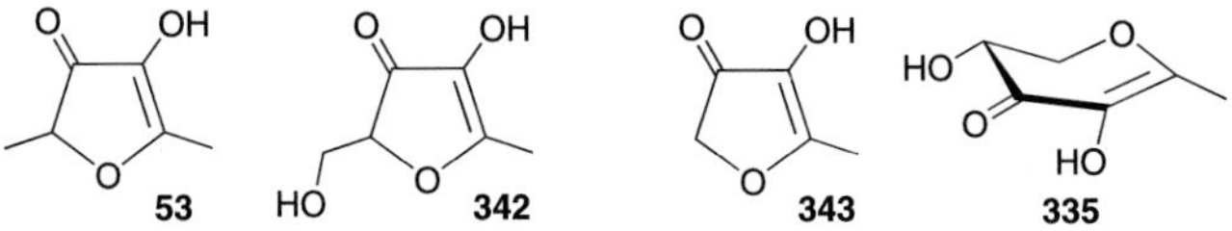

◻ Figure 9
Maillard products having strand-breaking activities to DNA

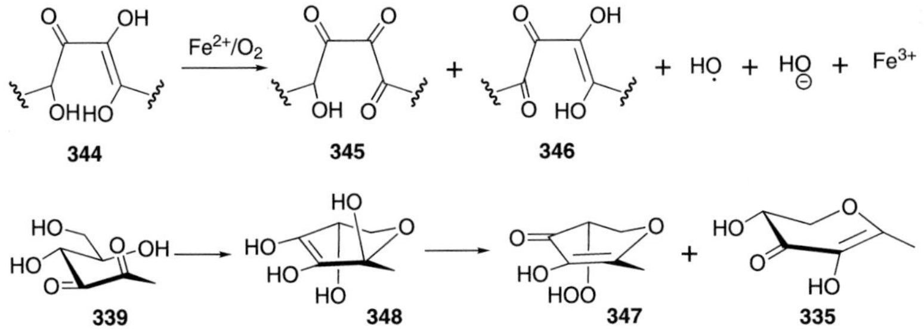

◻ Scheme 67

The major intermediate of the Maillard reaction **38**, having an allylic system, seems to furnish a cytotoxic ester on metabolic sulfonation [298]. In contrast to the above findings, some Amadori products, such as pyrazines have antimutagenicity [299,300,301]. Enkastines, the Amadori products of **33** and dipeptides, beneficially prolong the action of enkephaline by inhibiting enkephalinase [302].

Reactive small sugars and related acids appear to play a role in forming AGEs including cross-linked proteins in the aged body as well as inactivation of human Cu,Zn-superoxide dismutase [282,303,304]. Reaction of **33** with *n*-propylamine in phosphate-buffered, neutral solution generates several derivatives of small sugars [305,306], namely, C_2 and C_3 sugar derivatives. The 3-deoxyulose **349**, a hemiacetal form of **338**, yields **351**, the hydrate of methylglyoxal (**352**), as well as the Schiff's base **350** which is thought to be the precursor of the C_3-products [307,308,309] (**◉** *Scheme 68*).

N^ε-(carboxymethyl)lysine (**357**), is a main AGE product found in vivo [310,311]. About 50% of **357** seems to be formed via oxidative degradation of the Amadori product **356**. The reduced

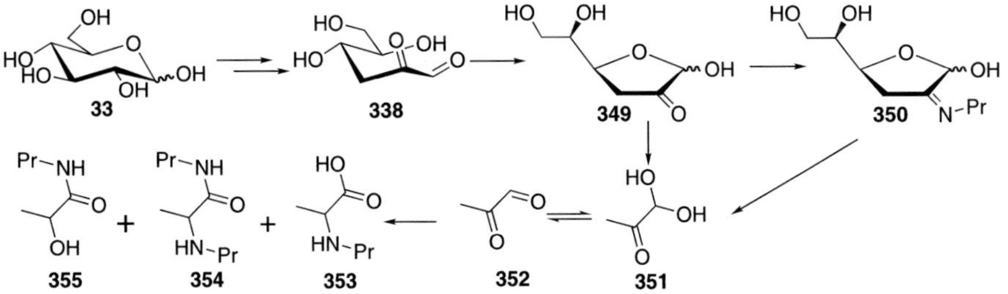

◻ Scheme 68

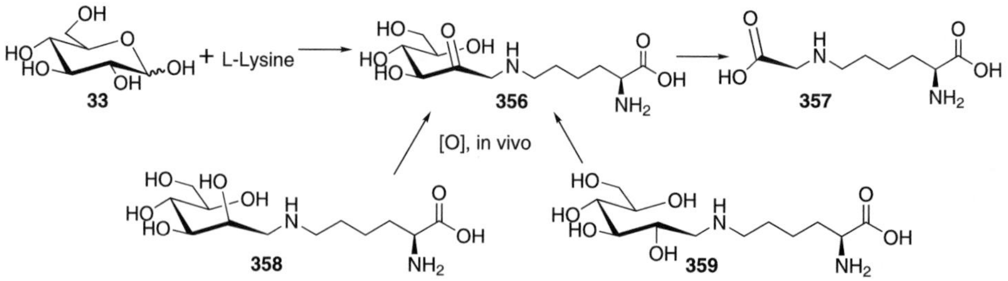

□ Scheme 69

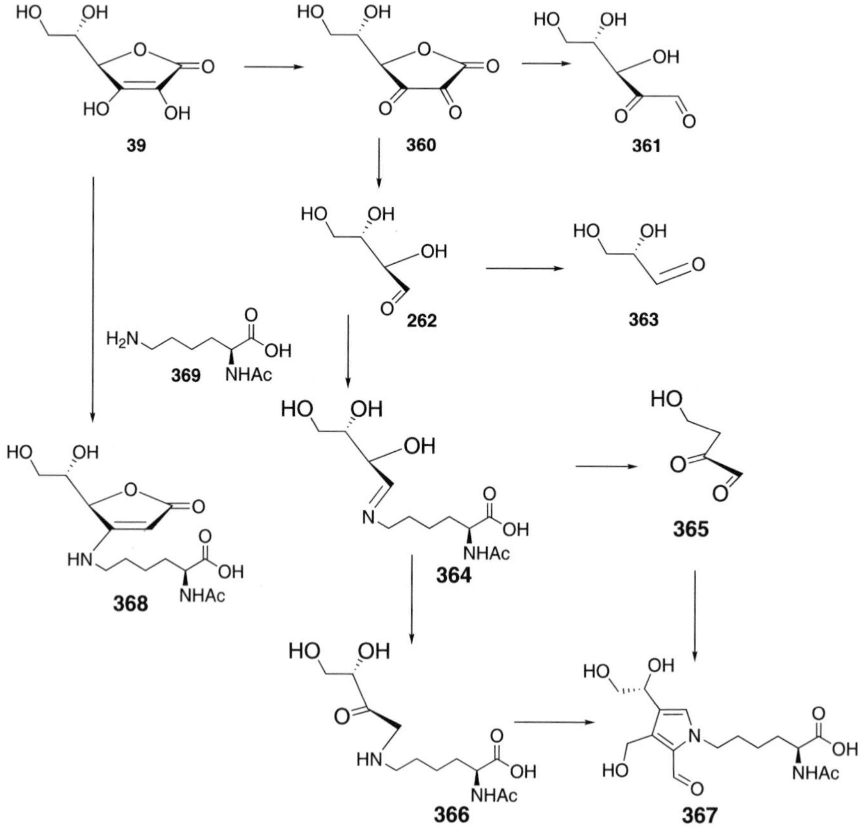

□ Scheme 70

compounds **358** and **359** also form **356** under aerobic physiological conditions [282]. Reactive **352** combines reversibly with lysine and cysteine residues and irreversibly with arginine residue [312] (❷ *Scheme 69*).

L-threose (**362**), the degradation product of **360** degrades in the presence of N^{α}-acetyl-L-lysine (**369**) at pH 7 into 3-deoxy-tetros-2-ulose (**365**) [313,314]. Only at pH 7 does retroaldolization of **362** occur to give glyceraldehyde (**363**). Under physiological conditions, the AGE product **364** is formed from **362** and **369**, apparently via condensation of the Amadori compounds **365** and **366** [315,316,317]. On heating at 100 °C, a hood-processing temperature, in methanol in a sealed vessel, **39** and **369** form an amine **368** [318] (❯ *Scheme 70*).

Some of the heterocyclic compounds among the Maillard products, for example, pentosidine **370** and pyrraline **371**, are AGEs in the skin of diabetic patients as well as in the brain of Alzheimer patients [319,320,321,322,323,324,325,326,327,328,329]. The observation that the aldehyde **372**, the Maillard product of **33** and *n*-propylamine, reacts with the amine and **369** to give **373** and **374**, respectively, led to the assumption that pyrrole aldehydes might also be precursors of the lysine side chain of proteins [320,330,331,332] (❯ *Scheme 71*).

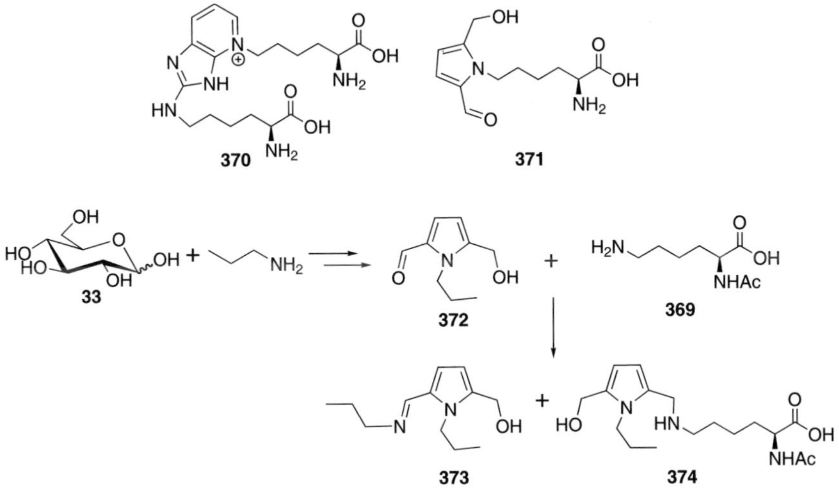

□ Scheme 71

References

1. Solomons TWG, Fryhle CB (2002) Organic chemistry. Wiley, New York
2. Bruice PY (2004) Organic chemistry. Pearson/Prentice Hall, Upper Saddle River, NJ
3. Stütz AE (2001) Glycoscience: epimerisation, isomerisation and rearrangement reactions of carbohydrates. Springer, Berlin, Heidelberg, New York
4. Lindhorst TK (2003) Essentials of carbohydrate chemistry and biochemistry. Wiley-VCH, Weinheim
5. Levy DE, Fügedi P (2006) The organic chemistry of sugars. Taylor & Francis, Boca Raton, FL
6. Sun Y, Cheng J (2002) Bioresource Technology 83:1
7. Wolfenden R, Lu X, Young G (1998) J Am Chem Soc 120:6814
8. Chenault HK, Chafin LF (1998) J Org Chem 63:833
9. Kirby AJ, Stromberg R (1994) J Chem Soc-Chem Comm: 709
10. Camilleri P, Jones RFD, Kirby AJ, Stromberg R (1994) J Chem Soc-Perkin Trans 2:2085
11. Sidelmann UG, Hansen SH, Gavaghan C, Carless HAJ, Lindon J, Farrant R, Wilson ID, Nicholson JK (1996) Anal Chem 68:2564

12. ManleyHarris M, Richards GN (1996) Carbohydr Res 287:183
13. Sinnott ML (1990) Chem Rev 90:1171
14. Kaczmarek J, Kaczynski Z, Trumpakaj Z, Szafranek J, Bogalecka M, Lonnberg H (2000) Carbohydr Res 325:16
15. Horenstein BA, Bruner M (1996) J Am Chem Soc 118:10371
16. Chong AK, Pegg MS, Taylor NR, von Itzstein M (1992) Eur J Biochem 207:335
17. Davies GJ, Mackenzie L, Varrot A, Dauter M, Brzozowski AM, Schulein M, Withers SG (1998) Biochemistry 37:11707
18. Nagano N, Noguchi T, Akiyama Y (2007) Proteins-Structure Function and Bioinformatics 66:147
19. Crennell SJ, Garman EF, Laver WG, Vimr ER, Taylor GL (1993) Proc Natl Acad Sci USA 90:9852
20. Guo X, Sinnott ML (1993) Biochem J 296 (Pt 2):291
21. Ashwell M, Sinnott ML, Zhang Y (1994) J Org Chem 59:7539
22. Ashwell M, Guo X, Sinnott ML (1992) J Am Chem Soc 114:10158
23. Pitson SM, Mutter M, van den Broek LA, Voragen AG, Beldman G (1998) Biochem Biophys Res Commun 242:552
24. Biely P, Benen J, Heinrichova K, Kester HC, Visser J (1996) FEBS Lett 382:249
25. Schultz PG, Lerner RA (1995) Science 269:1835
26. Yu J, Choi SY, Moon KD, Chung HH, Youn HJ, Jeong S, Park H, Schultz PG (1998) Proc Natl Acad Sci USA 95:2880
27. Yu S, Ahmad T, Kenne L, Pedersen M (1995) Biochim Biophys Acta 1244:1
28. Richard GT, Yu S, Monsan P, Remaud-Simeon M, Morel S (2005) Carbohydr Res 340:395
29. Yu S, Kenne L, Pedersen M (1993) Biochim Biophys Acta 1156:313
30. Saddler JN, Penner MH (1995) Am Chem Soc Meeting. Enzymatic degradation of insoluble carbohydrates. American Chemical Society, Washington, DC
31. Horn SJ, Sikorski P, Cederkvist JB, Vaaje-Kolstad G, Sorlie M, Synstad B, Vriend G, Varum KM, Eijsink VG (2006) Proc Natl Acad Sci USA 103:18089
32. National Research Council (US) Committee on Bioprocess Engineering (1992) Putting biotechnology to work: bioprocess engineering. National Academy of Sciences, Washington, DC
33. Mulimani VH, Ramalingam R (1995) Biochem Mol Biol Int 36:897
34. Kulkarni DS, Kapanoor SS, Girigouda K, Kote NV, Mulimani VH (2006) Biotechnol Appl Biochem 45:51
35. Hollingsworth RI, Wang G (2000) Chem Rev 100:4267
36. Nolasco J, De Massaguer PR (2006) J Food Proc Eng 29:462
37. Hollnagel A, Kroh LW (2000) J Agricult Food Chem 48:6219
38. Miyazawa T, Ohtsu S, Nakagawa Y, Funazukuri T (2006) J Mat Sci 41:1489
39. Srokol Z, Bouche AG, van Estrik A, Strik RCJ, Maschmeyer T, Peters JA (2004) Carbohydr Res 339:1717
40. Griebl A, Lange T, Weber H, Milacher W, Sixta H (2006) Macromolecular Symposia 232:107
41. Kruse A, Gawlik A (2003) Ind Eng Chem Res 42:267
42. Hashaikeh R, Fang Z, Butler IS, Kozinski JA (2005) Proc Combustion Institute 30:2231
43. Chheda JN, Roman-Leshkov Y, Dumesic JA (2007) Green Chem 9:342
44. Sanderson K (2006) Nature 444:673
45. Hayes MH (2006) Nature 443:144
46. Kruse A, Maniam P, Spieler F (2007) Ind Eng Chem Res 46:87
47. Mark J, Pollien P, Lindinger C, Blank I, Mark T (2006) J Agricult Food Chem 54:2786
48. Fan XT (2005) J Agricult Food Chem 53:7826
49. Ginz M, Balzer HH, Bradbury AGW, Maier HG (2000) Eur Food Res Technol 211:404
50. Arya M, Rao LJM (2007) Crit Rev Food Sci Nutrit 47:51
51. Burde RDL, Crayton F, Bavley A (1962) Nature 196:166
52. Qian X, Nimlos MR, Johnson DK, Himmel ME (2005) Appl Biochem Biotechnol 121–124:989
53. Buskas T, Konradsson P (2000) J Carbohydr Chem 19:25
54. Antal MJ, Jr., Mok WS, Richards GN (1990) Carbohydr Res 199:91
55. Halliday GA, Young RJ, Jr., Grushin VV (2003) Org Lett 5:2003
56. Hauck T, Landmann C, Bruhlmann F, Schwab W (2003) J Agric Food Chem 51:1410
57. deGoede ATJW, vanDeurzen MPJ, vanderLeij IG, vanderHeijden AM, Baas JMA, vanRantwijk F, vanBekkum H (1996) J Carbohydr Chem 15:331
58. Cancilla MT, Penn SG, Lebrilla CB (1998) Anal Chem 70:663

59. Yang BY, Montgomery R (2007) Bioresour Technol 98:3084
60. T'i VTU, Pisarnitskii AF (1998) Appl Biochem Microbiol 34:106
61. Manini P, La Pietra P, Panzella L, Napolitano A, d'Ischia M (2006) Carbohydr Res 341:1828
62. Hourdin GL, Germain A, Moreau C, Fajula F (2002) J Catalysis 209:217
63. van den Berg R, Peters JA, van Bekkum H (1995) Carbohydr Res 267:65
64. Rangappa KS, Raghavendra MP, Mahadevappa DS, Gowda DC (1998) Carbohydr Res 306:57
65. Hollingsworth RI (1996) Biotechnol Annu Rev 2:281
66. Hollingsworth RI (1999) J Org Chem 64:7633
67. Stapley JA, BeMiller JN (2007) Carbohyd Res 342:407
68. Hourdin G, Germain A, Moreau C, Fajula F (2000) Catal Lett 69:241
69. Fischer K, Bipp HP (2005) Bioresour Technol 96:831
70. Niu W, Molefe MN, Frost JW (2003) J Am Chem Soc 125:12998
71. Service RF (2007) Science 315:1488
72. Champagne P (2007) Resour Conserv Recy 50:211
73. Kakinuma K, Li HY (1989) Tetrahedron Lett 30:4157
74. Taillefumier C, Chapleur Y (2004) Chem Rev 104:263
75. Sasaki M, Higashi M, Masu H, Yamaguchi K, Takeda K (2005) Org Lett 7:5913
76. Bertrand P, Gesson JP, Renoux B, Tranoy I (1995) Tetrahedron Lett 36:4073
77. Halcomb RL, Boyer SH, Wittman MD, Olson SH, Denhart DJ, Liu KKC, Danishefsky SJ (1995) J Am Chem Soc 117:5720
78. Modica E, Compostella F, Colombo D, Franchini L, Cavallari M, Mori L, De Libero G, Panza L, Ronchetti F (2006) Org Lett 8:3255
79. Jaunzeme I, Jirgensons A (2005) Synlett: 2984
80. Montero A, Mann E, Herradon B (2005) Tetrahedron Lett 46:401
81. Banaszek A, Pakulski Z, Zamojski A (1995) Carbohyd Res 279:173
82. Ichikawa Y, Kobayashi C, Isobe M (1996) J Chem Soc-Perkin Trans 1:377
83. Ichikawa Y, Kobayashi C, Isobe M (1994) Synlett: 919
84. Oishi T, Ando K, Inomiya K, Sato H, Iida M, Chida N (2002) Org Lett 4:151
85. Werschkun B, Thiem J (2001) Glycoscience: Epimerisation, Isomerisation and Rearrangement Reactions of Carbohydrates. Topics in Current Chemistry Vol 215, Springer-Verlag, Berlin, Heidelberg 215:293
86. Sudha AVRL, Nagarajan M (1998) Chem Comm 925
87. Balasubramanian KK, Ramesh NG, Pramanik A, Chandrasekhar J (1994) J Chem Soc-Perkin Trans 2:1399
88. Kuhn C, Legouadec G, Skaltsounis AL, Florent JC (1995) Tetrahedron Lett 36:3137
89. Overman LE, Shim J (1993) J Org Chem 58:4662
90. Deng W, Overman LE (1994) J Am Chem Soc 116:11241
91. Knight SD, Overman LE, Pairaudeau G (1995) J Am Chem Soc 117:5776
92. Agami C, Couty F, Lin J, Mikaeloff A (1993) Synlett: 349
93. Agami C, Couty F, Puchot-Kadouri C (1998) Synlett: 449
94. Oberdorfer F, Haeckel R, Lauer G (1998) Synthesis-Stuttgart: 201
95. Mundorff EC, Hanson MA, Varvak A, Ulrich H, Schultz PG, Stevens RC (2000) Biochemistry 39:627
96. Black KA, Leach AG, Kalani MY, Houk KN (2004) J Am Chem Soc 126:9695
97. Yokoyama H, Otaya K, Yamaguchi S, Hirai Y (1998) Tetrahedron Lett 39:5971
98. Zamojski A (2002) Polish J Chem 76:1053
99. Yokoyama H, Ejiri H, Miyazawa M, Yamaguchi S, Hirai Y (2007) Tetrahedron: Asymmetry 18:852
100. Muzart J (2005) Tetrahedron 61:4179
101. Mcdonald FE, Gleason MM (1995) Angew Chem Int Ed Engl 34:350
102. Castro S, Peczuh MW (2005) J Org Chem 70:3312
103. Redlich H (1994) Angew Chem 106:1407
104. Collins P, Ferrier RJ (1995) Monosaccharides: their chemistry and their roles in natural products. Wiley, New York
105. Tatsuta K, Hosokawa S (2006) Science and Technology of Advanced Materials 7:397
106. Ponten F, Magnusson G (1994) Acta Chemica Scandinavica 48:566
107. Ito H (2003) Yakugaku Zasshi-J Pharmaceut Soc Japan 123:933
108. Ito H, Motoki Y, Taguchi T, Hanzawa Y (1993) J Am Chem Soc 115:8835

109. Jenkins DJ, Riley AM, Potter BVL (1996) J Org Chem 61:7719
110. Bettelli E, D'Andrea P, Mascanzoni S, Passacantilli P, Piancatelli G (1998) Carbohydr Res 306:221
111. Estevez JC, Saunders J, Besra GS, Brennan PJ, Bash RJ, Fleet GWJ (1996) Tetrahedron: Asymmetry 7:383
112. Bichard CJF, Brandstetter TW, Estevez JC, Fleet GWJ, Hughes DJ, Wheatley JR (1996) J Chem Soc, Perkin Trans 1: Organic and Bio-Organic Chemistry: 2151
113. Binkley RW, Ambrose MG (1984) J Carbohydr Chem 3:1
114. Binkley RW (1992) J Org Chem 57:2353
115. Wang YF, Fleet GWJ, Zhao LX (1998) Carbohydr Res 307:159
116. Fuzier M, Le Merrer Y, Depezay J-C (1995) Tetrahedron Lett 36:6443
117. Poitout L, LeMerrer Y, Depezay JC (1996) Tetrahedron Lett 37:1613
118. Charette AB, Cote B (1993) J Org Chem 58:933
119. Jeong LS, Moon HR, Yoo SJ, Lee SN, Chun MW, Lim YH (1998) Tetrahedron Lett 39:5201
120. Welch JT, Svahn B-M, Eswarakrishnan S, Hutchinson JP, Zubieta J (1984) Carbohydrate Research 132:221
121. Baudry M, Barberousse V, Descotes G, Faure R, Pires J, Praly JP (1998) Tetrahedron 54:7431
122. Goebel M, Nothofer HG, Ross G, Ugi I (1997) Tetrahedron 53:3123
123. Holton RA, Somoza C, Kim HB, Liang F, Biediger RJ, Boatman PD, Shindo M, Smith CC, Kim SC, Nadizadeh H, Suzuki Y, Tao CL, Vu P, Tang SH, Zhang PS, Murthi KK, Gentile LN, Liu JH (1994) J Am Chem Soc 116:1597
124. Lee SD, Chan TH, Kwon KS (1984) Tetrahedron Lett 25:3399
125. Csuk R, Fuerstner A, Weidmann H (1986) J Carbohydr Chem 5:271
126. Curran DP, Suh Y-G (1987) Carbohydr Res 171:161
127. Fredenhagen A, Peter HH (1996) Tetrahedron 52:1235
128. Stoltz BM, Wood JL (1996) Tetrahedron Lett 37:3929
129. Avalos M, Babiano R, Cabanillas A, Cintas P, Higes FJ, Jimenez JL, Palacios JC (1996) J Org Chem 61:3738
130. Hoberg JO, Bozell JJ (1995) Tetrahedron Lett 36:6831
131. Hoberg JO (1997) J Org Chem 62:6615
132. Boyer FD, Lallemand JY (1994) Tetrahedron 50:10443
133. Ramana CV, Murali R, Nagarajan M (1997) J Org Chem 62:7694
134. Wender PA, Jesudason CD, Nakahira H, Tamura N, Tebbe AL, Ueno Y (1997) J Am Chem Soc 119:12976
135. Chen XT, Gutteridge CE, Bhattacharya SK, Zhou BS, Pettus TRR, Hascall T, Danishefsky SJ (1998) Angew Chem Int Ed 37:185
136. Xu YM, Zhou WS (1997) J Chem Soc-Perkin Trans 1:741
137. Marshall JA, Tang Y (1994) J Org Chem 59:1457
138. Kim DK, Kim GH, Kim YW (1996) J Chem Soc-Perkin Trans 1:803
139. Tatsuta K (1998) J Synth Org Chem Japan 56:714
140. Boyer FD, Pancrazi A, Lallemand JY (1995) Synth Comm 25:1099
141. Chen YW, Vogel P (1994) J Org Chem 59:2487
142. Shilvock JP, Fleet GWJ (1998) Synlett: 554
143. Beacham AR, Smelt KH, Biggadike K, Britten CJ, Hackett L, Winchester BG, Nash RJ, Griffiths RC, Fleet GWJ (1998) Tetrahedron Lett 39:151
144. Poitout L, LeMerrer Y, Depezay JC (1996) Tetrahedron Lett 37:1609
145. Ernholt BV, Thomsen IB, Jensen KB, Bols M (1999) Synlett: 701
146. Lohse A, Jensen KB, Bols M (1999) Tetrahedron Lett 40:3033
147. Andreassen V, Svensson B, Bols M (2001) Synthesis-Stuttgart: 339
148. Moriyama H, Tsukida T, Inoue Y, Kondo H, Yoshino K, Nishimura SI (2003) Bioorg Med Chem Lett 13:2737
149. Tschamber T, Gessier F, Neuburger M, Gurcha SS, Besra GS, Streith J (2003) Eur J Org Chem: 2792
150. Fuentes J, Sayago FJ, Illangua JM, Gasch C, Angulo M, Pradera MA (2004) Tetrahedron-Asymmetry 15:603
151. Tsukida T, Moriyama H, Inoue Y, Kondo H, Yoshino K, Nishimura SI (2004) Bioorg Med Chem Lett 14:1569
152. Ichikawa Y, Osada M, Ohtani II, Isobe M (1997) J Chem Soc-Perkin Trans 1:1449
153. Blanco JLJ, Diaz Prez VM, Mellet CO, Fuentes J, Garcia Fernandez JM, Diaz Arribas JC, Canada FJ (1997) Chem Comm (Cambridge): 1969

154. Goti A, Cacciarini M, Cardona F, Cordero FM, Brandi A (2001) Org Lett 3:1367

155. Overkleeft HS, Bruggeman P, Pandit UK (1998) Tetrahedron Lett 39:3869

156. Haakansson AE, Palmelund A, Holm H, Madsen R (2006) Chemistry-A Eur J 12:3243

157. Kotha S, Mandal K, Tiwari A, Mobin SM (2006) Chemistry 12: 8024

158. Andresen TL, Skytte DM, Madsen R (2004) Org Biomol Chem 2: 2951

159. Ovaa H, Lastdrager B, Codee JDC, van der Marel GA, Overkleeft HS, van Boom JH (2002) J Chem Soc-Perkin Trans 1:2370

160. Yamazaki T, Mizutani K, Kitazume T (1996) ACS Symp Series 639:105

161. Dondoni A, Catozzi N, Marra A (2004) J Org Chem 69:5023

162. Pradera MA, Sayago FJ, Illangua JM, Gasch C, Fuentes J (2003) Tetrahedron Lett 44:6605

163. Fuentes J, Illangua JM, Sayago FJ, Angulo M, Gasch C, Pradera MA (2004) Tetrahedron-Asymmetry 15:3783

164. Dhavale DD, Jachak SM, Karche NP, Trombini C (2004) Tetrahedron 60:3009

165. Sun H, Abboud KA, Horenstein NA (2005) Tetrahedron 61:10462

166. Parr IB, Horenstein BA (1997) J Org Chem 62:7489

167. Castro S, Peczuh Mark W (2005) J Org Chem 70:3312

168. Foulard G, Brigaud T, Portella C (1997) J Org Chem 62:9107

169. Momotake A, Mito J, Yamaguchi K, Togo H, Yokoyama M (1998) J Org Chem 63: 7207

170. Zou W (2005) Curr Topic Med Chem 5:1363

171. Lin C-H, Lin H-C, Yang W-B (2005) Curr Topic Med Chem (Sharjah, United Arab Emirates) 5:1431

172. Lee DYW, He MS (2005) Curr Topic Med Chem 5:1333

173. Bililign T, Griffith BR, Thorson JS (2005) Natural Product Reports 22:742

174. Sharma GVM, Krishna PR (2004) Curr Org Chem 8:1187

175. Gascon-Lopez M, Motevalli M, Paloumbis G, Bladon P, Wyatt PB (2003) Tetrahedron 59:9349

176. Cipolla L, La Ferla B, Nicotra F (1998) Carbohydr Polym 37:291

177. Cipolla L, Peri F, La Ferla B, Redaelli C, Nicotra F (2005) Curr Org Synth 2:153

178. Johnson CR, Johns BA (1997) J Org Chem 62:6046

179. Hashimoto M, Terashima S (1994) Chem Lett: 1001

180. Brochard L, Lorin C, Spiess N, Rollin P (1998) Tetrahedron Lett 39:4267

181. Jun JG, Lee DW, Mundy BP (1998) Synth Comm 28:2499

182. Wei BG, Chen J, Huang PQ (2006) Tetrahedron 62:190

183. Pelyvas IF, Madi-Puskas M, Toth ZG, Varga Z, Hornyak M, Batta G, Sztaricskai F (1995) J Antibiotics 48:683

184. Zhou J, Wang G, Zhang L-H, Ye X-S (2006) Curr Org Chem 10:625

185. Ferrier RJ, Hoberg JO (2003) Advances in Carbohydrate Chemistry and Biochemistry, Vol 58, Academic Press, San Diego, London, 58:55

186. Chida N, Ohtsuka M, Ogura K, Ogawa S (1991) Bull Chem Soc Japan 64:2118

187. Wang A, Auzanneau FI (2007) J Org Chem 72:3585

188. Chida N, Takeoka J, Ando K, Tsutsumi N, Ogawa S (1996) Tennen Yuki Kagobutsu Toronkai Koen Yoshishu 38th: 259

189. Letellier P, Ralainirira R, Beaupere D, Uzan R (1994) Tetrahedron Lett 35:4555

190. Letellier P, Ralainairina R, Beaupere D, Uzan R (1997) Synthesis-Stuttgart: 925

191. Letellier P, ElMeslouti A, Beaupere D, Uzan R (1996) Synthesis-Stuttgart: 1435

192. Iimori T, Takahashi H, Ikegami S (1996) Tetrahedron Lett 37:649

193. Takahashi H (2001) Yuki Gosei Kagaku Kyokaishi 59:484

194. Takahashi H (2002) Yakugaku Zasshi-J Pharmaceut Soc Japan 122:755

195. Miyazaki H, Kobayashi Y, Shiozaki M, Ando O, Nakajima M, Hanzawa H, Haruyama H (1995) J Org Chem 60:6103

196. Friestad GK, Branchaud BP (1997) Tetrahedron Lett 38:5933

197. Das SK, Mallet JM, Sinay P (1997) Angew Chem Int Ed Engl 36:493

198. Sollogoub M, Mallet JM, Sinay P (2000) Angew Chem Int Ed Engl 39:362

199. Sollogoub M, Sinay P (2006) Organic Chemistry of Sugars: 349

200. Petasis NA, Yao X (2001) Abstracts of Papers of the American Chemical Society 222: U99

201. Chung S-K, Yu S-H (1996) Bioorg Med Chem Lett 6:1461

202. Jia C, Pearce AJ, Bleriot Y, Zhang Y, Zhang L-H, Sollogoub M, Sinay P (2004) Tetrahedron: Asymmetry 15:699

203. Strukul G (1998) Angew Chem Int Ed 37:1199
204. Mereyala HB, Guntha S (1995) Tetrahedron 51:1741
205. Zhou H, Wang GN, Zhang LH, Ye XS (2006) Curr Org Chem 10:625
206. Bennett SM, Biboutou RK, Zhou ZH, Pion R (1998) Tetrahedron 54:4761
207. Grove JJC, Holzapfel CW (1997) Tetrahedron Lett 38:7429
208. CronjeGrove JJ, Holzapfel CW, Williams DBG (1996) Tetrahedron Lett 37:1305
209. Chiara JL, Martinez S, Bernabe M (1996) J Org Chem 61:6488
210. Sano H, Sugai S (1995) Tetrahedron-Asymmetry 6:1143
211. deArmas P, GarciaTellado F, MarreroTellado JJ, Robles J (1997) Tetrahedron Lett 38:8081
212. Lewis BE, Choytun N, Schramm VL, Bennet AJ (2006) J Am Chem Soc 128:5049
213. Liu FW, Zhang YB, Liu HM, Song XP (2005) Carbohydr Res 340:489
214. Pacsu E (1930) J Am Chem Soc 52:2563
215. Pacsu J (1928) Berichte der Deutschen Chemischen Gesellschaft [Abteilung] B: Abhandlungen 61B:1508
216. Capozzi G, Mannocci F, Menichetti S, Nativi C, Paoletti S (1997) Chem Comm: 2291
217. Ellervik U, Jansson K, Magnusson G (1998) J Carbohydr Chem 17:777
218. Csuk R, Schaade M (1994) Tetrahedron 50:3333
219. Boons GJ, Stauch T (1996) Synlett: 906
220. Kim JM, Roy R (1997) J Carbohydr Chem 16:1281
221. Maunier V, Boullanger P, Lafont D (1997) J Carbohydr Chem 16:231
222. Gervay J, Nguyen TN, Hadd MJ (1997) Carbohydr Res 300:119
223. Fabian MA, Perrin CL, Sinnott ML (1994) J Am Chem Soc 116:8398
224. Vaino AR, Szarek WA (2001) J Org Chem 66:1097
225. Grundberg H, Eriksson-Bajtner J, Bergquist KE, Sundin A, Ellervik U (2006) J Org Chem 71:5892
226. Vaino AR, Chan SSC, Szarek WA, Thatcher GRJ (1996) J Org Chem 61:4514
227. Skorupowa E, Dmochowska B, Madaj J, Kasprzykowski F, Sokolowski J, Wisniewski A (1998) J Carbohydr Chem 17:49
228. Ren RXF, Chaudhuri NC, Paris PL, Rumney S, Kool ET (1996) J Am Chem Soc 118:7671
229. Yokoyama M, Nomura M, Togo H, Seki H (1996) J Chem Soc-Perkin Trans 1:2145
230. Di Florio R, Rizzacasa MA (1998) J Org Chem 63:8595
231. Brandstetter TW, Kim YH, Son JC, Taylor HM, Lilley PMD, Watkin DJ, Johnson LN, Oikonomakos NG, Fleet GWJ (1995) Tetrahedron Lett 36:2149
232. Leeuwenburgh MA, Timmers CM, vanderMarel GA, vanBoom JH, Mallet JM, Sinay PG (1997) Tetrahedron Lett 38:6251
233. Gervay J, Hadd MJ (1997) J Org Chem 62:6961
234. Vedso P, Chauvin R, Li Z, Bernet B, Vasella A (1994) Helvetica Chimica Acta 77:1631
235. Barker JL, Frost JW (2001) Biotechnol Bioeng 76:376
236. Zhu LZ, Talukdar A, Zhang GS, Kedenburg JP, Wang PG (2005) Synlett: 1547
237. Rowe D (2004) Chem Biodiversity 1:2034
238. vanDeurzen MPJ, vanRantwijk F, Sheldon RA (1997) J Carbohydr Chem 16:299
239. Mathews WB, Zajac WW (1995) J Carbohydr Chem 14:287
240. Devianne G, Escudier JM, Baltas M, Gorrichon L (1995) J Org Chem 60:7343
241. Lee CK, Kim EJ, Lee ISH (1998) Carbohydr Res 309:243
242. Diehl V, Cuny E, Lichtenthaler FW (1998) Heterocycles 48:1193
243. Oikawa N, Muller C, Kunz M, Lichtenthaler FW (1998) Carbohydr Res 309:269
244. Sun XL, Kai T, Takayanagi H, Furuhata K (1997) J Carbohydr Chem 16:541
245. Sano H, Mio S, Kitagawa J, Sugai S (1994) Tetrahedron-Asymmetry 5:2233
246. Chaudhari VD, Kumar KSA, Dhavale DD (2006) Tetrahedron 62:4349
247. Kozlowski JS, Marzabadi CH, Rath NP, Spilling CD (1997) Carbohydr Res 300:301
248. Mauron J (1981) Prog Food Nutr Sci 5:5
249. Ledl F, Beck J, Sengl M, Osiander H, Estendorfer S, Severin T, Huber B (1989) Prog Clin Biol Res 304:23
250. John WG, Lamb EJ (1993) Eye 7 (Pt 2): 230
251. Gerrard J (2005) The Maillard reaction: chemistry, biochemistry and implications by Harry Nursten. Royal Society of Chemistry
252. Erbersdobler HF, Somoza V (2007) Mol Nutr Food Res 51:423
253. Charissou A, Ait-Ameur L, Birlouez-Aragon I (2007) J Agric Food Chem 55:4532

254. van Boekel MA (2006) Biotechnol Adv 24:230

255. Kato A (2002) Food Sci Technol Res 8:193

256. Chen CQ, Robbins E (2000) ACS Symp Series 754:286

257. Gerrard JA (2006) Trends Food Sci Technol 17:324

258. Tuohy KM, Hinton DJ, Davies SJ, Crabbe MJ, Gibson GR, Ames JM (2006) Mol Nutr Food Res 50:847

259. Sun Y, Hayakawa S, Chuamanochan M, Fujimoto M, Innun A, Izumori K (2006) Biosci Biotechnol Biochem 70:598

260. Saraiva MA, Borges CM, Florencio MH (2006) J Mass Spectrom 41:755

261. Robert L, Labat-Robert J (2006) Pathologie Biologie 54:371

262. Reddy VP, Beyaz A (2006) Drug Discovery Today 11:646

263. Peyroux J, Sternberg M (2006) Pathol Biol (Paris) 54:405

264. Somoza V (2005) Mol Nutrit Food Res 49:663

265. Baynes JW (2000) Biogerontology 1:235

266. Colaco CA (1993) J R Soc Med 86:243

267. Wirth DD, Baertschi SW, Johnson RA, Maple SR, Miller MS, Hallenbeck DK, Gregg SM (1998) J Pharm Sci 87:31

268. Howard MJ, Smales CM (2005) J Biol Chem 280:22582

269. Rahbar S (2005) Ann N Y Acad Sci 1043:9

270. Monnier VM, Stevens VJ, Cerami A (1981) Prog Food Nutr Sci 5:315

271. Gottlieb Sheldon H (2002) Diabetes Forecast 55:34

272. Stitt Alan W (2005) Ann N Y Acad Sci 1043:582

273. Mossine VV, Glinsky GV, Barnes CL, Feather MS (1995) Carbohydr Res 266:5

274. Mossine VV, Barnes CL, Mawhinney TP (2007) Carbohydr Res 342:131

275. Yaylayan VA, Huyghues-Despointes A (1996) Carbohydr Res 286:179

276. Ge SJ, Lee TC (1996) J Agric Food Chem 44:1053

277. Yaylayan VA, Huyghues-Despointes A (1994) Crit Rev Food Sci Nutrit 34:321

278. Niwa T, Takeda N, Miyazaki T, Yoshizumi H, Tatematsu A, Maeda K, Ohara M, Tomiyama S, Niimura K (1995) Nephron 69:438

279. Yamada H, Miyata S, Igaki N, Yatabe H, Miyauchi Y, Ohara T, Sakai M, Shoda H, Oimomi M, Kasuga M (1994) J Biol Chem 269:20275

280. Hirsch J, Petrakova E, Feather MS (1992) Carbohydr Res 232:125

281. Hirsch J, Petrakova E, Feather MS, Barnes CL (1995) Carbohydr Res 267:17

282. Zyzak DV, Richardson JM, Thorpe SR, Baynes JW (1995) Arch Biochem Biophys 316:547

283. Hayase F, Shibuya T, Sato J, Yamamoto M (1996) Biosci Biotechnol Biochem 60: 1820

284. Hayase F, Nagaraj RH, Miyata S, Njoroge FG, Monnier VM (1989) J Biol Chem 264:3758

285. Yaylayan VA, Keyhani A (1999) J Agric Food Chem 47:3280

286. Kim SW, Rogers QR, Morris JG (1996) J Nutr 126:195

287. Hiramoto K, Aso-o R, Ni-iyama H, Hikage S, Kato T, Kikugawa K (1996) Mutat Res 359:17

288. Hiramoto K, Ishihara A, Sakui N, Daishima S, Kikugawa K (1998) Biol Pharm Bull 21:102

289. Hiramoto K, Kato T, Kikugawa K (1993) Mutat Res 285:191

290. Hiramoto K, Nasuhara A, Michikoshi K, Kato T, Kikugawa K (1997) Mutat Res 395:47

291. Hiramoto K, Sekiguchi K, Aso OR, Ayuha K, Ni-Iyama H, Kato T, Kikugawa K (1995) Food Chem Toxicol 33:803

292. Hiramoto S, Itoh K, Shizuuchi S, Kawachi Y, Morishita Y, Nagase M, Suzuki Y, Nobuta Y, Sudou Y, Nakamura O, Kagaya I, Goshima H, Kodama Y, Icatro Faustino C, Koizumi W, Saigenji K, Miura S, Sugiyama T, Kimura N (2004) Helicobacter 9:429

293. Lertsiri S, Fujimoto K, Miyazawa T (1995) Biochim Biophys Acta 1245:278

294. Lertsiri S, Maungma R, Assavanig A, Bhumiratana A (2001) J Food Process Preserv 25:149

295. Lertsiri S, Oak JH, Nakagawa K, Miyazawa T (2002) Biochim Biophys Acta 1573:48

296. Lertsiri S, Oak J-H, Nakagawa K, Miyazawa T (2002) Biochimica et Biophysica Acta, General Subjects 1573:48

297. Lertsiri S, Shiraishi M, Miyazawa T (1998) Biosci Biotechnol Biochem 62:893

298. Surh YJ, Liem A, Miller JA, Tannenbaum SR (1994) Carcinogenesis 15:2375

299. Tressi R, Piechotta CT, Rewicki D, Krause E (2002) Int Congr Series 1245:203

300. Jenq SN, Tsai SJ, Lee H (1994) Mutagenesis 9:483

301. Friedman M (1996) J Agric Food Chem 44:631

302. Vertesy L, Fehlhaber HW, Kogler H, Schindler PW (1996) Liebigs Annalen: 121
303. Ukeda H, Hasegawa Y, Ishi T, Sawamura M (1997) Biosci Biotechnol Biochem 61:2039
304. Ukeda H, Shimamura T, Tsubouchi M, Harada Y, Nakai Y, Sawamura M (2002) Anal Sci 18:1151
305. Buttner U, Gerum F, Severin T (1997) Carbohydr Res 300:265
306. Buttner U, Ochs S, Severin T (1996) Carbohydr Res 291:175
307. Ahmed MU, Dunn JA, Walla MD, Thorpe SR, Baynes JW (1988) J Biol Chem 263:8816
308. Ahmed MU, Thorpe SR, Baynes JW (1986) J Biol Chem 261:4889
309. Ahmed N, Babaei-Jadidi R, Howell SK, Thornalley PJ, Beisswenger PJ (2005) Diabetes Care 28:2465
310. Glomb MA, Pfahler C (2001) J Biol Chem 276:41638
311. Glomb MA, Tschirnich R (2001) J Agricult Food Chem 49:5543
312. Lo TW, Westwood ME, McLellan AC, Selwood T, Thornalley PJ (1994) J Biol Chem 269:32299
313. Lopez MG, Mancilla-Margalli NA (2000) Frontiers of Flavour Science. Proceedings of the 9th Weurman Flavour Research Symposium, Freising, Germany, June 22–25
314. Li EY, Feather MS (1994) Carbohydr Res 256:41
315. Nagaraj RH, Monnier VM (1995) Biochim Biophys Acta 1253:75
316. Nagaraj RH, Sady C (1996) FEBS Lett 382:234
317. Nagaraj RH, Sarkar P, Mally A, Biemel KM, Lederer MO, Padayatti PS (2002) Arch Biochem Biophys 402:110
318. Pischetsrieder M, Larisch B, Seidel W (1997) J Agric Food Chem 45:2070
319. Dyer DG, Blackledge JA, Katz BM, Hull CJ, Adkisson HD, Thorpe SR, Lyons TJ, Baynes JW (1991) Z Ernahrungswiss 30:29
320. Dyer DG, Blackledge JA, Thorpe SR, Baynes JW (1991) J Biol Chem 266:11654
321. Smith PR, Somani HH, Thornalley PJ, Benn J, Sonksen PH (1993) Clin Sci (Lond) 84:87
322. Smith PR, Thornalley PJ (1992) Eur J Biochem 210:729
323. Nissimov J, Elchalal U, Bakala H, Brownlee M, Berry E, Phillip M, Milner Y (2007) J Immunolog Method 320:1
324. Meerwaldt R, Lutgers HL, Links TP, Graaff R, Baynes JW, Gans ROB, Smit AJ (2007) Diabetes Care 30:107
325. Yu Y, Thorpe SR, Jenkins AJ, Shaw JN, Sochaski MA, Mcgee D, Aston CE, Orchard TJ, Silvers N, Peng YG, McKnight JA, Baynes JW, Lyons TJ, Grp DER (2006) Diabetologia 49:2488
326. Lutgers HL, Graaff R, Links TP, Ubink-Veltmaat LJ, Bilo HJ, Gans RO, Smit AJ (2006) Diabetes Care 29:2654
327. Stitt AW, Jenkins AJ, Cooper ME (2002) Expert Opin Inv Drug 11:1205
328. Raj DSC, Choudhury D, Welbourne TC, Levi M (2000) Am J Kidney Dis 35:365
329. Chiarelli F, Catino M, Tumini S, Cipollone F, Mezzetti A, Vanelli M, Verrotti A (2000) Pediatr Nephrol 14:841
330. DeGroot J (2004) Curr Opin Pharmacol 4:301
331. Chellan P, Nagaraj RH (2001) J Biol Chem 276:3895
332. Biemel KM, Reihl O, Conrad J, Lederer MO (2001) J Biol Chem 276:23405

Part 3
Chemical Glycosylation Reactions

3.1 Glycosyl Halides

Kazunobu Toshima
Department of Applied Chemistry, Faculty of Science and Technology,
Keio University, Yokohama 223–8522, Japan
toshima@applc.keio.ac.jp

1 Chemical Glycosylation of Glycosyl Bromide and Chloride 430
2 Chemical Glycosylation of Glycosyl Iodide 433
3 Preparation and Chemical Glycosylation of Glycosyl Fluoride 433
3.1 Preparation of Glycosyl Fluoride 433
3.2 Chemical Glycosylation of Glycosyl Fluoride 439

Abstract

This chapter describes the preparations and chemical glycosylation reactions of glycosyl halides as glycosyl donors including glycosyl bromides, chlorides, iodides, and fluorides. For a survey on the general current methodological advances, glycosyl halide donors are classified into four groups based on the type of anomeric functional group and their activating methods. Among them, glycosyl fluorides, which are frequently and widely used in current glycosylation reactions, are particularly emphasized and discussed in this chapter.

Keywords

Glycosylation; Glycosyl donor; Glycosyl halide; Glycosyl bromide; Glycosyl chloride; Glycosyl iodide; Glycosyl fluoride

Abbreviations

Ac	acetyl
acac	acetylacetone
Ar	aryl
Bn	benzyl
BTF	benzotrifluoride
t-Bu	*tert*-butyl
Bz	benzoyl
Cp	cyclopentadienyl
DAST	diethylaminosulfur trifluoride
DDQ	2,3-dichloro-5,6-dicyano-1,4-benzoquinone
DEAD	diethyl azodicarboxylate
DIEA	*N,N*-diisopropylethylamine

In: *Glycoscience*. Fraser-Reid B, Tatsuta K, Thiem J (eds)
Chapter-DOI 10-1007/978-3-540-30429-6_10: © Springer-Verlag Berlin Heidelberg 2008

Et	ethyl
Me	methyl
MS	molecular sieves
NBS	N-bromosuccinimide
NIS	N-iodosuccinimide
Ph	phenyl
***i*-Pr**	isopropyl
PPTS	pyridinium p-toluenesulfonate
Pv	pivaloyl
TBAF	tetrabutylammonium fluoride
TBAI	tetrabutylammonium iodide
TEA	triethylamine
Tf	trifluoromethanesulfonyl
THF	tetrahydrofuran
TMS	trimethylsilyl
TMU	1,1,3,3-tetramethylurea
Tr	triphenylmethyl
Ts	toluenesulfonyl

1 Chemical Glycosylation of Glycosyl Bromide and Chloride

The use of glycosyl bromide or chloride as an effective glycosyl donor in the glycosylation reaction was first introduced by Koenigs and Knorr in 1901 [1]. In relation to the anomeric stereochemistry of the glycosylation reaction, three significant basic methods, (1) the neighboring group assisted method for construction of 1,2-*trans* glycosides such as β-gluco- or α-manno type glycoside, (2) the in-situ anomerization method for synthesis of α-gluco or α-manno type glycoside [2], and (3) the heterogenic catalyst method for preparation of β-mannoglycoside [3], were developed in this area [4d, f]. The well-known classical Koenigs–Knorr method used heavy metal salts (mainly silver and mercury salts) as activating reagents as summarized in ❷ *Table 1*. A variety of heavy metal salts such as AgClO$_4$, AgOTf, AgNO$_3$, Ag$_2$CO$_3$, Ag$_2$O, Ag-silicate, Hg(CN)$_2$, HgBr$_2$, HgCl$_2$, and HgI$_2$ [5] and their combined use were employed in this area [4a–h, j, l]. The order of reactivity of some representative catalysts was generally confirmed [4d, f]. Furthermore, Ag$_2$CO$_3$, Ag$_2$O, HgO, CdCO$_3$, S-collidine, and TMU were frequently used as an acid scavenger, and water was generally removed by Drierite® and molecular sieves during these glycosylation reactions [4a–h, j, l].

On the other hand, other glycosylation methods using glycosyl bromide or chloride in the absence of any metal were also widely studied. Lemieux et al. [6] introduced a mild glycosylation method in the presence of Bu$_4$NBr. Also, the glycosylation protocols which involved a transformation of glycosyl bromide into the corresponding onium salts by Et$_3$N, Ph$_3$P, and Me$_2$S were developed by Schuerch et al. [7]. Furthermore, the uses of other activating reagents without heavy metals and including several Lewis acids such as SnCl$_4$ [8], BF$_3$Et$_2$O [8], Sn(OTf)$_2$-collidine [9a], Sn(OTf)$_2$-TMU [9b], TrCl-ZnCl$_2$ [10], LiClO$_4$ [11], I$_2$-DDQ [12],

☐ **Table 1**

Glycosylation of glycosyl bromide or chloride using heavy metal

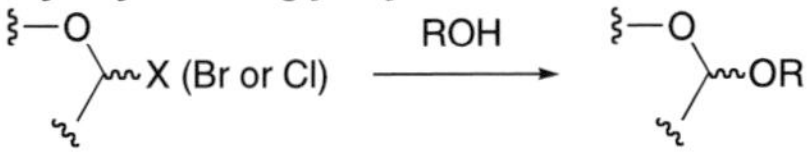

Activator	Acid scavenger	Drying agent	Ref.
AgClO$_4$	Ag$_2$CO$_3$	Drierite	[4a–h, j, l,5]
AgOTf	Ag$_2$O	molecular sieves	
AgNO$_3$	HgO		
Ag$_2$CO$_3$	CdCO$_3$		
Ag$_2$O	s-collidine		
Ag-silicate	TMU		
Hg(CN)			
HgBr$_2$			
HgCl$_2$			
HgI$_2$			

☐ **Table 2**

Glycosylation of glycosyl bromide or chloride using Lewis acid

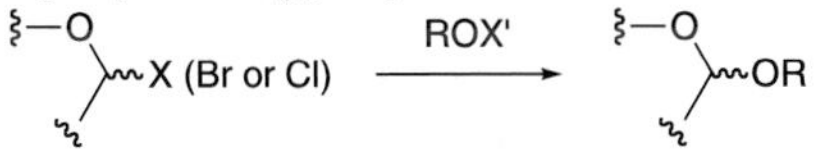

Activator	X′	Ref.
SnCl$_4$	SnBu$_3$	[8]
BF$_3$·OEt$_2$	SnBu$_3$	[8]
Sn(OTf)$_2$-collidine	H	[9a]
Sn(OTf)$_2$-TMU	H	[9b]
TrCl-ZnCl$_2$	H	[10]
LiClO$_4$	H	[11]
I$_2$-DDQ	H	[12]
IBr	H	[13]
NIS	H	[14]
In	H	[15]
InCl$_3$	H	[16]
TMSOTf	SnBu$_3$	[17]
$\left(\bigcirc\text{N} \right)_3 \text{P=O}$		
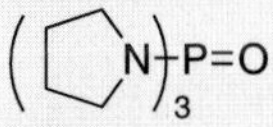 Zn(acac)$_2$	H	[18,19]
Zn(acac)$_2$	H	[20]
$\left(t\text{-Bu} - \bigcirc - \text{COO} \right)_2 \text{Zn}$		
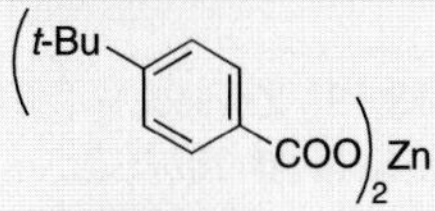 Cu(OTf)$_2$-BTF	H	[21]
Cu(OTf)$_2$-BTF	H	[22]

☐ Table 3

Glycosylation of glycosyl bromide or chloride using phase transfer catalyst

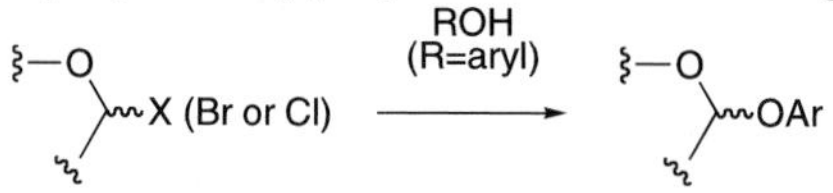

Catalyst	Conditions	Ref.
Et$_3$N$^+$BnBr$^-$	CHCl$_3$/H$_2$O/NaOH	[23]
Et$_3$N$^+$BnCl$^-$	CH$_2$Cl$_2$/H$_2$O/NaOH or KOH	[24]
Me(CH$_2$)$_{15}$N$^+$Me$_3$Br$^-$	CH$_2$Cl$_2$/H$_2$O/NaOH	[25,26]
Bu$_4$N$^+$Br$^-$	CH$_2$Cl$_2$/H$_2$O/NaOH	[27]
Bu$_4$N$^+$HSO$_4^-$	CH$_2$Cl$_2$/H$_2$O/NaOH or NaHCO$_3$	[28]
Aliquat®336 (R = alkyl)	CH$_2$Cl$_2$/H$_2$O/K$_2$CO$_3$	[29]
BnN$^+$Bu$_3$Cl$^-$	CHCl$_3$/H$_2$O/K$_2$CO$_3$	[30]

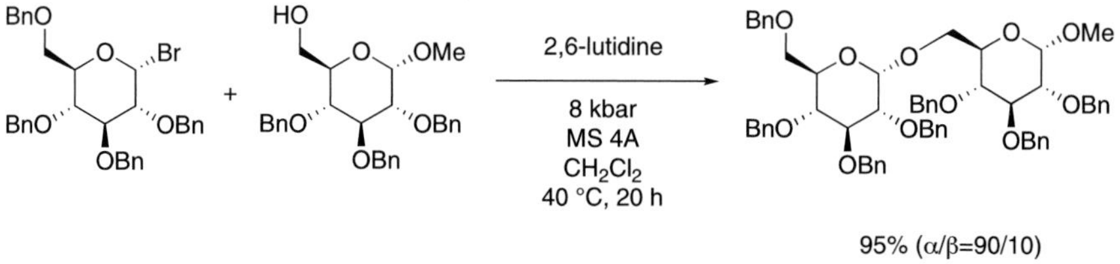

95% (α/β=90/10)

☐ Scheme 1

98% (α/β=28/72)

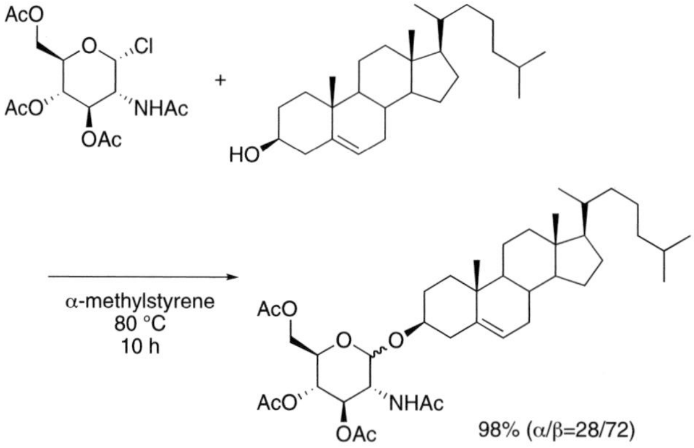

☐ Scheme 2

IBr [13], NIS [14], In [15], InCl$_3$ [16], TMSOTf [17], (C$_4$H$_8$N)$_3$P = O [18,19], Zn(acac)$_2$ [20], (t-BuC$_6$H$_4$CO$_2$)$_2$Zn [21], and Cu(OTf)$_2$-BTF [22] were reported in this field (❂ *Table 2*). The glycosylations of aryl alcohols using a phase transfer catalyst such as Et$_3$N$^+$BnBr$^-$ [23a, b], Et$_3$N$^+$BnCl$^-$ [24], Me(CH$_2$)$_{15}$N$^+$Me$_3$Br$^-$ [25,26], Bu$_4$N$^+$Br$^-$ [27], Bu$_4$NH$^+$SO$_4^-$ [28], Aliquat®336 [29], and BnN$^+$Bu$_3$Cl$^-$ [30] were also developed (❂ *Table 3*).

Alternatively, Sasaki et al. [31] provided a glycosylation method using glycosyl bromide in the presence of hindered amines such as 2,6-lutidine or TMU under high-pressure conditions (❷ *Scheme 1*). In addition, Nishizawa et al. [32] developed a thermal glycosylation of glycosyl chloride in the presence of α-methylstyrene or TMU as an acid scavenger without any metal salts (❷ *Scheme 2*).

2 Chemical Glycosylation of Glycosyl Iodide

Among the glycosyl halides, only a few examples of the use of glycosyl iodide as an isolatable glycosyl donor are demonstrated due to their instability (❷ *Table 4*). Thus, LiClO$_4$ [33], TBAI-DIEA [34], FeCl$_3$-I$_2$ [35], CuCl-I$_2$ [35], and NIS-I$_2$-TMSOTf [35] were reported as the activators of the glycosyl iodides (❷ *Scheme 3*).

◻ **Table 4**
Glycosylation of glycosyl iodide

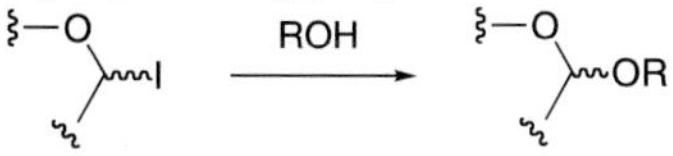

Activator	X	Ref.
LiClO$_4$	H	[33]
TBAI-DIPEA	H	[34]
NIS-I$_2$-TMSOTf	H	[35]
FeCl$_3$-I$_2$	H	[35]
CuCl-I$_2$	H	[35]

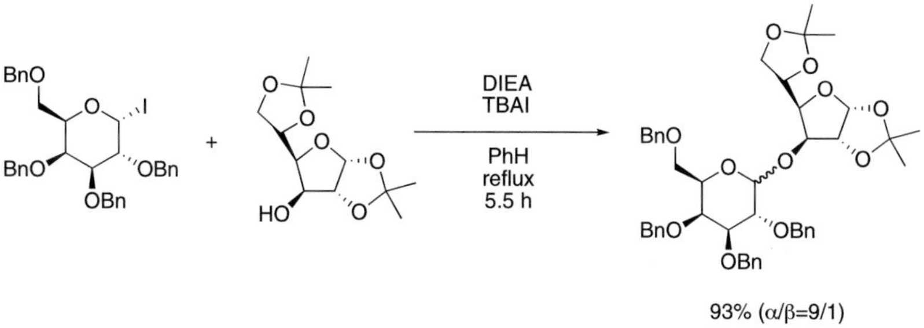

93% (α/β=9/1)

◻ **Scheme 3**

3 Preparation and Chemical Glycosylation of Glycosyl Fluoride

3.1 Preparation of Glycosyl Fluoride

Glycosyl fluorides have been widely and effectively used for glycosylation reactions. One of the notable advantages of the glycosyl fluoride as a glycosyl donor is its higher thermal and chemical stability as compared to other glycosyl halides such as glycosyl chlorides, bromides,

☐ **Table 5**
Preparation of glycosyl fluoride from aldose

Reagent	Ref.
2-F-C$_5$H$_4$NMe·OTs-Et$_3$N	[37]
DAST	[38,39]
HF/Py	[40,41]
TiF$_4$	[42]
CF$_3$ZnBr·2MeCN-TiF$_4$	[42]
M$_2$C=C(F)NMe$_2$	[43]
DEAD-PPh$_3$-Et$_3$OBF$_3$	[44]

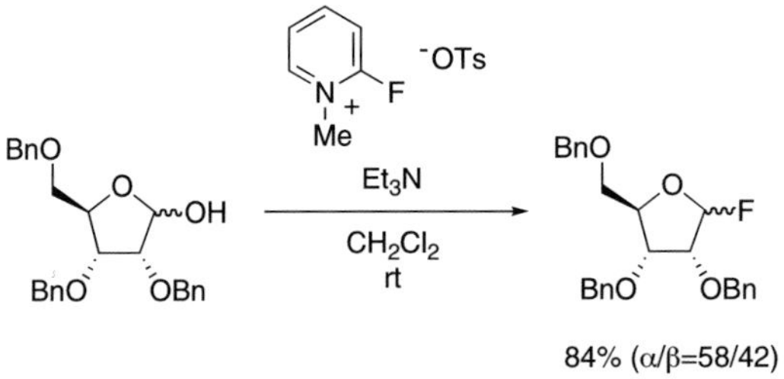

84% (α/β=58/42)

☐ **Scheme 4**

and iodides. Therefore, the glycosyl fluoride can be generally purified by the appropriate distillation and even by column chromatography with silica gel. The practical use of glycosyl fluoride as a glycosyl donor was first introduced by Mukaiyama et al. in 1981 [36]. After the significant advance in this field, a number of effective methods for fluorinating the anomeric center of several types of sugars have been developed. The representative methods for the conversion of a free anomeric hydroxyl group of sugars into the fluoride are summarized in ❯ *Table 5*.

Mukaiyama and coworkers reported that totally benzylated ribofuranose smoothly reacted with 2-fluoro-1-methylpyridinium tosylate in the presence of triethylamine at room temperature to afford an anomeric mixture of the corresponding ribofuranosyl fluorides in high yield (❯ *Scheme 4*) [37].

Diethylaminosulfur trifluoride (DAST) is now widely used as the most preferable fluorinating reagent of 1-glycoses (❯ *Scheme 5*) [38,39]. Pyridinium poly(hydrogen fluoride) (HF/Py) was also found to be an effective reagent for the conversion of various furanose and pyranose hemiacetals into the corresponding glycosyl fluorides (❯ *Scheme 6*) [40,41].

On the other hand, use of metal fluorides such as TiF$_4$ and the complex CF$_3$ZnBr·2MeCN-TiF$_4$ also afforded the glycosyl fluorides from the corresponding glycoses (❯ *Scheme 7*) [42].

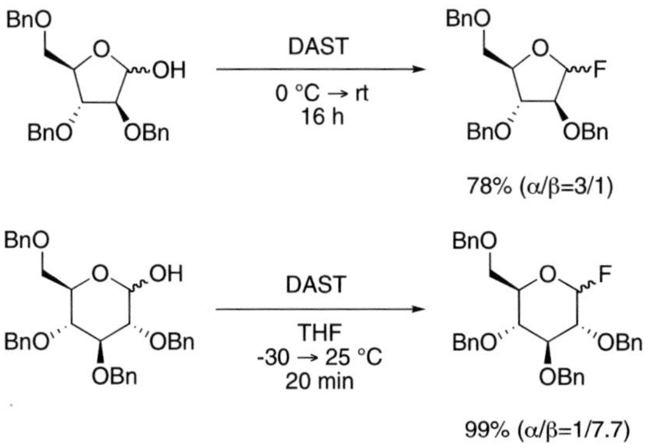

78% (α/β=3/1)

99% (α/β=1/7.7)

■ Scheme 5

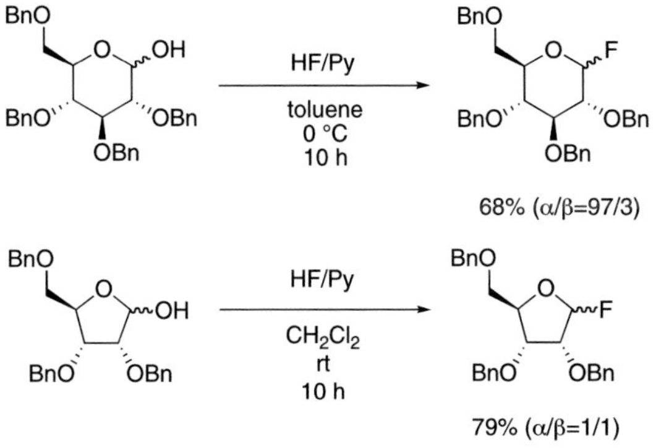

68% (α/β=97/3)

79% (α/β=1/1)

■ Scheme 6

83% (α/β=4/6)

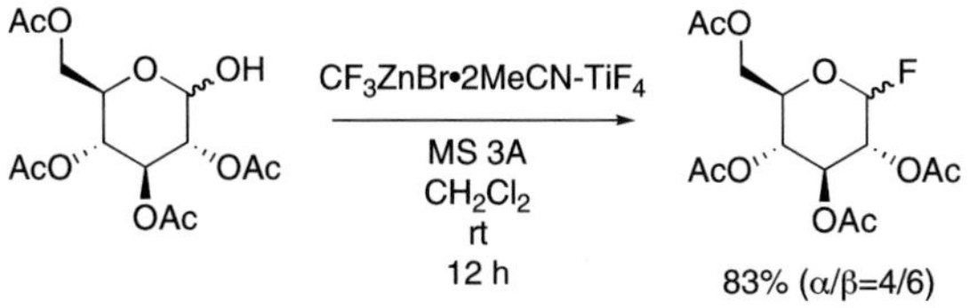

■ Scheme 7

Furthermore, the anomeric hydroxyl group of various furanose and pyranose hemiacetals can be replaced by a fluorine under neutral conditions using an α-fluoroenamine (❷ *Scheme 8*) [43].

Glycosyl fluorides can also be obtained by the modified Mitsunobu reaction using diethyl azodicarboxylate (DEAD)-PPh$_3$-Et$_3$O$^+$BF$_4^-$ (❷ *Scheme 9*) [44]. By this method, the acid sensitive isopropylidene-protected mannofuranosyl fluoride was prepared in moderate yield.

Alternatively, 1-*O*-acylated sugars can be easily transformed into the corresponding glycosyl fluorides. Thus, treatment of 1-*O*-acetylated sugars with HF/Py gave the corresponding glyco-

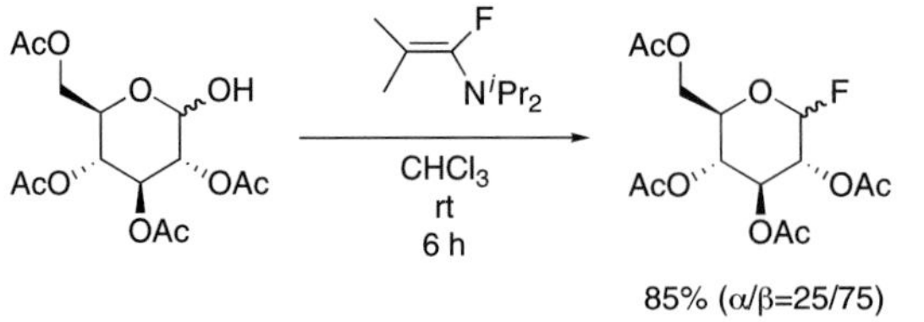

◻ Scheme 8

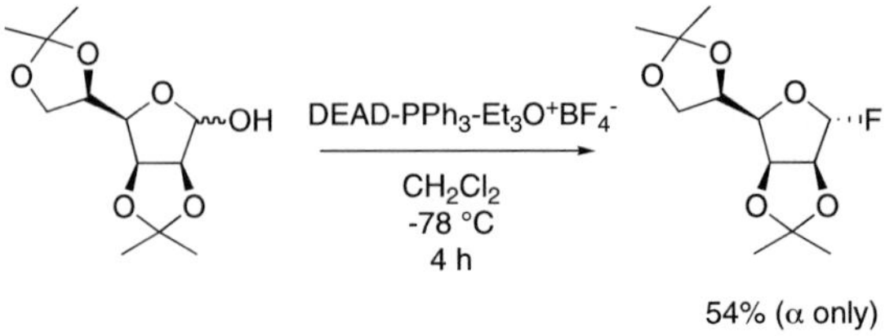

◻ Scheme 9

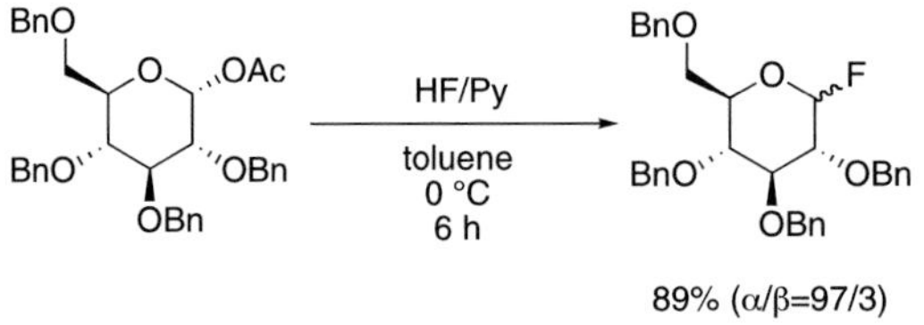

◻ Scheme 10

◻ Table 6
Preparation of glycosyl fluoride from glycosyl bromide

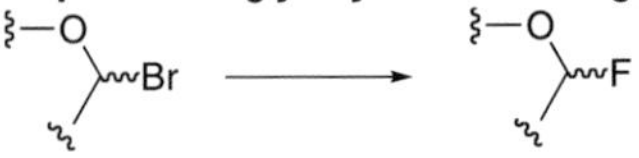

Reagent	Ref.
AgF	[45,46]
ZnF$_2$-2,2′-bipyridine	[47]
CF$_3$ZnBr·2MeCN	[42]

syl fluorides in high yields. Compared to the unprotected analogue, the 1-*O*-acetylated derivatives preferentially underwent fluorination by this reagent (❯ *Scheme 10*) [40]. Furthermore, it was found that the reaction of penta-*O*-acetyl-β-D-glucopyranose, which had a participating acetoxy group at the C-2 position, gave a mixture of the α-and β-fluorides in which the latter predominated.

The preparation of glycosyl fluorides from other glycosyl halides is summarized in ❯ *Table 6*. Glycosyl bromides were effectively reacted with metal fluorides such as AgF (❯ *Scheme 11*)

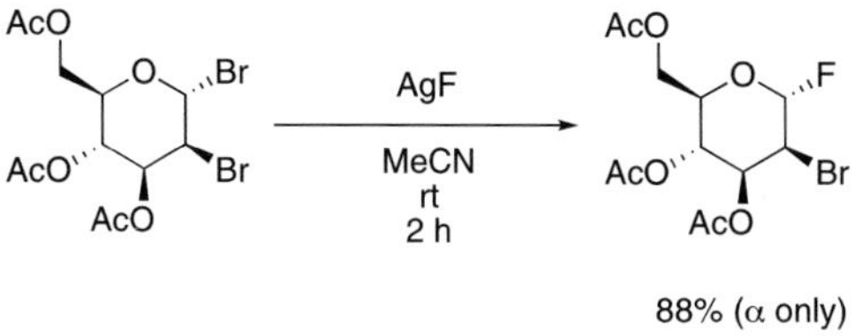

88% (α only)

◻ Scheme 11

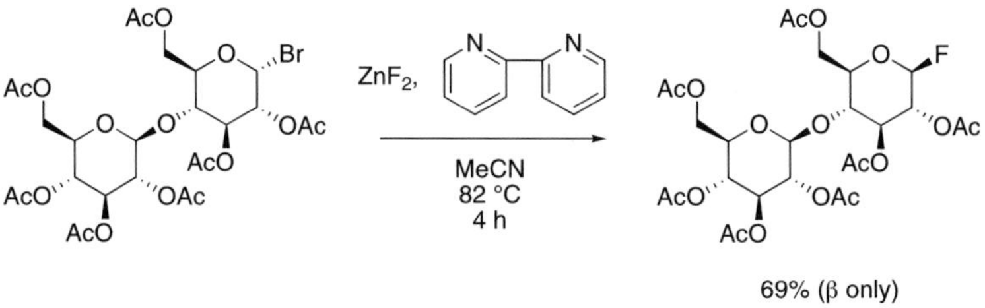

69% (β only)

◻ Scheme 12

83% (β only)

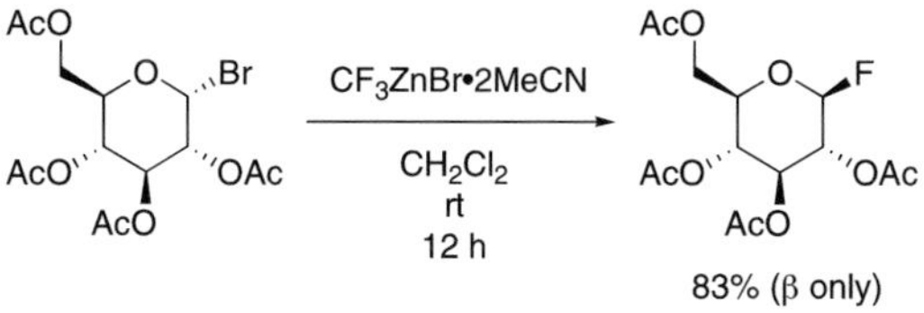

◻ Scheme 13

◻ Table 7
Preparation of glycosyl fluoride from thioglycoside

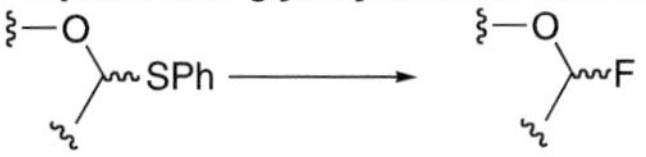

Reagent	Ref.
NBS-DAST	[48]
NBS-HF/Py	[48]
4-MeC$_6$H$_4$IF$_2$	[49]

[45,46], ZnF$_2$−2,2′-bipyridine (❷ *Scheme 12*) [47] or CF$_3$ZnBr·2MeCN (❷ *Scheme 13*) [42] to afford the corresponding glycosyl fluorides in high yields via the nucleophilic halide exchange reaction. Therefore, in general, the glycosyl fluorides were obtained with stereo-inversion of the anomeric center of the glycosyl bromides.

On the other hand, the glycosyl fluorides can be prepared from the corresponding thioglyco-sides (❷ *Table 7*). Thus, the reactions of thioglycosides with DAST-NBS or HF/Py-NBS gave the corresponding glycosyl fluorides in high yields (❷ *Scheme 14*) [48]. Furthermore, the thio-

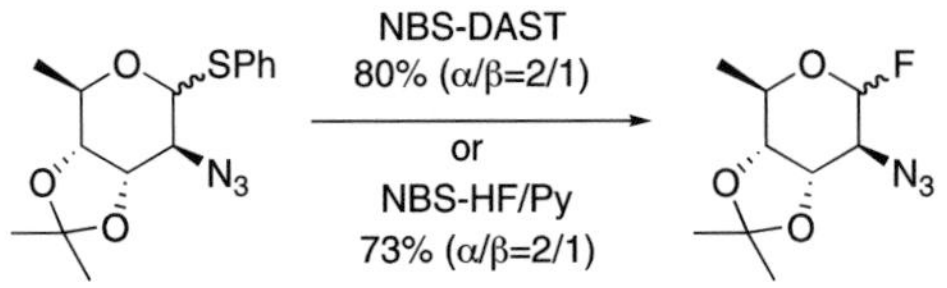

□ Scheme 14

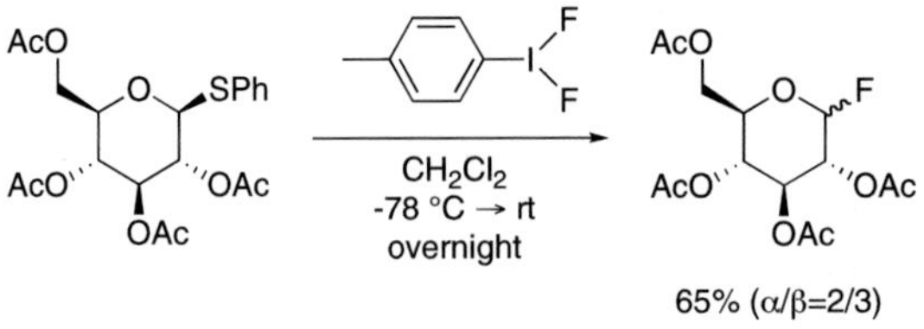

65% (α/β=2/3)

□ Scheme 15

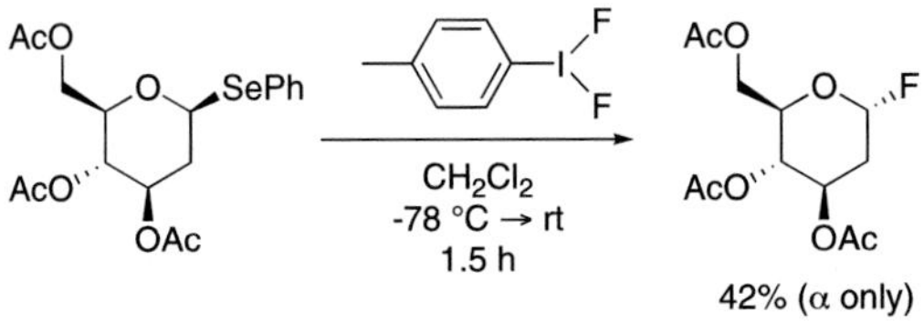

42% (α only)

□ Scheme 16

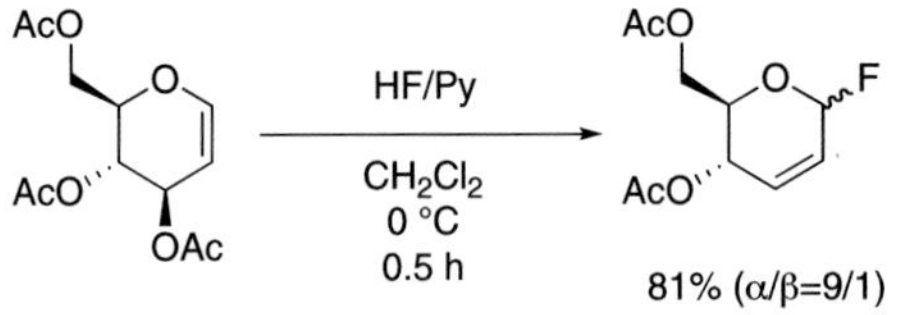

81% (α/β=9/1)

□ Scheme 17

53% (β only)

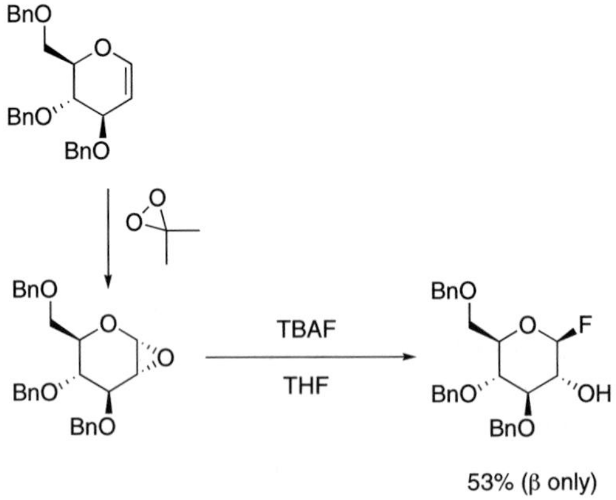

□ Scheme 18

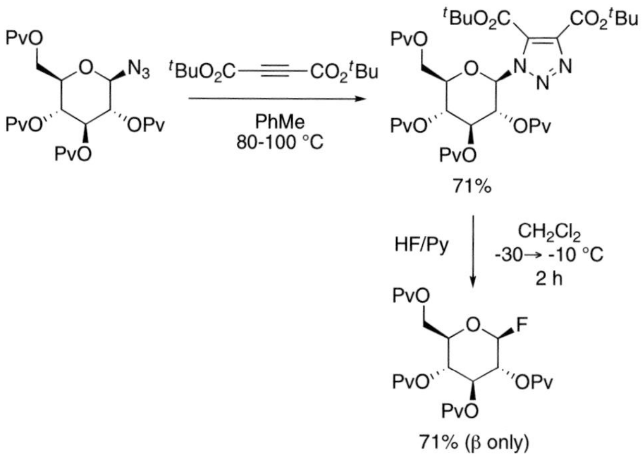

□ **Scheme 19**

glycosides may also react with the hypervalent iodoarene, 4-methyl(difluoroiodo)benzene, to yield the corresponding glycosyl fluorides (❯ *Scheme 15*) [49,50]. It is noteworthy that the yields of the produced glycosyl fluorides are significantly improved using the corresponding *p*-chloro-phenylthioglycosides as glycosyl donors in these reactions.

The selenoglycosides were also found to be transformed into the glycosyl fluorides using 4-methyl(difluoroiodo)benzene via an S_N2 inversion of the anomeric centers (❯ *Scheme 16*) [50].

Glycals and 1,2-anhydro sugars, the latter of which are readily obtained from glycals, were prepared and then transformed into the corresponding glycosyl fluorides using HF/Py (❯ *Scheme 17*) [51] and tetrabutylammonium fluoride (TBAF) (❯ *Scheme 18*) [52], respectively. In the former case, the corresponding 2,3-unsaturated glycosyl fluorides are exclusively produced via the Ferrier rearrangement [53] while 1,2-trans glycosyl fluorides are stereoselectively obtained in the latter case.

Furthermore, *N*-glycosyl triazole derivatives (❯ *Scheme 19*) [54], prepared from the corresponding glycosyl azides, and (1-phenyl-1*H*-tetrazol-5-yl)glycosides synthesized from the corresponding glycoses and 5-chloro-1-phenyl-1*H*-tetrazole (❯ *Scheme 20*) [55] both smoothly react with HF/Py to furnish the corresponding glycosyl fluorides in good yields.

3.2 Chemical Glycosylation of Glycosyl Fluoride

Since 1981 when Mukaiyama introduced the practical use of glycosyl fluoride with a fluorophilic activator, $SnCl_2$-$AgClO_4$, as a glycosyl donor (❯ *Scheme 21*) [36], a number of specific fluorophilic reagents have been developed for the effective glycosylation reactions (❯ *Table 8*). Following the $SnCl_2$-$AgClO_4$ promoter, the combined use of $SnCl_2$-$TrClO_4$ (❯ *Scheme 22*) [37] and $SnCl_2$-$AgOTf$ (❯ *Scheme 23*) [56] were reported by Mukaiyama and Ogawa, respectively. In these cases, 1,2-*cis*-α-glycosides were obtained predominantly in high yields.

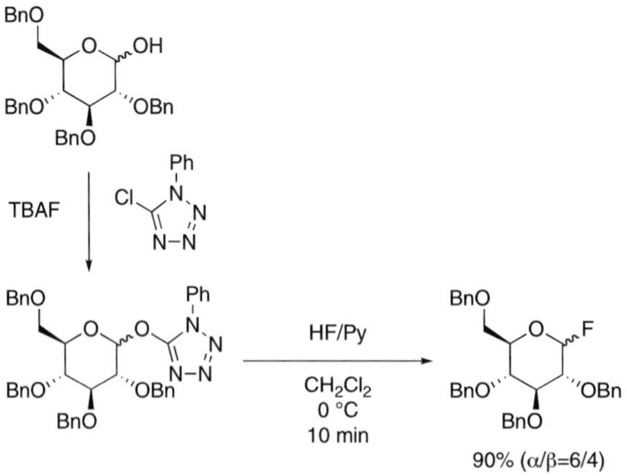

☐ **Scheme 20**

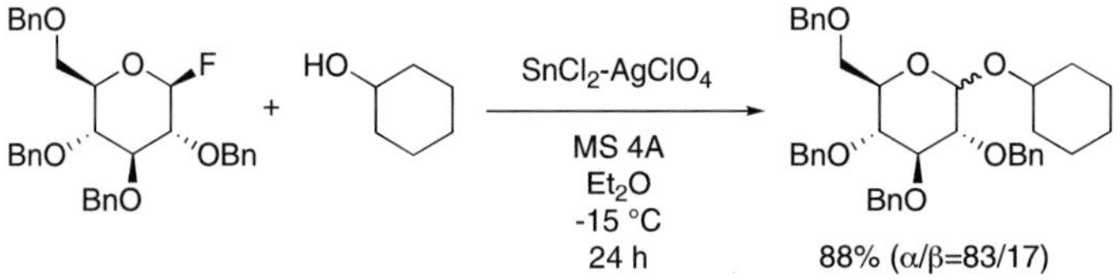

☐ **Scheme 21**

In 1984, Noyori and coworkers announced that the silyl compounds, both SiF$_4$ and trimethylsilyl trifluoromethanesulfonate (TMSOTf) were very effective for activation of the glycosyl fluorides (❷ *Scheme 24*) [57]. Furthermore, it was found that the stereoselectivity of the glycosylation was highly dependent on the reaction solvent. Thus, the glycosylation in MeCN exclusively gave the β-glycoside while the glycosylation in Et$_2$O predominately afforded the α-glycoside.

Furthermore, Nicolaou, Kunz, and Vozny independently reported that glycosyl fluorides effectively reacted with a variety of free alcohols and silyl ethers using BF$_3$Et$_2$O as an activator to give the corresponding *O*-glycosides in good yields (❷ *Scheme 25*) [58,59,60,61].

On the other hand, metal fluorides such as TiF$_4$ and SnF$_4$ were also used as effective promoters of glycosyl fluorides by Thiem et al., and the stereoselective glycosylation of 2-deoxy glycosyl fluoride was carried out with use of TiF$_4$. In the case of 2-deoxy-β-glycosyl fluoride, when hexane was used as the solvent, the α-glycoside was selectively produced with the inversion of the anomeric center. On the other hand, when the reaction was performed in Et$_2$O, the β-glycoside was obtained as the major product via "double S$_N$2" mechanism which involved the formation of the oxonium cation-ether complex (❷ *Scheme 26*) [62].

Suzuki and coworkers developed new and quite effective methods in which the combined activators including the group IV$_B$ metallocenes such as Cp$_2$MCl$_2$-AgClO$_4$ (M = Zr, Hf) (❷ *Scheme 27*) [63], Cp$_2$ZrCl$_2$-AgBF$_4$ (❷ *Scheme 28*) [64], and Cp$_2$HfCl$_2$-AgOTf [64,65] were used as milder reagents for promoting the glycosylations of glycosyl fluorides. Fur-

⬛ Table 8
Glycosylation of glycosyl fluoride

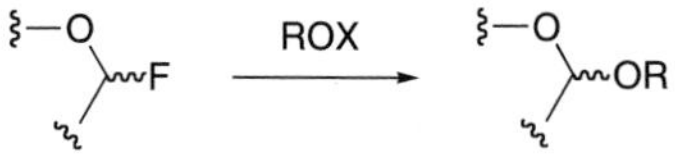

Activator	X	Ref.
$SnCl_2$-$AgClO_4$	H	[36]
$SnCl_2$-$TrClO_4$	H	[37]
$SnCl_2$-$AgOTf$	H	[56]
TMSOTf (cat.)	TMS	[57]
SiF_4 (cat.)	TMS	[58]
$BF_3 \cdot Et_2O$	H	[58,59,60,61]
TiF_4	H	[62]
SnF_4	H	[62]
Cp_2MCl_2-$AgClO_4$ (M = Zr or Hf)	H	[63]
Cp_2ZrCl_2-$AgBF_4$	H	[64]
Cp_2HfCl_2-$AgOTf$	H	[64,65]
$Bu_2Sn(ClO_4)_2$	H	[66]
Me_2GaCl	H	[67]
Tf_2O	H	[68]
$LiClO_4$	H	[69]
$Yb(OTf)_3$	H	[70]
$La(ClO_4)_3 \cdot nH_2O$ (cat.)	TMS	[71]
$La(ClO_4)_3 \cdot nH_2O$-$Sn(OTf)_2$	H	[72]
Yb-Amberlyst 15	H	[73]
SO_4/ZrO_2	H	[74]
Nafion-H	H	[74]
montmorillonite K-10	H	[74]
$TrB(C_6F_5)_4$ (cat.)	H	[75]
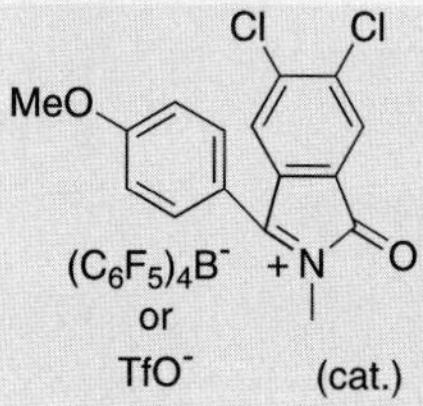	H	[75]
$SnCl_4$-$AgB(C_6F_5)_4$ (cat.)	H	[76]
TfOH (cat.)	H	[77]
$HB(C_6F_5)_4$ (cat.)	H	[78]
$HClO_4$ (cat.)	H	[78]
$HNTf_2$ (cat.)	H	[78]
$Yb(NTf_2)_3$	H	[79]
$ZrCl_4$	TMS	[80]
$Cu(OTf)_2$	H	[22]

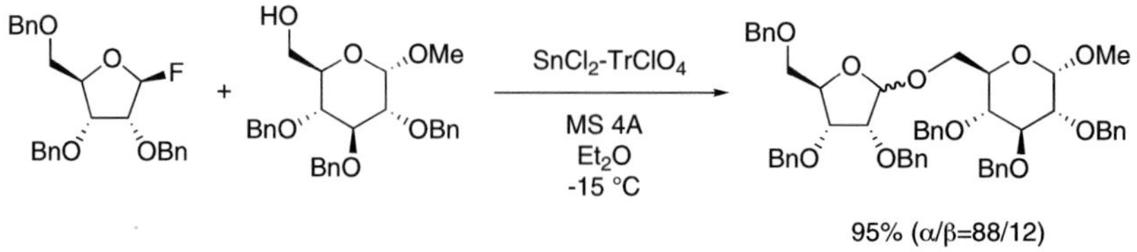

□ **Scheme 22**

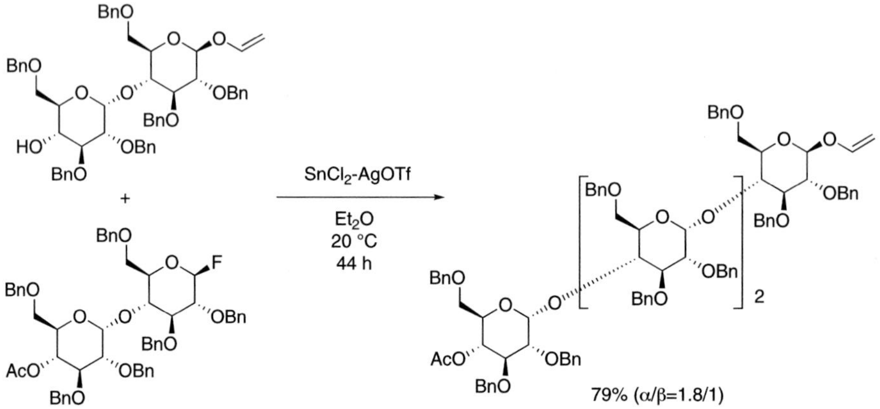

□ **Scheme 23**

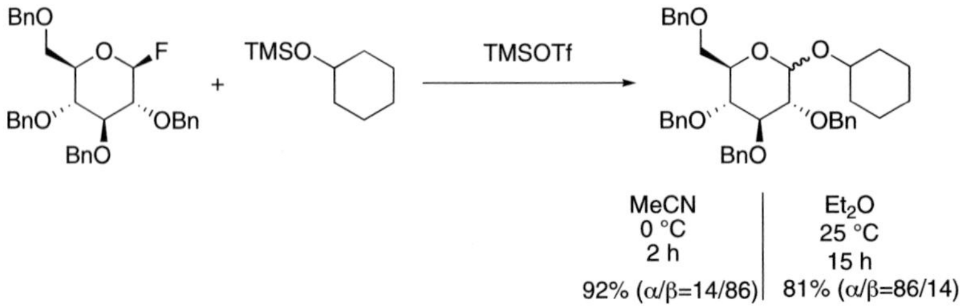

□ **Scheme 24**

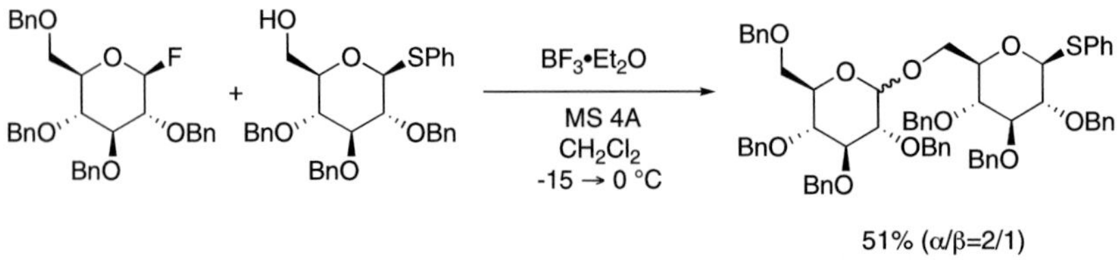

□ **Scheme 25**

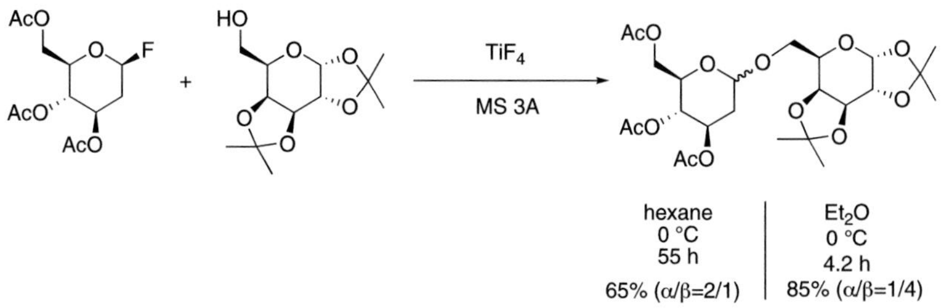

Scheme 26

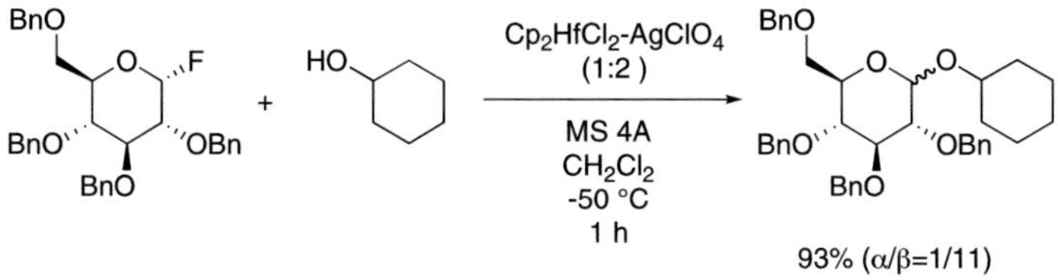

Scheme 27

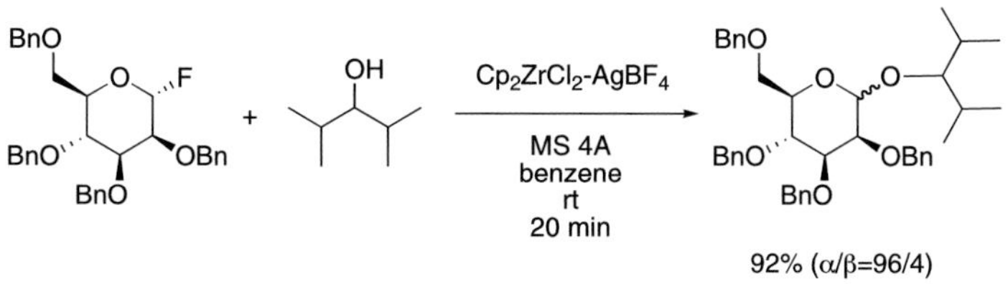

Scheme 28

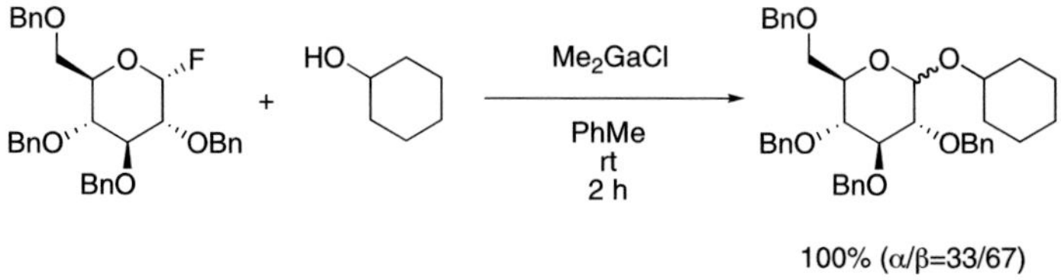

Scheme 29

thermore, the novel combined use of Bu_2SnCl_2–$2AgClO_4$ was found to show promise as an effective promoter of the glycosyl fluoride [66].

Me_2GaCl and Me_2GaOTf were introduced as new promoters of the glycosyl fluorides by Kobayashi due to their strong affinity for fluoride. In this study, it was found that the readily available Me_2GaCl was more effective for the glycosylation reactions of glycosyl fluorides (◉ *Scheme 29*) [67].

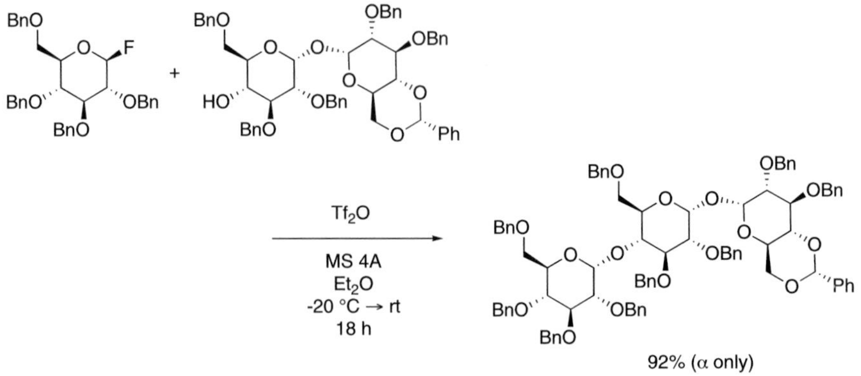

■ Scheme 30

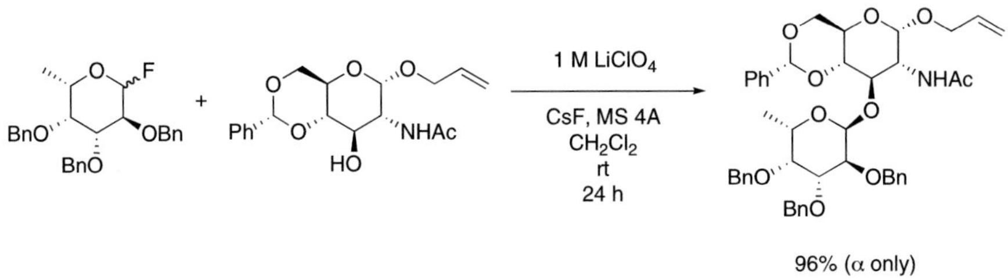

■ Scheme 31

Furthermore, Wessel et al. announced that Tf_2O was a highly reactive activator during the glycosylation of glycosyl fluorides (❯ *Scheme 30*) [68]. In this report, it was suggested that a sequence of relative reactivity of the examined catalysts was TMSOTf < $SnCl_2$-AgOTf < TiF_4 < Tf_2O.

On the other hand, Waldmann et al. reported the use of $LiClO_4$, a milder Lewis acid, for the glycosylation of fucosyl fluoride under neutral conditions (❯ *Scheme 31*) [69].

Shibasaki and coworkers developed the rare earth metal salts, such as $La(ClO_4)_3 \cdot nH_2O$ and $Yb(OTf)_3$, catalyzed glycosylations of glycosyl fluorides [70,71,72]. The use of either $Yb(OTf)_3$ or $YbCl_3$ in the presence of $CaCO_3$ and MS 4A in Et_2O was found to be effective for α-stereoselective glucosylation. On the other hand, for β-stereoselective glucosylation, the utilization of $Yb(OTf)_3$ containing K_2CO_3 and MS 4A in MeCN gave the best result (❯ *Scheme 32*) [70]. Furthermore, it was found that the combined use of $La(ClO_4)_3 \cdot nH_2O$ and $Sn(OTf)_2$ was very useful for β-stereoselective mannosylation (❯ *Scheme 33*) [72]. Along this line, Wang et al. reported the use of lanthanide (III) catalysts supported on ion exchange resins for simple glycosylations of glycosyl fluorides with methanol [73].

Toshima and coworkers demonstrated that environmentally friendly heterogeneous catalysts such as montmorillonite K-10, Nafion-H®, and SO_4/ZrO_2 were very effective for the glycosylations of glycosyl fluorides [74]. This is the first report on the use of a protic acid for activation of glycosyl fluoride. Among them, SO_4/ZrO_2 was shown to be superior to the

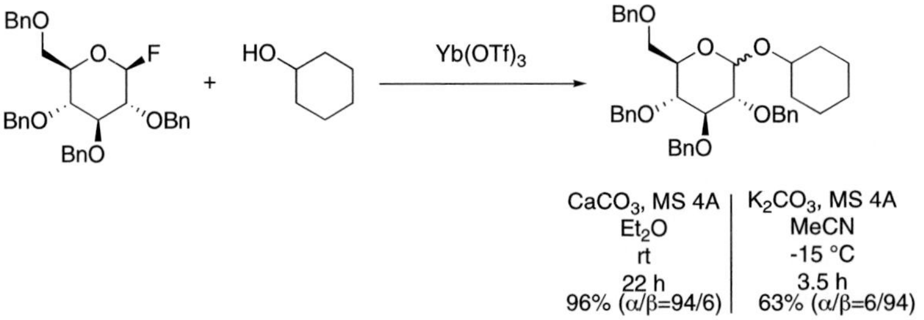
BnO
F
HO
Yb(OTf)3
BnO
OBn
BnO
CaCO3, MS 4A
Et2O
rt
22 h
96% (α/β=94/6)
K2CO3, MS 4A
MeCN
-15 °C
3.5 h
63% (α/β=6/94)

Scheme 32

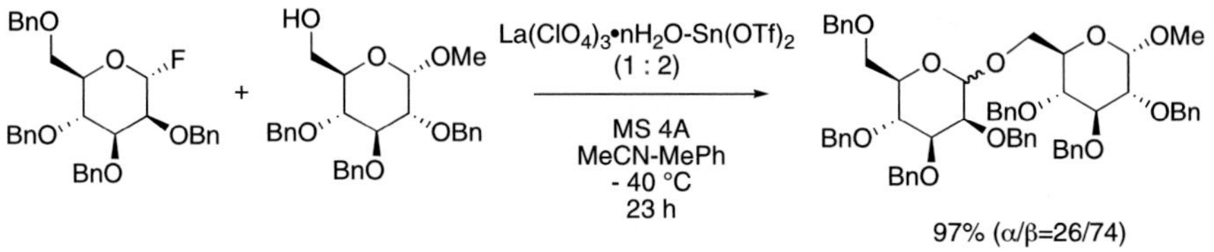
BnO
F
HO
OMe
BnO
OBn
BnO
La(ClO4)3•nH2O-Sn(OTf)2
(1 : 2)
MS 4A
MeCN-MePh
- 40 °C
23 h
97% (α/β=26/74)

Scheme 33

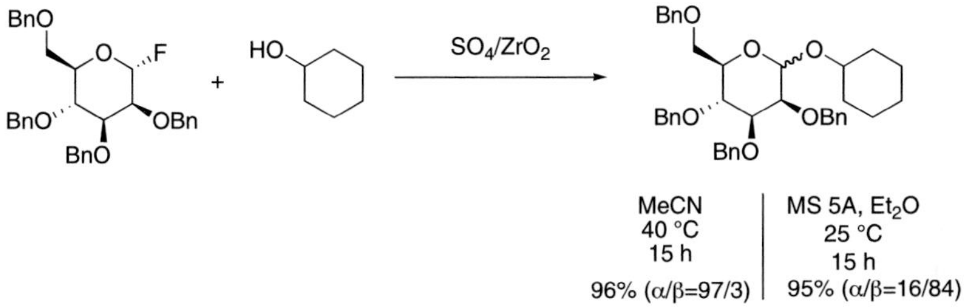
BnO
F
HO
SO4/ZrO2
BnO
OBn
BnO
MeCN
40 °C
15 h
96% (α/β=97/3)
MS 5A, Et2O
25 °C
15 h
95% (α/β=16/84)

Scheme 34

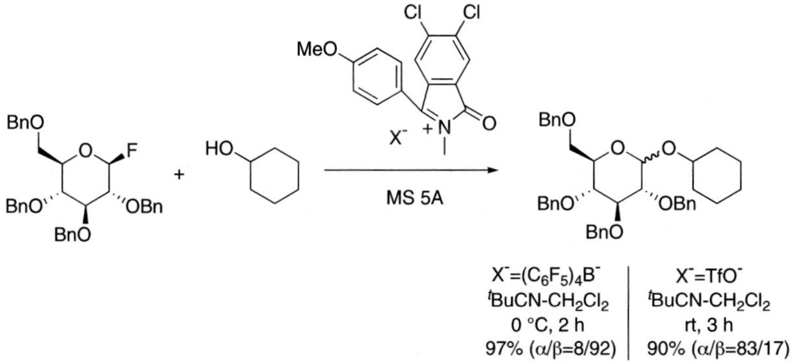
Cl
Cl
MeO
X-
O
MS 5A
BnO
F
HO
BnO
OBn
BnO
X-=(C6F5)4B-
tBuCN-CH2Cl2
0 °C, 2 h
97% (α/β=8/92)
X-=TfO-
tBuCN-CH2Cl2
rt, 3 h
90% (α/β=83/17)

Scheme 35

others for the stereocontrolled glycosylation with α-mannopyranosyl fluoride. Thus, the glycosylations of perbenzylated α-mannopyranosyl and and alcohols using SO_4/ZrO_2 in MeCN exclusively gave the corresponding α-glycosides. On the other hand, the corresponding β-glycosides were selectively obtained by the glycosylations employing SO_4/ZrO_2 in the presence of MS 5A in Et_2O (**◉** *Scheme 34*). Furthermore, this method could be applied to the stereocontrolled glycosylation of 2-deoxy-α-glucopyranosyl fluoride.

Mukaiyama et al. developed the glycosylations of glycosyl fluorides using carbocationic species paired with tetrakis(pentafluorophenyl) borate or trifluoromethanesulfonate [75]. In this study, it was found that the glycosylation using the former catalyst in CH_2Cl_2 containing t-BuCN gave the β-glycoside selectively while the α-glycoside was predominately obtained by the glycosylation using the latter catalyst in CH_2Cl_2 containing Et_2O (**◉** *Scheme 35*). In addition, they found that the use of a 1:2 combination of $SnCl_4$ and $AgB(C_6F_5)_4$ was very effective for the glycosylations of glycosyl fluorides possessing phthaloyl or dichlorophthaloyl protecting group for the C-2 amino function [76].

In contrast to Toshima's glycosylations of glycosyl fluorides using heterogeneous protic acids such as montmorillonite K-10, Nafion-H®, and SO_4/ZrO_2 [74], Mukaiyama and coworkers demonstrated the catalytic glycosylations using various homogeneous protic acids such as TfOH [77], $HClO_4$, $HB(C_6F_5)_4$, and $HNTf_2$ [78]. Among them, when the glycosylation was performed using $HClO_4$ in Et_2O, the α-glycoside was major product while the β-stereoselectivity was observed when $HB(C_6F_5)_4$ was employed in a mixture of BTF and t-BuCN (**◉** *Scheme 36*).

$Yb(NTf_2)_3$ [79] and $ZrCl_4$ [80] were also employed by Switzer et al. and Inazu et al., respectively. Furthermore, Yamada reported that $Cu(OTf)_2$ was a promoter not only for glycosyl bromide but also for glycosyl fluoride [22] (**◉** *Scheme 37*).

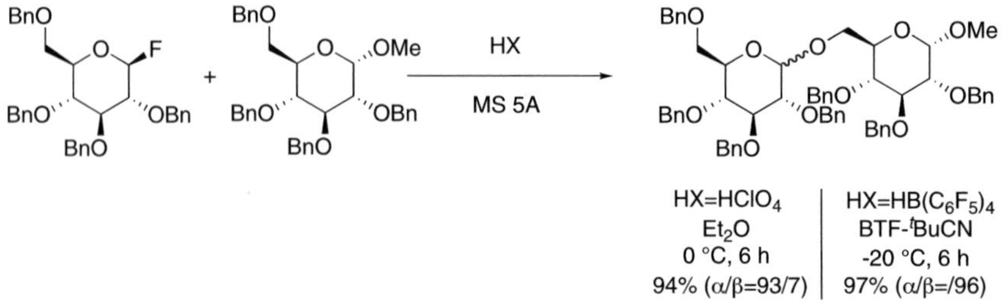

HX=HClO$_4$	HX=HB(C$_6$F$_5$)$_4$
Et$_2$O	BTF-^{t}BuCN
0 °C, 6 h	-20 °C, 6 h
94% (α/β=93/7)	97% (α/β=/96)

◻ Scheme 36

88% (α/β=80/20)

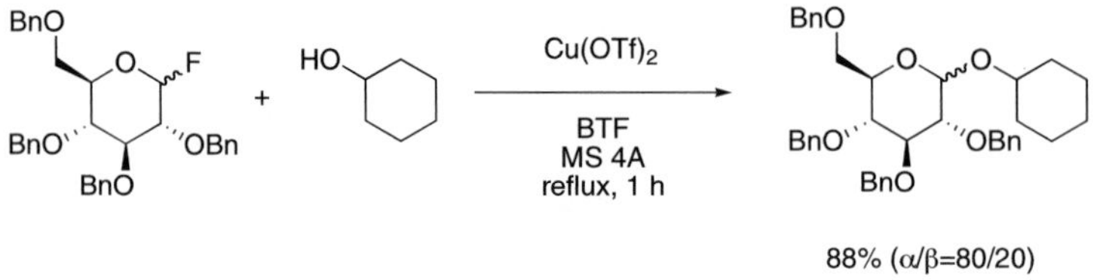

◻ Scheme 37

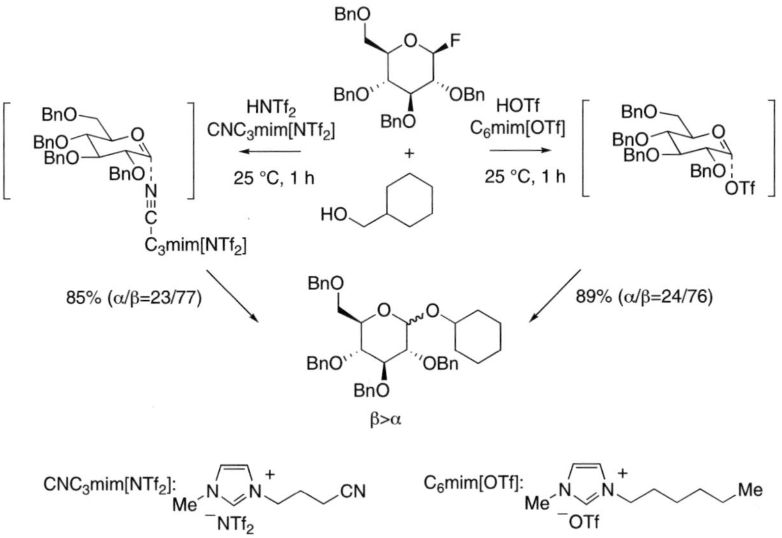

■ **Scheme 38**

Following the extensive studies using a heterogeneous and reusable solid catalyst in a glyco-sylation reaction by Toshima et al. for greening the chemical glycosylation method [74], the use of an ionic liquid as an environmentally benign reaction media for the glycosylation of glycosyl fluoride was reported by Toshima et al. [81]. In this study, it was demonstrated not only the reusability of the ionic liquid in the glycosylation reaction but also the potency of the ionic liquid for stereocontrol of the glycosylation reaction (❏ *Scheme 38*).

References

1. Koenigs W, Knorr E (1901) Chem Ber 34:957
2. Lemieux RU, Hayami JI (1965) Can J Chem 43:2162
3. (a) Paulsen H, Lockhoff O (1981) Chem Ber 114:3102; (b) Paulsen H, Kutschker W, Lockhoff O (1981) Chem Ber 114:3233
4. (a) Wulff G, Röhle G (1974) Angew Chem Int Ed Engl 13:157; (b) Bochkov A-F, Zaikov GE (1979) (eds) Chemistry of the O-Glycosidic Bond: Formation and Cleavage. Pergamon Press, Oxford; (c) Tsutsumi H, Ishido Y (1980) J Synth Org Chem Jpn 38:473; (d) Paulsen H (1982) Angew Chem Int Ed Engl 21:155; (e) Koto S, Morishima N, Zen S (1983) J Synth Org Chem Jpn 41:701; (f) Paulsen H (1984) Chem Soc Rev 13:15; (g) Schmidt RR (1986) Angew Chem Int Ed Engl 25:212; (h) Krohn K (1987) Nachr Chem Tech Lab 35:930; (i) Kunz H (1987) Angew Chem Int Ed Engl 26:294; (j) Schmidt RR (1989) Pure Appl Chem 61:1257; (k) Hashimoto S, Ikegami S (1991) Farmacia 27:50; (l) Schmidt RR (1991) In: Trost BM (ed) Comprehensive Organic Synthesis. Perga-mon Press, Oxford, vol 6, p 33; (m) Sinaÿ P (1991) Pure Appl Chem 63:519; (n) Suzuki K, Nagasawa T (1992) J Synth Org Chem Jpn 50:378; (o) Ito Y, Ogawa T (1992) In: Iguchi Y (ed) Jikkenkagakukoza. Maruzen, Tokyo, vol 26, p 267; (p) Banoub J (1992) Chem Rev 92:1167; (q) Toshima K, Tatsuta K (1993) Chem Rev 93:1503; (r) Boons G-J (1996) Tetrahedron 52:1095; (s) Davis BG (2000) J Chem Soc Perkin Trans 1 2137; (t) Pellissier H (2005) Tetrahedron 61:2947; (u) Fügedi P (2006) In: Levy DE, Füge-di P (eds) The Organic Chemistry of Sugars. CRC Press, Boca Raton, Fl, p 89

5. Bock K, Meldal M (1983) Acta Chem Scand Ser B B37:775

6. Lemieux RU, Hendriks KB, Stick RV, James K (1975) J Am Chem Soc 97:4056

7. (a) West AC, Schuerch C (1973) J Am Chem Soc 95:1333; (b) Kronzer FJ, Schuerch C (1974) Carbohydr Res 33:273; (c) Eby R, Schuerch C (1975) Carbohydr Res 39:33

8. Ogawa T, Matsui M (1976) Carbohydr Res 51:C13

9. (a) Lubineau A, Malleron A (1985) Tetrahedron Lett 26:1713; (b) Lubineau A, Le Gallic J, Malleron A (1987) Tetrahedron Lett 28:5041

10. Higashi K, Nakayama K, Soga T, Shioya E, Uoto K, Kusama T (1990) Chem Pharm Bull 38:3280

11. Waldmann H, Böehm G, Schmid U, Röettele H (1994) Angew Chem Int Ed Engl 33:1994

12. Kartha KPR, Aloui M, Field RA (1996) Tetrahedron Lett 37:8807

13. Kartha KPR, Aloui M, Field RA (1997) Tetrahedron Lett 87:8233

14. Stachulski AV (2001) Tetrahedron Lett 42:6611

15. Banik BK, Samajdar S, Banik I, Zegrocka O, Becker FF (2001) Heterocycles 55:227

16. Mukherjee D, Kumar RP, Sankar Chowdhury U (2001) Tetrahedron 57:7701

17. Yamago S, Yamada T, Nishimura R, Ito H, Mino Y, Yoshida J-I (2002) Chem Lett 152

18. Mukaiyama T, Kobashi Y (2004) Chem Lett 33:10

19. Kobashi Y, Mukaiyama T (2005) Bull Chem Soc Jpn 78:910

20. Nishizawa M, Garcia DM, Yamada H (1992) Synlett 797

21. Nishizawa M, Garcia DM, Shin T, Yamada H (1993) Chem Pharm Bull 41:784

22. Yamada H, Hayashi T (2002) Carbohydr Res 337:581

23. (a) Kleine HP, Weinberg DV, Kaufman RJ, Sidhu RS (1985) Carbohydr Res 142:333; (b) Dess D, Kleine HP, Weinberg DV, Kaufman RJ, Sidhu RS (1981) Synthesis 883

24. Brewster K, Harrison JM, Inch TD (1979) Tetrahedron Lett 20:5051

25. Loganathan D, Trivedi GK (1987) Carbohydr Res 162:117

26. Wang Y, Li L, Wang Q, Li Y (2001) Synth Commun 31:3423

27. Roy R, Tropper F (1990) Synth Commun 20:2097

28. Roy R, Tropper FD, Grand-Maître C (1991) Can J Chem 69:1462

29. Bliard C, Massiot G, Nazabadioko S (1994) Tetrahedron Lett 35:6107

30. Hongu M, Saito K, Tsujihara K (1999) Synth Commun 29:2775

31. Sasaki M, Gama Y, Yasumoto M, Ishigami Y (1990) Tetrahedron Lett 31:6549

32. (a) Nishizawa M, Kan Y, Yamada H (1988) Tetrahedron Lett 29:4597; (b) Nishizawa M, Kan Y, Yamada H (1989) Chem Pharm Bull 37:565; (c) Nishizawa M, Kan Y, Shimomoto W, Yamada H (1990) Tetrahedron Lett 31:2431; (d) Nishizawa M, Imagawa H, Kan Y, Yamada H (1991) Tetrahedron Lett 32:5551; (e) Nishizawa M, Shimomoto W, Momii F, Yamada H (1992) Tetrahedron Lett 33:1907

33. Schmid U, Waldmann H (1997) Liebigs Ann Chem 2573

34. (a) Hadd MJ, Gervay J (1999) Carbohydr Res 320:61; (b) Lam SN, Gervay-Hague J (2002) Org Lett 4:2039; (c) Lam SN, Gervay-Hague J (2002) Carbohydr Res 337:1953; (d) Dabideen DR, Gervay-Hague J (2004) Org Lett 6:973; (e) Lam SN, Gervay-Hague J (2005) J Org Chem 70:2387; Du W, Gervay-Hague J (2005) Org Lett 7:2063

35. (a) Perrie JA, Harding JR, King C, Sinnott D, Stachulski AV (2003) Org Lett 5:4545; (b) Harding JR, King CD, Perrie JA, Sinnott D, Stachulski AV (2005) Org Biomol Chem 3:1501

36. Mukaiyama T, Murai Y, Shoda S (1981) Chem Lett 431

37. Mukaiyama T, Hashimoto Y, Shoda S (1983) Chem Lett 935

38. Rosenbrook JrW, Riley DA, Lartey, PA (1985) Tetrahedron Lett 26:3

39. Posner GH, Haines SR (1985) Tetrahedron Lett 26:5

40. Hayashi M, Hashimoto S, Noyori R (1984) Chem Lett 1747

41. Szarek WA, Grynkiewicz G, Doboszewski B, Hay GW (1984) Chem Lett 1751

42. Miethchen R, Hager C, Hein M (1997) Synthesis 159

43. Ernst B, Winkler T (1989) Tetrahedron Lett 30:3081

44. Kunz H, Sanger W (1985) Helv Chim Acta 68:283

45. Teichmann M, Descotes G, Lafont D (1993) Synthesis 889

46. Lichtenthaler FW, Kläres U, Lergenmüller M, Schwidetzky S (1992) Synthesis 179

47. Goggin KD, Lambert JF, Walinsky SW (1994) Synlett 162
48. Nicolaou KC, Dolle RE, Papahatjis DP, Randall JL (1984) J Am Chem Soc 106:4189
49. Caddick S, Motherwell WB, Wilkinson JA (1991) J Chem Soc Chem Commun 674
50. Caddick S, Gazzard L, Montherwell WB, Wilkinson JA (1996) Tetrahedron 52:149
51. MacDonald SJF, McKenzie TC (1988) Tetrahedron Lett 29:1363
52. Gordon DM, Danishefsky SJ (1990) Carbohydr Res 206:361
53. Ferrier RJ, Prasad N (1969) J Chem Soc C:570
54. Bröder W, Kunz H (1993) Carbohydr Res 249:221
55. Palme M, Vasella A (1995) Helv Chim Acta 78:959
56. Ogawa T, Takahashi Y (1985) Carbohydr Res 138:C5; Takahashi Y, Ogawa T (1987) Carbohydr Res 164:277
57. Hashimoto S, Hayashi M, Noyori R (1984) Tetrahedron Lett 25:1379
58. Nicolaou KC, Chucholowski A, Dolle RE, Randall JL (1984) J Chem Soc Chem Commun 1155
59. Kunz H, Sager W (1985) Helv Chim Acta 68:283
60. Kunz H, Waldmann H (1985) J Chem Soc Chem Commun 638
61. Vozny YA, Galoyan AA, Chizhov OS (1985) Bioorg Khim 11:276
62. (a) Kreuzer M, Thiem J (1986) Carbohydr Res 149:347; (b) Jünnemann J, Lundt I, Thiem J (1991) Liebigs Ann Chem 759
63. (a) Matsumoto T, Maeta H, Suzuki K, Tsuchihashi G (1988) Tetrahedron Lett 29:3567; (b) Suzuki K, Maeta H, Matsumoto T, Tsuchihashi G (1988) Tetrahedron Lett 29:3571; (c) Matsumoto T, Maeta H, Suzuki K, Tsuchihashi G (1988) Tetrahedron Lett 29:3575; (d) Matsumoto T, Katsuki M, Suzuki K (1989) Chem Lett 437
64. Suzuki K, Maeta H, Suzuki T, Matsumoto T (1989) Tetrahedron Lett 30:6879
65. (a) Nicolaou KC, Caulfield TJ, Kataoka H, Stylianides NA (1990) J Am Chem Soc 112:3693; (b) Nicolaou KC, Hummel CW, Iwabuchi Y (1992) J Am Chem Soc 114: 3126
66. Maeta H, Matsumot T, Suzuki K (1993) Carbohydr Res 249:49
67. Kobayashi S, Koide K, Ohno M (1990) Tetrahedron Lett 31:2435
68. (a) Wessel HP (1990) Tetrahedron Lett 31: 6863; (b) Wessel HP, Ruiz N (1991) J Carbohydr Chem 10:901
69. Böhm G, Waldmann H (1995) Tetrahedron Lett 36:3843
70. Hosono S, Kim W-S, Sasai H, Shibasaki M (1995) J Org Chem 60:4
71. Kim W-S, Hosono S, Sasai H, Shibasaki M (1995) Tetrahedron Lett 36:4443
72. Kim W-S, Sasai H, Shibasaki M (1996) Tetrahedron Lett 37:7797
73. Yu L, Chen D, Li J, Wang PG (1997) J Org Chem 62:3575
74. (a) Toshima K, Kasumi K, Matsumura S (1998) Synlett 643; (b) Toshima K, Kasumi K, Matsumura S (1999) Synlett 1332
75. (a) Takeuchi K, Mukaiyama T (1998) Chem Lett 555; (b) Yanagisawa M, Mukaiyama T (2001) Chem Lett 224
76. Jona H, Maeshima H, Mukaiyama T (2001) Chem Lett 726–727
77. (a) Mukaiyama T, Jona H, Takeuchi K (2000) Chem Lett 696; (b) Jona H, Takeuchi K, Mukaiyama T (2000) Chem Lett 1278
78. (a) Jona H, Mandai H, Mukaiyama T (2001) Chem Lett 426; (b) Jona H, Mandai H, Chavasiri W, Takeuchi K, Mukaiyama T (2002) Bull Chem Soc Jpn 291
79. Yamanoi T, Nagayama S, Ishida H-K, Nishikido J, Inazu T (2001) Synth Commun 31:899
80. Pikul S, Switzer AG (1997) Tetrahedron: Asymmetr 8:1165
81. Sasaki K, Matsumura S, Toshima K (2004) Tetrahedron Lett 45:7043

3.2 Glycosyl Trichloroacetimidates

Richard R. Schmidt[*1], *Xiangming Zhu*[2]
[1] Fachbereich Chemie, Fach M 725, Universität Konstanz,
78457 Konstanz, Germany
[2] School of Chemistry and Chemical Biology, University College Dublin,
Belfield, Dublin 4, Ireland
richard.schmidt@uni-konstanz.de, xiangming@ucd.ie

Abstract

In the first part of this review, the basic principles of chemical glycosylation reactions are discussed; this way the advantages of *O*-glycosyl trichloroacetimidates and related systems as glycosyl donors become obvious. Many new methods for the generation of *O*-glycosyl trichloroacetimidates and for their use as glycosyl donors have been introduced which are

In: *Glycoscience*. Fraser-Reid B, Tatsuta K, Thiem J (eds)
Chapter-DOI 10-1007/978-3-540-30429-6_11: © Springer-Verlag Berlin Heidelberg 2008

compiled as well as their use in solid-phase oligosaccharide synthesis. The power of these gly-cosyl donors is demonstrated by their application in complex oligosaccharide and glycoconju-gate synthesis, as outlined in the second part of this review. Recent applications in glycolipid, glycosyl amino acid and glycopeptide, nucleoside and nucleotide glycosidation, glycosamino glycan, cell wall constituent, and GPI anchor synthesis, glycosylation of various natural prod-ucts and their metabolites, and finally cyclooligosaccharide generation are compiled. In the last section, related glycosyl donors are briefly discussed.

Abbreviations

ADMB	4-acetoxy-2,2-dimethylbutanoyl
CAc	chloroacetyl
CDs	cyclodextrins
CSs	chondroitin sulfates
DBU	1,8-diazabicyclo[5.4.0]undec-7-ene
DDQ	2,3-dichloro-5,6-dicyno-benzoquinone
DMM	*N*-dimethylmaleoyl
DPM	diphenylmethyl
FGFs	fibroblast growth factors
Fmoc	9-fluorenylmethyl carbamate
GAGs	glycosaminoglycans
GPI	glycosylphosphatidyl inositol
HS	heparan sulfate
ILs	ionic liquids
IP	inverse procedure
IPG	inositolphosphoglycan
LTAs	lipoteichoic acids
MPEG	mono-methyl polyethyleneglycol
MPLC	medium pressure liquid chromatography
NAP	naphthylmethyl
PLL	photolabile linker
PNBP	*p*-nitrobenzylpyridine
PPTS	pyridinium *p*-toluenesulfonate
PTBD	polymer-supported 1,5,7-triazabicyclo[4.4.0]dec-5-ene
PS-DBU	polystyrene-supported DBU
SPOS	solid-phase oligosaccharide syntheses
TBS	*tert*-butyl-dimethylsilyl
TMSOTf	trimethylsilyl trifluoromethanesulfonate

1 Introduction

The biological significance of oligosaccharides and glycoconjugates has stimulated many activities in carbohydrate chemistry. Most of these activities have been devoted to the devel-opment of methods for glycoside bond formation as the assembly of monosaccharide building blocks to complex oligosaccharides and glycoconjugates is the most difficult task in this endeavor.

2 Glycosyl Donor Generation Through Anomeric-Oxygen Exchange Reactions

For a long time the methods developed for glycosylation have essentially favored approaches which require for glycosyl donor generation an *anomeric oxygen exchange reaction* on the half-acetal moiety of pyranoses and furanoses [❷ *Fig. 1*, methods (A) and (B)] [1,2,3,4]. The *Fischer–Helferich method* (A), as a direct acid-catalyzed anomeric-oxygen replacement reaction has been very successfully applied to the synthesis of simple alkyl glycosides. However, because of its reversibility, it is of limited usefulness in the synthesis of complex oligosaccharides and glycoconjugates. For this endeavor, irreversible methods are required, which can be attained by preactivation of the anomeric center by introducing good leaving groups which can be released by leaving group specific promoters or—even better—by catalytic amounts of an activator.

The best-known of these methods is the *Koenigs–Knorr method* (B) in which the anomeric hydroxy group is replaced by chlorine or bromine. Thus, an α-halogen-ether is generated which in the glycosylation step can be readily activated by halophilic promoters frequently incorrectly termed "catalyst." Generally, from one up to four equivalents of heavy metal salts are employed resulting in an irreversible transfer of the glycosyl donor moiety to the acceptor. On the basis of this general method many valuable techniques for the synthesis of complex oligosaccharides and glycoconjugates have been introduced which have been extensively reviewed. However, the obvious limitations of this method were the reason for the search for alternative methods [1,2,3,4,5,6,7].

Therefore, other anomeric oxygen exchange reactions closely related to the *Koenigs–Knorr method* have been extensively investigated. Particularly, the introduction of fluorine and alkyl- and arylthio groups as leaving groups gained great interest because these groups also tolerate manipulations of orthogonal protecting groups. Differences in thio leaving group tendencies (in combination with the "armed"/"disarmed" principle) could be even employed for one-pot consecutive glycosylations leading to more or less pure oligosaccharide products. However, the basic drawbacks of the *Koenigs–Knorr method* are also associated with these promoter systems. For instance, the large amounts of promoter required (and often additional reagents) limit their usefulness particularly in large-scale oligosaccharide and glycoconjugate synthesis.

3 Direct Anomeric Oxygen Alkylation

For more than a century glycosylations were essentially based on methods where the anomeric carbon of the sugar residue to be coupled served as the electrophile (the glycosyl donor) and the alcohol (the glycosyl acceptor) as the nucleophile [❷ *Fig. 1*, (A), (B)]. Alternatively, base-mediated deprotonation of the anomeric hydroxy group generating at first an anomeric

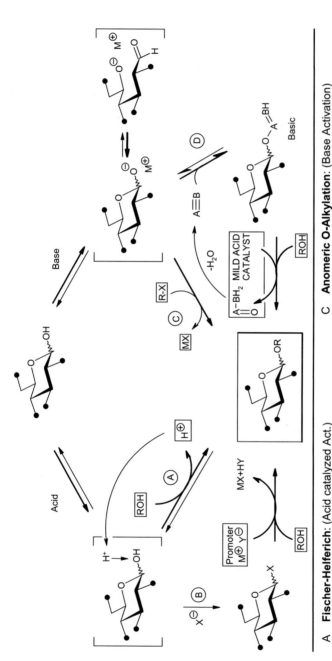

A   **Fischer-Helferich**: (Acid catalyzed Act.)
B   **Koenigs-Knorr**:   X = Cl, Br, (I) Activation
                          X = F-Activation
                          X = S-R-Activation (+ Hetarylthio)

Glycosylation: "X-Philic" Promoter
(Generally: Heavy metal salts: Ag⁺, Hg²⁺, etc.)

C   **Anomeric O-Alkylation**: (Base Activation)
D   **Trichloroacetimidate Activation**: A≡B = CCl₃–CN
    Related Systems (?):
        PO(OR)₂-, P(OR)₂-Activation
        SO₂(OR)-, SO(OR)-, SO₂R-Activation

    Glycosylation: Mild Acid catalyst
    (Generally: BF₃ · OEt₂, TMSOTf, etc.)

◻ **Figure 1**
Synthesis of glycosides and saccharides

oxide structure from a pyranose or a furanose and then anomeric *O*-alkylation leading direct-ly and irreversibly to glycosides should be also available [❯ *Fig. 1*, (C)]. Surprisingly, no studies employing this simple '*anomeric O-alkylation method*' as termed by us [1,2,3,4], for the synthesis of complex glycosides and glycoconjugates had been reported before our work. Only a few scattered examples with simple alkylating agents, for instance, excess methyl iodide or dimethyl sulfate, have been found in the literature [1]. However, in our hands, direct *anomeric O-alkylation* of variously protected and totally unprotected sugars in the presence of a base and triflates or Michael acceptors, respectively, as alkylating agents has become a very convenient method for glycoside bond formation [8,9,10,11]. Base-promoted decomposition reactions, particularly of the acyclic form, were practically not observed. Often even high anomeric diastereocontrol was available. The high diastereocontrol in pyranoses is based on the enhanced nucleophilicity of equatorial oxygen atoms (due to steric effects and the stereo-electronic *kinetic anomeric effect*) [1,2,3,4] and on the higher stability of axial oxygen atom-derived products (due to the *thermodynamic anomeric effect*). Chelation effects can be also employed to design anomeric stereocontrol. The availability and to some extent the stability of the carbohydrate-derived alkylating agents precluded the general applicability of this simple method to the synthesis of complex oligosaccharides and glycoconjugates.

4 Glycosyl Donor Generation Through Retention of the Anomeric Oxygen

The requirements for an efficient glycosylation method are the following:

- high chemical and stereochemical yield,
- applicability to large-scale preparations, and
- avoidance of large amounts of waste materials, i.e. activation of the donor by catalytic amounts of reagent.

These demands were not met by any of the above-described methods for the synthesis of complex oligosaccharides and glycoconjugates. However, the general strategy for glycoside bond formation seems to be correct:

- The first step (*activation step*) should consist of an activation of the anomeric center under formation of a stable glycosyl donor—best by a catalyzed attachment of a leaving group to the anomeric hydroxy group.
- The second step (*glycosylation step*) should consist of a sterically uniform high-yielding glycosyl transfer to the glycosyl acceptor based on glycosyl donor activation with catalytic amounts of promoter, i.e. a catalyst. Obviously, this catalytic procedure has to be orthogo-nal to the glycosyl donor preparation procedure. Diastereocontrol in the glycosylation step may be derived from the anomeric configuration of the glycosyl donor (by inversion or retention), by anchimeric assistance, by the solvent influence, by thermodynamic and/or stereoelectronic effects, or by any other effects.

The experience with the direct anomeric *O*-alkylation exhibited that these demands can be fulfilled by a simple base-catalyzed anomeric *O*-transformation into a leaving group and its acid-catalyzed activation in the glycosylation step. This approach should also satisfy the demand for simplicity in combination with efficiency which is decisive for general acceptance.

Obviously, for achieving stereocontrolled activation of the anomeric oxygen atom, the anomerization of the anomeric hydroxy group or the anomeric oxide ion, respectively, has to be considered. Thus, in a reversible activation process and with the help of kinetic and thermodynamic reaction control, possibly both activated anomers should be accessible. From these considerations it was concluded that suitable triple-bond systems A≡B (or compounds containing cumulative double bond systems A=B=C) might be found that add pyranoses and furanoses under base catalysis directly and reversibly in a stereocontrolled manner [❍ Fig. 1, (D)].

Electron-deficient nitriles, such as for instance trichloroacetonitrile (and trifluoroacetonitrile [11]), are known to undergo direct and reversible, base-catalyzed addition of alcohols providing O-alkyl trichloroacetimidates (or O-alkyl trifluoroacetimidate, respectively) [1,12,13] [❍ Fig. 2, (1)]. This imidate synthesis has the advantage that the imidates can be isolated as stable adducts which are less sensitive to hydrolysis than the corresponding salts. On acid addition leading to imidate activation [❍ Fig. 2, (2)], hydrolysis with water (R^2–OH = H_2O) is a fast reaction furnishing amide CCl_3–$CONH_2$ and alcohol R^1–OH [❍ Fig. 2, (3A)]; mechanistically this is an acylative attack of the imidate at the nucleophile. With other nucleophiles, after acylation by the imidate, other transformations are possible, as for instance ortho-ester formation. However, the basic question is: are these activated imidates also good alkylating agents [❍ Fig. 2, (3B)] as required for being effective in glycosidation reactions? On consider-

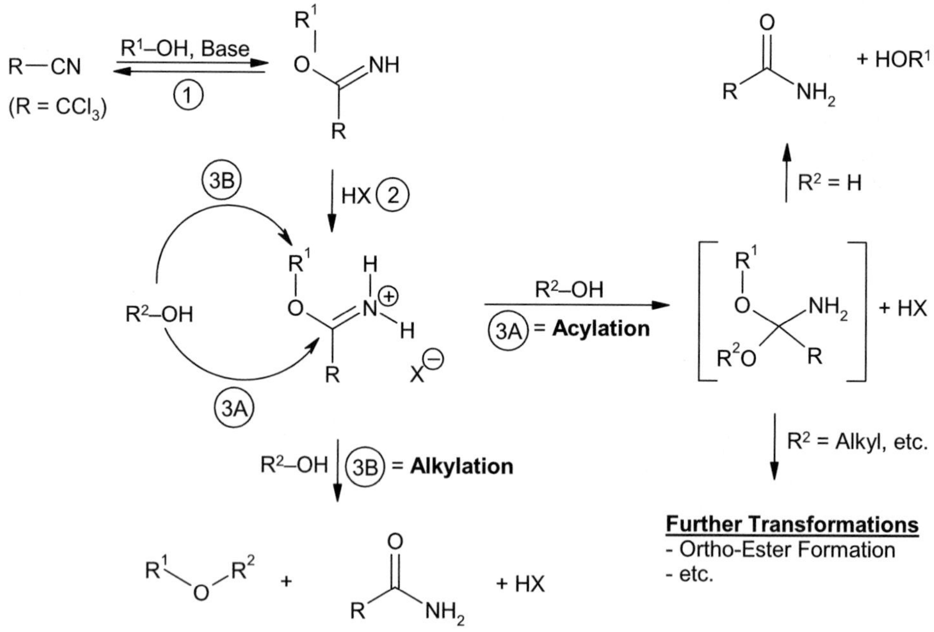

❑ Figure 2
Trichloroacetimidate formation and its acid-catalyzed transformations

ation of the influence of the substituents R and R^1 on these two competing reactions an attack of R^2–OH leads to the following expectations:

- Acylation [reaction (3A)] is supported by R being a small electron-withdrawing group, R^1 destabilizing carbenium ion formation.
- Alkylation [reaction (3B)] is supported by R being a sterically demanding electron-withdrawing group, R^1 supporting carbenium ion formation.

From these considerations it can be deduced that the bulky and strongly electron-withdrawing trichloromethyl group as R and the glycosyl group as R^1, which through the α-oxygen atom supports oxocarbenium ion formation at the anomeric center, should provide excellent alkylating agents; hence, O-glycosyl trichloroacetimidates should on acid activation exhibit excellent glycosyl donor properties.

As expected, on base-catalyzed addition of the anomeric hydroxy group to trichloroacetonitrile the O-glycosyl trichloroacetimidate is formed for which, due to the reversibility of the addition, the different nucleophilicities of the anomeric oxides, and the different thermodynamic stabilities of the O-glycosyl trichloroacetimidates, anomeric stereocontrol is possible. Thus, the weak base potassium carbonate could be employed for preferential or exclusive formation of the β-anomer and the strong base sodium hydride could be employed for the thermodynamically more stable α-anomer (❷ *Fig. 3*). This stereocontrol could be successfully extended to S_N2-type glycosidation reactions in solvents of low donicity and under low temperature conditions and particularly to glycosylations of O=X–OH nucleophiles with X being RC or R_2P (see below). However, for many cases anomeric stereocontrol is based on other effects such as for instance neighboring group participation and/or steric effects, stereoelectronic effects, solvent participation effects, etc.

The experience with the *trichloroacetimidate method* exhibited that the demands on a new glycosylation methodology are fulfilled:

(i) The O-glycosyl trichloroacetimidates are readily formed and generally stable under room temperature conditions. However, on acid catalysis they exhibit extraordinary high glycosyl donor properties.

(ii) The release of nonbasic trichloroacetamide fulfills the criteria for acid catalysis: The acid is not consumed by the leaving group, therefore generally only catalytic amounts of (Lewis) acid are required ($\sim$0.001 to 0.1 equivalents).

(iii) The released trichloroacetamide is also not acidic, therefore the acidity provided by the catalyst amount is maintained throughout the reaction course. Hence, negative effects of increasing acidity in the reaction mixture (found for instance for phosphate, sulfate, sulfonate leaving groups, ❷ *Fig. 1*, D) are avoided.

(iv) Glycosidation is basically a condensation reaction. In this procedure, water is bound to trichloroacetonitrile under trichloroacetamide formation. Hence, drying agents are not required. Often molecular sieves are used in glycosidation reactions. Because they may affect the acidity of the reaction in an unpredictable fashion, their use is not even recommended in this method.

(v) Trichloroacetamide can be removed from the reaction mixture and transformed back to trichloroacetonitrile, thus exhibiting the cost-effectiveness and ecofriendliness of this method in large-scale preparations.

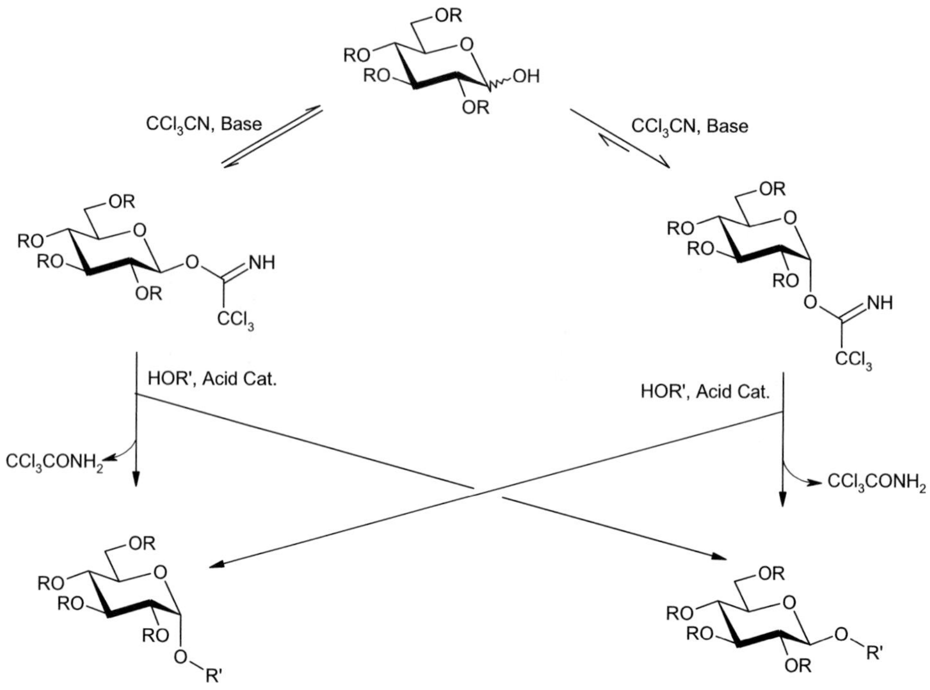

☐ Figure 3
O-Glycosyl trichloroacetimidate formation and acid-catalyzed glycosylation of acceptors HOR′

(vi) Neither in the formation of the *O*-glycosyl trichloroacetimidates nor in the glycosidation reactions are equivalent or even higher amounts of salts produced—a disadvantage of most of the above-mentioned methods, particularly in large-scale preparations. Also, highly expensive sterically hindered bases are not required.

This discussion exhibits that *O*-glycosyl *N*-methyl-acetimidates, introduced by the Sinaÿ group [14,15], are poor glycosyl donors: they lack the strongly electron-withdrawing sterically demanding trichloromethyl group. Hence, they are rather better acylating agents than alkylating agents. In addition, their formation via *O*-glycosylation of *N*-methyl-acetamide with excess amounts of silver oxide is quite cumbersome.

These favorable aspects of *O*-glycosyl trichloroacetimidates led to their frequent use as glycosyl donors for various types of glycosyl acceptors such as nucleophiles (❷ *Fig. 4*). Besides hydroxy groups (of standard alcohols, phenols, sugars, etc.), also carboxylic acids, phosphorous acids, and sulfonic and sulfuric acids, respectively, were successfully employed as acceptors; due to the acidity of these acidic acceptors, generally catalysts are not required for the activation of the glycosyl donors and—presumably via an eight-membered transition state—often the inversion product at the anomeric center is generated. However, also various *C*-, *N*-, *S*-, and *P*-nucleophiles have been successfully glycosylated by *O*-glycosyl trichloroacetimidates [1,2].

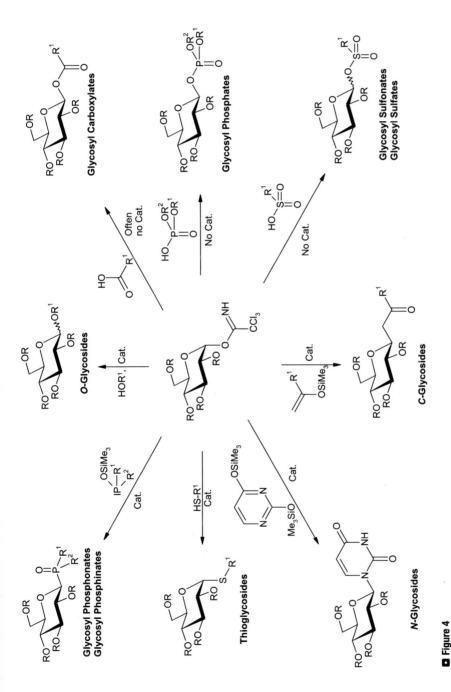

Figure 4

Reaction of *O*-glycosyl trichloroacetimidates with *O*-, *C*-, *N*-, *S*-, and *P*-nucleophiles—activation by (Lewis)acid catalysis

4.1 Methodological Aspects

As outlined above, the anomeric hydroxy groups add under base catalysis readily to the electron-deficient nitrile group of trichloroacetonitrile to furnish generally via equilibration at the anomeric center O-glycosyl trichloroacetimidates. As bases mainly potassium carbonate, sodium hydride, DBU, and cesium carbonate are employed. Weak bases, such as for instance potassium carbonate, permit kinetic product control (formation of the equatorial product) whereas strong bases, as for instance sodium hydride, lead to the thermodynamically more stable product (generally the axially oriented trichloroacetimidate group). Because of convenience, DBU has become the most popular base, which also favors axial product formation. Therefore, polystyrene-supported DBU (PS-DBU) [16,17] and various other solid-supported nitrogen bases were successfully employed in this reaction as shown for the particularly difficult trichloroacetimidate formation of 4-dimethylamino-2,4,6-trideoxy sugar kedarosamine (**1a** → **1b**) and its use in the glycosylation of the ansamacrolide substructure (**1b** + **1c** → **1d**) of the kedarcidin chromophore (❷ *Scheme 1*) [17]. Also Dowex 1-X8 OH⁻ is a good catalyst for the trichloroacetimidate formation of a glucosamine derivative [18,19]. Polymer-supported 1,5,7-triazabicyclo[4.4.0]dec-5-ene (PTBD) turned out to be a powerful catalyst for trichloroacetimidate formation as well [17,19]; this base catalyst in combination with Nafion®-SAC resin (a nanocomposite of Nafion® resin with porous silica) could be successfully employed in one-pot trichloroacetimidate formation and following glycosylation reactions [19].

O-Glycosyl trichloroacetimidates are quite stable under neutral but also under basic conditions. Hence, trichloroacetimidate formation is tolerated by standard O- and N-protecting groups, such as O-acyl, O-benzyl, O,O-alkylidene, O-silyl, N-acyl and N,N-diacyl, N-phthaloyl (Phth), N-dimethylmaleoyl (DMM), and the latent amino functionality azido. It is particularly worth mentioning that trichloroacetimidate formation is also tolerated by Fmoc-protected hydroxy groups; this group has become an important temporary protecting group which is orthogonal to other temporary protecting groups, thus permitting in oligosaccharide synthesis not only regioselective chain extension but also branching [20,21]. The useful 3-O-Fmoc-protected galactosyl donor **2d** can be readily prepared from compound **2a** via intermediates **2b** and **2c** (❷ *Scheme 2*) [20]. Additional protecting groups have been recently probed particularly at 2-O in order to influence the stereochemical outcome, such as for instance the diphenylmethyl (DPM), the 9-fluorenyl (Fl) group [22], and the 4-acetoxy-2,2-dimethylbutanoyl (ADMB) group [23]. The ADMB group is a useful alternative to the pivaloyl group; it combines the strong 1,2-*trans*-selection in glycosylation reactions of the pivaloyl group with the ease of removal of more reactive acyl groups. Other important new groups are the O-trifluoroethylsulfonate group which could be successfully employed to release the sulfate group and the replacement of the 6-hydroxy group by a phenylthio group which is employed for uronate formation [25].

O-Glycosyl trichloroacetimidates are under acidic conditions very powerful glycosyl donors. Reaction with nonacidic nucleophiles is generally performed with catalytic amounts of Brønsted or Lewis acids. TMSOTf and BF₃OEt₂ are the most frequently employed catalysts. TMSOTf is the catalyst of first choice and it is generally used in 0.001 to 0.1 equivalents based on the glycosyl donor. Commonly 1.0 to 1.5 equivalents of glycosyl acceptor are added. Dichloromethane is the solvent of first choice. Because of the high reactivity of O-glycosyl

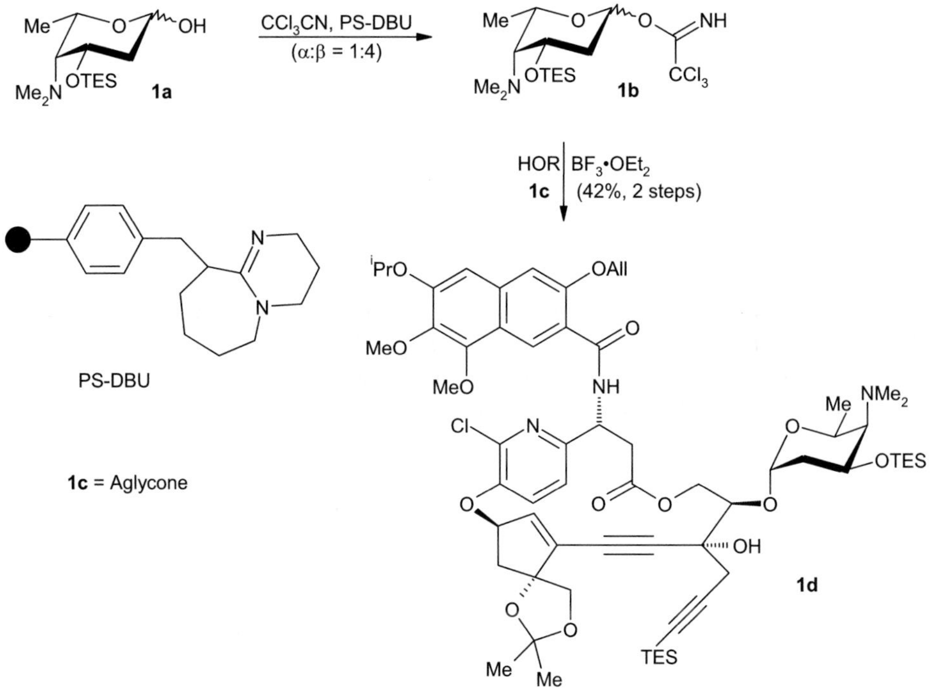

Scheme 1

trichloroacetimidates often reaction temperatures between 0 and −40 °C are selected. For highly sensitive glycosyl donors and for the anomeric stereocontrol many variations have been probed which are shortly summarized.

Numerous acid catalysts have been investigated. Brønsted acids, such as for instance p-TsOH and TfOH, can lead—as previously discussed [1,2,3,4,26]—to O-glycosyl sulfonates as reactive intermediates which are in most cases not the glycosyl donors but rather some type of glycosyl oxocarbenium ion intermediates (S_N1-type reactions). Similarly, pyridinium p-toluenesulfonate (PPTS) [27,28] and perchloric acid [29,30] are working as catalysts. Metal triflates such as AgOTf [31,32], Cu(OTf)$_2$ [33], Sn(OTf)$_2$ [34,35], Sm(OTf)$_3$ [36], and Yb(OTf)$_3$ [37], are particularly valuable in glycosylations with acid sensitive glycosyl donors and/or acceptors. On Cu(OTf)$_2$-activation of various donors the trichloroacetimidates have proven to be the most potent glycosyl donors [33]. For instance, AgOTf has proven to be useful, as a catalyst for highly reactive deoxysugars such as glycosyl donors, as shown in ❷ Scheme 3 for the reaction of fucosyl donor 3a with fucosyl acceptor 3b giving the α-oligosaccharide 3c [32]. Sn(OTf)$_2$ permitted successful glycosylations of acid sensitive glycal derivatives as acceptors as shown for the synthesis of 4c from 4a and 4b (❷ Scheme 4) [35]. Various other catalyst systems have proven to be useful such as for instance HB(C$_6$F$_5$)$_4$ [29,30], N-acyl-sulfonamides, phenols [38], I$_2$ [39], I$_2$/Et$_3$SiH [40], and electrophilic carbonyl compounds [41], such as for instance chloral. Silica-supported perchloric acid [42,43], acid-washed molecular sieves (MS

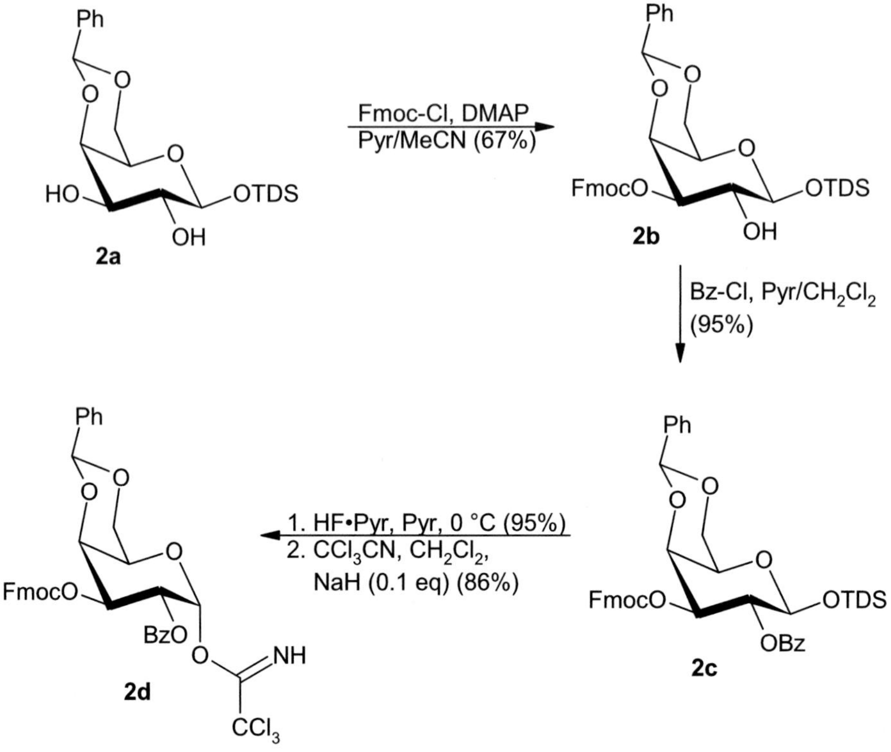

Scheme 2

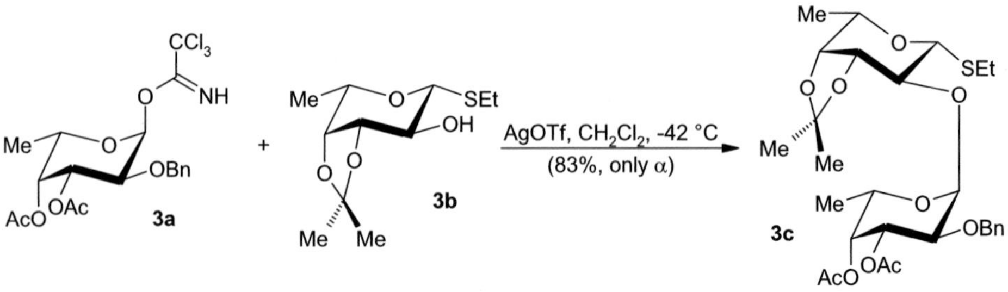

Scheme 3

AW 300) [44], and Nafion®-SAC resin [19] (see above) are highly valuable solid-supported acid catalysts. This is nicely demonstrated in an avermectin B_{1a} analog synthesis [43]: Rhamnosylation of intermediate **5b** with donor **5a** gave target molecule **5c** in quantitative yield (**Scheme 5**).

The solvent of choice in the glycosylation reactions is dichloromethane. In combination with a solvent of low polarity and/or of low donicity (such as for instance cyclohexane, petroleum ether, etc.), BF_3OEt_2 as a mild catalyst and at low temperature, S_N2-type reactions could be

BnO OBn
BnO
AcO O NH
4a CCl₃

+

HO OTIPS
O
HO

(a) or (b) or (c) or (d)

BnO OBn HO OTIPS
O O
BnO O
AcO
4c

4b

(a) 0.01 eq TMSOTf, -78°C   –   (**4c**)
(b) 0.01 eq BF₃·OEt₂, -78 °C   –   (**4c**)
(c) 0.5 eq ZnCl₂, 20 °C   44%   (**4c**)
(d) 0.02 eq Sn(OTf)₂, -20 °C   75%   (**4c**)

❑ **Scheme 4**

carried out [1,2,3,4]. Solvents of high donicity (such as for instance ethers and nitriles) permit—based on the exo-anomeric effect—a different anomeric stereocontrol: As found and previously explained, in ethers at room temperature generally the axial product (for most cases the α-product) is favored whereas in nitriles at low temperature, based on the nitrile effect, the equatorial product (for most cases the β-product) is preferentially or exclusively obtained [3,4,45,46]. Hence, solvent and temperature selection play also an important role in anomeric stereocontrol.

Another interesting solvent effect is associated with the application of the 'inverse procedure' (IP) which consists of the addition of the glycosyl donor to a mixture of acceptor and acid catalyst [47,48]. In this way, particularly with highly reactive glycosyl donors often dramatic glycosylation yield improvements could be obtained [49]. This result seems to be based on a cluster effect between the catalyst and acceptor molecules; hence, on penetration of the donor into this cluster, activation of the donor and following glycoside bond formation takes place within the cluster in a practically intramolecular fashion. Obviously, this effect is critically dependent on the acceptor type, the solvent, and the temperature; therefore it requires often some experimentation for its successful application.

A dramatic polarity increase of the solvent can be reached by adding LiClO₄ to organic solvents (for instance, ether, dichloromethane, etc.) which enabled O-glycosyl trichloroacetimidate activation under essentially neutral conditions [50,51]. High polarity solvents are also ionic liquids (ILs) which have gained a lot of interest as solvents for organic reactions [52]. ILs have been probed in glycosylation reactions with O-glycosyl trichloroacetimidates and excellent glycosylation yields have been obtained (❷ *Scheme 6*; **6a + 6b** → **6c**) [53]; for reactive systems again no acid activation is required [53]. Microwave heating has been employed for O-glycosyl trichloroacetimidate activation as well, furnishing glycosides in high yields [54]; however, presumably due to the temperature effects only modest anomeric stereocontrol was available.

Besides solvent, temperature, and (kinetic and thermodynamic) stereoelectronic effects, anchimeric assistance by neighboring acyl groups and/or by steric shielding is decisive for the anomeric stereocontrol [1,2,3,4,5,6,7,22]. This way 1,2-*trans*- (β-*gluco*, α-*manno*) and 1,2-*cis*-type (α-*gluco*) glycosides have been generally obtained at wish in good to excellent stereochemical yields. However, a 2-O-acyl group does not guarantee β-glucopyranoside or α-mannopyranoside formation. Besides undesired ortho-ester formation, which can often be

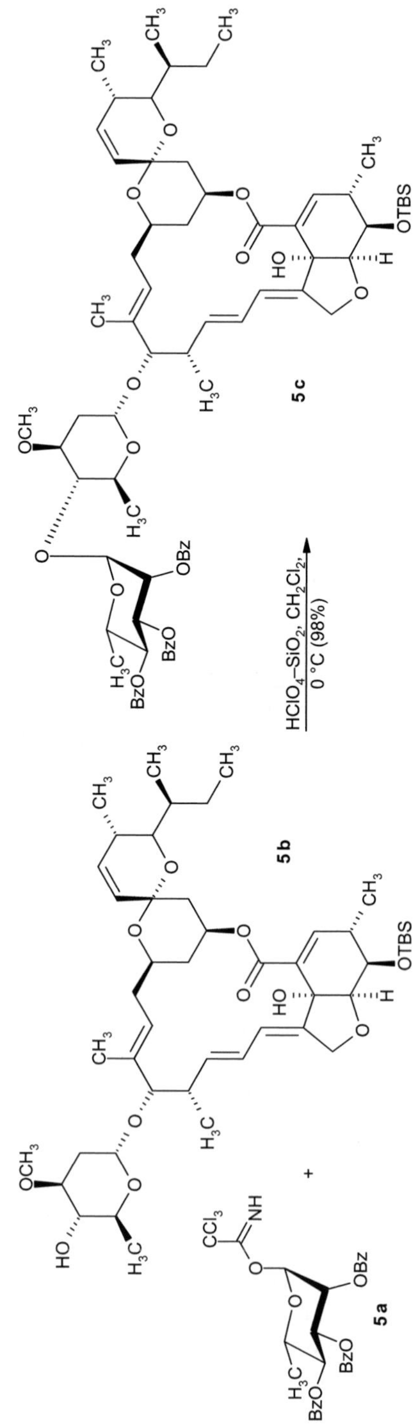

■ Scheme 5

◻ **Scheme 6**

overcome by using more catalyst or by varying the acyl group, once in a while still the α-glu-copyranoside or even some β-mannopyranoside, respectively, is obtained. Interesting cases of α-linkage on attempted 2-O-acyl supported $\beta(1\text{–}3)$-glucan synthesis were recently reported which were explained by 'remote control' [55,56,57]. **7aβ + 7bβ** gave **7cα** whereas **7aα + 7bα** gave **7cβ** (❷ *Scheme 7*). However, this problem could be readily overcome by employing the ADMB group [23]. The influence of the structure of the glycosyl acceptors on the anomeric stereocontrol has often been discussed and since the work of van Boeckel et al. [58,59] the potential importance of matched and mismatched donor-acceptor pairs on the stereochemical outcome of glycosylation reactions has become evident. Recent work on 2-azido-2-deoxy-glucopyranosyl trichloroacetimidates exhibited that the structure of the glycosyl donor has generally the major influence [60].

Other anchimerically assisting groups at C-2, such as–SPh, -SePh, -Br, and -I, have been investigated [61,62]. Particularly worth mentioning is the work on equatorial iodo substitution of glucopyranosyl trichloroacetimidates which strongly favored 1,2-*trans*- i. e. β-glycosidation. This result could be due to generation of an iodonium intermediate or due to the steric demand of the iodo group, thus favoring a twist-boat type oxocarbenium ion intermediate which—as previously discussed in β-mannopyranoside synthesis (see below) [26]—for steric and stereo-electronic reasons favors nucleophilic attack from the β-side [63,64].

An interesting addition to the repertoire of anchimerically assisting groups is the chiral 1-phenyl-2-(phenylsulfanyl)ethyl group [65]. Coupled as (*S*)-isomer **8a** to the 2-hydroxy group of the glucopyranosyl trichloroacetimidates with acceptor **8b** 1,2-*cis*- i. e. α-glucopyra-nosides **8c** have been obtained in excellent yields and stereoselectivities (❷ *Scheme 8*). The formation of a cyclic β-linked sulfonium ion intermediate **8a***, having a *trans*-decalin type structure has been confirmed by NMR experiments. This intermediate seems to be preferentially or exclusively attacked by the nucleophile from the α-side.

A particularly difficult problem is 1,2-*cis*- i. e. β-linkage formation in β-mannopyranoside synthesis. The presence of β-linked mannopyranosides in various natural products [66], particularly in the N-glycan core structure of glycoproteins [66,67], led to the search for efficient methodologies for generating this target structure. Investigation of mannopyranosyl donors with nonparticipating protecting groups and different leaving groups led in general mainly or exclusively to α-products [1,66,67,68]. Several specific methods led to some success in this endeavor [69,70,71,72,73,74], however finally epimerization of β-glucopyranosides at 2-O [74,75,76,77,78,79] and intramolecular aglycone delivery [80,81,82,83,84] have led

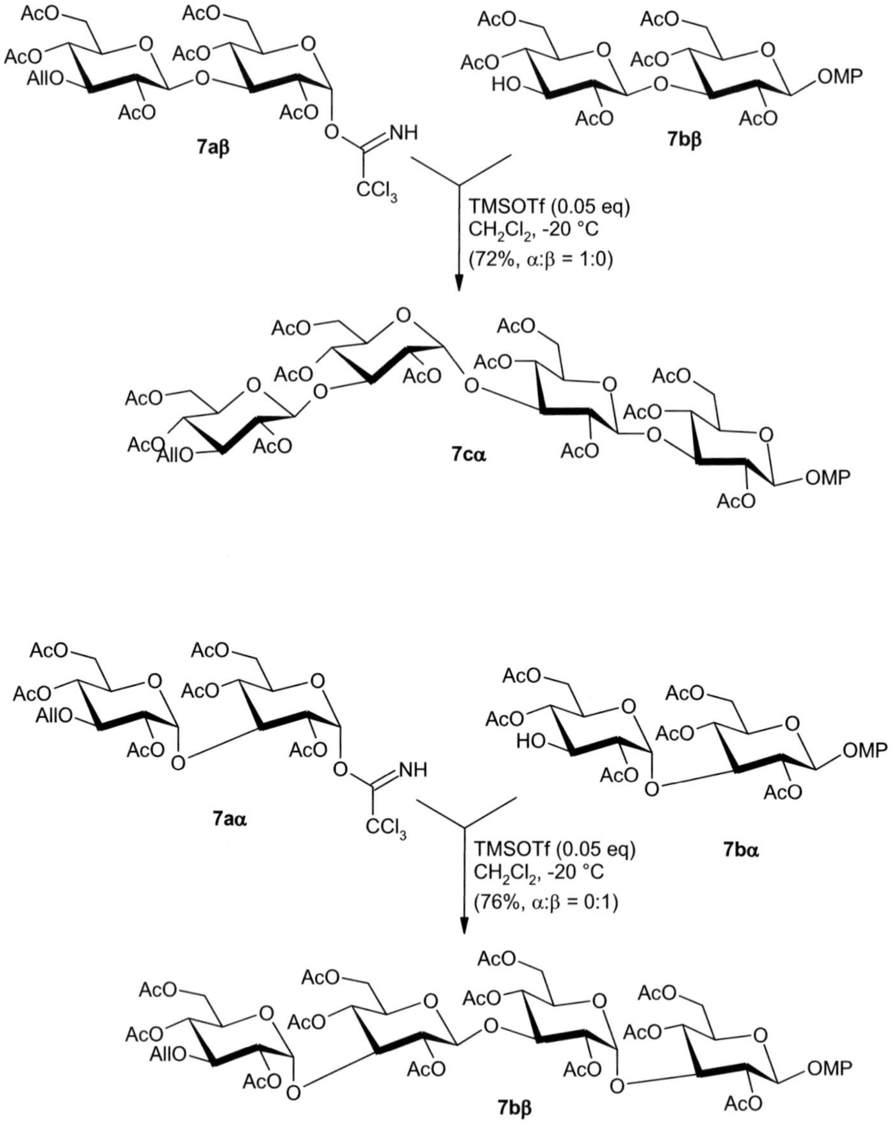

□ Scheme 7

to very successful results. Mannopyranosyl donors with diol O-protecting groups leading to ring annelation had already been investigated but again with limited success [85]. Therefore, it was a big surprise that 2,3-di-O-alkyl-4,6-O-benzylidene-protected mannopyranosyl sulfoxides as glycosyl donors gave preferentially β-products with various acceptors at low temperatures [86,87]. The same result is more conveniently obtained with the corresponding trichloroacetimidates as shown in ❷ Scheme 9 with glycosyl donor 9a furnishing with accep-

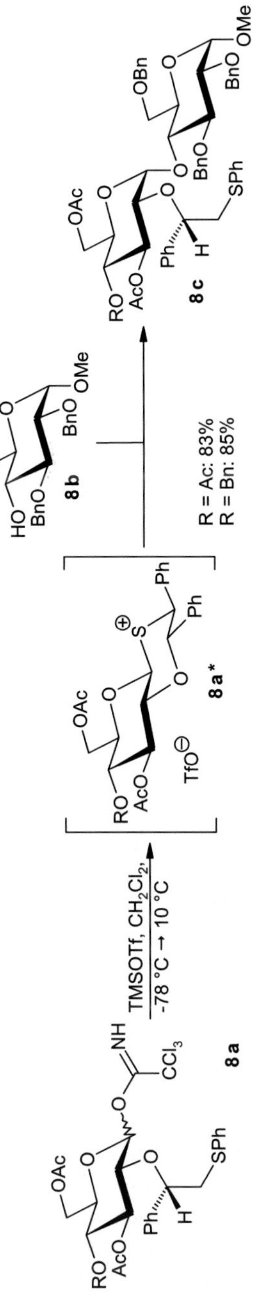

■ Scheme 8

X = SOEt: DTBMP (2 eq), Tf$_2$O (1 eq), CH$_2$Cl$_2$, -78 °C → 0 °C (18%; α:β = 1:2.7)
X = O⌇⌇⌇NH: TMSOTf (0.15 eq), CH$_2$Cl$_2$, -50 °C, IP (71%; α:β = 1:3.6)
CCl$_3$

◻ Scheme 9

tor **9b** preferentially the β-disaccharide **9c** [26]. The results obtained on varying the reaction parameters were not compatible with the reaction mechanism proposed for sulfoxide activation, in which an α-triflate intermediate is thought to play the decisive role [87]. Rather the anomeric stereocontrol is caused by a conformational effect enforced by the 4,6-*O*-benzylidene group on the pyranosyl ring, which favors the generation of a flattened twist-boat conformation as the intermediate. For stereoelectronic and steric reasons, this twist-boat intermediate will be preferentially attacked from the β-side, thus forming after equilibration the $^4C_1$-conformer of the β-mannopyranoside [26]. This mechanistic proposal is also confirmed by L-rhamnosylation reactions with bulky 3-*O*- and 4-*O*-protecting groups which enforce the $^4C_1$-conformation and this way favor β-L-rhamnoside formation [88]. On the basis of these mechanistic considerations, β-mannopyranoside formation should be facilitated by nonparticipating, strongly electron-withdrawing groups at the 2-*O* atom because generation of the twist-boat intermediate would gain from a strong dipole effect. This expectation could be confirmed [89].

Regioselectivity in glycosylation reactions is generally based on a sequence consisting first of regioselective functional group protection, making use of various principles, second glycosylation of the *O*-unprotected hydroxy group, and finally of *O*-deprotection. Direct regioselective glycosylation using the difference in reactivity of sugar hydroxy groups is of great interest as it often avoids cumbersome protection and deprotection steps. Many regioselective glycosylations of partly unprotected carbohydrate acceptors have been reported in the literature [1,2,3,4,5,6,7]. Generally, they make use of the higher reactivity of primary hydroxy groups over secondary hydroxy groups (particularly those in axial orientation) and the higher reactivity of equatorial 3-hydroxy groups, for instance in galacto- and glucopyranosides. This was again confirmed in the glycosylation of 4,6-*O*-benzylidene-glucopyranosides which gave with various glycosyl trichloroacetimidates as donors the (1–3)-linked disaccharides in very good yields [90]. Another interesting alternative is one-pot regioselective protection and following *O*-glycosylation with *O*-glycosyl trichloroacetimidates which, as shown in
❯ *Scheme 10* in the glycosylation of **10a** with donors **10b** and **10c** affording disaccharides **10d** and **10e**, respectively, makes use of this reactivity difference [91].

Separation of the target glycosides from the reaction mixture generally requires column chromatography purification. Alternatively, highly fluorinated compounds are readily separated from nonfluorinated compounds by a simple phase separation. Therefore, organic synthesis

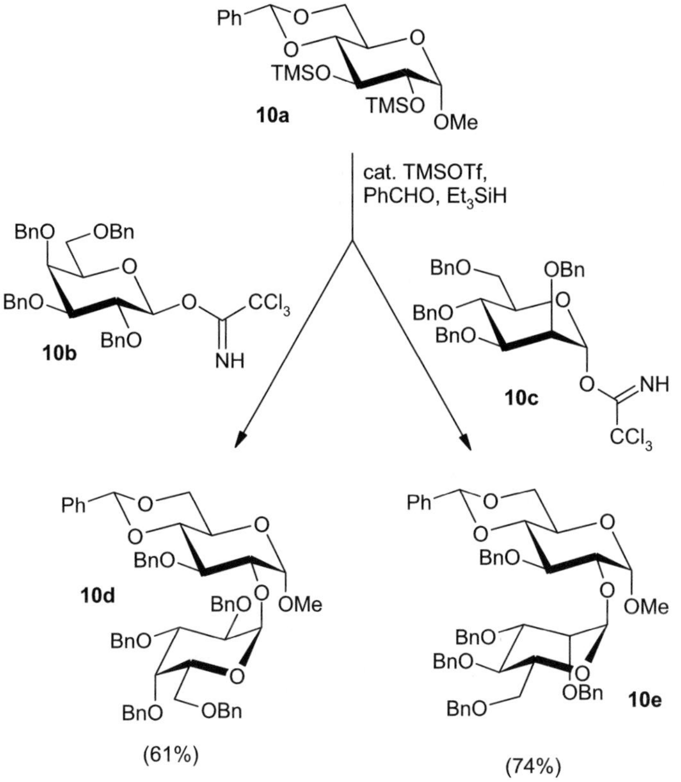

■ Scheme 10

with substrates having fluorinated tags has become a valuable alternative to solid-phase synthesis (see below). This concept has been already applied to oligosaccharide synthesis [92]. On the basis of the novel fluorous acyl protecting group Bfp (**b**is **f**luorous chain type **p**ropanoyl) highly successful oligosaccharide syntheses could be performed [93], which were extended to the synthesis of globo-triaosyl ceramide **11e** from lactoside **11a** as shown in ❯ *Scheme 11* [94]. All fluorous intermediates (**11b** and **11d**) were extracted with FC-72 (perfluorohexane isomers) and an organic solvent (toluene or methanol) and were purified without silica gel chromatography. After removal of the Bfp groups by simple base-catalyzed methanolysis the pure allyl trisaccharide **11e** was obtained by only one silica gel column chromatography in 34% total yield after five steps (~81% per step).

4.2 Glycoside Synthesis on Polymer Supports

Successful solid-phase oligosaccharide syntheses (SPOS) have been developed by several research groups [95,96,97,98,99,100,101,102,103,104,105], which exhibit the inherent advantages over solution phase synthesis, such as (i) higher reaction yields due to the use

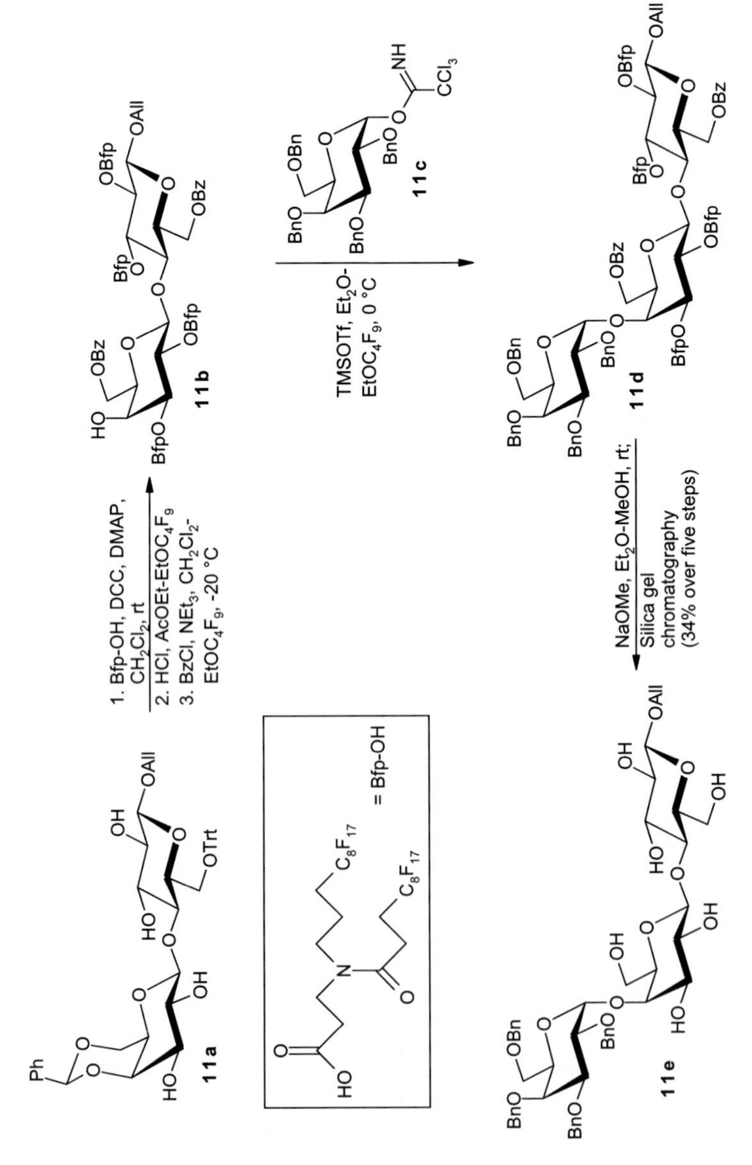

□ **Scheme 11**

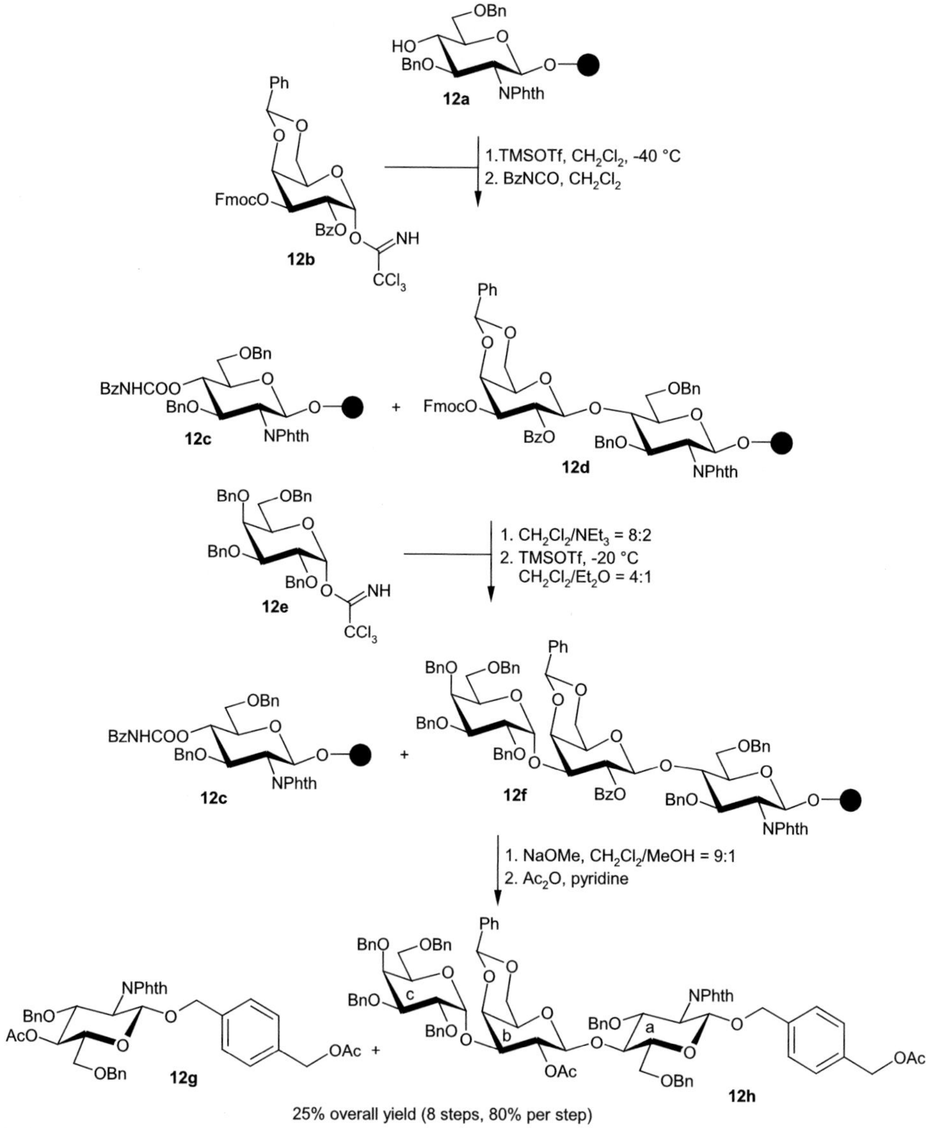

25% overall yield (8 steps, 80% per step)

Scheme 12

of excess building blocks, (ii) shorter reaction times for the completion of total syntheses, and (iii) convenient purification procedures by just washing the resin. In addition, methods to avoid undesired byproduct formation in the synthesis of the target molecules have been introduced [106,107,108,109], such as for instance capping procedures of unreacted intermediates (❯ *Scheme 12*) [108]. Low reactive acceptor **12a** after glycosylation with donor **12b** is capped with benzoyl isoyanate yielding **12c** and **12d**. Chain extension with **12e** (→ **12c + 12f**) and cleavage led to readily separable **12g** and **12h**. So far, no generally accepted strategy has yet appeared for the efficient construction of various complex oligosaccharides on polymer supports, thus limiting the commercialization of automated synthesizers. However, it seems that O-glycosyl trichloroacetimidates and the less reactive O-glycosyl phosphates have the potential to become the glycosyl donors of choice because the requirement of catalytic amounts of just one activator is a major advantage over the other glycosylation methods.

After various attempts of the Schmidt group with different linkers between solid support and the carbohydrate groups, such as for instance

- thioglycosides and their cleavage by thiophilic reagents,
- silyl glycosides and their cleavage by fluorides,
- pentenyl-type glycosides and their cleavage by electrophilic reagents, and
- allyl-type glycosides and their cleavage by cross metathesis or by ring-closing metathesis, respectively [21,110,111],

finally the ester-based SPOS design was introduced which gave excellent results in chain extension and product cleavage from the resin [110,111]. However, to cope with complex oligosaccharide synthesis, besides the ester-linker and -spacer system, three types of glycosyl donor building blocks for controlled chain extension (suffix **e**), branching (suffix **b**), and chain termination (suffix **t**) are required. An efficient solution to these requirements for SPOS of a small library of N-glycans is exhibited in ❷ *Scheme 13* which shows the retrosynthesis to the required building blocks and the required reaction steps. A particularly important role played the selection of the N-protecting group for the glucosamine residue.

The ester-based SPOS methodology (❷ *Scheme 13*) comprises (i) different types of esters, that is, the benzoate group as a linker and for chain termination and the Fmoc and PA (phenoxyacetyl) group as temporary protecting groups for chain extension and branching which can be chemoselectively cleaved (in the sequence Fmoc and then PA), (ii) the benzyl group for permanent O-protection and for the spacer between the anomeric center at the reducing end sugar, thus providing after final product cleavage from the resin a structurally defined target molecule, (iii) O-glycosyl trichloroacetimidates of type **e**, **b**, or **t** (for chain extension, branching, or termination) as powerful glycosyl donors, which can be readily activated by catalytic amounts of (Lewis) acid, and (iv) benzoic acid residues on the Merrifield resin for the linkage of the hydroxymethylbenzyl spacer. Hence, retrosynthesis of a typical N-glycan molecule containing the core pentasaccharide and some antennae leads to spacer-linker connected Merrifield resin SLP and to glycosyl donors G^e, G^t, Gal^t, Gal^b, M^b, M^t, and MG^b which can be selectively converted into acceptors on resin (**e** and **b**-type donor building blocks). Thus, as indicated in ❷ *Scheme 13*, only four simple procedures are required for successful SPOS: (a) glycosidation under TMSOTf catalysis; (b) product cleavage under transesterification conditions; (c) selective Fmoc cleavage under basic conditions; and (d) selective PA cleavage under milder transesterification conditions.

On the basis of this concept, as shown in ❷ *Scheme 14a–c*, a small library of N-glycans (compounds **14a** to **14q**) was successfully synthesized: (i) The Merrifield resin as solid support exhibited excellent results during all stages of the assembly; (ii) all glycosylations, including those with N-DMM protected glycosyl donors, gave high yields; (iii) the methodology presented herein shows the desired versatility in terms of efficient chain extension and branching requiring only two standard (in one direction) orthogonal protecting

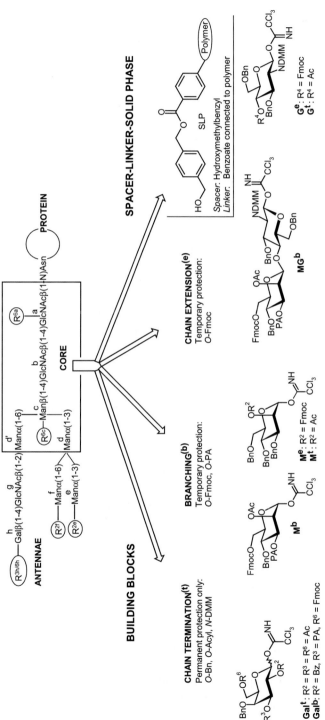

Scheme 13
Solid phase synthesis of high mannose, complex, and hybrid type *N*-glycans: increased donor and acceptor reactivity

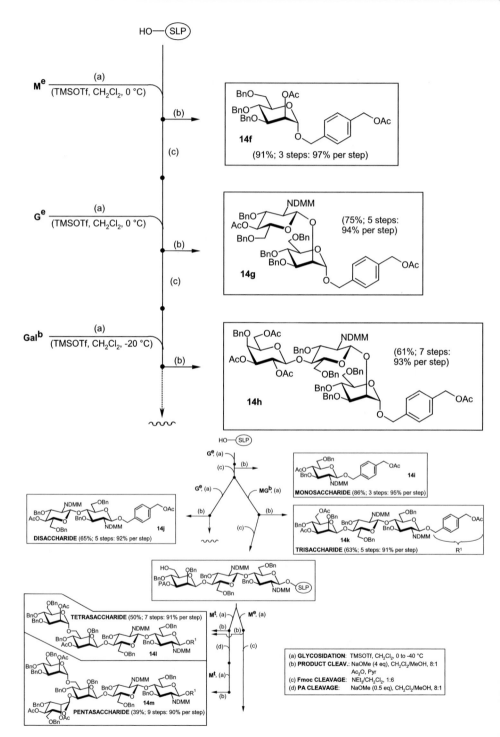

HO—(SLP)

M^e (a) (TMSOTf, CH$_2$Cl$_2$, 0 °C)
(b)
(c)

14f
(91%; 3 steps: 97% per step)

G^e (a) (TMSOTf, CH$_2$Cl$_2$, 0 °C)
(b)
(c)

14g
(75%; 5 steps: 94% per step)

Gal^b (a) (TMSOTf, CH$_2$Cl$_2$, -20 °C)
(b)

14h
(61%; 7 steps: 93% per step)

HO—(SLP)

G^e, (a)
(c)   (b)
G^v, (a)
(b)
MG^b, (a)
(b)
(c)

14i
MONOSACCHARIDE (86%; 3 steps: 95% per step)

14j
DISACCHARIDE (65%; 5 steps: 92% per step)

14k
TRISACCHARIDE (63%; 5 steps: 91% per step)
R^1

TETRASACCHARIDE (50%; 7 steps: 91% per step)

14l

M^t, (a)
(b)
(d)
M^e, (a)
(b)
(c)

M^t, (a)
(b)

14m
PENTASACCHARIDE (39%; 9 steps: 90% per step)

(a) **GLYCOSIDATION**: TMSOTf, CH$_2$Cl$_2$, 0 to -40 °C
(b) **PRODUCT CLEAV.**: NaOMe (4 eq), CH$_2$Cl$_2$/MeOH, 8:1
  Ac$_2$O, Pyr
(c) **Fmoc CLEAVAGE**: NEt$_3$/CH$_2$Cl$_2$, 1:6
(d) **PA CLEAVAGE**: NaOMe (0.5 eq), CH$_2$Cl$_2$/MeOH, 8:1

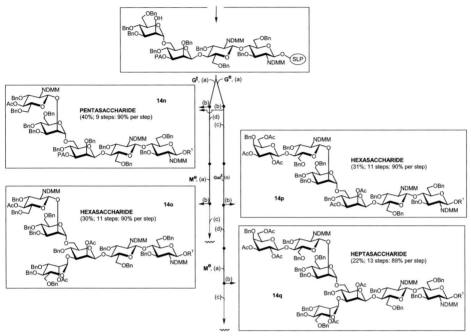

☐ **Scheme 14 (left page and above)**
Solid phase synthesis of a small library of *N*-glycans: typical constituents

groups; (iv) cleavage of the product from the resin was feasible leading to stable 1-*O*-ben-zyl type products with only benzyl, DMM and, after acetylation, acetyl protection; (v) the crude products were already of high purity; therefore, standard silica gel chromatography and MPLC were sufficient for purification; (vi) yields of isolated products were high, ranging from 97% per step (after three steps) to 89% per step (after 13 steps) on solid-phase; (vii) the methodology is technically simple, thus lending itself available to automa-tion.

Very good SPOS results have also been obtained by the Seeberger group with *O*-glyco-syl phosphates and *O*-glycosyl trichloroacetimidates, respectively, as glycosyl donors and (Z)-4-octene-1,8-diol as the linker between the Merrifield resin and the sugar residue. The linker was cleaved by cross-metathesis with ethylene, thus providing pentenyl glycosides as cleavage products (Schemes 15 and 16) [112]. ❷ *Scheme 15* shows the synthesis of pentenyl cellotrioside **15g** with **15a** as the donor over seven steps with **15b–f** as intermediates in 53% overall yield. As a temporary protecting group in the glycosyl donor the TBS group in the 4-position was employed. The β-selectivity was supported by a 2-*O*-pivaloyl group. This method was also employed in an automated SPOS which was based on a modified peptide syn-thesizer and *O*-glycosyl trichloroacetimidates as glycosyl donors (❷ *Scheme 16*) [96,113,114]. This way a glycosylphosphatidyl inositol (GPI) (**16n**) was synthesized which functions as a malarial toxin. Starting from three different *O*-glycosyl trichloroacetimidates (**16a, d, f**)

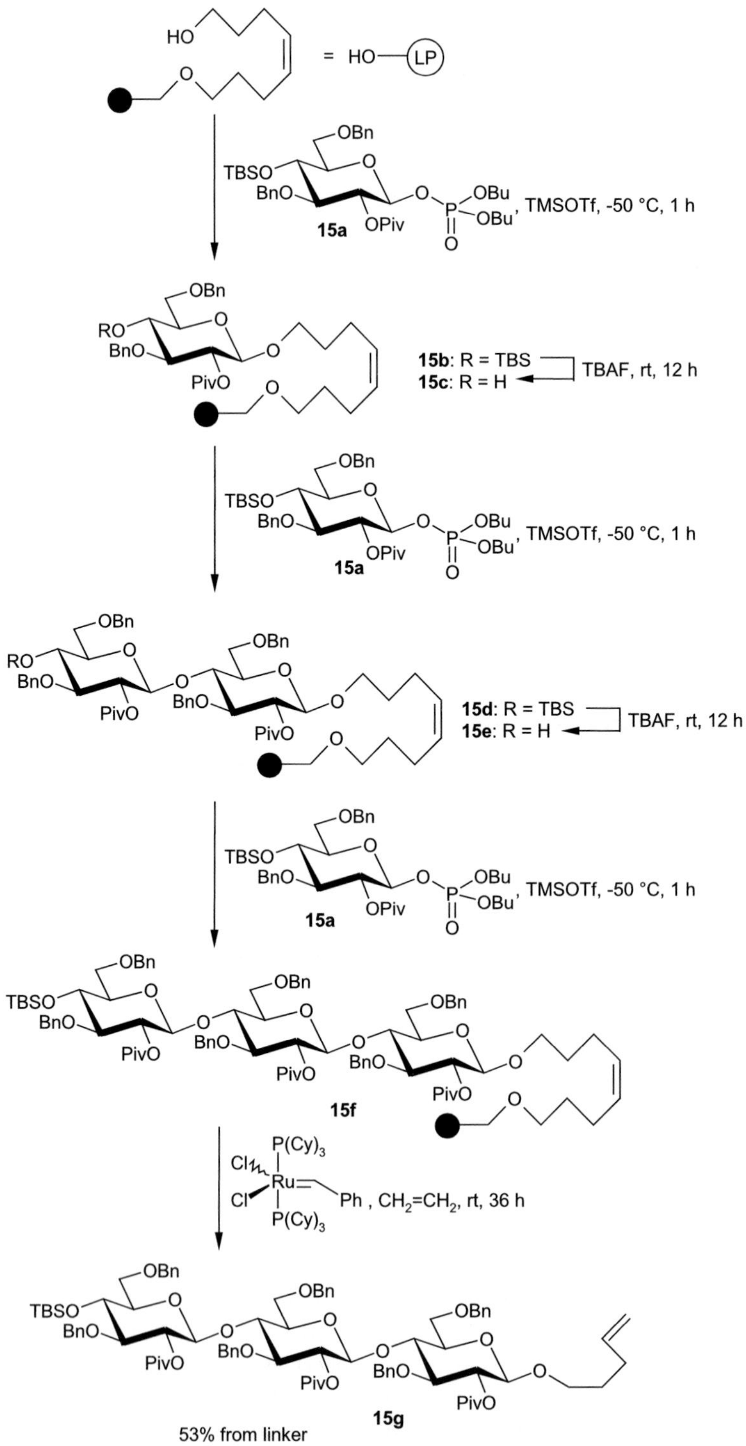

□ **Scheme 15**

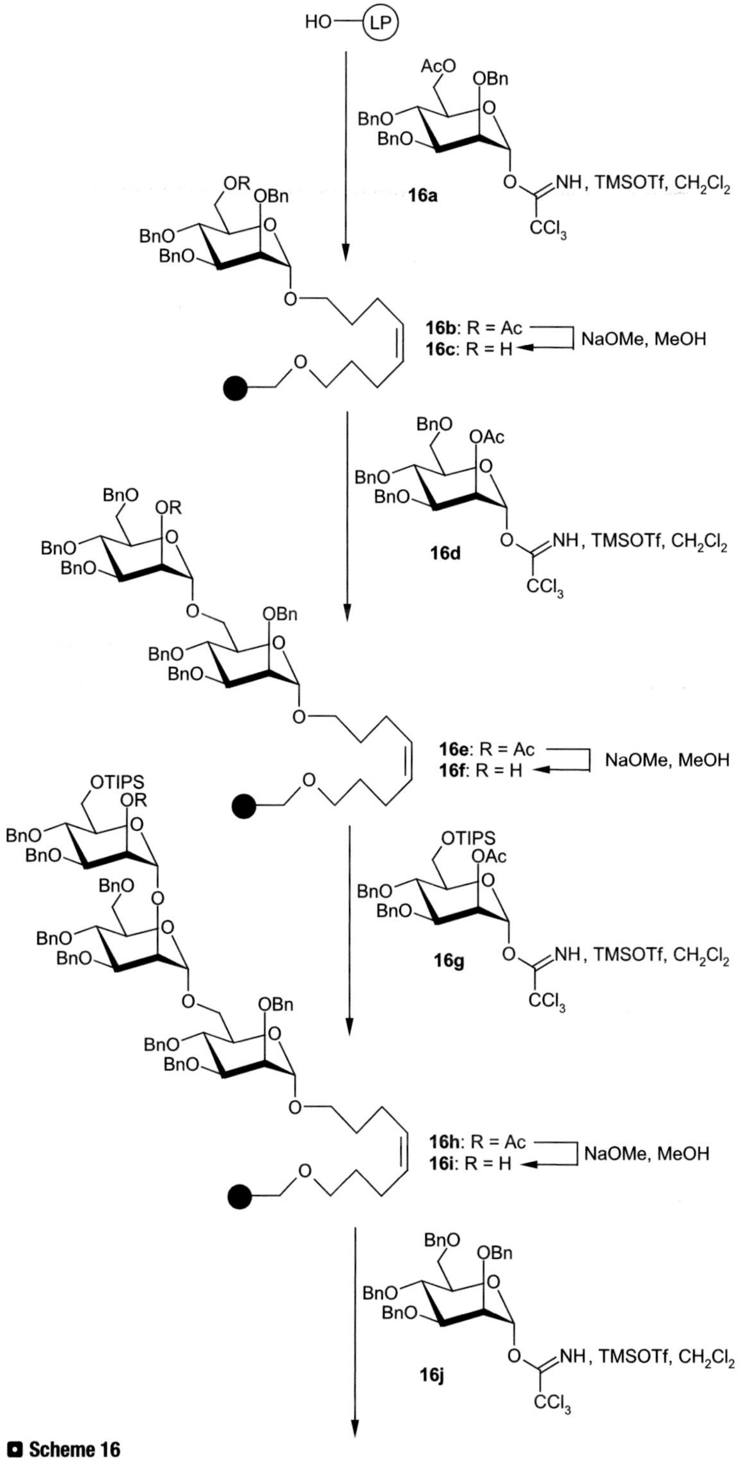

Scheme 16

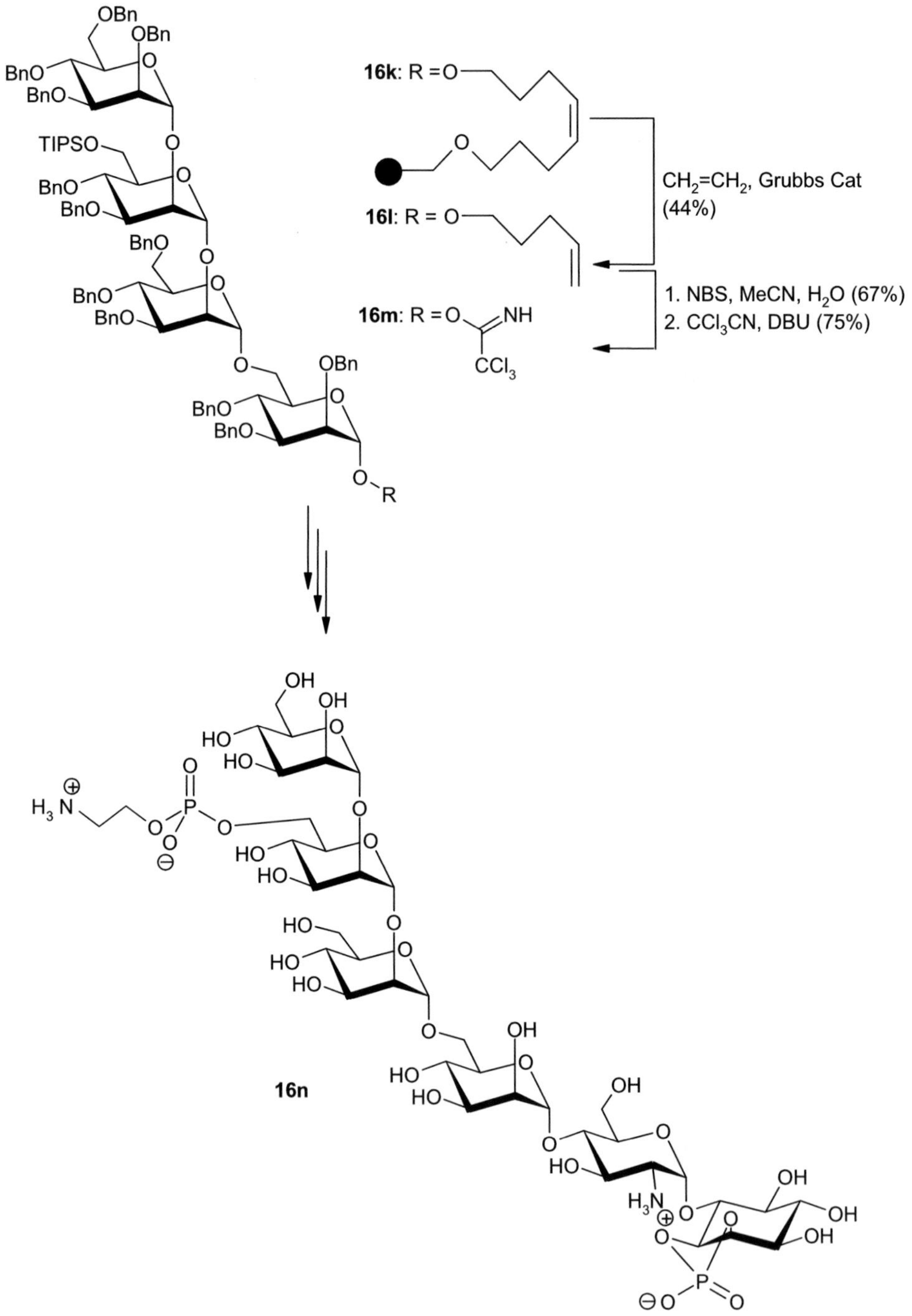

16k: R = O—

16l: R = O—

16m: R = O   NH
               |
              CCl₃

CH₂=CH₂, Grubbs Cat
(44%)

1. NBS, MeCN, H₂O (67%)
2. CCl₃CN, DBU (75%)

16n

☐ Scheme 16
(continued)

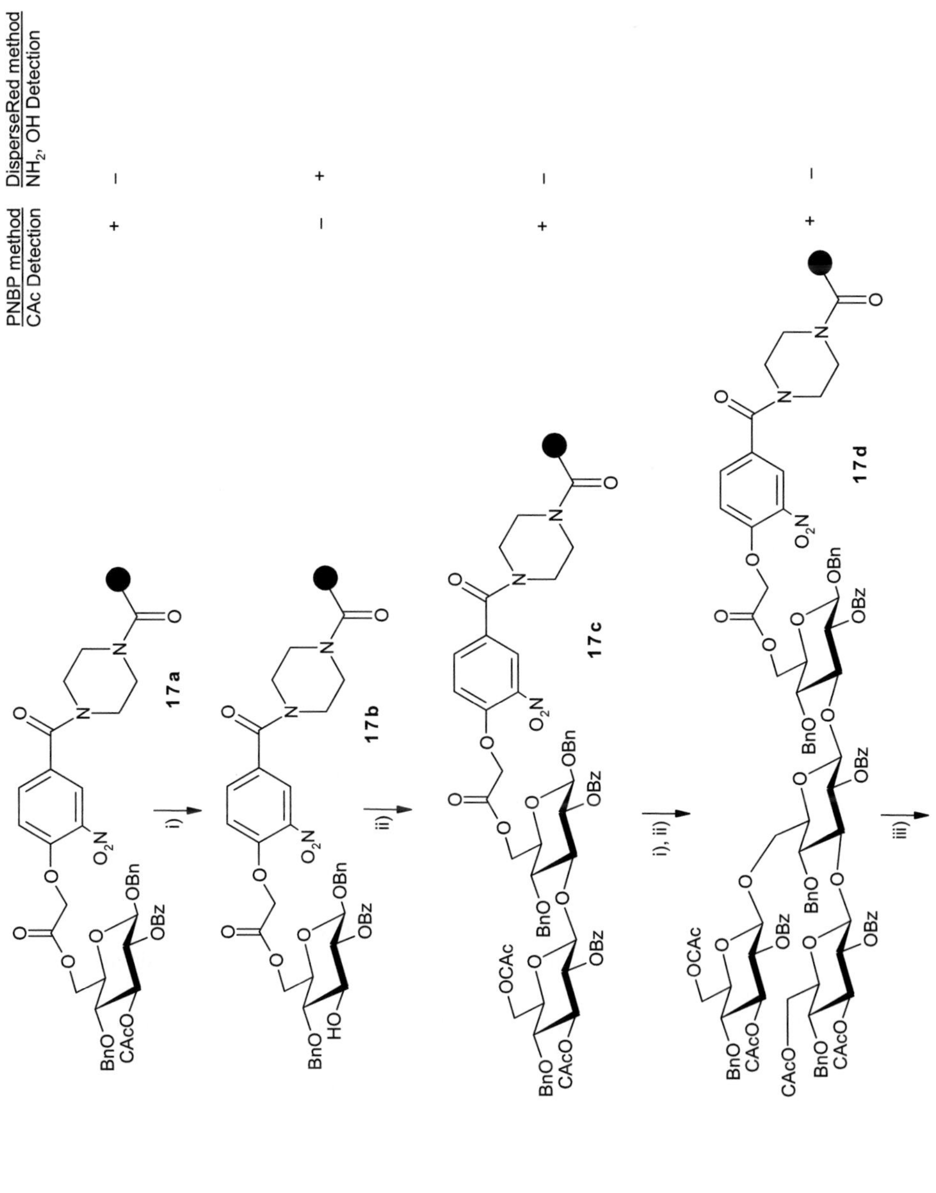

■ Scheme 17

i) (a) Capping: Ac$_2$O, ^{i}Pr$_2$NEt, CH$_2$Cl$_2$
   (b) CAc-Cleavage: HDTC, DMF
ii) Glycosylation: [structure] , BF$_3$·OEt$_2$, CH$_2$Cl$_2$ **17f**

iii) Cleavage from resin: NaOMe, MeOH, THF (30% overall yield)

◻ **Scheme 17**
(continued)

having *O*-acetyl groups as temporary protection for chain extension and one per-*O*-benzylated *O*-mannopyranosyl trichloroacetimidate (**16j**) for chain termination a tetrasaccharide **16k** was obtained which was cleaved from the resin (→ **16l**) by cross metathesis. Transformation into a trichloroacetimidate **16m** and glycosylation of a pseudodisaccharide led after deprotection to the target molecule **16n**.

For the real-time monitoring of the glycosylation result by the Ito group sensitive color tests were introduced with p-nitrobenzylpyridine (PNBP), reacting with chloroacetyl (CAc) as the temporary protecting group under red color formation, and with cyanuric chloride-Disperse Red conjugate, which readily reacts with hydroxy or amino groups to give a red color on the resin [115]. This way, as shown in ❯ *Scheme 17*, the repeating unit **17e** of the immuno-active oligosaccharide schizophyllan was synthesized based on a partly CAc-protected *O*-glucosyl trichloroacetimidate donor **17f** with **17a** as starting material and **17b** to **17d** as intermediates; the presence of the CAc groups could be readily monitored with PNBP and the presence or absence of hydroxy groups was monitored by the Disperse-Red method. As a linker to piperazine-modified TentaGel, a 2-nitrophenyloxyacetyl group [99] was employed that can be readily cleaved under transesterification conditions. This led also to loss of other ester protecting groups furnishing the target molecule **17f** in very good overall yield.

The synthesis of oligosaccharides on soluble supports has been further investigated as well. Particularly worth mentioning is the synthesis of heparin-like oligosaccharides on polyethylene glycol ω-monomethyl ether (MPEG) resin with a succinoyl ester linker and *O*-glycosyl trichloroacetimidates as glycosyl donors [116,117]. A typical example is shown in ❯ *Scheme 18* [117]. After the glycosylation steps of **18a** with polymer bound **18b** and **18c** with **18d**, respectively, catalyzed by TMSOTf, a capping step was included in which unreacted MPEG-bound acceptor was captured onto the solid-phase of succinate-functionalized Merri-

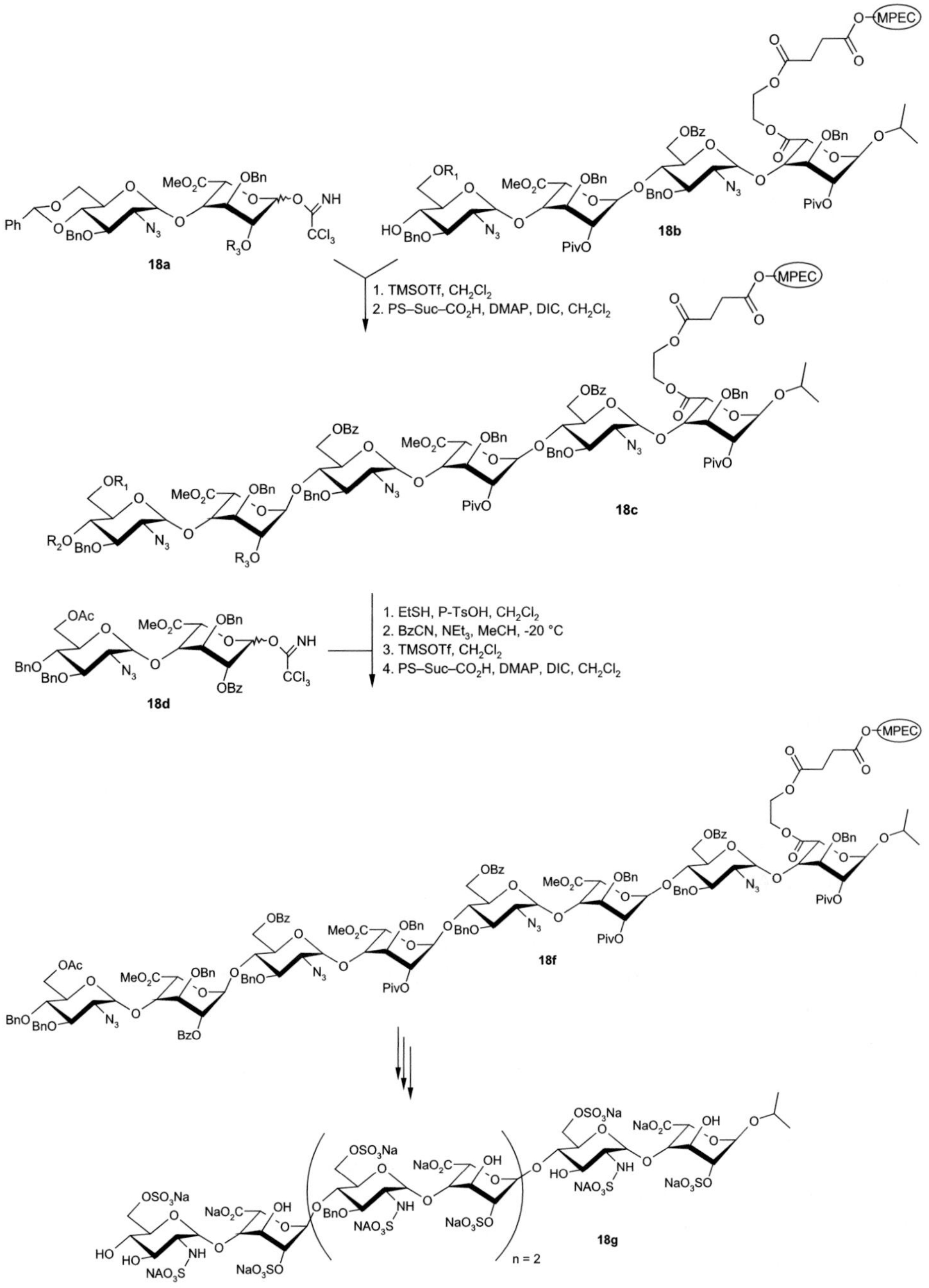

■ Scheme 18

field resin (PS–Suc–CO$_2$H) by a chemoselective ester formation (→ **18c** and **18f**). Alkaline hydrolysis released the octasaccharide from the resin and also all other ester groups were cleaved. By further known transformations [118] the octasaccharide fragment **18g**, containing the structural motif of the regular region of heparin, was obtained.

An interesting comparative study between solution phase synthesis and synthesis of the same oligosaccharides on MPEG as soluble support was carried out by the Furneaux group [119]. This study clearly reflected the incompatibility of MPEG as a soluble support with some common reagents, reaction conditions, and choice of standard protecting groups. Hence, the shortcomings associated with this support seem to outbalance the merits [119,120].

To avoid some limitations of PEGs, as for instance low loading, hyperbranched polyester such as Boltorn H40 and H50 have been successfully investigated in glycosylation reactions with *O*-glycosyl trichloroacetimidates [121]. For simple mono- and disaccharide synthesis optimization of the reaction conditions led to practically quantitative transformations.

5 Recent Applications of *O*-Glycosyl Trichloroacetimidates in Complex Oligosaccharide and Glycoconjugate Synthesis

Several comprehensive reviews have been devoted to the use of *O*-glycosyl trichloroacetimidates in glycosylation reactions. In this overview, so far the basic principles of the glycosylation methods, the strengths of the *trichloroacetimidate method*, differences to other methods, and methodological variations have been discussed. In this chapter recent applications of this method in complex glycoconjugate synthesis will be highlighted.

5.1 Glycolipids

Glycosphingolipid synthesis has remained a target of great interest because of the biological importance of these compounds. Assembly of the sugar residues with *O*-glycosyl trichloroacetimidates as donors and attachment of the ceramide residue via the azidosphingosine *glycosylation procedure* [122,123] has become the method of choice. On the basis of this approach, aminodeoxy analogs of globotriosyl ceramide [124], globo- and isoglobotriosides bearing cinnamoylphenyl tags [125], and a quite practical globotriose synthesis [126], which is essentially based on *O*-acyl protection, have been carried out. An efficient glycosylation protocol for the attachment of an α-linked galactosamine residue for asialo GM2 synthesis has been reported [127]. A novel ether-bridged GM3 lactone analog has been successfully prepared and used in antibody-based cancer therapy studies [128]. Also ganglioside GD3 was synthesized [129]; comparison with bovine-brain derived GD3 showed that the effects in GD3-triggered uncoupling of mitochondrial respiration and induction of apoptosis in oligodendrocytes are very similar. Also ganglioside mimics for binding studies with myelin-associated glycoprotein were prepared [130].

The *lacto-* and *neolacto-*series of glycosphingolipids also attracted further attention. The synthesis of the lacto-*N*-neotetraose and lacto-*N*-tetraose building blocks has been improved by using the *N*-dimethylmaleoyl protecting group [131]. A total synthesis of the natural antigen **19k** involved in a hyperacute rejection response to xenotransplants has been carried out as

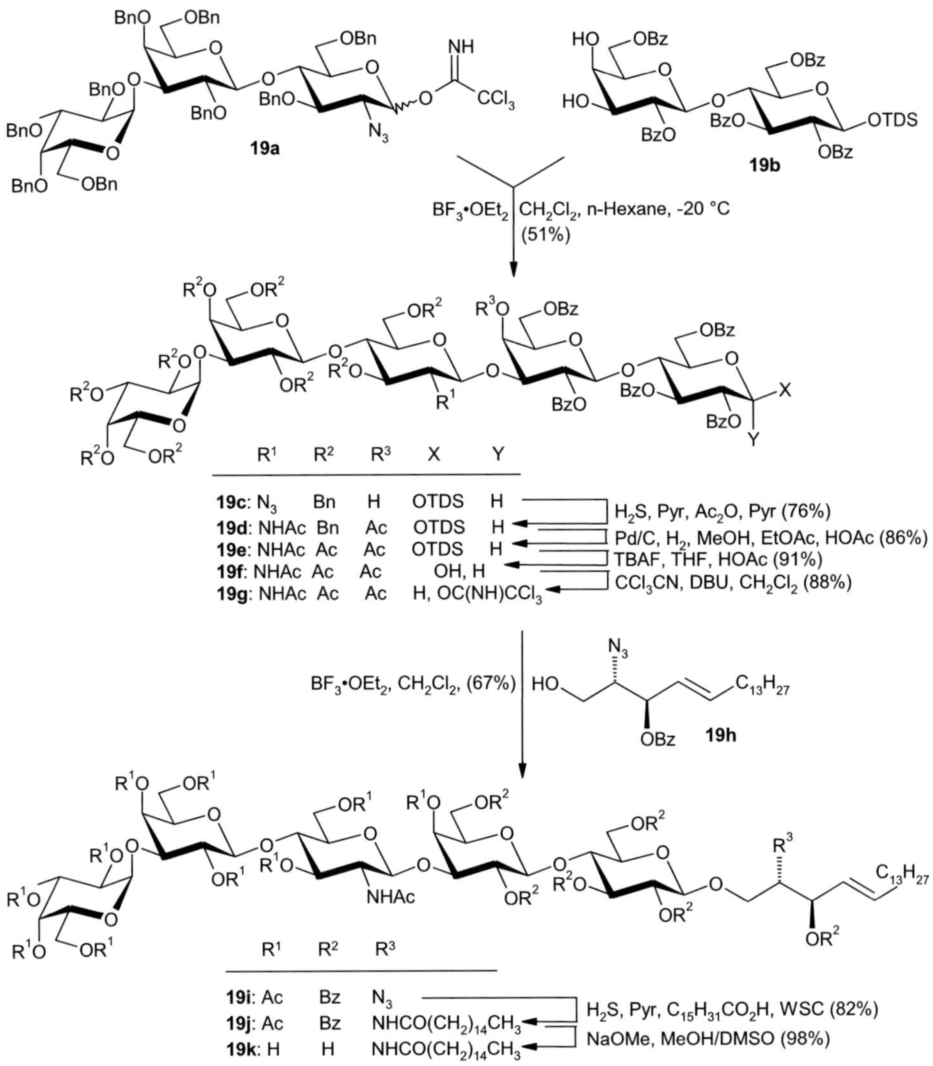

Scheme 19

shown in ❯ *Scheme 19* [132]. Glycosyl donor **19a**, also obtained via the *trichloroacetimidate method*, gave with acceptor **19b** pentasaccharide **19c**. Standard transformation via **19d**, **19e**, and **19f** led to pentasaccharide donor **19g**. Application of the *azidosphingosine glycosylation procedure* with **19h** as acceptor furnished **19i** which was transformed via **19j** into the target molecule **19k**.

Further, for microdomain formation studies fluorescence labeled sialyl Lewis X [133], for cluster effect studies dimeric sialyl Lewis X [134], and for carbohydrate–carbohydrate recognition studies a dimer of Lewis X [135] were synthesized (❯ *Scheme 20*, **20k**). To this aim,

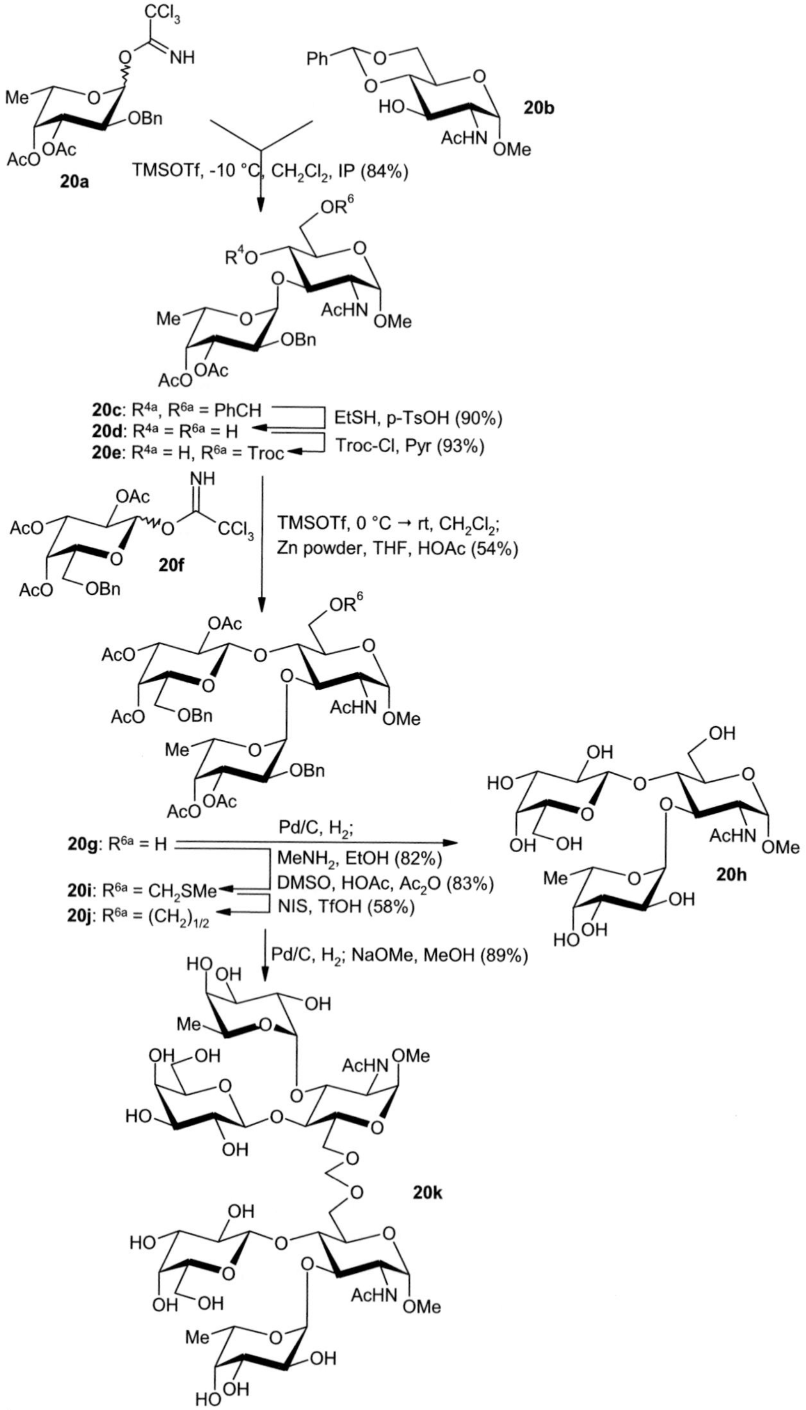

The synthetic scheme shows the following reagents and intermediates:

20a (glycosyl trichloroacetimidate donor) + **20b** → TMSOTf, -10 °C, CH₂Cl₂, IP (84%)

Product disaccharide with R⁴O, OR⁶ groups:
- **20c**: R⁴ᵃ, R⁶ᵃ = PhCH ┐ EtSH, p-TsOH (90%)
- **20d**: R⁴ᵃ = R⁶ᵃ = H ◄┘
- **20e**: R⁴ᵃ = H, R⁶ᵃ = Troc ◄── Troc-Cl, Pyr (93%)

20f (trichloroacetimidate donor) → TMSOTf, 0 °C → rt, CH₂Cl₂; Zn powder, THF, HOAc (54%)

Trisaccharide intermediate:
- **20g**: R⁶ᵃ = H ── Pd/C, H₂; MeNH₂, EtOH (82%) → **20h**
- **20i**: R⁶ᵃ = CH₂SMe ◄── DMSO, HOAc, Ac₂O (83%)
- **20j**: R⁶ᵃ = (CH₂)₁/₂ ◄── NIS, TfOH (58%)

20j → Pd/C, H₂; NaOMe, MeOH (89%) → **20k**

☐ **Scheme 20**

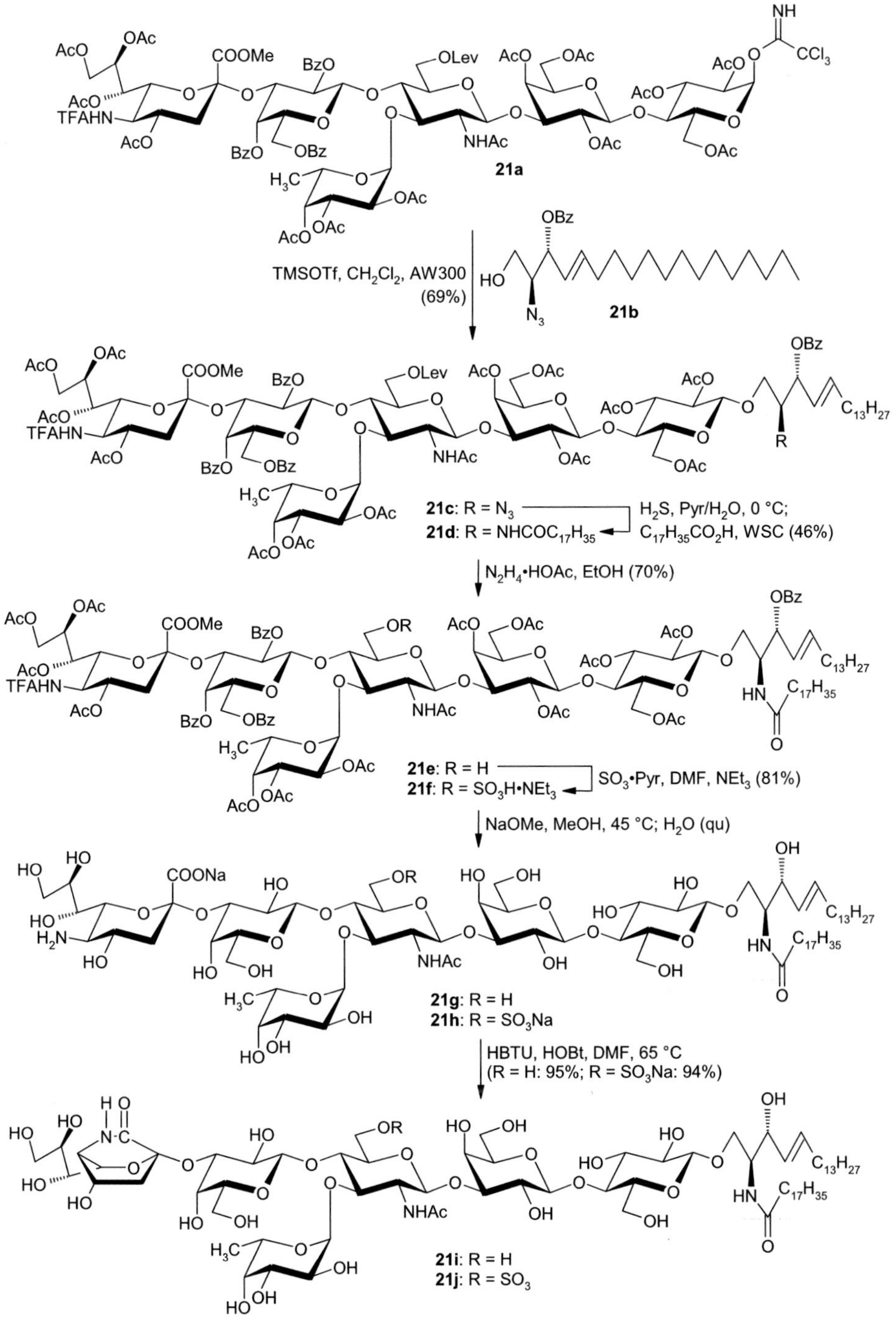

Scheme 21

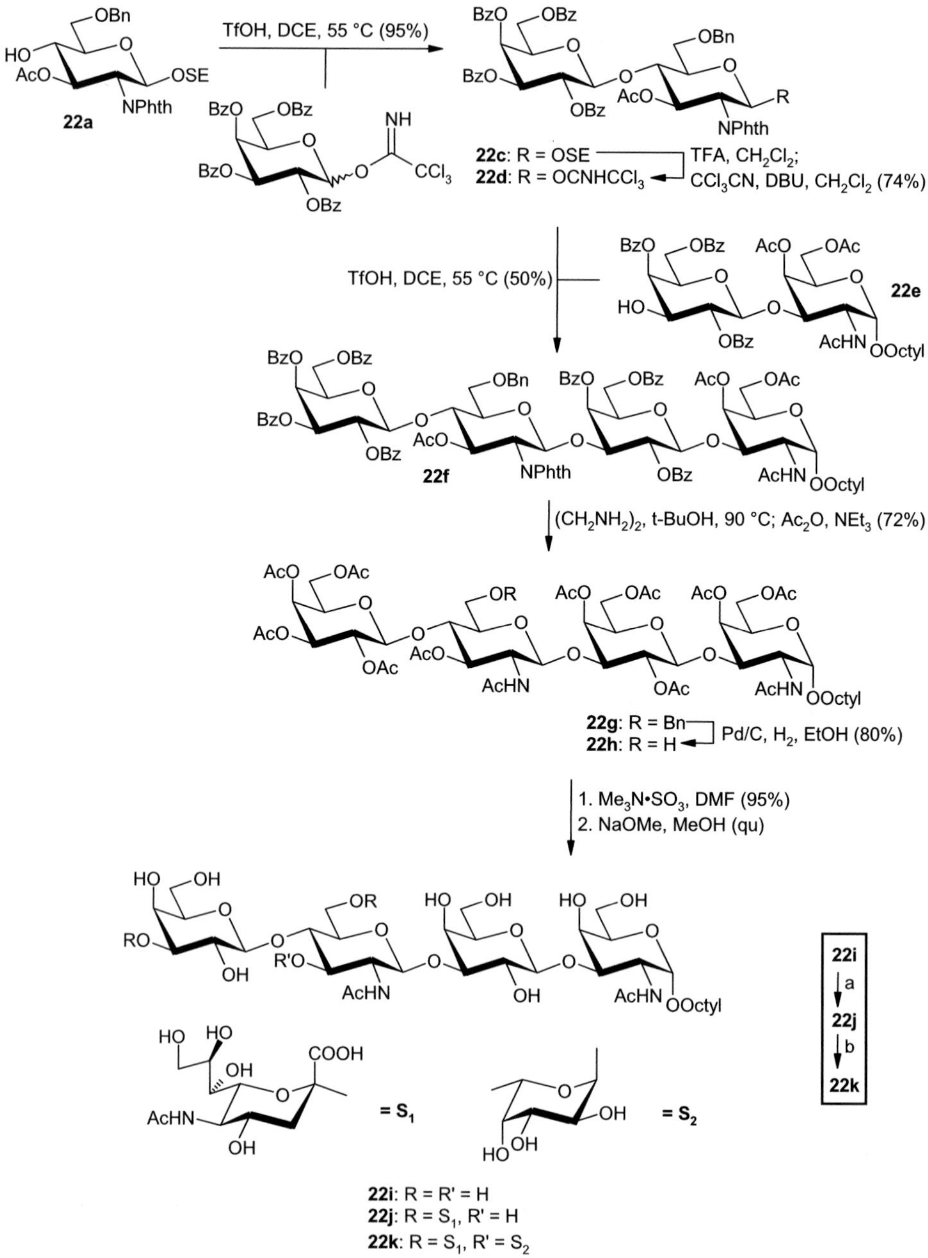

□ Scheme 22

Reagents and conditions: (a) SAT(N), CMP-NANA, MOPS (pH 7.4), Triton S4, BSA, H_2O, CIP, 55%; (b) FucT(V), GDP-Fuc, $MnCl_2$, DTT, Tris-HCl (pH 7.4), qu

from building blocks **20a**, **20b**, and **20f** via intermediates **20c**, **d**, **e** protected Lewis X inter-mediate **20g** was obtained which gave monomer **20h**. For the dimerization, via **20i**, **j**, the target molecule **20k** was obtained. The synthesis of sialyl Lewis X containing glycolipids with different core structures [136] exhibited the importance of the spacer in selectin-binding

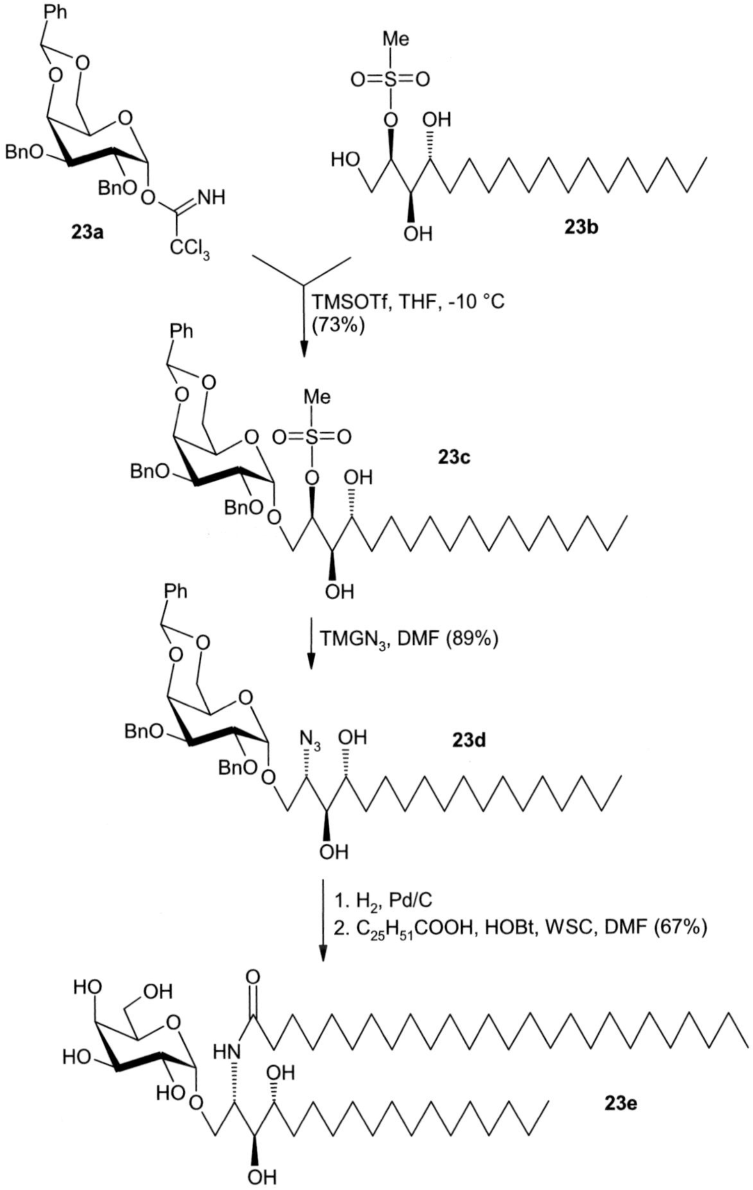

■ Scheme 23

studies [137]. Sulfated sialyl Lewis X variants, containing a lactamized sialyl residue, were obtained by the Kiso-Ishida group (❷ *Scheme 21*) [138]; these compounds were found to be potent antigenic determinants. Following the standard procedure **21a** and **21b** gave **21c**, which was transformed via **21d–h** into lactamized target molecules **21i** and **21j**. Also chemoenzymatic synthesis was successfully engaged in the synthesis of sulfated sialyl Lewis X connected to a core 1 mucin (T antigen) structure (❷ *Scheme 22*) [139]. From **22a, b** disaccharide **22c** was obtained, which was transformed into donor **22d** giving with acceptor **22e** tetrasaccharide **22f**. Partial deprotection (→ **22g, h**), regioselective sulfation and final deprotection gave **22i**; enzymatic sialylation and fucosylation led to target molecules **22j** and **22k**.

Some α-galactosyl ceramides with a phytosphingosine residue, isolated from the marine sponge *Agelas mauritianus*, were found to possess strong in vivo activities against several murine tumor cells [140,141]. The synthesis of analogs exhibited that compound **23e** (❷ *Scheme 23*) is a potent candidate for clinical trials [142]; **23e** was also found to have immunostimulatory activity [143]. These findings promoted a great demand for this compound, therefore, several syntheses of **23e** and analogs have been reported [144]. The Schmidt group reported an efficient synthesis which is based on galactosyl trichloroacetimidate **23a** as donor and phytosphingosine derivative **23b** as acceptor which led to the α-linked intermediate **23c**. Azide introduction (→ **23d**), hydrogenolysis of protecting groups and *N*-acylation furnished target molecule **23e** very efficiently.

Lipid-linked T and T_n antigens [145] and glycosidated phosphoglycerolipids [146] were also synthesized based on *O*-glycosyl trichloroacetimidates donors. In a comparative study glucopyranosylation of methyl ω-hydroxy-hexadecanoate with different glucosyl donors was carried out which demonstrated the advantageous properties of the trichloroacetimidate donor; this way the desired β-linked target molecule for biological studies was readily obtained [147]. 3,4,5-Tris(alkyloxy)benzyl glycosides were prepared with standard *O*-glycosyl trichloroacetimidates as donors [148]. The three lipid chains permitted the immobilization of these compounds on a hydrophobic surface and lectin affinity studies by surface plasmon resonance.

5.2 Glycosyl Amino Acids and Glycopeptides

The availability of *O*- or *N*-glycosyl amino acids as building blocks for glycopeptide or eventually glycoprotein synthesis is of great importance. Therefore, various synthetic methods have been reported [149,150]. Direct *O*-glycosylations of serine and threonine derivatives with *O*-glycosyl trichloroacetimidates as donors have been reported [151,152]. The straightforward hexafluoroacetone *O,N*-protection of Ser, Thr, Pyp, and Tyr gave generally the best results with trichloroacetimidate donors as shown in ❷ *Scheme 24* [153] on glucosylation of **24a, d, g** with per-*O*-acetylated glucopyranosyl donors affording **24b, e, h**. As this protection also leads to an activated carboxylate group, protecting group cleavage and peptide chain extension can be combined furnishing dipeptides **24c, f, i**.

The presence of the Glcα(1–2)Galβ(1-O)Hyl moiety in collagen was reason to synthesize this building block [154]. Replacement of the ω-amino group of Hyl by an azido group and protection of the α-amino group by the Z group and the carboxylate group by a *t*-butyl group led cleanly to the target molecule after two glycosylation steps. Glycosylation of Fmoc-protected Ser, Hse, and Thr with T_N, T, and ST-derived glycosyl donors were also success-

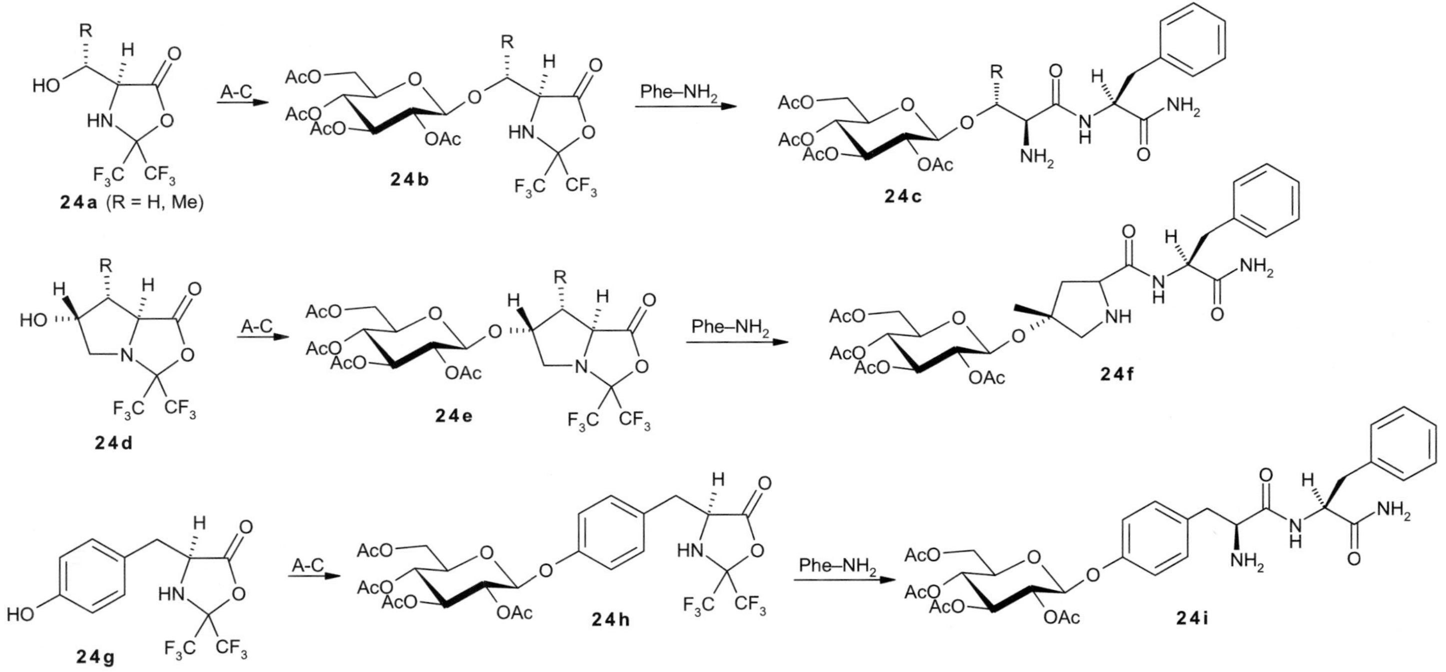

Compound			Method[a]/Yield [%]		
			A	B	C
24a (R = H)	HFA-Ser(Glc)		57	61	78
24a (R = Me)	HFA-Thr(Glc)		86	70	82
24d:	HFA-Hyp(Glc)		21	40	27
24g:	HFA-Tyr(Glc)		-	83	93

[a] Method A: Helferich variant of the Koenigs-Knorr reaction;
method B: Procedure according to Paulsen;
method C: Procedure according to Schmidt

◻ Scheme 24

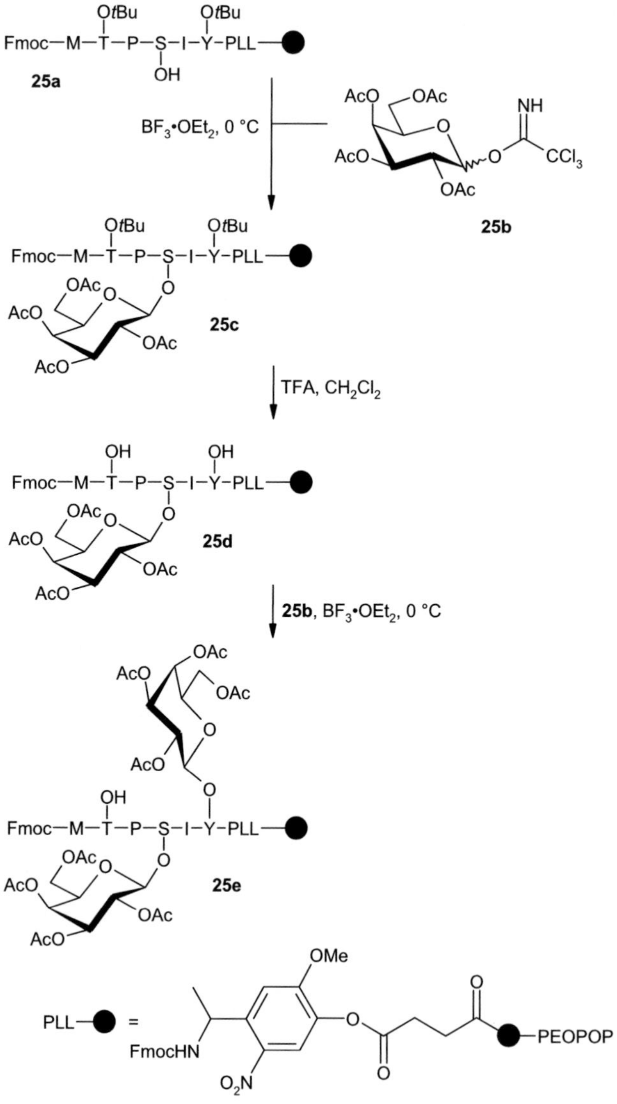

Scheme 25

fully carried out, partly by following known procedures [155,156]. Also an important Galβ (1–4)GlcNAcβ(1–3)-L-Fuc moiety was prepared, which is part of O-linked chains of human clotting factor IX [157].

More demanding is direct glycosylation of peptides, which was so far not very successful because of solubility problems and side reactions with the functional groups [158,159,160]. The Meldal group expected that a solid-phase approach would suffer less from these drawbacks, therefore, they undertook a study with Ser, Thr, and Tyr containing hexapeptides, hav-

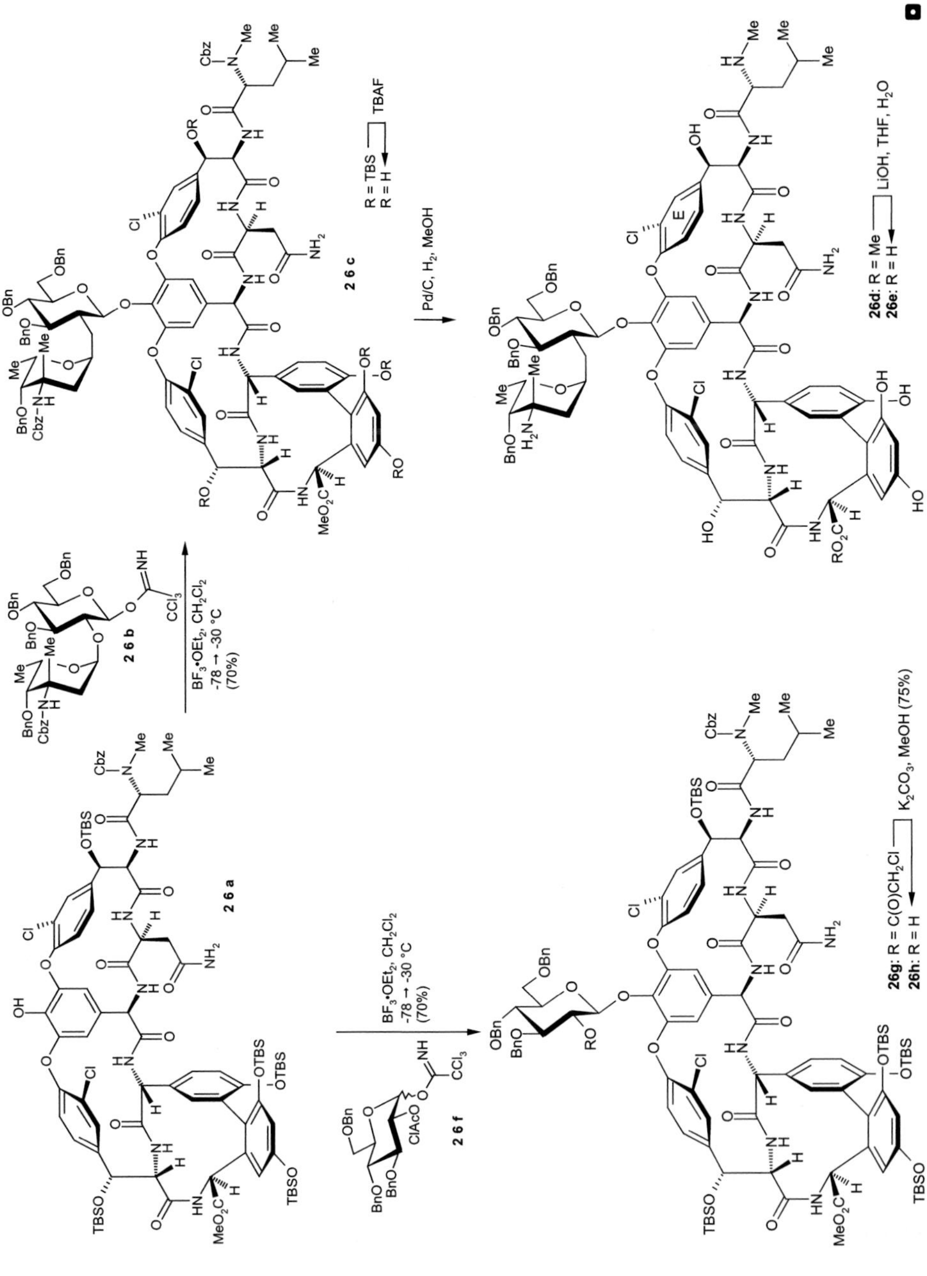

■ Scheme 26

ing no other functional side chains (!), leading to direct reaction in the order Tyr > Ser > Thr, as shown in ❷ *Scheme 25* [161]. The peptide was linked via a photolabile linker (PLL) to the resin (→ **25a**). Galactosylation with **25b** (→ **25c**), then *O*-deprotection (→ **25d**), and galactosylation with **25b** gave **25e**; threonine did not undergo reaction in this case.

A particularly interesting case is the glycosylation of the vancomycin aglycone to achieve the vancomycin total synthesis as successfully investigated by the Nicolaou group [162,163]. They obtained very good results on glycosylation of aglycone **26a** with glycosyl donors **26b** and **26g** affording intermediates **26c** and **26g** (❷ *Scheme 26*), which were further transformed, thus furnishing via **26d** target molecule **26e**. Similarly, vancomycin analogs were prepared by

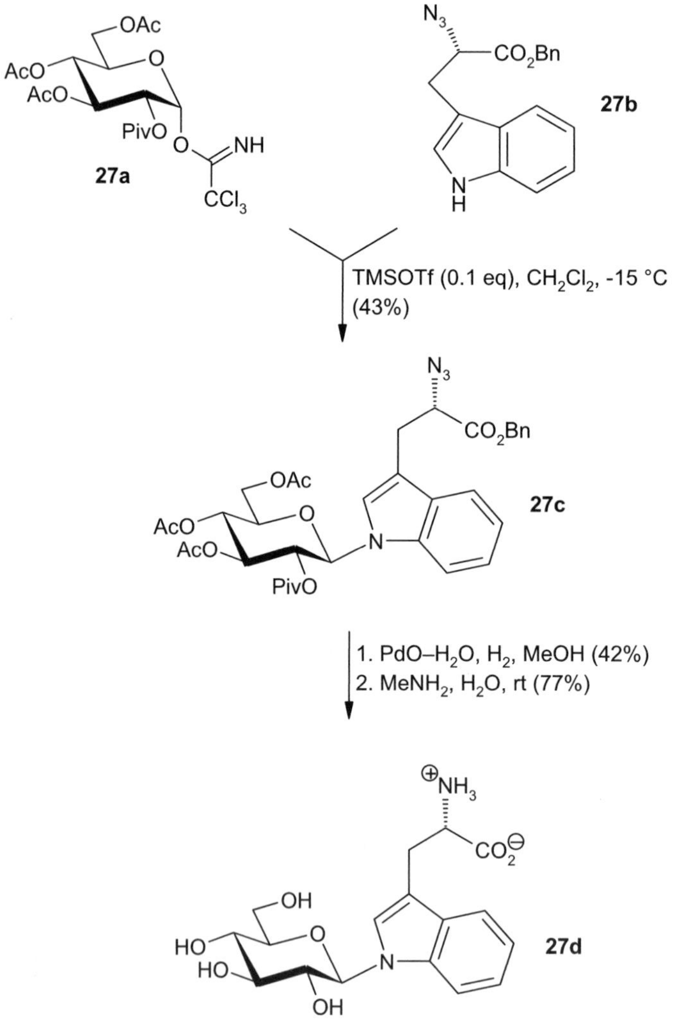

❏ Scheme 27

the Wong group [164]; biological studies exhibited growth inhibition of vancomycin sensitive bacteria by several of these compounds.

N-Glycopeptides are generally obtained by treatment of reducing sugars with ammonium bicarbonate [165] or reduction of glycosyl azides [166] with activated aspartic acid. This way a heptasaccharide synthesized with O-glycosyl trichloroacetimidates as donors was linked to Asp [167,168,169,170]; further work along these lines was carried out [171]. Direct formation of the N-glycosidic linkage with Asn by chemical glycosylation is still an important task. An interesting method was investigated by Ito et al. [172] based on hydroxyamination of the Asp side chain and then glycosylation with a glycosyl donor. The results show, for standard glycosyl donors direct N- glycosylation is possible, however so far yields and anomeric stereoselection are not yet satisfactory.

A number of C- and N-linked tryptophan glycoconjugates were discovered as constituents of natural products [173,174,175,176]. Also tryptophan N-glucoside **27d** has been detected for which Unverzagt et al. [177] reported a successful synthesis (❷ *Scheme 27*). With the help of 2-O-pivaloyl protection in glucosyl donor **27a** the undesired acetal formation could be overcome; thus with acceptor **27b** N-glucoside **27c** could be obtained and then transformed into target molecule **27d**.

5.3 Nucleoside and Nucleotide Glycosidation

The presence of carbohydrate moieties in biomolecules influences many biological functions, thus also modulation of the functions of the aglycone. Therefore, studies have been undertaken to glycosylate nucleosides [178] and nucleotides [179]. Particularly interesting is the direct glycosylation of CPG-bound oligo-deoxynucleotides containing pyrimidine residues leading to reaction preferentially at the 5′-end.

The occurrence at 5-(β-D-glucopyranosyloxymethyl-2′-deoxyuridine (βdJ) in DNA, for instance of *Trypanosoma brucei* [180,181,182], led to the development of an efficient route for the synthesis of βdJ (❷ *Scheme 28*) and its phosphoramidite as building blocks for DNA synthesis [183]. The synthesis was based on glucosyl donor **28a** (R = R′ = Bz) which gave with acceptor **28b** intermediate **28cβ** and after deprotection βdJ. Similarly, from **28a** (R = Bn, R′ = Ac) and **28b** α-linked **28cα** was obtained which led to αdJ.

5.4 Synthesis of Glycosaminoglycans

Glycosaminoglycans (GAGs) are bioactive oligosaccharides that are highly functionalized, linear, and anionically charged. Because of their complexity their chemical synthesis is a demanding task. The chondroitin sulfates (CSs) are a member of the GAG family. They are found in the extracellular matrix of connective tissues at the surface of many cells and in intracellular secretory granules [184,185]. They are linear heteroglycans built mainly from D-GlcA and GalNAc dimers which are β(1–3)-linked; chain extension employs a β(1–4)-linkage. In addition, the sugar residues are sulfated at various positions (→ A- E- and K-type CSs). Stereocontrolled shark cartilage CS (D-type) synthesis has been successfully carried out by Jacquinet et al. (❷ *Scheme 29*) [186,187]. The GlcA building block **29a** possesses as temporary protecting group for chain extension at 4-O the chloroacetyl group, and for sulfation at 2-O the benzoyl

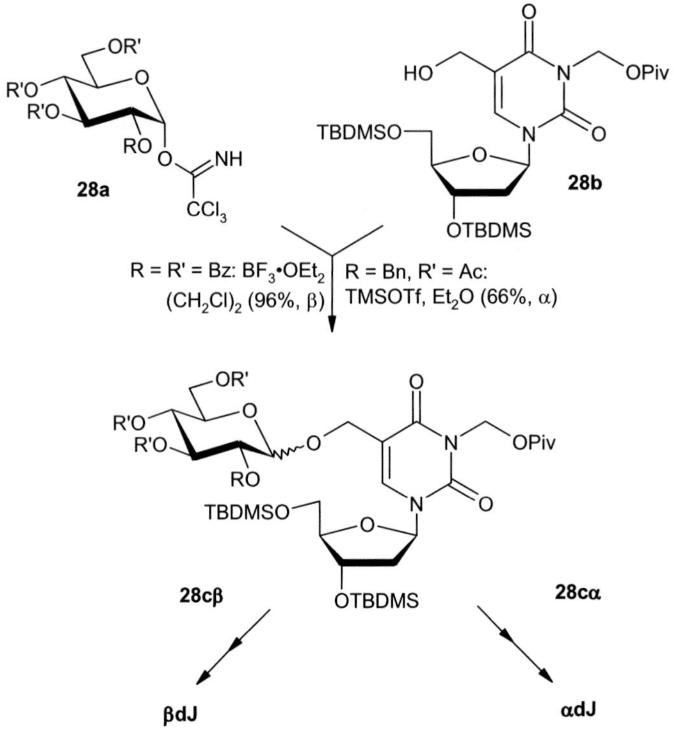

☐ Scheme 28

group. Similarly, the galactosamine residue **29b** carries for chain extension at 1-*O* an MP group and for sulfation at *O*-6 the TBDMS group; the amino group is protected by the trichloroacetyl group. Glycosylation and further transformations furnish donor **29c** which on chain extension with **29d** led to the tetrasaccharide **29e** (n = 1) and higher oligomers which could be deprotected to yield, for instance, target molecule **29f**. On the basis of a similar strategy E-type CS was obtained which was found to stimulate neuronal outgrowth [188]. A hexameric E-type CS was also synthesized by Tamura et al. [189]; in this synthetic approach the amino group of the galactosamine residue was replaced by an azido group.

Hyaluronic acids possess a repeating unit consisting of GlcA β(1–3)-linked to GlcNAc; chain extension of this dimer employs a β(1–4)-linkage. The dimeric unit has been successfully synthesized based on GlcA trichloroacetimidate as the donor [190]. A trimer consisting of two GlcNAc residues and one GlcA residue was also successfully obtained [191]. In this synthesis design the carboxylate function was introduced at a late stage, after construction of the trisaccharide backbone.

Fibroblast growth factors (FGFs) display high binding affinities for GAGs [185,192,193,194,195] such as heparan sulfate (HS) and heparin. Heparin is a linear, heterogeneously sulfated, anionic polysaccharide composed of alternating L-iduronic acid and D-glucosamine residues which is found in almost all animal tissues. The biological roles of several members of the FGFs

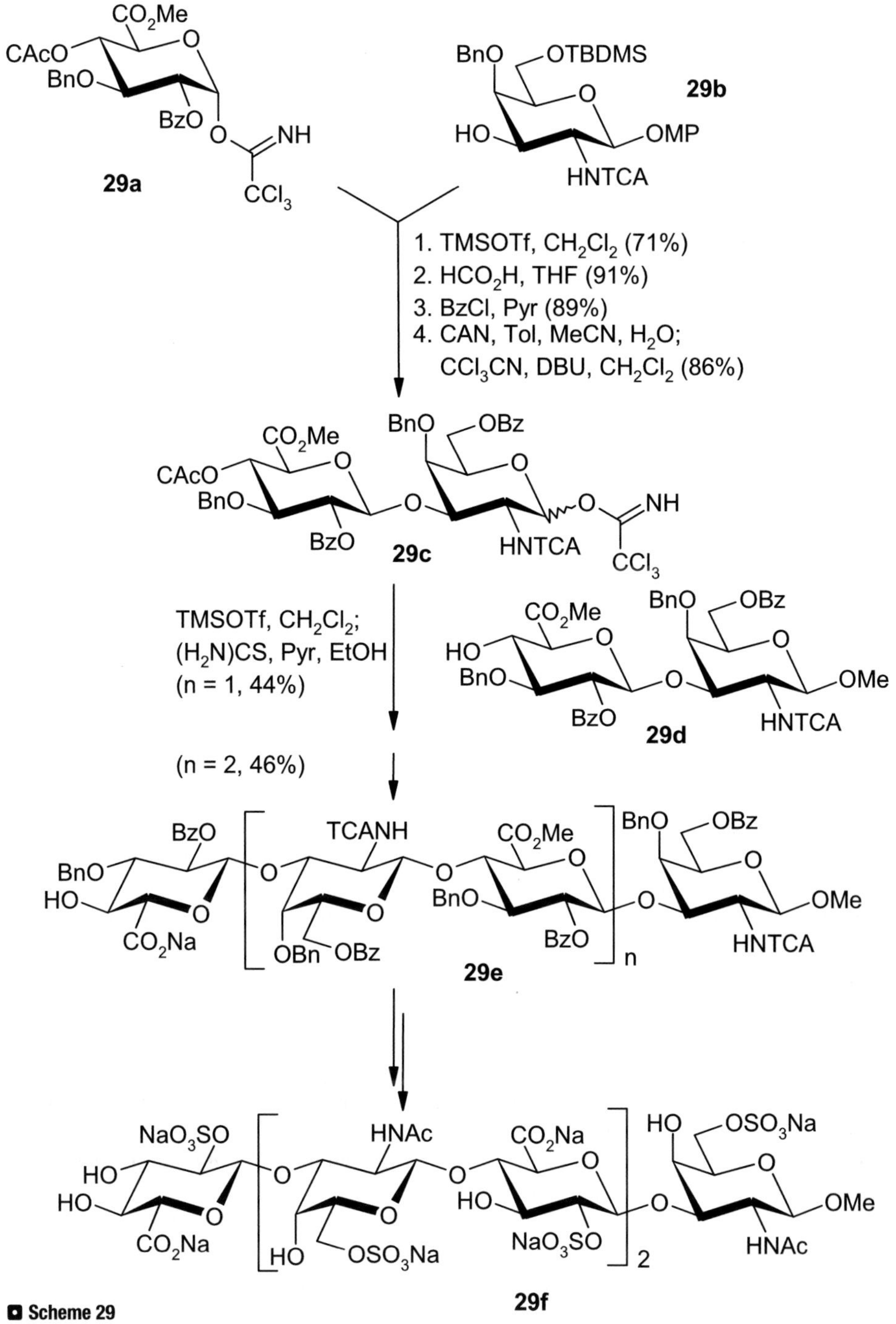

Scheme 29

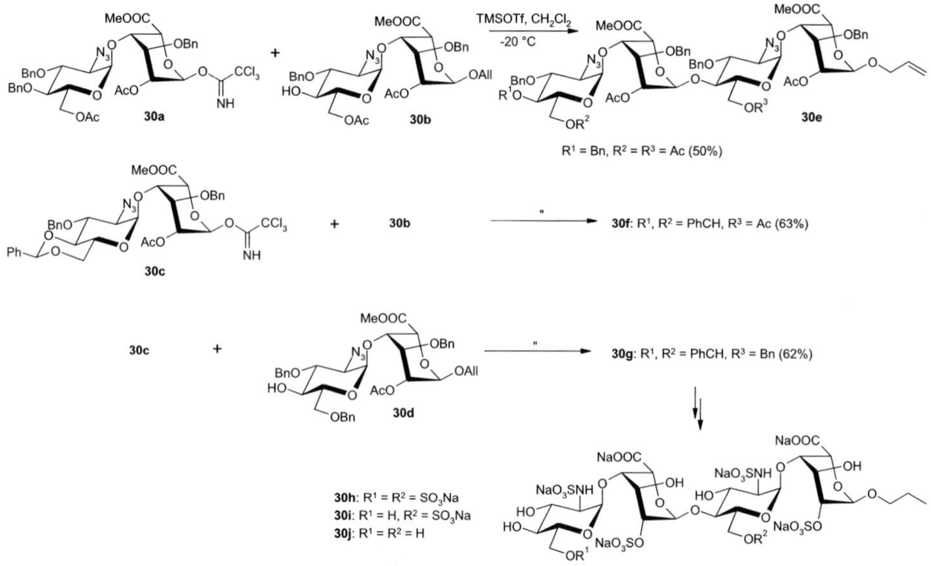

■ Scheme 30

have been extensively investigated [196,197], which also called for synthetic endeavors; they are still ongoing. The synthesis of a highly sulfated tetrasaccharide has been reported by Lay et al. (❷ *Scheme 30*) [198]. It is based on a versatile $\alpha(1\text{--}4)$-linked GalN$_3$-IdoA disaccharide building block from which four differently protected disaccharide acceptors and donors (**30a–d**) are obtained, thus leading to **30e, f, g**. The deprotection scheme is shown for **30g**, leading to target molecules **30h, i, j**. An interesting study was also reported by Bonnaffé et al. [199]; for successful glycosylation a remote *N*-acetyl group had to be replaced by an azido group. Tri-, penta-, hepta-, and nonasaccharides of heparin have been synthesized by Hung et al. (❷ *Scheme 31*) [200]. In this synthetic approach, the carboxylate function is again introduced at a late stage by oxidation of the hydroxymethyl group of the idose moiety. Glycosylation of Ido derivative **31b** with donor **31a** led via further transformations to disaccharide donor **31c** which was chain extended with glucosamine acceptor **31d** to yield **31e** with n = 1. Further chain extension is based on **31c** having naphthylmethyl (NAP) temporary protection; chemoselective NAP cleavage is readily performed with DDQ. Thus via the higher oligomers of **31e** after deprotection target molecules **31f** (n = 1, 2, 3, 4) are obtained. Heparin-like oligosaccharide syntheses were also reported by Martin-Lomas et al. [201], by Seeberger et al. [202], and by Yu et al. [203]. The low reactivity of the axial 4-hydroxy group in the iduronate residue in the Seeberger approach led to trichloroacetimidate rearrangement because the donor and acceptor reactivity did not match.

Proteoglycans are other important glycoconjugates with various roles [204], such as lubrication and blood anticoagulation. Several proteoglycans possess a highly conserved tetrasaccharide linkage region joining a GAG to a core protein. This GlcAβ(1--3)Galβ(1--3)Galβ(1--4)Xylβ(1-O)Me tetrasaccharide was also successfully synthesized based on the *trichloroacetimidate method* [205].

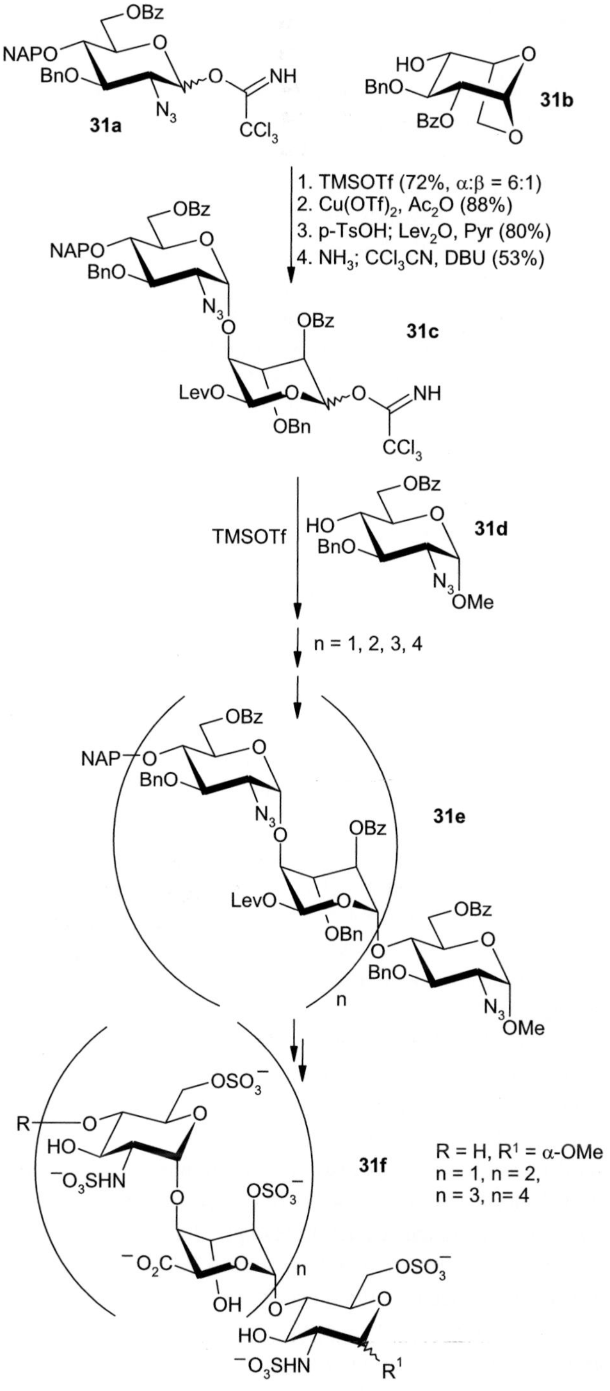

□ Scheme 31

5.5 Cell Wall Constituents

Bacterial cell wall peptidoglycan is known as a strong immunopotentiator which induces various mediators such as cytokines, prostaglandins, platelet activation factor, and NO, thus stimulating the immune system [206]. The receptor for peptidoglycans was shown to be TLR2 [207], the same as for lipoteichoic acids (LTAs) and lipoproteins. Therefore, the synthesis of peptidoglycan fragments is of continuing interest. Recently up to an octasaccharide fragment was obtained as shown in ❷ *Scheme 32* [208,209]. Starting material **32a** was transformed into donor **32b** and acceptor **32c** based on the temporary protection at *O*-1 with allyl and at *O*-4' with benzylidene. Glycosylation afforded tetrasaccharide **32d** which by similar transformation to **32e** and **32f** led to octasaccharide **32g** which could be deprotected to yield target molecule **32h**. This procedure of chain extension could be even extended to the hexadecasaccharide, however the Troc deprotection of this compound failed so far. To the octasaccharide **32h** also small peptides were attached via the lactate residue and the biological activity of the derived compounds was tested.

Many bacterial and fungal cell walls contain homo- and heteroglycans. The synthesis of fragments of these cell wall constituents has gained increasing interest in the last years. The Manα(1–3)Manα(1–2)Manα(1–6)Manα(1–2)Man heptasaccharide obtained by mild acetolysis from *C. glabrata* IFO 0622 has been successfully synthesized by Kong et al. [210]; they synthesized also a related sulfated pentasaccharide [211]. Also mannan repeating units of *Trichophytan mentagrophytes*, *T. rubrum* [212], *Saccharomyces cerevisiae* X 2180–1A [213], and *Candida kefyr* IFO 0586 [214] have been prepared. Another important class of glycans are β(1–3)- and β(1–6)-linked glucans which are frequently branched. Glucans consisting of β(1–3)-linked glucose residues having β(1–6)-linked branches have been found in many plants and fungi [215,216]. Since these glucans show antitumor activity, the synthesis of at least minimal structural units for biological activity studies has gained great interest [217]. Recently, several successful syntheses of such molecules have been reported mainly by Kong et al. [218,219,220,221,222,223]. Also a 2-branched β(1–3)-glucan was found and dimers of the trisaccharide repeating unit were prepared [224]. 3-Branched β(1–6)-linked glucans were found to be phytoalexin elicitors; they possess antitumor activities as well. A highly efficient synthesis of tetradecasaccharide **33t** was reported again by the Kong group [225] (❷ *Scheme 33*). The synthesis is based on readily available glucose-derived building blocks **33a**, **33b**, and **33e** and leads via disaccharides **33c** and **33d**, trisaccharides **33f–33i**, and **33n**, hexasaccharides **33j**, **k**, **o**, **p**, heptasaccharides **33l**, **m**, **q**, **r** to tetradecasaccharide **33s** which was deprotected to furnish target molecule **33t** in good overall yield. Such compounds were also prepared in large scale and with different aglycones [226,227,228].

Mannoglucan from *Microellabosperia grisea* possesses also high antitumor activity, therefore the repeating unit {[Glcα(1–3)][Glcα(1–6)]Glcβ(1–4)Glcβ(1–4)}$_n$ has been synthesized via an efficient route [229]. β(1–6)-Linked galactofuranosyl oligosaccharides are constituents of the cell wall of bacteria and fungi including some clinically significant pathogens [230,231,232]. The highly immunogenic arabinogalactans contain arabinofuranosyl and galactofuranosyl residues. β(1–6)-Linked galactofuranose oligomers, also found in the cell wall of the fungus *Fusarium*, exhibited in plants elicitor activity. Therefore, a β(1–6)-linked *O*-galactofuranosyl hexasaccharide was prepared based on the *trichloroacetimidate method* in high yield [233,234]. The Kong group also synthesized α(1–5)-linked L-arabinofu-

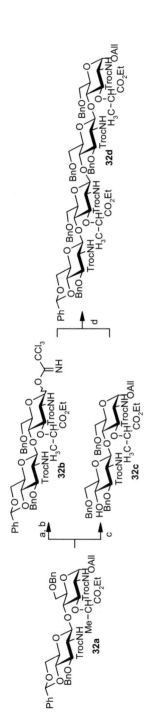

32a

32b

32c

32d

32e

32f

32g

32h

n = 3

R = L-Ala-D-isoGln

(a) Ir complex, H₂, THF; I₂, H₂O
(b) CCl₃CN, Cs₂CO₃; CH₂Cl₂
(c) NMe₃•BH₃, BF₃•OEt₂, MeCN
(d) TMSOTf (0.1 eq), CH₂Cl₂, −15 °C

■ Scheme 32

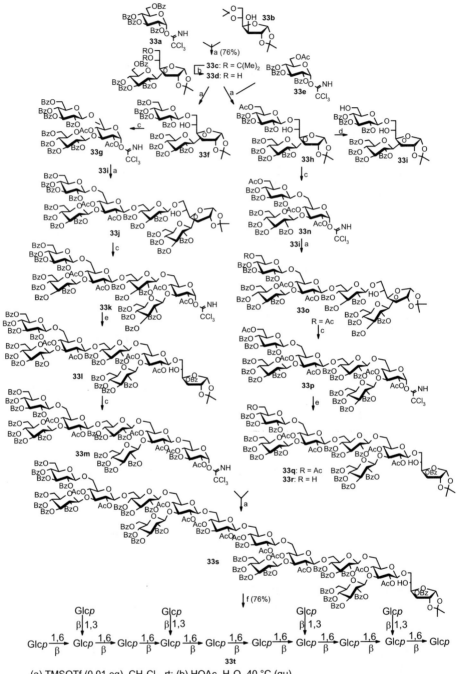

Glc$p$
β|1,3

Glc$p$
β|1,3

Glc$p$
β|1,3

Glc$p$
β|1,3

Glc$p$ $\xrightarrow[β]{1,6}$ Glc$p$ $\xrightarrow[β]{1,6}$ Glc$p$ $\xrightarrow[β]{1,6}$ Glc$p$ $\xrightarrow[β]{1,6}$ Glc$p$ $\xrightarrow[β]{1,6}$ Glc$p$ $\xrightarrow[β]{1,6}$ Glc$p$ $\xrightarrow[β]{1,6}$ Glc$p$ $\xrightarrow[β]{1,6}$ Glc$p$ $\xrightarrow[β]{1,6}$ Glc$p$

33t

(a) TMSOTf (0.01 eq), CH$_2$Cl$_2$, rt; (b) HOAc, H$_2$O, 40 °C (qu)

(c) HOAc, H$_2$O, refl.; Ac$_2$O, Pyr, rt; NH$_3$, THF, MeOH; CCl$_3$CN, K$_2$CO$_3$, CH$_2$Cl$_2$;

(d) HCl, CH$_2$Cl$_2$, MeOH; (e) 3-O-Bz-1,2-O-Ip-glucofuranose, TMSOTf, CH$_2$Cl$_2$, rt;

(f) HOAc, H$_2$O, refl.; NH$_3$, MeOH, CH$_2$Cl$_2$.

☐ **Scheme 33**

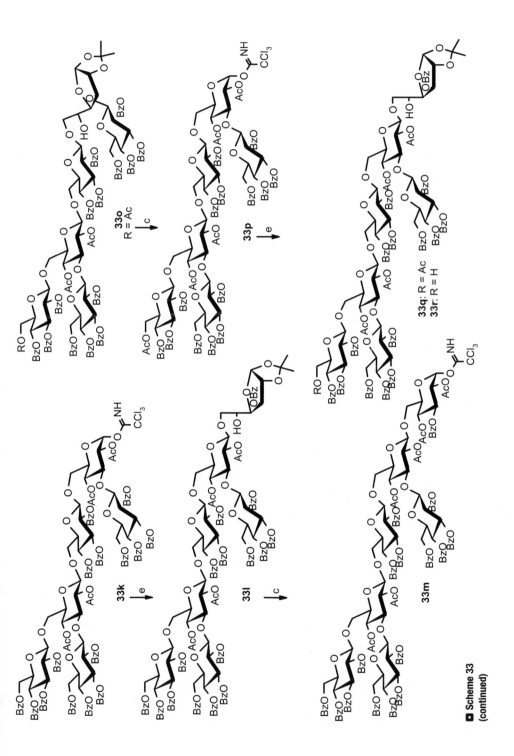

■ Scheme 33
(continued)

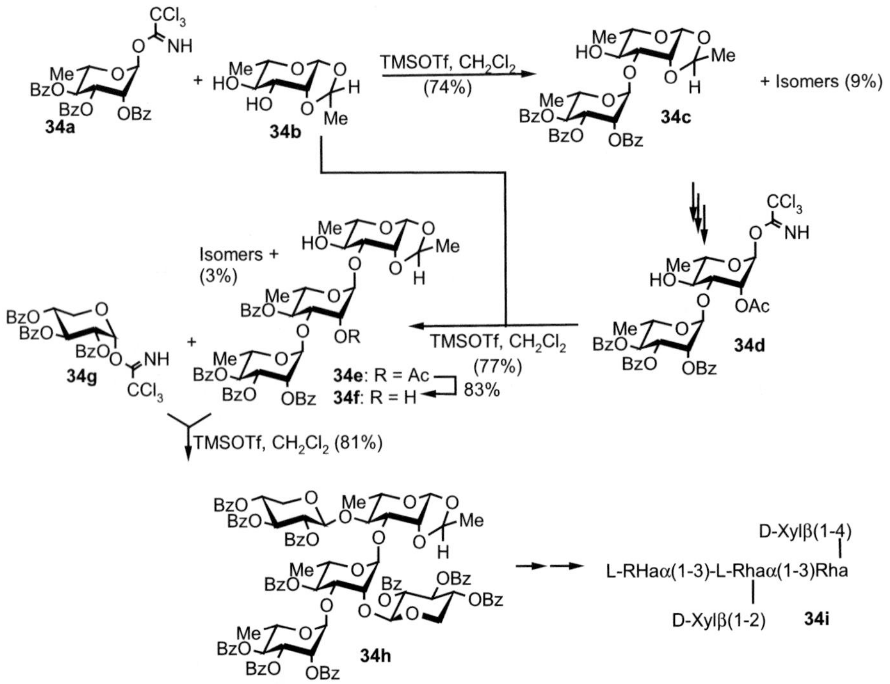

□ Scheme 34

ranosyl oligosaccharides up to the octamer based on related methodologies [235]. 2-*O*-Arabi-nofuranosylated β(1–6)-linked galactopyranosyl oligosaccharides, found in arabiogalactans, could be obtained also by Kong et al. [236] who also reported the successful synthesis of other arabinogalactan linkage types based on the *trichloroacetimidate method* [237,238,239,240]. Lipoarabinomannans have attracted great interest as well, because they are part of the cell surface oligosaccharide of mycobacterial species that cause many diseases, including tuberculosis and leprosy. The largest heteroglycan synthesis for this type of compound was recently reported in which the *trichloroacetimidate method* plays an important role [241].

In algal cell wall polysaccharides β(1–3)-linked xylans were found. Fragments up to the hexamer were successfully synthesized [242]. The specific *O*-chain of the lipopolysaccharide of many bacteria contains repeating units composed of various sugar residues. Several of these repeating units or fragments thereof have been synthesized based on the *trichloroacetimidate method*. Rhamnosylated rhamnan [243], the D-Rha*p*α(1–3)[L-Xyl*p*β(1–2)]D-Rha*p*α(1–3) [L-Xyl*p*β(1–2)]D-Rha unit [244], (❷ *Scheme 34*, **34i**), a more complex xylorhamnan [245], a xylosylated GlcNAc rhamnan [246], a rhamnan with GlcNAc in the branch [247,248], the pentasaccharide D-Glcβ(1–2)-[D-Rib*f*β(1–3)]L-Rha*p*α(1–3)L-Rha*p*α(1–3)L-Rha*p*α(1–2) L-Rha*p* [249], the glucurono xylomannan hexasaccharide repeating unit of *C. neoformans* serotype A (❷ *Scheme 35*) [250], a heptasaccharide fragment of *C. neoformans* serotype C [251], a repeating unit of *lactosillan* [252], the *Shigella flexneri* serotype 2a [253,254,255,256,25 and serotype 5a [258] have been efficiently obtained (❷ *Scheme 36*). The straight forward syn-

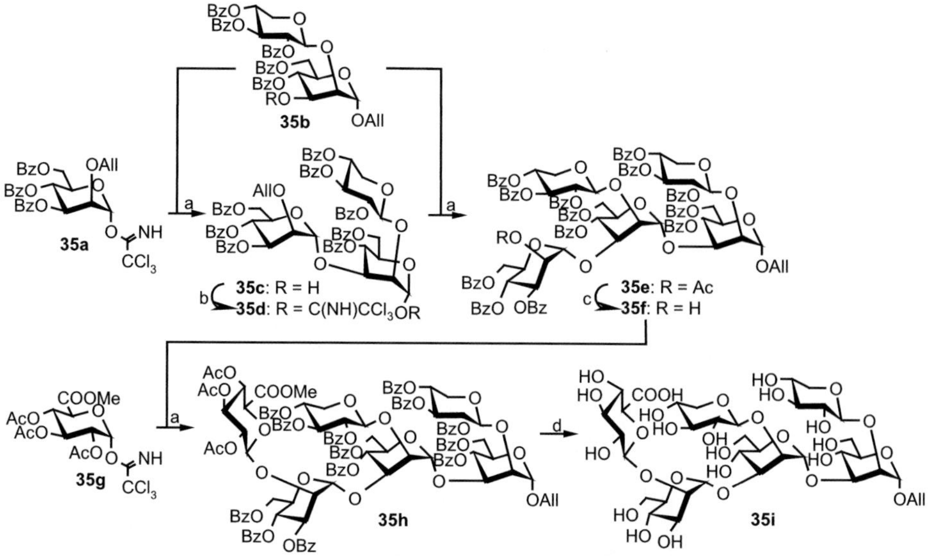

(a) TMSOTf, CH_2Cl_2, -10 °C → rt (dry); (b) $PdCl_2$, 90% acetic acid-NaOAc, rt, 12 h; then Gl_3CN, DBU, CH_2Cl_2, 2 h; (c) 2% CH_3COCl–CH_3OH, 0 °C → rt; (d) sat. NH_3–MeOH, rt, 36 h; then H_2O, rt, 5 h.

☐ Scheme 35

thesis of heptasaccharide **34i** is outlined in ❷ *Scheme 34* [244]. Rhamnosyl donor **34a** readily reacts with rhamnose acceptor **34b** to give mainly disaccharide **34c** which is transformed into donor **34d**. Chain extension with **34b** leads to **34e** and then to acceptor **34f** which on reaction with xylosyl donor **34g** furnishes pentasaccharide **34h**; deprotection leads to target molecule **34i**. The hexasaccharide repeating unit synthesis of *C. neoformans* type A is outlined in ❷ *Scheme 35* [29]. Mannosyl donor **35a** reacts with acceptor **35b** to afford trisaccharide **35c** leading to donor **35d** which on reaction with **35b** gives pentasaccharide **35e** and acceptor **35f**. Glucuronidation with donor **35g** furnished hexasaccharide **35h** which gave on deprotection target molecule **35i**. The *Shigella flexneri* serotype 2a pentasaccharide fragment is outlined in ❷ *Scheme 36* [36]. Rhamnosyl donor **36b** and acceptor **36a** yield disaccharide **36c** which is transformed via **36d** into donor **36e**. On reaction with acceptor **36f** tetrasaccharide **36g** is obtained which is again transformed via **36h** into donor **36i**. Reaction with acceptor **36j** pentasaccharide **36k** is obtained which led to target molecule **36m** having an aminoethyl group at *O*-1 for glycoconjugate synthesis. Linkage to the PADRE sequence, acting as a universal T-cell epitope, was successfully performed and immunogenicity studies were carried out. Also successful investigations towards the synthesis of a tetrasaccharide rhamnogalacturo-nan related to an antiulcer pectic polysaccharide have been reported [259]. The *Neisseria* lipooligosaccharide contains two heptose residues within the conserved core structure; one heptose residue is 3,4-branched. On the basis of the *trichloroacetimidate method* a success-ful synthesis of a branched tetrasaccharide unit was successfully performed by Yamasaki et al. [260,261].

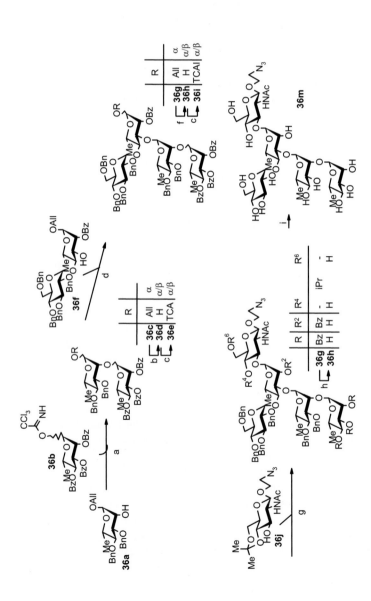

(a) cat. TMSOTf, Et₂O, -70 °C → rt, 8 h; (b) i. cat. [Ir(COD){PCH₃(C₆H₅)₂}₂]⁺PF6⁻, THF, rt, 16 h; ii. HgO, HgBr₂, acetone/water, rt, 1h;
(c) CCl₃CN, DBU, CH₂Cl₂, rt, 2 h; (d) cat. TMSOTf, Et₂O, -60 - -30 °C, 2 h; (f) i. cat. [Ir(COD){PCH₃(C₆H₅)₂}₂]⁺PF6⁻, THF, rt, 16 h,
ii. HgO, HgCl₂, acetone/water, rt, 1h; (g) cat. TMSOTf, 4 Å-MS, CH₂Cl₂, rt, 3 h; (h) i. 50% aq. TFA, CH₂Cl₂, 0 °C, 2 h; ii. cat. MeONa,
MeOH, 55 °C, 2 h; (i) 10% Pd/C, 1 M aq. HCl, EtOH/EtOAc, rt, 2 h.

◻ Scheme 36

5.6 Synthesis of Glycosylphosphatidyl Inositol Anchors

Glycosylphosphatidyl inositol anchors constitute a class of glycolipids that link proteins and glycoproteins via their *C*-terminus to eukaryotic cell membranes. The first structure of a GPI anchor, that of *Trypanosoma brucei*, was published by Ferguson et al. [262]. Since then quite a few examples of GPI anchors were described, allowing the definition of the core structure depicted in ❷ *Scheme 37* [263].

The diversity within GPI anchors is mainly reflected in the location and nature of the branching groups of the glycan residue (R2, R3, R4). Additional ethanolamine phosphates (R1) seem to be specific for higher eukaryotes [264]. Concerning the lipid residue, many of the structures of GPI anchors contain a diacylglycerol moiety but alkylacylglycerol residues are not uncommon and ceramide structures have also been identified [263]. These modifications of the evolutionary conserved structure give rise to species-, stage-, and tissue-specific GPI structures.

The function of GPI anchors has been extensively discussed. A controversial aspect of GPI anchors is their ability to mediate signaling mechanisms or to function as second messengers, e. g. in insulin-mediated signal transduction processes [265]. Therefore, to perform biological studies elucidating the functions of GPI anchors, it seems to be an important objective to have access to structurally homogeneous GPI anchors and their derivatives. For the total synthesis of GPI anchors, a combination of lipid, phosphate, and oligosaccharide chemistry is required. A highly versatile strategy has been successfully followed for a ceramide-containing GPI anchor of yeast [266,267]. Similarly obtained were the acylglycerol-containing GPI anchors of *Trypanosoma brucei* [268,269] and *rat brain* Thy-1 [270].

On the basis of earlier work [266,267], the development of a highly variable synthetic strategy, which is also applicable to the preparation of branched GPI anchors, was reported by Schmidt et al. [271,272]. This strategy allows also for the attachment of peptide or protein residues to the GPI anchor. The focus was on 4,6-branched mannose residues as there are sev-

Natural Source	R⁴	R³	R²	R¹	Lipid
S. cerevisiae	Manα(1-2)	H	H	H	Ceramide/DAG
T. brucei VSG	H	H	Gal₂₋₄α(1-3)	H	DAG
T. gondii A	H	GalNACβ(1-4)	H	H	DAG
B	H	Glcα(1-4)GalNAcβ(1-4)	H	H	DAG
Rat brain *Thy-1*	Manα(1-2)	GalNAcβ(1-4)	H	EA-P	Acylalkylglycerol

◻ **Scheme 37**
General structure of GPI anchors (EA, ethanolamine; P, phosphate; DAG, diacylglycerol)

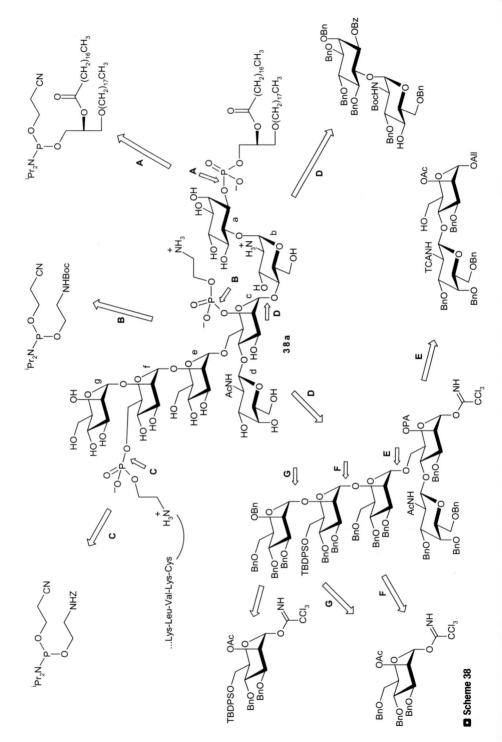

Scheme 38

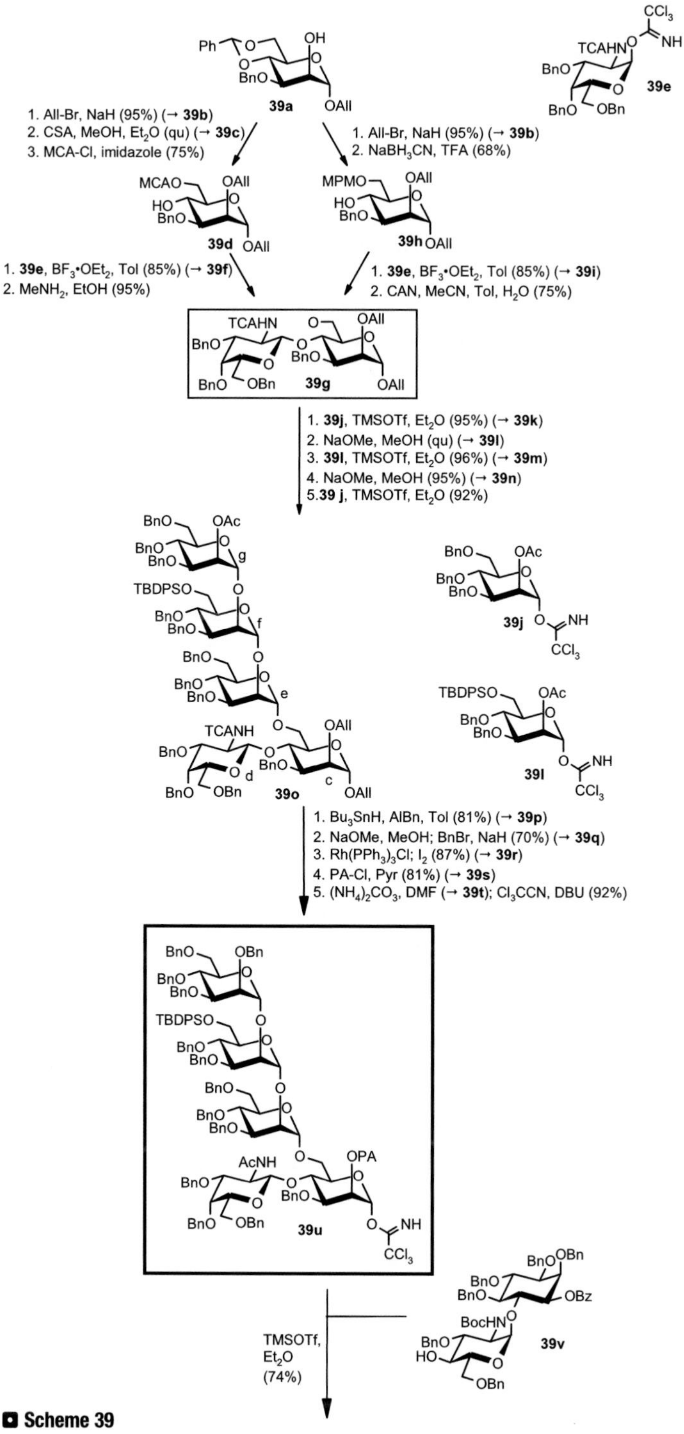

1. All-Br, NaH (95%) (→ **39b**)
2. CSA, MeOH, Et₂O (qu) (→ **39c**)
3. MCA-Cl, imidazole (75%)

1. All-Br, NaH (95%) (→ **39b**)
2. NaBH₃CN, TFA (68%)

1. **39e**, BF₃·OEt₂, Tol (85%) (→ **39f**)
2. MeNH₂, EtOH (95%)

1. **39e**, BF₃·OEt₂, Tol (85%) (→ **39i**)
2. CAN, MeCN, Tol, H₂O (75%)

1. **39j**, TMSOTf, Et₂O (95%) (→ **39k**)
2. NaOMe, MeOH (qu) (→ **39l**)
3. **39l**, TMSOTf, Et₂O (96%) (→ **39m**)
4. NaOMe, MeOH (95%) (→ **39n**)
5. **39 j**, TMSOTf, Et₂O (92%)

1. Bu₃SnH, AlBn, Tol (81%) (→ **39p**)
2. NaOMe, MeOH; BnBr, NaH (70%) (→ **39q**)
3. Rh(PPh₃)₃Cl; I₂ (87%) (→ **39r**)
4. PA-Cl, Pyr (81%) (→ **39s**)
5. (NH₄)₂CO₃, DMF (→ **39t**); Cl₃CCN, DBU (92%)

TMSOTf,
Et₂O
(74%)

□ **Scheme 39**

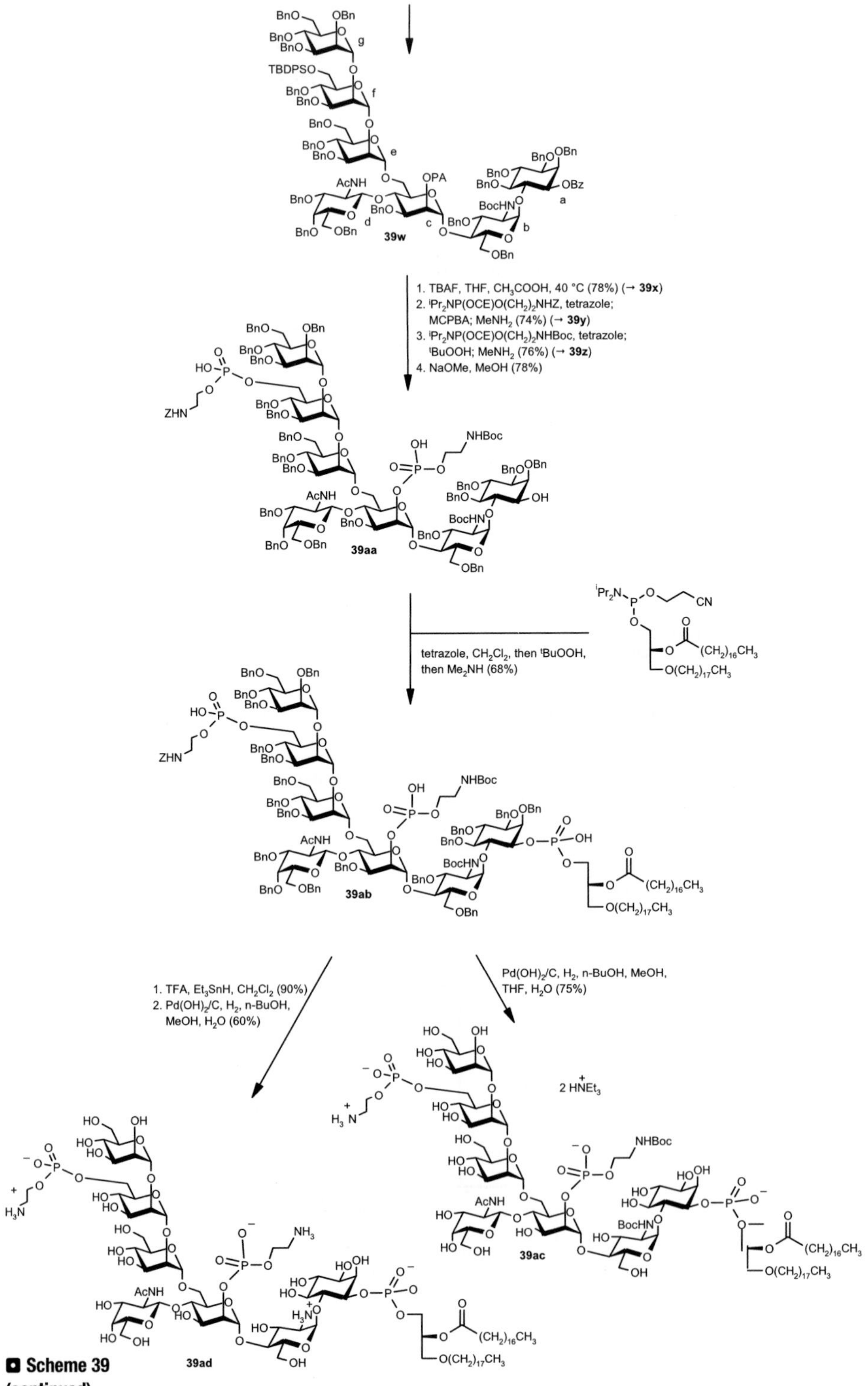
39w
1. TBAF, THF, CH3COOH, 40 °C (78%) (→ 39x)
2. iPr2NP(OCE)O(CH2)2NHZ, tetrazole; MCPBA; MeNH2 (74%) (→ 39y)
3. iPr2NP(OCE)O(CH2)2NHBoc, tetrazole; tBuOOH; MeNH2 (76%) (→ 39z)
4. NaOMe, MeOH (78%)
39aa
tetrazole, CH2Cl2, then tBuOOH, then Me2NH (68%)
39ab
1. TFA, Et3SnH, CH2Cl2 (90%)
2. Pd(OH)2/C, H2, n-BuOH, MeOH, H2O (60%)
Pd(OH)2/C, H2, n-BuOH, MeOH, THF, H2O (75%)
2 HNEt3
39ac
39ad

Scheme 39 (continued)

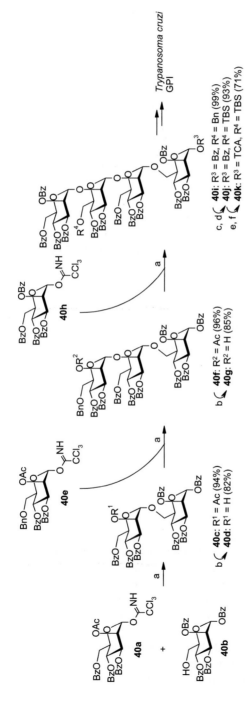

Reagents: (a) TMSOSO$_2$CF$_3$, CH$_2$Cl$_2$; (b) 2% HCl, MeOH/CH$_2$Cl$_2$; (c) H$_2$, Pd(OH)$_2$/C, THF; (d) TBSOSO$_2$CF$_3$, Et$_3$N, CH$_2$Cl$_2$; (e) NH$_2$CH$_2$CH$_2$NH$_2$ • HOAc, THF; (f) CCl$_3$CN, Cs$_2$CO$_3$, CH$_2$Cl$_2$, Bn = benzyl, THF = tetrahydrofuran, TMS = trimethylsilyl.

□ Scheme 40

eral prominent examples in nature (❯ *Scheme 37*). Therefore, the fully phosphorylated GPI anchor of *Toxoplasma gondii* was prepared [271]. Also a new route to rat-brain Thy-1 [272] was developed which is outlined in ❯ *Scheme 38*, showing important disconnections and building blocks; it is totally based on trichloroacetimidate donors. Two efficient routes starting from mannosyl acceptor **39a** and glucosamine-derived donor **39e** leading to disaccharide **39g** were developed (❯ *Scheme 39*). With variously protected mannosyl donors **39j** and **39l**, chain extension to pentasaccharide **39o** could be efficiently performed. Transformation into donor **39n** and reaction with disaccharide acceptor **39v** led to pseudoheptasaccharide **39w**. Introduction of two orthogonally protected aminoethyl phosphate residues led to **39aa** to which the phospholipid was attached (→ **39ab**). Debenzylation afforded *N*-Boc-protected **39ac** which permits regioselective attachment of peptide residues to one aminoethylphosphate moiety and complete deprotection led to target molecule **39ad**. This way a highly variable concept for the synthesis of branched GPI anchors could be established. It is based on versatile building blocks which are readily accessible and provide the desired regio- and stereocontrol. The concept further allows for the regioselective attachment of peptides or proteins.

The synthesis of a partial structure of a branched pseudohexasaccharide of an inositolphosphoglycan (IPG) has been reported by Martin-Lomas et al. [273]. Linear oligosaccharides of GPIs were also synthesized based on the *trichloroacetimidate method* [274,275]. The GPI anchor of *Trypanosoma cruzi* containing an unsaturated fatty acid was obtained without any *O*-benzyl protection in the decisive intermediate (❯ *Scheme 40*) [276]. Acyl-protected donor **40a** and acyl-protected acceptor **40b** gave disaccharide **40c** which was transformed into acceptor **40d** which on reaction with donor **40e** gave trisaccharide **40f** and then acceptor **40g**. Glycosylation with **40h** led to tetrasaccharide **40i** which was transformed via **40j** into donor **40k**. From this compound the desired target molecule was prepared.

5.7 Glycosylation of Various Natural Products and Their Metabolites

Many antibiotics are glycosylated, however the structural complexity of many antibiotics and their glycosylation is a demanding task. On the basis of the *trichloroacetimidate method* a successful synthesis of landamycin A hexasaccharide was reported [277]. Novobiocin was glycosylated with three different *O*-glycosyl trichloroacetimidates as donors at the 7-hydroxy group [278]. Also the quite sensitive (+)-Neo-carzinostatin aglycone **41a** (❯ *Scheme 41*) could be successfully glycosylated [279]. Reaction of **41a** with glycosyl donor **41b** furnished the desired glycoside **41c** in good yield which on deprotection afforded target molecule **41d**. Hydroquinone glycosylation leading to α-arbutin was also performed [280]; this compound is an inhibitor of human tyrosinase and therefore of interest in the cosmetic industry.

Aminoglycoside antibiotics are of special interest as glycosylation targets. Interesting neomycin trisaccharide mimetics have been prepared by Boons et al. [281]. Glycosidase inhibitors have also been combined with additional carbohydrate residues in order to fine tune their biological properties. Thus, deoxynojirimycin has been combined with β(1–3)- and β(1–6)-linked glucan residues [282].

Many metabolites are glucuronidated, therefore the introduction of the glucuronic acid residue is a major task because of the low reactivity of the derived donors and their tendency to 4,5-elimination. Therefore, often compounds are first glycosidated and at a later stage of the

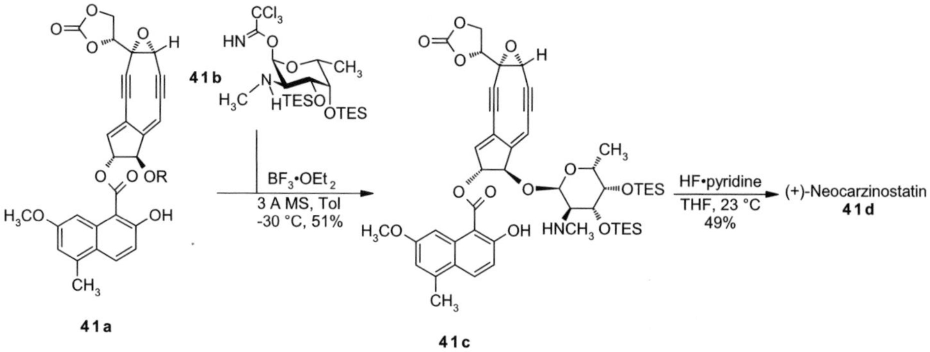

■ Scheme 41

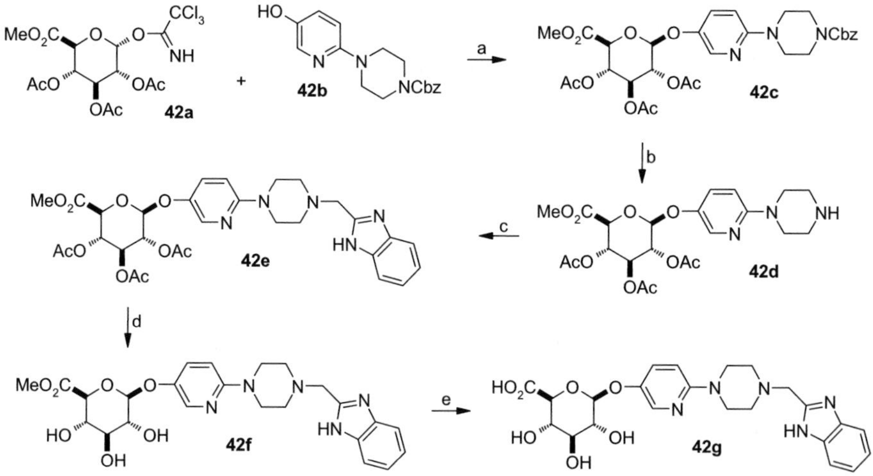

(a) 2.0 eq **42a**, 1.0 eq **42b**, BF$_3$•OEt$_2$, CH$_2$Cl$_2$, -60 °C → rt, 41 h, 75%; (b) 5% Pd/C, H$_2$, MeOH, rt, 83%;
(c) (1) benzimidazolylmethyl chloride, NEt$_3$, NMP, 0 °C → rt, (2) EtOAc, rt, 62%; (d) 0.1 eq LiOH, MeOH, -10 °C → rt;
(e) 1.9 eq LiOH, MeOH, rt; 2.0 eq AcOH, H$_2$O, CHCl$_3$ wash; Dowex hydroxide resin, filtration, wash, then AcOH$_{aq}$, 87%.

■ Scheme 42

synthesis the hydroxymethyl group is selectively oxidized to the carboxylate group. Excellent results in direct glucuronidation based on *O*-isobutyryl-protected trichloroacetimidate have been obtained by Scheinmann and Stachulski et al., who glucuronidated for instance an anti-estrogenic steroid [283,284] and morphine [285]. Also MS 209, a quinoline derivative exhibiting multidrug resistance in cancer therapy was directly glucuronidated based on an *O*-pivaloyl-protected donor [286]. 5-Hydroxypyridine derivative **42b** (❷ *Scheme 42*) which is part of ABT-724, a potent D4 dopamine receptor agonist, could be directly glucuronidated [287] leading to **42c**. Further transformations via intermediates **42e** and **42f** gave the desired glycosylated metabolite **42g**.

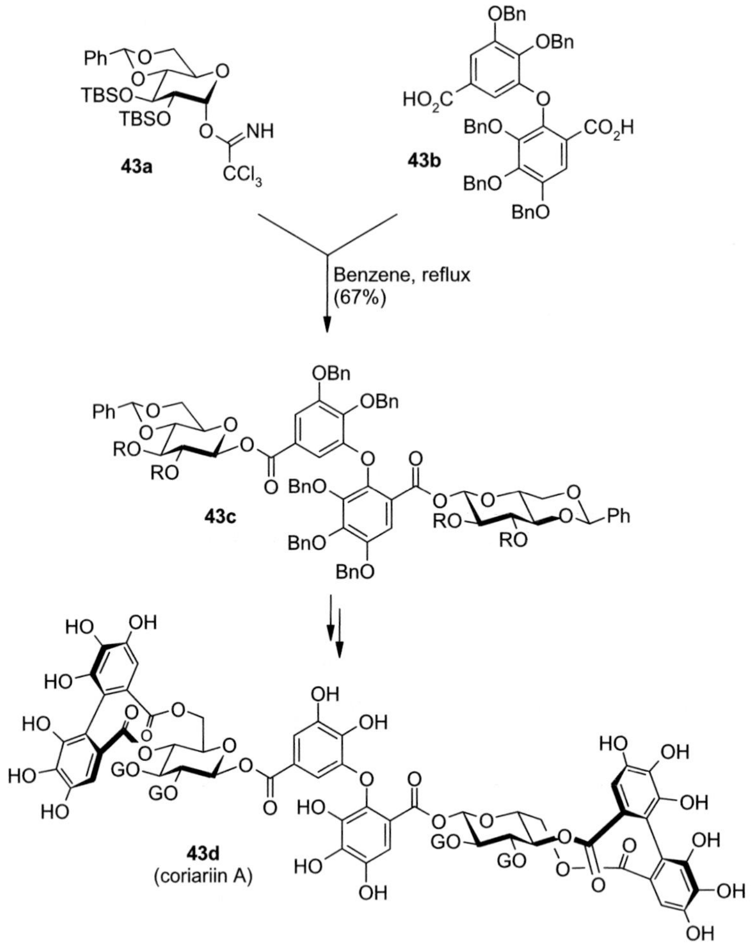

☐ **Scheme 43**

Dihydroxyphenyl glycosides are a family of plant components. Some representatives have also been synthesized based on the *trichloroacetimidate method*, such as for instance the conandroside [288]. Other plant metabolites, such as for instance the ellagitannins, contain benzoyl groups at the anomeric oxygen. The derived coriariin A (**43d**, ☐ *Scheme 43*) was obtained as discussed above just by heating **43b** with glycosyl donor **43a**, leading to **43c**, which was transformed into target molecule **43d** [289]. Macrophylloside D [290] and buprestin A [291] were similarly prepared.

N-Glycosylation of nitrogen-containing heterocycles has also been investigated. An interesting synthesis of *N*-indigoglycosides was undertaken which, as analogs of akashines, exhibit activity against various human tumor cell lines [292].

Unique glycolipids produced by plants are the resin glycosides for which the first synthesis was based on the *trichloroacetimidate method* [293,294]. Other members of this family have been

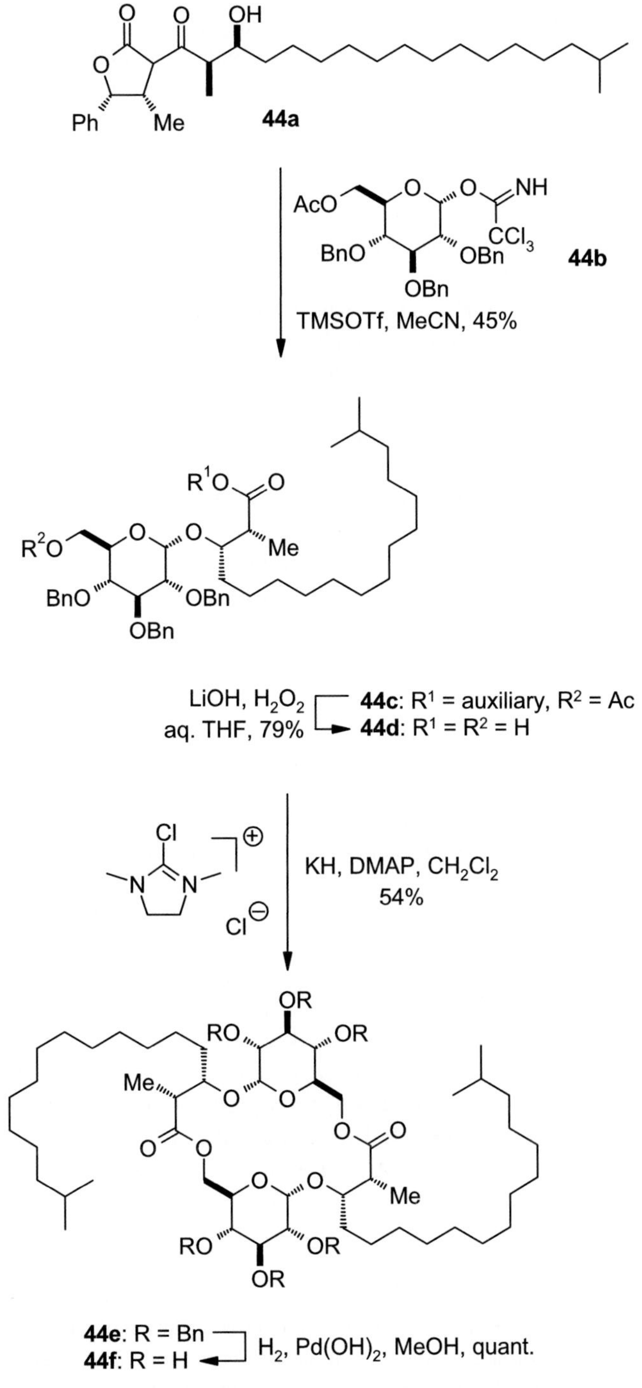

44a

44b

TMSOTf, MeCN, 45%

LiOH, H₂O₂ — 44c: R¹ = auxiliary, R² = Ac
aq. THF, 79% — 44d: R¹ = R² = H

KH, DMAP, CH₂Cl₂
54%

44e: R = Bn
44f: R = H — H₂, Pd(OH)₂, MeOH, quant.

◻ Scheme 44

prepared with the help of closely related trichloroacetimidate building blocks and a similar strategy, such as for instance for woodrosin I [295] and tricolorin F [296]. Structurally related, highly bioactive antiviral compounds, such as for instance cycloviracin B_1 or glucolipsin A were isolated and successfully synthesized [297,298,299]. The approach to the synthesis of glucolipsin A [299] is exhibited in ❷ *Scheme 44*. Glucosylation of fatty acid derivative **44a** with *O*-glycosyl trichloroacetimidate **44b** gave the desired glycoside **44c** in good yield. Ester hydrolysis ($\rightarrow$ **44d**) and dimerization to **44e** and then hydrogenolysis of the *O*-benzyl groups led to target molecule **44f** which had physical data in accordance with the natural product (see [299]).

Saponins are steroid- or triterpenoid-based glycolipids which are found in terrestrial and marine plants. They possess various biological activities [300]. Several papers reported on successful glycosylations at the 3-hydroxy group [301,302,303,304,305,306]. When this hydroxy group is sterically hindered due to disubstitution at C-4, this glycosylation is a difficult task. However, with *O*-glycosyl trichloroacetimidate donors good yields and anomeric selectivities have also been obtained for this glycosylation step [307,308,309,310,311,312]. Particularly worth mentioning is the total synthesis of QS 21 A published by Gin et al. [25] where the decisive step is carried out with a trisaccharide donor.

Most carbohydrates contain ethane-1,2-diol fragments, therefore they were also successfully included into crown ether macrocycles [313]. Polyvalency of carbohydrates was studied based on dendrimeric structures which were generated on successful glycosylations of highly branched polyethylene glycol units [314].

5.8 Cyclooligosaccharides

Cyclodextrins (CDs) have gained a lot of interest because they provide useful cavities for the generation of inclusion complexes. Therefore, branched CDs and CDs with various carbohydrate residues have been synthetic targets. The *trichloroacetimidate method* was successfully employed to reach such goals. Mannosyl, galactosyl, lactosyl, and the Galβ(1–4')-lactosyl residues were selectively introduced at *O*-6 of β- and γ-CD, respectively [315,316]. Also building blocks for oligomannoside attachment to CDs were prepared [317]. Most remarkable is the synthesis of a *O*-6 branched cyclo(1–3)-glucohexaose and octaose [318,319]. The ring closure of the prepared nona- and dodecasaccharide worked extremely well based on *O*-glycosyl trichloroacetimidate donors (❷ *Scheme 45*) [5]. Dodecasaccharide **45a** was transformed via **45b** into trichloroacetimidate **45c** having a 3-hydroxy group at the nonreducing end. Acid-catalyzed activation led to CD derivative **45d** which gave after deacylation target molecule **45e**.

5.9 *C*-Glycoside Synthesis

The synthesis of "*C*-glycosides" has been of great interest [320]. *C*-Glycosylation of electron-rich compounds with *O*-glycosyl trichloroacetimidates continues to be a method of choice. This work is not discussed here in detail. A few references give an entry to the work in this field [321,322,323,324].

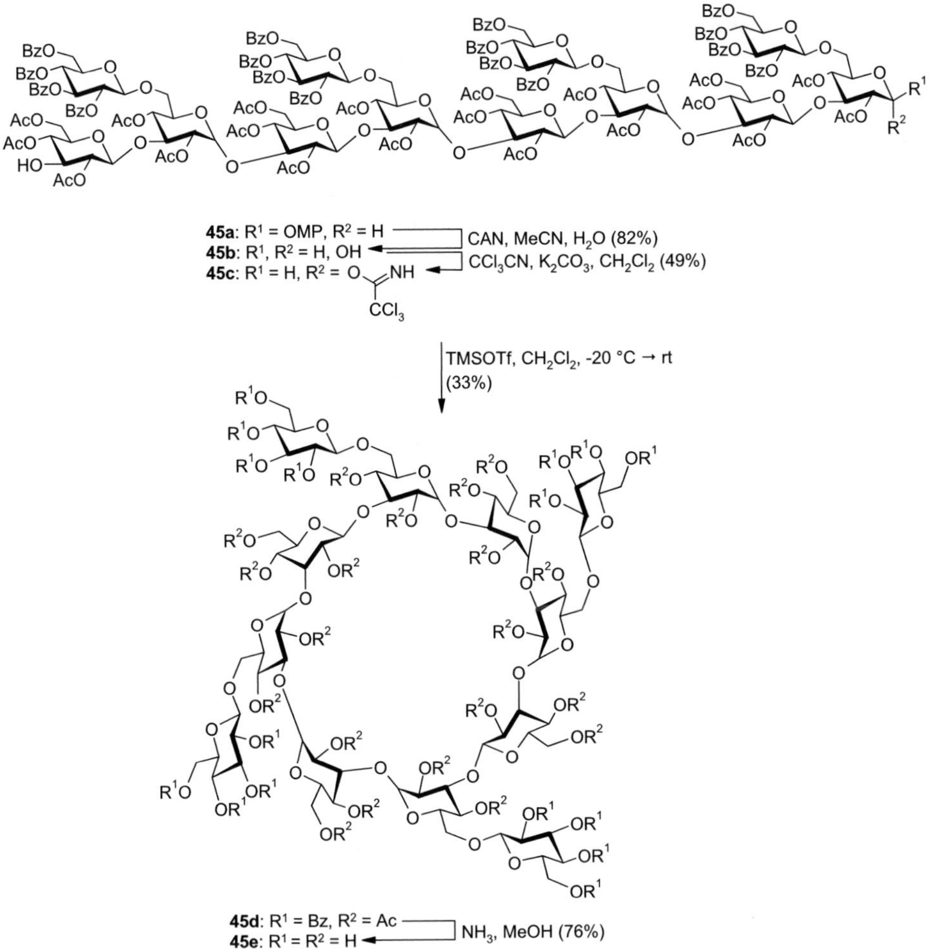

45a: R^1 = OMP, R^2 = H — CAN, MeCN, H_2O (82%)
45b: R^1, R^2 = H, OH ◄— CCl$_3$CN, K_2CO_3, CH_2Cl_2 (49%)
45c: R^1 = H, R^2 = O—NH / CCl$_3$ ◄

TMSOTf, CH_2Cl_2, -20 °C → rt
(33%)

45d: R^1 = Bz, R^2 = Ac — NH$_3$, MeOH (76%)
45e: R^1 = R^2 = H ◄

☐ **Scheme 45**

5.10 O-Glycosyl Trichloroacetimidates of N,O- and S,O-Halfacetals

Not only O,O-halfacetals but also N,O-, and S,O-halfacetals have successfully been transformed into trichloroacetimidate derivatives [325,326]. Thus carbohydrates with sulfur in the ring have been successfully transformed into O-glycosyl trichloroacetimidates and employed in glycosylation reactions by M. Hashimoto et al. [327,328] and by H. Hashimoto et al. [329,330]. Thy glycosyl donors could be readily obtained by base-catalyzed trichloroacetonitrile addition to the anomeric hydroxy group and glycosylation yields under standard conditions were generally very good.

6 Related Activation Systems

O-Glycosyl trichloroacetimidates have become popular because they are easily available, powerful glycosyl donors. Therefore, investigations were undertaken to provide glycosyl donors which follow the principle of *O*-glycosyl trichloroacetimidate formation and activation. A short outline of this work is attached here.

Already previously trifluoroacetonitrile has been investigated for the generation of glycosyl donors [12,331]. However, because trifluoroacetonitrile is a gas, its use is not as convenient as the use of trichloroacetonitrile. In addition, preliminary glycosylation results were inferior to those obtained with *O*-glycosyl trichloroacetimidates. Much more promising were studies with dichloromalonitrile which is a suitable reagent for the base-catalyzed generation of *O*-glycosyl dichloro-cyanoacetimidates [332,333]. These compounds exhibited glycosyl donor properties closely related to those of *O*-glycosyl trichloroacetimidates (❷ *Scheme 46*). A further important class of compounds is ketene imines, which should readily lead to *O*-glycosyl imidates. However, so far only a few examples were investigated [12,13,330], therefore, the potential of these compounds has not been elucidated yet.

Another interesting class of compounds are imide halides having electron-withdrawing carbon substituents and their heterocyclic equivalents. After some earlier work [12,334,335,336,337, 338], recently imide halides have gained increased interest and excellent glycosylation results have been reported [45,339,340]. However, these systems have the disadvantage of furnishing equimolar amounts of salt on glycosyl donor generation. Additionally, glycosyl donor generation is not reversible, therefore α/β-stereocontrol is more difficult or even impossible. Hence, matching or even surpassing the properties of *O*-glycosyl trichloroacetimidates remains a demanding task.

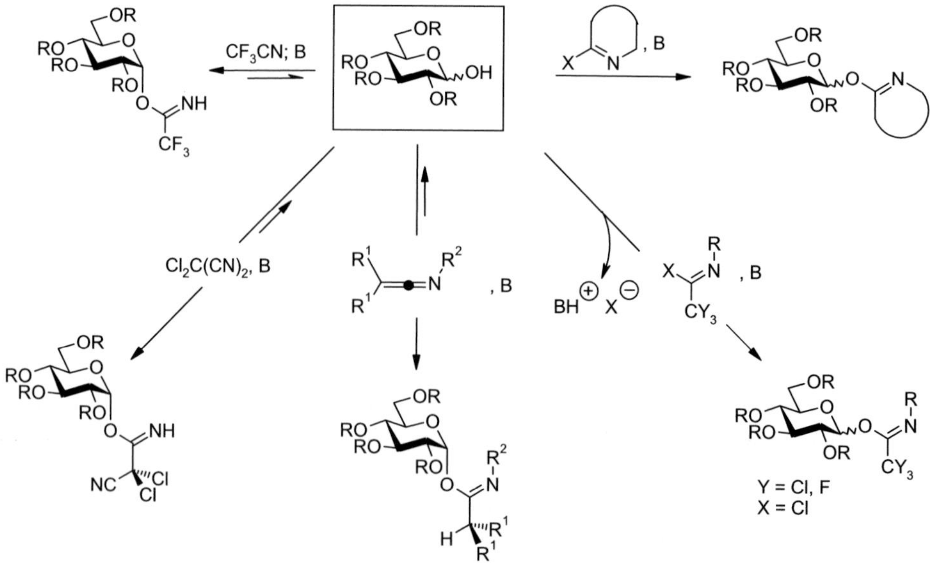

◻ Scheme 46

7 Conclusions

The requirement for efficient glycosylation methods, as outlined at the beginning of this chapter, namely convenient diastereocontrolled anomeric *O*-activation (first step) and subsequent efficient diastereocontrolled glycosylation promoted by catalytic amounts of a promoter (second step) are essentially completely fulfilled by the *trichloroacetimidate method*. In terms of reactivity and applicability toward different acceptors, the *O*-glycosyl trichloroacetimidates have generally proven to be outstanding glycosyl donors, which resemble in various respects the nucleoside diphosphate sugar derivatives used by nature as glycosyl donors. Thus, base-catalyzed generation of *O*-glycosyl trichloroacetimidates followed by acid-catalyzed glycosylation have become a very competitive alternative to other methods mainly requiring anomeric oxygen exchange reactions for glycosyl donor generation and at least equimolar amounts of a promoter system for the glycosylation step. Hence, the *trichloroacetimidate method* can readily be adapted for large-scale preparations. Recently, related activation systems have attracted increasing interest, thus widening the scope of this glycosylation method.

Acknowledgement

Our work reported in this chapter was supported by the Deutsche Forschungsgemeinschaft and the Fonds der Chemischen Industrie.

References

1. Schmidt RR (1986) Angew Chem Int Ed Engl 25:89
2. Schmidt RR, Kinzy W (1994) Adv Carbohydr Chem Biochem 50:21
3. Schmidt RR, Jung KH (1997) Oligosaccharide Synthesis via Trichloroacetimidates. In: Hannessian S (ed) Preparative Carbohydrate Chemistry. Marcel Dekker, p 283
4. Schmidt RR, Jung KH (2000) Trichloroacetimidates. In: Ernst B, Hart GW, Sinaÿ P (eds) Carbohydrates in Chemistry and Biology, Part I: Chemistry of Saccharides, vol 1. Wiley-VCH, Weinheim, p 5
5. Paulsen H (1982) Angew Chem Int Ed Engl 21:155
6. Kunz H (1987) Angew Chem Int Ed Engl 26:294; (1993) Pure Appl Chem 65:1223
7. Toshima K, Tatsuta K (1993) Chem Rev 93:1503
8. Schmidt RR, Reichrath M (1979) Angew Chem Int Ed Engl 18:466
9. Klotz W, Schmidt RR (1994) J Carbohydr Chem 13:1093
10. Schmidt RR, Klotz W (1991) Synlett 168
11. Klotz W, Schmidt RR (1993) Liebigs Ann Chem 683
12. Michel J (1983) PhD thesis, Universität Konstanz
13. Schmidt RR, Michel J (1980) Angew Chem Int Ed Engl 19:731
14. Pougny JR, Sinaÿ P (1976) Tetrahedron Lett 4073
15. Sinaÿ P (1978) Pure Appl Chem 50:1437
16. Ohashi I, Lear MJ, Yoshimura F, Dirama M (2004) Org Lett 6:719
17. Chiara JL, Encinas L, Díaz B (2005) Tetrahedron Lett 46:2445
18. Yoshizaki H, Fukuda N, Sato K, Oikawa M, Fukase K, Suda Y, Kusumoto S (2001) Angew Chem Int Ed 40:1475
19. Oikawa M, Tanaka T, Fukuda, N, Kusumoto S (2004) Tetrahedron Lett 45:4039
20. Roussel F, Knerr L, Grathwohl M, Schmidt RR (2000) Org Lett 2:3043
21. Knerr L, Schmidt RR (2001) The Use of *O*-Glycosyl Trichloroacetimidates for the Polymer Supported Synthesis of Oligosaccharides. In: Seeberger PH (ed) Solid Support Oligosaccha-

ride Synthesis and Combinatorial Carbohydrate Libraries. Wiley, New York, pp 67

22. Ali IAI, El-Ashry ESH, Schmidt RR (2003) Eur J Org Chem 4121

23. Yu H, Williams DL, Ensley HE (2005) Tetrahedron Lett 46:3417

24. Karst NA, Islam TF, Avci FY, Linhardt RJ (2004) Tetrahedron Lett 45:6433

25. Yu B, Zhu X, Hui Y (2001) Tetrahedron 57:9403

26. Weingart R, Schmidt RR (2000) Tetrahedron Lett 41:8753

27. Nicolaou KC, Daines RA, Ogawa Y, Chakraborty TK (1987) J Am Chem Soc 109:2821

28. Nicolaou KC, Daines RA, Ogawa Y, Chakraborty TK (1988) J Am Chem Soc 110:4696

29. Jona H, Mandai H, Chavasiri W, Takeuchi K, Mukaiyama T (2002) Bull Chem Soc Jpn 75:291

30. Jona H, Mandai H, Mukaiyama T (2001) Chem Lett 426

31. Douglas SP, Whitfield DM, Krepinsky JJ (1993) J Carbohydr Chem 12:131

32. Wei G, Gu G, Du Y (2003) J Carbohydr Chem 22:385

33. Yamada H, Hayashi T (2002) Carbohydr Res 337:581

34. Castro-Palomino JC, Schmidt RR (1995) Tetrahedron Lett 36:5343

35. Geiger J, Barroca N, Schmidt RR (2004) Synlett 836

36. Adinolfi M, Barone G, Guariniello L, Iadonisi A (2000) Tetrahedron Lett 41:9005

37. Adinolfi M, Barone G, Iadonisi A, Mangoni L, Schiattarella M (2001) Tetrahedron Lett 42:5967

38. Griswold KS, Horstmann TE, Miller SJ (2003) Synlett 1923

39. Kartha KPR, Karkkainen TS, March SJ, Field RA (2001) Synlett 260

40. Adinolfi M, Barone G, Iadonisi A, Mangoni L, Schiattarella M (2002) Synlett 269

41. Schmidt RR, Gaden H, Jatzke H (1990) Tetrahedron Lett 31:327

42. Mukhopadyay B, Maurer SV, Rudolph N, van Well RM, Russell DA, Field RA (2005) J Org Chem 70:9059

43. Du Y, Wei, G, Cheng S, Hua Y, Linhardt RJ (2006) Tetrahedron Lett 47:307

44. Adinolfi M, Barone G, Iadonisi A, Mangoni L, Schiattarella M (2003) Org Lett 5:987

45. Schmidt RR, Rücker E (1980) Tetrahedron Lett 21:1421

46. Schmidt RR, Behrendt M, Toepfer A (1990) Synlett 694

47. Schmidt RR (1992) New Aspects of Glycosylation Reactions. In: Ogura H, Hasegawa A, Suami T (eds) Carbohydrates—Synthetic Methods and Applications in Medicinal Chemistry. Kodanasha Ltd., Tokyo, 68

48. Toepfer A, Schmidt RR (1991) Tetrahedron Lett 32:3353

49. Bommer R, Kinzy W, Schmidt RR (1991) Liebigs Ann Chem 425

50. Böhm G, Waldmann H (1995) Tetrahedron Lett 36:3843

51. Schmid U, Waldmann H (1997) Liebigs Ann Chem 2573

52. Jain N, Kumar A, Chauhan S, Chauhan SMS (2005) Tetrahedron 61:1015

53. Rencurosi A, Lay L, Russo G, Caneva E, Poletti L (2005) J Org Chem 70:7765

54. Larsen K, Worm-Leonhard k, Olsen P, Hoel A, Jensen KJ (2005) Org Biomol Chem 3:3966

55. Zeng Y, Ning J, Kong F (2002) Tetrahedron Lett 43:3729

56. Yang F, He H, Du Y, Lü M (2002) Carbohydr Res 337:1165

57. Zeng Y, Ning J, Kong F (2003) Carbohydr Res 338:307

58. Spijker NM, van Boeckel CAA (1991) Angew Chem Int Ed Engl 30:180

59. Masamuni S, Choy W, Petersen JC, Sita LR (1985) Angew Chem Int Ed Engl 24:1

60. Cid MB, Alfonso F, Martín-Lomas M (2005) Chem Eur J 11:928

61. Castro-Palomino JC, Schmidt RR (1998) Synlett 501

62. Marzabadi CH, Franck RW (2000) Tetrahedron 56:8385

63. Roush WR, Gung BW, Bennett CE (1999) Org Lett 1:891

64. Chong PY, Roush WR (2002) Org Lett 4:4523

65. Kim JH, Yang H, Park J, Boons GJ (2005) J Am Chem Soc 127:12090

66. Gridley JJ, Osborn HMI (200) J Chem Soc Perkin Trans 1:1471

67. Dwek RA (1996) Chem Rev 96:683

68. Barresi F, Hindsgaul O (1996) In: Khan SH, O'Neill RA (eds) Modern Methods in Carbohydrate Synthesis. Harwood Academic Publishers, Amsterdam, p 251

69. Paulsen H, Lockhoff O (1981) Chem Ber 114:3102

70. Garegg PJ, Ossowski P (1983) Acta Chem Scand 337:249

71. Danishefsky SJ, Hu S, Cirilo PF, Eckhardt M, Seeberger PH (1997) Chem Eur J 3:1617
72. Lichtenthaler FW, Schneider-Adams T (1994) J Org Chem 59:6728
73. Schmidt RR, Moering U, Reichrath M (1982) Chem Ber 115:39
74. Hodosi G, Kovác P (1997) J Am Chem Soc 119:2335
75. Ekborg G, Lindberg B, Lonngren J (1972) Acta Chem Scand B 26:3287
76. Matsuo I, Isomura M, Walton R, Ajisaka K (1996) Tetrahedron Lett 37:8795
77. Kunz H, Günther W (1988) Angew Chem Int Ed Engl 27:1086
78. Weiler S, Schmidt RR (1998) Tetrahedron Lett 39:2299
79. Twaddle GWJ, Yashunsky DV, Nikolaev AV (2003) Org Biomol Chem 1:623
80. Barresi F, Hindsgaul O (1991) J Am Chem Soc 113:9376
81. Stork G, LaChair JJ (1996) J Am Chem Soc 118:247
82. Ito Y, Ogawa T (1994) Angew Chem Int Ed Engl 33:1765
83. Jung KH, Müller M, Schmidt RR (2000) Chem Rev 100:4423
84. Ziegler T, Lemanski G, Rakoczy A (1995) Tetrahedron Lett 36:8973
85. Caregg PJ, Iversen T, Johansson R (1980) Acta Chem Scand B 34:505
86. Crich D, Sun S (1996) J Org Chem 61:4506
87. Crich D, Smith M (2001) J Am Chem Soc 123:9015
88. Ikeda T, Yamada H (2000) Carbohydr Res 329:889
89. Abdel-Rahman AAH, Jonke S, El Ashry ESH, Schmidt RR (2002) Angew Chem Int Ed 41:2972
90. Zhou FY, Huang JY, Yuan Q, Wang YG (2005) Chem Lett 34:878
91. Wang CC, Lee JC, Luo SY, Fan HF, Pai CL, Yang WC Lu LD Hung SC (2002) Angew Chem Int Ed 41:2360
92. Curran DP, Ferritto R, Hua Y (1998) Tetrahedron Lett 39:4937
93. Miura T, Hirose Y, Ohmae M, Inazu T (2001) Org Lett 3:3947
94. Miura T, Inazu T (2003) Tetrahedron Lett 44:1819
95. Haase WC, Seeberger PH (2000) Chem Rev 100:4349
96. Plante OJ, Palmacci ER, Seeberger PH (2001) Science 291:1523
97. Roussel F, Takhi M, Schmidt RR (2001) J Org Chem 66:8540
98. Roussel F, Knerr L, Schmidt RR (2001) Eur J Org Chem 2066
99. Wu X, Grathwohl M, Schmidt RR (2001) Org Lett 3:747
100. Rademann J, Geyer A, Schmidt RR (1998) Angew Chem Int Ed 37:1241
101. Rademann J, Schmidt RR (1997) J Org Chem 62:3650
102. Zhu T, Boons GJ (2001) Chem Eur J 7:2382
103. Zhu T, Boons GJ (2000) J Am Chem Soc 122:10222
104. Nicolaou KC, Watanabe N, Li J, Pastor J, Winssinger N (1998) Angew Chem Int Ed 37:1559
105. Mogemark M, Elofsson M, Kihlberg J (2003) J Org Chem 68:7281
106. Ando H, Manabe S, Nakahara Y, Ito Y (2001) Angew Chem Int Ed 40:4725
107. Pamacci ER, Hewitt MC, Seeberger PH (2001) Angew Chem Int Ed 40:4433
108. Wu X, Schmidt RR (2004) J Org Chem 69:1853
109. Bauer J, Rademann (2005) J Am Chem Soc 127:7296
110. Wu X, Grathwohl M, Schmidt RR (2002) Angew Chem Int Ed 41:4489
111. Jonke S, Liu K, Schmidt RR (2006) Chem Eur J 12:1274
112. Andrade RB, Plante OJ, Melean LG, Seeberger PH (1999) Org Lett 1:1811
113. Ratner DM, Swanson ER, Seeberger PH (2003) Org Lett 5:4717
114. Hewitt MC, Snyder DA, Seeberger PH (2002) J Am Chem Soc 124:13434
115. Manabe S, Ito Y (2002) J Am Chem Soc 124:12638
116. Ojeda R, de Paz JL, Martín-Lomas M (2003) Chem Commun 2486
117. Ojeda R, Terentí O, de Paz JL, Martín-Lomas M (2004) Glycoconjugate J 21:179
118. de Paz JL, Angulo J, Lassaletta JM, Nieto PM, Ridondo-Horcajo M, Lozano RM, Gimenez-Gallego G, Martín-Lomas M (2001) ChemBioChem 2:673
119. Blattner R, Furneaux RH, Ludewig M (2006) Carbohydr Res 341:299
120. Knust B (1996) Diploma thesis, Universität Konstanz
121. Kantchev EAB, Parquette JR (2005) Synlett 1567
122. Schmidt RR, Zimmermann P (1986) Angew Chem Int Ed Engl 25:725

123. Zimmermann P, Schmidt RR (1988) Liebigs Ann Chem 663
124. Hansen HC, Magnusson G (1999) Carbohydr Res 322:190
125. Aly MRE, Rochaix P, Amessou M, Johannes L, Florent JC (2006) Carbohydr Res 341:2026
126. Chen L, Zhao XE, Lai D, Song Z, Kong F (2006) Carbohydr Res 341:1174
127. Sun B, Pukin AV, Visser GM, Zuilhof H (2006) Tetrahedron Lett 14:7371
128. Tietze LF, Keim K, Janßen CO, Tappertzhofen C, Olschimke J (2000) Chem Eur J 6:2801
129. Castro-Palomino JC, Simon B, Speer O, Leist M, Schmidt RR (2001) Chem Eur J 7:2178
130. Janssen S, Schmidt RR (2005) J Carbohydr Chem 24:611
131. Aly MRE, Ibrahim ESI, El Ashry ESH, Schmidt RR (1999) Carbohydr Res 316:121
132. Gege C, Kinzy W, Schmidt RR (2000) Carbohydr Res 328:459
133. Gege C, Oscarson S, Schmidt RR (2001) Tetrahedron Lett 42:377
134. Gege C, Schmidt RR (2002) Carbohydr Res 337:1089
135. Gege C, Geyer A, Schmidt RR (2002) Eur J Org Chem 2475
136. Gege C, Vogel J, Bendas G, Rothe U, Schmidt RR (2000) Chem Eur J 6:111
137. Gege C, Geyer A, Schmidt RR (2002) Chem Eur J 8:2454
138. Yamaguchi M, Ishida H, Kanamori A, Kannagi R, Kiso M (2003) Carbohydr Res 338: 2793
139. Bélot F, Rabuka D, Fukuda M, Hindsgaul O (2002) Tetrahedron Lett 43:7743
140. Natori T, Morita M, Akimoto K, Koezuka Y (1994) Tetrahedron 50:2771
141. Natori T, Koezuka Y, Higa T (1993) Tetrahedron Lett 34:5591
142. Morita M, Motoki K, Akimoto K, Natori T, Sakai T, Sawa E, Yamaji K, Koezuka Y Kobayashi E, Fukushima H (1995) J Med Chem 38:2176
143. Kawano T, Cui J, Koezuka Y, Toura I, Kaneko Y, Motoki K, Ueno H, Nakagawa R, Sato H, Kondo E, Koseki H, Taniguchi M (1997) Science 278:1626
144. Figueroa-Pérez S, Schmidt RR (2000) Carbohydr Res 328:95
145. Laurent N, lafont D, Buillange P (2006) Carbohydr Res 341:823
146. Bartolmäs T, Heyn T, Mickeleit M, Fischer A, Reutter W, Danker K (2005) J Med Chem 48:6750
147. Gouin SG, Pilgrim W, Porter RK, Murphy PV (2005) Carbohydr Res 340:1547
148. Sato R, Toma K, Nomura K, Takagi M, Yoshida T, Azefu Y, Tamiaki H (2004) J Carbohydr Chem 23:375
149. Herzner H, Reipen T, Schultz M, Kunz H (2000) Chem Rev 100:4495
150. Röhrig CH, Retz OA, Hareng L, Hartung T, Schmidt RR (2005) ChemBioChem 6:1805
151. Saha UK, Schmidt RR (1997) J Chem Soc Perkin Trans 1:1855
152. Saha UK, Griffith LS, Rademann J, Geyer A, Schmidt RR (1997) Carbohydr Res 304:21
153. Burger K, Kluge M, Fehn S, Koksch B, Hennig L, Müller G (1999) Angew Chem Int Ed 38:1414
154. Allevi P, Anastasia M, Paroni R, Ragusa A (2004) Bioorg Med Chem Lett 14:3319
155. Komba S, Meldal M, Werdelin O, Jensen T, Bock K (1999) J Chem Soc Perkin Trans 1:415
156. St. Hilaire PM, Cipolla L, Franco A, Tedebark U, Tilly DA, Meldal M (1999) J Chem Soc Perkin Trans 1:3559
157. Xue J, Khaja SK, Locke RD, Matta KL (2004) Synlett 861
158. Paulsen H, Paal M, Schultz M (1983) Tetrahedron Lett 24:1759
159. Paulsen H, Paal M, (1984) Carbohydr Res 135:53
160. Kessler H, Kottenhahn M, Kling A, Kolar C (1987) Angew Chem Int Ed Engl 99:919
161. Halkes KM, Gotfredsen CH, Grøtly M, Miranda LP, Duus JØ, Meldal M (2001) Chem Eur J 7:3584
162. Nicolaou KC, Mitchell HJ, Jain NF, Bando T, Hughes R, Winssinger N, Natarajan S, Koumbis AE (1999) Chem Eur J 5:2648
163. Nicolaou KC, Cho SY, Hughes R, Winssinger N, Smethurst C, Labischinski H, Endermann R (2001) Chem Eur J 7:3798
164. Ritter TK, Mong KKT, Liu H, Nakatani T, Wong CH (2003) Angew Chem Int Ed 42:4657
165. Meinjohanns E, Meldal M, Paulsen H, Dwek RA, Bock K (1998) J Chem Soc Perkin Trans 1: 549
166. Matsuo I, Nakahara Y, Ito Y, Nukada T, Nakahara Y, Ogawa T (1995) Bioorg Med Chem 3:1455
167. Chiesa MV, Schmidt RR (2000) Eur J Org Chem 3541

168. Aly MRE, Ibrahim ESI, El Ashry ESH, Schmidt RR (2001) Carbohydr Res 331:129

169. Zhu Y, Chen L, Kong F (2002) Carbohydr Res 337:207

170. Ratner DM, Plante OJ, Seeberger PH (2002) Eur J Org Chem 826

171. Mendoza VM, Agusti R, Gallo-Rodriguez C, de Lederkremer RM (2006) Carbohydr Res 341:1488

172. Nakano J, Ichiyanagi T, Ohta H, Ito Y (2003) Tetrahedron Lett 44:1742

173. Hofsteenge J, Müller DR, de Beer T, Lüffler A, Richter WJ, Vliegenthart JFG (1994) Biochemistry 33:13524

174. Gäde G, Kellner R, Rinehart KL, Proefke ML (1992) Biochem Biophys Res Commun 189:1303

175. Hofsteenge J, Blommers M, Hess D, Furmanek A, Miroshnichenko O (1999) J Biol Chem 274:32786

176. Diem S, Albert J, Herderich M (2001) Eur Food Res Technol 213:439

177. Schnabel M, Römpp B, Ruckdeschel D, Unverzagt C (2004) Tetrahedron Lett 45:295

178. Adinolfi M, Barone G, De Napoli L, Guariniello L, Iadonisi A, Piccialli G (1999) Tetrahedron Lett 40:2607

179. Adinolfi M, De Napoli L, di Fabio G, Guariniello L, Iadonisi A, Messere A, Montesachio D, Piccialli G (2001) Synlett 745

180. Bernards A, van Harten-Loosbroek N, Borst P (1984) Nucleic Acids Res 12:4153

181. Gommers-Ampt JH, Lugterink J, Borst P (1991) Nucleic Acids Res 19:1745

182. Gommers-Ampt JH, van Leeuwen F, de Beer ALJ, Vliegenthart JFG, Dizdaroglu M, Kowalak JA, Crain PF, Borst P (1993) Cell 75:1129

183. de Kort M, Ebrahimi E, Wijsman ER, van der Marel GA, van Boom JH (1999) Eur J Org Chem 2337

184. Fransson LÅ (1987) Trends Biochem Sci 12:406

185. Kjellen L, Lindahl U (1991) Annu Rev Biochem 60:443

186. Karst N, Jacquinet JC (2000) J Chem Soc Perkin Trans 1:2709

187. Karst N, Jacquinet JC (2002) Eur J Org Chem 815

188. Tully SE, Mabon R, Gama CI, Tsai SM, Liu X, Hsieh-Wilson LC (2004) J Am Chem Soc 126:7736

189. Tamura J, Tokuyoshi M (2004) Biosci Biotechnol Biochem 68:2436

190. Soliman Se, Bassily RW, El-Sokkary RI, Nashed MA (2003) Carbohydr Res 338:2337

191. Yeung BKS, Hill DC, Janicka M, Petillo PA (2000) Org Lett 2:1279

192. Casu B (1985) Adv Carbohydr Chem 43:51 and references cited therein

193. Faham S, Hileman RE, Fromm JR, Linhardt RJ, Rees DC (1996) Science 271:1116

194. Hileman RE, Fromm JR, Wiler JM, Linhardt RJ (1998) BioEssays 20:156

195. Pellegrine L, Burke DF, von Delft F, Mulloy B, Blundell TL (2000) Nature 107:1029

196. Mach H, Volkin DB, Burke CJ, Middaugh CR, Linhardt RJ, Fromm JR, Loganathan D, Mattsson L (1993) Biochemistry 32:5480

197. Guimond S, Maccarana M, Olwin BB, Lindahl U, Rapraeger AC (1993) J Biol Chem 268:23906

198. Poletti L, Fleischer M, Vogel C, Guerrini M, Torri G, Lay L (2001) Eur J Org Chem 2727

199. Lucas R, Hamza D, Lubineau A, Bonnaffé D (2004) Eur J Org Chem 2107

200. Lee JC, Lu XA, Kulkarni SS, Wen YS, Hung SC (2004) J Am Chem Soc 126:476

201. de Paz JL, Ojeda R, Reichardt N, Martín-Lomas M (2003) Eur J Org Chem 3308

202. Lohman GJS, Seeberger PH (2004) J Org Chem 69:4081

203. Zhou Y, Lin F, Chen J, Yu B (2006) Carbohydr Res 341:1619

204. Bernfield M, Gotte M, Park PW, Reizes O, Fitzgerald ML, Lincecum J, Zako M (1999) Annu Rev Biochem 68:729

205. Chen L, Kong F (2002) Carbohydr Res 337:1373

206. Rietschel ET, Schletter J, Weidemann B, El-Samalouti V, Mattern T, Zähringer U, Seydel U, Brade H, Flad HD, Kusumoto S, Gupta D, Dziarski R, Ulmer AJ (1998) Microb Drug Resist 4:37

207. Takeuchi O, Hoshino K, Kawai T, Sanjo H, Takada H, Ogawa T, Takeda K, Akira S (1999) Immunity 11:443

208. Inamura S, Fukase K, Kusumoto S (2001) Tetrahedron Lett 42:7613

209. Inamura S, Fujimoto Y, Kawasake A, Shiokawa Z, Woelk V, Heine H, Lindner B, Inohara N, Kusumoto S, Fukase K (2006) Org Biomol Chem 4:232

210. Zeng Y, Zhang J, Kong F (2002) Carbohydr Res 337:1367

211. Gu G, Wie G, Du Y (2004) Carbohydr Res 339:1155

212. Ning J, Heng L, Kong F (2002) Carbohydr Res 337:1159
213. Zeng Y, Zhang J, Ning J, Kong F (2003) Carbohydr Res 338:5
214. Xing Y, Ning J (2003) Tetrahedron Asymmetry 14:1275
215. Bruneteau M, Fabre I, Perret J, Michel G, Ricci P, Joseleau J, Kraus J, Schneider M, Blaschek W, Franz G (1988) Carbohydr Res 175:137
216. Johnson J, Kirkwood S, Misaki A, Nelson TE, Scalleti JD, Smith F (1963) Chem Ind (London) 820
217. He H, Yang,F, Du Y (2002) Carbohydr Res 337:1673
218. Zhu Y, Kong F (2000) Synlett 663
219. Zhang G, Fu M, Ning J (2005) Carbohydr Res 340:597
220. Zeng Y, Kong F (2003) Carbohydr Res 338:2359
221. Ning J, Zhang W, Yi Y, Yang G, Wu Z, Yi J, Kong F (2003) Bioorg Med Chem 11:2193
222. Wu Z, Kong F (2004) Carbohydr Res 339:2761
223. Wu Z, Kong F (2004) Carbohydr Res 339:377
224. Li A, Kong F (2004) Carbohydr Res 339:2499
225. Ning J, Yi Y, Kong F (2002) Tetrahedron Lett 43:5545
226. Ning J, Kong F, Lin B, Lei H (2003) J Agric Food Chem 51:987
227. Yi Y, Zhou Z, Ning J, Kong F, Li J (2003) Synthesis 491
228. Huang GL, Mei XY, Liu MX (2005) Carbohydr Res 340:603
229. Zhu Y, Kong F (2000) Carbohydr Res 329:199
230. McNeil M, Wallner SJ, Hunter SW, Brennan PJ (1987) Carbohydr Res 166:299
231. Nita-Lazar M, Chevolot L, Iwahara S, Takegawa K, Furmanek A, Lienart Y (2002) Acta Biochim Pol 49:1019
232. Ramli N, Shinohara H, Takegawa K, Iwahara SJ (1994) Ferment Bioeng 78:341
233. Zhang G, Fu M, Ning J (2005) Carbohydr Res 340:155
234. Gandolfi-Donadio L, Gallo-Rodriguez C, de Lederkremer RM (2003) J Org Chem 68:6928
235. Du Y, Pan Q, Kong F (2000) Carbohydr Res 329:17
236. Ning J, Yi Y, Yao Z (2003) Synlett 2208
237. Li A, Zeng Y, Kong F (2004) Carbohydr Res 339:673
238. Ma Z, Zhang J, Kong F (2004) Carbohydr Res 339:1761
239. Li A, Kong F (2004) Carbohydr Res 339:1847
240. Li A, Kong F (2005) Carbohydr Res 340:1949
241. Borman S (2006) C&EN 84:80
242. Chen L, Kong F (2002) Carbohydr Res 337:2335
243. Bedini E, Carabellese A, Corsaro MM, De Castro C, Parrilli M (2004) Carbohydr Res 339:1907
244. Zhang J, Kong F (2002) Carbohydr Res 337:391
245. Zhang J, Ning J, Kong F (2003) Carbohydr Res 338:1023
246. Zhang J, Kong F (2003) Carbohydr Res 338:19
247. Zhang J, Kong F (2003) Tetrahedron 59:1429
248. Ma Z, Zhang J, Kong F (2004) Carbohydr Res 339:43
249. Zhang J, Kong F (2002) J Carbohydr Chem 21:579
250. Zhang J, Kong F (2003) Carbohydr Res 338:1719
251. Zhao W, Kong F (2005) Carbohydr Res 340:1673
252. Hua Y, Xiao J, Huang Y, Du Y (2006) Carbohydr Res 341:191
253. Bélot F, Costachel C, Wright K, Phalipon A, Mulard LA (2002) Tetrahedron Lett 43:8215
254. Segat-Dioury F, Mulard LA (2002) Tetrahedron Asymmetry 13:2211
255. Mulard LA, Guerreiro C (2004) Tetrahedron 60:2475
256. Bélot F, Wright K, Costachel C, Phalipon A, Mulard LA (2004) J Org Chem 69:1060
257. Wright K, Guerreiro C, Laurent I, Baleux F, Mulard LA (2004) Org Biomol Chem 2:1518
258. Mulard LA, Clément MJ, Imberty A, Delepierre M (2002) Eur J Org Chem 2486
259. Maruyama M, Takeda T, Shimizu N, Hada N, Yamada H (2000) Carbohydr Res 325:83
260. Ishii K, Kubo H, Yamasaki R (2002) Carbohydr Res 337:11
261. Kubo H, Ishii K, Koshino H, Toubetto K, Naruchi K, Yamasaki R (2004) Eur J Org Chem 1202
262. Ferguson MAJ, Homans SW, Dwek RA, Rademacher TW (1988) Science 239:753
263. Frankhauser C, Homas SW, Oates JE, McConville MJ, Desponds C, Conzelmann A, Ferguson MAJ (1993) J Biol Chem 268:26365
264. Homans SW, Ferguson MAJ, Dwek RA, Rademacher TW, Anad R, Williams AF (1988) Nature 333:269
265. Caro HH, Martín-Lomas M, Bernabé M (1993) Carbohydr Res 240:119
266. Mayer TG, Kratzer B, Schmidt RR (1994) Angew Chem Int Ed Engl 33:2177
267. Mayer TG, Schmidt RR (1999) Eur J Org Chem 1153

268. Murakata C, Ogawa T (1992) Carbohydr Res 234:75 and 235:95

269. Baeschlin DK, Chaperon AR, Green LG, Hahn ME, Ince SJ, Ley SV (2000) Chem Eur J 6:172

270. Campbell AS, Fraser-Reid B (1995) J Am Chem Soc 117:10387

271. Pekari K, Tailler D, Weingart R, Schmidt RR (2001) J Org Chem 66:7432

272. Pekari K, Schmidt RR (2003) J Org Chem 68:1295

273. Martín-Lomas M, Flores-Mosquera M, Chiara JL (2000) Eur J Org Chem 1547

274. Ma Z, Zhang J, Kong F (2004) Carbohydr Res 339:29

275. Gandolfi-Donadio L, Gallo-Rodriguez C, de Lederkremer RM (2002) J Org Chem 67:4430

276. Yashunsky DV, Borodkin VS, Ferguson MAJ, Nikolaev AV (2006) Angew Chem 118:482

277. Roush WR, Bennett CE (2000) J Am Chem Soc 122:6124

278. Ješelnik M, Plavec J, Polanc S, Kočevar M (2000) Carbohydr Res 328:591

279. Myers AG, Liang J, Hammond M, Harrington PM, Wu Y, Kuo EY (1998) J Am Chem Soc 120:5319

280. Wang ZX, Shi XX, Chen GR, Ren ZH, Luo L, Yan J (2006) Carbohydr Res 341:1945

281. Rao Y, Venot A, Swayze EE, Griffey RH, Boons GJ (2006) Org Biomol Chem 4:1328

282. Blattner R, Furneaux RH, Pakulski Z (2006) Carbohydr Res 341:2115

283. Ferguson JR, Harding JR, Lumbard KW, Scheinmann F, Stachulski AV (2000) Tetrahedron Lett 41:389

284. Ferguson JR, Harding JR, Kilick DA, Lumbard KW, Scheinmann F, Stachulski AV (2001) J Chem: Soc Perkin Trans 1:3037

285. Brown RT, Carter NE, Mayalarp SP, Scheinmann F (2000) Tetrahedron 56:7591

286. Suzuki T, Mabuchi K, Fukazawa N (1999) Bioorg Med Chem Lett 9:659

287. Engstrom KM, Daanen JF, Wagaw S, Stewart AO (2006) J Org Chem 71:8378

288. Kawada T, Asano R, Makino K, Sakuno T (2000) Eur J Org Chem 2723

289. Feldman KS, Lawlor MD (2000) J Am Chem Soc 122:7396

290. Nicolaou KC, Pfefferkorn JA, Cao GQ (2000) Angew Chem Int Ed 39:734

291. Schramm S, Dettner K, Unverzagt C (2006) Tetrahedron Lett 47:7741

292. Hein M, Michalik D, Langer P (2005) Synthesis 3531

293. Jiang ZH, Geyer A, Schmidt RR (1995) Angew Chem Int Ed Engl 34:2520

294. Jiang ZH, Schmidt RR (1994) Liebigs Ann Chem 645

295. Fürstner A, Jeanjean F, Razon P (2002) Angew Chem Int Ed 41:2097

296. Brito-Arias M, Pereda-Miranda R, Heathcock CH (2004) J Org Chem 69:4567

297. Fürstner A, Albert M, Mlynarski J, Matheu M (2002) J Am Chem Soc 124:1168

298. Fürstner A, Mlynarski J, Albert M (2002) J Am Chem Soc 124:10274

299. Fürstner A, Ruiz-Caro J, Prinz H, Waldmann H (2004) J Org Chem 69:459

300. Hostettmann K, Marston A (1995) Saponins, Chemistry and Pharmacology of Natural Products, University Press, Cambridge

301. Chwalek M, Plé K, Voutquenne-Nazabadioko L (2004) Chem Pharm Bull 52:965

302. Deng S, Yu B, Xie J, Hui Y (1999) J Org Chem 36:7265

303. Plé K, Chwalek M, Voutquenne-Nazabadioki L (2004) Eur J Org Chem 1588

304. Thompson MJ, Hutchinson EJ, Stratford TH, Bowler WB, Blackburn GM (2004) Tetrahedron Lett 45:1207

305. Ikeda T, Yamauchi K, Nakano D, Naknishi K, Miyashita H, Ito S, Nohara T (2006) Tetrahedron Lett 47:4355

306. Ikeda T, Miyashita H, Kajimoto T, Nohara T (2001) Tetrahedron Lett 42:2353

307. Zhu X, Yu B, Hui Y, Schmidt RR (2004) Eur J Org Chem 965

308. Schimmel J, Passos Eleutério MI, Ritter G, Schmidt RR (2006) Eur J Org Chem 1701

309. Passos Eleutério MI, Schimmel J, Ritter G, do Céu Costa M, Schmidt RR (2006) Eur J Org Chem 5293

310. Wang P, Kim YJ, Navarro-Villalobos M, Rohde BD, Gin DY (2005) J Am Chem Soc 127:3256

311. Kim YJ, Wang P, Navarro-Villalobos M, Rohde BD, Derryberry JM, Gin DY (2006) J Am Chem Soc 128:11906

312. Sawada N (2004) Synthesis of Deoxy-QS-21A, Universität Konstanz, to be published

313. Dumont-Hornebeck B, Joly JP, Coulon J, Chapleur Y (1999) Carbohydr Res 320:147

314. Rele SM, Cui W, Wang L, Hou S, Barr-Zarse G, Tatton D, Gnanou Y, Esko JD, Chaikof EL (2005) J Am Chem Soc 127:10132

315. Ikuta A, Tanimoto T, Koizumi K (2003) J Carbohydr Chem 22:297
316. Ikuta A, Mizuta N, Kitahata S, Murata T, Usui T, Koizumi K, Tanimoto T (2004) Chem Pharm Bull 52:51
317. Smiljanic N, Halila S, Moreau V, Djedaïni-Pilard F (2003) Tetrahedron Lett 44:8999
318. Damager I, Olsen CE, Møller BL, Motawia MS (1999) Carbohydr Res 320:19
319. Wu Z, Kong F (2004) Synlett 2594
320. Levy DE, Tang C (1995) In: The Chemistry of C-Glycosides, Pergamon, Elmsford NY
321. Dondoni A, Marra A, Massi A (1999) J Org Chem 64:933
322. Armitt DJ, Banwell MG, Freeman C, Parish CR (2002) J Chem Soc Perkin Trans 1:1743
323. Herzner H, Palmacci ER, Seeberger PH (2002) Org Lett 4:2965
324. Furuta T, Kumura T, Kondo S, Mihara H, Wakimoto T, Nukaya H, Tsuji K, Tanaka K (2004) Tetrahedron 60:9375
325. Ali IAI, El-Ashry ESH, Schmidt RR (2004) Tetrahedron 40:4773; and references therein
326. Fuchss T, Schmidt RR (2000) J Carbohydr Chem 19:677; and references therein
327. Tsuruta O, Yuasa H, Hashimoto H, Kuromo S, Yazawa S (1999) Bioorg Med Chem Lett 9:1019
328. Izumi M, Tsuruta O, Kajihara Y, Yazawa S, Yuasa H, Hashimoto H (2005) Chem Eur J 11:3032
329. Ohara K, Matsuda H, Hashimoto M, Miyairi K, Okuno T (2002) Chem Lett 626
330. Morii Y, Matsuda H, Ohara K, Hashimoto M, Miyairi K, Okuno T (2005) Bioorg Med Chem 13:5113
331. Schmidt RR, Michel J, Roos M (1984) Liebigs Ann Chem 1343
332. Schmelzer U (1997) PhD thesis, Universität Konstanz
333. Schmelzer U, Zhang Z, Schmidt RR (2007) J Carbohydr Chem (submitted)
334. Huchel U (1998) PhD thesis, Universität Konstanz
335. Huchel U, Schmidt C, Schmidt RR (1995) Tetrahedron Lett 36:9457
336. Huchel U, Schmidt C, Schmidt RR (1998) Eur J Org Chem 1353
337. Hanessian S, Condé JJ, Lion B (1995) Tetrahedron Lett 36:5865
338. Hanessian S (1997) In: Hanessian S (ed) Preparative Carbohydrate Chemistry. Marcel Dekker, New York, p 381–388; and references therein
339. Yu B, Tao H (2001) Tetrahedron Lett 42:2405
340. Yu B, Tao H (2002) J Org Chem 67:9099

3.3 Further Anomeric Esters

Kwan Soo Kim, Heung Bae Jeon
Center for Bioactive Molecular Hybrids and Department of Chemistry, Yonsei University, Seoul 120–749, Korea
kwan@yonsei.ac.kr

Abstract

The most representative anomeric ester glycosyl donor is glycosyl acetate, which seems to be a promising and useful glycosylating reagent due to a variety of advantageous properties. Several other anomeric acyl and carbonate-type leaving groups have been devised as new glycosyl donors. In particular, the glycosyl acetate can be the precursor for glycosylating agents of other types, for example, glycosyl halides, 1-thioglycosides, and glycosyl trichloroacetimidates. Therefore, use of glycosyl acetate can make glycosylation reactions easier and less expensive. Furthermore, the intramolecular glycosylation method through the 1-*O*-acyl linkage and the glycosylation with glycosyl donors having remote acyl groups from the anomeric center have been developed to increase the efficiency and the stereoselectivity.

Keywords

Glycosylation; Glycosyl donor; Anomeric leaving group; Anomeric ester; Glycosyl acetate; Glycosyl carbonate; 1-*O*-acyl linkage; Remote acyl group

Abbreviations

CB	2′-carboxybenzyl
DCC	1,3-dicyclohexylcarbodiimide
DIC	diisopropylcarbodiimide
Dmob	*N*-2,4-dimethoxybenzyl
IDCP	iodonium dicollidine perchlorate
LPS	lipopolysaccharide
NIS	*N*-iodosuccinimide
TBSOTf	*tert*-butyldimethylsilyl trifluoromethanesulfonate
TMSOTf	trimethylsilyl trifluoromethanesulfonate

In: *Glycoscience*. Fraser-Reid B, Tatsuta K, Thiem J (eds)
Chapter-DOI 10-1007/978-3-540-30429-6_12: © Springer-Verlag Berlin Heidelberg 2008

1 Introduction

An advantage of the 1-*O*-acylated glycosyl donors in the glycosylation method is undoubtedly the ease of their preparation. The most representative anomeric functional group in this area is the acetyl group. Glycosyl acetates seem to be promising and useful glycosylating reagents due to a variety of advantageous properties. Among these, the most important are their availability and chemical stability, which allows one to obtain almost any desired amounts of glycosyl acetates which can be stored for unlimited time without special precautions. Furthermore, glycosyl acetates can be precursors for glycosylating agents of other types, for example, glycosyl halides, 1-thioglycosides, and glycosyl trichloroacetimidates. Therefore, the use of glycosyl acetates can make glycosylation reactions easier and less expensive.

Because of the relatively low reactivity of glycosyl acetates they used to occasionally bring about difficulty in controlling stereoselectivity. To overcome the problem associated with glycosyl acetates, several other anomeric acyl leaving groups, for example, ester-type leaving groups such as haloacetyl, (2-methoxyethoxy)acetyl, benzoyl, *p*-nitrobenzoyl, 2-pyridinecarbonyl, pivaloyl, 4-pentenoyl, and 5-hexynoyl, and carbonate-type leaving groups such as phenoxycarbonyl, isopropenyloxycarbonyl, 1-imidazolylcarbonyl, *N*-allyl carbamoyl, sulfonylcarbamoyl, and 2-pyridyl thiocarbonyl, have been devised as new glycosyl donors. Furthermore, the methods of intramolecular glycosylation through the 1-*O*-acyl linkage and glycosylation with glycosyl donors having acyl groups remote from the anomeric center have been developed to increase efficiency and stereoselectivity.

2 Glycosylation with Glycosyl Acetates

Generally the activation of glycosyl acetates is achieved in the presence of a Lewis acid. Since Helferich et al. introduced the use of glycosyl acetates as glycosyl donors in the reaction with phenol in the presence of *p*-toluenesulfonic acid (TsOH) or ZnCl$_2$ in 1933 [1], several Lewis acids have been utilized as effective promoters in the glycosylation with glycosyl acetates. The representative promoters used for the activation of glycosyl acetate donors for the glycosylation are summarized in ❷ *Table 1*. TsOH was also used as a promoter of peracetylated 2-deoxy-L-fucose for the glycosylation of digitoxigenin (❷ *Scheme 1*) [2].

60% (α/β = 1/1.4)

☐ Scheme 1
(Ref. [2])

■ **Table 1**
Glycosylation with 1-*O*-acetyl sugar

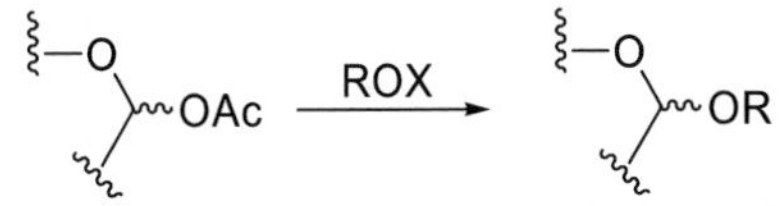

Promoters	X	References
TsOH	H	[1,2]
ZnCl₂	H	[1]
SnCl₄	H	[3,4,5,6,7,8]
FeCl₃	H	[9,10,11,12]
BF₃·Et₂O	H	[13,14,15,16,17,18,19,20,21]
BF₃·Et₂O/Bi(OTf)₃	H	[22]
TMSOTf	H	[23,24,25,26,27,28,29,30,31,32,33,34,35,36,37,40,41,42]
TBSOTf	H	[35,38,39]
TrClO₄	H	[43,44]
SnCl₄/Sn(OTf)₂/LiClO₄	TMS	[45]
SnCl₄/AgClO₄	TMS	[46,47,49]
GaCl₃/AgClO₄	TMS	[47]
Me₂SiCl₂/AgClO₄	H	[48]
K-10 montmorillonite	H	[50]
Yb[N(O₂SC₄F₉)₂]₃	H	[51]
TMSCl/Zn(OTf)₂	H	[52,53]

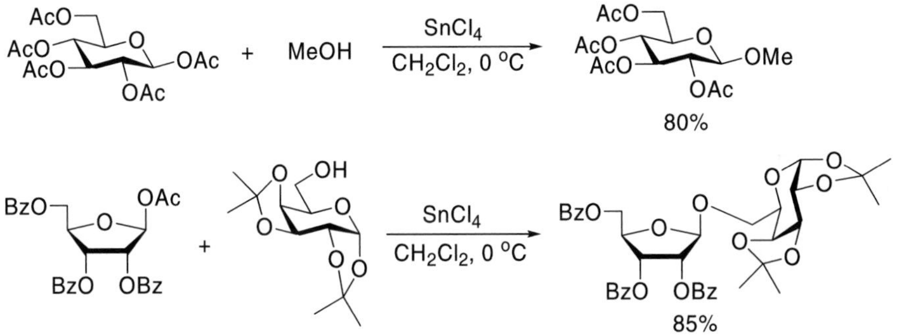

■ **Scheme 2**
(Ref. [3,4])

Lemieux and Shyluk reported that the reaction of peracetylated glucose and methanol in the presence of SnCl₄ gave methyl tetraacetyl-β-glucopyranoside in moderate yield [3]. Later, Hanessian and Banoub reported that the glycosylation using SnCl₄ for the activation of glycosyl acetate donors at low temperature afforded the corresponding β-*O*-glycosides in high yield (❯ *Scheme 2*) [4]. Furthermore, SnCl₄ was employed as a promoter for *O*-glucuronylation of some phenols and alcohols including estradiol 17-acetate and estradiol 16,17-diacetate with methyl 1,2,3,4-tetra-*O*-acetyl-β-D-glucopyranuronate (❯ *Scheme 3*) [5]. And, Williams et al. developed a mild and regiocontrolled method for the *O*-glucuronylation of the highly reactive 3-hydroxyl group of 5β-cholestane-3α,7α,12α,25-tetrol using SnCl₄ in order to study the transformations of bile alcohols and evaluate the biological effects on the blood-brain barrier (● *Scheme 4*) [6]. Krausz et al. reported the synthesis of bioactive nucleoside analogues, which have a spacer arm between the sugar and the base moieties, by SnCl₄-mediated glycosylation of 3-alkyl N^4-(3-hydroxylpropyl) 2-piperazinones with protected 1-O-acetyl ribofura-

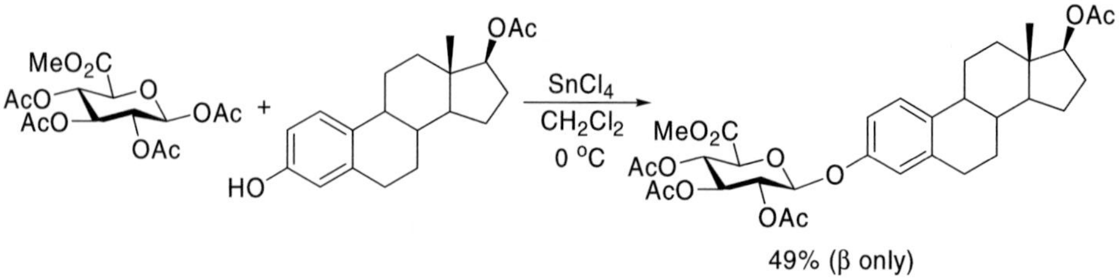

Scheme 3
(Ref. [5])

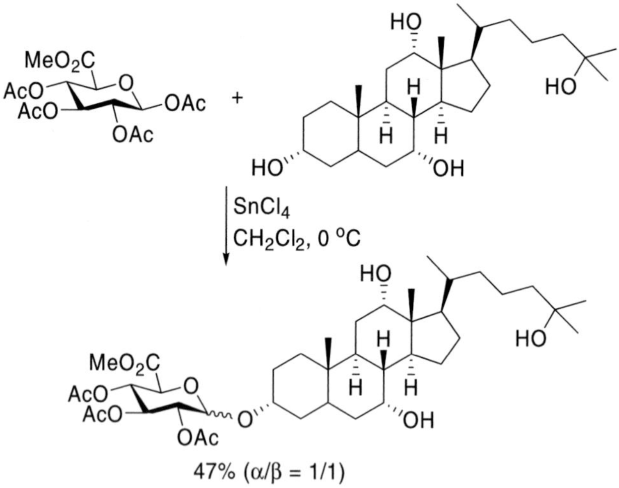

Scheme 4
(Ref. [6])

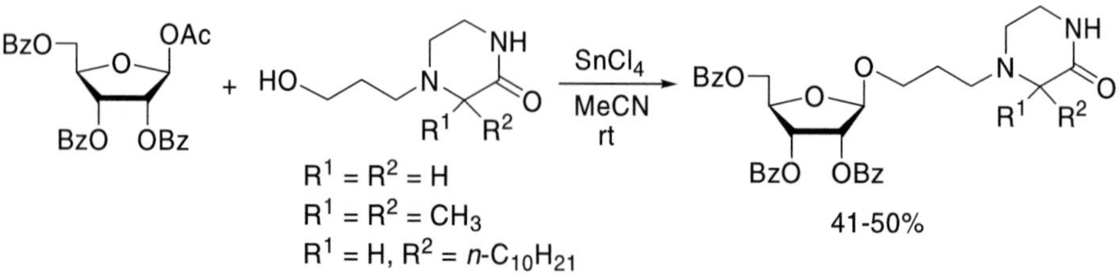

$R^1 = R^2 = H$
$R^1 = R^2 = CH_3$
$R^1 = H, R^2 = n\text{-}C_{10}H_{21}$

Scheme 5
(Ref. [7])

noses (❂ *Scheme 5*) [7] and the synthesis of novel *meso*-glycosylarylporphyrins by the similar methodology [8].

Iron(III) chloride, FeCl$_3$, was introduced for glycosylation with 1-*O*-acetyl-2-acylamido-2-deoxy-β-D-glucopyranose by Kiso and Enderson (❂ *Scheme 6*) [9], and later, Lerner used FeCl$_3$ in combination with molecular sieves (3Å) for the dimerization of 1-*O*-acetyl-2,3,5-tri-*O*-benzoyl-β-D-ribofuranose to afford a nonreducing disaccharide derivative (❂ *Scheme 7*) [10]. Uryu and coworkers reported the use of FeCl$_3$ as a promoter for the glycosylation of aliphatic alcohols with malto-hexaoside and malto-heptaoside peracetates (❂ *Scheme 8*) [11]. On the other hand, while peracetylated sugar donors are typically

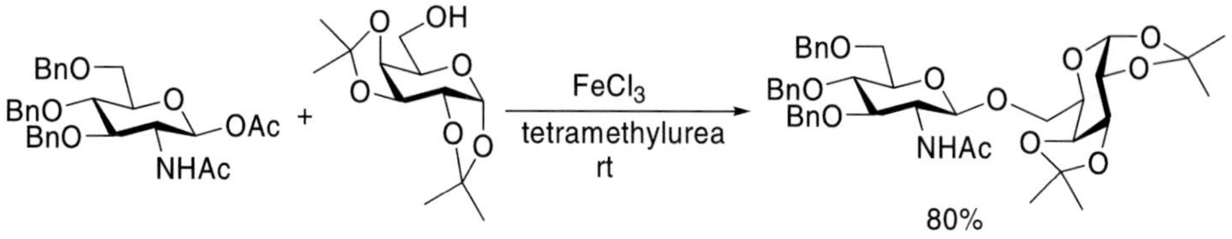

☐ Scheme 6
(Ref. [9])

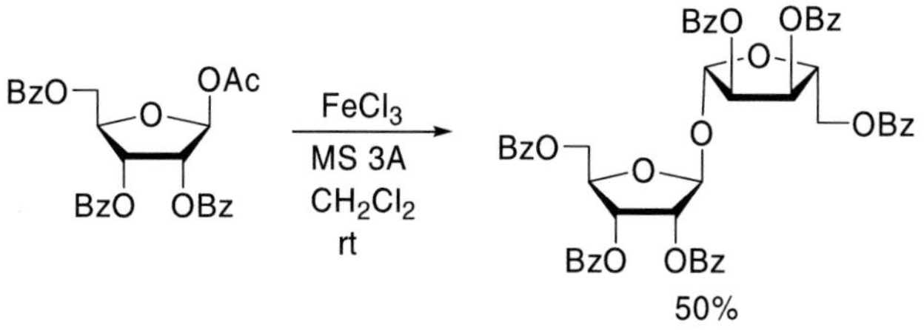

☐ Scheme 7
(Ref. [10])

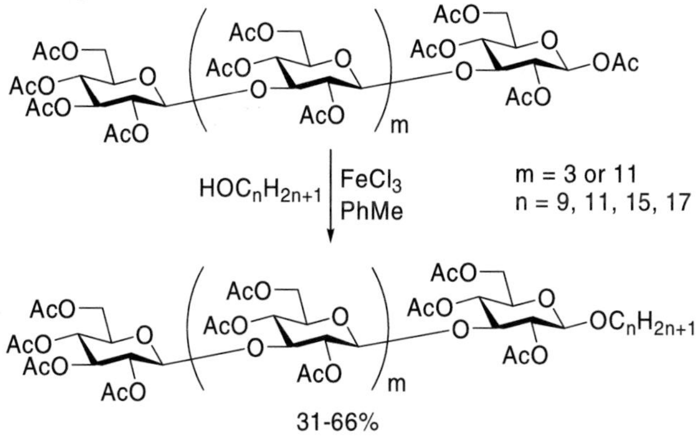

HOC_nH_{2n+1} FeCl$_3$ m = 3 or 11
PhMe   n = 9, 11, 15, 17

31–66%

☐ Scheme 8
(Ref. [11])

employed for the 1,2-*trans* glycosylation, it was reported that the glycosylation of a series of alcohol acceptors with peracetylated sugars in the presence of FeCl₃ provided stereoselectively 1,2-*cis*-α-glycosides, which might be generated by the anomerization of the 1,2-*trans*-β-glycopyranoside by FeCl₃ (❷ *Scheme 9*) [12].

Magnusson et al. introduced BF₃·Et₂O as a promoter for the formation of alkyl glycosides from 1-*O*-acetyl-2-deoxy-2-phthalimidoglucopyranose (❷ *Scheme 10*) [13]. And then, BF₃·Et₂O was applied for the synthesis of a number of oligosaccharides as an activator of glycosyl acetate donors. Gurjar et al. applied BF₃·Et₂O for the synthesis of the oligo-

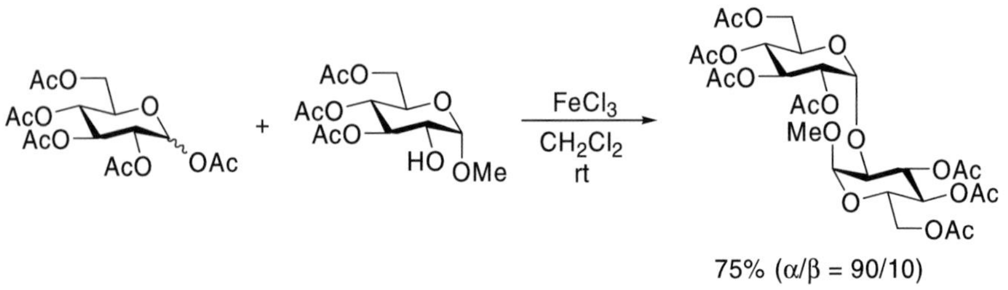

■ Scheme 9
(Ref. [12])

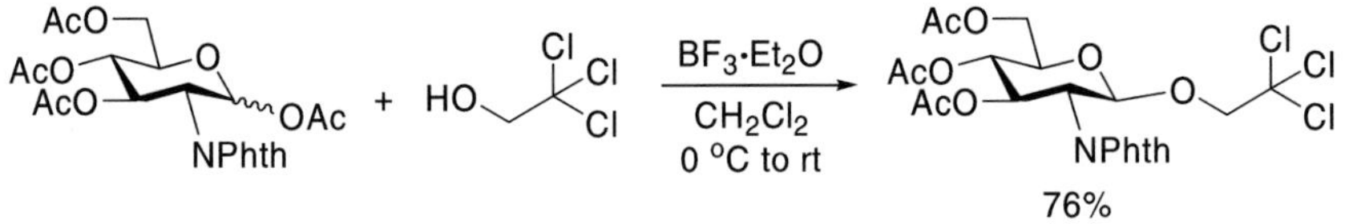

■ Scheme 10
(Ref. [13])

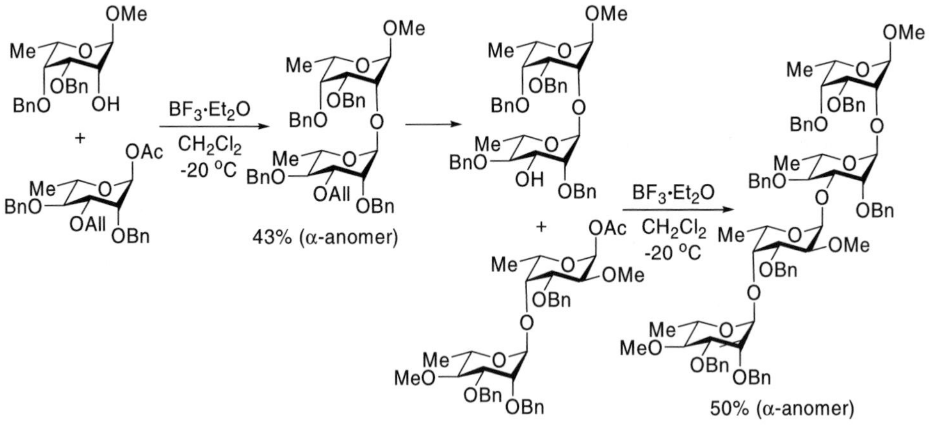

■ Scheme 11
(Ref. [14])

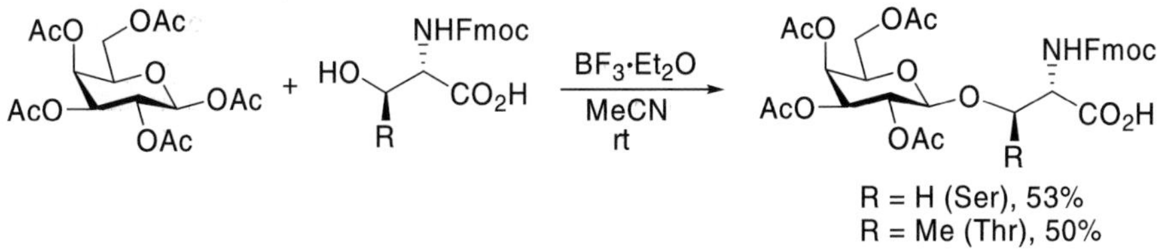

**Scheme 12
(Ref. [15])**

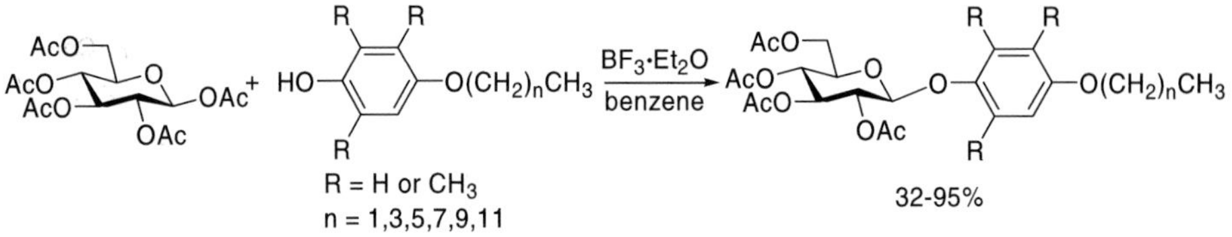

R = H (Ser), 53%
R = Me (Thr), 50%

R = H or CH$_3$
n = 1,3,5,7,9,11

32-95%

**Scheme 13
(Ref. [18])**

saccharide segment of the glycopeptidolipid antigen of mycobacterium avium serotype 4
(**Scheme 11**) [14]. Elofsson et al. discovered that glycosyl acetate donors could be used for
the glycosylation of amino acids, in which the carboxylic acid was unprotected, in the presence
of BF$_3$·Et$_2$O (**Scheme 12**) [15], while Steffan et al. showed that coupling of peracetylated
glucose with N-Fmoc-protected serine by the same methodology provided the corresponding
glycosylated amino acid [16]. This was a major breakthrough, because the approach circumvents the protection-deprotection procedure of the carboxylic acid prior to incorporation into
a peptide. Peracetylated glucose and BF$_3$·Et$_2$O were also utilized for the glycosylation of
protected serine and tyrosine to make O-glycopeptides [17]. On the other hand, Satoh et al.
used BF$_3$·Et$_2$O as an activator of peracetylated glucose in order to prepare 4-alkoxyphenyl
β-D-glucopyranosides, showing strong inhibitory effects on the histamine release from rat
peritoneal mast cells induced by concanavalin A (**Scheme 13**) [18]. A similar methodology was applied for the preparation of sannamycin-type aminoglycoside antibiotics from the
purpurosamine C-type glycosyl acetate donor and the sannaime-type acceptor by Prinzbach
et al. (**Scheme 14**) [19]. Gervay-Hague and Lam also reported the use of BF$_3$·Et$_2$O for the
glycosylation of methyl paraben with protected mannose 1-O-acetate to obtain the corresponding α-mannopyranoside (**Scheme 15**) [20]. And, BF$_3$·Et$_2$O was used for the glycosylation
with the N-2,4-dimethoxybenzyl (Dmob) protected 2-acetamido-1-O-acetate glucosyl donor,
which gave higher glycosylation yields than the corresponding 2-acetamido glucosyl donor
without Dmob protection (**Scheme 16**) [21].
Ikeda et al. reported that a combined system of bismuth triflate [Bi(OTf)$_3$] and BF$_3$·Et$_2$O in
dichloromethane is an efficient promoter for the glycosylation of sialyl acetates among various Lewis acids, for example, BiCl$_3$, Sc(OTf)$_2$, Yb(OTf)$_3$, Zn(OTf)$_2$, and TMSCl, and their
combinations, which showed the first example of utilization of Bi(OTf)$_3$ in the glycosylation
of common sialic acid derivatives (**Scheme 17**) [22].

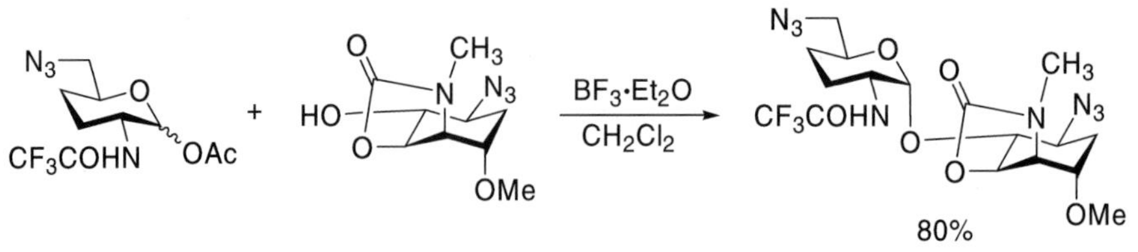

◘ Scheme 14
(Ref. [19])

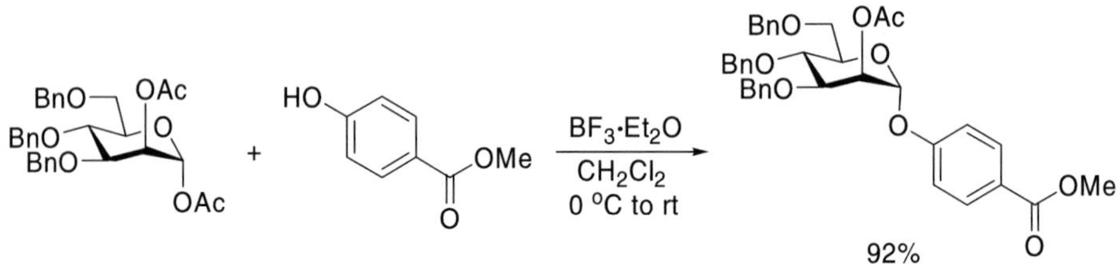

◘ Scheme 15
(Ref. [20])

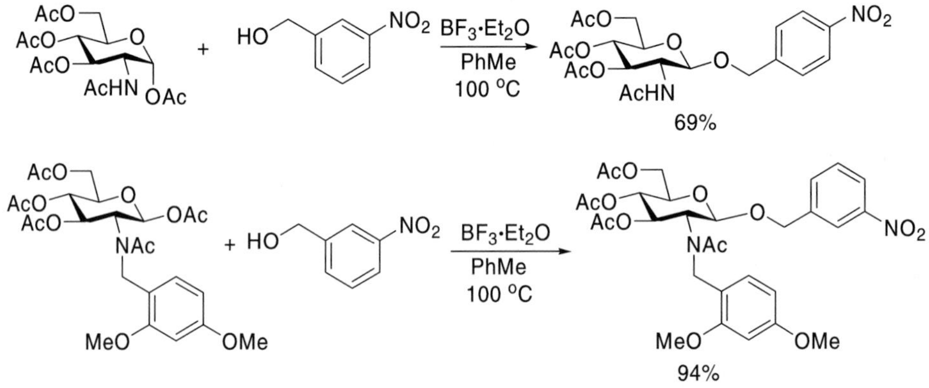

◘ Scheme 16
(Ref. [21])

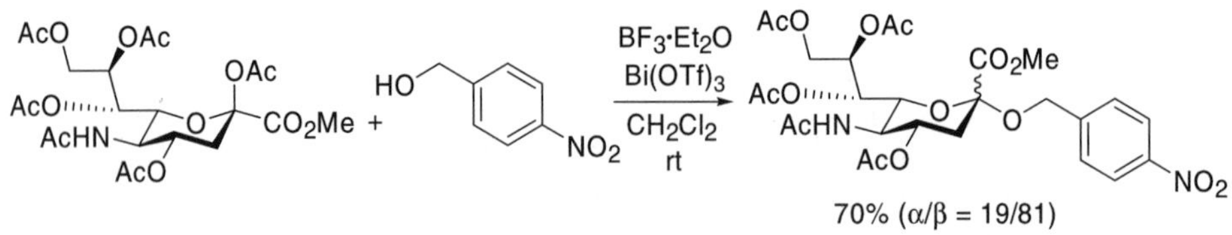

◘ Scheme 17
(Ref. [22])

Ogawa and coworkers reported that trimethylsilyl trifluoromethanesulfonate (TMSOTf) was a very effective promoter for the glycosylation of a glycerolipid alcohol with octa-*O*-acetyl-β-lactose [23] and with 1,3,4,6-tetra-*O*-acetyl-2-deoxy-2-phthalimido-β-D-galactopyranose [24] to synthesize model glycoglycerolipids (❯ *Scheme 18*). However, glycosylations with glycosyl acetates in the presence of TMSOTf sometimes accompany side reactions such as transfer of an acyl group from the glycosyl donor to the glycosyl acceptor [25],

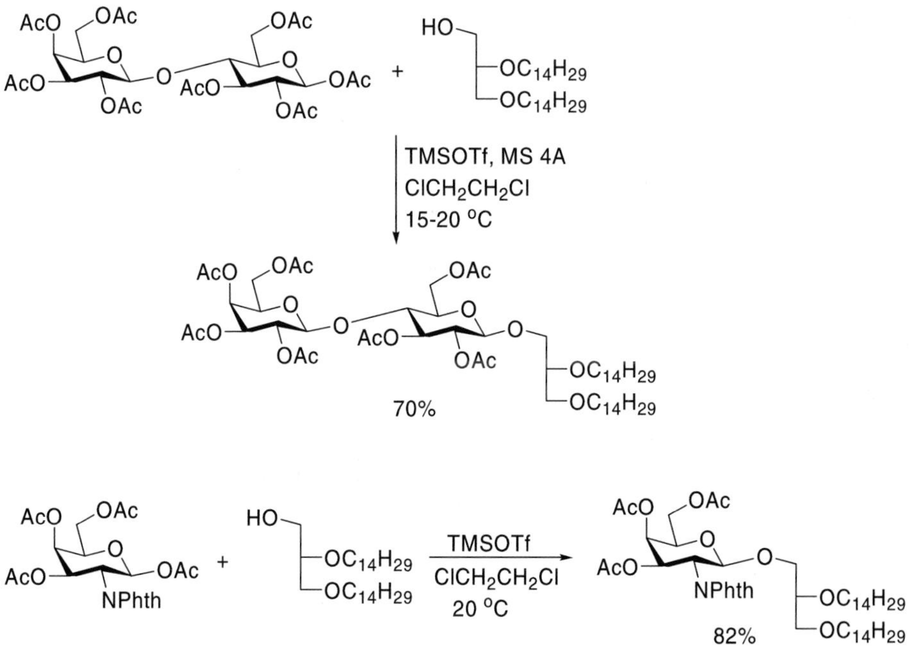

❑ Scheme 18
(Ref. [23,24])

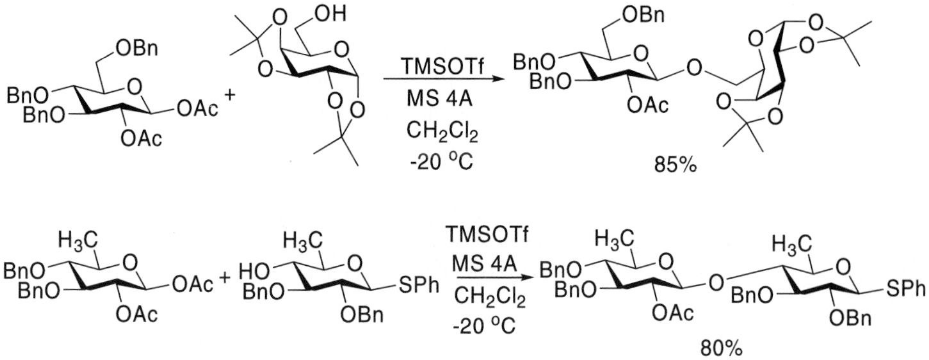

❑ Scheme 19
(Ref. [28])

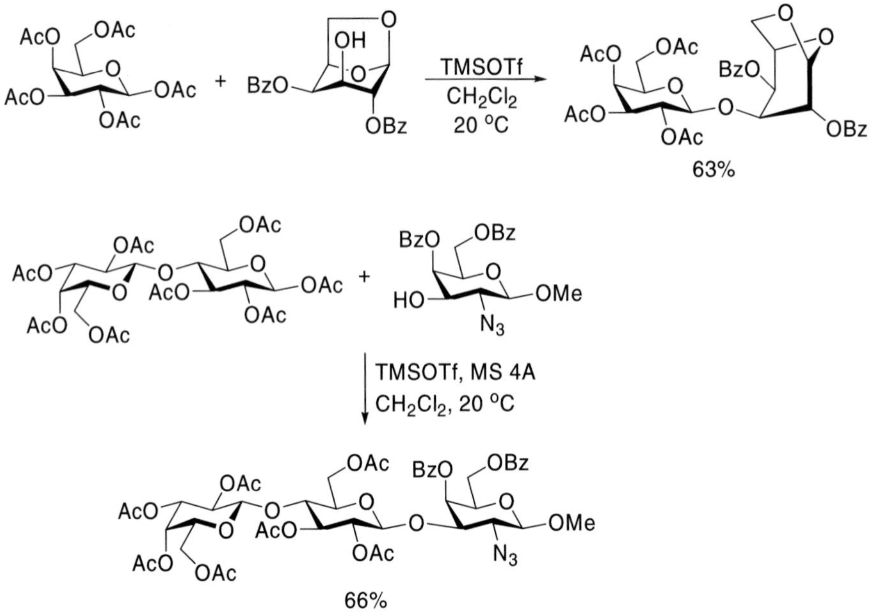

□ Scheme 20
(Ref. [29])

deacetylation [25b], and silylation of the glycosyl acceptor [26]. Thus, Nifant'ev et al. studied the scope and limitation of the use of β-glycosyl acetates as glycosyl donors in the presence of TMSOTf [27]. In spite of these problems, since its first introduction by Ogawa, TMSOTf has been widely used as the most preferred activator of glycosyl acetates for glycoside synthesis. Sinaÿ et al. reported that 1,2-*trans*-diacetates as glycosyl donors with various nonacylated glycosyl acceptors in the presence of TMSOTf gave the corresponding β-disaccharides in high yields (❍ *Scheme 19*) [28]. And, Paulsen and Paal reported the use of TMSOTf in reactions of various glycosyl acetate donors with secondary hydroxyl groups of saccharides having low reactivity to obtain the corresponding di- and oligosaccharide in good yields (❍ *Scheme 20*) [29]. They also used TMSOTf in the final coupling reaction for the synthesis of *N*-acetylneuraminic acid-containing trisaccharide (❍ *Scheme 21*) [30]. TMSOTf was also employed as an efficient activator for the rhamnosylation reaction with 1,2-di-*O*-acetyl-3,4-di-*O*-benzyl-α-L-rhamnopyranose in synthesizing the rhamnotriose moiety of the common antigen among oligosaccharides related to Group B Streptococcal polysaccharides by Jennings et al. (❍ *Scheme 22*) [25b]. They also reported the synthesis of a tri- and a tetrasaccharide fragment of the capsular polysaccharide of type III Group B *Streptococcus* by a similar methodology (❍ *Scheme 23*) [25c], [31]. On the other hand, Scharf et al. showed another example of the TMSOTf-mediated rhamnosylation with 1,4-di-*O*-acetyl-3-*O*-methyl-2-*O*-pivaloyl-α-L-rhamnopyranose for the synthesis of a tetradeoxydisaccharide found in the avermectin family (❍ *Scheme 24*) [32], and Lafont et al. reported the TMSOTf-mediated glycosylation of 1,6-anhydro-β-D-mannopyranose derivatives with peracetylated *N*-allyloxycarbonyl-β-D-glucosamine for the synthesis of glycan fragments of glycoproteins (❍ *Scheme 25*) [33].

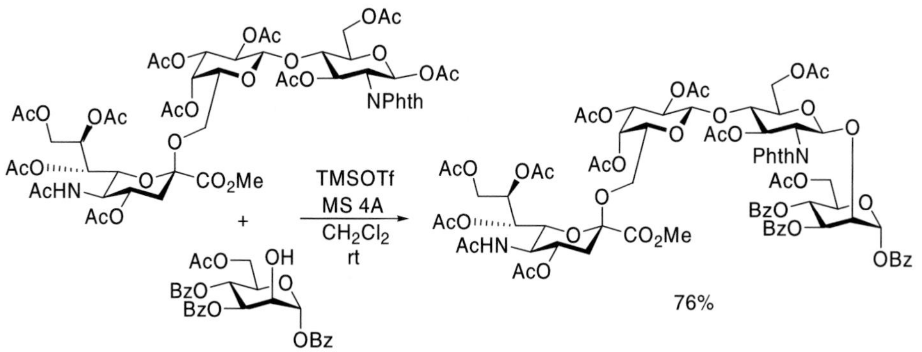

Scheme 21
(Ref. [30])

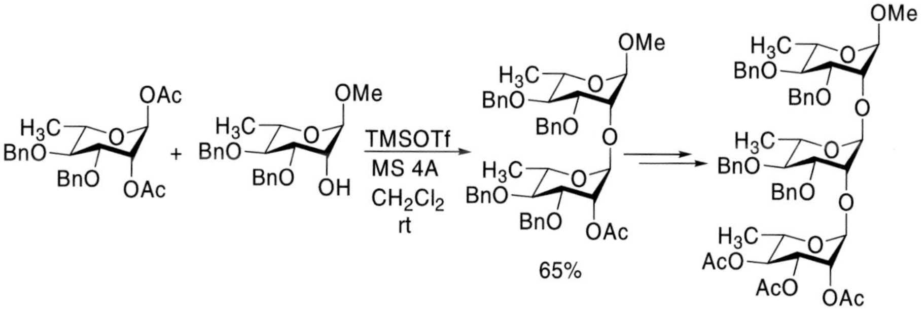

Scheme 22
(Ref. [25b])

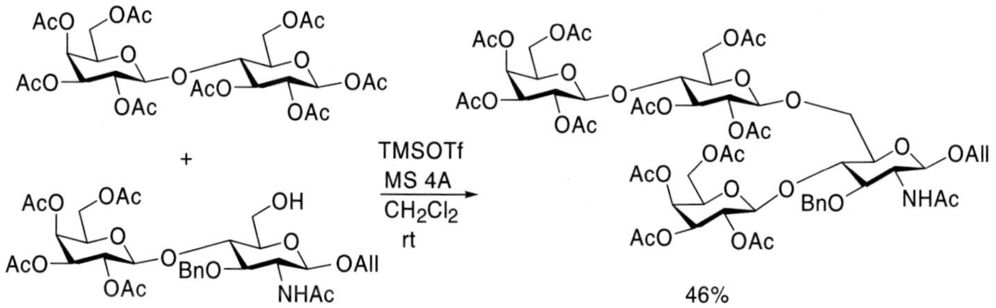

Scheme 23
(Ref. [25c], [31])

On the other hand, Roush and coworkers reported the synthesis of the functionalized C-D-E trisaccharide precursor of Olivomycin A, in which, particularly, the 2-deoxy-2-iodo-3-methyl-α-L-mannopyranosyl linkage was stereoselectively introduced by using the glycosyl acetate donor (❍ *Scheme 26*) [34]. This new α-glycosylation protocol was applied to the synthesis

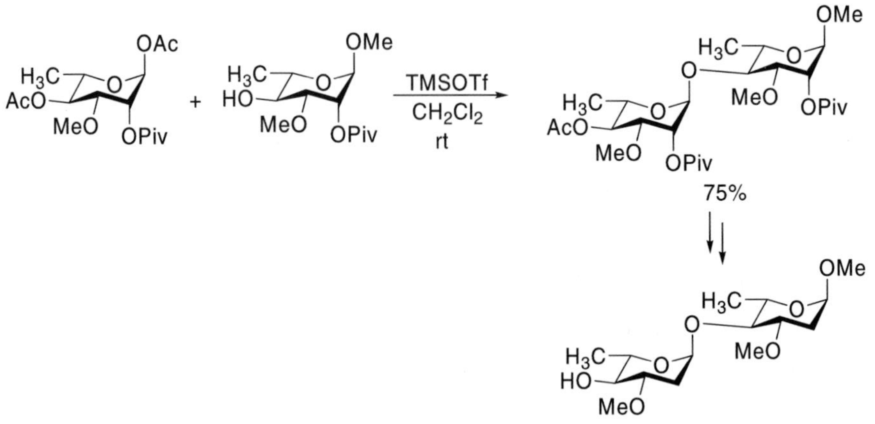

□ Scheme 24
(Ref. [32])

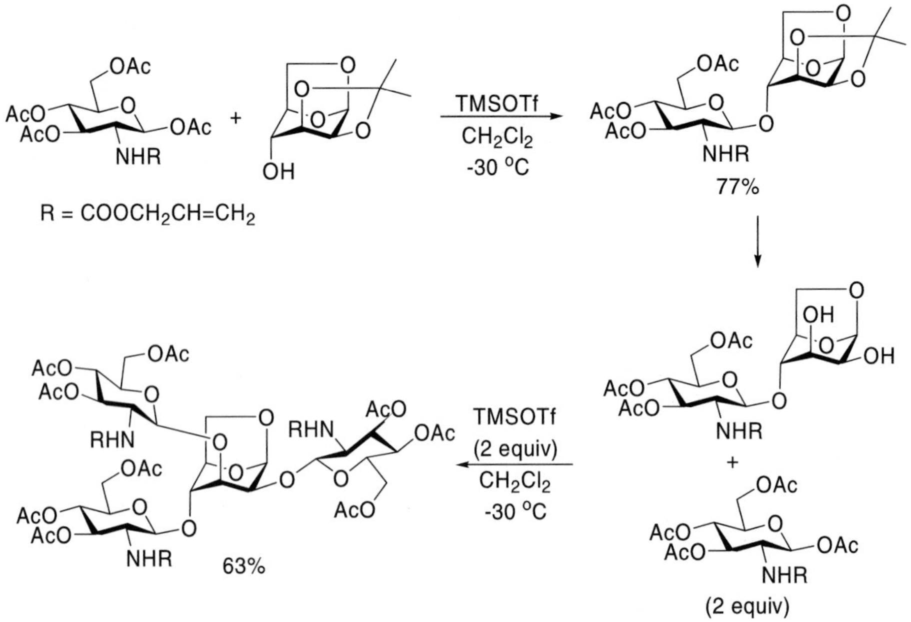

□ Scheme 25
(Ref. [33])

of 2-deoxy-α-D-mannosides and -talosides by reduction of 2-deoxy-2-iodo-α-glycosides generated from the α-stereoselective glycosylation with 2-deoxy-2-iodo glycosyl acetates in the presence of TMSOTf or TBSOTf as the promoter (❖ Scheme 27) [35]. They also demonstrated the stereoselective synthesis of 2-deoxy-β-galactosides using 2-deoxy-2-iodo-galactopyra-

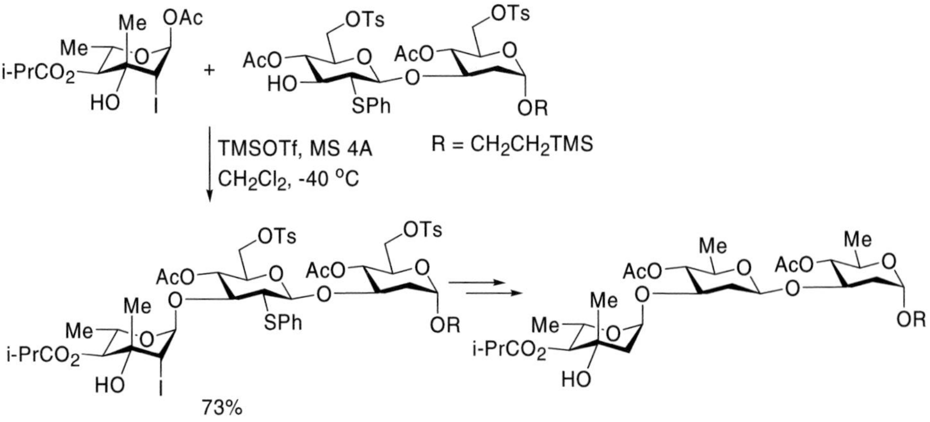

■ Scheme 26
(Ref. [34])

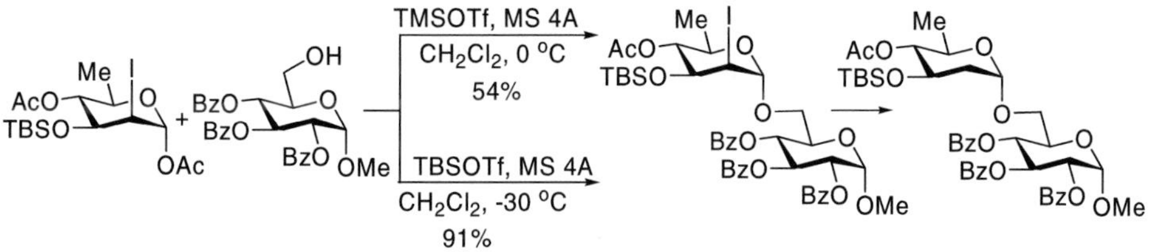

■ Scheme 27
(Ref. [35])

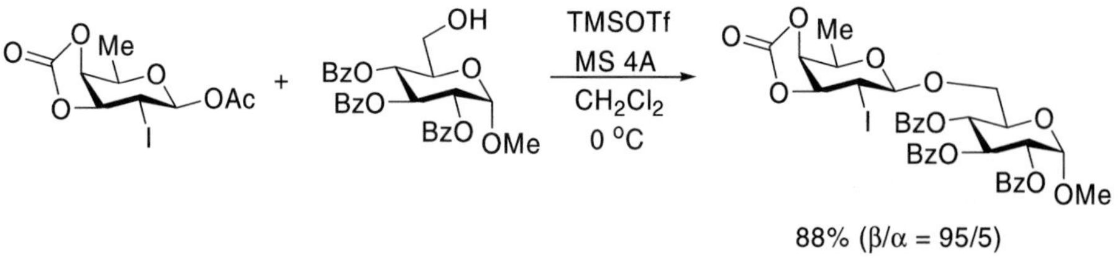

■ Scheme 28
(Ref. [36])

nosyl acetate donors by a similar methodology (❯ Scheme 28) [36]. Kaneko et al. reported an approach to the stereoselective semi-synthesis of GM-237354, a potent inhibitor of fungal elongation factor 2 (EF-2), by employing a highly β-selective glycosylation using 2-deoxy-2-iodo-glycopyranosyl acetate donor developed by the Roush group (❯ Scheme 29) [37].

Roush et al. also reported the use of TBSOTf as an activator for the glycosylation reactions with L-rhodinosyl (2,3,6-trideoxy-L-hexosyl) acetate derivatives in synthesizing the repeat A-B-C trisaccharide unit of Landomycin A (❯ Scheme 30) [38], and for the glycosylation reaction with α-L-rhamnopyranosyl acetate in the early stage of the synthesis of (−)-spinosyn A (❯ Scheme 31) [39].

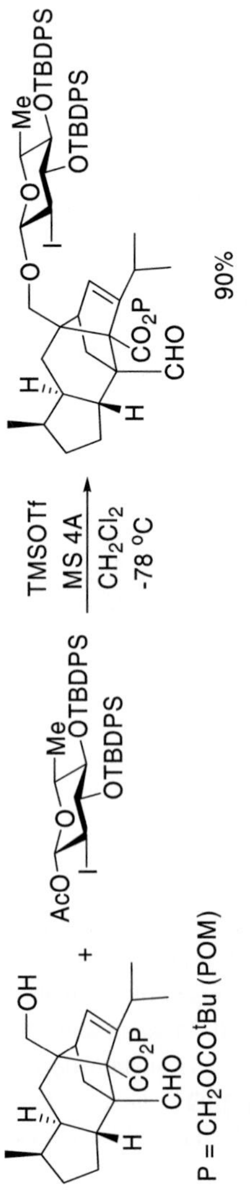

P = CH$_2$OCO^{t}Bu (POM)

■ Scheme 29
(Ref. [37])

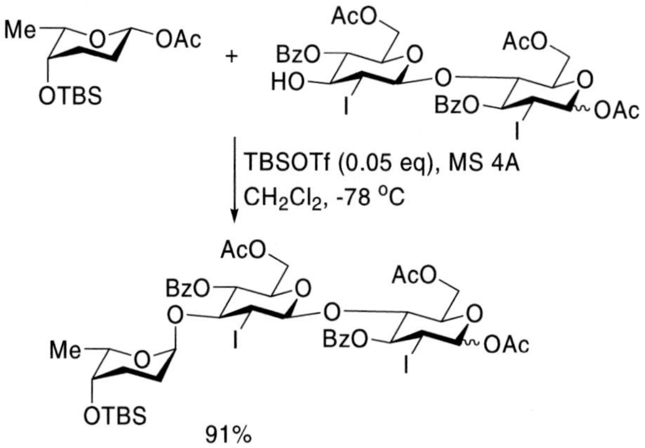

■ Scheme 30
(Ref. [38])

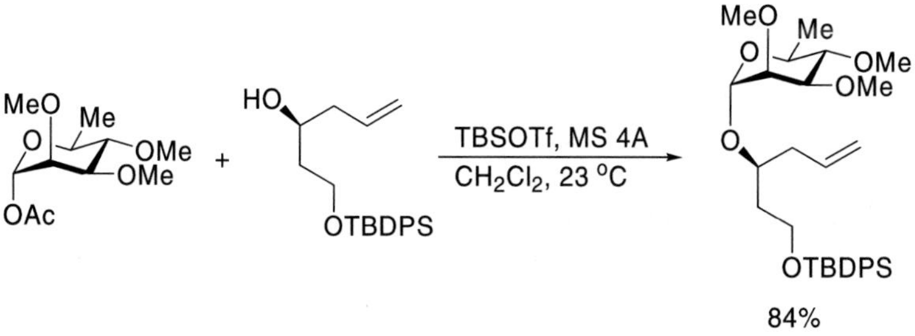

■ Scheme 31
(Ref. [39])

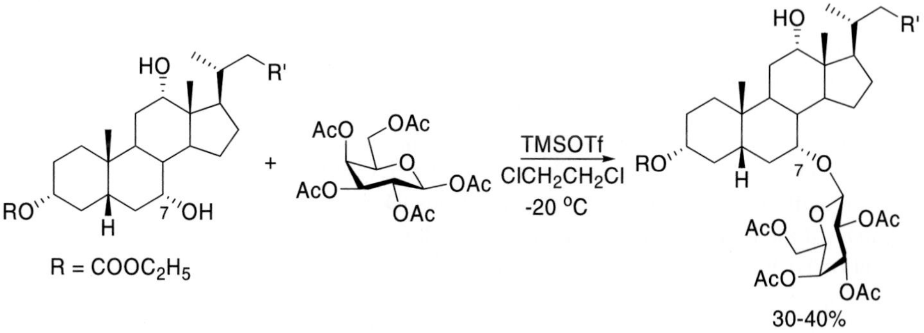

■ Scheme 32
(Ref. [40])

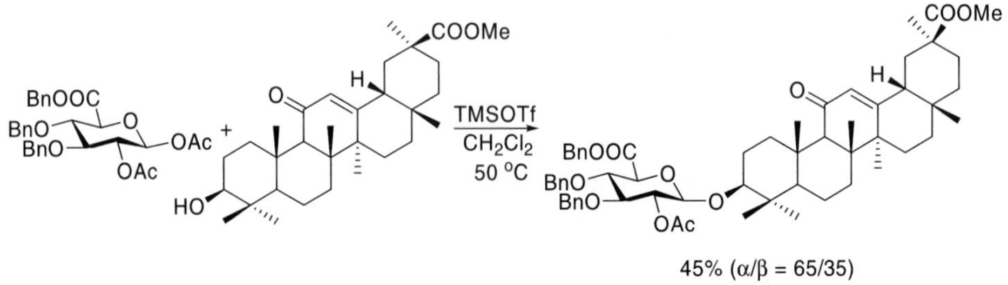

☐ Scheme 33
(Ref. [41])

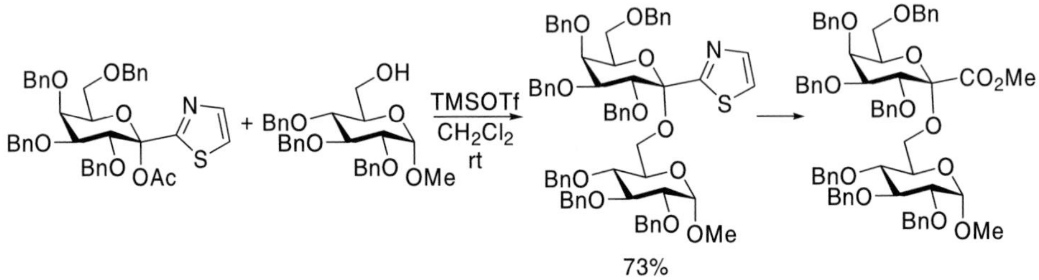

☐ Scheme 34
(Ref. [42])

Nakanishi and coworkers also reported the TMSOTf-mediated glycosylation of a cholic acid derivative with galactose pentaacetate for the synthesis of mosesin-4, a naturally occurring steroid saponin with shark repellent activity, in which the severely hindered 7α-position was selectively glycosylated (❂ *Scheme 32*) [40]. Satto et al. reported that the glycosylation of methyl glycyrrhetinate with methyl D-glucuronatopyranose 1-*O*-acetate in the presence of TMSOTf afforded the corresponding glycoside, a precursor of the glycyrrhetic acid glyco-side, although the stereoselectivity was not satisfactory (❂ *Scheme 33*) [41]. On the oth-er hand, Dondoni et al. introduced the TMSOTf-promoted glycosylation with thiazolyketol acetates as glycosyl donors for the stereoselective synthesis of α-linked ketodisaccharides (❂ *Scheme 34*) [42].

Trityl perchlorate (TrClO₄) was found to be an effective promoter for the anomeric acetates by Mukaiyama and coworkers. They reported that the TrClO₄-promoted glycosylations of alco-hols with 1-*O*-acetyl-β-D-ribose in ether afforded the corresponding β-ribosides exclusively, while the same reaction in the presence of molecular sieves (4Å) and additives such as LiClO₄ provided α-ribosides predominantly in good yields (❂ *Scheme 35*) [43]. Evans and cowork-ers modified the Mukaiyama's stoichiometric TrClO₄ glycosylation by employing a catalytic amount of TrClO₄ [44] and applied the modified method to the glycosylation with glycosyl acetates for the total synthesis of Cytovaricin [44a] and Lepicidin A [44b]. Thus, the catalyt-ic TrClO₄ (ca. 5%)-promoted glycosylation with 2,3,4-tri-*O*-methyl-α-D-rhamnosyl acetate in toluene provided the α-glycoside in 87% yield along with 5% of the undesired β-anomer, and the α-anomer was further transformed into the complex macrolide insecticide, Lepicidin A (❂ *Scheme 36*) [44b].

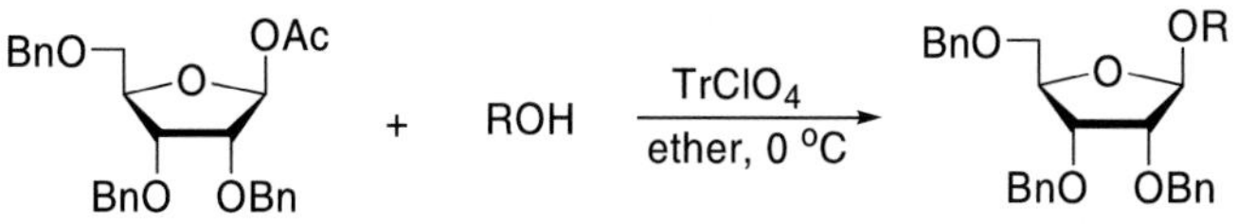

ROH = 1-octadecanol, 95% (β only)
cyclohexanol, 83% (β only)
3β-cholestanol, 88% (β only)

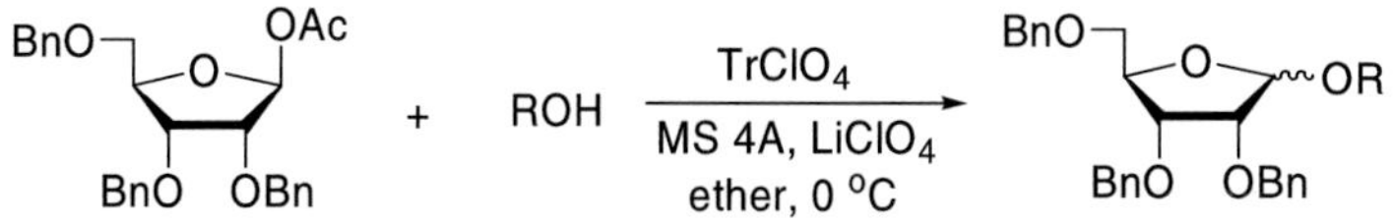

ROH = 2-propanol, 91% (α/β = 73/27)
cyclohexanol, 86% (α/β = 70/30)
3β-cholestanol, 75% (α/β = 79/21)

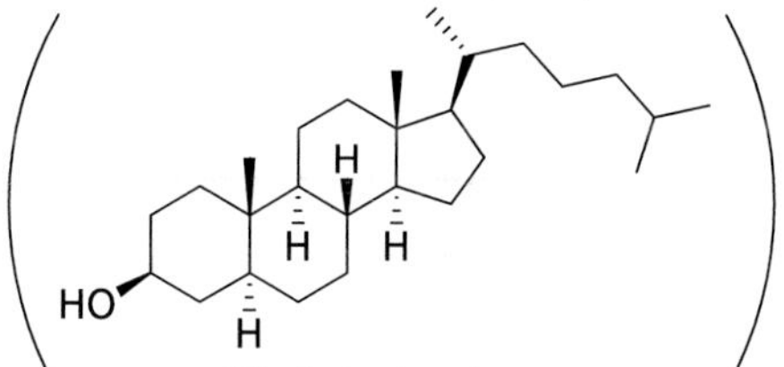

3β-cholestanol

☐ Scheme 35
(Ref. [43])

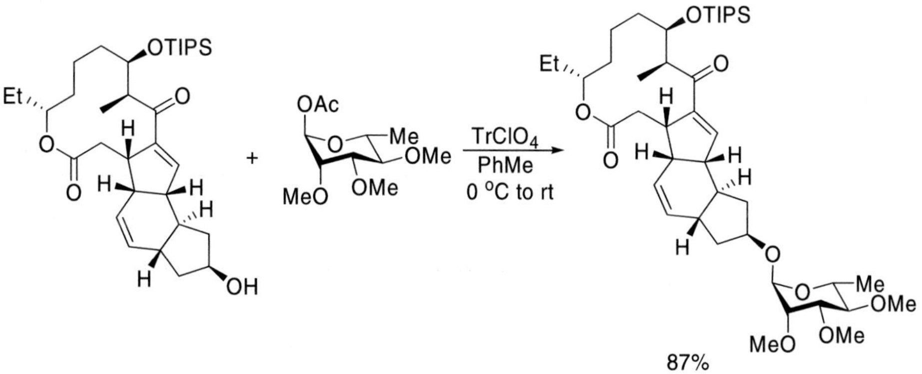

87%
(+ 5% β-anomer)

☐ Scheme 36
(Ref. [44b])

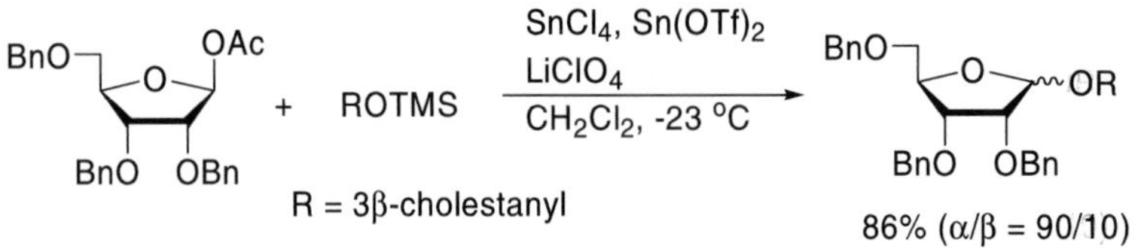

■ Scheme 37
(Ref. [45])

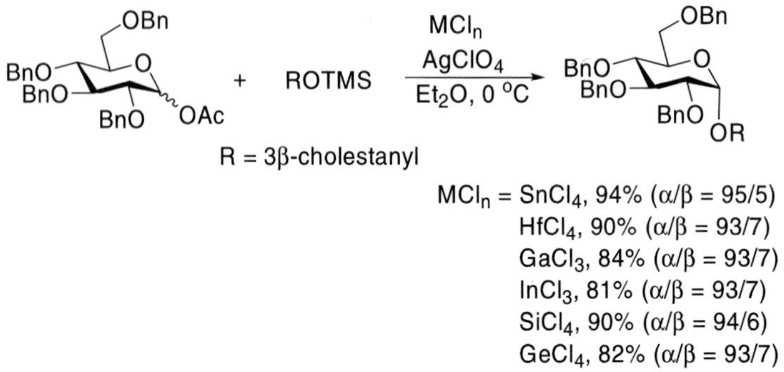

$MCl_n = SnCl_4$, 94% ($\alpha/\beta = 95/5$)
$HfCl_4$, 90% ($\alpha/\beta = 93/7$)
$GaCl_3$, 84% ($\alpha/\beta = 93/7$)
$InCl_3$, 81% ($\alpha/\beta = 93/7$)
$SiCl_4$, 90% ($\alpha/\beta = 94/6$)
$GeCl_4$, 82% ($\alpha/\beta = 93/7$)

■ Scheme 38
(Ref. [46,47])

Mukaiyama and coworkers also introduced tin-based Lewis acids such as $SnCl_4/Sn(OTf)_2/$ $LiClO_4$ and $SnCl_4/AgClO_4$ for the glycosylation of trimethylsilylated alcohols with the 1-O-acetyl sugar. For example, 1,2-cis-α-ribosides were predominantly obtained from 1-O-acetyl-riboses and silylated 3β-cholestanol (● Scheme 37) [45]. On the basis of the effectiveness of the $SnCl_4/AgClO_4$ combination as a promoter [46], they also investigated the glycosylation of silylated alcohols with 1-O-acetyl-2,3,4,6-tetra-O-benzyl-D-glucopyranose using a series of combinations of other Lewis acids and $AgClO_4$. Among them, the catalysts generated from $HfCl_4$, $GaCl_3$, $InCl_3$, $SiCl_4$, or $GeCl_4$, with $AgClO_4$ led to the formation of the corresponding glycosides effectively in high diastereomeric ratios (● Scheme 38) [47].

76% ($\alpha/\beta = 83/17$)

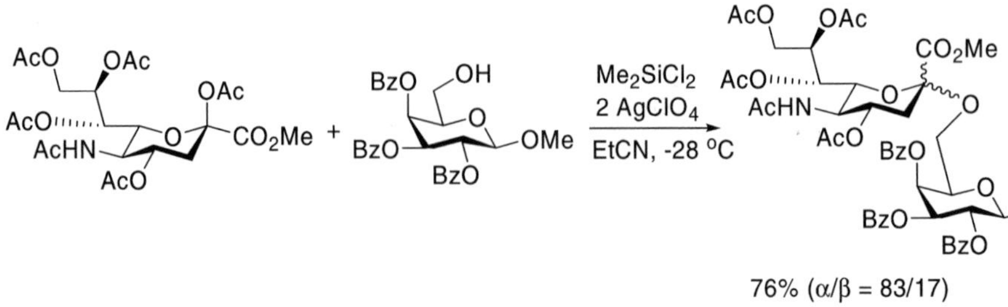

■ Scheme 39
(Ref. [48])

Furthermore, Mukaiyama et al. investigated the glycosylation with *N*-acetyl-α-neuraminosyl acetate for the stereoselective synthesis of *N*-acetyl-α-neuraminosyl-galactose disaccharide by using a variety of combinations of Lewis acids, for example, SiCl₄, TiCl₄, SnCl₄, MeSiCl₃, Me₂SiCl₂, or PhSiCl₃, with AgClO₄, in which the combination of Me₂SiCl₂ and AgClO₄ proved to be the most suitable promoter system (❿ *Scheme 39*) [48]. Kakinuma et al. applied the SnCl₄/AgClO₄ combination as the promoter to the glycosylation of an *O*-TMS-aglycone with the appropriately protected 1-*O*-acetyl aminosugar for the total synthesis of Vicenistatin, a novel 20-membered macrocyclic lactam antitumor antibiotic (❿ *Scheme 40*) [49].

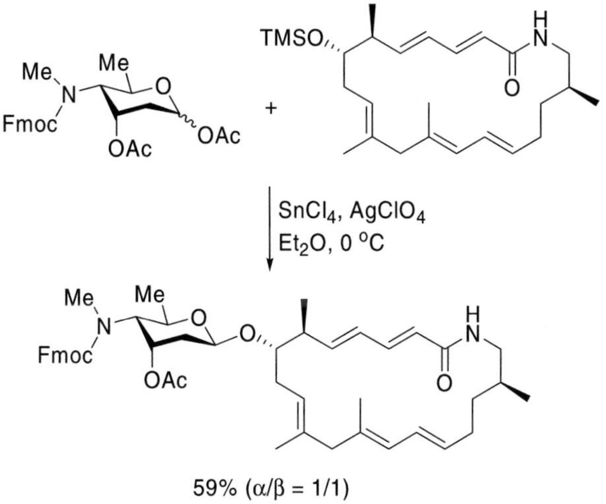

❏ Scheme 40
(Ref. [49])

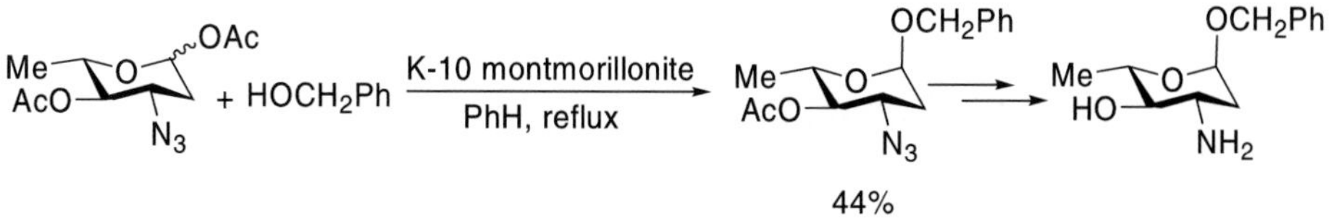

❏ Scheme 41
(Ref. [50])

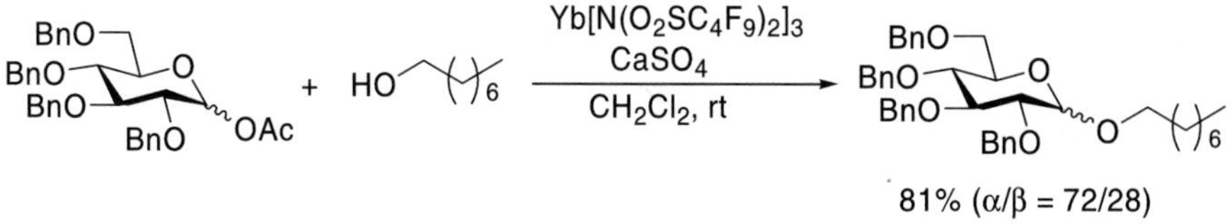

❏ Scheme 42
(Ref. [51])

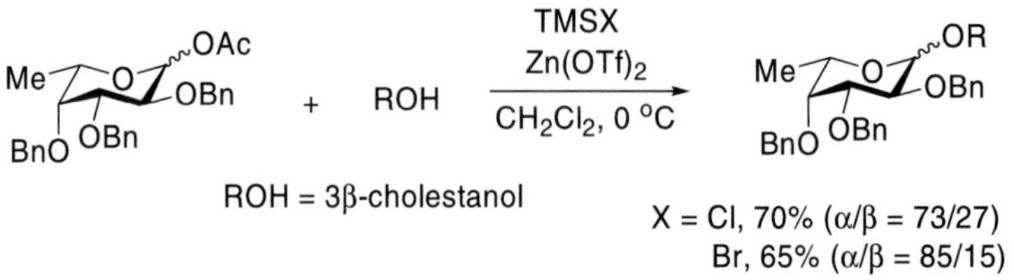

ROH = 3β-cholestanol

X = Cl, 70% (α/β = 73/27)
Br, 65% (α/β = 85/15)

□ Scheme 43
(Ref. [52])

92% (α/β = 97/3)

□ Scheme 44
(Ref. [53])

Besides these promoters for the activation of glycosyl acetate donors, K-10 montmorillonite was used as a new inexpensive catalyst in the glycosylation of the simple alcohol such as benzyl alcohol or methanol to give benzyl or methyl acosaminide with high stereoselectivity (❍ Scheme 41) [50], and a catalytic amount of ytterbium(III) tris[bis(perfluorobutylsulfonyl)amide] $(Yb[N(O_2SC_4F_9)_2]_3)$ was introduced as a new promoter for the glycosylation with glucosyl 1-acetate (❍ Scheme 42) [51]. And, the combination of trimethylsilyl halides and $Zn(OTf)_2$ was also employed to give the corresponding glycoside in good yield from benzyl-protected fucosyl acetate (❍ Scheme 43) [52]. In particular, a combination of TMSCl and $Zn(OTf)_2$ afforded α-mannosides stereoselectively from benzyl-protected mannosyl acetate without the participating group at C-2 (❍ Scheme 44) [53].

3 Glycosylation with Donors Having Other Anomeric Ester Groups

Besides glycosyl acetate, a variety of other acyl groups were employed as good anomeric leaving groups. The representative acyl leaving groups used in glycosyl donors for the glycosylation are summarized in ❍ Table 2.
Mukaiyama and coworkers prepared some 1-O-haloacetyl-β-D-glucopyranoses by treating 2,3,4,6-tetra-O-benzyl-D-glucopyranose with haloacetyl chloride in the presence of cesium fluoride or potassium fluoride and performed the glycosylation reactions with the 1-O-haloacetyl glucoses in the presence of $TrClO_4$ to obtain the corresponding glycosides in good yields (❍ Scheme 45) [43]. Furthermore, they investigated the reactivity of various glucoses having substituted acetyl groups at the anomeric center, such as $-COCH_2OMe$, $-COCH_2CH_2OMe$, $-COCH_2SMe$, and $-COCH_2OCH_2CH_2OMe$, in the glycosylation reaction in the presence of $SnCl_4$ and $AgClO_4$ as a promoter system and they reported that the glucopyranose having the (2-methoxyethoxy)acetyl $(-COCH_2OCH_2CH_2OMe)$ group at the anomeric

Table 2
Glycosylation with 1-O-ester sugars

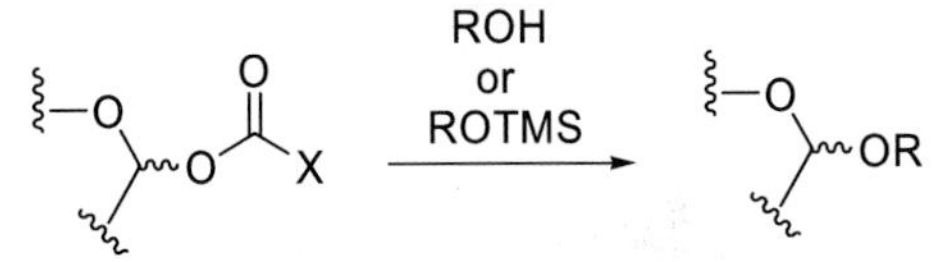

X	Activator	References
–CHCl$_2$, –CH$_2$Br, –CH$_2$I	TrClO$_4$, SnCl$_4$/AgClO$_4$	[43,54]
–CH$_2$OCH$_2$CH$_2$OMe	SnCl$_4$/AgClO$_4$	[54]
–CF$_3$, –CCl$_3$	TMSOTf, BF$_3$·OEt$_2$	[55,56,57,58]
–Ph	FeCl$_3$, TMSOTf	[10,59]
–Ph-p-NO$_2$	TMSOTf, BF$_3$·OEt$_2$, TMSCl/Zn(OTf)$_2$	[52,53,60,61,62]
–2-pyridine	Cu(OTf)$_2$, Sn(OTf)$_2$	[63]
–C(CH$_3$)$_3$	SnCl$_4$, TMSOTf	[64,65,66]
–CH$_2$CH$_2$CH=CH$_2$	IDCP, NIS, PhSeOTf	[67,68,69,70,71]
–CH$_2$CH$_2$CH$_2$C≡CH	Hg(OTf)$_2$	[72]

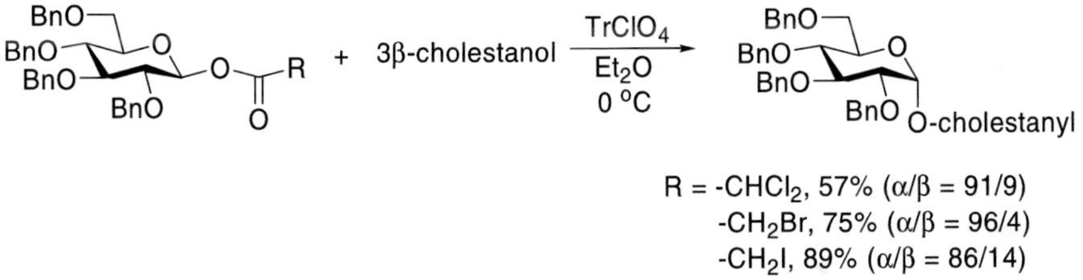

R = -CHCl$_2$, 57% (α/β = 91/9)
-CH$_2$Br, 75% (α/β = 96/4)
-CH$_2$I, 89% (α/β = 86/14)

Scheme 45
(Ref. [43])

center was the most efficient donor among them (❯ Scheme 46) [54]. The trihaloacetyl group was introduced as a new leaving group at the anomeric center in the glycosylation reaction by Cai and coworkers. 2,3,4,6-Tetra-O-acetyl-1-O-trifluoroacetyl-α-D-glucopyranose was found to be a good glycosyl donor in the presence of TMSOTf [55] and 1-O-trichloroacetyl 2,3,4-tri-O-benzoyl-α-L-rhamnopyranose was employed as an efficient donor to prepare disaccharide and trisaccharide residues, the sugar cores of phenylpropanoid glycosides (❯ Scheme 47) [56]. They also reported the glycosylation with 2,3,4,6-O-tetrabenzyl-1-α-D-galactopyranosyl trichloroacetate [57] and with 2,3,4,6-tetra-O-benzoyl-α-D-mannopyranosyl trichloroacetate [58] in the presence of BF$_3$·OEt$_2$ or TMSOTf.

On the other hand, benzoyl and p-nitrobenzoyl groups were employed by Lerner as good anomeric leaving groups and could be activated by FeCl$_3$, TMSOTf, or BF$_3$–OEt$_2$ like the acetyl group [10]. Charette et al. reported that the catalytic use of TMSOTf promoted the glycosylation of the trimethylsilyl ether of acceptor alcohols with 1-O-benzoyl sugar (❯ Scheme 48) [59]. Terashima and coworkers used the 1-O-(p-nitrobenzoyl)glycosyl donor and TMSOTf for the synthesis of anthracycline antibiotics (❯ Scheme 49) [60] while Scharf

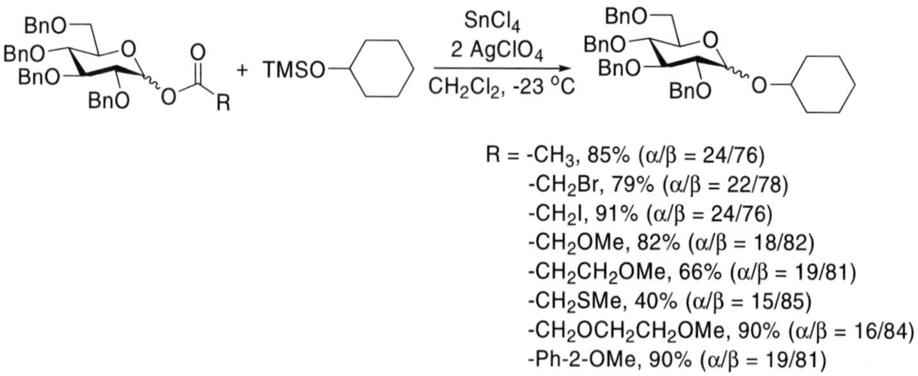

R = -CH₃, 85% (α/β = 24/76)
-CH₂Br, 79% (α/β = 22/78)
-CH₂I, 91% (α/β = 24/76)
-CH₂OMe, 82% (α/β = 18/82)
-CH₂CH₂OMe, 66% (α/β = 19/81)
-CH₂SMe, 40% (α/β = 15/85)
-CH₂OCH₂CH₂OMe, 90% (α/β = 16/84)
-Ph-2-OMe, 90% (α/β = 19/81)

☐ **Scheme 46**
(Ref. [54])

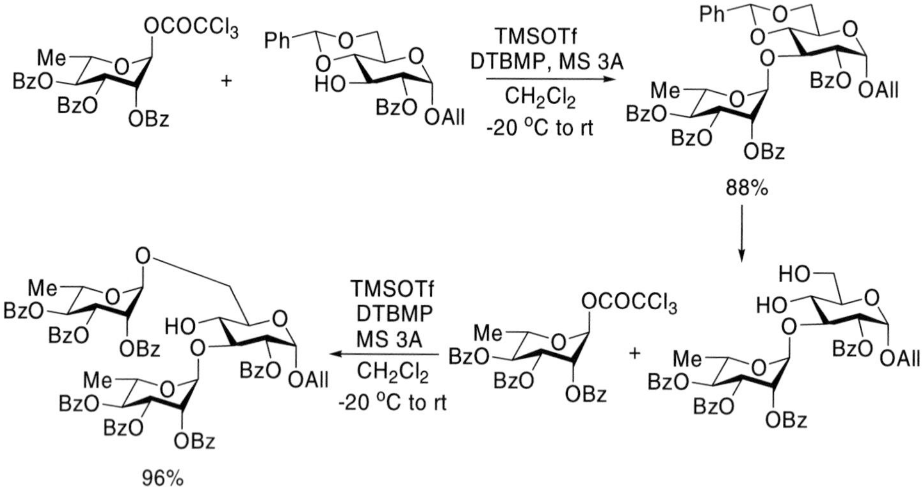

☐ **Scheme 47**
(Ref. [56])

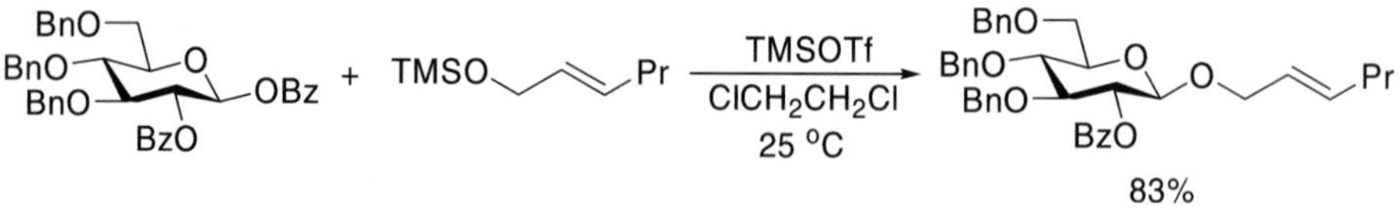

☐ **Scheme 48**
(Ref. [59])

and coworkers applied Terashima's method to their synthetic study on everninomicin antibiotics (❯ *Scheme 50*) [61]. Rosso et al. reported the use of the 1-*O*-(*p*-nitrobenzoyl)glycosyl donor for the practical synthesis of Disaccharide H, 2-*O*-(α-L-fucopyranosyl)-D-galactopyranose—which is one of the biologically important L-fucose-containing oligosaccharides—in

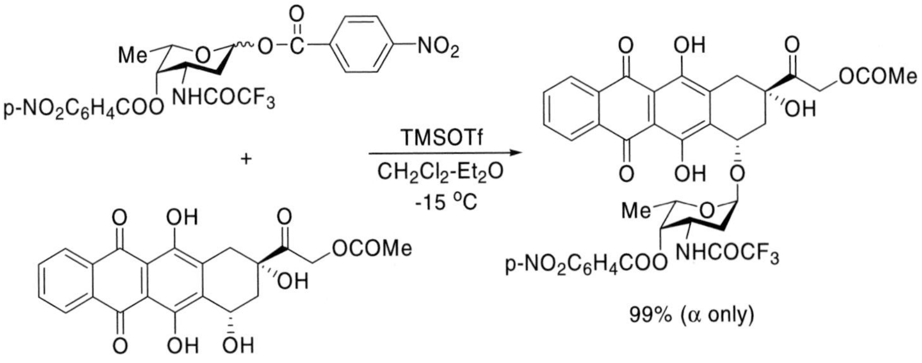

Scheme 49
(Ref. [60])

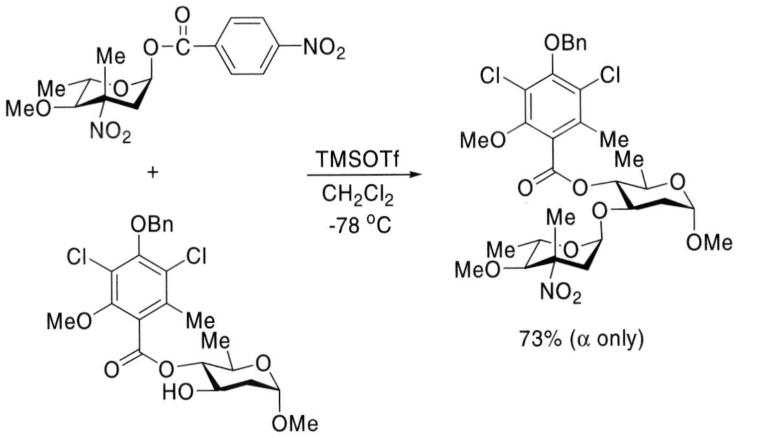

Scheme 50
(Ref. [61])

the presence of BF$_3$·OEt$_2$ (◉ *Scheme 51*) [62]. The p-nitrobenzoate as a leaving group was also activated by the combination of TMSCl and Zn(OTf)$_2$ and thus α-mannosides were obtained from benzyl-protected 1-O-(p-nitrobenzoyl) mannopyranose (◉ *Scheme 52*) [52,53]. Kobayashi et al. reported that glycosyl 2-pyridinecarboxylate, as a glycosyl donor, could be activated by Cu(OTf)$_2$ in Et$_2$O or Sn(OTf)$_2$ in MeCN to predominantly produce the corresponding α- or β-glucoside, respectively (◉ *Scheme 53*) [63].

In addition, the pivaloyl group has been employed as an anomeric leaving group. Thus, the fully O-pivaloyl-protected galactofuranose glycosyl donor reacted with alkyl primary alcohol acceptors in the presence of SnCl$_4$ as the promoter (◉ *Scheme 54*) [64]. Kunz et al. also reported that the fully O-pivaloyl-protected galactopyranosyl donor reacted with 6-O-protected 2-azido-galactose in the presence of TMSOTf to give the precursor structure of the Thomsen–Friedenreich antigen disaccharide, β(1–3)-linked galactosyl galactosamine, with high regioselectivity, although in low yield (◉ *Scheme 55*) [65]. Very recently, pivaloylated glucosamin-

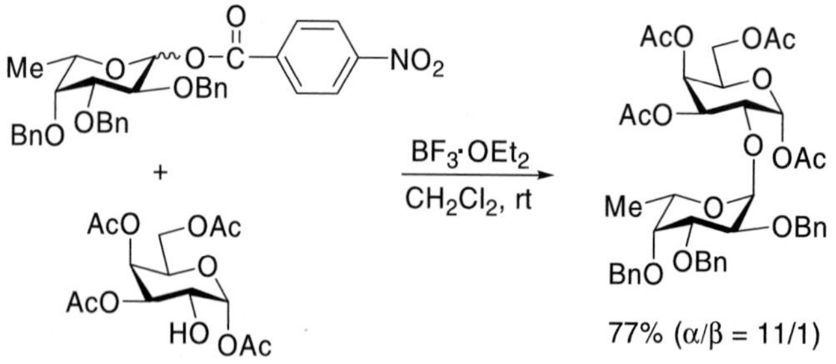

Scheme 51
(Ref. [62])

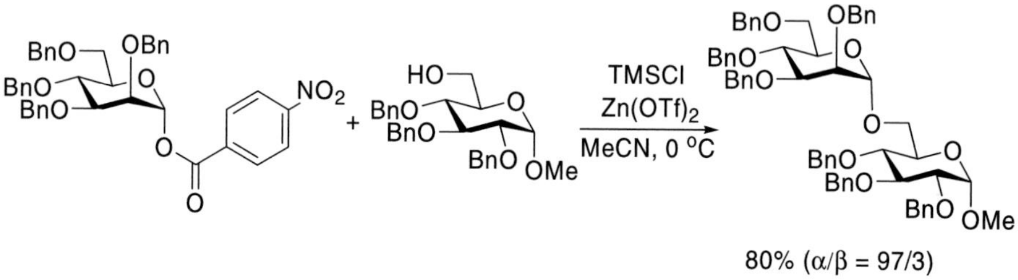

77% (α/β = 11/1)

Scheme 52
(Ref. [52,53])

80% (α/β = 97/3)

MX_n = Cu(OTf)$_2$ in Et$_2$O, 97% (α/β = 83/17)
Sn(OTf)$_2$ in MeCN/CH$_2$Cl$_2$, 77% (α/β = 9/91)

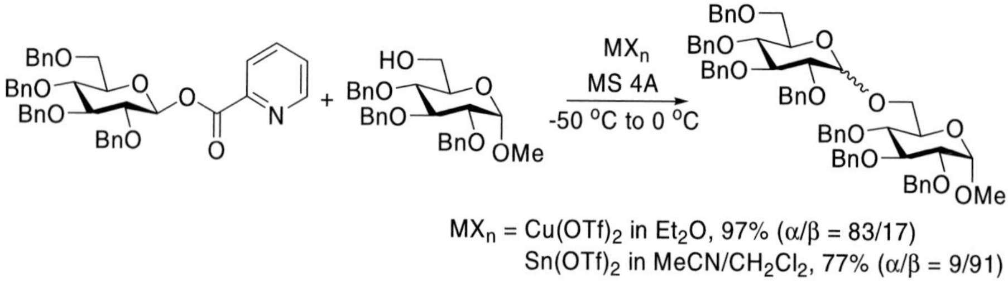

Scheme 53
(Ref. [63])

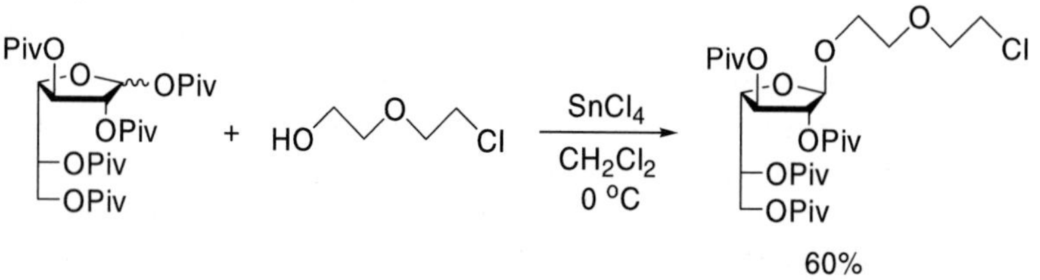

Scheme 54
(Ref. [64])

60%

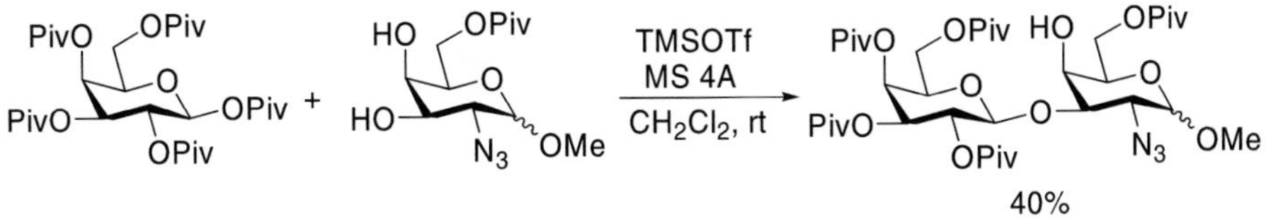

Scheme 55
(Ref. [65])

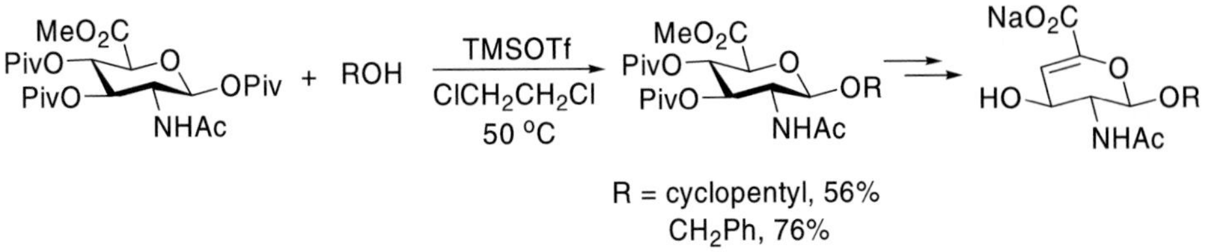

R = cyclopentyl, 56%
CH$_2$Ph, 76%

Scheme 56
(Ref. [66])

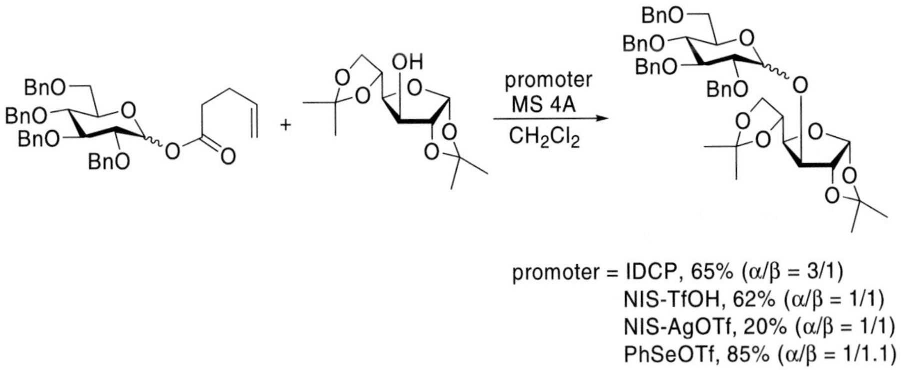

promoter = IDCP, 65% (α/β = 3/1)
NIS-TfOH, 62% (α/β = 1/1)
NIS-AgOTf, 20% (α/β = 1/1)
PhSeOTf, 85% (α/β = 1/1.1)

Scheme 57
(Ref. [67,69])

uronates as glycosyl donors were employed with TMSOTf to provide a series of novel β-*O*-glycosides, which are the precursors of unsaturated *N*-acetyl-D-glucosaminuronic acid glycosides as inhibitors of the influenza virus sialidase (**❯** *Scheme 56*) [66].

Kunz [67] and Fraser-Reid [68] et al. introduced glucosyl 4-pentenoates as new glycosyl donors. These glucosyl 4-pentenoates were prepared by the condensation of the corresponding glucopyranoses with 4-pentenoic acid using 1,3-dicyclohexylcarbodiimide (DCC) or diisopropylcarbodiimide (DIC) and activated with soft electrophiles, for example, iodonium compounds, iodonium dicollidine perchlorate (IDCP) or *N*-iodosuccinimide (NIS) (**❯** *Scheme 57*), or 1,3-dithian-2-ylium tetrafluoroborate (**❯** *Scheme 58*). However, the glycosylations with the pentenoates using these promoters were not very efficient. Very recently, Kim and coworkers reported that phenylselenyl triflate (PhSeOTf) was found to be a much more efficient promoter than IDCP, NIS-TfOH, and 1,3-dithian-2-yl tetrafluoroborate for glycosylations with glyco-

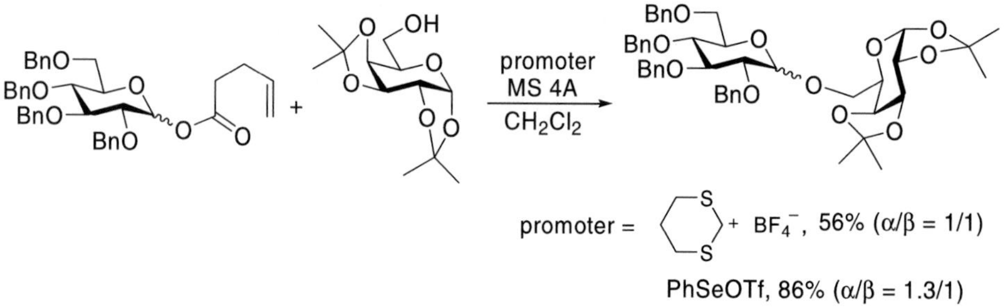

Scheme 58
(Ref. [68,69])

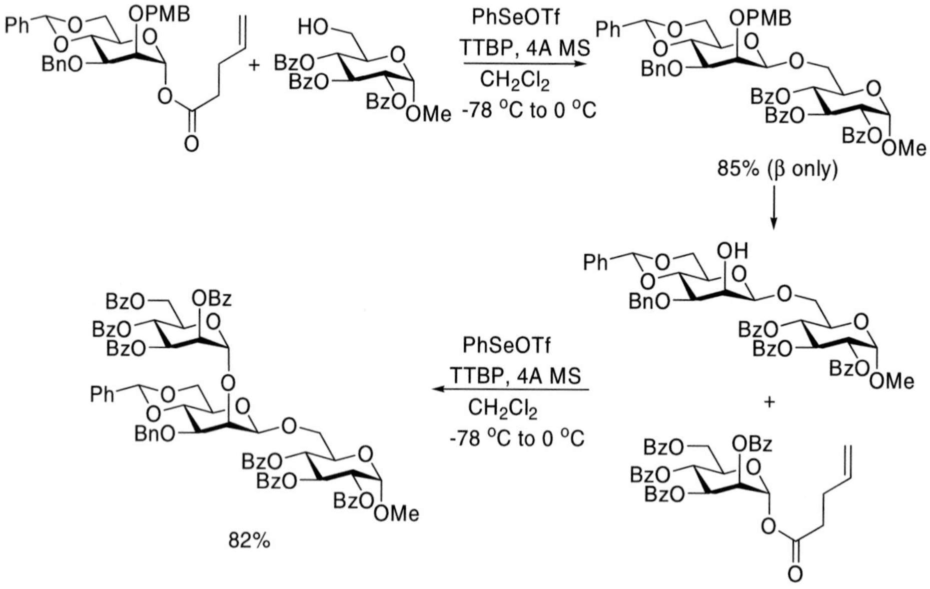

Scheme 59
(Ref. [70])

syl 4-pentenoates as donors (❷ *Scheme 57* and ❷ *Scheme 58*) [69]. Moreover, they described a highly reactive and stereoselective procedure for the β-mannopyranosylation employing 4,6-*O*-benzylidene mannopyranosyl pentenoate and PhSeOTf (❷ *Scheme 59*), and this method was proved to be comparable to and even better than other methods for the β-mannosylation of the simple reactive primary alcohols [70]. On the other hand, the reactivity and stereoselectivity of 4-pentenoic acid ester of *N*-acetylneuraminic acid as a sialyl donor were also investigated with a series of promoters, such as NIS-TESOTf, dimethyl(methylthio)sulfonium triflate (DMTST), and IDCP, but the glycosylation with the sialyl pentenoate gave the *O*-sialosides in low yield and low stereoselectivity (❷ *Scheme 60*) [71].

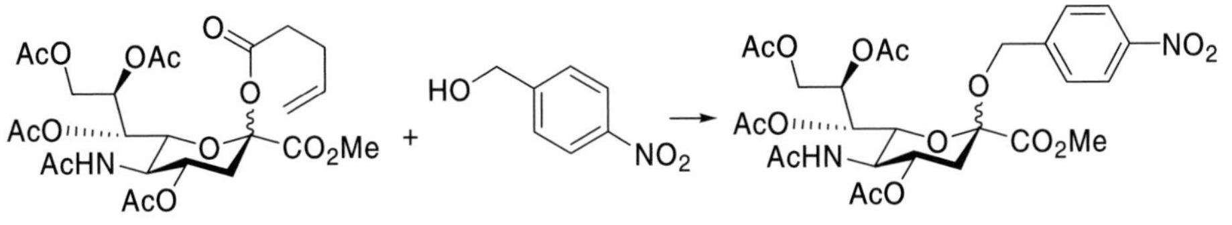

NIS-TESOTf in MeCN, 68% ($\alpha/\beta = 2/1$)
DMTST in CH$_2$Cl$_2$, 48% ($\alpha/\beta = 1/4$)
IDCP in MeCN, trace

☐ Scheme 60
(Ref. [71])

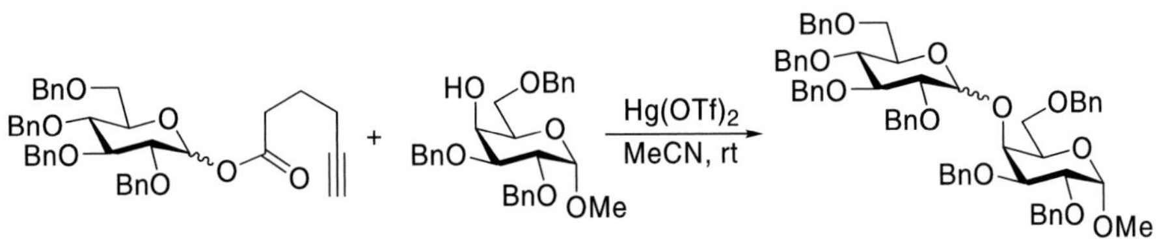

80% ($\alpha/\beta = 24/76$)

☐ Scheme 61
(Ref. [72])

Recently, Nishizawa et al. reported a mercuric triflate [Hg(OTf)$_2$]-catalyzed glycosylation using the alkynoate as the leaving group (❯ *Scheme 61*) [72].

4 Glycosylation with Glycosyl Carbonates and Related Donors

Some representative methods using 1-*O*-carbonate derivatives as glycosyl donors are summarized in ❯ *Table 3*.

Mukaiyama et al. reported the glycosylation with the *N*-acetylneuraminic acid donor having an easily accessible phenoxycarbonyloxy leaving group at the anomeric center using a combination of Me$_2$SiCl$_2$ and AgClO$_4$ as the promoter for the stereoselective synthesis of *N*-acetyl-α-neuraminosyl-galactose disaccharide (❯ *Scheme 62*) [48]. They also reported the highly stereoselective synthesis of 1,2-*trans*-glycosides using *p*-chlorobenzylated glycosyl carbonates as glycosyl donors in the presence of trityl tetrakis(pentafluorophenyl)borate [TrB(C$_6$F$_5$)$_4$], in which several β-D-glucopyranosides were prepared in good yields with high stereoselectivities even in the absence of the anchimeric assistance by the neighboring group at C-2 position (❯ *Scheme 63*) [73,74]. Furthermore, they applied the same method using a galactosyl phenylcarbonate as the glycosyl donor to the convergent total synthesis of the F1α antigen, a member of the tumor-associated *O*-linked mucin glycosyl amino acids (❯ *Scheme 64*) [75]. On the other hand, Sinaÿ and Descotes independently introduced the isopropenyloxycarbonyloxy leaving group at the anomeric center for the glycosylation reaction with TMSOTf. Sinaÿ

◻ **Table 3**
Glycosylation with 1-*O*-carbonates and related sugars

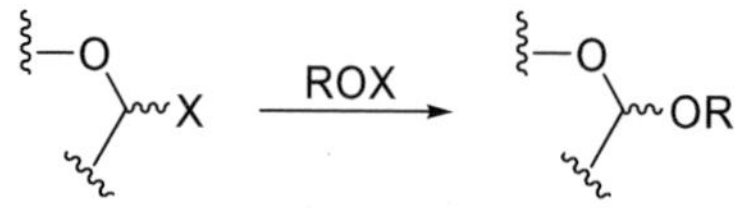

References	X	Activator
—O–C(=O)–OPh	Me₂SiCl₂/AgClO₄, TrB(C₆F₅)₄	[48,73,74,75]
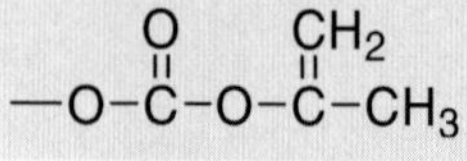 —O–C(=O)–O–C(=CH₂)–CH₃	TMSOTf	[76]
—O–C(=O)–N (imidazolyl)	ZnBr₂	[77]
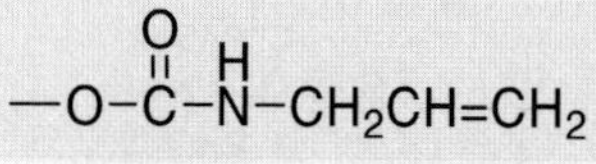 —O–C(=O)–N(H)–CH₂CH=CH₂	IDCP	[78]
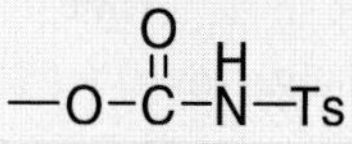 —O–C(=O)–N(H)–Ts	TMSOTf	[79]
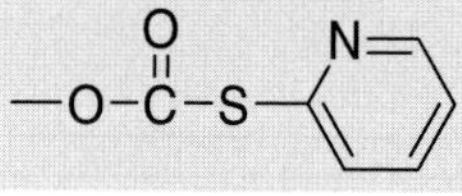 —O–C(=O)–S–(2-pyridyl)	AgOTf	[80,81]

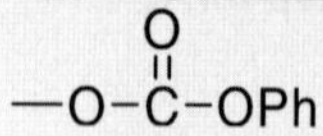

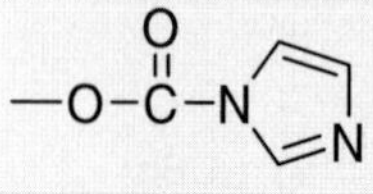

86% (α/β = 86/14)

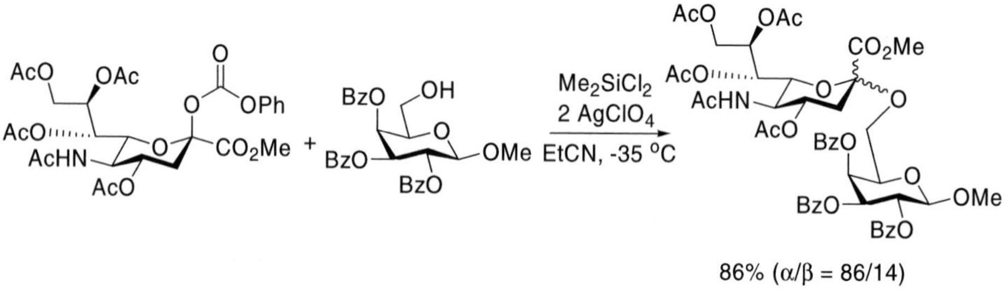

◻ **Scheme 62**
(Ref. [48])

and coworkers showed the high-yielding β-stereoselective glycosylation employing glucosyl isopropenylcarbonate as the glycosyl donor (❸ *Scheme 65*) [76].
Ley and Ford reported the C-1 (1-imidazolylcarbonyl) glycosides as new glycosyl donors, which reacted with a series of alcohol acceptors in the presence of ZnBr₂ to give the corresponding *O*-glycosides (❸ *Scheme 66*) [77]. Kunz and Zimmer reported the *N*-allyl carbamate as an anomeric leaving group with soft electrophiles, for example, IDCP, DMTST, and methyl bis-methylthiosulfonium hexachloroantimonate, as promoters for the *O*-glycoside synthesis (❸ *Scheme 67*) [78]. Kiessling and Hinklin reported glycosyl sulfonylcarbamates

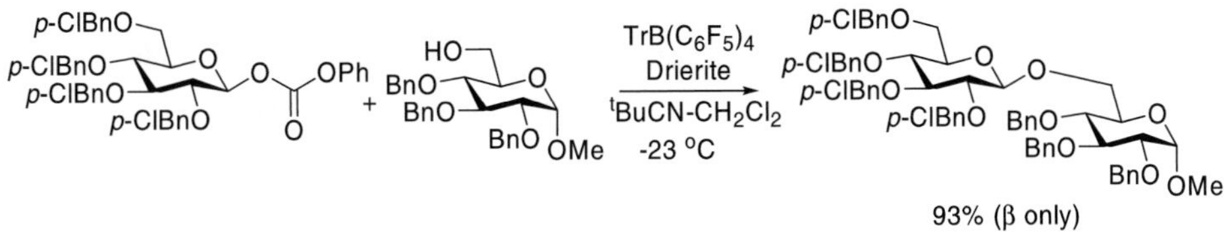

Scheme 63
(Ref. [73,74])

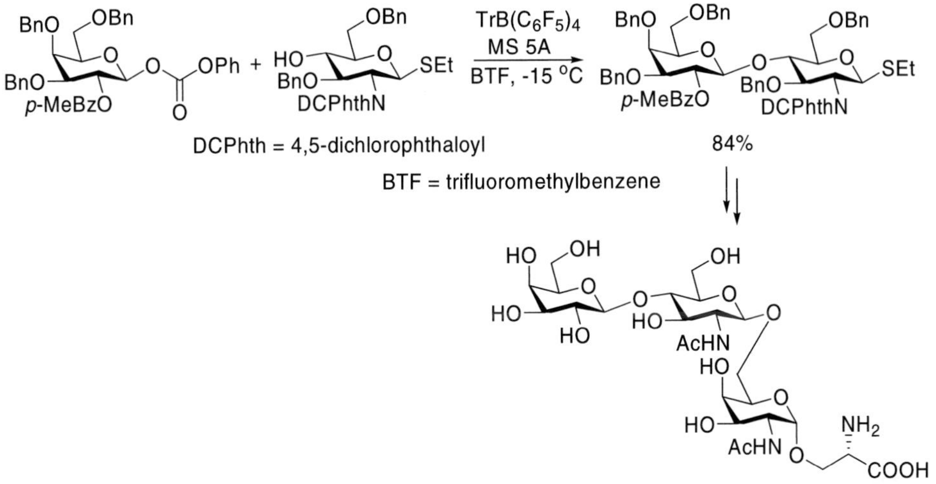

DCPhth = 4,5-dichlorophthaloyl

BTF = trifluoromethylbenzene

Scheme 64
(Ref. [75])

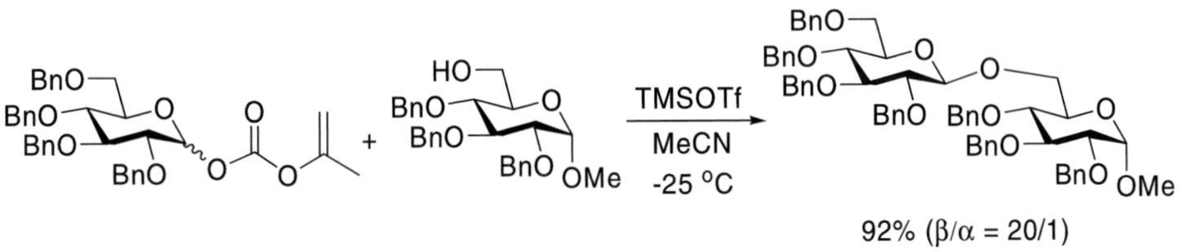

Scheme 65
(Ref. [76])

as new glycosyl donors with tunable reactivity by introducing different *N*-alkyl substituents on the sulfonylcarbamate group. For example, the glycosylation of a glucose with tetrabenzylgalactosyl *N*-tosylcarbamate was more efficient than that with tetrabenzylgalactosyl *N,N*-(cyanomethyl)tosylcarbamate while the glycosylation of a poor nucleophilic phenol acceptor with tetrabenzylgalactosyl *N,N*-(cyanomethyl)tosylcarbamate turned out to be more efficient and stereoselective than that with tetrabenzylgalactosyl *N*-tosylcarbamate (**◉** *Scheme 68*) [79].

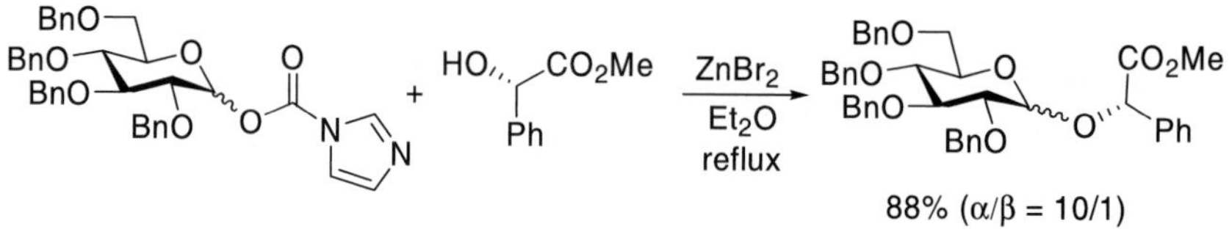

BnO
BnO
BnO
BnO
O
O
O
N
N
+
HO
CO2Me
Ph
ZnBr2
Et2O
reflux
CO2Me
Ph
88% (α/β = 10/1)

Scheme 66
(Ref. [77])

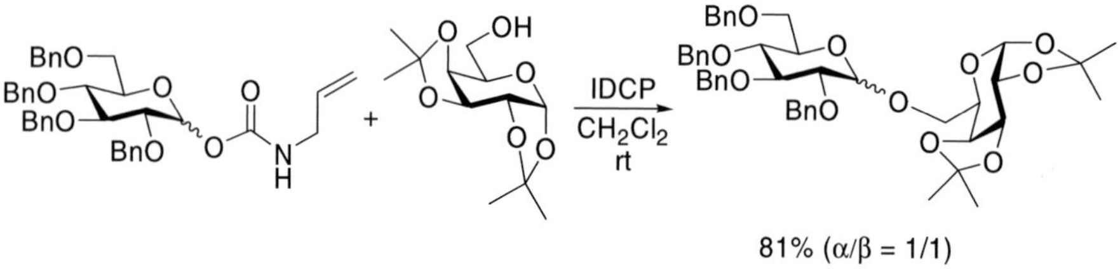

BnO
BnO
BnO
BnO
N
H
+
OH
IDCP
CH2Cl2
rt
81% (α/β = 1/1)

Scheme 67
(Ref. [78])

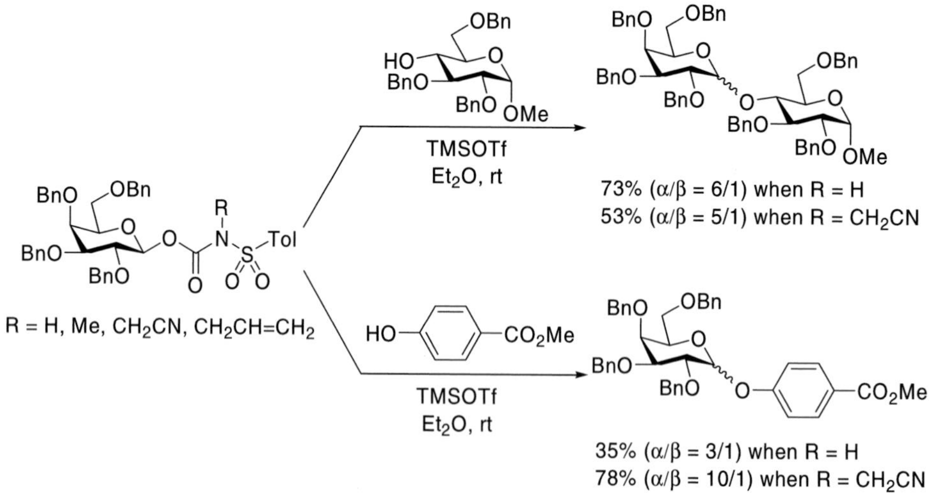

BnO
OBn
R
N
S
Tol
R = H, Me, CH2CN, CH2CH=CH2
HO
BnO
OMe
TMSOTf
Et2O, rt
73% (α/β = 6/1) when R = H
53% (α/β = 5/1) when R = CH2CN
CO2Me
35% (α/β = 3/1) when R = H
78% (α/β = 10/1) when R = CH2CN

Scheme 68
(Ref. [79])

On the other hand, in order to explore a new efficient α-galactopyranosylation method, Hanessian et al. introduced a crystalline glycosyl 2-pyridyl thiocarbonate donor, which gave the pseudo-disaccharide in good yield using AgOTf as the promoter (➋ *Scheme 69*) [80]. Furthermore, they successfully utilized the same method to prepare a carbohydrate-based antagonist of E-selectin, and they found that pretreatment of the hindered disaccharide acceptor with sodium hydride, followed by addition of AgOTf, and finally adding the glycosyl donor led to the target trisaccharide (➋ *Scheme 70*) [81].

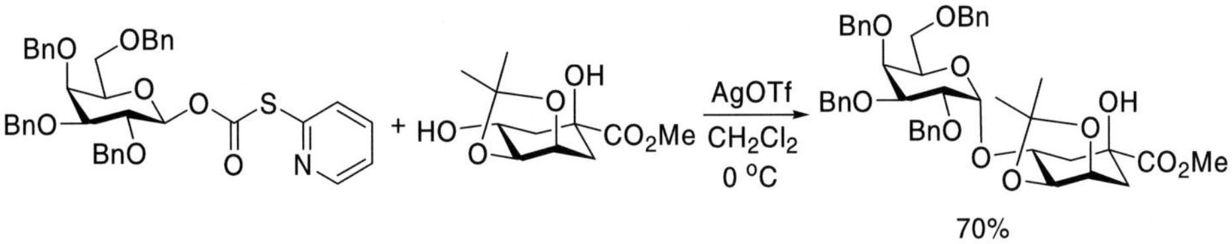

□ Scheme 69
(Ref. [80])

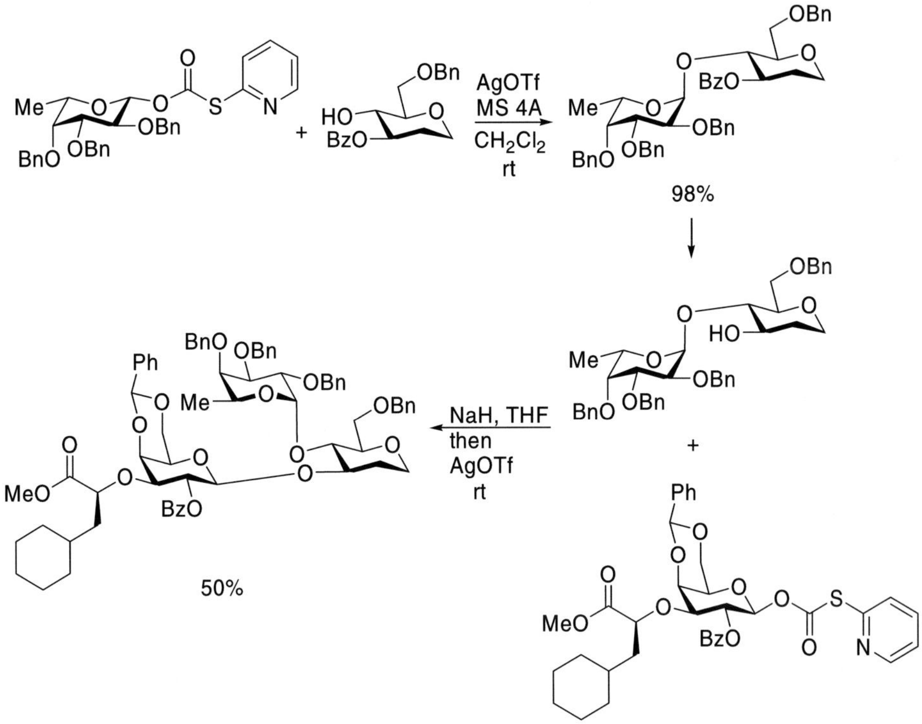

□ Scheme 70
(Ref. [81])

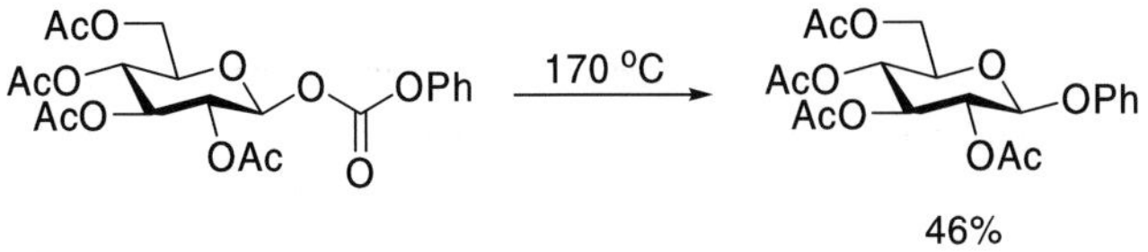

□ Scheme 71
(Ref. [82,83])

5 Intramolecular Glycosylation Through 1-*O*-Acyl Linkages

Ishido and coworkers reported a new method for the synthesis of *O*-glycosides by the pyrolysis of 1-*O*-aryloxycarbonyl sugar derivatives: for example, the pyrolysis of 2,3,4,6-tetra-*O*-acetyl-1-*O*-phenoxycarbonyl-β-D-glucopyranose at 170°C afforded phenyl 2,3,4,6-tetra-*O*-acetyl-β-D-glycopyranoside in moderate yield (❷ *Scheme 71*) [82,83]. Ikegami [84] and Schmidt [85] et al. developed a novel intramolecular decarboxylative glycosylation via mixed carbonate as a two-step glycosylation procedure, which involves linking two sugars by using the carbonate as a connector and subsequent extrusion of carbon dioxide to form a glycosyl bond by the aid of Lewis acid, TMSOTf or TBSOTf (❷ *Scheme 72*). Ikegami et al. also reported that

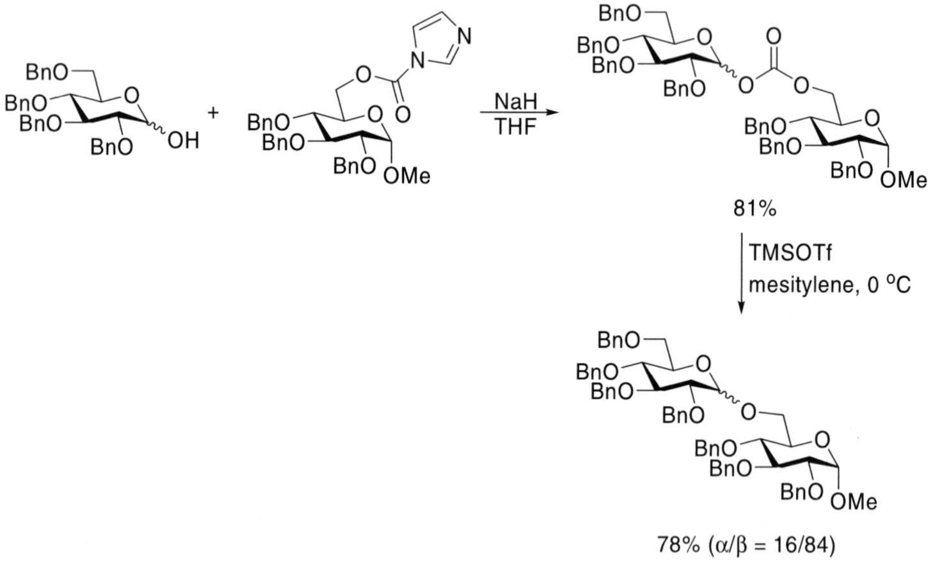

81%

TMSOTf
mesitylene, 0 °C

78% (α/β = 16/84)

❏ Scheme 72
(Ref. [84,85])

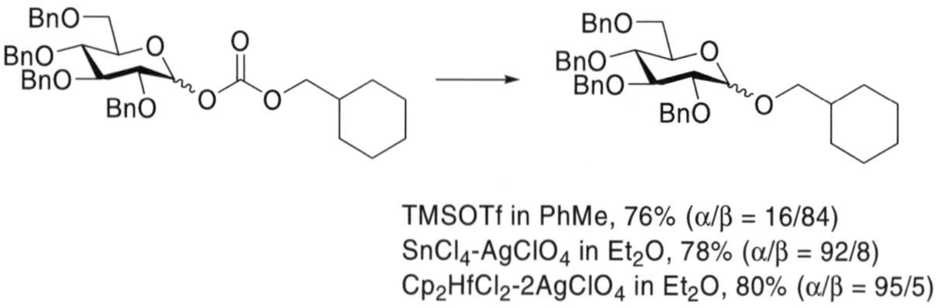

TMSOTf in PhMe, 76% (α/β = 16/84)
SnCl$_4$-AgClO$_4$ in Et$_2$O, 78% (α/β = 92/8)
Cp$_2$HfCl$_2$-2AgClO$_4$ in Et$_2$O, 80% (α/β = 95/5)

❏ Scheme 73
(Ref. [86,87,88])

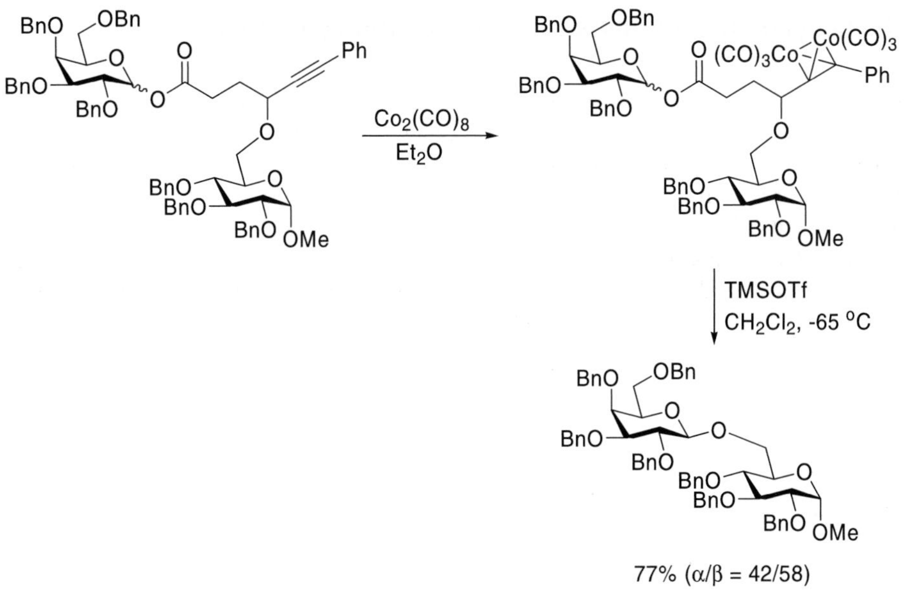

□ Scheme 74
(Ref. [89])

a catalytic amount of SnCl$_4$-AgClO$_4$ or Cp$_2$HfCl$_2$-2AgClO$_4$ promoted decarboxylation of an 1-O-carbonate sugar to afford the α-glycoside stereoselectively (❍ Scheme 73) [86]. At the same time, they reported the synthesis of mixed β-carbonates of acyl-protected sugar and their decarboxylative glycosylations promoted by TMSOTf [87] or Hf(OTf)$_4$ [88] and obtained the β-glycosides stereoselectively.

On the other hand, Mukai et al. reported a glycosylation reaction based on the alkyne-Co$_2$(CO)$_6$ complex, thus O-protected glycopyranose possessing 4-(O-protected-glycosyl)-6-phenyl-5-hexynoate residue at an anomeric position was converted to the corresponding cobalt complex, which on exposure to TMSOTf afforded the product disaccharide in good yield via the internal delivery of the glycosyl acceptor (❍ Scheme 74) [89].

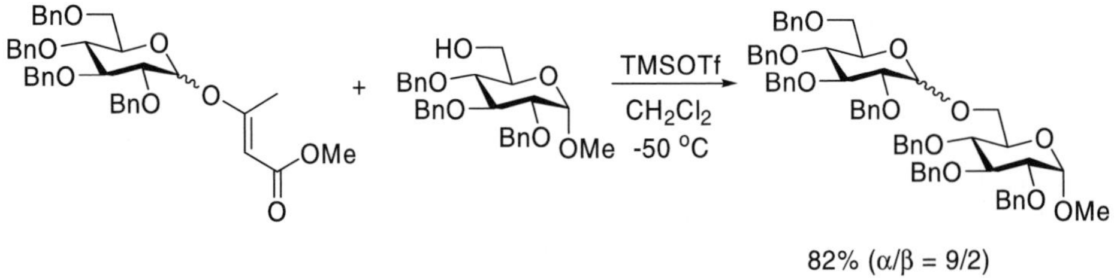

□ Scheme 75
(Ref. [90])

6 Glycosylation with Donors Having Remote Acyl Groups

Takeda et al. reported a glycosylation method using glycosyl donors having the enol ether conjugated with the ester functionality as a leaving group and TMSOTf as the promoter (❷ *Scheme 75*) [90]. Kim et al. introduced a stereocontrolled glycosylation method employing 2'-carboxybenzyl (CB) glycosides as glycosyl donors, which was found to be a useful tool for the stereoselective β-mannopyranosylation using CB 4,6-*O*-benzylidene-2,3-di-*O*-benzyl-D-mannoside and triflic anhydride (Tf$_2$O) as the promoter, and for the stereoselective α-glucopyranosylation using CB 4,6-*O*-benzylidene-2,3-di-*O*-benzyl-D-glucoside (❷ *Scheme 76*) [91]. Next, they showed a highly α- and β-stereoselective (dual stereoselective)

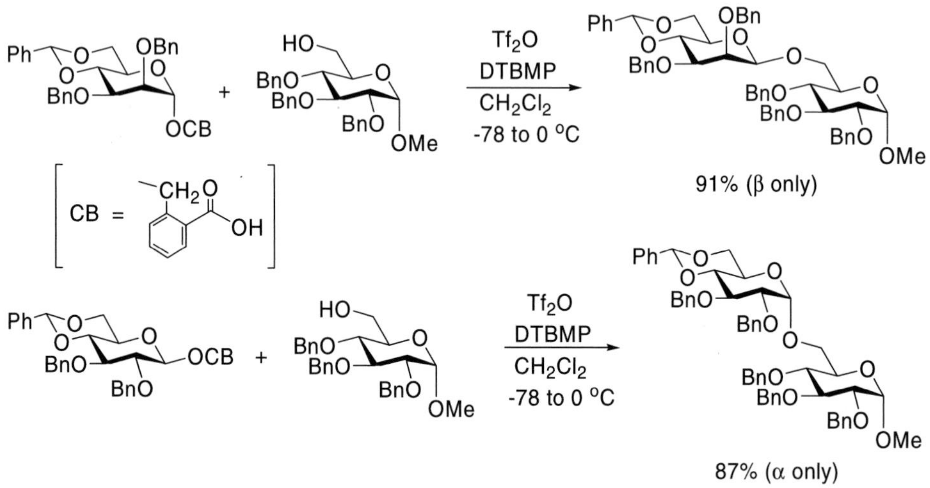

■ Scheme 76
(Ref. [91])

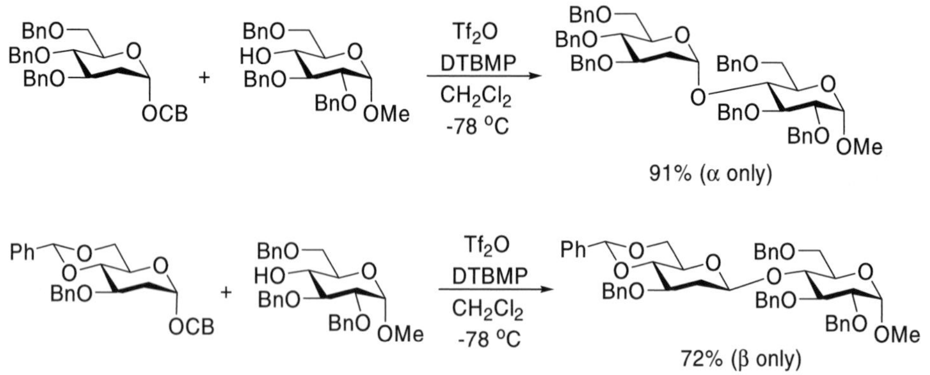

■ Scheme 77
(Ref. [92])

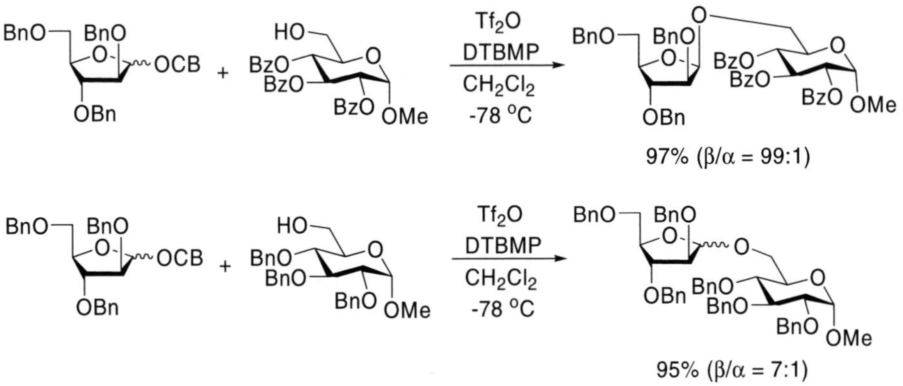

97% (β/α = 99:1)

95% (β/α = 7:1)

□ Scheme 78
(Ref. [93])

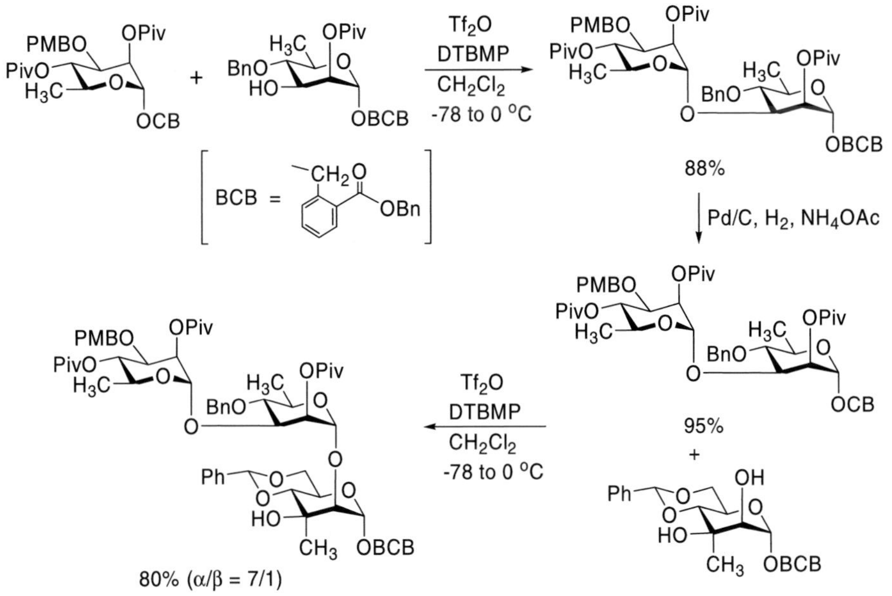

BCB =

88%

Pd/C, H₂, NH₄OAc

80% (α/β = 7/1)

95%

□ Scheme 79
(Ref. [94])

method for the synthesis of 2-deoxyglycosides by employing CB 2-deoxyglycosides as glycosyl donors, in which glycosylation of the 4,6-O-benzylidene-protected glycosyl donor with secondary alcohols afforded predominantly β-glycosides whereas glycosylation of the benzyl-protected glycosyl donor with secondary alcohols afforded α-glycosides (**○** *Scheme 77*) [92]. They also established a reliable and generally applicable direct method for the stereoselective β-arabinofuranosylation employing CB tri-O-benzylarabinoside as the glycosyl donor, in

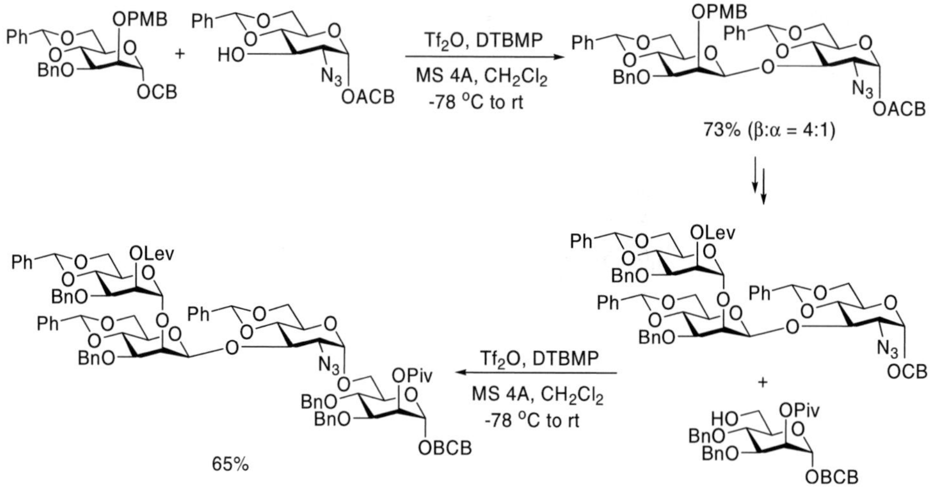

□ Scheme 80
(Ref. [95])

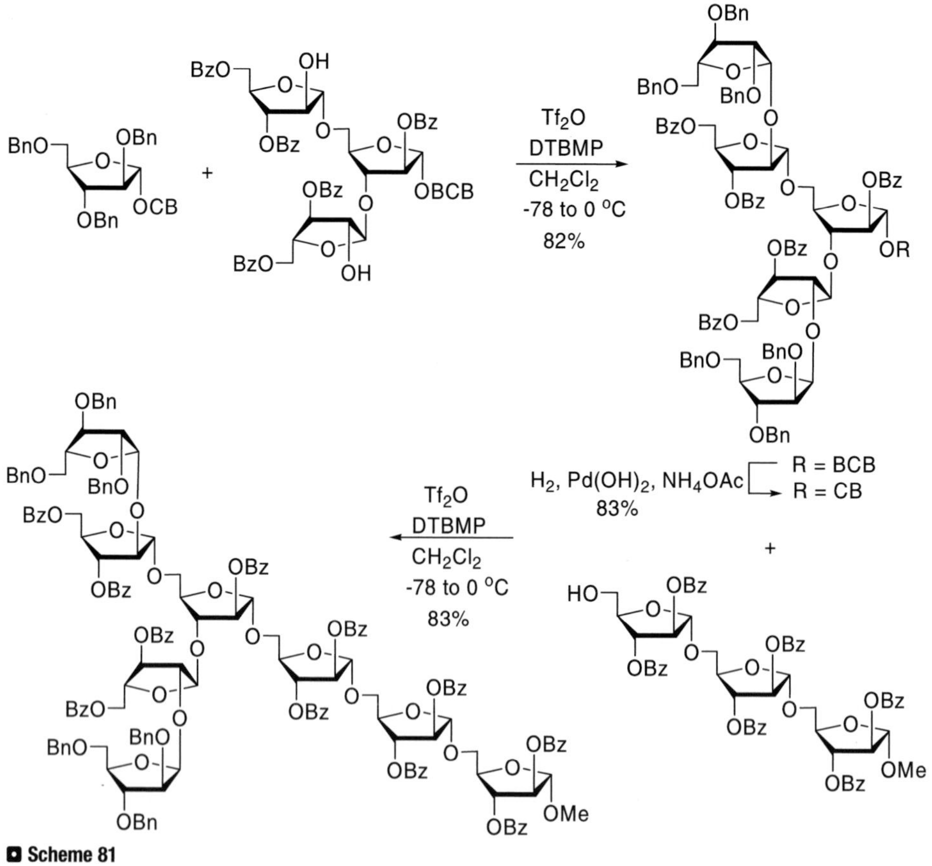

□ Scheme 81
(Ref. [93])

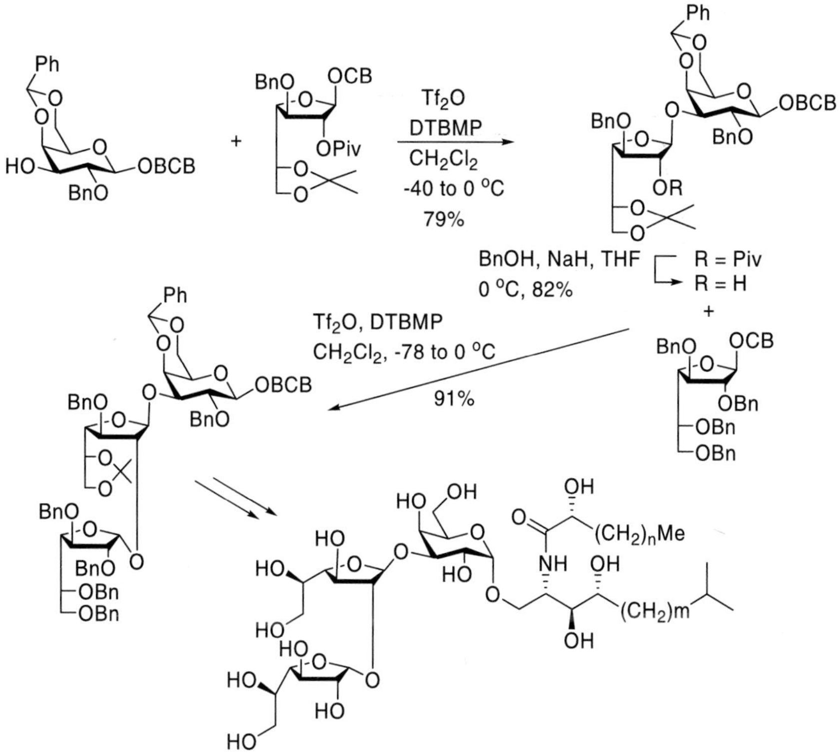

Scheme 82
(Ref. [97])

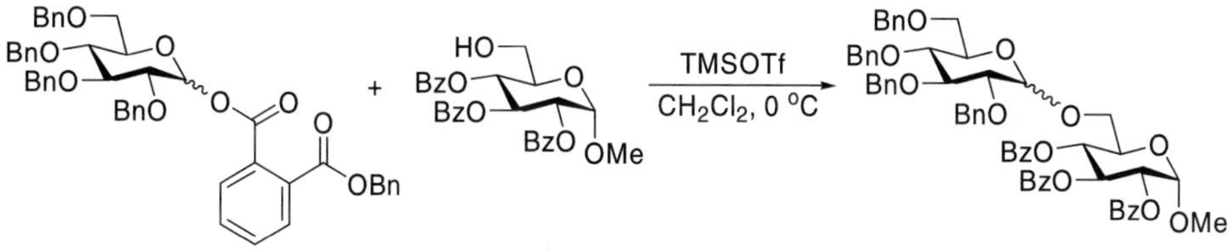

91% (α/β = 2.4/1)

Scheme 83
(Ref. [98])

which the acyl-protective group on glycosyl acceptors was essential for the β-stereoselectivity (● *Scheme 78*) [93]. Furthermore, they applied this CB glycoside methodology to the synthesis of the repeat unit of the O-antigen polysaccharide of the lipopolysaccharide (LPS) from Gram-negative bacteria; a protected form of the trisaccharide repeat unit of the atypical O-polysaccharide from Danish *Helicobacter pylori* strains (● *Scheme 79*) [94], and a protected form of the tetrasaccharide repeat unit of the O-antigen polysaccharide from the *E. coli* lipopolysaccharide (● *Scheme 80*) [95]. An octaarabinofuranoside in arabinogalactan and lipoarabinomannan, found in the mycobacterial cell wall, was also synthesized by employing the latent (BCB)-active (CB) glycosylation method (❯ *Scheme 81*) [93]. Very recently, they reported the total synthesis of agelagalastatin, which is a trisaccharide sphingolipid and displays significant in vitro inhibitory activities against human cancer cell growth [96]. Among three glycosyl linkages in agelagalastatin, α-galactofuranosyl and β-galactofuranosyl linkages were stereoselectively constructed employing the CB glycoside method (❯ *Scheme 82*) [97]. On the other hand, they reported the glycosylation of various acceptors with glycosyl benzyl phthalates as glycosyl donors using TMSOTf as the promoter (❯ *Scheme 83*) [98].

References

1. Helferich B, Shimitz-Hillebrecht E (1933) Chem Ber 66:378
2. Boivin J, Monneret C, Pais M (1978) Tetrahedron Lett 19:1111
3. Lemieux RU, Shyluk WP (1953) Can J Chem 31:528
4. Hanessian S, Banoub J (1977) Carbohydr Res 59:261
5. Honma K, Nakazima K, Uematsu T, Hamada A (1976) Chem Pharm Bull 24:394
6. Dayal B, Salen G, Padia J, Shefer S, Tint GS, Sasso G, Williams TH (1993) Carbohydr Res 240:133
7. Benjahad A, Benhaddou R, Granet R, Kaouadji M, Krausz P, Piekarski S, Thomasson F, Bosgiraud C, Delebassée S (1994) Tetrahedron Lett 35:9545
8. Gaud O, Granet R, Kaouadji M, Krausz P, Blais JC, Bolbach G (1996) Can J Chem 74:481
9. Kiso M, Anderson L (1979) Carbohydr Res 72:C15
10. Lerner LM (1990) Carbohydr Res 207:138
11. Katsuraya K, Ikushima N, Takahashi N, Shoji T, Nakashima H, Yamamoto N, Yoshida T, Uryu T (1994) Carbohydr Res 260:51
12. Chatterjee SK, Nuhn P (1998) Chem Commun 1729
13. Dahmén J, Frejd T, Magnusson G, Noori G (1983) Carbohydr Res 114:328
14. Gurjar MK, Viswanadham G (1991) Tetrahedron Lett 32:6191
15. Elofsson M, Walse B, Kihlberg J (1991) Tetrahedron Lett 32:7613
16. Steffan W, Schutkowski M, Fischer G (1996) J Chem Soc Chem Commun 313
17. Gangadhar BP, Jois SDS, Balasubramaniam A (2004) Tetrahedron Lett 45:355
18. Wang TC, Furukawa H, Nihro Y, Kakegawa H, Matsumoto H, Satoh T (1994) Chem Pharm Bull 42:570
19. (a) Ludin C, Weller T, Seitz B, Meier W, Erbeck S, Hoenke C, Krieger R, Keller M, Knothe L, Pelz K, Wittmer A, Prinzbach H (1995) Liebigs Ann 291;
(b) Erbeck S, Prinzbach H (1997) Tetrahedron Lett 38:2653;
(c) Erbeck S, Liang X, Hunkler D, Krieger R, Prinzbach H (1998) Eur J Org Chem 1935
20. Lam SN, Gervay-Hague J (2005) J Org Chem 70:8772
21. Kelly NM, Jensen KJ (2001) J Carbohydr Chem 20:537
22. Ikeda K, Torisawa Y, Nishi T, Minamikawa J, Tanaka, K, Sato M (2003) Bioorg Med Chem 11:3073
23. Ogawa T, Beppu K, Nakabayashi S (1981) Carbohydr Res 93:C6
24. Ogawa T, Beppu K (1982) Carbohydr Res 101:271
25. (a) Kovac P (1986) Carbohydr Res 153:237;
(b) PozsgayV, Brisson JR, Jennings HJ (1987) Can J Chem 65:2764;
(c) Pozsgay V, Brisson JR, Jennings HJ (1990) Carbohydr Res 205:133;
(d) Nifant'ev NE, Lipkind GM, Shashkov AS, Kochetkov NK (1992) Carbohydr Res 223:109
26. Nifant'ev NE, Backinowsky LV, Kochetkov NK (1988) Carbohydr Res 174:61
27. Nifant'ev NE, Khatuntseva EA, Shashkov AS, Bock K (1996) Carbohydr Lett 1:399
28. Trumtel M, Tavecchia P, Veyriéres A, Sinaÿ P (1989) Carbohydr Res 191:29
29. Paulsen H, Paal M (1984) Carbohydr Res 135:53
30. Paulsen H, Tietz H (1985) Angew Chem Int Ed Engl 24:128
31. Pozsgay V, Brisson JR, Jennings HJ (1991) J Org Chem 56:3377
32. Rainer H, Scharf HD, Runsink J (1992) Liebigs Ann Chem 103

33. Lafont D, Boullanger P, Banoub J, Descotes G (1990) Can J Chem 68:828
34. (a) Sebesta DP, Roush WR (1992) J Org Chem 57:4799;
    (b) Roush WR, Briner K, Sebesta DP (1993) Synlett 264
35. Roush WR, Narayan S (1999) Org Lett 1:899
36. Durham TB, Roush WR (2003) Org Lett 5:1871
37. Arai M, Kaneko S, Konosu T (2002) Tetrahedron Lett 43:6705
38. Roush WR, Bennett CE, Roberts SE (2001) J Org Chem 66:6389
39. Mergott DJ, Frank SA, Roush WR (2004) Proc Natl Acad Sci USA 101:11955
40. Gargiulo D, Blizzard TA, Nakanishi K (1989) Tetrahedron 45:5423
41. Satto S, Kuroda K, Hayashi Y, Sasaki Y, Nagamura Y, Nishida K, Ishiguro I (1991) Chem Pharm Bull 39:2333
42. Dondoni A, Marra A, Rojo I, Scherrmann MC (1996) Tetrahedron 52:3057
43. Mukaiyama T, Kobayashi S, Shoda S (1984) Chem Lett 907
44. (a) Evans DA, Kaldor SW, Jones TK, Clardy J, Stout TJ (1990) J Am Chem Soc 112:7001;
    (b) Evans DA, Black WC (1993) J Am Chem Soc 115:4497
45. Mukaiyama T, Shimpuku T, Takashima T, Kobayashi S (1989) Chem Lett 145
46. Mukaiyama T, Takashima T, Katsurada M, Aizawa H (1991) Chem Lett 533
47. Mukaiyama T, Katsurada M, Takashima T (1991) Chem Lett 985
48. Mukaiyama T, Sasaki T, Iwashita E, Matsubara K (1995) Chem Lett 455
49. Matsushima Y, Itoh H, Nakayama T, Horiuchi S, Eguchi T, Kakinuma K (2002) J Chem Soc Perkin Trans 1 949
50. Florent JC, Monneret C (1987) J Chem Soc Chem Commun 1171
51. Yamanoi T, Nagayama S, Ishida H, Nishikido J, Inazu T (2001) Synth Commun 31:899
52. Higashi K, Susaki H (1992) Chem Pharm Bull 40:2019
53. Susaki H, Higashi K (1993) Chem Pharm Bull 41:201
54. Matsubara K, Sasaki T, Mukaiyama T (1993) Chem Lett 1373
55. Yu CF, Li ZJ, Cai MS (1990) Synth Commun 20:943
56. Li ZJ, Huang HQ, Cai MS (1994) Carbohydr Res 265:227
57. Mao J, Chen H, Zhang J, Cai M (1995) Synth Commun 25:1563
58. Li ZJ, Huang HQ, Cai M (1996) J Carbohydr Chem 15:501
59. Charette AB, Marcoux JF, Côté B (1991) Tetrahedron Lett 32:7215
60. Kimura Y, Suzuki M, Matsumoto T, Abe R, Terashima S (1984) Chem Lett 501
61. Jütten P, Scharf HD, Raabe G (1991) J Org Chem 56:7144
62. Nicotra F, Panza L, Romanò A, Russo G (1992) J Carbohydr Chem 11:397
63. Koide K, Ohno M, Kobayashi S (1991) Tetrahedron Lett 32:7065
64. Carcía-Barrientos A, García-López JJ, Isac-García J, Ortega-Caballero F, Uriel C, Vargas-Berenguel A, Santoyo-González F (2001) Synlett 1057
65. Oßwald M, Lang U, Friedrich-Bochnitschek S, Pfrengle W, Kunz H (2003) Z Naturforsch 58b:764
66. Mann MC, Islam T, Dyason JC, Florio P, Trower CJ, Thomson RJ, von Itzstein M (2006) Glycoconj J 23:127
67. Kunz H, Wernig P, Schultz M (1990) Synlett 631
68. Lopez JC, Fraser-Reid B (1991) J Chem Soc Chem Commun 159
69. Choi TJ, Baek JY, Jeon HB, Kim KS (2006) Tetrahedron Lett 47:9191
70. Baek JY, Choi TJ, Jeon HB, Kim KS (2006) Angew Chem Int Ed 45:7436
71. Ikeda K, Fukuyo J, Sato K, Sato M (2005) Chem Pharm Bull 53:1490
72. Imagawa H, Kinoshita A, Fukuyama T, Yamamoto H, Nishizawa M (2006) Tetrahedron Lett 47:4729
73. Mukaiyama T, Miyazaki K, Uchiro H (1998) Chem Lett 635
74. Mukaiyama T, Wakiyama Y, Miyazaki K, Takeuchi K (1999) Chem Lett 933
75. Mukaiyama T, Ikegai K, Jona H, Hashihayata T, Takeuchi K (2001) Chem Lett 840
76. (a) Boursier M, Descotes G (1989) C R Acad Sci Ser II 308:919;
    (b) Sinaÿ P (1991) Pure Appl Chem 63:519;
    (c) Marra A, Esnault J, Veyrières A, Sinaÿ P (1992) J Am Chem Soc 114:6354
77. Ford MJ, Ley SV (1990) Synlett 255
78. Kunz H, Zimmer J (1993) Tetrahedron Lett 34:2907
79. Hinklin RJ, Kiessling LL (2001) J Am Chem Soc 123:3379

80. Hanessian S, Huynh HK, Reddy GV, Duthaler RO, Katopodis A, Streiff MB, Kinzy W, Oehrlein R (2001) Tetrahedron 57:3281

81. Hanessian S, Mascitti V, Rogel O (2002) J Org Chem 67:3346

82. Inaba S, Yamada M, Yoshino T, Ishido Y (1973) J Am Chem Soc 95:2062

83. Ishido Y, Inaba S, Matsuno A, Yoshino T, Umezawa H (1977) J Chem Soc Perkin Trans 1 1382

84. Iimori T, Shibazaki T, Ikegami S (1996) Tetrahedron Lett 37:2267

85. Scheffler G, Schmidt RR (1997) Tetrahedron Lett 38:2943

86. Iimori T, Azumaya I, Shibazaki T, Ikegami S (1997) Heterocycles 46:221

87. Azumaya I, Niwa T, Kotani M, Iimori T, Ikegami S (1999) Tetrahedron Lett 40:4683

88. Azumaya I, Kotani M, Ikegami S (2004) Synlett 959

89. Mukai C, Itoh T, Hanaoka M (1997) Tetrahedron Lett 38:4595

90. Osa Y, Takeda K, Sato, T, Kaji E, Mizuno Y, Takayanagi H (1999) Tetrahedron Lett 40:1531

91. Kim KS, Kim JH, Lee YJ, Lee YJ, Park J (2001) J Am Chem Soc 123:8477

92. Kim KS, Park J, Lee YJ, Seo YS (2003) Angew Chem Int Ed 42:459

93. Lee YJ, Lee K, Jung EH, Jeon HB, Kim KS (2005) Org Lett 7:3263

94. Kwon YT, Lee YJ, Lee K, Kim KS (2004) Org Lett 6:3901

95. Lee BR, Jeon JM, Jung JH, Jeon HB, Kim KS (2006) Can J Chem 84:506

96. Pettit GR, Xu JP, Gingrich DE, Williams MD, Doubek DL, Chapius JC, Schmidt JM (1999) Chem Commun 915

97. Lee YJ, Lee BY, Jeon HB, Kim KS (2006) Org Lett 8:3971

98. (a) Kim KS, Lee YJ, Kim HY, Kang SS, Kwon SY (2004) Org Biomol Chem 2:2408;
(b) Kwon SY, Lee BY, Jeon HB, Kim KS (2005) Bull Korean Chem Soc 26:815

3.4 *O*-Glycosyl Donors

J. Cristóbal López
Instituto de Química Orgánica General, CSIC,
Juan de la Cierva 3, 28006 Madrid, Spain
clopez@iqog.csic.es

In: *Glycoscience*. Fraser-Reid B, Tatsuta K, Thiem J (eds)
Chapter-DOI 10-1007/978-3-540-30429-6_13: © Springer-Verlag Berlin Heidelberg 2008

Abstract

O-Glycosyl donors, despite being one of the last successful donors to appear, have developed themselves into a burgeoning class of glycosyl donors. They can be classified in two main types: *O*-alkyl and *O*-aryl (or hetaryl) glycosyl donors. They share, however, many characteristics, they can be (1) synthesized from aldoses, either by modified Fisher glycosidation (*O*-alkyl) or by nucleophilic aromatic substitution (*O*-aryl or *O*-hetaryl), (2) stable to diverse chemical manipulations, (3) directly used for saccharide coupling, and (4) chemoselectively activated. Among these, *n*-pentenyl glycosides stand apart. They were the first *O*-alkyl glycosyl donors to be described and have paved the way to many conceptual developments in oligosaccharide synthesis. The development of the chemoselectivity-based "armed-disarmed" approach for saccharide coupling, including its stereoelectronic or torsional variants, now extended to other kinds of glycosyl donors, was first recognized in *n*-pentenyl glycosides. The chemical manipulation of the anomeric substituent in the glycosyl donor to induce reactivity differences between related species (sidetracking) was also introduced in *n*-pentenyl glycosides. An evolution of this concept, the "latent-active" strategy for glycosyl couplings, first described in thioglycosyl donors (vide infra), has been elegantly applied to *O*-glycosyl donors. Thus, allyl and vinyl glycosides, 2-(benzyloxycarbonyl)benzyl (BCB) glycosides and 2′-carboxybenzyl (CB) glycosides are useful "latent-active" glycosyl pairs. Finally, unprotected 3-methoxy-2-pyridyl (MOP) glycosides have been used in glycosylation processes with moderate success.

Keywords

2′-carboxybenzyl (CB) glycosides; 3-methoxy-2-pyridyl glycosides; Armed–disarmed; DISAL glycosyl donors; Halonium ion transfer; Latent-active glycosylation; n-Pentenyl glycosides; O-heteroaryl glycosyl donors; Oligosaccharide synthesis; Vinyl glycosides

Abbreviations

BCB	2-(benzyloxycarbonyl)benzyl
CAN	cerium ammonium nitrate
CB	2′-carboxybenzyl
DAST	(diethylamino)-sulfur trifluoride
DDQ	2,3-dichloro-5,6-dicyano-p-benzoquinone
DISAL	a dinitrosalicylic acid glycoside derivative
DMAP	4-(N,N-dimethylamino)pyridine
DTBMP	di-tert-butylmethylpyridine
IDCP	iodonium di-sym-collidine perchlorate
MOP	3-methoxy-2-pyridyl
NBS	N-bromosuccinimide
NIS	N-iodosuccinimide
NMP	1-methylpyrrolidin-2-one
NPG	n-pentenyl glycosides
NPhth	N-phthaloyl
NPOE	n-pentenyl orthoester
PMB	p-methoxybenzyl
PMP	p-methoxyphenyl
TBAF	tetra-n-butylammonium bromide
TBAI	tetra-n-butylammonium iodide
TBSOTf	tert-butyldimethylsilyl trifluoromethanesulfonate
TESOTf	triethylsilyl trifluoromethanesulfonate
Tf₂O	trifluoromethanesulfonic anhydride
TfOH	trifluoromethanesulfonic acid
THF	tetrahydrofuran
TMSOTf	trimethylsilyl trifluoromethanesulfonate
TPSOTf	tert-butyldiphenylsilyl trifluoromethane sulfonate
Troc	N-trichloroethoxycarbonyl
TTCP	N-tetrachlorophthaloyl

1 Introduction

From the early days, chemists involved in chemical glycosylation have been trying to develop successful glycosyl donors. In general, the characteristics of a successful glycosyl donor might include: (*a*) preparation under mild reaction conditions, (*b*) selective activation by reagents that would not interfere with the protecting and functional groups present in the donor and the glycosyl acceptor, and (*c*) good reactivity [1,2,3,4]. More recently, the advent of convergent block synthesis to tackle complex oligosaccharide preparations have also demanded that glycosyl donor building blocks (*d*) are sufficiently stable to be purified and stored for considerable periods of time, and (*e*) are resistant towards a wide range of reaction conditions [5,6,7,8,9]. According to this, *O*-glycosyl donors (the topic of this chapter), because of their remarkable "shelf-life" and stability (conditions *d*, *e*) will be attractive candidates for oligosaccharide block synthesis, provided that conditions *a*, *b*, and *c* are also met.

A chronology, displayed in ➋ *Fig. 1*, highlights the relatively recent arrival of *O*-glycosyl donors to the assortment of relevant glycosyl donors. In fact, the first *O*-alkyl glycosyl donor (*n*-pentenyl glycoside) [10], was introduced more than a century after the first glycosylation was described (synthesis of aryl glycosides from glycosyl chlorides) [11].

This late arrival is understandable on the basis of the outline in ➋ *Scheme 1a*, which makes it obvious that in situ transformation of one alkyl glycoside donor into a disaccharide (or into another alkyl glycoside) could be problematic. The acidic conditions normally used to cleave alkyl glycosides (**1**), generating oxocarbenium ion **2**, could tamper with the newly formed intersaccharidic linkage in **3**, notwithstanding the liberation of alkanol that might compete for glycosylation with the sought glycosyl acceptor, thus regenerating **1** (➋ *Scheme 1b*). The successful implementation of the strategy represented in ➋ *Scheme 1a* would imply that: (*a*) the alkanols have to be released under a non-nucleophilic form, and (*b*) the newly formed glycosidic linkage must be compatible with the promoter employed.

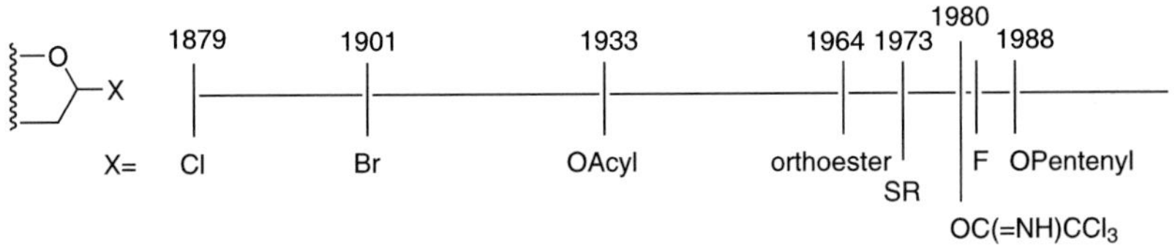

■ Figure 1
Chronology of selected glycosyl donors

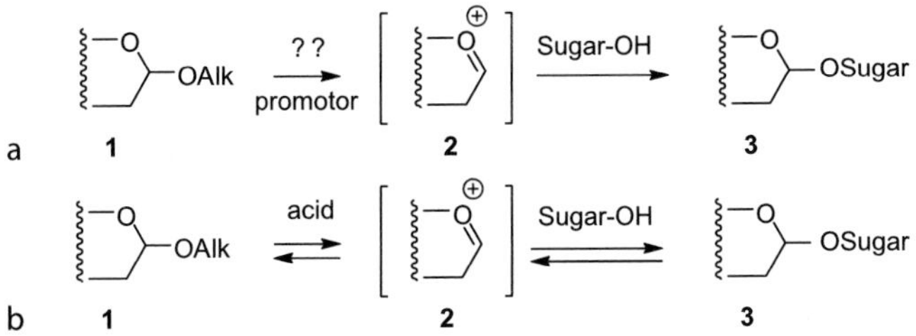

■ Scheme 1
O-Alkyl-glycosyl donors in glycosylation

2 *n*-Pentenyl Glycosides

2.1 Introduction

2.1.1 The Origin of *n*-Pentenyl Glycosides (NPGs)

The discovery of *n*-pentenyl glycosides (NPGs) [12], was derived from an observation made by Mootoo and Fraser-Reid in a completely unrelated project [13]. Attempted formation of bromohydrin **5** by reaction of **4** with NBS in 1% aqueous acetonitrile led, instead to bromomethyl tetrahydrofuran **6** (❷ *Scheme 2*) [14]. To rationalize this transformation (**4**→**6**), the authors invoked a 5-*exo*-cyclization [15] of the pyranosidic oxygen in **7** leading to cationic intermediate **8**, and thence to oxocarbenium ion **9**, that upon capture of H_2O led to hemiacetal **6**. The overall result of the process had been a, nonhydrolytic, electrophilic unravelling of the glycosidic-type bond in **4**.

The overlap between structures **4** and **10** permitted the authors to design structure **11**, as a candidate for testing electrophilic deprotection at the anomeric center of a pyranose (❷ *Scheme 3*). It has now become clear, 20 years after this observation, that the correlation shown in ❷ *Scheme 3* led to a breakthrough in glycoside synthesis.

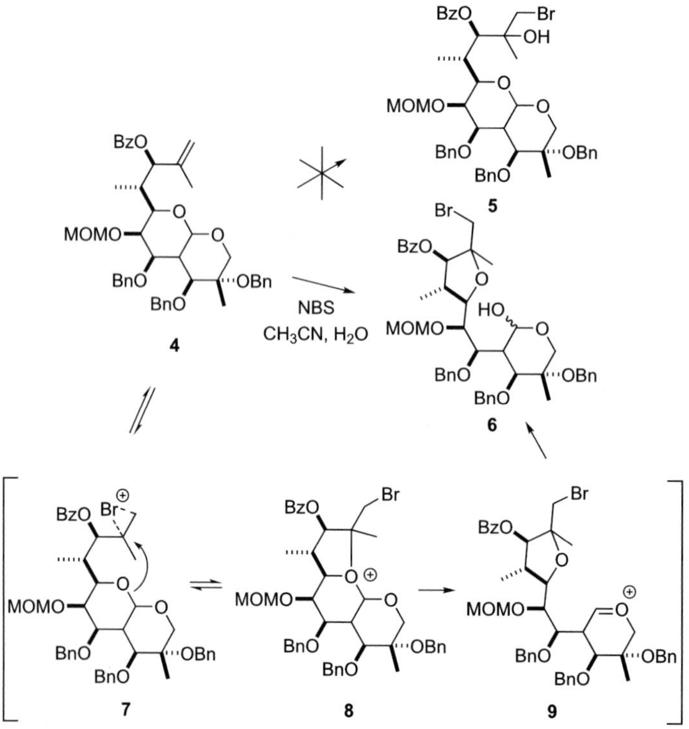

❑ **Scheme 2**
The origin of *n*-pentenyl glycosides

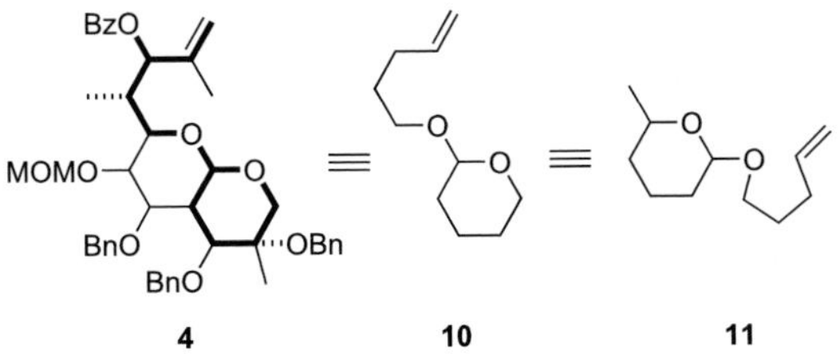

□ Scheme 3
The design of *n*-pentenyl glycosides

2.1.2 Chemoselective Liberation of the Anomeric Group in NPGs

To test the validity of their assumption, Mootoo and Fraser-Reid prepared NPGs **12–18** and treated them with NBS in 1% aqueous acetonitrile [16]. Their results, summarized in ❯ *Table 1*, showed that differently substituted NPGs could be chemoselectively liberated at the anomeric center to yield hemiacetals **19–24**. Furthermore, benzylidene, silyl, p-methoxybenzyl (PMB), ethoxyethyl, and allyl protecting groups proved to be compatible with the conditions employed in the deprotection of the anomeric pent-4-enyl group. Diol **18**, however, furnished a complex reaction mixture probably related to competing glycosylation processes, vide infra.

2.2 NPGs as Glycosyl Donors

To test the potential of NPGs as glycosyl donors, Fraser-Reid et al. first examined the reaction of compound **12** with NBS in MeCN-MeOH [10]. The reaction took place in 3 h, at room temperature yielding methyl glucoside **25** in 85% yield as a 1:3 (α:β) anomeric mixture (❯ *Table 2*, entry i). The utilization of iodonium di-*sym*-collidine perchlorate (IDCP) [17] as a promoter resulted in a faster reaction (0.5 h), which maintained the previous anomeric mixture (❯ *Table 2*, entry ii). The use of a 4:1 mixture of Et$_2$O-CH$_2$Cl$_2$ as solvent, to favor α-glucoside formation while solubilizing IDCP, resulted in a 3:1 (α:β) anomeric mixture of **25** (❯ *Table 2*, entry iii). When CH$_2$Cl$_2$ was used as a solvent a 1.2:1 (α:β) anomeric mixture was obtained.

NPGs were next tested in the elaboration of disaccharides by glycosylation of monosaccharide acceptors. Gluco- (**12**), *manno*- (**26**), and 2-deoxy- (**27**) NPGs reacted with sterically demanding methyl glucoside **28**, to give disaccharides **29–31** (❯ *Table 3*). Gluco-derivatives gave the best α versus β solvent dependence, MeCN favoring β, and Et$_2$O favoring α (the same trend as noted for MeOH, ❯ *Table 2*). For the *manno*- and 2-deoxy donors **26** and **27**, no consistent pattern of solvent dependence was noticed. 2-Deoxy-donor **27**, gave appreciable α-selectivities with secondary acceptors (❯ *Table 3*, entries vii–ix). MeCN gave the lowest overall yield of disaccharide products (❯ *Table 3*, entries i, iv, vii). Reactions of 2-deoxy NPG **27** (❯ *Table 3*, entries vii–ix) were generally much faster than those of either **12** or **26**, an observation that parallels the observed trends in acid lability of the three donors (❯ *Table 3*, entries i–vi). With Et$_2$O as solvent, reactions with primary alcohol acceptors displayed more stereoselectivity than reactions with secondary hydroxyl acceptors.

▫ **Table 1**
Oxidative hydrolysis of some NPGs with NBS in 1% aqueous acetonitrile

Substrate	Product	Yield (%)
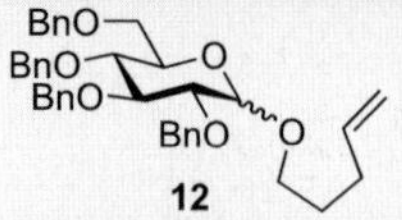 **12**	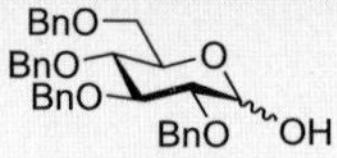 **19**	85
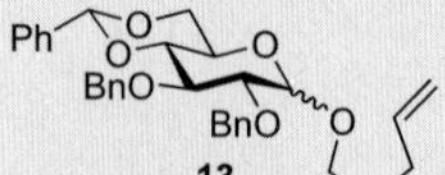 **13**	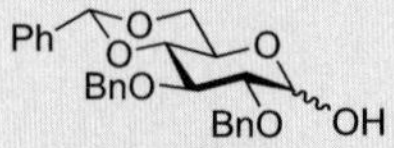 **20**	70
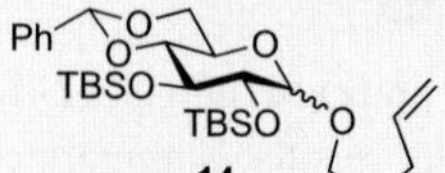 **14**	 **21**	90
 **15**	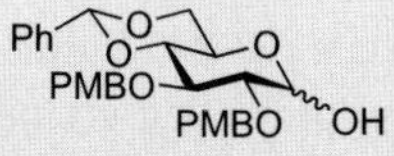 **22**	68
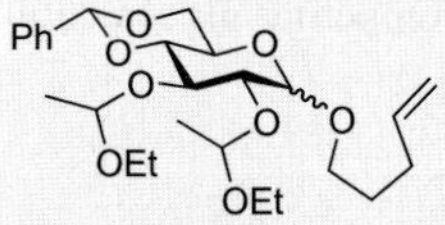 **16**	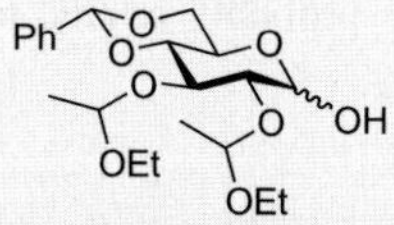 **23**	63
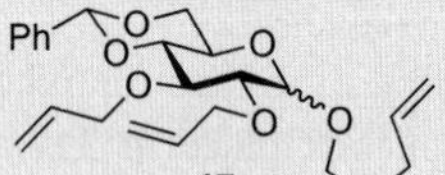 **17**	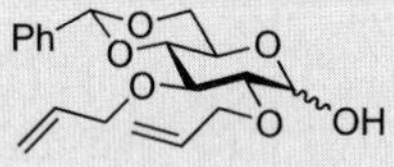 **24**	72
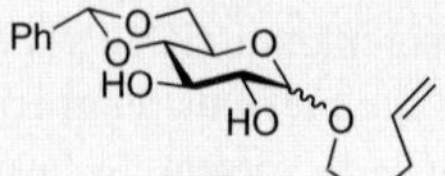 **18**	complex mixture	

▫ **Table 2**
Reaction of NPG 12 with methanol in the presence of halonium ions

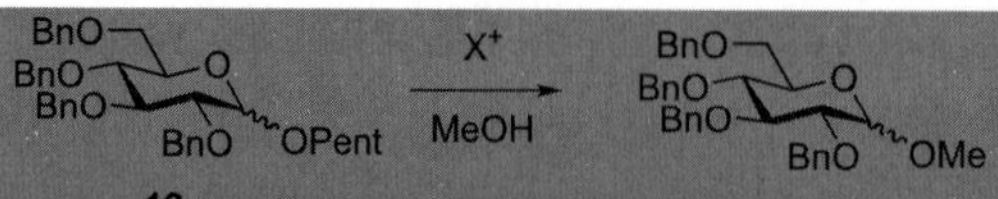

Entry	Promoter	Time (h)	Solvent	$\alpha:\beta$	Yield (%)
i	NBS	3	MeCN	1:3	85
ii	I(collidine)$_2$ClO$_4$	0,5	MeCN	1:3	75
iii	I(collidine)$_2$ClO$_4$	0,5	CH$_2$Cl$_2$	1.2:1	85
iv	I(collidine)$_2$ClO$_4$	24	CH$_2$Cl$_2$-Et$_2$O	3:1	75

◻ **Table 3**
Direct elaboration of NPGs into disaccharides on treatment with IDCP

Entry	NPG	Acceptor	Time (h)	Solvent α:β	Product	Yield (%)
i	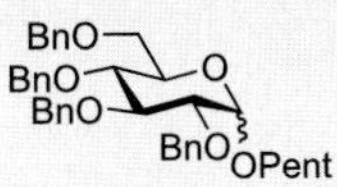 **12**	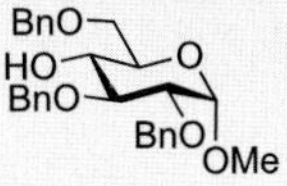 **28**	1–2	MeCN 1:2	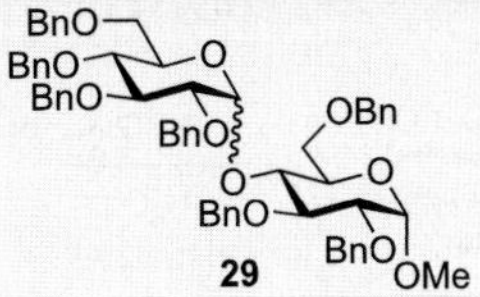 **29**	20
ii	12	28	1–2	CH$_2$Cl$_2$ 1.2:1	29	75
iii	12	28	16–24	Et$_2$O/CH$_2$Cl$_2$ 3:1	29	95
iv	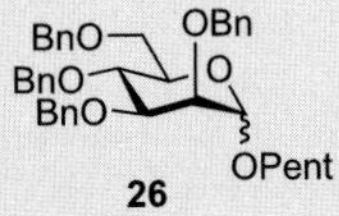 **26**	28	2–5	MeCN 6:1	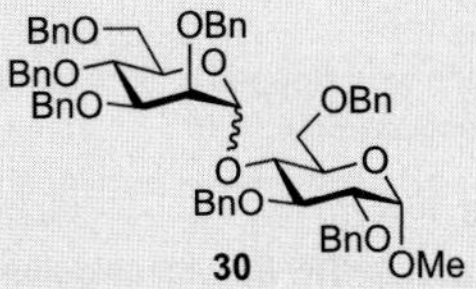 **30**	36
v	26	28	2–5	CH$_2$Cl$_2$ 9:1	30	76
vi	26	28	24	Et$_2$O/CH$_2$Cl$_2$ α only	30	92
vii	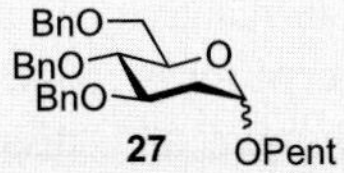 **27**	28	0,5	MeCN 7:3	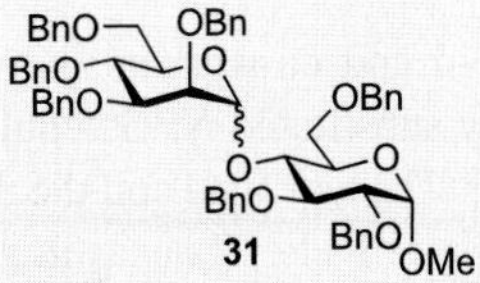 **31**	51
viii	27	28	0,5	CH$_2$Cl$_2$ 4:1	31	57
ix	27	28	4–6	Et$_2$O/CH$_2$Cl$_2$ 4:1	31	61

2.2.1 Acyl-Substituted NPG Donors

The results in ❷ *Table 3* made it clear that the use of different solvents to induce (α versus β) stereoselectivity in glycosyl couplings of NPGs was only moderately successful, generally leading to anomeric mixtures [10]. As it has been established, good stereocontrol in the formation of 1,2-*trans* glycosidic linkages can be conveniently obtained with the assistance of a neighboring participating group, generally an acyl moiety [18]. In this context, Fraser-Reid and co-workers examined the glycosylation of NPG 32 for the preparation of 1,2-*trans* glycoside 33 (❷ *Scheme 4*). Unfortunately, the reaction did not lead to glycoside 33, but to compounds resulting from addition across the terminal double bond of the pent-4-enyl moiety. Along this line, the authors had previously noticed that hydrolysis of acyl derivative 34 was considerably slower than that of 13 [16]. These results paralleled previous observations by Paulsen [1] on the deactivating effect of esters versus ether protecting groups upon differently substituted glycosyl halides. These observations, however, according to the state-of-the-art in glycosylation in 1988, only meant that acyl-NPGs would not be useful as glycosyl donors.

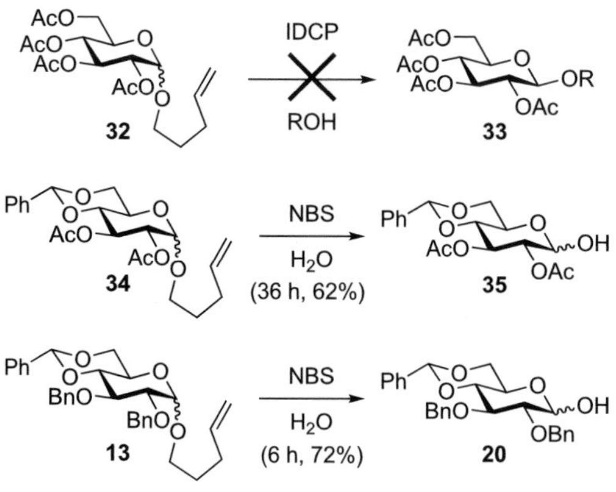

^a The reaction was stopped at 50% conversion, the yield is based on recovered **34**.

⬛ Scheme 4
Reaction of acylated NPGs with halonium ions

2.3 *Armed–Disarmed* Strategy for Glycosyl Coupling

Fraser-Reid and co-workers, however, anticipated that the difference in reactivity found in differently substituted NPGs could be applied in a chemoselective protocol for glycosyl coupling [19]. The activated and the deactivated NPGs were termed "armed sugar" and "disarmed sugar", respectively. Thus, as illustrated in ❷ *Scheme 5a*, coupling of **12** and **36**, mediated by IDCP afforded a 62% yield of disaccharide **37** [19]. Therefore, the acyl groups of **36** indeed "disarmed" the NPG, thereby ensuring that **12** served as the only glycosyl donor. No evidence for a hexaacetyl disaccharide **38**, arising from self-condensation of **12** was found, nor of further reaction of (disarmed) disaccharide **37** with the acceptor **36**.

The chemoselective coupling, however, is not the only quality of the armed-disarmed strategy for glycosyl assembly. An additional aspect of this strategy is the ability to "rearm" disarmed glycosyl donors for further glycosyl coupling. Thus, "disarmed" **37** was converted to "armed" disaccharide **39**, (by replacing the acetyl groups with benzyl substituents) which could then glycosylate galacto- derivative **40**, to yield trisaccharide **41**, in 60% yield (❷ *Scheme 5b*). An alternative way of "rearming" NPGs by increasing the potency of the promoter used for glycosylation was also introduced by the same authors [20]. According to that, an iodonium ion generated in situ from NIS and TfOH was able to promote the coupling of "disarmed" pent-4-enyl glycosides (e. g. α-**32**, ❷ *Scheme 5c*) with acceptors to give 1,2-*trans* disaccharides, e. g. **43**, via neighboring group participation [18].

The armed-disarmed concept takes advantage of reactivity differences induced by the ring substituents on the anomeric leaving group and, although originally described for NPG donors, it has been extended to various types of glycosyl donors. These include thioglycosides [21,22], glycals [23], glycosyl fluorides [24], selenoglycosides [25], glycosyl phosphoroamidates [26,27], glycosyl thioformimidates [28,29,30], and S-benzoxazolyl glycosides [31,32,33].

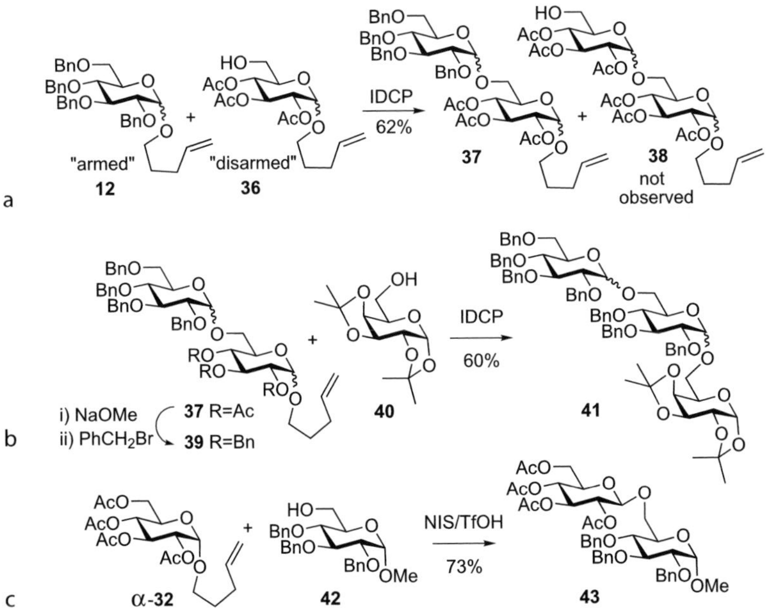

■ Scheme 5
Armed-disarmed strategy for chemoselective glycosyl couplings

2.3.1 Mechanistic Aspects of the Oxidative Hydrolysis of NPGs

The currently accepted mechanism for the reaction of NPGs (e. g. **11**) with halonium ions is outlined in ❯ *Scheme 6a*. The oxygen in the acetal function participates in a favored *5-exo-tet* ring opening of an intermediate cyclic halonium ion, **44** [15]. The ensuing furanilium ion **45**, evolves by splitting off a non-nucleophilic halotetrahydrofuran **47** [34], thus leading to oxo-carbenium ion **46**, that can trap the nucleophile (ROH). The overall result is the cleavage of the acetal moiety with the formation of a new glycosyl derivative, **48**. Madsen and Fraser-Reid have demonstrated that even when the system NIS/TESOTf [20] is used to promote the cleavage of NPGs, the reaction is not acid catalyzed but still halonium ion catalyzed [35].

In this connection, the question of why the reaction of an NPG in the presence of water leads to an aldose **48** (R=H), rather than to a halohydrin **49** (R=H) was raised (❯ *Scheme 6b*). In fact, the successful cleavage observed for NPGs rests on two issues: (*a*) the concentration of nucleophile (water in the case of hydrolysis), and (*b*) the rate of the *5-exo-tet* cyclization, **44**→**45**. Pertinent to question *a* the intramolecular reaction **44**→**45**, is preferred to the bimolecular reaction with water leading to a halohydrin, **44**→**49**, under the conditions used by the authors. An increase in the concentration of water would enhance the rate of the bimolecular reaction, without affecting the intramolecular process. Indeed added water led to the formation of bromohydrin **49** (R=H) [36]. Related to the second issue, Rodebaugh and Fraser-Reid examined the same reaction with allyl, butenyl, and hexenyl glycosides **50** (*n* = 1, 2, 4), differing on the rate of cyclization compared to NPGs [37,38]. They found that, unlike NPGs, they all gave rise to isomeric halohydrins **51** and **52**, upon treatment with NBS in aqueous MeCN (❯ *Scheme 6c*).

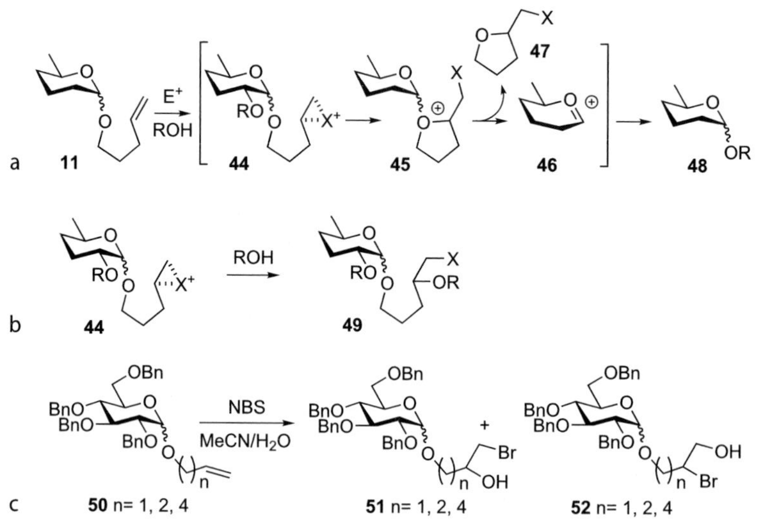

■ Scheme 6
Oxidative cleavage of NPGs

2.3.2 Evidence for Intermolecular Halonium-Ion Transfer

In a related experiment, hexenyl glycoside **50** ($n = 4$), that has been found to react 2.3-times slower than NPG **12** [37,38], was made to compete with **12** for 1 equiv. of NBS. The hexenyl glycoside **50** ($n = 4$) was recovered unchanged together with hemiacetal **19**, arising from the hydrolysis of **12** (❏ *Scheme 7a*). Rodebaugh and Fraser-Reid proved that this phenomenon was due to a diffusion-controlled intermolecular halonium-ion transfer (a similar process has been previously noted by Brown and co-workers for "sterically encumbered olefins" [39]). Accordingly, bromonium species **53**, obtained by irreversible reaction of **50** with NBS [37,38], would undergo a fast bromonium ion transfer to NPG **12** leading to halonium **54** (i. e. **44**, ❏ *Scheme 6a*) which will undergo a fast transformation to **19**. The general process, a classic example of Le Chatelier's principle, is represented in ❏ *Scheme 7b*. When two alkenes are made to compete for one equivalent of halonium ion X^+, a steady-state regime can be envisaged whereby the faster (F) (e. g. NPG **12**) reacts completely, and the slower (S) (e. g. **50**, $n = 4$) is recovered completely.

2.3.3 Intermolecular Halonium-Ion Transfer: A Key Factor in the Implementation of the Armed-Disarmed Protocol

This intermolecular halonium ion transfer had indeed been postulated earlier as the key factor to account for the absence of self-coupling product **38**, when armed and disarmed NPGs, **12** and **36** respectively, were made to compete for one equivalent of NBS (❏ *Scheme 5a*). The observed 6-fold difference in the hydrolysis rates of **36** and **12** should have resulted in the presence of **38**, in at least 10% [40]. Irreversible reaction of NPGs **55** and **59** with NBS leads to halonium ions **56** and **60**, respectively (❏ *Scheme 8*). The transfer of halonium (e. g. **60**→**55**) is reversible and rapid compared with the subsequent steps leading to glycoside formation

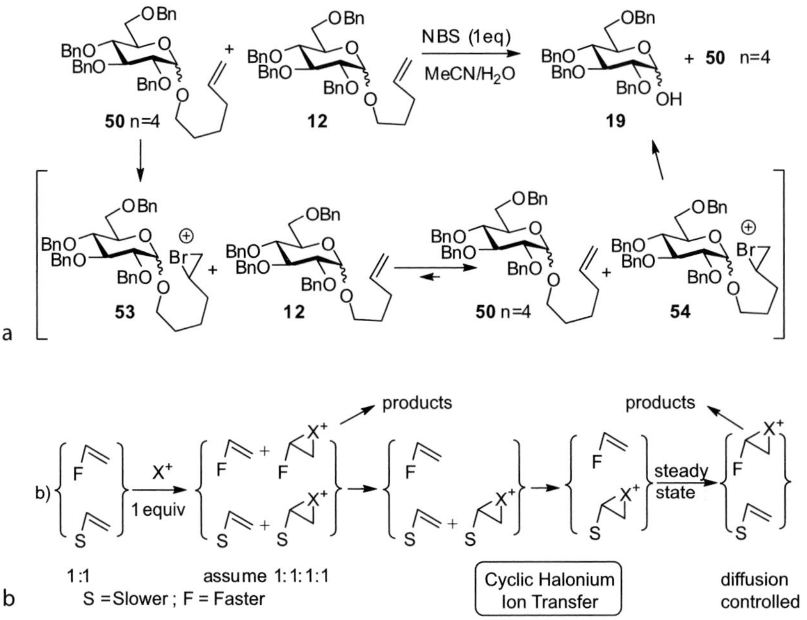

Scheme 7
Intermolecular halonium-ion transfer

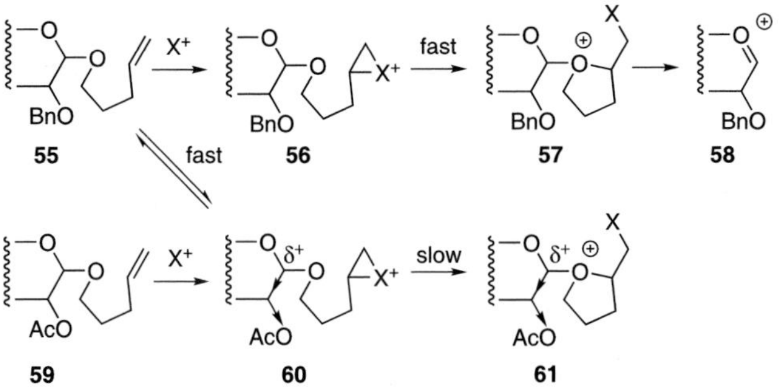

Scheme 8
Halonium ion transfer: a key factor in armed-disarmed couplings

($56 \rightarrow 57 \rightarrow 58$). By corollary, the inherent reactivity of the glycosyl donors is thus revealed in the final product distribution. If the acceptor functionality is located in the less reactive component, selective glycosylations take place leading to a specific disaccharide.

2.3.4 Torsional Disarming of NPGs

Fraser-Reid and co-workers found that widely used cyclic acetals also affected anomeric reactivity [41]. They showed that these reactivity differences could be applied to an armed-

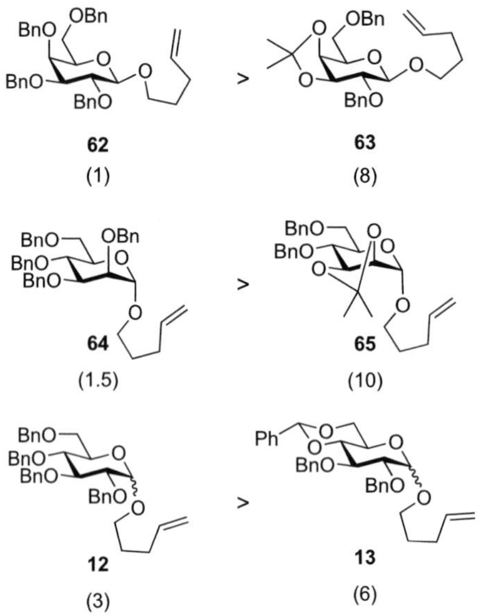

□ **Figure 2**
Relative rates of oxidative hydrolysis for acetal-protected NPGs

disarmed protocol based on torsional, rather that electronic, effects. The measured experimental relative rates of oxidative hydrolysis for some pairs of *galacto*- (**62, 63**), *manno*- (**64, 65**), and *gluco*- (**12, 13**), acetalated and nonacetalated NPGs are displayed in ❷ *Fig. 2*. From these data, the authors were able to design the successful armed-disarmed couplings shown in ❷ *Scheme 9*. These reactivity differences were ascribed to the fact that *trans*-fused protection restricts the molecule from ring flexibility, thereby making it increasingly difficult to reach a half-chair transition state from a chair ground state.

2.3.5 Sidetracking of NPGs: A Reversal for the Armed-Disarmed Strategy

In the armed-disarmed protocol, the reactivity differences induced by the ring substituents upon the anomeric center were exploited for chemoselective couplings. In these protocols, the more reactive "armed" NPG always glycosylates the "disarmed" NPG. On the basis of chemical manipulation of the pent-4-enyl moiety, rather than the ring substituents, Fraser-Reid and co-workers showed that "disarmed" NPGs could be used to glycosylate "armed" NPGs [40]. Treatment of NPG **72** with bromine and tetra-*n*-butylammonium bromide (TBAF), in a bimolecular reaction (e. g. **44**→**48**, ❷ *Scheme 6b*) yielded dibromo-derivative (❷ *Scheme 10*) **73**. Glycosylation of the latter with "disarmed" **74**, under the agency of NIS/TESOTf, yielded disaccharide **75a**, which could be transformed to the pentenyl disaccharide **75b**. Several methods proved to be successful for the restoration of the double bond from the dibromoderivative including, (a) Zn/TBAI in sonicating EtOH, (b) NaI in methyl ethyl ketone, and (c) SmI$_2$ in THF [42]. The choice of the reagent will vary with the

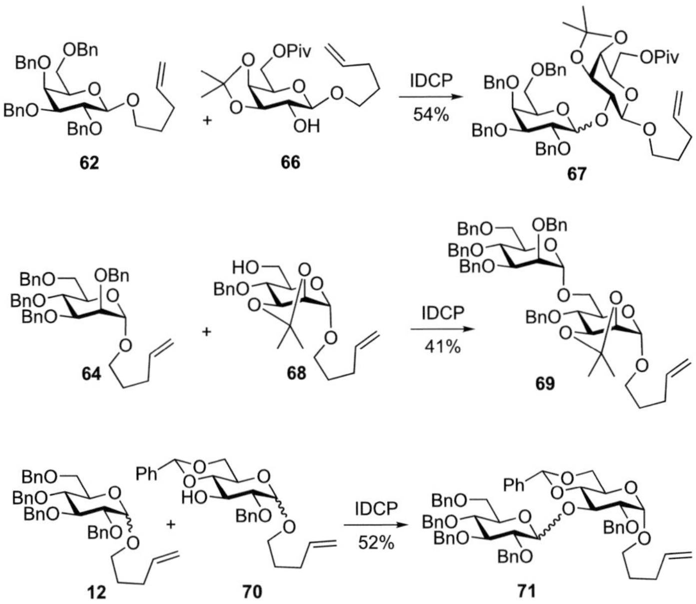

Scheme 9
Armed-disarmed couplings based on torsional effects

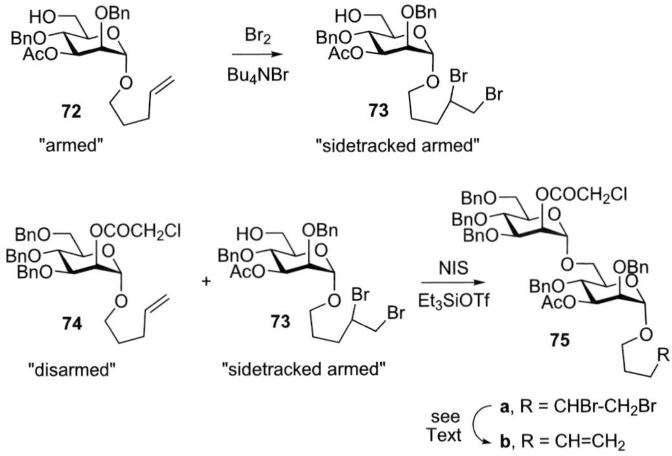

Scheme 10
Sidetracking of NPGs in saccharide synthesis

reactivity of the substrate, as well as the protecting groups thereon. More recently, a milder brominating system, the combination of CuBr$_2$ and LiBr in MeCN:THF (3:1), has been used to brominate *n*-pentenyl glycosides containing *O*-benzyl, *O*-*p*-methoxybenzyl, *N*-phthaloyl, and *N*-tetrachlorophthaloyl protecting groups [43].

2.4 Conversion of NPGs to Other Glycosyl Donors

NPGs have been converted into different glycosyl donors.

2.4.1 Conversion to Glycosyl Bromides

Konradsson and Fraser-Reid [44] reported the conversion of NPGs into glycosyl bromides, e. g. **76**, by treatment of the corresponding pentenyl glycoside with a dilute dichloromethane solution of bromine, conditions that favor unimolecular reaction. The reaction was shown to be compatible with acetals, benzyl, silyl, and allyl protecting groups in the NPG (❷ *Scheme 11a*).

2.4.2 Conversion to Glycosyl Phosphates

Pale and Whitesides [45] described the synthesis of glycosyl phosphates **77** [46,47], by reaction of dibenzyl phosphate with NPG **12**, with the use of either IDCP or NBS as promoters. The authors noted the influence of the solvent (MeCN, Et_2O, CH_2Cl_2) and the promoter in the α/β selectivity of the glycosyl phosphates formed (❷ *Scheme 11b*).

2.4.3 Conversion to Glycosyl Fluorides

Clausen and Madsen [48] reported the transformation of NPG **78** into glycosyl fluoride **79** by treatment with NBS and (diethylamino)-sulfur trifluoride (DAST) (❷ *Scheme 11c*).
López et al. [49] described the preparation of glycosyl fluorides **80**, by reaction of NPGs with bis(pyridinium) iodonium (I) tetrafluoroborate (IPy_2BF_4) in the presence of tetrafluoroboric acid (❷ *Scheme 11d*). The process was compatible with the presence of silyl and benzyl groups in the NPG.

2.4.4 Chemoselective Liberation Followed by Anomeric Activation

The ability to chemoselectively deprotect pent-4-enyl glycosides opens an avenue for a two-step transformation of NPGs into different glycosyl donors. In this context, NPGs can be transformed [50] into thioglycosides **81** [51,52,53], glycosyl trichloroacetimidates **82** [54], and glycosyl chlorides **83** [55] (❷ *Scheme 11e*).

2.5 NPGs in the Stereocontrolled Assembly of α- and β- Glycoproteins

2.5.1 Pyranosylacetonitrilium Ions from NPGs

Ratcliffe and Fraser-Reid found that acetonitrile was able to trap glycosyl oxocarbenium ions (e. g. **46**), arising from NPGs, to give acetonitrilium ions, e. g. **84** (❷ *Scheme 12a*) [56]. The latter reacted with water to produce intermediate **85** that evolves to α-amide **86**, in a Ritter-type reaction [57] (❷ *Scheme 12a*).
This transformation was significant from a mechanistic standpoint. The formation of the α-acetonitrilium ion was not expected on the basis of the reverse anomeric effect (originally defined as the tendency of positively charged substituents at C-1 of a pyranose ring to adopt the equatorial orientation [58]). The authors, however, unambiguously established the α-orientation of

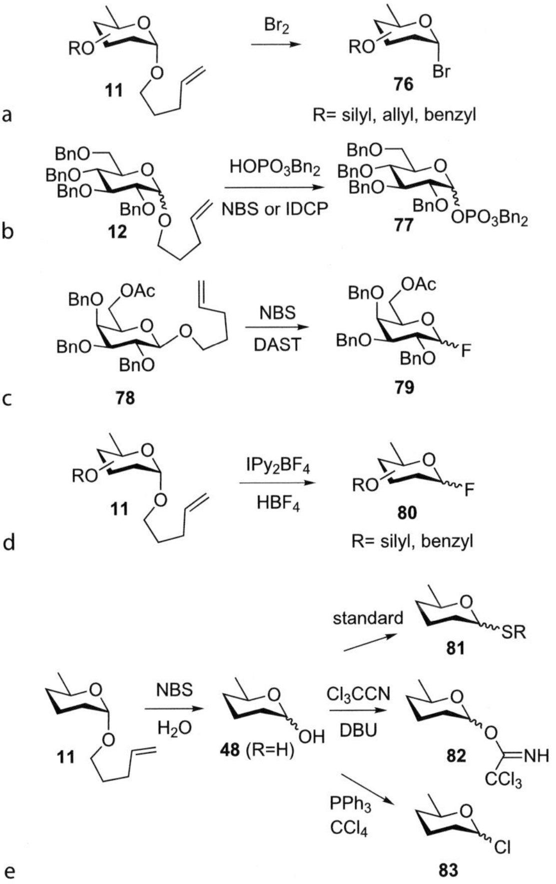

a

b

c

d

e

□ Scheme 11
Conversion of NPGs into different glycosyl donors

the amide and explained this result assuming the formation of the kinetically favored α-D-glu-copyranosylacetonitrilium ion, **84** [59].

2.5.2 Synthesis of *N*-α-Linked Glycoproteins from Pyranosylacetonitrilium Ions

The synthetic value of the above-mentioned transformation was considerably enhanced when a carboxylic acid, rather than water, was used to trap the pyranosylacetonitrilium ion (❷ *Scheme 12b*) [59]. Reaction of **87** with aspartic acid derivative **88**, in dry acetonitrile containing NBS, led to α-imide **89** in 61% yield [60,61]. The acetonitrilium ion **90** was trapped by carboxylic acid **88**, to give an imidic anhydride **91**, which rearranged in situ to give the *N*,*N*-diacyl derivative **89**. The route to 2-acetamido-1-*N*-(L-aspart-4-oyl)-2-deoxy-β-D-glycopyranosyl-amine **92**, was completed by selective *N*-deacetylation of **89** with piperidine (❷ *Scheme 12c*) [59].

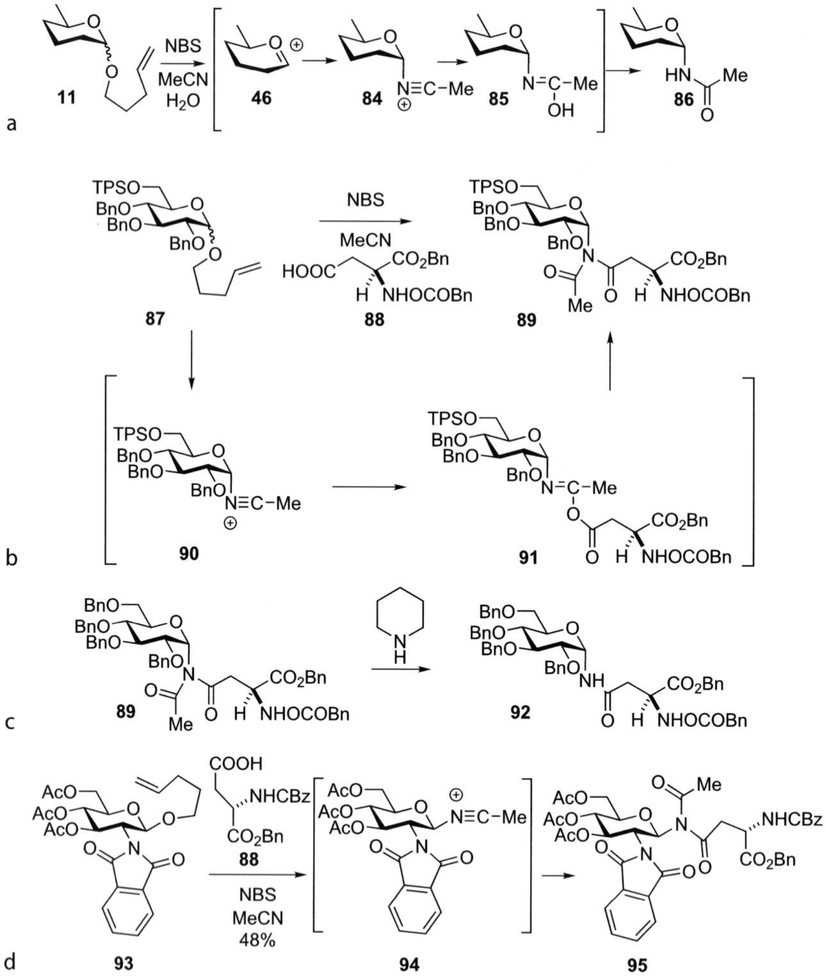

■ Scheme 12
Reactions of pyranosylacetonitrilium ions arising from NPGs

2.5.3 Synthesis of *N*-β-Linked Glycoproteins from Pyranosylacetonitrilium Ions

More interestingly, the presence of a neighboring participating group at C2 induces the formation of β-nitrilium ion intermediates, e. g. **94** (● *Scheme 12d*), thus paving the way to β-linked glycoproteins [62]. Accordingly, phthalimido NPG **93**, reacted with aspartic acid derivative **88**, in acetonitrile using NBS as promoter, via the β-nitrilium intermediate **94**, to give the β-asparagine-linked product **95** in 48% yield.

More recently, this method has been elaborated in a three-component-reaction (NPG, acetonitrile, carboxylic acid) route to *N*-glycosylamines [63].

2.6 *n*-Pentenyl 2-Amino-2-Deoxy Glycoside Derivatives as Glycosyl Donors

Several pent-4-enyl 2-amino-2-deoxy glycoside derivatives were evaluated as glycosyl donors for the synthesis of 2-amino-2-deoxy oligosaccharides [64]. 2-Deoxy-2-phthalimido **96**, and 2-anisylimino-2-deoxy-D-glucopyranosides **98**, underwent IDCP-induced coupling with

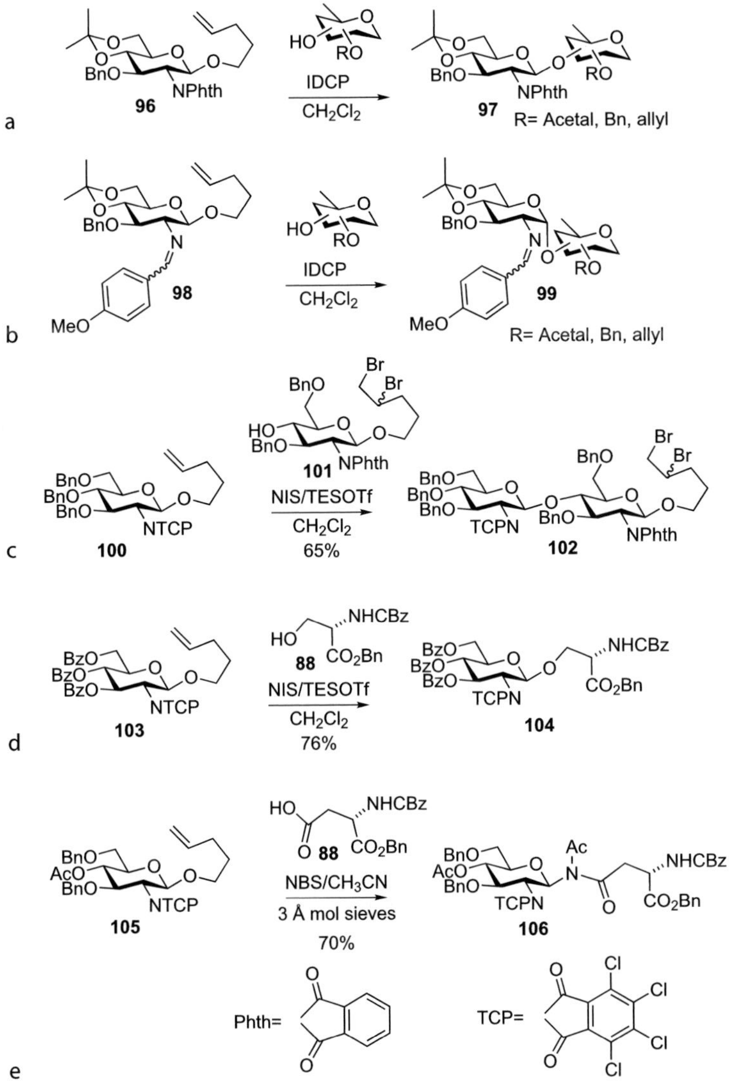

Scheme 13
Glycoside formation from pent-4-enyl 2-amino-2-deoxy glycosides

a variety of sugar alcohols to give β and α disaccharides **97** and **99**, respectively, in moderate to good yields (❷ *Scheme 13a,b*) [65,66]. 2-Deoxy-2-*N*-tetrachlorophthaloyl NPGs, e. g. **100**, **103**, [67,68,69] are useful donors for the stereocontrolled access to 1,2-*trans* glycosides as exemplified in ❷ *Scheme 13c,d*. Good yields of disaccharide **102**, and aminoacid **104** were obtained by the use of NIS/TESOTf as promoter. β-*N*-Linked glycopeptide **106** was prepared by treatment of **105** with acid **88** in dry MeCN containing NBS (❷ *Scheme 13e*).

Controversial results have been reported when 2-deoxy-2-azido NPGs were used as glycosyl donors (❷ *Scheme 14*). Fraser-Reid and co-workers reported that **107** failed to give pseudo-disaccharide **109** upon reaction with acceptor **108** under the agency of NIS/TESOTf

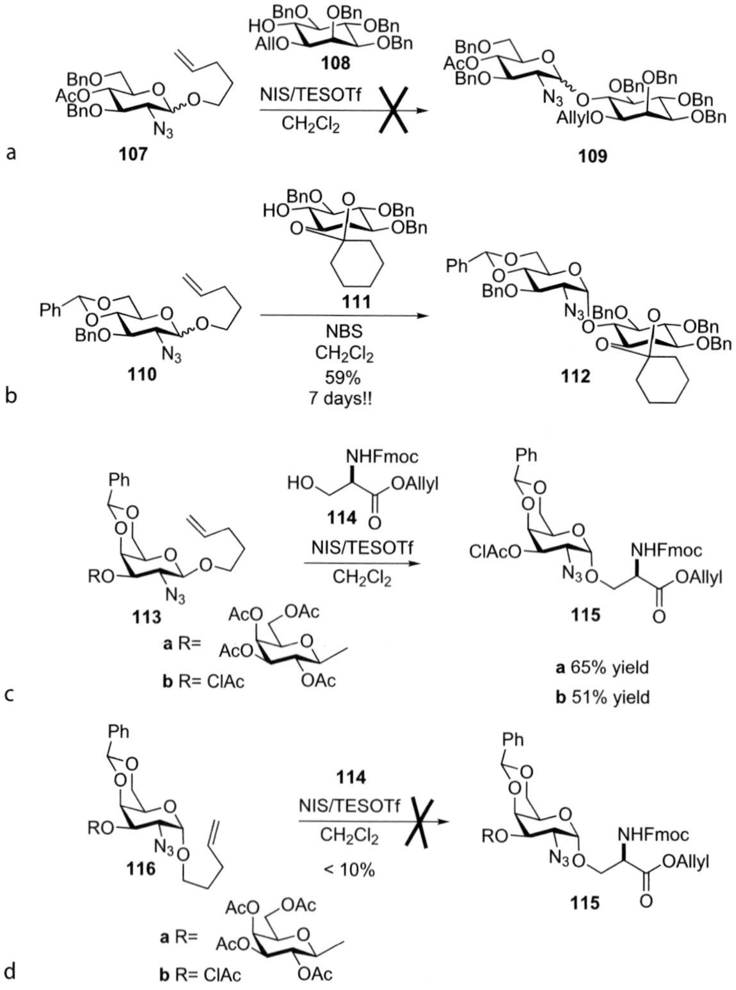

❏ **Scheme 14**
Glycoside formation from pent-4-enyl 2-deoxy-2-azido NPGs

(❸ *Scheme 14a*) [70]. By contrast, good results were obtained with the benzylidenated deriva-tive **110** (❸ *Scheme 14b*). The authors ascribed this result to the conformational constraint imposed by the benzylidene ring, in keeping with their precedents [41].

Svarovsky and Barchi [71] observed a striking reactivity difference between pent-4-enyl β- and α-2-azido-2-deoxy galactosides **113** and **116**, respectively (❸ *Scheme 14c*-X). Thus, where-

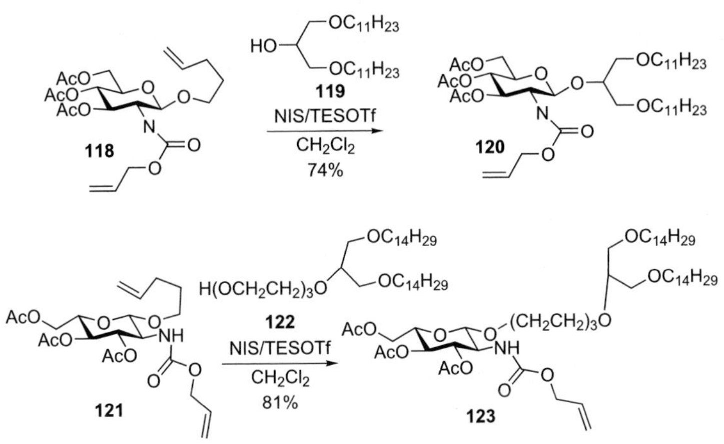

a

b

■ Scheme 15
Pent-4-enyl 2-allyloxycarbonyl-2-deoxy-D- and L-glucopyranosides

■ Table 4
Glycosidation of (−)-menthol (125) by oxazolidinone protected NPGs, 124

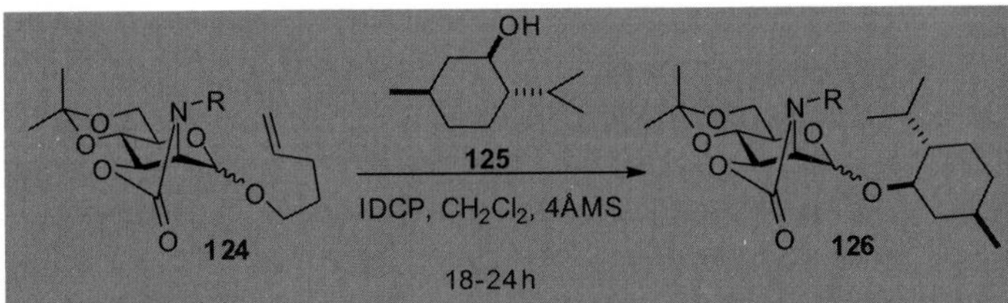

Entry	Oxazolidinone	R[a]	IDCP (equiv)	α:β	Yield (%)
i	α	H	2	4:1	53
ii	β	H	2	5:1	24
iii	α	CBz	4	α only	71
iv	β	CBz	4	α only	4
v	α	TCBoc	4	α only	41
vi	α	Troc	4	α only	50
vii	α	Boc	4	α only	63

[a] CBz = (benzyloxy)carbonyl, TCBoc = (2,2,2-trichloro-1,1-dimethylethoxy)carbonyl, Troc = (2,2,2-trichlroethoxy)carbonyl, Boc = *tert*-butoxycarbonyl

as β-NPGs **113a** and **113b** reacted with serine derivative **114** to give the sought α-glycosyl aminoacids **115a,b**, with complete stereocontrol, the corresponding α-anomers **116a,b** gave very poor yields of **115a,b** ($<10\%$) and much slower reaction rates.

Fraser-Reid's group advanced the synperiplanar lone-pair hypothesis (SLPH), to account for the fact that β-D-glycopyranosides hydrolyze $\approx$2–3 times faster [72] than the corresponding α-anomers [73]. This theory advocates that as the reaction progresses synperiplanar lone-pair interactions in the energetically accessible half-chair conformation of the β-anomer are equivalent to the antiperiplanar interactions in the half-chair of the α-anomer (antiperiplanar lone pair hypothesis, ALPH) [74]. On the other hand, the torsional effects associated with the conformational restraint imposed by the presence of the benzylidene ring might enhance this β/α reactivity difference to the point that the α-anomer hardly reacts [75].

2-Allyloxycarbonylamino-2-deoxy D- and L-glucopyranosides **118** and **121**, respectively, have been reported by Lafont and Boullanger [76], to successfully glycosylate 10-tetradecyloxymethyl-3,6,9,12-tetraoxahexacosanol (**119**) and 1,3-bis(undecyloxy)propan-2-ol (**122**), in the course on their studies on neoglycolipids for monolayers (◐ *Scheme 15*). In this case, the chemoselectivity in the reaction of the anomeric pent-4-enyl moiety in the presence of the

◻ **Table 5**
Stereocontrolled glycosylation using *N*-CBz NPG donor 127

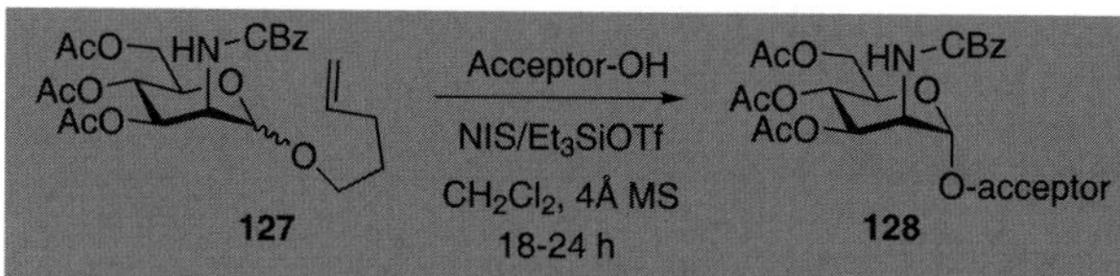

Entry	Donor	Acceptor-OH	α:β	Yield (%)
i	α	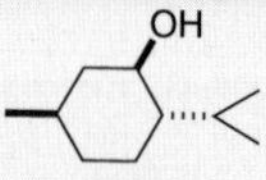 **125**	α only	63
ii	β	125	α only	60
iii	α	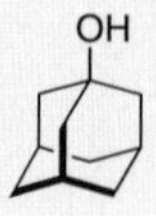 **129**	α only	66
iv	β	129	α only	60
v	α	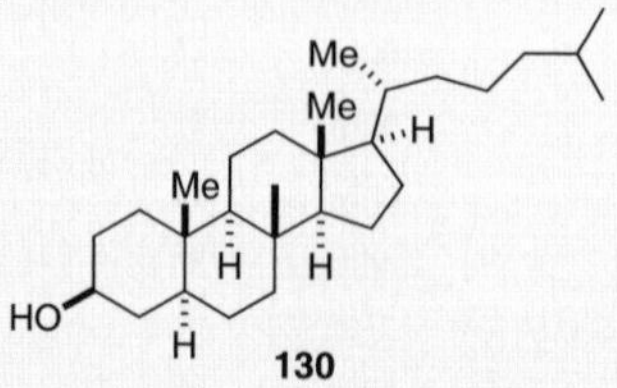 **130**	α only	68
vi	β	130	α only	63

[a] CBz = (benzyloxy)carbonyl

allyloxycarbonyl group is noteworthy. Related *n*-pentenoyl derivatives have been reported as efficient protecting groups for amines by its mild deprotection with iodine in THF-water [77]. Rojas and co-workers [78] described a novel synthetic route to α-linked 2-deoxy-2-man-nosamine derivatives, which involved a stereocontrolled glycosidation step of NPG oxazo-lidinones (e. g. **124**, ❷ *Table 4*) and *N*-CBz NPGs (e. g. **127**, ❷ *Table 5*). The authors found a striking difference in reactivity between α- and β-anomers of oxazolidinones **124**. α-NPG oxazolidinones served as highly stereoselective donors (❷ *Table 4*, entries iii, v–vii), where-as the β-anomer was nearly inert (❷ *Table 4*, entries ii, iv). However, regioselective *N*-CBz oxazolidinone ring opening to **127**, prior to glycosylation permitted elaboration of either NPG anomer to the desired α-Man-NCBz products **128** (❷ *Table 5*).

2.7 Semi-Orthogonal Couplings of NPGs

The term "semi-orthogonality" between glycosyl donors (e. g. **A** and **B**, ❷ *Scheme 16*) was introduced by Demchenko [79]. It indicates that whereas selective activation of armed and dis-armed glycosyl donor **A** can be effected in the presence of either armed or disarmed donor **B** (❷ *Scheme 16a*), the opposite is not feasible. Thus, in semi-orthogonal donors, the selective activation of disarmed glycosyl donor **B** in the presence of glycosyl donor **A** can not be accom-plished (❷ *Scheme 16c*).

2.7.1 Semi-Orthogonality of *O*-Pentenyl and *S*-Ethyl Glycosides

Demchenko and De Meo found conditions for the selective activation of NPGs and ethyl 1-thioglycosides (❷ *Scheme 17*) [79]. They demonstrated that armed NPGs (e. g. β-**12**)

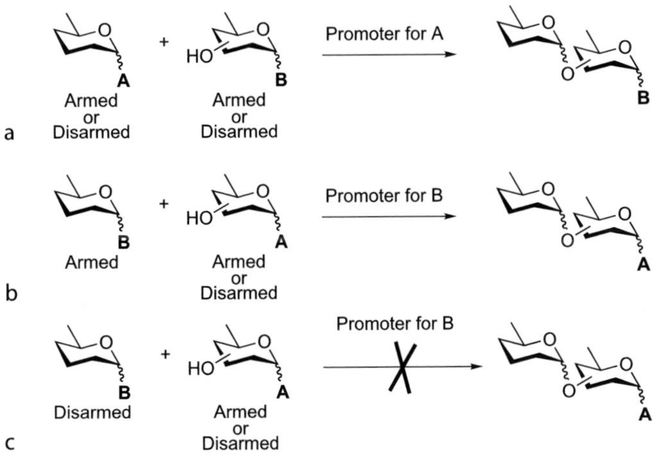

❏ **Scheme 16**
Semi-orthogonality of glycosyl donors

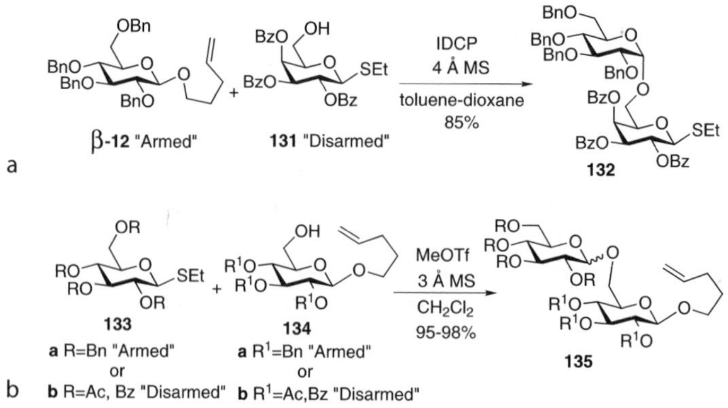

■ Scheme 17
Semi-orthogonality of *O*-pentenyl and *S*-ethyl-glycosides

could be activated in the presence of thioglycosides (e. g. **131**) with IDCP as the promoter
(❯ *Scheme 17a*). On the other hand, the use of methyl triflate (MeOTf) permitted the activation of disarmed thioglycosides (e. g. **133b**) in the presence of armed or disarmed NPGs, **134**
(❯ *Scheme 17b*).

2.7.2 Semi-Orthogonality of NPGs and Glycosyl Fluorides

López et al. reported the selective activation of armed NPGs (e. g. **136**) in the presence of
armed glycosyl fluorides (e. g. **137**) on treatment with IDCP (❯ *Scheme 18a*) [49]. On the
other hand, armed and disarmed glycosyl fluorides **139**, could be activated in the presence of
armed NPGs (e. g. **140**) on treatment with ytterbium triflate (Yb(OTf)$_3$) (❯ *Scheme 18b*).

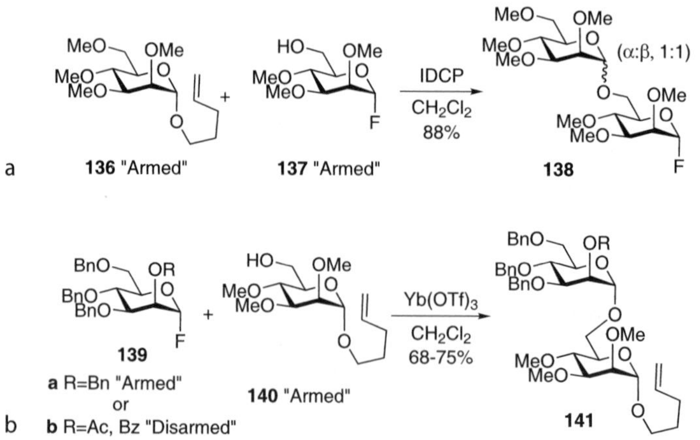

■ Scheme 18
Semi-orthogonality of NPGs and glycosyl fluorides

2.8 n-Pentenyl Furanoside Donors

2.8.1 Chemoselective Deprotection of the Anomeric Center

Unlike *n*-pentenyl pyranosides, the corresponding furanosides have attracted comparatively little attention. Sharma and Rao reported the preparation of *n*-pentenyl D-allo-, and D-gulo-furanosides **143** and **146**, respectively (❯ *Scheme 19*) [55]. They made use of an efficient acid-induced rearrangement of diacetonides **142** and **145**, in the presence of *n*-pentenyl alcohol. The ensuing pent-4-enyl diacetonides **143** and **146**, were chemoselectively cleaved to hemiacetals **144** and **147**.

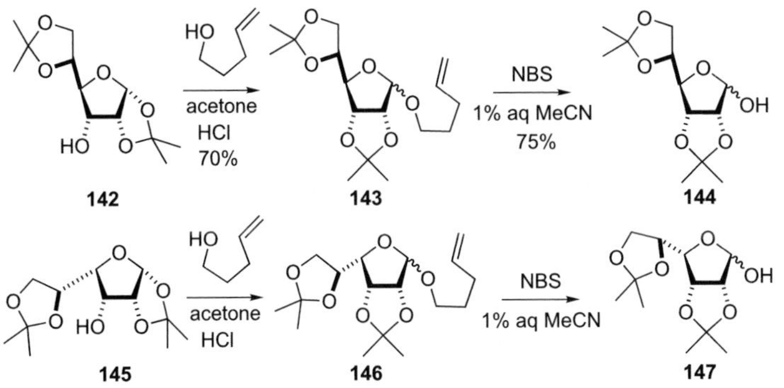

◻ Scheme 19
Pent-4-enyl furanosides from D-allose and D-gulose diacetonides

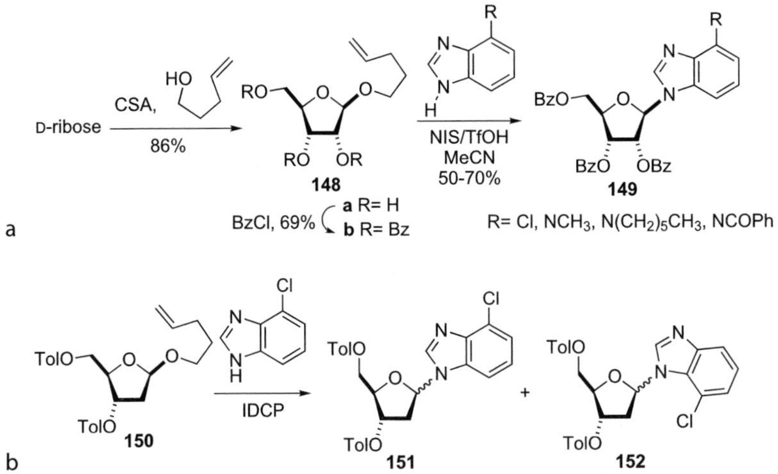

◻ Scheme 20
Pentenyl ribofuranosides in the synthesis of purine nucleosides

2.8.2 Application to the Synthesis of Nucleosides

Chapeau and Marnett developed a synthetic route to purine nucleosides from *n*-pentenyl ribo-sides (❷ *Scheme 20*) [80]. The authors used a Fischer glycosylation of D-ribose with 4-pentenol to produce pent-4-enyl β-D-erythro-pentofuranoside **148a**, in 86% yield. Reaction of the latter with benzoyl chloride produced their glycosyl donor **148b**, in 69% yield. Addition of TfOH to acetonitrile solutions containing **148b**, the selected purine, and NIS, resulted in a rapid coupling to form the desired nucleosides **149**, in a stereocontrolled manner with yields ranging from 50 to 70% (❷ *Scheme 20a*). The absence of an acyl group at *O2* in 2-deoxy NPG **150**, enhanced its reactivity to iodonium sources so IDCP could be used as the promoter. Thus, reaction of **150** with 6-chloropurine in acetonitrile was neither regio- nor stereo-selective, yielding four coupling products **151α,β** and **152α,β**, in similar amounts (❷ *Scheme 20b*).

2.8.3 *n*-Pentenyl Furanosides as Glycosyl Donors

Arasappan and Fraser-Reid described the preparation of *n*-pentenyl galactofuranosides and evaluated their prospects as glycosyl donors (❷ *Scheme 21*) [81]. Fischer glycosidation of D-galactose under kinetic conditions using *n*-pentenyl alcohol and DMSO as co-solvent [82]

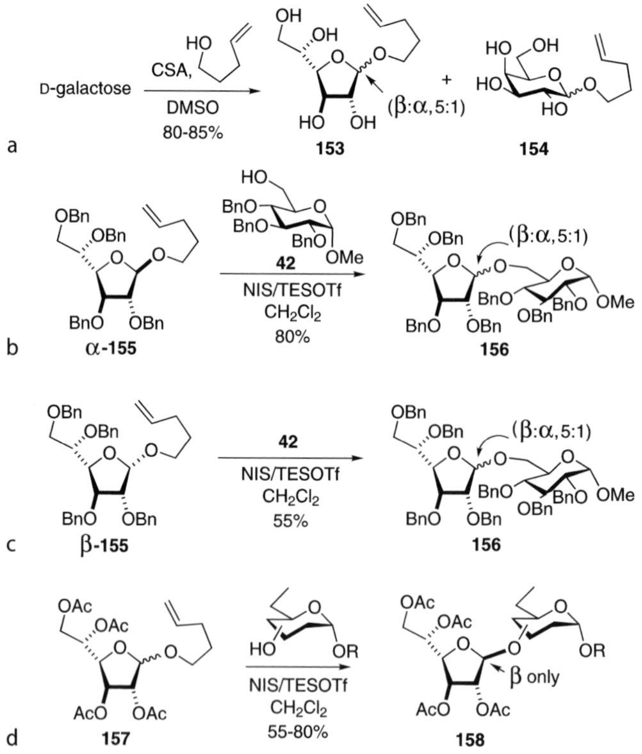

❏ Scheme 21
Pentenyl galactofuranosides

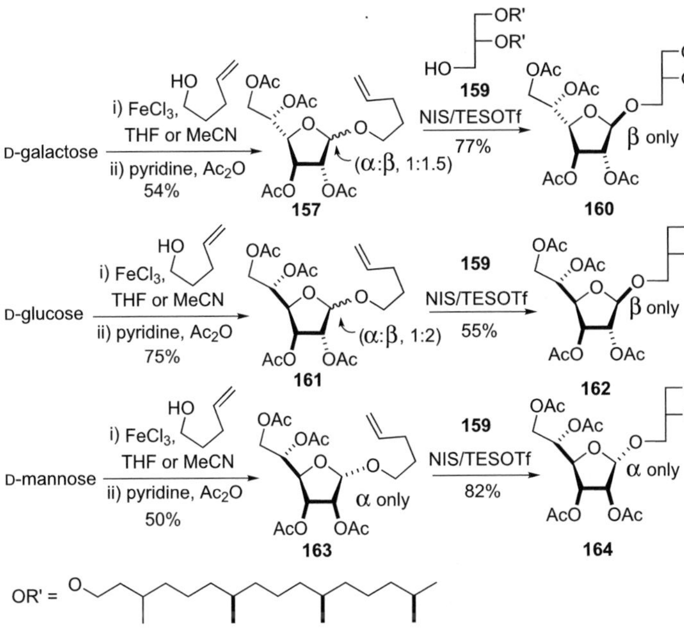

☐ Scheme 22
Synthesis of archaeol glycolipid analogues from pent-4-enyl furanosides

afforded an anomeric α/β (1:3) mixture of *n*-pentenyl galactofuranosides **153**, ($\approx$80–85% yield), contaminated with small amounts of the corresponding *n*-pentenyl galactopyranosides, **154** (❍ *Scheme 21a*). Glycosylation with α- or β- pentenyl glycosides **155** was irrelevant to the product β/α ratio, both favoring the β-furanoside β-**156** (❍ *Scheme 21b,c*). Reactions with donors α- or β-**157** resulted in the β-linkage product **158**, exclusively (saccharide acceptors with free hydroxyl groups at C-2, C-4, and C-6 were assayed), presumably due to the neighboring group participation of the C-2 ester functionality (❍ *Scheme 21d*).

Plusquellec and co-workers reported an improved method for the preparation of *n*-pentenyl furanosides [83] based on their previously described use of FeCl$_3$ as a catalyst in Fischer-type glycosylations [84]. Accordingly, D-glucose, D-galactose, and D-mannose upon treatment with FeCl$_3$ and *n*-pentenyl alcohol followed by in situ acetylation, yielded pent-4-enyl D-gluco-, D-galacto-, and D-mannofuranoside derivatives **157**, **161**, and **163**, respectively in yields ranging from 50 to 75%. Glycosylation of glycerol diether **159** with these donors, promoted by NIS/TESOTf yielded glycolipids **160**, **162**, and **164** in high yields and with excellent 1,2-*trans* stereoselectivity (❍ *Scheme 22*).

2.8.4 *n*-Pentenyl Arabinofuranosides in the Assembly of Oligoarabinans of *Mycobacterium tuberculosis*

Recent interest in oligoarabinans, have been triggered by their presence in the lipoarabinomannan polysaccharide component of the cell wall complex of mycobacteria [85]. Several research groups have employed *n*-pentenyl arabinofuranosides in their approaches to oligoarabinans.

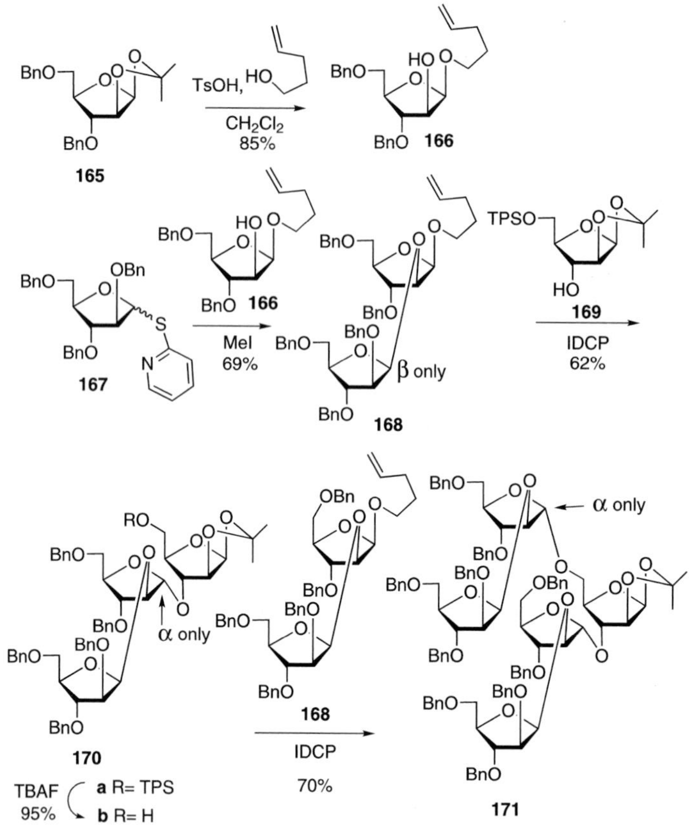

■ Scheme 23
Synthesis of a pentaarabinofuranosyl structure motif of *Mycobacterium tuberculosis*

n-Pentenyl β-D-arabinofuranoside **166**, readily prepared from **165**, was employed as acceptor/donor in Gurjar's approach to arabinosyl pentasaccharide **171** (❷ *Scheme 23*) [86]. Accordingly, **166** was glycosylated with *S*-(2-pyridyl)-1-thiofuranose **167** to yield, in a stereoselective manner, β-disaccharide **168**. The latter, itself a pentenyl donor, was then used as a glycosyl donor in two glycosylation events. First, IDCP-promoted glycosylation of silyl derivative **169** yielded trisaccharide **170a** in 62% yield, and in a stereocontrolled manner. Desilylation of **170a** furnished **170b**, which then functioned as the acceptor in the second IDCP-induced glycosylation with **168** to produce pentasaccharide **171**.

More recent studies by Fraser-Reid's group, have focused on the use of NPOEs both as arabinofuranosyl donors, and as convenient starting materials for the preparation of *n*-pentenyl arabinofuranosyl acceptors [87]. TPSOTf-induced rearrangement of NPOE **172**, followed by desilylation afforded pentenyl glycoside **173b** (❷ *Scheme 24*). Glycosylation of pentenyl glycoside **173b** with NPOE **172** was carried out using NIS/Yb(OTf)₃, a chemospecific promoter for NPOEs [88,89]. Iteration of the sequence permitted the preparation of the α-1,5-linked arabinan segment of the complex lipoarabinomanan cell wall array of *Mycobacterium tuberculosis*, **175**.

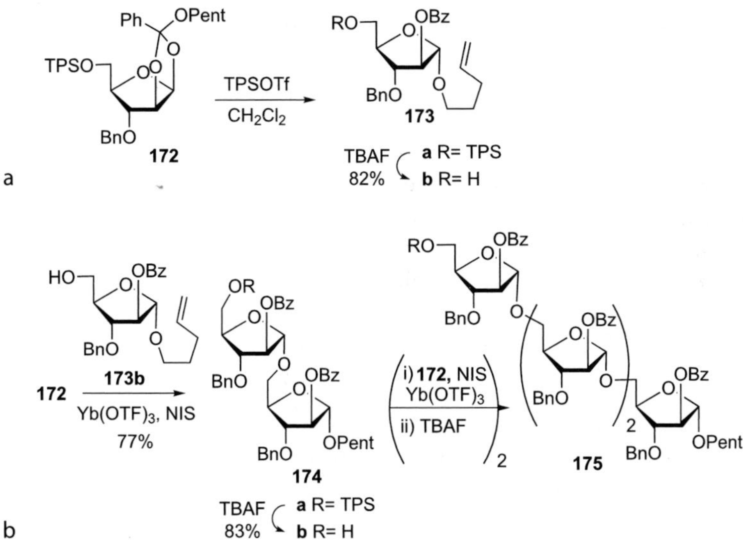

a

b

☐ Scheme 24
NPOEs in the synthesis of arabinofuranosyl donors

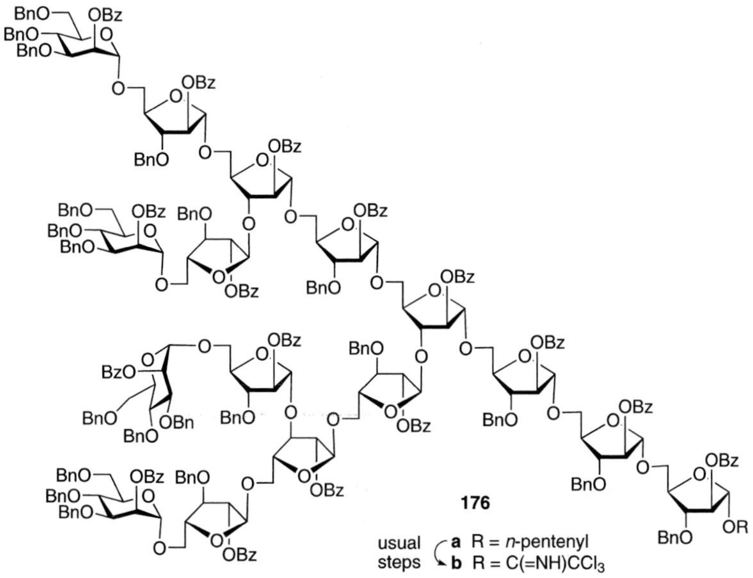

☐ Figure 3
Mannose-capped multibranched dodecafuranoarabinan of *Mycobacterium* species

The potency of this strategy relies ultimately in the sturdiness, and yet the possibility for chemoselective cleavage, of pentenyl arabinofuranosides (e. g. **175**) [90]. Its value has been demonstrated recently with the synthesis of the pentenyl glycoside of mannose-capped dode-

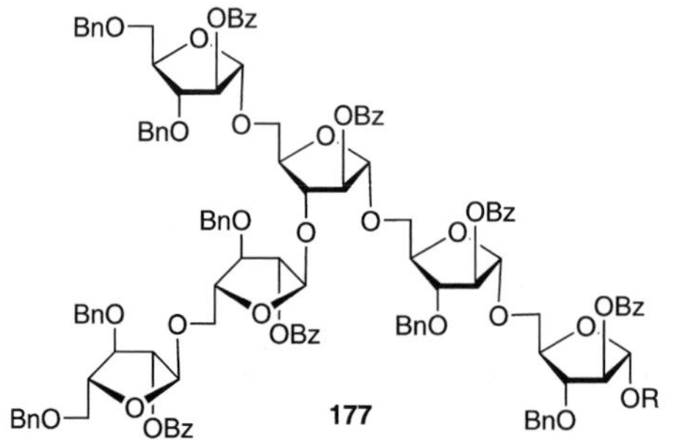

□ Figure 4
Multibranched hexafuranoarabinan of *Mycobacterium* species

cafuranoarabinan of *Mycobacterium* species, **176a** (❷ *Fig. 3*). The final NPG→trichloro-acetimidate transformation (**176a→176b**) made possible the coupling of this arabinan segment to an oligomannan acceptor, thus resulting in the synthesis of the largest heterooligosaccharide to date, a 28-mer arabinomannan [91].

n-Pentenyl arabinofuranosides have also been used by Seeberger and co-workers in the final stages of their synthesis of a 12-mer component of *Mycobacterium tuberculosis* [92]. *n*-Pentenyl arabinan hexasaccharide **177a**, was transformed to the corresponding trichloroacetimidate **177b** and coupled with a mannan hexasaccharide acceptor to yield the sought arabinomannan dodecasaccharide (❷ *Fig. 4*).

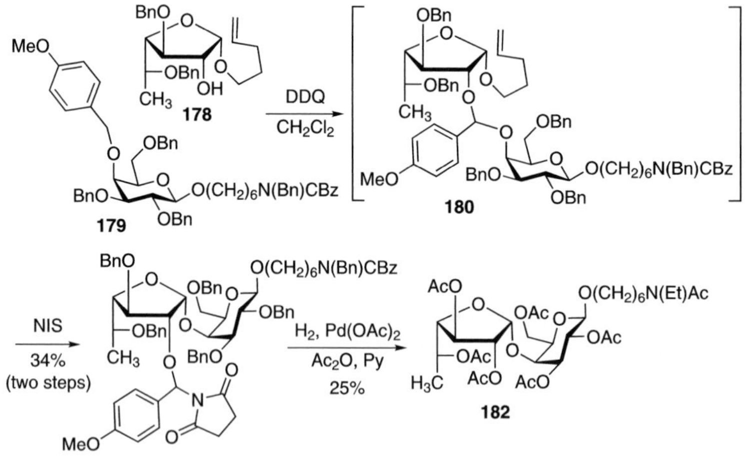

□ Scheme 25
NPGs as glycosyl donors in intramolecular aglycon delivery

2.8.5 Intramolecular Aglycon Delivery from *n*-Pentenyl Glycofuranosides

An approach to α-D-fucofuranosyl glycosides that makes use of the intramolecular aglycon delivery (IMAD) [93,94,95,96] starting from an *n*-pentenyl fucofuranoside has been described (❍ *Scheme 25*) [97]. *n*-Pentenyl fucofuranoside **178** bearing a free 2-OH group was attached to a 4-*O*-PMB-protected galactopyranoside **179**, upon treatment of the mixture with DDQ. The unstable tethered compound **180**, could be activated with NIS, in the absence of even catalytic amounts of acid, to undergo an efficient *p*-methoxybenzyl-assisted aglycon delivery [95] leading to the desired glycoside **181**. The unusual structure **181**, resulting from quenching of the benzylic cation with *N*-succinimide, was then processed to α-D-fucofuranoside **182**.

2.8.6 Intramolecular *C*-Glycosylation of NPGs

The intramolecular *C*-glycosylation of NPGs has been studied by Martin's group in the course of their approaches to bergenin [98] and related natural products [99,100]. The treatment of pentenyl β-D-glucopyranose **183** with IDCP promoted an internal, Friedel–Crafts type, *C*-arylation reaction in excellent yield (❍ *Scheme 26a*). The resulting product was exclusively the kinetically favored, *cis*-fused tricyclic system **184**. Treatment of the latter with an oxophilic Lewis acid (BF₃·OEt₂) led to the *trans*-fused (β-linked) **185**. On the contrary, analogous reac-

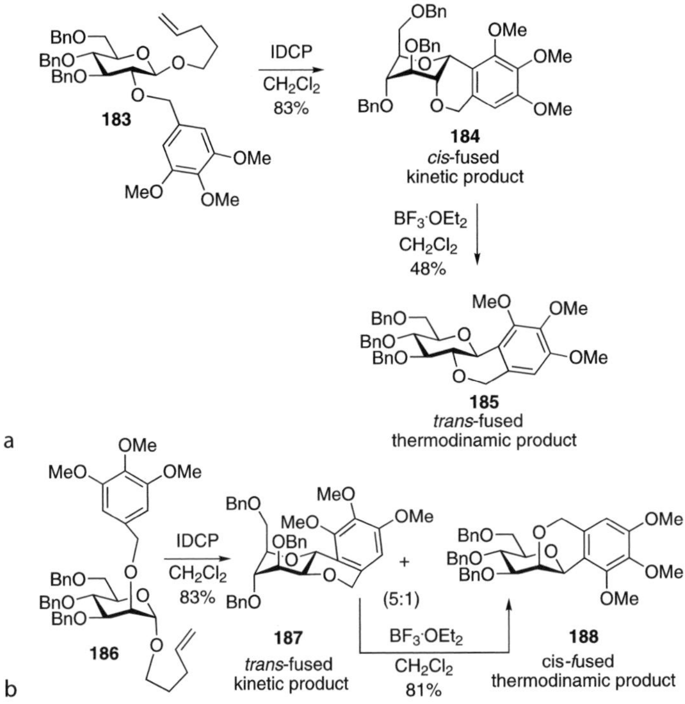

a

b

□ Scheme 26
Intramolecular *C*-glycosylation of NPGs

tion of α-D-mannopyranoside **186** led to a mixture of *trans*- and *cis*-fused compounds **187**, and **188** (5:1), where the major *trans*-fused (α-linked) product **187**, was this time the kinetic product (❯ *Scheme 26b*). Treatment of the latter with $BF_3 \cdot OEt_2$ promoted the epimerization to the, more stable, 1,2-*cis* epimer **188**, in 81% yield.

2.8.7 NPGs of *N*-Acetylneuraminic Acids (Neu5Ac)

One report describing *O*-sialylation of 4-pentenyl glycosides of Neu5OAc, e. g. **189**, has appeared (❯ *Table 6*) [101]. Good α/β selectivity (11:1) was attained in the glycosylation of primary acceptor **40** in MeCN using NIS/TfOH as the promoter (❯ *Table 6*, entry i), however, with the secondary acceptor **191** the α/β selectivity dropped to 4:1 (entry iii). The use of Et_2O as solvent produced a 1:1 mixture of anomers (entry ii).

2.8.8 NPGs of L-Iduronic Acid as Glycosyl Donors

In their studies on heparin/heparin sulfate, and dermatan sulfate, Petitou, Sinaÿ and co-workers found that *n*-pentenyl glycosides of L-Iduronic acid, e. g. **192**, were efficient glycosyl donors [102]. In contrast, the corresponding thioglycosides, and glycosyl fluorides did not give the expected disaccharides. Reaction of *n*-pentenyl glycosyl donors **192** (α or β) with acceptors **194–197** (❯ *Table 7*) was carried out in CH_2Cl_2 with NIS/TfOH to furnish the corresponding α-disaccharides **193**, in good yields.

More recently, Reichardt and Martín-Lomas have evaluated *n*-pentenyl glycosides of glucosamine $\alpha1{\rightarrow}4$L-iduronic acid disaccharide, as substrates for autocondensation in their approach to heparin oligosaccharide fragments. However, the NIS used as promoter, being

◻ **Table 6**
n-Pentenyl glycosides of Neu5Ac in glycosylation

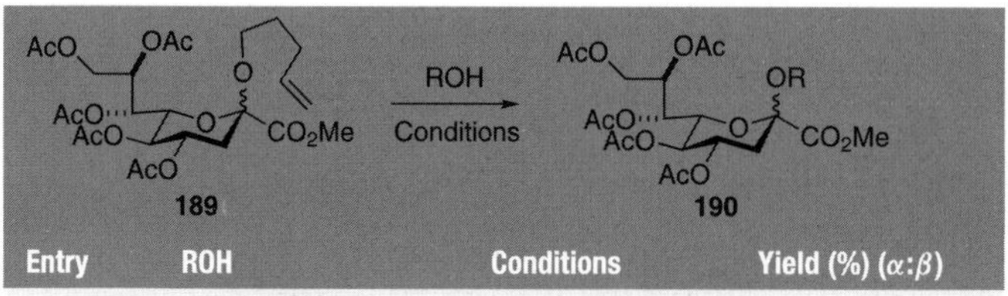

Entry	ROH	Conditions	Yield (%) (α:β)
i	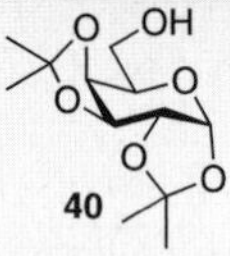 **40**	NIS/TfOH MeCN, −40 °C	60 (11:1)
ii	40	NIS/TfOH Et_2O, −40 °C	33 (1:1)
iii	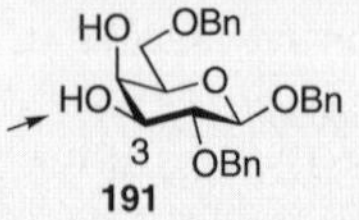 **191**	NIS/TfOH MeCN, −40 °C	37[a] (4:1)

[a] The glycosylation is regioselective at *O*-3

◼ **Table 7**
n-Pentenyl glycosides of L-iduronic acids as glycosyl donors

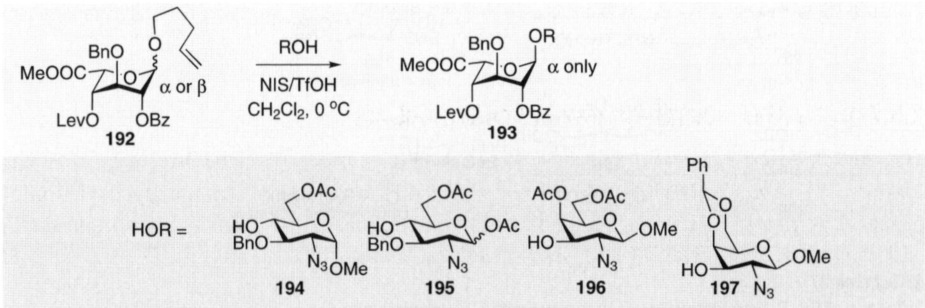

HOR =				
	194	**195**	**196**	**197**
Yields of disaccharide	80% from α-192 83% from β-192	75% from α-192 77% from β-192	72% from α-192	90% from α-192

itself a nucleophile, competes with the acceptor disaccharide in the polycondensation process, which results in fast chain-reaction termination and a low yield and degree of polymerization [103].

2.9 NPGs in Regioselective Couplings

The synthesis of branched saccharides by multiple glycosylations onto a central monosaccharide normally requires the use of orthogonal protecting groups in the acceptor. In this context, regioselective glycosylation of diols or polyols would ease the number of protection-deprotection steps in these synthetic protocols.

2.9.1 The Role of the *O*-2 Substituent in Regioselective Couplings

In their studies on *myo*-inositol glycosylation, Fraser-Reid and co-workers made the observations summarized in ❯ *Scheme 27* [104]. In the hope of achieving selective glycosidation of the equatorial-OH, they treated diol **198** with the armed *n*-pentenyl donor **64** (❯ *Scheme 27a*). However, the major product was the mixture of α/β glycosides **199** from glycosidation at the axial-OH (❯ *Scheme 27a*). In order to improve α anomeric stereoselectivity they selected the corresponding disarmed NPG **200**, as the donor (❯ *Scheme 27b*). Surprisingly, the only product obtained was the disaccharide **201** from glycosidation at the equatorial-OH.

In a series of subsequent papers Fraser-Reid and co-workers confirmed these discrepancies, and showed that the *O*-2 substituent in glycosyl donors, besides its recognized role for stereocontrol, exerts a profound influence in eliciting regioselective glycosyl couplings [105,106,107]. In most cases, 2-*O*-acyl NPGs and NPOEs shared the same regiopreferences, which were usually different from the ones displayed by 2-*O*-alkyl NPGs. The regiopreferences of the former were generally more pronounced or even exclusive.

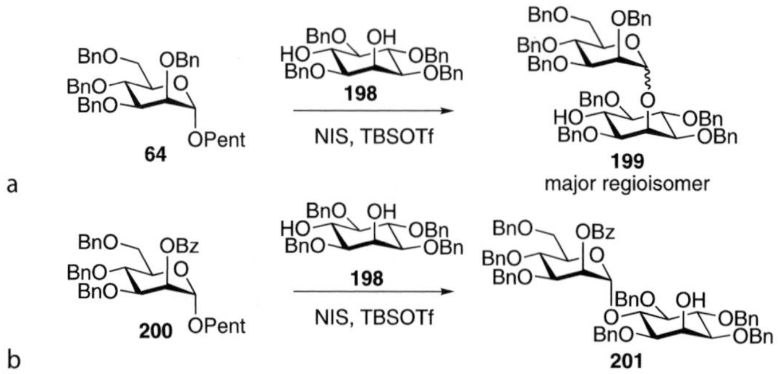

a

b

□ **Scheme 27**
Influence of the *O*-2 protecting group in regioselective glycosylations

2.9.2 Reciprocal Donor Acceptor Selectivity (RDAS)

The influence of the *O*-2 substituent in regioselective couplings is not limited to pentenyl glycoside donors. Thioglycoside and trichloroacetimidate donors have shown the same tendency [108]. The glycosylation of allose diol **203** with donors **64** and **202a–f** (❍ *Table 8*) illustrates this point. NPOE **202a** (that shows the same regiopreferences as disarmed NPGS), disarmed thiomannoside **202c**, and disarmed trichloroacetimidate **202e**, exhibited the same preference for the *O*3 of allose acceptor **203** (❍ *Table 8*, entries i, iii, v). On the contrary, armed donors **64**, **202c**, and **202e** furnished a 2:1 mixture of disaccharides **204b** and **205**. The above-mentioned examples indicate that each donor expresses preference for one of the diol–OHs in the acceptor and vice versa. The authors coined the term *Reciprocal Donor Acceptor Selectivity* (RDAS) [109] to account for these findings.

□ **Table 8**
Influence of the *O*-2 substituent in the regioselective coupling of various glycosyl donors with allose diol 203

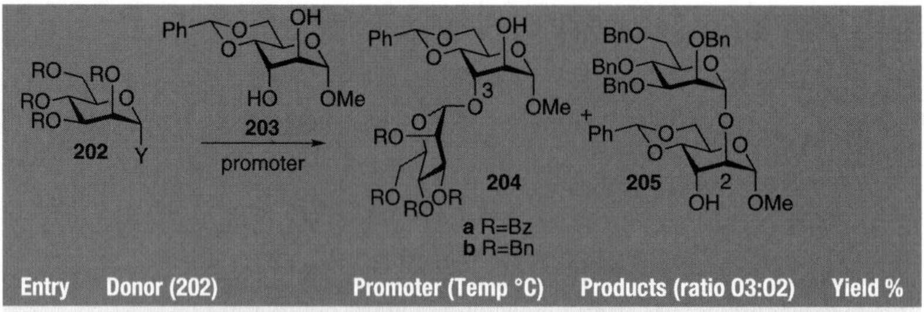

Entry	Donor (202)	Promoter (Temp °C)	Products (ratio O3:O2)	Yield %
i	a Y=orthoester; R=Bz	NIS/BF$_3$·Et$_2$O −30	204a only	92
ii	64 Y=OPent; R=Bn	NIS/BF$_3$·Et$_2$O −30	204b + 205 (2:1)	37
iii	b Y=SPh; R=Bz	NIS/BF$_3$·Et$_2$O −30	204a only	58
iv	c Y=SPh; R=Bn	NIS/BF$_3$·Et$_2$O −30	204b + 205 (2:1)	66
v	d Y=OC(NH)CCl$_3$; R=Bz	BF$_3$·Et$_2$O −78	204a only	65
vi	e Y=OC(NH)CCl$_3$; R=Bn	BF$_3$·Et$_2$O −78	204b + 205 (2:1)	54

2.9.3 In Situ Double Differential Glycosylations of Two Donors with One Acceptor

The practical utility of this concept was further demonstrated when diol acceptor **203**, NPOE
202a, and armed-NPG **64** were treated with NIS/BF$_3$·Et$_2$O to give one single trisaccharide
206, in 57% yield (❂ *Scheme 28*) [110]. It seemed that in the formation of the trisaccharide
the regiopreferences of the NPOE **202a** and the armed NPG **64**, displayed in ❂ *Table 8*, have
been followed.
Analogously, the regiopreferences (RDAS) of disarmed NPG **200**, and armed NPG **64** vis a vis
mannose diol **207**, were evaluated (❂ *Scheme 29a,b*). With the disarmed donor **200** mannosylation occurred at the (C6)-OH only to give **208** in 53% yield, and also the symmetrical
trisaccharide **209** in 13% yield, but with no evidence for the dimannan resulting from glycosylation of the (C3)-OH (❂ *Scheme 29a*). By contrast, the armed donor **64** gave a 38% yield
of the *O*-6 product, **210**, but also 11% of the *O*3 regioisomer **211** (❂ *Scheme 29b*). Analysis
of these results according to conventional wisdom, dictates that the preference of both donors,
200 and **64**, for the primary −OH was to be expected [1] but raised the question of the possible outcome of a three-components double glycosylation when **200** and **64** compete for diol
207. Previous calculations had shown that the relative reactivity of these donors (k_{64}/k_{200}) is
3.2 [111]. Hence, it was expected that *O*6 mannosylation by the armed donor, **64**, would pre-
dominate in any trimannan produced. Surprisingly, a single trimannan **212**, in which the *less
reactive* donor **200** ended up at *O*6 was obtained, even in the presence of 2 equiv. of the "*more
reactive*" **64**.

2.10 The Origin of Regioselectivity in Three-Component Couplings

In searching for the origin of the regioselectivity observed in the formation of trisaccharides
206 and **212** (❂ *Scheme 28* and ❂ *Scheme 29*) several factors were considered. The reactions in
❂ *Scheme 28* and ❂ *Scheme 29c* were carried out with excess NIS promoter, conditions under
which the intermolecular halonium ion transfer (responsible for the armed-disarmed effect) is
not operative. A study of the three types of *n*-pentenyl donors indicated that their relative reactivities were in the order NPOE > armed > disarmed (e. g. **202a** > **64** > **200**) [111]. Therefore,
the most and the least reactive donors have "chosen" their preferred −OH in the final trisac-

❏ Scheme 28
In situ three-component double differential glycosylation of two donors and one diol acceptor

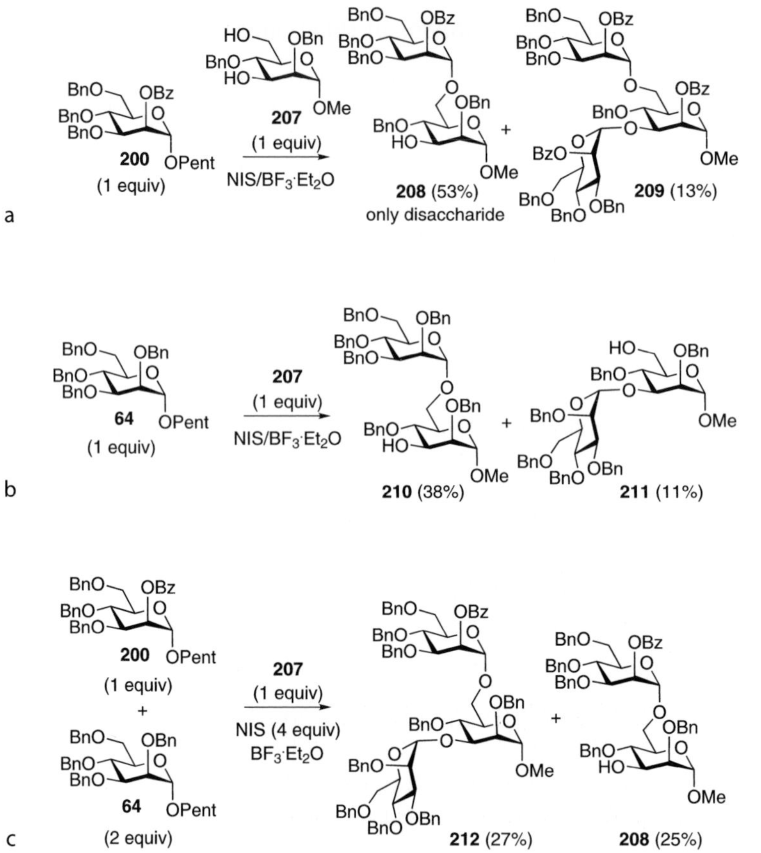

■ Scheme 29
In situ three-component double differential glycosylation of two donors and one diol acceptor

charide. On the other hand, the most and the least reactive donors give rise to the highly delo-calized, more stable intermediate **213**, while the armed donor gives the less stable oxocarbe-nium ion **214** (**◉** *Scheme 30*) [112]. The conclusion was that in *competitive* glycosylations the more *stable* donor/intermediate (not the most *reactive* donor) controls regioselectivity, result-ing in the formation of the single trisaccharides **206** and **212** and the single disaccharides **204a** and **208**.

In order to confirm this assumption the authors performed the experiments in **◉** *Table 9* [113]. Equimolar amounts of armed and disarmed donors **64**, and **200** or **202b** were allowed to com-pete for one equivalent of acceptor **215** under the agency of NIS. When one equivalent of NIS was used the major product obtained was that of glycosylation of armed NPG **216**, thus in agreement with a process of intermolecular halonium transfer and preferred reaction of the more reactive donor (**◉** *Table 9*, entries i, iii). When the amount of NIS was increased to three equivalents, the observed ratio of compounds **216** and **217** indicated enhanced coupling of the disarmed donor (**◉** *Table 9*, entries ii, iv), thus in agreement with the proposed rationalization for the regiopreferences observed in the three-component reactions.

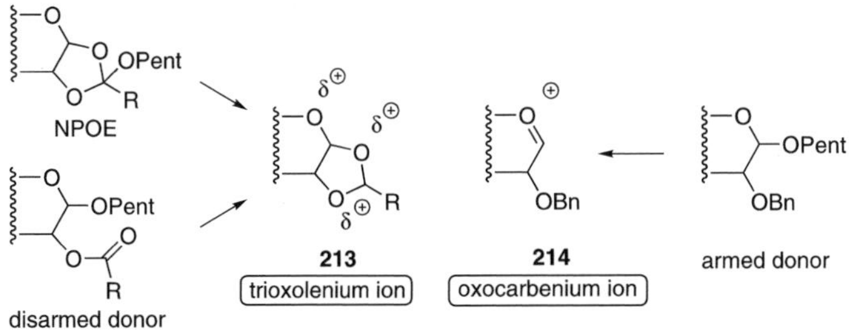

Scheme 30
Reactive intermediates from different glycosyl donors

Table 9
Competition studies on the glycosylation of acceptor 215 in the presence of glycosyl donors (64 and 200 or 202b) with variable amounts of NIS

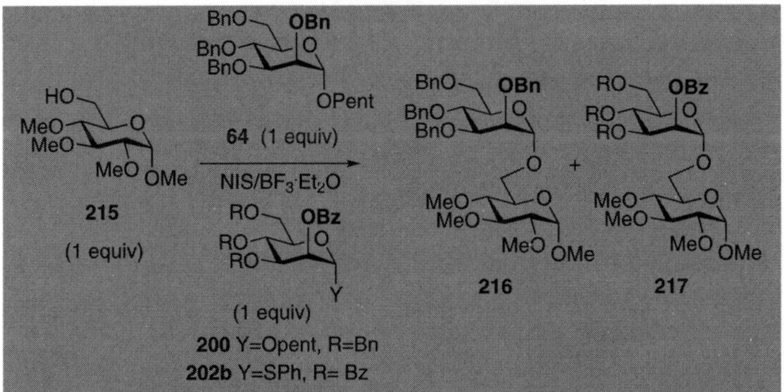

Entry	Donors armed/disarmed		NIS (equiv)	Product ratio 216	217	Yield (%)
i	64	200	1	2	1.1	57
ii	64	200	3	1.1	1	69
iii	64	202b	1	1	0	52
iv	64	202b	3	1	2.2	83

2.11 NPGs in Oligosaccharide Synthesis

Since their discovery, the unique properties of NPGs have allowed the preparation of several oligosaccharides. The pentenyl moiety may be installed early in the synthetic sequence and can survive many types of protecting group manipulations. Some selected syntheses of oligosaccharides are briefly discussed below.

2.11.1 The Pentasaccharide Core of the Protein Membrane Anchor Found in *Trypanosoma brucei*

Fraser-Reid and co-workers described a block (i. e., convergent), and a linear approach to the title compound, **226** (❱ *Scheme 31*) [114]. The convergent approach, outlined in ❱ *Scheme 31*,

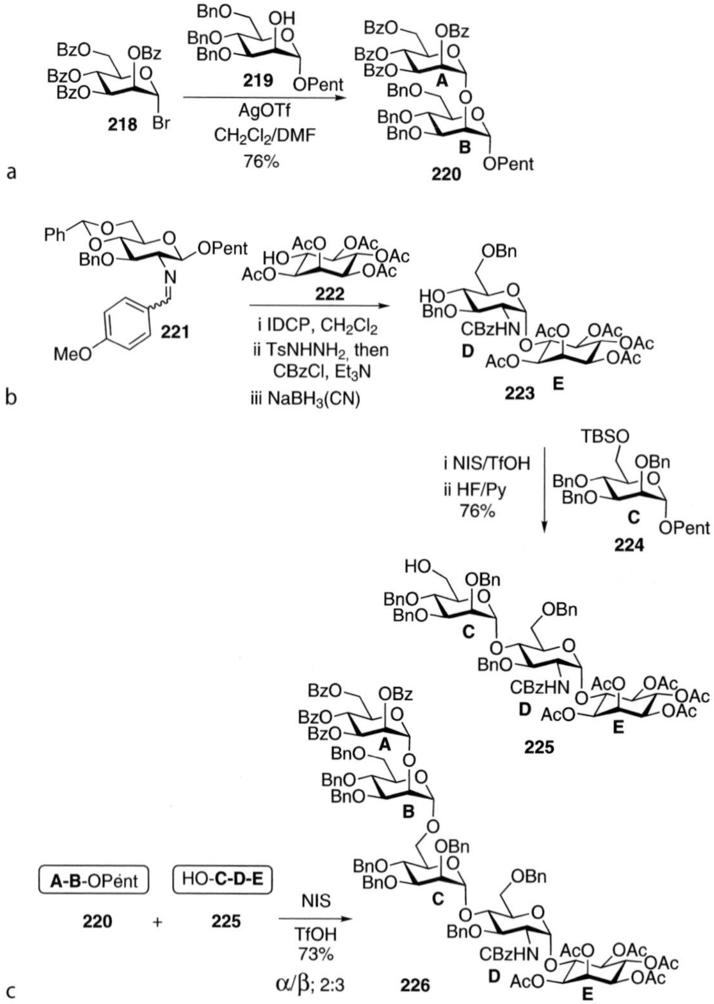

Scheme 31
The pentasaccharide core of the protein membrane anchor of *Trypanosoma brucei*

makes use of the stereocontrolled glycosylation of inositol derivative **222** with 2-deoxy-2-imino NPG **221** (❷ *Scheme 31b*). Protecting group manipulations led to acceptor **223** that was glycosylated with NPG **224**, to furnish, after desilylation, the acceptor **CDE** block, **225**. The donor counterpart **220**, had been readily prepared by Koenigs–Knorr [115] coupling of NPG **219** with glycosyl bromide **218** (❷ *Scheme 31a*). Finally, coupling of fragments AB (**220**) and CDE (**225**) promoted by NIS/TfOH led to pentasaccharide **226**, in 73% as an α/β (2:3) mixture (❷ *Scheme 31c*).

2.11.2 The Nonamannan Component of High Mannose Glycoproteins

The concise approach to nonamannan **227** (❷ *Scheme 32*), was greatly simplified with the sidetracking of NPGs [116,117] that allows the same NPG synthon to function as glycosyl donor or as glycosyl acceptor.

In the retrosynthesis of **227**, the authors identified three types of elements depending on the number of sugar units attached to them. Two components carried sugars at *O*3 and *O*6, four held substituents at *O*2, and the last three had no monosaccharides attached. According to that, the nonasaccharide target could be correlated with only two mannopyranose precursors **228** and **229**, since synthon **228** could be used to access the last two kinds of sugars. The approach featured the final link of a pentasaccharide donor with a tetrasaccharide acceptor, as outlined in ❷ *Scheme 32*.

The synthesis of pentasaccharide donor **233** started with the mannosylation of sidetracked NPG **230** with disarmed NPG donor **228** (❷ *Scheme 33*). The ensuing, stereoselectively formed disaccharide **231a**, after unveiling of its C6-OH, underwent a second mannosyla-tion with **228**. Removal of the acetates in **232a** with NH$_3$/MeOH led to diol trisaccharide **232b**, which was bis-mannosylated with **228** to give pentasaccharide **233a** in 59% yield. Regeneration of the pentenyl moiety in sidetracked **233a**, with Zn/nBu$_4$NI, granted access to pentasaccharide NPG donor, **233b**.

The lowest antenna of **227** was built from **234** (❷ *Scheme 34*). By taking advantage of the sidetracking concept, compound **228** could be used as a glycosyl donor or, after dibromina-tion and deacetylation as the glycosyl acceptor **234**, thereby facilitating the rapid assembly of trisaccharide fragment **235**. Thus, coupling of **234** and **228** afforded the expected disaccharide in 73% yield, deacetylation and additional coupling with **228** led to trisaccharide **235a** in 62% yield. The latter was transformed in glycosyl donor **235b** by reductive elimination, and cou-pled with **230** to give sidetracked tetrasaccharide **236a**. Dechloroacetylation of the latter led to **236b** that was glycosylated by pentasaccharide donor **233b** to give nonasaccharide **227** in 57% yield.

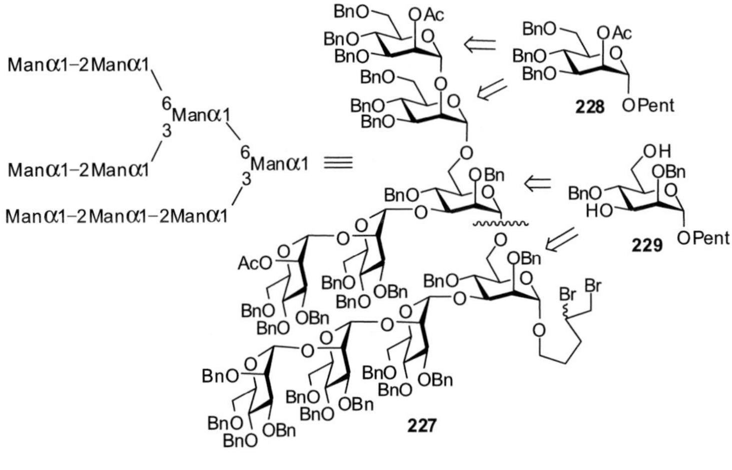

■ Scheme 32
The nonamannan component of high mannose glycoproteins

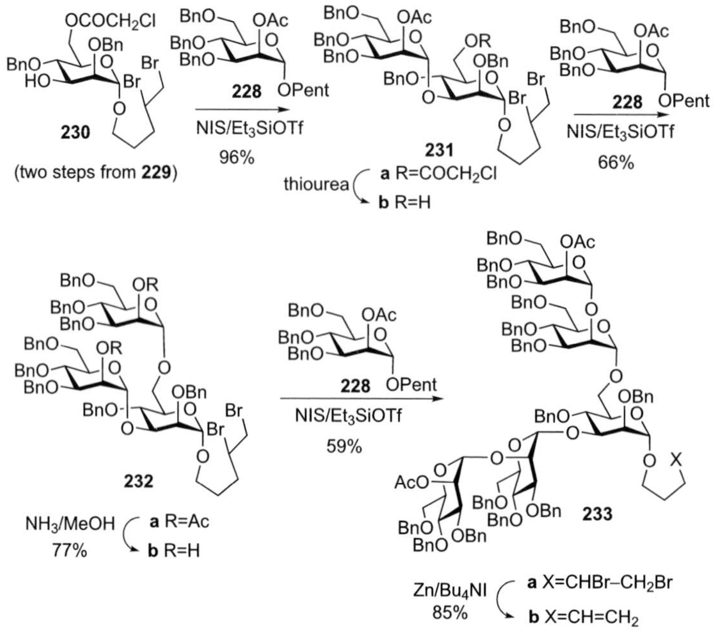

Scheme 33
Convergent synthesis of nonamannan 227. Synthesis of the pentasaccharide donor 233

2.11.3 Synthesis of NodRf-III (C18:1) (MeFuc)

Nodulation factors comprise a family of unique oligosaccharides composed substantially of glucosamine (2-amino-2-deoxy-D-glucose) units that are N-acylated with acetic acid and fatty acids residues, the latter residing at the nonreducing terminus [118]. The block synthesis of NodRf-III (C18:1) (MeFuc) 237, a nod factor produced by *Rhizobium fredii*, is an illustrative example of the chemistry developed around 2-amino-2-deoxy-NPGs (❷ *Scheme 35*) [119]. The key elements in this stereocontrolled synthesis are: (a) the use of the TCP protecting group, which provides a facile method for N-differentiation in the glucosamine oligomer, (b) the assistance of the sidetracking methodology, (c) a solvent-assisted stereoselective α-fucosylation, (d) a β-selective, neighboring group assisted, glycosidation, and (e) the use of FeCl₃ for late-stage debenzylation of the oligosaccharide moiety [120,121]. In the retrosynthesis (❷ *Scheme 35*), the authors selected a TCP as protecting group for the nitrogen atom that would bear the unique fatty acid, while the repeating unit would be a 2-deoxy-2-N-phthaloyl NPG capable of acting as a glycosyl donor (e. g. 238). The reducing end retron was identified with benzyl glycoside 239.

The disaccharide acceptor 239 was prepared in 85% yield by coupling (NIS/TESOTf) of acceptor 241 with n-pentenyl fucoside 240 in Et₂O:CH₂Cl₂, (5:1) (❷ *Scheme 36*). The disaccharide donor 238 was assembled in 71% by coupling of NPG 242 with sidetracked acceptor 243 (NIS/TESOTf), followed by reinstating of the pent-4-enyl moiety from the dibromo pentenyl residue in 244. Final coupling (NIS/TESOTf) of donor 238 with acceptor 239 yielded

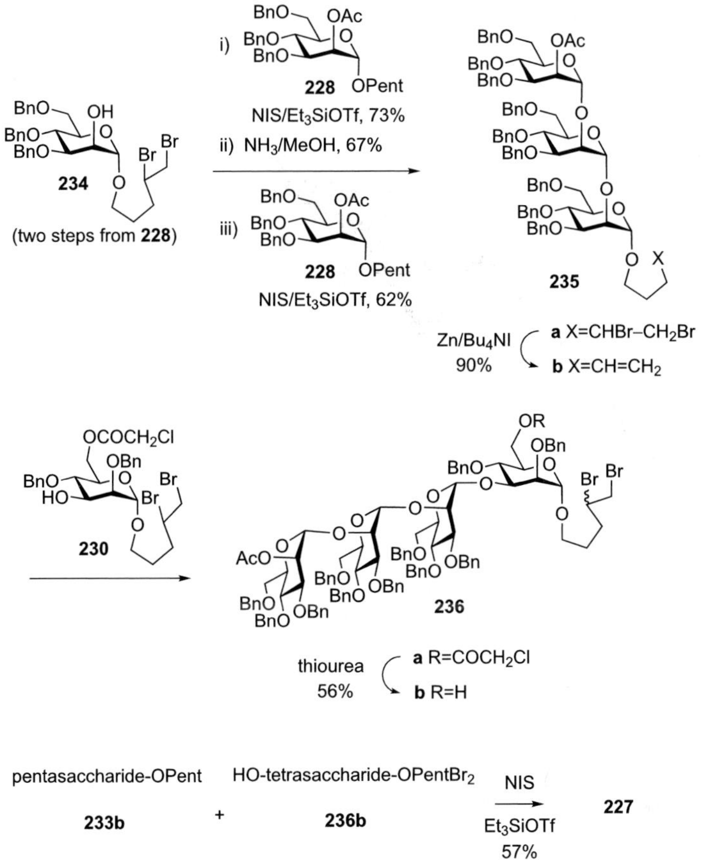

Scheme 34
Convergent synthesis of nonamannan 227. Synthesis of the tetrasaccharide acceptor 236b and final assembly

tetrasaccharide **245** in 65% yield. The final stages in the preparation of **237** involved: (i) FeCl₃ debenzylation, (ii) silylation of the resulting free –OHs, (iii) deprotection of the TCP and condensation with an activated fatty acid, (iv) removal of the phthalimido protecting groups, and (v) acylation, saponification, and desilylation.

2.11.4 Synthesis of Phosphorylated Rat Brain Thy-1 Glycosylphosphatidylinositol Anchor

Glycosylphosphatidylinositol (GPI) membrane anchors constitute a class of glycolipids that covalently link certain proteins to cell and virion surfaces [122,123]. A boost in their chemistry occurred in 1988 when Ferguson et al. reported the first covalent structure of a member of this family [124,125]. The first synthesis of a fully phosphorylated GPI, compound **246** (❍ *Fig. 5*), was accomplished by Fraser-Reid's group based entirely on NPG chemistry [126,127,128,129,130].

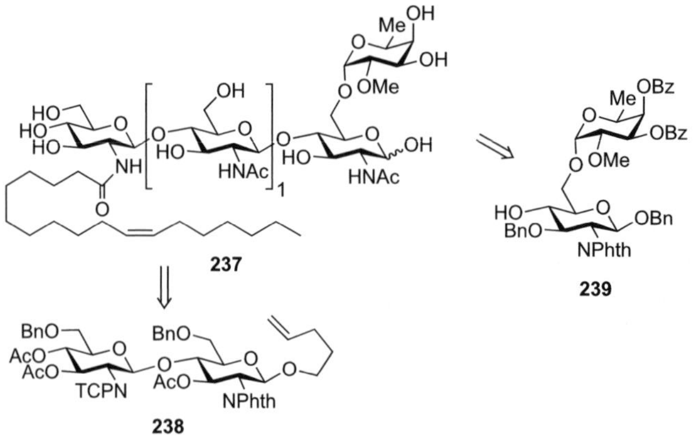

■ Scheme 35
Retrosynthesis of nodulation factor NodRf-III (C18:1) (MeFuc)

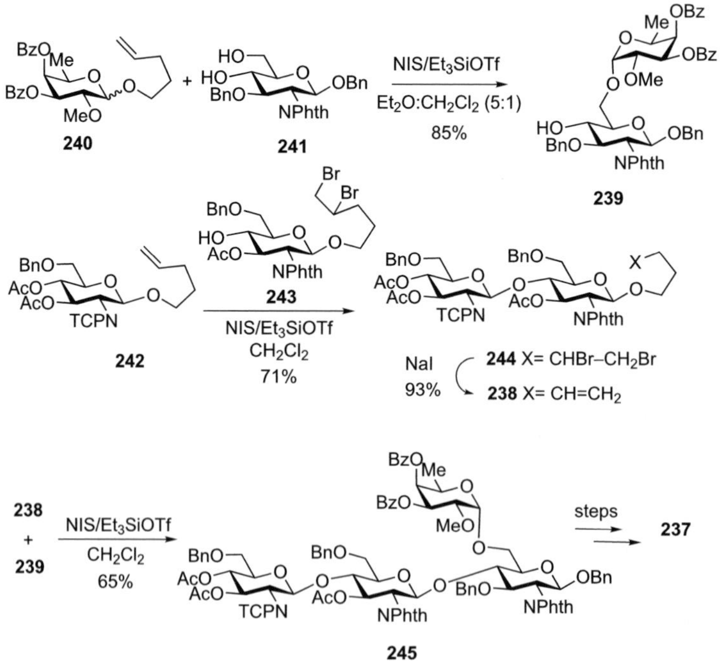

■ Scheme 36
Synthesis of nodulation factor NodRf-III (C18:1) (MeFuc)

The retrosynthetic analysis dictated a heptasaccharide **251**, with all free hydroxyls in the final product benzylated, and the three sites of phosphorylation all differentially protected, so that all three can be manipulated separately for maximal flexibility. The free amine of glucosamine is protected as an azide, which can be taken through multiple transformations and will only be unmasked at the end of the synthesis (❯ *Scheme 37*). The heptasaccharide is in turn put

246
Site for protein attachment

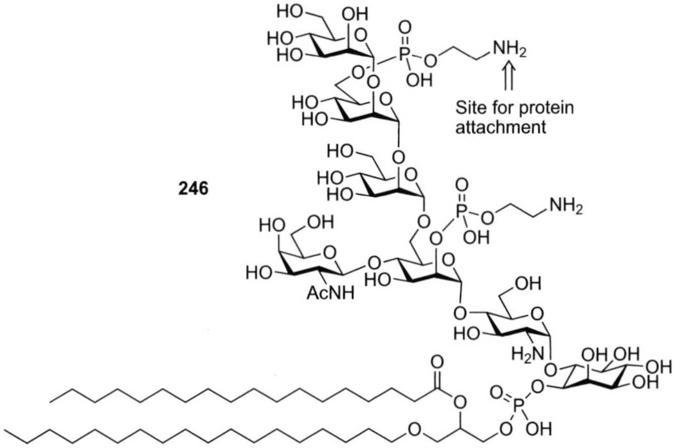

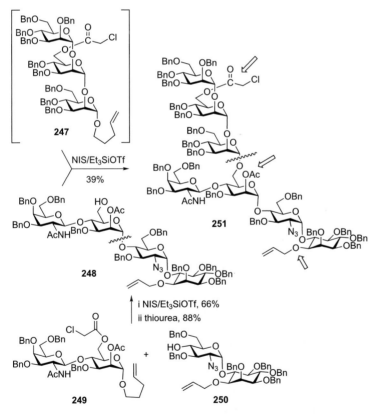

Figure 5
Rat brain Thy-1 GPI anchor 1, 246

247
NIS/Et3SiOTf
39%
251
248
i NIS/Et3SiOTf, 66%
ii thiourea, 88%
249
250

Scheme 37
Synthesis of rat brain Thy-1 GPI anchor 1

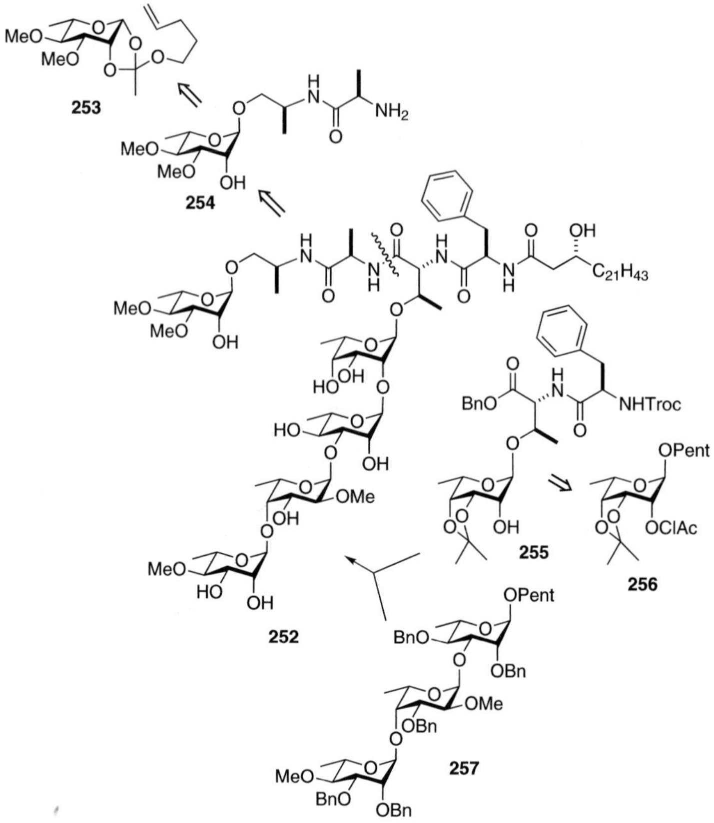

☐ **Scheme 38**
Synthesis of the glycopeptidolipid of *Micobacterium avium* Serovar 4, 252

together in three portions, a galatosaminylmannose **249** being coupled to azidoglucosylinosi-tol **250** and then a trimannose, **247**, being coupled to that moiety. Coupling of glycosyl donor **249** with the disaccharide acceptor **250** was carried out with NIS/TESOTf to give α-linked tetrasaccharide **248** in 66% yield. Notably, the allyl protecting group survived the treatment with NIS. The *O*6 of the mannose residue was deprotected by removal of the chloroacetate moiety with thiourea and glycosylated with pentenyl trimannoside **247** to give the fully pro-tected heptasaccharide **251** in 39% yield. The three positions (marked with arrows) were then deprotected and phosphorylated according to the following sequence: dechloroacetylation with thiourea, saponification of the acetate with methoxide, and deallylation with $PdCl_2$. Complete debenzylation, then culminated the synthesis of **246**.

2.11.5 Synthesis of the Glycopeptidolipid of *Micobacterium avium* Serovar 4

Heidelberg and Martin described the first synthesis of the "polar mycoside C" **252** (❷ *Scheme 38*) [131]. The synthesis was based on the disconnection of the final structure into three saccharidic building blocks, an L-rhamnosyl pseudodipeptide **254**, a 6-deoxy-L-talosyl dipeptide **255**, and a pentenyl trisaccharide donor **257**. The key steps were the creation of the glycosidic linkage between the trisaccharide donor **257**, and the 6-deoxy-L-talose unit **255** (IDCP, 60% yield), and the final coupling of the two glycopeptide fragments. Other pentenyl mediated couplings were the glycosylation of orthoester **253** leading to **254** (NIS/TMSOTf, 81%) and the stereoselective α-coupling of disarmed NPG **256** to give glycodipeptide **255** (NIS/TMSOTf, 70%).

2.11.6 Synthesis of Oligogalacturonates Based on NPGs

Madsen and co-workers described a concise approach to oligogalacturonates (e. g. **258**, ❯ *Scheme 39*) conjugated to bovine serum albumin (BSA) based on NPGs. They synthesized several oligogalacturonates, which were linked to the BSA by reductive amination via an aldehyde spacer at the reducing end (❯ *Scheme 39*) [48,132,133]. Their strategy called for two orthogonal protecting groups (P^1 and P^2), and three different monomeric building blocks: a spacer galactoside **C** to serve as glycosyl acceptor for the reducing end, and two glycosyl donors **A** and **B**, the former for the nonreducing end and the latter for the galacturonic

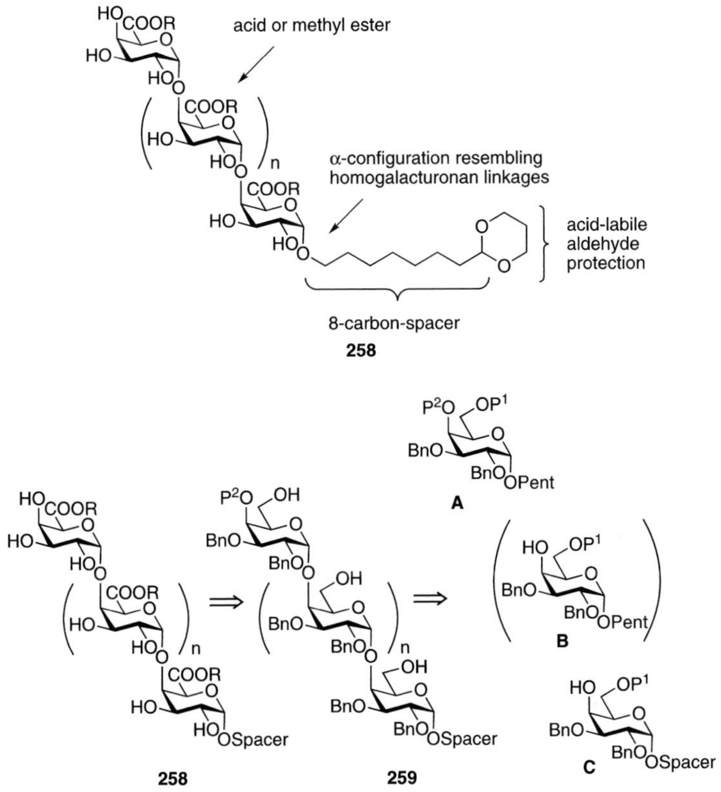

◻ Scheme 39
Retrosynthesis of oligogalacturonates for conjugation to bovine serum albumin

repeating unit. *p*-Methoxyphenyl (PMP) and acetyl groups were used as protecting groups. The methodology was then based on the repeated coupling of galactose donors onto galactose acceptors followed by deprotection at *O*6, as in **259**, which permitted the oxidation of these primary positions to either the carboxylic acids or methyl esters.

The attaching of the spacer to galactose (i.e. **261**, building block **C**) was carried out by glycosylation, under Lemieux conditions [134], of a glycosyl bromide readily obtained from NPG, **260** (❷ *Scheme 40*). Coupling of NPG **260** with galactosyl bromide **262** (AgOTf, 71%) led to pentenyl disaccharide **263** that glycosylated acceptor **261** (NIS/TESOTf, 71%) to give trisaccharide **264a**. Deprotection of the latter to **264b**, and glycosylation with glycosyl donor **263** (NIS/TESOTf, 90%) led to pentasaccharide **265a**. Glycosyl assembly to hexasaccharide **267**, included the glycosylation of pentenyl donor **266** with acceptor **265b** (NIS/TESOTf, 69%), obtained by NaOMe treatment of **265a**. The final stages of the synthesis included CAN-mediated deprotection of the *p*-methoxyphenyl groups, Dess–Martin oxidation and esterification.

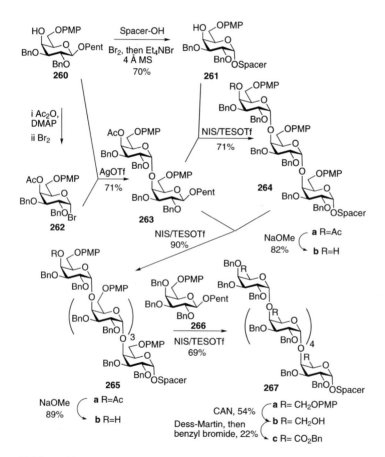

❏ Scheme 40
Synthesis of oligogalacturonates

2.11.7 Miscellaneous

Arasappan and Fraser-Reid reported an NPG-based methodology for the stereoselective construction of the tetrasaccharyl cap portion of Leishmania Lipophosphoglycan [135].

Kuzuhara and co-workers have reported the use of a NPG disaccharidic synthon as the chain elongating unit in the synthesis of amphiphilic chitopentaose and chitoheptaose derivatives [136].

Toshida and co-workers have described the synthesis of a set of di- and tri-sulfated galabioses by using an *n*-pentenyl galactoside donor and IDCP as the catalyst [137].

2.12 NPGs in Solid-Phase Oligosaccharide Synthesis

Solid-phase oligosaccharide synthesis has received considerable attention in the last years [138]. Some of the approaches that involve NPGs are discussed below.

2.12.1 Glycosylation of Supported Alcohol Acceptors with NPG Donors

Fraser-Reid and co-workers designed a photolabile *o*-nitrobenzylic linker **268**, which was used in the synthesis of a branched trimannan oligosaccharide **271** (❷ *Scheme 41*) [139]. Differentially protected NPG **269** was coupled to the resin via linker **268**. Selective removal of the C6 chloroacetyl and C3 acetyl groups, followed each by mannosylation (NIS/TESOTf) with NPG **64**, afforded trimannan **271**, in 42% overall yield after photolytic cleavage.

In a related approach, Fraser-Reid and co-workers used Chiron's polystyrene-grafted "crowns" with Rich's photocleavable o-nitrobenzyl linker [140] and NPG donors in the synthesis of trisaccharide **277** (❷ *Scheme 42*) [141]. After attachment of the first aminoglucosyl moiety to

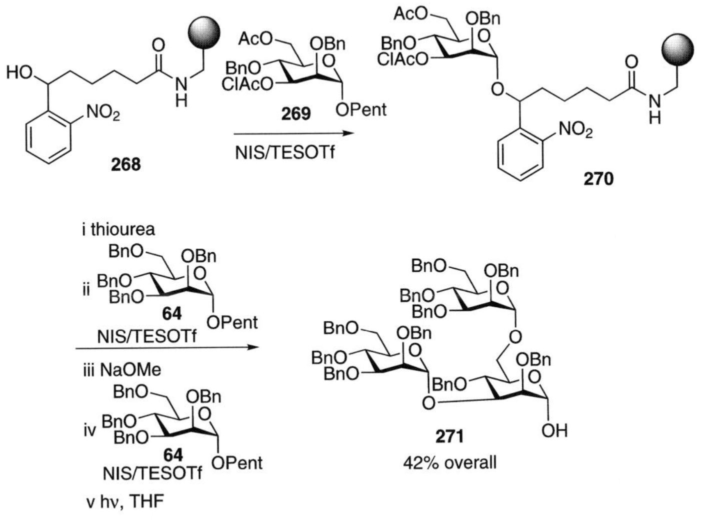

❏ Scheme 41
NPGs in solid-phase oligosaccharide synthesis

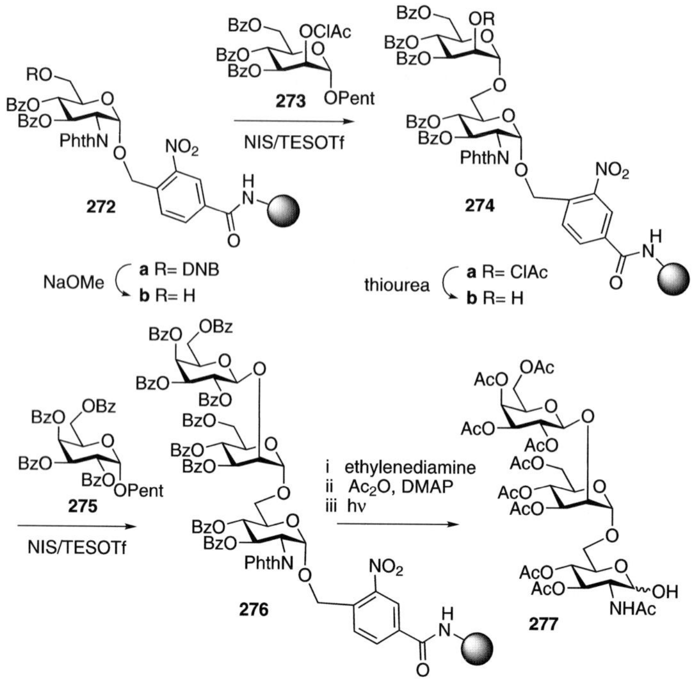

■ Scheme 42
NPGs in solid-phase oligosaccharide synthesis

the linker, via its corresponding NPG, the C6 dinitrobenzoyl (DNB) group was removed to give **272b**. Coupling with mannose donor **273**, deprotection of the O2 chloroacetyl group, and galactosylation with NPG **275** furnished trisaccharide **276**. Global deprotection followed by peracetylation and photolytic cleavage from the solid support provided trisaccharide **277**.

2.12.2 Pentenyl Glycoside-Based Linkers

Seeberger and co-workers developed a new linker concept in solid-phase oligosaccharide synthesis. They designed a new NPG-based linker that upon deprotection rendered an oligosaccharide NPG suitable for further glycosylations in fragment couplings (❷ *Scheme 43*) [142]. The first carbohydrate moiety (e. g. **278**) was connected via a glycosidic bond to octenediol-functionalized Merrifield's resin, **279**. Resins with loadings of up to 0.65 mmol/g were obtained and employed in oligosaccharide synthesis. Glycosylation events can now take place in deprotected saccharide **280** to yield oligosaccharide **282**. The octenediol linker was then cleaved by olefin cross metathesis using Grubbs' catalyst under an atmosphere of ethylene to afford fully protected oligosaccharides in the form of NPGs, e. g. **283**. A further refinement of this strategy is that it can be made compatible with glycosyl donors that require electrophiles as activators by sidetracking the linker to the corresponding dibromooctane derivative (e. g. **284**) [143]. Seeberger and co-workers have illustrated the potency of this strategy with several oligosaccharide syntheses [144,145,146,147].

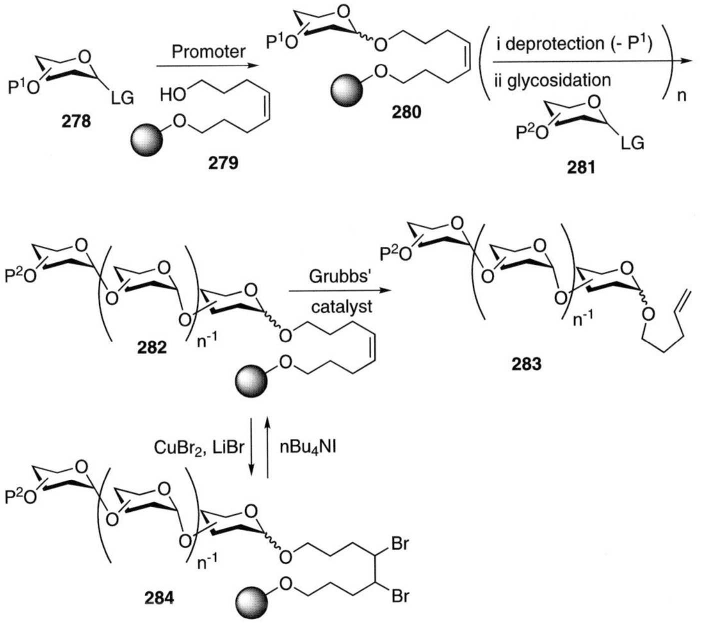

◻ Scheme 43
Seeberger's NPG-based linkers for oligosaccharide synthesis

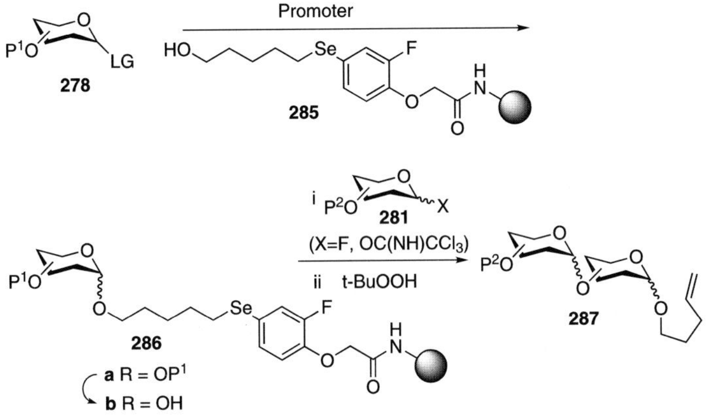

◻ Scheme 44
A fluorinated selenide linker for solid-phase synthesis of NPGs

Mogemark et al. described a fluorinated selenide linker **285**, for solid-phase synthesis of NPGs (**◉** *Scheme 44*) [148]. The resin-bound linker could be glycosylated both with trichloroace-timidates and glycosyl fluorides to give anchored saccharides, e. g. **286**, that can be submitted to glycosylation once deprotected. After oxidation to a selenoxide with t-BuOOH the linker undergoes *β*-elimination upon heating, and releases the NPG **287**, in excellent yield.

2.13 Miscellaneous Uses of NPGs

The versatility of NPGs has been further enhanced by chemical modifications of the pent-4-enyl moiety itself. In this context, the pentenyl moiety has been transformed in many spacer functionalities [149,150], used as a handle to incorporate amino-acid moieties [151,152,153, 154], used as a monomer in copolymerization strategies [155,156,157], used in the formation of dendrimers [158], and converted to dimeric and trimeric structures for multivalent presentations [159,160]. These applications fall beyond the scope of this chapter.

2.14 Preparation of NPGs

NPGs, being normal O-glycosides, can be readily obtained by application of the standard procedures for preparing such derivatives [12]. They can be obtained by Fischer glycosida-

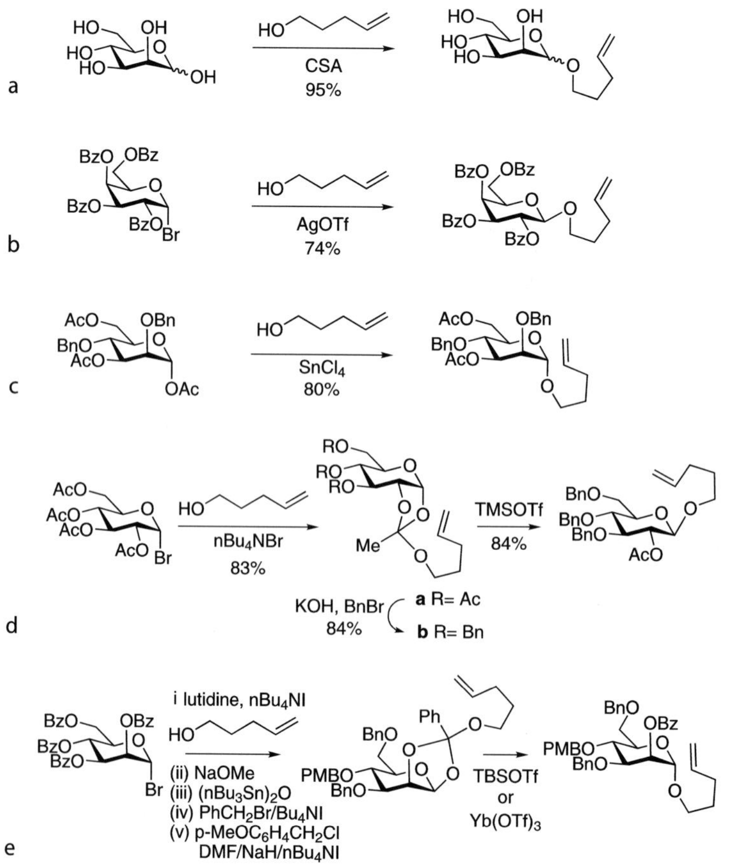

◘ Scheme 45
Methods for the preparation of NPGs

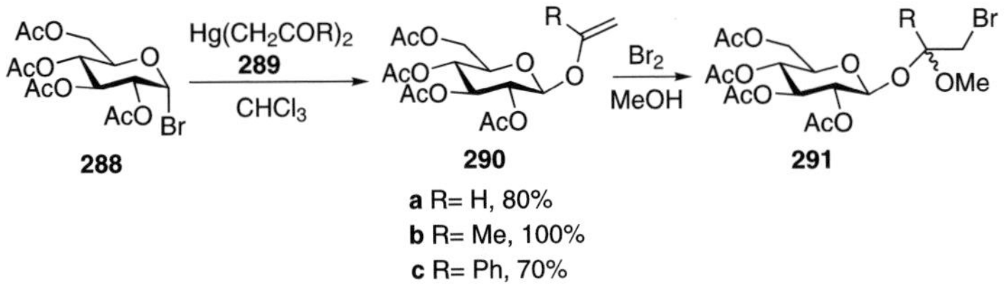

■ Scheme 46
Synthesis of 1′-substituted vinyl glucosides

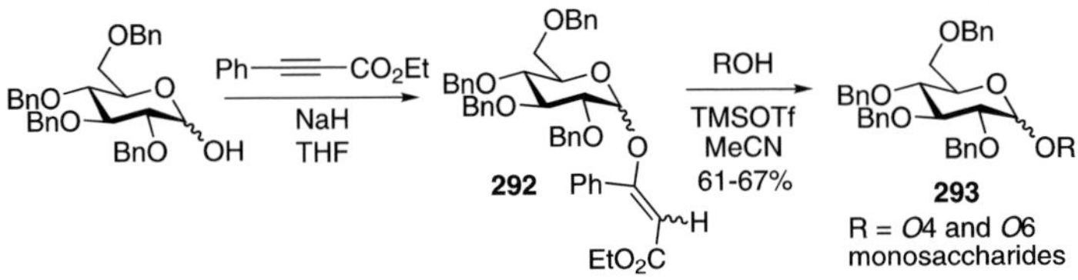

■ Scheme 47
Synthesis of 1′-substituted vinyl glucosides

tion (❍ *Scheme 45a*) [82,161,162]. An obvious advantage of this procedure is that the *n*-pentenyl group can be installed right at the outset of the synthesis; however, the formation of *α*/*β* anomers might sometimes be a drawback. Use of the Koenigs–Knorr coupling [115] permits the stereocontrolled preparation of NPGs (❍ *Scheme 45b*). SnCl₄ facilitates the formation of NPGs from acetyl mannosides (❍ *Scheme 45c*) [117]. The most useful method for the preparation of NPGs is, arguably, the acid-catalyzed rearrangement of NPOEs, prepared under Lemieux–Morgan conditions (❍ *Scheme 45d,e*) [163]. This method permits the stereoselective synthesis of NPGs with different protecting groups [50,164]. Rousseau and Martin described the rearrangement of acetyl NPOEs with TMSOTf (❍ *Scheme 45d*) [99], and Fraser-Reid and co-workers have used TBSOTf [165] or ytterbium triflate [166] to rearrange benzoyl-substituted NPOEs (❍ *Scheme 45e*).

3 Enol Ether-Type Glycosides

3.1 Early Contributions

De Raadt and Ferrier were the first to report the preparation and attempted glycosylation of 1′-substituted-vinyl glycosides [167]. Reaction of tetra-*O*-acetyl-*α*-D-glucopyranosyl bromide (**288**) with bis(acetonyl)mercury derivatives **289a–c** in refluxing chloroform afforded vinyl-, isopropenyl-, and styryl-*β*-D-glucosides **290a–c** in excellent yields (❍ *Scheme 46*). However, when **290a–c** were each treated with either NBS or bromine/AgClO₄ in the presence of methanol no glycosides were formed, the products in each case being mixed stereoisomers of the glycosyl acetals **291a–c**.

Schmidt and co-workers described the preparation of vinyl glucosides **292** from the reaction of tetra-*O*-benzyl glucose with ethyl phenyl propiolate under the agency of sodium hydride

(❍ *Scheme 47*) [168]. The reaction of **292**, as an anomeric mixture, with various acceptors was examined in acetonitrile at −40 °C in the presence of TMSOTf as catalyst. Reaction of **2** with 6-OH and 4-OH methyl glucosides as acceptors gave the corresponding disaccharides in 61 and 67% yield and as 85:15 and 75:25 β/α mixtures, respectively. Similar results were obtained for tetra-*O*-benzyl galactose.

3.2 Isopropenyl Glycosides

Sinaÿ and co-workers described the synthesis of isopropenyl glycosides [169] by reaction of the corresponding anomeric acetates with the Tebbe reagent [170]. Reaction of 1-*O*-acetyl-2,3,4,6-tetra-*O*-benzyl-D-glucopyranose (**294**) with a solution of Tebbe reagent in toluene gave the isopropenyl glycosides **295**, in 87–90% yields (❍ *Scheme 48a*). Likewise, isopropenyl galactoside **297** ($\alpha:\beta \approx 1:1$) was prepared from the corresponding acetate **296** by Tebbe methylenation in 88% yield (❍ *Scheme 48b*). Treatment of **295** ($\alpha:\beta = 4:1$) in MeCN at −25 °C with the primary hydroxyl acceptor **298**, in the presence of TMSOTf gave the disaccharides **299** (68%) with an excellent β-selectivity (20:1) (❍ *Scheme 48c*). The condensation of **295** with the secondary alcohol **300** in MeCN at −25 °C in the presence of BF$_3$·Et$_2$O afforded the disaccharide **301** in good yield, albeit with reduced stereoselectivity ($\beta:\alpha = 5:1$) (❍ *Scheme 48d*). When the same glycosylation was carried out in CH$_2$Cl$_2$ instead of MeCN the disaccharide **301** was obtained in limited yield (❍ *Scheme 48e*). The successful glycosylation of phenyl 1-thio-glycoside **302** with **295** in the presence of TMSOTf illustrates the usefulness of isopropenyl glycosides in the synthesis of thiophenyl disaccharides (e. g. **303**, ❍ *Scheme 48f*). The authors found no significant variations on yield or stereoselectivity by the use of either mainly α or mainly β isopropenyl derivatives. The best results for galactosylation were achieved in CH$_2$Cl$_2$ with TMSOTf as promoter (❍ *Scheme 48g*).

Chenault and co-workers reported the use of *O*-isopropenyl glycosides bearing ester protecting groups [171,172]. These compounds are stable at room temperature and can be readily purified by column chromatography on silicagel, moreover their glycosylation would proceed to give β-glycosides via neighboring group participation. The reaction of bis(acetonyl)mercury [173] with glycopyranosyl halides proved to be a good method for the preparation of isopropenyl β-glycopyranosides (e. g. **305**, ❍ *Scheme 49a*). The authors described routes to *O*-isopropenyl 2,3,4,6-tetra-*O*-pivaloyl-α, and β-D-glucopyranosides α-**307** and β-**307**, respectively. Reaction of 2,3,4,6-tetra-*O*-pivaloyl-α-D-glucopyranosyl bromide (**306**) with diacetonyl mercury led to β-**307** (❍ *Scheme 49b*), whereas regioselective methylidenation [174] of **309** (prepared stereoselectively by acid-catalyzed exchange of the anomeric pivaloyloxy group of penta-*O*-pivaloyl-β-D-glucopyranose, **308**) generated α-**307** as the only product (❍ *Scheme 49c*). The β-isomer, however, exhibited greater shelf life than the latter.

On the basis of the reaction of NPGs with electrophiles, Chenault et al. considered the possible activation of isopropenyl glycosides with electrophiles. The mechanism of activation was expected to involve initial capture of the electrophile (E$^+$) by the vinyl ether double bond of **310** leading to the formation of cation **311** or **312** (❍ *Scheme 50*). Collapse of **311** or **312** to form glycosyl oxocarbenium ion **313** and acetone derivative **314** would be followed by nucleophilic attack on **313** to generate glycoside **315**. An alternative reaction would involve direct nucleophilic attack on **311** or **312** to generate the addition product **316**.

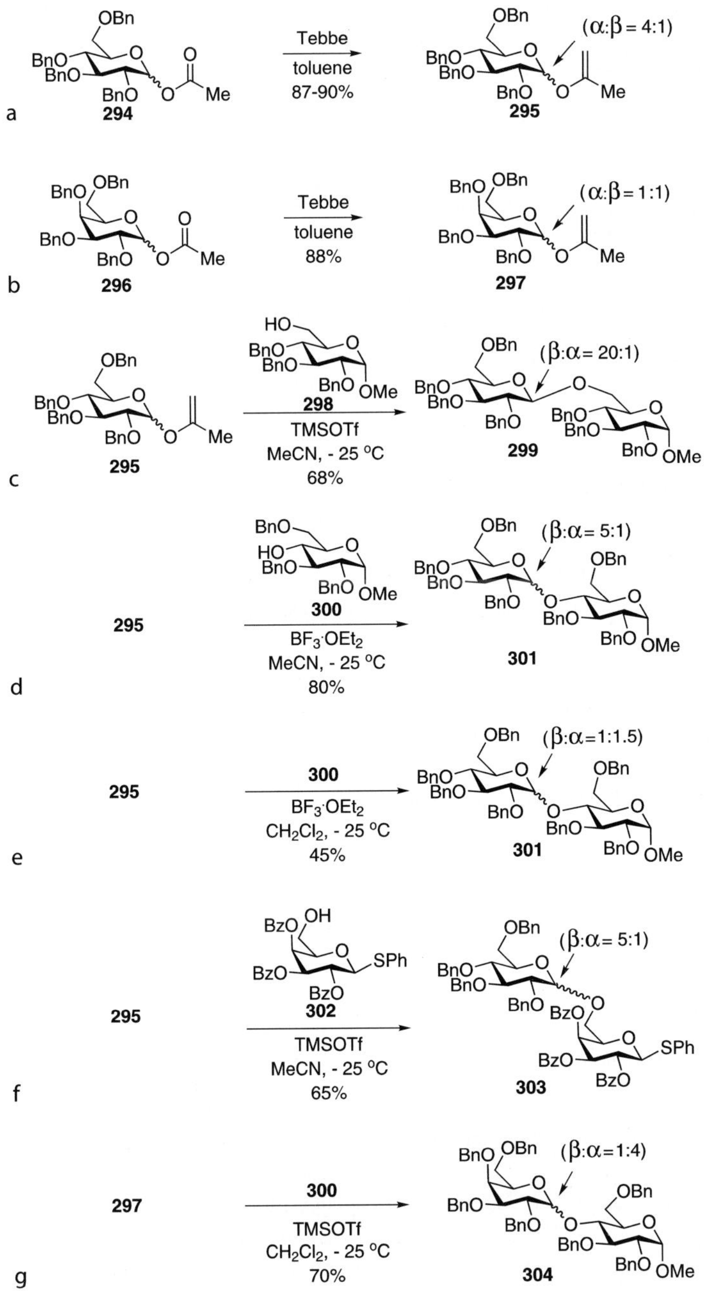

◘ Scheme 48
Synthesis and glycosylation reactions of isopropenyl glycosides

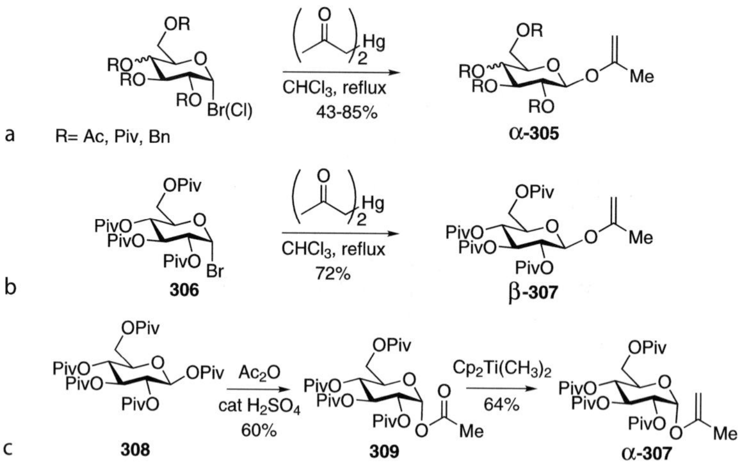

Scheme 49
Synthesis of pivaloyl isopropenyl glycosides

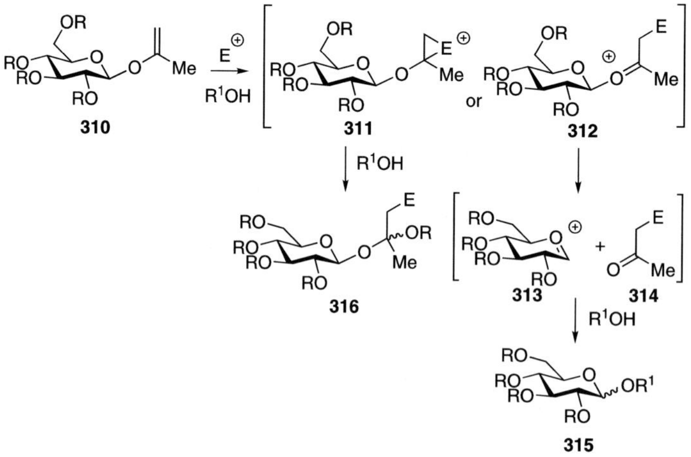

Scheme 50
Activation of isopropenyl glycopyranosides

The authors found that "armed" and "disarmed" isopropenyl glycosides displayed different behavior towards electrophiles (**Scheme 51**). Armed isopropenyl glycoside **β-294**, glycosylated acceptor **40** to give disaccharide **317** under the agency of IDCP, a relatively weak electrophile, in a nonpolar solvent (CH$_2$Cl$_2$) (**Scheme 51a**). On the other hand, disarmed glycoside **β-307**, led under the same conditions to the electrophilic addition product **318**. Use of a more potent electrophile (NIS/TfOH) in CH$_2$Cl$_2$ also resulted in the formation of the addition product **318** (**Scheme 51b**). However, NIS/TfOH in more polar MeCN successfully promoted the glycosidic coupling (**Scheme 51c**). Apparently, the relatively electron-releasing ethereal protecting groups lower the energy barrier to oxocarbenium ion formation from

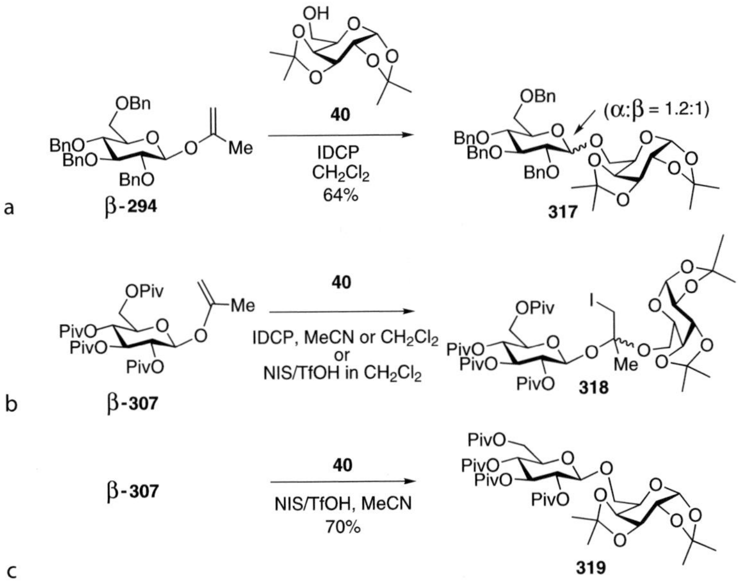

Scheme 51
Reaction of armed and disarmed isopropenyl glycopyranosides

armed **β-294** relative to that from disarmed glycoside **β-307**. In general, factors which favor the formation of the glycosyl oxocarbenium ion (strong electrophile, polar solvent, electron-releasing protecting groups on the glycosyl donor) lead to transglycosylation. Factors which retard the formation of the glycosyl cation (weak electrophile, nonpolar solvent, electron-withdrawing protecting groups on the glycosyl donor) lead to addition across the isopropenyl ether double bond.

The ability of various electrophiles to promote transglycosylation of disarmed isopropenyl glycosides is outlined in ❖ *Table 10*. NIS/TfOH, TMSOTf, and Tf₂O in MeCN, all led to the formation of disaccharide **319** in good yield (❖ *Table 10*, entries i–iii). Reactions were carried out at 0 °C and were complete within 2–5 min. With silver triflate (AgOTf) the reaction was slower and gave a lower yield of disaccharide **319** (❖ *Table 10*, entry iv). When TfOH, NIS, or NBS were used alone **β-307** failed to react and the glycosyl donor was recovered unchanged (❖ *Table 10*, entries vi, viii). Thus, neither NIS/TfOH, TMSOTf, Tf₂O, nor AgOTf seem to activate isopropenyl glycosides by acting as a source of TfOH. Dimethyl(methylthio)-sulfonium triflate (DMTST) was the only promoter that led exclusively to the formation of disaccharide **319** from **β-307** when CH₂Cl₂ was used as the solvent.

In terms of glycosyl donors, either **α-307** or **β-307** gave the same results in terms of yields. Likewise, isopropenyl galactopyranosides reacted in a similar manner to glucopyranosides. Acylated isopropenyl donors gave lower yields than pivaloyl analogs, presumably because of complications due to orthoester formation [175].

Isopropenyl glycosides could be activated selectively in the presence of armed NPGs, and that allowed a one-pot synthesis of trisaccharide **322** involving the successive glycosyl coupling of a vinyl glycoside **β-307**, and an NPG, **321** (❖ *Scheme 52*).

◻ **Table 10**
Evaluation of promoters for transglycosylation of β-307

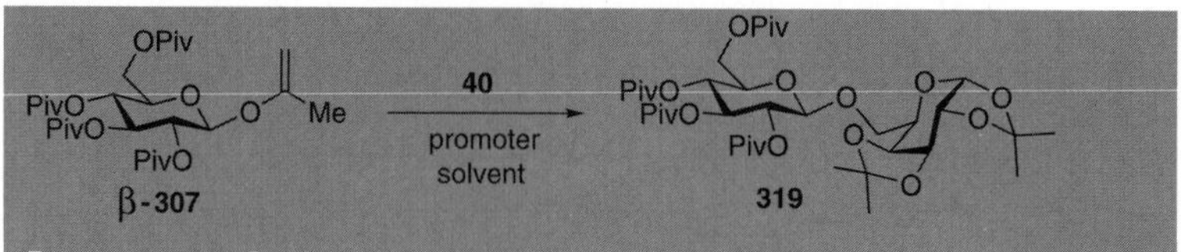

Entry	Promoter	Solvent	Yield
i	NIS/TfOH	MeCN	70%
ii	TMSOTf	MeCN	69%
iii	Tf$_2$O	MeCN	65%
iv	silver triflate	MeCN	24% (24h)
v	DMTST	CH$_2$Cl$_2$	48%
vi	TfOH	MeCN	no reaction
vii	trimethylsilyl iodide	MeCN	no reaction
viii	NIS or NBS	MeCN	no reaction

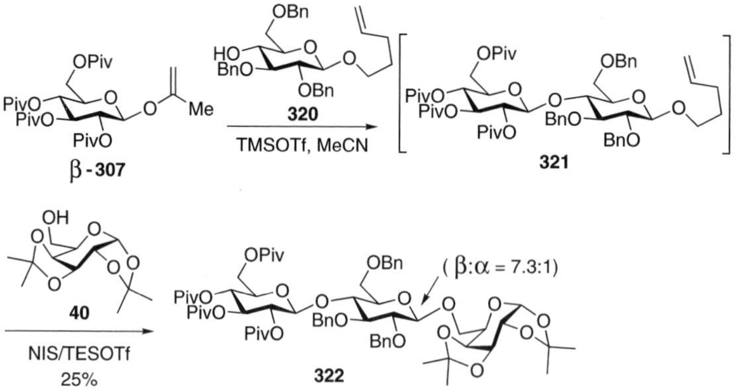

◻ **Scheme 52**
One-pot synthesis of trisaccharide 322

3.3 3-Butene-2-yl Glycosides as Precursors for Vinyl Glycosides

Boons and co-workers introduced stable allyl glycosides (e. g. **323**, ◗ *Scheme 53*), which are converted to the enol ether-type glycosides **324**, prior to glycosylation [176].

3.3.1 Latent-Active Glycosylation Strategy

The allyl glycoside **323**, can be considered a "latent" [177] form of a glycosyl donor which can be efficiently isomerized to the "active" vinyl glycoside, **324**. The isomerization reaction was performed by a rhodium catalyst obtained by treating the Wilkinson's catalyst, (Ph$_3$)P$_3$RhCl,

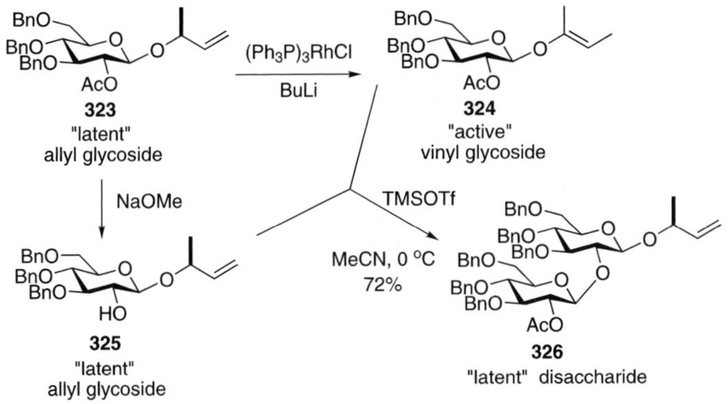

Scheme 53
Vinyl glycoside-based latent-active strategy for glycosyl coupling

with BuLi [178]. Base labile functionalities in the molecule are compatible with these isomerization conditions [179]. The "active" vinyl glycoside **324**, undergoes Lewis acid-catalyzed glycosylation reactions with "latent" allyl glycoside **325**, to give "latent" disaccharide **326** (● *Scheme 53*). Unlike isopropenyl glycosides, which require stoichiometric amounts of Lewis acids for activation [169], the reaction of Boons' vinyl glycosides only demands catalytic amounts of TMSOTf. The higher reactivity of the substituted vinyl glycoside was ascribed to the additional methyl substituent of the vinyl moiety that makes the double bond more electron rich. Although racemic 3-buten-2-ol could be used for the preparation of **323** without affecting its reactivity, the use of diastereomeric allyl glycosides can be avoided with the use of optically pure 3-buten-2-ol, easily obtainable in multigram amounts.

The use of neighboring participating groups permits the formation of 1,2-*trans* glycosides (e. g. **326**, ● *Scheme 53*). The choice of solvent and, to some extent, the choice of activator, was used to control the α/β ratio in glycosyl donors without participating groups at *O*2. TMSOTf-

Table 11
Effect of the solvent and promotor in α/β selectivity

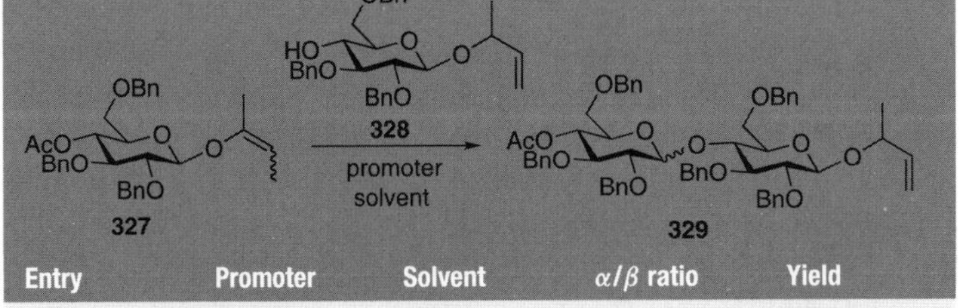

Entry	Promoter	Solvent	α/β ratio	Yield
i	TMSOTf	MeCN	1:8	78%
ii	TMSOTf	CH$_2$Cl$_2$	1.5:1	69%
iii	BF$_3$·OEt$_2$	Et$_2$O	1.5:1	69%
iv	NIS/TfOH	Et$_2$O/dichloroeth	3:1	73%

promoted condensation of **327** with **328** in MeCN gave disaccharide **329** as the β-anomer mainly (α/β = 1:8) (◉ *Table 11*, entry i). An improved α-selectivity was obtained (73%, α/β = 3:1) when the coupling was performed in ether/dichloroethane (◉ *Table 11*, entry iv).

3.3.2 Preparation of Trisaccharide Libraries

Linear Trisaccharide Libraries   Boons et al. described an approach to combinatorial synthesis of trisaccharide libraries based on their latent-active glycosylation strategy [180]. One major building block, **330** (i. e. **B^1**, ◉ *Scheme 54*) can be converted into a glycosyl donor **331** (i. e. **D^1**) and a glycosyl acceptor **332** (i. e. **A^1**). Coupling of compounds **331** and **332** gives disaccharide **333a** in excellent yield (the anomeric ratio can be greatly influenced by changes in the temperature: α/β = 1:20 at low temperature; α/β = 1:1 at ambient temperature). The latter can be converted into a glycosyl acceptor **333b** (i. e. **D^{1}A^1**) by removing the acetyl protecting

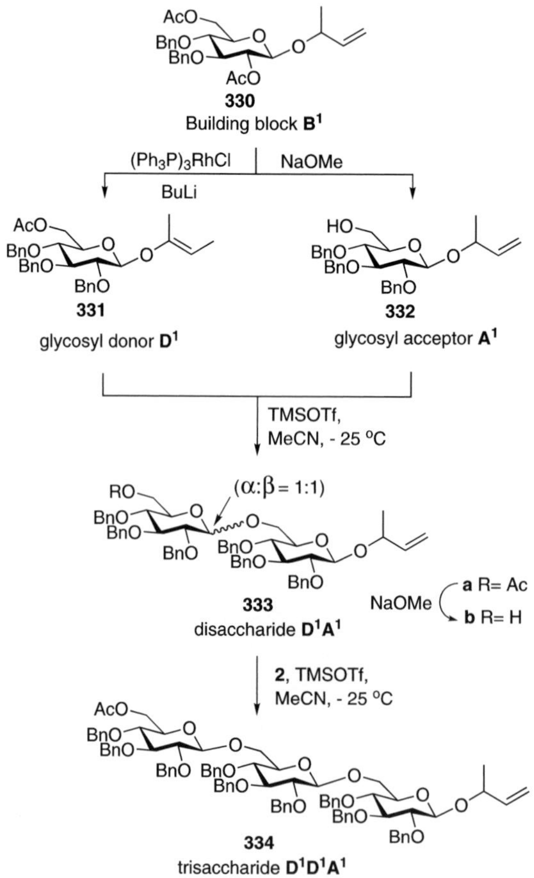

◻ Scheme 54
Preparation of linear trisaccharide libraries

group and into a glycosyl donor by isomerizing the allyl moiety. These compounds can be used in oligosaccharide synthesis, as outlined in ❖ *Scheme 54*, for example by coupling **333b** with **332** to give trisaccharide **334** (i. e. $D^1 D^1 A^1$). Application of this strategy to four allyl building blocks ($B^{1 \to 4}$) would lead to four vinyl glycosyl donors ($D^{1 \to 4}$) and four allyl glycosyl acceptors ($A^{1 \to 4}$). Individual glycosylations of each donor with each acceptor will furnish 16 disaccharides ($D^{1 \to 4} A^{1 \to 4}$) (if glycosylations are stereoselective, or 32 disaccharides if conditions are met for 1:1 anomeric selectivity). Next, the disaccharides can be mixed, and removal

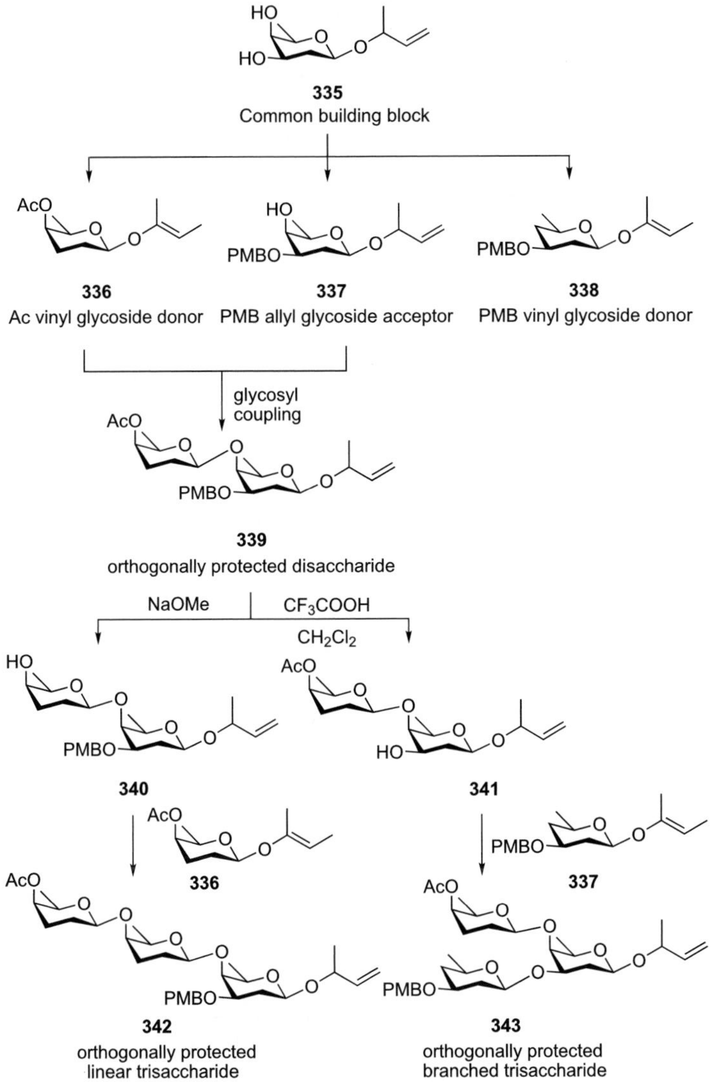

Scheme 55
Preparation of branched trisaccharide libraries

of the acetyl groups will give an assortment of acceptors. The pool of compounds can be split, and in combinatorial steps each pool of glycosyl acceptors can be coupled with a particular glycosyl donor ($\mathbf{D}^{1\rightarrow4}$) resulting in four libraries of 32 (or 64, as above) trisaccharides each.

Branched Trisaccharide Libraries Biologically important oligosaccharides often contain more complex features such as branching points and further functional groups. In this context, Boons and co-workers, using the latent-active strategy, designed a synthetic method to create orthogonally protected saccharides (acetyl and p-methoxybenzyl groups were used as orthogonal protecting groups) that could be easily further derivatized [181]. Thus, a common allyl glycoside building block (e. g. **335**, ◆ *Scheme 55*) can be converted to two vinyl glycoside donors bearing orthogonal protecting groups (e. g. **336** and **338**), and to an allyl glycosyl acceptor **337**, bearing one free hydroxyl and one selectively removable PMB ether. The latter will be coupled with donor **336** bearing an acetyl protecting group to give an orthogonally

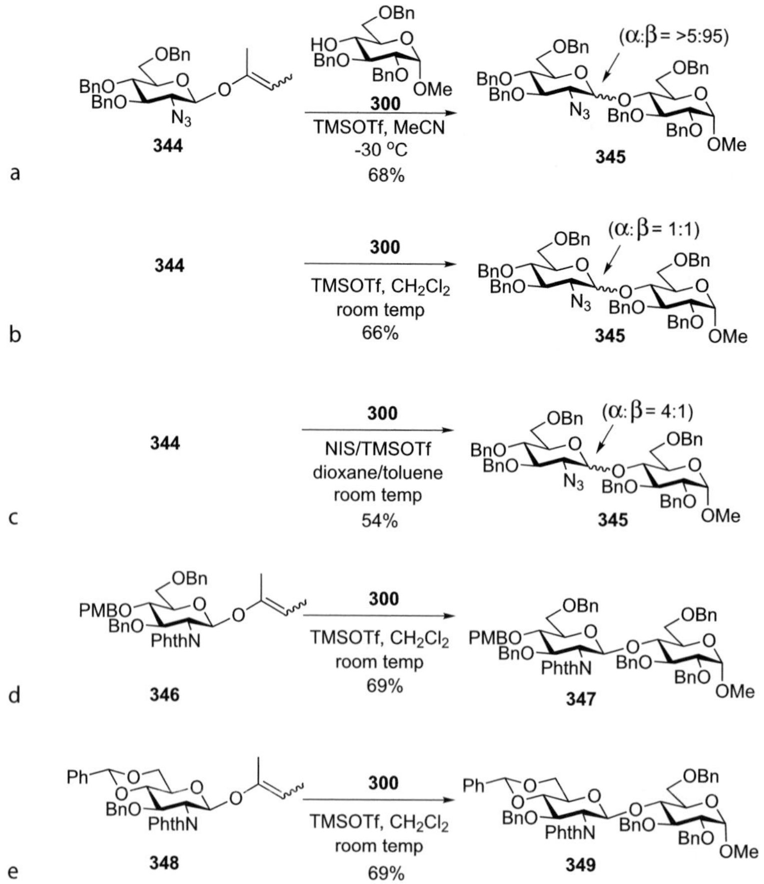

◆ **Scheme 56**
3-Buten-2-yl derivatives for the synthesis of amino sugar containing disaccharides

protected disaccharide, **339**. Compound **339** can be elaborated into linear or branched trisac-charides **342** and **343**. Thus, deprotection of the acetyl group in **339**, and glycosylation with vinyl donor **336** will yield linear trisaccharide **342**, whereas removal of the PMB group and coupling with **337** will produce orthogonally protected branched trisaccharide **343**.

3.3.3 3-Buten-2-yl 2-amino-2-deoxy Glycosides as Glycosyl Donors

Boons and co-workers studied the use of 3-buten-2-yl 2-azido-2-deoxy, and 2-deoxy-2-phthalimido glycosides, as building blocks for the preparation of sugar containing oligosac-charides [182]. Vinyl glycoside donors **344**, **346**, and **348**, were uneventfully prepared by isomerization of the corresponding 3-buten-2-yl glycosides with $(Ph_3)P_3RhCl/BuLi$ in yields exceeding 90%. Several glycosyl acceptors were used in the study, although representative data in ❷ *Scheme 56* refer solely to acceptor **300**. The glycosylation with azido donor **344**, in MeCN using TMSOTf as the promoter at $-30\,°C$, proceeded with high β-selectivity (❷ *Scheme 56a*), whereas NIS/TMSOTf in a dioxane/toluene mixture gave good α-selectiv-ities (❷ *Scheme 56c*). 2-Buten-2-yl 2-deoxy-2-phthalimido glycosides **346** and **348** reacted in CH_2Cl_2 in the presence of a catalytic amount of TMSOTf to give only the β-linked disaccharides **347** and **349**, respectively.

3.3.4 An Approach for Heparin Synthesis Based on 3-Buten-2-yl Glycosides

Haller and Boons described an approach fully based on 3-buten-2-yl glycosides for the synthe-sis of trisaccharide **350** and sulfated disaccharide **351** (❷ *Scheme 57*) [183]. In their strategy the glucuronic acid moieties were introduced at a late stage of the synthetic sequence by selec-tive oxidation of primary hydroxyl groups with TEMPO and NaOCl. "Latent" allyl glycoside **353** functioned as an acceptor for the reducing end in compounds **350** and **351**, and was also transformed to "active" vinyl glycoside **352**, for the nonreducing unit of **350**. The 2-aceta-

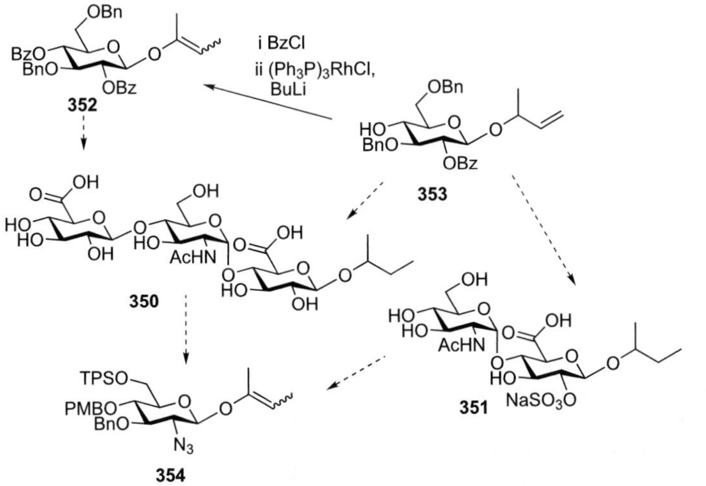

❏ Scheme 57
3-Buten-2-yl derivatives for the synthesis of amino sugar containing oligosaccharides

mido-2-deoxy unit in **350**, was retrosynthetically correlated with 2-azido-2-deoxy glycosyl donor **354**.

3.3.5 Conversion of 2-Buten-2-yl Glycosides to Other Glycosyl Donors

Treatment of "active" vinyl glycosides with NIS/TMSOTf in CH_2Cl_2 in the presence of dibenzyl phosphate gives good yield of glycosyl phosphates [184].

2-Buten-2-yl glycosides can also be transformed to glycosyl fluorides and trichloroacetimidates by hydrolysis to the corresponding hemiacetal (HgO, $HgBr_2$, aq. acetone) followed by standard treatment (CCl_3CN, DBU, CH_2Cl_2 or DAST, THF, respectively) [183].

3.3.6 Synthesis of 3-Buten-2-yl Glycosides

Being normal alkyl glycopyranosides, 3-buten-2-yl glycosides can be prepared by standard glycosylation methods, as previously mentioned for NPGs.

3.4 Oxathiines: Vinyl Glycosyl Donors for the Synthesis of 2-Deoxy Glycosides

Cycloadduct **357**, readily available by cycloaddition of tri-*O*-benzyl glucal (**355**) with the electron-poor 3-thioxopentane-2,4-dione (**356**) [185] has been used by two research groups as precursor glycosyl for vinyl glycosyl donors **358**, **361**, and **363** (❷ *Scheme 58*). Franck and co-workers showed that glycoside **358**, prepared by methylenation of **357**, underwent β-selective glycosylation with a variety of glycosyl acceptors in the presence of TfOH to give glycosides **359**, in good yields [186]. Moreover, Raney nickel desulfurization of **359** granted access to 2-deoxy-glycosides **360** [187]. Capozzi and co-workers reported that acetyl [188], and silyl [189] derivatives **361** and **363**, also functioned as glycosyl donors in reactions catalyzed by MeOTf in nitromethane and TMSOTf in CH_2Cl_2, respectively. The timing in the quenching of the reactions is crucial for obtaining completely selective β-glycosylations, and prolonged reaction times led to α/β anomeric mixtures. The total β-stereoselectivity of the coupling was ascribed by Capozzi and co-workers to an S_N2 type reaction (**361**→**362**, ❷ *Scheme 59*) that induces β-stereospecific glycosylation [188]. The observed subsequent α/β–equilibration presumably proceeds through an oxonium intermediate **364** (❷ *Scheme 59*).

4 DISAL Glycosyl Donors

4.1 Synthesis and Glycosylation Reactions

Petersen and Jensen reasoned that glycosides of phenols (e. g. **366**) carrying sufficiently electron-withdrawing substituents could possibly serve as *O*-glycosyl donors under neutral or mildly basic conditions (❷ *Scheme 60*) [190,191]. Carbohydrate hemiacetals have been used as nucleophiles in aromatic substitutions using activated fluoroarenes [192,193]. Accordingly, glycosides of methyl 2-hydroxy-3,5-dinitrobenzoate (DISAL, a *DI*nitro*SAL*icylic acid derivative), e. g. **367** (❷ *Scheme 61*), and methyl 4-hydroxy-3,5-dinitrobenzoate (*para*-isomer)

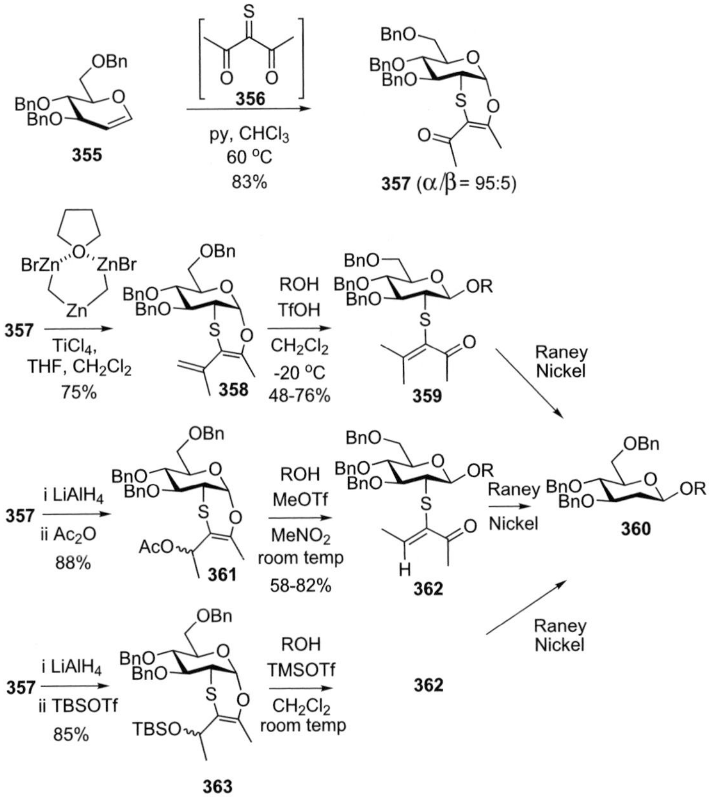

◻ Scheme 58
Oxathiines: vinyl glycoside donors for the synthesis of 2-deoxy glycosides

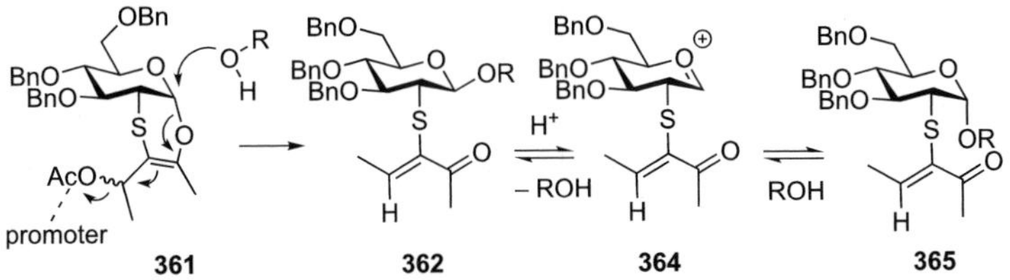

◻ Scheme 59
Stereocontrol in the formation of 2-deoxy glycosides from oxathiines

were prepared by reaction of carbohydrate hemiacetals with the corresponding activated fluoroarenes in the presence of a base. The use of 4-(*N,N*-dimethylamino)pyridine (DMAP) gave an α/β ratio similar to the starting 1-OH, i. e. predominantly α. In contrast, the formation of β-DISAL donors was favored using 1,4-dimethylpiperazine as base. The fluoroarenes were prepared by nitration of 2-fluoro- or 4-fluoro-benzoic acid.

The preparation of disaccharides, from benzyl-protected DISAL donors (e. g. **367**, ❷ *Scheme 61*), was best carried out in 1-methylpyrrolidin-2-one (NMP), a high polar, aprotic solvent,

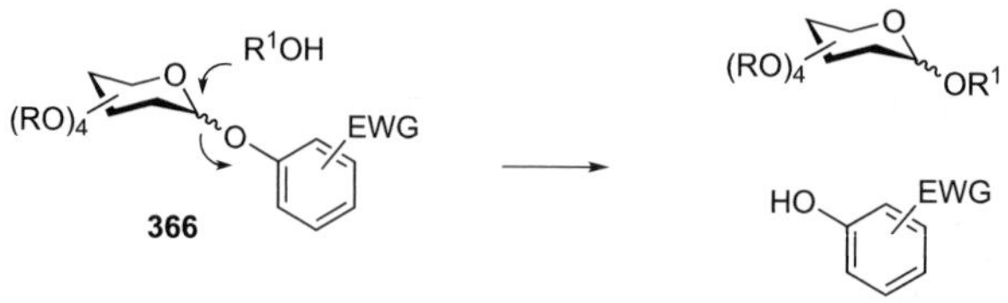

□ **Scheme 60**
Glycosides of phenols with electron-withdrawing substituents as glycosyl donors

at 40 °C, in the absence of Lewis acids (❷ *Scheme 61a,b*). The fact that glycosylations also occurred in the presence of base (e. g. Et₃N, 2,6-lutidine) indicated that the glycosylations were not auto-catalytically promoted by the released phenol. Under these conditions, galactose derivative **40** was glycosylated with DISAL donor **367** (1.5 equiv.) to give disaccharide **368** in 90% yield (α/β = 2.4:1). Glycosylation of a secondary hydroxyl group with DISAL donor **367** required increasing the temperature to 60 °C, and resulted in the formation of disaccharide **370** as the α-glycoside in 74% yield (❷ *Scheme 61b*). The *para*-glycosyl donor, (vide supra) also proved effective in analogous glycosylations.

Unlike benzyl-protected DISAL donors, benzoyl-protected donors, e. g. **371**, did not give the expected glycosides under these neutral conditions, in part due to trapping of intermediates as the orthoesters (❷ *Scheme 61c*). Lewis acids, such as BF₃·Et₂O or TMSOTf, activated the acylated DISAL donor **367**, albeit diisopropylidene acceptors **40** and **369** were not stable in the reaction media [194]. More robust benzyl-protected acceptors were glycosylated with alkylated and acylated DISAL donors in the presence of BF₃·Et₂O to give disaccharides **372** and **373** in 82 and 46% yield, respectively (❷ *Scheme 61d,e*). Interestingly, LiClO₄ was found to be an efficient additive for activation of DISAL donors in nonpolar solvents, giving significantly higher yields of disaccharides than BF₃·Et₂O (❷ *Scheme 61f*). Acylated DISAL donor **371** did not give good yield of disaccharides when reacting with secondary hydroxyl acceptors (❷ *Scheme 61g*). More recently, Jensen and co-workers have shown that high-temperature glycosylation of DISAL donors using precise microwave heating results in improved yield of disaccharides (❷ *Scheme 61h*) [195].

4.2 DISAL Donors in Solid-Phase Synthesis

This approach was extended to solid-phase glycosylation of D-glucosamine derivatives anchored by the 2-amino group through a Backbone Amide Linker to a solid support [196].

4.3 Intramolecular Glycosylation Approach to the Synthesis of 1,4-Linked Disaccharides

The DISAL donor concept was developed further to allow intramolecular glycosylations [197]. The glycosyl donor and acceptor were linked through the DISAL leaving group positioned to facilitate intramolecular glycosyl transfer to 4-OH by a 1,9-glycosyl shift (❷ *Scheme 62a*).

□ **Scheme 61** ❷
Glycosylation with DISAL glycosyl donors

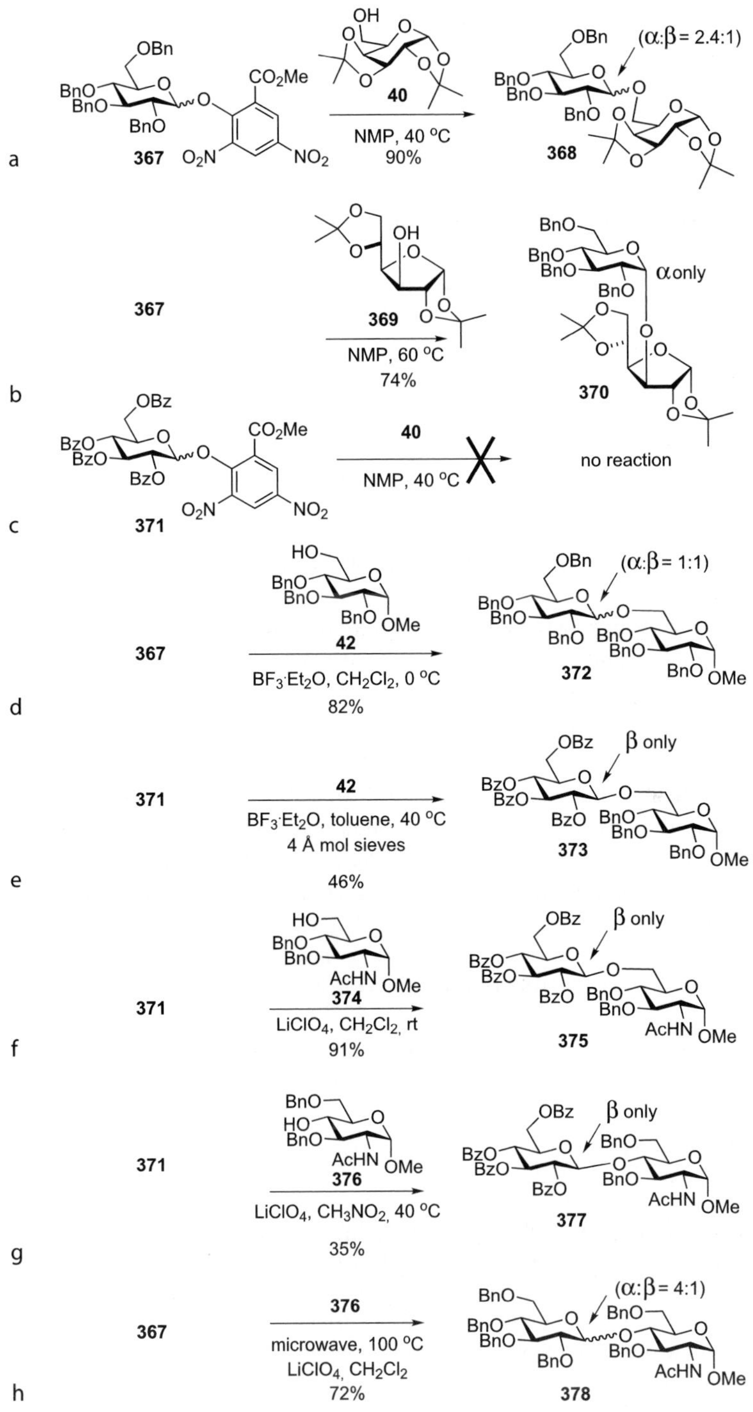

a

367 **40** NMP, 40 °C 90% **368** (α:β = 2.4:1)

b

367 **369** NMP, 60 °C 74% **370** α only

c

371 **40** NMP, 40 °C no reaction

d

367 **42** BF₃·Et₂O, CH₂Cl₂, 0 °C 82% **372** (α:β = 1:1)

e

371 **42** BF₃·Et₂O, toluene, 40 °C 4 Å mol sieves 46% **373** β only

f

371 **374** LiClO₄, CH₂Cl₂, rt 91% **375** β only

g

371 **376** LiClO₄, CH₃NO₂, 40 °C 35% **377** β only

h

367 **376** microwave, 100 °C LiClO₄, CH₂Cl₂ 72% **378** (α:β = 4:1)

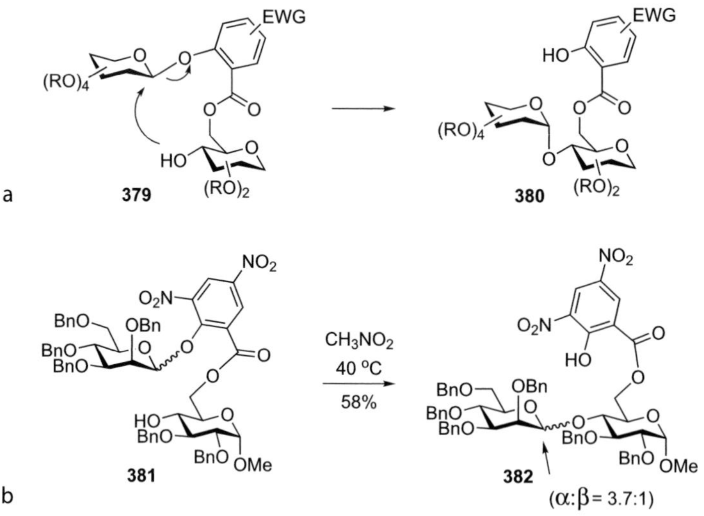

a

b

Scheme 62
DISAL-Based intramolecular glycosylation approach to 1,4-linked disaccharides

The tethered glycoside **381** underwent intramolecular transglycosylation to form the 1,4-linked mannoside **382** as an anomeric mixture (α/β = 3.7:1) in moderate yield (❷ *Scheme 62b*).

4.4 Application of DISAL Donors to Oligosaccharide Synthesis

Jensen and co-workers reported the synthesis of hexasaccharide **383**, a starch-related hexasaccharide (❷ *Scheme 63*) [198]. Their approach was based on the use of DISAL disaccharides **384** and **385**, readily obtained from the corresponding disaccharide hemiacetals, for sequential glycosylations. Glycosylation of phenyl 1-thio disaccharide **386** with DISAL donor **385** took place with good yield and excellent α-selectivity in CH_3NO_2 in the presence $LiClO_4$ and Li_2CO_3. The trityl group that have survived the coupling, was next removed and the ensuing tetrasaccharide glycosylated with DISAL donor **384** ($LiClO_4$, Li_2CO_3, $(CH_2Cl_2)_2$, 35 °C, 38% yield, α/β = 3:2).

DISAL donors have also been used in the preparation of phenazine natural products and analogs [199].

4.5 2-Deoxy-2-amino Derivatives as DISAL Donors

Jensen and co-workers evaluated the behavior of different glucosamine-derived DISAL donors in glycosylation reactions [200]. N-tetrachlorophthaloyl (TCP), N-trifluoroacetyl (TFAc), and N-trichloroethoxycarbonyl (Troc) DISAL donors **387**, **388**, and **389** and **390**, respectively, were prepared from the corresponding hemiacetals (❷ *Fig. 6*). Glycosylation of cyclohexanol, in NMP at 60 °C, with these donors took place with yields ranging from 35 to 76%. The N-TCP protected donor **387**, was the least reactive. N-Troc protected donors **389** and **390**, gave the highest glycosylation yields with monosaccharides (63–71% yield), although they displayed

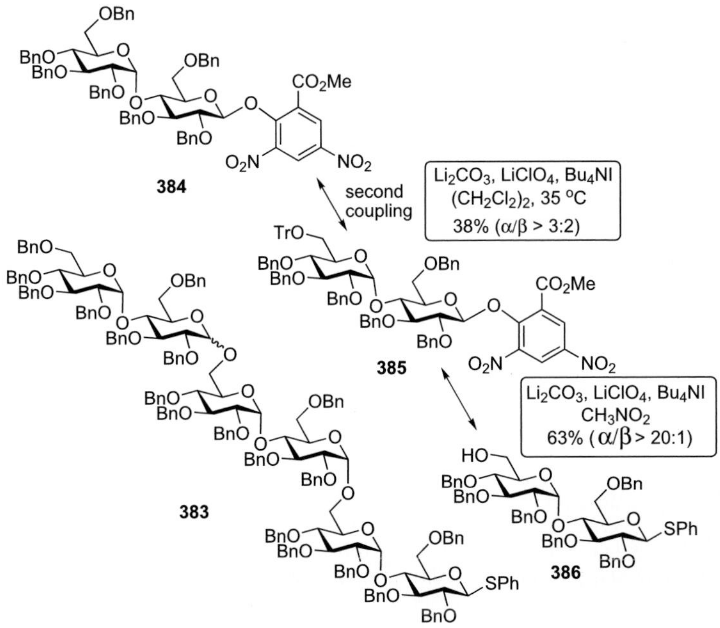

■ Scheme 63
Synthesis of hexasaccharide 383 based on DISAL donors

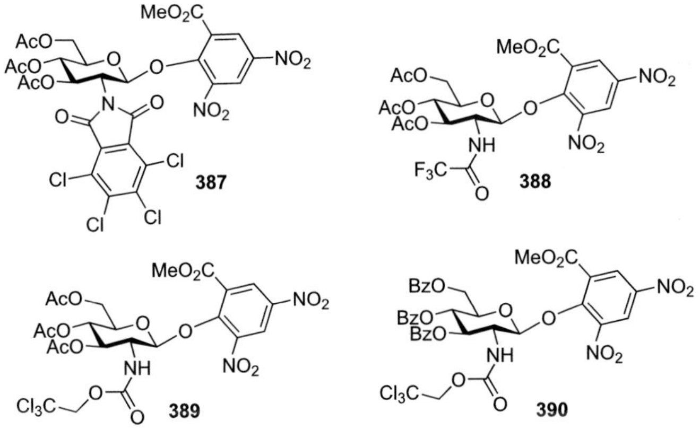

■ Figure 6
Glucosamine-derived DISAL donors

lower selectivities with primary hydroxyl acceptors (α/β ratio, from 1:1 to 1:7). A secondary hydroxyl acceptor was glycosylated with *N*-Troc DISAL donor **389** under microwave heating (130 °C, CH$_3$NO$_2$, LiClO$_4$) to give the corresponding disaccharide in 38% yield (β-anomer only). *N*-TFAc DISAL donor **388** gave even lower yields on coupling reactions with primary hydroxyl acceptors (35–45%) although β-disaccharides were obtained exclusively.

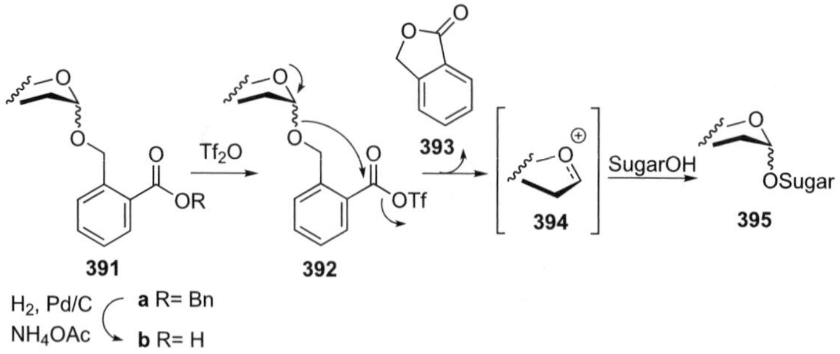

Scheme 64
2′-Carboxybenzyl (CB) glycosides

Table 12
β-Mannopyranosylation with 2′-carboxybenzyl glycosides

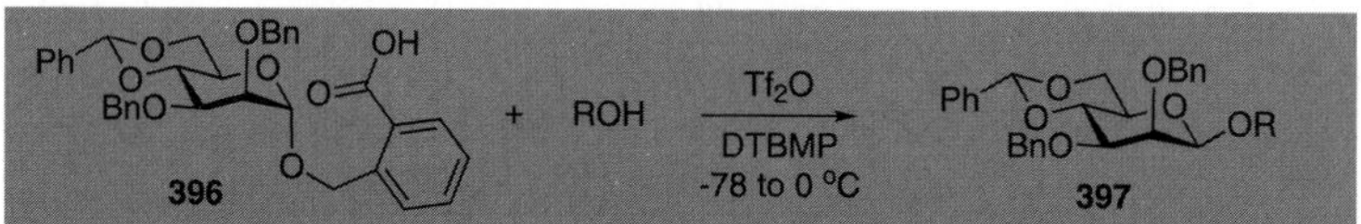

Entry	Glycosyl acceptor (ROH)	Solvent	β/α ratio	Yield (%)
i	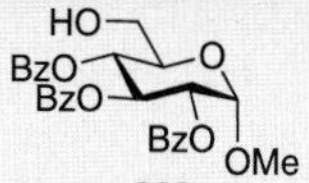 HO–BzO–BzO–BzO–OMe **398**	CH_2Cl_2	β only	91
ii	**389**	toluene	20:1	95
iii	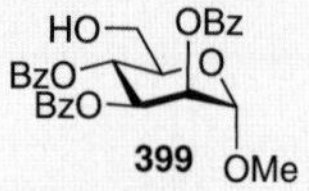 HO–OBz BzO–BzO–OMe **399**	CH_2Cl_2	β only	91
iv	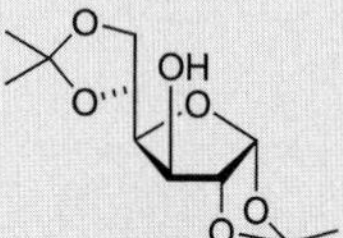 **369**	CH_2Cl_2	> 20:1	91
v	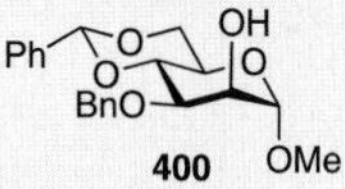 Ph–O–BnO–OH OMe **400**	CH_2Cl_2	17:1	96
vi	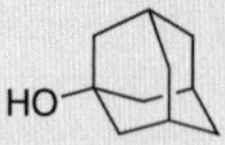 HO– **129**	CH_2Cl_2	> 20:1	90

5 2′-Carboxybenzyl (CB) Glycosides

Kim and co-workers introduced a novel type of *O*-glycosyl donor, the 2′-carboxybenzyl (CB) glycoside **391b**, readily available by selective hydrogenolysis of the benzyl ester functionality of 2-(benzyloxycarbonyl)benzyl (BCB) glycosides, **391a** [201,202,203]. Lactonization of the glycosyl triflate **392**, which was derived from the CB glycoside **391b**, is the driving force for the facile generation of the oxocarbenium ion **394** (❷ *Scheme 64*). Reaction of **394** with the glycosyl acceptor (Sugar–OH) would give the desired saccharide **395**. In the course of the transformation, a non-nucleophilic phthalide **393** is extruded. Treatment of CB glycosides with Tf$_2$O in the presence of di-tert-butylmethylpyridine (DTBMP) at −78 °C and subsequent addition of the glycosyl acceptor afforded the expected disaccharides in excellent yields.

5.1 β-D-Mannosylation Employing 2′-Carboxybenzyl Glycosyl Donors

The stereospecific formation of β-mannopyranosyl linkages is a challenging task in oligosaccharide synthesis [204]. Crich and co-workers found that 4,6-*O*-benzylidene-protected glycosyl sulfoxides or thioglycosides are useful donors in the construction of β-mannopyranosyl linkages [205,206,207,208,209]. Kim and co-workers have shown that CB glycosides with a 4,6-benzylidene group can also be applied for stereoselective β-mannopyranosylation. Glycosylations of primary alcohol acceptors, **398** and **399**, in CH$_2$Cl$_2$ were completed in 1 h at −78 °C to afford only β-mannosides in high yields (❷ *Table 12* entries i, iii). Toluene was also found to be a good solvent (❷ *Table 12*, entry ii). This high β-selective mannosylation was also achieved with secondary alcohols, e. g. **369**, **400**, and with hindered tertiary alcohol **129** (❷ *Table 12*, entries iv–vi). Glucosyl CB donors possessing the 4,6-benzylidene group gave high yields of α-glucosides.

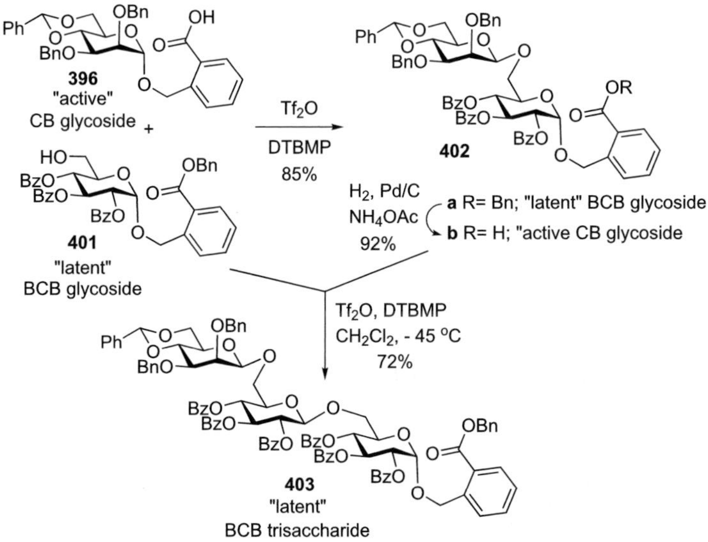

■ Scheme 65
2′-Carboxybenzyl glycoside-based "latent-active" strategy for glycosyl coupling

◘ Table 13
β-Selective glycosylation of secondary hydroxyl acceptors

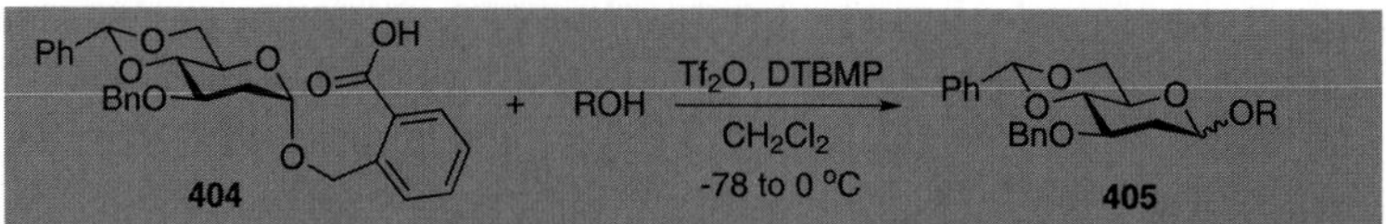

Entry	Glycosyl acceptor (ROH)	α/β ratio	Yield (%)
i	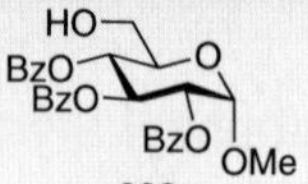 **398**	1:1	92
ii	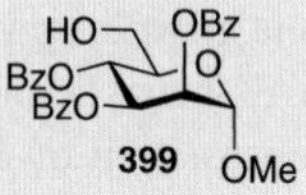 **399**	1:1.2	80
iii	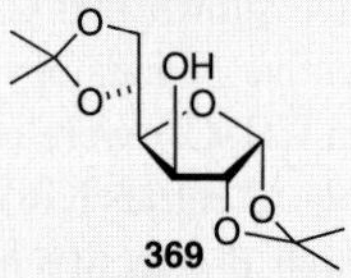 **369**	1:10	76
iv	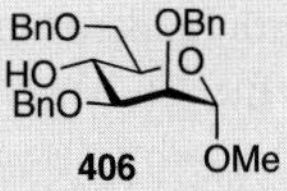 **406**	1:10	78
v	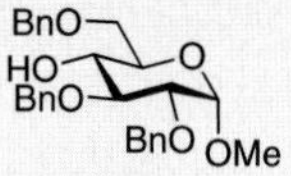 **300**	β only	72

5.2 Latent-Active Glycosylation Strategy

A remarkable feature of 2'-carboxybenzyl glycosides (e. g. **391b, ◐** *Scheme 64*) is that they can be used as a latent-active pair, together with their synthetic precursors 2-(benzyloxycarbonyl)benzyl (BCB) glycosides (e. g. **391a, ◐** *Scheme 64*). The successful mannosylation of "latent" BCB-glycoside **401** with "active" CB glycoside **396**, to give disaccharide **402a** indicated that a sequential glycosylation strategy for oligosaccharide synthesis would be possible (◐ *Scheme 65*). Thus, BCB disaccharide **402a** was readily converted into the active CB disaccharide **402b** by selective hydrogenolysis (92%, in the presence of benzyl and benzylidene groups), which upon treatment with Tf$_2$O/DTBMP glycosylated the latent BCB glycoside **401** to yield trisaccharide **403** in 72% yield.

5.3 Stereoselective Construction of 2-Deoxyglycosyl Linkages

Kim and co-workers have developed a highly α- and β- stereoselective (dual stereoselective) [210] method for the synthesis of 2-deoxyglycosides by employing CB 2-deoxyglycosides as glycosyl donors. Glycosylation of the 4,6-*O*-benzylidene-protected glycosyl donor

◘ **Table 14**
α-Selective glycosylation of secondary hydroxyl acceptors

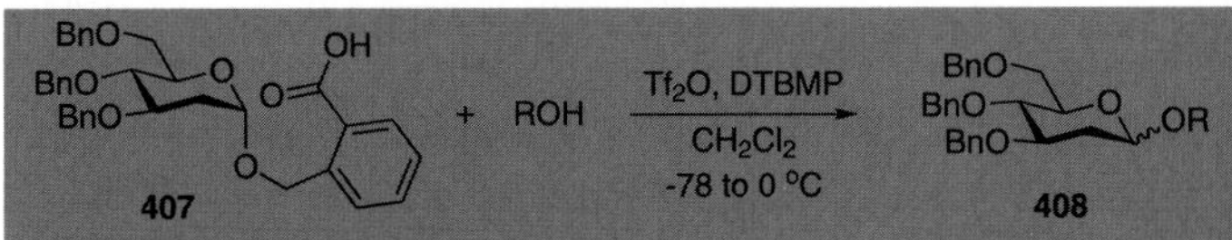

Entry	Glycosyl acceptor (ROH)	β/α ratio	Yield (%)
i	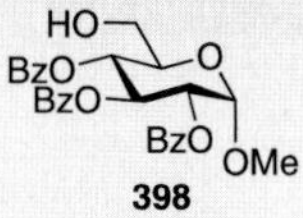 398	1:1	98
ii	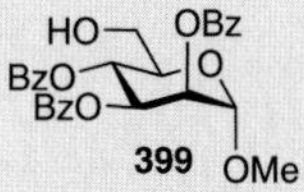 399	1:1.2	93
iii	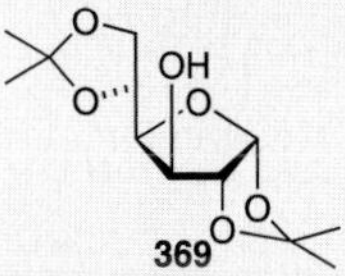 369	α only	91
iv	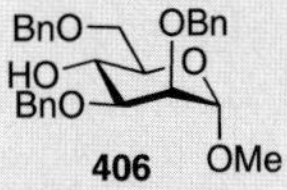 406	α only	91
v	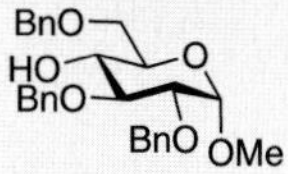 300	α only	88

404 with secondary alcohols afforded predominantly β-glycosides (❷ *Table 13*, entries iii–v). Complete reversal of the stereoselectivity, from β to α, was observed in the glycosylation of secondary alcohols with benzyl-protected glycosyl donor **406** (❷ *Table 14*, entries iii–v). On the other hand, glycosylation of primary hydroxyl acceptors with both donors did not show appreciable stereoselectivity (❷ *Table 13* and ❷ *Table 14*, entries i, ii). The authors suggested that the secondary hydroxyl acceptors formed β-disaccharides by S_N2-like displacement of an α-triflate favored in 4,6-O-benzylidene derivatives, as previously mentioned in the formation of β-mannosides of 4,6-O-benzylidene derivatives. No or poor β-selectivity in the reaction of **404** with primary alcohols was interpreted assuming that the more reactive primary alcohols reacted both with the α-triflate and an oxocarbenium ion.

5.4 2'-Carboxybenzyl Furanosyl Donors. Acceptor-Dependent Stereoselective β-D-Arabinofuranosylation

Kim and co-workers reported recently that CB tribenzyl-D-arabino furanoside **409** (easily available from methyl tribenzyl-D-arabinofuranoside) could be efficiently applied in stereos-

☐ **Table 15**
β-Selective arabinofuranosylation of acyl-protected acceptors

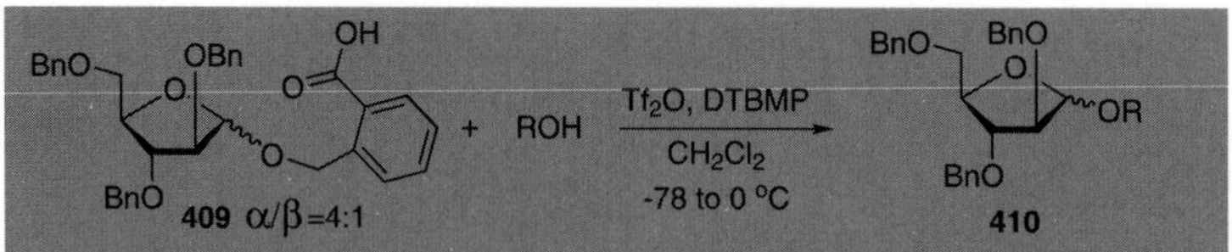

Reaction scheme: **409** (α/β = 4:1) + ROH →[Tf₂O, DTBMP; CH₂Cl₂; -78 to 0 °C]→ **410**

Entry	Glycosyl acceptor (ROH)	β/α ratio	Yield β/α ratio (%)
i	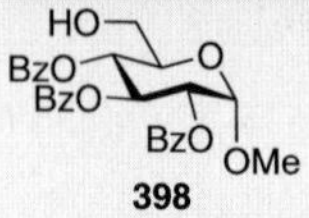 398	99:1	97
ii	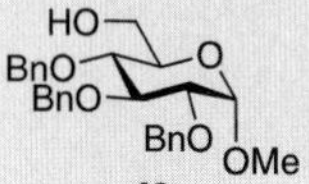 42	7:1	95
iii	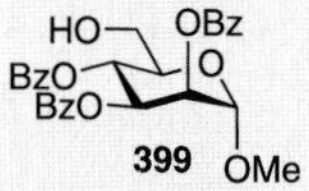 399	β only	95
iv	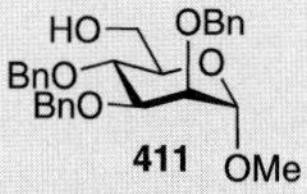 411	4:1	95
v	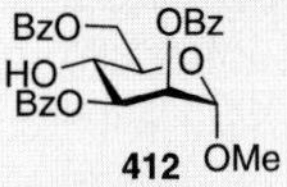 412	20:1	86
vi	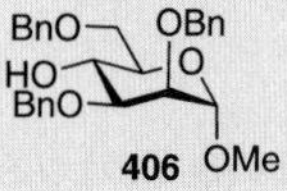 406	4:1	95
vii	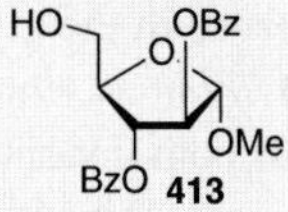 413	β only	92
viii	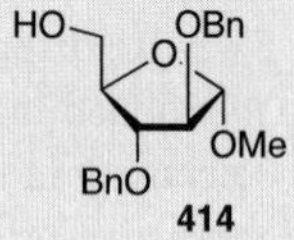 414	2.2:1	95

elective β-arabinofuranosylation processes [211]. They found that the presence of acyl-protective groups on the glycosyl acceptors was essential for attaining β-stereoselective glycosyl couplings. Thus, reaction of donor **409** with acceptor **398** having benzoyl-protective groups afforded a β-disaccharide almost exclusively (β/α = 99:1) in 97% yield (❯ *Table 15*, entry i), while the same reaction with acceptor **42** having benzyl-protective groups gave a mixture of α- and β-disaccharides (β/α = 7:1) (❯ *Table 15*, entry ii). Further examples in ❯ *Table 15* clearly showed that the protective groups in the acceptors, regardless of pyranoses or furanoses and of primary alcohols or secondary alcohols, were the crucial factor for the outcome of the stereochemistry in glycosylations with **409**. This observed stereoselectivity was also donor dependent, since glycosylation with 2-benzyl-3,5-dibenzoyl CB arabinofuranoside was not as stereoselective [211].

5.4.1 Synthesis of an Octaarabinofuranoside Based on Stereoselective β-D-Arabinofuranosylation

The authors applied this acceptor-dependent β-arabinofuranosylation method to the synthesis of octaarabinofuranoside **417**. Their retrosynthesis of compound **417** led to three components, a linear methyl trisaccharide **416**, a branched BCB trisaccharide **415**, and to CB furanosyl donor **409**. Levulinyl protective groups were chosen in fragments **415** and **416** for selective deprotection prior to furanosyl coupling (❷ *Scheme 66*). Three arabinose building blocks were used in the assembly.

Arabinofuranosyl donor **418** glycosylated acceptor **419**, to yield after levulinyl-deprotection and repetitive glycosylation with **419**, the linear trisaccharide **420** (❷ *Scheme 67a*). Coupling of latent BCB donor **422**, with active CB donor **421**, led after deprotection of the levulinyl groups to diol **423** (❷ *Scheme 67b*). The crucial double β-arabinofuranosylation of diol **423**

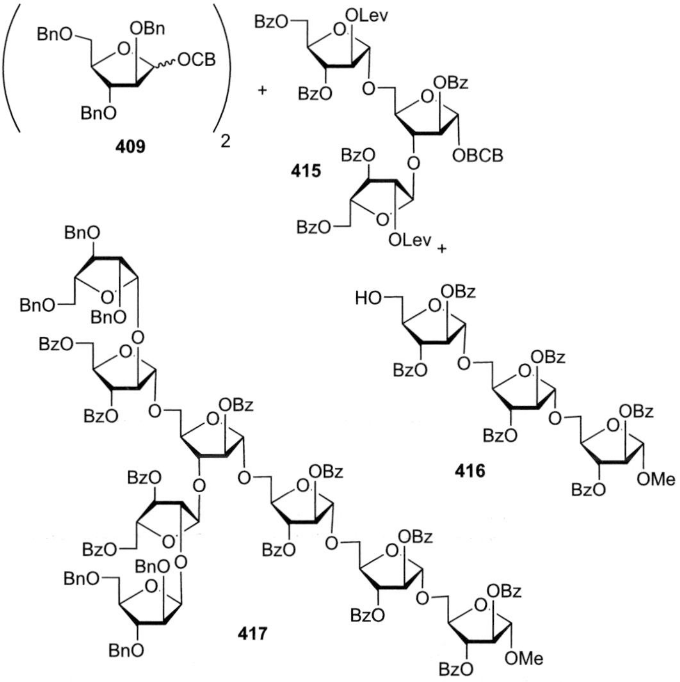

❑ Scheme 66
Retrosynthesis of octaarabinose 417. A component of lipoarabinomannan in mycobacterial cell wall

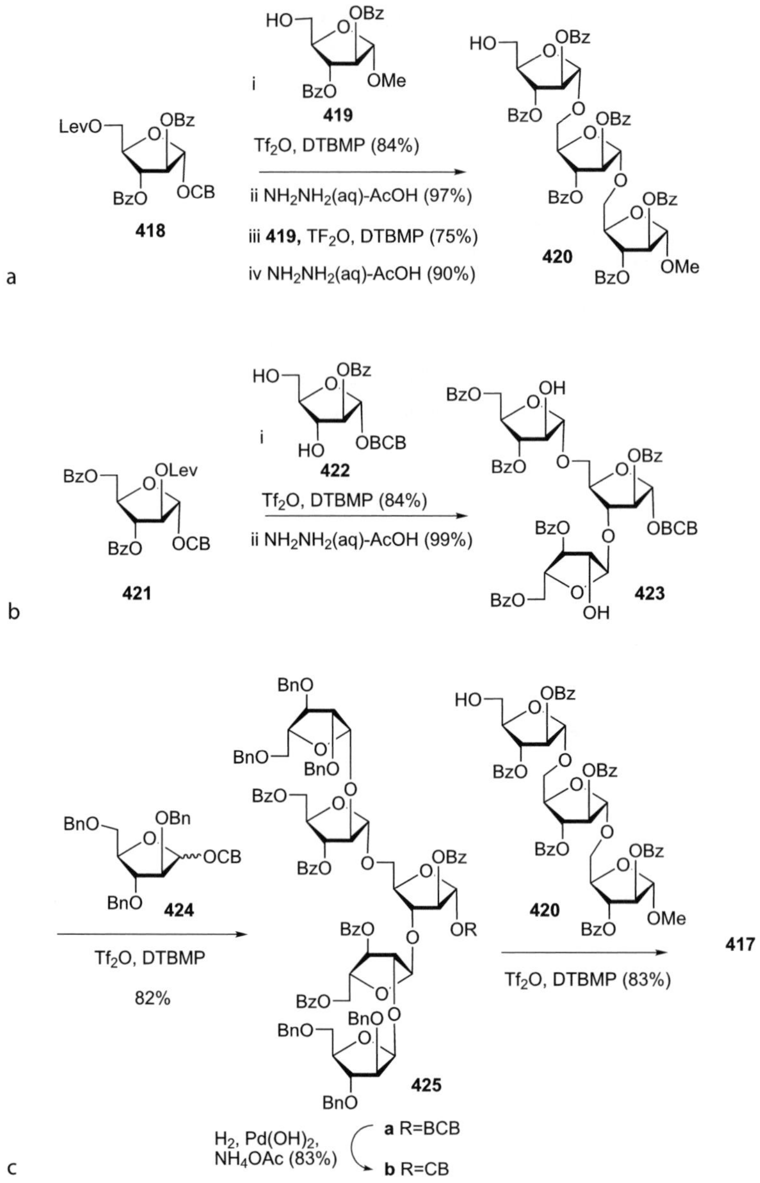

◻ Scheme 67
Synthesis of octaarabinose 417

with 3.7 equiv. of the arabinofuranosyl donor **424**, paved the way to pentaarabinofuranoside **425a** (82% yield) with complete β-selectivity. The latent BCB arabinofuranoside **425a** was converted into the active CB arabinoside **425b**. Finally, coupling of the latter with triarabinofuranosyl acceptor **420**, afforded octaarabinofuranoside **417**, in 83% yield.

5.5 2′-(Allyloxycarbonyl)benzyl (ACB) Glycosides: New "Latent" Donor for the Preparation of "Active" 2-Azido-2-deoxy BC Glycosyl Donors

Kim and co-workers introduced 2′-(allyloxycarbonyl)benzyl (ACB) glycosides, e. g. **426a**, as new "latent" glycosyl donors for 2-azido-2-deoxy-glucosides [212]. Introduction of the new ACB group in the place of the previously used BCB group was necessary because the azide functionality at C-2 was also reduced during the conversion of the BCB group into the CB group under the normally used hydrogenolysis conditions (Pd/C, H_2, NH_4OAc, MeOH). 2-Azido-2-deoxy ACB glycosides could be converted into active CB glycosyl donors (e. g. **426b**, ❷ *Scheme 68*) without affecting the azide functionality on treatment with a catalytic amount of Pd(Ph₃P)₄ in the presence of morpholine [213].

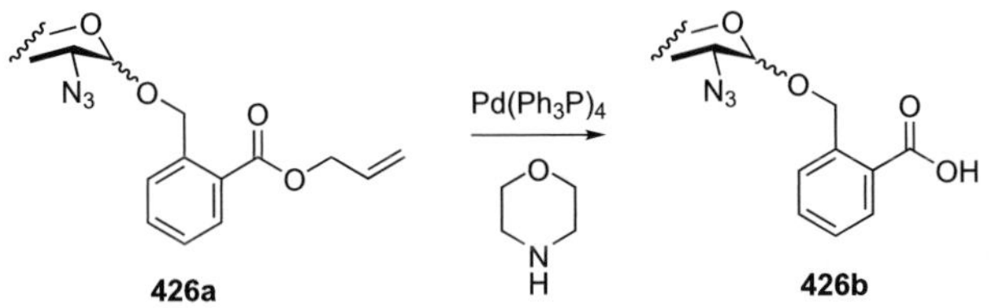

426a
"latent" ACB glycosides

426b
"active" CB glycosides

❑ **Scheme 68**
2′-(Allyloxycarbonyl)benzyl (ACB) glycosides

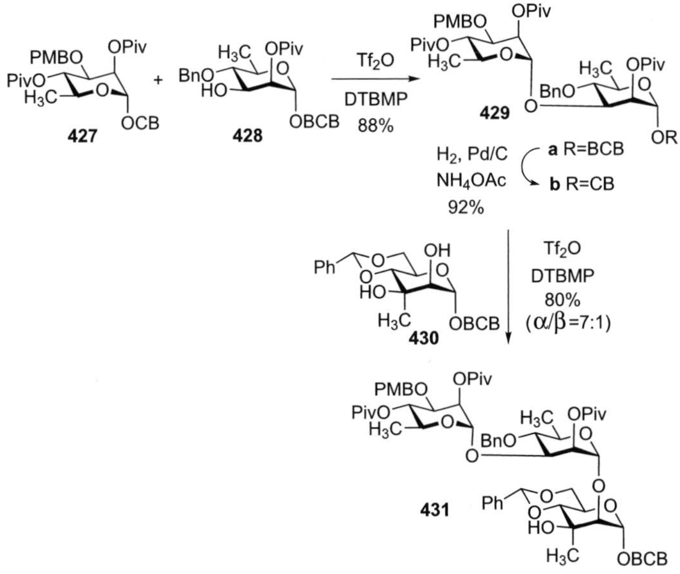

❑ **Scheme 69**
Synthesis of trisaccharide 431

5.6 Synthesis of Oligosaccharides Based on BC Glycosyl Donors

The CB glycoside methodology by means of the "latent" BCB (or ACB) glycoside and the "active" CB glycoside has proved itself as a reliable method for the synthesis of complex oligosaccharides.

5.6.1 Synthesis of Trisaccharide 431, the Repeat Unit of the *O*-Antigen Polysaccharide from Danish *Helicobacter pylori* Strains

Kim and co-workers synthesized the repeat unit of the *O*-antigen polysaccharide from Danish *Helicobacter pylori* strains, **431** (❷ *Scheme 69*) [214]. Coupling of donor CB L-rhamnoside **427** and acceptor BCB D-rhamnoside **428** gave α-disaccharide **429a** in 88% yield. Selective hydrogenolysis of "latent" BCB disaccharide afforded "active" CB disaccharide **429b** in

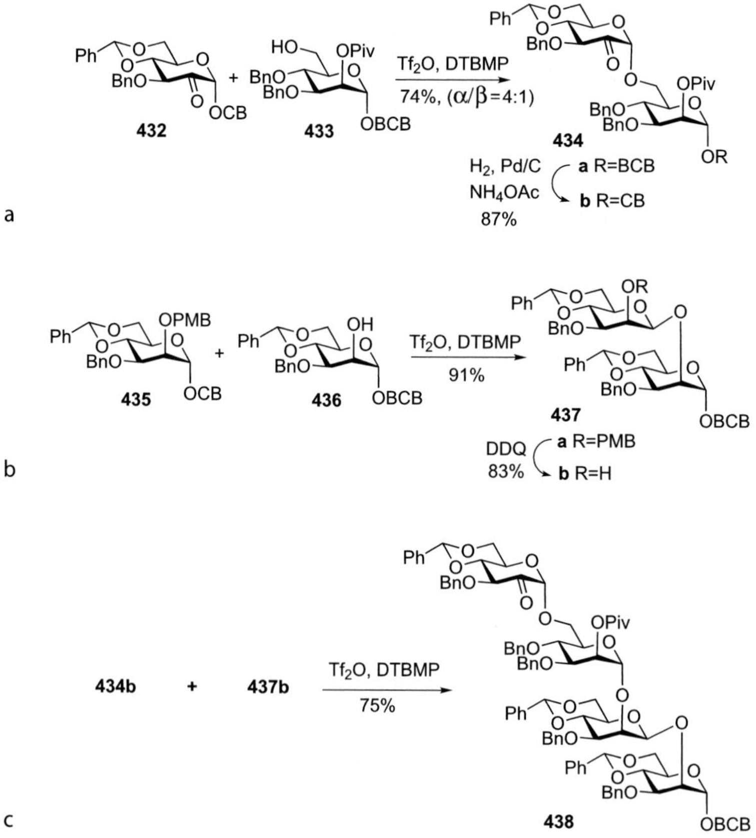

❑ Scheme 70
Synthesis of tetrasaccharide 438

92% yield. Finally, glycosylation of 3-*C*-methyl mannoside, **430**, with **429b** yielded the target α-trisaccharide **431**, along with its β-anomer in 7:1 ratio in 80% yield. A result that indicated that neighboring group participation is operative in CB glycosides.

5.6.2 Synthesis of Tetrasaccharide 438

The CB methodology was also applied to the synthesis of protected tetrasaccharide **438**, an analogue of the tetrasaccharide repeat unit of the *O*-antigen polysaccharide from the *E. coli* lipopolysaccharide (❷ *Scheme 70*) [215]. Coupling of "latent" BCB acceptor **433** with "active" CB glycosyl donor **432** gave a mixture of α-disaccharide **434a** along with its β-isomer (4:1) in 74% yield (❷ *Scheme 70a*). Glycosylation of acceptor **436** with donor **435** gave β-mannoside **437a**, that after removal of the PMB protecting group led to **437b** (❷ *Scheme 70b*). Finally, coupling of the latter with active donor **434b**, prepared from latent **434a**, yielded tetrasaccharide **438**, in 75% yield (❷ *Scheme 70c*).

5.6.3 Synthesis of Tetrasaccharide Repeat Unit from *E. coli* 077

A route to a tetrasaccharide **439** was reported, which made use of the previously mentioned "latent" 2'-(allyloxycarbonyl)benzyl (ACB) glycosides in combination with "latent" and "active" BCB and CB glycosides, respectively [212]. The retrosynthesis is outlined in

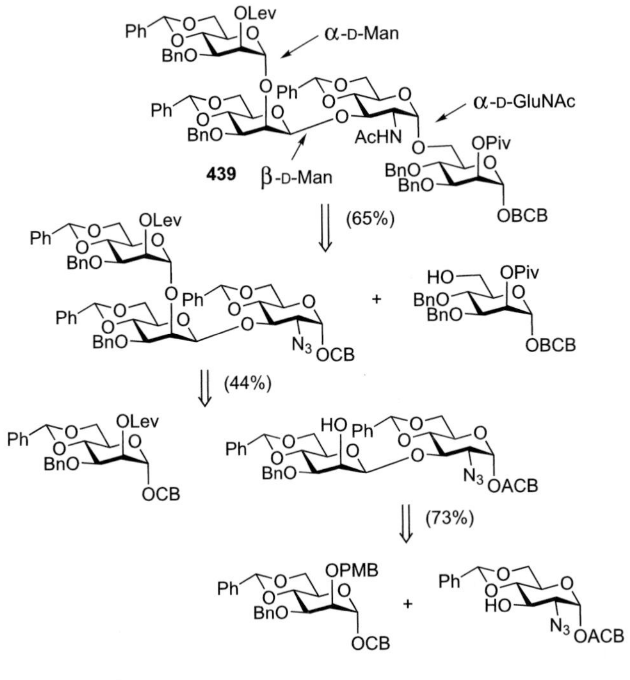

■ **Scheme 71**
Retrosynthesis of tetrasaccharide 439

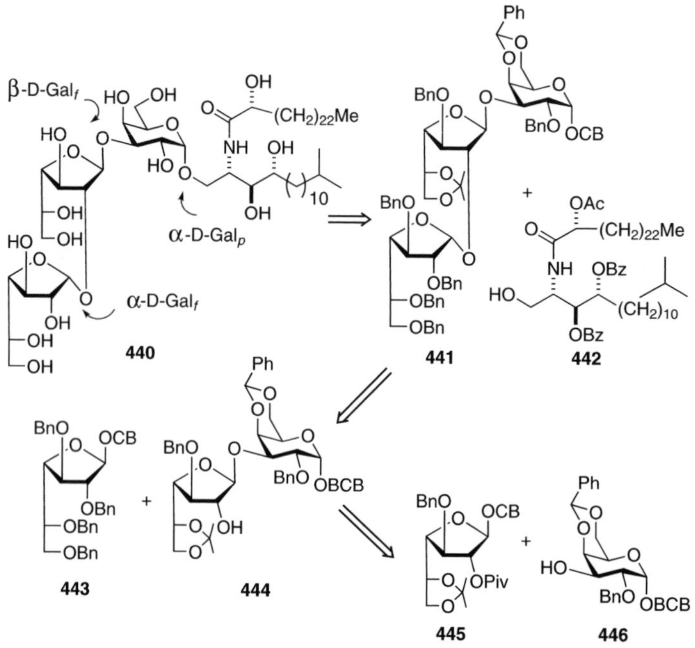

☐ Scheme 72
Retrosynthesis of Agelagalastatin 440

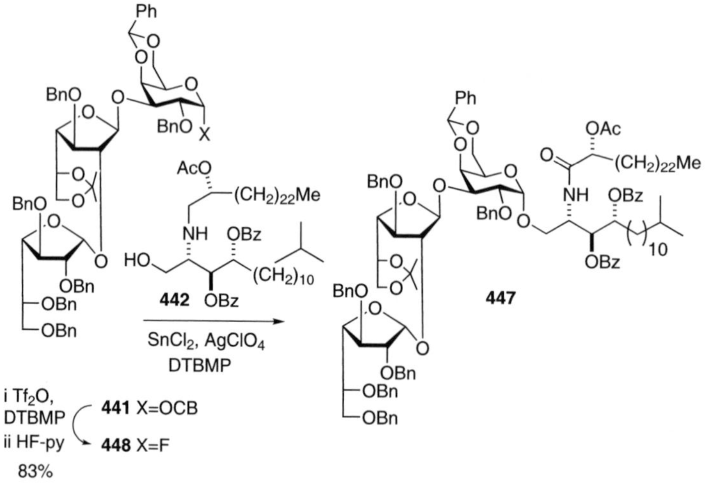

☐ Scheme 73
Transformation of CB glycoside 441 into glycosyl fluoride 448 and final coupling of Agelagalastatin

❯ *Scheme 71.* All glycosyl couplings were based on the "latent-active" methodology, and all were stereoselective. A slightly modified synthesis of **439** has been reported including one CB mediated coupling [216].

5.6.4 Total Synthesis of Agelagalastatin

The total synthesis of agelagalastatin, an antineoplastic glycosphingolipid, has been described by Kim and co-workers [217]. The retrosynthesis, outlined in ❷ *Scheme 72*, involved a β-D-galactofuranosylation, an α-D-galactofuranosylation, and a final α-D-galactopyranosylation. The β-D-galactofuranosylation was achieved in 79% yield via neighboring group participation of the pivaloyl group at *O*2 in compound **445**. The α-D-galactofuranosylation to **441**, took place with 91% yield with a nonparticipant benzyl group at *O*2 in donor **443**. The final α-D-galactopyranosylation (❷ *Scheme 73*) was carried out with CB trisaccharide donor **441** furnishing compound **447** in 77% yield as a 1.4:1 (α/β) mixture of saccharides. The efficiency of this coupling was improved by conversion of the CB trisaccharide donor to glycosyl fluoride **448**. Treatment of **441** with TF$_2$O/DTBMP followed by HF-pyridine, as a source of fluoride, yielded glycosyl fluoride **448**. The glycosylation of acceptor **442** with glycosyl fluoride **448**, then gave the target protected agelagalastatin **447**, in 72% yield as the pure α-isomer.

5.7 Conversion of 2′-Carboxybenzyl Glycosides into Other Glycosyl Donors

CB glycosides have been converted to phenyl 1-thio glycosides and glycosyl fluorides in one-pot operations [218]. Thus, treatment of CB glycosyl donors with TF$_2$O/DTBMP in CH$_2$Cl$_2$ at −78 °C for 10 min followed by addition of PhSH furnished thioglycosides, e. g. **451**, **453** (❷ *Scheme 74*), whereas treatment with DAST or HF-pyridine (see ❷ *Scheme 73*) yielded the corresponding glycosyl fluorides, e. g. **449**, **452** (❷ *Scheme 74*). The high β-selectivity observed in the formation of glycosyl fluoride **452** and thioglycoside **453** from 4,6-*O*-benzylidenemannopyranoside **396** was ascribed to the presence of a highly reactive 4,6-*O*-benzylidenemannopyranosyl α-triflate, in keeping with previously mentioned findings.

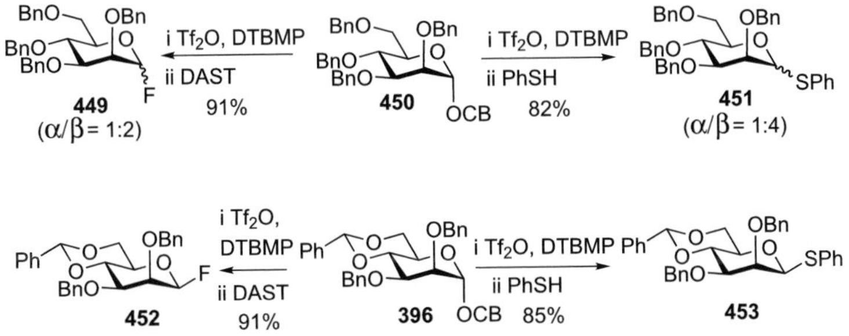

❏ **Scheme 74**
Conversion of CB glycosides to glycosyl fluorides and thiophenyl glycosides

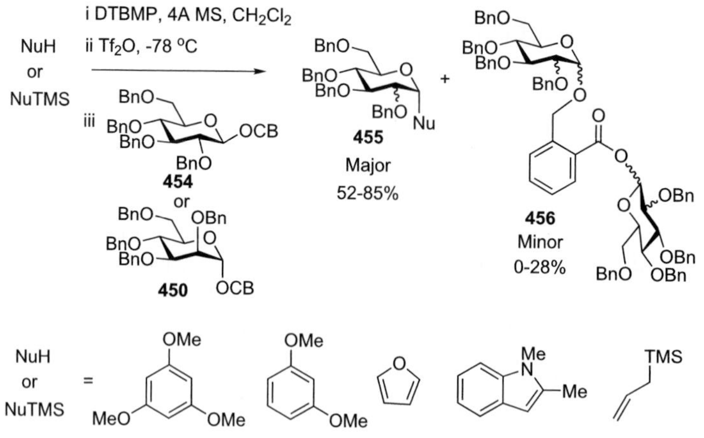

☐ Scheme 75
CB glycosides in the synthesis of α-C-glycosides

5.8 2'-Carboxybenzyl Glycosides as Glycosyl Donors for C-Glycosylation

Glycosylation of various glycosyl acceptors (NuH or NuTMS, ❷ *Scheme 75*) with *manno-* and *gluco-* CB glycosyl donors **450** and **454**, respectively afforded α-C-glycosides **455**, exclusively or predominantly in good yields [218]. Experimentally these reactions were carried out by addition of the donor to a solution of the acceptor, DTBMP, and Tf$_2$O in CH$_2$Cl$_2$ at −78 °C. These modified conditions led to increased yields of C-glycosides and minimized the amount of self-condensed esters **456**.

6 O-Heteroaryl Glycosyl Donors

Glycosides of some heterocycles have also been investigated as glycosyl donors.

6.1 2-Pyridyl 2,3,4,6-tetra-O-benzyl-D-glucosides

The first example, reported by Nikolaev and Kochetkov [219], dealt with the use of 2-pyridyl 2,3,4,6-tetra-O-benzyl-β-D-glucoside in glycosylation. This heteroaryl glycoside was prepared by glycosylation of 2(1H)-pyridinone by the corresponding sugar chloride, and was activated by electrophiles, such as MeOTf and Et$_3$O·BF$_4$, to give mixtures of *cis-* and *trans-*glycosides.

6.2 O-Hetaryl Glycosides by Schmidt's Group

Schmidt and co-workers [168,220,221] reported the preparation, and use in glycosylation reactions of several O-hetaryl glycosides, e. g. **458**, conveniently prepared by anomeric

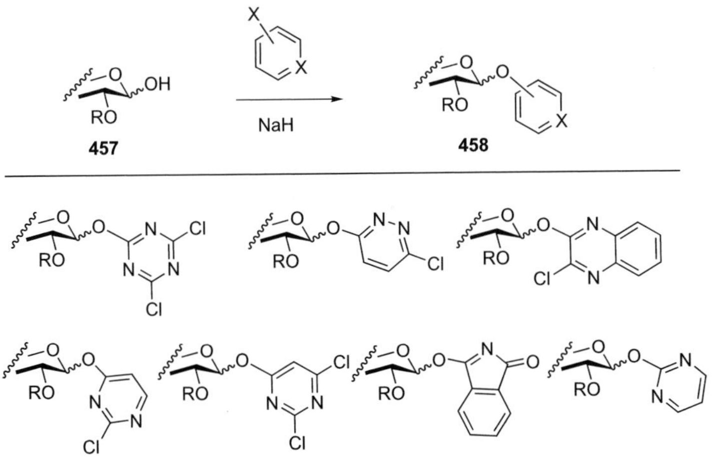

■ Scheme 76
O-Hetaryl glycosides synthesized by Schmidt's group

O-hetarylation of hexoses, e. g. **457**, with the corresponding electron-deficient heteroaromatic/heterocyclic systems (❷ *Scheme 76*). The best results in terms of glycosylation were obtained with tetrafluoropyridyl glycosides **460** and **462**, obtained by reaction of hexoses **459** and **461** with 2,3,4,5,6-pentafluoro pyridine (❷ *Scheme 77a,b*). Under TMSOTf catalysis, in CH$_2$Cl$_2$ at room temperature, they furnished the corresponding α- and β-disaccharides **463** and **464** in 98 and 74% yield, respectively (❷ *Scheme 77c,d*).

6.3 3-Methoxy-2-pyridyl (MOP) Glycosides

On the basis of the concept of remote activation [222], first applied to pyridine thioglycosides [223], Hanessian and co-workers introduced 3-methoxy-2-pyridyl (MOP) glycosides [224,225]. They first reported the usefulness of ribofuranosyl MOP donor **465** in the coupling with silylated pyrimidine bases, by activation with TMSOTf, to give 1,2-*cis* nucleosides, **466**, with high selectivity (❷ *Scheme 78*) [226]. These glycosides react in MeOTf-, Cu(OTf)$_2$-, TfOH-, or Yb(OTf)$_3$-promoted reactions to give disaccharides [227].

6.3.1 Coupling of Unprotected MOP Glycosyl Donors

Interestingly, unprotected MOP glycosides could also be used as donors. In fact, when using an excess of glycosyl acceptor (≈10 equiv.), disaccharides are obtained in reasonable yields, as illustrated in ❷ *Scheme 79*.
It was also found that introduction of any protecting group on the unprotected MOP glycosyl donors resulted in a significant decrease of the reactivity. This deactivation was considerable when *p*-fluorobenzoates (FBz) were used as protecting groups, and it was applied to the synthesis of disaccharides, and to iterative oligosaccharide synthesis (❷ *Scheme 80*).

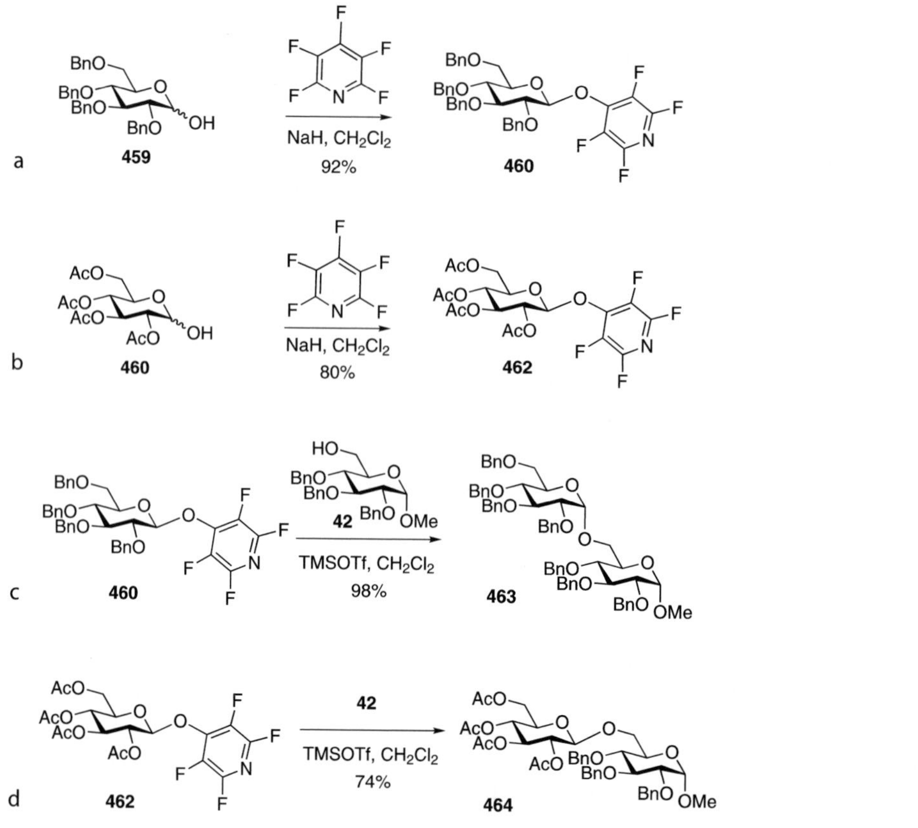

Scheme 77
2,3,5,6-Tetrafluoro-4-(2,3,4,6-tetra-*O*-acetyl or *O*-benzyl-β-D-glucopyranosyloxy) pyridine as glycosyl donors

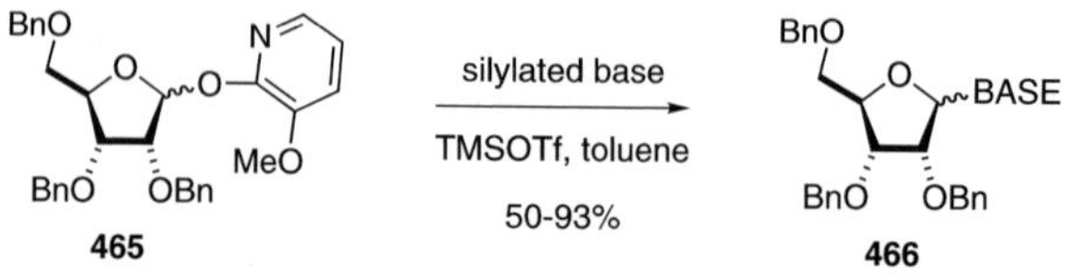

Scheme 78
MOP Ribofuranosyl donors in the synthesis of 1,2-*cis* furanosyl nucleosides

6.3.2 Esterification and Phosphorylation of Unprotected MOP Glycosides

MOP glycosyl donors have been used in stereocontrolled esterification and phosphorylation, leading to glycosyl 1,2-*cis*-1-carboxylates or glycosyl 1,2-*cis*-glycosyl-1-phosphates in one step. Treatment of MOP donor **475** in acetonitrile or DMP with an excess (20–200 equiv.) of a carboxylic acid under anhydrous conditions led to the corresponding D-glycosyl carboxylate **477**, in excellent yields [228]. Moreover, treatment of 6-*O*-tert-butyldiphenylsilyl MOP donor **478** with only 1.5 equiv. of the corresponding carboxylic acid in CH$_2$Cl$_2$ resulted in the formation of 1,2-*cis*-glycosyl carboxylate **479** (◯ *Scheme 81a*). Treatment of β-D-galactopyranosyl, and 2-azido-2-deoxy-α-D-galactopyranosyl MOP donors **475** and **471**, respectively with

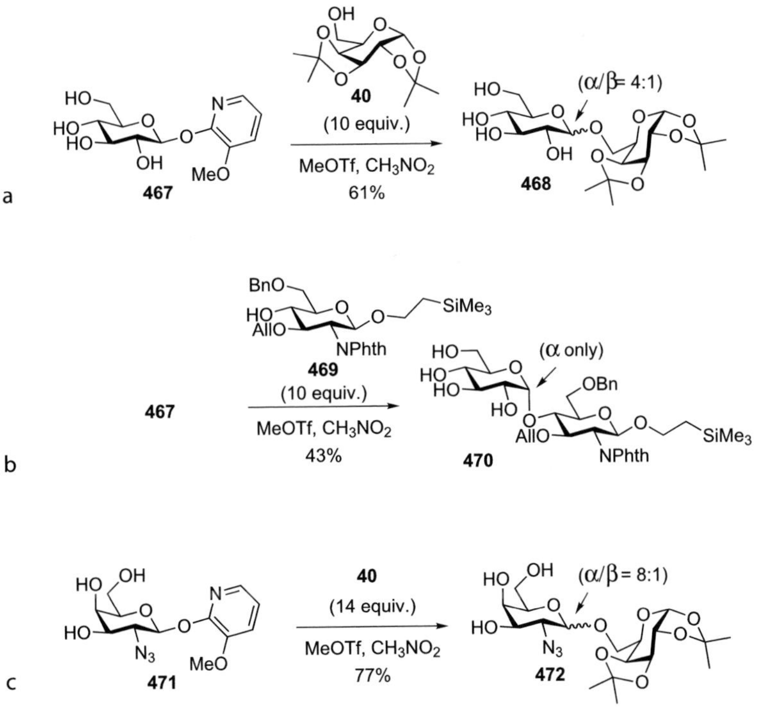

Scheme 79
Disaccharide synthesis with unprotected MOP donors

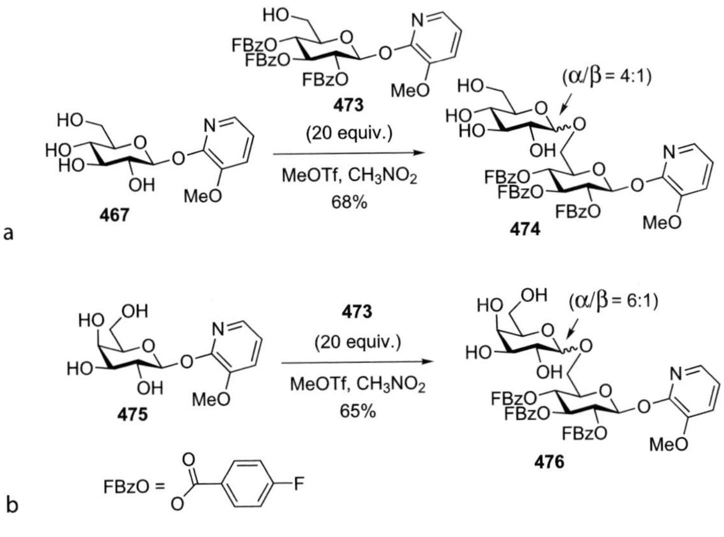

Scheme 80
Selective activation of unprotected- versus protected-MOP glycosides

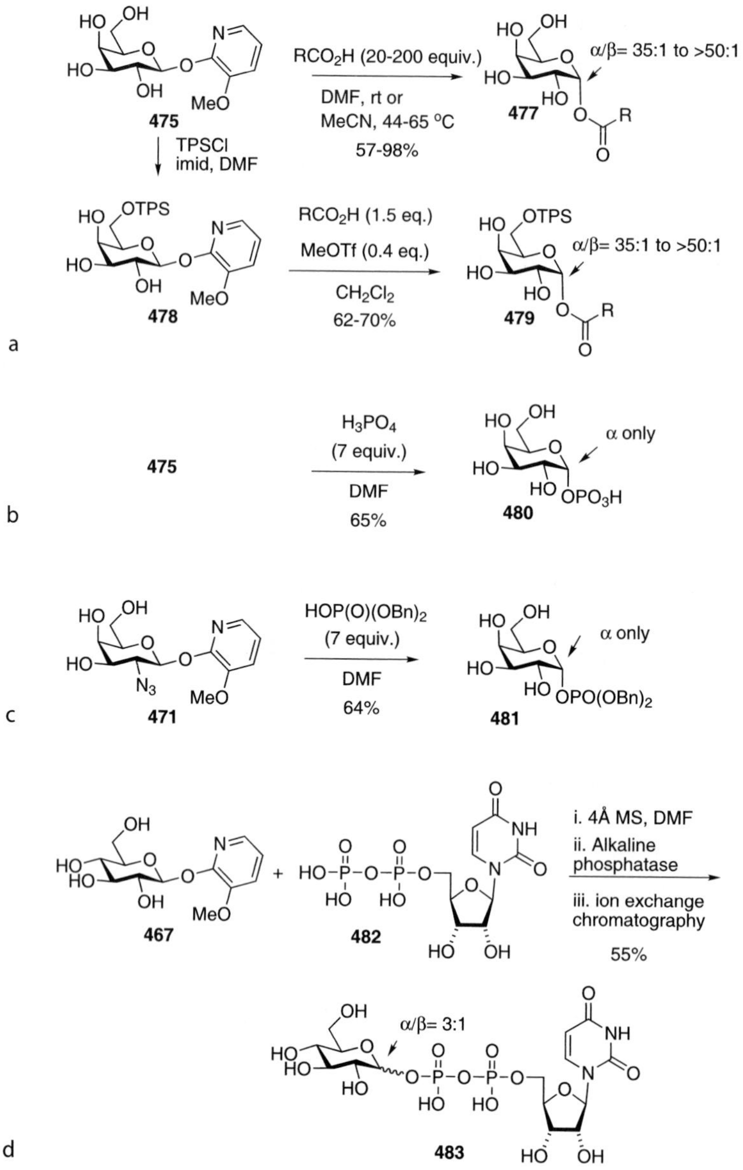

□ Scheme 81
Stereocontrolled synthesis of glycosyl 1-carboxylates and 1-phosphates

7 equiv. of phosphoric acid or dibenzyl phosphate in DMF led to the corresponding α-glyco-syl phosphates **480** and **481**, respectively. The same donors are also capable of transferring glucopyranosyl (e. g. **467**, ❷ *Scheme 81d*) and galactopyranosyl units to UDP-free acid **482**, to afford the corresponding uridine 5′ diphosphosugars (e. g. **483**) in one step [229].

6.3.3 MOP Glycosides in Oligosaccharide Synthesis

A solid-phase oligosaccharide synthesis based on the MOP donor/acceptor methodology was developed by Hanessian and co-workers [224]. Thus, an *O*-unprotected polymer-phase bound MOP donor is coupled with an excess of a partially esterified MOP acceptor. Selective removal of the ester (or related protecting groups) from the new saccharides generates a new *O*-unprotected MOP donor to engage in a subsequent iteration.

Hanessian and co-workers also illustrated the usefulness of the MOP methodology with some syntheses of oligosaccharides [230,231]. They reported a concise synthesis of a Galα1→ 3Galβ1→4GlcNAcOR trisaccharide **489**, outlined in ❷ *Scheme 82* [232]. Treatment of MOP donor **484** with 3-benzyloxycarbonylamino 1-propanol in the presence of HBF₄·Et₂O in CH₂Cl₂ led to the expected β-glycoside **485**. Protecting group manipulation and glycosylation with MOP galactopyranosyl **486** in the presence of Cu(TfO)₂ as activator gave the intended β-disaccharide **487**. Final glycosylation of **487** with β-galacto MOP donor **488** (Cu(OTf)₂) or Yb(OTf)₃, as promoters) led to protected trisaccharide **489**.

6.4 6-Nitro-2-benzothiazolyl Glycosides

Mukaiyama et al. described glycosyl 6-nitro-2-benzothiazoates (e. g. **490**) as useful glycosyl donors [233]. They are prepared by reaction of glucose derived hemiacetals (e. g. **459**) with 2-chloro-6-nitro-2-benzothiazoate (❷ *Scheme 83a*). The purified α-isomer **α-490**, react-

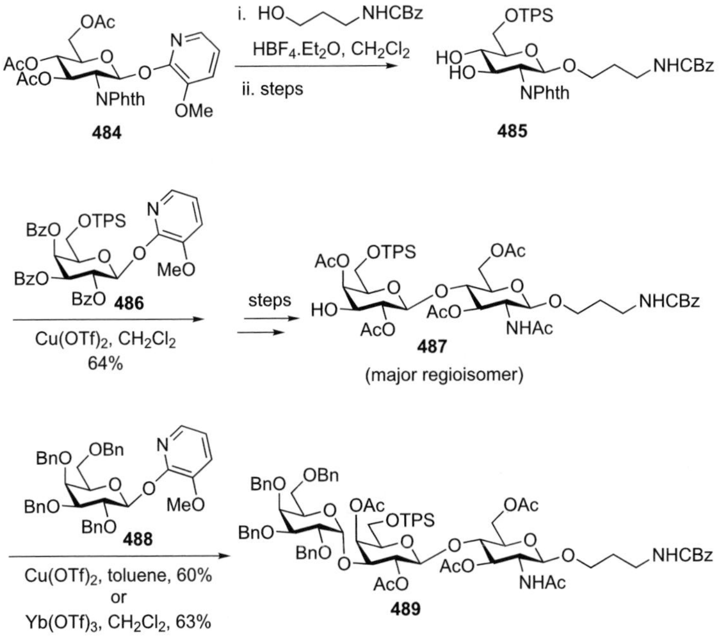

❏ Scheme 82
Synthesis of a trisaccharide using MOP glycosyl donors

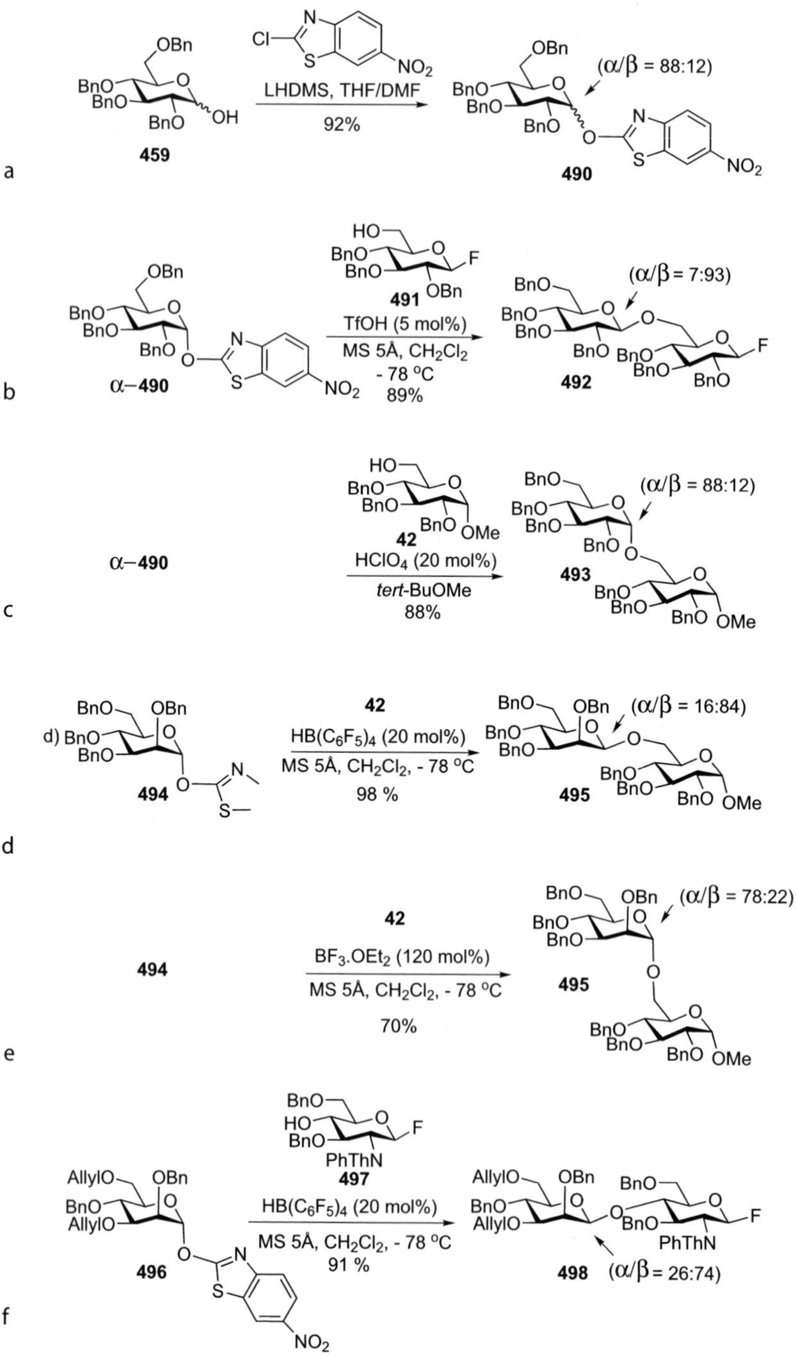

☐ **Scheme 83**
Glycosyl 6-nitro-2-benzothiazoate as a glycosyl donor

ed with primary hydroxyl acceptors in the presence of catalytic TfOH at $-78\,°C$ to give mainly β-glucosides, e. g. **492** (❷ *Scheme 83b*). Although, a highly stereoselective α-glucosyla-
tion ($\alpha/\beta = 88{:}12$) was carried out in high yield using 20 mol% of $HClO_4$ in *tert*-BuOMe
(❷ *Scheme 83c*). 6-Nitro-2-benzothiazolyl α-mannosides (e. g. **494**) effected stereoselective
β-mannosylation with several glycosyl acceptors [234,235]. The highest β-stereoselectivity
was achieved when tetrakis(pentafluorophenyl)boric acid [$HB(C_6F_5)_4$] [236] was employed
as catalyst (❷ *Scheme 83d*). $BF_3{\cdot}Et_2O$, a weaker Lewis acid, showed a reversed stereoselectivity [237] (❷ *Scheme 83e*). The β-selective coupling was employed by Mukaiyama and co-
workers in the formation of the β-Man(1→4)GlcN linkage, e. g. **498**, that exists in *N*-linked
glycans (❷ *Scheme 83f*) [238,239].

7 Miscellaneous *O*-Glycosyl Donors

Noyori and Kurimoto [240] described that hydroxyl-protected and -unprotected glycosyl aryloxides reacted with alcohols under mild electrolytic conditions to give the corresponding
glycosides. They hypothesized that the glycosylation reaction proceeded via oxocarbenium
ion intermediates generated from the radical cation of the easily oxidizable aryloxy substrate
(❷ *Scheme 84*).
A combination of trimethylsilyl bromide and zinc triflate promoted the glycosylation of
benzyl-, isopropyl-, and methyl glycosides with several glycosyl acceptors in moderate to
good yields [241,242].
2-Deoxyglycosides, e. g. **501**, were obtained by DDQ oxidation of 3,4-dimethoxybenzyl
glycosides **500**, in MeCN in the presence of primary, secondary, and tertiary alcohols
(❷ *Scheme 85*) [243].
Davis and co-workers [244] examined the self-activating properties of unprotected and acetylated bromobutyl glycosides **505** and **508**, respectively (❷ *Scheme 86*). These readily available
compounds reacted with galactose acceptor **40** (1 equiv.) in the presence of a halophilic Lewis

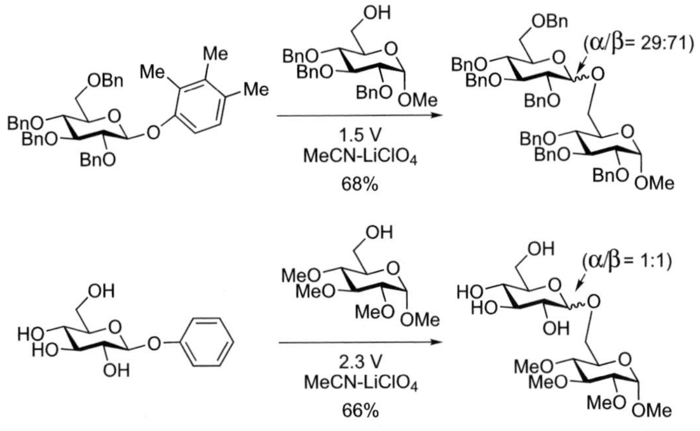

❑ Scheme 84
Electrochemical glycosylation of glycosyl aryloxides

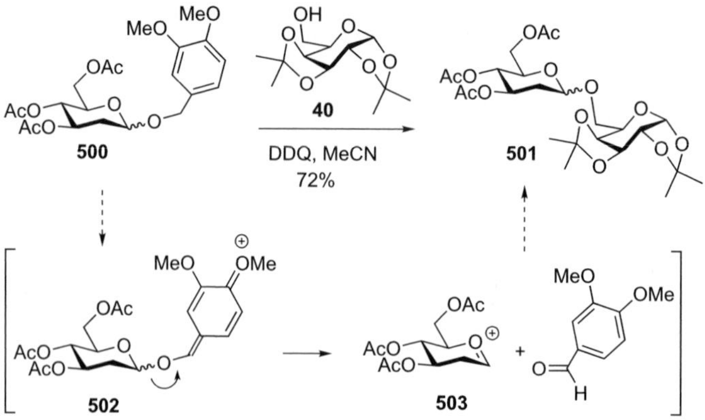

☐ Scheme 85
3,4-Dimethoxybenzyl 2-deoxy glycosides as glycosyl donors

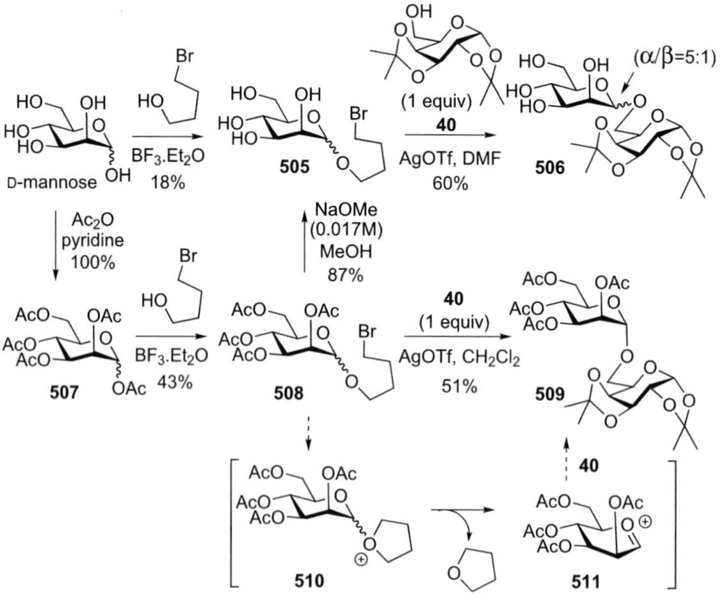

☐ Scheme 86
Bromobutyl glycosides as glycosyl donors

acid promoter (AgOTg) to give disaccharides **506** and **509** in moderate yields. The suggested reaction pathway involved a spontaneous, or acid, triggered, 5-*exo-tet* cyclization of the bromobutyl glycoside, e. g. **508**→**510**, to form an anomeric furanosyl cation **510**, which would evolve to give non-nucleophilic volatile tetrahydrofuran, and oxocarbenium ion **511**. The latter will then react with acceptor **40** to furnish the disaccharide.

Hung and co-workers have reported on the use of 2-allyloxyphenyl mannoside **512** as a useful glycosyl donor [245]. Mannoside **512** reacted in the presence of NIS/TfOH in CH$_2$Cl$_2$ at room

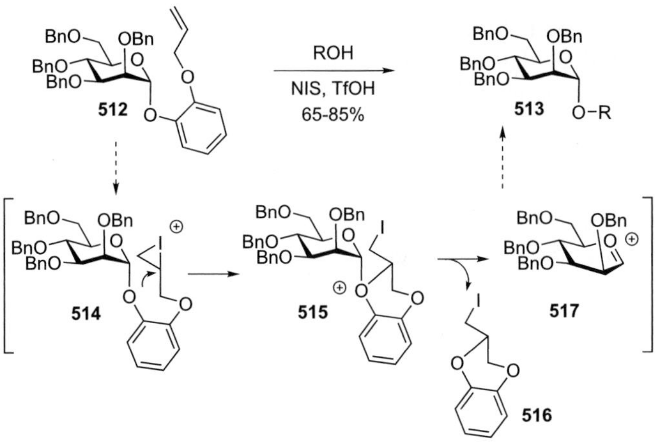

■ Scheme 87
2-Allyloxyphenyl mannosides as glycosyl donors

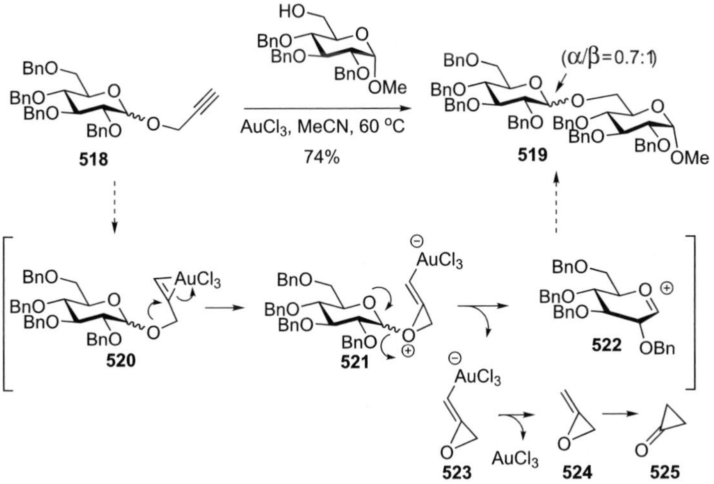

■ Scheme 88
Propargyl glycosides as glycosyl donors

temperature, with a series of primary and secondary hydroxyl acceptors to give α-mannosides, e. g. **513**, in good yields (❖ *Scheme 87*). The proposed mechanism for the formation of the oxocarbenium ion **517**, outlined in ❖ *Scheme 87*, implies a 6-*exo-tet* cyclization on halonium ion **514**, and the ejection of the non-nucleophilic species **516**.

Hotha and Kashyap have identified propargyl glycosides **518**, as new glycosyl donors (❖ *Scheme 88*) [246,247]. Various aglycones, including primary and secondary alcohols, reacted with propargyl glycosides in the presence of 3 mol% of AuCl$_3$ in MeCN at 60 °C, to give α/β-mixtures of glycosides and disaccharides in good yields. The α,β-ratio of the transglycosylation products was found to be independent of the anomeric ratio of the donor. per-*O*-Acylated propargyl glycosides did not give transglycosylation products. A possible reaction pathway to the generation of an intermediate oxocarbenium ion, based on the alkynophilicity of gold catalysts, was advanced by the authors (❷ *Scheme 88*). Coordination of AuCl₃ to the glycosyl donor **518** would be followed by formation of the cyclopropyl gold carbene intermediate (**520**) that could evolve to intermediate **521**, which would lead to oxocarbenium ion **522**, and alkenyl gold complex **523**. The latter upon protodemetalation will generate AuCl₃ and cyclopropanone **525** via intermediate **524**.

Acknowledgements

The author thanks the *Dirección General de Enseñanza Superior* (Grant CTQ2006-15279-C03-02) for financial support.

References

1. Paulsen H (1982) Angew Chem Int Ed Engl 21:155
2. Schmidt RR (1986) Angew Chem Int Ed Engl 25:212
3. Paulsen H (1990) Angew Chem Int Ed Engl 29:823
4. Toshima K, Tatsuta K (1993) Chem Rev 93:1503
5. Schmidt RR (1986) Angew Chem Int Ed Engl 25:212
6. Boons GJ (1996) Tetrahedron 52:1095
7. Boons GJ (1996) Contemp Org Synth 3:173
8. Davis BG (2000) J Chem Soc Perkin Trans 1 2137
9. Demchenko AV (2005) Lett Org Chem 2:580
10. Fraser-Reid B, Konradsson P, Mootoo DR, Udodong U (1988) J Chem Soc Chem Commun 823
11. Michael A (1879) Am Chem J 1:305
12. Fraser-Reid B, Udodong UE, Wu Z, Ottoson H, Merrit JR, Rao CS, Roberts C, Madsen R (1992) Synlett 927
13. Mootoo DR, Date V, Fraser-Reid B (1988) J Chem Soc Chem Commun 1462
14. Mootoo DR, Fraser-Reid B (1986) J Chem Soc Chem Commun 1570
15. Baldwin JE (1976) J Chem Soc Chem Commun 734
16. Mootoo DR, Date V, Fraser-Reid B (1988) J Am Chem Soc 110:2662
17. Lemieux RU, Morgan AR (1965) Can J Chem 43:2190
18. Goodman L (1967) Adv Carbohydr Chem Biochem 22:109
19. Mootoo DR, Konradsson P, Udodong U, Fraser-Reid B (1988) J Am Chem Soc 110: 5583
20. Konradsson P, Mootoo DR, McDevitt RE, Fraser-Reid B (1990) J Chem Soc Chem Commun 270
21. Veeneman GH, van Boom JH (1990) Tetrahedron Lett 31:275
22. Veeneman GH, van Leeuwen SH, van Boom JH (1990) Tetrahedron Lett 31:1331
23. Friesen RW, Danishefsky SJ (1989) J Am Chem Soc 111:6656
24. Barrena MI, Echarri R, Castillón S (1996) Synlett 675
25. Cheung MK, Douglas NL, Hinzen B, Ley SV, Pannecoucke X (1997) Synlett 257
26. Hashimoto S, Sakamoto H, Honda T, Ikegami S (1997) Tetrahedron Lett 38:5181
27. Hashimoto S, Sakamoto H, Honda T, Abe H, Nakamura S, Ikegami S (1997) Tetrahedron Lett 38:8969
28. Mukaiyama T, Chiba H, Funasaka S (2002) Chem Lett 392
29. Chiba H, Funasaka S, Kiyota K, Mukaiyama T (2002) Chem Lett 746
30. Chiba H, Funasaka S, Mukaiyama T (2003) Bull Chem Soc Jpn 76:1629
31. Demchenko AV, Kamat MN, De Meo C (2003) Synlett 1287
32. Kamat MN, Demchenko AV (2005) Org Lett 7:3215
33. Bongat AFG, Kamat MN, Demchenko AV (2007) J Org Chem 72:1480

34. Llera JM, López JC, Fraser-Reid B (1990) J Org Chem 55:2997
35. Madsen R, Fraser-Reid B (1995) J Org Chem 60:772
36. Fraser-Reid B, Merrit RJ, Handlon AJ, Andrews CW (1993) Pure Appl Chem 65:779
37. Rodebaugh R, Fraser-Reid B (1994) J Am Chem Soc 116:3155
38. Rodebaugh R, Fraser-Reid B (1996) Tetrahedron 52:7663
39. Brown RS (1997) Acc Chem Res 30:131
40. Fraser-Reid B, Wu Z, Udodong UE, Ottosson, H (1990) J Org Chem 55:6068
41. Fraser-Reid B, Wu Z, Andrews CW, Skowronski E, Bowen JP (1991) J Am Chem Soc 113:1434
42. Merrit JR, Debenham JS, Fraser-Reid B (1996) J Carbohydr Chem 15:65
43. Rodebaugh R, Debenham JS, Fraser-Reid B, Snyder JP (1999) J Org Chem 64:1758
44. Konradsson P, Fraser-Reid B (1988) J Chem Soc Chem Commun 1124
45. Pale P, Whitesides GM (1991) J Org Chem 56:4547
46. Plante OJ, Andrade RB, Seeberger PH (1999) Org Lett 1:211
47. Hashimoto S, Honda T, Ikegami S (1999) J Chem Soc Chem Commun 685
48. Clausen MH, Madsen R (2003) Chem Eur J 9:3281
49. López JC, Uriel C, Guillamón-Martín A, Valverde S, Gómez AM (2007) Org Lett 9:2759
50. Fraser-Reid B, Lu J, Jayaprakash KN, López JC (2006) Tetrahedron Asymmetry 17:2449
51. Ferrier RJ, Hay RW, Vethaviyasar N (1973) Carbohydr Res 27:55
52. Fügedi P, Garegg PJ, Lönn H, Norberg T (1987) Glycoconjugate J 4:97
53. Codée JDC, Litjens REJN, van den Bos LJ, Overkleeft HS, van der Marel GA (2005) Chem Soc Rev 34:769
54. Schmidt RR, Michel J (1980) Angew Chem Int Ed Engl 19:731
55. Sharma GVM, Rao SM (1992) Tetrahedron Lett 33:2365
56. Ratcliffe AJ, Fraser-Reid B (1989) J Chem Soc Perkin Trans 1 1805
57. Ritter JJ, Minieri PP (1948) J Am Chem Soc 70:4045
58. Lemieux RU, Morgan AR (1965) Can J Chem 43:2205
59. Ratcliffe AJ, Fraser-Reid B (1990) J Chem Soc Perkin Trans 1 747
60. Ratcliffe AJ, Konradsson P, Fraser-Reid B (1990) J Am Chem Soc 112:5665
61. Ratcliffe AJ, Konradsson P, Fraser-Reid B (1991) Carbohydr Res 216:323
62. Handlon AL, Fraser-Reid B (1993) J Am Chem Soc 115:3796
63. Nair LG, Fraser-Reid B, Szardenings AK (2001) Org Lett 3:317
64. Banoub J, Boullanger P, Lafont D (1992) Chem Rev 92:1167
65. Mootoo DR, Fraser-Reid B (1989) Tetrahedron Lett 30:2363
66. Klaffke W, Warren CD, Jeanloz RW (1992) Carbohydr Res 244:171
67. Debenham JS, Fraser-Reid B (1994) XVIIth International Carbohydrate Symposium, Ottawa, Canada
68. Debenham JS, Madsen R, Roberts C, Fraser-Reid B (1995) J Am Chem Soc 117:3302
69. Debenham JS, Fraser-Reid B (1996) J Org Chem 61:432
70. Fraser-Reid B, Anilkumar G, Nair LG, Olsson L, García Martín M, Daniels JK (2000) Isr J Chem 40:255
71. Svarovsky SA, Barchi JJ (2003) Carbohydr Res 338:1925
72. Feather MS, Harris JF (1965) J Org Chem 30:153
73. Andrews CW, Fraser-Reid B, Bowen JP (1993) Involvement of nσ* interactions in glycoside cleavage. In: Thatcher GRJ (ed) The anomeric effect and associated stereoelectronic effects. ACS Symposium Series 539, American Chemical Society, Washington, DC, p 114
74. Deslongchamps P (1983) Stereoelectronic effects in organic chemistry. Pergamon Press, New York, pp 30–35
75. Ratcliffe AJ, Mootoo DR, Andrews CW, Fraser-Reid B (1989) J Am Chem Soc 111:7661
76. Lafont D, Boullanger P (2006) Tetrahedron Assymmetry 17:3368
77. Madsen R, Roberts C, Fraser-Reid B (1995) J Org Chem 60:7920
78. Bodner R, Marcellino BK, Severino A, Smenton AL, Rojas CM (2005) J Org Chem 70:3988
79. Demchenko, AV, De Meo C (2002) Tetrahedron Lett 43:8819
80. Chapeau M-C, Marnett LJ (1993) J Org Chem 58:7258
81. Arasappan A, Fraser-Reid B (1995) Tetrahedron Lett 36:7967
82. Konradsson P, Roberts C, Fraser-Reid B (1991) Recl Trav Chim Pays-Bas 110:23

83. Velty R, Benvegnu T, Plusquellec D (1996) Synlett 817
84. Ferrières V, Bertho J-N, Plusquellec D (1995) Tetrahedron Lett 36:2749
85. Brennan PJ, Nikaido H (1995) Annu Rev Biochem 64:29
86. Mereyala HB, Hotha S, Gurjar MK (1998) 685
87. Lu J, Fraser-Reid B (2004) Org Lett 6:3051
88. Jayaprakash KN, Radhakrishnan KV, Fraser-Reid B (2002) Tetrahedron Lett 43:6955
89. Jayaprakash KN, Fraser-Reid B (2004) Synlett 301
90. Lu J, Fraser-Reid B (2005) Chem Commun 862
91. Fraser-Reid B, Lu J, Jayaprakash KN, López JC (2006) Tetrahedron Asymmetry 17:2449
92. Hölemann A, Stocket BL, Seeberger PH (2006) J Org Chem 71:8071
93. Barresi F, Hindsgaul O (1991) J Am Chem Soc 113:9337
94. Stork G, Kim G (1992) J Am Chem Soc 114:1087
95. Ito Y, Ogawa T (1994) Angew Chem Int Ed Engl 33:1765
96. Krog-Jensen C, Oscarson S (1996) J Org Chem 61:4512
97. Gelin M, Ferrières V, Lefeuvre M, Plusquellec D (2003) Eur J Org Chem 1285
98. Hay JE, Haynes LJ (1958) J Chem Soc 2231
99. Rousseau C, Martin OR (2000) Tetrahedron Asymmetry 11:409
100. Girard N, Rousseau C, Martin OR (2003) Tetrahedron Lett 44:8971
101. Ikeda K, Fukuyo J, Sato K, Sato M (2005) Chem Pharm Bull 53:1490
102. Tabeur C, Machetto F, Mallet J-M, Duchaussoy P, Petitou M, Sinaÿ P (1996) Carbohydr Res 281:253
103. Reichardt N-C, Martín-Lomas M (2005) Arkivoc ix:133
104. Anilkumar G, Nair LG, Fraser-Reid B (2000) Org Lett 2587–2589
105. Fraser-Reid B, López JC, Radhakrishnan KV, Mach M, Schlueter U, Gómez AM, Uriel C (2002) J Am Chem Soc 124:3198
106. Fraser-Reid B, Anilkumar GN, Nair LG, Radhakrishnan KV, López JC, Gómez AM, Uriel C (2002) Aust J Chem 55:123
107. Fraser-Reid B, López JC, Radhakrishnan KV, Mach M, Schlueter U, Gómez AM, Uriel C (2002) Can J Chem 80:1075
108. López JC, Gómez AM, Uriel C, Fraser-Reid B (2003) Tetrahedron Lett 44:1417
109. Fraser-Reid B, López JC, Gómez AM, Uriel C (2004) Eur J Org Chem 1387
110. Fraser-Reid B, López JC, Radhakrishnan KV, Nandakumar MV, Gómez AM, Uriel C (2002) Chem Común 2104
111. Wilson BG, Fraser-Reid B (1995) J Org Chem 60:317
112. Fraser-Reid B, Grimme S, Piacenza M, Mach M, Schlueter U (2003) Chem Eur J 9:4687
113. Uriel C, Gómez AM, López JC, Fraser-Reid B (2003) Synlett 2203
114. Mootoo DR, Konradsson P, Fraser-Reid B (1989) J Am Chem Soc 111:8540
115. Koenigs W, Knorr E (1901) Ber 34:957
116. Merritt JR, Fraser-Reid B (1992) J Am Chem Soc 114:8334
117. Merritt JR, Naisang E, Fraser-Reid B (1994) J Org Chem 59:4443
118. Lerouge P (1994) Glycobiology 4:127
119. Debenham JS, Rodebaugh R, Fraser-Reid B (1996) J Org Chem 61:6478
120. Rodebaugh R, Debenham JS, Fraser-Reid B (1996) Tetrahedron Lett 37:5477
121. Park MH, Takeda R, Nakanishi K (1987) Tetrahedron Lett 28:3823
122. Ferguson MAJ (1991) Curr Opin Struct Biol 1:52
123. Ferguson MAJ, Low MG, Cross GAM, (1985) J Biol Chem 260:14547
124. Ferguson MAJ, Homans SW, Dwek RA, Rademacher TW (1988) Science 239:753
125. Homans SW, Ferguson MAJ, Dwek RA, Rademacher TW, Anand R, Williams AF (1988) Nature 333:269
126. Campbell AS, Fraser-Reid B (1995) J Am Chem Soc 117:10387
127. Roberts C, Madsen R, Fraser-Reid B (1995) J Am Chem Soc 117:1546
128. Madsen R, Udodong UE, Roberts C, Mootoo DR, Konradsson P, Fraser-Reid B (1995) J Am Chem Soc 117:1554
129. Udodong UE, Madsen R, Roberts C, Fraser-Reid B (1993) J Am Chem Soc 115:7886
130. Bridges AJ (1996) Chemtracts Org Chem 9:215
131. Heidelberg T, Martin OR (2004) J Org Chem 69:2290
132. Clausen MH, Madsen R (2004) Carbohydr Res 339:2159
133. Clausen MH, Jørgensen MR, Thorsen J, Madsen R (2001) 543
134. Lemieux RU, Hendricks KB, Stick RV, James K (1975) J Am Chem Soc 97:4056

135. Arasappan A, Fraser-Reid B 1996 J Org Chem 61:2401
136. Hinou H, Umino A, Matsuoka K, Terunuma D, Takahashi S, Esumi Y, Kuzuhara H (2000) Bull Chem Soc Jpn 73:163
137. Yoshida T, Chiba T, Yokochi T, Onozaki K, Sugiyama T, Nakashima I (2001) Carbohydr Res 335:167
138. Seeberger PH, Haase WH (2000) Chem Rev 100:4349
139. Rodebaugh R, Fraser-Reid B, Geysen HM (1997) Tetrahedron Lett 44:7653
140. Rich DH, Gurwara SK (1975) J Am Chem Soc 97:1575
141. Rodebaugh R, Joshi S, Fraser-Reid B, Geysen HM (1997) J Org Chem 62:5660
142. Andrade RB, Plante OJ, Melean LG, Seeberger PH (1999) Org Lett 1:1811
143. Melean LG, Haase WC, Seeberger PH (2000) Tetrahedron Lett 41:4329
144. Hewitt MC, Seeberger PH (2001) J Org Chem 66:4233
145. Ratner DM, Plante OJ, Seeberger PH (2002) Eur J Org Chem 826
146. Hewitt MC, Zinder DA, Seeberger PH (2002) J Am Chem Soc 124:13434
147. Palmacci ER, Plante OJ, Hewitt MC, Seeberger PH (2003) Helv Chim Acta 86:3975
148. Mogemark M, Gustafsson L, Bengtsson C, Elofsson M, Kihlberg J (2004) Org Lett 6:4885
149. Buskas T, Söderberg E, Konradsson P, Fraser-Reid B (2000) J Org Chem 65:958
150. Bindschädler P, Noti C, Castagnetti E, Seeberger PH (2006) Helv Chim Acta 89:2591
151. Allen JR, Allen JG, Zhang X-F, Williams LJ, Zatorski A, Ragupathi G, Livingston PO, Danishefsky SJ (2000) Chem Eur J 6:1366
152. Allen JR, Danishefsky SJ (2000) J Prakt Chem 342:736
153. Allen JR, Ragupathi G, Livingston PO, Danishefsky SJ (1999) J Am Chem Soc 121:10875
154. Allen JR, Harris CR, Danishefsky SJ (2001) J Am Chem Soc 123:1890
155. Nishimura SI, Matsuoka K, Furuike T, Ishii S, Kurita K, Nishimura KM (1991) Macromolecules 24:4236
156. Nishimura SI, Furuike T, Matsuoka K, Maruyama K, Nagata K, Kurita K, Nishi N, Tokura S (1994) Macromolecules 27:4876
157. Miyagawa A, Kurosawa H, Watanabe T, Koyama T, Terunuma D, Matsuoka K (2004) Carbohydr Polymers 57:441
158. Yamada A, Hatano K, Koyama T, Matsuoka K, Esumi Y, Terunuma D (2006) Carbohydr Res 341:467
159. Rele SM, Iyer SS, Baskaran S, Chaikof EL (2004) J Org Chem 69:9159
160. Lu J, Fraser-Reid B, Chowda G (2005) Org Lett 7:3841
161. Fischer E (1893) Ber 26:2400
162. Fischer E (1895) Ber 28:1145
163. Lemieux RU, Morgan AR (1965) Can J Chem 43:2199
164. Lu J, Jayaprakash KN, Schlueter U, Fraser-Reid B (2004) J Am Chem Soc 126:7540
165. Mach M, Schlueter U, Mathew F, Fraser-Reid B, Hazen KC (2002) Tetrahedron 58:7345
166. Jayaprakash KN, Fraser-Reid B (2004) Org Lett 6:4211
167. de Raadt A, Ferrier RJ (1991) Carbohydr Res 216:93
168. Vankar YD, Vankar PS, Behrendt M, Schmidt RR (1991) Tetrahedron 47:9985
169. Marra A, Esnault J, Veyrières A, Sinaÿ P (1992) J Am Chem Soc 114:6354
170. Tebbe FN, Parshall GW, Reddy GS (1978) J Am Chem Soc 100:3611
171. Chenault HK, Castro A (1994) Tetrahedron Lett 35:9145
172. Chenault HK, Castro A, Chafin LF, Yang Y (1996) J Org Chem 61:5024
173. Lutsenko IF, Khomutov RV (1955) Doklady Akad Nauk SSSR 102:97
174. Petasis NA, Bzowej EI (1990) J Am Chem Soc 112:6392
175. Kunz H, Harreus A (1982) Liebigs Ann Chem 41
176. Boons GJ, Isles S (1994) Tetrahedron Lett 35:3593
177. Roy R, Andersson FO, Letellier M (2002) Tetrahedron Lett 33:6053
178. Boons GJ, Burton A, Isles S (1996) J Chem Soc Chem Commun 141
179. Boons GJ, Isles S (1996) J Org Chem 61:4262
180. Boons GJ, Heskamp B, Hout F (1996) Angew Chem Int Ed Engl 35:2845
181. Johnson M, Aries C, Boons GJ (1998) Tetrahedron Lett 39:9801
182. Bai Y, Boons GJ, Burton A, Johnson M, Haller M (2000) J Carbohydr Chem 19:939
183. Haller M, Boons GJ (2001) J Chem Soc Perkin Trans 1 814
184. Boons GJ, Burton A, Wyatt P (1996) Synlett 310

185. Capozzi G, Dios A, Franck RW, Geer A, Marzabadi C, Menichetti S, Nativi C, Tamarez M (1996) Angew Chem Int Ed Engl 35:777
186. Marzabadi CH, Franck RW (1996) Chem Commun 2651
187. Marzabadi CH, Franck RW (2000) Tetrahedron 56:8385
188. Capozzi G, Mannocci F, Menichetti S, Nativi C, Paoletti S (1997) Chem Commun 2291
189. Bartolozzi A, Capozzi G, Menichetti S, Nativi C (2001) Eur J Org Chem 2083
190. Petersen L, Jensen KJ (2001) J Org Chem 66:6268
191. Jensen KJ (2002) J Chem Soc Perkin Trans 1 2219
192. Koeners HJ, de Kok AJ, Romers C, van Boom JH (1980) Recl Trav Chim Pays-Bas 99:355
193. Sharma SK, Corrales G, Penadés S (1995) Tetrahedron Lett 36:5627
194. Petersen L, Jensen KJ (2001) J Chem Soc Perkin Trans 1 2175
195. Larsen K, Worm-Leonhard K, Olsen P, Hoel A, Jensen KJ (2005) Org Biomol Chem 3:3966
196. Tolborg JF, Jensen KJ (2000) Chem Commun 147
197. Laursen JB, Petersen L, Jensen KJ (2001) Org Lett 3:687
198. Petersen L, Laursen JB, Larsen K, Motawia MS, Jensen KJ (2003) Org Lett 5:1309
199. Laursen JB, Petersen L, Jensen KJ, Nielsen J (2003) Org Biomol Chem 1:3147
200. Grathe S, Thygessen MB, Larsen K, Petersen L, Jensen KJ (2005) Tetrahedron Asymmetry 16:1439
201. Kim KS, Lee YJ, Kim HY, Kang SS, Kwon SY (2001) J Am Chem Soc 123:8477
202. Kim KS, Lee ME, Cho JW (2004) Bull Korean Chem Soc 25:139
203. Kim KS, Jeon HB (2007) In: Demchenko AV (ed) Frontiers in modern carbohydrate chemistry. ACS Symposium Series, vol 960. ACS, Washington, DC, Ch 9
204. Demchenko AV (2003) Curr. Org. Chem. 7:35
205. Crich D, Sun S (1997) J Am Chem Soc 119:11217
206. Crich D, Sun S (1996) J Org Chem 61:4506
207. Crich D, Sun S (1997) J Org Chem 62:1198
208. Crich D, Sun S (1998) Tetrahedron 54:8321
209. Crich D, Smith M (2000) Org Lett 2:4067
210. Kim KS, Park J, Lee YJ, Seo YS (2003) Angew Chem Int Ed 42:459
211. Lee YJ, Lee K, Jung EH, Jeon HB, Kim KS (2005) Org Lett 7:3263
212. Lee R, Jeon JM, Jung JH, Jeon HB, Kim KS (2006) Can J Chem 84:506
213. Kunz H, Waldmann H (1984) Angew Chem Int Ed Engl 23:71
214. Kwon YT, Lee YJ, Lee K, Kim KS (2004) Org Lett 6:3901
215. Kim KS, Kang SS, Seo YS, Kim HJ, Lee YJ, Jeong KS (2003) Synlett 1311
216. Lee BY, Baek JY, Jeon HB, Kim KS (2007) Bull Korean Chem Soc 28:257
217. Lee YJ, Lee BY, Jeon HB, Kim KS (2006) Org Lett 8:3971
218. Lee YJ, Baek JY, Lee BY, Kang SS, Park HS, Jeon HB, Kim KS (2006) Carbohydr Res 341:1708
219. Nikolaev AV, Kochetkov NK (1986) Izv Akad Nauk SSSR Ser Khim 2556
220. Huchel U, Schmidt C, Schmidt RR (1995) Tetrahedron Lett 36:9457
221. Huchel U, Schmidt C, Schmidt RR (1998) Eur J Org Chem 1353
222. Hanessian S (1997) In: Hanessian S (ed) Preparative carbohydrate chemistry. Marcel Dekker Inc., New York, p 381
223. Hanessian S, Bacquet C, Lehong N (1980) Carbohydr Res 80:C17
224. Hanessian S, Lou B (2000) Chem Rev 100:4443
225. Lou B, Reddy GV, Wang H, Hanessian S (1997) In: Hanessian S (ed) Preparative carbohydrate chemistry. Marcel Dekker Inc., New York, p 389
226. Hanessian S, Condé JJ, Lou B (1995) Tetrahedron Lett 36:5865
227. Lou B, Huynh HK, Hanessian S (1997) In: Hanessian S (ed) Preparative carbohydrate chemistry. Marcel Dekker Inc., New York, p 413
228. Hanessian S, Mascitti V, Lu PP, Ishida H (2002) Synthesis 1959
229. Hanessian S, Lu PP, Ishida H (1998) J Am Chem Soc 120:13296
230. Lou B, Eckhardt E, Hanessian S (1997) In: Hanessian S (ed) Preparative carbohydrate chemistry, Marcel Dekker Inc., New York, p 449
231. Hanessian S, Huynh HK, Reddy GV, Duthaler RO, Katopodis A, Streiff MB, Kinzy W, Oehrlein R (2001) Tetrahedron 57:3281
232. Hanessian S, Saavedra OM, Mascitti V, Marterer W, Oehrlein R, Mak CP (2001) Tetrahedron 57:3267
233. Mukaiyama T, Hashihayata T, Mandai H (2003) Chem Lett 32:340

234. Hashihayata T, Mandai H, Mukaiyama T (2003) Chem Lett 32:442

235. Hashihayata T, Mukaiyama T (2003) Heterocycles 61:51

236. Jona H, Mandai H, Chavasri W, Takeuchi K, Mukaiyama T (2002) Bull Chem Soc Jpn 75:291

237. Hashihayata T, Mandai H, Mukaiyama T (2004) Bull Chem Soc Jpn 77:169

238. Mandai H, Mukaiyama T (2005) Chem Lett 34:702

239. Mandai H, Mukaiyama T (2006) Bull Chem Soc Jpn 79:479

240. Noyori R, Kurimoto I (1986) J Org Chem 51:4320

241. Higashi K, Nakayama K, Shioya E, Kusama T (1992) Chem Pharm Bull 40:1042

242. Higashi, K, Susaki H (1992) Chem Pharm Bull 40:2019

243. Inanaga J, Yokoyama Y, Hanamoto T (1993) Chem Lett 85

244. Davis BG, Word, SD, Maughan MAT (2002) Can J Chem 80:555

245. Lee JC, Pan GR, Kulkarni SS, Luo SY, Liao CC, Hung SC (2006) Tetrahedron Lett 47:1621

246. Hotha S, Kashyap S (2006) J Am Chem Soc 128:9620

247. Hotha S, Kashyap S (2006) J Am Chem Soc 128:17153

3.5 *S*-Glycosylation

Stefan Oscarson
Department of Organic Chemistry, Arrhenius Laboratory,
Stockholm University, Stockholm 106 91, Sweden
stefan.oscarson@ucd.ie

In: *Glycoscience.* Fraser-Reid B, Tatsuta K, Thiem J (eds)
Chapter-DOI 10-1007/978-3-540-30429-6_14: © Springer-Verlag Berlin Heidelberg 2008

Abstract

Published methods to synthesize thioglycosides have been reviewed. Two major pathways are recognized, either a displacement of an anomeric leaving group with a thiol acceptor to give the thioglycoside directly, or an initial formation of an anomeric thiol, which then is alkylated to give the thioglycoside. For the direct approach a listing is made based on the diverse anomeric leaving groups that have been utilized, acetates and halides being the most common. For the second pathway, various ways to form an anomeric thiol in a stereoselective manner are summarized, hydrolysis of anomeric thioacetates or pseudothiouronium salts being the most frequent, as well as different methodologies to perform the subsequent alkylation or arylation. Some miscellaneous pathways to thioglycosides, mainly through rearrangement of special O-glycosyl derivatives, are also discussed. In the last part an overview of the use of the discussed methodologies in the synthesis of thiooligosaccharides, i. e. oligosaccharides with interglycosidic thiolinkages, is presented.

Keywords

Thioglycosylation; Thioglycosides; Thiosaccharides; Synthesis

Abbreviations

AIBN	azobisisobutyronitrile
CAN	ceric ammonium nitrate
DMF	dimethylformamide
DTT	dithiothreitol
HMPA	hexamethylphosphoramide
IDCP	iodonium dicollidine perchlorate
Kdo	3-deoxy-D-*manno*-2-octulosonic acid
mCPBA	*meta*-chloroperoxybenzoic acid
NIS	N-iodosuccinimide
OTMS	octadecyltrimethoxysilane
TFA	trifluoroacetic acid
TMSO	N-trimethylsilyl-2-oxazolidinone
UDP	uridine diphosphate

1 Introduction

Carbohydrate derivatives with anomeric sulfur are not very common in nature, the only examples found are the various glucosinolates ("mustard oil glycosides") mainly from the Brassicaceae family [1,2,3], the simple alkyl thioglycosides of lincomycin and structurally related antibiotics found in *Streptomyces* species [4,5,6] and quite recently the methylthiomethyl and methylsulfonylmethyl thioglucosides from the Huaceae family [7] (❯ *Fig. 1*). However, there is an increasing interest in the use of thioglycosides, and other derivatives with anomeric sulfur, as efficient glycosyl donors, as ligands in chiral catalysts [8,9] and as stable *O*-glycoside analogues to be used in various biological context, e. g., as glycosidase inhibitors or vaccines [10], and, thus, their efficient syntheses are of great importance. The chemistry of thiosugars [1], their use as glycosyl donors [11,12,13,14], and the synthesis and biological application of thiooligosaccharides [15,16,17,18] have been reviewed extensively.

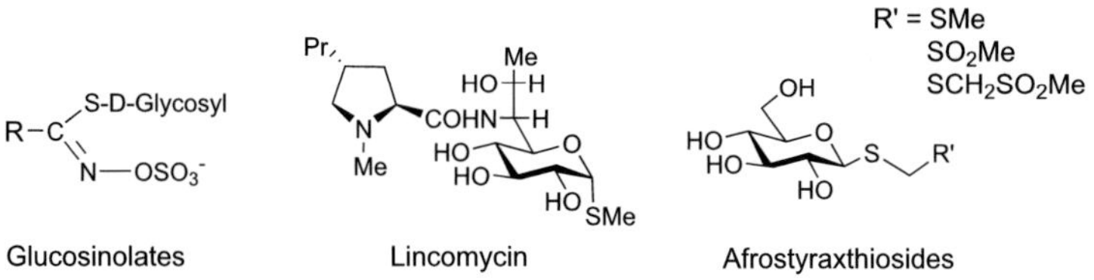

❏ Figure 1
Naturally occurring thioglycosides

Two general approaches to thioglycosides are used: either a direct introduction of a mercaptan through a displacement reaction of an anomeric leaving group, sometimes aided by a promoter, in a way similar to an *O*-glycosylation reaction (❯ *Scheme 1*), or a two (or more)-step procedure in which a non-thiol anomeric thio group is first introduced as above and then cleaved or rearranged in different fashions to give either directly a thioglycoside or an anomeric thiol or thiolate, which is then reacted with an alkyl electrophile to give the target thioglycoside (❯ *Scheme 2*). In this chapter the synthesis of thiomono- and thiooligosaccharides using these general pathways and also some other miscellaneous methods are discussed.

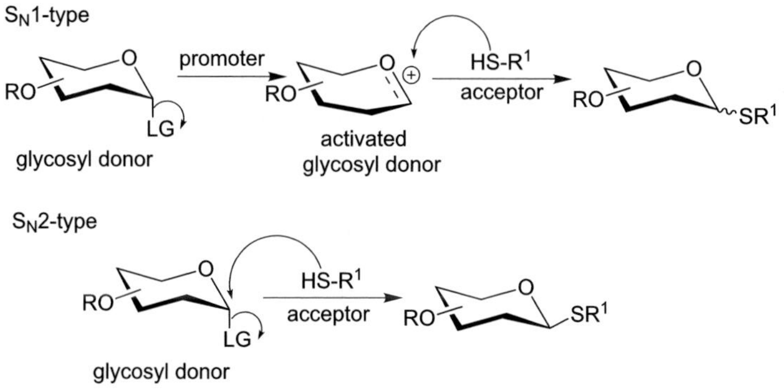

❏ Scheme 1

Two other types of glycosides worth mentioning in this context are glycosyl sulfoxides and selenoglycosides. Glycosyl sulfoxides were introduced as glycosyl donors by Kahne [19] and are often reviewed together with thioglycosides [11,12,14,20]. They are prepared by oxidation of the corresponding thioglycosides. Various reagents have been used, the most common

R^1 = e.g., Ac, CN, C(S)OMe

☐ **Scheme 2**

being *m*CPBA [19,21], also chiral glycosyl sulfoxides have been synthesized [22]. The interest in selenoglycosides, i. e., the equivalent of thioglycosides but with a selenium atom instead of a sulfur atom, has also increased recently, partly because they as well have been found to be good glycosyl donors, but with a different and complementary reactivity as compared to thioglycosides [23,24,25]. Thus, selenoglycosides can be activated in the presence of thioglycosides giving the possibility to build up thioglycoside building block donors. Also, selenoglycosides have proven to be very effective in helping and simplifying the phasing of protein-carbohydrate complex crystals in X-ray crystallography [26,27]. The syntheses of selenoglycosides follow the same general routes as are discussed for thioglycosides in this chapter [24,28]. Other thiosugars apart from thioglycosides (1-thiosugars), i. e., with the sulfur atom substituting another oxygen atom than the anomeric one and especially the ring oxygen, are also of interest as carbohydrate mimics [29]. 2-, 3-, 4- and 6-thiosugars are often synthesized as precursors (acceptors) for the synthesis of oligosaccharides containing a sulfur atom in the interglycosidic linkage (see ❷ *Sect. 3.1*).

2 Synthesis of Monothioglycosides

In the synthesis of *S*-monosaccharides both the above-mentioned pathways (❷ *Scheme 1*, ❷ *Scheme 2*) have been used extensively. The first step is the same in both routes: the displacement of an anomeric leaving group by a thionucleophile. Various leaving groups as well as thionucleophiles have been used. In ❷ *Sect. 2.1* is a listing organized by the nature of the leaving group. The emphasis will be on reactions with thiols and derivatives thereof as acceptors to give thioglycosides directly, but examples of other thio compounds, e. g., thioacetic acid, as acceptors are also included. The formation of the latter compounds and their transformation into proper thioglycosides, i. e., with a thiol aglycone part, will be predominantly discussed in ❷ *Sect. 2.2*.

2.1 Direct Synthesis

2.1.1 From Anomeric Acetates

An anomeric acetate is efficiently displaced by a thiol under the influence of an acidic catalyst, and this is probably presently the most used and efficient way to produce thioglycosides of simple mercaptans, especially on a large scale. The standard procedure is to react a peracetylated aldose dissolved in dichloromethane with a slight excess of thiol using a hard Lewis acid as promoter, which generally gives a high yield of mainly the 1,2-*trans* product (❷ *Scheme 3*) [11].

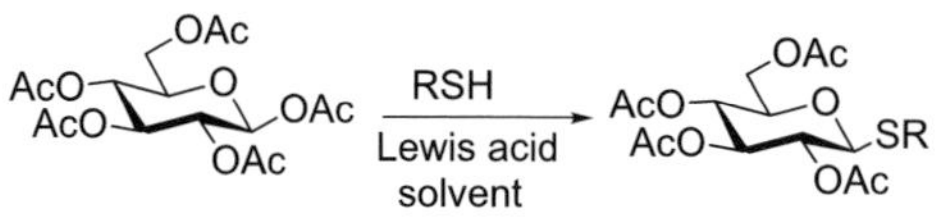

■ Scheme 3

Early studies employing hydrogen chloride as catalyst on perbenzoylated sugars showed complicated product mixtures [1,30], while Lewis acid-catalyzed mercaptolysis of peracetylated monosaccharides afforded the corresponding thioglycosides in good yields [31,32,33]. However, from a preparative point of view (smell, work-up), the present use of a solvent and only a small excess of thiol is definitely an advantage.

Various Lewis acids have been employed, most commonly BF$_3$-etherate [34,35], but also zinc chloride, stannic chloride, ferric chloride [36], zirconium chloride [37] and titanium chloride/indium chloride [38], with similar results. Recently 18 different oxometallic species were investigated as catalysts, and of these MoO$_2$Cl$_2$ (3 mol%) was found to be the most efficient affording thioglycosides in 75–94% yield [39].

β-Acetates react faster than α-acetates, which sometimes are difficult to induce to react at all [31,40]. With simple mercaptans the reaction can be performed almost stereospecifically to give the 1,2-*trans* product due to neighboring group participation, minor 1,2-*cis* impurities are often possible to remove by crystallization. However, if less reactive mercaptans [41,42,43,44] or sugar precursors (e. g., α-acetates or glucuronic acids [45,46]) are used, and, thus, prolonged reaction times have to be employed, more of the 1,2-*cis*-α product is generally formed, because of an anomerization of the kinetically favored β-thioglycoside to the more thermodynamically stable α-glycoside under the reaction conditions [47]. High stereoselectivity is also observed in the synthesis of thioglycosides of Kdo [48,49] and neuraminic acid [50] according to this approach, in spite of their 3-deoxy functions and thus lack of neighboring group participation. Both sugars yield preferentially the β-thioglycoside.

The approach works equally well with thio reagents other than mercaptans, e. g., thioacetic acid [37], thiourea [51] or O,O-dialkylphosphorodithioic acid derivatives [52], to yield the corresponding derivatives with an anomeric sulfur in high yields and stereoselectivity.

With the more reactive furanosides the per-benzoylated derivatives can also effectively be used as precursors (● Scheme 4) [53].

With more elaborate protecting group patterns, introduced to allow consecutive oligosaccharide synthesis, optimal conditions for the specific transformation often have to be worked out. Obviously, the conditions used are not compatible with acid-labile groups. Problems encountered might be cleavage of ether protecting groups, e. g. benzyl groups [54]. With the levulinoyl

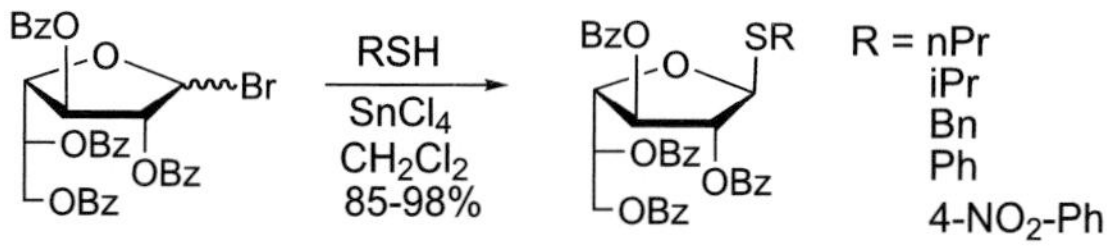

■ Scheme 4

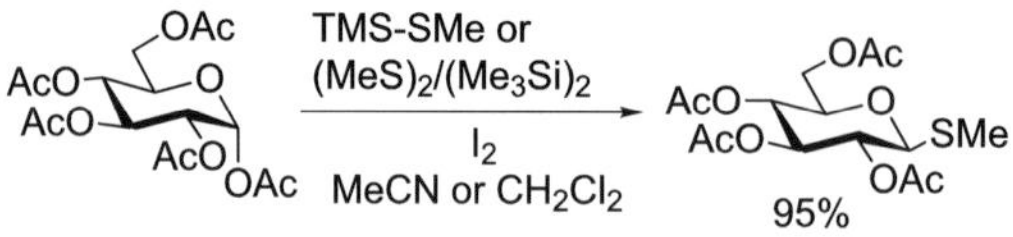

□ Scheme 5

protecting group, the keto function is transformed into a dithioacetal as expected, but this can be chemoselectively hydrolyzed in the presence of the thioglycoside by following treatment with $AgNO_3/Ag_2O$ [55]. However, as mentioned above, in most oligosaccharide synthetic pathways the thioglycoside is formed first using the per-acetylated derivative and subsequently the desired elaborate protecting group pattern is introduced.

One-pot syntheses, via per-acetylated intermediates, of acetylated thioglycosides from unprotected reducing sugars have been reported. Either the same (BF_3-Et_2O [56]) or two different Lewis acids ($Cu(OTf)_2$ or $LiClO_4$ followed by BF_3-Et_2O [57,58]) were used for the initial acetylation and the succeeding phenylthioglycosylation step. No solvent other than Ac_2O was necessary and the yields were in the 70–75% range. One-pot synthesis via acetylated halide sugar intermediates has also been described (see ❷ Sect. 2.1.2).

Variants of the method are the use of tributylstannyl [59] or trimethylsilyl [60] derivatives of the thiol, which give less odor and sometimes higher yields, due to less formation of 1,2-cis-products and dithioacetals, but these are also more expensive reagents. With the latter type of reagents α-acetates have also successfully been used as precursors using either zinc iodide [61] or iodine [62] as promoter. In the latter publication the authors could not detect any difference in reactivity between the α- and the β-acetate precursor under these reaction conditions. Furthermore, a cheaper version utilizing dimethyl sulfide/hexamethyldisilane as reagent was found to be equally efficient (❷ Scheme 5).

Also perbenzylated anomeric acetates have been employed as precursors using the stannylated thiol approach to give good yields of products, but with, as expected, low stereoselectivity [54,63]. A more preferred way to perbenzylated thioglycosides is to first construct the peracetylated thioglycoside stereospecifically and then change the acetyl groups for benzyl groups.

Modern laboratory techniques have been used to further improve the efficacy of these reactions. Thus, the use of standard conditions (peracetylated monosaccharides, thiophenol, BF_3-Et_2O, CH_2Cl_2) in combination with sonication gave the target 1,2-trans-thioglycosides in almost quantitative yield after less than 15 min, even when starting from the α-acetate [64]. With 2-acetamido-2-deoxy-sugars the reaction time was 2 h, but still only 1,2-trans-products were obtain in high yields (83–87%). Also microwave heating has been utilized but then with thiouronium salt precursors (see ❷ Sect. 2.2.2).

2.1.2 From Halogen Sugars

A classical route to thioglycosides is the reaction between an acetobromo sugar and a thiolate anion (❷ Scheme 6). The first thioglycoside ever synthesized was prepared by this method [65]. However, it is also a general method, and a good alternative to the use of peracetylated donors, and numerous thioglycosides, both alkyl and aryl have been prepared

☐ Scheme 6

☐ Scheme 7

accordingly [66,67,68]. Thioglycosides containing a heterocyclic aglycon, e. g. various nucleoside analogues, are preferentially synthesized from halogen sugars, although other precursors are also utilized [69]. The high nucleophilicity of the sulfur towards the anomeric position combined with its rather low basicity makes it possible to perform the reaction in acetone or methanol or even acetone-water. Since, especially with alkyl thiolates, de-*O*-acetylation is a frequent concomitant side-reaction, reacetylation prior to work-up is often necessary.

Usually a 1,2-*trans* product is obtained, possibly through participation of the 2-*O*-acyl group, but if the conditions are carefully selected a direct S_N2 displacement reaction can take place, to give, e. g., α-thioglucosides from β-halogen precursors or β-thiomannosides from α-halogen precursors, in spite of the presence of 2-*O*-participating groups. Thus, using an aprotic polar solvent (HMPA), the 1,2-*cis* *p*-nitrophenyl thioglycosides of mannose, galactose, and glucose were prepared (❷ *Scheme 7*) [70,71].

Instead of thiolate anions, stannyl sulfides have been utilized, which necessitate elevated temperature during the formation of the thioglycoside, due to the lower reactivity of the stannyl derivatives. However, if a promoter ($SnCl_4$) is used, reactions at room temperature or below can be performed, this, though, also causes anomerization and α/β product mixtures [59].

A variant is to use phase-transfer conditions, which allows the use of free thiols and non-polar solvents [72,73,74]. These conditions gave, starting from acetylated bromosugars, high yields of the corresponding 1,2-*trans*-thioglycoside both with various alkylthiols and phenylthiols, although ethyl mercaptan gave considerable deacetylation, which could be circumvented by using the perbenzoylated derivative instead. Interestingly, the β-chloroglucosyl donor gave exclusively the 1,2-thioorthoester under these conditions, which might indicate that an acyloxonium ion is not an intermediate in the reaction with α-bromo precursors [72]. If chlorinated solvents were used, competing displacement reactions with the solvents occurred, making toluene and ethyl acetate better choices as solvents [73].

In addition, thioglycosylations using halide-assisted conditions (❷ *Scheme 8*) [75] and Koenigs–Knorr conditions (❷ *Scheme 9*) [76], the latter both with stannyl sulfides and free thiols, have been described.

One-pot syntheses from unprotected reducing sugars via acetylated halide sugars have been described, too. Using iodine and iodine/HMDS as catalyst the iodosugar was prepared in situ and subsequently treated with dimethyldisulfide (compare ❷ *Scheme 5*) to give the acetylated

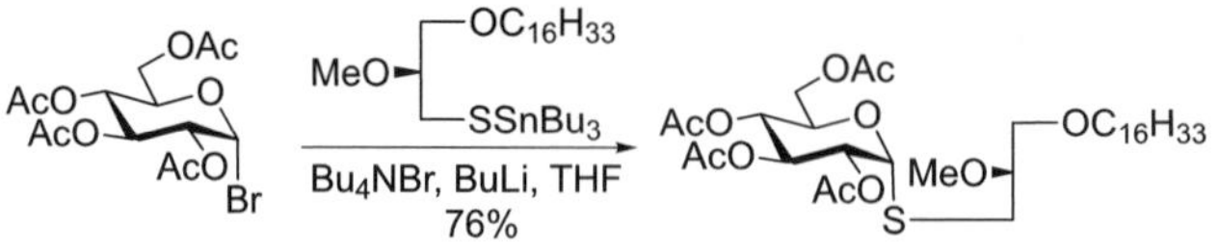

☐ Scheme 8

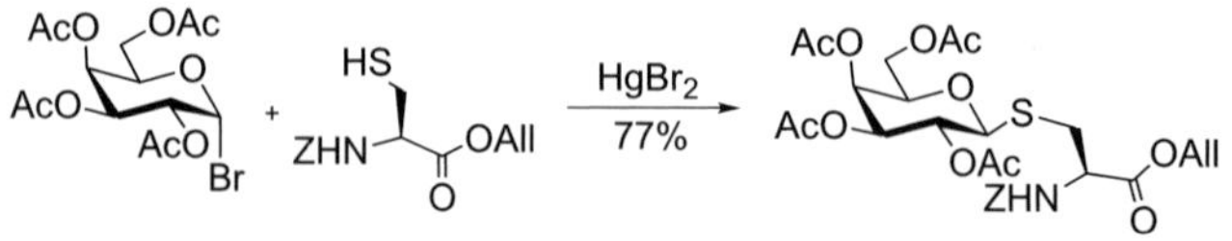

☐ Scheme 9

methyl thioglycoside in high yield (~80%) [77]. When HBr/HOAc were used to affect both acetylation and glycosyl bromide formation, the following thioglycosylation was performed under phase transfer conditions with various mercaptans (ethyl, phenyl, toluyl) to give the target thioglycosides in comparable high yields [78].

2.1.3 From Hemiacetals

The problem when making thioglycosides from hemiacetals, either a free sugar or a protected derivative with a free anomeric position, is that the higher nucleophilicity of sulfur, as compared to oxygen, facilitates the consecutive reaction with an additional thiol to give the dithiomercaptal. Thus, under mercaptolytic conditions (thiol, strong acid), most free sugars directly form the open type dithioacetals, some of these (especially mannose) after standing and at elevated temperature then producing thioglycosides in low yields. The isolated dithioacetal can sometimes be transformed in fair yield to the thioglycoside, but usually an equilibrium is established containing both the dithioacetal and the furanosidic and pyranosidic thioglycosides (❯ Scheme 10) [1]. However, with partially protected derivatives these ring-closures, promoted by standard thiophilic promoters (usually NIS), are often quite high-yielding and is a reaction now and then used as final step in de novo synthesis of thioglycoside monosaccharides (❯ Scheme 11 and ❯ Scheme 12) [79,80,81].

Another route to thioglycosides from hemiacetals involving a cyclization of an open-chain sulfide has been described recently. A Wittig–Horner reaction afforded the α,β-unsaturated thiophenyl derivative, which then was treated with iodonium reagents to induce cyclization to afford 2-iodo-1-thioglycosides, which are good precursors for 2-deoxy-sugars (❯ Scheme 13) [167].

In addition, a number of methods to synthesize thioglycosides direct from hemiacetals (especially perbenzylated) have been reported. Hence, treatment of fully protected hemiacetal monosaccharides (both furanoses and pyranoses) with a trialkylphosphine and a diaryl disulfide gave high yields of the corresponding aryl thioglycosides, predominantly with converted stereochemistry at the anomeric center, indicating a S_N2 type of reaction (❯ Scheme 14) [82,83,84].

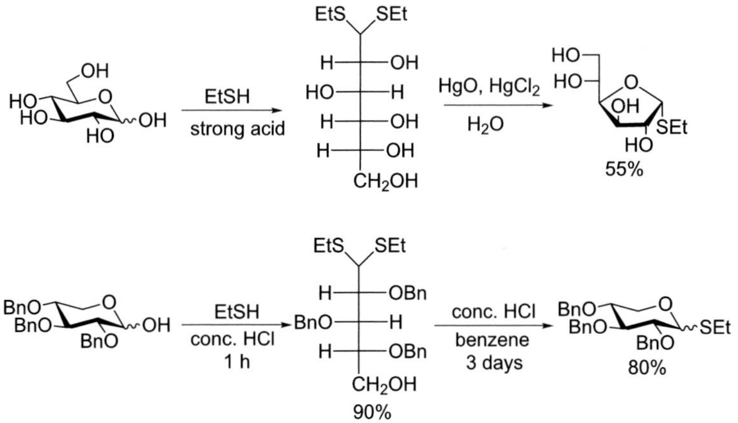

■ Scheme 10

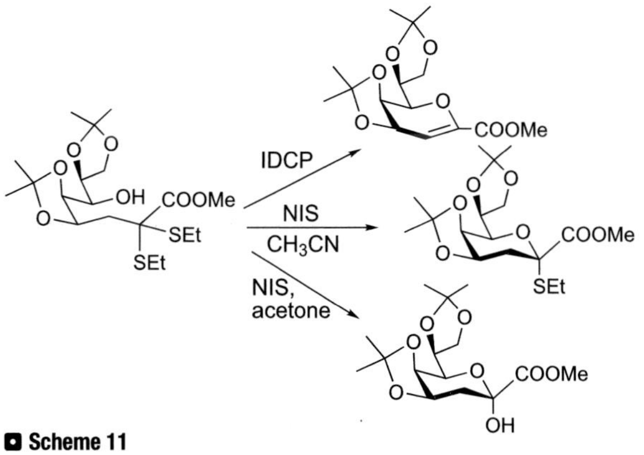

■ Scheme 11

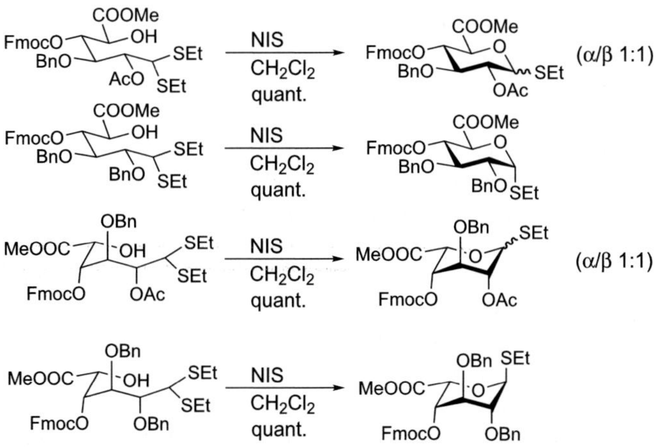

■ Scheme 12

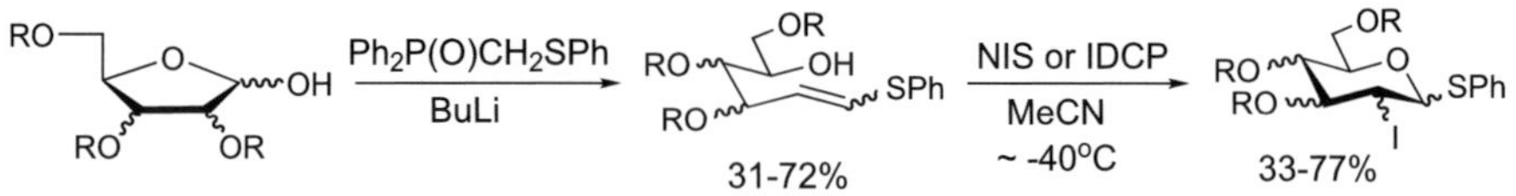

◻ Scheme 13

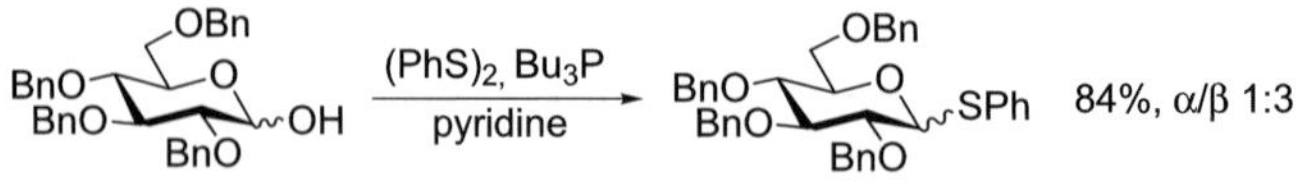

◻ Scheme 14

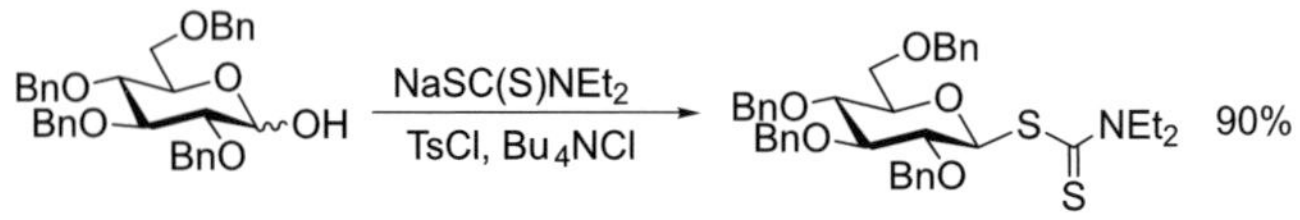

◻ Scheme 15

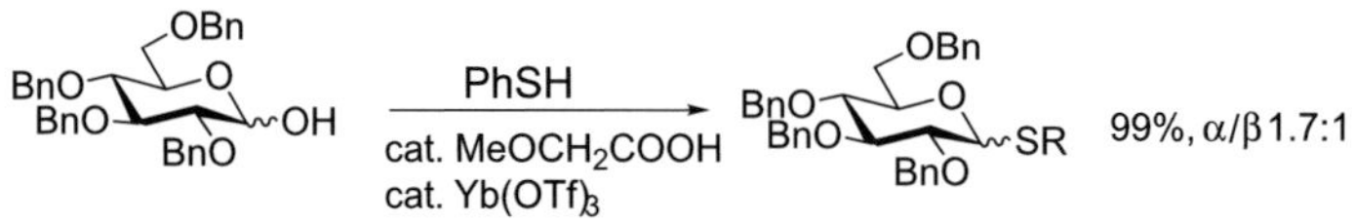

◻ Scheme 16

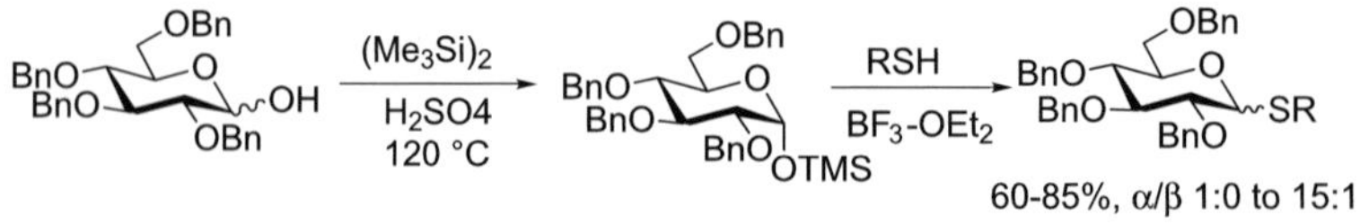

◻ Scheme 17

The use of diphenyl phosphorochloridate or tosyl chloride and sodium N,N-diethyldithiocarba-mates under phase-transfer conditions yields the S-glycosyl diethyldithiocarbamates through intermediate 1-O-diphenylphosphates or tosylates, respectively [85] (❷ Scheme 15).

Similarly, perbenzylated glucopyranose and ribofuranose almost quantitatively yield the phenyl thioglycoside when reacted with thiophenol and catalytical amounts of methoxyacetic acid and ytterbium triflate (❷ Scheme 16) [86].

An alternative is, of course, to convert the hemiacetals to isolable derivatives with bet-ter anomeric leaving groups, e. g., acetates (❷ Sect. 2.1.1), halogens (❷ Sect. 2.1.2), or trichloroacetimidates (❷ Sect. 2.1.4). 1-O-Trimethylsilyl derivatives can also be smoothly converted to the corresponding thioglycoside through the reaction with a thiol and BF_3-ether-ate (❷ Scheme 17) [83]. Since no participating groups were involved predominantly α-thio-glycosides were produced.

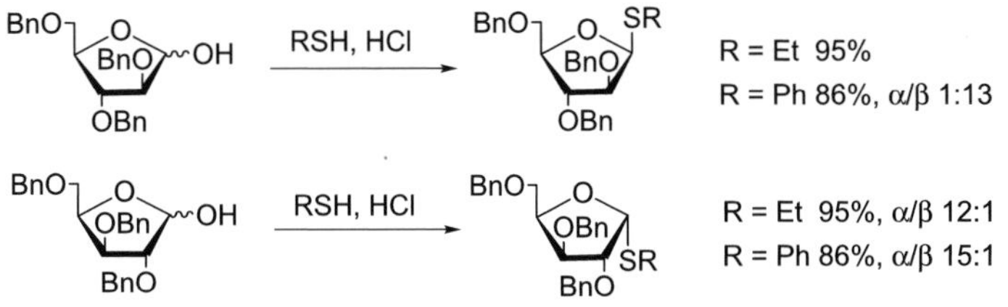

Scheme 18

In special cases also the use of a strong proton acid as catalyst can result in an efficient synthesis of the thioglycoside. Thus, perbenzylated pentofuranoses, when treated with ethyl or phenyl mercaptan and a catalytic amount of HCl, gave predominantly β-linked thioglycosides in high yields (Scheme 18), whereas perbenzylated arabinopyranose (as did unprotected arabinose) with ethyl mercaptan gave exclusively the dithioacetal, which was slowly converted to the ethyl thioarabinopyranoside upon standing (see Scheme 10) [87]. Perbenzylated hexopyranoses (glucose and galactose) were inert under these conditions.

2.1.4 From Trichloroacetimidates

Since the activation of trichloroacetimidates involves hard Lewis acids like BF_3-etherate or TMS-triflate, promoters which usually do not activate thioglycosides, it is possible to effectively prepare thioglycosides from this type of donor [88].

Both acetylated and benzylated precursors have been employed. In the *gluco*-series using the α-imidates, an alkylthiol, and BF_3-activation, the acetylated derivative gave, as expected, the β-thioglycoside, whereas the benzylated donor gave, surprisingly, exclusively the α-linked product (Scheme 19). Not even at low temperatures ($-42\,°C$) was any β-linked product observed, giving evidence that this reaction does not proceed through an S_N2-mechanism, which is the case using alcohol nucleophiles, and probably not through an S_N1-mechanism either, since this would give an α/β-mixture. The authors suggest a mechanism involving an intramolecular reaction of tight ion-pairs to give the retention of configuration. However, the use of a L-cysteine derivative as thionucleophile under identical coupling conditions gave a 5:1 α/β-mixture (Scheme 20) [89] (see also Sect. 3.1.2).

2.1.5 From Glycosides

Mercaptolysis (excess thiol (often ethyl mercaptan) in the presence of a strong proton acid) of a glycosidic linkage generally yields the open dithioacetal as the main product [1,90]. *O*-Glycosides can, however, be converted to *S*-glycosides in quite acceptable yields by treatment with a thiol or a thiotrimethylsilane and a Lewis acid. The conditions needed are still rather harsh and acid-sensitive protecting groups, including methyl (primary) and benzyl ethers, are often concomitantly cleaved [91,92,93]. Treatment of permethylated or persilylated methyl α-D-glucopyranoside with phenylthio(trimethyl) silane and zinc iodide gave the permethylated and the unprotected phenyl thioglycoside, respectively, as α/β-mixtures with the α-anomer predominating (Scheme 21) [91]. TMS-triflate has also been used as Lewis acid [92], as well as zinc bromide in the thiolysis of permethylated cyclodextrins [94].

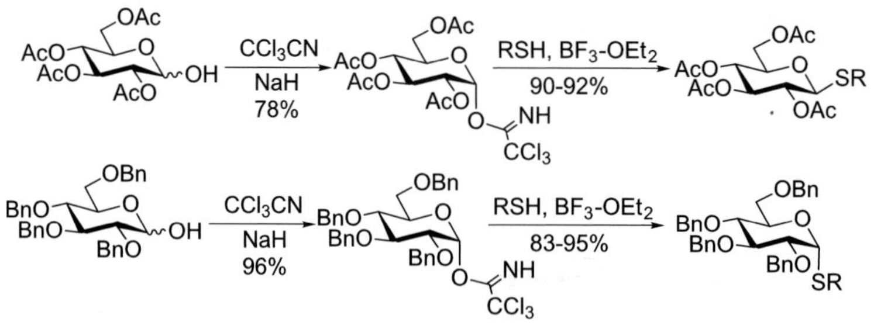

R = Pr, iPr, $(CH_2)_3Ph$, $(CH_2)_2COOMe$

Scheme 19

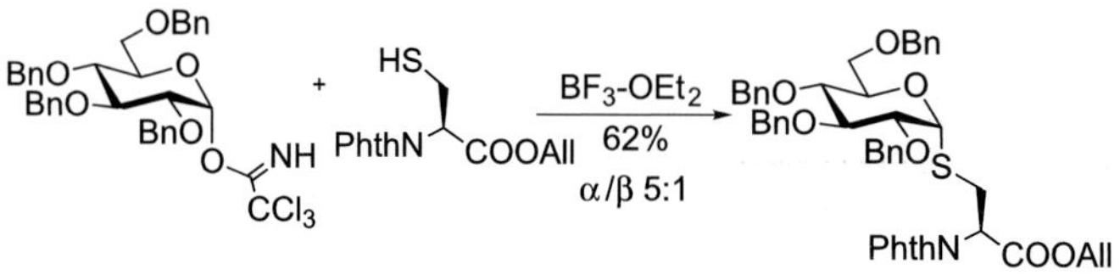

Scheme 20

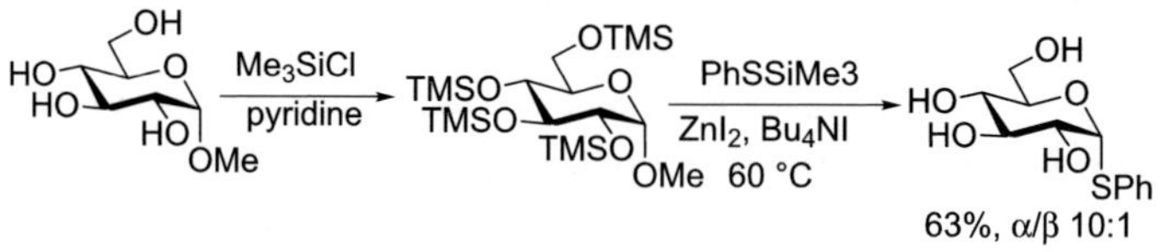

Scheme 21

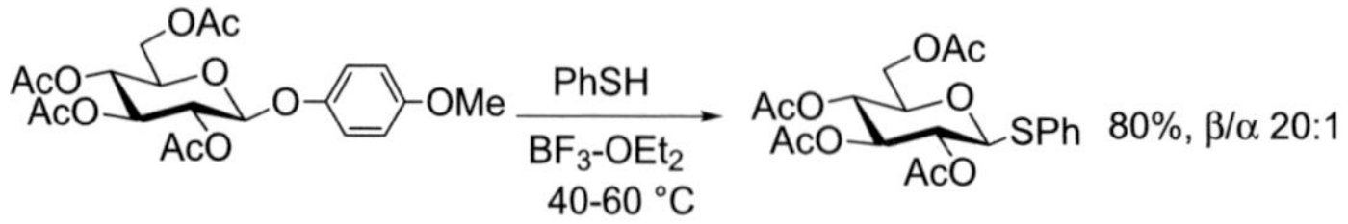

Scheme 22

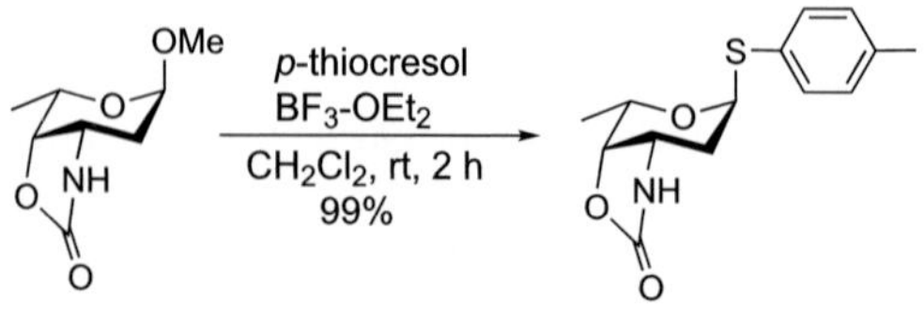

Scheme 23

With a more acid-labile *O*-glycoside, the transformation into a thioglycoside can be performed more efficiently. Accordingly, variously protected *p*-methoxyphenyl mono- and disaccharides have been smoothly converted into the corresponding phenyl thioglycosides using thiophenol and BF$_3$-etherate (● *Scheme 22*) [95]. With a 2-participating group the β-selectivity of the reaction is high, but, as above, with a non-participating group in the 2-position the α-anomer heavily prevails in the product thioglycoside.

Deoxy functions in the sugar ring also increase the acid lability of the glycoside. Thus, was the methyl glycoside of daunosamine (2,3,6-trideoxy-3-amino-L-*lyxo*-hexopyranose) efficiently transformed into the corresponding aryl thioglycoside by treatment with a thiol and BF$_3$-etherate (● *Scheme 23*) [96]. The reaction was highly α-selective.

2.1.6 From Glycals

Radical addition to peracetylated 2-hydroxyglucals proceeds efficiently using either photochemical initiation and thiols or *t*-butyl or cumene peroxides and thioacetic acid to give the 1-thio-α-D-*gluco* derivative (● *Scheme 24*) [97,98].

Radical addition to peracetylated glycals, however, yields the opposite regioselectivity and poorer stereoselectivity and 2-thio-D-glucitols and mannitols are here the normal products (● *Scheme 25*) [97,99,100].

Acid catalyzed reactions of glycals with thiols on the other hand, involve an allylic displacement reaction and a product mixture of the 2,3-dideoxy-1-thioglycoside (Ferrier rearrangement product) and the 3-*S*-alkyl-3-thioglycal is obtained (● *Scheme 26*) [101,102]. The ratio between these products is dependent both on the substrate, the thio compound, and the acid used. The thioglycoside is the kinetic product and prolonged treatment with SnCl$_4$ gives a thermodynamic equilibrium where the 3-thioglycal is dominant (about 95%). Shorter reaction time and the use of, e. g., trimethylsilylated thiols, BF$_3$-etherate, Sc(OTf)$_3$ [103] or ZrCl$_4$ [104] as Lewis acid gave almost exclusively the 1-thio compound (● *Scheme 27*). With thiols (also trimethylsilylated) preferentially α-thioglycosides were formed, but (thionoace-

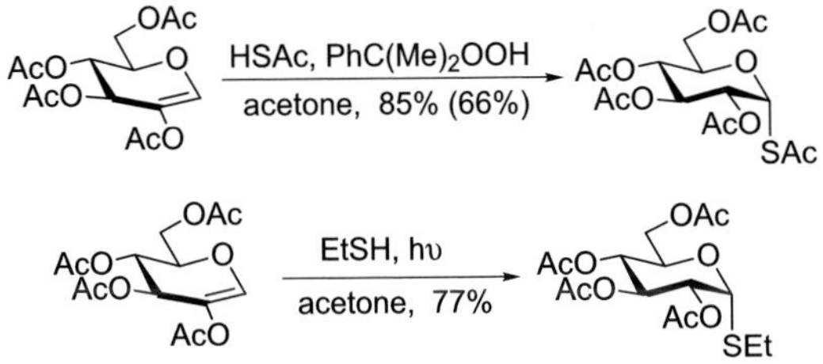

■ Scheme 24

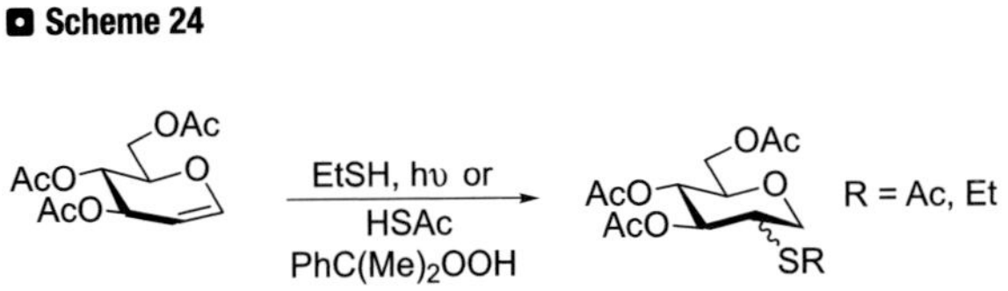

■ Scheme 25

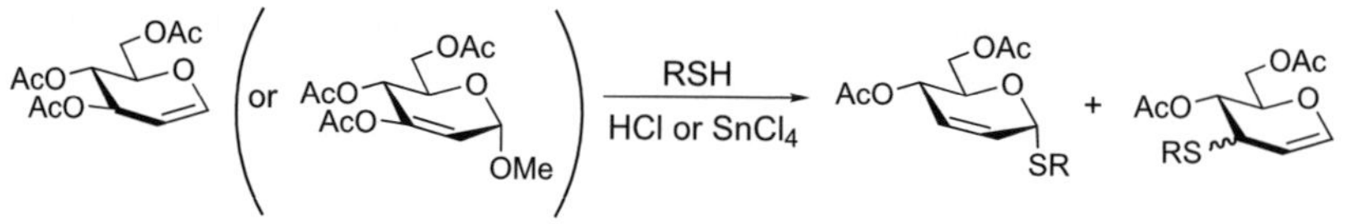

Scheme 26

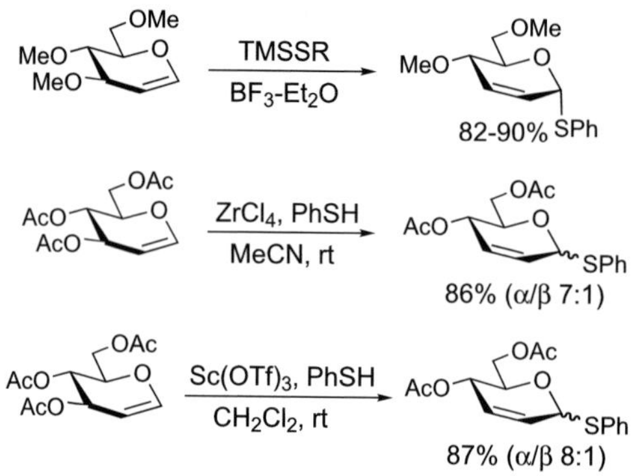

82-90% SPh

86% (α/β 7:1)

87% (α/β 8:1)

Scheme 27

72%, α/β 1:1.5

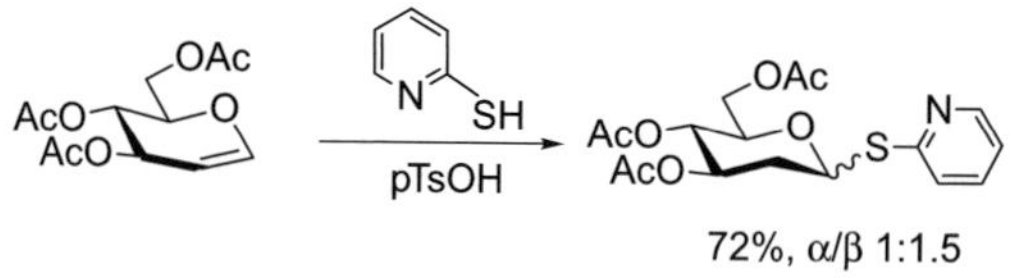

Scheme 28

toxy)trimethylsilane (Me$_3$SiOC(S)Me) gave good β-selectivity, especially with the poorer OMe leaving group (as compared to OAc).

Under some conditions the formation of the intermediate allylic cation can be prevented and direct addition to the double bond is the dominant reaction to give 2-deoxy thioglycosides. The treatment of tri-O-acetylglucal (and -galactal and -rhamnal) with 2-mercaptopyridine and p-toluenesulfonic acid resulted in an addition reaction and gave the 2-deoxythioglycoside in good yield (❂ Scheme 28) [105]. Use of BF$_3$-etherate as catalyst, however, gave the 2,3-dideoxy-1-thio-α-D-glycoside. Recently, use of CAN, a Re(V)-oxo complex, or GaCl$_3$ were reported to give good selectivity for the 2-deoxythioglycoside (❂ Scheme 29, ❂ Scheme 30 and ❂ Scheme 31) [106,107,108], although, as expected, the outcome is also dependent on the glycal and the thiol used.

Thioglycosides from glycals can also be obtained in a two-step procedure via a 1,2-anhydro intermediate (❂ Sect. 2.1.7).

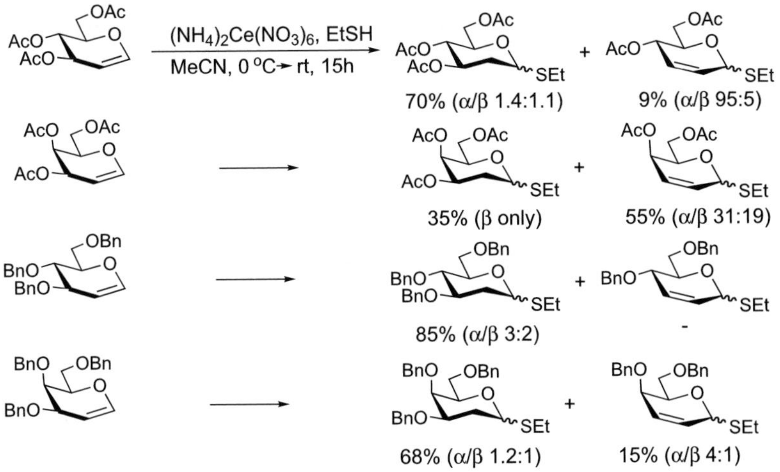

Scheme 29

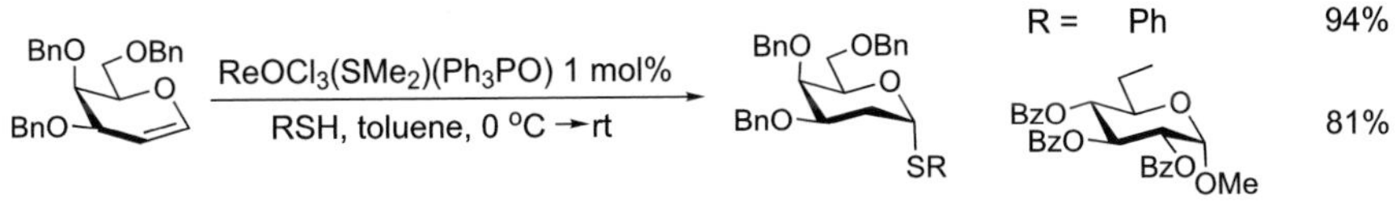

Scheme 30

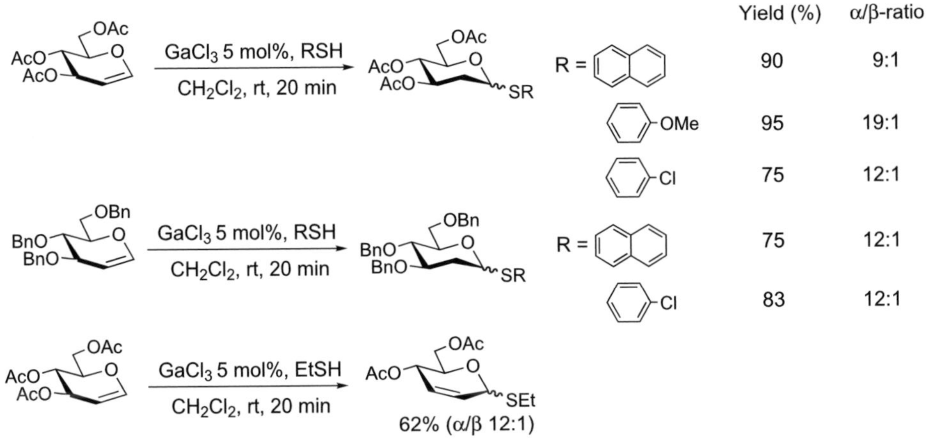

		Yield (%)	α/β-ratio

R =		Yield (%)	α/β-ratio
naphthyl		90	9:1
phenyl-OMe		95	19:1
phenyl-Cl		75	12:1
R = naphthyl		75	12:1
phenyl-Cl		83	12:1

Scheme 31

2.1.7 From Anhydro Sugars

Similar to glycosides, anhydro derivatives involving the anomeric oxygen, which can be considered as internal glycosides, can also be cleaved by a Lewis acid in the presence of a thiol, or a TMS-derivative thereof, to give thioglycosides. Examples with 1,2- and 1,6-anhydro sugars have been published (❯ *Scheme 32* and ❯ *Scheme 33*). The 1,2-anhydro precursors, obtained, e. g., from glycals, gave exclusively the 1,2-*trans* products [109,110], whereas the 1,6-anhydro precursors gave 1,2-*trans* products with a 2-participating group, but α/β-mixtures otherwise [111,112,113]. The ratio obtained was dependent both on the anhydro derivative, the thiol, and the promoter used and the proportion between these. Starting from a *gluco*-derivative the α-anomer was always predominant, but the use of TMSOTf as promoter gave much higher α/β-ratios (about 20:1) than ZnI_2 (2–5/1) [112].

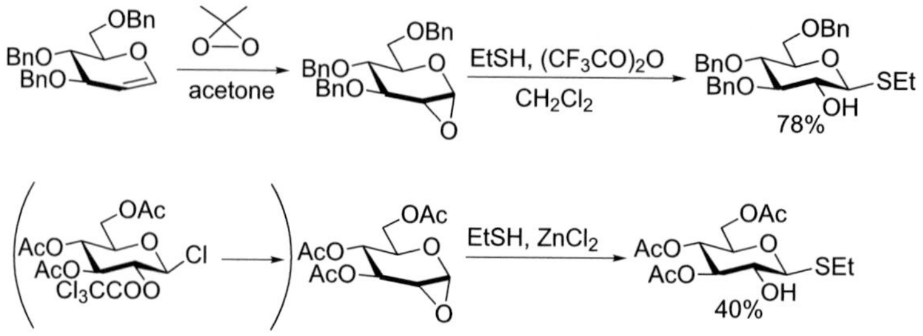

□ Scheme 32

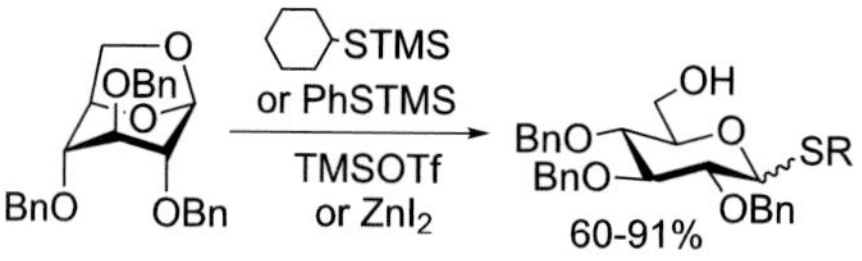

□ Scheme 33

The corresponding thio derivative, 1,6-dideoxy-1,6-epithio sugars, are not good precursors for thioglycosides, since the sulfur normally ends up in the 6-position when the sulfur bridge is opened [114]. However, with soft nucleophiles, like thiols, attack on the 6-position is observed to afford 1,6-dithio-glycosides (compare ❯ *Scheme 60*, 3.2.1) [115]. Also, using a carbene-mediated reaction a thioglycoside was produced in low yield via hydrolysis of an intermediate sulfur ylid (❯ *Scheme 34*) [116].

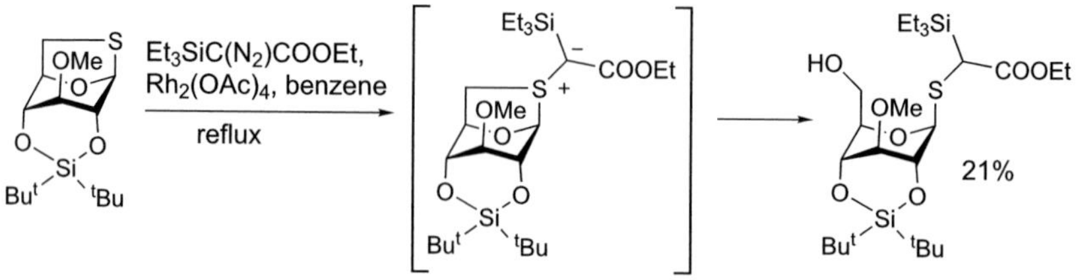

□ Scheme 34

2.2 Indirect Synthesis

The second main pathway to thioglycosides involves, as mentioned in the introduction, at least two steps: first the introduction of the anomeric sulfur using one of the ways described above, most often displacement of an anomeric halogen or acetate with a thionucleophile to give a 1-thio derivative, which then is directly transformed into a thioglycoside or cleaved to give a thioaldose or its thiolate salt, which finally is alkylated (❯ *Scheme 2*). Recently it was also shown that anomeric thiols can be produced directly from protected hemiacetal saccharides by treatment with Lawesson's reagent [197]. Sometimes the uncovered thiol is protected, e. g., by a xanthenyl [117] or a trityl [118] group, to allow various manipulations prior to re-exposure of the thiol and eventual formation of a thioglycoside. The 1-thio precursors mainly employed are glycosyl thioacetates, thiouronium salts, thiocyanates, and xanthates. Their syntheses, as well as their transformations into thioglycosides, are discussed below.

2.2.1 Via Glycosyl Thioacetates

Glycosyl thioacetates are prepared from glycopyranosyl halides or acetates employing either thioacetic acid or potassium thioacetate as reagent under various conditions as discussed above [1,37,119]. The higher nucleophilicity of sulfur, as compared to oxygen, is evident since only 1-thioacetates are obtained. As with mercaptans under controlled conditions, a S_N2-displacement reaction can be performed, e. g., to give the α-product from a β-chloro sugar (❯ *Scheme 35*) [70,120].

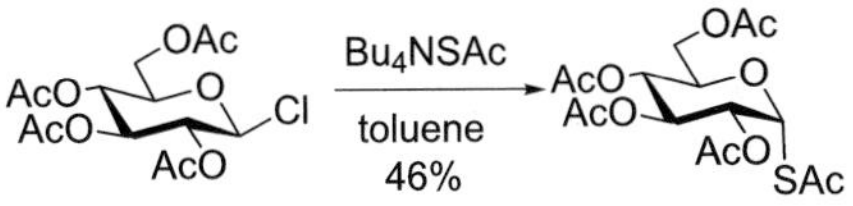

■ Scheme 35

Chemoselective cleavage of anomeric thioacetates to the thioaldose in the presence of *O*-acetyl groups is possible under various conditions, e. g., methanolic sodium methoxide at low temperatures [121], phenylmercury acetate followed by demercuration [122], 2-aminoethanethiol [123], or hydrazinium acetate [124], and these methods have therefore often been used in the synthesis of thiooligosaccharides (❯ *Sect. 3.2.1*). As mentioned, glycosyl thioacetates can also be prepared from the 2-acetoxyglycal derivative (see ❯ *Scheme 24*).

2.2.2 Via Glycosyl Pseudothiouronium Salts

These derivatives are also obtained by a displacement reaction of a sugar halide, this time with thiourea as nucleophile [125,126]. The higher nucleophilicity of sulfur as compared to nitrogen towards the (soft) electrophilic anomeric center ensures the sole formation of the 2-glycosyl-2-pseudothiourea salt (i. e., with the sulfur attached to the glycosyl moiety) which often can be isolated by direct crystallization. An advantage with these derivatives is their very mild conversion into thioaldoses (❯ *Scheme 36*). Thus, the amidino group can be cleaved with potassium carbonate or pyrosulfite or by simply heating the bicarbonate salt avoiding de-*O*-acetylation, which is often a concomitant reaction in the fission of the other anomeric

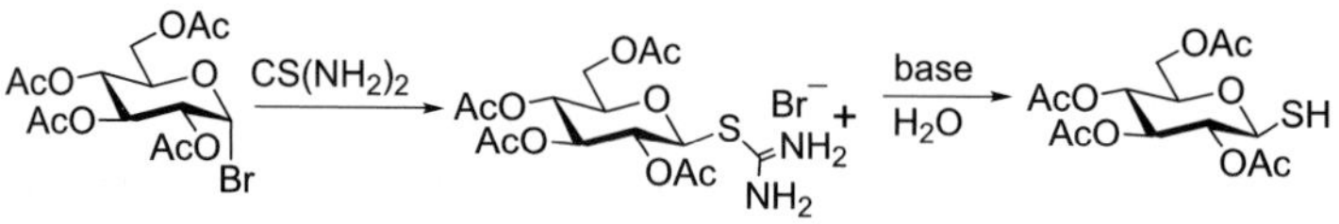

3

Scheme 36

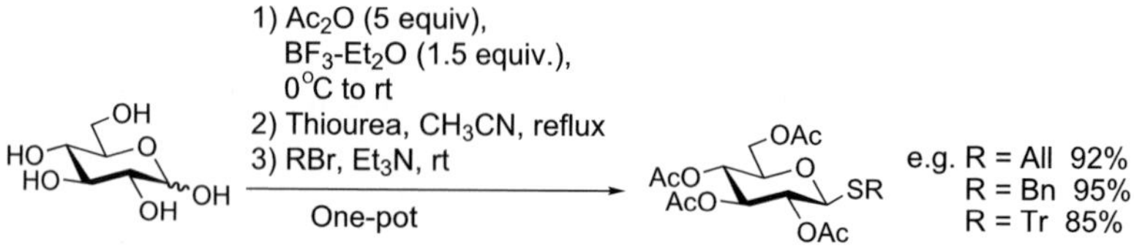

1) Ac$_2$O (5 equiv),
   BF$_3$-Et$_2$O (1.5 equiv.),
   0°C to rt
2) Thiourea, CH$_3$CN, reflux
3) RBr, Et$_3$N, rt

One-pot

e.g. R = All 92%
     R = Bn 95%
     R = Tr 85%

Scheme 37

thio groups discussed. This allows the easy and non-smelling formation of various acetylated alkyl thioglycosides on a large scale [125], a drawback is the use of the probably carcinogenic substances thiourea and alkyl iodide.

Recently, Et$_3$N/CH$_3$CN has been introduced as an efficient reagent for the conversion into the thiol. This procedure, furthermore, allows the in situ formation of an array of thioglycosides [127]. A one-pot protocol for the transformation of unprotected reducing sugars into thioglycosides via pseudothiouronium salts has been described [128]. In this protocol the thiouronium salt is formed directly from the in situ formed (BF$_3$-etherate/Ac$_2$O) peracetate by treatment with thiourea in refluxing acetonitrile (❯ Scheme 37). Also, microwave heating has been used to accelerate the formation of the thiouronium salt as well as the subsequent transformations into free anomeric thiol and thioglycoside [129,130].

2.2.3 Via Glycosyl Thiocyanates

Glycosyl thiocyanates are prepared from the corresponding bromo sugar through a displacement reaction using alkali thiocyanates as nucleophiles [131,132] (❯ Scheme 38). Peracylated bromo sugars give the 1,2-*trans* linked product, whereas perbenzylated derivatives in the *gluco*- and *galacto*-series give predominantly the 1,2-*cis* derivatives. However, if the reaction is carefully monitored the 1,2-*trans* product can be obtained in good yield even without a participating group [134]. A competing reaction is always the formation of the 1-isothiocyanates. If the silver [133,134] or lead salt [135], or potassium salt under phase-transfer [136] or solvent-free conditions [137] are used, the product will be the isothiocyanate, which also can be obtained by rearrangement of the thiocyanate by heating.

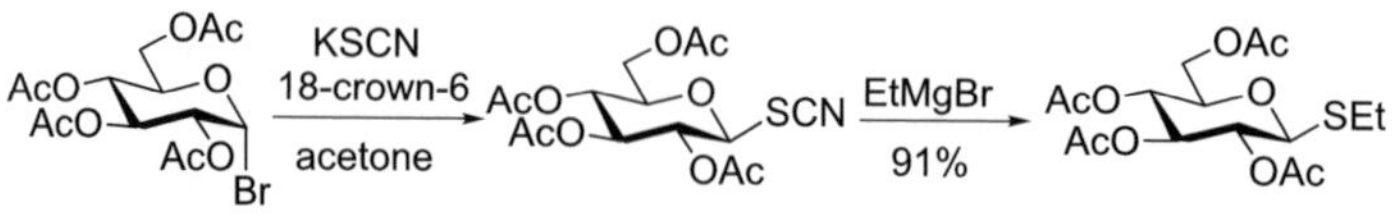

Scheme 38

The glycosyl thiocyanates can be transformed into thioglycosides through cleavage to the thioaldose by treatment with sodium methoxide under mild conditions followed by alkylation, or by a Grignard reaction which directly gives the thioglycoside (mannose-α-D-thiocyanate was inert) (● *Scheme 38*) [138].

2.2.4 Via Glycosyl Xanthates

Glycosyl xanthates (dithiocarbonates) are synthesized by the reaction of an acylated glycopyranosyl halide with a potassium alkyl (or benzyl) xanthate in an alcohol [139] or using phase-transfer conditions [140,141] (● *Scheme 39*). Also, glycosyl alkyl *N*-xanthates (dithiocarbamates) have been produced utilizing the same procedure with a bromo sugar and a sodium alkyl *N*-xanthate in DMF [142]. The glycosyl xanthates are converted to 1-thioaldoses by treatment with a base, generally ammonia or sodium methoxide, to allow later alkylation to afford the thioglycosides.

However, if the primary formation of the xanthate is performed with acetone instead of alcohol as solvent and at elevated temperature, the initially formed glycosyl xanthate will react further to give the bis(glycoside) sulfide and the alkyl thioglycoside as main products [143]. Thus, under these conditions the xanthate is cleaved to give an alkoxythiocarbonyl cation and the glycosyl thiolate, which then can react with the bromo sugar, to give the bis(glycosyl) sulfide, or an alkyl cation, obtained through decomposition of the alkoxythiocarbonyl cation, which then produces the thioglycoside (● *Scheme 40*, compare also ● *Scheme 51*) [144]. This rearrangement is most effective in polar solvents and with an added salt, e. g., NaI, also thermal decomposition is effective but slightly slower. To avoid the formation of bis(glycosyl) sulfides, the xanthate can first be isolated and then rearranged to give high yields of thioglycosides [45,144].

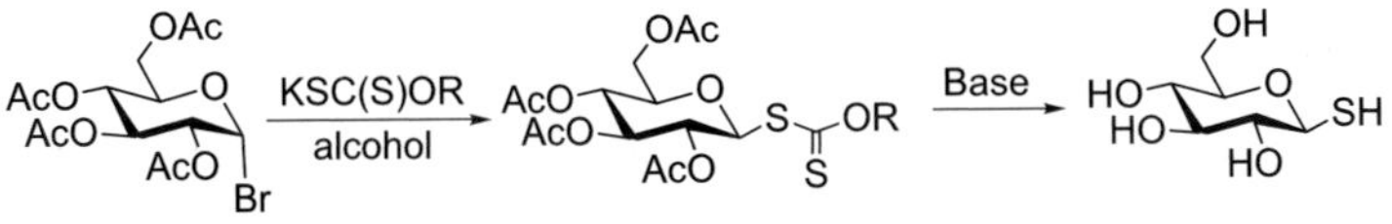

☐ Scheme 39

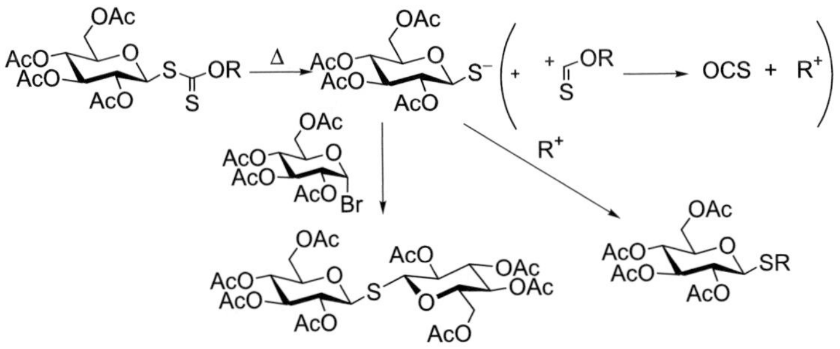

☐ Scheme 40

2.2.5 Via 1,2-Thiazoline Derivatives (2-Acetamido Donors)

Treatment of peracetylated 2-acetamido-β-D-gluco- or galacto-pyranose with the Lawesson's reagent gave the corresponding 1,2-thiazoline derivatives in quantitative yield (❷ Scheme 41). Consecutive hydrolysis using TFA in wet methanol yielded the corresponding anomeric α-thiols, which were then acetylated to the α-thioacetate or alkylated in various ways (compare ❷ Sect. 2.3 below) to give a number of different α-thioglycosides [145,146,147].

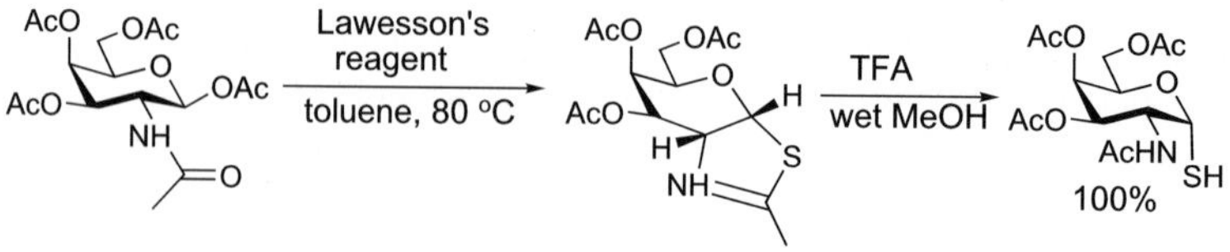

❏ Scheme 41

Another efficient way to 2-acetamido thiosugars is described from a 2-azido-1-chloro precursor. The β-thioacetate of 2-acetamido-2-deoxy-lactose was obtained by treatment of the 2-azido-2-deoxy α-chloride with a large excess of thioacetic acid, which both displaced the chloride and concomitantly reduced and acetylated the azido group (❷ Scheme 42) [148].

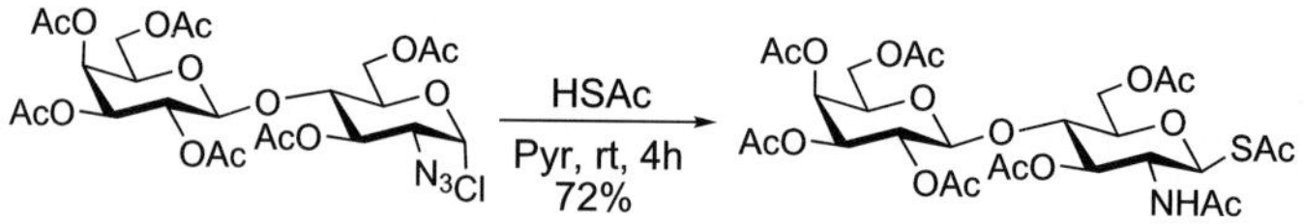

❏ Scheme 42

2.2.6 Via Glycosyl Thioimidocarbonates and Diglycosyl Disulfides

Other miscellaneous 1-thio derivatives convertible to thioaldoses are glycosyl thioimidocarbonates and diglycosyl disulfides [1]. Their syntheses are similar to the ones described above, i.e., displacement reactions of an acylated glycopyranosyl halide with a thionucleophile, an

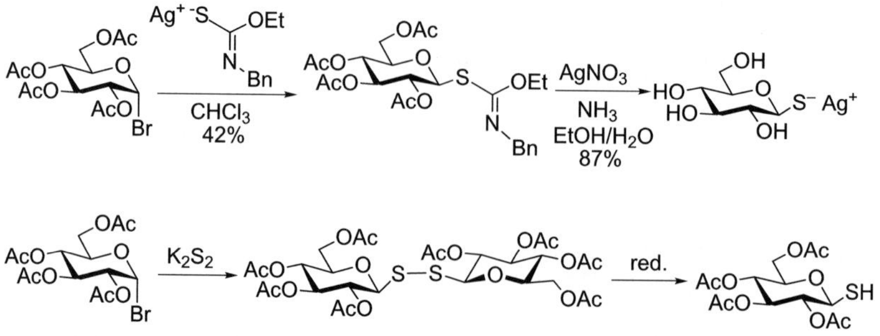

❏ Scheme 43

O-ethylbenzyl thioimidocarbonate, or potassium disulfide, respectively (❯ *Scheme 43*). Digly-cosyl disulfides have also been prepared using tetrathiomolybdate as the sulfur transferring reagent [149]. The former product is saponified, whereas in the latter product the disulfide bridge is reduced to give the thioaldose. With the tetrathiomolybdate reagent this could be performed in a tandem sulfur transfer/reduction/Michael reaction to afford thioglycosides directly in a one-pot reaction (❯ *Scheme 44*) [150].

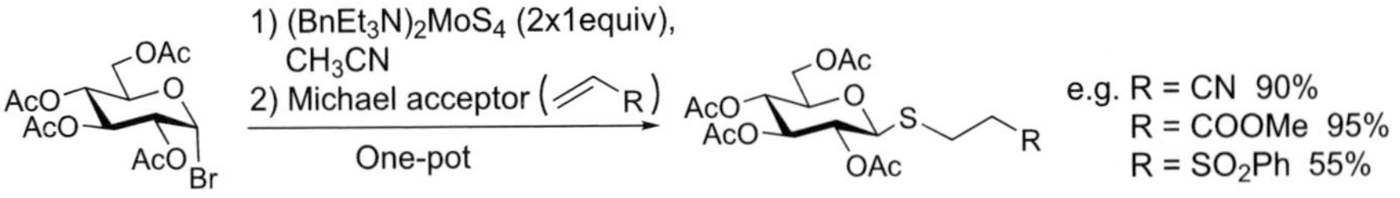

❐ Scheme 44

2.3 Alkylation of Thioaldoses

Thialdoses are easily *S*-alkylated to give the corresponding alkyl thioglycoside using alkyl iodides or bromides under basic conditions [1]. The pseudothiouronium salt derivative (❯ *Scheme 36*) can be used as precursor in a one-pot procedure, since the thioaldose is obtained in situ under these conditions [1,127]. Other leaving groups, like tosylates [151], have been used as well as a solid phase approach with the thioaldose attached to the resin (❯ *Scheme 45*, compare ❯ *Scheme 61*) [152]. Furthermore, thioalkylation using an alcohol and Mitsunobu conditions has been reported [153]. Arylations have been performed with aryl halides susceptible to nucleophilic displacement, e. g., 4-nitro- and 2,4-dinitrophenyl fluoride, in the presence of, e. g., potassium carbonate [127,154,155]. Also reactions with triazene substituted aryl iodides have been carried out using copper mediated arylation [156]. Unsubstituted iodobenzene did not react properly under the former conditions, but the thiophenyl glucoside can be obtained by refluxing the peracetylated thioaldose in DMF in the presence of phenylboronic acid and Cu(OAc)$_2$ [157].

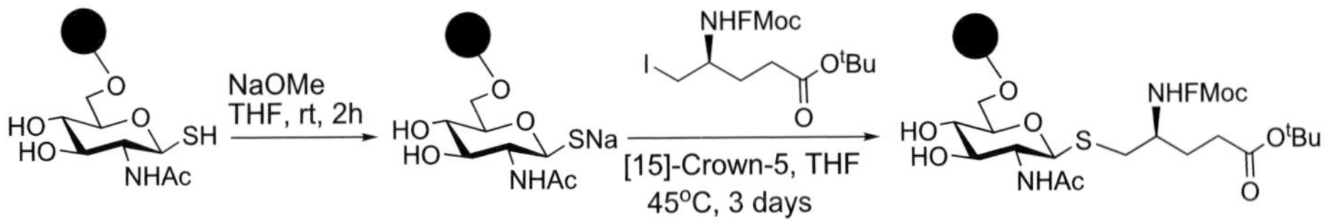

❐ Scheme 45

2.3.1 Via Aryldiazonium Derivatives

Apart from direct arylation using aryl halides, aryl thioglycosides can be prepared most efficiently by treatment of the thioaldose with an aryldiazonium salt followed by thermal decomposition of the produced diazo derivative (❯ *Scheme 46*) [158]. The procedure is usually very

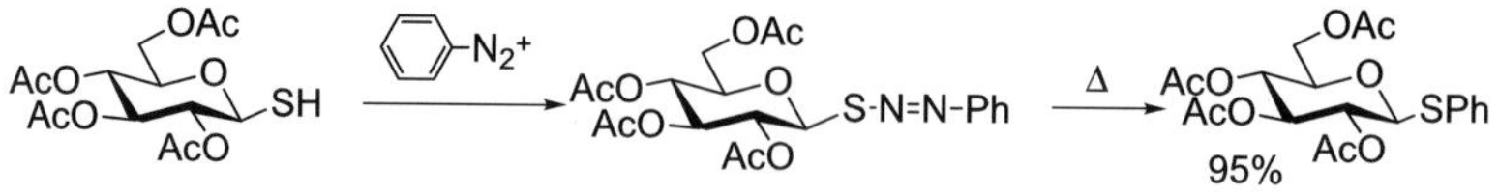

□ Scheme 46

high-yielding and is for many aryl groups (e. g., not containing electron-withdrawing groups) superior to the direct arylation pathway.

2.3.2 By Addition to Alkenes

The well-known reaction of radical thiol addition to unsaturated compounds has been utilized to produce thioglycosides from thioaldoses. A number of alkenes, including bifunctional ones, were coupled to various acetylated thioaldoses using AIBN as radical initiator to give alkyl and spacer thioglycosides in good to high yields (50–93%) (● Scheme 47) [159].
Addition of thiols to Michael acceptors is another classical reaction that has been used extensively to produce thioglycosides from the corresponding thioaldose nucleophile (● Scheme 44 and ● Scheme 48) [150,160]. Both monosaccharide and oligosaccharide thioglycosides (compare ● Scheme 59) have been prepared.

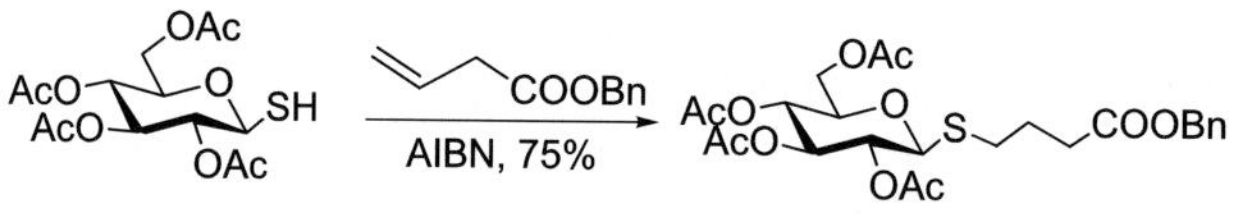

□ Scheme 47

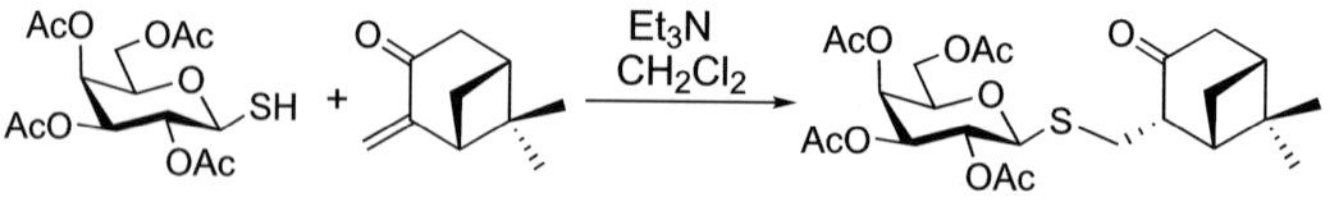

80% (2'S/2'R 9:1 mixture)

□ Scheme 48

2.4 Miscellaneous Methods

A few syntheses of thioglycosides by rearrangement of non-1-thio compounds have been reported.

2.4.1 Rearrangement of 1-Sulfenates

1-Sulfenates, prepared from the hemiacetal by treatment with sulfenyl chlorides and triethylamine, can be rearranged in the presence of a phosphorus(III) reagent, trialkyl phosphites or trialkylphosphines, to give the corresponding thioglycoside with inversed configuration (● Scheme 49) [161,162].

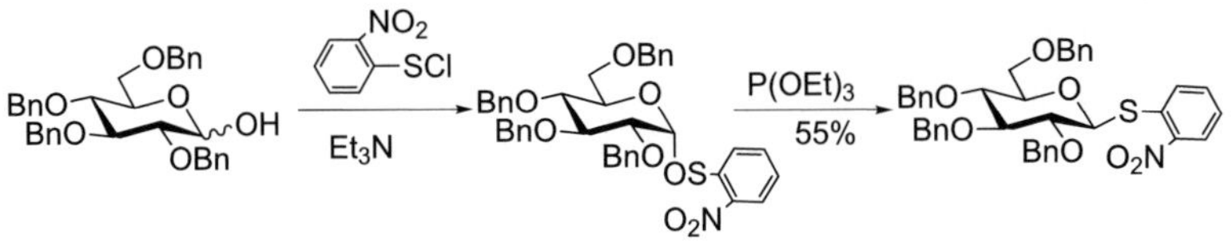

Scheme 49

The suggested mechanism involves an attack of the nucleophilic phosphorus reagent on the anomeric oxygen to afford an anomeric phosphonium ion and a thiolate counterion, which then attacks in a S_N2 reaction the anomeric position to give the thioglycoside and trialkyl phosphate or trialkylphosphine oxide.

2.4.2 Rearrangement of Thioorthoesters

Thioorthoesters, easily prepared from the corresponding acetobromo sugar by treatment with a thiol and collidine, can be rearranged to the 1,2-*trans*-thioglycoside (➋ *Scheme 50*). Deactivated Raney-nickel or TMSOTf have been used as promoters to give excellent yields of the thioglycosides of glucose, galactose, xylose, lactose, and glucuronic acid [163,164].

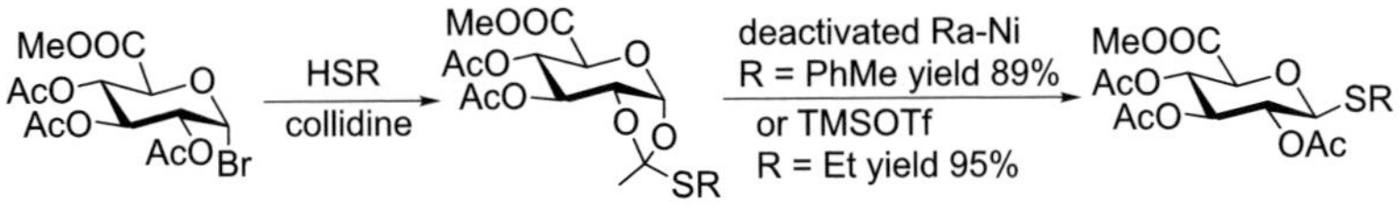

Scheme 50

2.4.3 Rearrangement of *p*-Nitrobenzoylxanthates

In a one-pot, two-step procedure anomeric thio *p*-nitrobenzoyl esters were formed from the corresponding free hemiacetal precursor via the rearrangement of in situ formed 1-*O*-(*S*-*p*-nitrobenzoyl) dithiocarbonate glycopyranose intermediates (➋ *Scheme 51*) [165,166].

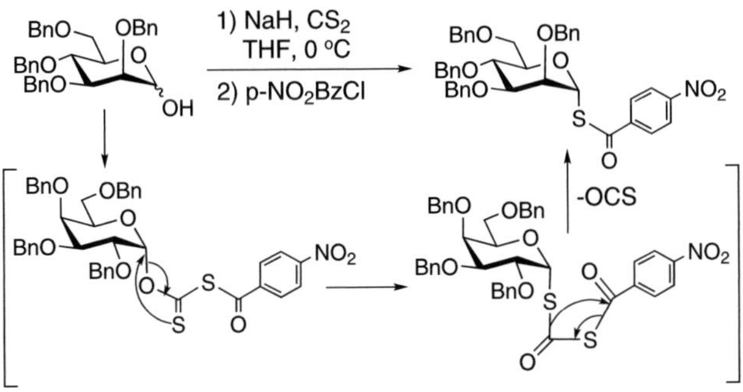

Scheme 51

The mechanism involves the rearrangement of the intermediate into the 1-*S*-(*S*-thio-*p*-nitro-benzoyl)dithiocarbonyl glycopyranose followed by a subsequent rearrangement with loss of carbon oxide sulphide to give the product (compare ❷ *Scheme 40*).

3 Synthesis of Thiooligosaccharides

As mentioned in the introduction, thiooligosaccharides (i. e., oligosaccharides in which at least one interglycosidic oxygen atom is substituted by a sulfur atom) are of great value as glycosyl-hydrolase stable analogues of *O*-glycosides and an increasing number of syntheses of complex thiooligosaccharides is continuously being reported [15,16,17,18,168].

The two general approaches to thiooligosaccharides are identical to the ones discussed above in the synthesis of thiomonosaccharides. Either a "normal" glycosylation is performed, but with a saccharide thiol acceptor, or the anomeric thio function is introduced first to yield an anomeric thiol or thiolate which then is reacted with a saccharide electrophile to form the inter-*S*-glycosidic linkage through a S$_N$2 displacement reaction. Both approaches are commonly used, but perhaps the second approach, where the anomeric conformation is more easily controlled, since it is constructed during the formation of the anomeric thiol function and not during the coupling reaction, has been the most abundant.

General problems, as compared to the formation of *O*-glycosides, are the incompatibility between catalytic hydrogenolysis and sulfur functions, which complicates the use of benzyl ethers as protecting groups, although Birch reduction might be an alternative, and the easy formation of disulfides from thiols, irrespective of if they are used as donors or acceptors.

An interesting new possibility is to use enzymes in the synthesis of thiooligosaccharides, a methodology that has been made possible by the development of "thioglycoligases" by point mutation of glycosidases (See ❷ *Sect. 3.3*).

3.1 Glycosylations with Thiol Acceptors

Under this heading methods according to ❷ *Scheme 1* will be discussed, i. e., with thiol saccharide acceptors. As compared to the synthesis of monosaccharides along this pathway, the formation of thiooligosaccharides is complicated by the fact that now the acceptor moiety is often the most laborious to construct and thus cannot, if necessary, be used in large excess to effect the glycosidic bond formation. Only glycosylation methods using promoters that are not thiophilic can obviously be used, and this disqualifies many *O*-glycosylation methods. More or less only the trichloroacetimidate method has been generally applied together with a few examples utilizing anhydro sugars and glycals as donors. Also thioglycosylation reactions under basic conditions have been employed quite frequently.

3.1.1 Anhydro Sugars as Donors

Benzylated 1,6-anhydroglucose has been used as donor in the construction of maltose and maltotriose structures with high α-selectivity (❷ *Scheme 52*) [112]. Interestingly, 1,6-anhydrodisaccharides were found to be unreactive as donors with 4-thiosaccharide acceptors, and could consequently be used as alternative acceptors.

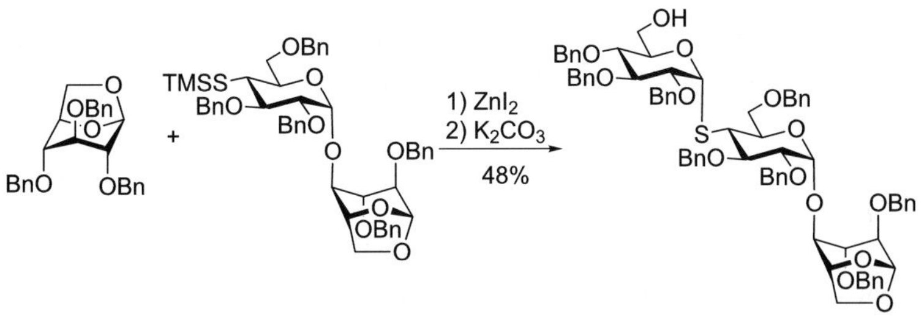

■ Scheme 52

3.1.2 Trichloroacetimidates as Donors

Trichloroacetimidates function well as donors also with saccharide thiol acceptors. A number of thiodisaccharides have accordingly been prepared using this method [169,170,171,172,173]. Possible by-products are various elimination products and orthoester formation. The stereochemical outcome is not easy to predict or control and, as so often in carbohydrate chemistry, an optimization of conditions is usually necessary for each glycosylation. For example, the use of an acetylated α-trichloroacetimidate glucose donor can yield not only an almost exclusively β-linked disaccharide product but also preferentially an α-linked product, depending on the acceptor and conditions used (❯ *Scheme 53*) (see also ❯ *Sect. 3.3*).

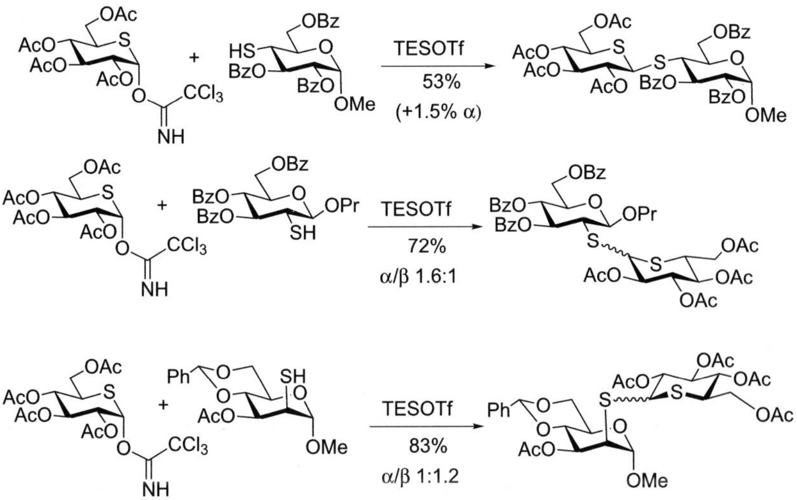

■ Scheme 53

3.1.3 Glycals as Donors

Addition reactions of saccharide thiols to glucals have been performed (❯ *Scheme 54*, compare with ❯ *Scheme 26*). The α-linked thiooligosaccharide was the main product but the β-anomer and the product resulting from a displacement of the 3-*O*-acetate were also isolated (ratio 5 : 1 : 1) [174].

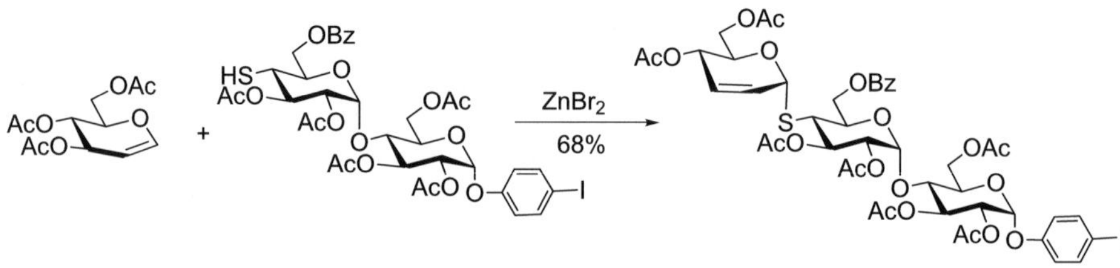

□ Scheme 54

3.1.4 Base-Promoted Glycosylations

Also the S_N2-displacement of an anomeric halogen by saccharide thiolates, even secondary ones, is quite feasible. Using this approach and α-acetobromo sugars for construction of β-linkages and β-acetochloro sugars for the construction of α-linkages, the thiodisaccha-ride analogues of gentiobiose, maltose, lactose, cellobiose, and galabios, have been synthe-sized in yields of around 50% [70,175,176,177]. Treatment of the β-chloro sugar of per-acetylated neuraminic acid with a 3-thiogalactopyranoside derivative in the presence of NaH and a crown ether yielded the α-thio-linked disaccharide in high yield and stereospecificity (❷ Scheme 55) [178].

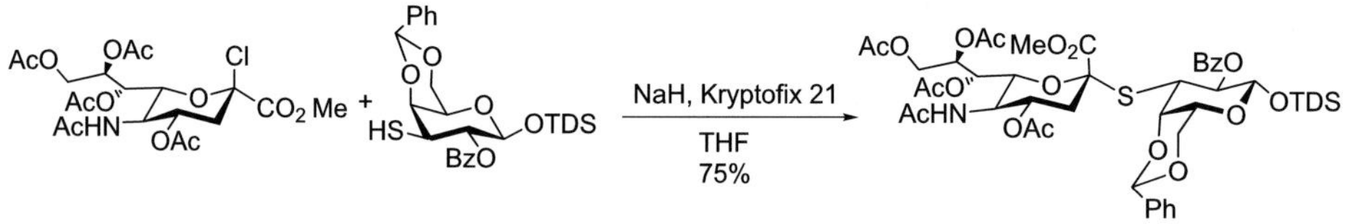

□ Scheme 55

3.2 Displacement Reactions with 1-Thioglycoses

The second approach to thiooligosaccharides follows that outlined in ❷ Scheme 2, i.e., an anomeric thiol (or thiolate) is first constructed and then reacted with a saccharide electrophile to give the thiooligosaccharide through an S_N2 displacement reaction. Problems to consid-er are the stereospecific introduction of the anomeric sulfur group, its (preferably selective) conversion into a thiol, and the nature of the electrophile. Side-reactions encountered in the couplings are elimination of the electrophile and anomerization of the thiol. In a model study of the synthesis of methyl 4-thiocellobioside, this approach was found to be superior to the one using thiol acceptors [179].

3.2.1 Formation of Anomeric Thiols

Of all the possible routes to anomeric thiols discussed in Sects. 1 and 2, almost exclusively thioacetates and thiourea salts have been used as intermediates in the synthesis of thiooligosac-charides, above all, since they can be selectively cleaved into the thiol in the presence of acyl protecting groups (❷ Sects. 2.2.1 and ❷ 2.2.2). The synthesis of the 1,2-trans-1-thio derivatives is usually straight-forward using participating protecting groups. However, also the construction of 1,2-*cis* compounds is usually performed using acetyl protecting groups, but then starting from 1,2-*trans* halogeno sugars employing conditions favoring S_N2-type reactions to give the 1,2-*cis* products. The anomeric thiol can either be isolated (with the concomitant problem of disulfide formation) and is then often activated as the thiolate during the coupling with the electrophile using, e. g., sodium hydride, or the cleavage and activation can be accomplished in situ. The displacement reaction is performed using an aprotic polar solvent or a phase-transfer reagent and non-polar solvents. Activation through complete deacetylation

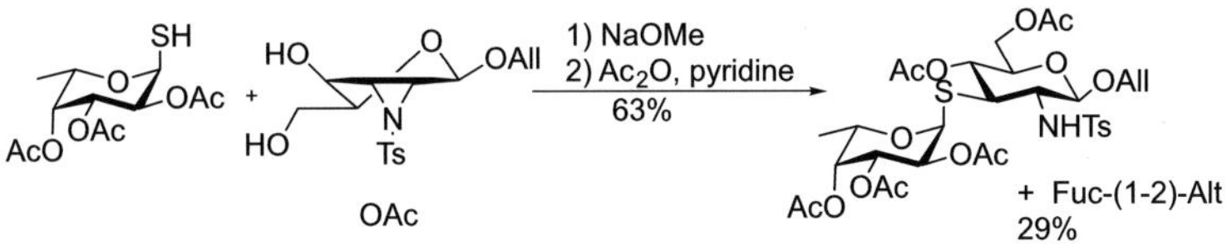

■ Scheme 56

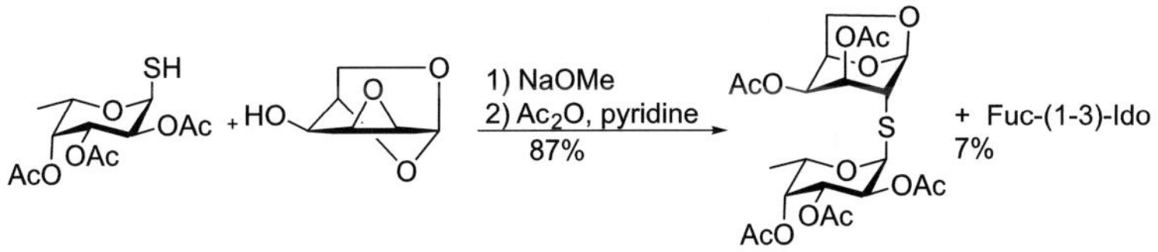

■ Scheme 57

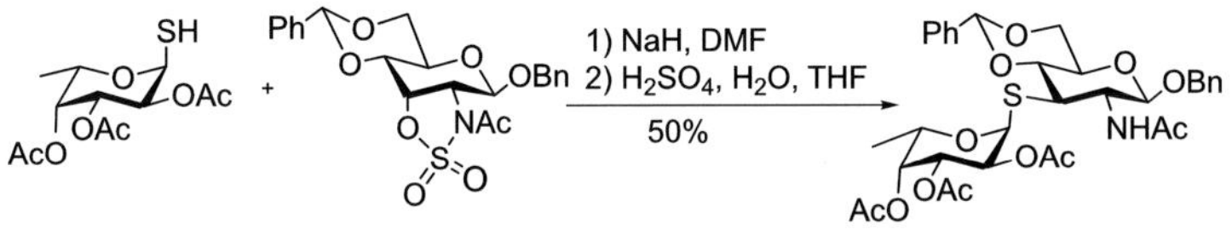

■ Scheme 58

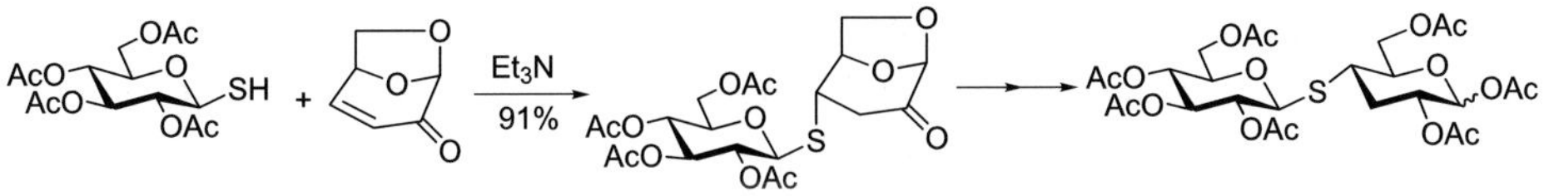

■ Scheme 59

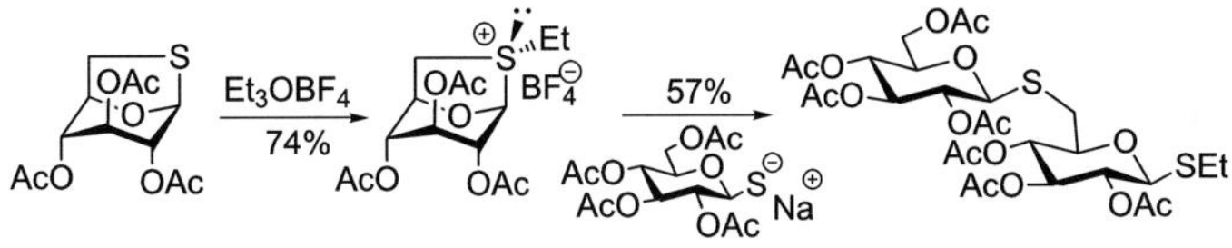

■ Scheme 60

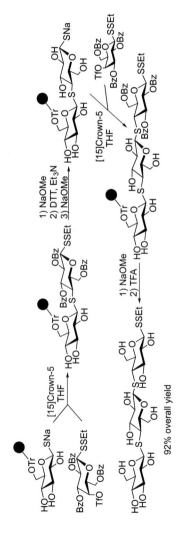

⬛ **Scheme 61**

and formation of the unprotected anomeric thiolate, utilizing the higher nucleophilicity (and lower basicity) of sulfur in the consecutive substitution reaction is also an often used alternative.

3.2.2 The Nature of the Electrophile (Leaving Group)

In the synthesis of (1→6)-linked structures, with a primary leaving group, even less effective leaving groups can be used. Bromides, iodides, tosylates, and triflates have been employed, bromides and tosylates requiring harsher conditions (sodium salt, high temperature) for reaction. Obviously there is also no problem with stereochemistry, neither in the construction of the electrophile nor in the displacement reaction. However, at secondary positions only triflates are good enough leaving groups to allow the displacement by anomeric thiolates and the construction of thiooligosaccharides. The substitution reactions yield the inverted configuration, which might pose a problem if the required electrophile has an unusual configuration. Accordingly, (1→4)-linked galactose and glucose derivatives are conveniently converted into each other, as is the case with (1→2)-linked glucose and mannose derivatives. For other oligosaccharides, for instance (1→3)-linked, a more laborious construction of the electrophile is often necessary, for example through a double inversion pathway.

Various other electrophiles have also been utilized, e. g., when their construction is more convenient than that of the corresponding triflate or when the triflate has failed to give the right product. Ring opening of a 2,3-*allo*-tosylaziridine, easily available from a GlcNAc precursor, with a 1-thiofucopyranosyl derivative gave with high regioselectivity the (1→3)-linked disaccharide (❯ *Scheme 56*) [180,181].

Similarly, ring opening of a 2,3-*talo*-oxirane with the same thionucleophile gave mainly the (1→2)-linked product (❯ *Scheme 57*) [180,181].

After failures in the synthesis and in the use of the 3-*O-allo*-triflate, the corresponding cyclic sulfamidate was constructed and reacted with the same 1-thiofucopyranosyl derivative to yield the desired (1→3)-linked thiodisaccharide (❯ *Scheme 58*) [182].

Other saccharide electrophiles utilized have been an α,β-unsaturated keto derivative and ethylated thiolevoglucosan. The former yielded in a Michael addition with tetra-*O*-acetyl-1-thio-β-D-glucopyranose the (1→4)-linked disaccharide in an excellent yield (❯ *Scheme 59*) [183,184]. The latter, when reacted with the sodium salt of the above glucose thiol, gave the (1→6)-linked disaccharide as its thioethyl glycoside (❯ *Scheme 60*) [185].

◻ Scheme 62

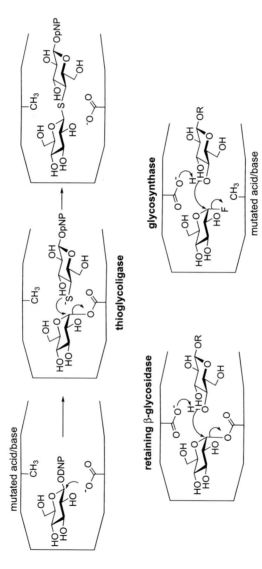

⬛ **Scheme 63**

3.2.3 Solid-Phase Synthesis of Thiooligosaccharides

Solid-phase syntheses of thiooligosaccharides in excellent yields have been described [186]. Unprotected monosaccharides with an anomeric ethyl disulfide group were attached to a solid phase via their primary hydroxy group. Activation through the formation of the sodium thiolate and subsequent reaction with a saccharide triflate gave thiodisaccharides, which can be cleaved from the resin by treatment with TFA. If the electrophile contains an anomeric ethyl disulfide group, the reaction sequence can be repeated to give thiooligosaccharides (❯ *Scheme 61*).

3.2.4 Synthesis of Non-Reducing Sugars

The syntheses of thio analogues of non-reducing disaccharides could be categorized under any of the above two headings since they involve both an anomeric thiol and an anomeric electrophile. Both halo sugars and hemiacetals have been used as electrophiles (❯ *Scheme 62*) [70,187].

3.2.5 Enzymatic Synthesis

Attempts to use glucosidases to form interglycosidic thioglycoside linkages from thiol precursors have not been successful. Neither was the use of glycosynthases, mutant glycosidases optimized to catalyze the formation and not the hydrolysis of *O*-glycosidic bonds where one (the nucleophilic) of the two carboxylate residues in the active site of the glycosidases has been changed to a non-nucleophilic (methyl) group. However, when the other carboxylic acid

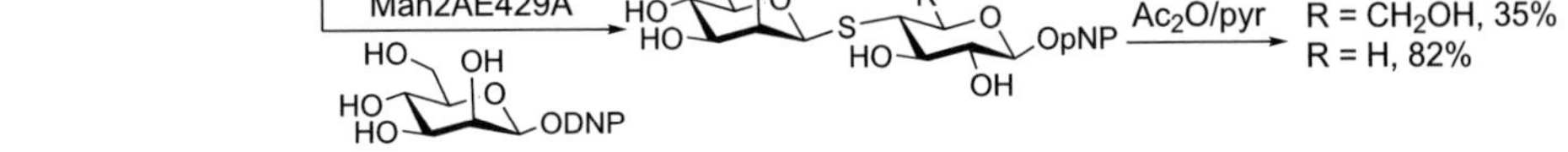

◻ Scheme 64

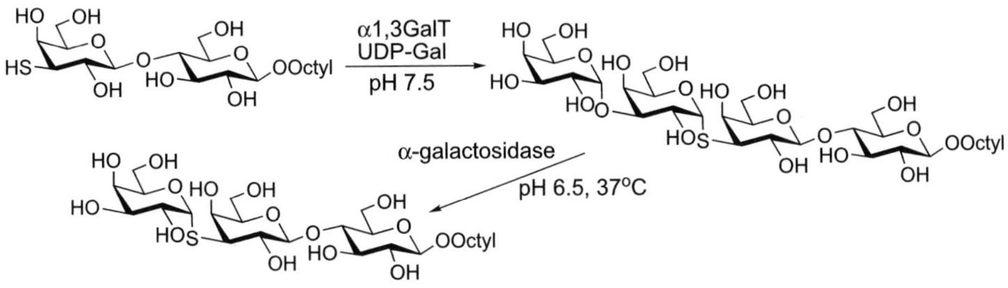

◻ Scheme 65

residue (the general acid/base) was mutated, a new type of mutant glycosidases was created, named thioglycoligases by the developers, which were able to catalyze the reaction between a glycosyl donor and a thiol acceptor to efficiently give thioglycosidically linked oligosaccharides (❷ *Scheme 63*) [188]. Further studies showed that also the "double mutant" (called thioglycosynthases), where both the carboxylates were removed could catalyze this type of reaction [189]. So far only β-(1→4)-linkages have been produced using glucose and xylose acceptors and glucose and mannose donors (❷ *Scheme 64*) [190,191,192,193].

Interestingly, glycosyl transferases seem to tolerate thiol acceptors. Thus, using an α1,3-Gal-transferase and an octyl 3′-thio-lactoside acceptor a thio-interglycosidic linkage was formed in a very high yield. Surprisingly, the product was not the expected trisaccharide, but a tetrasaccharide with an extra (1→3)-*O*-linked galactose residue added. Here, the thiolinkages stability towards glycosidases could be utilized to selectively cleave the Gal-(1→3)-Gal *O*-glycosidic bond to obtain the desired trisaccharide product (❷ *Scheme 65*). In addition, MS-data were reported to show that the 3′-thiolactoside was also accepted by a β1,3GlcNac-transferase [194].

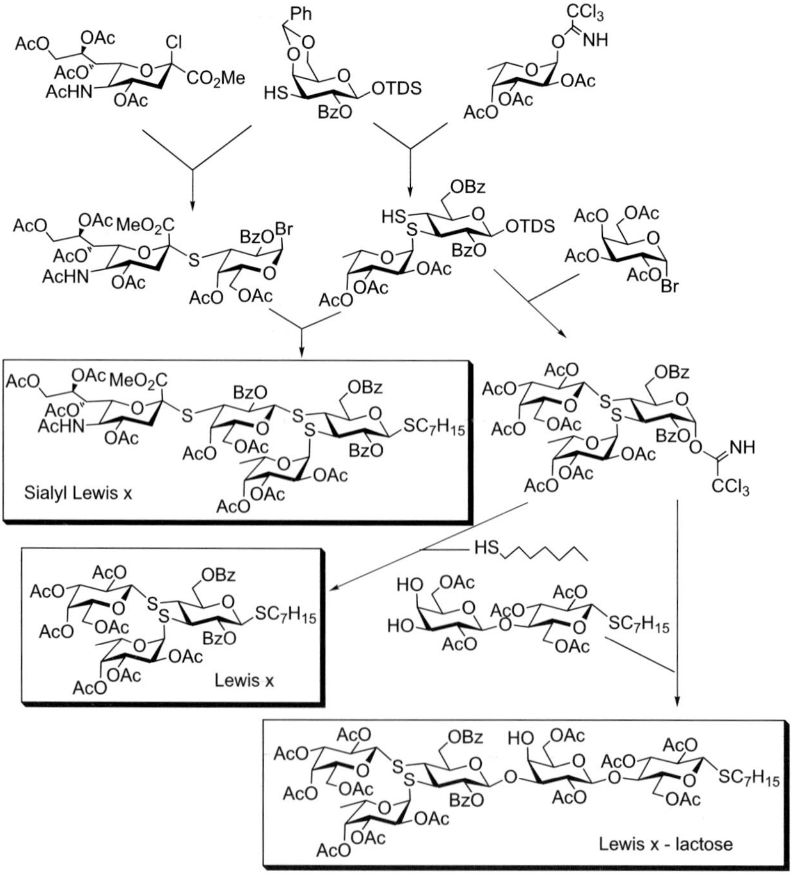

❑ **Scheme 66**

3.3 Examples

With the continuous development of new methods and improvement of old methods, the construction of more and more complex thiooligosaccharides has become possible. Two recently published examples of this are shown in ❷ *Scheme 66* and ❷ *Scheme 67*. In ❷ *Scheme 66* the synthesis of thio analogues of Lewis x and sialyl-Lewis x, using exclusively non-anomeric thiols in both acid-and base-catalyzed glycosylation reactions, is summarized [195].

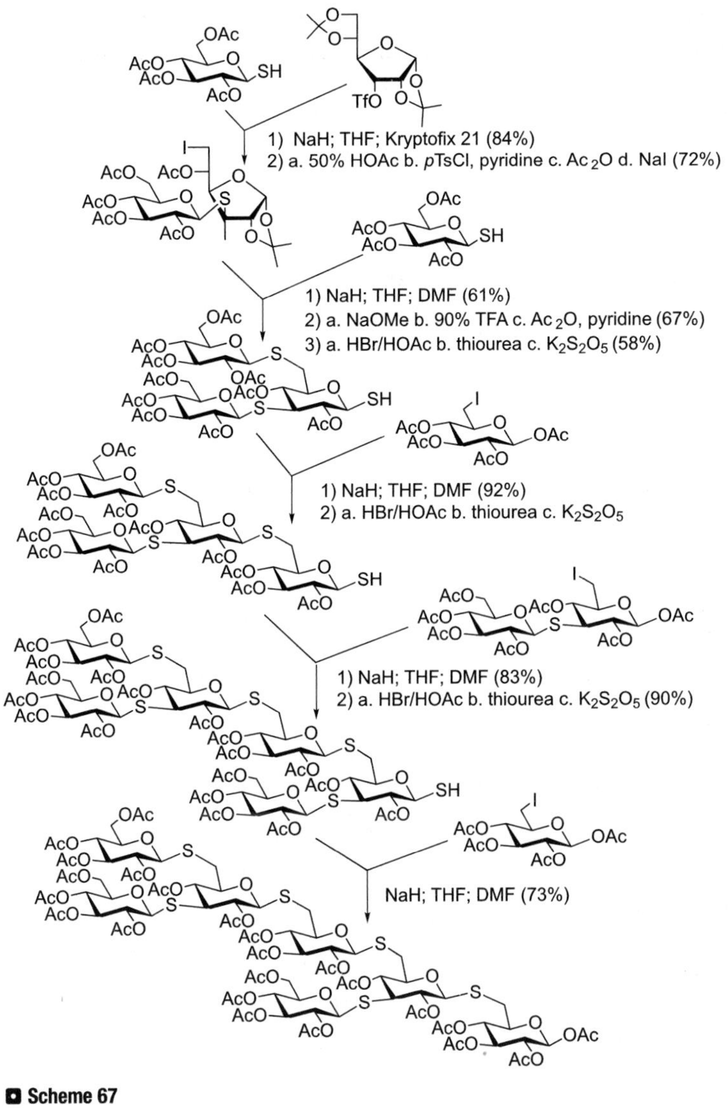

❏ Scheme 67

In ❷ *Scheme 67* is shown the assembly of the thio analogue of a phytoelicitor active heptasaccharide. In contrast to the above example, here all the thio-interglycosidic linkages are formed through the reaction of anomeric thiols with non-anomeric electrophiles [196].

In spite of the difference in methodology between these syntheses, both are impressive and excellent representatives of the state of the art of thiooligosaccharide synthesis.

References

1. Horton D, Hutson DH (1963) Adv Carbohydr Chem 18:123
2. Gil V, MacLeod AJ (1980) Phytochemistry 19:2071
3. Fahey JW, Zalcmann AT, Talalay P (2001) Phytochemistry 56:5
4. Herr RR (1967) J Am Chem Soc 89:2444
5. Hoeksema H, Bannister B, Birkenmeyer RD, Kagan F, Magerlein BJ, MacKellar FA, Schroeder W, Slomp G, Herr RR (1964) J Am Chem Soc 86:4223
6. Hoeksema H (1964) J Am Chem Soc 86:4224
7. Ngane AN, Lavault M, Seraphin D, Landreau A, Richomme P (2006) Carbohydr Res 341:2799
8. Khiar N, Suarez B, Valdivia V, Fernandez I (2005) Synlett 19:2963
9. Khiar N, Araujo CS, Suarez B, Fernandez I (2006) Eur J Org Chem 7:1685
10. Bundle DR, Rich JR, Jacques S, Yu HN, Nitz M, Ling C-C (2005) Angew Chem Int Ed 44:7725
11. Norberg T (1996) In: Khan SH, O'Neill RA (eds) Modern Methods in Carbohydrate Synthesis. Harwood, London, p 82
12. Garegg PJ (1997) Adv Carbohydr Chem Biochem 52:179
13. Oscarson S (2000) In: Ernst B, Hart G, Sinaÿ P (eds) Oligosaccharides in Chemistry and Biology: A Comprehensive Handbook. Wiley-VCH, Weinheim
14. Kartha KPR, Field RA (2003) In: Osborn HMI (ed) Carbohydrates. Elsevier Science Ltd, Oxford, p 121
15. Defaye J, Gelas J (1991) In: Atta-ur-Rahman (ed) Studies in Natural Products Chemistry, vol 8 E. Elsevier, Amsterdam, p 315
16. Defaye J (1995) In: Petersen SB, Svensson B, Petersen S (eds) Progress in Biotechnology, vol 10. Elsevier, Amsterdam, p 113
17. Driguez H (1997) Topics Curr Chem 187:85
18. Fairweather JK, Driguez H (2000) In: Ernst B, Hart G, Sinaÿ P (eds) Oligosaccharides in Chemistry and Biology: A Comprehensive Handbook. Wiley-VCH, Weinheim, p 531
19. Kahne D, Walker S, Cheng Y, van Engen D (1989) J Am Chem Soc 111:6881
20. Crich D, Lim LBL (2004) Org React 64:115
21. Agnihotri G, Misra AK (2005) Tetrahedron Lett 46:8113
22. Fernandez I, Khiar N (2003) Chem Rev 103:3651
23. Mehta S, Pinto BM (1991) Tetrahedron Lett 32:4435
24. Mehta S, Pinto BM (1996) In: Khan SH, O'Neill RA (eds) Modern Methods in Carbohydrate Synthesis. Harwood, London, p 107
25. van Well RM, Kaerkkaeinen TS, Kartha KPR, Field RA (2006) Carbohydr Res 341:1391
26. Buts L, Bouckaert J, De Genst E, Loris R, Oscarson S, Lahmann M, Messens J, Brosens E, Wyns L, De Greve H (2003) Acta Cryst D 59:1012
27. Cioci G, Mitchell EP, Chazalet V, Debray H, Oscarson S, Lahmann M, Breton C, Gautier C, Perez S, Imberty A (2006) J Mol Biol 357:1575
28. Witczak ZJ, Czernecki S (1998) Adv Carbohydr Chem Biochem 53:143
29. Robina I, Vogel P, Witczak ZJ (2001) Curr Org Chem 5:1177
30. Brigl P, Schinle R (1932) Ber dtsch chem Ges 65:1890
31. Lemieux RU (1951) Can J Chem 29:1079
32. Lemieux RU, Brice C (1955) Can J Chem 33:109
33. Hough L, Taha MI (1956) J Chem Soc 2042
34. Ferrier RJ, Furneaux RH (1976) Carbohydr Res 52:63
35. Das SK, Roy N (1996) Carbohydr Res 296:275
36. Dasgupta F, Garegg PJ (1989) Acta Chem Scand 43:471
37. Contour M-O, Defaye J, Little M, Wong E (1989) Carbohydr Res 193:283
38. Das SK, Roy J, Reddy KA, Abbineni C (2003) Carbohydr Res 338:2237
39. Weng S-S, Lin Y-D, Chen C-T (2006) Org Lett 8:5633
40. Vesely J, Ledvina M, Jindrich J, Saman D, Trnka T (2003) Collect Czech Chem Comm 68: 1264

41. Dahlhoff WV (1990) Liebigs Ann Chem 1025
42. Doren HAV, Geest RVD, Kellogg RM, Wynberg H (1989) Carbohydr Res 194:71
43. Galema SA, Engberts JBFN, van Doren HA (1997) Carbohydr Res 303:423
44. Boons GJ, Geurtsen R, Holmes D (1995) Tetrahedron Lett 36:6325
45. Garegg PJ, Olsson L, Oscarson S (1995) J Org Chem 60:2200
46. Nakano T, Ito Y, Ogawa T (1990) Tetrahedron Lett 31:1597
47. Lindberg B, Erbing B (1976) Acta Chem Scand Ser B 30:611
48. Boons GJPH, Delft FL, vanKlein PAM, van der Marel GA, van Boom JH (1992) Tetrahedron 48:885
49. Mannerstedt K, Ekelöf K, Oscarson S (2006) Carbohydr Res 342:631
50. Marra A (1989) Carbohydr Res 187:35
51. Ibatullin FM, Shabalin KA, Jänis JV, Shavva AG (2003) Tetrahedron Lett 44:7961
52. Kudelska W, Michalska M (1995) Synthesis 1539
53. Marino C, Marino K, Miletti L, Alves MJM, Colli W, de Lederkremer RM (1998) Glycobiology 8:901
54. Helander A, Kenne L, Oscarson S, Peters T, Brisson JR (1992) Carbohydr Res 230:299.
55. Adamo R, Kovac P (2006) Eur J Org Chem 2803
56. Agnihotri G, Tiwari P, Misra AK (2005) Carbohydr Res 340:1393
57. Tai CA, Kulkarni SS, Hung SC (2003) J Org Chem 68:8719
58. Lin CC, Huang LC, Liang PH, Liu CY (2006) J Carbohydr Chem 25:303
59. Ogawa T, Matsui M (1977) Carbohydr Res 54:C17
60. Pozsgay V, Jennings HJ (1987) Tetrahedron Lett 28:1375
61. Takahashi S, Terayama H, Kuzuhara H (1992) Tetrahedron Lett 33:7565
62. Kartha KPR, Field RA (1998) J Carbohydr Chem 17:693
63. Sato T, Fujita Y, Otera J, Nozaki H (1992) Tetrahedron Lett 33:239
64. Deng S, Gangadharmath U, Chang CWT (2006) J Org Chem 71:5179
65. Fischer E, Delbrück K (1909) Ber dtsch chem Ges 42:1476
66. Schneider W, Sepp J, Stiehler O (1918) Ber dtsch chem Ges 51:220
67. Helferich B, Grünewald H, Langenhoff F (1953) Chem Ber 86:873
68. Pedretti V, Veyrières A, Sinaÿ P (1990) Tetrahedron 46:77
69. El Ashry ESH, Awad LF, Atta AI (2006) Tetrahedron 62:2943
70. Blanc-Muesser M, Defaye J, Driguez H (1978) Carbohydr Res 67:305
71. Apparu M, Blanc-Muesser M, Defaye J, Driguez H (1981) Can J Chem 59:314
72. Bogusiak J, Szeja W (1985) Pol J Chem 59:293
73. Tropper FD, Andersson FO, Grand-Maître C, Roy R (1991) Synthesis 734
74. Meunier SJ, Andersson FO, Letellier M, Roy R (1994) Tetrahedron Asymm 5:2303
75. Gerz M, Matter H, Kessler H (1993) Angew Chem Int Ed Engl 32:269
76. Byun H-S, Bittman R (1995) Tetrahedron Lett 36:5143
77. Mukhopadhyay B, Kartha KPK, Russell DA, Field RA (2004) J Org Chem 69:7758
78. Kumar R, Tiwari P, Maulik PR, Misra AK (2006) Eur J Org Chem 74
79. van der Klein PAM, Boons GJPH, Veeneman GH, van der Marel GA, van Boom JH (1990) Synlett 6:311
80. Reimer M, Schmidt RR (2000) Tetrahedron Asymm 11:319
81. Timmer MSM, Adibekian A, Seeberger PH (2005) Angew Chem Int Ed 44:7605
82. Stewart AO, Williams RM (1985) J Am Chem Soc 107:4289
83. Li ZJ, Liu PL, Qui DX (1990) Synth Commun 20:2169
84. Fürstner A (1993) Liebigs Ann Chem 1211
85. Szeja W, Bogusiak J (1988) Synthesis 224
86. Inanaga J, Yokoyama Y, Hanamoto T (1993) J Chem Soc Chem Commun 1090
87. Hiranuma S, Kajimoto T, Wong C-H (1994) Tetrahedron Lett 35:5257
88. Schmidt RR, Stumpp M (1983) Liebigs Ann Chem 1249
89. Käsbeck L, Kessler H (1996) Liebigs Ann/Recueil 165
90. Wolfrom ML, Thompson A (1963) Methods Carbohydr Chem 3:150
91. Hannesian S, Guindon Y (1980) Carbohydr Res 86:C3
92. Nicolau KC, Seitz SP, Papahatjis DP (1983) J Am Chem Soc 105:2430
93. Liu D, Chen R, Hong L, Sofia MJ (1998) Tetrahedron Lett 39:4951
94. Sakairi N, Kuzuhara H (1996) Carbohydr Res 280:139

95. Zhang Z, Magnusson G (1996) Carbohydr Res 295:41

96. Mendlik TM, Tao P, Hadad CM, Coleman RS, Lowary TL (2006) J Org Chem 71: 8059

97. Araki Y, Matsuura K, Ishido Y, Kushida K (1973) Chem Lett 383

98. Gadelle A, Defaye J, Pedersen C (1990) Carbohydr Res 200:497

99. Lehmann J (1966) Carbohydr Res 2:486

100. Igarashi K, Honma T(1970) J Org Chem 35:606

101. Priebe W, Zamojski A (1980) Tetrahedron 36:274

102. Dunkerton LV, Adair NA, Euske JM, Brady KT, Robinson PD (1988) J Org Chem 53:845

103. Yadav JS, Reddy BVS, Geetha V (2003) Synth Comm 33:717

104. Smitha G, Reddy S (2004) Synthesis 834

105. Mereyala HB (1987) Carbohydr Res 168:136

106. Paul S, Jayaraman N (2004) Carbohydr Res 339:2197

107. Sherry BD, Loy RN, Toste FD (2004) J Am Chem Soc 126:4510

108. Yadav JS, Reddy BVS, Bhasker EV, Raghavendra S, Narsaiah AV (2007) Tetrahedron Lett 48:677

109. Weygand F, Ziemann H (1962) Liebigs Ann Chem 657:179

110. Seeberger PH, Eckhardt M, Gutteridge CE, Danishefsky SJ (1997) J Am Chem Soc 119:10064

111. Sakairi N, Hayashida M, Kuzuhara H (1987) Tetrahedron Lett 28:2871

112. Wang LX, Sakairi N, Kuzuhara H (1990) J Chem Soc Perkin Trans 1 1677

113. Garegg PJ, Oscarson S, Tedebark U (1998) J Carbohydr Chem 17:587

114. Skelton BW, Stick RV, Tilbrook DMG, White AH, Williams, SJ (2000) Austr J Chem 53:389

115. Lundt I, Skelbaeck-Pedersen B (1981) Acta Chem Scand Ser B 35:637

116. Plet JRH, Porter MJ (2006) Chem Comm 1197

117. Falconer RA (2002) Tetrahedron Lett 43:8503

118. Zhu X (2006) Tetrahedron Lett 47:7935

119. Gehrke M, Kohler W (1931) Ber dtsch chem Ges 64:2696

120. Blanc-Muesser M, Defaye J, Driguez H (1982) J Chem Soc Perkin Trans 1:15

121. Hasegawa A, Nakamura J, Kiso M (1986) J Carbohydr Chem 5:11

122. Ferrier RJ, Furneaux RH (1977) Carbohydr Res 57:63

123. Defaye J, Guillot JM (1994) Carbohydr Res 253:185

124. Park WKC, Meunier SJ, Zanini D, Roy R (1995) Carbohydr Lett 1:179

125. Horton D (1963) Methods Carbohydr Chem 2:433

126. Cerny M, Stanek J, Pacák J (1963) Monatsh Chem 94:290

127. Ibatullin FM, Selivanov SI, Shavva AG (2001) Synthesis 419

128. Tiwari P, Agnihotri G, Misra AK (2005) J Carbohydr Chem 24:723

129. El Ashry ESH, Awad LF, Hamid HMA, Atta AI (2005) J Carbohydr Chem 24:745

130. El Ashry ESH, Awad LF, Hamid HMA, Atta AI (2006) Synth Comm 36:2769

131. Fischer E, Helferich B, Ostman P (1920) Ber dtsch chem Ges 53:873

132. Müller A, Wilhelms A (1941) Ber dtsch chem Ges 74:698

133. Kochetkov NK, Klimov E, Malysheva NN, Demchenko AV (1991) Carbohydr Res 212:77

134. Fischer E (1914) Ber dtsch chem Ges 47:1378

135. Ogura H, Takahashi H (1982) Heterocycles 17:87

136. Camarasa MJ, Fernández-Resa P, López MTG, Heras FGdL, Méndez-Castrillón PP, Felix AS (1984) Synthesis 509

137. Lindhorst TK, Kieburg C (1995) Synthesis 1228

138. Pakulski Z, Pierozynski D, Zamojski A (1994) Tetrahedron 50:2975

139. Schneider W, Gille R, Eisfeld K (1928) Ber dtsch chem Ges 61:528, 1244

140. Tropper FD, Andersson FO, Cao S, Roy R (1992) J Carbohydr Chem 11:741

141. Bogusiak J, Wandzik I, Szeja W (1996) Carbohydr Lett 1:411

142. Fügedi P, Garegg PJ, Oscarson S, Rosén G, Silwanis BA (1991) Carbohydr Res 211:157

143. Sakata M, Haga M, Tejima S, Akagi M (1963) Chem Pharm Bull 11:1081

144. Sakata M, M MH, Tejima S (1970) Carbohydr Res 13:379

145. Knapp S, Gonzales S, Myers DS, Eckman LL, Bewley CA (2002) Org Lett 4:4337

146. Knapp S, Myers DS (2001) J Org Chem 66:3636

147. Knapp S, Myers DS (2002) J Org Chem 67:2995

148. Matsuoka K, Ohtawa T, Hinou H, Koyama T, Esumi Y, Nishimura SI, Hatano K, Terunuma D (2003) Tetrahedron Lett 44:3617

149. Bhar D, Chandrasekaran S (1997) Carbohydr Res 301:221

150. Sridhar PR, Prabhu KR, Chandrasekaran S (2004) Eur J Org Chem 4809

151. McDougall JM, Zhang XD, Polgar WE, Khroyan TV, Toll L, Cashman JR (2004) J Med Chem 47:5809

152. Jobron L, Hummel G (2000) Org Lett 2:2265

153. Falconer RA, Jablonkai I, Toth I (1999) Tetrahedron Lett 40:8663

154. Falco EA, Hitchings GH, Russell PB (1949) J Am Chem Soc 71:362

155. Driguez H, Szeja W (1994) Synthesis 1413

156. Naus P, Leseticky L, Smrcek S, Tislerova I, Sticha M (2003) Synlett 2117

157. Herradura PS, Pendola KA, Guy RK (2000) Org Lett 2:2019

158. Cerny M, Zachystalova D, Pacák J (1961) Coll Czech Chem Commun 26:2206

159. Lacombe JM, Rakotomanomana N, Pavia AA (1988) Tetrahedron Lett 29:4293

160. Bourgeois MJ, Gueyrard D, Montaudon E, Rollin P (2005) Lett Org Chem 2:665

161. Fokt I, Szeja W (1992) Carbohydr Res 232:169

162. Fokt I, Bogusiak J, Szeja W (1998) Carbohydr Lett 3:191

163. Magnusson G (1977) J Org Chem 42:913

164. Krog-Jensen C, Oscarson S (1998) Carbohydr Res 308:287

165. Josse S, Le Gal J, Pipelier M, Cleophax J, Olesker A, Pradere JP, Dubreuil D (2002) Tetrahedron Lett 43:237

166. Ane A, Josse S, Naud S, Lacone L, Vidot S, Fournial A, Kar A, Pipelier M, Dubreuil D (2006) Tetrahedron 62:4784

167. Rodriguez MA, Boutureira O, Arnes X, Matheu MI, Diaz Y, Castillon S (2005) J Org Chem 70:10297

168. Driguez H (2001) Chem Bio Chem 2:311

169. Mehta S, Andrews JS, Johnston BD, Pinto BM (1994) J Am Chem Soc 116:1569

170. Mehta S, Andrews JS, Johnston BD, Svensson B, Pinto BM (1995) J Am Chem Soc 117:9783

171. Andrews JS, Pinto BM (1995) Carbohydr Res 270:51

172. Andrews JS, Johnston BD, Pinto BM (1998) Carbohydr Res 310:27

173. Johnston BD, Pinto BM (1998) Carbohydr Res 310:17

174. Blanc-Meusser M, Driguez H (1988) J Chem Soc Perkin Trans 1 3345

175. Hutson DH (1967) J Chem Soc (C) 442

176. Reed LA, Goodman L (1981) Carbohydr Res 94:91

177. Nilsson U, Johansson R, Magnusson G (1996) Chem Eur J 2:295

178. Eisele T, Toepfer A, Kretzschmar G, Schmidt RR (1996) Tetrahedron Lett 37:1389

179. Moreau V, Norrild J C, Driguez H (1997) Carbohydr Res 300:271

180. Hashimoto H, Shimada K, Horito S (1993) Tetrahedron Lett 31:4953

181. Hashimoto H, Shimada K, Horito S (1994) Tetrahedron Asymm 5:2351

182. Aguilera B, Fernández-Mayoralas A (1996) J Chem Soc Chem Commun 127

183. Witczak ZJ, Sun J, Mielguj R (1995) Bioorg Med Chem Lett 5:2169

184. Witczak ZJ, Chhabra R, Chen H, Xie X-Q (1997) Carbohydr Res 301:167

185. Lundt I, Skelbaeck-Pedersen B (1981) Acta Chem Scand Ser B 35:637

186. Hummel G, Hindsgaul O (1999) Angew Chem Int Ed 38:1782

187. Defaye J, Driguez H, Poncet S, Chambert R, Petit-Glatron M-F (1984) Carbohydr Res 130:300

188. Jahn M, Marles J, Warren RAJ, Withers SG (2003) Angew Chem Int Ed 42:352

189. Jahn M, Chen H, Mullegger J, Marles J, Warren RAJ, Withers SG (2004) Chem Comm 274

190. Jahn M, Marles J, Warren RAJ, Withers SG (2003) Chem Comm 1327

191. Jahn M, Withers SG (2003) Biocat Biotrans 21:159

192. Mullegger J, Jahn M, Chen HM, Warren RAJ, Withers SG (2005) Protein Eng Design Select 18:33

193. Stick RV, Stubbs KA (2005) Tetrahedron Asymm 16:321

194. Rich JR, Szpacenko A, Palcic MM, Bundle DR (2004) Angew Chem Int Ed 43: 613

195. Eisele T, Schmidt RR (1996) Liebigs Ann/Recueil 1303

196. Ding Y, Contour-Galcera M-O, Ebel J, Ortiz-Mellet C, Defaye J (1999) Eur J Org Chem 1143

197. Bernardes JLG, Gamblin DP, Davis BG (2006) Angew Chem Int Ed 45:4007

3.6　Glycal Derivatives

Waldemar Priebe[*][1], *Izabela Fokt*[1], *Grzegorz Grynkiewicz*[2]
[1] M. D. Anderson Cancer Center, The University of Texas,
　Houston, TX 77030, USA
[2] Pharmaceutical Research Institute, Rydgiera 8, 01-793 Warsaw, Poland
wpriebe@mdanderson.org, wp@wt.net, g.grynkiewicz@ifarm.waw.pl

Abstract

Glycals are unsaturated sugar derivatives in which the double bond engages the anomeric carbon atom. Such cyclic vinyl ethers are characterized by high reactivity, allowing for regio- and stereoselective transformations, directly or indirectly related to glycosylation, as well as to formation of carbon–carbon and carbon–heteroatom bonds at the anomeric center. This review provides a systematic survey of chemical synthetic methods, by which the carbon–carbon double bond is introduced next to the ring oxygen, in an endo- or exocyclic position. Some mechanistic aspects are discussed in relation to traditional methods of glycal preparation, which rely on elimination reactions. Glycal-to-glycal rearrangement and applications of organometallic chemistry and heteroatom-induced transformations for syntheses and activation of glycals are also highlighted.

Keywords

Glycal; Synthesis; Carbohydrate; Monosaccharides

In: *Glycoscience*. Fraser-Reid B, Tatsuta K, Thiem J (eds)
Chapter-DOI 10-1007/978-3-540-30429-6_15: © Springer-Verlag Berlin Heidelberg 2008

Abbreviations

AIBN	azobisisobutyronitrile
DBU	1,8-diazabicyclo[5.4.0]undec-7-ene
HMPA	hexamethylphosphoramide
HSAB	hard and soft acids and bases
LAH	lithium aluminum hydride
LDA	lithium diisopropylamide
m-CPBA	3-chloroperbenzoic acid
NIS	*N*-iodosuccinimide
THF	tetrahydrofuran
TMSBr	trimethylsilyl bromide

1 Introduction

Monosaccharides possessing a double bond between C1 and C2 are known under their trivial name as glycals. The name glycals was coined after the product of the first synthesis of 1,2-unsaturated sugar from D-glucose **1** which was named D-glucal [1]. Consequently, the generic name "glycals" was adopted for this class of unsaturated sugars, which are in fact extensively substituted cyclic vinyl ethers. Such cyclic vinyl ethers are also encountered as structural motifs in total synthesis of some complex natural products [2,3].

Systematic naming [4,5] uses alditol descriptors and not the parent aldose, as is the case in commonly used trivial names. Thus, **4** should be named 1,5-anhydro-2-deoxy-D-*arabino*-hex-1-enitol. However, trivial names are being widely used because of their simplicity. Advances in the synthesis of glycals have been summarized regularly in comprehensive reviews [2,6,7,8] and new reactions were regularly summarized in annual reports [9]. Recent developments in the chemistry of glycals have expanded their importance as useful organic substrates and intermediates far beyond the limits set by the traditional placement of somewhat esoteric unsaturated sugars. Glycals are extensively used as glycosylating agents, [10,11,12] especially in the synthesis of 2-deoxy sugars [13,14,15,16] and oligosaccharides [17], and also as precursors of *C*-glycosyl compounds [18,19,20]. They are also widely recognized as versatile chiral building blocks [21,22,23]. Selectively protected glycals are being used in chemical and enzymatic synthesis of complex carbohydrates and in the synthesis of *C*-oligosaccharides and glycoproteins [24,25]. There is an increased interest in exploring carbohydrate scaffolds in drug discovery as a consequence of identifying the role of glycoconjugates in biological recognition and signal transduction processes [26]. After genomics and proteomics, glycomics appears to be the next significant frontier for the development of new medical therapeutics, where progress will require concerted multidisciplinary efforts. Specifically, chemical glycomics will require the maturation of synthetic carbohydrate methodologies that would allow the preparation of glycoconjugates in a parallel, automated synthesis [27,28]. It now appears that unsaturated sugar derivatives, like glycals, can be used as useful substrates for such a purpose [29,30].

Various aspects of essential glycal chemistry connected to Ferrier rearrangement and anomeric carbanions have been extensively reviewed [31,32,33,34].

2 Reductive Elimination of Glycosyl Halides

Because of their facile preparation and reactivity, glycosyl halides are commonly used in carbohydrate chemistry. They are primarily used as glycosyl donors [35,36] and are the most important substrates in the synthesis of glycals from monosaccharides. Especially useful are glycosyl bromides **3**; they are conveniently prepared from glycosyl esters **2** [37,38] (one-pot procedures starting from reducing sugars are also available [39,40] and are often used without purification in the next step of synthesis (❷ *Scheme 1*)). Reductive elimination of glycosyl bromides, prepared in situ with zinc generates glycals with good to excellent yields [41]. The mechanism of reductive elimination is believed to involve a two-electron reduction process localized at the anomeric center, followed by elimination of an acetoxonium anion from the vicinal position [42].

☐ Scheme 1

Fischer's original procedure [1] as well as subsequent improvements dating from the early 1900s, are practically unavailable; however, they are summarized and in some cases specifically described in *Methods in Carbohydrate Chemistry* [43,44,45,46]. All these procedures for reduction of glycosyl bromides into acetylated glycals used zinc powder alone or in combination with salts of other metals in aqueous acetic acid, and several generations of chemists used them as standard procedures for the preparation of glycals. Specifically, in the synthesis of 3,4-di-*O*-acetyl-D-xylal, the recommended reductive agent was zinc dust alone in 50% acetic acid (at a temperature below −10 °C) [47]. However, heterogeneous reaction conditions can sometimes create problems of reproducibility, and although the quality of the zinc powder used was not as critical as it is in typical metaloorganic transformations [48], it was often mentioned as a factor affecting yield [49]. To overcome this problem, pre-washing with diluted hydrochloric acid or the addition of platinum [50] and copper [51] salts were proposed for more effective reduction on a metal surface. Such improvements are now part of the standard protocols for preparation of L-arabinal [43] and D-glucal [44], where zinc is activated with cupric sulfate pentahydrate. On the other hand, in the protocol for preparation of D-galactal, zinc is activated with platinic chloride [46]. In general, the yields for most glycals are good to excellent, and the final product can be purified by distillation or crystallization.

Another approach (and one more important for assessing the mechanisms of glycal formation than for practical use) involved the electro-reduction of 2,3,4,6-tetra-*O*-acetyl-D-glucopyranosyl bromide **3** on mercury electrode in dipolar aprotic solvents containing tetraethylammonium perchlorate to support a mechanistic two-electron reduction hypothesis [52]. However, 3,4,6-tri-*O*-acetyl-D-glucal **4** was obtained in only 55% yield. The main product **4** was accompanied by penta-*O*-acetyl-D-glucopyranose (35%) and 2-acetoxy-3,4,6-tri-*O*-acetyl-D-glucal (10%), indicating the extent of secondary solvolytic processes in this method.

The composition of reaction mixtures obtained under classical Fischer–Zach conditions was determined by capillary gas chromatography [53] of per-*O*-silylated and per-*O*-acetylated components of the crude reaction mixtures, and some progress was made in explaining the mechanism behind typical side-product formation. The compounds identified included products of solvolytic displacement of an anomeric bromine by acetic acid and water as well as 2,3-unsaturated derivatives. Interestingly, there was a difference in the reaction outcome of pentopyranosyl versus hexopyranosyl substrates. In the former case, glycals were accompanied by acetylated 1,5-anhydropentitols; in the latter, hex-1-enitols contained admixtures of peracetylated 2-deoxyhexopyranosides.

One failed attempt to convert peracetylated glycosyl chlorides into the corresponding glycals used zinc dust in aqueous acetic acid [54]. However, when an excess of chromium(II) diacetate dimer monohydrate was used together with 1,2-ethanediamine in dimethylformamide to generate α-pyranosyl chlorides of D-glucose, D-xylose, and D-arabinose, the corresponding glycals were obtained in yields of 82–97%.

Aluminum amalgam has been shown to be an effective reducing-eliminating reagent for glycosyl bromides when used in aqueous tetrahydrofuran [55], producing 60–85% yields of product. From the design of alternative procedures employing aprotic reaction conditions have come synthetic methodologies for the preparation of glycals from pyranose derivatives bearing acid-sensitive protecting groups and even from furanose derivatives. The first attempts in this direction, consisting of treating pentofuranosyl bromides bearing good leaving groups at C-2 with sodium iodide in dry acetone [56], generated a hitherto unknown class of furanosyl glycals. However, it also became immediately obvious that, to prevent Ferrier rearrangement, the oxygen at C-3 would require protection of the ether type [57]. Later, 2,3:5,6-di-*O*-isopropylidene-D-mannofuranosyl bromide **5** was shown to react smoothly with an excess of sodium or potassium in dry tetrahydrofuran to give corresponding glycal **6** in 59% yield [58], whereas lithium and magnesium failed to produce glycals (❯ *Scheme 2*).

Sodium naphtalide was found to be a good replacement for the sodium and to afford good yields of pyranoid and furanoid glycals from glycosyl halides having alkali-stable protecting groups [59]. The furanoid glycal **6** was subsequently prepared using modifications of this pro-

5   **6**

■ Scheme 2

cedure [60,61]. Meanwhile, Ireland's innovative application of lithium in liquid ammonia as a reducing agent for furanosyl halides [62,63] became a widespread procedure [64]. However, it has been demonstrated that, in special cases (i. e., when there are benzyl protective groups present and there is no heteroatom substitution at C-3), even furanoid glycals can be prepared by the zinc/acetic acid method [65].

Considerable attention has been given to reactive metal–graphite combinations, which were found to reduce both O-acylated pyranosyl and O-alkylidene furanosyl halide substrates under neutral conditions and to give excellent yields of high-purity glycals. Zinc/silver-graphite in tetrahydrofuran [66,67] and potassium graphite laminate [68] in the same solvent are very reactive, which allows the reaction work-up to be reduced to simple filtration and evaporation. The potassium graphite reagent is unique in effectively eliminating the 2-benzyloxy substituent from per-O-benzylated glycosyl halides, thus affording direct access to 3,4,6-tri-O-benzyl-D-glucal [69,70] and circumventing, in this case, the need for benzylation of preformed glucal derivatives.

Two other interesting reagents are samarium iodide and dimeric cyclopentadiene-titanium complex (Cp$_2$TiCl)$_2$. When used at 6 equivalents, samarium iodide can convert acetylated glucopyranosyl bromide into D-glucal 4 at yields of 90% [71]. Dimeric cyclopentadiene-titanium complex (Cp$_2$TiCl)$_2$, which is known to react with activated alkyl halides via halogen atom abstraction forming alkyl radicals, also reacts smoothly with glycosyl bromides in tetrahydrofuran at room temperature, affording corresponding glycals in very good yields [72]. The proposed reaction scheme, involving repeated transformation of Ti(III) into Ti(IV) in intermediate titanocene species [73], is shown on ❷ Scheme 3.

R = OAc

Cp$_2$Ti = titanocene

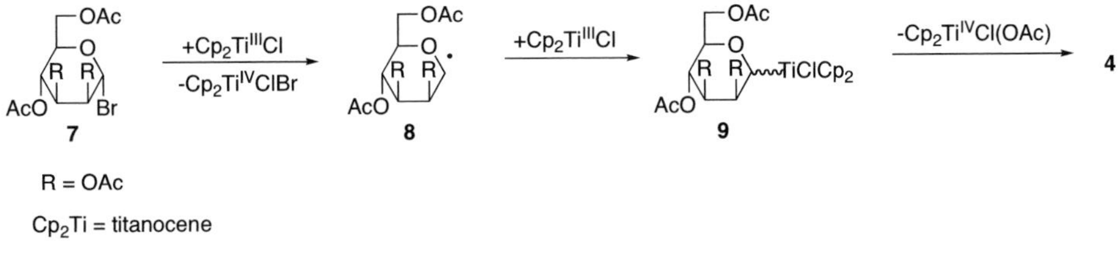

◘ Scheme 3

The intermediate glycosylic-titanium complex is likely to be a mixture of anomers. But, when taking into account the facile formation of acetylated D-glucal from both -D-glucopyranosyl and -D-mannopyranosyl bromides under the conditions discussed above, it seems that both syn and anti elimination of Cp$_2$TiCl(OAc) is possible. Glycosyl radical intermediates, on the other hand, show a strong preference for alpha orientation, as evidenced by the configuration of C-glycosyl compounds obtained by trapping with unsaturated species [74].

In the case of organochromium(III) complex intermediates obtained from glycosyl halides, work by Somsak and coworkers showed that their decay into glycals [75] is marked by the remarkable stability of the organometalic compounds in aqueous solutions. From the same group came the radically new concept of using zinc metal under aprotic conditions [76]. In brief, they proposed that acetylated glycosyl bromides be treated with zinc dust in the presence of N-heterocyclic bases (e. g., 4-methylpyridine or 1-methylimidazole) in a variety of aprotic solvents (i. e., benzene, ethyl acetate, tetrahydrofuran, acetone, dichloromethane) to afford the corresponding glycals. The beneficial influence of triethylamine and pyridine addition on zinc-mediated reductive elimination was first noticed in the case of acylated 1-bromo-D-glycosyl cyanides [77] and was further noticed in other monosaccharide derivatives substituted with an electron-withdrawing group and bromine atom at the anomeric center [78,79]. To identify the reaction mechanism operating under aprotic solvent conditions, two model reactions were chosen and compositions of corresponding reaction mixtures were examined using capillary gas chromatography and a standard co-injection procedure [80]. The postulated reaction scheme, based on isolable product distribution as well as free radical inhibition and trapping experiments, involved glycosyl and 2-deoxyglycosyl radicals in addition to a glycosyl carbanion (1,5-anhydro-2,3,4,6-tetra-O-acetyl-D-glucitol-1-ide) [5]. This postulated carbanion, however, resisted attempts to trap it in the presence of a benzaldehyde serving as an electrophile (❷ *Scheme 4*).

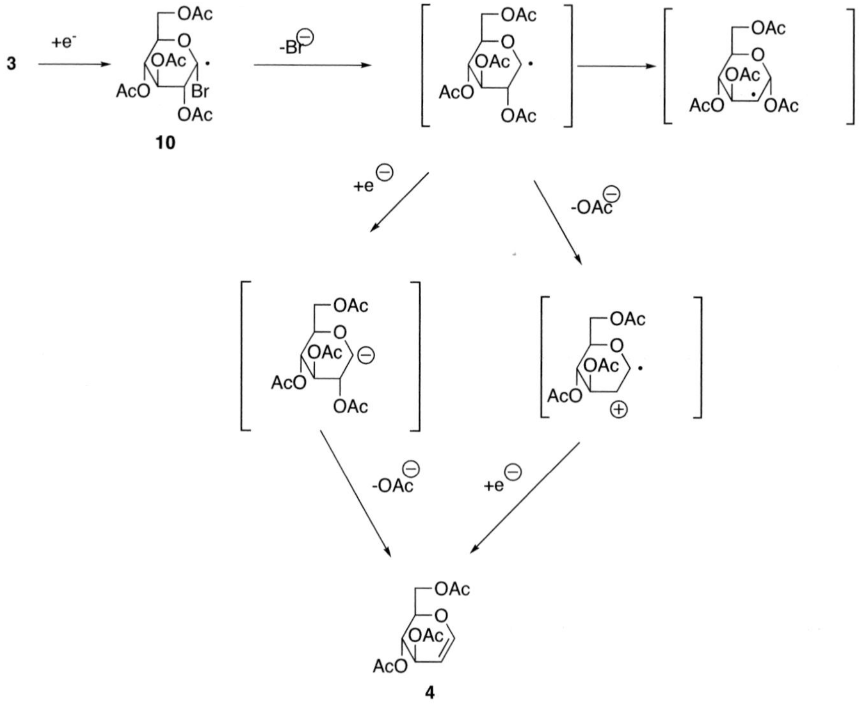

□ Scheme 4

Anomeric substituents, other than halides, are also capable of undergoing 1,2-elimination under a variety of reduction conditions. Anomeric 2-pyridylsulfonyl substituents can undergo 1,2-elimination, yielding hex-2-enitols, in reaction with samarium diiodide, particularly in the presence of HMPA [81]. Less obvious, but apparently general transformation leading to glycals has been discovered by Spanish researchers [82]. 2-Deoxy-2-iodo-1-phenylthioglycoside undergoes 1,2-elimination under various reducing conditions, preferably in the presence of metallic zinc or zinc-copper couple. Substrates for this reaction are easily accessible from

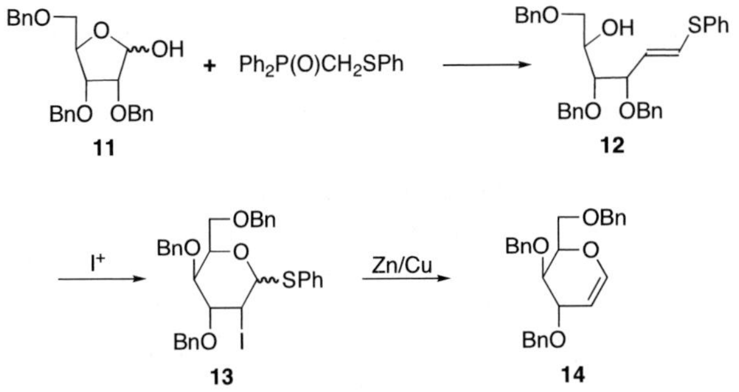

■ Scheme 5

1-OH pentofuranoses, as shown in ❥ *Scheme 5*. Thus, otherwise scarcely available D-allal and D-gulal derivatives can be readily prepared from D-ribose and D-arabinose, respectively [82]. It has recently been demonstrated that reductive elimination of β-hydroxycarboxylic acid can be applied for preparation of hex-1-enitols. The procedure involves refluxing appropriate substrate in toluene with *N,N*-dimethylformamide dineopentyl acetal.

By using such a transformation, uronic acid obtained from methyl D-glucopyranoside was converted into intermediate 4-deoxypentenoside and then via epoxidation followed by a nucleophile attack, into a variety of pyranosides of the L-configuration series, as shown in ❥ *Scheme 6* [83].

It should be noted that, under radical-generating conditions, a number of vicinal 1,2-substituents in cyclic sugar derivatives will tend to undergo elimination. This tendency has been demonstrated in a series of pyranosyl 1-phenylselenide-2-azides, which form corresponding glycals in yields close to quantitative upon treatment with tributyltin hydride [84].

Under the same conditions, glycals are also formed from 1-thioglycoside-2-xanthates. Phenyl thioglycopyranosides easily undergo reductive lithiation with lithium naphtalenide in tetrahydrofuran at low temperatures and subsequent elimination of C-2 substituents [85,86]. Such reaction conditions are compatible with labile acid- and base-protecting groups and yields

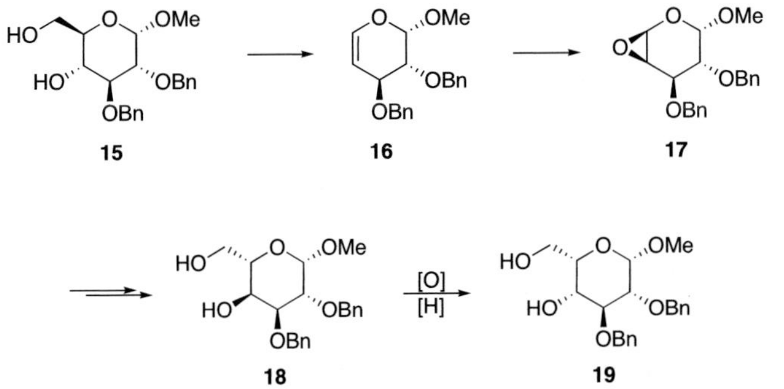

■ Scheme 6

of isolated glycals are high. Corresponding phenyl sulfones follow the same reaction pathway [87]. Anomeric sulfoxides form glycals upon treatment with 3 equivalents of butyllithium in THF at −78 °C. The yield of products generated from pyranosyl derivatives protected with acetal and ether groups ranges from 51 to 86% [88].

Although there is much new evidence testifying to the potential complexity of glycosyl halide reductive transformations, this evidence has not significantly influenced laboratory practices for the preparation of glycals from simple sugars. One paper [89] outlined a general method amenable to large-scale synthesis; the method involved a one-pot, three-step procedure involving peracetylation of an aldose, synthesis of a glycosyl bromide, and bromide reductive elimination. For D-glucose, the last step uses a Zn/AcOH/AcONa/CuSO$_4$ mixture that yields 98% 3,4,6-tri-O-acetyl-D-glucal. Another improved version of glycal synthesis, devised by Franck [90], uses vitamin B-12 as a catalyst for reductive eliminations [91]. With methanol as solvent, acetylated glycals were conveniently prepared by this method in the presence of metallic zinc and ammonium chloride in yields exceeding 90%. In the case of an L-rhamnal derivative, however, column chromatography had to be applied to isolate product in a 45% yield. Both the original and modified procedures of glycal preparation are reasonably tolerant of the inclusion of commonly applied functionalities, including glycosidic bonds. In fact, a number of glycals have already been prepared by different procedures from peracetylated disaccharides and trisaccharides [92,93,94,95,96,97,98,99,100].

3 Base-Catalyzed Eliminations of Glycosyl Halides to Yield 2-Oxy-hex-1-enitols

Peracylated glycosyl halides 3 are easily dehydrohalogenated to per-O-acyl-2-hydroxy-hex-1-enitols 20 (acylated 2-hydroxyglycals) under a variety of basic conditions (◉ Scheme 7) [101]. The first transformation of this kind was recorded during the attempted N-glycosylation of secondary amines. As a result, diethylamine in benzene [102,103] or in chloroform [104,105] was used routinely in early preparations to effect such elimination. Since then, dehydrobromination has emerged as one of the competing reactions, and factors influencing their relative rates have been assessed [106]. For example, formation of 2-O-acyl-hex-1-enitols was often observed during attempted glycosylation of phenolates or carboxylates under basic (including phase-transfer) conditions. A reaction of 2,3,4-tri-O-acetyl-L-rhamnopyranosyl bromide with 4-methylbenzenethiol in the presence of 2,4,6-collidine in nitromethane resulted in a mixture containing not only the desired thioglycosides but also the elimination product (2,3,4-tri-O-acetyl-1,5-anhydro-6-deoxy-L-*arabino*-hex-1-enitol) and two isomeric thiophenyl

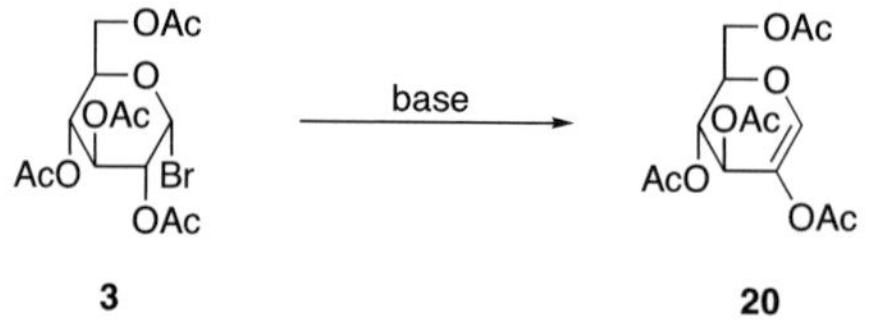

□ Scheme 7

orthoesters [107]. In the case of ethanethiol, glycosylation in the presence of triethylamine in acetonitrile resulted in elimination [108].

Other studies led to the successful yet sometimes unintentional isolation of glycals. For instance, heating acetylated glucopyranosyl bromide in nitromethane with nicotinamide gave peracetyl-2-hydroxy-D-glucal 21 along with both anomers of quaternary *N*-glycosides [109]. Studies of reactions of peracetylated -D-hexopyranosyl bromides with sodium 4-nitrophenoxide in dimethylformamide [110] revealed a strong correlation between sugar configuration and elimination product. Depending on the sugar, the yields of 2-acetoxyglycals diminished from 61 to 8% (*allo* > *gulo* > *gluco* > *galacto*), while the effectiveness of substitution with phenoxide anion followed the reverse order. A dispute over the scope of phase-transfer glycosylating reactions employing acylated glycosyl bromides and protic acid substrates led to the finding that application of 2,3,4,6-tetra-*O*-acetyl-D-glucopyranosyl bromide [111] or benzoylated glycosyl bromide [112] can give acylated glycals as common side products. For example, in phase-transfer reactions with a series of 44 phenolic substrates, acetylated α-D-galactopyranosyl bromide afforded the expected aryl β-galactopyranosides in yields ranging from 29 to 87% but no elimination product [113]; in contrast, during attempted glycosylation of a sugar substrate in the presence of silver trifluoromethanesulfonate and 2,4,6-trimethylpyridine, the benzoylated analog formed as much as 40% of the corresponding 2-*O*-benzoyl-D-galactal derivative [114]. When reacted with 1,2:5,6-di-*O*-isopropylidene-D-glucofuranose in the presence of diisopropylethylamine and tetrabutylammonium iodide in refluxing benzene, 2,3,4,6-tetra-*O*-benzyl-D-glucopyranosyl iodide 21 gave equimolar amounts of the expected product 22 (linked disaccharide) and the glycal 23 (❷ *Scheme 8*) [115].

R = 1,2:5,6-di-O-isopropylidene-α-D-glucofuranos-3-yl

☐ **Scheme 8**

Such isolations of 2-*O*-substituted glycals were, however, unintentional. Therefore, more purposeful and effective methods and conditions for glycal preparation were sought. Excellent yields of disaccharide 2-*O*-benzoylated glycals were obtained by supplementing diethylamine with sodium iodide and carrying out the elimination reaction in acetone at ambient temperature [116]. Application of tetrabutylammonium bromide, as an addition to diethylamine, proved somewhat less effective [96,117,118]. When tetra-*O*-benzoyl-D-glucopyranosyl bromide was subjected to treatment with NaH in dimethylformamide, the corresponding glycal was isolated in yields of 52% [119]. Sodium hydride-promoted glycosylation of an enolic substrate gave a similar yield [120]. Cesium fluoride in acetonitrile was applied successfully for efficient elimination of acetylated glycosyl bromides with D- and L-*manno* configuration [121]. Introduction of 1,5-diazabicyclo bases as reagents for dehydrohalogenation [122] led to the establishment [123] of 1,8-diazabicyclo[5.4.0.]undec-7-ene (DBU) as a very effective agent for the elimination of hydrogen bromide from acetylated [124], benzoylated [125], and pival-

oylated [126] glycosyl bromides. In the case of 2-*O*-acetyl-3,4,6-tri-*O*-benzyl-D-glucopyra-
nosyl bromide, DBU was applied, together with sodium iodide, in acetone [117].

Obviously, the tendency to eliminate HBr from glycosyl bromides, or other protic acids from
a sugar derivative bearing a good leaving group at an anomeric position, can be considerably
enhanced by exploiting the possibility of forming a conjugated double bond system. Such
an arrangement of functional groups exists in natural products (for example, in members of
the neuraminic acid family [127,128,129], and it is well known that corresponding sialic acid
glycals [130,131,132,133] can be prepared by treating peracetylated derivatives with substoi-
chiometric amounts of trimethylsilyl triflate.

On the other hand, an electron-withdrawing functional group can be introduced in position C-3
of the sugar ring. A good example of this approach is the preparation of chiral pyranone synthon **26** from di-*O*-isopropylidene-D-glucofuranose **24** as shown below (❷ *Scheme 9*) [134].
Other benzoylated 3-enones were obtained using the same approach [135].

Analogously, various 2-aminoglycals were obtained through oxidation of a C-3 hydroxyl
group in 4,6-benzylidene derivatives of *N*-protected 2-amino-D-pyranosides [136]. 2-Acetami-
do-*O*-acetylated glycals, on the other hand, were obtained under acidic conditions, during
reaction of protected 2-amino-2-deoxypyranosyl chlorides with propenyl acetate [137,138].
Attempted displacement of mannoside 2-triflate **27** with an azide anion resulted in the formation of enone **28**, while attempted displacement of the corresponding anomer gave, as
expected, 2-azide **30** in the D-gluco configuration (❷ *Scheme 10*) [139].

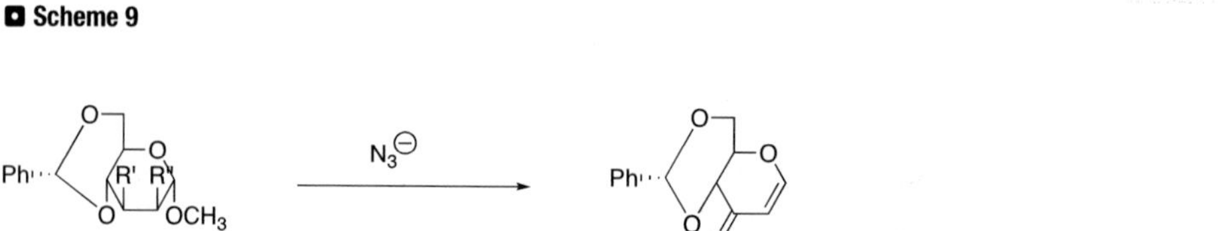

■ Scheme 9

R' = OAc ; R" = OTf

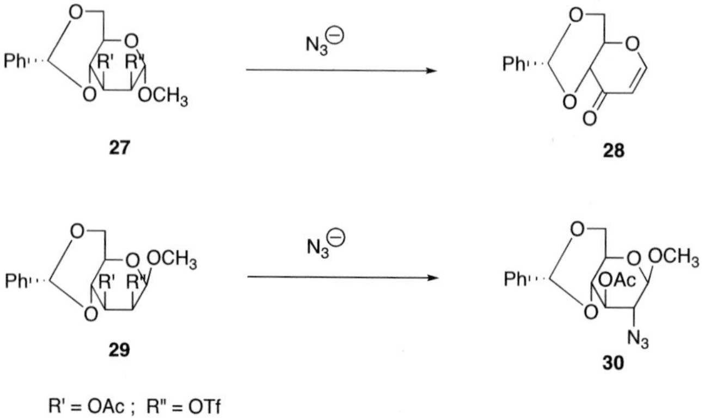

■ Scheme 10

In one interesting case, hexopyranoses **31** and pentofuranoses **33** protected with nonparticipating groups were formally dehydrated in a two-step process involving preparation of anomeric mesylate and palladium catalysis [140]. The researchers who observed this case, however, ruled out the simple explanation of bases assisting in Bronsted acid removal. Instead, they claimed that treatment of an intermediate anomeric mesylate with 0.9–5% Pd(Ph$_3$P)$_4$ at 50 °C (in a solution of methylene dichloride and *s*-collidine) resulted in oxidative addition followed by hydride elimination, consequently affording *O*-protected glycals **32** and **34** (❍ *Scheme 11*).

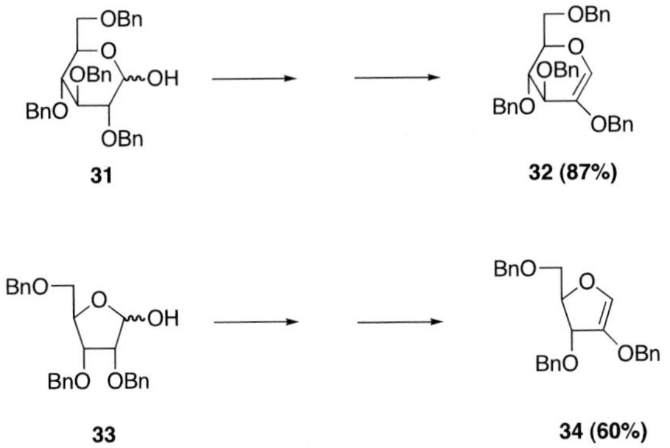

☐ **Scheme 11**

4 Eliminations Induced by Alkyllithium Reagents

Deprotonation of heterosubstituted carbon in protected sugar derivatives, as well as deprotonation of a hydroxyl group, can start anion fragmentation that, as illustrated below, can lead to unsaturated products. Subsequent deprotonation of glycals at C-1 is also possible and has already been expertly covered in a monograph [141]. This section, in addition to discussing the synthesis leading to glycals, will also discuss deprotonation at C-1 as a general approach to obtaining a great variety of C-1 heterosubstituted and C–C linked products.

During a study devoted to fragmentation of sugar acetals in the presence of butyllithium [142], the formation of unsaturated pyranoses (**36–38**) from L-rhamnose derivative **35** was observed and explained in terms of anionic species rearrangement (❍ *Scheme 12*).

In a related process, D-allal derivative **40** was first obtained [143] from the *allo* epoxide **39** by treatment with methyllithium. Subsequently, it was demonstrated that 2-iodo-D-altroside [144] **41** and its deprotected analog [145] **42** gave even higher yields of the allal (**40** and **43**, respectively) when treated with an excess of alkyllithium (❍ *Scheme 13*). Incidentally, benzylidene derivative **40** could not be deprotected to **43** without decomposition.

2-Deoxy-2-iodopyranosides constitute a class of synthetic sugar derivatives that are useful intermediates for preparation of 2-deoxy glycosides, which in turn are natural compounds of considerable medicinal importance [15]. These iodopyranosides can be conveniently prepared by using *N*-iodosuccinimide (NIS)-assisted addition of hydroxylic substrates to glycals or using oxirane ring opening by iodine anion of 2,3-anhydro pyranosides [12]. The resulting

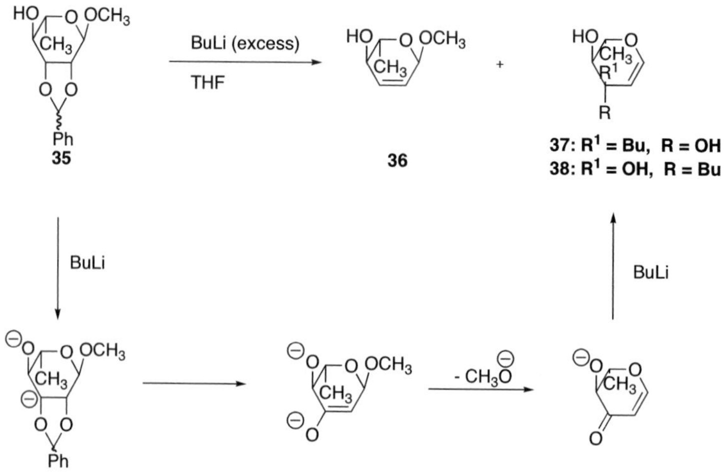

☐ Scheme 12

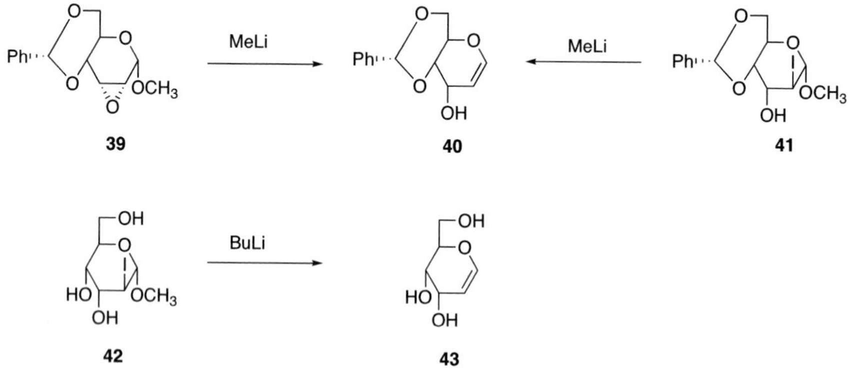

☐ Scheme 13

sugar iodohydrines and their acylated derivatives can be transformed into unsaturated sugars using methyllithium [146,147]. The formal reversibility of NIS-promoted addition to glycals can be illustrated by the transformation of D-digitose-derived disaccharide synthons **44** and **45** (❏ *Scheme 14*).

Only one-step is required to derive glycals from antibiotic sugars. For example, L-mycarose and L-olivose can be derived in one step from methyl 2,3-benzylidene-L-rhamnoside by applying the same principle of acetal fragmentation in reaction with a 6-fold excess of methyllithium [148]. Glycosyl sulfoxides with nonparticipating protecting groups require only 3 equivalents of butyllithium (at −78 °C in THF) for smooth conversion to the corresponding glycals [88].

As these examples indicate, introduction of a double bond via deprotonation of sugar derivatives having nonparticipating protecting groups can be synthetically useful. In practical syn-

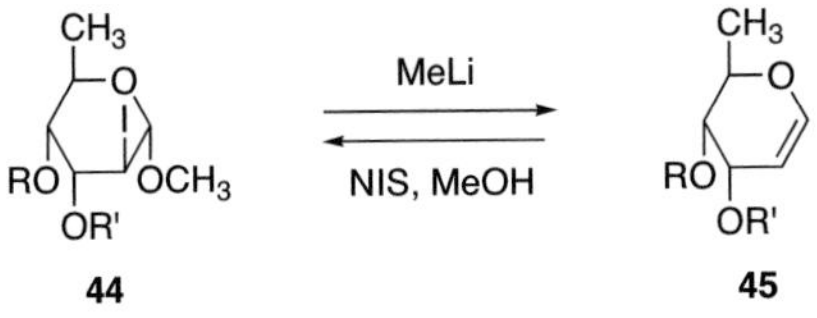

Scheme 14

R - monosacharide
R' - H, Ac

theses of glycals, deprotonation is facilitated by placing a strong electron-withdrawing group at the anomeric position. For example, C-glycosyl derivative **46** undergoing acetic acid elimination upon treatment with pyridine at room temperature easily affords **47** in a 75% yield (● *Scheme 15*) [149].

Easily available anomeric sulfones, which have already been mentioned as glycal precursors, readily undergo base-induced eliminations resulting in formation of 1,2-unsaturated pyranose derivatives in which the sulfonyl group has been retained [150,151]. Such compounds have considerable synthetic potential, particularly for C–C bond formation. For example, compound **49** can be lithiated with LDA/THF at 80 °C to afford C-2 branched glycals **50** (● *Scheme 16*) [152].

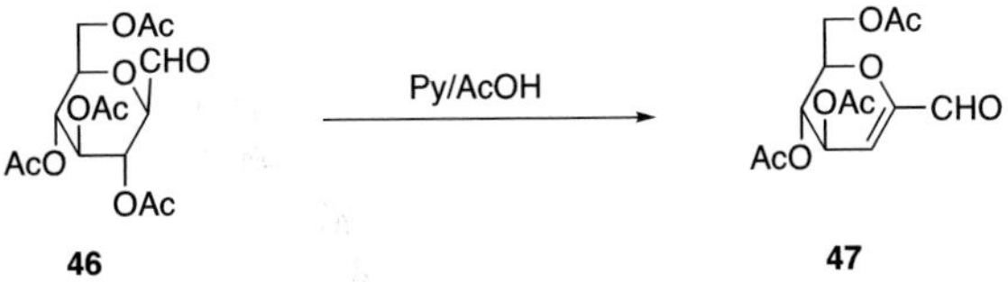

Scheme 15

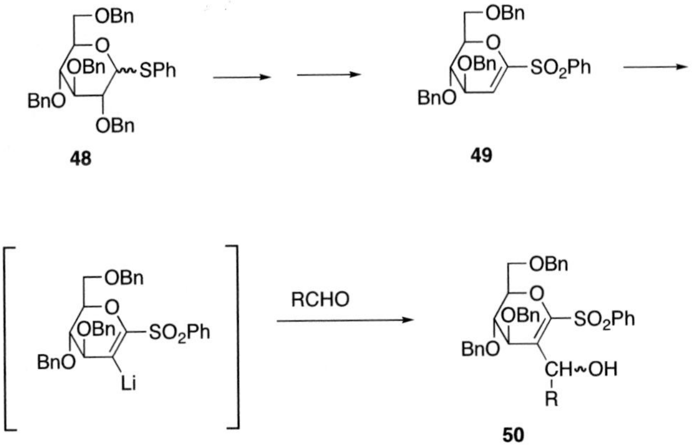

Scheme 16

Sulfonylated glycals are also useful precursors of 1-unsaturated *C*-glycosyl compounds. For example, treatment of **52** with an excess of tri-*n*-butyltin hydride in the presence of a free radical initiator (AIBN) in refluxing toluene leads to exchange of the anomeric sulfonyl group resulting in 1-stannyl glycal **53**, which in turn can be coupled with bromoaryl substrates in a palladium-catalyzed reaction to **54** (❷ *Scheme 17*) [153].

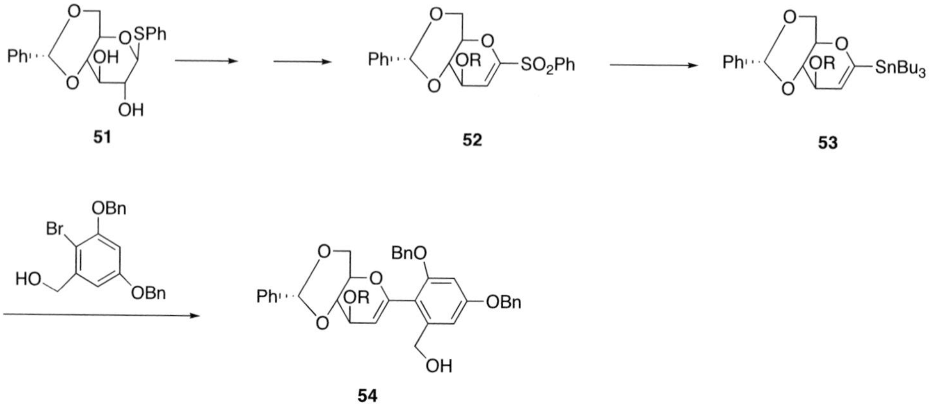

R = TBDMS

❑ **Scheme 17**

Since stannylated glycals are of particular interest as prospective components of Pd(0)-catalyzed cross-couplings (i. e. Stille reactions) [154], their preparation deserves some attention. Reactions involving the direct deprotonation of glycal derivatives, which require *tert*-butyl lithium, were first reported in 1986 [155,156]. Use of silyl ethers as protective groups is considered critical for the success of this transformation. In fact, one report recommends using triisopropylsilyl protective groups and 4 equivalents of *tert*-butyllithium for best results [157]. As shown below, lithiated glycals can be trapped by a variety of electrophiles, including tributylstannyl chloride. The stannyl derivatives **56** thus formed can be used to generate reactive intermediates that are more convenient than primary lithiation products, via exchange of protective groups for more generally compatible benzyl ethers (❷ *Scheme 18*) [141].

Lithiated glycals can also be transformed into more stable and more synthetically useful organozinc reagents [158]. Nevertheless, as has already been shown, even the primary lithiation products of the furanoid glycals **59** can be efficiently trapped by appropriate electrophiles to **60** (❷ *Scheme 19*) [159].

5 Thermal Elimination of Sulfoxides and Selenoxides to Glycals

Sulfoxides and selenoxides are known to easily undergo beta-elimination reactions. Sugar sulfoxides obtained by oxidation of respective thioglycosides were efficiently transformed to glycals upon heating in toluene (❷ *Scheme 20*) [160].

Analogous oxidation of selenoglycosides has been used to directly synthesize glycals due to the low temperature spontaneous *syn*-elimination of selenoxides [161,162].

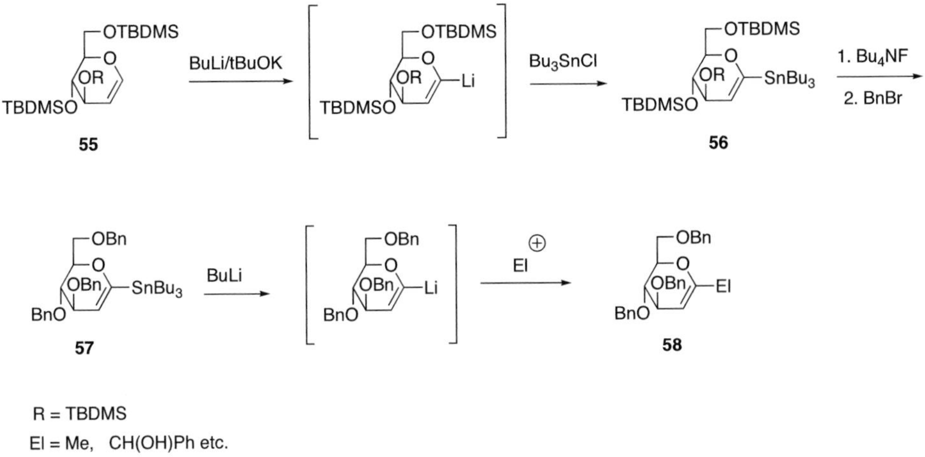

R = TBDMS
El = Me, CH(OH)Ph etc.

⬛ **Scheme 18**

Furanoid glycals utilizing elimination of selenoxides were prepared from unusual substrates, the 4-phenylselenyltetrahydrofurans. These intermediates were obtained with good yields from D-glyceraldehyde (❷ *Scheme 21*). Their oxidations have led to selenoxides, however, the elimination reaction required heating. The best yields were noted for reactions refluxed in 1,2-dichloroethane and yields ranged from 62 to 95%.

6 Glycals from 2-Deoxysaccharides

Methods for the preparation of glycals from 2-deoxy sugars, although very efficient, are limited by the availability of 2-deoxy sugars. Very often the best substrates for 2-deoxy sugars are glycals themselves. Therefore, such methods are not practical for the preparation of simple glycals, which can be easily obtained directly from commercially available monosaccharides. However, such approaches are very useful in the preparation of glycals of rare sugars or glycals of naturally occurring 2-deoxy sugars. Modified glycals are often needed for the preparation of biologically important glycosides, oligosaccharides, and sugars modified at C-2.

1-*O*-esters of 2-deoxyhexopyranoses or respective glycosyl halides appear to be useful substrates in the synthesis of glycals. When added to a silica gel suspended in dry xylene and upon heating, 1,4-di-O acetyl-*N*-trifluoroacetyl-α,β-daunosamine **68**, formed the respective glycal **69** (4-*O*-acetyl 1,5-anhydro-2,3,6-trideoxy-3-trifluoroacetamido-L-lyxo-hex-1-enitol) in 60% yield (❷ *Scheme 22*) [163]. In fact, the whole spectrum of 3-aminoglycals was prepared from 2-deoxy sugars starting from either esters or glycosyl halides. These transformations along with other approaches to 3-aminoglycals are described in detail by Pelyvas et al. [164].

Biologically interesting unsaturated derivatives of *N*-acetylneuraminic acid (Neu5Ac), a potent inhibitors of sialidase, were also obtained from peracetylated substrates. Unsaturated peracetate methyl ester of Neu5Ac2en **71** was obtained in over 90% yield via the CF$_3$SO$_3$SiMe$_3$-mediated elimination of acetic acid from **70** (❷ *Scheme 23*). Other unsaturated products were similarly prepared [133].

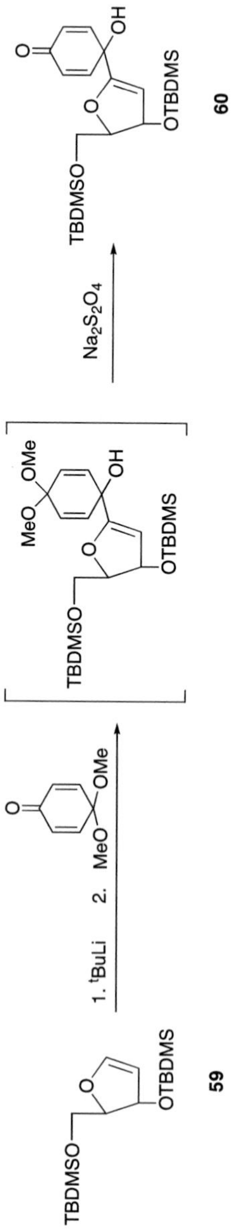

■ Scheme 19

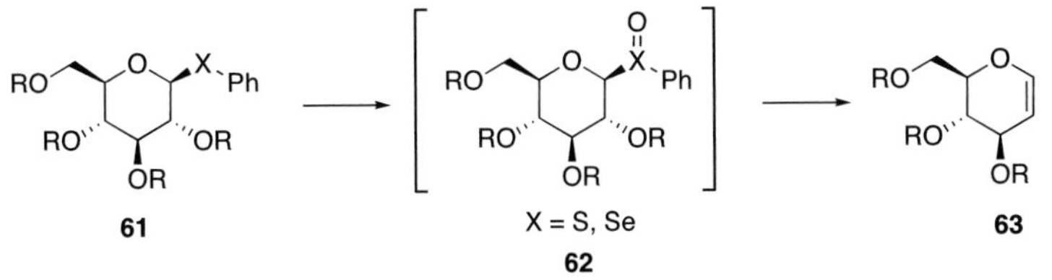

◻ Scheme 20

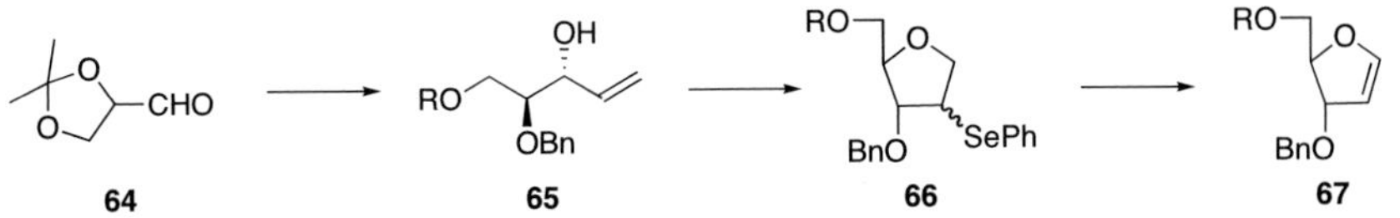

◻ Scheme 21

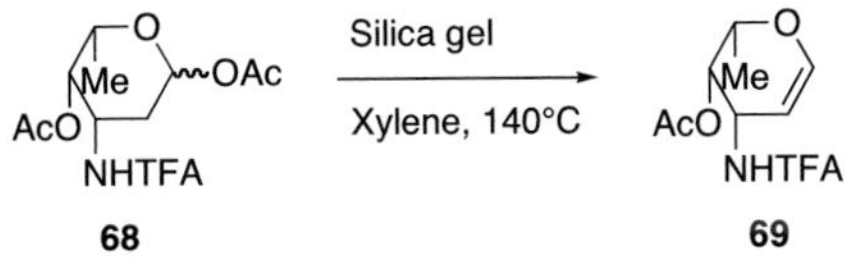

◻ Scheme 22

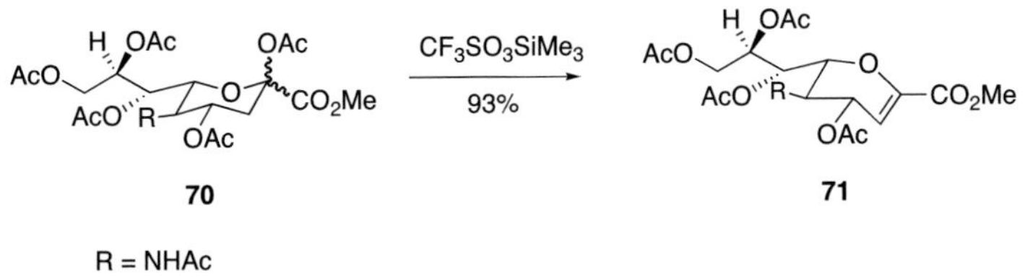

R = NHAc

◻ Scheme 23

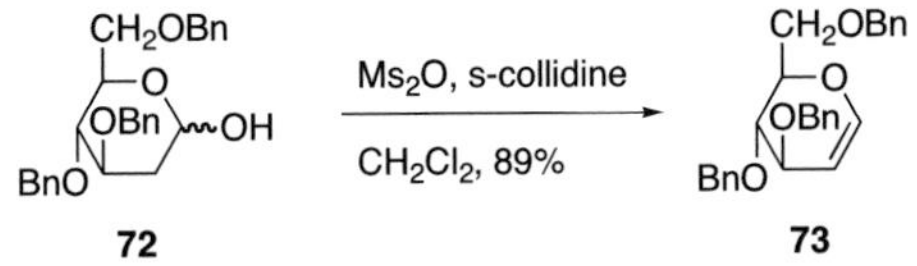

◻ Scheme 24

Glycals can also be obtained in a one-pot reaction from 2-deoxyhexopyranoses having a free hydroxyl at C-1. An example of such a transformation is high-yield (89%) synthesis of tri-O-benzyl-D-glucal **73** from 2-deoxy-3,4,6-tri-O-benzyl-D-glucopyranose **72** via 1-O-mesylate (❯ *Scheme 24*) [165].

An interesting demonstration of base-catalyzed *cis* elimination leading to 3,4,6-tri-O-acetyl-glucal (**4**) used O,O-dimethyl-S-(3,4,6-tri-O-acetyl-2(R)-2-deoxy-2-deuterio-α-D-arabinohex-opyranosyl)phosphorodithioate **72** (❯ *Scheme 25*) [166].

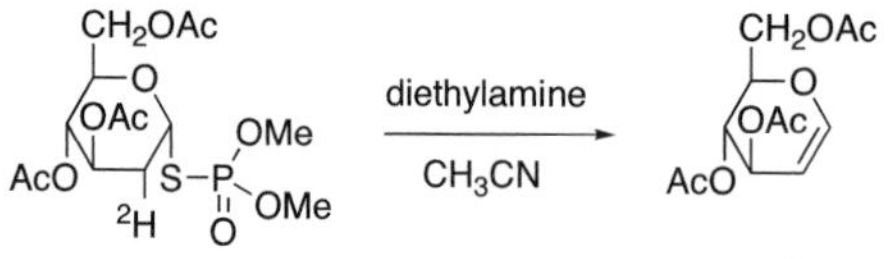

72   **4**

□ Scheme 25

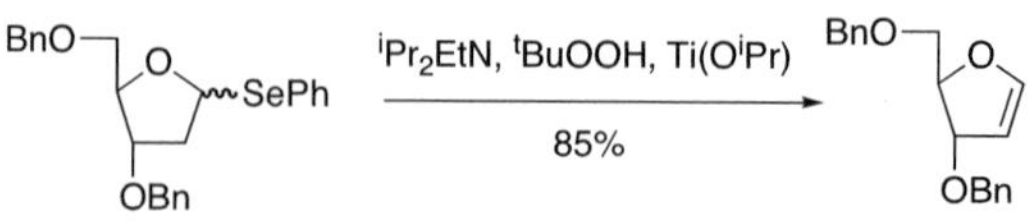

75   **76**

□ Scheme 26

77   **78**

79   **80**

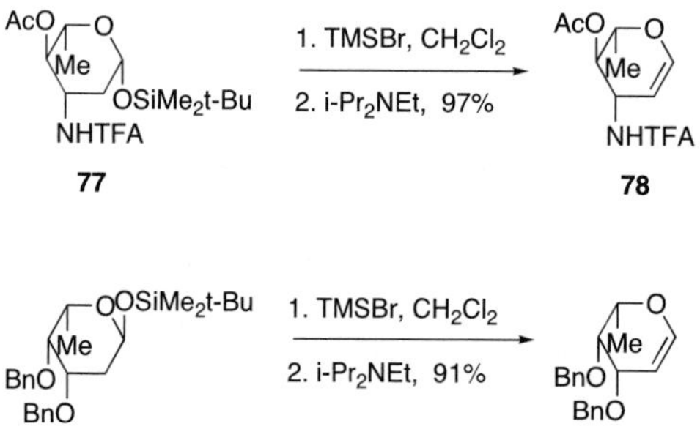

□ Scheme 27

An elimination reaction was also successfully used in the synthesis of furanoid glycals from 2-deoxyribose. The substrates used were 2-deoxy-1-seleno-furanosides **75**, which were transformed to their respective glycals **76** via selenoxide elimination in yields ranging from 71 to 85% (❷ Scheme 26) [167].

A one-pot, two-step reaction leading to glycals from 2-deoxyhexopyranoses silylated or acetylated at the anomeric center or from 2-deoxyhexopyranosides can be considered a method of choice because of its high yield and simplicity. The two-step reaction includes addition of trimethylbromosilane (TMSBr) to **77** or **79** (generation of glycosyl bromide) and subsequent addition of N,N-diisopropylethylamine to cause elimination to glycals **78** and **80**, respectively(❷ Scheme 27). In all cases, yields of isolated product exceeded 90% [163,168,169].

7 Reactions Involving 2,3-Unsaturated Saccharide-to-Glycal and Glycal-to-Glycal Rearrangements

Esterified glycals **81** easily undergo reaction with nucleophiles, typically under Lewis acid catalysis to 2,3-unsaturated pyranoses **82**. This reaction is known as Ferrier rearrangement [7].

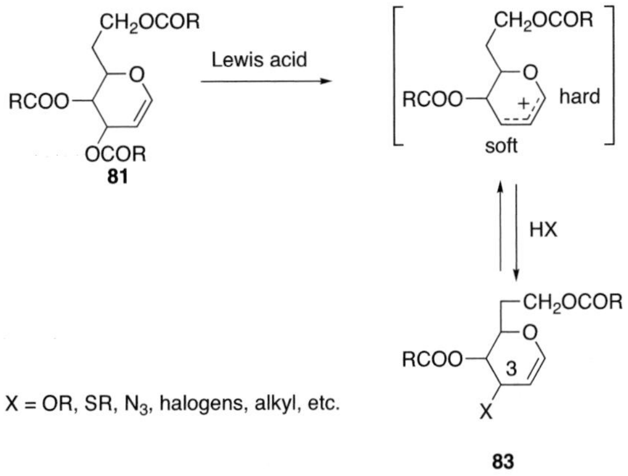

Scheme 28

The reaction initially leads to the formation of carbenium ions leaving open the possibility of attack at C-1 or C-3. Even though initial studies showed that reaction of glycals with alcohols led to 2,3-unsaturated products but no 3-*O*-alkylated glycals **83**, later studies explained the regioselectivity of the Ferrier rearrangement and clearly demonstrated the possibility of selecting attack at the C-3 carbocation to form C-3-modified glycals with good yields [170]. Use of the reverse Ferrier rearrangement of 2,3-unsaturated pyranose derivatives to produce glycals **82** to **83** is not widespread but should nevertheless be considered along with glycal-to-glycal transformation **81** to **83** (❂ *Scheme 28*).

By testing nucleophilic substrates categorized according to Pearson's Theory of Hard and Soft Acids and Bases (HSAB) [171,172] Priebe and Zamojski proposed that the regioselectivity of the Ferrier rearrangement reaction could be rationalized in terms of the HSAB principle, with the carbenium ion center at C-1 being hard and C-3 being soft [170]. They demonstrated that hard bases preferentially form 2,3-unsaturated products (via attack at C-1) following typical

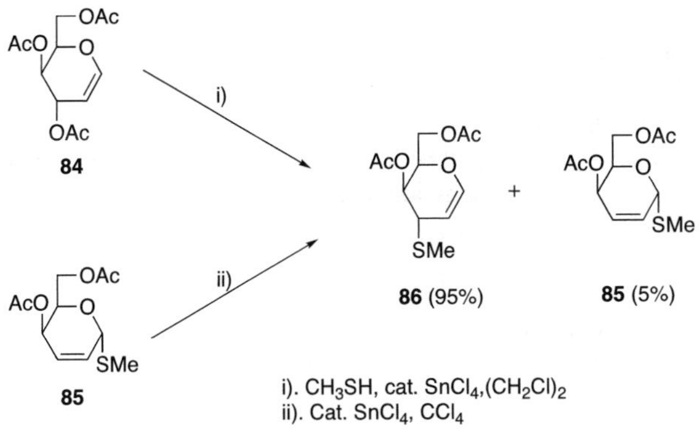

i). CH₃SH, cat. SnCl₄,(CH₂Cl)₂
ii). Cat. SnCl₄, CCl₄

Scheme 29

Ferrier rearrangement prediction, that soft bases preferentially form bonds with a carbenium ion at C-3, leading to novel glycals modified at C-3 (see below the synthesis of 3-thiomethyl glycal **86**), and that borderline bases can form both C-1 and C-3 substituted products [170]. Priebe and Zamojski also clearly demonstrated that reaction of glycals with nucleophiles is reversible and that substrates, as well as kinetically controlled products, are prone to generate intermediate carbenium ions before attaining thermodynamic equilibrium [170]. Thus, alcohols and other *O*-nucleophiles end up as anomeric (a hard acidic center) substituents, while *S*-(**86**) [170], *N*- [173,174], and *C*- [175] anions tend to form a new bond at C-3 (soft center), which leads to substituted glycals (● *Scheme 29*).

Further support for the HSAB-based rationalization comes also from later studies demonstrating that making sulfur harder (*S*-alkyl to *S*-acetyl) dramatically alters regioselectivity and leads to 2,3-unsaturated products [176,177]. In the reaction from 3,4-di-*O*-acetyl-L-rhamnal **87** with thioacetic acid or potassium thioacetate in the presence of $BF_3 \cdot Et_2O$, only products substituted at C-1 **88** and **89** were isolated (● *Scheme 30*) [177].

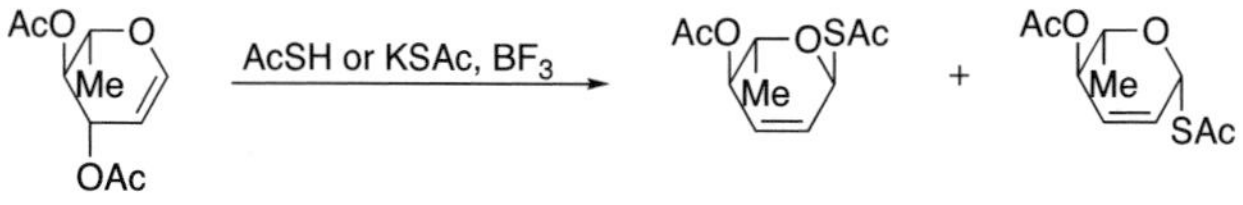

■ Scheme 30

In the 1970s and 1980s, unsaturated pyranosyl azides were studied extensively because of their potential use in synthesis of aminated antibiotic sugars [175,178,179,180,181,182]. These studies showed that both regioisomers were being formed. This can be explained by the borderline nature of azides according to the principles of HSAB. It is interesting to note that, according to an HASB-based rationalization, a hydride anion will, as a soft base in reaction with 2,3-unsaturated pyranosides, produce 3-deoxyglycals [183,184,185].

Once the stereospecific character of hydride delivery was demonstrated in experiments with deuterated LAH, the reductive removal of glycosidic functions were subsequently used to prepare pyranosyl dienes [186], which are precursors of *C*-disaccharides. The usefulness of this transformation even extends to tertiary C-4 branched hex-2-enopyranosides, provided the leaving oxygen substituents of the substrate are in a relative 1,4-*cis* configuration [187]. A similar type of reductive rearrangement was observed [188] during catalytic hydrogenation of acetylated hex-2-enopyranose **90** over platinum in the presence of amine. Under carefully selected conditions, the 3-deoxy glycal **91** could be isolated in 83% yield using a substrate obtained by $ZnCl_2$-catalyzed rearrangement of **4** in acetic anhydride (● *Scheme 31*).

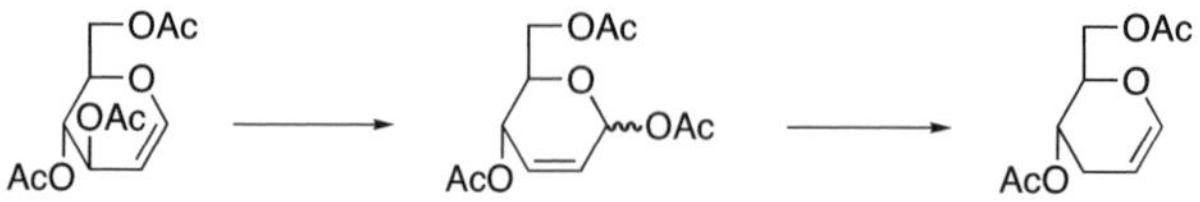

■ Scheme 31

Subsequently, it has been demonstrated that, under aprotic reaction conditions, various metal chlorides can catalyze not only double bond migration but also C-3 epimerization [189].

Some caution is required in predicting regioselectivity in analyzing Ferrier type rearrangements that are not catalyzed by Lewis acid or that occur under conditions that favor kinetically controlled products. For example, nucleophilic agents can sometimes react with glycals in the absence of Lewis catalysts to produce 2,3-unsaturated products [2]. Thus, as observed in one study, condensation of 3,4-bis-O-(4-nitrobenzoyl)-D-xylal with purines and pyrimidines without an acid catalyst led to anomeric mixtures of 2′,3′-unsaturated nucleosides [190]. Moreover, **4** can react with thiophenol to give kinetically favored phenyl 4,6-di-O-acetyl-2,3-dideoxy-1-thio-D-*erythro*-hex-2-enopyranoside **92** [191] and only a minor admixture of the expected 3-regioisomer. The tendency of this unsaturated thioglycoside to undergo rearrangement was exploited in order to synthesize 1-C-substituted glycals and related bicyclic spiroketals [192]. Also, oxidation of the phenyl thioglycoside **92** with m-CPBA to **93**, followed by treatment with piperidine, led to the allal derivative **95**, apparently via sulfenate **94** as intermediate (❯ *Scheme 32*) [193]. However, because m-CPBA can give rise to complex mixtures upon oxidation of unsaturated thioglycosides, use of 3,3-dimethyldioxirane, which cleanly affords the desired sulfoxides, is recommended [194].

The same synthetic approach was applied in obtaining a unique thiosugar and constituent of ene-diyne antitumor antibiotics; in contrast, unsaturated glycosides **96** were treated with [195] phenylselenol at −90 °C in the presence of BF$_3$·Et$_2$O to give mainly unsaturated glycoside **97** [196,197], which in turn was rearranged into glycal **98** after oxidation with m-CPBA (❯ *Scheme 33*). An analogous sequence of reactions employing a dinitrophenyl thiol protective group has been used to obtain 2,3-unsaturated thiophenyl glycosides [198].

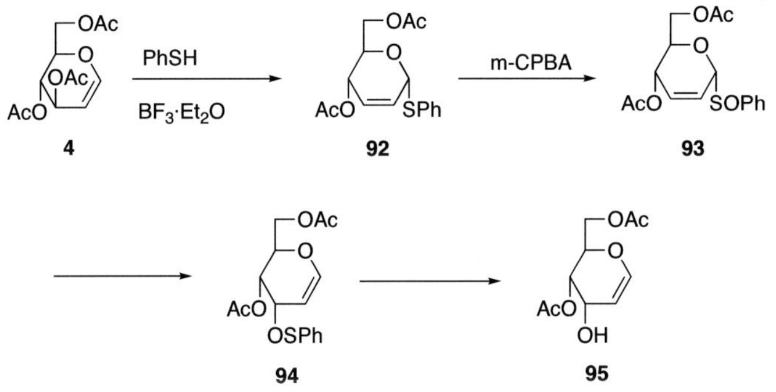

❑ Scheme 32

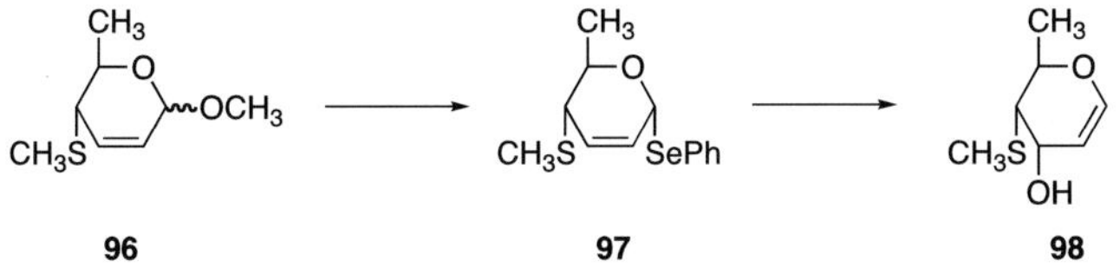

❑ Scheme 33

The above remarks concerning the regioselectivity of the Ferrier rearrangement apply also to reactions with *C*-nucleophiles but not to reactions with C-1-substituted kinetic products since such kinetic products are not susceptible to equilibration and therefore become final products. Anomeric allylation of glycals [199] and their reactions with enol ethers [200] are good examples of this. On the other hand, reaction of **4** with dimethylcuprate afforded mainly the S_N2 substitution product [201]. However, reaction with a number of *C*-nucleophiles allowed formation of all four possible alkylation products [202,203]. 2,3-Unsaturated vinyl glycosides were shown to exhibit regioselective *C*-alkylating properties depending on the conditions under which they were applied. Lewis acid catalysis afforded anomeric mixtures of *C*-glycosides, while thermal Claisen rearrangement [204] afforded stereospecific 3-*C*-branched glycals [205,206,207]. An analogous transformation of aryl glycosides (**99** and **101**) was achieved by heating substrates in *N*,*N*-diethylaniline [208]; as a result and as expected, the aryl substituents remained on the same face of the pyranoid ring (**100** and **102**) (❯ *Scheme 34*). Reaction rates differed markedly between the stereoisomers. For example, 1,4-*cis* substrates took almost 100-times longer to complete their migration.

In other experiments, a range of nucleophilic agents were reacted with sulfonyl hex-2-enopyranoside **103**, and corresponding glycals **104** and **105** were obtained stereoselectively (❯ *Scheme 35*) [209]. In this case, C-3 stereochemistry was regulated by the nature of the incoming anion.

The stereochemistry of 3-*C*-nitro glycals has been studied in some detail [210]. For example, when nitroanhydroglucitol **106** was subjected to reaction with triethylamine, it gave the elimination product **107**, which was then rearranged to an equilibrated mixture of glycals **108** and **109** (❯ *Scheme 36*) [211,212]. Some researchers have tried to explain this equilibrium shift by arguing that the quasi-equatorial anomeric proton is made more acidic by the stereoelectronic effect [213].

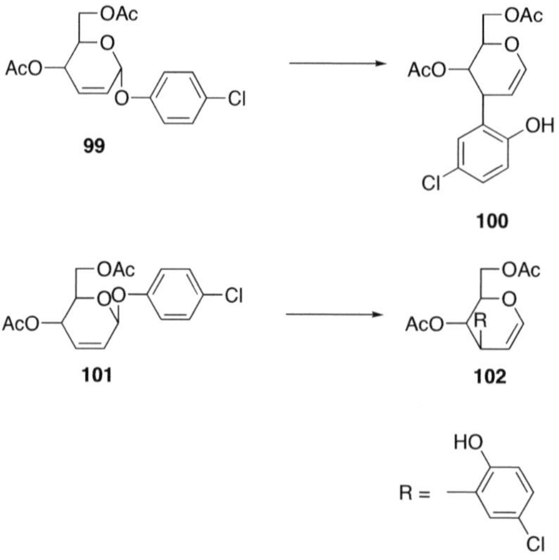

❏ Scheme 34

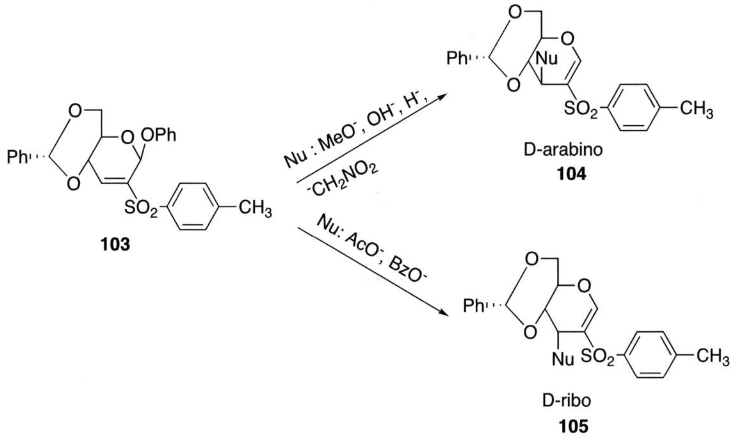

■ Scheme 35

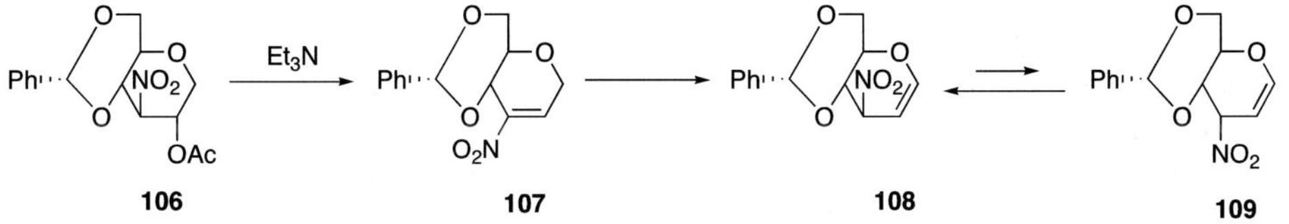

■ Scheme 36

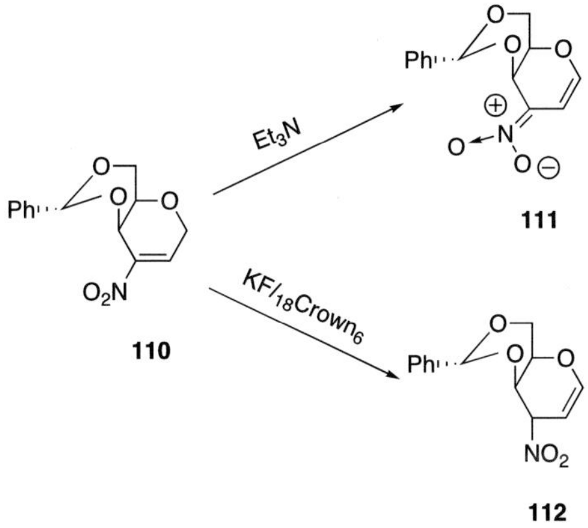

■ Scheme 37

In the case of the D-*threo* isomer **110**, nitronate **111** was isolated under the same conditions, while 3-nitroglycal **112** was isolated under conditions that included treatment with KF in the presence of 18-crown-6-ether (❷ *Scheme 37*).

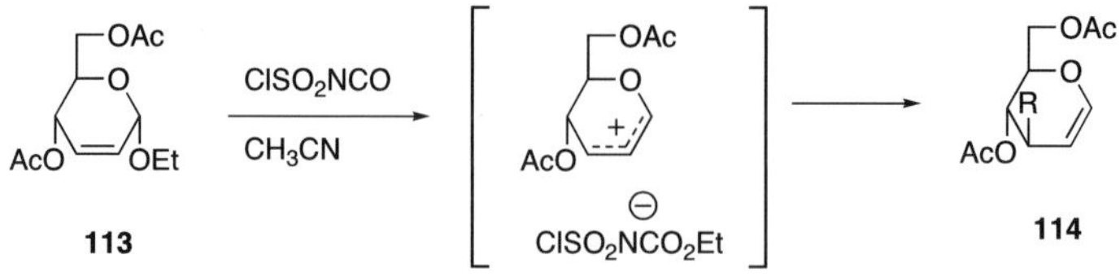

■ Scheme 38

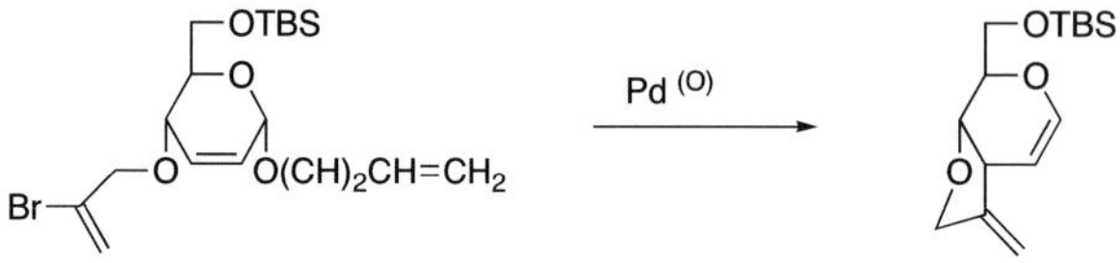

■ Scheme 39

3-Aminoglycal derivatives are of special interest in connection with the synthesis of sugar constituents of antibiotics [164], and their preparation is covered above in ● *Sect. 6*. However, reactions of hex-2-enopyranosides with chlorosulfonyl isocyanate [214,215], which involve double bond migration, (**113** to **114**), should also be mentioned since they are one-step procedures of considerable synthetic potential (● *Scheme 38*).

In recent years unsaturated sugars have evoked much interest as chiral templates for assembling polycyclic systems in a stereocontrolled way (e. g., by palladium-catalyzed transformation). An example of palladium-catalyzed transformation involving the formation of a 1,2-double bond **116** from a 2,3-unsaturated precursor **115** is given below (● *Scheme 39*) [216,217]. Many more "annulated sugars" of the same general structure have been obtained by using an analogous methodology [218].

8 Cycloadditions and Cyclizations

The potential of the hetero Diels–Alder reaction as a means of synthesizing sugar derivatives has long been recognized and explored [219,220]. More recently, introduction of a new generation of electron-rich dienes [221] and stereoselective catalysts of cycloaddition [222] have expanded the usefulness and practicality of this approach. The reaction of 1-alkoxy-3-silyloxy-1,3-butadienes with aldehydes shown below has been studied in great detail with respect to the influence exerted by the structural features of the reagents, as well as by the catalysts (● *Scheme 40*). Two general pathways—cycloaddition and aldol-like condensation followed by heterocyclization—have been identified [223,224], and both are relevant to glycal chemistry.

Indeed, pent-1-enitols [225], glucal derivatives [226], and galactal and fucal analogs have all been obtained via 1-en-3-ones by using the above-mentioned approach. Thermal cyclocondensation of aldehyde **117** with diene **118** gave racemic enone **119** (● *Scheme 41*), which is used as a synthon in the preparation of the rifamycin S chain [227].

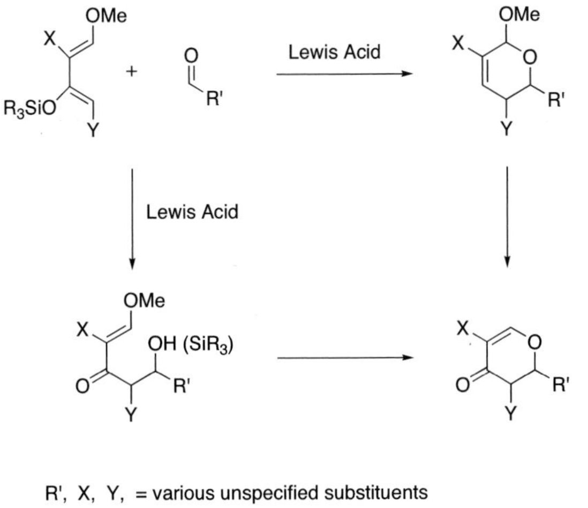

R', X, Y, = various unspecified substituents

■ Scheme 40

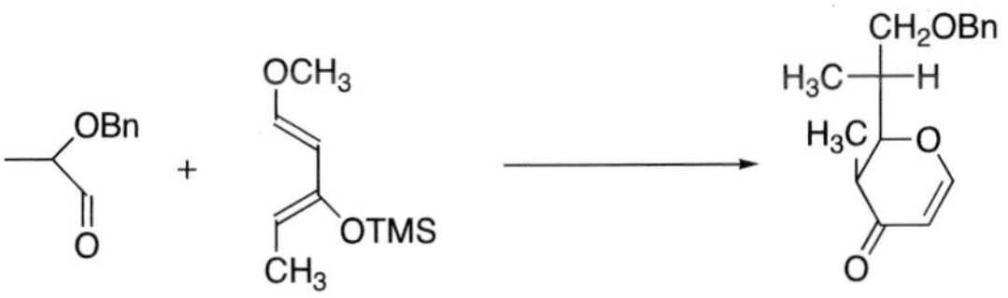

117   118   119

■ Scheme 41

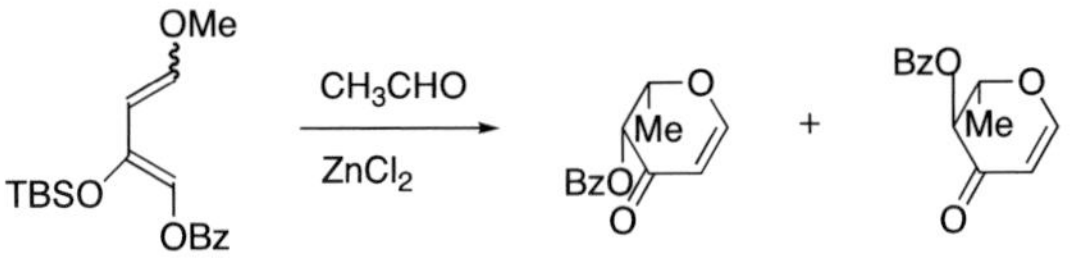

120   121   122

■ Scheme 42

Reaction of isomeric dienes **120** with acetaldehyde in the presence of zinc chloride gave a mixture of C-4 epimeric enones **121**, **122** (❷ *Scheme 42*) [224], which are convenient substrates for synthesis of all four racemic 3-amino-2,3,6-trideoxy pyranoses.

Similarly, reaction of 1,4-diacetoxy-1,3-butadiene with methyl glyoxalate afforded glucuronate glycal [228] and a 2,3-unsaturated isomer. Cycloaddition reactions performed on substrates with inverse electronic properties (e. g., enol ether as a dienophile and unsaturated carbonyl compound as a heterodiene) afford not only the expected products but also ones having high *endo* selectivity [229,230] as exemplified below by olivose **126** synthesis (❷ *Scheme 43*).

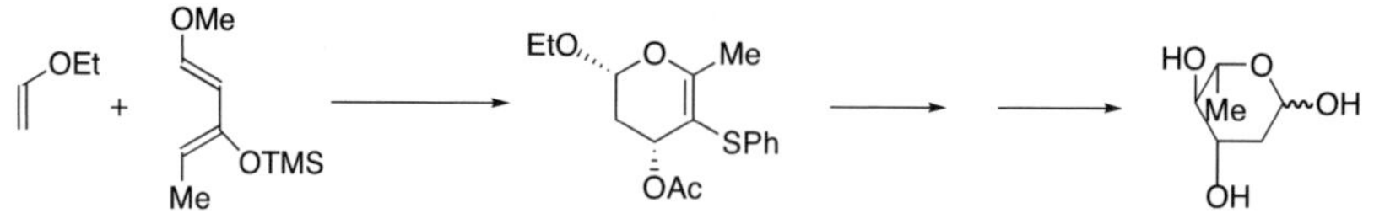

123   124   125   126

endo : exo > 10 : 1

◻ Scheme 43

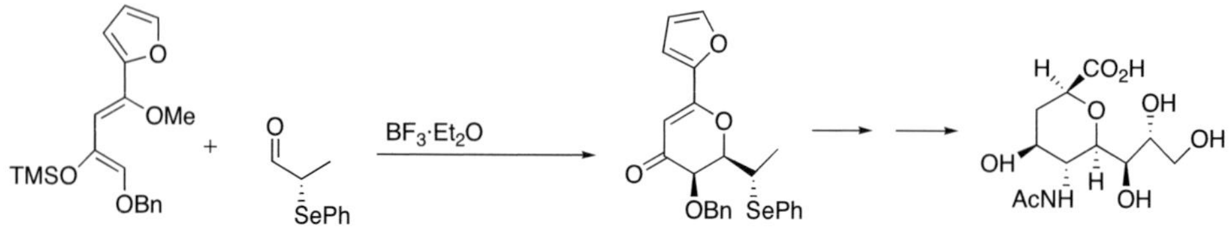

127   128   129   130

◻ Scheme 44

Studies of hetero Diels–Alder condensation that used enantiopure substituted aldehydes found that the catalyzed additions had a high and predictable stereoselectivity [231]. The synthesis of *N*-acetylneuraminic acid **130** [232] from diene **127** and (S)-selenoaldehyde **128**, which is shown below, illustrates this point (**❷** *Scheme 44*).

A brand-new methodology for synthesizing glycals from noncarbohydrate precursors, one based on cyclization of acetylenic alcohols, has emerged from the field of metalorganics. Molybdenum pentacarbonyl-trialkylamine complexes have been found to efficiently catalyze cyclization of 1-alkyn-4-ols to substituted dihydrofurans [233,234]. This same transformation has been successfully carried out on asymmetrically substituted alcohols; the furanoid glycals **132**, **134**, and **135** (**❷** *Scheme 45*) so obtained have in turn been used as intermediates in the synthesis of nucleosides [235].

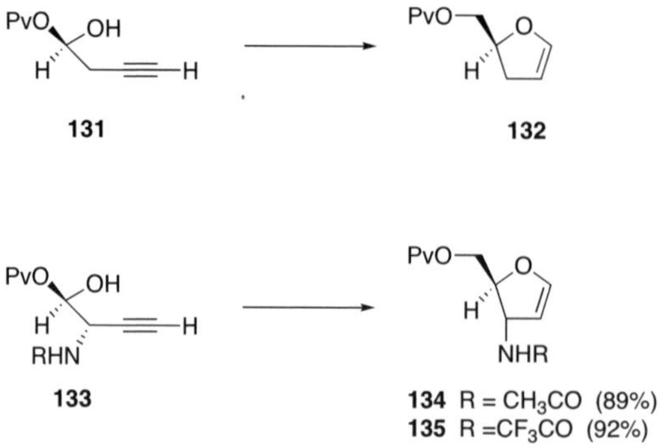

131   132

133   134 R = CH₃CO (89%)
                       135 R = CF₃CO (92%)

◻ Scheme 45

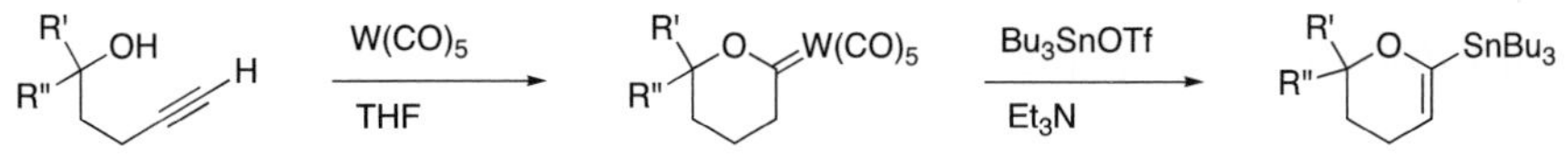

136 **137** **138**

Scheme 46

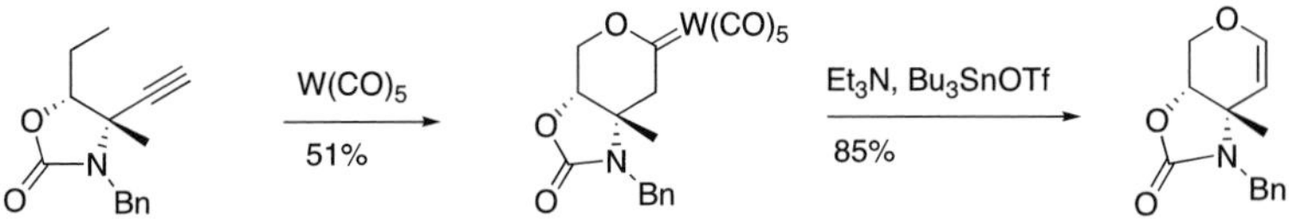

139 **140** **141**

Scheme 47

The transformation has been extended even further to pyran derivatives by using a tungsten pentacarbonyl-tetrahydrofuran complex. Primary, secondary, and tertiary 1-alkyn-5-ols **136** undergo cyclization to dihydropyranylidene carbenes, which can then be converted into the corresponding stannyl dihydropyranes **138** upon treatment with tributyltin triflate and triethylamine (➧ *Scheme 46*) [236].

3-Aminosugar-derived glycal **141** was obtained by this method from an appropriate chiral precursor **139** [237]. Unexpectedly, however, reaction of tungsten intermediate **140** with tri-*n*-butyltin triflate and triethylamine in ether gave the glycal **141** (in 85% yield) instead of the 1-stannyl derivative (➧ *Scheme 47*). The molybdenum carbonyl-triethylamine complex afforded the same product, although in a much lower yield.

Glycals can also be obtained from suitable substrates, by ring-closing olefin metathesis reactions. In a general approach to variously linked *C*-disaccharides, illustrated in ➧ *Scheme 48*, nonreducing-end glycals are typical intermediates, finally subjected to hydroboration or dihydroxylation to afford the desired *C*-analogs of *O*-disaccharides [238,239].

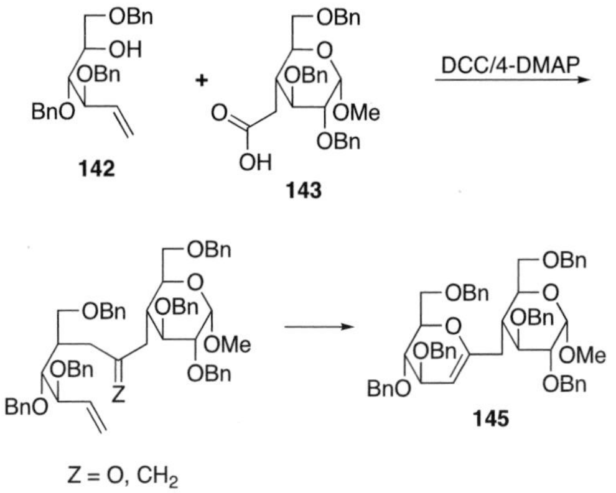

142 **143** **144** **145**

Scheme 48

9 Preparation of exo-Glycals

Exomethylene monosaccharide derivatives became of interest in connection with increased application of *C*-glycosyl synthons in the preparation of glycomimetics intended as glycosidase inhibitors and glycoconjugates. Their syntheses and uses have been comprehensively reviewed [240].

Sugar lactones are the most convenient substrates in the synthesis of exo-glycals. Their methylenation using Tebbe's reagent has been efficiently used but application of the Wittig reaction also allowed for the preparation of a series of exo-glycals (❯ *Scheme 49*) [241,242]. Suitably protected sugar lactones can also easily undergo addition of nucleophiles, with creation of a ketosyl anomeric center from which the hydroxyl group can be easily eliminated to exo-glycals by treatment with trifluoroacetic anhydride in the presence of pyridine [243]. The same group of authors described simple and general synthesis of conjugated *exo*-glycals. In the key step β-C-1 carbonyl sugar derivative was reacted with Wittig reagent followed by phenylselenyl chloride to place the substituent which undergoes easy elimination after oxidation to sulfoxide [244]. Glycosyl cyanides (2,5-anhydroaldonitriles) are easily converted into corresponding aldehyde tosylhydrazones, which in turn undergo base-catalyzed decomposition to *C*-glycosylmethylene carbenes and eventually to *exo*-glycals (❯ *Scheme 50*) [245,246]. These reaction conditions are compatible with the presence of acyl protecting groups.

Application of Ramberg–Bäcklund rearrangement provides a very effective approach to exo-glycals, in which α-halogenated sulfones are converted into olefins, with concomitant expulsion of sulfur dioxide. Since thioglycosides are easily available and there are several methods for their efficient oxidation to sulfones, this method seems to offer a wide scope, frequently

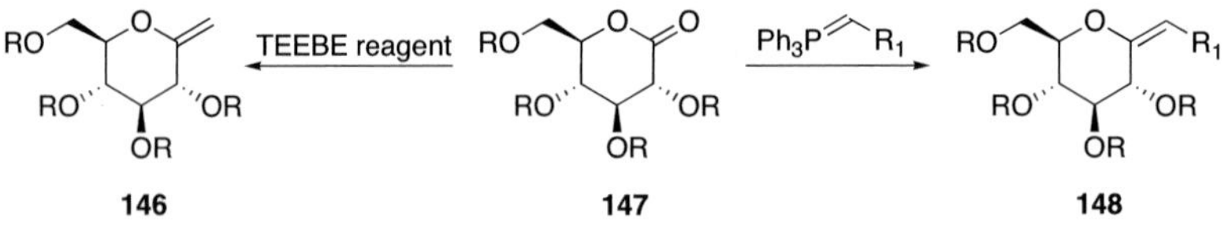

□ Scheme 49

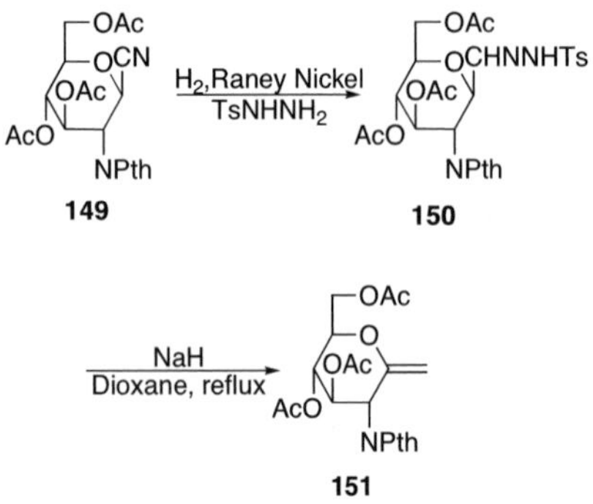

□ Scheme 50

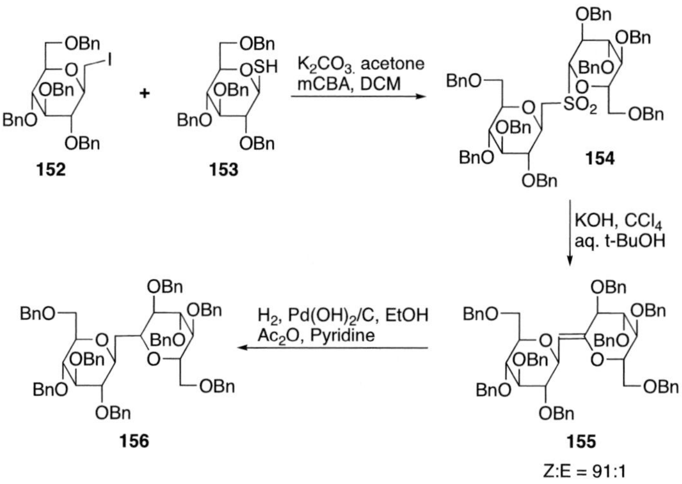

□ **Scheme 51**

lacking in the case of previously described methods. In particular, this method is suitable for the synthesis of *C*-linked disaccharides and their analogs via reaction of 1-thio-sugars with iodo-sugars as shown in ❍ *Scheme 51*.

The application of the Ramberg–Bäcklund reaction for the synthesis of *C*-glycosides and related compounds has been recently discussed in detail [247].

10 Miscellaneous Methods

Specifically substituted glycals, which are often used as chiral synthons or key substrates, can be prepared in many different ways. Double bond formation frequently precedes the assembly of the complete carbon framework designed for a cyclic enol ether intermediate. Of particular interest here are glycals substituted at C-1 and C-2, which are obtained by functionalization of the existing double bond. For example, C-2-acylated glycals offer a variety of ways to generate synthetically useful transformations. C-2 formyl glycals were first obtained by syntheses based on dithiane [248] and enol ether [249]. Exhaustive hydrogenation of the enone, which leads to C-2-methyl pyrane, has been applied as the final step in the total synthesis of the antibiotic restricitin [250]. In an alternative approach, direct formylation of methyl **157** or benzyl ethers of glycals with Vilsmeier–Haack reagent has been achieved [251], and it has been shown that these products **158** can undergo BF₃/Et₂O catalyzed C-3 arylation to **159** (❍ *Scheme 52*) [252]. The C-2 formyl glycals have an obvious connection, through Wittig chemistry, with C-2 vinyl glycals. The C-2 vinyl glycals, in their turn, have been studied as the diene components of Diels–Alder cycloaddition reactions [253]. Interestingly, C-2 formylation can also be achieved by free radical-promoted cyclization-fragmentation of suitably arranged pyranose derivatives. For example, in one study, 2-deoxy-2-iodo– mannoside **160** was converted into formylated glycal **161** [254] in 65% yield (❍ *Scheme 53*), while the corresponding benzylated glycoside afforded saturated C-2 branched products.

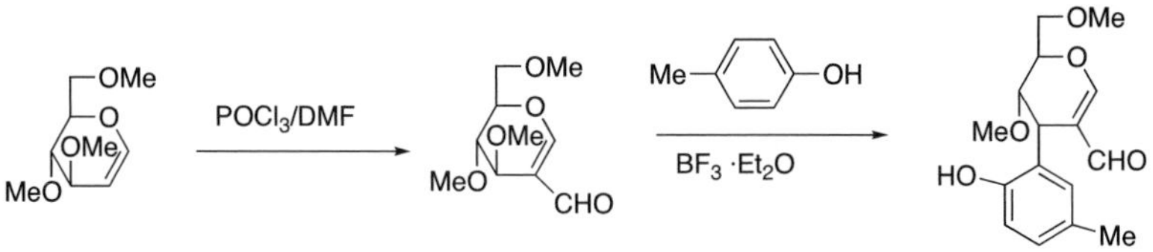

157 158 159

Scheme 52

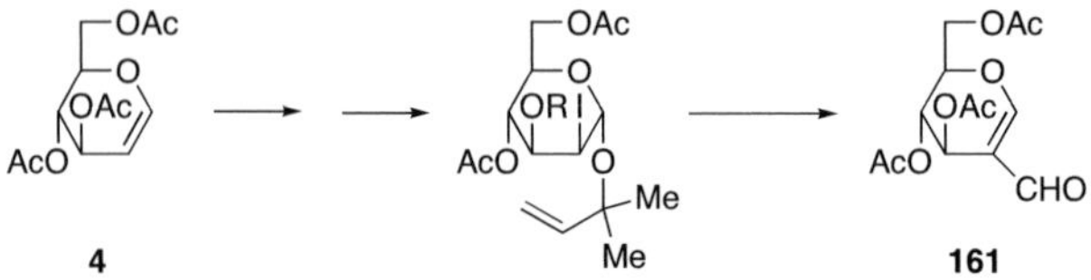

4 160: R = Ac 161

Scheme 53

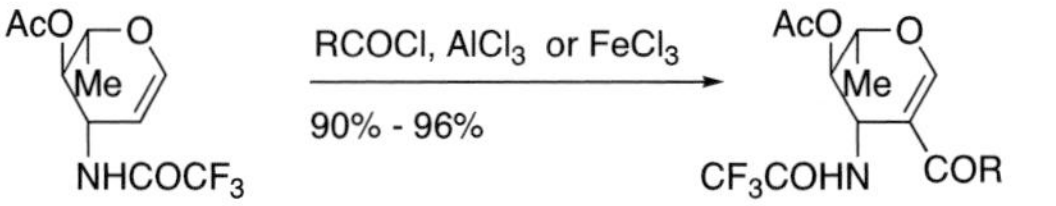

162 163

R = alkyl; C1 - C17

Scheme 54

Although the general scope of direct acylation of olefins under Friedel–Crafts conditions is rather limited, it has been demonstrated that such reactions can be applied to glycals like **162** to produce C-2 acylated glycals **163** in excellent yields and selectivity (➋ *Scheme 54*). Moreover, as reagents, the long-chain fatty acid based acylating species have been shown to be as effective as acetyl chloride [255].

Interestingly, such C-acylated glycals **163** can also be obtained in a one-pot reaction from 2-deoxyhexopyranoses **164** and **165** in surprisingly good yield (➋ *Scheme 55*). Friedel–Crafts acylation is only one of several sequentially occurring reactions [255].

This approach also allowed preparation, in a one-pot reaction, of the 2-C and 4-O acylated glycal **167** [255] starting from 2-deoxyhexopyranose **166** having a free hydroxyl at C-4 (➋ *Scheme 56*).

The addition of isocyanates to glycals, which has been studied thoroughly in connection with new methods of synthesizing lactams [256], usually affords a mixture of [2+2] and [4+2] cycloadducts. These primary reaction products are rather unstable and slowly rearrange to unsaturated amides, which can often be isolated directly from reaction mixtures [257,258]. In the case of simple dihydropyran derivatives **168**, unsaturated *N*-substituted C-2 formamides **170** are the only isolable products (➋ *Scheme 57*) [259,260].

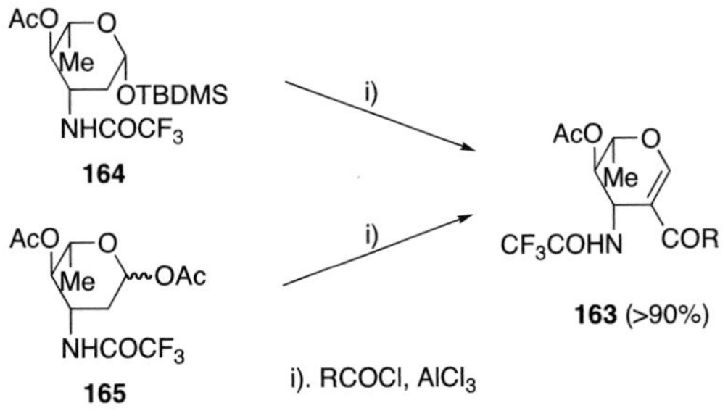

Scheme 55

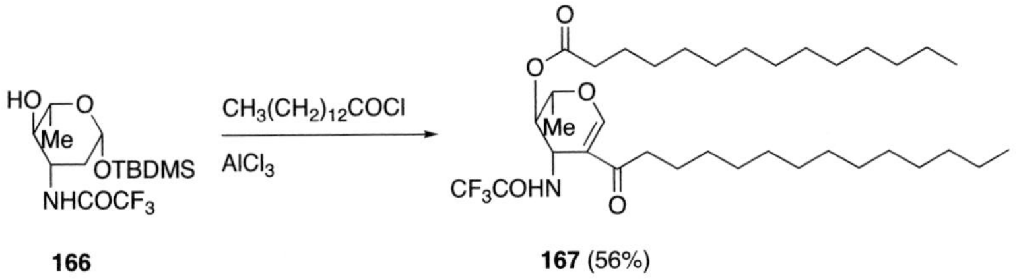

Scheme 56

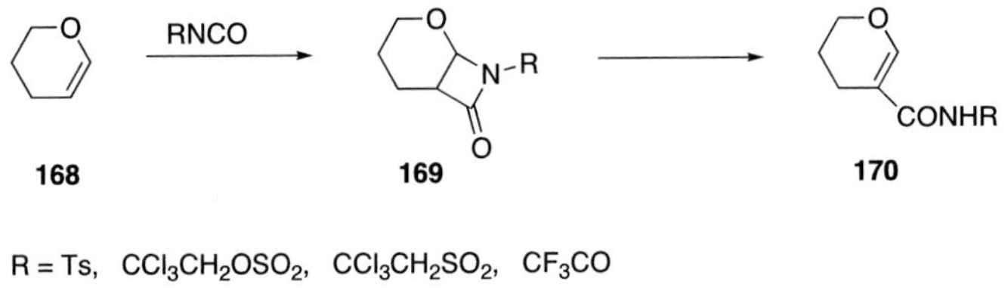

R = Ts, CCl₃CH₂OSO₂, CCl₃CH₂SO₂, CF₃CO

$R = Ts,\ CCl_3CH_2OSO_2,\ CCl_3CH_2SO_2,\ CF_3CO$

Scheme 57

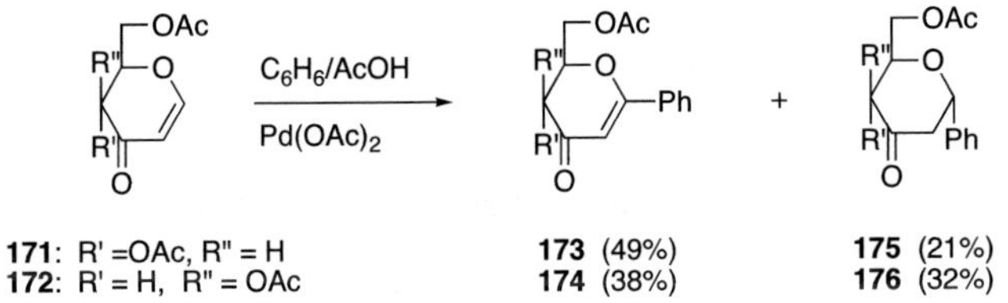

171: R' =OAc, R" = H
172: R' = H, R" = OAc

173 (49%)
174 (38%)

175 (21%)
176 (32%)

Scheme 58

Glycals have been used extensively as substrates in various approaches to *C*-glycosyl compounds. Among the studies devoted to the Heck-type catalytic coupling [261] between carbohydrate derivatives and aromatic compounds, there are several examples of the glycal double bond being preserved during such reaction. For example, in reaction with benzene, peracylated pyranosyl eneones **171** and **172** mainly form unsaturated 1-*C*-phenyl substituted products **173–176** (◉ *Scheme 58*) [262].

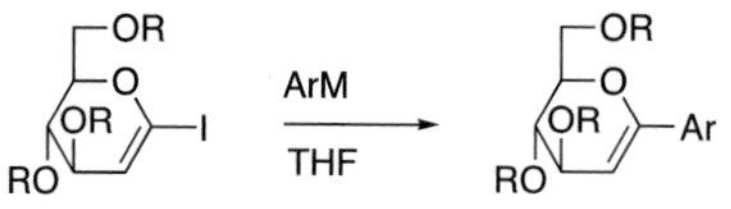

○ **Scheme 59**

The discovery (already mentioned) that benzylated or silylated 1-tri-n-butylstannyl derivatives of glucal undergo palladium-catalyzed cross-couplings with aryl or alkenyl halides [263,264] facilitated synthesis of naturally occurring *C*-glycosyl compounds [264]. Interestingly, silylated 1-iodo D-glucal **177**, obtained by a two-step procedure consisting of stannylation and tin-iodine exchange, can also be coupled with a variety of metalated aromatic derivatives to afford a high yield of 1-*C*-arylated compounds **178** (❯ *Scheme 59*) (the best results were obtained using arylboronic acids and ArZnCl derivatives) [265]. 1-*C*-arylated compounds, in turn, are useful precursors of more complex *C*-glycosyl natural products. In an analogous way, iodoglucal can be obtained in unexpectedly high yield (75%) upon attempted coupling of C-1 stannylated derivative to 3-iodo-2-propyn-1-ol [263].

2-Cyanoglycals [266], which have found application in the synthesis of branched sugars [267], can be obtained in moderate yields by addition of chlorosulfonyl isocyanate to acylated glycals and treatment of intermediate *N*-chlorosulfonamides with triethylamine. Potentially useful *S*-heterocyclic analogs of glycals have been obtained predominantly by adaptation of the reductive elimination method described in the first part of this chapter [268,269].

References

1. Fischer E, Zach K (1913) Sitzber Kgl Preuss Akad Wiss 16:311
2. Helferich B (1952) Adv Carbohydr Chem Biochem 7
3. Nicolau KC, Sorensen EJ (1966) Classics in Total Synthesis. VCH, Weinheim
4. McNaught AD (1997) Adv Carbohydr Chem Biochem 52
5. McNaught AD (1997) J Carbohydr Chem 16:1191
6. Blair MG (1954) Adv Carbohydr Chem 9:97
7. Ferrier RJ (1965) Adv Carbohydr Chem 20:67
8. Ferrier RJ (1969) Adv Carbohydr Chem 24:199
9. Ferrier RJ (1967–1999) In: Ferrier RJ (ed) (1967–1999) A Specialist Periodical Report; Carbohydr Chem 1–33
10. Danishefsky SJ, Bilodeau MT (1996) Angew Chem Int Ed Engl 35:1380
11. Seeberger PH, Bilodeau MT, Danishefsky SJ (1997) Aldrichimica Acta 30:75
12. Di Busolo V, Kim YJ, Gin DY (1998) J Am Chem Soc 120:13515
13. Thiem J, Karl H, Schwentner J (1978) Synthesis: 696
14. Gervay J, Danishefsky SJ (1991) J Org Chem 56:5448
15. Thiem J, Klaffke W (1990) Top Curr Chem 154:285
16. Grewal G, Kaila N, Franck RW (1992) J Org Chem 57:2084
17. Seeberger PH, Danishefsky SJ (1998) Acc Chem Res 31:685
18. Daves JGD (1990) Acc Chem Res 23:201
19. Postema MHD (1995) C-Glycoside Synthesis. CRC Press, Boca Raton, FL
20. Levy DE, Tang C (1995) The Chemistry of C-glycosides. Pergamon Press, Oxford

21. Bols M (1996) Carbohydrate building blocks. Wiley, New York
22. Hanessian S (1993) Total Synthesis of Natural Products: The "Chiron" Approach. Pergamon Press, Oxford
23. Klingler FD, Psiorz M (1992) Chimica Oggi 3:47
24. Yuan X, Linhardt RJ (2005) Curr Top Med Chem 5:1393
25. Hang HC, Bertozzi CR (2001) Accounts Chem Res 34:727
26. Meutermans W, Le GT, Becker B (2006) Chem Med Chem 1:1164
27. Sofia MJ (1998) Mol Divers 3:75
28. Chakraborty TK, Gosh S, Jayaprakash S (2002) Curr Med Chem 102:49
29. Seeberger PH (2003) Chem Comm: 1115
30. Palmacci ER, Plante OJ, Hewitt MC, Seeberger PH (2003) Helvetica Chimica Acta 86:3975
31. Ferrier RJ (2001) Topics Curr Chem 215:153
32. Ferrier RJ, Hoberg JO (2003) Adv Carbohydr Chem Biochem 58:55
33. Somsak L (2001) Chem Rev 101:81
34. Ferrier RJ, Zubkov OA (2003) Org React 62:569
35. Wulff G, Rohle G (1973) Angew Chem Int Ed Engl 13:157
36. Paulsen H (1982) Angew Chem Int Ed Engl 21:155
37. Jeanloz RW, Stoffyn PJ (1962) In: Whistler RL (ed) Methods in Carbohydrate Chemistry, vol I. Academic Press, New York, p 221
38. Wolfrom ML, Thompson A (1963) In: Whistler RL (ed) Methods in Carbohydrate Chemistry, vol II. Academic Press, New York, p 211
39. Redemann CE, Niemann C (1942) Org Synth 22:1
40. Kartha KPR, Jennings HJ (1990) J Carbohydr Chem 9:777
41. Helferich B, Mulcahy EN, Zeigler H (1954) Chem Ber 87:233
42. Collins P, Ferrier RJ (1995) Monosaccharides: Their Chemistry and Their Roles in Natural Products. Wiley, Chichester, p 316
43. Humoller FL (1962) Methods in Carbohydrate Chemistry, vol 1. Academic Press, New York, p 83
44. Roth W, Pigman W (1963) In: Whistler RL (ed) Methods in Carbohydrate Chemistry, vol II. Academic Press, New York, p 405
45. Blair MG (1963) Methods in Carbohydrate Chemistry, vol II. Academic Press, New York, p 411
46. Shafizadeh F (1963) Methods in Carbohydrate Chemistry, vol II. Academic Press, New York, p 409
47. Weygand F (1962) Methods in Carbohydrate Chemistry, vol 1. Academic Press, New York, p 182
48. Erdik E (1987) Tetrahedron 43:2203
49. Hurd CD, H. J (1966) Carbohydr Res 2:240
50. Deriaz RE, Overend WG, Stacey M, Teece EG, Wiggins LF (1949) J Chem Soc: 1879
51. Iselin B, Reichstein T (1944) Helv Chim Acta: 1146
52. Maran FEV, Catelani G, D'Angeli F (1989) Electrochimica Acta 34:587
53. Wisniewski A, Skorupowa ERW, Glod D (1991) Pol J Chem 65:875
54. Pollon JHP, Llewellyn G, Williams JM (1989) Synthesis: 758
55. Jain S, Suryawaneshi SN, Bhakuni DS (1987) Indian J Chem 26B:866
56. Ness RK, Fletcher JHG (1963) J Org Chem 28:435
57. Haga M, Ness RK (1965) J Org Chem 30:158
58. Eitelman SJ, Jordaan A (1977) J Chem Soc Chem Commun: 552
59. Eitelman SJ, Hall RH, Jordaan A (1978) J Chem Soc Perkin Trans I: 595
60. Ireland RE, Thaisrivongs S, Vanier N, Wilcox CS (1980) J Org Chem 45:48
61. Armstrong PL, Coull IC, Hewson AT, Slater MJ (1995) Tetrahedron Lett 36:4311
62. Ireland RE, Wilcox CS (1977) Tetrahedron Lett: 2389
63. Ireland RE, Wilcox CS, Thaisrivongs S (1978) J Org Chem 43:786
64. Cheng JCY, Hacksell U, Daves Jr GD (1985) J Org Chem 50:2778
65. Holzapfel CW, Koekemoer JM, Verdoorn GH (1986) S Afr J Chem 39:151
66. Csuk R, Furstner A, Glanzer BI, Weidmann H (1986) J Chem Soc Chem Commun: 1149
67. Csuk R, Glanzer BI, Furstner A, Weidmann H, Formacek V (1986) Carbohydr Res 157:235
68. Furstner A, Weidmann H (1988) J Carbohydr Chem 7:773
69. Boullanger P, Martin JC, Descotes G (1975) J Heterocycl Chem 12:91
70. Chmielewski M, Fokt I, Grodner J, Grynkiewicz G, Szeja W (1989) J Carbohydr Chem 8:735

71. de Pouilly P, Vauzeilles B, Mallet JM, Sinay P (1991) CR Acad Sci Paris 313:1391

72. Cavallaro CL, Schwartz J (1995) J Org Chem 60:7055

73. Spencer RP, Schwartz J (1996) Tetrahedron Lett 37:4357

74. Spencer RP, Schwartz J (1997) J Org Chem 62:4204

75. Kovacs G, Gyarmati J, Somsak L, Micskei K (1996) Tetrahedron Lett 37:1293

76. Somsak L, Nemeth I (1993) J Carbohydr Chem 12:679

77. Somsak L, Bajza I, Batta G (1990) Liebids Ann Chem: 1265

78. Mahmoud SH, Somsak L, Farkas I (1994) Carbohydr Res 254:91

79. Kiss L, Somsak L (1996) Carbohydr Res 291:43

80. Somsak L, Madaj J, Wisniewski A (1997) J Carbohydr Chem 16:1075

81. Chemla F (2002) J Chem Soc Perkin Trans 1:275

82. Boutureira O, Rodriguez MA, Matheu MI, Diaz Y, Castillion S (2006) Org Lett 8:673

83. Boulineau FP, Wei A (2002) Org Lett 4:2281

84. Santoyo-Gonzales F, Calvo-Flores FG, Hernandez-Mateo F, Garcia-Mendoza P, Isac-Garcia J, Perez-Alvarez D (1994) Synlett: 454

85. Lancelin JM, Morin-Allory L, Sinay P (1984) J Chem Soc Chem Commun: 355

86. Fernandez-Mayorales A, Marra A, Trumtel M, Veyriers A, Sinay P (1989) Tetrahedron Lett 30:2537

87. Fernandez-Mayorales A, Marra A, Trumtel M, Veyriers A, Sinay P (1989) Carbohydr Res 188:81

88. Casillas M, Gomez AM, Lopez CJ, Valverde S (1996) Synlett: 628

89. Shull BK, Wu Z, Koreeda M (1996) J Carbohydr Chem 15:955

90. Forbes CL, Franck RW (1999) J Org Chem 64:1424

91. Scheffold R, Rytz G, Walder R (1983) In: Scheffold R (ed) Modern Synthetic Methods, vol 3. Springer, Berlin Heidelberg New York, p 355

92. Bergmann M (1923) Justus Liebigs Ann Chem 434:79

93. Bergmann M, Freudenberg K (1929) Ber Dtsch Chem Ges 62

94. Haworth WN (1930) J Chem Soc: 2644

95. Dauben WL, Evans X (1938) J Am Chem Soc 60:886

96. Rolland N, Vass G, Cleophax J, Sepulchre AM, Gero SD, Cier A (1982) Helv Chim Acta 65:1627

97. Wang LX, Sakairi N, Kuzuhara H (1991) Carbohydr Res 219:133

98. Spohr U, Bach M, Spiro RG (1993) Can J Chem 71:1919

99. Ziegler T (1995) Liebigs Ann Org Bioorg Chem 6:949

100. Kanemitsu T, Ogihara Y, Takeda T (1997) Chem Pharm Bull 45:643

101. Ferrier RJ (1980) In: Pigman W (ed) The Carbohydrates, Chemistry and Biochemistry, vol IB. Academic Press, New York, p 843

102. Maurer K, Mahn H (1927) Ber Dtsch Chem Ges 60:1316

103. Bredereck H, Wagner A, Faber G, Huber W, Immel G (1958) Chem Ber 91:2891

104. Zemplen G, Bruckner X (1928) Ber Dtsch Chem Ges 61:2481

105. Maurer K (1930) Ber Dtsch Chem Ges 63:25

106. Lemieux RU, Lineback DR (1965) Can J Chem 43:94

107. Backinovski LV, Tsvetkov YE, Bairamova NE, Balan NF, Kochetkov NK (1980) Izv Akad Nauk SSSR, Ser Khim 8:1905

108. Tsvetkov YE, Byramova NE, Backinovski LV (1983) Carbohydr Res 115:254

109. Honda T, Inoue M, Kato M, Shima K, Shimamoto T (1987) Chem Pharm Bull 35:3975

110. Shah RH, Bahl OP (1979) Carbohydr Res 74:105

111. Dess D, Kleine HP, Weinberg DV, Kaufman RJ, Sidhu RS (1981) Synthesis: 883

112. Loganathan D, Trivedi GK (1987) Carbohydr Res 162:117

113. Kleine HP, Weinberg DV, Kaufman RJ, Sidhu RS (1985) Carbohydr Res 142:333

114. Kovac P, Taylor RB (1987) Carbohydr Res 167:153

115. Hadd MJ, Gervay J (1999) Carbohydr Res 320:61

116. Lichtenthaler FW, Kaji E, Weprek S (1985) J Org Chem 50:3505

117. Kaji E, Osa Y, Takahashi K, Hirooka M, Zen S, Lichtenthaler FW (1994) Bull Chem Soc Jpn 67:1130

118. Konda Y, Iwasaki Y, Takahata S, Arima S, Toida T, Kaji E, Takeda K, Harigaya Y (1997) Chem Pharm Bull 45:626

119. Chretien F (1989) Synth Commun 19:1015

120. Marais C, Steenkamp JA, Ferreira D (1996) J Chem Soc Perkin Trans 1:2915

121. Jain S, Suryawanshi SN, Misra S, Bhakuni DS (1988) Indian J Chem, Sect B 27:866
122. Oedinger H, Moeller F (1967) Angew Chem Int Ed Engl 6:76
123. Rao DR, Lerner LM (1972) Carbohydr Res 22:345
124. Lain V, Coste-Sarguet A, Gadelle A, Defay J, Perly B, Djedaini-Pilard F (1995) J Chem Soc Perkin Trans 2:1479
125. Varela O, de Fina MG, de Lederkremer RM (1987) Carbohydr Res 167:187
126. Lichtenthaler FW, Klaeres U, Lergenmueller M, Schwidetzky S (1992) Synthesis: 179
127. Schauer R (1982) Adv Carbohydr Chem Biochem 40:131
128. Reutter PW, Kottgen E, Bauer C, Gerok W (1982) Sialic Acids: Chemistry, Metabolism and Function. In: Cell Biology Monographs, vol 10. Springer, Berlin Heidelberg New York, p 263
129. von Itzstein M, Thomson RJ (1986) Glycoscience: Synthesis of Glycoconjugates and Oligosaccharides, vol 186. Springer, Berlin Heidelberg New York
130. Zbiral E, Schreiner E, Christian R, Kleineidam RG, Schauer R (1989) Liebids Ann Chem: 159
131. Kumar V, Tanenbaum SW, Flashner M (1982) Carbohydr Res 101:155
132. Kumar V, Kessler J, Scott ME, Patwardhan BH, Tanenbaum SW, Flashner M (1981) Carbohydr Res 94:123
133. Kok GB, Mackey BL, von Itzstein M (1996) Carbohydr Res 289:67
134. Lichtenthaler FW (1992) Modern Synthetic Methods 1992, vol 6. Verlag Helvetica Chimica Acta, Verlag Chemie, Weinheim, p 273
135. Lichtenthaler FW, Nishiyama S, Weimer T (1989) Liebigs Ann Chem: 1163
136. Iglesias-Guerra F, Candela JI, Espareto JL, Vega-Perez JM (1994) Tetrahedron Lett 35:5031
137. Pravdic N, Fletcher JHG (1967) J Org Chem 32:1806
138. Pravdic N, Franjic-Mihalic I, Danilov B (1975) Carbohydr Res 45:302
139. Vos JN, van Boom JH, van Boeckel CAA, Beetz T (1984) J Carbohydr Chem 3:117
140. Jones GS, Scott WJ (1992) J Am Chem Soc 114:1491
141. Beau JM, Gallagher T (1997) Top Curr Chem, Glycoscience, vol 187. Springer, Berlin Heidelberg New York, 2
142. Clode DM, Horton D, Weckerle W (1976) Carbohydr Res 49:305
143. Feast AA, Overend WG, Williams NR (1965) J Chem Soc: 7378
144. Lemieux RU, Frage E, Watanabe KA (1968) Can J Chem 46:61
145. Guthrie RD, Irvine RW (1979) Carbohydr Res 72:285
146. Thiem J, Ossowski P, Schwentner J (1979) Angew Chem Int Ed Engl 18:222
147. Thiem J, Ossowski P, Schwentner J (1980) Chem Ber 113:955
148. Jung G, Klemer A (1981) Chem Ber 114:740
149. Dettinger HM, Kurz G, Lehmann J (1979) Carbohydr Res 74:301
150. Lesimple P, Beau JM, Jaurand G, Sinay P (1986) Tetrahedron Lett 27:6201
151. Qiu D, Schmidt RR (1990) Synthesis: 875
152. Qiu D, Schmidt RR (1995) Carbohydr Lett 1:291
153. Dubois E, Beau JM (1990) Tetrahedron Lett 31:5165
154. Stille JK (1986) Angew Chem Int Ed Engl 25:508
155. Nicolau KC, Hwang CK, Duggan ME (1986) J Chem Soc Chem Commun: 925
156. Hanessian S, Martin M, Desai RC (1986) J Chem Soc Chem Commun: 926
157. Friesen RW, Sturino CF, Daljeet AK, Kolaczewska AE (1991) J Org Chem 56:1944
158. Tius MA, Gu Y, Gomez-Galeno J (1990) J Am Chem Soc 112:8188
159. Parker KA, Su DS (1996) J Org Chem 61:2191
160. Liu J, Huang CH, Wong CH (2002) Tetrahedron Lett 43:3447
161. Chambers DJ, Evans GR, Fairbanks AJ (2003) Tetrahedron Lett 44:5221
162. Chambers DJ, Evans GR, Fairbanks AJ, Fairbanks AJ (2004) Tetrahedron 60:8411
163. Horton D, Priebe W, Sznaidman M (1989) Carbohydr Res 187:145
164. Pelyvas IF, Monneret C, Herczegh P (1988) Synthetic Aspects of Aminodeoxy Sugars of Antibiotics. Springer, Berlin Heidelberg New York
165. Charette AB, Cote B (1993) J Org Chem 58:933
166. Borowiecka J, Lipka P, Michalska M (1988) Tetrahedron 44:2067
167. Kassou M, Castillon S (1994) Tetrahedron Lett 30:5513
168. Priebe W, Grynkiewicz G, Krawczyk M, Fokt I (1994) Abstr Papers. Am Chem Soc 207:78-CARB
169. Priebe W, Grynkiewicz G, Krawczyk M, Fokt I (1994) Proc XVII IUPAC Int Carbohyd Symp, Ottawa, Canada, B2.55

170. Priebe W, Zamojski A (1980) Tetrahedron 36:287
171. Pearson J (1963) J Am Chem Soc 85:3533
172. Ho TL (1975) Chem Rev 75:1
173. Leutzinger EE, Meguro T, Towwnsend LB, Shuman DA, Schweizer MP, Stewart CM, Robins RK (1972) J Org Chem 37:3695
174. De las Heras FG, Stud M (1977) Tetrahedron 33:1513
175. Heyns K, Park JI (1976) Chem Ber 109:3262
176. Dunkerton LV, Adair NK, Euske JM, Brady KT, Robinson PD (1988) J Org Chem 53:845
177. Priebe W, Grynkiewicz G, Neamati N (1991) Tetrahedron Lett 32:3313
178. Heyns K, Lim MT, Park JI (1976) Tetrahedron Lett: 1477
179. Guthrie RD, Irvine RW (1980) Carbohydr Res 82:207
180. Guthrie RD, Irvine RW (1980) Carbohydr Res 82:225
181. Boivin J, Pais M, Monneret C (1980) Carbohydr Res 79:193
182. Thiem J, Springer D (1985) Carbohydr Res 136:325
183. Fraser-Reid B, Radatus B (1970) J Am Chem Soc 92:6661
184. Achmatowicz O, Szechner B (1972) Tetrahedron Lett: 1205
185. Martin A, Pais M, Monneret C (1983) Carbohydr Res 113:21
186. Fraser-Reid B, Radatus B (1975) Acc Chem Res 8:192
187. Achmatowicz O, Szechner B (1997) Tetrahedron Lett 38:4701
188. Habus I, Sunjic V (1985) Croatica Chem Acta 58:321
189. Inaba K, Matsumura S, Yoshikawa S (1991) Chem Lett: 485
190. Doboszewski B, Blaton N, Herdewijn P (1995) J Org Chem 60:7909
191. Valverde S, Garcia-Ochoa S, Martin-Lomas M (1987) J Chem Soc Chem Commun: 383
192. Gomez AM, Valverde S, Fraser-Reid B (1991) J Chem Soc Chem Commun: 1207
193. Wittman MD, Halcomb RL, Danishefsky SJ, Golik J, Vyas D (1990) J Org Chem 55:1979
194. Griffith DA, Danishefsky SJ (1996) J Am Chem Soc 118:9526
195. Halcomb RL, Boyer SH, Wittman MD, Olson SH, Denhart DJ, Liu KKC, Danishefsky SJ (1995) J Am Chem Soc 117:5720
196. Dupradeau FY, Allaire S, Prandi J, Beau JM (1993) Tetrahedron Lett 34:4513
197. Dupradeau FY, Prandi J, Beau JM (1995) Tetrahedron 51:3205
198. Danishefsky SJ, Shair MD (1996) J Org Chem 61:16
199. Dawe RD, Fraser-Reid B (1981) J Chem Soc Chem Commun: 1180
200. Grynkiewicz G, BeMiller JN (1982) J Carbohydr Chem 1:121
201. Kawauchi N, Hashimoto H (1987) Bull Chem Soc Jpn 60:1441
202. Marco-Contelles JL, Fernandez C, Gomez A, Martin-Leon N (1990) Tetrahedron Lett 31:1467
203. Priebe W, Grynkiewicz G, Neamati N (1991) Monatsh Chem 121:419
204. Wipf P (1997) In: Trost BM (ed) Comprehensive Organic Synthesis, vol 5. Pergamon Press, Oxford, p 827
205. Heyns K, Hohlweg R (1978) Chem Ber 111:1632
206. Cottier L, Remy G, Descotes G (1979) Synthesis: 711
207. de Raadt A, Ferrier RJ (1991) Carbohydr Res 216:93
208. Balasubramanian KK, Rhamesh NG, Pramanik A, Chandrasekar J (1994) J Chem Soc Perkin Trans 2:1399
209. Takai I, Yamamoto A, Ishido Y, Sakakibara T, Yagi E (1991) Carbohydr Res 220:195
210. Seta A, Nagano C, Ito S, Tokuda K, Tamura T, Kamitani T, Sakakibara T (1998) Tetrahedron Lett 39:591
211. Sakakibara T, Nishitani A, Seta A, Nakagawa T (1991) Tetrahedron Lett 32:5809
212. Sakakibara T, Nomura Y, Sudoh R (1983) Carbohydr Res 124:53
213. Sakakibara T, Ito S, Ikegawa H, Matsuo I, Seta A (1993) Tetrahedron Lett 34:3429
214. Hall RH, Jordaan A, Lourens GJ (1973) J Chem Soc Perkin Trans 1:38
215. Hall RH, Jordaan A, Villiers OG (1975) J Chem Soc Perkin Trans 1:626
216. Nguefack JF, Bolitt V, Sinou D (1996) Tetrahedron Lett 37:59
217. Nguefack JF, Bolitt V, Sinou D (1997) J Org Chem 62:6827
218. Holzapfel CW, Engelbrecht GJ, Marais L, Toerien F (1997) Tetrahedron 53:3975
219. Zamojski A, Banaszek A, Grynkiewicz G (1982) Adv Carbohydr Chem Biochem 40:1
220. Zamojski A, Grynkiewicz G (1984) ApSimon J (ed) Total Synthesis of Natural Products. Wiley, New York 6:141

221. Danishefsky S (1980) Acc Chem Res 14:400
222. Carruthers W (1990) Cycloaddition Reactions in Organic Synhesis. Pergamon Press, Oxford
223. Larson E, Danishefsky S (1982) J Am Chem Soc 104:6458
224. Danishefsky S, Larson E, Askin D, Kato N (1985) J Am Chem Soc 107:1246
225. Danishefsky SJ, Webb RR (1984) J Org Chem 49:1955
226. Danishefsky SJ, Bednarski M (1985) Tetrahedron Lett 26:3411
227. Danishefsky SJ, Myles DC, Harvey DF (1987) J Am Chem Soc 109:862
228. Schmidt RR, Angerbauer R (1981) Carbohydr Res 89:159
229. Schmidt RR, Maier M (1985) Tetrahedron Lett 26:2065
230. Schmidt RR (1986) Acc Chem Res 19:250
231. Danishefsky SJ, Kobayashi S, Kerwin JF (1982) J Org Chem 104:358
232. Danishefsky SJ, De Ninno MP, Chen S (1988) J Am Chem Soc 110:3929
233. McDonald FE, Schultz CC, Chatterjee AK (1995) Organomtallics 14:3628
234. McDonald FE, Gleason MM (1995) Angew Chem Int Ed Engl 34:350
235. McDonald FE, Gleason MM (1996) J Am Chem Soc 118:6648
236. McDonald FE, Bowman JL (1996) Tetrahedron Lett 37:4675
237. McDonald, Zhu HYH (1997) Tetrahedron 53:11061
238. Postema MHD, Piper JL, Liu L, Shen J, Faust M, Andreana P (2003) J Org Chem 68:4748
239. Postema MHD, Piper JL, Betts RL, Valeriote FA, Pietraszkiewicz (2005) J Org Chem 70:829
240. Taillefumier C, Chapleur Y (2004) Chem Rev 104:263
241. Ali MH, Collins PM, Overend WG (1990) Carbohydr Res 205:428
242. Gascon-Lopez M, Motevalli M, Paloumbis G, Bladon P, Wyatt PB (2003) Tetrahedron 59:9349
243. Yang WB, Wu CY, Chang CC, Wang SH, Teo CF, Lin CH (2001) Tetrahedron Lett 42:6907
244. Yang WB, Yang YY, Gu YF, Wang SH, Chang CC, Lin CH (2002) J Org Chem 67:3773
245. Toth M, Somsak L (2001) J Chem Soc Perkin Trans 1:942
246. Toth M, Kover KE, Benyei A, Somsak L (2003) Org Biomol Chem 1:4039
247. Taylor RJK, Mc Allister GD, Franck RW (2006) Carbohydr Res 341:1298
248. Lopez JC, Lameignere E, Lukacs G (1988) J Chem Soc Chem Commun: 514
249. Burnouf C, Lopez JC, Laborde MA, Olesker A, Lukacs G (1988) Tetrahedron Lett 28:5533
250. Jedrzejewski S, Ermann P (1993) Tetrahedron Lett 34:615
251. Ramesh NG, Balasubramanian KK (1991) Tetrahedron Lett 32:3875
252. Booma C, Balasubramanian KK (1991) Tetrahedron Lett 32:3875
253. Lopez CJ, Lukacs G (eds) (1992) Cycloaddition Reactions in Carbohydrate Chemistry, vol 494. American Chemical Society, Washington, DC
254. Jung ME, Choe SWT (1993) Tetrahedron Lett 34:6247
255. Priebe W, Grynkiewicz G, Neamati N (1992) Tetrahedron Lett 33:7681
256. Chmielewski M, Kaluza Z, Grodner J, Urbanski R (1992) In: Guiliano RM (ed) Cycloaddition Reactions in Carbohydrate Chemistry, vol 494. American Chemical Society, Washington, DC
257. Chmielewski M, Kaluza Z, Belzecki C, Salanski P, Jurczak J, Adamowicz H (1985) Tetrahedron 41:2441
258. Chmielewski M, Kaluza Z, Mostowicz D, Belzecki C, Baranowska E, Jacobsen JP, Salanski P, Jurczak J (1987) Tetrahedron 43:4563
259. Barret AGM, Fenwick A (1983) J Chem Soc Chem Commun: 299
260. Chan JH, Hall SS (1984) J Org Chem 49:195
261. Heck RF (1991) In: Semmelhack MF (ed) Comprehensive Organic Synthesis, vol 4: Additions to and substitutions at C-C π-bonds. Pergamon Press, Oxford, p 833
262. Benhaddou R, Czernecki S, Ville G (1992) J Org Chem 57:4612
263. Dubois E, Beau JM (1990) J Chem Soc Chem Commun: 1191
264. Friesen RW, Sturino CF, Daljeet AK, Kolaczewska AE (1990) J Org Chem 55:5808
265. Friesen RW, Loo RW (1991) J Org Chem 56:4821
266. Hall RH, Jordaan A (1973) J Chem Soc Perkin Trans 1:1059
267. Bischofberger K, Hall RH, Jordaan A, Woolard GR (1980) S Afr J Chem 33:92
268. Korytnyk W, Angelino N, Dodson-Simmons O, Hanchak M, Madson M, Valentkovic-Horvath S (1983) Carbohydr Res 113:166
269. Bozo E, Boros S, Kuszmann J (1997) Carbohydr Res 299:59

3.7 Anomeric Anhydro Sugars

*Nathan W. McGill, Spencer J. Williams**
School of Chemistry, The University of Melbourne,
Parkville, VIC 3052, Australia
n.mcgill@pgrad.unimelb.edu.au, sjwill@unimelb.edu.au

Abstract

Anomeric anhydro sugars are sugar derivatives where the anomeric carbon participates in an acetal linkage with two of the hydroxyl groups of the sugar. They are essentially intramolecular glycosides, and their bicyclic nature provides a powerful conformational constraint that greatly influences their reactivity. This chapter reviews the occurrence, properties, formation, and reactions of anomeric anhydro sugars. Particular emphasis is placed on 1,2- and 1,6-anhydropyranoses, including conformational aspects and ring-opening reactions. Epoxide-containing 1,6-anhydro sugars (Černý epoxides) are briefly reviewed, and the formation and some reactions of the 1,6-anhydro sugar enone, levoglucosenone, is covered. An overview is given of the use of 1,2-anhydro sugar as glycosyl donors. Also discussed are the formation and reactions of anomeric anhydro sugars containing nitrogen, sulfur, or selenium.

Keywords

Anhydro sugars; Dianhydro sugars; Epoxide; Glucosan; Glycosan; Levoglucosan; Levoglucosenone; Seleno sugar; Selenolevoglucosan; Thiolevoglucosan

Abbreviations

CSA   camphorsulfonic acid
DMDO   3,3-dimethyldioxirane

In: *Glycoscience*. Fraser-Reid B, Tatsuta K, Thiem J (eds)
Chapter-DOI 10-1007/978-3-540-30429-6_16: © Springer-Verlag Berlin Heidelberg 2008

mw microwave
NIS *N*-iodosuccinimide
TESOTf triethylsilyl triflate

1 Introduction

Anomeric anhydro sugars are sugar derivatives in which the anomeric carbon participates in an acetal linkage with two of the hydroxyl groups of the sugar. They are essentially intramolecular glycosides, which may be formally derived from the parent sugar by the loss of a molecule of water. Importantly, the intramolecular acetalation of the anomeric center acts to protect this position and two other hydroxyl groups. The bicyclic structure of anomeric anhydro sugars ensures that they are restricted to a limited conformational coordinate that strongly influences the reactivity of the remaining hydroxy groups. Many anhydro sugars are highly crystalline and this feature makes them easy compounds to work with and to purify. This chapter focuses on anomeric anhydro sugars of synthetic utility and on the more common reactions known for this class of sugars.

1.1 General Remarks

Solutions of aldohexoses in aqueous acid at equilibrium contain varying amounts of anhydro sugars, particularly the 1,6-forms. The composition of the equilibrium mixture has been determined for a number of aldohexoses in the presence of acidic ion-exchange resin or in 0.25-M H_2SO_4 at 100 °C [1]. For D-glucose and D-galactose, the proportion of anhydro sugars present at equilibrium is small; for D-altrose, D-gulose, and D-idose the proportion of the 1,6-anhydropyranose present is greater than 50%. The formation of 1,6-anhydrofuranoses under these conditions is much less favorable. Of the eight aldohexoses, D-talose forms an anhydrofuranose to the greatest degree and, in this case, only to the extent of 2.5%. As a rule, as the number of hydroxy groups oriented in the axial direction in the 'usual' $^4C_1$ conformation of an aldose increases, so does its tendency to form 1,6-anhydro derivatives. The acetal of anomeric anhydro sugars may be cleaved by solvolysis under acidic conditions. In the absence of a nucleophilic solvent, Lewis acid-catalyzed polycondensation of anomeric anhydro sugar monomers affords glycans. This approach provides an efficient method for the preparation of certain stereoregular polysaccharides [2].

2 1,6-Anhydrohexopyranoses

2.1 Occurrence and Formation

Of all the anhydro sugars, 1,6-anhydrohexopyranoses are the most studied and are of the greatest synthetic utility. 1,6-Anhydrohexopyranoses have been the subject of several comprehensive reviews, which the interested reader should consult for more details [3,4,5,6]. 1,6-Anhydro-β-D-glucopyranose (**1**; ❷ *Fig. 1*) (also known as 'glucosan' or 'levoglucosan' on account

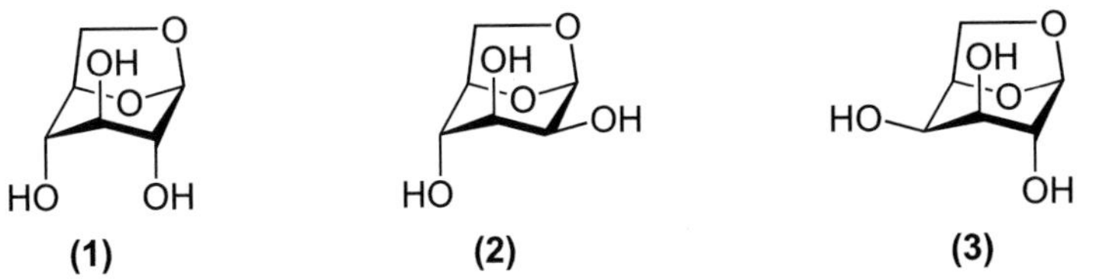

(1)   **(2)**   **(3)**

Figure 1
Common 1,6-anhydro sugars: D-glucosan (1), D-mannosan (2) and D-galactosan (3)

of its significantly negative optical rotation) was first isolated in 1894 by Tanret upon treatment of the naturally occurring phenolic glycosides, picein, salicin, and coniferin, with barium hydroxide [7]. On the basis of the molecular formula of $C_6H_{10}O_5$, Tanret correctly described this crystalline material as 'glucose anhydride,' which is often shortened to 'glucosan.' The structure of this anhydride remained unknown until 1920, when Pictet proposed the existence of a 1,6-anhydro bridge. As a result of the 1,6-anhydro bridge, levoglucosan adopts a $^1C_4$ conformation in which the hydroxyl groups are axially disposed, and opposite to the normal $^4C_1$ conformation observed for D-glucopyranose. The crystal structure [8] and ^{1}H-NMR spectral data [9] are in agreement with a $^1C_4$ conformation, which is common to all eight 1,6-anhydropyranoses. Despite the obvious structural differences, many of the chemical properties of levoglucosan are similar to methyl β-D-glucopyranoside: it is stable to alkali, but is hydrolyzed by acid.

In 1918, Pictet and Sarasin isolated 1,6-anhydro-β-D-glycopyranose from the distillate obtained from the pyrolysis of cellulose [10]. Indeed, the pyrolysis of oligo- and polysaccharide-rich biomass remains an effective method for the preparation of simple 1,6-anhydro sugars. Levoglucosan has been produced on a large scale by the uncatalyzed pyrolysis of powdered corn starch [11] or cellulose [12]. The pyrolysis of cellulosic-biomass during bushfires or residential wood combustion produces large amounts of smoke aerosol, of which levoglucosan is a significant component. Consequently, levoglucosan has been used as a molecular marker for quantifying the contribution of woodsmoke to atmospheric pollution; levoglucosan can be quantified by GC-MS or LC-ESI-MS analysis [13]. Mannosan (2) and galactosan (3) (❷ Fig. 1) may be produced by pyrolysis of the seeds of the ivory nut palm, Phytelephas macrocarpa [14], and of α-lactose monohydrate [15], respectively (❷ Scheme 1). In both cases, purification of the anhydro sugars from other pyrolyzates is facilitated by conversion of each into the corresponding isopropylidene acetal. Extraction of this product into an organic solvent and subsequent removal of the acetonide group yields the parent glycosans.

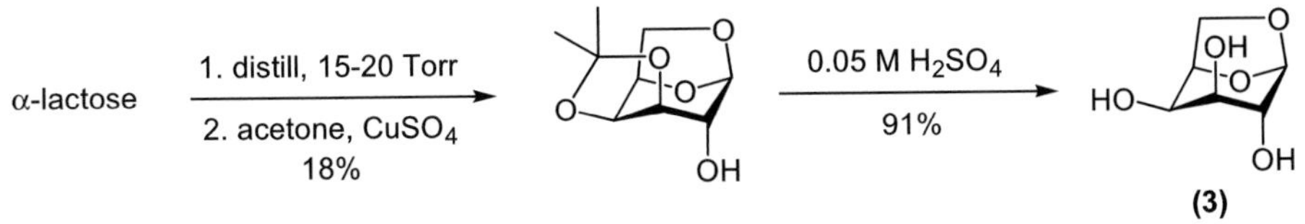

α-lactose
1. distill, 15-20 Torr
2. acetone, CuSO$_4$
18%

0.05 M H$_2$SO$_4$
91%

(3)

Scheme 1

Pyrolysis of a variety of monosaccharides has been shown to afford the corresponding 1,6-anhydro sugars; however, the yields are lower than that seen when using the glycan precursors. It has been proposed that the pyrolysis of simple sugars proceeds by way of an initial condensation to intermediate polysaccharides, which then thermally depolymerize, affording the 1,6-anhydro sugars [3].

In addition to the pyrolytic procedures described above, there are three main synthetic routes to 1,6-anhydro sugars. The first method relies on selective activation of C-1 or C-6 with a suitable leaving group, such as a halide or sulfonate, and displacement of this group by an oxyanion, generated with base, at the alternate position. Large scale, one-pot syntheses of tri-O-acetyl-1,6-anhydro-β-D-glucopyranose and -β-D-mannopyranose, involving the selective tosylation of D-glucose and D-mannose at O-6, and subsequent treatment of the intermediate sulfonates with sodium hydroxide, have been described by Fraser-Reid [16]. More recently, Cleophax reported a high yielding, solvent-free synthesis of tri-O-acetyl-1,6-anhydro-β-D-glycopyranose. In an analogous approach to that described by Fraser-Reid, selective tosylation of D-galactose, D-glucose, and D-mannose at O-6, followed by microwave (mw) irradiation of the sulfonate intermediates in the presence of basic alumina, afforded the desired 1,6-anhydrohexopyranoses (● Scheme 2) [17].

In analogy to Tanret's original observation, treatment of a number of 1,2-trans-glycosides with base affords 1,6-anhydropyranoses, by way of the 1,2-anhydropyranose intermediate. In this way, a large-scale synthesis of tri-O-acetyl-1,6-anhydro-β-D-glucopyranose, utilizing the basic hydrolysis of phenyl tetra-O-acetyl-β-D-glucopyranoside as the key step, has been described [18]. More recently, Boons and co-workers have shown that treatment of pentabromophenyl β-D-glycosides with Amberlite IRA-400 (OH⁻) resin affords 1,6-anhydro derivatives in high yields [19].

In the second approach to 1,6-anhydro sugars, anomeric esters, particularly glycosyl acetates, are activated by Lewis acids and displaced by 'labile' ethers at O-6. Rao has reported an efficient synthesis of tri-O-acetyl-1,6-anhydro-β-D-glucopyranose by treatment of the trityl ether (4) with titanium chloride (● Scheme 3) [20].

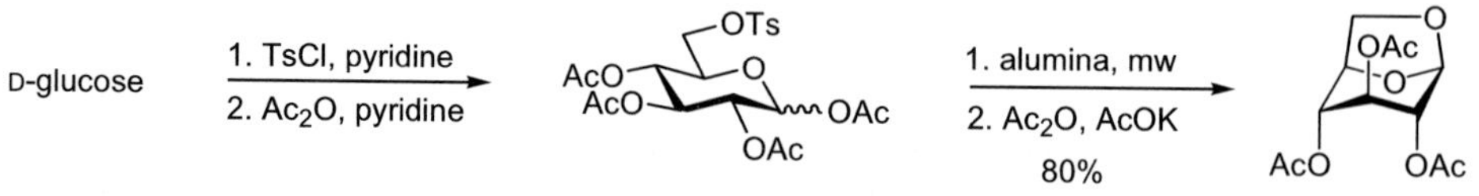

■ Scheme 2

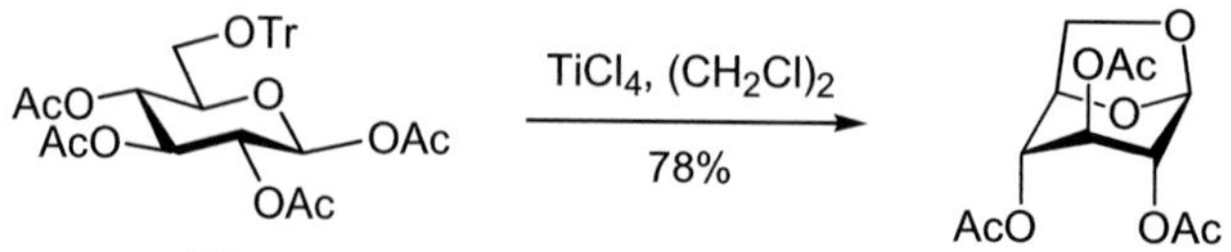

■ Scheme 3

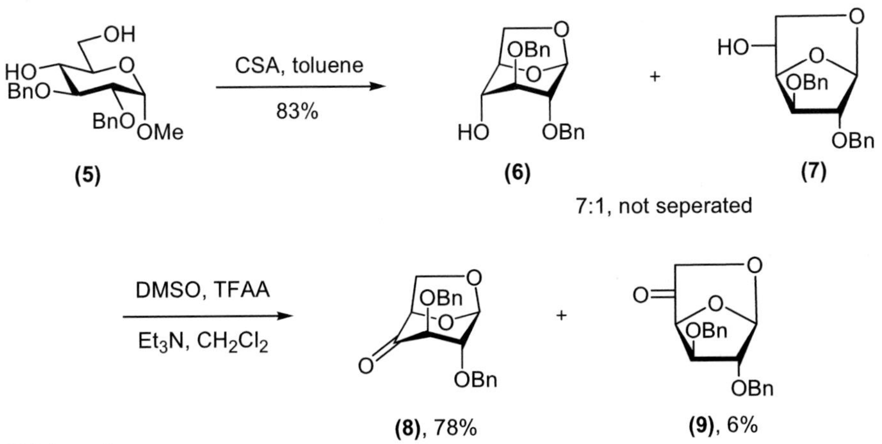

Scheme 4

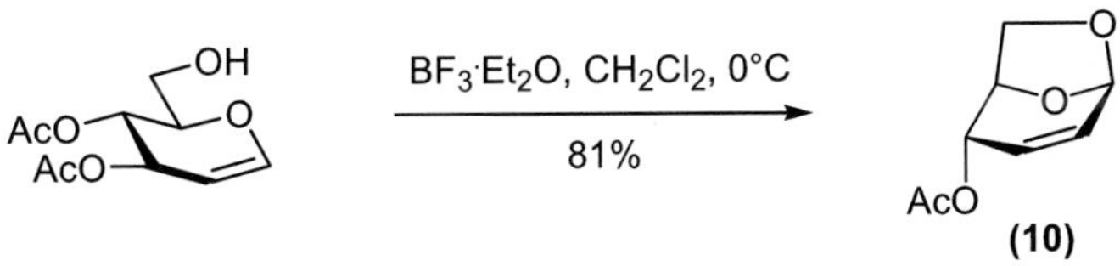

Scheme 5

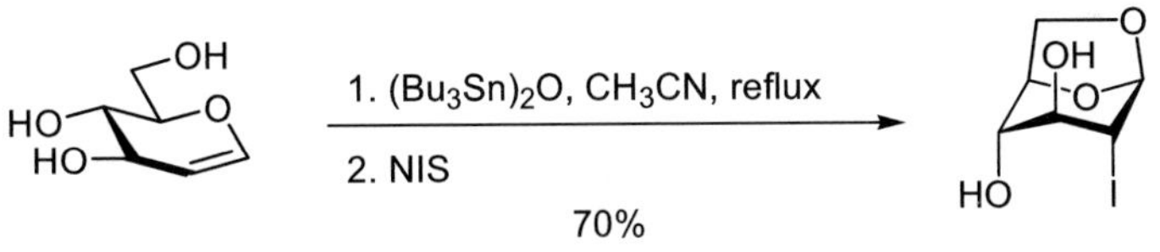

Scheme 6

In a similar approach, treatment of methyl glycosides with camphorsulfonic acid (CSA) in toluene under Dean–Stark conditions provides a good route to functionalized 1,6-anhydro sugars. Heathcock and co-workers showed that oxidation of the difficult-to-separate mixture of the 1,6-anhydro-β-D-glucopyranose (**6**) and 1,6-anhydro-β-D-glucofuranose (**7**) allowed for easy isolation of the desired ketone (**8**), by either chromatography or direct crystallization, depending on the reaction scale (❍ *Scheme 4*) [21].

In the final method, glycals are transformed into various 1,6-anhydro sugar derivatives. Glycals may be efficiently cyclized to yield 1,6-anhydro-2,3-dehydro sugars, such as (**10**), by the action of Lewis or Brønsted acids; this reaction is an intramolecular Ferrier glycosylation (❍ *Scheme 5*) [22,23]. 1,6-Anhydro-2-deoxy-2-halo sugars are accessed through 'halocyclization' reactions of *O*-stannylated glycals with a source of halogen such as iodine, bromine, or NIS (❍ *Scheme 6*) [22,24,25,26].

O NH-L-Ala-D-Glu-*meso*-Dap-D-Ala

peptidoglycan → lytic transglycosylases / endopeptidases

◻ **Scheme 7**

While the aforementioned methodologies constitute the most useful and general synthetic procedures for the preparation of 1,6-anhydropyranoses, a number of alternative routes have been described [27]. As mentioned previously, solutions of a number of sugars in aqueous acid at equilibrium contain considerable amounts of 1,6-anhydro sugars. A simple synthesis of tri-*O*-acetyl-1,6-anhydro-β-L-idopyranose has been reported by Stoffyn and Jeanloz that utilizes the acid-catalyzed dehydration of L-idose to form the anhydro ring [28].

A 1,6-anhydromuropeptide is formed during the recycling of cell wall peptidoglycan in *E. coli* (❷ *Scheme 7*) [29,30]. Upon cleavage of the GlcNAc and peptide moieties, the liberated 1,6-anhydro-*N*-acetylmuramic acid residue is converted in two steps to *N*-acetyl-D-glucosamine-6-phosphate which enters an established pathway for amino sugar metabolism [31]. Interestingly, a 1,6-anhydromuropeptide is formed during the recycling of cell wall peptidoglycan in *E. coli* (❷ *Scheme 7*) [29,30]. Upon cleavage of the GlcNAc and peptide moieties, the liberated 1,6-anhydro-*N*-acetylmuramic acid residue is converted in two steps to *N*-acetyl-D-glucosamine-6-phosphate [31].

2.2 Reactions

Aside from the protection afforded to the anomeric and 6 positions of 1,6-anhydro sugars, considerable differences in the reactivities at the remaining ring-positions are observed. The reactivities of the hydroxy groups of a number of 1,6-anhydrohexopyranoses towards sulfonylation with tosyl chloride have been determined and were shown to depend upon both the configuration, i. e. axial or equatorial, and the position around the ring [3]. In one case, Pulido has reported a regioselective 4-*O*-acylation of levoglucosan by action of a lipase isolated from *Candida antarctica* [32]; under similar conditions D-glucose is acylated at the primary hydroxyl group [33]. Epoxides of 1,6-anhydrohexopyranoses, commonly called 'Černý' epoxides [25], are useful compounds for the selective introduction of substituents onto the sugar ring. Owing to the conformational restraint in a $^1C_4$ conformation imposed on the sugar ring by the 1,6-anhydro bridge, these reactions generally proceed with reversed regioselectivity when compared to the more usual $^4C_1$ conformation of most hexopyranoses. This is in accordance with the Fürst–Plattner rule that predicts diaxial opening of constrained epoxides. The preparation of such dianhydro sugars from a number of 1,6-anhydropyranoses is relatively straightforward. For example, treatment of 1,6-anhydro-β-D-glucopyranose with tosyl chloride affords the ditosylate (**11**) that, upon treatment with base, is converted in high yield into the 1,6:3,4-dianhydro-β-D-galactoside (**12**) (❷ *Scheme 8*). In another approach, the 1,6-anhydro-2,3-didehydro sugar (**13**) upon treatment with peracid under Prilezhaev conditions fur-

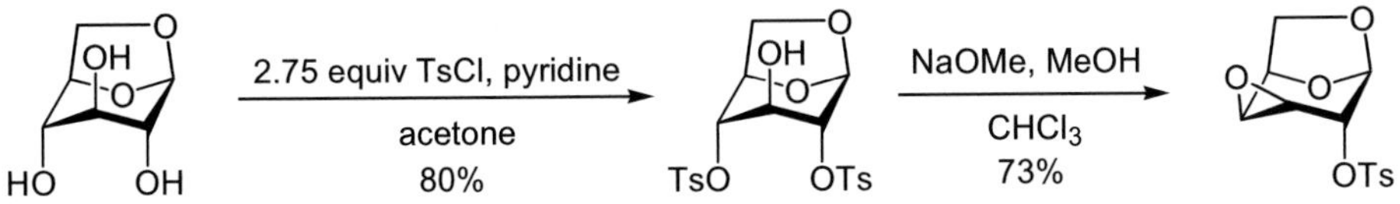

Scheme 8

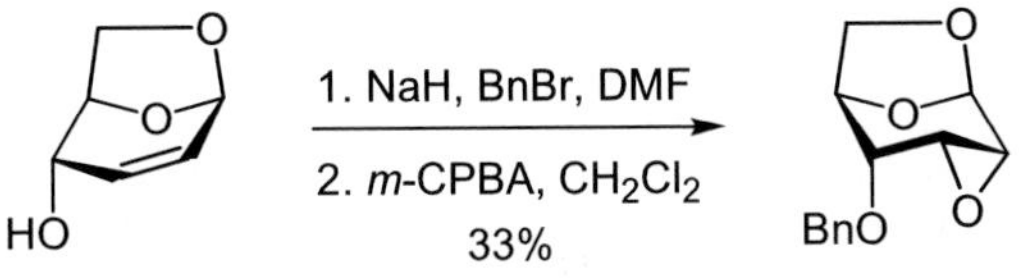

Scheme 9

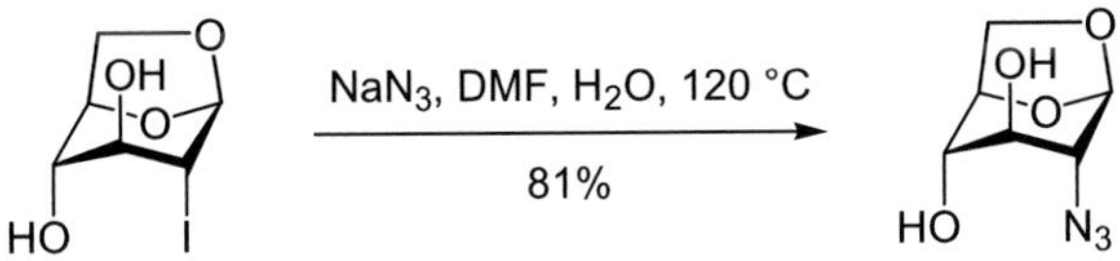

Scheme 10

nishes the 1,6:2,3-dianhydro-2,3-β-D-allopyranose (**14**) (❷ *Scheme 9*) [24]. The presence of the 1,6-anhydro bridge directs formation of the oxirane ring to the *exo*-face.

Treatment of the 2-iodo sugar (**15**) (derived by iodocyclization of D-glucal) with sodium azide affords the 2-azido-2-deoxy sugar with retention of stereochemistry, presumably by way of the 1,6:2,3-dianhydro sugar (❷ *Scheme 10*) [26].

Acylated 1,6-anhydro sugars readily undergo photobromination with high regio- and stereoselectivity, affording *exo*-6-bromo sugars (❷ *Scheme 11*) [34,35], and prolonged reaction times yield small amounts of reactive 6,6-dibromo anhydro sugars [36]. The 6-bromo anhydro sugars can be reduced with tributyltin deuteride to the 6-deutero compounds with retention of stereochemistry (❷ *Scheme 11*) [35]. After cleavage of the anhydro ring, this method provides a good route to D-glucose stereoselectively labeled with deuterium at C-6.

6-Bromo anhydro sugars react with nucleophiles, generally with inversion, affording the corresponding *endo*-products [35]; however, many carbon nucleophiles react with high *exo*-selectivity [37]. The reaction of 6-bromo sugars with carbon nucleophiles allows for stereoselective carbon-chain elongation at C-6. This is a useful alternative to the reaction of hexopyranose 6-aldehydes with Grignard reagents. A procedure has been devised for the hydrolysis of the 6-bromo anhydro sugar (**16**); oxidation of the intermediate hemiacetals affords a urono-6,1-lactone (❷ *Scheme 12*) [38].

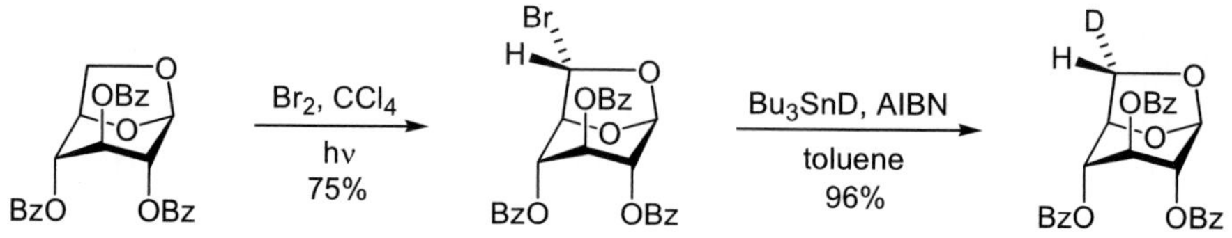

◘ Scheme 11

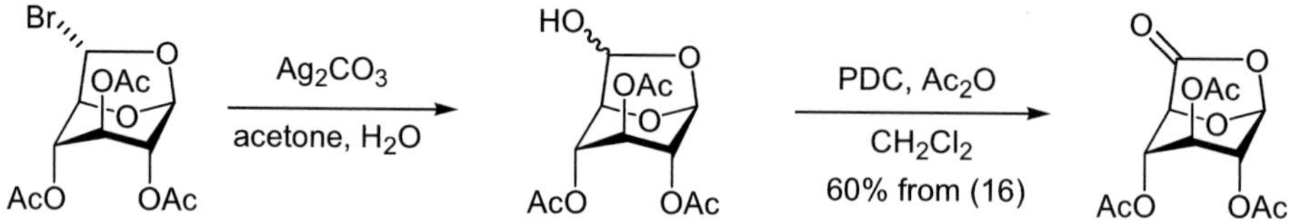

(16)

◘ Scheme 12

2.3 Cleavage

Cleavage of the 1,6-anhydro bridge can be effected by aqueous mineral acids, such as hydrochloric or sulfuric acid. Hydrolysis of 1,6-anhydro-β-D-glucopyranose is achieved in 5 h by 0.5-M sulfuric acid at 100 °C [7]. Glycosans are also susceptible to acetolysis by the action of acetic anhydride and catalytic Lewis acid. The treatment of a number of glycosans with triethylsilyl triflate (TESOTf) in acetic acid provides a rapid cleavage of the 1,6-anhydro bridge in good to excellent yields [39]. Treatment of various protected 1,6-anhydropyranoses with catalytic scandium triflate and a minimum of acetic anhydride leads to the expected products in crude yields exceeding 95% (◉ Scheme 13) [40]. This method is particularly attractive as it circumvents the use of moisture sensitive Lewis acids.

The rate of acetolysis of 1,6-anhydro sugars is dependent upon the electron-withdrawing ability of the protecting groups and, in particular, of the group at C-2. Thus, ester protecting groups lower the rate of acetolysis compared to their benzylated counterparts [41].

Methods for the direct conversion of 1,6-anhydropyranoses to thioglycosides have been described using either substituted thio(trimethyl)silanes or ethanethiol in the presence of Lewis acid [42,43]; the resulting thioglycosides are furnished with a free primary hydroxyl group that is available for further reaction (◉ Scheme 14).

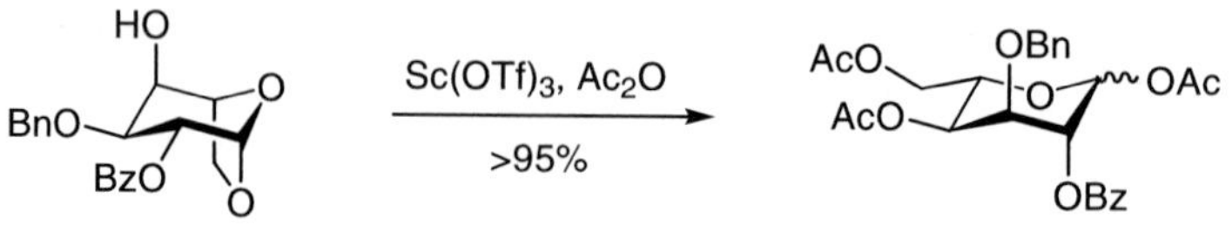

◘ Scheme 13

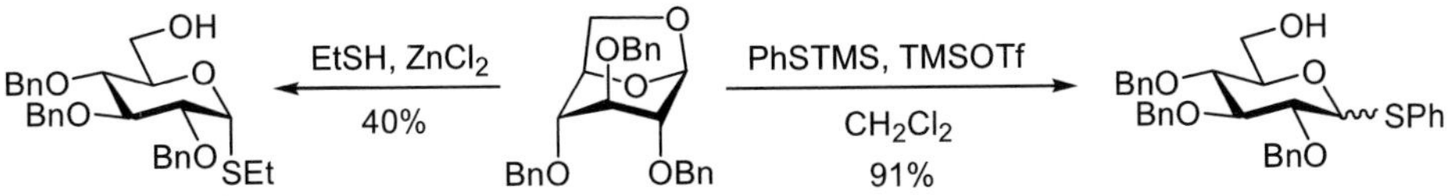

■ Scheme 14

Ring-opening polymerization of 1,6-anhydro sugars provides an efficient route to high molecular weight, stereoregular glycans. For example, the first reported synthesis of stereoregular (1→6)-α-D-glucopyranan (dextran) was achieved through the ring-opening polymerization of benzylated levoglucosan [44]. More recently, this approach has been employed in the synthesis of dextran sulfonates that possess anti-coagulant activity and inhibitory effects on HIV infection in vitro [45].

3 1,2-Anhydro Sugars

1,2-Anhydro sugars have proven to be versatile glycosyl donors for the preparation of many glycoconjugates. The 1,2-anhydro bridge is an epoxide that is also part of an acetal and is responsible for the unique reactivity of 1,2-anhydrohexopyranoses. The first well-defined member of this class was prepared by treatment of the partially functionalized pyranose (**17**) with ammonia in benzene, and is commonly referred to as Brigl's anhydride (❷ *Scheme 15*) [46].
Brigl's anhydride was utilized by Lemieux and Huber in the first chemical synthesis of sucrose; a milestone achievement in synthetic organic chemistry [47]. Non-protected derivatives of Brigl's anhydride have not been isolated on account of their high reactivity; however, they have been implicated as intermediates in the alkali-mediated formation of 1,6-anhydro sugars. There exists three main routes to 1,2-anhydropyranoses. The first uses the approach of Brigl: treatment of 2-hydroxy, C-1 activated sugars with base results in an intermediate oxyanion that displaces the leaving group at C-1, thus leading to 1,2-anhydro sugars directly. This method necessitates a *trans*-arrangement of the groups at the C-1 and C-2 positions of the pyranose ring. Alternatively, treatment of 2-*O*-tosyl or 2-*O*-bromo-2-deoxy derivatives of reducing sugars with base yields 1,2-anhydro sugars through *trans*-cyclization [47]. The third method for synthesis of 1,2-anhydro sugars involves epoxidation of glycals. In pioneering work, Halcomb and Danishefsky reported that epoxidation of glycals with 3,3-dimethyldioxirane (DMDO) directly affords 1,2-anhydropyranoses (❷ *Scheme 16*) [48]. DMDO epoxidation enjoys a sim-

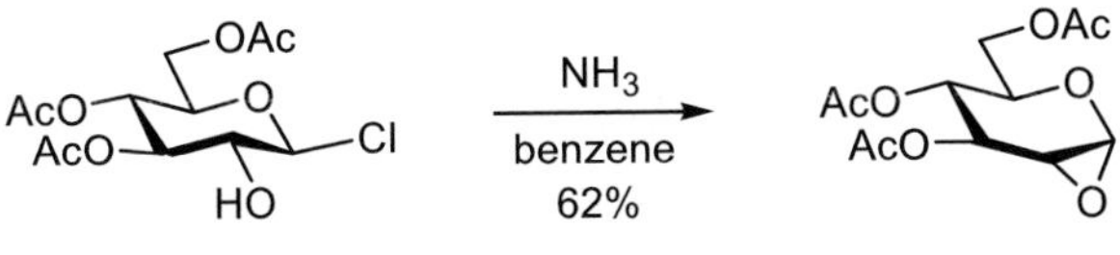

(17)

■ Scheme 15

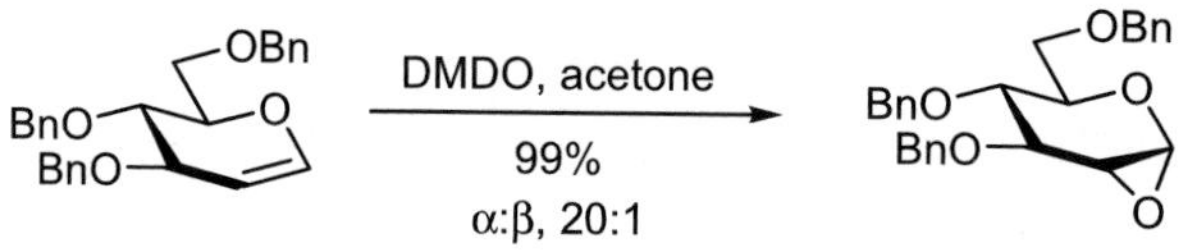

Scheme 16

ple work-up, which requires only evaporation. The stereochemistry of the resulting epoxide is usually *trans*- to that of the substituent at C-3 and, as a result, the first two procedures may be necessary for access to certain 1,2-anhydro sugars. It is noteworthy that the use of other epoxidation reagents, such as peracids, usually fail owing to the susceptibility of the anomeric oxirane ring to nucleophilic attack by the solvent or carboxylic acid by-products of the peracid reagents [48].

Interest in 1,2-anhydro sugars stems largely from their ability to function as glycosyl donors under neutral or mildly acidic conditions. Danishefsky and Bilodeau have reported routes to several complex oligosaccharides in which glycosylation steps were carried out using 1,2-anhydro sugars as glycosyl donors [49]. In each case, the epoxide donors were activated with Lewis acid and treated with suitably protected alcohols to afford glycosides in good to excellent yields. For reactions of 1,2-anhydro sugars with alcohol acceptors, including

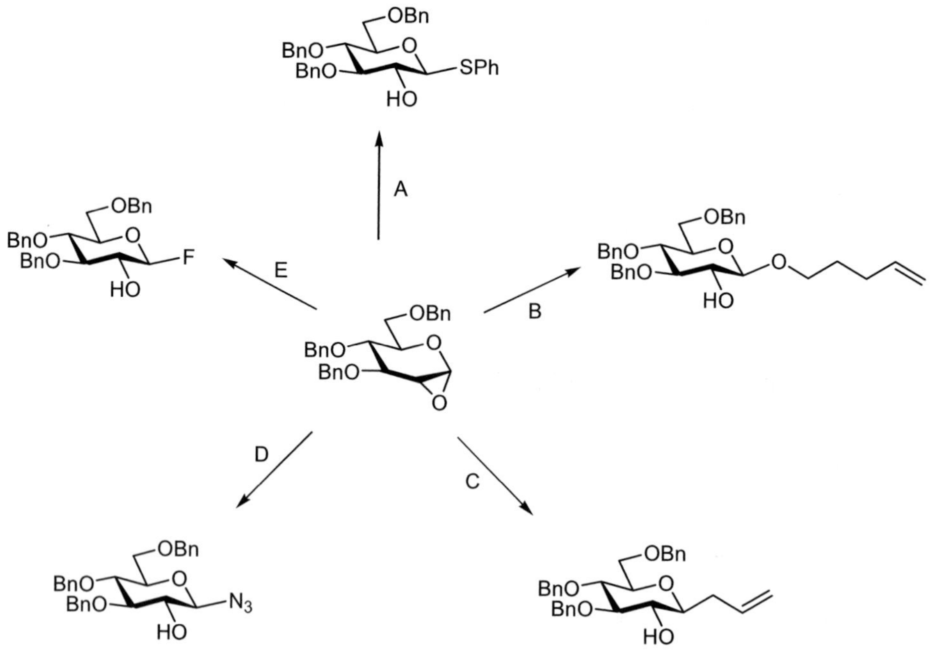

Scheme 17
Reagents: (A) Bu₄NSPh, α:β=1:6, 50%; (B) pent-4-en-1-ol, ZnCl₂, 63%; (C) AllMgBr, 75%; (D) DIBAL-LiN₃, 73%; (E) TBAF, 53%

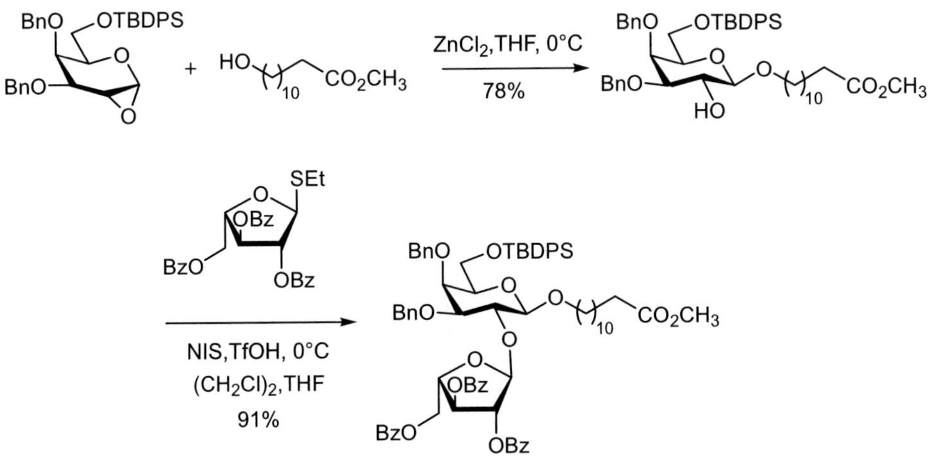

■ Scheme 18

those in which the hydroxyl function is part of a saccharide, the catalyst most frequently employed is anhydrous zinc chloride [49]. Importantly, glycal epoxides can be easily converted into a number of glycosides or glycosylating agents such as thioglycosides, *n*-pentenyl glycosides, glycosyl fluorides, glycosyl azides, and *C*-glycosides (❷ *Scheme 17*) [49,50,51]. A propitious feature of these conversions is that *O*-2 is rendered free of protecting groups and this free hydroxyl group is amenable to glycosylation in a subsequent step, allowing for the ready preparation of 1,2-glycosides (❷ *Scheme 18*) [52]. The use of glycals as precursors for 1,2-anhydro sugar glycosyl donors has been developed by the Danishefsky group and termed the 'glycal approach' or 'glycal assembly method.' This approach can be used in an iterative fashion for the stepwise elongation of saccharides; thus, a glycal is converted to a 1,2-anhydro sugar and used as a glycosyl donor to glycosylate a selectively protected glycal acceptor; the resultant glycoside is a glycal and sequential epoxidation/glycosylations can be performed. The 'glycal approach' has been utilized on a solid phase [53].

4 Miscellaneous Anhydro Sugars

4.1 1,3-Anhydro Sugars

Interest in 1,3-anhydrosugars stems predominately from their analogy to the dioxabicyclo[3.1.1]heptane ring system present in the blood-platelet aggregation factors, the thromboxanes. The first synthesis of a thromboxane-like nucleus was achieved by the action of sodium hydride on the glycosyl chloride (**18**) (❷ *Scheme 19*) [54]. In general, 1,3-anhydrosugars are prepared in an analogous manner to 1,6-anhydro sugars through treatment of activated sugars (at C-1 or C-3) with base. Still and co-workers have described a procedure for the synthesis of 1,3-anhydro sugars using a modified Mitsunobu reaction on a sugar hemiacetal unprotected at C-3 [55]. 1,3-Anhydrosugars have been utilized in the synthesis of $(1\rightarrow3)$-β-D-glucans by

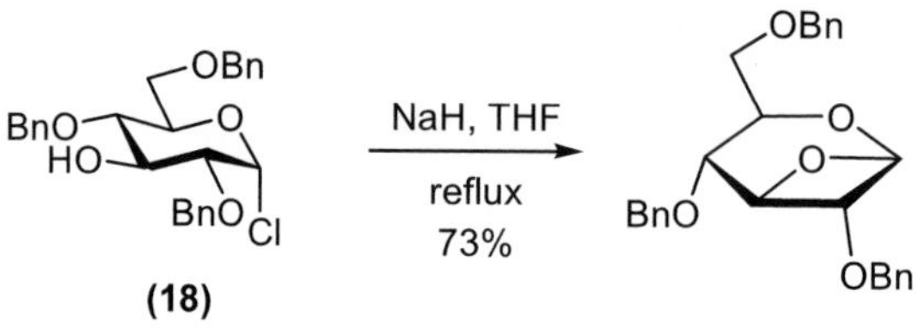

⬛ Scheme 19

stereoregular polymerization [54]; polymeric glycoforms of this type are commonly found in plants, bacteria, yeast, and fungi.

4.2 1,4-Anhydropyranoses (1,5-Anhydrofuranoses)

Relatively few studies have been undertaken on 1,4-anhydro sugars. 1,4-Anhydropyranoses (which may be considered as 1,5-anhydrofuranoses) have been synthesized by treatment of glycosyl fluorides possessing an unprotected C-4 hydroxy group with Lewis acid, or by treatment of 4-O-sulfonyl (usually mesyl) derivatives with bases such as sodium azide [56,57,58]. Reaction times may be shortened by mw irradiation (❷ Scheme 20) [59], however, as with conventional heating, the anhydro bridge remains susceptible to ring-opening and formation of self-polymerization by-products.

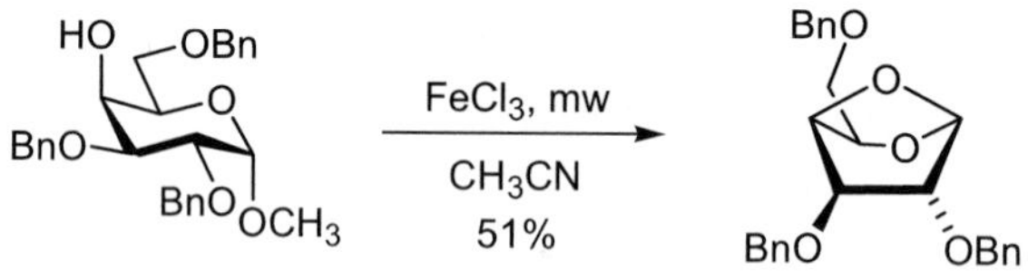

⬛ Scheme 20

4.3 1,6-Anhydrohexofuranoses

1,6-Anhydrohexofuranoses are present to a small extent in equilibrium mixtures of aldohexoses in water [1] and have been isolated from the pyrolysis distillates of sugars. 1,6-Anhydro-β-D-glucofuranose and 1,6-anhydro-α-D-galactofuranose were first isolated from pyrolysis residues after removal of the bulk of the more abundant 1,6-anhydrohexopyranoses [60,61]. These two sugars have been the subject of review [3,62]. Specific methods for the synthesis of 1,6-anhydrohexofuranoses are somewhat limited; however, Angyal and Beveridge have reported that treatment of dilute solutions of a number of aldohexoses in DMF/benzene with 4-toluenesulfonic acid under Dean–Stark conditions affords moderate yields of 1,6-anhydrohexofuranoses such as (**19**) (❷ Scheme 21) [63].

As described earlier, treatment of methyl α-D-glucoside (**5**) with camphorsulfonic acid under Dean–Stark conditions yields a 7:1 mixture of 1,6-anhydro-β-D-glucopyranose (**6**)

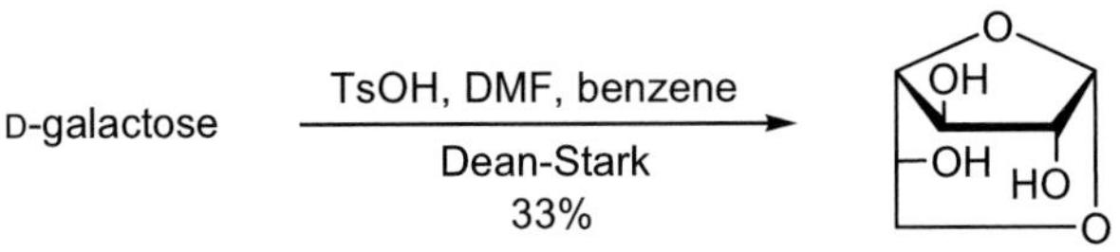

D-galactose $\xrightarrow[\substack{\text{Dean-Stark} \\ 33\%}]{\text{TsOH, DMF, benzene}}$

(19)

□ Scheme 21

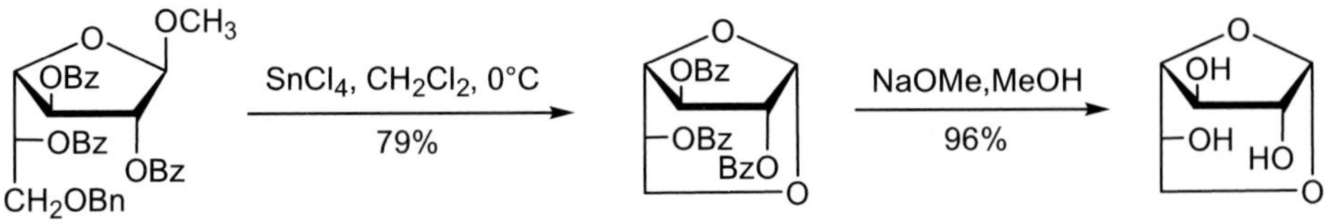

(20)   **(19)**

□ Scheme 22

and 1,6-anhydro-β-D-glucofuranose (**7**) (❷ *Scheme 4*). Dehydration of methyl α-D-galac-topyranoside with CSA yields an equimolar mixture of the furanose and pyranose anhydro derivatives [21]. In an alternative approach, synthesis of 1,6-anhydro-α-D-galactofuranose (**19**) through stannic chloride catalyzed ring-closure of the fully protected galactofuranose (**20**) was achieved in two steps in excellent yield (❷ *Scheme 22*) [64]. It has been observed that the *trans*-oriented 2,3-diols of these bicyclic compounds possess a remarkable resistance to both periodate and lead tetraacetate oxidation, a consequence of the rigidity imposed upon these molecules by the 1,6-anhydro bridge.

4.4 Levoglucosenone

Levoglucosenone is an enone-containing 1,6-anhydro sugar, the considerable differences of which merit a separate section to 1,6-anhydro sugars, although only a brief overview of the extensive chemistry pertaining to this compound is possible here [65]. The presence of the acetal and enone moieties of levoglucosenone provides ready access to a wide range of func-tionalized sugars from a single chemical feedstock. Halpern and co-workers properly char-acterized levoglucosenone in 1973, correcting several structures proposed by other work-ers [66]. Levoglucosenone may be prepared by the Brønsted-acid catalyzed pyrolysis of cel-lulosic materials and is easily isolated in multi-gram quantities from the pyrolyzate of waste paper [67]. The 1,6-anhydro bridge sterically hinders the *endo*-face of levoglucosenone and, as a consequence, nucleophilic addition is directed to the non-hindered *exo*-face. For exam-ple, reduction of the carbonyl group with sodium borohydride or lithium aluminum hydride leads to the expected allylic alcohol in good yield [68,69]. Carbon nucleophiles derived from Grignard or Gilman reagents add to the β-carbon yielding the corresponding Michael adducts. In this way, stereospecific addition of methylmagnesium iodide or lithium dimethylcuprate

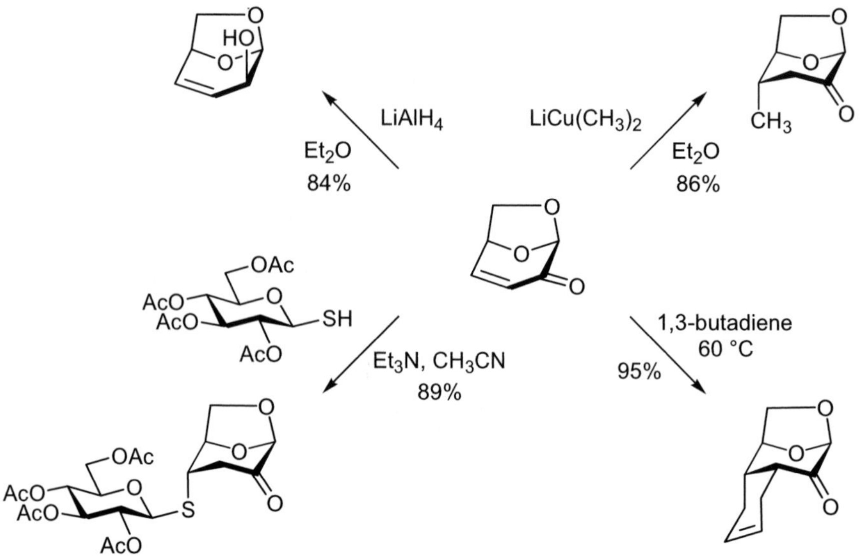

■ Scheme 23

to the enone of levoglucosenone leads to C-4 methyl-substituted carbohydrates [69,70]. Furthermore, conjugate addition of glycosyl nitronates [71] or thioglycosides [72] with levoglucosenone affords *C*- or *S*-linked disaccharides, respectively. The alkene of levoglucosenone acts as a dienophile in Diels–Alder reactions [73]. Some transformations of levoglucosenone are summarized in ❷ *Scheme 23*.

5 Anhydro Sugars Containing Nitrogen, Sulfur, or Selenium

A number of anhydro sugars are known that contain nitrogen, sulfur, or selenium. Tri-*O*-acetyl thioglucosan was first prepared in 1963 by Akagi and co-workers by treatment of *S*-glucosyl dithiocarbonate (**21**; ❷ *Fig. 2*) with sodium methoxide [74]. Since that time, numerous syntheses have been reported that rely on the introduction of a thioester at C-1 or C-6 and the activation of the alternate position with a halide or sulfonate group [75,76]. Treatment of a 'doubly activated' hexopyranose such as (**24**) with hydrogen sulfide and triethylamine

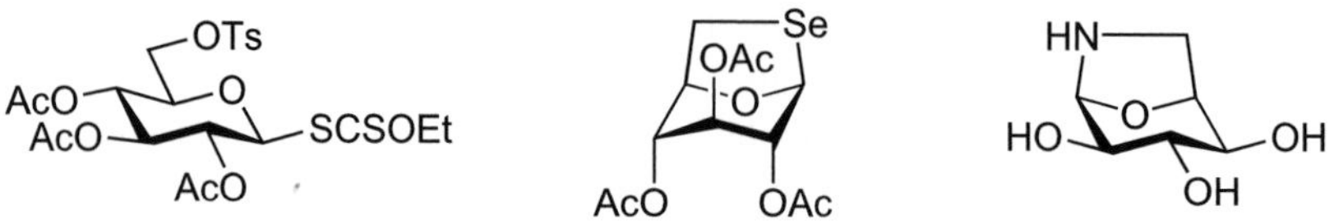

(21)   (22)   (23)

■ Figure 2

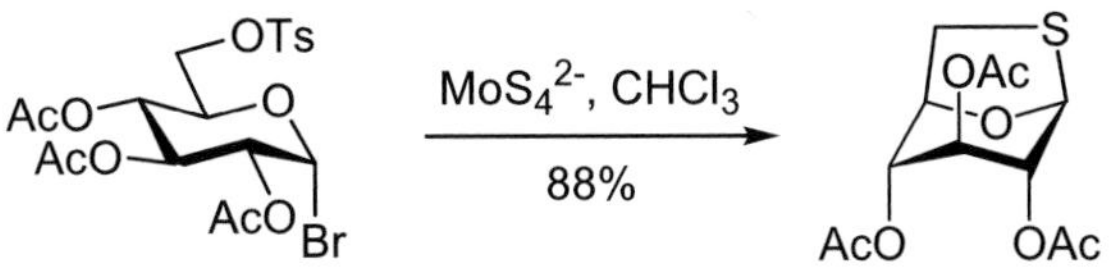

Scheme 24

leads to thioglycosans in good yields [77]. Treatment of doubly activated sugars with the sulfur-transfer reagent benzyltriethylammonium tetrathiomolybdate [78] now represents the most expeditious route to thioglycosans (❷ *Scheme 24*) [77,79].

Thioglycosans are stable under basic conditions but are cleaved either by sulfuric acid in acetic anhydride to afford a 6-*S*-acetyl anomeric acetate [80] or by dilute hydrochloric acid in water to yield the 6-thioglycose [74]. The sulfur atom of thiolevoglucosan is susceptible to oxidation and efficient routes to the corresponding sulfoxides and sulfones have been reported [81]. The sulfoxides undergo stereospecific Pummerer rearrangement to *exo*-α-acetoxy sulfides, which can be oxidized in turn to α-acetoxy sulfoxides and sulfones [81]. Thioglycosans are intramolecular thioglycosides and this observation has led to their investigation as glycosyl donors [75]. Stick and co-workers have shown that thioglucosans are effective glycosyl donors when activated with the *N*-iodosuccinimide/TfOH couple [82]. The initial products are disulfide-linked disaccharides and desulfurization (with Raney nickel) leads to 6-deoxy glycosides. A selenium counterpart, tri-*O*-acetyl selenoglucosan (**22**; ❷ *Fig. 2*) was prepared through the agency of sodium hydrogen selenide (prepared from sodium borohydride and selenium in anhydrous ethanol) and the doubly activated sugar (**24**) [77]. Selenoglycosans can also act as glycosyl donors in the presence of NIS/TfOH, and the weak nature of the carbon–selenium bond allows a mild tributylstannane-mediated reduction of intermediate 6-seleno sugars to afford the 6-deoxy glycosides [83]. Detailed structural analyses of a number of sulfur and selenium-containing 1,6-anhydro sugars have been reported [81,84].

Several nitrogen-containing anhydro sugars are known. These compounds, while formally anhydro amino sugars, are better considered aza sugars. In alkaline media, equilibrated solutions of 6-amino-6-deoxy aldoses contain varying amounts of the corresponding aza sugars, such as the amine (**23**; ❷ *Fig. 2*) [3,85]. Treatment of C-6 activated anomeric azides with triphenylphosphine yields anomeric iminophosphoranes that rearrange in situ by elimination of the *O*-6 sulfonate affording, after ion-exchange chromatography and *N*-acylation, 6-amino-1,6-anhydro-6-deoxy-β-D-glycopyranoses (❷ *Scheme 25*) [86].

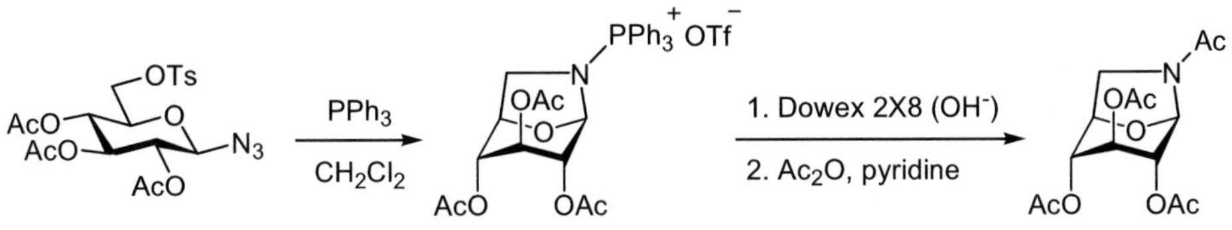

Scheme 25

References

1. Angyal SJ, Dawes K (1968) Aust J Chem 21:2747
2. Kochetkov NK (1987) Tetrahedron 43:2389
3. Černý M, Stanek J Jr (1977) Adv Carbohydr Chem Biochem 34:23
4. Bols M (1996) 1,6-Anhydro sugars. Carbohydrate building blocks. Wiley, New York
5. Peat S (1946) Adv Carbohydr Chem 2:37
6. Witczak ZJ (1994) Selective protection of levoglucosan derivatives. In: Witczak ZJ (ed) Frontiers in biomedicine and biotechnology. Levoglucosenone and levoglucosans, chemistry and applications. ATL Press, Mount Prospect, 2:165
7. Tanret M (1894) Bull Soc Chim Fr 211:944
8. Park YJ, Kim HS, Jeffrey GA (1971) Acta Cryst B 27:220
9. Heyns K, Weyer J (1968) Liebigs Ann Chem 718:224
10. Pictet A, Sarasin J (1918) Helv Chim Acta 1:87
11. Ward RB (1963) Methods Carbohydr Chem 2:394
12. Shafizadeh F, Furneaux RH, Stevenson TT, Cochran TG (1978) Carbohydr Res 61:519
13. Schkolnik G, Rudich Y (2006) Anal Bioanal Chem 385:26
14. Knauf AE, Hann RM, Hudson CS (1941) J Am Chem Soc 63:1447
15. Hann RM, Hudson CS (1942) J Am Chem Soc 64:925
16. Zottola MA, Alonso R, Vite GD, Fraser-Reid B (1989) J Org Chem 54:6123
17. Cleophax J (2003) Synthesis 1015
18. Coleman GH (1963) Methods Carbohydr Chem 2:397
19. Boons G-J, Isles S, Setälä P (1995) Synlett 755
20. Rao MV, Nagarajan M (1987) Carbohydr Res 162:141
21. Caron S, McDonald A, Heathcock CH (1996) Carbohydr Res 281:179
22. Haeckel R, Lauer G, Oberdorfer F (1996) Synlett 21
23. Mereyala HB, Venkataramanaiah KC, Dalvoy VS (1992) Carbohydr Res 225:151
24. Lauer G, Oberdorfer F (1993) Angew Chem Int Ed Engl 32:272
25. Černý M (1994) 1,6:2,3- and 1,6:3,4-Dianhydro-β-D-hexopyranoses. Synthesis and preparative applications. In: Witczak ZJ (ed) Frontiers in biomedicine and biotechnology. Levoglucosenone and levoglucosans, chemistry and applications. ATL Press, Mount Prospect, 2:121
26. Tailler D, Jacquinet JC, Noirot AM, Beau JM (1992) J Chem Soc Perkin 1 3163
27. Černý M (2003) Adv Carbohydr Chem Biochem 58:122
28. Stoffyn PJ, Jeanloz RW (1960) J Biol Chem 235:2507
29. Hölte JV (1998) Microbiol Mol Biol R 62:181
30. Hölte JV, Mirelman D, Sharon N, Schwarz U (1975) J Bacteriol 124:1067
31. Uehara T, Suefuji K, Valbuena N, Meehan B, Donegan M, Park J (2005) J Bacteriol 187:3643
32. Pulido R, Gotor V (1994) Carbohydr Res 252:55
33. Ljunger G, Adlercreutz P, Mattiasson B (1994) Biotechnol Lett 16:1167
34. Somsák L, Ferrier RJ (1991) Adv Carbohydr Chem Biochem 49:37
35. Ohrui H, Horiki H, Kishi H, Meguro H (1983) Agric Biol Chem 47:1101
36. Ferrier RJ, Furneaux RH (1980) Aust J Chem 33:1025
37. Nishikawa T, Mishima Y, Ohyabu N, Isobe M (2004) Tetrahedron Lett 45:175
38. Vogel C, Liebelt B, Steffan W, Kristen H (1992) J Carbohydr Chem 11:287
39. Zottola M, Rao BV, Fraser-Reid B (1991) J Chem Soc Chem Commun 969
40. Lee JC, Tai CA, Hung SC (2002) Tetrahedron Lett 43:851
41. Burgey CS, Vollerthun R, Fraser-Reid B (1994) Tetrahedron Lett 35:2637
42. Wang L-X, Sakairi N, Kuzuhara H (1990) J Chem Soc Perkin 1 1677
43. Koto S, Uchida T, Zen S (1972) Chem Lett 1049
44. Ruckel ER, Schuerch C (1966) J Org Chem 31:2233
45. Yoshida T, Nakashima H, Yamamoto N, Uryu T (1993) Polymer J 25:1069
46. Brigl P (1922) Hoppe-Seyler's Z Physiol Chem 122:245
47. Lemieux RU, Huber G (1956) J Am Chem Soc 78:4117
48. Halcomb RL, Danishefsky SJ (1989) J Am Chem Soc 111:6661
49. Danishefsky SJ, Bilodeau MT (1996) Angew Chem Int Ed Engl 35:1380
50. Collins P, Ferrier R (1995) Monosaccharides: Their Chemistry and Their Roles in Natural Products. Wiley, Chichester

51. Leeuwenburgh MA, van der Marel GA, Overkleeft HS, van Boom JH (2003) J Carbohydr Chem 22:549
52. Timmers CM, Wigchert SCM, Leeuwenburgh MA, van der Marel GA, van Boom JH (1998) Eur J Org Chem 1998:91
53. Seeberger PH, Danishefsky SJ (1998) Acc Chem Res 31:685
54. Ito H, Eby R, Kramer S, Schuerch C (1980) Carbohydr Res 86:193
55. Bhagwat SS, Hamam PR, Still WC (1985) J Am Chem Soc 107:6372
56. Kops J, Schuerch C (1965) J Org Chem 30:3951
57. Bullock C, Hough L, Richardson AC (1990) Carbohydr Res 197:131
58. Thiem J, Wiesner M (1993) Carbohydr Res 249:197
59. Nokami T, Werz DB, Seeberger PH (2005) Helv Chim Acta 88:2823
60. Dimler RJ, Davis HA, Hilbert GE (1946) J Am Chem Soc 68:1377
61. Alexander BH, Dimler RJ, Mehltretter CL (1951) J Am Chem Soc 73:4658
62. Dimler RJ (1952) Adv Carbohydr Chem 7:37
63. Angyal SJ, Beveridge RJ (1978) Aust J Chem 31:1151
64. Sarkar SK, Choudhury AK, Mukhopadhyay B, Roy N (1999) J Carbohydr Chem 18:1121
65. Witczak ZJ (1994) Levoglucosenone; past, present and further applications In: Witczak ZJ (ed) Frontiers in biomedicine and biotechnology. Levoglucosenone and levoglucosans, chemistry and applications. ATL Press, Mount Prospect, 2:3
66. Halpern Y, Riffer R, Broido A (1973) J Org Chem 38:204
67. Shafizadeh F, Furneaux RH, Stevenson TT (1979) Carbohydr Res 71:169
68. Brimacombe JS, Hunedy F, Tucker LCN (1978) Carbohydr Res 60:C11
69. Shafizadeh F, Chin PPS (1977) Carbohydr Res 58:79
70. Mori M, Chuman T, Kato K (1984) Carbohydr Res 129:73
71. Witczak ZJ (1994) Pure Appl Chem 66:2189
72. Witczak ZJ, Chhabra R, Chen H, Xie X-Q (1997) Carbohydr Res 301:167
73. Ward DD, Shafizadeh F (1981) Carbohydr Res 95:155
74. Akagi M, Tejima S, Haga M (1963) Chem Pharm Bull 11:58
75. Lundt I, Skelbæk-Pedersen B (1981) Acta Chem Scand B35:637
76. Whistler RL, Seib PA (1966) Carbohydr Res 2:93
77. Driguez H, McAuliffe JC, Stick RV, Tilbrook DMG, Williams SJ (1996) Aust J Chem 49:343
78. Ramesha AR, Chandrasekaran S (1992) Synth Commun 22:3227
79. Sridhar PR, Saravanan V, Chandrasekaran S (2005) Pure Appl Chem 77:145
80. Yamamoto K, Haga M, Tejima S (1975) Chem Pharm Bull 23:233
81. Skelton BW, Stick RV, Tilbrook DMG, White AH, Williams SJ (2000) Aust J Chem 53:389
82. Stick RV, Tilbrook DMG, Williams SJ (1997) Tetrahedron Lett 38:2741
83. Stick RV, Tilbrook DMG, Williams SJ (1999) Aust J Chem 52:685
84. Buděšínský M, Poláková J, Hamerníková M, Císařová I, Trnka TS, Černý M (2006) Coll Czech Chem Commun 71:311
85. Paulsen H, Todt K (1967) Chem Ber 100:512
86. Lafont D, Wollny A, Boullanger P (1998) Carbohydr Res 310:9

3.8 *C*-Glycosylation

Toshio Nishikawa, Masaatsu Adachi, Minoru Isobe
Graduate School of Bioagricultural Sciences, Nagoya University,
464-8601 Nagoya, Japan
nisikawa@agr.nagoya-u.ac.jp, madachi@agr.nagoya-u.ac.jp,
isobem@agr.nagoya-u.ac.jp

Abstract

This chapter deals with *C*-glycosylation, a carbon–carbon bond-forming reaction at an anomeric carbon of carbohydrate and its derivatives. Since synthesis of natural products containing the *C*-glycosidic linkage was one of the major issues in organic synthesis in the 1970s, extensive efforts have been devoted to developing stereoselective synthesis of *C*-glycoside, which includes a variety of C–C bond formations by using anomeric cations (oxocarbenium cations), anomeric radicals, and anomeric anions, as well as sigmatropic

In: *Glycoscience*. Fraser-Reid B, Tatsuta K, Thiem J (eds)
Chapter-DOI 10-1007/978-3-540-30429-6_17: © Springer-Verlag Berlin Heidelberg 2008

rearrangement and transition metal-catalyzed reactions. Among these reactions, useful C-glycosylations with high stereoselectivities are reviewed in this chapter with an emphasis on the reaction mechanisms.

Keywords

C-Glycosylation; Oxocarbenium cation; Anomeric radical; Anomeric anion; Stereoelectronic effect

Abbreviations

HMPA   hexamethylphosphoramide
MCPBA m-chloroperoxybenzoic acid
TIPS   triisopropylsilyl
TMSCN trimethylsilyl cyanide

1 Introduction

C-Glycosylation, a carbon–carbon bond-forming reaction at an anomeric center of carbohydrates and related compounds, has been the subject of great attention due to the importance of C-glycoside-containing compounds such as C-glycoside mimics of glycolipids, oligosaccharides, and glycoproteins, and natural products in chemistry and biology.

In 1993, researchers of the Kirin Brewery Corporation, Japan, reported the isolation and structure of agelasphines, novel anticancer sphingolipids isolated from an Okinawan marine sponge, *Agelas mauritianus* (❷ *Fig. 1*) [1]. Since these compounds are the first examples of α-galactosyl ceramide found in nature, and a novel mode of action for their anticancer activity was expected, development of an antitumor drug candidate based on agelasphines led to KRN7000, a synthetic sphingolipid. Extensive studies regarding the mode of action revealed that KRN7000 activated natural killer T-cells by binding CD1d, an antigen presentation protein, thus releasing cytokines including Interferon-γ (IFN-γ) causing tumor rejection and Interleukin-4 (IL-4) suppressing autoimmune diseases. Interestingly, OCH, a truncated synthetic analog of KRN7000 induced a predominant production of IL-4 over IFN-γ, while KRN7000 induced production of both cytokines. In 2003, Franck and co-workers reported the synthesis of a C-glycoside analog of KRN7000, surprisingly, this compound exhibited 1000-fold more activity than KRN7000 in a mouse malaria assay as well as a 100-fold activity increase in a mouse melanoma model. Significant efforts have been devoted towards the synthesis of C-glycoside that specifically releases one of these cytokines in order to yield compounds for chemotherapeutic application. This recent example demonstrates the importance of C-glycoside analogs of biologically active glycoconjugates in biology and pharmacy. These results are stimulating development of stereoselective and efficient methods for synthesizing C-glycoside analogues of KRN7000 (❷ *Fig. 1*).

The C-glycosidic linkage is also found in many biologically important natural products such as marine natural toxins and antibiotics as shown in ❷ *Fig. 2*. Since synthesis of these structurally complex natural products has been a major subject in modern organic synthesis, stere-

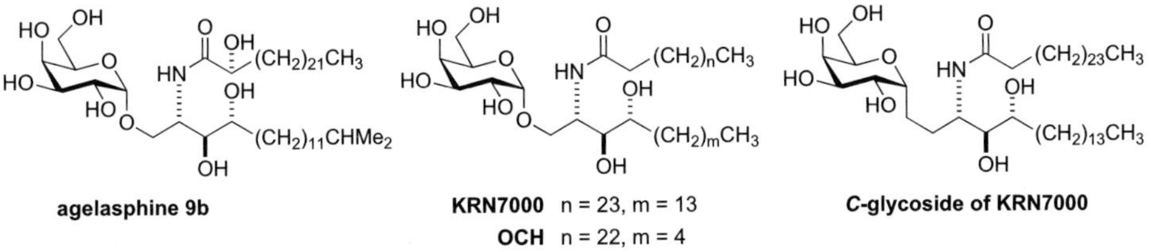

Figure 1
Agelasphines, novel anticancer sphingolipids and its related synthetic compounds

oselective *C*-glycosylation has been extensively studied. In the synthesis of these molecules, carbohydrates have been employed as chiral starting materials (chiral pool), because a variety of carbohydrates with multi-functionality are readily available from nature [2]. In ❯ *Fig. 2*, the arrows indicate the C–C bonds that were synthesized by "*C*-glycosylation". It is worth noting that *C*-glycosylation has been also employed for the synthesis of hidden *C*-glycosidic bonds (see the structure of brevetoxin A).

This chapter will survey *C*-glycosylation by classifying the reactions mainly according to the type of reaction and in some cases the structure of the substrate with focus on the selectivity and its applicability [3].

2 Nucleophilic Addition to Electrophilic Carbohydrate Derivatives

Addition reaction of carbon nucleophile to electrophilic carbohydrate derivative is one of the most widely used methods for *C*-glycosylation. Since there are many carbohydrate derivatives that have been reported for *O*-glycosylation, these materials have also been employed for *C*-glycosylations. The difference between *O* and *C*-glycosylation is lack of the thermodynamic anomeric effect to control stereochemistry, and neighboring group participation has been rarely operated in *C*-glycosylation. In this section, the reactions are classified by the structure of the electrophilic carbohydrate such as (a) lactol, (b) lactone, (c) glycoside, (d) glycal, and (e) anhydrosugar.

2.1 Lactol as an Electrophilic Carbohydrate

Reaction of lactol with stabilized nucleophilic reagents such as 1,3-diketone and the Wittig reagent is a classic method and is still a useful method for the synthesis of *C*-glycosides. The reaction undergoes formation of olefin with an electron-withdrawing group followed by intramolecular conjugate addition with the resulting alkoxide anion (❯ *Eq. 1*). As the process is reversible, thermodynamically stable *C*-glycoside would be obtained as the major product. Among many of these reactions reported to date, several recent examples with special characters are shown below.

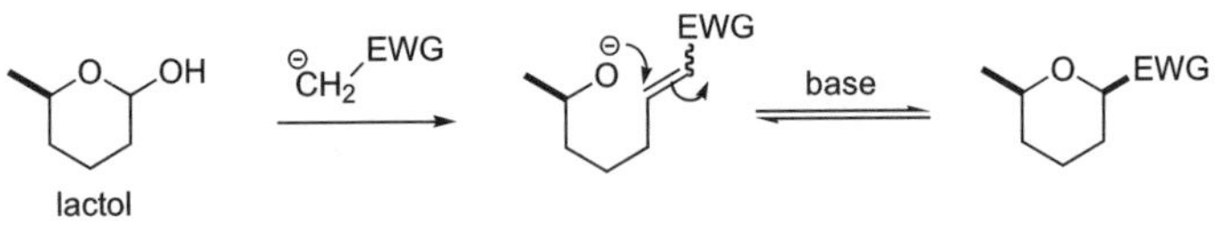

(1)

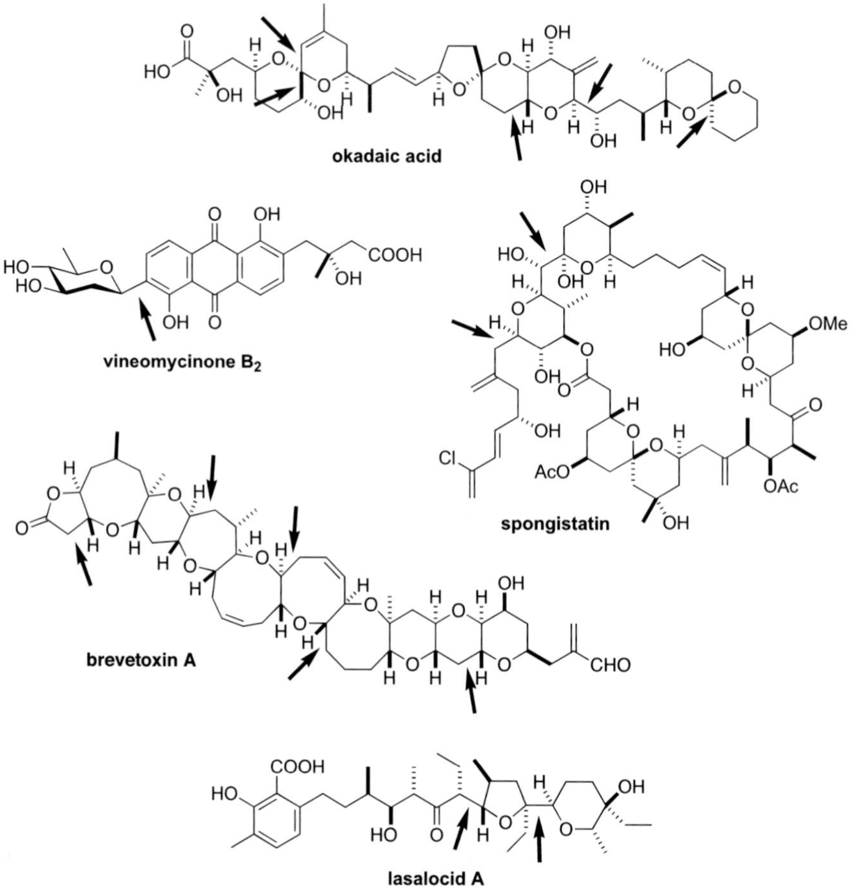

okadaic acid

vineomycinone B₂

spongistatin

brevetoxin A

lasalocid A

☐ Figure 2
Naturally occurring compounds containing the *C*-glycosidic linkage

One-step synthesis of β-glycosyl ketone from unprotected carbohydrate in aqueous media was reported by Rubineau and co-workers (❯ *Scheme 1*) [4]. Unprotected D-glucose, mannose, and galactose were treated with pentane-2,4-dione at 90 °C in aqueous sodium bicarbonate to afford β-glycosyl ketones in high yields. The reaction began with Knoevenagel condensation to give the unsaturated ketone, which underwent an intramolecular conjugate addition of the alkoxide. The reaction conducted at room temperature gave a mixture of the four possible furanosides and pyranosides. It is worth noting that epimerization at the C-2 position did not occur in the reaction with glucose and mannose, while the same reactions with N-acetyl-gluco-, *manno*-, and galactosamine afforded a mixture of the C-2 epimeric products. The β-glycosyl ketone was transformed into β-1-formyl sugar, a versatile synthetic intermediate for a variety of *C*-glycosidic compounds.

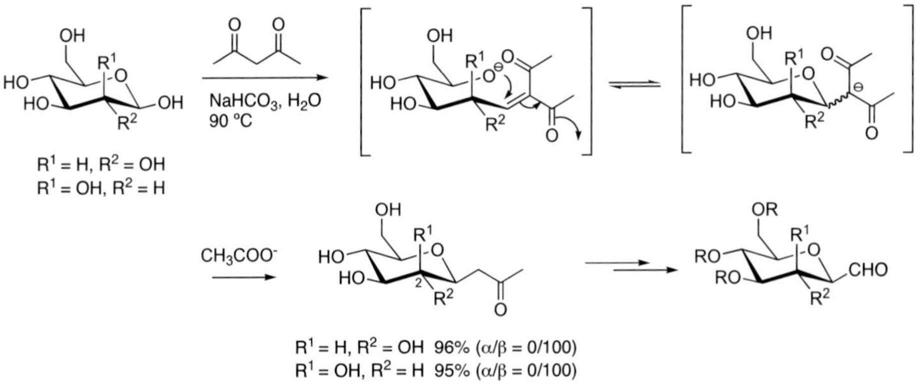
R1 = H, R2 = OH
R1 = OH, R2 = H
NaHCO3, H2O
90 °C
CH3COO-
R1 = H, R2 = OH 96% (α/β = 0/100)
R1 = OH, R2 = H 95% (α/β = 0/100)
CHO

Scheme 1

This reaction generally gives *β-C*-hexopyranose, however, *α-C*-glycoside could be obtained in a special case as shown in *Scheme 2* [5]. Reaction of 2-deoxylactol with a Wittig reagent yielded an unsaturated ester, which was treated with a catalytic amount of NaOMe to afford *α-C*-glycoside as a single product in good yield. Preferential formation of the *C*-glycoside with *α*-configuration is rationalized by the transition state conformers shown in *Scheme 2*. The right conformer with the two axial substituents is involved in the recyclization step, affording *α-C*-glycoside, because the left conformer is much destabilized by A-strain between the exo-olefin and the two equatorial substituents. The product was further transformed into luminacins, angiogenesis inhibitors isolated from a *Streptomyces*.

Epoxy lactol was employed for the *C*-glycosylation depicted in *Scheme 3* [6]. The lactol was treated with a stabilized Wittig reagent to preferentially give the *Z*-unsaturated ester, which upon exposure to NaH gave the thermodynamically stable *β-C*-pyranoside through conjugate addition of the resulting alkoxide. On the other hand, the same unsaturated epoxide was treated with a palladium catalyst to give *β-C*-furanoside as a single product via a *π*-allyl palladium complex with net retention of configuration.

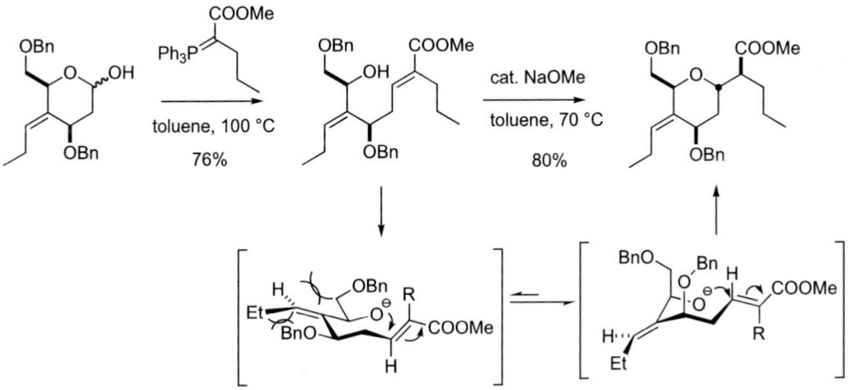
toluene, 100 °C
76%
cat. NaOMe
toluene, 70 °C
80%
COOMe

Scheme 2

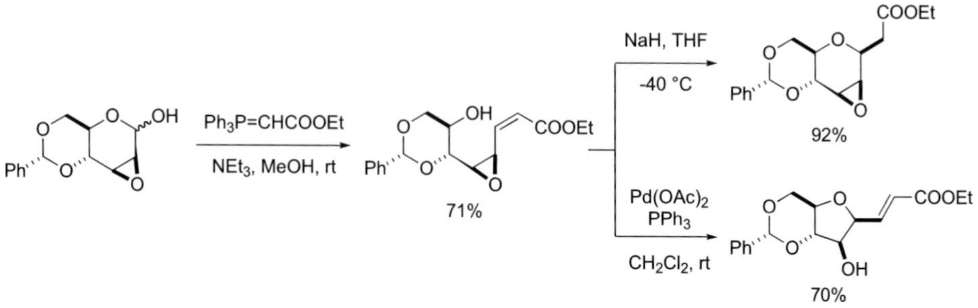

◘ Scheme 3

2.2 Lactone as an Electrophilic Carbohydrate

Glyconolactones (Sugar lactones) have been widely employed as electrophilic donors for C-glycosylation, because of their easy availability from oxidation of the corresponding hemi-acetal. Reaction of the lactone with a carbon nucleophile gives lactol, which can be further manipulated into C-glycoside or olefin at the anomeric position. This section deals with C-gly-cosylations starting from sugar lactone classified by the nucleophiles; (1) enolate and its relat-ed anion stabilized by carbonyl, sulfonyl group, (2) Wittig and its related reagents [7], and (3) organometallic reagents such as Grignard and organolithium reagents followed by reduc-tion.

2.2.1 Reaction with Enolate and Related Anions

Reaction of sugar lactone with ester enolate (the Claisen condensation) gives β-keto ester, the resulting 1,3-dicarbonyl compound immediately reacts with the hydroxyl group present in the same molecule to afford a lactol with an axial hydroxyl group. In a recent example shown in ❏ Scheme 4 [8], the product of the reaction was further transformed to the corresponding nitrile or amide by addition of cyanide or a Ritter reaction with benzonitrile in the presence of TMSOTf as a Lewis acid.

In the reaction with an enolate of α-substituted ester, stereocontrol of the α-position is gener-ally difficult because of the easy epimerization. However, Nakata reported a highly stereose-

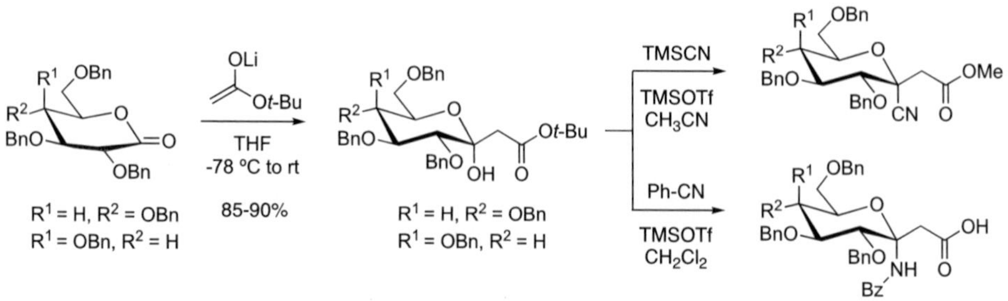

◘ Scheme 4

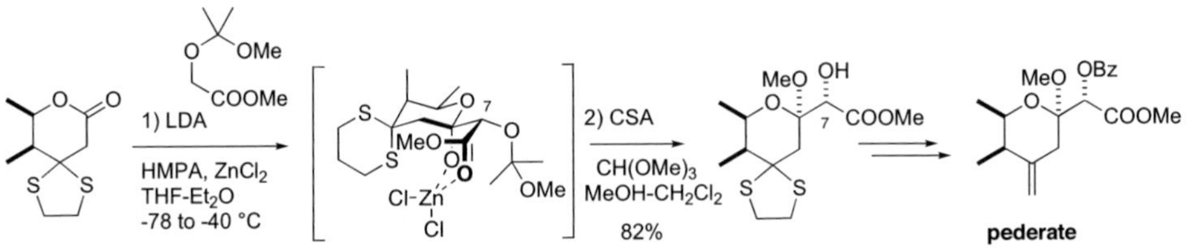

□ Scheme 5

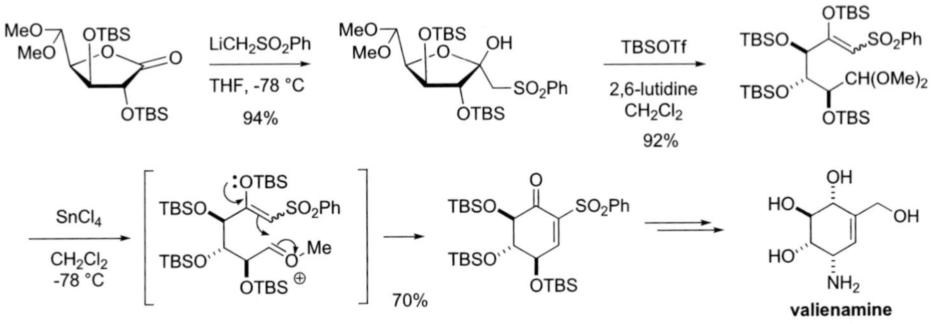

□ Scheme 6

lective reaction of lactone with enolate derived from glycolate in the presence of ZnCl₂ and HMPA (❍ *Scheme 5*) [9]. The high stereoselectivity was explained by formation of the thermodynamically stable chelate intermediate, which has an equatorial hydroxyl group at the C-7 position. The product was further transformed to pederate, the left part of pederine, an insect toxin.

Many carbanions stabilized by other electron-withdrawing groups have also been employed as nucleophiles. For example, the lithium anion of methyl phenyl sulfone reacted with furanolactone to give a hemiacetal, which was transformed to silyl enol ether (❍ *Scheme 6*) [10]. Upon treatment with SnCl₄ as a Lewis acid, an intramolecular reaction of the silyl enol ether (the so-called Mukaiyama reaction) takes place followed by elimination of methanol, giving the highly substituted cyclohexenone, an important intermediate for valienamine, a glycosidase inhibitor.

2.2.2 Reaction with Wittig Reagents and Related Reagents

Reaction of ester with Wittig reagents has rarely been employed, because of the low reactivity compared with aldehyde and ketone. However, sugar lactone with suitable protecting groups reacts with (carbomethoxymethylene)tributylphosphorane at elevated temperature to give *exo*-glycal (also known as glycosylidene) in good yield. Examples of Wittig reactions with glucono-, galactono-, and mannolactone are shown in ❍ *Scheme 7* [11]. Lactones derived from amino sugars such as GalNAc and GlcNAc also underwent the Wittig reaction giving the corresponding *exo*-glycals in good yields [12]. The carbon–carbon double bond can be stereoselectively reduced by hydrogenation to give β-*C*-glycoside.

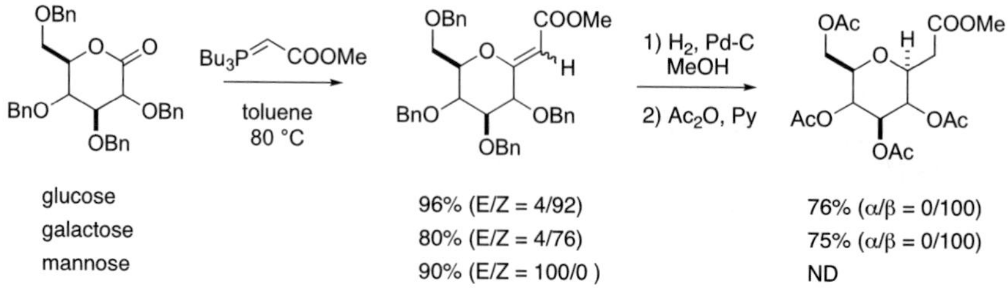

glucose   96% (E/Z = 4/92)   76% (α/β = 0/100)
galactose   80% (E/Z = 4/76)   75% (α/β = 0/100)
mannose   90% (E/Z = 100/0)   ND

◘ Scheme 7

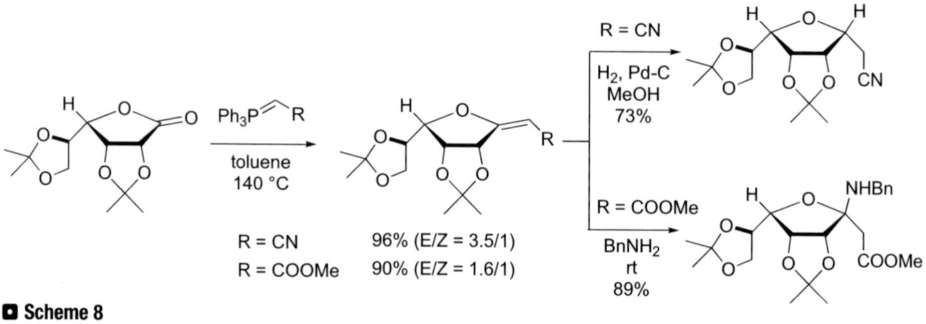

R = CN   96% (E/Z = 3.5/1)
R = COOMe   90% (E/Z = 1.6/1)

◘ Scheme 8

Wittig reactions of furanolactone have also been reported (❯ *Scheme 8*) [13]. Use of microwave irradiation reduces the reaction time due to the effect of microwave flash heating. The resulting *exo*-methylene moiety can be utilized as a stepping stone for further modifications. Catalytic hydrogenation gave α-*C*-glycoside, while conjugate addition of benzylamine afforded β-amino acid derivatives in a stereoselective manner [14].

Sugar lactone was treated with tris(dimethylamino)phosphine-carbon tetrachloride or triphenylphosphine-carbon tetrachloride to give *exo*-dichloromethylene as shown in ❯ *Scheme 9* [15]. A wide range of sugar lactones including furanolactone and pyranolactone with a variety of protecting groups have been employed as a substrate for this reaction. The *exo*-methylene moiety was transformed to an amino acid precursor by oxidation with MCPBA followed by addition of azide to the resulting chloride [16].

Simple methylenation of sugar lactone was carried out by a titanium-based reagent such as Tebbe and Petasis reagent (Me$_2$TiCp$_2$) (❯ *Table 1*) [17]. Since the carbonyl group of ester also reacts with these reagents, ester should be avoided as a protecting group. The Tebbe reagent reacted with the lactone at a low temperature (entry 1), while reaction with the Petasis

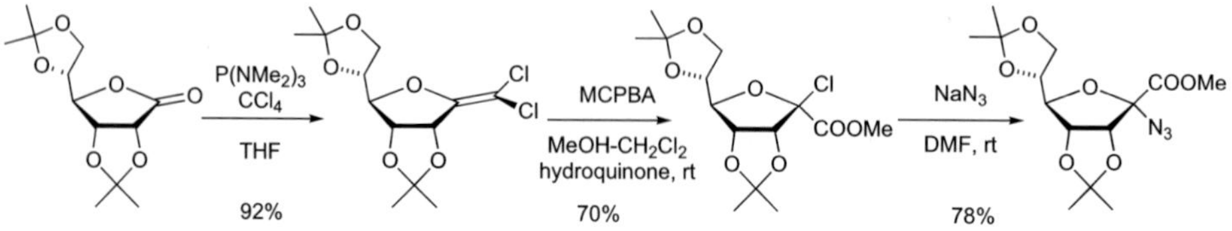

92%   70%   78%

◘ Scheme 9

◻ **Table 1**
[17]

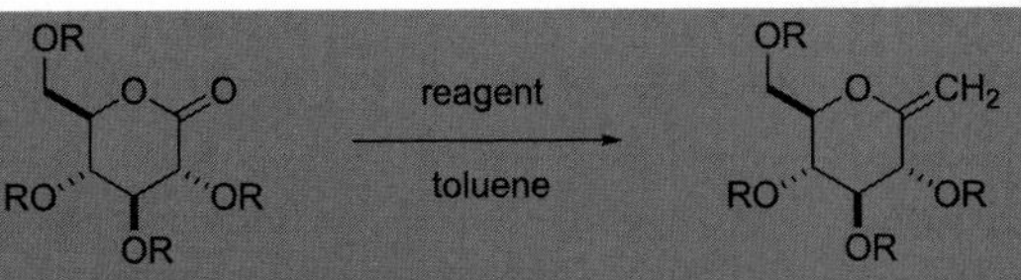

Entry	R	Reagent	Temp	Yield (%)	Ref.
1	Bn	Cp₂ Cl Me / Ti Al (Tebbe) / Cp Me	−40 to 0 °C	82	[17a]
2	Bn	Me₂TiCp₂ (Petasis)	70 °C	94	[17b]
3	TMS	Cp₂ Cl Me / Ti Al (Tebbe) / Cp Me	−40 to 0 °C	54	[17a]
4	MOM	Me₂TiCp₂ (Petasis)	70 °C	79	[17c]
5	TES	i) 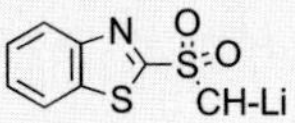 [benzothiazole-S–S(=O)O–CH-Li]; ii) DBU, rt	−78 °C	74	[17d]

reagent required heating at 60–70 °C. Recently, methylenation of the lactone by Julia coupling was also reported (entry 5). The product, *exo*-glycal is a versatile synthetic intermediate for a variety of *C*-glycosides by means of Suzuki–Miyaura coupling (❷ *Scheme 10*). For example, alkenyl iodide **2** prepared from **1** was coupled with an arylboron reagent to **3** in the presence of a palladium catalyst [17b]. On the other hand, hydroboration of *exo*-glycal **1** with 9-BBN gave alkylborane intermediate **4** [18], which was coupled with alkenyl bromide in the presence of a palladium catalyst to give β-*C*-glycoside **5** [17c]. This synthetic methodology has been extensively used for the synthesis of marine polyether toxins. The related reactions will be discussed in ❷ *Sect. 6*.

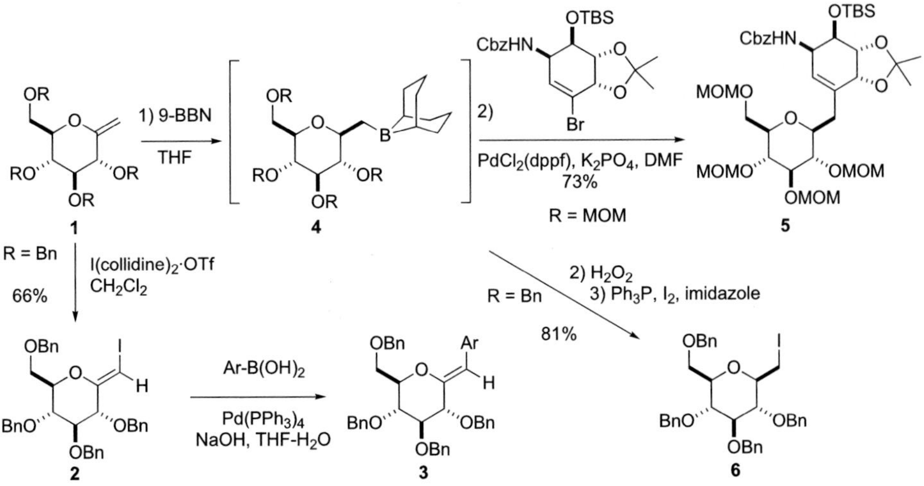

◻ **Scheme 10**

2.2.3 Reaction with Organometallic Reagents Followed by Reduction of the Newly Formed Hemiacetal

Reaction of glyconolactone with organometallic reagents such as Grignard and organolithium reagents gives hemiacetal (ketose). The resulting hemiacetal is reduced with Et_3SiH in the presence of a Lewis acid such as TMSOTf or $BF_3 \cdot OEt_2$ to afford C-glycoside (❷ *Eq. 2*). In the reduction of a hexopyranose intermediate, the hydride attacks the oxocarbenium cation generated from the reaction between the hemiacetal and the Lewis acid from the axial position, because of the stereoelectronic effect, affording β-C-glycoside. This two-step procedure established by Kishi is one of the most widely used reactions for β-C-hexopyranoside (❷ *Table 2*) [19].

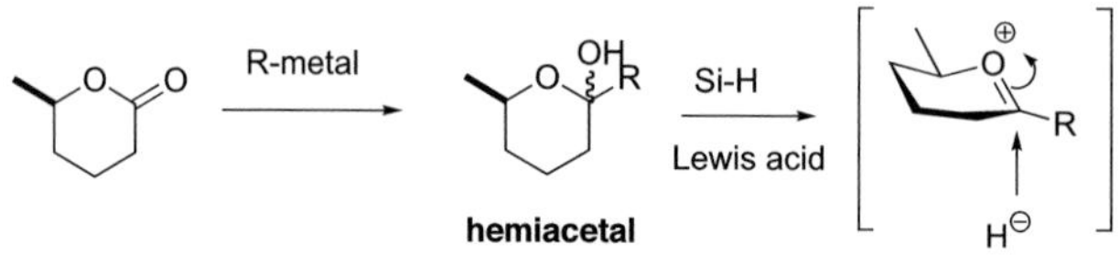

$$\text{(2)}$$

As shown in ❷ *Table 2*, this procedure enables the synthesis of a variety of β-C-glucosides and β-C-galactosides with substituents such as alkyl, allyl, vinyl, alkynyl, propargyl, aromatic, heteroaromatic [20], and acetyl groups.

In contrast, the reduction of mannose-derived hemiacetals under similar conditions gave lower selectivity (❷ *Table 2*) [19]. In order to improve the stereoselectivity, several methods have been proposed for special substrates; use of sterically hindered reducing agents such as TIPS-H (i-Pr_3SiH) or $(TMS)_3SiH$ (entry 21 in ❷ *Table 2*), restricted conformation of pyranose with 4,6-benzylidene acetal [21], and neighboring group participation of an acetyl group at the C-2 position (❷ *Scheme 11*) [22].

The same procedure was applied to C-glycosylations of furanose-derived lactones (❷ *Table 3*) [23]. The lactone reacted with organolithium reagents such as alkyl, alkynyl and heteroaromatic lithium to give the corresponding hemiacetals, which were reduced with triethylsilane in the presence of a Lewis acid. Since the reaction proceeds through an oxocarbenium cation intermediate, the hydride attack determines the stereochemistry of the product. However, rational explanation of these selectivities is difficult at present, because the selectivity is influenced by many factors such as steric and stereoelectronic effects together with the kind of sugar used in the reaction, the substituent, and the protective group.

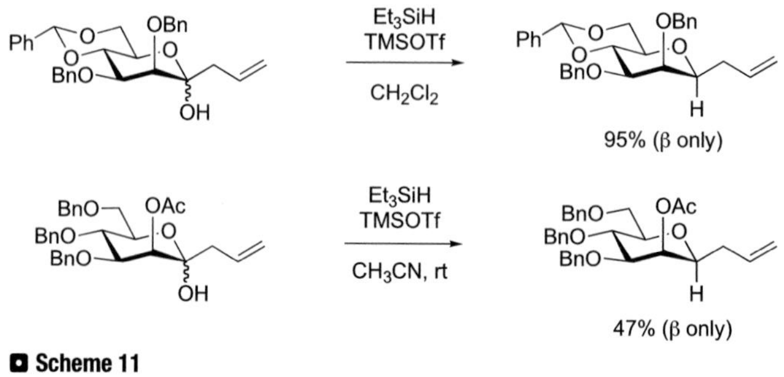

❑ **Scheme 11**

□ Table 2
[19]

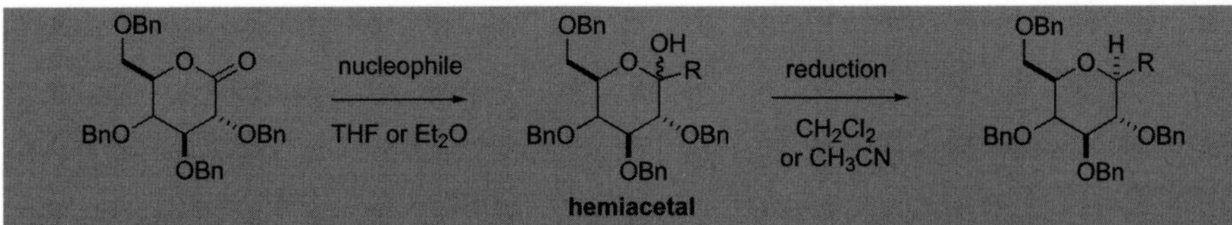

hemiacetal

Entry	Sugar	Nucleophile	Hemiacetal	Reduction Silane, Lewis acid	Product Yield (α/β)	Ref.
1	Glucose	MeMgBr	90%	Et$_3$SiH, TMSOTf	81% (0/100)	[19a,b]
2		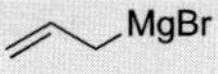 ⟋⟍MgBr	ND	Et$_3$SiH, BF$_3$·OEt$_2$	85% (10/90)[a]	[19c]
3		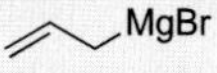 ⟋⟍MgBr	82%	Et$_3$SiH, TMSOTf	79% (10/90)	[19a,b]
4		 ⟋⟍MgBr	ND	Et$_3$SiH, BF$_3$·OEt$_2$	49% (0/100)[a]	[19d]
5		TMS—≡—Li	ND	Et$_3$SiH, BF$_3$·OEt$_2$	74% (0/100)	[19e]
6		BnO(CH$_2$)$_2$—≡—Li	Quant	Et$_3$SiH, BF$_3$·OEt$_2$	72% (0/100)	[19f]
7		PhLi	85%	Et$_3$SiH, BF$_3$·OEt$_2$	80% (1/4)	[19g,h]
8		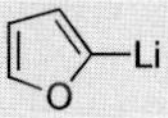 furyl-Li	ND	Et$_3$SiH, BF$_3$·OEt$_2$	77% (0/100)[a]	[19g]
9		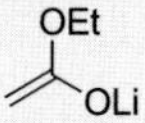 OEt / OLi ketene acetal	ND	Et$_3$SiH, BF$_3$·OEt$_2$	72% (0/100)[a]	[19c]
10		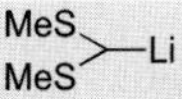 MeS / MeS —Li	74%	Et$_3$SiH, BF$_3$·OEt$_2$	99% (0/100)	[19i]
11	Galactose	 ⟋⟍⟋MgBr	ND	Et$_3$SiH, BF$_3$·OEt$_2$	81% (0/100)[a]	[19j]
12		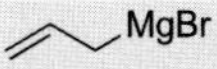 ⟋⟍MgBr	ND	Et$_3$SiH, BF$_3$·OEt$_2$	76% (10/90)[a]	[19c]
13		 ⟋⟍MgBr	ND	Et$_3$SiH, BF$_3$·OEt$_2$	50% (0/100)[a]	[19d]
14		TMS—≡—Li, CeCl$_3$	91%	Et$_3$SiH, BF$_3$·OEt$_2$	71% (0/100)[b]	[19k]
15		(i) 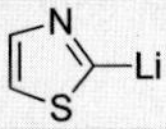 thiazolyl-Li (ii) Ac$_2$O	75%[c]	Et$_3$SiH, TMSOTf	96% (0/100)	[19l]
16		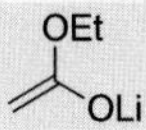 OEt / OLi ketene acetal	ND	Et$_3$SiH, BF$_3$·OEt$_2$	79% (0/100)[a]	[19c]
17	Mannose	MeMgBr	86%	Et$_3$SiH, TMSOTf	68% (22/78)	[19a,b]
18		 ⟋⟍MgBr	ND	Et$_3$SiH, BF$_3$·OEt$_2$	67% (1/1)[a]	[19c]
19			ND	Et$_3$SiH, TMSOTf	61% (33/67)[a]	[19a,b]
20		TMS—≡—Li, CeCl$_3$	93%	Et$_3$SiH, BF$_3$·OEt$_2$	77% (1/2.5)[b]	[19k]
21		TMS—≡—Li	–	(TMS)$_3$SiH, BF$_3$·OEt$_2$	76% (0/100)[b]	[19m]
22		PhLi	85%	Et$_3$SiH, TMSOTf	53% (0/100)	[19b]
23		(i) 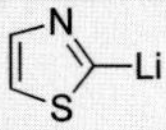 thiazolyl-Li (ii) Ac$_2$O	78%[c]	Et$_3$SiH, TMSOTf	97% (0/100)	[19l]
24		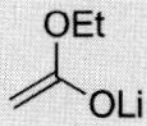 OEt / OLi ketene acetal	ND	Et$_3$SiH, BF$_3$·OEt$_2$	66% (3/1)[a]	[19c]

[a]: 2-step yield; [b]: after removal of TMS with TBAF; [c]: product was the corresponding acetate

☐ **Table 3**
[23]

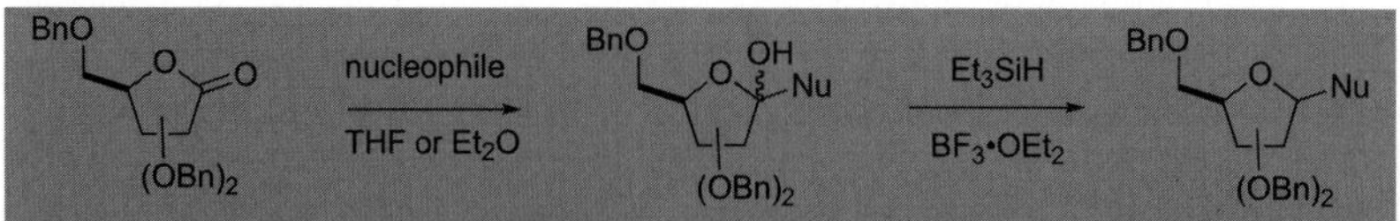

BnO ... O, O → nucleophile (THF or Et₂O) → BnO ... O, OH, Nu (OBn)₂ → Et₃SiH, BF₃·OEt₂ → BnO ... O, Nu (OBn)₂

Entry	Furanolactone	Nucleophiles	Hemiacetal	Product Yield (α/β)	Ref.
1	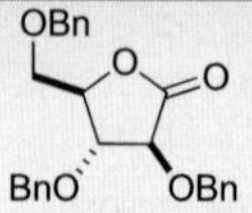 OBn ... BnO OBn (O, O)	MeLi	ND	53% (100/0)*	[23a]
2		≡—Li	ND	81% (5/1)*	[23a]
3		LiCH₂PO(OMe)₂	93%	95% (100/0)	[23b]
4	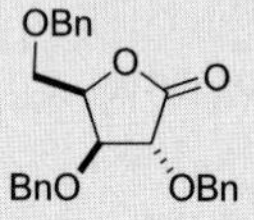 OBn ... BnO OBn (O, O)	MeLi	81%	76% (1/5)	[23a]
5		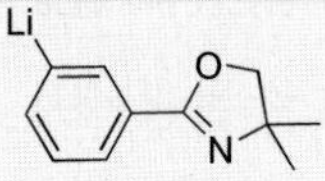 Li, aryl oxazoline	ND	87% (0/100)*	[23c]
6	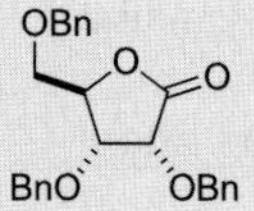 OBn ... BnO OBn (O, O)	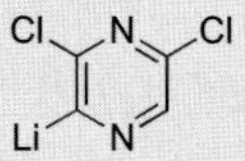 Cl N Cl, Li N (dichloropyrazine)	ND	72% (0/100)*	[23c]

*: yield over 2 steps

Synthesis of aryl-*C*-glycoside has been reported by the same methodologies. One recent example is shown in ❷ *Scheme 12*. 2-Deoxysugar reacts with napthtyl lithium to give the corresponding hemiacetal, which was reduced with NaBH₃CN in an acidic aqueous medium to afford α-*C*-glycoside exclusively [24].

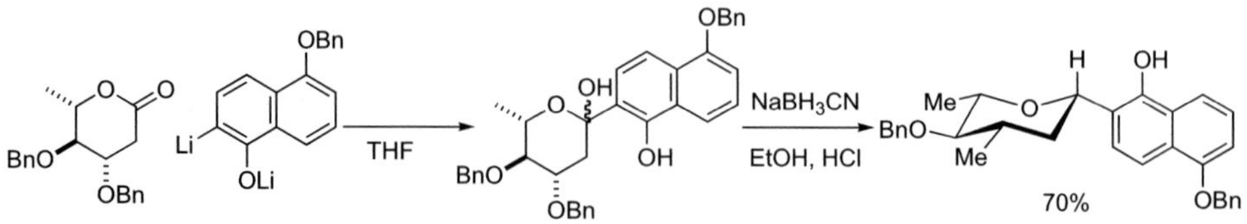

Scheme 12 reaction: BnO ... (O, O) OBn, OBn + Li naphthyl OLi → THF → hemiacetal → NaBH₃CN, EtOH, HCl → product (70%)

☐ **Scheme 12**

2.3 Glycosides (Halides, Oxygen, and others) as Electrophilic Carbohydrates

C-Glycosylation reactions of glycosyl-halide, -ester and -ether with carbon nucleophiles are divided into two types: the S$_N$2 type of substitution with carbon nucleophiles such as Grignard reagents, and addition of carbon nucleophiles to oxocarbenium cations generated from the glycoside upon reaction with Lewis acid. In the latter case, the stereochemistry of the addition is controlled by steric hindrance and/or stereoelectronic effects. This section deals with these types of reactions classified by nucleophiles such as (i) Grignard reagents and organoaluminum reagents, (ii) organosilane and tin reagents, (iii) aromatic nucleophiles (for Friedel–Crafts reaction).

2.3.1 Reaction with Grignard Reagents and Organoaluminum Reagents

Addition of a Grignard reagent to glycosyl halide has been employed as one of the classic methods for *C*-glycoside synthesis. However, the use of this method is limited to the reaction in which the desired product is easily separable. ❷ *Scheme 13* shows one such example [25]; reaction of a commercially available peracetyl α-galactosyl bromide (or chloride) with a large excess of allyl magnesium bromide (ca. 10 equiv.) gave a mixture of the products depicted in ❷ *Scheme 13*. Since the acetyl groups were deprotected under the reaction conditions, the crude products were reacetylated and then separated by column chromatography. The major *C*-glycoside product was found to have β-configuration indicating preferential inversion of configuration at the anomeric position.

Despite instability of glycosyl halides having benzyl protective groups, the preparation of α-galactosyl iodide was recently reported from the corresponding acetate with TMS-I at a low temperature (❷ *Scheme 14*) [26]. Reaction of galactosyl iodide with vinyl Grignard reagent in the presence of TBAI (*n*-Bu₄NI) in toluene at 110 °C gave α-vinyl-*C*-galactoside, stereoselectively. Under these conditions, vinyl magnesium bromide preferentially reacts with the more reactive β-iodo galactoside equilibrated with the corresponding α-iodide through in situ anomerization, to afford α-*C*-galactoside in an S_N2 manner. It is noted that this isomerization under thermal conditions would be much faster than *C*-glycosylation. The vinyl product was further transformed into a *C*-analogue of the antitumor glycolipid, KRN7000 through olefin cross metathesis.

Glycosyl fluoride has been widely used as a stable donor compared with the corresponding glycosyl bromide and chloride in *O*-glycosylation. Two examples of reactions between glycosyl fluoride and Grignard reagents are shown in ❷ *Scheme 15*, indicating that the reactions take place through an oxocarbenium cation intermediate [27].

❏ Scheme 13

❏ Scheme 14

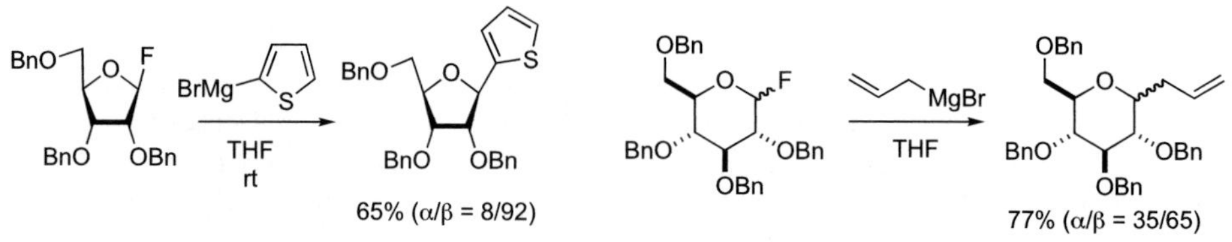
BnO
F
BrMg
S
THF
rt
BnO
OBn
65% (α/β = 8/92)
OBn
MgBr
THF
77% (α/β = 35/65)

Scheme 15

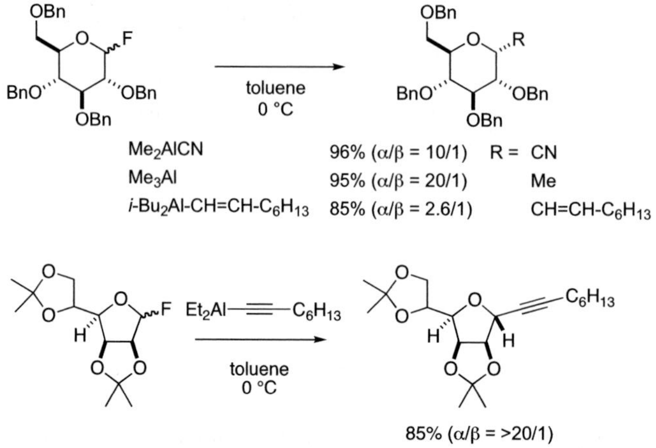
OBn
F
BnO
R
toluene
0 °C
Me2AlCN 96% (α/β = 10/1) R = CN
Me3Al 95% (α/β = 20/1) Me
i-Bu2Al-CH=CH-C6H13 85% (α/β = 2.6/1) CH=CH-C6H13
Et2Al—≡—C6H13
toluene
0 °C
C6H13
H
85% (α/β = >20/1)

Scheme 16

Glycosyl fluoride reacts with a variety of organoaluminum reagents having alkyl, alkenyl, alkynyl, and aryl groups without the use of a promoter, this is due to high affinity between aluminum and fluoride. The reactions shown in ➋ *Scheme 16* indicate that the reaction proceeded via an oxocarbenium cation, thereby giving *α-C*-glycosides as the major product from hexopyranose, and *α-C*-glycoside was obtained as the major product from the corresponding furanose [28].

2.3.2 Reaction with Organosilane and Organotin Reagents

Reaction of glycosyl donors with organosilane or organotin reagents as nucleophiles in the presence of a Lewis acid has been widely employed as one of the most important *C*-glycosidation methods. The reaction takes place through addition of the nucleophiles to an oxocarbenium cation generated from the glycoside by using a Lewis acid (➋ *Eq. 3*). In the case of the hexopyranose derivative, attack of the nucleophile occurs preferentially from the axial direction, due to stereoelectronic effects, leading to *α-C*-glycoside in a highly stereoselective manner.

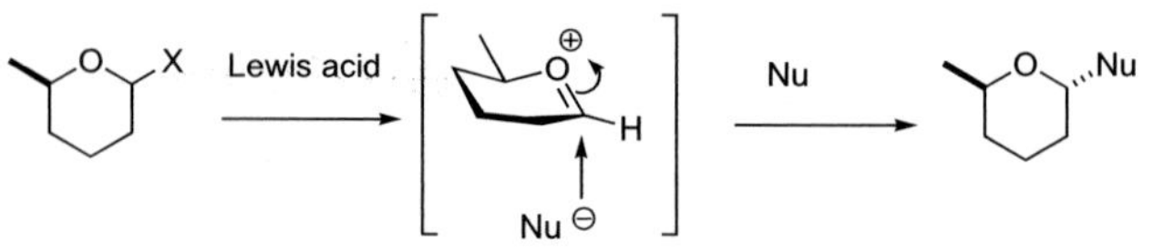
X
Lewis acid
H
Nu
Nu

(3)

□ Table 4
[29]

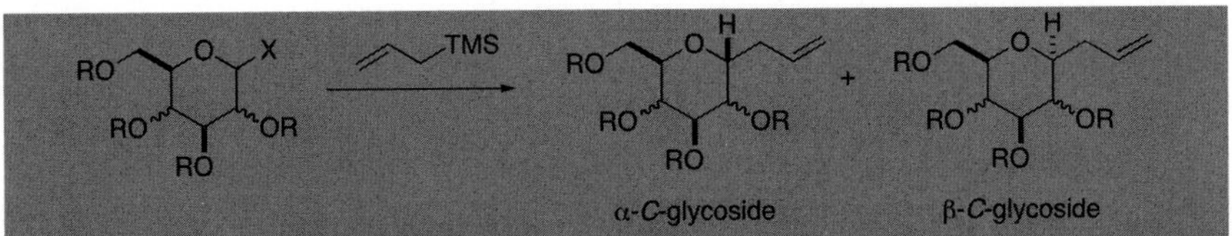

Entry	Carbohydrate			Conditions		Products		Ref.
		R	X (α/β)	Solvent,	Lewis acid	Yield (%),	α/β	
1	Glc	Bn	*p*-NBz (α)	CH$_3$CN,	BF$_3\cdot$OEt$_2$	55	10/1	[19c]
2		Bn	OMe (α)	CH$_3$CN,	TMSOTf	86	91/9	[29a]
3		Bn	F (1/1)	CH$_2$Cl$_2$,	BF$_3\cdot$OEt$_2$	95	>20/1	[29b]
4		Bn	Ac (mix)	CH$_3$CN,	BF$_3\cdot$OEt$_2$	89	α	[29e]
5		Ac	Ac (β)	CH$_3$CN,	BF$_3\cdot$OEt$_2$	81	95/5	[29c]
6	Gal	Bn	*p*-NBz (α)	CH$_3$CN,	BF$_3\cdot$OEt$_2$	79	10/1	[19c]
7		Bn	OMe (α)	CH$_3$CN,	TMSOTf	91	α	[29f]
8		Ac	Ac (β)	CH$_3$CN,	BF$_3\cdot$OEt$_2$	80	95/5	[29c]
9	Man	Bn	*p*-NBz (α)	CH$_3$CN,	BF$_3\cdot$OEt$_2$	79	>10/1	[19c]
10		Bn	OMe (α)	CH$_3$CN,	TMSOTf	87	α	[29a]
11		Bn	PO(OBn)$_2$ (α)	CH$_2$Cl$_2$,	TMSOTf	93	α	[29d]
12		Ac	Ac (β)	CH$_3$CN,	BF$_3\cdot$OEt$_2$	68	4/1	[29c]

Since Kishi reported allylation of glycosyl *p*-nitrobenzoate with allyltrimethylsilane in the presence of a Lewis acid (entry 1 in ❷ *Table 4*) [19c], a variety of leaving groups such as halogen, ester, ether, imidate, and phosphate have been employed. Typical examples are shown in ❷ *Table 4*, indicating the generality of this method for the synthesis of α-allyl-*C*-glycosides [29].

Under similar conditions, allenyl and ketone could be installed by reaction with propargyl silane and silyl enol ether. Methyl galactoside was treated with propargylsilane in the presence of TMSOTf to give α-allenyl-*C*-glycoside, which was directly treated with acetic anhydride to afford the corresponding acetate at the 6-position (❷ *Scheme 17*) [30]. Glucosyl chloride reacted with silyl enol ether in the presence of silver salt to yield α-glycosyl ketone in a highly stereoselective manner (❷ *Scheme 18*) [31].

The acetylene unit is also introduced by the reaction with silyl or tin acetylene in the presence of a Lewis acid. Tin acetylene was used in order to introduce acetylene into glucose, galactose, and mannose (❷ *Table 5*) [32], while *C*-glycosylation of deoxyhexopyranose was carried out with silylacetylene (❷ *Table 6*) [33]. These examples indicate an easy and reliable route for the preparation of α-glycosylacetylene (α-sugar acetylene).

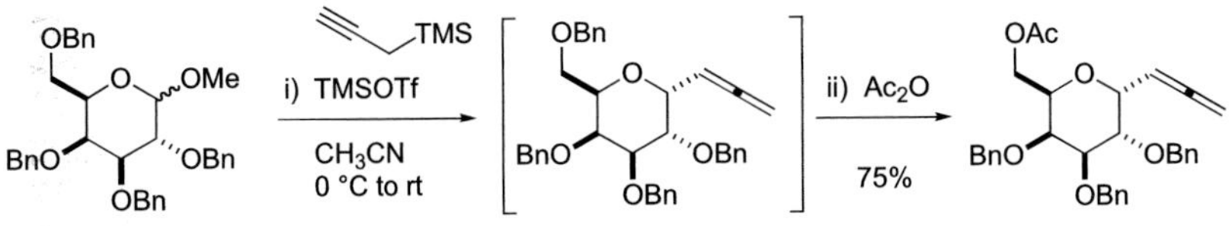

□ Scheme 17

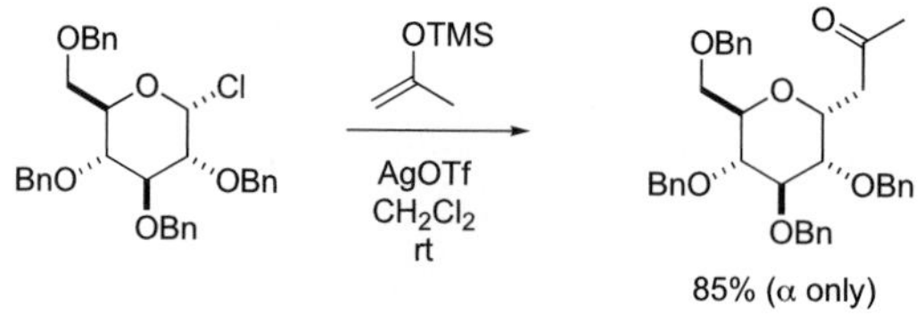

○ Scheme 18

○ Table 5
[32]

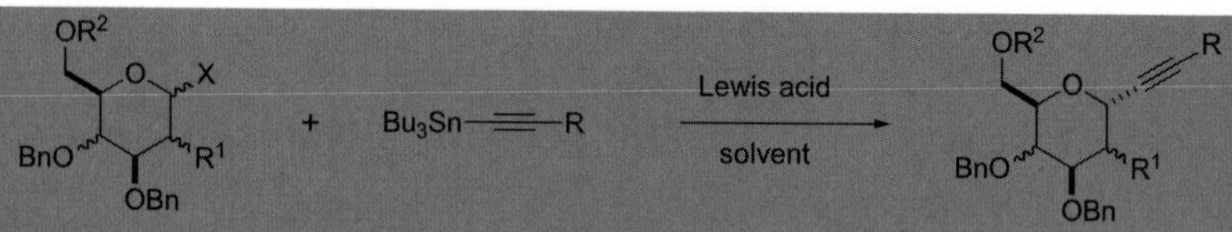

	Carbohydrate			Acetylene	Conditions	Products		Ref.
	R^1	R^2	X	R	Lewis acid, solvent	R	Yield (%)	
Glc	OBn	Bn	Br	Ph	$ZnCl_2$, CCl_4	Ph	61	[32a]
	OBn	Ac	Cl	Ph	$AgBF_4$, $(CH_2Cl)_2$	Ph	73	[32b]
	OBn	Bn	OAc	TMS	TMSOTf, CH_2Cl_2	H	71[a]	[32c]
	N_3	Bn	OAc	TMS	TMSOTf, CH_2Cl_2	TMS	43	[32d]
	N_3	Ac	Br	C_6H_{13}	$AgBF_4$, $(CH_2Cl)_2$	C_6H_{13}	76	[32b]
Man	OBn	Ac	OAc	TMS	TMSOTf, CH_2Cl_2	H	83[a]	[32c]
	OBn	Bn	OAc	TMS	TMSOTf, CH_2Cl_2	H	65[a]	[32c]
	N_3	Bn	OAc	TMS	TMSOTf, CH_2Cl_2	TMS	35	[32d]
Gal	OBn	Bn	OAc	TMS	TMSOTf, CH_2Cl_2	H	54[a]	[32c]
	N_3	Bn	OAc	TMS	TMSOTf, CH_2Cl_2	TMS	63	[32d]

[a]: The yields after desilylation with TBAF in THF or K_2CO_3 in MeOH

○ Table 6
[33]

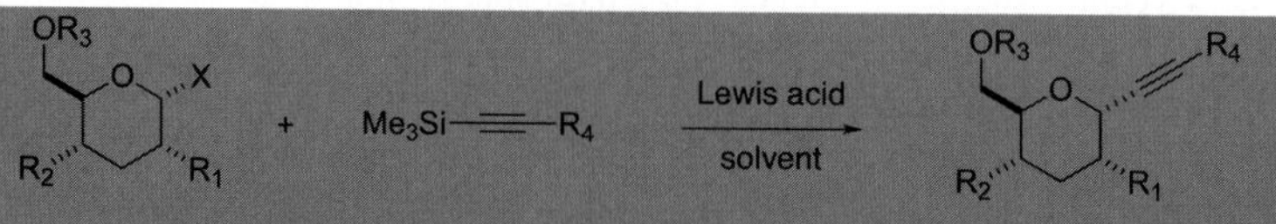

Entry	Carbohydrate				Acetylene	Conditions	Yield (%)	Ref.
	R_1	R_2	R_3	X	R_4	Lewis acid, solvent		
1	OAc	OAc	Ac	OAc	TMS	$SnCl_4$, CH_2Cl_2	73	[33a]
2	H	OAc	Ac	OAc	SPh	$BF_3 \cdot OEt_2$, CH_3CN	72	[33b]
3	OAc	OAc	Ac	OAc	≡—TMS	$SnCl_4$, CH_2Cl_2	71	[33a]
4	OAc	OAc	Ac	OAc	 Cl (image)	$SnCl_4$, CH_2Cl_2	83	[33a]
5	Me	H	TBDPS	Oi-Pr	SPh	$BF_3 \cdot OEt_2$, CH_3CN	61	[33c]
6	H	Me	H	OEt	SPh	$BF_3 \cdot OEt_2$, CH_3CN	62	[33d]

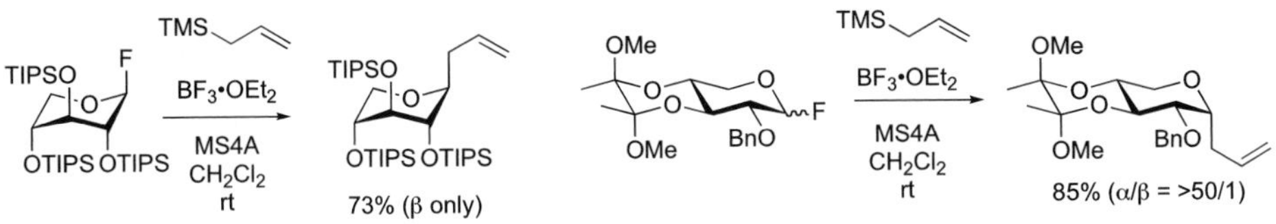

Scheme 19

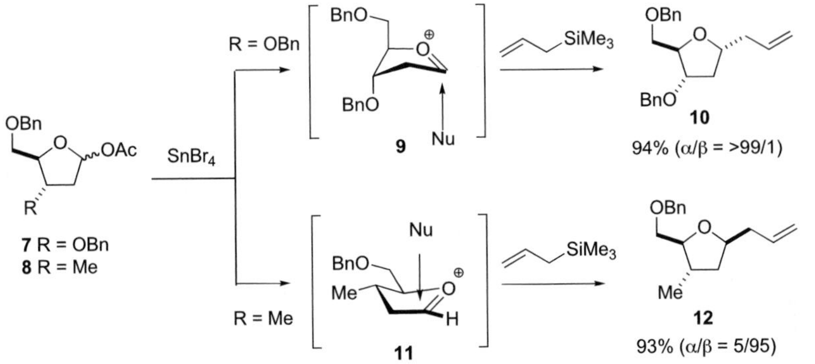

Scheme 20

Since the stereoelectronic effect determines the stereochemical outcomes in the above-mentioned *C*-glycosidation of pyranose derivatives, application of the same method for the synthesis of β-*C*-glycosylation seems to be difficult. However, when the conformation of the substrate can be inverted, axial attack of a nucleophile from the β face would be predominant for stereoelectronic reasons, yielding β-*C*-glycoside. On the basis of the above outlined hypothesis, Shuto and co-workers demonstrated β-*C*-allylation of pentoxylose protected with TIPS (tri-isopropylsilyl) (❷ *Scheme 19*) [34]. Installation of the sterically hindered protecting group fixed the inverted conformation ($^1C_4$) to avoid the gauche repulsion, thereby axial attack of allylsilane from the β face resulted in the β-*C*-glycoside, while allylation of 1-fluoroxylose protected with butane 2,3-diacetal (BDA) gave α-*C*-glycoside exclusively.

The stereochemical outcome of the similar *C*-allylation of pentofuranose does not depend on the substituent at the C-4 position of the carbohydrate, but rather depends on the type of substituent at the C-3 position as shown in ❷ *Scheme 20*. Woerpel explained the stereochemical outcome by stereoelectronically controlled attack of allylsilane to the lower energy envelope conformer as shown below [35]. The reaction of **7** gave α-*C*-products through the transition state shown for intermediate **9**, while the reaction of **8** afforded β-allyl product **12** through a different transition state **11**, in which the methyl substituent occupied the equatorial position.

2.3.3 Aryl-*C*-Glycosylation with Aromatic Nucleophiles

Stereoselective formation of the aryl-*C*-glycosidic linkage has been developed for synthesis of biologically important antibiotics and plant constituents. The most direct and widely used synthetic method is the Friedel–Crafts type reaction between electrophilic carbohydrate and electron-rich aromatic compounds. Carbohydrates with an appropriate leaving group react with an aromatic derivative in the presence of various Lewis acids to afford aryl-C-glycoside through addition of aromatic compounds to an oxocarbenium ion generated by a Lewis acid (❯ *Eq. 4*). Generally, the thermodynamically stable product is preferentially obtained with an equatorial C-glycosidic bond.

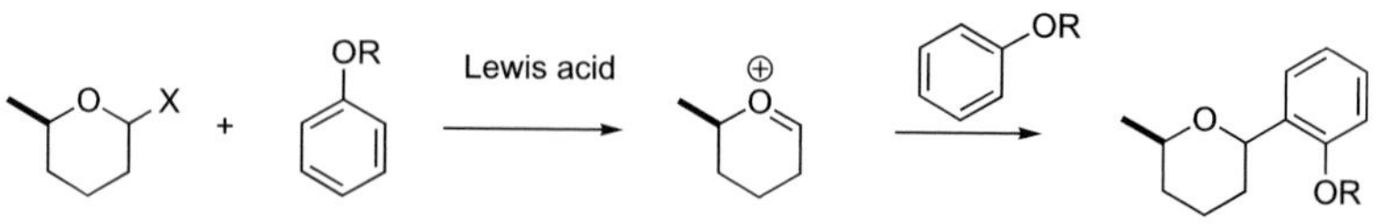

(4)

Typical examples are shown below. Combination of the leaving group with Lewis acid is critical in some special cases. Acetate, trifluoroacetate, and imidate were employed as leaving groups of the glycosyl donor, which were activated with a Lewis acid such as $BF_3 \cdot OEt_2$, TMSOTf, and $SnCl_4$, etc. In all cases, products possessing an equatorial C-glycosidic linkage were exclusively obtained and transformed to naturally occurring aryl-C-glycosides such as carminic acid (❯ *Scheme 21*) [36], an analogue of medermycin (❯ *Scheme 22*) [37], and a flavone from oolong tea (❯ *Scheme 23*) [38]. Combination of AgOTFA (AgOCOCF$_3$) and $SnCl_4$ exhibited a superior reactivity for these types of reaction, and by such means it was possible to form α-C-arylation of N-acetyl nueraminic acid, which had not been reported previously (❯ *Scheme 24*) [39]. C-Glycosylation of a complete unprotected 2,6-deoxysugar with naphthalenediol has also been reported for the synthesis of urdamycinone B (❯ *Scheme 25*) [40]. Among the examples shown above, reaction with a phenol derivative may take place through O- to C-glycoside rearrangement (vide infra).

The general problems in the above-mentioned Friedel–Crafts arylations are the regioselectivity of the aromatic substitution and stereoselectivity of the aryl-C-glycoside. Since many naturally occurring aryl-C-glycoside compounds contain the C-glycosidic linkage at the phenol *ortho*-position, an O- to C-glycoside rearrangement strategy has been developed by Suzuki and co-workers in order to solve the problems (❯ *Scheme 26*) [41]. Lewis acid-promoted O-glycosylation of a phenol derivative is followed by rearrangement to C-glycoside affording aryl-C-glycoside. The mechanism of the rearrangement is not as simple as it seems. The reaction proceeds through an ion pair generated from O-glycoside, followed by re-combination, resulting in *ortho*-substituted C-glycoside. Under the Lewis acid conditions, anomerization via a quinone methide occurs to afford the thermodynamically stable product.

$Sc(OTf)_3$ was reported as an efficient catalyst for the O- to C-glycoside rearrangement (❯ *Scheme 27*) [42]. The conditions enabled an efficient coupling between glycosyl acetate and an electron-deficient phenol derivative that reduced the nucleophilicity of the donor [❯ *Scheme 27* (a)], while the same reaction with $BF_3 \cdot OEt_2$ gave a 38% combined yield of the products with a poor selectivity ($\alpha/\beta = 2/3$). Total synthesis of ravidomycin demonstrated the power of the O- to C-glycoside rearrangement methodology for the synthesis of structurally complex aryl-C-glycoside antibiotics [❯ *Scheme 27* (b)] [43]. O-Glycosylation of glycosyl fluoride with phenol was carried out upon treatment with Cp_2HfCl_2 and $AgClO_4$, which promoted the subsequent rearrangement to C-glycoside.

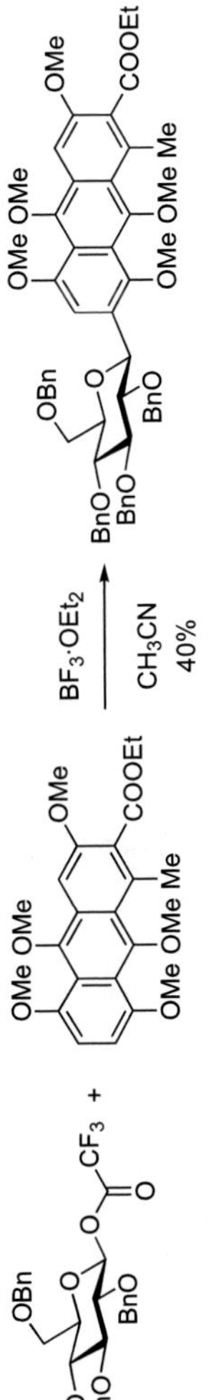

□ Scheme 21

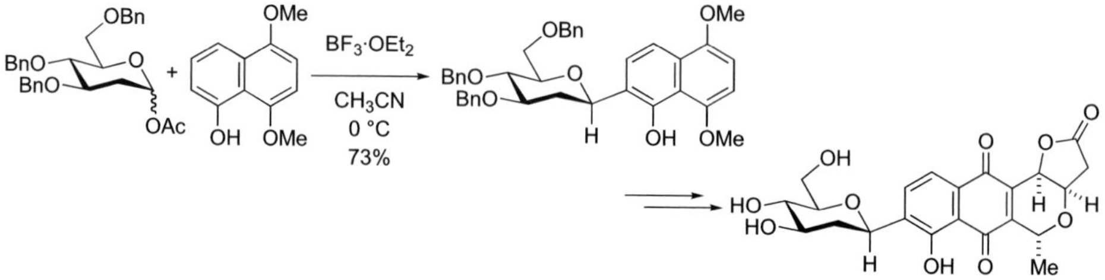

⬛ Scheme 22

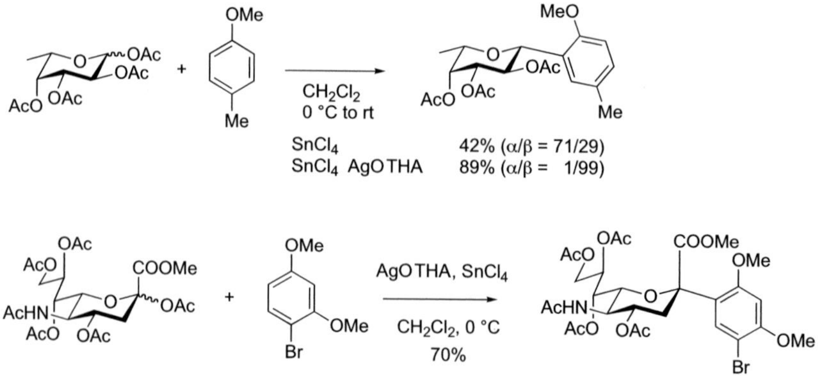

⬛ Scheme 23

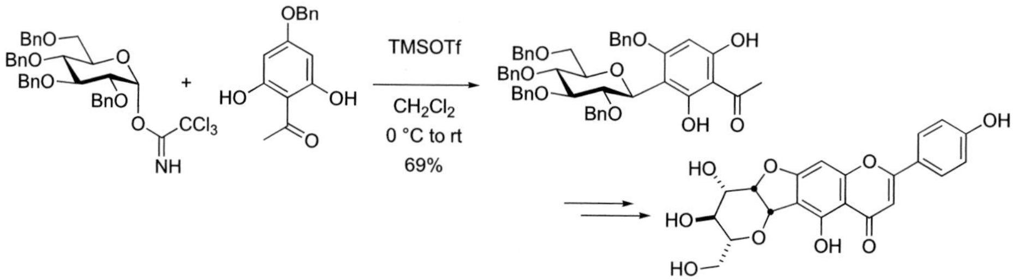

⬛ Scheme 24

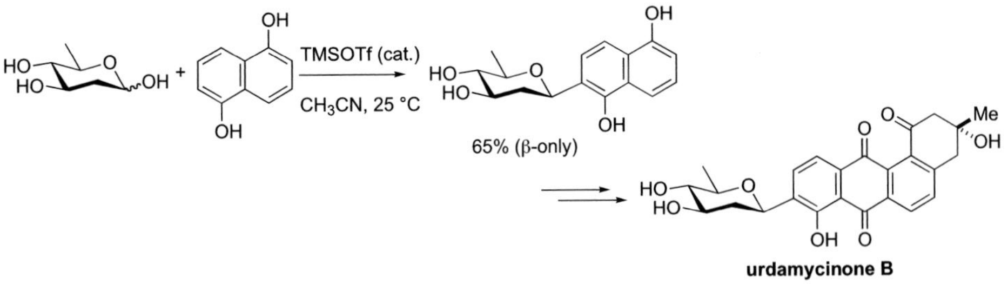

◻ Scheme 25

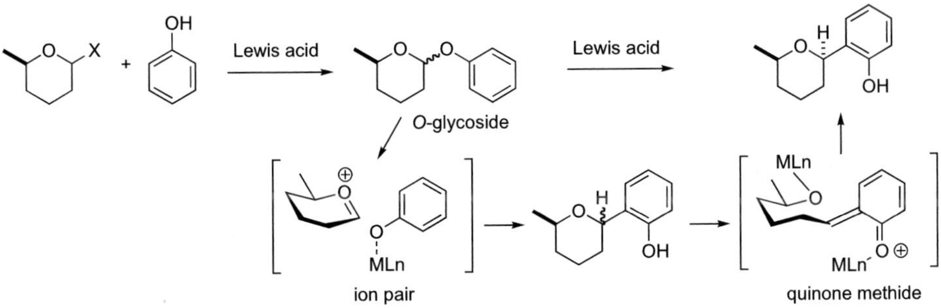

◻ Scheme 26

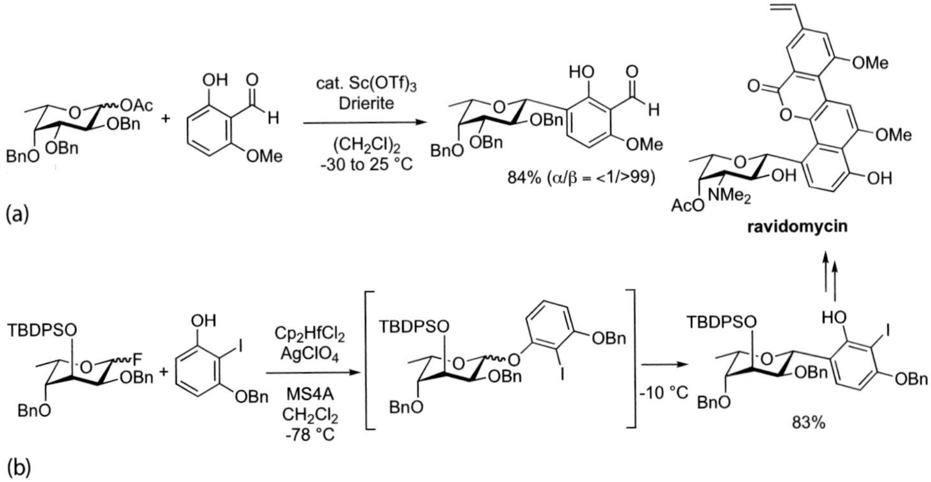

(a)

(b)

◻ Scheme 27

2.4 Glycal as an Electrophilic Carbohydrate

Glycals, easily available electrophilic carbohydrates with great diversity of functionality, react with a variety of carbon nucleophiles in the presence of Lewis acid to give unsaturated C-glycosides. The reaction seemingly proceeds in S_N2' manner with migration of the double bond (so-called Ferrier rearrangement). However, in the reaction mechanism the Lewis acid activates the acyl group at the C-3 position to generate an oxocarbenium cation, which is attacked by a carbon nucleophile (⊙ *Scheme 28*). The cation is an ambident electrophile that will react at the C-1 or C-3 position with nucleophiles. Most nucleophiles such as organozinc reagent, silane, and stannane react at the C-1 position to give C-glycoside [44], while organocopper reagents react at the C-3 position to afford a carbon-branched carbohydrate [45]. This section deals with the reactions with organozinc reagents, silanes, and stannanes in the mentioned order.

The stereochemical outcome is determined by the nucleophilic attack of a carbon nucleophile to an oxocarbenium cation; in the case of hexopyranose as tri-*O*-acetyl-D-glucal, axial attack of the nucleophile takes place at the anomeric position due to stereoelectronic effects, giving α-C-glycoside as the major product.

On the other hand, in the case of pentopyranose, the 1,4-*anti*-adduct was preferentially obtained (⊙ *Scheme 29*), due to the preferred conformation of the intermediate oxonium cation derived from the glycal [46,51a].

Aryl- and alkylzinc reagents having functional groups reacted with glycals without the addition of Lewis acid to give the corresponding C-glycosides in high α-stereoselectivity (⊙ *Scheme 30*) [47].

All of the above-mentioned reactions using glycals derived from hexopyranoses gave stereoelectronically controlled α-C-glycoside products. On the other hand, it is difficult to synthesize the corresponding β-C-glycoside by using the same method. Recently, reaction of

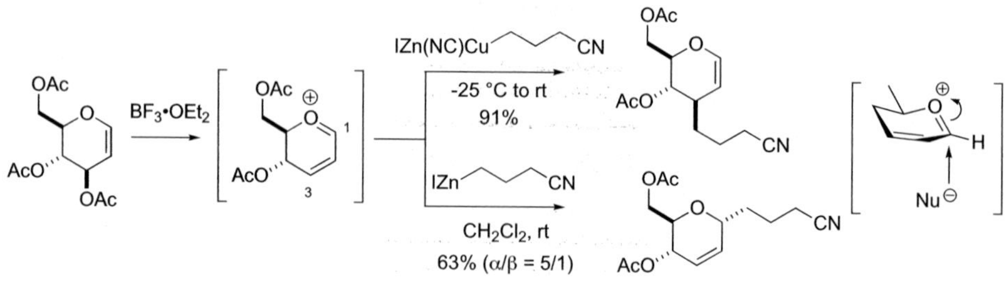

◻ Scheme 28

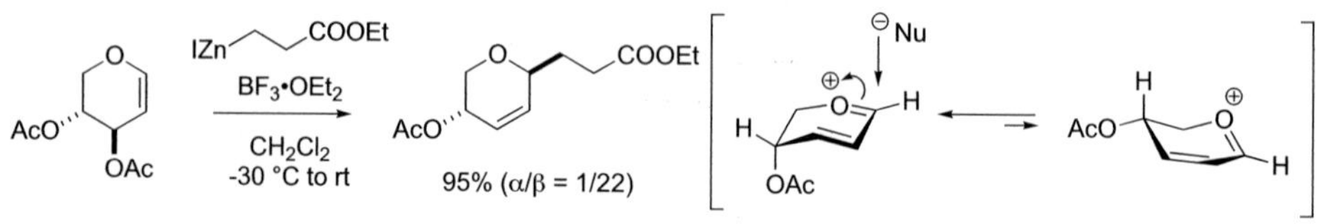

◻ Scheme 29

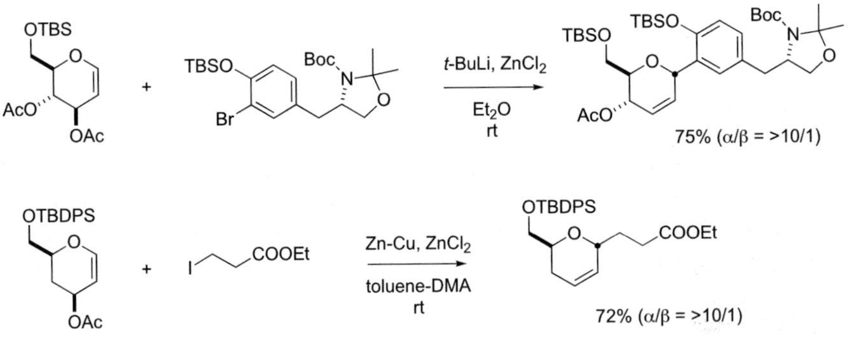

OTBS ... AcO''' OAc   +   TBSO Boc N O Br t-BuLi, ZnCl$_2$ / Et$_2$O / rt   →   TBSO TBSO Boc N O AcO'''   75% (α/β = >10/1)

OTBDPS ... OAc   +   I COOEt   Zn-Cu, ZnCl$_2$ / toluene-DMA / rt   →   OTBDPS ... COOEt   72% (α/β = >10/1)

◘ **Scheme 30**

OBn MsO''' OH t-BuOK / Et$_2$O   →   OBn O   RLi (3eq.) / Et$_2$O, 0 °C   [O Li R OBn O]   →   OBn HO R   78-93%

R = Me, n-Bu, i-Pr, Ph

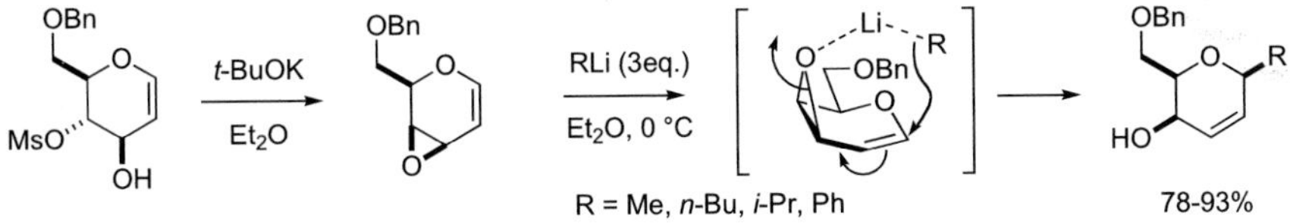

◘ **Scheme 31**

3,4-β-epoxyglycal with alkyl lithium was reported to give β-*C*-glycoside (❷ *Scheme 31*) [48]. This reaction proceeded intramolecularly through chelation between the epoxide oxygen and the lithium reagent. The same mechanism also worked in the reaction of the corresponding 3,4-α-epoxyglycal with alkyl lithium, giving α-*C*-glycoside, exclusively.

The most widely employed nucleophiles in the *C*-glycosylation of glycals are silicon-based reagents, such as allylsilane, silylacetylene [49], silylenol ether and its related reagents. ❷ *Table 7* lists typical examples of such nucleophiles, indicating the versatility of the reaction [50]. Reactions between glycal and these nucleophiles take place in the presence of conventional Lewis acids to give the corresponding *C*-glycosidic products generally with very high stereoselectivities. Recently, many other Lewis acids including Sc(OTf)$_3$, Yb(OTf)$_3$, InX$_3$ (X = Cl, Br), and I$_2$ were reported as catalysts for the reaction. The observed stereochemistries of the products are rational in light of the stereoelectronic effect. The reactions with TMSCN and silyl enol ether exhibited lower selectivities.

The methodology has been extended to glycal derivatives of pentopyranose. ❷ *Table 8* [51] and ❷ *Table 9* [51a] show typical examples of *C*-glycosylation of arabinal and xylal, indicating the reliability of the reaction.

These reactions have been extensively employed for the synthesis of complex natural products due to easy installation of various substituents with high stereoselectivities. In particular, the resulting sugar acetylenes are attractive for further transformation by means of Co-mediated reactions developed by Isobe and co-workers [33a].

The configuration of the acetylene group was inverted through an acetylene-cobalt complex prepared by treatment of the acetylene with Co$_2$(CO)$_8$ (❷ *Scheme 32*) [52]. Upon treatment of the complex with acid, epimerization occurred via a propargylic cation intermediate stabilized by the cobalt complex to afford the thermodynamically more stable β-*C*-glycoside as the

Table 7
[50]

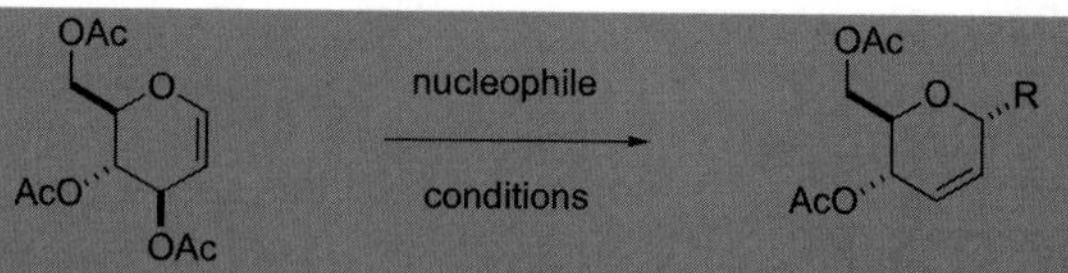

Entry	Nucleophile	Conditions Lewis acid, solvent	Products R	Yield (α/β)	Ref.
1	 ⟋⟋—TMS	TiCl$_4$, CH$_2$Cl$_2$	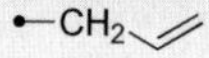 —CH$_2$⟋⟋	85% (16/1)	[50a]
2		Yb(OTf)$_3$ (10 mol%), CH$_2$Cl$_2$		94% (1/0)	[50b]
3	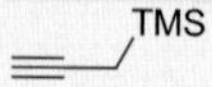 ≡—/—TMS	TiCl$_4$, CH$_2$Cl$_2$		89%	[50c]
4	TMS-CN	I$_2$ (5 mol%), CH$_2$Cl$_2$	—CN	80% (6/4)	[50d]
5	TMS—≡—TMS	SnCl$_4$, CH$_2$Cl$_2$	—≡—TMS	99%	[50e]
6	TMS—≡—≡—TMS	SnCl$_4$, CH$_2$Cl$_2$	—≡—≡—TMS	96%	[50e]
7	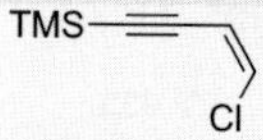 TMS—≡—⟋⟋Cl	SnCl$_4$, CH$_2$Cl$_2$	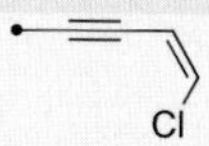 —≡—⟋⟋Cl	81%	[50e]
8	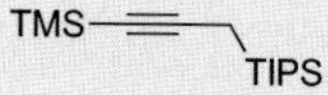 TMS—≡—⟍TIPS	I$_2$, CH$_2$Cl$_2$	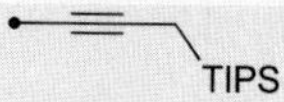 —≡—⟍TIPS	53%	[50f]
9	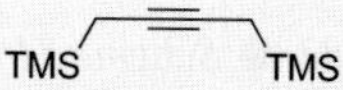 TMS—⟋≡⟍TMS	BF$_3$·OEt$_2$, CH$_2$Cl$_2$	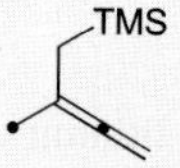 (TMS)	82%	[50g]
10	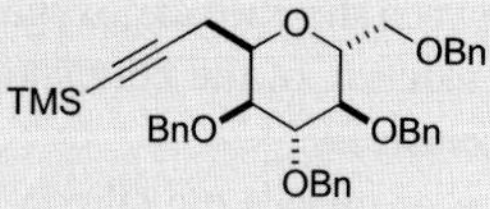 TMS—≡—(sugar, BnO/OBn/OBn/OBn)	I$_2$, CH$_2$Cl$_2$	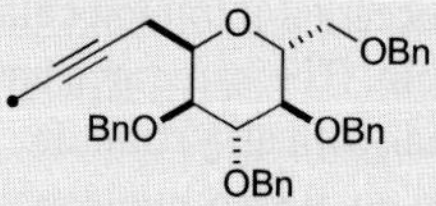	88%	[50f]
11	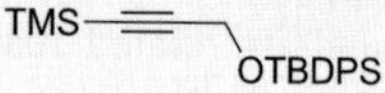 TMS—≡—⟍OTBDPS	SnCl$_4$, CH$_2$Cl$_2$	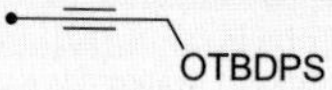 —≡—⟍OTBDPS	84%	[50g]

major product. Decomplexation with iodine afforded β-sugar acetylene. The acetylene cobalt complex was transformed into a variety of structures such as *cis*-olefin and vinylsilane by treatment with Bu$_3$SnH and Et$_3$SiH, respectively [53].

An acetylene cobalt complex stabilizes the propargyl cation, which can be captured by a variety of nucleophiles (Nicholas reaction). By using the unique properties of these complexes, an efficient synthetic methodology for the preparation of medium-sized cyclic ether has been developed (**◐** *Scheme 33*) [54]. An acetylene-cobalt complex prepared from sugar acetylene **13** was transformed to the acyclic cobalt complex **14**, which was exposed to pivalic anhydride under acidic conditions to give **16**. Upon treatment with a Lewis acid, the seven-membered cyclic ether was efficiently obtained through a stabilized propargyl cation **15** by the Co-complex. Hydrogenation with Wilkinson catalyst gave the ABC ring model compound **17** of ciguatoxin, a highly toxic marine polyether natural product.

Sugar acetylene reacted with a Fisher chromium carbene complex to give a phenol derivative, a possible intermediate for naturally occurring aryl-*C*-glycoside antibiotics (**◐** *Scheme 34*) [55].

◘ Table 8
[51]

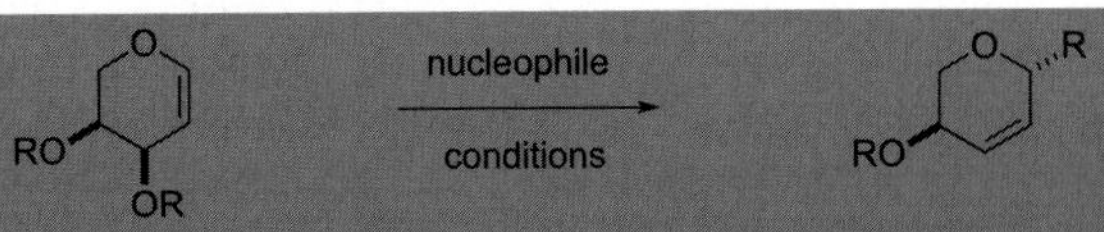

Entry	R	Nucleophile	Conditions Lewis acid, solvent	Products R	Yield (α/β)	Ref.
1	Ac	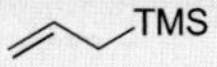 ⟋⟋TMS	BF$_3$·OEt$_2$, CH$_2$Cl$_2$	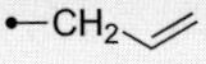 •—CH$_2$⟋	95% (95/5)	[51a]
2	Ac	TMS-CN	InCl$_3$(20mol%), CH$_3$CN	•—CN	72% (10/1)	[51b]
3	Ac	TMS—≡—TMS	TiCl$_4$, CH$_2$Cl$_2$	•—≡—TMS	97% (95/5)	[51a]
4	Ac	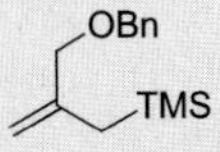 OBn ⟋⟍TMS	Yb(OTf)$_3$(10mol%), CH$_2$Cl$_2$	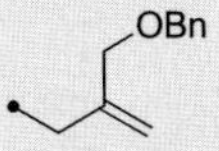 OBn	85%	[51c]
5	Piv	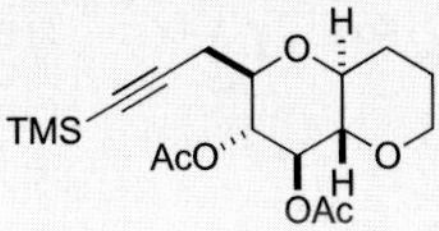 TMS≡ AcO OAc	SnCl$_4$, CH$_2$Cl$_2$	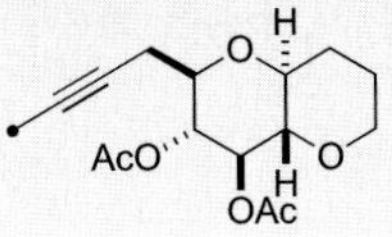 AcO OAc	83% (95/5)	[51a]

◘ Table 9
[51a]

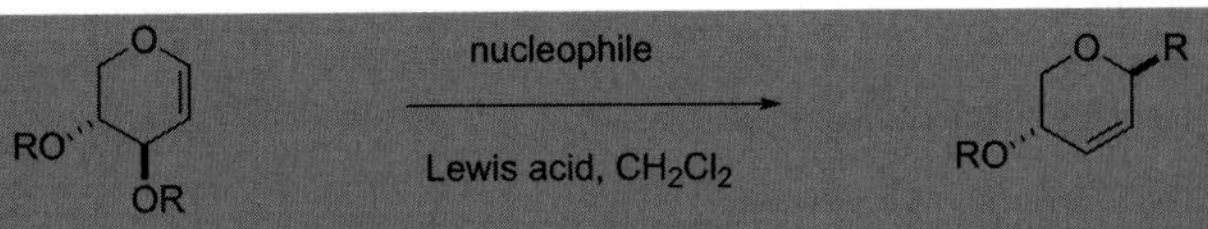

Entry	R	Nucleophile	Conditions Lewis acid	Products R	Yield (α/β)	Ref.
1	Ac	 ⟋⟋TMS	BF$_3$·OEt$_2$	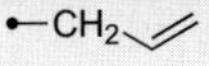 •—CH$_2$⟋	99% (5/95)	[51a]
2	Ac	TMS—≡—TMS	TiCl$_4$	•—≡—TMS	73% (5/95)	[51a]
3	Ac	TMS—≡—SPh	BF$_3$·OEt$_2$	•—≡—SPh	88% (5/95)	[51a]
4	Ac	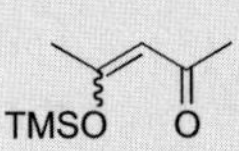 TMSO O	BF$_3$·OEt$_2$	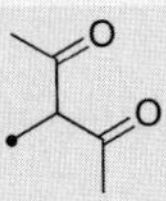 O O	65% (15/85)	[51a]
5	Piv	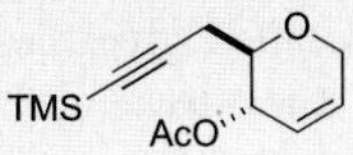 TMS≡ AcO	TiCl$_4$	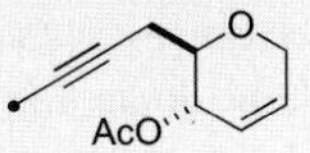 AcO	87% (5/95)	[51a]

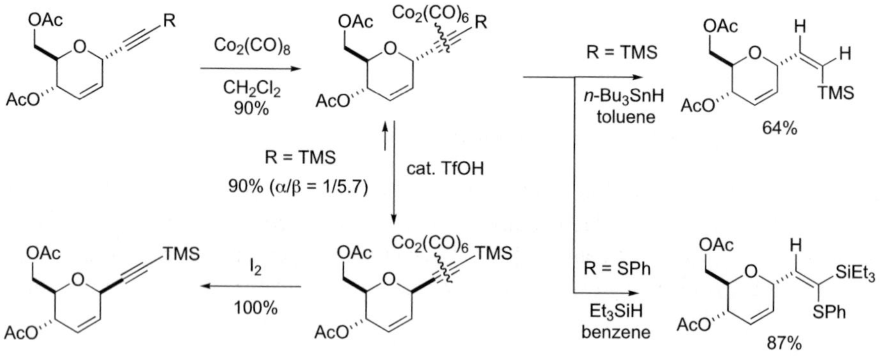

☐ Scheme 32

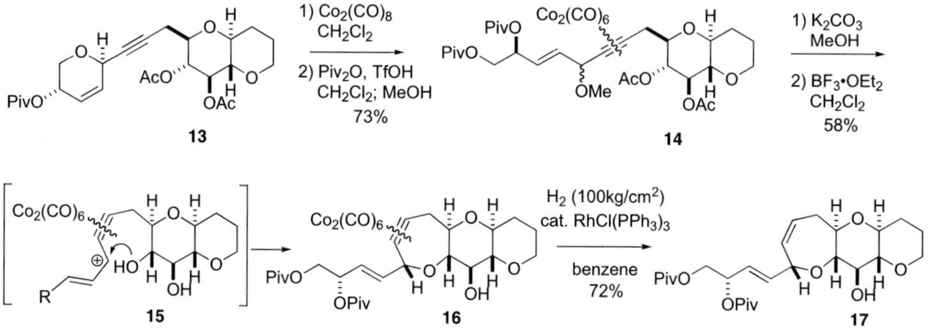

☐ Scheme 33

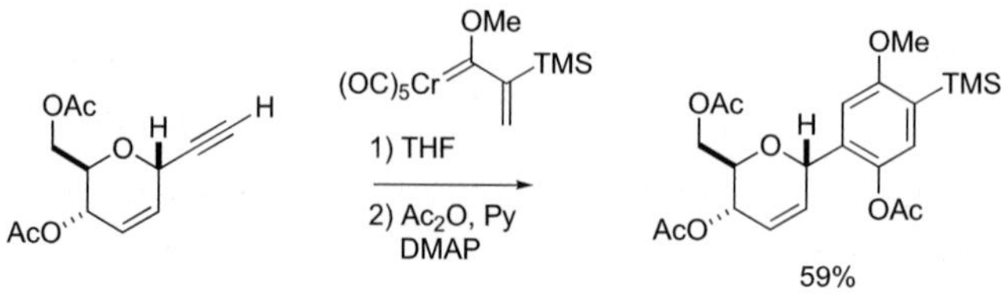

☐ Scheme 34

2.5 Anhydrosugar as an Electrophilic Carbohydrate

Among anhydrosugar derivatives, 1,2-anhydrosugar (so-called glycal epoxide) has been employed as the most useful donor for C-glycosylation. Since these unstable materials are easily prepared by epoxidation of glycal with DMDO (dimethyldioxorane) under very mild conditions, C-glycosylations of the glycal epoxide with various organometallic reagents have been extensively explored (❿ Eq. 5).

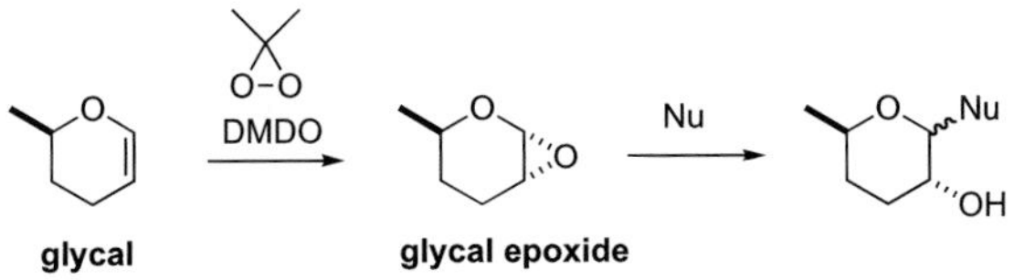

$$(5)$$

glycal   **glycal epoxide**

The stereoselectivity of the reaction with organometallic reagents is strongly dependent on the Lewis acidity as well as the nucleophilicity of the reagents. ❍ *Table 10* shows reactions of glucal epoxide with a variety of reagents [56]. When organocopper reagents such as dimethyl and diphenyl cuprates were employed as nucleophiles, opening of the epoxide occurred with inversion of configuration to give β-*C*-glycosides (entries 1 and 2). Reactions with allyl and propargyl Grignard reagents also gave the corresponding β-*C*-glycosides in high stereoselectivities (entries 3 and 4). In the case of vinyl Grignard reagents, reaction temperature affected the stereoselectivity; the reaction with vinyl Grignard reagents at 0 °C showed no selectivity, while the same reaction at −40 °C gave β-*C*-glycoside exclusively (entry 5). Allylation was also carried out by allylstannane with a catalytic amount of tributyltin triflate (entry 6). Sodi-

❑ **Table 10**
[56]

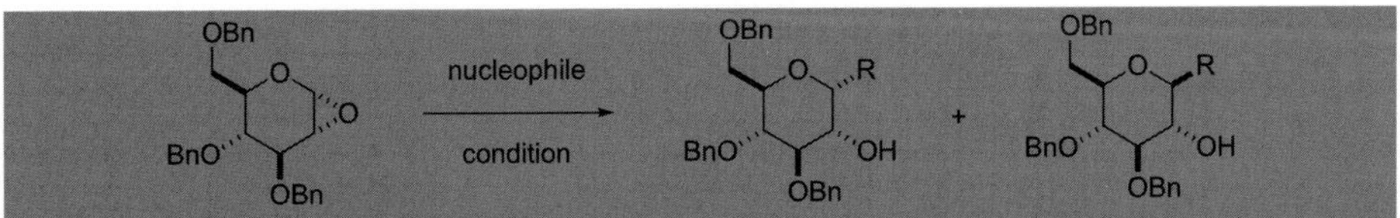

Entry	Nucleophile	Conditions		Products		Ref.
		Solvent	Temp.	R	Yield (α/β)	
1	Me₂CuLi	THF	0 °C	Me	82% (0/100)	[56a]
2	Ph₂CuLi	Et₂O	0 °C	Ph	84% (0/100)	[56a]
3	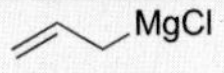 ⌇MgCl	THF	0 °C	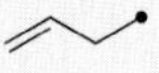 ⌇	82% (0/100)	[56a]
4	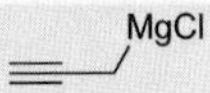 ⌇MgCl	CH₂Cl₂	0 °C	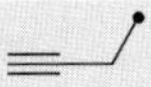 ⌇	78% (0/100)	[56a]
5	 ⌇MgBr	CH₂Cl₂	−40 °C	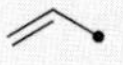 ⌇	57% (0/100)	[56a]
6	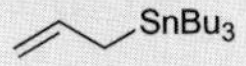 ⌇SnBu₃	Bu₃SnOTf, CH₂Cl₂	−78 °C	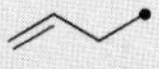 ⌇	57% (5/95)	[56c]
7	NaCH(COOEt)₂	ZnCl₂, THF	ND	CH(COOEt)₂	76% (0/100)	[56d]
8	AlMe₃	CH₂Cl₂	−95 °C	Me	82% (1/0)	[56a]
9	Ph₃Al	CH₂Cl₂	−65 °C to rt	Ph	79%	[56a]
10	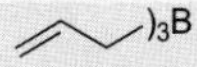 (⌇)₃B	CH₂Cl₂	−60 °C	 ⌇	70% (13/1)	[56b]
11	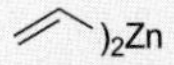 (⌇)₂Zn	THF	0 °C	 ⌇	50% (1/0)	[56e]
12	 TMS–≡–AlMe₂	THF	−95 °C	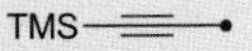 TMS–≡	80% (1/0)	[56a]
13	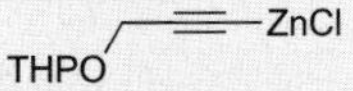 ≡–ZnCl, THPO	THF	−50 °C to rt	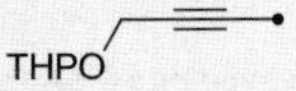 THPO	72% (1/0)	[56e]
14	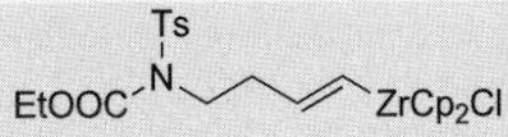 EtOOC–N(Ts)⌇ZrCp₂Cl	AgClO₄, CH₂Cl₂	rt	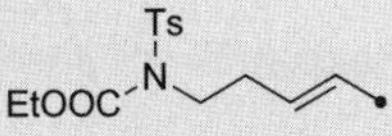 EtOOC–N(Ts)⌇	70% (1/0)	[56f]

um malonate ester as the nucleophile reacted with the glucal epoxide in the presence of ZnCl$_2$ to afford β-C-glycoside (entry 7). In this case, the thermodynamically stable β-C-glycoside was preferentially obtained after equilibrium through a retro-Michael reaction followed by conjugate addition.

On the other hand, organoaluminum reagents such as trialkyl, triaryl, and trivinyl aluminum except allyl aluminum gave high α-stereoselectivities (entries 8 and 9). For the synthesis of α-allyl-C-glycoside, allyl borane has been shown to be the best reagent (entry 10). Vinyl zinc and vinyl-zirconium reagents prepared by hydrozirconation of alkyne served as good nucleophiles for the synthesis of alkenyl-α-C-glycoside (entries 11 and 14). Reaction with aluminum and zinc acetylides afforded α-C-glycosides in high stereoselectivities (entries 12 and 13).

The stereochemical outcomes were explained by the following mechanisms (❷ *Scheme 35*); in the reactions with copper reagents, Grignard reagents, or a combination of allyl stannanes and Bu$_3$SnOTf, a mild Lewis acid, a S$_N$2-like reaction with the nucleophile took place at the anomeric position to afford the β-C-glycoside. On the other hand, aluminum reagents promoted formation of the oxocarbenium ion due to the strong Lewis acidity, followed by intramolecular transfer of the nucleophile to the cation through chelation giving the α-C-glycoside.

Since some of the above-mentioned C-glycosylations are mild enough and compatible with some functionalities, the reaction has been widely utilized in the synthesis of complex natural products. In the total synthesis of a complex marine macrolide, altohyrtin (spongistatin), Kishi and Evans independently employed the reaction between glycal epoxide and an organometallic reagent for the synthesis of the substituted tetrahydropyrane structure. An allyl cuprate prepared from allylstannane and cyanocuprate, reacted with the glycal epoxide to give 1,5-*syn*-pyranose (❷ *Scheme 36*) [57]. Reaction of glycal epoxide with allylstannane in the presence of tributyltin triflate gave the 1,5-*syn*-pyranose product in good yield with very high stereoselectivity (❷ *Scheme 37*) [58].

α-C-glycoside   glycal epoxide   β-C-glycoside

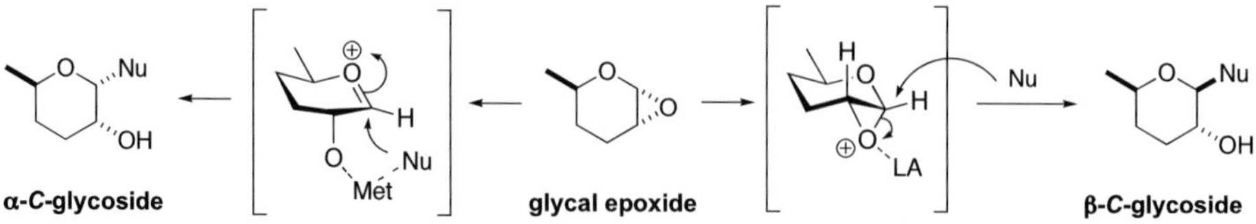

◻ Scheme 35

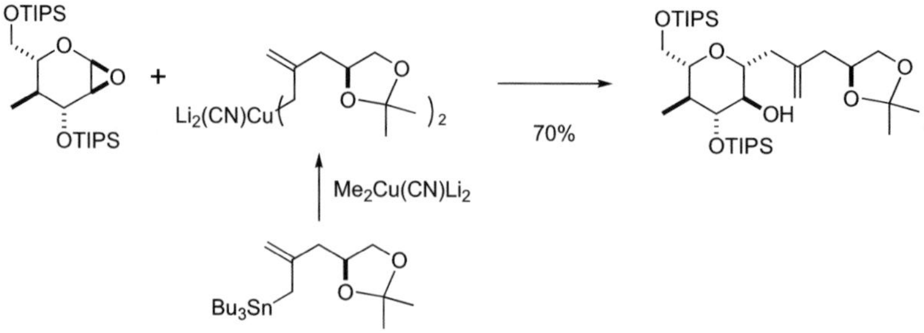

◻ Scheme 36

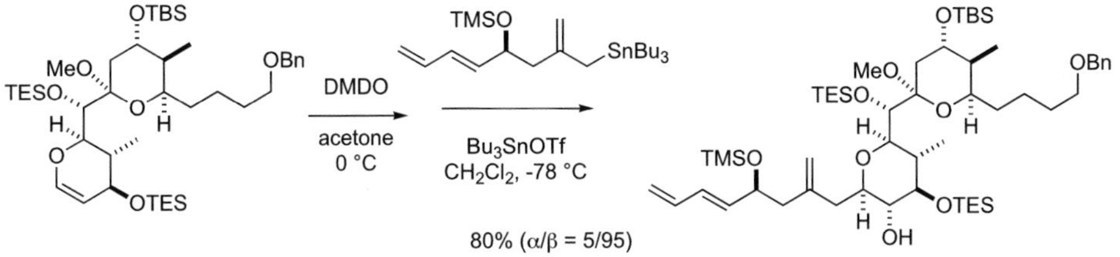

Scheme 37

Reaction of 1,2-anhydro-mannose with lithiated indole in the presence of $BF_3 \cdot OEt_2$ as Lewis acid yielded α-C-mannosylindole, which was further transformed into C-mannosyltryptophan, a naturally occurring C-glycosyl amino acid found in some proteins (● *Scheme 38*) [59].

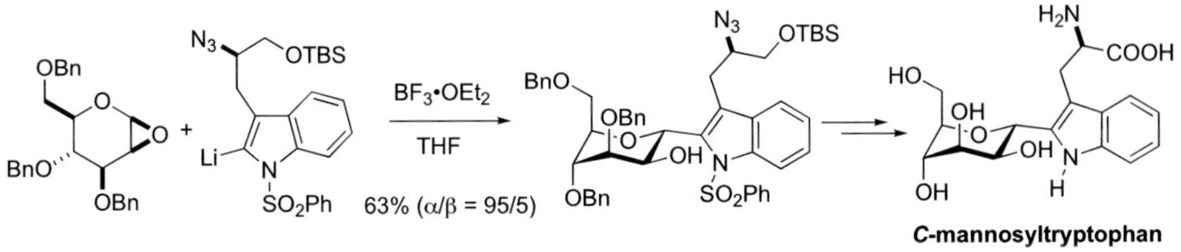

Scheme 38

In contrast to the reaction of 1,2-anhydrosugar, C-glycosylation of 1,6-anhydropyranose requires a strong Lewis acid in order to open the anhydro bridge, generating the corresponding oxocarbenium cation (● *Eq. 6*). Addition of nucleophiles to the cation could be controlled by the stereoelectronic effect, chelation between the nucleophile and the Lewis acid coordinated to the hydroxyl group, or intramolecular delivery of nucleophiles as shown below.

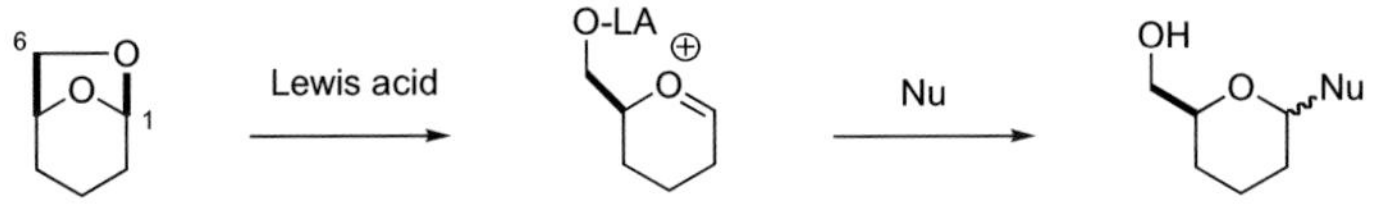

$$\text{(6)}$$

For example, reaction of 1,6-anhydroglucopyranose with allylsilane in the presence of $BF_3 \cdot OEt_2$ gave α-C-glycoside in a high stereoselective manner (● *Scheme 39*) [19c]. In this reaction, an initially formed oxocarbenium cation changed the conformation of the substrate to a more stable one ($^4C_1$), followed by attack of the allylsilane to the cation from the axial direction, affording α-C-glycoside.

1,6-Anhydromannopyranose (R = TIPS) reacted with lithium acetylide in the presence of Me_3Al to afford α-C-glycoside (● *Scheme 40*) [60]. On the other hand, when a similar substrate (R = H) was treated with lithium acetylide in the presence of $AlCl_3$ instead of Me_3Al, β-C-glycoside was exclusively obtained [61]. The mechanism for the reaction was proposed as follows; the acetylide reacted with $AlCl_3$ to form the alkynylaluminum species, which opened the anhydro bridge by participating with the hydroxyl group at the C-3 position. Intramolecular transfer of the acetylene to the resulting oxocarbenium cation afforded β-C-glycoside stereoselectively.

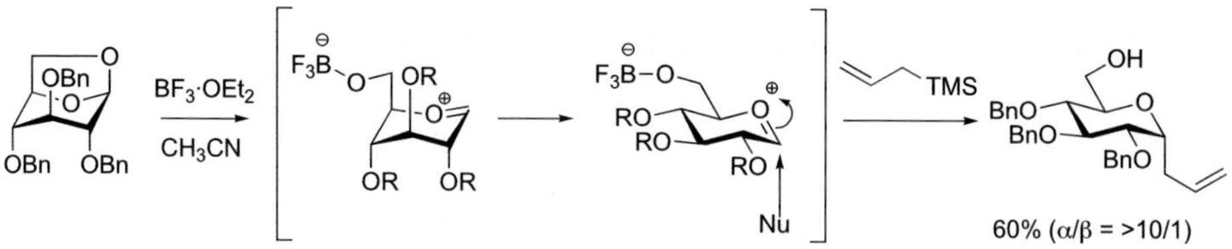

◘ Scheme 39

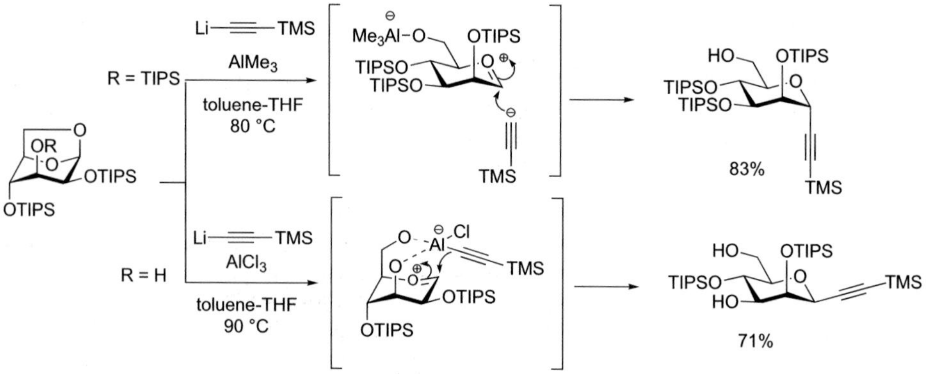

◘ Scheme 40

There are a few reports of *C*-glycosylation by using 1,4-anhydrosugar as a glycosyl donor. The reaction proceeds through an oxocarbenium cation as in the case for 1,6-anhydrosugar. Under similar conditions, reaction of 1,4-anhydrosuger with allylsilane did not give the allyl-*C*-pyranoside, instead allyl-*C*-furanoside was exclusively obtained (◑ *Scheme 41*) [62]. The stereochemistry was explained by attack of allylsilane to the oxocarbenium cation from the less hindered face.

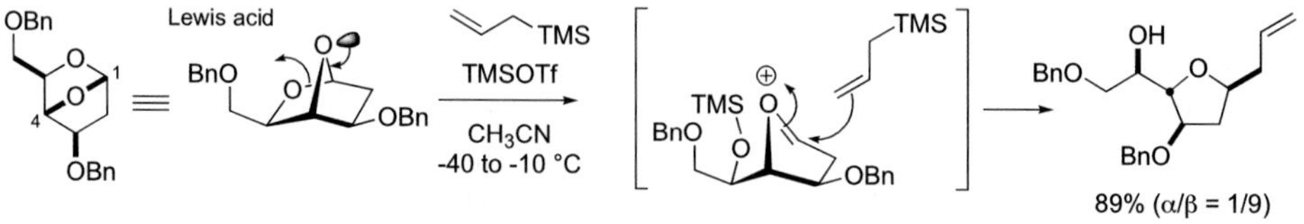

◘ Scheme 41

3 Anomeric Radical Intermediate

Free radical reactions proceed under essentially neutral conditions that are compatible with various functional and protective groups used for saccharides. C-Glycosylation by means of radical reaction has been extensively developed [63]. This section will discuss intermolecular and intramolecular reactions for control of the stereochemistry of the C-glycoside linkage.

3.1 Intermolecular Radical Reaction

C-Glycosylation by means of anomeric radical species was first reported by Giese and co-workers in 1983 [64]. In general, an anomeric radical is readily generated from glycosyl halides (Cl or Br) or phenylseleno-glycoside by using tributyltin hydride and AIBN or irradiation as initiator. The resulting anomeric radical possesses a nucleophilic character and reacts with activated olefins such as acrylonitrile, acrylate ester, and allyltin reagents to afford chain extension products (C-glycosides). The general mechanism is outlined in ❯ Eq. 7. The tin radical generated by an initiator abstracts the X group to give an anomeric radical, which reacts with unsaturated olefin or allyltin reagent to form the C-glycosidic bond. Radical-mediated C-glycosylation of hexopyranose predominantly affords the α-C-glycosidic bond, independent of configuration at the anomeric position of the starting material. In the transition state, the anomeric radical is pyramidal (sp^3-like) and the axial radical preferentially reacts with olefine to afford α-C-glycoside [65].

$$(7)$$

α-Glycosyl bromides have been widely studied as glycosyl donors for the radical C-glycosylation (❯ Table 11) [66]. A variety of terminal olefins with electron-withdrawing groups can be installed into glucose, galactose, and mannose derivatives in high α-selectivities by using this methodology. It is also possible to conduct this chemistry without the use of a tin reagent (entries 2 and 9).

Radical reactions have also been employed for the synthesis of C-glycoside of 2-amino sugars as shown in ❯ Table 12 [67] in which most of the reaction gave α-C-glycosides. However, when a sterically hindered protective group such as phtaloyl or tetrachlorophtaloyl was installed in order to protect the amino functionality at the C-2 position, the reverse stereoselectivity was observed (entry 2). In the reaction with an activated olefin such as acrylate ester or styrene, a milder reaction by using phenylseleno glycoside with Et$_3$B as the radical initiator was required (entry 5). The conventional conditions resulted in simple reduction of the glycosyl halide as a serious side reaction. Et$_3$B in the presence of traces of oxygen generates a radical species at room temperature. This method could be applied for C-glycosylation of

⬛ **Table 11**
[66]

Entry	Sugar	Olefin	Conditions	Products R	Yield (α/β)	Ref.
1	Glc	CH$_2$=CH-CN	Bu$_3$SnH, benzene, $h\nu$	CH$_2$CH$_2$CN	75% (93/7)	[66a]
2		CH$_2$=CH-CN	vitamine B$_{12}$, Zn, NH$_4$Cl/DMF	CH$_2$CH$_2$CN	68% (13/1)	[66b]
3		CH$_2$=CH$_2$-CH$_2$-SnBu$_3$	AIBN, benzene, 80 °C	CH$_2$-CH=CH$_2$	64% (ND)	[66c]
4		(SnBu$_3$ / COOEt olefin)	AIBN, benzene, 80 °C	CH$_2$ / COOEt	84% (ND)	[66a]
5		(vinyl sulfonyl benzothiazole)	n-Bu$_3$SnH, AIBN Et$_2$O, reflux	(product)	66%	[66d]
6	Gal	CH$_2$=CH-CN	Bu$_3$SnH, Et$_2$O, $h\nu$	CH$_2$CH$_2$CN	70%	[66a]
7		NHBoc / COOBn olefin	Bu$_3$SnH, AIBN, toluene	CH$_2$ / NHBoc / COOBn	61%	[66e]
8		CH$_2$=CH$_2$-CH$_2$-SPh	(Bu$_3$Sn)$_2$Sn, benzene, $h\nu$	CH$_2$CH=CH$_2$	73%	[66f]
9		(P(OMe)$_2$ / COOMe olefin)	(TMS)$_3$SiH Et$_2$O, $h\nu$	(P(OMe)$_2$ / COOMe product)	80% (98/2)	[66g]
10	Man	CH$_2$=CH-CN	Bu$_3$SnH, Et$_2$O, $h\nu$	CH$_2$CH$_2$CN	65%	[66a]
11		COOEt / COOEt olefin	Bu$_3$SnH, AIBN toluene, 67 °C	CH$_2$ / COOEt / COOEt	99%	[66a]

other 2-acetamide sugars such as GalNAc and ManNAc. The use of tristrimethylsilylsilane as an alternative to tin hydride has also been reported (entry 4).

The above-mentioned radical C-glycosylation has been applied for the synthesis of complex compounds. ❷ *Scheme 42* shows two such examples of the radical C-glycosylation of mannosyl bromide with complex olefins such as unsaturated ketone and unsaturated lactone. This highlights the mildness of the reaction and these conditions have compatibility with many functional groups [68]. It is noteworthy that an acceptor bearing seleno and chloro groups can be employed without interference under the radical conditions. In both reactions, the α-C-glycosidic products were exclusively obtained, although in moderate yields.

Since the conformation of hexopyranose is fixed by the presence of the equatorial hydroxymethyl group at the C-5 position, the corresponding radical intermediate possesses the axial radical in the α-direction, which reacted with the olefine to form α-C-glycoside. On the other hand, the conformation of pentopyranose is rather flexible, and easily inverted by several modifications. The anomeric radical adopting the $^1C_4$ conformation reacts with the olefin to give β-C-glycoside. Shuto and co-workers reported formation of β-C-glycosides by using

■ Table 12
[67]

Entry	Carbohydrate		Olefine	Condition	Products		Ref.
	R_1	X			R	yield (α/β)	
1	GlcNAc	Ac	Cl	⌁SnBu$_3$	AIBN, toluene	CH$_2$-CH=CH$_2$	73% (12/1)	[67a]
2		TCP	Br		AIBN, benzene[a]		77% (1/20)	[67b]
3		Ac	SePh	⌁COO-t-Bu	n-Bu$_3$SnH, AIBN, benzene[a]	CH$_2$CH$_2$COO-t-Bu	76% (α)	[67c]
4		Ac	SePh		(TMS)$_3$SiH, AIBN, benzene[a]		93% (α)	[67c]
5		Ac	SePh		Et$_3$B, AIBN, benzene, rt		71% (α)	[67c]
6		Ac	SePh	⌁Ph	n-Bu$_3$SnH, AIBN, benzene[a]	CH$_2$CH$_2$Ph	30% (α)	[67c]
7	GalNAc	Ac	Br	⌁SO$_2$Ph	(Bu$_3$Sn)$_2$, hv, benzene	CH$_2$-CH=CH$_2$	42% (α)	[67a]
8		Ac	SePh	⌁COO-t-Bu	n-Bu$_3$SnH, AIBN, benzene[a]	CH$_2$CH$_2$COO-t-Bu	68% (α)	[67c]
9		Ac	SePh		Et$_3$B, n-Bu$_3$SnH, toluene, hv		90% (α)	[67d]
10		Ac	SePh	⌁Ph	n-Bu$_3$SnH, AIBN, benzene[a]	CH$_2$CH$_2$Ph	41% (α)	[67a]
11	ManNAc	Ac	SePh	⌁COO-t-Bu	Et$_3$B, n-Bu$_3$SnH, toluene, hv	CH$_2$CH$_2$COO-t-Bu	64% (α)	[67c]
12		Ac	SePh	⌁Ph	Et$_3$B, n-Bu$_3$SnH, toluene, hv	CH$_2$CH$_2$Ph	28% (α)	[67c]
13		Ac	SePh	SnBu$_3$⌁COOMe	Ph$_3$SnH, toluene	CH$_2$⌁COOMe	80% (α)[b]	[67e]

[a]: The reaction was carried out at a reflux. [b]: The yield of ketoester after ozonolysis of the product

substrates with an inverting conformer (❸ *Scheme 43*) [65]. When xylose was protected with a sterically hindered TIPS group, the conformation was fixed to be $^1C_4$, due to gauche repulsion between the TIPS groups. By utilizing a cyclic boronate, it is also possible to fix the $^1C_4$ conformation. Radical C-glycosylation of these substrates afforded β-C-glycoside in high stereoselectivities through the transition state as shown in the parentheses in ❸ *Scheme 43*.

The radical C-glycosylation of N-acetylneuraminic acid was independently reported by Paulsen and Bednarski (❸ *Scheme 44*) [69]. The products were obtained as an anomeric mixture, which were easily separated after deprotection. The radical allylation is one of a few synthetic methods for formation of C-glycoside with neuraminc acid derivatives.

☐ **Scheme 42**

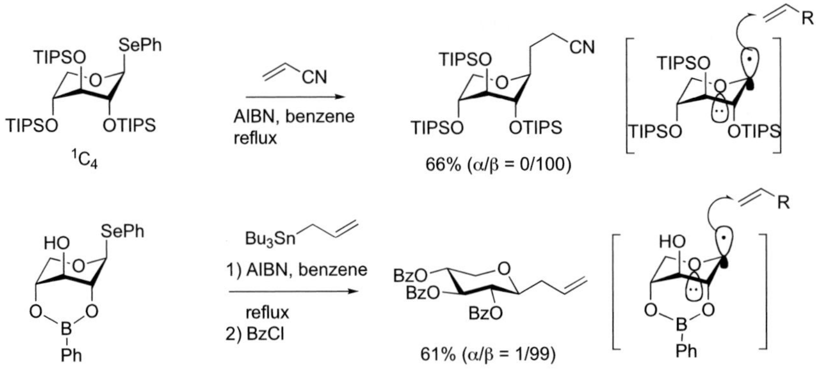

Scheme 43

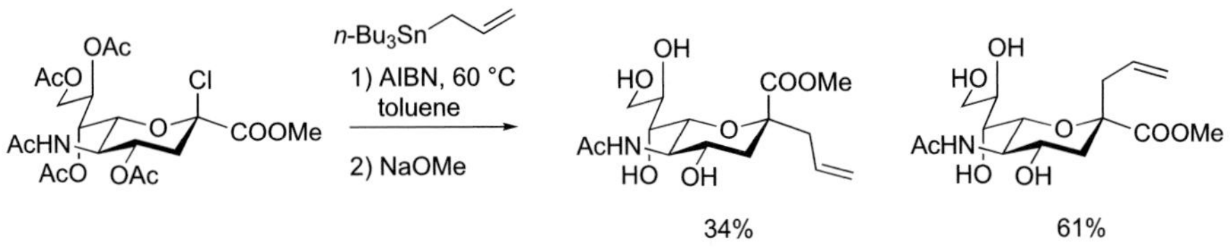

Scheme 44

3.2 Intramolecular Radical Reaction

Intermolecular *C*-Glycosylation of hexopyranoses inherently provides α-*C*-glycoside in high stereoselectivity, while intramolecular radical *C*-glycosylation controls the stereochemistry at the anomeric position by the configuration of the hydroxyl group that is connected to the radical acceptor through a tether. Therefore, the intramolecular reaction is of significance despite the fact that several additional steps are necessary in order to synthesize the precursor and for removal of the tether after conducting the radical reactions. Phenylselenoglycoside and silicon tethers have been widely employed for this purpose [70]. After *C*-glycoside bond formation, the silicon tether was removed to give the product with the desired configuration (❷ *Eq. 8*).

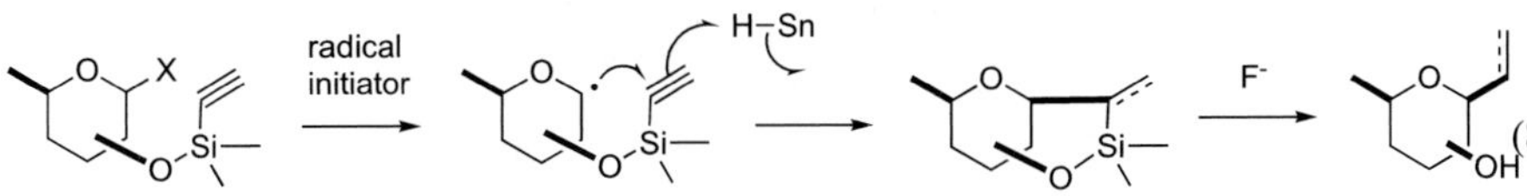

β-*C*-Mannoside is one of the most difficult to synthesize among *C*-glycosylations, however, intramolecular radical *C*-glycosylation of mannose by using a silicon tether at the C-2 position, afforded β-*C*-mannoside, exclusively (❷ *Scheme 45*) [71]. Phenylsulfonyl- or pyridylsulfonyl groups can be employed as glycosyl donors for the same purpose by using SmI_2. This method avoids the use of toxic tin hydride and the phenylseleno group [72].

Hydroxyl groups at the C-3 and -6 position of glucoside were employed for the introduction of the silicon tether, leading to β-*C*-glycoside (❷ *Scheme 46*). The resulting styryl group can be transferred to a variety of functional groups. The strategy was also applied to the synthesis of *C*-furanosides.

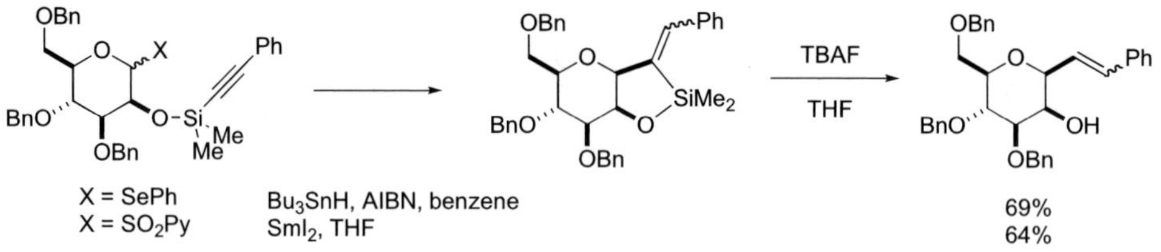

X = SePh   Bu₃SnH, AIBN, benzene
X = SO₂Py   SmI₂, THF

69%
64%

□ **Scheme 45**

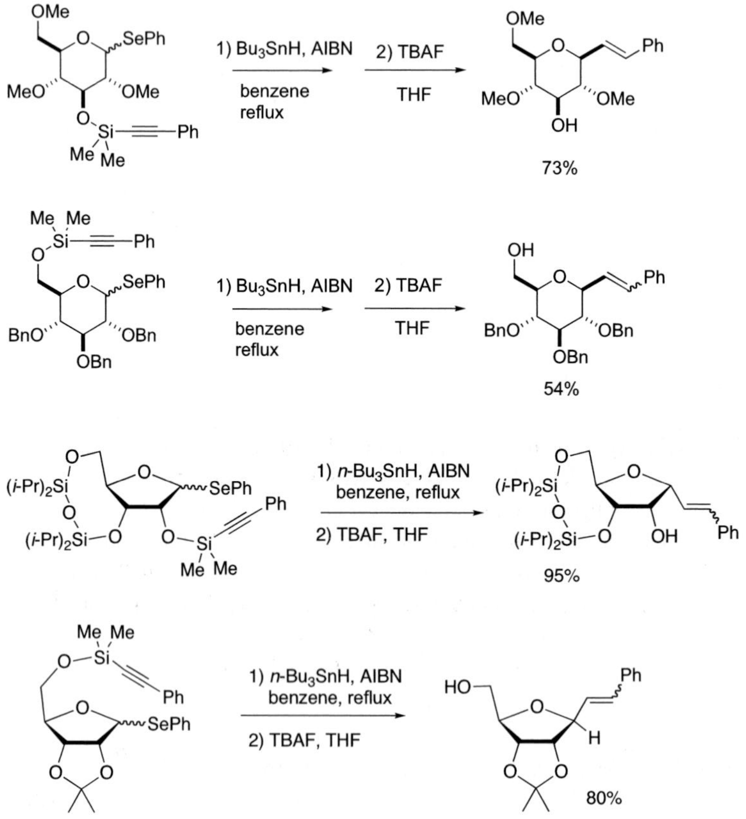

73%

54%

95%

80%

□ **Scheme 46**

Carbohydrates with a hydrazono ester at the C-2 position underwent intramolecular radical C-glycosylation upon reaction with tin hydride and AIBN, to afford the corresponding lactone in good yield (◉ *Scheme 47*). Although the stereoselectivities of the amino acid moiety were generally low, the methods have been used for the synthesis of a variety of C-glycosyl amino acids [73]. The following reaction was a key step in the total synthesis of furanomycin, an antibiotic.

◻ Scheme 47

4 Anomeric Anion

In contrast to C-glycosylation by nucleophilic addition to an electrophilic carbohydrate, the use of an anomeric anion has found limited use for such a reaction due to instability. However, C-glycosylation by means of anomeric anion species is of significance in some special cases [74]. The anomeric anion intermediate can be stereoselectively prepared by (i) reductive metalation of anomeric halides or sulfones, (ii) transmetalation of glycosyl stannanes, and (iii) direct deprotonation of the anomeric proton. This section deals with three types of the reaction using different anomeric anions.

4.1 Lithium Anomeric Anion

Anomeric anion (sp³) species without a stabilizing group can be employed in stereoselective C-glycosylation, because the anions are configurationally stable at low temperature, and react with a variety of electrophiles such as alkyl halides and compounds containing carbonyl groups. However, the method has not been well exploited probably due to strict anhydrous conditions that are required in order to conduct these reactions. A typical example of N-acetyl-glucosamine is shown in ❷ Scheme 48 [75]. In this case, dianion chemistry was employed to prevent elimination of the substituent at the C-2 position; N-acetyl-glucosyl chloride was treated with n-butyl lithium to abstract the amide proton, followed by reduction with lithium naphthalenide at −78 °C to generate the α-anomeric anion with retention of configuration. On the other hand, the same glycosyl chloride was treated with 1 equivalent of n-butyl lithium followed by 1 equivalent of tributyltin lithium to give β-stannane with inversion of configuration. Upon treatment with 2 equivalents of n-butyl lithium, deprotonation and transmetalation took place to generate the β-anomeric anion with retention of configuration. These stereoselectively prepared anomeric anions reacted with aldehydes, CO_2, and alkyl iodide to afford the corresponding α- and β-C-glycosides, respectively. This method is applied to C-glycosylation of other carbohydrates such as 2-hydroxy and 2-deoxy sugars. It is worth noting that a carbohydrate with a protected hydroxyl group at the C-2 position undergoes β-elimination to give the corresponding glycal under these conditions.

C-Glycosylation by means of reaction with an anomeric anion containing stabilizing groups such as nitro, cyano, and sulfonyl groups has been extensively studied. The anomeric anion is usually prepared by proton abstraction with a base, and reacts with electrophiles to afford the C-glycoside product. However, due to the potential β-elimination to the corresponding glycal, the use of the stabilizing anomeric anion for C-glycosylation is limited. Since comprehensive

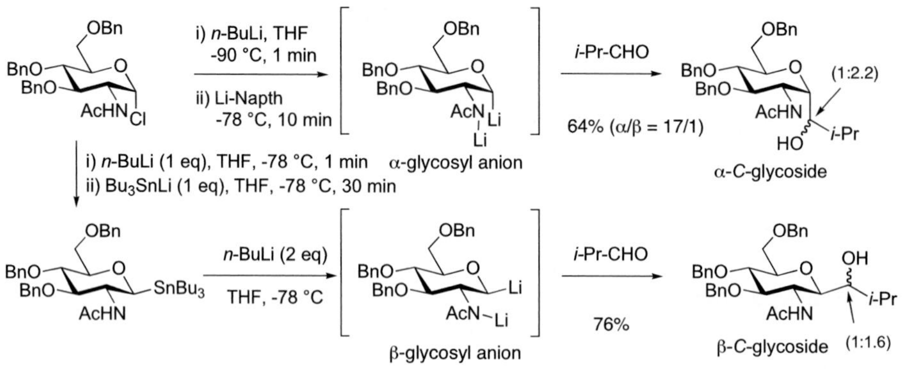

□ Scheme 48

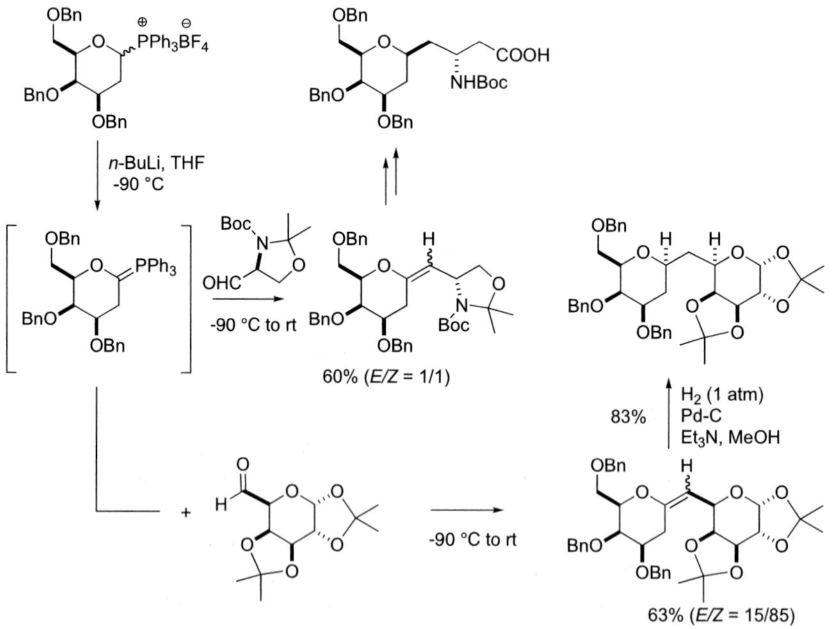

□ Scheme 49

reviews for this type of *C*-glycosylation have been published [74], this section will only deal with a recent example utilizing glycosyl ylide. The ylide prepared from a galactose derivative reacted with Garner aldehyde and sugar aldehyde to give the corresponding *exo*-olefinic products as a mixture of the geometric isomers (❷ *Scheme 49*) [76]. Catalytic hydrogenation gave a single isomer of β-*C*-glycosyl amino acid and *C*-disaccharide.

Lithio glycal is prepared by direct abstraction of a vinylic proton with butyl lithium or Schlosser's base (*n*-BuLi and *t*-BuOK). Alternatively, transmetalation of 1-stannyl-glycal with butyl

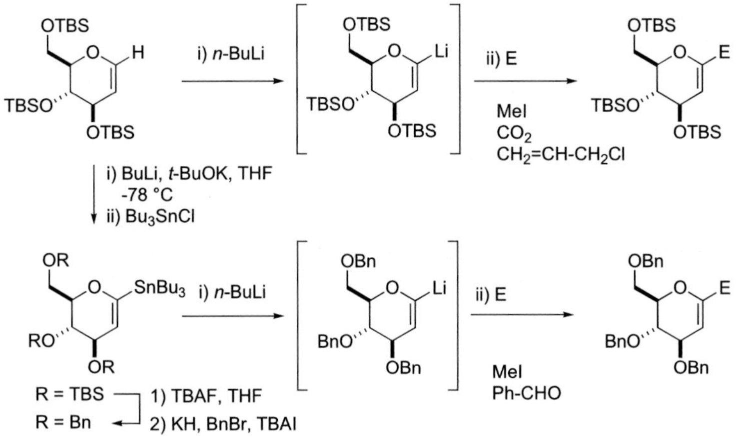

■ Scheme 50

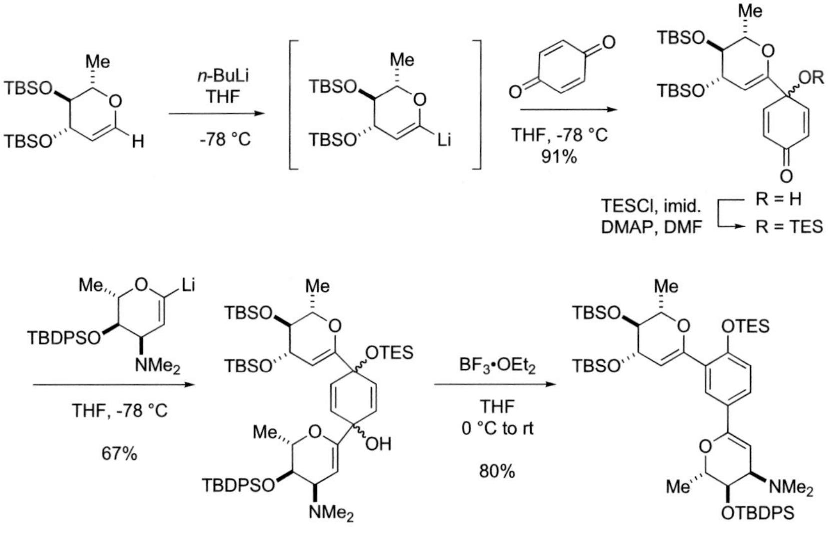

■ Scheme 51

lithium has been frequently employed as exemplified in ❯ *Scheme 50* [77] due to the inertness of the resulting tetrabutyltin, the side product. The resulting alkenyl anion reacts with a wide variety of electrophiles such as alkyl iodide, carbon dioxide, allyl halide, and quinone. A recent application of this methodology is shown in ❯ *Scheme 51* [78]. Iterative addition of 1-lithioglycal to *p*-quinone was followed by 1,2-migration of one of the glycals (dienol phenol-type rearrangement) by treatment with a Lewis acid to afford a model compound of pluramycin, an aryl-*C*-glycoside antibiotic.

4.2 Ramberg–Bäcklund Reaction

Another valuable type of *C*-glycosylation by using an anomeric anion is known as the Ramberg–Bäcklund reaction, which is classified as an intramolecular glycosylation. The reaction is, in general, a transformation of α-halosulfone to olefin under alkaline conditions. A typical example of *C*-glycosylation by the use of this reaction is shown in ❷ *Scheme 52* [79]. When glucosyl sulfone was treated with pulverized KOH suspended in a mixture of CCl₄ and *t*-BuOH or alumina-supported KOH in CBr₂F₂/*t*-BuOH, halogenation of the sulfone moiety occurred, which resulted in the formation of *epi*-sulfone. Elimination of SO₂ gave the *exo*-glycal in good to moderate yield. The product was, in most cases, obtained as a mixture of regioisomers. The selectivity is influenced by the stereochemistry of the *S*-glycoside. It is worthwhile to note that β-alkoxy elimination dose not occur even though an anomeric anion is generated as the intermediate. Since preparation of glycosyl sulfone, a precursor for the Ramberg–Bäcklund reaction, is easy, the method has been employed for a wide range of applications, giving a variety of *C*-glycosides. ❷ *Scheme 53* shows an example of the synthesis of *C*-disaccharide commencing with the synthesis of the corresponding sulfide [80].

4.3 Samarium Anion Mediated Reactions

Samarium diiodide (SmI₂)-reduction of glycosyl pyridyl sulfone with aldehydes or ketones under Barbier conditions provides an easily accessible method for the formation of *C*-glycosides. Since the conditions are very mild compared to the conditions used for lithium anomeric anions, a variety of substrates could be used [81]. Stereochemistry and yield of the *C*-glycosylation highly depends on the kind of carbohydrate and the protective groups used. Typical examples are listed in ❷ *Table 13*. Samarium-mediated *C*-glycosylation of mannosyl pyridyl sulfone with cyclohexanone gave α-*C*-glycoside in good yield without giving glycal as a by-

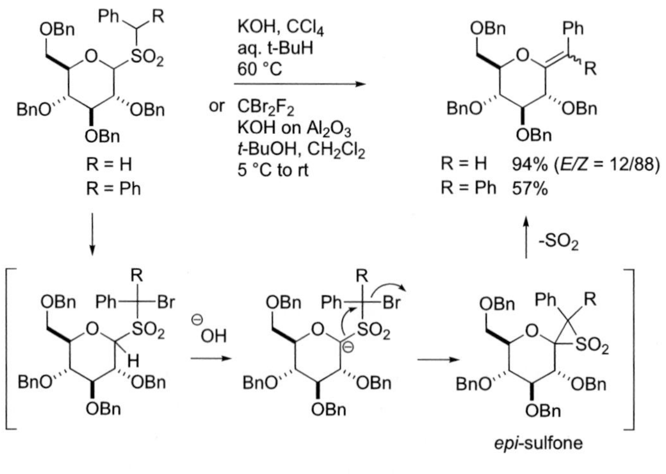

■ Scheme 52

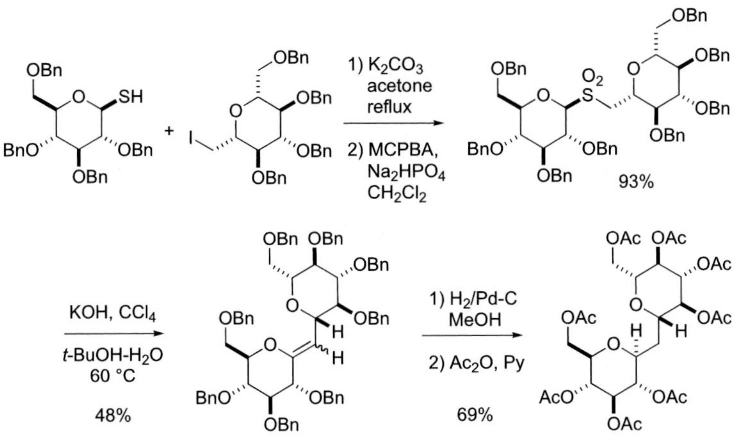

◻ Scheme 53

◻ Table 13
[81]

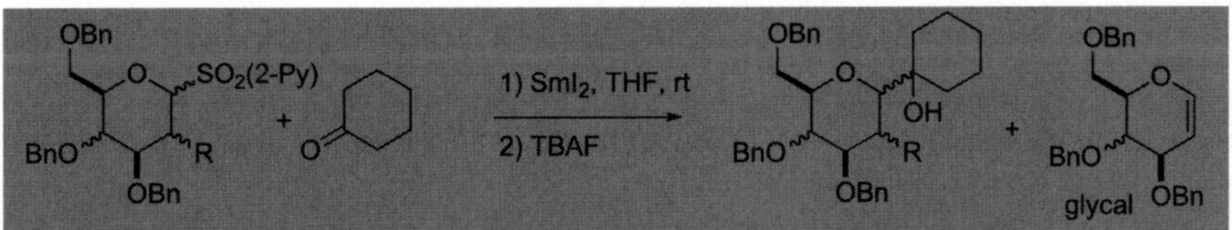

Entry	Glycosyl pyridyl sulfone			Product		
		Anomer	R	C-Glycoside	α/β	Glycal
1	Man	α	OTBS	80%[a]	1/0	0
2	Glc	β	OTBS	57%	0/1	21%
3	Gal	β	OTBS	22%	0/1	32%
4	2-Deoxy Gal	Mixture	H	86%	1/1	0

[a]: The yield was before desilylation

product (entry 1). On the other hand, *C*-glycosylation of glucosyl and galactosyl pyridyl sulfone under the same reaction conditions gave β-*C*-glycosylated products in lower yields with a considerable amount of glycal formation (entries 2 and 3). The configuration of the glycosyl pyridyl sulfone substrate does not influence the stereoselectivity of the reaction.

The stereochemical outcome of this *C*-glycosylation is explained by the following mechanism. Single electron transfer from SmI₂ to glycosyl sulfone leads to the formation of the thermodynamically more stable α-anomeric radical, which is further reduced by SmI₂ to give the α-samarium anion (◐ *Scheme 54*). In order to avoid a repulsive interaction between the Sm-anion and the lone-pair electron of the ring oxygen in the case of mannose, conformation of the samarium anion flips to a boat-like conformer, which reacts with aldehyde to afford the α-*C*-mannoside. In the case of glucose or galactose derivatives, the conformation of the anion is equilibrated with a boat-like conformation, which leads to formation of glycal via a facile

BnO OBn   SmI$_2$ BnO OBn   SmI$_2$ BnO OBn   slow   BnO OBn
BnO O → BnO O → BnO O ------ BnO O—SmI$_2$
BnO   BnO   BnO   BnO
SO$_2$(2-Pyr)

Man

SmI$_2$

fast ↕

BnO
BnO O—SmI$_2$
BnO OBn → α-C-mannoside

syn-elimination

glycal

BnO OBn → syn-elimination → glycal

BnO SmI$_2$ BnO   SmI$_2$ BnO   fast   BnO
BnO O → BnO O → BnO O → BnO O—SmI$_2$
BnO SO$_2$(2-Pyr)   BnO   BnO   BnO
BnO   BnO   BnO SmI$_2$ BnO

Glc, Gal

β-C-glycoside

fast ↕

BnO
BnO O—SmI$_2$
BnO
BnO → glycal

syn-elimination

▯ Scheme 54

syn-elimination. On the other hand, the configurational change (inversion of the anomeric configuration) gives a more stable β-samarium anion, which reacts with carbonyl compounds to form β-C-glycosides. It is worth noting that C-glycosylation of 2-deoxysugar gave a good yield of the product, however no stereoselectivity was observed (entry 4). Glycosyl phosphates instead of glycosyl pyridyl sulfone can also be employed as the substrate for samarium-mediated C-glycosylation [82].

Samarium-mediated C-glycosylation in order to form N-acetyl-sugars such as N-acetyl-glucosamine and N-acetyl-galactosamine has also been reported (❷ Scheme 55) [83]. These C-glycosylations exhibited α-selectivity, presumably because the strong coordination of the N-acetyl group to samarium (III) in an intermediate fixed the α-configuration of the samarium anomeric anion. This chemistry has also been employed for the synthesis of C-glycosyl amino acid such as the mimic of the tumor-associated antigen depicted in ❷ Scheme 56 [84].

In addition to radical allylation of sialic acid, Sm-mediated C-glycosylation of sialic acids has been recognized as a reliable method. Sialic acid derivatives with a suitable leaving group at the C-2 position are reduced with SmI$_2$ in the presence of a ketone or an aldehyde, giving the α-C-glycoside in high yields [85]. A special characteristic of this reaction is the exclusive production of α-C-glycoside, because the samarium enolate intermediate chelates with ring oxygen resulting in the reaction of the carbonyl from the less hindered face. As leaving groups, phenylsulfone, 2-pyridyl sulfone, phenyl sulfide, acetate, and chloride can be employed (❷ Scheme 57). The method can be applied to the synthesis of C-glycoside of other ulosonic acids such as KDN and KDO. Many applications of this methodology for the synthesis of C-disaccharide have been reported, some of which are shown in ❷ Scheme 58 [86].

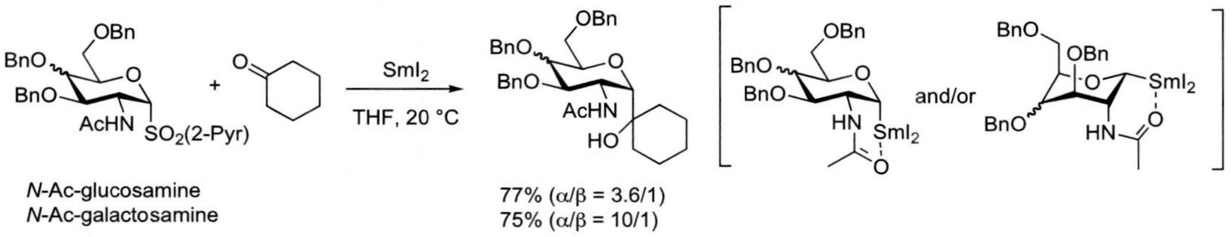

❏ Scheme 55

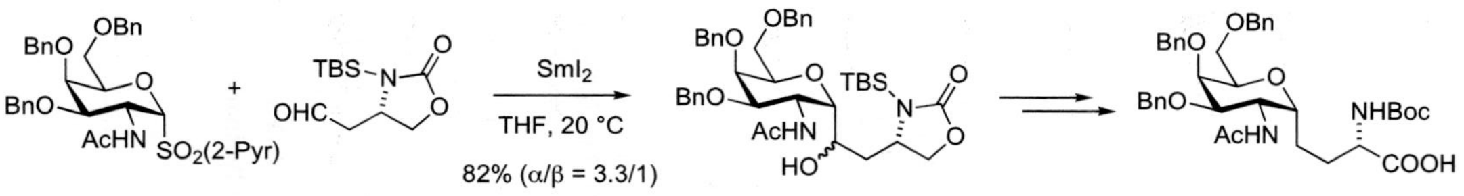

❏ Scheme 56

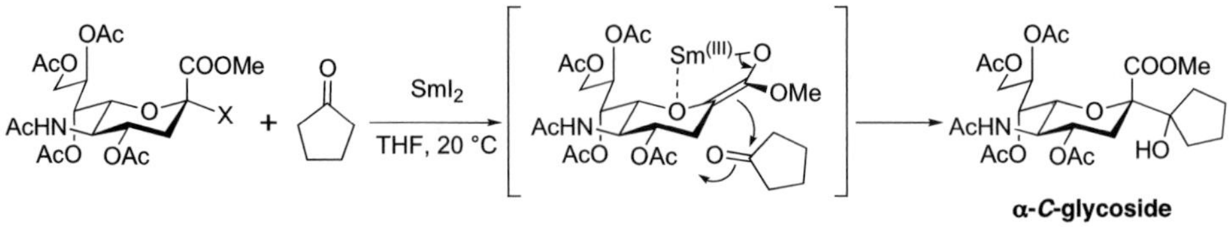

X = SO₂Ph (95%), SPy (76%), OAc (97%), Cl (95% in the reaction with acetone)

◻ **Scheme 57**

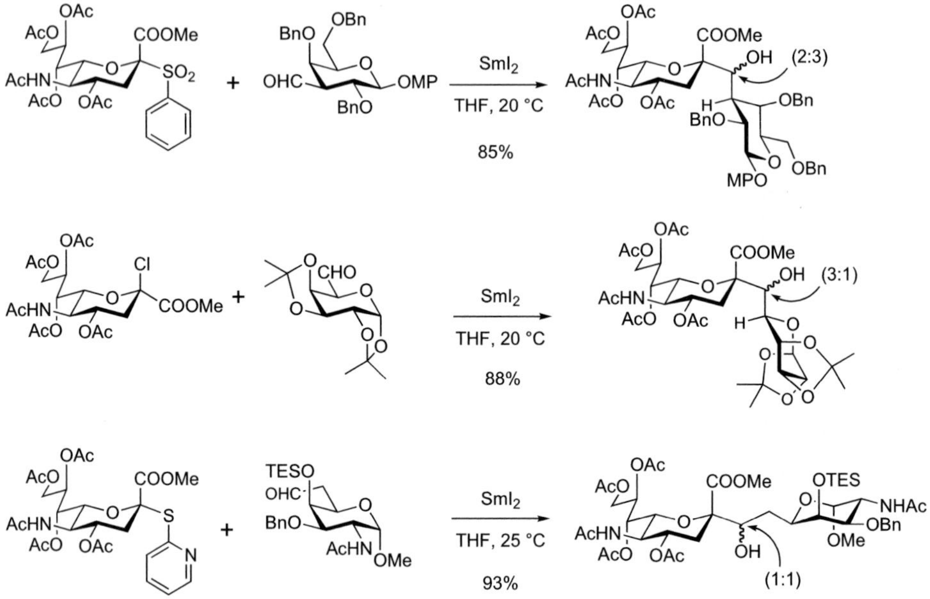

◻ **Scheme 58**

5 Sigmatropic Rearrangement

The Claisen rearrangement is the most important reaction among sigmatropic rearrangements for *C*-glycosylation. In the late 1970s, R. Ireland and B. Fraser-Reid reported Claisen rearrangement of glycal ester and its derivative [87] [88]. The Claisen rearrangement generally proceeds through a 6-membered chair-like transition-state, however, in the case of reaction with a glycal ester, a boat-like transition state was proposed as shown in ❷ *Eq. 9*. Since the reaction proceeds in a concerted manner, chirality at the C-3 position is transferred to the newly formed *C*-glycoside.

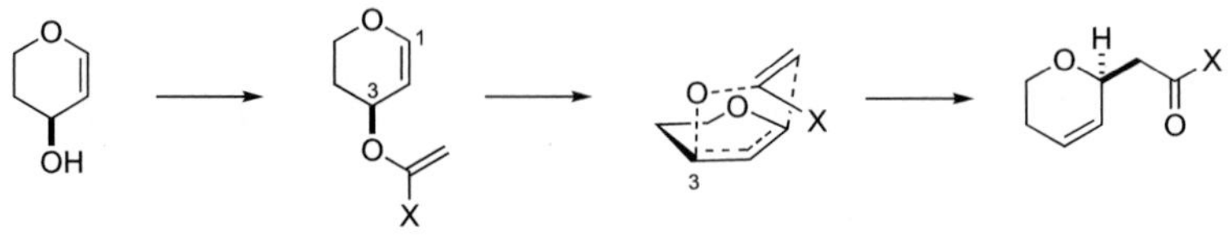

(9)

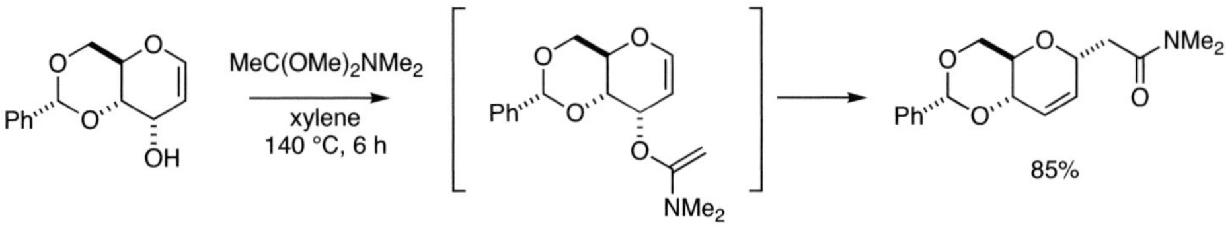

□ Scheme 59

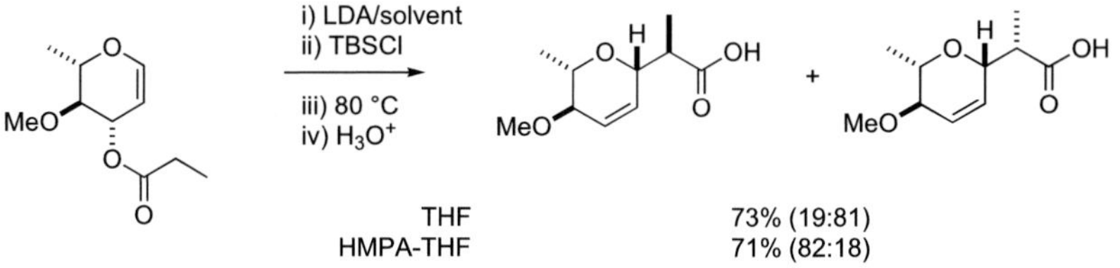

□ Scheme 60

The reaction outlined in ❯ *Scheme 59* is an example of a variant of the Claisen rearrangement of allyl ketene aminal (so-called Eschenmoser–Claisen rearrangement) [87]. The reaction dose not require an acid catalyst; glycal was just heated with dimethylacetamide dimethyl acetal to form ketene aminal, which underwent the sigmatropic rearrangement to form the corresponding γ,δ-unsaturated amide.

Claisen rearrangement of ketene silyl acetal derived from glycal ester, so-called Ireland–Claisen rearrangement, has been extensively developed for synthesis of polyether antibiotics. Glycals of pyranose or furanose with an enolizable ester can be employed as substrate for the reaction. However, it has been generally difficult to control the configuration at the α-position of the resulting acetic acid, although some improvement has been reported as shown in ❯ *Scheme 60* [88]. In some reactions, epimerization of the products was observed under the conditions.

Ireland demonstrated the usefulness of this reaction in the total synthesis of the complex polyether antibiotic, lasalocid A (X537A) [89]. Two Claisen rearrangements were used as key steps for the introduction of the carbon chain to the tetrahydrofuran and tetrahydropyrane rings (arrows indicate the bonds generated by the Claisen rearrangement in ❯ *Scheme 61*).

Claisen rearrangement of an oxazole derived from ester of alanine has also been reported (❯ *Scheme 62*) [90]. The product was further transformed into *C*-mannosyl-alanine.

Recently, Claisen rearrangement of allyl-vinyl ether prepared from glycal ester with Tebbe reagent was reported [91]. In contrast to the Ireland–Claisen rearrangement, in principle, a non-enolizable ester can be employed (❯ *Scheme 63*). This method was applied for the synthesis of *C*-disaccharide.

A few examples of [2,3] sigmatropic rearrangements to form *C*-glycoside have been reported [92]. On the other hand, [1,2] Wittig rearrangement has been extensively developed for the synthesis of *C*-glycoside. *O*-Benzyl, allyl, and propargyl glycoside is treated with base (BuLi) to generate a carbanion, which undergoes the [1,2] rearrangement to form *C*-glycoside with

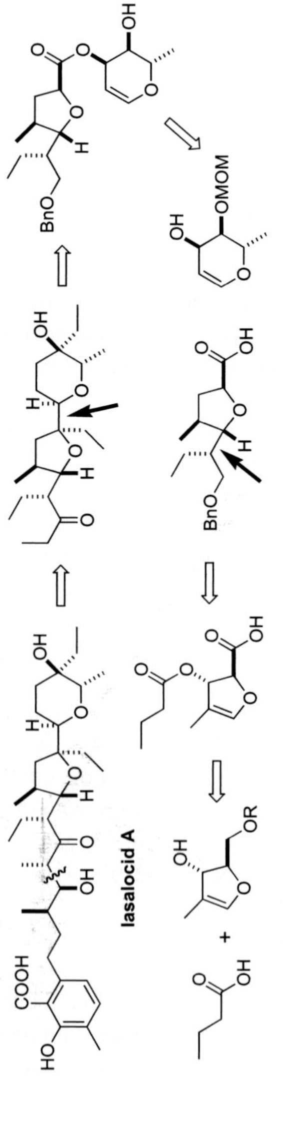

lasalocid A

⊡ Scheme 61

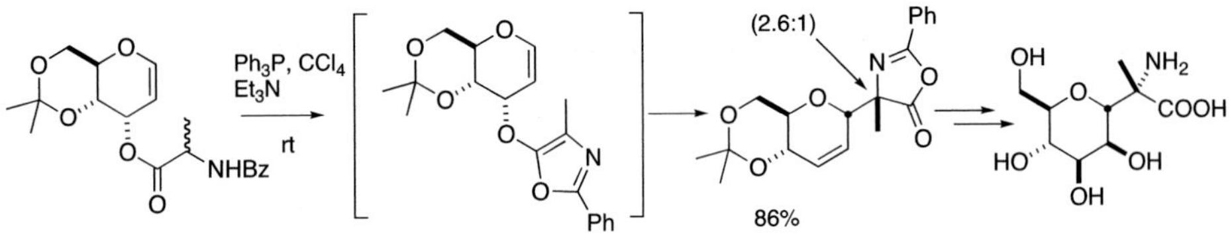

◻ Scheme 62

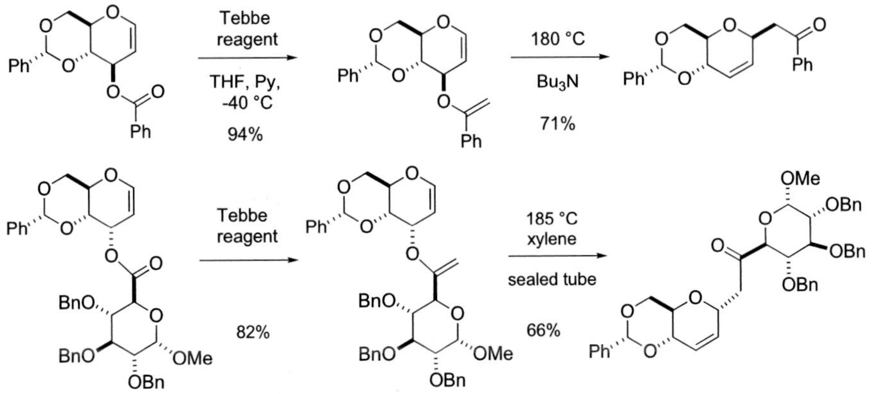

◻ Scheme 63

retention of configuration. The reaction mechanism was reported to be non-concerted but rad-ical-dissociation and recombination in a cage (❷ *Eq. 10*). However, this section will deal with the [1,2] Wittig reaction.

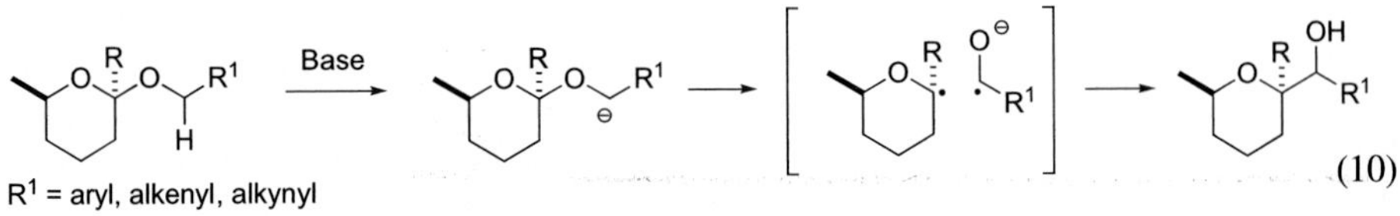

R^1 = aryl, alkenyl, alkynyl

(10)

For example, upon lithiation at the benzylic position of β-*O*-benzyl ribose with butyl lithium, the rearrangement takes place to give the β-*C*-glycosidic product [❷ *Scheme 64*(a)] [93]. The stereochemistry of the secondary alcohol on the side chain was controlled by chelation between lithium alkoxide and ring oxygen. In the total synthesis of zaragozic acid, a potent inhibitor of cholesterol biosynthesis, [1,2] Wittig rearrangement of bis-ethynylalcohol-α-*O*-glycoside was efficiently employed as one of the key steps for constructing the highly congested stereochemistries of zaragozic acid [❷ *Scheme 64*(b)].

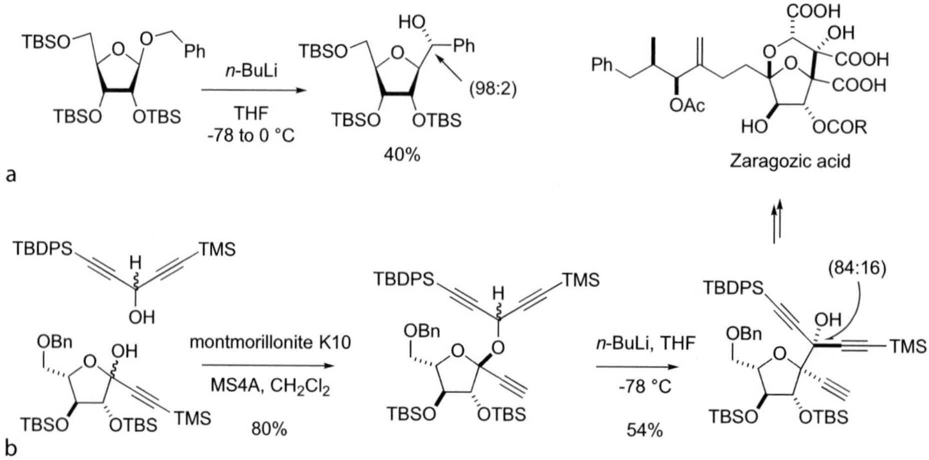

a

b

◘ Scheme 64

6 Transition Metal-Catalyzed Reactions

Since transition metal-catalyzed C–C bond formations have been extensively studied, a number of reliable reactions with palladium or nickel catalysts have been reported for the synthesis of *C*-glycosidic linkages due to the high functional group compatibility.

Reaction of the π-allyl-palladium complex with a stabilized anion (this reaction is known as the Tsuji–Trost reaction) has been applied for the stereoselective *C*-glycosylation of unsaturated sugars. 2,3-Unsaturated sugars are treated with a carbon nucleophile in the presence of a palladium(0) catalyst to give *C*-glycosides with migration of the double bond. Stereochemistry of the products mainly depends on the nature of the nucleophiles. The reaction mechanism involves formation of an electrophilic π-allyl-palladium complex by nucleophilic attack of palladium(0) catalyst to allyl ester with inversion of configuration (*anti* attack). When a stable anion such as malonate ester (i. e. soft nucleophile) was employed, the nucleophile attacked the anomeric carbon from the opposite side of the palladium complex to afford the C-1 substituted product with net retention of configuration (❂ *Scheme 65*) [94]. On the other hand, when an unstable anion nucleophile such as organo zinc reagent (i. e. hard nucleophile) was used, the anion attacked the palladium and then the nucleophile was transferred to the C-1 position to give the product with net inversion of configuration. Epimerization might occur when an unsubstituted malonate is employed as the nucleophile due to retro-Michael reaction followed by ring closing.

Acetyl glycal has not been employed for the Tsuji–Trost reaction due to the fact that the vinyl ether is inert under the conditions. However, trifluoroacetyl glycal is reactive enough for the reaction with the potassium salt of malonate ester in the presence of Pd(0) catalyst to form the product with migration of the double bond (❂ *Scheme 66*) [95]. According to the mechanisms mentioned above, the reaction proceeded with double inversion of configuration to give β-*C*-glycoside.

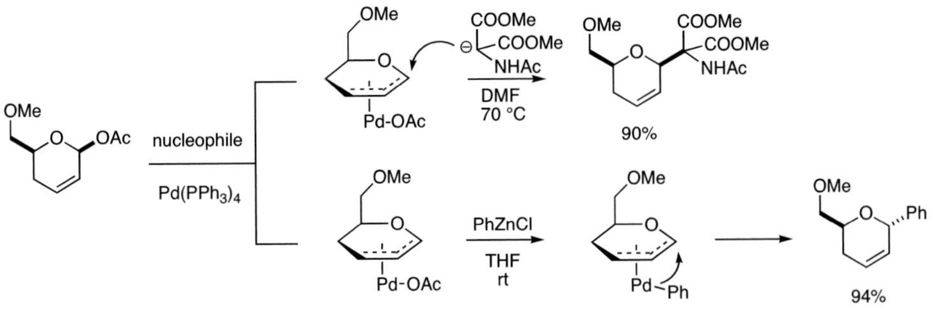

□ **Scheme 65**

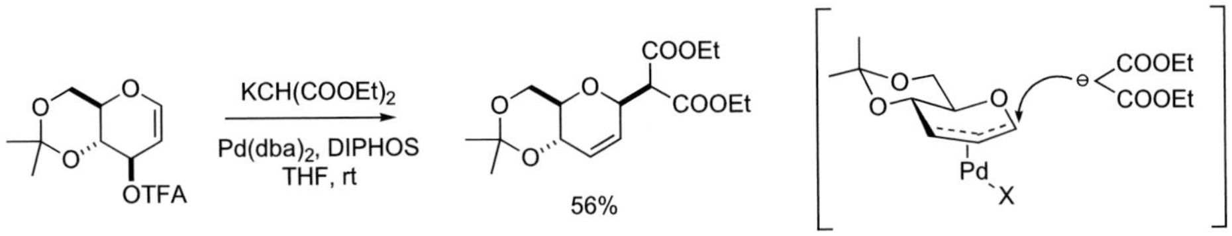

□ **Scheme 66**

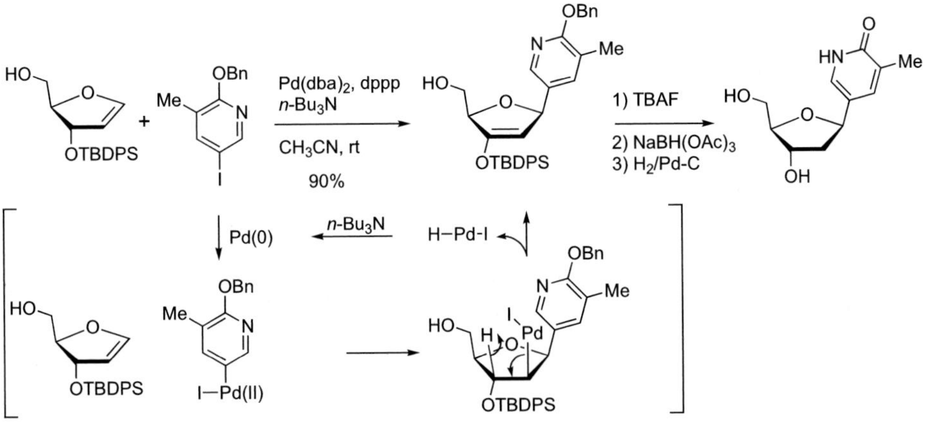

□ **Scheme 67**

Palladium-catalyzed arylation and vinylation of alkene is referred to as the Mizoroki–Heck reaction and is one of the most widely used Pd(0)-catalyzed C–C bond formations in organic synthesis. However, the reaction has not been extensively employed for *C*-glycosylation [96]. The example shown in ❷ *Scheme 67* outlines the reaction of iodopyridine and furanose glycal for the synthesis of *C*-nucleoside [97]. The mechanism began with the oxidative addition of iodopyridine to Pd(0) catalyst, and the resulting organo-palladium species was inserted by

⬛ Scheme 68

glycal double bond with *syn* fashion, followed by elimination of palladium hydride in stereospecific *syn* manner to give β-*C*-glycoside with migration of the double bond. The stereochemistry was explained by the preferential addition of the organo-palladium intermediate to the less hindered β-face of the double bond due to the siloxy group at the C-3 position. In the reaction, dppp (1,3-bis(diphenylphosphino)propane), a bidentate ligand, was critical in order to obtain the high yield of the product. The resulting vinylsilyl ether was desilylated upon treatment with fluoride and the ketone was stereoselectively reduced with triacetoxyborohydride to form the aryl-*C*-nucleoside.

Pd-catalyzed coupling between aryl or alkenyl halides and an organostannane is known as Migita–Stille coupling. Since organostannanes are stable to air and moisture, and the reaction proceeds under essentially neutral conditions, the Migita–Stille coupling has been employed in the synthesis of complex molecules, despite stoichiometric use of toxic organostannane as a coupling partner [98]. ❷ *Scheme 68* shows the coupling between the sterically hindered aryl group and stannylglucal [99]. The product was further transformed into the spiroketal core structure of papulacandin, an antifungal antibiotic.

On the other hand, similar coupling by using organozinc reagents as the nucleophilic components is called Negishi coupling. Since organozinc reagents with functional groups such as ester, amide, and carbamate can be prepared, a highly convergent synthesis of multi-functional compounds can be carried out. When both organotin and organozinc reagents with the same substituents are available, the Negishi coupling with the organozinc reagent generally gives better results than that of the Stille coupling (❷ *Scheme 69*) [100]. ❷ *Scheme 70* demonstrates the power of this reaction in the total synthesis of vineomycinone, a naturally occurring aryl-*C*-glycoside antibiotic [101].

In place of the aryl halide or alkenyl halide, the corresponding phosphate or triflate have been employed for the above cross-coupling reaction (❷ *Scheme 71*) [102]. These substrates were easily prepared from the corresponding lactone. The reaction of alkenyl phosphate or triflate has frequently been employed in the synthesis of marine polycyclic ethers such as brevetoxin etc (❷ *Scheme 72*) [103].

Recently, the reaction of glycosyl halides with organozinc reagent in the presence of a nickel catalyst was reported (❷ *Scheme 73*) [104]. This is a rare example of cross-coupling between sp^3-sp^3 carbons. Stereoselectivities of the reaction between glucosyl or galactosyl halide and the organozinc reagents are very low, while *C*-glycosylation of mannose under the same conditions gave α-*C*-glycosides in high selectivity, due to the steric hindrance of the protected hydroxyl group at the C-2 position.

Palladium-catalyzed cross-coupling reactions between organo borane compound and organic halides (or triflate, phosphate) provide a powerful and reliable method for C–C bond formation known as the Suzuki–Miyaura coupling. sp^3-Alkyl boranes prepared by hydroboration of alkane can be employed as the coupling partner for 1-iodoglucal as shown in ❷ *Scheme 74* [105]. When the reaction was conducted under a carbon monoxide atmosphere, carbonylative coupling occurred. This coupling has been frequently used as a key reaction in the synthesis of marine polyether toxins (❷ *Scheme 75*) [106]. For more examples, see ❷ *Sect. 2.2.2*.

A useful *C*-glycosylation reaction employing a cobalt catalyst is shown in ❷ *Scheme 76* [107]. Glycosyl acetate was treated with diethylmethylsilane and catalytic amounts of $Co_2(CO)_8$ under a carbon monoxide atmosphere, stereospecific hydroxymethylation occurred in 1,2-*trans* manner through neighboring group participation with the acetyl group.

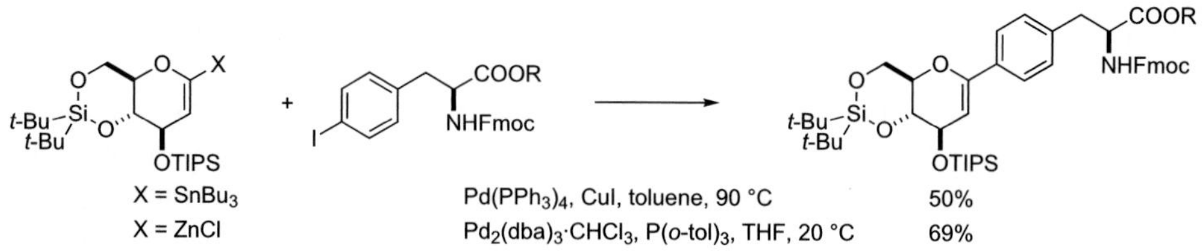

X
t-Bu
t-Bu
Si
OTIPS
X = SnBu3
X = ZnCl
COOR
NHFmoc
I
Pd(PPh3)4, CuI, toluene, 90 °C
Pd2(dba)3·CHCl3, P(o-tol)3, THF, 20 °C
50%
69%

Scheme 69

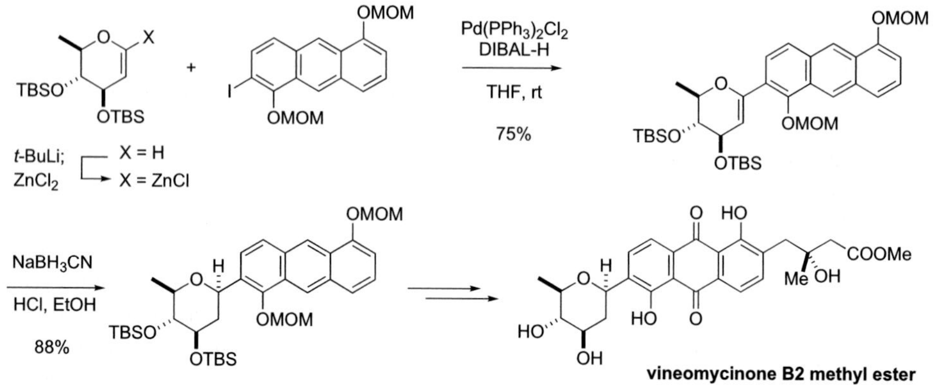

TBSO
OTBS
t-BuLi;
ZnCl2
X = H
X = ZnCl
OMOM
I
Pd(PPh3)2Cl2
DIBAL-H
THF, rt
75%
NaBH3CN
HCl, EtOH
88%
COOMe
Me OH
HO
OH
vineomycinone B2 methyl ester

Scheme 70

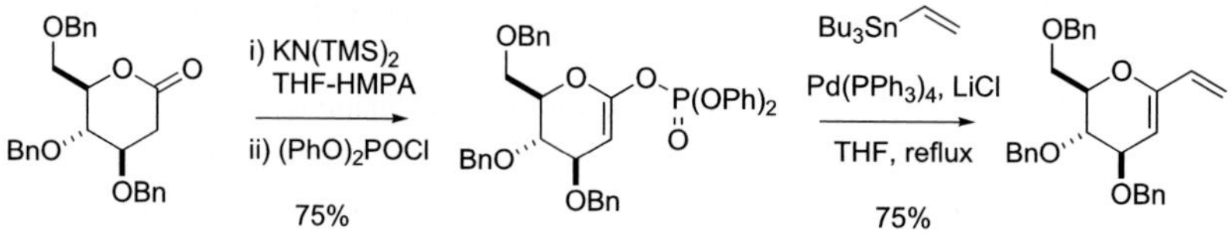

OBn
BnO
i) KN(TMS)2
THF-HMPA
ii) (PhO)2POCl
75%
P(OPh)2
Bu3Sn
Pd(PPh3)4, LiCl
THF, reflux
75%

Scheme 71

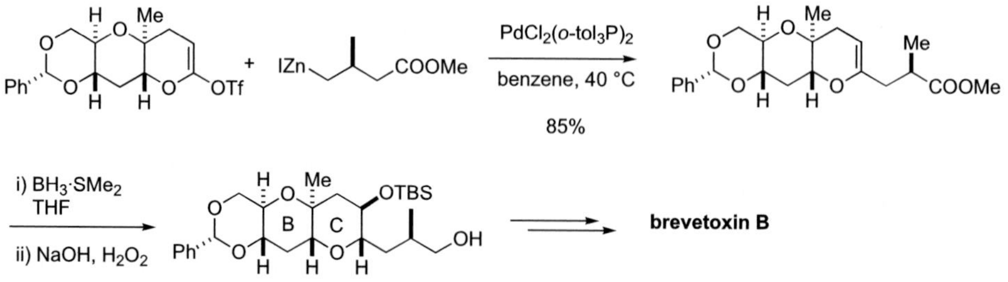

Me
Ph
OTf
IZn
COOMe
PdCl2(o-tol3P)2
benzene, 40 °C
85%
i) BH3·SMe2
THF
ii) NaOH, H2O2
OTBS
OH
brevetoxin B

Scheme 72

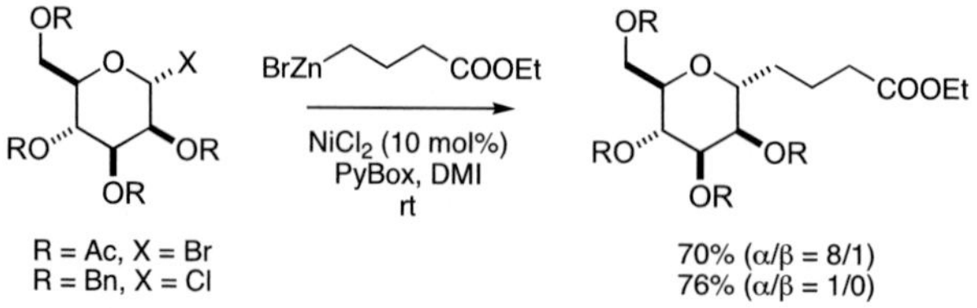

OR
RO
X
R = Ac, X = Br
R = Bn, X = Cl
BrZn
COOEt
NiCl2 (10 mol%)
PyBox, DMI
rt
70% (α/β = 8/1)
76% (α/β = 1/0)

Scheme 73

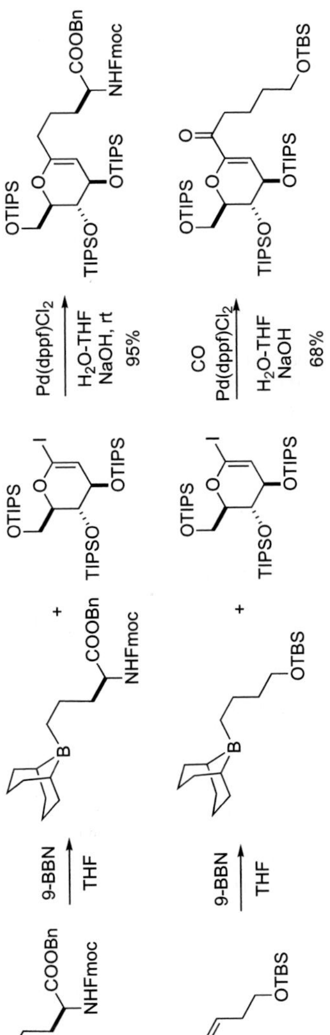

❏ Scheme 74

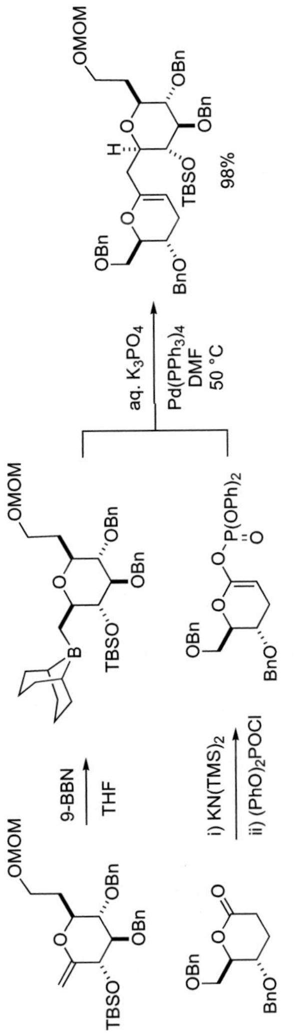

❏ Scheme 75

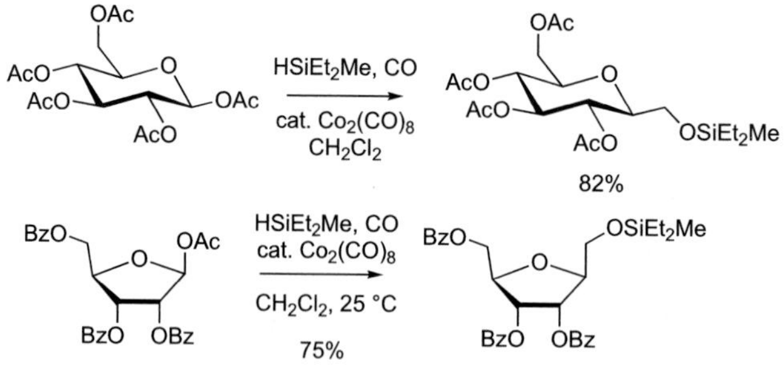

□ **Scheme 76**

References

1. Frank RW, Tsuji M (2006) Acc Chem Res 39:692
2. Okadaic acid: (a) Isobe M, Ichikawa Y, Bai DL, Masaki H, Goto T (1986) Tetrahedron 27:4767 and references cited therein; (b) Ley SV, Humphries AC, Eick H, Downham R, Ross AR, Boyce RJ, Pavey JBJ, Pietruszka J (1998) J Chem Soc Perkin Trans 1 3907; (c) Forsyth CJ, Sabes SF, Urbanek RA (1997) J Am Chem Soc 119:8381; Brevetoxin A: Nicolaou KC, Yang Z, Shi GQ, Gunzner JL, Agrios KA, Gärtner P (1998) Nature 392:264; Review of aryl-C-glycosides: Jaramillo C, Knapp S (1994) Synthesis 1
3. Previous reviews on synthesis of C-glycosides: (a) Postema MHD (1995) C-Glycoside Synthesis. CRC Press, London; (b) Levy DE, Tang C (1995) The Chemistry of C-Glycosides. Pergamon, Oxford; (c) Postema MHD (1992) Tetrahedron 48:8545; (d) Du Y, Linhardt RJ, Vlahov IR (1998) Tetrahedron 54:9913; (e) Meo P, Osborn HMI (2003) In: Osborn HMI (ed) Carbohydrate. Academic Press, Amsterdam p 337
4. (a) Rodrigues F, Canac Y, Lubineau A (2000) J Chem Soc Chem Commun 2049; (b) Bragnier N, Scherrmann MC (2005) Synthesis 814
5. Tatsuta K, Nakano S, Narazaki F, Nakamura Y (2001) Tetrahedron Lett 42:7625
6. Harvey JE, Raw SA, Taylor RJK (2004) Org Lett 6:2611
7. Review of exo-glycal: Taillefumier C, Chapleur Y (2004) Chem Rev 104:263
8. (a) Schweizer F, Otter A, Hindsgaul O (2001) Synlett 1743; (b) Penner M, Taylor D, Desautels D, Marat K, Schweizer F (2005) Synlett 212
9. Trotter NS, Takahashi S, Nakata T (1999) Org Lett 1:957
10. Tatsuta K, Mukai H, Takahashi M (2000) J Antibiotics 53:430
11. (a) Gascón-López M, Motevalli M, Paloumbis G, Bladon P, Wyatt PB (2003) Tetrahedron 59:0349; (b) Coumbarides GS, Motevalli M, Muse WA, Wyatt PB (2006) J Org Chem 71:7888
12. Molina A, Czernecki S, Xie J (1998) Tetrahedron Lett 39:7507
13. (a) Lakhrissi M, Chapleur Y (1996) Angew Chem Int Ed 35:750; (b) Lakhrissi M, Taillefumier C, Lakhrissi M, Chapleur Y (2000) Tetrahedron Asymmetry 11:417
14. Taillefumier C, Lakhrissi Y, Lakhrissi M, Chapleur Y (2002) Tetrahedron Asymmetry 13:1707
15. Lakhrissi M, Chapleur Y (1994) J Org Chem 59:5752
16. Lakhrissi M, Chapleur Y (1998) Tetrahedron Lett 39:4659
17. (a) RajanBabu TV, Reddy GS (1986) J Org Chem 51:5458; (b) Gómez AM, Pedregosa A, Valverde S, López JC (2003) Tetrahedron Lett 44:6111; (c) John BA, Pan YT, Elbein AD, Johnson CR (1997) J Am Chem Soc 119:4856; (d) Gueyrard D, Haddoub R, Salem A, Bacar NS, Goekjian PG (2005) Synlett 520
18. Alcaraz ML, Griffin FK, Paterson DE, Taylor RJK (1998) Tetrahedron Lett 39:8183
19. (a) Yang WB, Yang YY, Gu YF, Wang SH, Chang CC, Lin CH (2002) J Org Chem 67:3773; (b) Terauchi M, Abe H, Matsuda A, Shuto S (2004) Org Lett 6:3751; (c) Lewis MD, Cha JK, Kishi Y (1982) J Am Chem Soc 104:4976; (d) Xie J, Durrat F, Valéry JM (2003) J Org Chem 68:7896; (e) Pulley SR, Carey JP (1998) J Org Chem 63:5275; (f) Lancelin JM, Zollo

PHA, Sinaÿ P (1983) Tetrahedron Lett 24:4833; (g) Czernecki S, Ville G (1989) J Org Chem 54:610; (h) Ellsworth BA, Doyle AG, Patel M, Caceres-Cortes J, Meng W, Deshpande PP, Pullockaran A, Washburn WN (2003) Tetrahedron Asymmetry 14:3243; (i) Labéguère F, Lavergne JP, Martinez J (2002) Tetrahedron Lett 43:7271; (j) Wellner E, Gustafsson T, Bäcklund J, Holmdahl R, Kihlberg J (2000) ChemBioChem 1:272; (k) Lowary T, Meldal M, Helmboldt A, Vasella A, Bock K (1998) J Org Chem 63:9657; (l) Dondoni A, Scherrmann MC (1994) J Org Chem 59:6404; (m) Nishikawa T, Koide Y, Kanakubo A, Yoshimura H, Isobe M (2006) Org Biomol Chem 4:1268

20. Review of thiazole as a nucleophile: Dondoni A, Marra A (1999) J Chem Soc Chem Commun 2133

21. Beignet J, Tiernan J, Woo CH, Kariuki BM, Cox LR (2004) J Org Chem 69:6341

22. Terauchi M, Abe H, Matsuda A, Shuto S (2004) Org Lett 6:3751

23. (a) Calzada E, Clarke CA, Roussin-Bouchard C, Wightman RH (1995) J Chem Soc Perkin Trans 1 517; (b) Centrone CA, Lowary TL (2002) J Org Chem 67:8862; (c) Krohn K, Heins H, Wielckens K (1992) J Med Chem 35:511

24. (a) Boyd VA, Drake BE, Sulikowski GA (1993) J Org Chem 58:3191; (b) Kaelin DE Jr, Lopez OD, Martin SF (2001) J Am Chem Soc 123:6937

25. Uchiyama T, Woltering TJ, Wong W, Lin CC, Kajimoto T, Takebayashi M, Weitz-Schmidt G, Asakura T, Noda M, Wong CH (1996) Bioorg Med Chem 4:1149

26. Kulkarni SS, Gervay-Hague J (2006) Org Lett 8:5765

27. (a) Toshima K (2000) Carbohydr Res 327:15; (b) Yokoyama M, Toyoshima H, Shimizu M, Mito J, Togo H (1998) Synthesis 409

28. (a) Nicolaou KC, Chucholowski A, Dolle RE, Randall JL (1984) J Chem Soc Chem Commun 1155; (b) Posner GH, Haines SR (1985) Tetrahedron Lett 26:1823

29. (a) Hosomi A, Sakata Y, Sakurai H (1984) Tetrahedron Lett 25:2383; Hosomi A, Sakata Y, Sakurai H (1987) Carbohydr Res 171:223; (b) Nicolaou KC, Dolle RE, Chucholowski A, Randall JL J Chem Soc Chem Commun (1984) 1153; (c) Giannis A, Sandhoff K (1985) 26:1479; (d) Palmacci ER, Seeberger PH (2001) Org Lett 3:1547; Plante OJ, Palmacci ER, Andrade RB, Seeberger PH (2001) J Am Chem Soc

123:9545; (e) Brenna E, Fuganti C, Grasselli P, Serra S, Zambotti S (2002) Chem Eur J 8:1872; (f) Fletcher S, Jorgensen MR, Miller AD (2004) Org Lett 6:4245

30. Hung SC, Lin CC, Wong CH (1997) Tetrahedron Lett 38:5419

31. Allevi P, Anastasia M, Ciuffreda P, Fiecchi A, Scala A (1987) J Chem Soc Chem Commun 101

32. (a) Zhai D, Zhai W, Williams RM (1998) J Am Chem Soc 110:2501; (b) Jobron L, Leteux C, Veyrières A, Beau JM (1994) J Carbohydr Chem 13:507; Désiré J, Veyrières (1995) Carbohydr Res 268:177; (c) Nishikawa T, Ishikawa M, Isobe M (1999) Synlett 123; Nishikawa T, Koide Y, Kajii S, Wada K, Ishikawa M, Isobe M (2005) Org Biomol Chem 3:687; (d) Dondoni A, Mariotti G, Marra A (2002) J Org Chem 67:4475

33. (a) Isobe M, Nishizawa R, Hosokawa S, Nishikawa T (1998) J Chem Soc Chem Commun 2665; (b) Jiang Y, Isobe M (1996) Tetrahedron 52:2877; (c) Tsuboi K, Ichikawa Y, Isobe M (1997) Synlett 713; (d) Jiang Y, Ichikawa Y, Isobe M (1997) Tetrahedron 53:5103

34. Tamura S, Abe H, Matsuda A, Shuto S (2003) Angew Chem Int Ed 42:1021

35. (a) Smith DM, Woerpel KA (2006) Org Biomol Chem 4:1195; (b) Larsen CH, Ridgway BH, Shaw JT, Smith DM, Woerpel KA (2005) J Am Chem Soc 127:10879

36. Brimble MA, Brenstrum TJ (2000) Tetrahedron Lett 41:2991

37. Allevi P, Anastasia M, Ciuffreda P, Fiecchi A, Scala A, Bingham S, Muir M, Tyman J (1991) J Chem Soc Chem Commun 1319

38. Furuta T, Kimura T, Kondo S, Mihara H, Wakimoto T, Nukaya H, Tsuji K, Tanaka K (2004) Tetrahedron 60:9375

39. (a) Kuribayashi T, Ohkawa N, Satoh S (1998) Tetrahedron Lett 39:4537; (b) Kuribayashi T, Mizuno Y, Gohya S, Satoh S (1999) J Carbohydr Chem 18:371

40. Matsuo G, Miki Y, Nakata M, Matsumura S, Toshima K (1999) J Org Chem 64:7101

41. Matsumoto T, Katsuki M, Suzuki K (1988) Tetrahedron Lett 29:6935

42. Ben A, Yamauchi T, Matsumoto T, Suzuki K (2004) Synlett 225

43. Futagami S, Ohashi Y, Imura K, Hosoya T, Ohmori K, Matsumoto T, Suzuki K (2000) Tetrahedron Lett 41:1063

44. Thorn SN, Gallagher T (1996) Synlett 185

45. Thorn SN, Gallagher T (1996) Synlett 856

46. Cook MJ, Fletcher MJE, Gray D, Lovell PJ, Gallagher T (2004) Tetrahedron 60:5085

47. Steinhuebel DP, Fleming JJ, Du Bois J (2002) Org Lett 4:293

48. (a) Bussolo VD, Caselli M, Romano MR, Pineschi M, Crotti P (2004) J Org Chem 69:7383; (b) Bussolo VD, Caselli M, Romano R, Pineschi M, Crotti P (2004) J Org Chem 69:8702

49. Saeeng R, Isobe M (2006) Chem Lett Highlight Review 35:552

50. (a) Danishefsky SJ, Kerwin JF Jr (1982) J Org Chem 47:3805; (b) Takhi M, Rahman AA-HA, Schmidt RR (2001) Tetrahedron Lett 42:4053; (c) Zhu YH, Vogel P (2001) Synlett 82; (d) Yadav JS, Reddy BVS, Rao CV, Chand PK, Prasad AR (2001) Synlett 1638; (e) Tsukiyama T, Isobe M (1992) Tetrahedron Lett 33:7911; (f) Saeeng R, Sirion U, Sahakitpichan P, Isobe M (2003) Tetrahedron Lett 44:6211; (g) Saeeng R, Isobe M (2005) Org Lett 7:1585; (h) Isobe M, Saeeng R, Nishizawa R, Konobe M, Nishikawa T (1999) Chem Lett 467

51. (a) Hosokawa S, Kirschbaum B, Isobe M (1998) Tetrahedron Lett 39:1917; (b) Das SK, Reddy KA, Abbineni C, Roy J, Rao KVLN, Sachwani RH, Iqbal J (2003) Tetrahedron Lett 44:4507; (c) Shoji M, Akiyama N, Tsubone K, Lash LL, Sanders JM, Swanson GT, Sakai R, Shimamoto K, Oikawa M, Sasaki M (2006) J Org Chem 71:5201

52. Tanaka S, Tsukiyama T, Isobe M (1993) Tetrahedron Lett 34:5757

53. Hosokawa S, Isobe M (1998) Tetrahedron Lett 39:2609

54. Hosokawa S, Isobe M (1999) J Org Chem 64:37

55. Pulley SR, Carey JP (1998) J Org Chem 63:5257

56. (a) Allwein SP, Cox JM, Howard BE, Johnson HWB, Rainier JD (2002) Tetrahedron 58:1997; (b) Rainier JD, Cox JM (2000) Org Lett 2:2707; (c) Evans DA, Trotter BW, Côté B (1998) Tetrahedron Lett 39:1709; (d) Timmers CM, Dekker M, Buijsman RC, van der Marel GA, Ethell B, Burchell AB, Mulder GJ, van Boom JH (1997) Bioorg Med Chem Lett 7:1501; (e) Leeuwenburgh MA, van der Marel GA, Overkleeft HS, van Boom JH (2003) J Carbohydr Chem 22:549; (f) Wipf P, Pierce JG, Zhuang N (2005) Org Lett 7:483

57. Hayward MM, Roth RM, Duffy KJ, Dalko PI, Stevens KL, Guo J, Kishi Y (1998) Angew Chem Int Ed 37:192

58. Evans DA, Trotter BW, Coleman PJ, Côté B, Dias LC, Rajapakse HA, Taylor AN (1999) Tetrahedron 55:8671

59. Manabe S, Marui Y, Ito Y (2003) Chem Eur J 9:1435

60. Stichler-Bonaparte J, Vasella A (2001) Helv Chim Acta 84:2355

61. Stichler-Bonaparte J, Bernet B, Vasella A (2002) Helv Chim Acta 85:2235

62. Jaouen V, Jégou A, Veyrières A (1996) Synlett 1218

63. Review: Togo h, He W, Waki Y, Yokoyama M (1998) Synlett 700

64. Giese B (1989) Angew Chem Int Ed 28:969

65. Abe H, Shuto S, Matsuda A (2001) J Am Chem Soc 123:11870

66. (a) Giese B, Linker T, Muhn R (1989) Tetrahedron 45:935; Giese B, Dupuis J (1983) Angew Chem Int Ed 22:622; Giese B, Dupuis J, Nix M (1987) Org Synth 65:236; Giese B, Dupuis J, Leising M, Nix M, Lindner HJ (1987) Carbohydr Res 171:329; (b) Harenbrock M, Matzeit A, Schäfer HJ (1996) Liebigs Ann Chem 55; (c) Keck GE, Enholm EJ, Yates JB, Wiley MR (1985) Tetrahedron 41:4079; (d) Krishna PR, Lavanya B, Jyothi Y, Sharma GVM (2003) J Carbohydr Chem 22:423; (e) Kessler H, Wittmann V, Köck M, Kottenhahn M (1992) Angew Chem Int Ed 31:902; (f) Pontén F, Magnusson G (1996) J Org Chem 61:7463; (g) Junker HD, Phung N, Fessner WD (1999) Tetrahedron Lett 40:7063

67. (a) Bouvet VR, Ben RN (2006) J Org Chem 71:3619; (b) Roe BA, Boojamra CG, Griggs JL, Bertozzi CR (1996) J Org Chem 61:6442; (c) Grant L, Liu Y, Walsh KE, Walter DS, Gallagher T (2002) Org Lett 4:4623; (d) SanMartin R, Tavassoli B, Walsh KE, Walter DS, Gallagher T (2000) Org Lett 2:4051; (e) Abel S, Linker T, Giese B (1991) Synlett 171

68. (a) Giese B, Hoch M, Lamberth C, Schmidt RR (1988) Tetrahedron Lett 29:1375; (b) Pasquarello C, Demange R, Vogel P (1999) Bioorg Med Chem Lett 9:793

69. (a) Paulsen H, Matschulat P (1991) Liebigs Ann Chem 487; (b) Nagy JO, Bednarski MD (1991) Tetrahedron Lett 32:3953

70. Bois M, Skrydstrup T (1995) Chem Rev 95:1253

71. Stork G, Suh HS, Kim G (1991) J Am Chem Soc 113:7054

72. Mazéas D, Skrydstrup T, Doumeix O, Beau JM (1994) Angew Chem Int Ed 33:1383

73. (a) Zhang J, Clive DLJ (1999) J Org Chem 64:770; (b) Zhang J, Clive DLJ (1999) J Org Chem 64:1754

74. Review: (a) Somsák L (2001) Chem Rev 101:81; (b) Beau JM, Gallagher T (1997) In Topics in Current Chemistry 187:1. Springer, Berlin Heidelberg New York

75. (a) Hoffmann M, Kessler H (1994) Tetrahedron Lett 35:6067; (b) Burkhart F, Kessler H (1998) Tetrahedron Lett 39:255

76. (a) Lieberknecht A, Griesser H, Krämer B, Bravo RD, Colinas PA, Grigera RJ (1999) Tetrahedron 55:6475; (b) Colinas PA, Ponzinibbio A, Lieberknecht A, Bravo RD (2003) Tetrahedron Lett 44:7985

77. (a) Lesimple P, Beau JM, Jaurand G, Sinaÿ P (1986) Tetrahedron Lett 27:6201; (b) Hanessian S, Martin M, Desai RC (1986) J Chem Soc Chem Commun 926

78. Parker KA, Su DS (2005) J Carbohydr Chem 24:199

79. (a) Griffin FK, Murphy PV, Paterson DE, Taylor RJK (1998) Tetrahedron Lett 39:8179; (b) Belica PS, Franck RW (1998) Tetrahedron Lett 39:8225

80. Griffin FK, Paterson DE, Murphy PV, Taylor RJK (2002) Eur J Org Chem 1305

81. Skrydstrup T, Jarreton O, Mazéas D, Urban D, Beau JM (1998) Chem Eur J 4:655

82. Hung SH, Wong CH (1996) Angew Chem Int Ed 35:2671

83. (a) Andersen L, Mikkelsen LM, Beau JM, Skrydstrup T (1998) Synlett 139; (b) Urban D, Skrydstrup T, Beau JM (1998) J Org Chem 63:2507

84. Urban D, Skrydstrup T, Beau JM (1998) J Chem Soc Chem Commun 955

85. (a) Du Y, Linhardt RJ (1998) Carbohydr Res 308:161; (b) Polat T, Du Y, Linhardt RJ (1998) Synlett 1195; (c) Malapelle A, Abdallah Z, Doisneau G, Beau JM (2006) Angew Chem Int Ed 45:6016

86. (a) Bazin HG, Du Y, Polat T, Linhardt RJ (1999) J Org Chem 64:7254; (b) Abdallah Z, Doisneau G, Beau JM (2003) Angew Chem Int Ed 42:5209

87. Fraser-Reid B, Dawe RD, Tulshian DB (1979) Can J Chem 57:1746

88. Ireland RE, Wilcox CS, Thaisrivongs S, Vanier NR (1979) Can J Chem 57:1743

89. Ireland RE, Anderson RC, Badoud R, Fitzsimmons BJ, McGarvey GJ, Thaisrivongs S, Wilcox CS (1983) J Am Chem Soc 105:1988

90. Colombo L, Di Giacomo M, Ciceri P (2002) Tetrahedron 58:9381

91. (a) Godage HY, Fairbanks AJ (2000) Tetrahedron Lett 41:7589; (b) Chambers DJ, Evans GR, Fairbanks AJ (2005) Tetrahedron Asymmetry 16:45

92. Tulshian DB, Fraiser-Reid B (1984) J Org Chem 49:518

93. (a) Tomooka K, Kikuchi M, Igawa K, Keong PH, Nakai T (1999) Tetrahedron Lett 40:1917; (b) Tomooka K, Kikuchi M, Igawa K, Suzuki M, Keong PH, Nakai T (2000) Angew Chem Int Ed 39:4502

94. (a) Dunkerton LV, Serino AJ (1982) J Org Chem 47:2814; (b) Moineau C, Bolitt V, Sinou D (1998) J Org Chem 63:582

95. RajanBabu TV (1985) J Org Chem 50:3642

96. Daves GD Jr (1990) Acc Chem Res 23:201

97. Hsieh HP, McLaughlin LW (1995) J Org Chem 60:5356

98. Friesen RW (2001) J Chem Soc Perkin Trans 1 1969

99. Friesen RW, Sturino CF (1990) J Org Chem 55:5808

100. Boucard V, Larrieu K, Lubin-Germain N, Uziel J, Auge J (2003) Synlett 1834

101. Tius MA, Gu XQ, Gomez-Galeno J (1990) J Am Chem Soc 112:8188

102. Nicolaou KC, Shi GQ, Gunzner L, Gärtner P, Yang Z (1997) J Am Chem Soc 119:5467

103. Kadota I, Takamura H, Nishii H, Yamamoto Y (2005) J Am Chem Soc 127:9246

104. Gong H, Sinisi R, Gagné MR (2007) J Am Chem Soc 129:1908

105. Potuzak JS, Tan DS (2004) Tetrahedron Lett 45:1797

106. Sasaki M, Fuwa H, Ishikawa M, Tachibana K (1999) Org Lett 1:1075

107. Chatani N, Ikeda T, Sano T, Sonoda N, Kurosawa H, Kawasaki Y, Murai S (1988) J Org Chem 53:3387

Index